The Parabola $(a \neq 0)$

Equation	$y = ax^2 + bx + c$ $= a(x-h)^2 + k$	$x = ay^2 + by + c$ $= a(y-k)^2 + h$
Vertex	(h, k) where $h = -b/2a$	(h, k) where $k = -b/2a$
Axis	$x = h$	$y = k$
Opening	Upward, if $a > 0$ Downward, if $a < 0$	To the right, if $a > 0$ To the left, if $a < 0$

The Circle with Center (h, k) and Radius r

$$(x-h)^2 + (y-k)^2 = r^2$$

Ellipse with Center at (h, k)

$$\frac{(x-h)^2}{a^2} + \frac{(y-k)^2}{b^2} = 1$$

Vertical Line Test

If any vertical line intersects the graph at two or more points, the graph does *not* represent a function. If no vertical line intersects the graph at more than one point, the relation is a function.

Hyperbola with Center at (h, k)

Foci on x-axis $\quad \dfrac{(x-h)^2}{a^2} - \dfrac{(y-k)^2}{b^2} = 1$

Foci on y-axis $\quad \dfrac{(y-k)^2}{a^2} - \dfrac{(x-h)^2}{b^2} = 1$

Asymptotes of $f(x) = \dfrac{P(x)}{Q(x)}$

Vertical $\quad x = a$, if $Q(a) = 0$ and $P(a) \neq 0$

$$\text{Horizontal} \quad \begin{cases} y = 0, \text{ if } n < m \\ y = a_n/b_m, \text{ if } n = m \\ \text{none if } n > m \end{cases}$$

Properties of Logarithms

$\log_a a^x = x \qquad \log_a u = \log_a v$ if and only if $u = v$

$a^{\log_a x} = x \qquad \log_a(uv) = \log_a u + \log_a v$

$\log_a a = 1 \qquad \log_a(u/v) = \log_a u - \log_a v$

$\log_a 1 = 0 \qquad \log_a u^r = r \log_a u$

$$\log_b N = \frac{\log_a N}{\log_a b}$$

Graphing Functions

To graph

$y = cf(x)$, multiply each y-coordinate on the graph of $y = f(x)$ by c,

$y = -f(x)$, reflect the graph of $y = f(x)$ through the x-axis,

$$y = f(x) + c, \begin{cases} c > 0, \text{ shift the graph of } y = f(x) \text{ upward } c \text{ units,} \\ c < 0, \text{ shift the graph of } y = f(x) \text{ downward } |c| \text{ units,} \end{cases}$$

$$y = f(x - c), \begin{cases} c > 0, \text{ shift the graph of } y = f(x) \text{ to the right } c \text{ units,} \\ c < 0, \text{ shift the graph of } y = f(x) \text{ to the left } |c| \text{ units.} \end{cases}$$

Binomial Theorem

$$(a+b)^n = a^n + \binom{n}{1}a^{n-1}b + \binom{n}{2}a^{n-2}b^2 + \cdots + \binom{n}{r}a^{n-r}b^r + \cdots + \binom{n}{n-2}a^2b^{n-2} + \binom{n}{n-1}ab^{n-1} + b^n$$

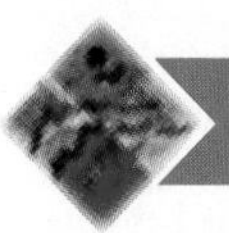

COLLEGE ALGEBRA with TRIGONOMETRY

Also Available from McGraw-Hill

Schaum's Outline Series in Mathematics & Statistics

Most outlines include basic theory, definitions and hundreds of example problems solved in step-by-step detail, and supplementary problems with answers.

Related titles on the current list include:

Advanced Calculus
Advanced Mathematics for Engineers & Scientists
Analytic Geometry
Basic Mathematics for Electricity & Electronics
Basic Mathematics with Applications to Science & Technology
Beginning Calculus
Boolean Algebra & Switching Circuits
Calculus
Calculus for Business, Economics, & the Social Sciences
College Algebra
College Mathematics
Complex Variables
Descriptive Geometry
Differential Equations
Differential Geometry
Discrete Mathematics
Elementary Algebra
Essential Computer Mathematics
Finite Differences & Difference Equations
Finite Mathematics
Fourier Analysis
General Topology
Geometry
Group Theory
Laplace Transforms
Linear Algebra
Mathematical Handbook of Formulas & Tables
Mathematical Methods for Business & Economics
Mathematics for Nurses
Matrix Operations
Modern Abstract Algebra
Numerical Analysis
Partial Differential Equations
Probability
Probability & Statistics
Real Variables

Review of Elementary Mathematics
Set Theory & Related Topics
Statistics
Technical Mathematics
Tensor Calculus
Trigonometry
Vector Analysis

Schaum's Solved Problems Series

Each title in this series is a complete and expert source of solved problems with solutions worked out in step-by-step detail.

Titles on the current list include:

3000 Solved Problems In Calculus
2500 Solved Problems in College Algebra and Trigonometry
2500 Solved Problems in Differential Equations
2000 Solved Problems in Discrete Mathematics
3000 Solved Problems in Linear Algebra
2000 Solved Problems in Numerical Analysis
3000 Solved Problems in Precalculus

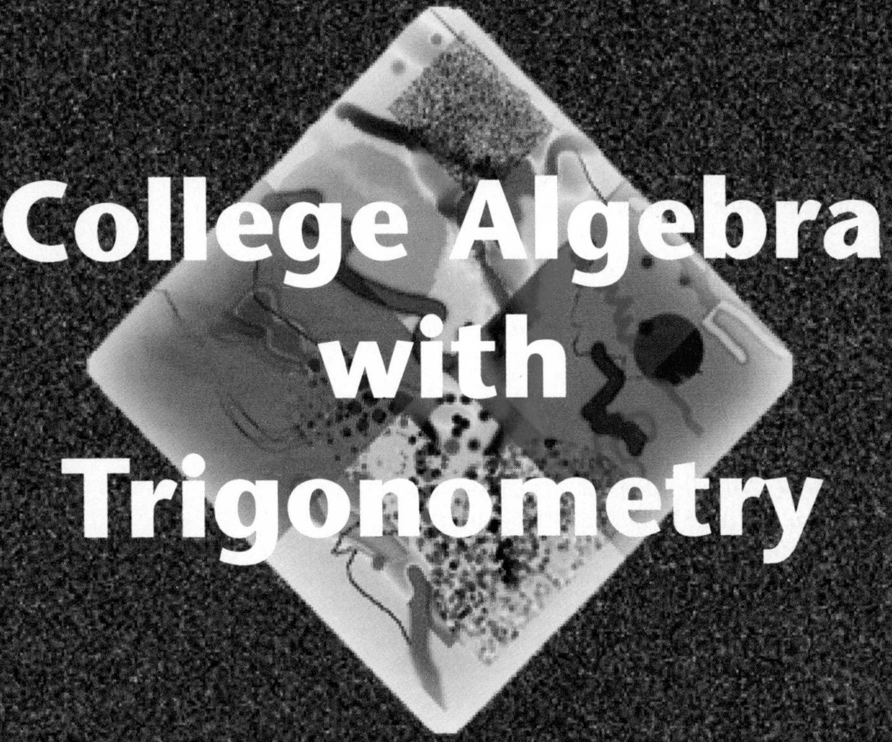

College Algebra with Trigonometry

Linda Gilbert and Jimmie Gilbert
University of South Carolina at Spartanburg

McGraw-Hill, Inc.

New York St. Louis San Francisco Auckland Bogotá Caracas
Lisbon London Madrid Mexico City Milan Montreal
New Delhi San Juan Singapore Sydney Tokyo Toronto

College Algebra with Trigonometry

This book is printed on acid-free paper.

1 2 3 4 5 6 7 8 9 0 VNH VNH 9 0 9 8 7 6 5 4

ISBN 0-07-023586-4

This book was set in Minion by York Graphic Services, Inc.
The editors were Michael Johnson and Margery Luhrs;
the design was done by A Good Thing, Inc.;
the production supervisor was Phil Galea.
The photo researcher was Elyse Rieder.
The cover art was done by Shane Canon.
Von Hoffmann Press, Inc., was printer and binder.

Library of Congress Cataloging-in-Publication Data

Gilbert, Linda.
College algebra with trigonometry / Linda Gilbert and Jimmie Gilbert.
p. cm.
Includes index.
ISBN 0-07-023586-4
1. Algebra. 2. Trigonometry. I. Gilbert, Jimmie, (date). II. Title.
QA154.2.G5224 1995
512'.13—dc20 94-13265

ABOUT THE AUTHORS

Linda and **Jimmie Gilbert** are professors of mathematics at the University of South Carolina at Spartanburg. Linda holds a Ph.D. in mathematics from Louisiana Tech University, and Jimmie received his Ph.D. in mathematics from Auburn University. Both taught at Louisiana Tech University before coming to USC-Spartanburg in 1986. They have written the companion text *College Algebra* (McGraw-Hill, 1995) and several other college mathematics textbooks, including *Elements of Modern Algebra* (PWS-Kent, 1992), now in its third edition, and a forthcoming *Advanced Linear Algebra* (Academic Press).

The Gilberts not only teach and write together, but they also enjoy their children, their garden, and their fishing.

ABOUT THE COVER

Glance at the front cover and you will see a pleasing design. To thoroughly appreciate what's there requires looking deeper, in keeping with the **critical thinking** theme of this text. This computer generated art, called an *auto-stereogram* or a *random dot stereogram,* lends itself well to this theme. Some familiar three-dimensional mathematical objects are embedded in this graphic as hidden images. The viewer who takes time to delve deeper into the third dimension is rewarded with a visual experience. Throughout this text, the student is enticed to delve deeper into new dimensions of mathematics through critical and logical thought and is rewarded with a deeper understanding.

The technology behind an autostereogram is surprisingly simple. The hidden image is formed by overlapping patterns of dots generated by a computer program. The brain detects the pattern and sends a message to the eye that allows it to focus in such a way as to reveal the hidden image. There are two popular viewing techniques. With the first, simply stare at a reflection (your own or some object positioned behind you) until the hidden image comes into focus. With the second method, hold the cover close (very close) to your face, let it blur, and slowly (very slowly) move it away from your face. At some point, the three-dimensional effect becomes evident and then the image itself. A black and white version of the image hidden in the front cover of this book appears below.

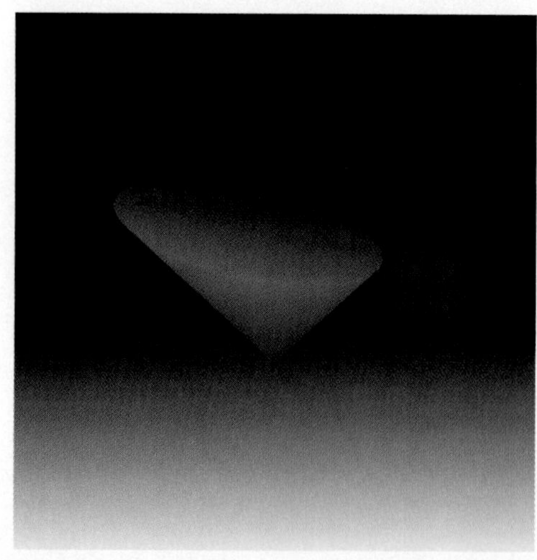

CONTENTS

Preface xiii

1 Fundamentals

1.1 Real Numbers 2
1.2 Ordering and Absolute Value 12
1.3 Integral Exponents and Scientific Notation 21
1.4 Polynomials 32
1.5 Factoring Polynomials 38
1.6 Rational Expressions 46
1.7 Radicals and Rational Exponents 57
1.8 Operations with Radicals 71
1.9 Complex Numbers 80

CHAPTER REVIEW **87**
Summary of Important Concepts and Formulas 87
Critical Thinking: Find the Errors 90
Review Problems for Chapter 1 92
Encore 94

2 Equations and Inequalities in One Variable

2.1 Solutions of Equations 96
2.2 Applications 104
2.3 Linear and Absolute Value Inequalities 115
2.4 Quadratic Equations 123
2.5 The Quadratic Formula 131
2.6 More on Quadratic Equations 139
2.7 Nonlinear Inequalities in One Real Variable 145

CHAPTER REVIEW **153**
Summary of Important Concepts and Formulas 153
Critical Thinking: Find the Errors 154
Review Problems for Chapter 2 155
Encore 157

3 Functions

3.1 Graphing in Two-Dimensional Coordinates 161
3.2 Functions 171
3.3 The Algebra of Functions 182
3.4 Linear Functions 187
3.5 Quadratic Functions 197

CHAPTER REVIEW **205**
Summary of Important Concepts and Formulas 205
Critical Thinking: Find the Errors 206
Review Problems for Chapter 3 206
Encore 209

Cumulative Test Chapters 1–3 211

4 Graphing Techniques

4.1 The Parabola and the Circle 214
4.2 Symmetry and Translations 224
4.3 The Ellipse 231
4.4 The Hyperbola 237
4.5 Translations of Ellipses and Hyperbolas 242
4.6 Inverse Functions 248
4.7 Variation 256

CHAPTER REVIEW **266**
Summary of Important Concepts and Formulas 266
Critical Thinking: Find the Errors 267
Review Problems for Chapter 4 268
Encore 270

5 Polynomials and Rational Functions

5.1 Synthetic Division 274
5.2 The Remainder and Factor Theorems 280
5.3 The Fundamental Theorem of Algebra and Descartes' Rule of Signs 287
5.4 Rational Zeros 295
5.5 Approximation of Real Zeros 307
5.6 Graphs of Polynomial Functions 311
5.7 Graphs of Rational Functions 321

CHAPTER REVIEW **335**
Summary of Important Theorems and Concepts 335
Critical Thinking: Find the Errors 336
Review Problems for Chapter 5 337
Encore 338

6 Exponential and Logarithmic Functions

6.1 Exponential Functions 341
6.2 The Natural Exponential Function 350
6.3 Logarithmic Functions 357
6.4 Logarithmic and Exponential Equations 365

CHAPTER REVIEW **373**
Summary of Important Concepts and Formulas 373
Critical Thinking: Find the Errors 373
Review Problems for Chapter 6 374
Encore 375

Cumulative Test Chapters 3–6 377

7 The Trigonometric Functions

7.1 Angles and Triangles 380
7.2 Radian Measure 388
7.3 Trigonometric Functions of Angles 398
7.4 Some Fundamental Properties 410
7.5 Values of Trigonometric Functions 414
7.6 Right Triangles 419
7.7 Trigonometric Functions of Real Numbers 434
7.8 Sinusoidal Graphs 441
7.9 Other Trigonometric Graphs 456

CHAPTER REVIEW **469**
Summary of Important Concepts and Formulas 469
Critical Thinking: Find the Errors 470
Review Problems for Chapter 7 472
Encore 475

8 Analytic Trigonometry

8.1 Verifying Identities 478
8.2 Cosine of the Sum or Difference of Two Angles 489
8.3 Identities for the Sine and Tangent 497
8.4 Double-Angle and Half-Angle Identities 504
8.5 Product and Sum Identities 513
8.6 Trigonometric Equations 517
8.7 Inverse Trigonometric Functions 527

CHAPTER REVIEW **538**
Summary of Important Concepts and Formulas 538
Critical Thinking: Find the Errors 539
Review Problems for Chapter 8 541
Encore 543

9 Additional Topics in Trigonometry

9.1 The Law of Sines 546
9.2 The Law of Cosines 558
9.3 The Area of a Triangle 567
9.4 Vectors 573

9.5 Trigonometric Form of Complex Numbers 587
9.6 Powers and Roots of Complex Numbers 596
9.7 Polar Coordinates 601

CHAPTER REVIEW 610
Summary of Important Concepts and Formulas 610
Critical Thinking: Find the Errors 610
Review Problems for Chapter 9 612
Encore 613

Cumulative Test Chapters 3–9 616

10 Systems of Equations and Inequalities

10.1 Systems of Linear Equations in Two Variables 618
10.2 Systems of Linear Equations in Three Variables 629
10.3 Systems Involving Nonlinear Equations 637
10.4 Systems of Inequalities 646
10.5 Linear Programming 653

CHAPTER REVIEW 661
Summary of Important Concepts and Formulas 661
Critical Thinking: Find the Errors 661
Review Problems for Chapter 10 663
Encore 665

11 Matrices and Determinants

11.1 Notation and Definitions 668
11.2 Matrix Multiplication 676
11.3 Solution of Linear Systems by Matrix Methods 684
11.4 Calculation of Inverses 693
11.5 Determinants 701
11.6 Evaluation of Determinants 709
11.7 Cramer's Rule 716

CHAPTER REVIEW 722
Summary of Important Concepts and Formulas 722
Critical Thinking: Find the Errors 723
Review Problems for Chapter 11 725
Encore 726

Cumulative Test Chapters 3–11 729

12 Further Topics

12.1 Sequences and Series 732
12.2 Arithmetic Sequences 738
12.3 Geometric Sequences 747
12.4 Infinite Geometric Sequences 753
12.5 The Binomial Theorem 759
12.6 Mathematical Induction 766
12.7 Permutations 772
12.8 Combinations 783
12.9 Probability 789

CHAPTER REVIEW 798
Summary of Important Concepts and Formulas 798
Critical Thinking: Find the Errors 799
Review Problems for Chapter 12 800
Encore 802

Appendix

A.1 Table Evaluation of Logarithms A-1

A.2 Table Evaluation of Trigonometric Functions A-6

Tables

I Trigonometric Functions—Degrees and Minutes or Radians A-11

II Trigonometric Functions—Decimal Degrees A-14

III Common Logarithms A-20

IV Natural Logarithms and Powers of *e* A-22

Answers to Selected Exercises AN-1

Index I-1

PREFACE

This book is written as a text for the traditional topics in college algebra and trigonometry at the freshman level. In response to the AMATYC and MAA guidelines and the NCTM standards, several features involving writing, critical thinking, discussion, exploration, and technology appear in the book. In order to make the material more relevant to students, an abundance and variety of applications have been placed in the examples as well as in the exercises. These applications are presented utilizing visualization techniques that involve a generous quantity of graphs, drawings, photographs, and tables. The knowledge and maturity that an average student has after two courses of algebra in high school are adequate for the study of this text. One year of geometry is a highly desirable prerequisite but is not absolutely necessary.

ORGANIZATION

The first two chapters furnish a review of fundamental topics from intermediate algebra such as the real number system, absolute value, factoring, radicals, and quadratic equations. This material is presented in a manner so that it can be covered thoroughly, briefly reviewed, or skipped completely, depending on the needs of the students. It seems desirable to include this material, even if it serves only as a reference source to refresh the students' memory.

The contents of Chapters 2–11 are fairly standard in college algebra and trigonometry courses. Matrices and determinants are presented in the same chapter, but either topic can be taught without the other. Inequalities, which always seem to be troublesome, are treated with exposures on four different occasions in Sections 1.2, 2.3, 2.7, and 10.4.

Inverse functions (Section 4.6) are introduced after conic sections (Sections 4.1–4.5), at a point where more knowledge of graphing is available. Experience in graphing halves of conics enables the student to more easily visualize the concept of an inverse function.

Since a familiarity with exponential and logarithmic functions is needed in almost all majors, Chapter 6 is devoted to these functions. Section 6.2 is a special section on the natural exponential function with applications so important to engineering, science, and business students. Chapter 12, *Further Topics,* contains several supplemental topics. Selections from these topics can be used to stimulate interest in other areas of mathematics such as probability, statistics, and calculus.

TECHNOLOGY

A reasonable sampling of problems that require the use of a scientific calculator or a graphing device is included, and these problems are flagged in the margin with either a calculator icon  or a graphing device icon 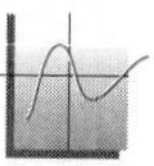. We have left the choice of incorporating graphing calculator or computer technology into the course entirely to the instructor. Exercises are included that will enable a student to develop skill and knowledge in the use of a graphing calculator or computer if the opportunity is available. Also we have indicated how this technology greatly enhances the study of mathematics through visualization and geometric interpretation. Notes that point out the usefulness of technology are placed in the margins of the book at appropriate points to invite further exploration. These marginal notes are also flagged with an icon 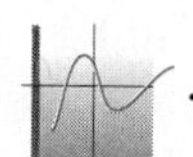.

The exercises requiring trigonometric function values or logarithms may be worked either with a calculator or with tables. Explanations are included in the body of the text for the calculator approach, and the use of tables is explained in the appendix. Tables are provided for common logarithms, natural logarithms, powers of *e*, and trigonometric functions in degrees and minutes, in decimal degrees, and in radians or real numbers. The choice of computational methods is left to the instructor.

ENRICHMENT

Critical Thinking Problems of several types are included throughout the book. A selection of these problems is found at the end of each exercise set and marked in the margin. Most require the student to *write* mathematics or to *discuss* topics presented in the text, and some encourage the student to *explore* beyond the text presentation. Some of the *Critical Thinking: Writing* problems involve writing a brief summary paragraph, while others involve library research and a short essay. Some of the *Critical Thinking: Discussion* problems can be writing assignments, while others are more suitable for group discussion. Some of the *Critical Thinking: Exploration* problems give the students an opportunity to think for themselves, while others require exploration with a graphing device. Problems labeled *Critical Thinking: Find the Errors* are found at the end of each chapter. Each is followed by a worked-out solution that contains one or more errors. The challenge for the student is to think critically and find the errors. The errors are marked and corrected in the answer section of the book. We've discovered that this type of critical thinking problem helps the student recognize and avoid common pitfalls.

In each set of exercises, **Core Problems** are marked with check marks ✓ in the margin. Assignment of these problems will ensure minimal coverage of the core concepts in that section. Additional problems may be assigned to provide more practice as needed. Problems that utilize technology or critical thinking skills may be assigned to enhance coverage.

Each chapter ends with an interesting, relevant, and often timely application that indicates the usefulness of the mathematics presented in the chapter. These concluding items are labeled **Encore.**

PEDAGOGY

The following features are intended to make the study of this book a successful and rewarding experience for the student.

- **Chapter Openers** convey the significance and applications of the material in each chapter.
- **Illustrations** in which equations and graphs match in color enhance the visual aspect of learning.
- **Four-Color Format** that highlights key equations, rules, and procedures for problem solving promotes readability and assimilation of the material.
- **Examples** are labeled for easy reference.
- **Exercises** (at the end of each section) and review problems (at the end of each chapter) are plentiful, with the exercises odd-even paired and arranged in order of increasing difficulty to allow a gradual growth in skills.
- **Labeled Applications** appear in abundance in the examples as well as the exercises in order to exhibit the relevance of the material to the students.
- **Practice Problems** are integrated in each section to provide an intermediate step between the examples and exercises. The student will find worked-out solutions for the practice problems at the end of each section.
- **Core Problems** are marked in each exercise set. When a student has worked these problems, all the basic concepts of that section have been encountered.
- **Technology** can be incorporated in the course by assigning problems that require the use of a graphing device or scientific calculator 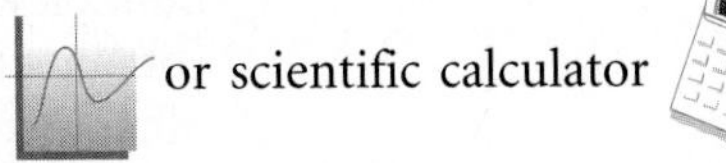 .
- **Marginal Notes** point out the usefulness of technology as a learning tool.
- **Critical Thinking Problems** of three types, *Writing, Discussion,* and *Exploration* (at the end of each section) and *Find the Errors* (in each chapter review) promote deeper understanding of the concepts.
- **A Chapter Review** at the end of each chapter includes a *Summary of Important Concepts and Formulas,* a selection of *Critical Thinking Problems: Find the Errors,*

and a generous supply of *Review Problems* to provide a valuable synopsis of the chapter.

- **An Encore** concludes each chapter to stimulate the student's interest.
- **Cumulative Tests** appear at the end of Chapters 3, 6, 9 and 11 to furnish continuous reinforcement.
- **Answers** for odd-numbered exercises, all review problems, and all cumulative test problems are found in the answer section at the back of the book to expedite student self-checking.

SUPPLEMENTS

Student Supplements

1. A STUDENT'S SOLUTIONS MANUAL is available through your bookstore. The manual contains worked-out solutions and answers to the odd-numbered end-of-section exercises and all review exercises.
2. The STUDENT TUTORIAL reinforces topics and provides opportunities to review concepts and to practice problem solving. It requires virtually *no* computer training and is available for IBM, IBM-compatible, and Macintosh computers.
3. The INTERACTIVE VIDEODISC TUTORIAL combines video and software to create illustrations of real-world applications, to model the problem-solving process, to present worked-out examples, and to provide an unlimited number of problems for students to solve.

Instructor Supplements

1. The INSTRUCTOR'S SOLUTIONS MANUAL contains an answer section as well as detailed solutions to all exercises and cumulative tests found in the text.
2. The INSTRUCTOR'S RESOURCE MANUAL contains multiple-choice and open-ended chapter tests, multiple-choice and open-ended cumulative tests, multiple-choice and open-ended final tests, and an answer section. It also contains teaching hints and suggestions, transparency masters, and various other elements to assist the instructor.
3. The INTERACTIVE VIDEODISC TUTORIAL combines video and software to create illustrations of real-world applications, to model the problem-solving process, to present worked-out examples, and to provide an unlimited number of problems for students to solve.
4. Course VIDEOTAPES are available for use through your institution.
5. THE PROFESSOR'S ASSISTANT is a unique computerized test generator available to instructors. This system allows the instructor to create tests

using algorithmically generated test questions or those from a standard test bank. This testing system enables the instructor to choose questions either manually or randomly by section, question type, difficulty level, and other criteria. This system is available for IBM, IBM-compatible, and Macintosh computers.

6. A PRINTED AND BOUND TEST BANK is also available. This is a hard-copy listing of the questions found in the standard test bank.

For further information about these supplements, please contact your local college division sales representative.

ACKNOWLEDGMENTS

We wish to acknowledge with thanks the helpful suggestions made in reviews by the following persons:

Karen Barker, Indiana University, South Bend
James Brasel, Phillips County Community College
Carl Cuneo, Essex Community College
Michael Ecker, Pennsylvania State University, Wilkes-Barre
Iris Fetta, Clemson University
Gilbert French, Manatee Community College
Rita Hussung, College of Mount Saint Joseph
Margaret Karpinski, Holy Name College
Debra Landre, San Joaquin Delta College
Barbara Sausen, Fresno City College
M. J. Still, Palm Beach Community College

A special thank you to Richard Semmler of Northern Virginia Community College for checking all the text exercises and for preparing the Student's Solutions Manual and also to all college algebra survey respondents. Thanks to Shane Canon for supplying the art for the cover.

We also wish to express our appreciation to Michael Johnson, mathematics editor at McGraw-Hill, for the valuable advice and encouragement that he gave us in planning this project. Thanks go to Nancy Evans for her helpful suggestions concerning the features incorporated in the book, to Karen Minette for all her efforts in coordinating the supplements package, and to Margery Luhrs for her excellent editorial and production supervision. Finally, we extend thanks to Charles Stavely, a colleague and friend, for his special insight that made all of us look beyond the surface.

Linda Gilbert
Jimmie Gilbert

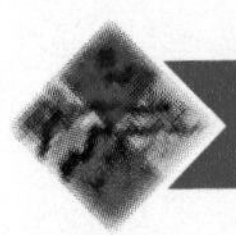

COLLEGE ALGEBRA with TRIGONOMETRY

Fundamentals

The importance of mathematics rests in large part on the fact that its language, formulas, and procedures can often be used to analyze a real-world problem and produce a solution to that problem. A mathematical description of a real-world problem is called a **mathematical model** of the problem. In order to work effectively with these mathematical models, it is necessary to possess some skill and knowledge of the basic tools used in building them: polynomials, rational expressions, radicals, exponents, and complex numbers. This material provides a basis for the solution of equations in Chapter 2.

1.1 REAL NUMBERS

Since much of our work in algebra involves sets of real numbers, it is frequently convenient to use some of the terminology and notation for sets. The word **set** refers to a collection of objects where it is possible to determine whether or not a certain object is in the set. The individual objects in the set are called the **members** of the set, or the **elements** of the set.

The simplest way to describe a set is by listing the members of the set. For example, we write

$$T = \{2, 4, 6, 8\}$$

to indicate that T is the set that contains the numbers 2, 4, 6, and 8 and has no other members.

The use of braces to indicate a set is standard notation even when all the elements cannot be listed. Thus the notation

$$E = \{2, 4, 6, 8, \ldots\}$$

indicates that E is the set of all positive even integers. Another standard way to describe a set is to use braces and indicate the property that is the qualification for membership in the set. In this manner, the same set E is indicated as

$$E = \{x \mid x \text{ is a positive even integer}\}.$$

This notation is called **set-builder notation** and is read as "E is the set of all x such that x is a positive even integer." The vertical slash is taken as shorthand for "such that."

There are other shorthand symbols that are convenient to use in connection with sets. We write $x \in S$ as shorthand for the phrase "x is an element of S" or "x is in S." If x is not an element of S, we write $x \notin S$. With T as above, we write $6 \in T$ and $10 \notin T$.

If it happens that every element of the set A is an element of the set B, then A is called a **subset** of B, and we write $A \subseteq B$ or $B \supseteq A$. With the sets T and E above, $T \subseteq E$. We write $A \not\subseteq B$ or $B \not\supseteq A$ to indicate that A is not a subset of B, and we write $A = B$ to mean that A and B are composed of exactly the same elements.

There are two operations on sets that are frequently used in mathematics. One of these operations is that of forming the union of two sets. If A and B are sets, the **union** of A and B is that set $A \cup B$ given by $\mathbf{A \cup B = \{x \mid x \in A \text{ or } x \in B\}}$. That is, $A \cup B$ consists of all those elements x that are an element of A, an element of B, or an element of both A and B.

Example 1 • The Union of Sets

a. If $A = \{2, 4, 6\}$ and $B = \{1, 3, 6\}$, then $A \cup B = \{1, 2, 3, 4, 6\}$.
b. If $A = \{1, 2, 3, 4\}$ and $B = \{2, 4\}$, then $A \cup B = \{1, 2, 3, 4\} = A$.
c. $\{1, 2, 3\} \cup \{4, 5, 6\} = \{1, 2, 3, 4, 5, 6\}$. □

The other operation frequently used is that of forming the intersection of sets. For sets A and B, the **intersection** of A and B is the set $A \cap B$ given by $\mathbf{A \cap B = \{x \mid x \in A \text{ and } x \in B\}}$. Thus $A \cap B$ consists of those elements that are in both A and B.

Example 2 • The Intersection of Sets

a. If $A = \{2, 4, 6, 8\}$ and $B = \{1, 3, 6, 2\}$, then $A \cap B = \{2, 6\}$.

b. If $A = \{1, 2, 3, 4\}$ and $B = \{2, 4\}$, then $A \cap B = \{2, 4\} = B$. □

The empty set is the set that has no members. The **empty set** is denoted by $\emptyset$ or { } and is regarded as a subset of every set.

The operations of union and intersection can be applied in repetition. For instance, one might form the intersection of A and B, obtaining $A \cap B$, and then form the union of this set with C. The resulting set is denoted by $(A \cap B) \cup C$, where the parentheses indicate the order in which the operations are to be performed.

Example 3 • Using Symbols of Grouping

Let $A = \{1, 2, 3, 4\}$, $B = \{3, 4, 5, 6\}$, and $C = \{2,4, 6\}$. Then

$$\begin{aligned}(A \cap B) \cup C &= \{3, 4\} \cup \{2, 4, 6\} && \text{since } A \cap B = \{3, 4\}\\ &= \{2, 3, 4, 6\}\end{aligned}$$

and

$$\begin{aligned}A \cap (B \cup C) &= \{1, 2, 3, 4\} \cap \{2, 3, 4, 5, 6\} && \text{since } B \cup C = \{2, 3, 4, 5, 6\}\\ &= \{2, 3, 4\}.\end{aligned}$$

This illustrates that the way we position the symbols of grouping in an expression may affect the resulting set, for we have $(A \cap B) \cup C \neq A \cap (B \cup C)$. □

Throughout this book the set of real numbers is denoted by $\mathcal{R}$. The special subsets N, W, Z, Q, and H are described in Figure 1.1 on the next page, and their relationships to each other are shown there.

◆ **Practice Problem 1** ◆

In which of the sets N, W, Z, Q, $\mathcal{R}$, and H does each of the following belong?

a. -2 b. $\frac{17}{3}$ c. $\sqrt{3}$ d. $0.353535\ldots$

One of the fundamental properties of the real numbers is that there is a relation of **equality** defined in $\mathcal{R}$. The familiar equality symbol $=$ for real numbers is used to make a statement that two expressions represent the same real number. Such a statement is called an **equation.**

In later work with equations, we will frequently use the substitution property

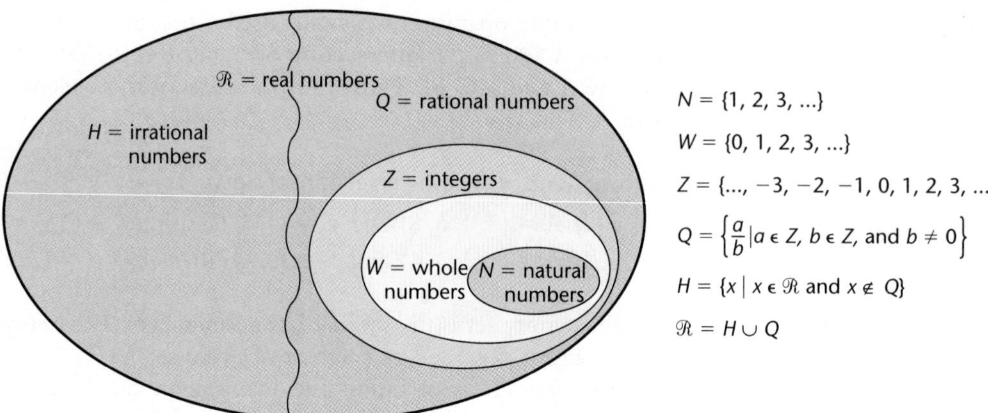

Figure 1.1 Special subsets of the real numbers

of equality. The **substitution property** states that if $a = b$, then the value of any computation involving a is unchanged when a is replaced by b.

Two special cases of the substitution property are important enough to require extra attention. In these two statements, a, b, and c represent arbitrary real numbers:

Addition Property of Equality

If $a = b$, then $a + c = b + c$.

Multiplication Property of Equality

If $a = b$, then $ac = bc$.

The addition property allows us to add the same real number to both sides of an equation. The multiplication property allows us to multiply both sides of an equation by the same real number.

In the study of arithmetic, skill is developed in performing the operations of addition, subtraction, multiplication, and division with real numbers. The computations performed with these operations are based on the following list of basic properties. These basic properties are so familiar that we normally use them without being consciously aware of their formal statements.

Basic Properties for the Real Numbers

1. Addition properties: The operation of addition defined in $\mathcal{R}$ has the following properties:
   a. Closure property: $a + b \in \mathcal{R}$ for all $a \in \mathcal{R}$ and $b \in \mathcal{R}$.
   b. Associative property: $a + (b + c) = (a + b) + c$ for all $a, b, c \in \mathcal{R}$.

c. Additive identity: There is a real number 0 such that $a + 0 = 0 + a = a$ for all $a \in \mathcal{R}$.

d. Additive inverses: For each $a \in \mathcal{R}$, there is an element $-a \in \mathcal{R}$ such that $a + (-a) = (-a) + a = 0$.

e. Commutative property: $a + b = b + a$ for all a and $b \in \mathcal{R}$.

2. Multiplication properties: The operation of multiplication defined in $\mathcal{R}$ has the following properties:

a. Closure property: $a \cdot b \in \mathcal{R}$ for all $a \in \mathcal{R}$ and $b \in \mathcal{R}$.

b. Associative property: $a \cdot (b \cdot c) = (a \cdot b) \cdot c$ for all $a, b, c \in \mathcal{R}$.

c. Multiplicative identity: There is a real number 1 such that $1 \neq 0$ and $a \cdot 1 = 1 \cdot a = a$ for all $a \in \mathcal{R}$.

d. Multiplicative inverses: For each *nonzero* $a \in \mathcal{R}$, there is an element

$$\frac{1}{a} \in \mathcal{R} \text{ such that } a\left(\frac{1}{a}\right) = \left(\frac{1}{a}\right)a = 1.$$

e. Commutative property: $a \cdot b = b \cdot a$ for all $a, b \in \mathcal{R}$.

3. Distributive property: $a \cdot (b + c) = (a \cdot b) + (a \cdot c)$ for all $a, b, c \in \mathcal{R}$.

Example 4 • Identifying the Basic Properties

Each of the following statements is a true statement concerning addition or multiplication of real numbers. To the right of each statement, basic properties that justify the statement are given. Letters represent arbitrary real numbers.

a. $2 + (3 + x) = 5 + x$	associative property, +
b. $5(\frac{1}{5}) = 1$	multiplicative inverse
c. $a(x + 1) = ax + a \cdot 1$	distributive property
$= ax + a$	multiplicative identity
d. $ax + ay = ay + ax$	commutative property, +

□

◆ **Practice Problem 2** ◆

Identify the basic property that justifies each statement.

a. $ax + ay = ax + ya$

b. $x + y = 1 \cdot x + 1 \cdot y$

c. $2 + \pi \in \mathcal{R}$

The **difference,** $a - b$, of real numbers a and b is defined by

$$a - b = a + (-b).$$

The operation that combines a and b to yield $a - b$ is called **subtraction.**

We note the different uses of the minus sign in the equation $a - b = a + (-b)$. In $a - b$, it represents the operation of subtraction; in $a + (-b)$, the term $-b$ denotes the real number that yields zero when added to b.

Just as the operation of addition is used to define subtraction, the operation of multiplication is used to define division. Let a and b be real numbers with $b \neq 0$. The quotient $\frac{a}{b}$ (or $a \div b$, or a/b) is defined by

$$\frac{a}{b} = a \cdot \left(\frac{1}{b}\right), \text{ where } b \neq 0.$$

The operation that combines a and b to yield a/b when $b \neq 0$ is called **division.**

Note that division by zero is excluded. We do not go into the details here, but any attempt to extend the operation of division to include division by zero leads invariably to a conflict with the basic properties.

The operations of addition, multiplication, subtraction, and division are frequently referred to as the **four fundamental operations.** In many instances, it is necessary to indicate a sequence of operations to be performed. Such a situation could give rise to ambiguity. For example, $3 + 5 \cdot 6$ could mean $(8)(6) = 48$, or it could mean $3 + 30 = 33$. Rules concerning the order of operations and the use of symbols of grouping have been established to govern these situations. The rules commonly used are stated here, but the algebraic operating systems of some computers and calculators may vary slightly from these.

Order of Operations and Symbols of Grouping

1. In an expression involving powers and fundamental operations we proceed as follows:
   a. All powers are evaluated first.
   b. All multiplications and divisions are performed next, from left to right.
   c. All additions and subtractions are then performed, from left to right.
2. Parentheses, ( ), brackets, [ ], or braces, { }, may be used to indicate the order of operations, with the understanding that the innermost symbols of grouping are removed first. That is, the innermost operations are performed first.

According to the first of these rules, the correct evaluation of $3 + 5 \cdot 6$ is

$$3 + 5 \cdot 6 = 3 + 30 = 33.$$

More generally, $3 + 5x$ indicates the number obtained by first multiplying x by 5 and then adding 3 to the product. The following example illustrates the removal of symbols of grouping.

Example 5 • Evaluation That Involves Symbols of Grouping

If $x = 8$ and $y = 2$, find the value of each of the following.

a. $(6x + 8y) \div (x - 3y)$

b. $\{[(6x + 8y) \div x] - 3\} \cdot y$

Solution

a. $$\begin{aligned}(6x + 8y) \div (x - 3y) &= (6 \cdot 8 + 8 \cdot 2) \div (8 - 3 \cdot 2)\\ &= (48 + 16) \div (8 - 6)\\ &= 64 \div 2\\ &= 32\end{aligned}$$

b. $$\begin{aligned}\{[(6x + 8y) \div x] - 3\} \cdot y &= \{[(6 \cdot 8 + 8 \cdot 2) \div 8] - 3\} \cdot 2\\ &= \{[(48 + 16) \div 8] - 3\} \cdot 2\\ &= \{[64 \div 8] - 3\} \cdot 2\\ &= \{8 - 3\} \cdot 2\\ &= 10\end{aligned}$$ □

A letter that is used to refer to an arbitrary element in a given set of numbers is called a **variable.** In the list of basic properties for real numbers, the letters a, b, and c are variables representing arbitrary elements from the set of all real numbers.

In the expression $4xy + 5$, the quantities $4xy$ and 5 are called the terms of the expression. In any sum the quantities that are to be added are referred to as the **terms** of the sum. A term that represents a single real number is called a **constant term,** or simply a **constant.** The 5 in $4xy + 5$ is a constant. In the term $4xy$, we refer to xy as the **variable part** of the term and to 4 as the **constant part** of the term. The constant part of a term involving a variable is called the **coefficient** of the term.

Terms that have exactly the same variable part are called **like terms,** and they can be added or subtracted by using the distributive property. For example, we can write

$$8xy + 6xy = (8 + 6)xy = 14xy$$

and

$$8xy - 6xy = (8 - 6)xy = 2xy.$$

In each of these uses of the distributive property, we say that we have **combined like terms.**

Example 6 • Simplification That Involves Symbols of Grouping

Simplify the following expression by removing symbols of grouping and combining like terms.

$$2[4 - 3(1 + x)] + 5x - \{[2(2 - x) + 7] - (x - 1)\}$$

Solution

$$\begin{aligned}&2[4 - 3(1 + x)] + 5x - \{[2(2 - x) + 7] - (x - 1)\}\\ &\quad = 2[4 - 3 - 3x] + 5x - \{[4 - 2x + 7] - (x - 1)\}\\ &\quad = 2[1 - 3x] + 5x - \{11 - 2x - x + 1\}\\ &\quad = 2 - 6x + 5x - 12 + 3x\\ &\quad = 2x - 10\end{aligned}$$ □

◆ Practice Problem 3 ◆

Simplify by removing symbols of grouping and combining like terms.

$$3 - 2[x + 3(1 - x)] - [2 - (x + 3) - 3x]$$

The dividing bar in a fraction has the effect of a grouping symbol. In such cases, the numerator and denominator should be evaluated first, with the division performed last. For example,

$$17 + \frac{30 - 18}{10 - 4} = 17 + \frac{12}{6} = 17 + 2 = 19.$$

The following list is a summary of the basic formulas for working with fractions, or quotients. All these properties should be familiar from earlier courses.

Basic Formulas for Fractions

Let a, b, c, and d be real numbers with $b \neq 0$ and $d \neq 0$.

1. $\frac{a}{b} = \frac{c}{d}$ if and only if $ad = bc$ — equality
2. $\frac{ad}{bd} = \frac{a}{b}$ — fundamental principle of fractions
3. $\frac{a}{b} + \frac{c}{b} = \frac{a + c}{b}$ — addition
4. $\frac{a}{b} + \frac{c}{d} = \frac{ad + bc}{bd}$ — addition
5. $\frac{a}{b} \cdot \frac{c}{d} = \frac{ac}{bd}$ — multiplication
6. If b, c, and d are all nonzero, then

$$\frac{a/b}{c/d} = \frac{a}{b} \div \frac{c}{d} = \frac{a}{b} \cdot \frac{d}{c} = \frac{ad}{bc}.$$ — division

The fundamental principle is the one that is used to reduce a quotient of integers to lowest terms. A quotient of integers is in **lowest terms** if the only common factors in the numerator and denominator are 1 and -1. A fraction is reduced to lowest terms by dividing out all common factors different from 1 and -1.

Example 7 • Evaluating a Fractional Expression

If $a = 3$ and $b = 5$, find the value of the following expression, reduced to lowest terms.

$$\frac{b - a\left(2 - \dfrac{b-3}{b-7}\right)}{(b-a)\left(\dfrac{4a}{-b-1}\right)}$$

Solution

$$\begin{aligned}\frac{b - a\left(2 - \dfrac{b-3}{b-7}\right)}{(b-a)\left(\dfrac{4a}{-b-1}\right)} &= \frac{5 - 3\left(2 - \dfrac{5-3}{5-7}\right)}{(5-3)\left(\dfrac{4\cdot 3}{-5-1}\right)} \\ &= \frac{5 - 3\left(2 - \dfrac{2}{-2}\right)}{2\left(\dfrac{12}{-6}\right)} \\ &= \frac{5-3(3)}{2(-2)} \\ &= \frac{-4}{-4} \\ &= 1\end{aligned}$$

□

EXERCISES 1.1

Let $A = \{0, 2, 3, 4, 6, 9\}$, $B = \{0, 2, 4, 6, 8, 10\}$, and $C = \{3, 4, 5, 6\}$. Find each of the following. (See Examples 1–3.)

1. $A \cup C$
2. $A \cup B$
3. $C \cap A$
4. $C \cap B$
5. $(A \cap B) \cup C$
6. $(A \cup B) \cap C$
7. $A \cap (B \cup C)$
8. $C \cap (A \cup C)$
9. $(C \cap A) \cup (C \cap B)$
10. $(A \cup B) \cap (A \cup C)$

Let N be the set of natural numbers, W the set of whole numbers, Z the set of integers, Q the set of rational numbers, $\mathcal{R}$ the set of real numbers, and H the set of irrational numbers. In which of these sets does each of the following belong? (See Practice Problem 1.)

11. 1
12. 2/3
13. $-\pi$
14. 1.3
15. $-\sqrt{2}$
16. 0

17. 0.6666 . . .
18. $-135/222$
19. $22/7$
20. -1

Determine each of the following.

21. $N \cap Z$
22. $Q \cap H$
23. $Q \cup H$
24. $\mathcal{R} \cup H$
25. $Z \cap Q$
26. $Z \cap H$
27. $W \cap N$
28. $W \cap Z$
29. $W \cup Z$
30. $W \cup N$
31. $N \cup Q$
32. $Z \cup Q$

In Exercises 33–42, a true statement is made concerning addition or multiplication of real numbers. For each statement, identify the basic property that justifies the statement. (See Example 4 and Practice Problem 2.)

33. $3 + [4 + (-4)] = 3 + 0$
34. $4 + (-4) = 0$
35. $ax + ay = a(x + y)$
36. $ax + ay = xa + ya$
37. $(1)(3) + 4 = 3 + 4$
38. $(1)(3) + (1)(5) = (1)(3 + 5)$
39. $0 + (1)(5) = (1)(5)$
40. $0 + (1)(5) = 0 + 5$
41. $(-\sqrt{2})\left(\frac{1}{-\sqrt{2}}\right) = 1$
42. $3 + (4 + 5) = (3 + 4) + 5$

In Exercises 43–52, fill in the blank so that the statement is an application of the stated property. Variables represent real numbers.

43. $5 + 7 =$ _____ commutative property, +
44. $(3)(8) =$ _____ commutative property, ·
45. $x \cdot (y \cdot z) =$ _____ associative property, ·
46. $x + (y + z) =$ _____ associative property, +
47. $a +$ _____ $= a$ additive identity
48. $a($_____$) = a$ multiplicative identity
49. $a($_____$) = 1$ multiplicative inverse
50. $a +$ _____ $= 0$ additive inverse
51. If $x = y$, then $ax =$ _____ multiplication property of equality
52. If $x = y$, then _____ $= y + b$ addition property of equality

If $x = 3$, $y = 5$, and $z = -2$, find the value of each of the following expressions and write the answer in lowest terms. (See Examples 5 and 7.)

53. $x - 3/x + 1$
54. $1 - x(y - 1)$
55. $[15(x + 1) - (x + 1)(2 + y)] \div (3 + y)$
56. $[(5x + 2)(4) - 4(12 - y)] \div (2y - 2)$
57. $\{8x[3y + 2 - 2(2y - 7)] - 4[12x - 2(2y)]\} \div [y + 2 + x(14 - x)]$
58. $\{12z[5x - 3 + z(y - 2)] - 4z(5x - 7)\} \div [x + 4 + x(2y + 1)]$

59. $\{z[x - 6(z - x)] + x[z - 2y(3x + y)]\} \div (2xz)$
60. $\{x[y - 3(z - y)] + 2x[x - 2z(x - y)]\} \div (2x + z)$

If $x = 1.2$, $y = 4.1$, and $z = -2.7$, find the value of each of the following expressions and round your answers to four decimal places.

61–64

61. $[2y(x - y) + 4xy(x - 2y - 3z)] \div [-yz(2xy + z)]$
62. $[z(x + y - z) - 30z(x + 5z + 6)] \div [2y(xyz + 2)]$
63. $y - 1 - x\left(\dfrac{11-6x}{7 - 2x}\right)$
64. $\dfrac{-6x - (y + 2)(z + 4)}{9z - (y - 1)(-2y - z)}$

Simplify the following expressions by removing symbols of grouping and combining like terms. (See Example 6 and Practice Problem 3.)

65. $5(2r - 3) - 7(4 - r)$
66. $6(3r - 2) - 5(2 - r)$
67. $-2(a - b + 1) - (3a - 4b - 3)$
68. $-3(a - 2b + 4) - (2a - 5b - 7)$
69. $2[7 - 3(q - 2) + 5q]$
70. $4[6 - 2(q - 3) + 4q]$
71. $3[-(2 - p) - (-p + 3)]$
72. $2[-(4 - p) - (-2p + 5)]$
73. $-3[-2(x - 4) - (5x - 1)]$
74. $-4[-3(x - 1) - (2x - 5)]$
75. $2[5 - 3(y + 2)] - 3[2 - 4(y - 2)]$
76. $4[6 - 2(y - 3)] - 5[2 - 3(y + 2)]$
77. $3[2 - 4(a - b) + 7a]$
78. $4[5a - 2(a - 3b) - 7b]$
79. $2[3(a - b) - 2(a + 2b)]$
80. $3[4(a - 2b) - 6(2a + b)]$
81. $4\{2[3 - 2(x + 2)] - 4\}$
82. $3\{2[6 - 8(x + 3)] + 40\}$
83. $5\{2 - 3[(2y + 1) - (5y - 2)]\}$
84. $2\{6 - 2[(3y - 2) - (-2y - 1)]\}$
85. $6\{5 - 2[(x - 2y) - 3(2x - y)]\}$
86. $3\{8 - 3[(2x - 3y) - 5(x - 2y)]\}$
87. $2[3(4a - b) - 2(5a - 3b)] - \{5[3a - (b - 4a)] + b\}$
88. $4[2(7a - 2b) + 3(2a + b)] - \{6[4a - 3(b - 5a)] + 2b\}$
89. $2\{6 - [2 + 3(2p - r)]\} - 3\{4 - 2[3 - 2(4p - 3r)]\}$
90. $4\{7 - [4 - 2(5p - 3r)]\} - 6\{3 - [2 - 3(p + 2r)]\}$

Critical Thinking: Exploration and Writing

91. Does subtraction of real numbers have the associative property? That is, does $a - (b - c) = (a - b) - c$ for all real numbers a, b, and c? Explain why, or why not.

Critical Thinking: Exploration and Writing

92. Does division of nonzero real numbers have the associative property? Explain why, or why not.

Critical Thinking: Exploration and Writing

93. Does subtraction of real numbers have the commutative property? Give an explanation and an example that illustrates your decision.

Critical Thinking: Exploration and Writing

94. Does division of nonzero real numbers have the commutative property? Give an explanation and an example that illustrates your decision.

Critical Thinking: Writing

95. Explain the difference between $\varnothing$ and $\{\varnothing\}$.

◆ Solutions for Practice Problems

1. a. $Z, Q, \mathcal{R}$ b. $Q, \mathcal{R}$ c. $\mathcal{R}, H$ d. $Q, \mathcal{R}$
2. a. commutative property, ·
   b. multiplicative identity
   c. closure property, +
3. $$\begin{aligned} 3 - 2[x + 3(1 - x)] - [2 - (x + 3) - 3x] &= 3 - 2[x + 3 - 3x] - [2 - x - 3 - 3x] \\ &= 3 - 2[3 - 2x] - [-1 - 4x] \\ &= 3 - 6 + 4x + 1 + 4x \\ &= 8x - 2 \end{aligned}$$

1.2 ORDERING AND ABSOLUTE VALUE

One of the most useful aids in working with real numbers is the association between real numbers and points on a straight line. We start with a horizontal line that extends indefinitely in each direction as indicated by the arrowheads in Figure 1.2. A point is chosen, labeled with 0, and referred to as the **origin.** A unit of measure is chosen, and points successively 1 unit apart are located on the line in both directions from the origin. It is conventional to label the points to the right of 0 in succession with the positive integers 1, 2, 3, . . . , and those to the left of 0 are labeled successively with the negative integers $-1, -2, -3, \ldots$. Since any rational number can be represented as the quotient of two integers, rational numbers are then located on the line by using appropriate portions of the chosen unit of measure. For example, $\frac{5}{2} = 2.5$ is located two and one-half units to the right of 0, $-\frac{7}{4} = -1.75$ is located one and three-fourths units to the left of 0, and so on. An irrational number is characterized by the fact that its decimal representation is nonterminating and nonrepeating. Although these irrational numbers cannot be located exactly by use of their decimal representations, their location can be approximated to any desired degree of accuracy by using decimal approximations. In Figure 1.2, $\sqrt{2}$ is located using $\sqrt{2} \approx 1.41$, and $-\pi$ is located using $-\pi \approx -3.14$. (The symbol $\approx$ means "approximately equals.")

Figure 1.2 Location of real numbers

The association that has thus been established between real numbers and the points on the line is a one-to-one correspondence: each real number corresponds to exactly one point on the line, and each point on the line corresponds to one real number. After this correspondence between real numbers and points on the line has been made, the line is commonly referred to as a **number line.**

Ordering between real numbers is indicated by using the inequality symbols $>$ and $<$, read "greater than" and "less than," respectively. The notation $a > 0$ means that a lies to the right of 0 on the number line, or that a is *positive.* Similarly, $a > b$ means that a lies to the right of b, or that $a - b$ is *positive* (see Figure 1.3).

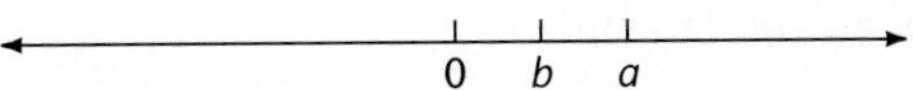

Figure 1.3 Ordering between real numbers

On the other hand, $a < b$ means that a lies to the left of b, or $a - b$ is *negative.* In particular, $a < 0$ means that a lies to the left of 0, or that a is *negative.* Whether we write $a < b$ or $b > a$ is often a matter of emphasis or personal preference.

The symbols $<$ and $>$ are often used in combination with equality to form compound statements. For example, either $x \leqq a$ or $x \leq a$ means that x is less than or equal to a. In this text, the symbol $x \leq a$ is used for this relation. Similarly, $x \geq a$ indicates that x is greater than or equal to a.

Example 1 • Use of Inequality Symbols

Some uses of $\leq$ and $\geq$ are illustrated by the following statements:

$-9 \leq -3$ since $-9 < -3$.

$8 \geq 8$ since $8 = 8$.

$x \leq 2$ means that x lies at 2 or to the left of 2 on the number line.

$x \geq 3$ means that x lies at 3 or to the right of 3 on the number line. □

The graph of $x \leq 2$ is obtained by placing a large dot at 2 and indicating a bold arrow to the left on the number line. This is illustrated in Figure 1.4. Similarly, $x \geq 3$ is represented as shown in Figure 1.5. The large dot at the end of the arrow is to indicate that the endpoint is included. If the endpoint is not to be included, we indicate this by an open circle at the end of the arrow. As an illustration, $x > 1$ is represented in Figure 1.6.

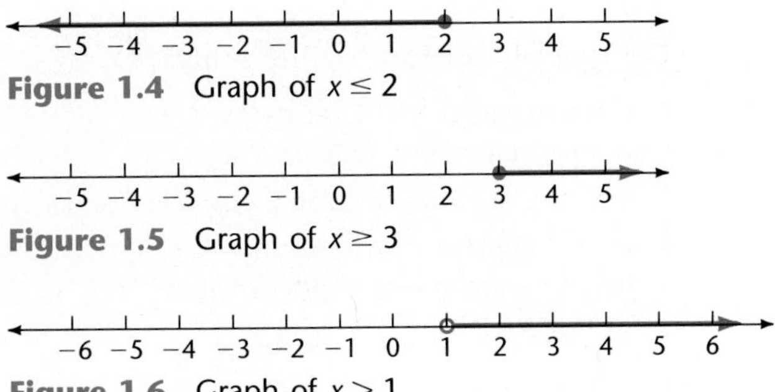

Figure 1.4 Graph of $x \leq 2$

Figure 1.5 Graph of $x \geq 3$

Figure 1.6 Graph of $x > 1$

Statements using the symbols $<$, $>$, $\leq$, and $\geq$ are called **inequalities.** At times it is convenient to use inequalities that involve two of these symbols. As an example, suppose we know that x is located either to the right of 2 or to the left of -1 on the number line. This can be indicated by

$$x > 2 \qquad \text{or} \qquad x < -1.$$

The graph is shown in Figure 1.7, where the open circles again indicate that the endpoints are not included.

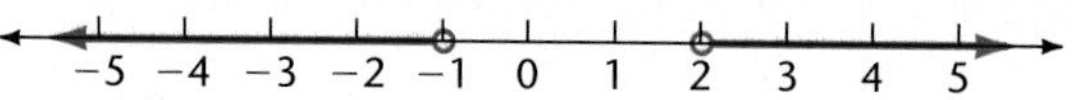

Figure 1.7 Graph of $x < -1$ or $x > 2$

As another illustration, suppose we wish to indicate that x lies between 2 and 4 on the number line. This can be indicated by

$$x > 2 \qquad \text{and} \qquad x < 4$$

or by

$$2 < x \qquad \text{and} \qquad x < 4.$$

This is represented in Figure 1.8 by a line segment from 2 to 4, not including the endpoints.

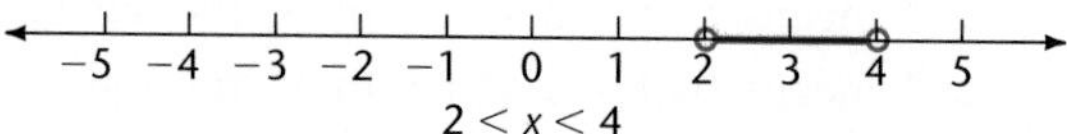

Figure 1.8 Graph of $2 < x$ and $x < 4$

Notice that in Figure 1.8 we have combined the compound statement $2 < x$ and $x < 4$ into the compact form $2 < x < 4$. This is conventional notation to indicate that x lies between 2 and 4. There are three important rules for use of this type of compact notation.

Compact Notation for Inequalities

1. Compact forms such as $a < x < b$ or $a > x > b$ are reserved for compound statements that use "and."
2. The two inequality symbols involved must have the same direction.
3. In a statement of the form $a < x < b$, the smaller number goes on the left. In a statement of the form $a > x > b$, the smaller number goes on the right.

Example 2 • Graphing Compound Inequalities

Figures 1.9 and 1.10 show the graphs of compound statements involving inequalities.

a. *Statement:* $-1 < x$ and $x \leq 2$
Compact form: $-1 < x \leq 2$
Representation:

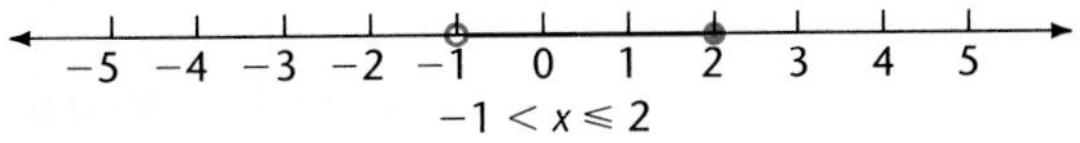

Figure 1.9 Graph of $-1 < x$ and $x \leq 2$

b. *Statement:* $x \leq 0$ or $x > 2$
No compact form: Not an "and" statement
Representation:

Figure 1.10 Graph of $x \leq 0$ or $x > 2$ □

◆ Practice Problem 1 ◆

a. Draw the graph of the statement $1 \leq x < 5$.
b. Write an inequality in compact form that has the graph shown in Figure 1.11.

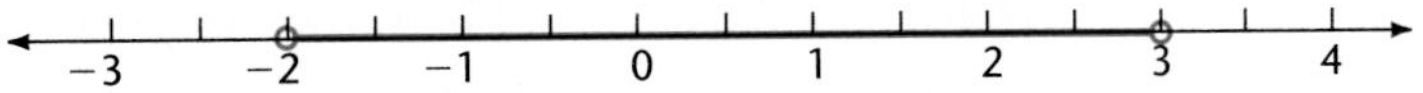

Figure 1.11 Graph for Practice Problem 1b

In later work, the following properties will be invaluable.

Addition and Multiplication Properties of Inequalities

For any real numbers a, b, and c,

1. If $a > b$, then $a + c > b + c$. — addition property
2. If $a > b$ and $c > 0$, then $ac > bc$.
3. If $a > b$ and $c < 0$, then $ac < bc$. — multiplication properties (2 and 3)

Similar statements can be made using the symbols $\geq$, $<$, or $\leq$.

Example 3 • Identifying Properties of Inequalities

Name the property of inequality that justifies the given statement.

a. If $-3x - 4 > 8$, then $-3x > 12$.
b. If $-3x > 12$, then $x < -4$.

Solution

a. The addition property allows us to add 4 to both sides of $-3x - 4 > 8$ to obtain $-3x > 12$.

b. Since $-\frac{1}{3} < 0$, the multiplication property allows us to multiply both sides of $-3x > 12$ by $-\frac{1}{3}$ and reverse the inequality to obtain $x < -4$. □

In describing distances between points on the number line, the concept of absolute value is essential.

Definition of Absolute Value

The **absolute value** of a real number a is denoted by $|a|$. It is defined by

$$|a| = \begin{cases} a, & \text{if } a \geq 0 \\ -a, & \text{if } a < 0 \end{cases}.$$

Note that $|a|$ is never negative. If $a \neq 0$, then $|a|$ is a positive number. It is also worth noting that in the case where $|a| = -a$, the value $-a$ signifies a positive number. That is, when a is a negative number, then $-a$ is a positive number. For example, if $a = -3$, then $-a = -(-3) = 3$ and $|a| = 3$.

Example 4 • Evaluation of Absolute Value Expressions

Express the following without absolute value symbols.

a. $|-5 + 4|$ b. $|-5| + |4|$ c. $|x - 1|$, if $x < 1$

Solution

a. $|-5 + 4| = |-1| = 1$

b. $|-5| + |4| = 5 + 4 = 9$

c. If $x < 1$, then $x - 1 < 0$ and $|x - 1| = -(x - 1) = 1 - x$. □

The following properties relate absolute value to the operations of multiplication, division, and addition.

Absolute Value and the Fundamental Operations

Let a and b be real numbers. Then

1. $|ab| = |a| \cdot |b|$, multiplication property
2. $\left|\dfrac{a}{b}\right| = \dfrac{|a|}{|b|}$, if $b \neq 0$, division property
3. $|a + b| \leq |a| + |b|$. the triangle inequality

The geometric interpretation of $|a|$ on the number line is that $|a|$ denotes the distance between 0 and a.

Example 5 • Absolute Value and Distance

a. $|2| = 2$ and 2 is located two units from 0.

b. $|-2| = 2$, and -2 is located two units from 0.

c. $|a| = |-a|$ for any real number a, since both a and $-a$ are located at distance $|a|$ from 0. □

When we think of absolute value in terms of distance on the number line, the following properties become evident.

Absolute Value and Distance

Let x and d be real numbers with $d > 0$. Then

1. $|x| < d$ if and only if $-d < x < d$,
2. $|x| > d$ if and only if either $x > d$ or $x < -d$.

Graphs that illustrate the connection between absolute value and distance are given in Figures 1.12 and 1.13.

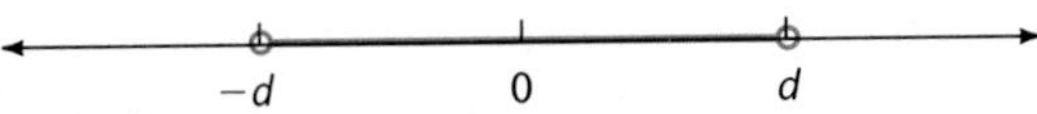

Figure 1.12 Graph of $|x| < d$

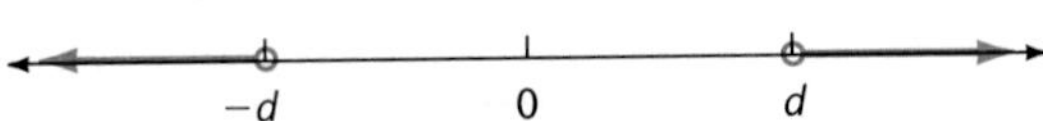

Figure 1.13 Graph of $|x| > d$

The properties of absolute value allow us to generalize the preceding discussion connecting absolute value and distance. For any positive real number d, the inequality $|x - a| < d$ can be written as follows.

$$|x - a| < d$$

$$-d < x - a < d \quad \text{rewriting without absolute value}$$

$$a - d < x < a + d \quad \text{adding } a \text{ to all three members}$$

This means that x lies between $a - d$ and $a + d$ (see Figure 1.14). Looking at it another way, $|x - a|$ represents the distance between x and a on the number line, and $|x - a| < d$ means that the distance between x and a is less than d.

$a - d$ $\quad$ a $\quad$ $a + d$

$|x - a| < d$

$a - d < x < a + d$

Figure 1.14 Graph of $|x - a| < d$

Similarly, $|x - a| > d$ means that the distance between x and a is greater than d. This is represented in Figure 1.15.

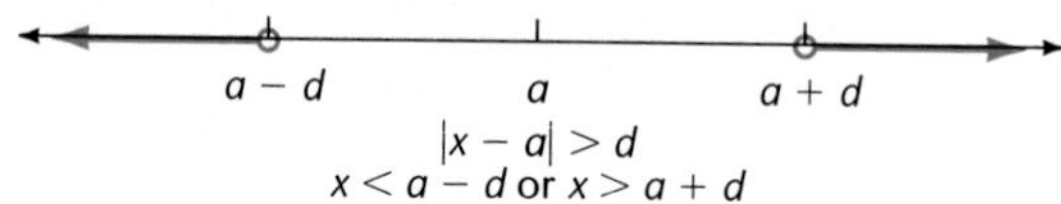

Figure 1.15 Graph of $|x - a| > d$

Example 6 • Graphing Absolute Value Inequalities

The foregoing discussion is illustrated in the following examples.

a. $|x - 2| < 3$ means that the distance between x and 2 is less than 3. Thus x lies between $2 - 3 = -1$ and $2 + 3 = 5$. This is represented in Figure 1.16.

−5 −4 −3 −2 −1 0 1 2 3 4 5

$|x - 2| < 3$

$-3 < x - 2 < 3$

$-1 < x < 5$

Figure 1.16 Graph of $|x - 2| < 3$

b. $|x + 2| > 1$ means that the distance between x and -2 is greater than 1, since $|x + 2| = |x - (-2)|$. On the number line, x is either to the right of $-2 + 1 = -1$, or x is to the left of $-2 - 1 = -3$. See Figure 1.17.

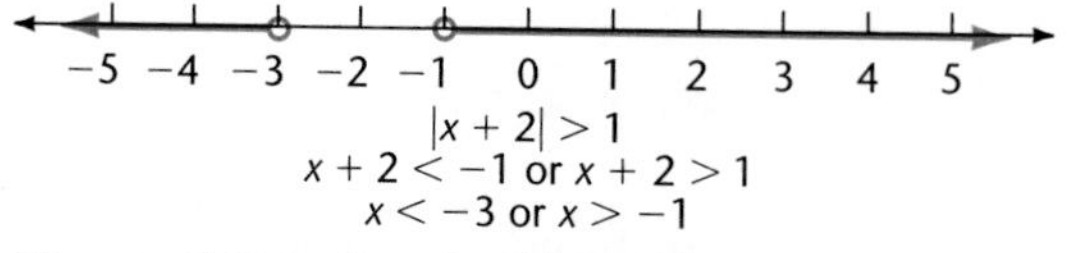

Figure 1.17 Graph of $|x + 2| > 1$ □

◆ Practice Problem 2 ◆

Draw the graph of each inequality.

a. $|x - 1| < 2$

b. $|x + 3| \geq 4$

EXERCISES 1.2

In Exercises 1–16, draw the graphs of the given statements. (See Example 2 and Practice Problem 1.)

1. $-7 \leq x < -2$
2. $-1 \leq x \leq 3$
3. $-4 < x \leq 0$
4. $-2 < x < 2$

5. $1 < x \leq 3$
6. $0 \leq x \leq 2$
7. $x > 1$ or $x < -1$
8. $x \leq 1$ or $x > 3$
9. $x > -1$ or $x < 1$
10. $x > 2$ or $x \leq 0$
11. $-2 < x \leq 0$ or $x > 1$
12. $1 \leq x < 3$ or $x < 0$
13. $2 \geq x > -1$ or $x < -3$
14. $3 > x > 1$ or $x > 5$
15. $2 \geq x \geq -1$ or $x < -4$
16. $3 \geq x \geq 1$ or $-3 < x < -1$

Write a statement that has the given graph. (See Practice Problem 1.)

17. −2
18. 1
19. 2 5
20. −3 0
21. −2 1
22. −1 2
23. 1 3 5
24. −2 3 8

In Exercises 25–34, name the property of inequalities that justifies each statement. (See Example 3.)

25. If $2x - 3 > 7$, then $2x > 10$.
26. If $2x > 10$, then $x > 5$.
27. If $-\dfrac{x}{2} \leq 6$, then $x \geq -12$.
28. If $3 + x > 2 + 2x$, then $3 > 2 + x$.
29. If $2x + 3 > 7$, then $2x > 4$.
30. If $x > 0$, then $5x > 0$.
31. If $5x > 0$, then $x > 0$.
32. If $-3x > 12$, then $x < -4$.
33. If $x < 0$ and $y < 0$, then $xy > 0$.
34. If $x > 0$ and $y < 0$, then $xy < 0$.

Arrange each set of numbers in order from smallest to largest.

35. $-|8 - 3|, -8 - |3|, |8 - 3|, |8| + 3, 0$
36. $2 - |5|, -|5 - 2|, |-2 - 5|, 0, -|2 - 5|$
37. $|-2|, -1, \sqrt{2}, -\sqrt{2}, -2, 1$

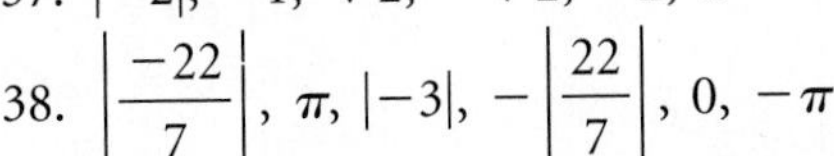

38. $\left|\dfrac{-22}{7}\right|, \pi, |-3|, -\left|\dfrac{22}{7}\right|, 0, -\pi$

37–40

39. $\sqrt{80}, -9, |-9|, -\sqrt{80}, \sqrt{80} - 9, |\sqrt{80} - 9|$
40. $\left|-\dfrac{11}{7}\right|, -\left|\dfrac{11}{7}\right|, \left|-\dfrac{9}{7}\right|, -\left|\dfrac{9}{7}\right|, \left|-\dfrac{7}{11}\right|, -\left|\dfrac{7}{11}\right|, \left|\dfrac{-7}{9}\right|, -\left|\dfrac{7}{9}\right|$

Express each of the following without using absolute value symbols. (See Example 4.)

41. $|-7|$
42. $\left|-\frac{3}{2}\right|$
43. $|-7+3|$
44. $|-8+2|$
45. $|-7|+3$
46. $-|7|-3$
47. $|9-7|$
48. $|9|-|7|$
49. $|9|+|-7|$
50. $|7-9|$
51. $|7|-|9|$
52. $|7|-|-9|$
53. $|y-4|$, if $y>4$
54. $|x-5|$, if $x<5$
55. $|y-4|$, if $y<4$
56. $|x-5|$, if $x>5$
57. $|9-\sqrt{80}|$
58. $\left|\frac{22}{7}-\pi\right|$
59. $|2x-10|$, if $x<5$
60. $|3x-12|$, if $x<4$
61. $|a-b|$, if $a<b$
62. $|2a-b|$, if $a>b/2$
63. $|2a-b|$, if $a<b/2$
64. $|a-2b|$, if $a<2b$
65. $\left|\frac{x}{5}\right|$ if $x<0$
66. $\frac{|x|}{5}$ if $x>0$
67. $|(-1)^2|$
68. $|(-4)|^2$
69. $|-a^2|$, where $a^2=a\cdot a$
70. $|a^2|$, where $a^2=a\cdot a$
71. $|-5-2a|$ if $a<-3$
72. $|3a-14|$ if $a<4$
73. $|3a-14|$ if $a>6$
74. $|15-6a|$ if $a>3$

Name the property of absolute value that justifies each statement.

75. $|x|\cdot|-5|=|-5x|$
76. $|-3x|=3|x|$
77. $\left|\frac{x}{5}\right|=\frac{|x|}{5}$
78. $\left|\frac{x}{-2}\right|=\frac{|x|}{2}$
79. $|x-y|\le|x|+|-y|$
80. $|-3+x|\le 3+|x|$

Draw the graph of each inequality. (See Example 6 and Practice Problem 2.)

81. $|x|<2$
82. $|x|<4$
83. $|x|>3$
84. $|x|>1$
85. $|x|\le 4$
86. $|x|\ge 3$
87. $|x-2|<1$
88. $|x-1|<3$
89. $|x+1|<2$
90. $|x+3|<1$
91. $|x+6|\le 4$
92. $|x+2|\le 5$
93. $|x-3|\ge 4$
94. $|x-4|>3$
95. $|x+4|>5$
96. $|x+2|>3$

Critical Thinking: Writing

97. Describe a way of deciding which of two numbers is greater. Use your procedure to decide which of π or $\frac{22}{7}$ is greater.

Critical Thinking: Writing

98. Use words instead of an equation to describe the absolute value of a number.

Critical Thinking: Writing

99. Explain why there is no real number x for which $|x| < -5$ is true.

Critical Thinking: Writing

100. A calculus student was overheard speaking of "a very large negative number." Write two possible meanings that might have been intended by this phrase.

◆ Solutions for Practice Problems

1. a. 

   b. $-2 < x < 3$

2. a. 

   b. 

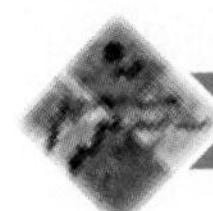

1.3 INTEGRAL EXPONENTS AND SCIENTIFIC NOTATION

In this section, we briefly review the properties of integral exponents. From earlier courses, we recall examples such as

$$x^2 = x \cdot x \qquad \text{and} \qquad 3^4 = 3 \cdot 3 \cdot 3 \cdot 3.$$

These examples illustrate the general definition of a positive integral exponent.

Definition of Positive Integral Exponents

If a is a real number and n is a positive integer, then

$$a^n = \underbrace{a \cdot a \cdot a \cdots a}_{n \text{ factors of } a}.$$

An expression of the form a^n is called an **exponential,** a is called the **base,** and n is called the **exponent,** or **power.**

The special case where $n = 1$ is worth noting. In this case, there is only one factor, and

$$a^1 = a.$$

In general, the exponent is understood to be 1 when no exponent is written.

The rules for order of operations stated in Section 1.1 apply to exponential expressions. For example, $2a^3 = 2 \cdot a \cdot a \cdot a$, but $(2a)^3 = (2a)(2a)(2a) = 8a^3$. Also, $-4a^2 = (-4)(a)(a)$, but $(-4a)^2 = (-4a)(-4a) = 16a^2$.

Several properties of exponentials follow from the fact that an exponential is a shorthand notation for a product. For example, if a is a real number and m and n are any positive integers, then

$$\begin{aligned} a^m \cdot a^n &= \underbrace{a \cdot a \cdot a \cdots a}_{m \text{ factors}} \cdot \underbrace{a \cdot a \cdot a \cdots a}_{n \text{ factors}} \\ &= \underbrace{(a \cdot a \cdot a \cdots a) \cdot (a \cdot a \cdot a \cdots a)}_{m + n \text{ factors}} \\ &= a^{m+n}. \end{aligned}$$

This property is called the **product rule** for exponents. It is one of the five **laws of exponents** that are extremely useful in simplifying expressions that involve exponentials.

$$a^m \cdot a^n = a^{m+n}$$ product rule

Another of these laws of exponents states that if a is a real number, then $(a^m)^n = a^{mn}$ for any positive integers m and n. This property is one of the power rules for exponents. It follows from the fact that

$$\begin{aligned} (a^m)^n &= \underbrace{a^m \cdot a^m \cdot a^m \cdots a^m}_{n \text{ factors of } a^m} \\ &= \underbrace{\underbrace{(a \cdot a \cdot a \cdots a)}_{m \text{ factors of } a} \underbrace{(a \cdot a \cdot a \cdots a)}_{m \text{ factors of } a} \cdots \underbrace{(a \cdot a \cdot a \cdots a)}_{m \text{ factors of } a}}_{n \text{ factors of } (a \cdot a \cdot a \cdots a)} \\ &= \underbrace{a \cdot a \cdot a \cdots a}_{mn \text{ factors of } a} \\ &= a^{mn}. \end{aligned}$$

$$(a^m)^n = a^{mn}$$ power of a power rule

Quotients of exponentials can be simplified by using the **quotient rule** for exponents. This rule can be stated as follows, where a is a nonzero real number and m and n are any positive integers.

$$\frac{a^m}{a^n} = \begin{cases} a^{m-n}, & \text{if } a \neq 0 \text{ and } m > n \\ \dfrac{1}{a^{n-m}}, & \text{if } a \neq 0 \text{ and } n > m \\ 1, & \text{if } a \neq 0 \text{ and } m = n \end{cases}$$ quotient rule

This rule can be established by explanations similar to those for the product and power rules.

In using the quotient rule, it is a nuisance to note repeatedly restrictions that must be made in order to avoid zero denominators. For this reason, *in the remaining examples and exercises of this chapter, we assume that all denominators are nonzero.*

The following example illustrates the three laws of exponents that we have considered to this point. To "simplify" an exponential expression means to rewrite it so that each variable occurs only once in the result.

Example 1 • Applying the Laws of Exponents

Simplify the following expressions.

a. $(2x^3y^2)(7x^4y^3)$ b. $(r^2)^3$ c. $\dfrac{2^7(1-t)^3}{2^3(1-t)^5}$

Solution

a. $(2x^3y^2)(7x^4y^3) = 2 \cdot 7 \cdot x^3 \cdot x^4 \cdot y^2 \cdot y^3 = 14x^7y^5$

b. $(r^2)^3 = r^6$

c. $\dfrac{2^7(1-t)^3}{2^3(1-t)^5} = \dfrac{2^7}{2^3} \cdot \dfrac{(1-t)^3}{(1-t)^5} = 2^4 \cdot \dfrac{1}{(1-t)^2} = \dfrac{16}{(1-t)^2}$ □

The other two laws of exponents involve two variables in the base. Consider a power that has a product ab as its base. Using the associative and commutative properties of multiplication freely, we have

$$\begin{aligned}(ab)^n &= \underbrace{(ab)(ab)(ab)\cdots(ab)}_{n \text{ factors of } ab}\\ &= \underbrace{(a \cdot a \cdot a \cdots a)}_{n \text{ factors of } a}\underbrace{(b \cdot b \cdot b \cdots b)}_{n \text{ factors of } b}\\ &= a^nb^n\end{aligned}$$

for any positive integer n. A similar property holds for quotients.

$(ab)^n = a^nb^n$	power of a product rule
$\left(\dfrac{a}{b}\right)^n = \dfrac{a^n}{b^n}$, if $b \neq 0$	power of a quotient rule

The following example illustrates how several of the laws of exponents may be used in simplifying a single expression.

Example 2 • Applying the Laws of Exponents

Simplify the following expressions.

a. $(2a^2b)^3 \cdot (3ab^4)^2$ b. $\left(\frac{2a^3}{b}\right)^3 \cdot \left(\frac{b}{4a^4}\right)^2$

Solution

a.
$$\begin{aligned}(2a^2b)^3 \cdot (3ab^4)^2 &= 2^3(a^2)^3b^3 \cdot 3^2a^2(b^4)^2 && \text{power of a product rule}\\ &= 8a^6b^3 \cdot 9a^2b^8 && \text{power of a power rule}\\ &= 72a^8b^{11} && \text{product rule}\end{aligned}$$

b.
$$\begin{aligned}\left(\frac{2a^3}{b}\right)^3 \cdot \left(\frac{b}{4a^4}\right)^2 &= \frac{(2a^3)^3}{b^3} \cdot \frac{b^2}{(4a^4)^2} && \text{power of a quotient rule}\\ &= \frac{8a^9}{b^3} \cdot \frac{b^2}{16a^8} && \text{power rules}\\ &= \frac{8}{16} \cdot \frac{a^9}{a^8} \cdot \frac{b^2}{b^3} && \text{rearranging factors}\\ &= \frac{1}{2} \cdot a \cdot \frac{1}{b} && \text{quotient rule}\\ &= \frac{a}{2b}\end{aligned}$$

□

Before summarizing the laws of exponents, we will extend the definition of a^n to include negative and zero exponents. Of course, this should be done so that all the laws of exponents hold for all integral exponents.

Consider first the case of a zero exponent. If we ignore the condition $m > n$ in the first part of the quotient rule, we have

$$\frac{a^m}{a^m} = a^{m-m} = a^0.$$

However,

$$\frac{a^m}{a^m} = 1,$$

and this leads to the conclusion that

$$a^0 = 1.$$

This case motivates us to make the following definition.

Definition of Zero Exponents

If a is a real number, then

$a^0 = 1$, if $a \neq 0$,

0^0 is undefined.

To guide us in making a definition for negative exponents, we drop the conditions on m and n in the quotient rule and use $m = 0$, with $a \neq 0$. This yields

$$\frac{a^0}{a^n} = a^{0-n} = a^{-n}.$$

Since $a^0 = 1$, we have

$$\frac{1}{a^n} = a^{-n}.$$

We are thus led to make the following definition.

Definition of Negative Integral Exponents

If n is any integer, then

$$a^{-n} = \frac{1}{a^n}, \qquad \text{for } a \neq 0.$$

We note that

$$a^{-1} = \frac{1}{a} \qquad \text{and} \qquad (a^n)^{-1} = \frac{1}{a^n}.$$

With these definitions, the quotient rule is valid for all integers m and n, and it may be rewritten as follows.

$$\frac{a^m}{a^n} = a^{m-n} = \frac{1}{a^{n-m}}, \text{ for } a \neq 0$$ quotient rule

Similar results hold for the other laws of exponents. Each of them is valid for all integers m and n, and these results are summarized in the following statement.

Laws of Exponents

Let a and b be real numbers. The following properties hold for all integers m and n for which the quantities involved are defined.

1. $a^m \cdot a^n = a^{m+n}$ — product rule
2. $(a^m)^n = a^{mn}$ — power of a power rule
3. $\dfrac{a^m}{a^n} = a^{m-n}$, for $a \neq 0$ — quotient rule
4. $(ab)^n = a^n b^n$ — power of a product rule
5. $\left(\dfrac{a}{b}\right)^n = \dfrac{a^n}{b^n}$, for $b \neq 0$ — power of a quotient rule

To avoid cumbersome instructions we adhere to the following convention.

Simplification of Exponential Expressions

To simplify an exponential expression is to rewrite it so that in each term, each variable occurs only once, and all negative and zero exponents have been eliminated.

There is usually more than one correct way to simplify an expression involving exponents. The steps may be different because of the order in which the laws of exponents are used.

Example 3 • Simplifying Exponential Expressions

Use the laws of exponents to simplify each of the following.

a. $[(x^{-1}y^2)^3]^{-2}$ b. $\left(\dfrac{(-r^2t^{-3})^{-1}s^3}{2r^2s^{-2}}\right)^{-2}$

Solution

a. $[(x^{-1}y^2)^3]^{-2} \overset{\text{Power rules}}{=} (x^{-3}y^6)^{-2} \overset{\text{Power rules}}{=} x^6y^{-12} = \dfrac{x^6}{y^{12}}$

b.
$$\left(\frac{(-r^2t^{-3})^{-1}s^3}{2r^2s^{-2}}\right)^{-2} = \left(\frac{-r^{-2}t^3s^3}{2r^2s^{-2}}\right)^{-2}$$ removing inner parentheses

$$= \frac{(-1)^{-2}r^4t^{-6}s^{-6}}{2^{-2}r^{-4}s^4}$$ applying the power rules

$$= \frac{2^2r^4r^4}{(-1)^2s^4s^6t^6}$$ eliminating negative exponents

$$= \frac{4r^8}{s^{10}t^6}$$ simplifying

□

◆ Practice Problem 1 ◆

Use the laws of exponents to simplify.

$$\left[\frac{(-2a)^3b^{-2}}{(a^2b)^4(2a^2)^0}\right]^{-1}$$

Scientific and technical work often involves very large numbers or very small numbers. In these cases, it is helpful to write the numbers in scientific notation. In scientific notation, a positive number N is written in the following form.

Scientific Notation

$$N = a \times 10^n, \qquad 1 \le a < 10 \text{ and } n \text{ is an integer}$$

The following examples illustrate changes between decimal notation and scientific notation.

Example 4 • Converting From Scientific to Decimal Notation

Write the following numbers in decimal notation.

a. 2.016×10^5 b. 5.9×10^{-6}

Solution

a. $2.016 \times 10^5 = 201{,}600$ moving the decimal 5 places

b. $5.9 \times 10^{-6} = 0.0000059$ moving the decimal 6 places □

Example 5 • Converting From Decimal to Scientific Notation

Write the following numbers in scientific notation.

a. 10,400,000 b. 0.0000000318

Solution

a. $10{,}400{,}000 = 1.04 \times 10^7$ (7 places)

b. $0.0000000318 = 3.18 \times 10^{-8}$ (8 places) □

Example 6 • The Hubble Space Telescope

A question with a mind-boggling answer: How far can the Hubble Space Telescope see? Translated from scientific notation into everyday language, the distance is seen to be enormous.

The Edwin P. Hubble Space Telescope is designed to view quasars and galaxies at distances up to 14 billion light-years. One light-year is approximately 5.88×10^{12} miles and 14 billion is 1.4×10^{10}, and so the viewing range is approximately

$$(5.88 \times 10^{12})(1.4 \times 10^{10}) = 8.232 \times 10^{22} \text{ miles.}$$

Written in decimal notation, this appears as

82,320,000,000,000,000,000,000 miles. □

The Hubble Space Telescope in orbit.

◆ **Practice Problem 2** ◆

The star Aldebaran in the constellation Taurus is 54 light-years away from the earth. Using 1 light-year as approximately 5.88×10^{12} miles, find the distance from the earth to Aldebaran and write this number of miles in both scientific notation and decimal notation.

The next example illustrates how the use of a calculator sometimes leads to a use of scientific notation.

Example 7 • Epidemiology

A cholera outbreak was reported in Peru near the end of January 1991. By the end of August that year, the disease had spread to eight other countries. Under favorable laboratory conditions, a population of cholera bacteria in a culture can double in 2 hours. At this rate, a single bacterium would grow to 2^{48} bacteria in 96 hours, or 4 days. Whenever a scientific calculator or graphing device is used to evaluate 2^{48}, the value is given in scientific notation. Depending on the accuracy of the device, this computation results in

$$2^{48} \approx 2.814749767 \times 10^{14}.$$

This may appear as 2.814749767E14 or some similar notation, depending on the device in use. Even in a situation as simple as this one, a knowledge of scientific notation is required to interpret the result. □

EXERCISES 1.3

Compute the value if possible; otherwise, simplify by using the laws of exponents. (See Examples 1–3.)

1. $(-3)^4$
2. $(-2)^6$
3. -5^2
4. -2^4
5. -2^{-4}
6. -3^{-2}
7. $\left(\frac{3}{2}\right)^{-2}$
8. $\left(-\frac{3}{4}\right)^{-4}$
9. $(2^2)^{-3}$
10. $(2^{-3})^{-2}$
11. $(4pq^2)^3$
12. $(-2r^2s^3)^4$
13. $\left(\frac{2x}{7}\right)^2$
14. $\left(\frac{2r}{3s}\right)^3$
15. $2y^{-3}$
16. $3z^{-2}$
17. $3^{-2}ab^{-2}$
18. $2^{-4}mn^{-3}$
19. $\frac{(5r^5)(6r^4)}{2r^3}$
20. $\frac{(xy)^2(x^2y^2)}{xy^2}$
21. $\frac{(6xy^4)^0}{2x^2y^{-3}}$
22. $\frac{(2y^2z)^3}{(16y^3)^0}$
23. $[(uv^{-1})^3]^{-2}$
24. $[(p^{-1}q^{-1})^{-2}]^{-1}$
25. $(4xy^2)^{-3}(x^{-2}y^3)^2$
26. $(3a^{-2}b^3)^{-1}(a^3b)^{-2}$
27. $\frac{r^{-3}}{s^{-5}}$
28. $\frac{u^{-6}}{2v^{-3}}$
29. $\frac{2t^{-3}}{(2r)^{-3}}$
30. $\frac{5a^{-2}}{bc^{-3}}$
31. $\frac{2^{-3}z^{-3}}{5^{-3}z^{-5}}$
32. $\frac{4p^{-2}q^3}{5^{-1}p^3q^{-2}}$
33. $\frac{4^{-2}a^{-4}}{3^{-2}a^{-4}}$
34. $\frac{7^{-2}x^4y^{-3}}{2x^{-3}y^{-3}}$
35. $\frac{3p^{-3}q^{-1}}{8^{-1}pq^3}$
36. $\frac{5m^{-2}n^3}{4^{-2}m^4n^{-2}}$
37. $\left(\frac{x^3}{3y^2}\right)^4$
38. $\left(-\frac{2x^2y}{z^3}\right)^3$
39. $\left(\frac{x^2}{y^{-1}z^3}\right)^{-3}$
40. $\left(\frac{yz^{-2}}{2x^{-1}w^2}\right)^{-2}$
41. $\left(\frac{(-x^2)^3z}{(2x^4y)^2}\right)^3$
42. $\left(\frac{(a^2b)^3}{(-ab^3)^2}\right)^2$
43. $\left(\frac{(ab)^{-1}c^2}{(ac^{-2})^{-1}b^2}\right)^{-2}$
44. $\left(\frac{-2x^{-3}z}{(xy)^{-2}z^{-4}}\right)^{-3}$

45. $\left(\dfrac{(x-y)^2(2z^4)}{[z(x-y)]^3}\right)^2$

46. $\left(\dfrac{(3-b)^3(a+c)^2}{[(a+c)(3-b)]^4}\right)^{-1}$

In Exercises 47–50, m and n denote positive integers.

47. $\left(\dfrac{3x^m}{y^n}\right)^2$

48. $\left(\dfrac{a^m b^n}{c^2}\right)^2$

49. $\dfrac{x^{2n-2}}{x^{n-1}}$

50. $\dfrac{x^{n+4}}{x^{2n-1}} \cdot \dfrac{x^n}{x^2}$

Write the following numbers in scientific notation. (See Examples 4 and 6.)

51. 1020

52. 40,300

53. 41,610,000

54. 940,200,000

55. 0.0035

56. 0.00073

57. $(8 \times 10^3)(7 \times 10^4)$

58. $(4 \times 10^5)(9 \times 10^2)$

59. $(1.2 \times 10^6)(5 \times 10^{-4})$

60. $(1.1 \times 10^7)(6 \times 10^{-3})$

61. $\dfrac{7.8 \times 10^{-9}}{1.3 \times 10^{-7}}$

62. $\dfrac{1.56 \times 10^8}{1.2 \times 10^6}$

Write the following numbers in decimal notation. (See Example 5.)

63. 8.22×10^3

64. 1.06×10^4

65. 6.6×10^{-1}

66. 7.9×10^{-2}

67. 2.87×10^{-5}

68. 5.1×10^{-4}

69. $\dfrac{4}{5 \times 10^3}$

70. $\dfrac{3}{2 \times 10^5}$

71. $\dfrac{68{,}000{,}000}{1{,}700{,}000}$

72. $\dfrac{90{,}000{,}000}{150{,}000}$

In Exercises 73–82, write the value of each expression in scientific notation, rounded to four digits.

73–82 73. $\dfrac{1{,}240{,}000 \times 67{,}100}{0.0082}$

74. $\dfrac{384{,}000 \times 715{,}900}{0.00061}$

75. $\dfrac{501{,}484 \times 0.000293}{61{,}941 \times 0.00275}$

76. $\dfrac{630{,}042 \times 0.000577}{56{,}229 \times 0.0000683}$

77. $\dfrac{0.004141 \times 73{,}290 \times 0.0619}{513{,}500 \times 0.00606}$

78. $\dfrac{649{,}200 \times 58{,}300 \times 0.00317}{2{,}374{,}000 \times 0.000565}$

79. $\dfrac{4^{-2} \cdot 3^4}{7(4.9)^{-2}}$

80. $\dfrac{5^{-7} \cdot 9^6}{8^{-1} \cdot 4^0}$

81. $\dfrac{(17^4 \times 8^{-1})^{-2}}{(0.19)^5}$

82. $\left(\dfrac{7^{12}(3.4)^{-6}}{(-2.9)^3}\right)^{-2}$

Planetary Orbits 83. The orbits of the earth and three other planets are shown in the accompanying figure. The orbit of the earth is a slightly flattened circle with the shortest possible distance between the earth and the sun approximately 91,110,000 miles. Write this distance in scientific notation.

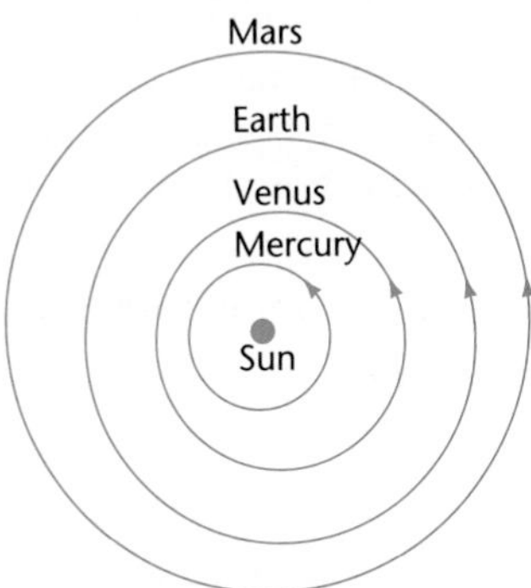

Figure for Exercise 83

Largest Animal 84. The largest animal on earth is the blue whale, which may weigh as much as 120,000,000 grams. Write this weight in scientific notation. (Note the contrast with the weight of the fairy fly in Exercise 85.)

Smallest Insect 85. The smallest insect on earth is the fairy fly, which may weigh as little as 5.01×10^{-6} gram. Write this weight in decimal notation.

Cave Paintings 86. Using carbon-14 dating, prehistoric cave paintings at Lascaux, France, have been estimated to be 1.55×10^4 years old. Write this estimate in decimal notation.

pH Scale 87. In chemistry, the pH scale is used to measure the acidity of all sorts of liquids, including hair shampoo. The pH value is defined in terms of the concentration of the hydronium ion (H_3O^+), and they are related by the equation

$$(H_3O^+) = 10^{-\text{pH}},$$

where (H_3O^+) is measured in moles per liter. Pert Plus brand of shampoo is pH-balanced with a pH of 6. Find the concentration of the hydronium ion in Pert Plus.

pH Scale 88. Distilled water has a pH of 7. Use the equation in Exercise 87 to find the decimal value of (H_3O^+) for distilled water.

Speed of Light

89–90

89. (See Exercise 83.) A **light-year** is the distance that a ray of light travels in 1 year, approximately 5.88×10^{12} miles. Use 91,110,000 miles as the distance from the sun to the earth and find the number of minutes that it takes a ray of light to travel from the sun to the earth.

Distance From the Sun to the Earth

90. (See Exercise 83.) The average distance from the sun to the earth is about 93,000,000 miles. Use 1 kilometer = 0.62 mile and express this average distance in kilometers, rounded to the nearest million kilometers. Write your answer in scientific notation.

Critical Thinking: Writing

91. The distributive property may lead to confusion between expressions of the form $a^n + b^n$ and $(a + b)^n$. Use the meaning of an exponent and write an explanation of why $3^2 + 4^2$ and $(3 + 4)^2$ are different values.

Critical Thinking: Writing

92. The distributive property also may lead to confusion between $a^m - a^n$ and a^{m-n}. Explain why $a^m - a^n$ and a^{m-n} are *never equal* if $a \neq 0$ and $m = n$.

◆ Solutions for Practice Problems

1. $$\left[\frac{(-2a)^3b^{-2}}{(a^2b)^4(2a^2)^0}\right]^{-1} = \frac{[(-2a)^3b^{-2}]^{-1}}{[(a^2b)^4(2a^2)^0]^{-1}} = \frac{(-2a)^{-3}b^2}{(a^2b)^{-4}(1)}$$
$$= \frac{(a^2b)^4b^2}{(-2a)^3} = \frac{a^8b^4b^2}{-8a^3} = -\frac{a^5b^6}{8}$$

2. $$54(5.88 \times 10^{12}) = 317.52 \times 10^{12} \text{ miles}$$
$$= 3.1752 \times 10^{14} \text{ miles, in scientific notation}$$
$$= 317{,}520{,}000{,}000{,}000 \text{ miles, in decimal notation}$$

1.4 POLYNOMIALS

An **algebraic expression** is an expression involving numbers and variables that represents a result obtained by performing the operations of addition, subtraction, multiplication (including constant powers), division, or extraction of roots. In this section, we assume that all variables represent real numbers.

A **polynomial** is an algebraic expression in which variables have only nonnegative integral exponents and no variable is in a denominator.

If only one variable is involved in a polynomial, we can give a more precise description.

Definition of a Polynomial

A **polynomial in the variable *x*** is an algebraic expression of the form

$$a_nx^n + a_{n-1}x^{n-1} + \cdots + a_1x + a_0,$$

where n is a nonnegative integer, and the coefficients $a_0, a_1, a_2, \ldots, a_n$ are constants. If $n = 0$, the polynomial has only one term, a_0.

One distinguishing feature of a polynomial is the number of terms that it contains. A polynomial that contains only one term is called a **monomial.** One with two terms is called a **binomial,** and one with three terms is a **trinomial.**

Another important feature of a polynomial is its degree. The **degree of a nonzero term** in a polynomial is the sum of the exponents on the variables that are involved in that term. The **degree of a nonzero polynomial** is the largest degree that occurs among the terms of the polynomial. The term $-xy^3$ in the polynomial $3x^2 - xy^3 + 5xy^2 + y^3$ has degree $1 + 3 = 4$, and the polynomial has degree 4 also. The coefficient in the term of highest degree is called the **leading coefficient,** and a **monic polynomial** is one in which the leading coefficient is 1.

Nonzero constants are polynomials of degree 0, and the real number 0 is the only polynomial that has no degree defined.†

Example 1 • Classifying Polynomials

The main points in the preceding discussion are illustrated in Table 1.1.

Table 1.1 Examples of Polynomials

Polynomial	Type	Degree
$-4x^3$	Monomial	3
$6x - 2$	Binomial	1
$x^4 - 7x^2 + 3$ (monic)	Trinomial	4
9	Monomial	0
0	Monomial	No degree
$8x^3y^2 - x^2y$	Binomial	5
$x^4 + 3y^2$ (monic)	Binomial	4

□

Since all variables in this section represent real numbers, polynomials also represent real numbers. Hence the properties of real numbers stated in Section 1.1 can be used freely in adding, subtracting, and multiplying polynomials. In particular, we can use the distributive property to combine like terms in a sum or difference.

A polynomial in x is in a **simplest form** if all like terms have been combined and written either in decreasing powers of x or in increasing powers of x. Addition, subtraction, and multiplication may be accomplished by the following procedures.

Addition, Subtraction, and Multiplication of Polynomials

1. To add or subtract polynomials, combine like terms and write the result in simplest form.
2. To find the product of two polynomials, use the distributive property to multiply each term of the first polynomial by each term of the second polynomial, combine like terms, and write the result in simplest form.

†With this exception made regarding the degree of the zero polynomial, it is always true that the degree of a product is the sum of the degrees of the factors.

Example 2 • Operations on Polynomials

Perform the indicated operations and write the results in simplest form.

a. $(4x^3 - 8x + 7) + (5x^3 + 3x - 6)$
b. $(6x^6 + 9x^2 - 5x) - (11x^2 - 3x - 8)$
c. $(2x^2 - 5x - 3)(4x^2 + 3x)$
d. $(3x^4 - 2x^2 + x)(-x^2 + 2x + 4)$
e. $(x - 2)(4x - 3)$

Solution

a. $$\begin{aligned}(4x^3 - 8x + 7) + (5x^3 + 3x - 6) &= (4 + 5)x^3 + (-8 + 3)x + (7 - 6)\\ &= 9x^3 - 5x + 1\end{aligned}$$

b. $$\begin{aligned}(6x^6 + 9x^2 - 5x) - (11x^2 - 3x - 8) &= 6x^6 + (9 - 11)x^2\\ &\quad + [-5 - (-3)]x - (-8)\\ &= 6x^6 - 2x^2 - 2x + 8\end{aligned}$$

c. The distributive property can be used systematically to write out all the products involved.

$$\begin{aligned}(2x^2 - 5x - 3)(4x^2 + 3x) &= (2x^2 - 5x - 3)(4x^2) + (2x^2 - 5x - 3)(3x)\\ &= 8x^4 - 20x^3 - 12x^2 + 6x^3 - 15x^2 - 9x\\ &= 8x^4 - 14x^3 - 27x^2 - 9x\end{aligned}$$

d. If both polynomials have several terms, multiplication may be easier using the **vertical method.** In this method, one polynomial is written under the other, with each of them in simplest form. The top polynomial is multiplied by each term in the lower one, with like terms written one under the other and then combined.

$$\begin{array}{rrrrrr} & & 3x^4 & & -\ 2x^2 & +\ x \\ & & & -\ x^2 & +\ 2x & +\ 4 \\ \hline -3x^6 & & +\ 2x^4 & -\ x^3 & & \\ & 6x^5 & & -\ 4x^3 & +\ 2x^2 & \\ & & 12x^4 & & -\ 8x^2 & +\ 4x \\ \hline -3x^6 & +\ 6x^5 & +\ 14x^4 & -\ 5x^3 & -\ 6x^2 & +\ 4x \end{array}$$

e. The product of two binomials is worthy of special note. The FOIL method can be used to multiply binomials where the word FOIL serves as a memory aid in ensuring that all products have been formed.

$$\begin{aligned}(x - 2)(4x - 3) &= \underset{\text{First terms}}{\overset{\textbf{F}}{x(4x)}} + \underset{\text{Outside terms}}{\overset{\textbf{O}}{x(-3)}} + \underset{\text{Inside terms}}{\overset{\textbf{I}}{(-2)(4x)}} + \underset{\text{Last terms}}{\overset{\textbf{L}}{(-2)(-3)}}\\ &= 4x^2 - 3x - 8x + 6\\ &= 4x^2 - 11x + 6\end{aligned}$$

□

Certain types of products are important enough to deserve special attention. These are stated in the following list of special product rules. Each of them can be verified by direct multiplication.

Special Product Rules

Let x and y represent arbitrary real numbers. Then the following special product rules are valid:

Difference of two squares: $(x - y)(x + y) = x^2 - y^2$

Square of a binomial: $\begin{cases}(x + y)^2 = x^2 + 2xy + y^2 \\ (x - y)^2 = x^2 - 2xy + y^2\end{cases}$

Cube of a binomial: $\begin{cases}(x + y)^3 = x^3 + 3x^2y + 3xy^2 + y^3 \\ (x - y)^3 = x^3 - 3x^2y + 3xy^2 - y^3\end{cases}$

Sum of two cubes: $(x + y)(x^2 - xy + y^2) = x^3 + y^3$

Difference of two cubes: $(x - y)(x^2 + xy + y^2) = x^3 - y^3$

Example 3 • Applying the Special Product Rules

One of the special product rules is illustrated in each of the following products:

a. $(3a - b^2)(3a + b^2) = (3a)^2 - (b^2)^2$
$= 9a^2 - b^4$ difference of two squares

b. $(3r + 4)^2 = (3r)^2 + 2(3r)(4) + 4^2$
$= 9r^2 + 24r + 16$ square of a binomial

c. $(5s - 2t^3)^2 = (5s)^2 - 2(5s)(2t^3) + (2t^3)^2$
$= 25s^2 - 20st^3 + 4t^6$ square of a binomial

d. $(p + 2q)^3 = p^3 + 3p^2(2q) + 3p(2q)^2 + (2q)^3$
$= p^3 + 6p^2q + 12pq^2 + 8q^3$ cube of a binomial

e. $(u - 2v)(u^2 + 2uv + 4v^2) = u^3 - (2v)^3$
$= u^3 - 8v^3$ difference of two cubes

□

Formulas that involve polynomials are used in manufacturing design, revenue or profit prediction, and other planning activities by industry and business. A simple illustration is presented in the next example.

Example 4 • Manufacturing Design

An open box without a top is to be made from a rectangular piece of tin by cutting equal squares from each corner and bending up the sides as shown in Figure 1.18. If the piece of tin measures 8 inches by 12 inches and the side of each square is x inches, the volume of the resulting box in cubic inches is given by $V = x(8 - 2x)(12 - 2x)$.

a. Find V when $x = 1$ and when $x = 3$.

b. Write V as a polynomial in simplest form.

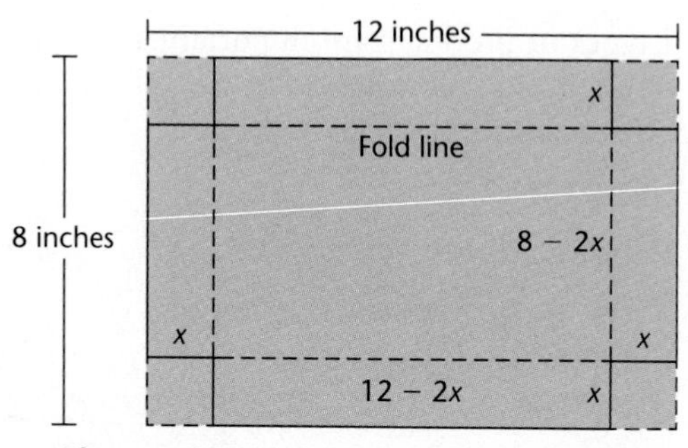

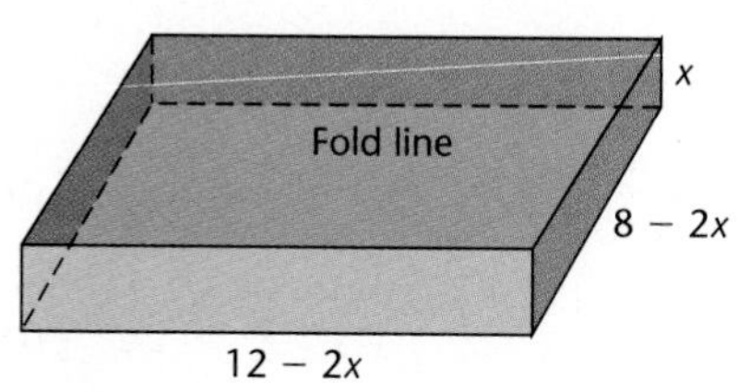

Figure 1.18 Constructing an open box

Solution

a. When $x = 1$, we have $V = 1(8 - 2)(12 - 2) = (1)(6)(10) = 60$. That is, $V = 60$ cubic inches. When $x = 3$, $V = 3(8 - 6)(12 - 6) = 36$ cubic inches.

b. $V = x(8 - 2x)(12 - 2x)$

$= x(96 - 16x - 24x + 4x^2)$ multiplying the binomials

$= x(4x^2 - 40x + 96)$ simplifying

$= 4x^3 - 40x^2 + 96x$ using the distributive property ◻

◆ Practice Problem 1 ◆

Perform the following multiplications by using the special product rules.

a. $(2a - 3b^2)^2$ b. $(x + y + z)(x + y - z)$

EXERCISES 1.4

Perform the indicated operations and write the results in simplest form. (See Examples 2 and 3.)

1. $(5x^4 - 8x^2 + 4) + (2x^3 + 7x^2 - 9)$
2. $(6x^5 - 14x^3 + 3x + 8) + (-8x^4 + 8x^3 - 5x)$
3. $(8r^4 - 3r^5 + 4) - (r^2 + 2r^4 - r^5)$
4. $(7p^5 - 4p^3 + p) - (5p^4 - 2p^5 + 4p)$
5. $(-3 + 8z + 4z^3) - (-5z^2 + 3z^3 - 12)$
6. $(2y^4 - 5y^2 - 7y^3 + 8) - (1 + 7y - 4y^2 - 6y^4)$
7. $5x^2(4x^3 - 3x + 8)$
8. $-y^3(6y^4 - 5y^2 - 2)$
9. $(4z + 3)(2z - 7)$
10. $(7t - 10)(3t - 2)$
11. $(5q + 3)(8q - 1)$
12. $(11r + 4)(r - 3)$
13. $(5a - b)(2a + 3b)$
14. $(8m + 3n)(2m - 5n)$
15. $(x - 4)(2x^3 + 3x^2 - 1)$
16. $(x^2 - 3)(x^3 - x + 5)$
17. $(2x^2 + 3x - 1)(x^2 - 2x + 4)$
18. $(9x^2 - 2x - 5)(3x^2 - x - 2)$
19. $(3y^2 - 7y - 2)(y^3 - y^2 - 1)$
20. $(4y^3 - 5y^2 + 1)(2y^4 - 3y^2 + 1)$

21. $(1 + x)[1 + x(1 - x^2)]$
22. $1 + x[1 + x(1 - x^2)]$
23. $(b - 3)(b + 3)$
24. $(y - 5)(y + 5)$
25. $(2r + s)(2r - s)$
26. $(4p + 3q)(4p - 3q)$
27. $(y^2 - 2)^2$
28. $(x^2 + 3)^2$
29. $(a + 3bc^2)^2$
30. $(m^2 - 2np)^2$
31. $[2(3m - 4n)]^2$
32. $[3(2p + 5q)]^2$
33. $(\frac{1}{3}x + \frac{1}{2}y)(\frac{1}{3}x - \frac{1}{2}y)$
34. $(\frac{2}{3}u + v)(\frac{2}{3}u - v)$
35. $(2x^3 - y)(2x^3 + y)$
36. $(p - 3q^2)(p + 3q^2)$
37. $(p - q)(p^2 + pq + q^2)$
38. $(a - 2b)(a^2 + 2ab + 4b^2)$
39. $(2x + 1)(4x^2 - 2x + 1)$
40. $(y + 3)(y^2 - 3y + 9)$
41. $(n + k)^3$
42. $(r - d)^3$
43. $(x - 2y)^3$
44. $(2m + n)^3$
45. $(3x^2 - 5y^2)(x^2 + 2y^2)$
46. $(4u^2 + v^2)(3u^2 - 5v^2)$
47. $(2x^3 + 3y)(x^3 - 4y)$
48. $(5x - 2y^3)(x + 2y^3)$
49. $(3a - 2b)(9a^2 + 6ab + 4b^2)$
50. $(4w + 3)(16w^2 - 12w + 9)$
51. $[(x + y) + 1][(x + y) - 1]$
52. $[2 + (a - b)][2 - (a - b)]$
53. $(a + 2b + c)^2$
54. $(x + y + 2)^2$
55. $(x + y + z)(x - y - z)$
56. $(a + b + c + 1)(a + b - c - 1)$

In Exercises 57–64, variable exponents represent positive integers.

57. $(2x^n - 1)(3x^n + 4)$
58. $(5y^m + 2)(3y^m - 4)$
59. $(2z^r + 9)(2z^r - 9)$
60. $(3t^k - 4)(3t^k + 4)$
61. $(5y^n + 3)^2$
62. $(3a^p + 4)^2$
63. $(2b^k - 5)^2$
64. $(4c^n - 3)^2$

Manufacturing Design

65. A closed box with a square bottom and top as shown in the accompanying figure is to be made using 800 square inches of plastic. If w is the length in inches of a bottom edge, the volume V of the box is given by $V = w(400 - w^2)/2$ cubic inches. Find the volume of the box when (a) $w = 4$ and (b) $w = 12$.

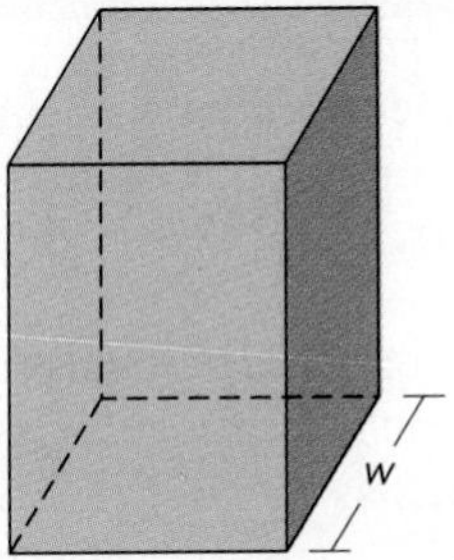

Figure for Exercise 65

Profit 66. Acme Anchor Company has a profit P (in thousands of dollars per day) given by $P = x(x - 4)(x - 6)$, where x is the daily production (in thousands of anchors). Find the profit when daily production is 2000 anchors.

Revenue 67. Acme Boat Company has found that the revenue R from producing x canoes is given (in dollars) by $R = 8x^2 - 0.02x^3$. Find the revenue from producing 100 canoes.

Demand 68. The demand D for Pedalex cars (in thousands of cars) is given by $D = x(2 + x)(8 - x)$, where x is the price (in thousands of dollars). Find the demand when the price is \$7000.

Find the value of the given polynomial for the given value of x. Round your answer to four decimal places.

69–72

69. $x^9 - 7x^6 + 2x^2 - 43,\ x = 2.3$
70. $x^{15} + 4x^{11} - 6x^3 + 721,\ x = -1.5$
71. $x^{11} - 5x^9 + 7x^6 - 345,\ x = -2.6$
72. $3x^{12} - 24x^7 - 6x^5 + 429,\ x = 1.4$

Critical Thinking: Writing 73. State the special product rule for the difference of two squares in words.

Critical Thinking: Writing 74. State in words the special product rule for the square of a binomial that involves a sum of two numbers.

◆ Solution for Practice Problem

1. a. $(2a - 3b^2)^2 = (2a)^2 - 2(2a)(3b^2) + (3b^2)^2 = 4a^2 - 12ab^2 + 9b^4$
   b. $(x + y + z)(x + y - z) = (x + y)^2 - z^2 = x^2 + 2xy + y^2 - z^2$

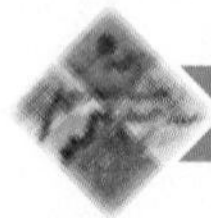

1.5 FACTORING POLYNOMIALS

In a product of polynomials, the polynomials being multiplied are called the **factors** of the product. To factor a given polynomial is to write the polynomial as a product of two or more factors. In this sense, factoring is the reverse process of multiplication.

In this section, we consider only polynomials that have integers as coefficients, and we require that all factors be polynomials whose coefficients are integers. We do *not* allow factorizations such as

$$x^2 - 1 = (2x - 2)(\tfrac{1}{2}x + \tfrac{1}{2}) \quad \text{and} \quad x - y = (\sqrt{x} - \sqrt{y})(\sqrt{x} + \sqrt{y}).$$

A polynomial with integral coefficients is **completely factored** when it is written as a product of polynomials with integral coefficients, and none of the nonconstant polynomial factors can be written as a product of two polynomials.

The first step in factoring is to use the distributive property to factor out a common factor from each term. This method is illustrated in the following example.

Example 1 • Common Factor

a. The terms in $4m^3 - 2m^2 - 8m$ have $2m$ as a common factor, so we can write

$$4m^3 - 2m^2 - 8m = 2m(2m^2 - m - 4).$$

b. The terms in $3x^2y - 4xy - 7xy^2$ have xy as a common factor, and

$$3x^2y - 4xy - 7xy^2 = xy(3x - 4 - 7y).$$ □

Since factoring is the reverse process of multiplication, the special product rules can be regarded as patterns for factoring. With this in mind, we state the following list.

Formulas for Factoring

Common Factor:	$ax + ay = a(x + y)$
Difference of two squares:	$x^2 - y^2 = (x + y)(x - y)$
Square of a binomial:	$\begin{cases} x^2 + 2xy + y^2 = (x + y)^2 \\ x^2 - 2xy + y^2 = (x - y)^2 \end{cases}$
Sum of two cubes:	$x^3 + y^3 = (x + y)(x^2 - xy + y^2)$
Difference of two cubes:	$x^3 - y^3 = (x - y)(x^2 + xy + y^2)$

The use of these formulas is illustrated in the following examples.

Example 2 • Difference of Two Squares

Factor each of the following polynomials.

a. $9a^2 - 25b^2$ b. $4x^3 - 100xy^2$

Solution

a. Since $9a^2 = (3a)^2$ and $25b^2 = (5b)^2$, we have the difference of two squares and

$$\begin{aligned} 9a^2 - 25b^2 &= (3a)^2 - (5b)^2 \\ &= (3a + 5b)(3a - 5b). \end{aligned}$$

b. Any common factor should be factored out of each term first. Thus

$$\begin{aligned} 4x^3 - 100xy^2 &= 4x(x^2 - 25y^2) \\ &= 4x(x + 5y)(x - 5y). \end{aligned}$$ □

Example 3 • Square of a Binomial

Factor completely.

a. $x^2 - 8xy + 16y^2$ b. $4x^2 + 12xy + 9y^2$

Solution

Each of these trinomials fits the form of the square of a binomial.

a. $x^2 - 8xy + 16y^2 = x^2 - 2(4xy) + (4y)^2$
$= (x - 4y)^2$

b. $4x^2 + 12xy + 9y^2 = (2x)^2 + 2(2x)(3y) + (3y)^2$
$= (2x + 3y)^2$ □

Example 4 • Sum or Difference of Two Cubes

Factor completely.

a. $x^3 + 8$ b. $2a^4b - 2ab^4$

Solution

a. The binomial $x^3 + 8$ can be expressed as the sum of two cubes. Hence we factor $x^3 + 8$ as

$$x^3 + 8 = x^3 + 2^3 = (x + 2)(x^2 - 2x + 4).$$

b. First we factor out the common factor $2ab$ and then use the formula for factoring the difference of two cubes.

$$2a^4b - 2ab^4 = 2ab(a^3 - b^3) = 2ab(a - b)(a^2 + ab + b^2)$$ □

◆ **Practice Problem 1** ◆

Factor completely.

a. $16a^2 - 1$ b. $4x^2a^3 + 24x^2a^2 + 36x^2a$

c. $25r^2 - 20rs + 4s^2$ d. $27 - b^3$

In some instances, a factorization of the type we are considering may not be possible. The polynomial $x^2 - 2$, for example, cannot be factored as a product of polynomials with integers as coefficients (except trivial factorizations, with one factor 1 or -1). The polynomial $x^2 + y^2$ furnishes another example. A polynomial with integral coefficients that cannot be factored as a product of polynomials, both different from 1 and -1, and both with integral coefficients, is called a **prime polynomial,** or an **irreducible polynomial.**†

†In Chapter 5, we consider other factorizations of polynomials.

The FOIL method of multiplying two binomials can be reversed to factor some trinomials. Since

$$(x + a)(x + b) = x^2 + bx + ax + ab = x^2 + (b + a)x + ab$$
$$= \mathbf{F} \quad \mathbf{O} \quad \mathbf{I} \quad \mathbf{L} = \mathbf{F} \quad \mathbf{O} + \mathbf{I} \quad \mathbf{L}$$

we can factor

$$x^2 + rx + s = (x + a)(x + b)$$

by choosing a and b so that $a + b = r$ and $ab = s$.

Example 5 • Reversing FOIL

Factor $12x^2 - 24x - 96$.

Solution

We first factor out the common factor 12 to obtain

$$12x^2 - 24x - 96 = 12(x^2 - 2x - 8).$$

The trinomial factor can then be factored as the product of two binomials

$$x^2 - 2x - 8 = (x + a)(x + b),$$

where a and b are chosen so that their sum is -2 and their product is -8. The possible choices are tabulated as follows:

Factors	Product	Sum	
$-8, 1$	-8	-7	
$8, -1$	-8	7	
$4, -2$	-8	2	
$-4, 2$	-8	-2	Correct choice

Thus we have

$$x^2 - 2x - 8 = (x - 4)(x + 2)$$

and

$$12x^2 - 24x - 96 = 12(x - 4)(x + 2).$$ □

Example 6 • Trial and Error

Factor $3x^2 + 5x - 2$.

Solution

Reversing the FOIL method of multiplication to factor a quadratic trinomial becomes somewhat more involved when the trinomial is not monic. The following diagram indicates how the product of two binomials might yield the given trinomial:

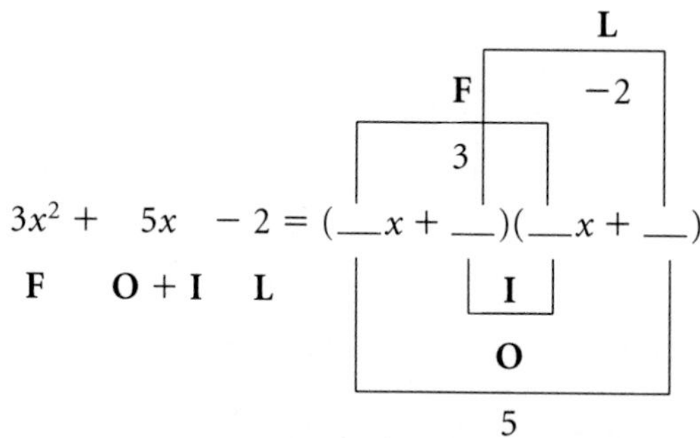

The trick in filling in the blanks in the diagram is to determine factors of 3 (the leading coefficient) and factors of -2 (the constant term), which when arranged properly will yield the correct middle coefficient, 5. For this reason, this method is called the **trial-and-error** method. The correct factorization is

$$3x^2 + 5x - 2 = (3x - 1)(x + 2).$$

□

◆ Practice Problem 2 ◆

Factor each expression completely.

a. $2x^2y - 7xy^2 + 6y^3$ b. $4x^2 - 12x + 5$

If a polynomial has more than three terms, it is usually necessary to group two or more terms together and then factor out a common factor in each group. This method is called **factoring by grouping.**

Example 7 • Grouping

Factor $4hx - 4bh - 8cx + 8bc$.

Solution

The first two terms contain the common factor $4h$, and the last two terms contain the common factor $-8c$. Thus we can write

$$4hx - 4bh = 4h(x - b)$$
$$-8cx + 8bc = -8c(x - b)$$

and we note that $4(x - b)$ is common to each group. Hence the polynomial can be factored by grouping as follows.

$$\begin{aligned} 4hx - 4bh - 8cx + 8bc &= 4h(x - b) - 8c(x - b) \\ &= 4(x - b)(h - 2c) \end{aligned}$$

□

Sometimes a polynomial of degree greater than 3 can be factored by using one of the special product formulas.

Example 8 • Square of a Binomial

Factor $x^4 - 2x^2 + 1$.

Solution

This polynomial fits the form of the square of a binomial.

$$\begin{aligned} x^4 - 2x^2 + 1 &= (x^2)^2 - 2x^2 + 1 \\ &= (x^2 - 1)^2 \\ &= [(x + 1)(x - 1)]^2 \\ &= (x + 1)^2(x - 1)^2 \end{aligned}$$

□

◆ Practice Problem 3 ◆

Factor completely.

a. $x^3 - x^2 - 3x + 3$ b. $x^4 - 1$

We conclude this section with a general procedure for factoring.

Factoring

1. Factor out any common factors.
2. If the polynomial is a binomial that fits the form of the difference of two squares or sum or difference of two cubes, then use the special factoring formulas.
3. If the polynomial is a quadratic trinomial that fits the form of a square of a binomial, use the special factoring formula. Otherwise use the trial-and-error method.
4. If the polynomial has more than three terms, factor by grouping.
5. If the polynomial has degree greater than 3, determine whether it will fit one of the special factoring formulas.
6. Finally, check to see if each factor is itself completely factored.

EXERCISES 1.5

Factor each expression completely, if possible.

Common Factor 1–10

(See Example 1.)

1. $3a - 6b$
2. $9y - 12z$
3. $4x^3 - 16xy$
4. $3a^5 - 12a^2$
5. $6m^3 - 24m^2 + 6mn$
6. $6xy^2 - 9x^2y^2 + 3x^3y$

7. $12x^2y - 4xy + 20xy^2$
8. $8m^2n^3 - 4m^3n^2 + 2m^2n^2$
9. $7(x + 2y) - 3(x + 2y)$
10. $a(x - y) - b(x - y)$

Difference of Two Squares 11–20

(See Example 2.)

11. $x^2 - 25$
12. $y^2 - 100$
13. $9y^2 - 1$
14. $49 - v^2$
15. $16x^2 - 25y^2$
16. $100a^2 - 9b^2$
17. $(x - 1)x^2 - (x - 1)y^2$
18. $(3a - 2)r^2 - (3a - 2)s^2$
19. $-y^3x^2 + 9y^3$
20. $-8a^2b + 2b$

Square of a Binomial 21–30

(See Example 3.)

21. $u^2 - 8u + 16$
22. $z^2 - 6z + 9$
23. $25y^2 + 10y + 1$
24. $49x^2 + 14x + 1$
25. $4a^2 - 20ab + 25b^2$
26. $9x^2 - 60xy + 100y^2$
27. $3a^3 + 6a^2 + 3a$
28. $2a^2b + 8ab + 8b$
29. $1 - 12rs + 36r^2s^2$
30. $1 + 14ab + 49a^2b^2$

Sum or Difference of Two Cubes 31–40

(See Example 4.)

31. $x^3 - z^3$
32. $1000 - x^3$
33. $8 + u^3$
34. $27 + a^3$
35. $8x^3 + 27$
36. $64x^3 + 27y^3$
37. $216x^3 - 8y^3$
38. $1000a^3 - 27b^3$
39. $-4a^4 + 256a$
40. $8a^2b^2 - 8a^2b^5$

Trial and Error 41–52

(See Examples 5 and 6 and Practice Problem 2.)

41. $x^2 - 2x - 24$
42. $x^2 + 8x + 15$
43. $x^2 + 10x + 21$
44. $12 - 7y + y^2$
45. $4y^2 - 3xy - x^2$
46. $5a^2 - 12ab + 7b^2$
47. $3x^2 + 7ax - 6a^2$
48. $45a^2 - 8ay - 4y^2$
49. $2x^3 + 4x^2 - 30x$
50. $4x^3 + 4x^2 - 24x$
51. $8y^3 - 20y^2 + 12y$
52. $-14y^3 - 26y^2 + 4y$

Grouping 53–62

(See Example 7.)

53. $3ac + 3bc + ad + bd$
54. $5ax + 2bx - 10ad - 4bd$
55. $ax^2 + bx^2 + ad^2 + bd^2$
56. $2ax^2 - bx^2 + 6a - 3b$
57. $9x^2 - y^2 + 2yz - z^2$
58. $16a^2 - 9b^2 + 6b - 1$
59. $2 + 4x - 10x^4 - 5x^3$
60. $3a^2 + a^3 - 15 - 5a$
61. $x^2 - 2x + 1 - y^2 - 2yz - z^2$
62. $x^2 + 2xy + y^2 - z^2 - 4z - 4$

Factor each expression completely, if possible. (See Example 8 and Practice Problems 1–3.)

63. $2x^2 - 13x + 21$
64. $3a^2 - 13a - 10$
65. $4a^2 + a - 3$
66. $6z^2 - 43z + 7$
67. $-2x(a + h) - 3y(a + h)$
68. $x^2(y - 2) + y^2(y - 2)$
69. $16a^4 + 64a^3 + 64a^2$
70. $3x^3y - 30x^2y^2 + 75xy^3$
71. $x^3 - x$
72. $y^4 - 4y^2$
73. $6y^2 + 11y - 10$
74. $20x^2 - 23x + 6$
75. $6a^2 + 13a - 15$
76. $21x^2 - 25x - 4$
77. $x^2 + 1$
78. $a^2 + 4b^2$
79. $3x^2 - 9$
80. $-6x^2 + 36$
81. $25x^2 + 10x + 1 - y^2$
82. $16x^2 + 8x + 1 - 4y^2$
83. $-x^6 - x^3$
84. $-64y - 8yx^3$
85. $ax^2 - 9ay^4$
86. $x^4 - 81y^4$
87. $(4x - 3y)^2 - 25$
88. $(2x + y)^2 - 9z^2$
89. $(4a - b)^2 - (2x - z)^2$
90. $(2x + y)^2 - (3z - w)^2$
91. $x^4 - 16x^2 + 64$
92. $y^4 - 6y^2 + 9$
93. $9x^4 - 30x^2 + 25$
94. $4x^4 - 28x^2 + 49$
95. $4a^4 + 12a^2b^2 + 9b^4$
96. $4x^4 + 28x^2y^2 + 49y^4$
97. $u(u^2 - v^2) + v(u^2 - v^2)$
98. $p(p + q)^2 + q(p + q)^2$
99. $9w^8 + 27w^5 - 9w^2$
100. $y^{50} + 6y^{30} + 9y^{10}$
101. $x^4 + 4$ (*Hint:* $x^4 + 4 = x^4 + 4x^2 + 4 - 4x^2$)
102. $y^4 + 64$ (*Hint:* Add and subtract $16y^2$.)
103. $z^4 + z^2 + 1$ (*Hint:* Add and subtract z^2.)
104. $x^4 - 3x^2 + 1$ (Hint: Add and subtract x^2.)

In Exercises 105–108, m and n represent positive integers.

105. $2x^{2n} + 10x^n + 12$
106. $2x^{2n} - 4x^n - 7$
107. $9z^{2m} - w^{2n}$
108. $x^{4m} - 16y^{4n}$

Critical Thinking: Exploration and Writing

109. Use areas of rectangles in the accompanying figure to explain why

$$x^2 + 2xy + y^2 = (x + y)^2.$$

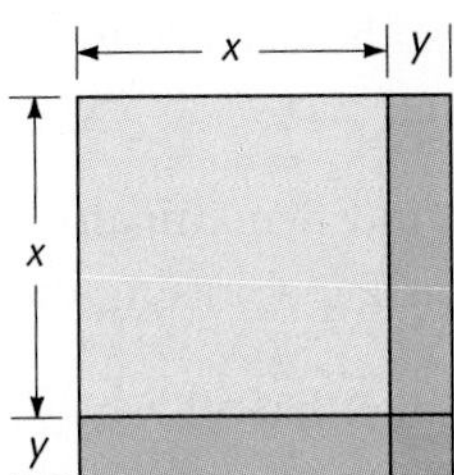

Figure for Exercise 109

Critical Thinking: Exploration and Writing

110. Explain how the formula for the square of a sum can be used to compute $(51)^2$ mentally.

◆ Solutions for Practice Problems

1. a. $16a^2 - 1 = (4a)^2 - 1^2 = (4a + 1)(4a - 1)$
   b. $4x^2a^3 + 24x^2a^2 + 36x^2a = 4x^2a(a^2 + 6a + 9) = 4x^2a(a + 3)^2$
   c. $25r^2 - 20rs + 4s^2 = (5r)^2 - 2(5r)(2s) + (2s)^2 = (5r - 2s)^2$
   d. $27 - b^3 = 3^3 - b^3 = (3 - b)(9 + 3b + b^2)$
2. a. $2x^2y - 7xy^2 + 6y^3 = y(2x^2 - 7xy + 6y^2) = y(2x - 3y)(x - 2y)$
   b. $4x^2 - 12x + 5 = (2x - 5)(2x - 1)$
3. a. $x^3 - x^2 - 3x + 3 = (x^3 - x^2) - (3x - 3)$
   $= x^2(x - 1) - 3(x - 1) = (x - 1)(x^2 - 3)$
   b. $x^4 - 1 = (x^2)^2 - 1 = (x^2 + 1)(x^2 - 1) = (x^2 + 1)(x + 1)(x - 1)$

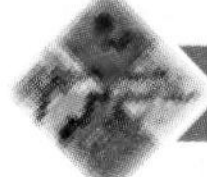

1.6 RATIONAL EXPRESSIONS

A **rational expression** is a quotient of polynomials that have integral coefficients—that is, a fraction formed by dividing one polynomial with integral coefficients by another. In all manipulations involving rational expressions, it is understood that *all values of the variables that make a zero denominator are excluded.*

We say that a rational expression is reduced to **lowest terms** if the only common factors in the numerator and denominator are 1 and −1. The fundamental principle of fractions stated in Section 1.1 for real numbers generalizes to rational expressions and can be used to reduce a rational expression to lowest terms.

The Fundamental Principle of Fractions

Let P, Q, and R be polynomials. Then

$$\frac{PR}{QR} = \frac{P}{Q}, \quad \text{if } Q \neq 0 \text{ and } R \neq 0.$$

Any rational expression can be reduced to lowest terms using the following procedure.

1. Factor the numerator and denominator completely.
2. Divide out all common factors using the fundamental principle of fractions.

In this procedure, we assume that all values of the variables that make a common factor zero are *excluded* in the resulting fraction as well as in the original fraction.

Example 1 • Applying the Fundamental Principle

Reduce $\dfrac{x^2 - 4xy + 4y^2}{x^2 - 4y^2}$ to lowest terms.

Solution

$$\frac{x^2 - 4xy + 4y^2}{x^2 - 4y^2} = \frac{(x - 2y)^2}{(x + 2y)(x - 2y)} = \frac{x - 2y}{x + 2y}$$

We assume all values of x and y are excluded for which $x - 2y = 0$ or $x + 2y = 0$. In other words,

$$\frac{x^2 - 4xy + 4y^2}{x^2 - 4y^2} = \frac{x - 2y}{x + 2y},$$

subject to the conditions that $x \neq 2y$ and $x \neq -2y$. □

The rules for multiplying and dividing fractions given in Chapter 1 also extend to rational expressions.

Multiplication and Division of Rational Expressions

Let P, Q, R, and S be polynomials. Then

$$\frac{P}{Q} \cdot \frac{R}{S} = \frac{PR}{QS} \quad \text{and} \quad \frac{P}{Q} \div \frac{R}{S} = \frac{P}{Q} \cdot \frac{S}{R} = \frac{PS}{QR},$$

where all denominators are nonzero.

When applying either of these rules, we factor both numerators and denominators completely and reduce the result to lowest terms by dividing out common factors. The following examples demonstrate the procedures.

Example 2 • Multiplying Rational Expressions

$$\frac{y^2 - 9}{y^2 + 6y + 9} \cdot \frac{y^3 - 1}{y^2 - 4y + 3}$$

$$= \frac{\overset{1}{\cancel{(y - 3)}}\overset{1}{\cancel{(y + 3)}}\overset{1}{\cancel{(y - 1)}}(y^2 + y + 1)}{(y + 3)^{\cancel{2}}\underset{1}{\cancel{(y - 1)}}\underset{1}{\cancel{(y - 3)}}} \quad \text{factoring all polynomials}$$

$$= \frac{y^2 + y + 1}{y + 3}, \quad \text{for } y \neq -3, 1, 3 \quad \text{reducing to lowest terms}$$ □

Except where otherwise noted, we will assume from now on that all denominators are not zero without explicit mention of the necessary restrictions.

Example 3 • Dividing Rational Expressions

$$\frac{25-9x^2}{x+3} \div (5x-3x^2) = \frac{25-9x^2}{x+3}\cdot\frac{1}{5x-3x^2}$$

$$= \frac{(5-3x)(5+3x)}{(x+3)(x)(5-3x)} \qquad \text{factoring}$$

$$= \frac{5+3x}{x(x+3)} \qquad \text{reducing to lowest terms}$$

□

Sometimes a quotient of rational expressions is written in fractional form. It is then referred to as a complex fraction. In general, a **complex fraction** is a fraction that has fractions in its numerator or its denominator. In one method of simplification, the complex fraction is rewritten as a division problem involving rational expressions. This method is illustrated in Example 4.

Example 4 • Simplifying a Complex Fraction

$$\frac{\dfrac{6y^2+5y-4}{2y^2+7y+3}}{\dfrac{2y^2+3y-2}{y^2-9}} = \frac{6y^2+5y-4}{2y^2+7y+3} \div \frac{2y^2+3y-2}{y^2-9} \qquad \text{rewriting}$$

$$= \frac{6y^2+5y-4}{2y^2+7y+3}\cdot\frac{y^2-9}{2y^2+3y-2} \qquad \text{inverting the divisor}$$

$$= \frac{(3y+4)(2y-1)(y-3)(y+3)}{(2y+1)(y+3)(2y-1)(y+2)} \qquad \text{factoring}$$

$$= \frac{(3y+4)(y-3)}{(2y+1)(y+2)} \qquad \text{reducing to lowest terms}$$

□

◆ **Practice Problem 1** ◆

Perform the indicated operations and reduce the results to lowest terms.

a. $\dfrac{3x^2-5xy-2y^2}{2x^2-5xy+2y^2}\cdot\dfrac{4x^2-4xy+y^2}{3x^2+4xy+y^2}$

b. $\dfrac{6x^2+7x-5}{1-9x^2} \div \dfrac{4x^2-4x+1}{3x^2-7x+2}$

To add or subtract fractions with the same denominator, we again extend one of the basic formulas for fractions to rational expressions.

Addition and Subtraction of Rational Expressions

Let P, Q, and R be polynomials with $Q \neq 0$. Then

$$\frac{P}{Q} + \frac{R}{Q} = \frac{P+R}{Q} \quad \text{and} \quad \frac{P}{Q} - \frac{R}{Q} = \frac{P-R}{Q}.$$

With different denominators, we first convert the fractions to fractions with the same denominator. The most efficient denominator to use is the **least common denominator** (abbreviated LCD).

Finding the Least Common Denominator

1. Factor each denominator completely (that is, into prime factors).
2. Form the product of all the different prime factors, with each prime factor raised to the highest exponent with which it appears in any of the given denominators. In the prime factors, the leading coefficients are chosen to be positive.

To form a sum or difference involving two or more rational expressions that are in lowest terms, we follow these steps:

1. Find the LCD of the expressions.
2. Change each fraction to a fraction having as denominator the LCD found in step 1.
3. Combine the new numerators and place the result over the LCD.
4. Reduce the result to lowest terms.

Example 5 • Subtracting Rational Expressions

Perform the indicated operation.

$$\frac{2}{6x-3} - \frac{x}{4x^2-1}$$

Solution

Since $6x - 3 = 3(2x-1)$ and $4x^2 - 1 = (2x-1)(2x+1)$, the LCD is $3(2x-1)(2x+1)$. Thus

$$\frac{2}{3(2x-1)} - \frac{x}{(2x-1)(2x+1)}$$
$$= \frac{2(2x+1)}{3(2x-1)(2x+1)} - \frac{3x}{3(2x-1)(2x+1)} \quad \text{using the LCD}$$

$$= \frac{4x + 2 - 3x}{3(2x - 1)(2x + 1)} \qquad \text{subtracting}$$

$$= \frac{x + 2}{3(2x - 1)(2x + 1)}. \qquad \text{simplifying} \quad \square$$

We use the result

$$b - a = -(a - b)$$

in the next example.

Example 6 • Adding Rational Expressions

Add $\dfrac{1}{b - a} + \dfrac{2a}{a^2 - b^2}$.

Solution

Since $a^2 - b^2 = (a - b)(a + b)$ and $b - a = -(a - b)$, the LCD is $(a - b)(a + b)$. Thus

$$\frac{1}{b - a} + \frac{2}{a^2 - b^2}$$

$$= \frac{1(-1)(a + b)}{(b - a)(-1)(a + b)} + \frac{2a}{(a - b)(a + b)} \qquad \text{using the LCD}$$

$$= \frac{-a - b}{(a - b)(a + b)} + \frac{2a}{(a - b)(a + b)} \qquad \text{simplifying}$$

$$= \frac{-a - b + 2a}{(a - b)(a + b)} \qquad \text{adding}$$

$$= \frac{a - b}{(a - b)(a + b)} \qquad \text{simplifying}$$

$$= \frac{1}{a + b}. \qquad \text{reducing to lowest terms} \quad \square$$

◆ Practice Problem 2 ◆

Perform the indicated addition and reduce the result to lowest terms.

$$\frac{p + 2}{3p^2 + 4p - 4} + \frac{2p - 1}{3p^2 - 5p + 2}$$

Some operations with rational expressions may involve both sums and products or quotients. In such cases, the operations are performed subject to the same rules as those given in Section 1.1 for real numbers.

Example 7 • Simplifying Rational Expressions

Express

$$\left(\frac{2}{y} + \frac{5}{y^2} - \frac{12}{y^3}\right) \div \left(4 - \frac{8}{y} + \frac{3}{y^2}\right)$$

as a single fraction in lowest terms.

Solution

$$\begin{aligned}\left(\frac{2}{y} + \frac{5}{y^2} - \frac{12}{y^3}\right) \div \left(4 - \frac{8}{y} + \frac{3}{y^2}\right) &= \left(\frac{2y^2 + 5y - 12}{y^3}\right) \div \left(\frac{4y^2 - 8y + 3}{y^2}\right)\\ &= \frac{2y^2 + 5y - 12}{y^3} \cdot \frac{y^2}{4y^2 - 8y + 3}\\ &= \frac{\overset{1}{\cancel{(2y-3)}}(y+4)}{\underset{y}{\cancel{y^3}}} \cdot \frac{\overset{1}{\cancel{y^2}}}{\underset{1}{\cancel{(2y-3)}}(2y-1)}\\ &= \frac{y+4}{y(2y-1)}\end{aligned}$$

□

The final example of this section illustrates another type of complex fraction: one involving negative exponents.

Example 8 • Simplifying Rational Expressions Involving Negative Exponents

Express the following as a rational expression in lowest terms.

$$\frac{x^{-2} + y^{-1}}{(xy)^{-2}}$$

Solution

When we eliminate the negative exponents, the resulting fraction is a complex fraction.

$$\frac{x^{-2} + y^{-1}}{(xy)^{-2}} = \frac{\dfrac{1}{x^2} + \dfrac{1}{y}}{\dfrac{1}{(xy)^2}} \qquad \text{eliminating negative exponents}$$

We could rewrite this complex fraction as a quotient of two fractions and then invert the divisor and multiply. Alternatively, we could simplify the complex function by multiplying its numerator and denominator by the least common denominator of all the fractions within the complex fraction.

$$\frac{x^{-2}+y^{-1}}{(xy)^{-2}} = \frac{\left(\frac{1}{x^2}+\frac{1}{y}\right)x^2y^2}{\left[\frac{1}{(xy)^2}\right]x^2y^2} \quad \text{fundamental principle of fractions}$$

$$= \frac{y^2+x^2y}{1} \quad \text{simplifying}$$

$$= y(y+x^2)$$

□

◆ Practice Problem 3 ◆

Write as a single fraction in lowest terms.

$(x^0 - x^{-1})^{-1}$

EXERCISES 1.6

Restrict the values of the variables so that each rational expression is defined. (See Example 1.)

1. $\frac{3}{x+2}$
2. $\frac{x}{2x-7}$
3. $\frac{4}{x(x+3)}$
4. $\frac{x}{(x-1)(x+2)}$
5. $\frac{x^2-y^2}{x-y}$
6. $\frac{x+1}{x^2-1}$
7. $\frac{3}{x^2+1}$
8. $\frac{x}{x^2+9}$

Reduce each expression to lowest terms. (See Example 1.)

9. $\frac{x^2-y^2}{x-y}$
10. $\frac{x^2-y^2}{x^3+y^3}$
11. $\frac{x^2-x-12}{x^2-2x-8}$
12. $\frac{2x^2+5x-3}{x^2+2x-3}$
13. $\frac{4(a-1)^2(a+2)^3}{(a+2)(a-1)^3}$
14. $\frac{12x^2(x-3)^2(3x-1)}{15x(x-3)(3x-1)^2}$
15. $\frac{(x^3-27y^3)(x^2-4y^2)}{x^2-5xy+6y^2}$
16. $\frac{(x+y)(x^2+3xy-4y^2)}{(x-y)(x^2+2xy+y^2)}$

Perform the indicated multiplications and divisions and reduce the results to lowest terms. (See Examples 2 and 3 and Practice Problem 1.)

17. $\frac{4pq^3}{p^2-q^2}\cdot\frac{p-q}{12p^2q^2}$
18. $\frac{3r^2s}{(r-s)^2}\cdot\frac{r^2-s^2}{15rs^2}$
19. $\frac{2b-3y}{3c+6d}\cdot\frac{2c+4d}{4b-6y}$
20. $\frac{3a-ay}{6z-xz}\cdot\frac{4z-xz}{3a-ya}$

21. $(x^2 - 4y^2) \cdot \dfrac{5a}{xy - 2y^2}$

22. $(9m^2 - n^2) \cdot \dfrac{3p}{3m^2 + mn}$

23. $(cy + dy) \div \dfrac{d^2 - c^2}{3y}$

24. $(a^2 - 4b^2) \div \dfrac{a^2 + 2ab}{a - 2b}$

25. $\dfrac{3x - 3y}{4x + 2y} \cdot \dfrac{4x^2 - y^2}{(x - y)^2}$

26. $\dfrac{u^2 - v^2}{u^2 + uv} \cdot \dfrac{uv^2 + u^2v}{uv^3 - u^3v}$

27. $\dfrac{y^2 - 2y - 15}{y^2 - 9} \cdot \dfrac{y^2 - 6y + 9}{12 - 4y}$

28. $\dfrac{2z^2 - 5z - 12}{2z^2 + 5z - 12} \cdot \dfrac{2z^2 + 3z - 9}{2z^2 + 9z + 9}$

29. $\dfrac{w + 3}{w^2 + 7w + 10} \div \dfrac{w^2 + 6w + 9}{w^2 + 5w + 6}$

30. $\dfrac{w + 2}{w^2 + 7w + 12} \div \dfrac{w^2 + 4w + 4}{w^2 + 5w + 6}$

31. $\dfrac{y^2 + 2y - 15}{y^2 + 11y + 30} \div \dfrac{y^2 - 8y + 15}{y^2 + 2y - 24}$

32. $\dfrac{t^2 + 5t - 14}{t^2 + 10t + 21} \div \dfrac{t^2 + 5t + 6}{t^2 + 7t + 12}$

33. $\dfrac{x^2 - 4y^2}{x^2 + xy - 6y^2} \cdot \dfrac{x^2 + 4xy + 3y^2}{x^2 + xy - 2y^2}$

34. $\dfrac{2y^2 + yz - z^2}{y^2 - z^2} \div \dfrac{3yz - 6y^2}{2y^2 + yz - 3z^2}$

35. $\dfrac{25 - 4x^2}{x^3 - 1} \cdot \dfrac{x^3 + x^2 + x}{15 - x - 2x^2}$

36. $\dfrac{16 - 9z^2}{z^3 + 8} \cdot \dfrac{z^2 - 2z + 4}{8 + 10z + 3z^2}$

37. $\dfrac{rs^4 + rs^2}{r^2s^2 + s^2} \div \dfrac{r^3s^2 - r^2s^3}{r^2 - rs}$

38. $\dfrac{x^3 + y^3}{x^2 - y^2} \div \dfrac{2x^2 + 3xy + y^2}{x^4 - y^4}$

39. $\dfrac{x^3 - 27}{2x^2 + 5} \div (3x - 9)$

40. $\dfrac{w^3 + 8}{w^2 + 2} \div (2w + 4)$

41. $\dfrac{w^3 - 4w^2 + 9w - 36}{w^4 - 81} \cdot \dfrac{2aw^2 - aw - 15a}{w^3 - 64}$

42. $\dfrac{x^3 - 3x^2 + 4x - 12}{x^4 - 16} \cdot \dfrac{3x^2y - xy - 10y}{x^3 - 27}$

43. $\dfrac{3a + 2b}{2ya - 2xyb} \cdot \dfrac{a^3 - x^3b^3}{xa + 2xb} \cdot \dfrac{a^2 + 4ab + 4b^2}{9a^2 - 4b^2}$

44. $\dfrac{2x + y}{3abx + 3by} \cdot \dfrac{a^3x^3 + y^3}{ax + 3ay} \cdot \dfrac{x^2 + 6xy + 9y^2}{4x^2 - y^2}$

45. $\dfrac{2x^2 + x - 6}{x^3 + 2x^2 + 4x} \cdot \dfrac{6x^3 + 24x}{3x^4 - 48} \div \dfrac{8x^2 - 18}{x^3 - 8}$

46. $\dfrac{3x^2 + 7x - 6}{x^3 + 3x^2 + 9x} \cdot \dfrac{6x^3 + 54x}{2x^4 - 162} \div \dfrac{27x^2 - 12}{x^3 - 27}$

Perform the indicated operations, and reduce the result to lowest terms. (See Examples 5–7 and Practice Problem 2.)

47. $\dfrac{5}{2 + x} - \dfrac{3}{2 + x}$

48. $\dfrac{6}{y + 7} + \dfrac{1}{y + 7} + \dfrac{y}{y + 7}$

49. $\dfrac{x}{x^2-1}+\dfrac{1}{x^2-1}$

50. $\dfrac{x}{x^3-1}-\dfrac{1}{x^3-1}$

51. $\dfrac{4}{x-1}+\dfrac{2}{1-x}$

52. $\dfrac{3r}{s-r}+\dfrac{2s}{r-s}$

✓53. $\dfrac{a}{a-b}-\dfrac{b}{b-a}$

54. $\dfrac{2x-y}{3x-2y}-\dfrac{3y-5x}{2y-3x}$

55. $\dfrac{y}{4x^2}-\dfrac{z}{3x}$

56. $\dfrac{2-y}{3xy^2}-\dfrac{3-4y}{2x^2y}$

57. $\dfrac{7}{2x}-\dfrac{1}{x-2}$

58. $\dfrac{3}{y-1}-\dfrac{5}{4y}$

59. $\dfrac{12}{x^2-9}-\dfrac{2}{x-3}$

60. $\dfrac{12}{x^2-4}+\dfrac{3}{x+2}$

✓61. $\dfrac{2y}{2x+2y}-\dfrac{3}{x^2-y^2}$

62. $\dfrac{2}{3x-6y}-\dfrac{3}{x^2-4y^2}$

63. $\dfrac{x-28}{x^2-x-6}+\dfrac{5}{x-3}$

64. $\dfrac{x-6}{x^2+3x-4}-\dfrac{2}{x+4}$

65. $\dfrac{x+5}{x^2+7x+10}-\dfrac{x+5}{2x^2+x-6}$

66. $\dfrac{x+5}{x^2+7x+10}-\dfrac{x-1}{x^2+5x+6}$

✓67. $\dfrac{w-1}{2w^2-13w+15}+\dfrac{w+3}{2w^2-15w+18}$

68. $\dfrac{2y+3}{3y^2+y-2}+\dfrac{4-3y}{2y^2-3y-5}$

69. $\dfrac{z-5}{z^2-9}+\dfrac{z-2}{12-4z}$

70. $\dfrac{3r-1}{r^2-16}+\dfrac{2r+1}{12-3r}$

71. $y-\dfrac{2y}{y^2-1}+\dfrac{3}{y+1}$

72. $t-\dfrac{4}{2-t}-\dfrac{t^3}{t^2-4}$

73. $(9x^2-25y^2)\div\left(1+\dfrac{5y}{3x}\right)$

74. $(16a^2-9b^2)\div\left(1-\dfrac{3b}{4a}\right)$

✓75. $\left(\dfrac{2}{x+2}-\dfrac{1}{x-2}\right)\left(\dfrac{x-2}{x+1}\right)$

76. $\left(\dfrac{2}{x-3}-\dfrac{1}{x+3}\right)\left(\dfrac{x+3}{x+9}\right)$

77. $\left(x-1-\dfrac{6}{x}\right)\div\left(1+\dfrac{2}{x}-\dfrac{15}{x^2}\right)$

78. $\left(1-\dfrac{8}{x^3}\right)\div\left(\dfrac{2}{x}-1\right)$

79. $\dfrac{2x+2y}{x^2+2xy-3y^2}-\dfrac{x-2y}{x^2+xy-6y^2}$

80. $\dfrac{2x+y}{4x^2-4xy-3y^2}+\dfrac{x-2y}{2x^2-5xy+3y^2}$

In Exercises 81–96, express the complex fraction as a rational expression in lowest terms. (See Examples 4 and 8.)

81. $\dfrac{\dfrac{3w^2 + w - 2}{4w^2 - w - 5}}{\dfrac{3w^2 - 11w + 6}{4w^2 - 7w - 15}}$

82. $\dfrac{\dfrac{2p^2 - 7p + 5}{3p^2 + p - 4}}{\dfrac{2p^2 - p - 10}{3p^2 + 2p - 8}}$

83. $\dfrac{\dfrac{x^3 + y^3}{2x - 3y}}{\dfrac{2x + 2y}{4x^2 - 9y^2}}$

84. $\dfrac{\dfrac{a^3 - b^3}{3a + 4b}}{\dfrac{3a - 3b}{9a^2 - 16b^2}}$

85. $\dfrac{1 + \dfrac{y}{x}}{1 - \dfrac{y}{x}}$

86. $\dfrac{4 - \dfrac{3}{x}}{3 + \dfrac{5}{x^2}}$

87. $\dfrac{9x^2 - 4y^2}{\dfrac{x - y}{y - 2x} - 1}$

88. $\dfrac{\dfrac{2xy}{x + 2y} - x}{\dfrac{3xy}{x + 3y} + x}$

89. $\dfrac{\dfrac{1}{x^2} - \dfrac{1}{y^2}}{\dfrac{1}{x} + \dfrac{1}{y}}$

90. $\dfrac{\dfrac{1}{a} + \dfrac{1}{b}}{\dfrac{1}{a + b}}$

91. $\dfrac{\dfrac{b^2}{a^2} - \dfrac{a}{b}}{\dfrac{1}{2b} - \dfrac{a}{2b^2}}$

92. $\dfrac{\dfrac{a + 3b}{xy} + \dfrac{b}{yz}}{\dfrac{b - a}{x} + \dfrac{b + a}{x}}$

93. $1 + \dfrac{1}{1 + \dfrac{1}{x}}$

94. $\dfrac{y}{2 + \dfrac{y}{1 + y}}$

95. $x - \dfrac{x}{2 - \dfrac{1}{x}}$

96. $1 - \dfrac{2}{1 - \dfrac{2}{1 - \dfrac{2}{x}}}$

Rewrite each as a rational expression in lowest terms. (See Example 8 and Practice Problem 3.)

√97. $xy^{-1} - \dfrac{x^{-1}}{y^{-1}}$

98. $(x^{-2} - x^{-1})^{-1}$

99. $\dfrac{(ab)^{-2}}{a^{-2} - b^{-2}}$

100. $\dfrac{x^{-1} + y^{-1}}{x^{-1} - y^{-1}}$

√101. $\dfrac{x^{-1} - y^{-1}}{x^{-2} - y^{-2}}$

102. $\dfrac{9x - x^{-1}}{3 + x^{-1}}$

103. $\dfrac{x^{-2} - y^{-2}}{(x + y)^2}$

104. $\dfrac{x - y}{x^{-2} - y^{-2}}$

105–106

105. Evaluate the following rational expression for $x = 2.17$ and round your result to four decimal places.

$$\frac{\dfrac{1}{x} - \dfrac{1}{3}}{\dfrac{1}{x - 3}}$$

106. Evaluate the following rational expression for $x = 3$.

$$1 - \cfrac{1}{1 - \cfrac{1}{1 - \cfrac{1}{x}}}$$

Critical Thinking: Discussion

107. Discuss the restrictions that are necessary on the rational expression

$$\frac{x^2 - 9}{x^3 - 5x^2 + 6x}$$

and explain the conditions on x that are understood when we write

$$\frac{x^2 - 9}{x^3 - 5x^2 + 6x} = \frac{x + 3}{x(x - 2)}.$$

Critical Thinking: Writing

108. a. State the rule for multiplication of rational expressions in words.
b. State the rule for division of rational expressions in words.

◆ Solutions for Practice Problems

1. a. $$\frac{3x^2-5xy-2y^2}{2x^2-5xy+2y^2}\cdot\frac{4x^2-4xy+y^2}{3x^2+4xy+y^2}=\frac{(3x+y)(x-2y)(2x-y)^2}{(2x-y)(x-2y)(3x+y)(x+y)}$$

$$=\frac{2x-y}{x+y}$$

b. $$\frac{6x^2+7x-5}{1-9x^2}\div\frac{4x^2-4x+1}{3x^2-7x+2}=\frac{(2x-1)(3x+5)}{(1-3x)(1+3x)}\cdot\frac{(3x-1)(x-2)}{(2x-1)^2}$$

$$=-\frac{(3x+5)(x-2)}{(3x+1)(2x-1)}$$

2. $$\frac{p+2}{3p^2+4p-4}+\frac{2p-1}{3p^2-5p+2}=\frac{p+2}{(3p-2)(p+2)}+\frac{2p-1}{(3p-2)(p-1)}$$

$$=\frac{p-1+2p-1}{(3p-2)(p-1)}$$

$$=\frac{3p-2}{(3p-2)(p-1)}$$

$$=\frac{1}{p-1}$$

3. $$(x^0-x^{-1})^{-1}=\left(1-\frac{1}{x}\right)^{-1}=\left(\frac{x-1}{x}\right)^{-1}=\frac{x}{x-1}$$

1.7 RADICALS AND RATIONAL EXPONENTS

If a is a *positive* real number and b is a number such that $b^2=a$, then b is called a **square root** of a. Since $(3)^2$ and $(-3)^2=9$, both 3 and -3 are square roots of 9. The positive square root, 3, is called the principal square root of 9, and the symbol $\sqrt{9}$ is used to denote it. We write

$$\sqrt{9}=3 \qquad \text{and} \qquad -\sqrt{9}=-3.$$

A *negative* real number a does not have a square root in the set of real numbers because the square of a real number is never negative. If $a=0$, its only square root is zero. For any *nonnegative* real number a, the **principal square root** of a is defined to be the *nonnegative* number $\sqrt{a}$ such that

$$(\sqrt{a})^2 = a.$$

The symbol $\sqrt{}$ is called the **radical sign.**

Other roots of numbers are defined in a manner similar to square roots. A **cube root** of a is a number b such that $b^3 = a$, and a **fourth root** of a is a number b such that $b^4 = a$. In general, if n is a positive integer, then b is an ***n*th root** of a if $b^n = a$. An nth root is sometimes referred to as a root of **order *n*.**

With cube roots, the situation is simpler than it is with square roots. Every real number a has exactly one real cube root. This cube root of a is denoted by $\sqrt[3]{a}$ and is called the **principal cube root** of a. For example,

$$\sqrt[3]{64} = 4 \text{ since } (4)^3 = 64,$$
$$\sqrt[3]{-27} = -3 \text{ since } (-3)^3 = -27.$$

In fact, for any odd n, every real number has exactly one real nth root, called the **principal *n*th root.**

The situation for fourth roots, sixth roots, and other roots of even order is much the same as for square roots. For example, 16 has two real fourth roots, 2 (the principal fourth root) and -2. However, a real fourth root of -16 does not exist. These examples illustrate the facts that

1. every positive number has two real roots of each even order, and
2. every negative number has no real roots of an even order.

This discussion leads to the following definition.

Definition of Principal *n*th root

Let n be a positive integer. The **principal *n*th root** of a real number a is the real number $\sqrt[n]{a}$ defined by the following statements.

1. If n is even and $a \geq 0$, then $\sqrt[n]{a}$ is the *nonnegative number* such that $(\sqrt[n]{a})^n = a$.
2. If n is odd, then $\sqrt[n]{a}$ is the real number such that $(\sqrt[n]{a})^n = a$.

The symbol $\sqrt[n]{a}$ is called a **radical,** a is called the **radicand,** and n is called the **index** or **order** of the radical.

We note that $\sqrt[n]{a}$ is left undefined for the present when n is even and a is negative. This point comes up again later in the study of complex numbers.

Example 1 • Evaluating *n*th Roots

Find the value of each of the following.

a. $\sqrt[4]{625}$ b. $\sqrt[3]{-8}$ c. $\sqrt[5]{32}$ d. $\sqrt{\frac{4}{9}}$ e. $\sqrt[4]{-1}$

Solution

We have

a. $\sqrt[4]{625} = 5$ since $5^4 = 625$ with both 5 and 625 positive.
b. $\sqrt[3]{-8} = -2$ since $(-2)^3 = -8$.
c. $\sqrt[5]{32} = 2$ since $2^5 = 32$.
d. $\sqrt{\frac{4}{9}} = \frac{2}{3}$ since $(\frac{2}{3})^2 = \frac{4}{9}$ with both $\frac{2}{3}$ and $\frac{4}{9}$ positive.
e. $\sqrt[4]{-1}$ is not defined because no real number raised to the fourth power yields a negative result. □

Since a^2 is always nonnegative, then $\sqrt{a^2}$ is always defined and designates the nonnegative number that yields a^2 when it is squared. Since a may be negative, $\sqrt{a^2}$ is not always the same as a. For example, if $a = -1$, then

$$\sqrt{a^2} = \sqrt{(-1)^2} = \sqrt{1} = 1 \neq a.$$

However, it is true that

$$\sqrt{a^2} = \sqrt{(-1)^2} = 1 = |-1| = |a|.$$

For any real a, we can write

$$\sqrt{a^2} = |a|.$$

This property of radicals can be extended for any positive integer n and any real a:

$$\sqrt[n]{a^n} = \begin{cases} a, & \text{if } n \text{ is odd} \\ |a|, & \text{if } n \text{ is even} \end{cases}.$$

Whenever $a \geq 0$, this reduces to

$$\sqrt[n]{a^n} = a, \qquad a \geq 0.$$

Example 2 • Using Absolute Value to Simplify *n*th Roots

Use absolute value, if necessary, to simplify the following expressions.

a. $\sqrt[3]{(-2x)^3}$ b. $\sqrt[4]{(-5x)^4}$ c. $\sqrt[6]{b^{12}}$ d. $\sqrt{x^2y^4}$

Solution

a. $\sqrt[3]{(-2x)^3} = -2x$ (no absolute values needed since 3 is odd)
b. $\sqrt[4]{(-5x)^4} = |-5x| = 5|x|$
c. $\sqrt[6]{b^{12}} = \sqrt[6]{(b^2)^6} = |b^2| = b^2$
d. $\sqrt{x^2y^4} = \sqrt{(xy^2)^2} = |xy^2| = |x|y^2$ □

The properties of radicals in the following list are useful in changing the form of expressions involving radicals.

Basic Properties of Radicals

Let n and m be positive integers, and let a and b be real numbers. Whenever each indicated root is defined, then:

1. $(\sqrt[n]{a})^n = a$
2. $\sqrt[n]{a^n} = \begin{cases} a, & \text{if } n \text{ is odd} \\ |a|, & \text{if } n \text{ is even} \end{cases}$ (if $a \geq 0$, $\sqrt[n]{a^n} = a$)
3. $\sqrt[n]{ab} = \sqrt[n]{a}\,\sqrt[n]{b}$
4. $\sqrt[n]{\dfrac{a}{b}} = \dfrac{\sqrt[n]{a}}{\sqrt[n]{b}}$, $b \neq 0$
5. $\sqrt[m]{\sqrt[n]{a}} = \sqrt[mn]{a}$

In the remainder of this section, we make the assumption that all variables represent positive real numbers, unless otherwise indicated. Thus we can be certain that all indicated roots are defined, that all denominators are not zero, and the basic property 2 occurs only in the form $\sqrt[n]{a^n} = a$.

Example 3 • Using the Basic Properties of Radicals

Find the value of each expression by using the basic properties of radicals.

a. $(\sqrt[4]{5})^4$ b. $\sqrt[3]{2x}\,\sqrt[3]{4x^2}$ c. $\sqrt[5]{\dfrac{-32a^5}{b^{10}}}$ d. $\sqrt[4]{16a^4}$ e. $\sqrt[3]{\sqrt{7}}$

Solution

a. $(\sqrt[4]{5})^4 \overset{\text{Property 1}}{=} 5$

b. $\sqrt[3]{2x}\,\sqrt[3]{4x^2} \overset{\text{Property 3}}{=} \sqrt[3]{8x^3} \overset{\text{Property 3}}{=} \sqrt[3]{2^3}\,\sqrt[3]{x^3} \overset{\text{Property 2}}{=} 2x$

c. $\sqrt[5]{\dfrac{-32a^5}{b^{10}}} \overset{\text{Property 4}}{=} \dfrac{\sqrt[5]{-32a^5}}{\sqrt[5]{b^{10}}} \overset{\text{Property 3}}{=} \dfrac{\sqrt[5]{(-2)^5}\,\sqrt[5]{a^5}}{\sqrt[5]{(b^2)^5}} \overset{\text{Property 2}}{=} \dfrac{-2a}{b^2}$

d. $\sqrt[4]{16a^4} \overset{\text{Property 3}}{=} \sqrt[4]{(2)^4}\,\sqrt[4]{a^4} \overset{\text{Property 2}}{=} 2a$ recall: $a > 0$

e. $\sqrt[3]{\sqrt{7}} \overset{\text{Property 5}}{=} \sqrt[6]{7}$ since $3 \cdot 2 = 6$

□

◆ Practice Problem 1 ◆

Find the value of each expression.

a. $\sqrt{16a^4b^{16}}$ b. $\sqrt[3]{\dfrac{-64}{a^3b^6}}$

Many formulas used in applications involve a square root or cube root. One such formula is Heron's Formula for calculating the area of a triangle as shown in Figure 1.19.

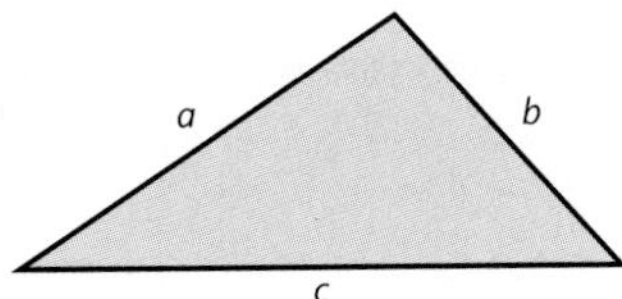

Figure 1.19 A triangle with sides a, b, and c

Heron's Formula

The area A of a triangle with sides a, b, and c is

$$A = \sqrt{s(s-a)(s-b)(s-c)},$$

where s is the semiperimeter of the triangle:

$$s = \tfrac{1}{2}(a + b + c).$$

Most calculators have a $\boxed{\sqrt{x}}$ button. Its use in conjunction with Heron's formula is demonstrated in the next example.

Example 4 • Boat Design

A triangular side window in a ski boat, as shown in Figure 1.20, measures 36 inches, 44 inches, and 20 inches along its three sides. How many square inches of glass does the side window contain?

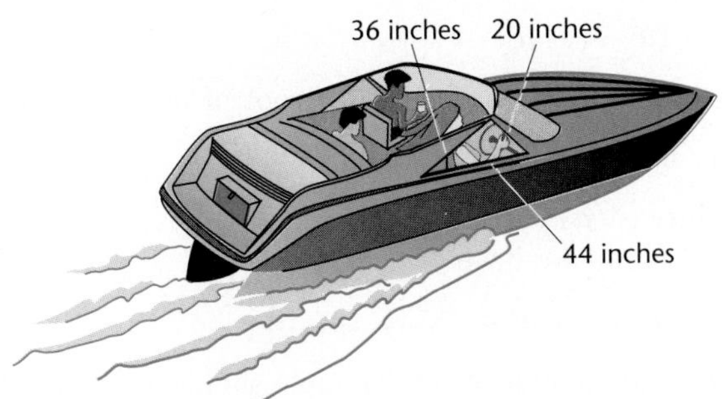

Figure 1.20 Dimensions of a boat window

Solution

We first compute the semiperimeter

$$s = \tfrac{1}{2}(a + b + c) = \tfrac{1}{2}(36 + 44 + 20) = 50.$$

Then Heron's formula gives

$$\begin{aligned} A &= \sqrt{s(s-a)(s-b)(s-c)} \\ &= \sqrt{50(50-36)(50-44)(50-20)} \\ &= \sqrt{50(14)(6)(30)} \\ &= \sqrt{126{,}000}. \end{aligned}$$

Using the $\boxed{\sqrt{x}}$ button and rounding to the nearest square inch, we find that there are approximately 355 square inches of glass in the side window. □

In Section 1.3, we defined what is meant by a^n when n is an integer. We wish now to extend this definition to a^n, where n is a rational number. At the same time, we wish to retain the properties that hold for integral exponents. This guides us in making our new definitions.

If rational exponents are to satisfy the power rule, for example, we must have

$$(a^{1/3})^3 = a.$$

That is, $a^{1/3}$ must be a cube root of a, or

$$a^{1/3} = \sqrt[3]{a}.$$

This leads us to define $a^{1/n}$ as follows:

Definition of $a^{1/n}$

Let n be a positive integer. Then

$$a^{1/n} = \sqrt[n]{a}.$$

A consequence of this definition is that $a^{1/n}$ is not defined if n is even and a is negative.

Again using the power rule as a guide, we need $a^{5/3}$ to satisfy

$$(a^{5/3})^3 = a^5.$$

That is, $a^{5/3}$ must be a cube root of a^5:

$$a^{5/3} = \sqrt[3]{a^5}.$$

Since $(\sqrt[3]{a})^5(\sqrt[3]{a})^5(\sqrt[3]{a})^5 = (\sqrt[3]{a}\,\sqrt[3]{a}\,\sqrt[3]{a})^5 = a^5$, then $(\sqrt[3]{a})^5$ is also a cube root of a^5 and we have

$$a^{5/3} = \sqrt[3]{a^5} = (\sqrt[3]{a})^5.$$

This discussion of our example should make the following definition plausible.

Definition of $a^{m/n}$

If m is an integer and n is a positive integer, then we define $a^{m/n}$ by

$$a^{m/n} = \sqrt[n]{a^m} = (\sqrt[n]{a})^m,$$

with the conditions that $a \geq 0$ if n is even and $a \neq 0$ if $m \leq 0$.

Under the conditions we have placed on m and n, each of the expressions $\sqrt[n]{a^m}$ and $(\sqrt[n]{a})^m$ is defined when the other is, and they are always equal. We note that $a^{m/n}$ is undefined when n is even and $a < 0$, or when $m \leq 0$ and $a = 0$. As in the last section, we make the assumption that all variables represent positive real numbers in order that all roots be defined.

Since rational exponents are defined in terms of radicals, any expression involving one of them can be changed to an expression involving the other. These kinds of changes are illustrated in the following example. The form $a^{m/n}$ involving a rational exponent is referred to as the **exponential form** of either of the **radical forms** $\sqrt[n]{a^m}$ or $(\sqrt[n]{a})^m$.

Example 5 • Conversions Between Radical and Exponential Form

Each of the following expressions is written in both radical form and exponential form.

a. $\sqrt[3]{a^5} = a^{5/3}$

b. $\left(\sqrt[4]{\dfrac{ab}{c^2}}\right)^3 = \left(\dfrac{ab}{c^2}\right)^{3/4}$

c. $\left(\dfrac{x^2y}{8}\right)^{2/5} = \sqrt[5]{\left(\dfrac{x^2y}{8}\right)^2}$ or $\left(\sqrt[5]{\dfrac{x^2y}{8}}\right)^2$

d. $a^{7/5} = \sqrt[5]{a^7}$ or $(\sqrt[5]{a})^7$ □

In the next example we illustrate the definition of $a^{m/n}$ by evaluating several expressions involving rational exponents. An evaluation can sometimes be made easier by an appropriate choice between the forms $\sqrt[n]{a^m}$ and $(\sqrt[n]{a})^m$.

Example 6 • Evaluating Expressions with Rational Exponents

a. $(16)^{1/4} = \sqrt[4]{16} = 2$

b. $\left(\dfrac{a^2}{16}\right)^{3/2} = \left(\sqrt{\dfrac{a^2}{16}}\right)^3 = \left(\dfrac{a}{4}\right)^3 = \dfrac{a^3}{64}$ recall: $a > 0$

c. $\left(-\dfrac{8}{27}\right)^{2/3} = \left(\sqrt[3]{-\dfrac{8}{27}}\right)^2 = \left(-\dfrac{2}{3}\right)^2 = \dfrac{4}{9}$

If the other radical form is chosen here, the evaluation is a little more difficult.

$$\sqrt[3]{\left(-\frac{8}{27}\right)^2} = \sqrt[3]{\frac{64}{729}} = \frac{4}{9}$$

d. $\left(\dfrac{-8a^6}{b^3}\right)^{4/3} = \left(\sqrt[3]{\dfrac{-8a^6}{b^3}}\right)^4 = \left(\dfrac{-2a^2}{b}\right)^4 = \dfrac{16a^8}{b^4}$

e. $(64)^{-5/6} = (\sqrt[6]{64})^{-5} = (2)^{-5} = \dfrac{1}{32}$ □

◆ Practice Problem 2 ◆

Assume that variables represent positive real numbers and evaluate each of the following expressions.

a. $(-32)^{-3/5}$ b. $\left(-\frac{x^6}{27y^9}\right)^{2/3}$

It can be shown that the laws of exponents hold for rational exponents as long as the radicals involved are defined. Although we do not prove them here, we will illustrate their use in the following example.

Example 7 • Simplifying Expressions with Rational Exponents

Use the laws of exponents to simplify.

a. $\frac{x^{1/2}x^{3/4}}{x^{1/3}}$ b. $(a^{1/4}b^{1/5})^{20}$ c. $(r^{-1/3}s^{2/3})^{-3/2}$

Solution

a. $\frac{x^{1/2}x^{3/4}}{x^{1/3}} = x^{1/2+3/4-1/3}$ product and quotient rules

$= x^{(6+9-4)/12}$ adding fractional exponents

$= x^{11/12}$

b. $(a^{1/4}b^{1/5})^{20} = (a^{1/4})^{20}(b^{1/5})^{20}$ using $(ab)^n = a^n b^n$

$= a^{20/4}b^{20/5}$ power rule

$= a^5b^4$ simplifying

c. $(r^{-1/3}s^{2/3})^{-3/2} = (r^{-1/3})^{-3/2}(s^{2/3})^{-3/2}$ using $(ab)^n = a^n b^n$

$= r^{1/2}s^{-1}$ power rule

$= \frac{r^{1/2}}{s}$ removing the negative exponent □

◆ Practice Problem 3 ◆

Use the laws of exponents to simplify.

a. $[(x^{1/2}y^{1/3})^6]^{1/2}$ b. $[(a^3b^{-4})^0]^{-3/5}$ c. $\left(\frac{x^{1/3}y^{4/3}}{x^{2/3}y^{2/3}}\right)^{-6}$

The $\boxed{y^x}$ key on a calculator can be used to evaluate any radical by changing to the corresponding exponential form as illustrated in the next example.

Example 8 • Evaluating Radicals with a Calculator

Evaluate each of the following and round the result to the nearest hundredth.

a. $\sqrt[3]{2.3}$ b. $\left(\dfrac{1}{\sqrt[5]{93.8}}\right)^2$

Solution

a. $\sqrt[3]{2.3} = (2.3)^{1/3}$ writing in exponential form

≈ 1.3200061 using the $\boxed{y^x}$ key

≈ 1.32 rounding to the nearest hundredth

b. $\left(\dfrac{1}{\sqrt[5]{93.8}}\right)^2 = (93.8)^{-2/5}$ writing in exponential form

≈ 0.1625994 using the $\boxed{y^x}$ key

≈ 0.16 rounding to the nearest hundredth

Alternatively, we can evaluate $\left(\dfrac{1}{\sqrt[5]{93.8}}\right)^2$ by first writing $\sqrt[5]{93.8}$ as $(93.8)^{1/5}$.

$$\left(\frac{1}{\sqrt[5]{93.8}}\right)^2 = \left(\frac{1}{(93.8)^{1/5}}\right)^2 \overset{\boxed{y^x}}{\approx} \left(\frac{1}{2.4799365}\right)^2$$

$$\overset{\boxed{1/x}}{\approx} (0.4032361)^2 \overset{\boxed{x^2}}{\approx} 0.1625994 \approx 0.16$$

□

Example 9 • Radius of a Balloon

If the inflated spherical balloon in Figure 1.21 holds 310 cubic inches of air, use the formula

$$r = \sqrt[3]{\frac{3V}{4\pi}}$$

to determine the radius r to the nearest tenth of an inch.

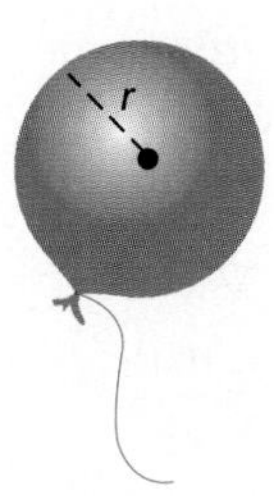

Figure 1.21 Spherical balloon with radius r

Solution

Replacing V with 310 and using the exponential form of the radical, we compute the radius to be

$$r = \sqrt[3]{\frac{3(310)}{4\pi}} = \left(\frac{3(310)}{4\pi}\right)^{1/3} \approx 4.2 \text{ inches.}$$

□

EXERCISES 1.7

Find the value of each expression. (See Example 1.)

1. $\sqrt{25}$ 2. $\sqrt{1.21}$ 3. $\sqrt[3]{64}$ 4. $\sqrt[3]{0.008}$

5. $(\sqrt[5]{73})^5$
6. $(\sqrt{29})^2$
7. $\sqrt[4]{\frac{256}{81}}$
8. $\sqrt[3]{-\frac{1}{8}}$
9. $\sqrt{49}$
10. $\sqrt[5]{0.00032}$
11. $\sqrt{\sqrt[5]{1024}}$
12. $\sqrt[3]{\sqrt{64}}$

In Exercises 13–20, variables may have negative values as well as positive values. Use absolute values as necessary to evaluate the following expression. (See Example 2.)

13. $\sqrt{25x^2}$
14. $\sqrt{(x+2)^2}$
15. $\sqrt[3]{a^3b^3}$
16. $\sqrt[5]{(x+y)^5}$
17. $\sqrt[4]{(-2x)^4}$
18. $\sqrt[6]{(4y)^6}$
19. $\sqrt{(r+s)^4}$
20. $\sqrt{(2a-b)^4}$

Find the value of the indicated root by using the basic properties of radicals. Assume that variables represent positive real numbers. (See Example 3 and Practice Problem 1.)

21. $\sqrt[3]{8b^6}$
22. $\sqrt{a^2}$
23. $\sqrt{y^4w^6}$
24. $\sqrt[4]{a^{12}}$
25. $\sqrt{0.16x^4}$
26. $\sqrt[3]{0.125x^3}$
27. $\sqrt[7]{-1}$
28. $\sqrt[3]{-8x^3y^6}$
29. $\sqrt[4]{\frac{16x^4}{81y^8}}$
30. $\sqrt[5]{\frac{-243x^{10}}{32y^5}}$
31. $\sqrt[3]{\frac{216x^3}{125}}$
32. $\sqrt{\frac{y^4}{w^6}}$
33. $\sqrt[4]{x^8y^{-12}}$
34. $\sqrt[5]{\frac{-1}{x^{15}}}$
35. $(\sqrt[4]{19y^5})^4$
36. $(\sqrt[3]{-2x})^3$
37. $\sqrt{\sqrt[3]{x^6y^{12}}}$
38. $\sqrt{\sqrt{x^4y^{12}}}$
39. $\sqrt[3]{\sqrt{\frac{x^{18}}{z^6y^0}}}$
40. $\sqrt{\sqrt[5]{\frac{1024}{y^{40}}}}$
41. $\sqrt{\left(\frac{81x^4}{25z^6}\right)^{-1}}$
42. $\sqrt[3]{\left(\frac{-8y^0}{x^6}\right)^{-2}}$

For each expression, change from radical form to exponential form, or from exponential form to radical form, whichever is appropriate. (See Example 5.)

43. $a^{1/5}$
44. $x^{5/7}$
45. $3a^{1/4}$
46. $bx^{4/5}$
47. $\sqrt[4]{a^5}$
48. $\sqrt[3]{x^8}$
49. $\sqrt[9]{x^{10}}$
50. $\sqrt[5]{x^{10}}$
51. $\sqrt[3]{a^6}$
52. $\sqrt{a^8}$
53. $(3b)^{1/4}$
54. $(5a^3)^{1/5}$
55. $(2xy^2)^{2/3}$
56. $(xy)^{5/3}$
57. $\sqrt[7]{bx^2}$
58. $\sqrt[5]{ay^3}$

Evaluate each of the following expressions. Assume that variables represent positive real numbers. (See Example 6 and Practice Problem 2.)

59. $(16)^{1/4}$
60. $(4)^{1/2}$
61. $(-64a^6)^{1/6}$
62. $(-27)^{1/3}$
63. $(64)^{-4/3}$
64. $(25)^{-3/2}$
65. $(81)^{3/4}$
66. $(16)^{5/4}$
67. $\left(\frac{1}{32}\right)^{-4/5}$
68. $\left(\frac{1}{16}\right)^{-3/4}$
69. $(8p^6)^{2/3}$
70. $(16m^8)^{3/4}$

71. $(a^2b^4)^{3/2}$
72. $(a^3b^6)^{2/3}$
73. $\left(-\dfrac{27}{64}\right)^{2/3}$
74. $\left(-\dfrac{8}{125}\right)^{2/3}$
75. $-\left(\dfrac{4m^2}{n^4}\right)^{3/2}$
76. $-\left(\dfrac{9p^4}{q^6}\right)^{3/2}$
77. $-\left(\dfrac{x^4}{16y^2}\right)^{-3/2}$
78. $-\left(\dfrac{m^6}{25p^4}\right)^{-3/2}$

Use the laws of exponents to simplify. (See Example 7 and Practice Problem 3.)

79. $x^{2/3}x^{7/3}$
80. $r^{1/5}r^{4/5}$
81. $(x^{1/3}y^{5/6})^{12}$
82. $(a^{4/9}b^{2/3})^{18}$
83. $(a^{-1/2}b^{3/2})^{-4/3}$
84. $(t^{20/9}r^{-10/3})^{-9/10}$
85. $[(a^{2/3}b^{-1/4})^{-2}]^{-6}$
86. $[(x^{1/3}y^{-2/3})^{-1/2}]^6$
87. $[(x^2y^{-2/3})^{-3}]^0$
88. $[(2c^9d^{5/4})^0]^{-3}$
89. $x^{1/3}(x^{2/3} - x^{-1/3})$
90. $y^{2/5}(y^{3/5} + y^{-2/5})$
91. $(a^{1/2} - b^{1/2})(a^{1/2} + b^{1/2})$
92. $(a^{-3/2} + b^{-3/2})(a^{-3/2} - b^{-3/2})$
93. $\dfrac{ab^0c^{2/5}}{a^{3/4}b^{-1/3}c^{1/5}}$
94. $\dfrac{x^{2/3}y^{7/4}z}{x^0y^{3/4}z^{1/2}}$

95–114

Use a calculator with a $\boxed{y^x}$ key to evaluate each of the following numbers. Round your results to the nearest hundredth. (See Example 8.)

95. $(2.5)^{1.4}$
96. $(0.015)^{0.003}$
97. $(27.1)^{2/5}$
98. $(4.12)^{-1/3}$
99. $\sqrt[5]{731}$
100. $(\sqrt[4]{49.3})^3$
101. $\left(\dfrac{1}{\sqrt[10]{90.2}}\right)^3$
102. $\left(\dfrac{1}{\sqrt[8]{1921}}\right)^3$

Windsurfer Sail

103. Use Heron's formula to find the area (to the nearest square foot) of the windsurfer sail in the accompanying photograph if the sides of the sail measure 114 inches, 108 inches, and 66 inches.

Christian Le Bozec/Agence Vandystadt Photo Researchers

Surveying 104. A triangular lot on a corner at a street intersection has sides with lengths 472 feet, 413 feet, and 161 feet. Use Heron's formula to find the area of the lot.

Pendulum 105. The time t (in seconds) required for one swing of the pendulum in the accompanying figure is given by $t = 0.555\sqrt{d}$, where d is the length (in feet) of the pendulum. Compute the time (to the nearest hundredth of a second) of one swing of a pendulum 2.3 feet long.

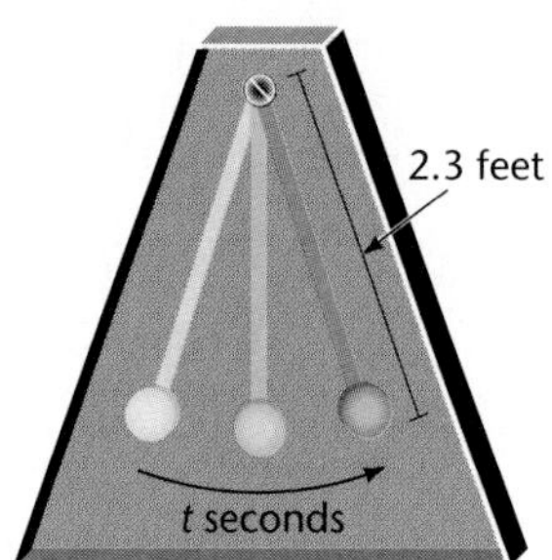

Figure for Exercise 105

Illumination 106. In the formula $d = 30/\sqrt{I}$, I represents the intensity (in foot-candles) of illumination of a light at a distance d (in feet) from the source of the light. Find the distance (to the nearest foot) from the street light to the park bench in the accompanying figure if the intensity of illumination at the bench is 7 foot-candles.

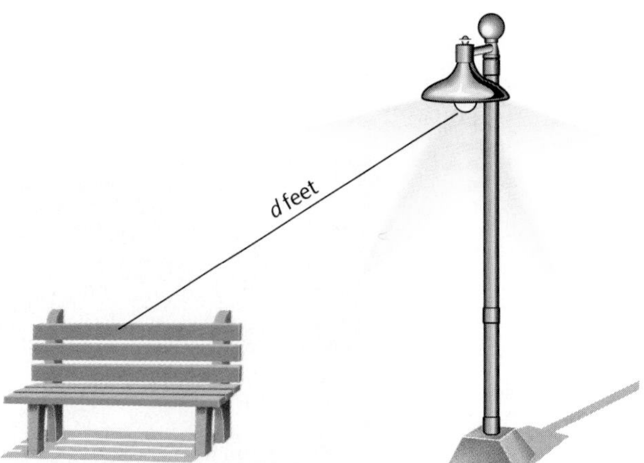

Figure for Exercise 106

Wind on a Sail 107. The wind speed s (in miles per hour) can be computed by the formula $s = 100\sqrt{P/A}$, where A is the area (in square feet) of a sail and P is the pressure (in pounds per square foot) of the wind on the sail. Find the wind speed (to the nearest mile per hour) that generates 10 pounds per square foot of pressure on a 50-square-foot sail.

Standard Deviation and Mean

108. Statisticians use the following formulas to compute the standard deviation, denoted by the Greek letter σ (sigma), of a population of four values x_1, x_2, x_3, x_4, where the Greek letter μ (mu) represents the mean or average of the four values.

$$\mu = \frac{x_1 + x_2 + x_3 + x_4}{4}$$

$$\sigma = \sqrt{\frac{(x_1 - \mu)^2 + (x_2 - \mu)^2 + (x_3 - \mu)^2 + (x_4 - \mu)^2}{4}}$$

Use these formulas to compute the standard deviation (to the nearest tenth) of the test scores 93, 87, 99, and 65.

For Exercises 109 and 110, if your calculator does not have a $\boxed{\pi}$ key, use 3.14 as an approximation of π.

Drinking Cup Design

109. Find the radius r (to the nearest tenth of an inch) of the conical drinking cup in the accompanying figure that holds 9.4 cubic inches of water and is 3.5 inches tall. Use the formula $r = \sqrt{\frac{3V}{\pi h}}$, where V represents the volume (in cubic inches) and h represents the height (in inches) of the cup.

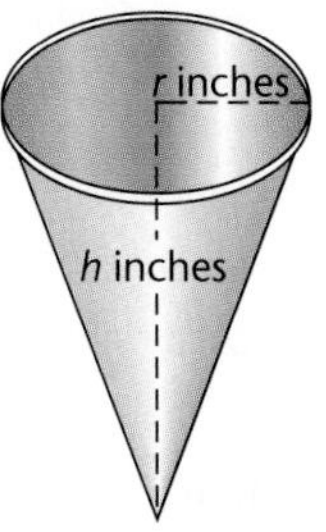

Figure for Exercise 109

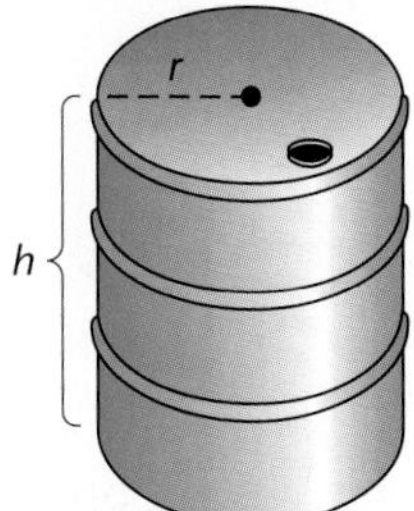

Figure for Exercise 110

Dimensions of an Oil Drum

110. Find the radius (to the nearest hundredth of a foot) of a 55-gallon oil drum that is 2.83 feet tall by using the formula $r = \sqrt{\frac{0.133V}{\pi h}}$, where V represents the volume (in gallons) and h represents the height (in feet) of the oil drum. (See the accompanying figure.)

Design of a Die

111. The solid cubical die in the accompanying figure was manufactured using 2.2 cubic centimeters of plastic (with no waste in construction). Find the length of the edge x to the nearest tenth of a centimeter.

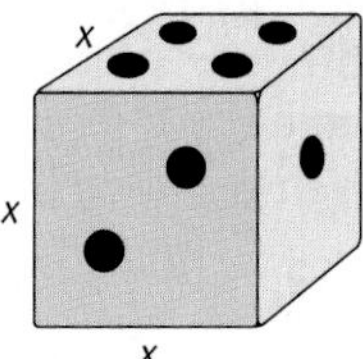

Figure for Exercises 111 and 112

Surface Area of a Die

112. The sum S of the areas of the six faces (the surface area) of the die in the accompanying figure can be computed by the formula $S = 6V^{2/3}$, where V is the volume of the die. Find the surface area (to the nearest tenth of a square centimeter) of the die with a volume of 5.8 cubic centimeters.

Propane Tank Design

113. The propane tank in the accompanying figure holds 65.5 cubic feet of fuel when completely full. Use the formula $V = \frac{4}{3}\pi r^3$ to find the radius r (to the nearest tenth of a foot), where V is the volume (in cubic feet) of the tank.

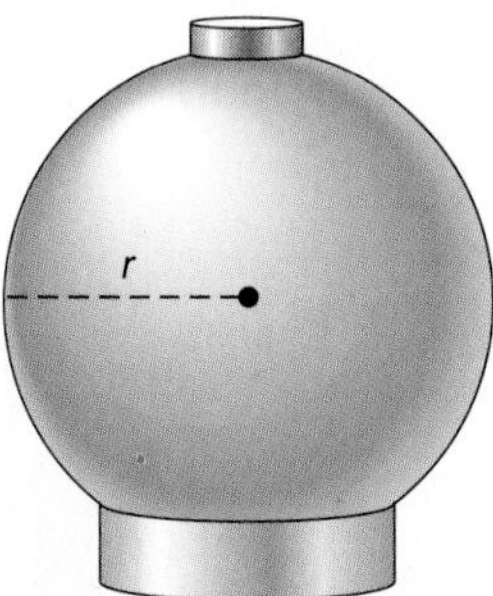

Figure for Exercises 113 and 114

Amount of Paint Needed

114. In order to repaint the propane tank in Exercise 113, the painter needs to know the surface area S of the spherical portion of the tank. Use the formula $S = (4\pi)^{1/3}(3V)^{2/3}$ to compute the surface area S (to the nearest tenth of a square foot) of the tank with a volume V of 65.5 cubic feet.

Critical Thinking: Exploration and Writing

115. According to one of the basic properties of radicals, it is true that $\sqrt{ab} = \sqrt{a}\sqrt{b}$ whenever all the square roots involved are defined. This sometimes leads to confusion between $\sqrt{a + b}$ and $\sqrt{a} + \sqrt{b}$. Use the definition of a square root to explain why $\sqrt{a + b}$ and $\sqrt{a} + \sqrt{b}$ are *never equal* whenever a and b are *both nonzero*.

Critical Thinking: Discussion

116. Use the fact that every real number has one cube root to explain why $\sqrt[3]{a^2}$ and $(\sqrt[3]{a})^2$ are the same number.

Critical Thinking: Exploration and Writing

117. We have seen that $\sqrt{a^2} = |a|$ for any real number a. Explain why the statement $\sqrt[3]{a^3} = |a|$ is not true.

Critical Thinking: Discussion

118. Explain why 0^{-2} is undefined.

Critical Thinking: Discussion

119. Explain why $a^{m/2}$ is undefined if a is negative.

◆ Solutions for Practice Problems

1. a. $\sqrt{16a^4b^{16}} = \sqrt{4^2}\sqrt{(a^2)^2}\sqrt{(b^8)^2} = 4a^2b^8$

   b. $\sqrt[3]{\dfrac{-64}{a^3b^6}} = \dfrac{\sqrt[3]{(-4)^3}}{\sqrt[3]{a^3}\sqrt[3]{(b^2)^3}} = \dfrac{-4}{ab^2}$

2. a. $(-32)^{-3/5} = (\sqrt[5]{-32})^{-3} = (-2)^{-3} = \dfrac{1}{(-2)^3} = -\dfrac{1}{8}$

   b. $\left(-\dfrac{x^6}{27y^9}\right)^{2/3} = \left(\sqrt[3]{-\dfrac{x^6}{27y^9}}\right)^2 = \left(-\dfrac{x^2}{3y^3}\right)^2 = \dfrac{x^4}{9y^6}$

3. a. $[(x^{1/2}y^{1/3})^6]^{1/2} = (x^3y^2)^{1/2} = x^{3/2}y$

   b. $[(a^3b^{-4})^0]^{-3/5} = (1)^{-3/5} = 1$

   c. $\left(\dfrac{x^{1/3}y^{4/3}}{x^{2/3}y^{2/3}}\right)^{-6} = \left(\dfrac{y^{2/3}}{x^{1/3}}\right)^{-6} = \dfrac{y^{-4}}{x^{-2}} = \dfrac{x^2}{y^4}$

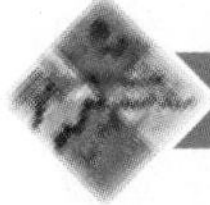

1.8 OPERATIONS WITH RADICALS

In this section, we consider ways in which an expression involving radicals or rational exponents may be changed. Frequently these changes lead to a simpler form of an expression. We will discuss and illustrate each of these ways. Again, we assume that all variables represent positive real numbers.

Procedure 1: Removal of Factors from the Radicand

A radical may be simplified by using these facts.

1. $\sqrt[n]{ab} = \sqrt[n]{a}\sqrt[n]{b}$
2. $\sqrt[n]{a^n} = a$

Example 1 • Removing Factors from the Radicand

Simplify the following radicals by removing as many factors as possible from the radicand.

a. $\sqrt{32x^2}$ b. $\sqrt[3]{32x^4y^5}$

Solution

a. We look for perfect-square factors under the radical.

$$\begin{aligned} \sqrt{32x^2} &= \sqrt{16x^2 \cdot 2} && \text{factoring the radicand} \\ &= \sqrt{(4x)^2}\sqrt{2} && \text{using } \sqrt{ab} = \sqrt{a}\sqrt{b} \\ &= 4x\sqrt{2} && \text{using } \sqrt{a^2} = a \end{aligned}$$

b. Since the index is 3, we look for perfect-cube factors in the radicand.

$$\begin{aligned} \sqrt[3]{32x^4y^5} &= \sqrt[3]{8x^3y^3 \cdot 4xy^2} && \text{factoring the radicand} \\ &= \sqrt[3]{(2xy)^3}\sqrt[3]{4xy^2} && \text{using } \sqrt[3]{ab} = \sqrt[3]{a}\sqrt[3]{b} \\ &= 2xy\sqrt[3]{4xy^2} && \text{using } \sqrt[3]{a^3} = a \end{aligned}$$

□

Procedure 2: Additions by Use of the Distributive Property

In a sum involving radicals, we follow this procedure:

1. First we simplify the terms by removing all factors possible from the radicand as described in Procedure 1.
2. Then, if two or more radicals have the same index and radicand, they can be combined by using the distributive property.

Example 2 • Addition and Subtraction of Radicals

The following simplifications illustrate Procedure 2.

a. $\sqrt{32} + \sqrt{18} = \sqrt{(16)(2)} + \sqrt{(9)(2)} = 4\sqrt{2} + 3\sqrt{2}$ Procedure 1
$= 7\sqrt{2}$ distributive property

This simplification illustrates the important fact that

$$\sqrt{x + y} \neq \sqrt{x} + \sqrt{y}.$$

For, in this case, we have

$$\sqrt{32 + 18} = \sqrt{50} = \sqrt{(25)(2)} = 5\sqrt{2},$$

whereas

$$\sqrt{32} + \sqrt{18} = 7\sqrt{2}.$$

b. $\sqrt{8} - b\sqrt{18a^4} = 2\sqrt{2} - 3a^2b\sqrt{2}$ Procedure 1
$= (2 - 3a^2b)\sqrt{2}$ distributive property

c. $\sqrt{3} + \sqrt{7} - 5\sqrt{2}$ cannot be simplified. □

Procedure 3: Multiplications or Divisions

The product or quotient of two radicals of the same index can be expressed as a single radical by using the properties

$$\sqrt[n]{a}\,\sqrt[n]{b} = \sqrt[n]{ab} \qquad \text{and} \qquad \frac{\sqrt[n]{a}}{\sqrt[n]{b}} = \sqrt[n]{\frac{a}{b}},$$

when $\sqrt[n]{a}$ and $\sqrt[n]{b}$ are defined.

Example 3 • Multiplication and Division of Radicals

Procedure 3 is illustrated by the following examples.

a. $(2\sqrt{3})(3\sqrt{2}) = 2 \cdot 3 \cdot \sqrt{3} \cdot \sqrt{2} = 6\sqrt{6}$

b. $\dfrac{\sqrt{4}}{\sqrt{6}} = \sqrt{\dfrac{4}{6}} = \sqrt{\dfrac{2}{3}}$, or $\dfrac{\sqrt{4}}{\sqrt{6}} = \dfrac{\sqrt{2}\sqrt{2}}{\sqrt{2}\sqrt{3}} = \dfrac{\sqrt{2}}{\sqrt{3}} = \sqrt{\dfrac{2}{3}}$

c. $\dfrac{\sqrt[3]{ab}}{\sqrt[3]{b^2}} = \sqrt[3]{\dfrac{ab}{b^2}} = \sqrt[3]{\dfrac{a}{b}}$

d. $\sqrt[3]{4a^2} \cdot \sqrt[3]{10a^2b^5} = \sqrt[3]{40a^4b^5}$

$= \sqrt[3]{8a^3b^3}\,\sqrt[3]{5ab^2}$

$= 2ab\sqrt[3]{5ab^2}$ □

◆ Practice Problem 1 ◆

Perform the multiplications and remove all factors possible from the radicands.

a. $\sqrt[3]{9x^2y^2}\,\sqrt[3]{12x^2y}$ b. $(\sqrt{5} - 3\sqrt{3})(\sqrt{15} - 4)$

Procedure 4: Reduction of the Index of a Radical

It may be possible to reduce the index of a radical if the radicand can be written as a power that has a factor in common with the index.

Example 4 • Reducing the Index of a Radical

The usefulness of rational exponents is brought out in the following illustrations of a reduction of the index.

a. $\sqrt[4]{25} = \sqrt[4]{5^2} = 5^{2/4} = 5^{1/2} = \sqrt{5}$

b. $\sqrt[9]{8} = \sqrt[9]{2^3} = 2^{3/9} = 2^{1/3} = \sqrt[3]{2}$ □

In some instances, it is desirable to change a quotient that has radicals in the denominator to an expression that has no radicals in the denominator. This process is called **rationalizing the denominator.**

Procedure 5: Rationalizing the Denominator

Radicals can be eliminated from the denominator of a quotient by multiplying the numerator and denominator of the fraction by an appropriate factor.

Example 5 • Rationalizing the Denominator

Rationalize the denominators of the following fractions.

a. $\dfrac{3}{\sqrt{7}}$ b. $\sqrt[3]{\dfrac{3}{4}}$ c. $\sqrt[3]{\dfrac{2a}{9b^2}}$

Solution

In each case, we multiply by a factor chosen so that it forms a power in the denominator which is the same as the index of the radical. It is efficient to make the multiplying factor as small as possible.

a. $\dfrac{3}{\sqrt{7}} = \dfrac{3\sqrt{7}}{\sqrt{7}\sqrt{7}} = \dfrac{3\sqrt{7}}{(\sqrt{7})^2} = \dfrac{3\sqrt{7}}{7}$

b. $\sqrt[3]{\dfrac{3}{4}} = \sqrt[3]{\dfrac{3 \cdot 2}{4 \cdot 2}} = \sqrt[3]{\dfrac{6}{(2)^3}} = \dfrac{\sqrt[3]{6}}{2}$

c. $\sqrt[3]{\dfrac{2a}{9b^2}} = \sqrt[3]{\dfrac{(2a)(3b)}{(9b^2)(3b)}} = \sqrt[3]{\dfrac{6ab}{(3)^3b^3}} = \dfrac{\sqrt[3]{6ab}}{3b}$ □

We frequently encounter quotients with a denominator that contains radicals and is more complex than those in Example 5. Often a denominator is of the form $a + \sqrt{b}$. To rationalize this type of denominator, we multiply the numerator and denominator of the fraction by $a - \sqrt{b}$. The expression obtained from a binomial by changing the sign on the second term is called the **conjugate** of the binomial. Thus $a + \sqrt{b}$ and $a - \sqrt{b}$ are conjugates of each other. The product of a binomial that contains square roots and its conjugate is always free of radicals. For example,

$$(a + \sqrt{b})(a - \sqrt{b}) = a^2 - (\sqrt{b})^2 = a^2 - b.$$

Example 6 • Using the Conjugate

Rationalize the denominator.

$$\frac{3}{3 - \sqrt{5}}$$

Solution

$$\frac{3}{3-\sqrt{5}} = \frac{3}{3-\sqrt{5}} \cdot \frac{3+\sqrt{5}}{3+\sqrt{5}}$$ multiplying numerator and denominator by the conjugate of $3 - \sqrt{5}$

$$= \frac{3(3+\sqrt{5})}{9-5}$$ multiplying the denominators

$$= \frac{3(3+\sqrt{5})}{4}$$ simplifying

□

◆ Practice Problem 2 ◆

Rationalize the denominator.

$$\frac{2}{4\sqrt{2}+\sqrt{3}}$$

There are some problems in which it is desirable to rationalize the numerator, that is, to eliminate all radicals in the numerator. (See Exercises 101–108.)

If a radical expression satisfies a certain list of conditions, it is said to be in simplest radical form. These conditions are given in the next definition.

Definition of Simplest Radical Form

A radical expression is in **simplest radical form** if the following conditions are satisfied:

1. The expression contains no negative or zero exponents.
2. The radicand contains no factor to a power equal to or greater than the index of the radical.
3. The denominator contains no radicals, and no fraction appears under a radical.
4. The index of the radical is as small as possible.

An expression involving radicals can be reduced to simplest form by the following procedure:

Reduction of a Radical Expression to Simplest Form

1. Eliminate all negative or zero exponents.
2. Express any power or root of a radical, or any product of radicals, as a single radical.
3. Reduce each radicand to a simple fraction in lowest terms and rationalize all denominators.

4. Remove from each radicand any factor that is a perfect nth power, where n is the index of the radical.
5. Reduce the index of the radical to the lowest possible order.
6. In a sum, combine any terms with the same index and radicand.

Example 7 • Simplest Radical Form

Reduce each of the following expressions to the simplest radical form. Assume that all variables represent positive real numbers.

a. $\sqrt[6]{\dfrac{a^2}{27c^{10}}}$ b. $\dfrac{3}{2-3\sqrt{2}} \cdot \dfrac{\sqrt{2}-1}{\sqrt{2}}$

Solution

a.
$$\sqrt[6]{\frac{a^2}{27c^{10}}} = \sqrt[6]{\frac{a^2}{27c^{10}} \cdot \frac{27c^2}{27c^2}} \quad \text{forcing perfect sixth powers in the denominator}$$
$$= \sqrt[6]{\frac{27a^2c^2}{(3)^6(c^2)^6}} \quad \text{factoring the denominator}$$
$$= \frac{\sqrt[6]{27a^2c^2}}{3c^2} \quad \text{simplifying the denominator}$$

b.
$$\frac{3}{2-3\sqrt{2}} \cdot \frac{\sqrt{2}-1}{\sqrt{2}} = \frac{3(\sqrt{2}-1)}{2\sqrt{2}-6} \quad \text{multiplying the denominators}$$
$$= \frac{3(\sqrt{2}-1)}{2\sqrt{2}-6} \cdot \frac{2\sqrt{2}+6}{2\sqrt{2}+6} \quad \text{rationalizing the denominator}$$
$$= \frac{3(4+6\sqrt{2}-2\sqrt{2}-6)}{8-36} \quad \text{multiplying}$$
$$= \frac{3(-2+4\sqrt{2})}{-28} \quad \text{simplifying}$$
$$= \frac{3(1-2\sqrt{2})}{14} \quad \text{reducing to lowest terms}$$
□

◆ **Practice Problem 3** ◆

Change each of the following to simplest radical form. All variables represent positive real numbers.

a. $\dfrac{\sqrt{5a^3b^5}}{\sqrt{10a^4b}}$ b. $\sqrt[3]{9x} - \sqrt[3]{243x^{-2}}$

EXERCISES 1.8

Simplify by removing all possible factors from the radicand. All variables represent positive real numbers. (See Example 1.)

1. $\sqrt{500}$
2. $\sqrt[3]{-16}$
3. $\sqrt{32a^3b^4}$
4. $\sqrt{8a^6b^5}$
5. $\sqrt[3]{128x^7}$
6. $\sqrt[3]{8x^8}$
7. $\sqrt[5]{x^{13}}$
8. $\sqrt[5]{-x^6y^7}$
9. $\sqrt[3]{-128a^9}$
10. $\sqrt[4]{16z^6}$
11. $\sqrt[4]{162(a+b)^5}$
12. $\sqrt[3]{108(x+y)^7}$

Combine whenever possible by using the distributive property. (See Example 2.)

13. $\sqrt{2}+\sqrt{3}$
14. $\sqrt{3}+\sqrt{5}$
15. $\sqrt{150}-\sqrt{24}$
16. $\sqrt{12}+\sqrt{27}$
17. $3\sqrt{50}-2\sqrt{18}$
18. $5\sqrt{72}+3\sqrt{98}$
19. $3\sqrt{2}-2\sqrt[3]{2}$
20. $5\sqrt[3]{4}-4\sqrt[5]{4}$
21. $\sqrt[3]{54a^3}+2b\sqrt[3]{16}$
22. $3a\sqrt[3]{16}-4b\sqrt[3]{2}$
23. $\sqrt[3]{375x^3y^4}+\sqrt[3]{-3x^3y^4}$
24. $\sqrt[3]{16a^3x}+\sqrt[3]{-54a^6x}$

Perform the indicated multiplications or divisions and then remove all factors possible from the radicand. All variables represent positive real numbers. (See Example 3 and Practice Problem 1.)

25. $\sqrt[3]{5}\cdot\sqrt[3]{50}$
26. $\sqrt[3]{4}\cdot\sqrt[3]{12}$
27. $\dfrac{\sqrt{15}}{\sqrt{3}}$
28. $\dfrac{\sqrt[3]{20}}{\sqrt[3]{5}}$
29. $\dfrac{\sqrt[3]{20x^4}}{\sqrt[3]{2x}}$
30. $\dfrac{\sqrt{15x}}{\sqrt{5x}}$
31. $\sqrt[3]{4x^2}\cdot\sqrt[3]{6x^2y^4}$
32. $\sqrt{3a}\cdot\sqrt{12a^3}$
33. $(2-\sqrt{3})(3+2\sqrt{3})$
34. $(3-2\sqrt{7})(2+\sqrt{7})$
35. $(\sqrt{5}+4\sqrt{2})(\sqrt{5}-4\sqrt{2})$
36. $(\sqrt{6}-3\sqrt{2})(\sqrt{6}+3\sqrt{2})$
37. $(\sqrt{5}-3\sqrt{2})(\sqrt{10}-3)$
38. $(\sqrt{7}+\sqrt{3})(2\sqrt{21}-1)$

Reduce the radical index and remove any factors possible from the radicand. All variables represent positive real numbers. (See Example 4.)

39. $\sqrt[9]{125}$
40. $\sqrt[4]{0.64}$
41. $\sqrt[6]{u^4}$
42. $\sqrt[9]{a^6}$
43. $\sqrt[12]{y^8}$
44. $\sqrt[12]{x^9}$
45. $\sqrt[10]{25a^6b^4}$
46. $\sqrt[12]{16x^4y^8}$

Rationalize the denominators and express the results in simplest radical form. Assume that all variables represent positive real numbers. (See Examples 5 and 6 and Practice Problem 2.)

47. $\sqrt{\dfrac{1}{5}}$
48. $\sqrt{\dfrac{3}{7}}$
49. $\sqrt{\dfrac{5}{8}}$
50. $\sqrt{\dfrac{2}{125}}$

51. $\frac{5}{\sqrt{7}}$ 52. $\frac{2}{\sqrt{11}}$ 53. $\sqrt[3]{\frac{1}{4}}$ 54. $\sqrt[3]{\frac{4}{9}}$

55. $\frac{3\sqrt{5}}{2\sqrt{7}}$ 56. $\frac{5\sqrt{2}}{3\sqrt{11}}$ 57. $\frac{\sqrt[3]{12}}{\sqrt[3]{4}}$ 58. $\frac{\sqrt[3]{18}}{\sqrt[3]{9}}$

59. $\sqrt{\frac{a}{3}}$ 60. $\sqrt{\frac{2a}{c}}$ 61. $\sqrt{\frac{5a}{2b}}$ 62. $\sqrt{\frac{3x}{5y}}$

63. $\sqrt[3]{\frac{2cd}{9x^2}}$ 64. $\sqrt[3]{\frac{3d}{4z^2}}$ 65. $\sqrt{\frac{1}{5u^5}}$ 66. $\sqrt{\frac{x^2}{z^5}}$

67. $\sqrt[4]{\frac{2a^2y^6}{x^3z^5}}$ 68. $\sqrt[4]{\frac{cu^6}{27v^7}}$ 69. $\frac{2\sqrt{7}-3}{\sqrt{7}-2}$ 70. $\frac{1-\sqrt{5}}{2+\sqrt{5}}$

71. $\frac{3}{5\sqrt{2}-\sqrt{7}}$ 72. $\frac{2}{2\sqrt{3}-\sqrt{5}}$ 73. $\frac{\sqrt{5}-\sqrt{3}}{3\sqrt{3}+\sqrt{2}}$ 74. $\frac{\sqrt{3}+2\sqrt{2}}{3\sqrt{2}+2\sqrt{3}}$

Change to simplest radical form. Assume that all variables represent positive real numbers. (See Example 7 and Practice Problem 3.)

75. $\sqrt[5]{32uv^7}$ 76. $\sqrt{40u^5}$ 77. $\sqrt[4]{3y^{-8}}$

78. $\sqrt[3]{5x^{-6}}$ 79. $\sqrt{x^9y^{-2}z^0}$ 80. $\sqrt{x^7y^0z^{-4}}$

81. $\sqrt{(a+2b)^6}$ 82. $\sqrt[3]{(x-y)^4}$ 83. $\sqrt[4]{\frac{5}{8}}$

84. $\sqrt[4]{\frac{5}{27}}$ 85. $\frac{\sqrt[3]{3c^2v^3}}{\sqrt[3]{15c^2v^2}}$ 86. $\frac{\sqrt{12y^3z}}{\sqrt{24y^4z^3}}$

87. $\frac{x}{\sqrt{4x^{-2}}}$ 88. $\sqrt{\frac{x^{-3}}{xy^{-6}}}$ 89. $\sqrt{\frac{27(a+b)^2}{(a+b)^{-3}}}$

90. $\sqrt{\frac{8(x-y)^{-3}}{(x-y)^2}}$

91. $(\sqrt{y}-3\sqrt{x})^2$ 92. $(\sqrt{3}+2\sqrt{b})^2$

93. $(1+\sqrt{2})^{-2}$ 94. $(\sqrt{a}-\sqrt{b})^{-1}$

95. $\sqrt[3]{32y^{-4}}-\sqrt[3]{-4y^2}$ 96. $\sqrt{3x^3+x^2}+\sqrt{12x+4}$

97. $\sqrt[3]{-2\sqrt{16x^8}}$ 98. $\sqrt{12\sqrt{9x^6}}$

99. $\sqrt{x^3\sqrt[3]{x^{-3}}}$ 100. $\sqrt{x\sqrt[3]{x^6}}$

Rationalize the numerator. Assume that all variables represent positive real numbers.

101. $\frac{\sqrt{3}}{6}$ 102. $\frac{\sqrt{2}}{10}$ 103. $\frac{1-\sqrt{3}}{2}$ 104. $\frac{\sqrt{2}-3}{2}$

105. $\frac{\sqrt{x}-2}{x}$ 106. $\frac{\sqrt{x}-\sqrt{3}}{x-3}$ 107. $\frac{\sqrt{a}+\sqrt{b}}{a-b}$ 108. $\frac{\sqrt{x+h}-\sqrt{x}}{h}$

Radicals with different indices can be multiplied or divided by using rational exponents. Write each of the following as a single radical in simplest form.

109. $\sqrt[3]{x^2}\sqrt[4]{x^3}$ 110. $\sqrt[5]{x^3y}\sqrt[4]{xy^3}$

The $\boxed{\sqrt{x}}$ key on a calculator can also be used to evaluate fourth roots (since $\sqrt[4]{x} = \sqrt{\sqrt{x}}$), eighth roots (since $\sqrt[8]{x} = \sqrt{\sqrt{\sqrt{x}}}$), and so on. Use the $\boxed{\sqrt{x}}$ key to evaluate each of the following. Round your answer to the nearest hundredth.

111–116

111. $\sqrt[4]{293}$ 112. $\sqrt[4]{11.56}$ 113. $\sqrt[8]{5691}$

114. $\sqrt[8]{528.1}$ 115. $\sqrt[16]{5772}$ 116. $\sqrt[16]{991.9}$

Critical Thinking: Discussion

117. One of the conditions for simplest radical form calls for no radical in the denominator. There is a very good reason for this condition if computation must be done by hand.

a. Compute, by hand, the values of

$$\frac{1}{\sqrt{2}} \approx \frac{1}{1.414} \quad \text{and} \quad \frac{\sqrt{2}}{2} \approx \frac{1.414}{2}.$$

Discuss and decide which is easier to evaluate.

b. Discuss and decide which of $3/\sqrt{2}$ or $3\sqrt{2}/2$ is easier to evaluate when a calculator is used.

Critical Thinking: Exploration and Discussion

118. The reduction of an index in a radical can be stated with symbols as follows: If p is a positive integer, then

$$\sqrt[pn]{a^{pm}} = \sqrt[n]{a^m},$$

provided $\sqrt[n]{a^m}$ *is defined.* Show that the condition $\sqrt[n]{a^m}$ be defined is necessary by finding a case with $n = 2$ and a negative where $\sqrt[pn]{a^{pm}}$ is defined but $\sqrt[n]{a^m}$ is not defined.

◆ Solutions for Practice Problems

1. a. $\sqrt[3]{9x^2y^2}\sqrt[3]{12x^2y} = \sqrt[3]{3^2 \cdot 3 \cdot 2^2x^4y^3} = \sqrt[3]{3^3x^3y^3}\sqrt[3]{2^2x} = 3xy\sqrt[3]{4x}$

   b. $(\sqrt{5} - 3\sqrt{3})(\sqrt{15} - 4) = \sqrt{75} - 4\sqrt{5} - 3\sqrt{45} + 12\sqrt{3}$

$$= 5\sqrt{3} - 4\sqrt{5} - 9\sqrt{5} + 12\sqrt{3}$$
$$= 17\sqrt{3} - 13\sqrt{5}$$

2. $$\frac{2}{4\sqrt{2} + \sqrt{3}} = \frac{2}{4\sqrt{2} + \sqrt{3}} \cdot \frac{4\sqrt{2} - \sqrt{3}}{4\sqrt{2} - \sqrt{3}}$$
$$= \frac{2(4\sqrt{2} - \sqrt{3})}{32 - 3} = \frac{2(4\sqrt{2} - \sqrt{3})}{29}$$

3. a. $\dfrac{\sqrt{5a^3b^5}}{\sqrt{10a^4b}} = \sqrt{\dfrac{5a^3b^5}{10a^4b}} = \sqrt{\dfrac{b^4}{2a}} = \sqrt{\dfrac{(b^2)^2(2a)}{(2a)^2}} = \dfrac{b^2\sqrt{2a}}{2a}$

b. $\sqrt[3]{9x} - \sqrt[3]{243x^{-2}} = \sqrt[3]{9x} - \sqrt[3]{\dfrac{3^3 \cdot 3^2}{x^2}} = \sqrt[3]{9x} - \dfrac{3\sqrt[3]{9x}}{x}$

$$= \frac{x\sqrt[3]{9x} - 3\sqrt[3]{9x}}{x} = \frac{(x-3)\sqrt[3]{9x}}{x}$$

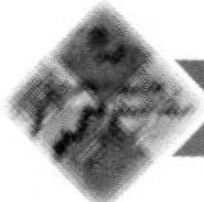

1.9 COMPLEX NUMBERS

In Section 1.7, $\sqrt[n]{a}$ is not defined when a is negative and n is even. This is a consequence of a fundamental deficiency in the set of real numbers: *A negative real number does not have a root of even order in the set of real numbers* since an even power of a real number is never negative.

This situation is very unsatisfactory to a mathematician. This inadequacy of the real numbers is the primary reason for construction of the system of complex numbers, which begins with the introduction of a number i such that $i^2 = -1$. The definition is as follows.

Definition of a Complex Number

The number i is, by definition, a number such that

$$i^2 = -1.$$

That is, $i = \sqrt{-1}$. A **complex number** is a number of the form

$$a + bi,$$

where a and b are real numbers. The real number a is called the **real part** of the complex number, and the real number b is called the **imaginary part** of the complex number. The set of all complex numbers is denoted by $\mathscr{C}$.

Example 1 • Examples of Complex Numbers

Examples of complex numbers are

$$3 + 7i, \quad \pi + 4i, \quad 1 + (-2)i, \quad -5 + (-\tfrac{3}{7})i, \quad 0 + 3i, \quad 4 + 0i.$$ □

If b is negative, as it is in $1 + (-2)i$, we drop the + sign and simply write $1 - 2i$. In this set $\mathscr{C}$, we identify each real number a as being the same as $a + 0i$. For example, we write 4 for $4 + 0i$. Numbers of the form $0 + bi$ are called **imaginary numbers** and are written as bi instead of $0 + bi$. The list of complex numbers in Example 1 can be written as

$$3 + 7i, \quad \pi + 4i, \quad 1 - 2i, \quad -5 - \tfrac{3}{7}i, \quad 3i, \quad 4.$$

Two complex numbers are said to be **equal** if they have equal real parts and equal imaginary parts.

Definition of Equality

For two complex numbers, $a + bi$ and $c + di$,

$$a + bi = c + di \text{ if and only if } a = c \text{ and } b = d.$$

Example 2 • Equality of Complex Numbers

If x and y are real numbers, then

$$x - 3i = 2 - yi,$$

if and only if $x = 2$ and $y = 3$. □

Our goal in this section is to define the operations of addition, subtraction, multiplication, and division in the set of complex numbers so that the calculations are consistent with those of the real numbers (numbers of the form $a + 0i$). It is appropriate to begin with the definition of addition and subtraction.

Definition of Addition and Subtraction

Let $a + bi$ and $c + di$ be arbitrary complex numbers. Then

$$(a + bi) + (c + di) = (a + c) + (b + d)i$$

and

$$(a + bi) - (c + di) = (a - c) + (b - d)i.$$

That is, to add two complex numbers, we simply add their real parts and add their imaginary parts. A similar statement can be made concerning subtraction. Addition and subtraction are illustrated by the following examples.

Example 3 • Adding Complex Numbers

a. $(2 - 3i) + (7 + i) = (2 + 7) + (-3 + 1)i = 9 - 2i$

b. $(3i) + (8 - 2i) = 8 + (3 - 2)i = 8 + i$ □

Example 4 • Subtracting Complex Numbers

a. $(9 - 5i) - (6 + 4i) = (9 - 6) + (-5 - 4)i = 3 - 9i$

b. $(-2 + i) - (-5i) = -2 + [1 - (-5)]i = -2 + 6i$ □

◆ **Practice Problem 1** ◆

Perform the following operations.

a. $(6 + 3i) + (-10 + 5i)$ b. $(-2 - i) - (3 - 2i)$

The work in Examples 3 and 4 illustrates that the same complex number can be written in different forms. The form where the complex number is written as $a + bi$, with a and b real numbers, is called the **standard form** of the complex number.

To understand the definition of multiplication of complex numbers, we consider the product of two binomials involving real numbers:

$$(a + bx)(c + dx) = \overset{\textbf{F}}{ac} + (\overset{\textbf{O}}{ad} + \overset{\textbf{I}}{bc})x + \overset{\textbf{L}}{bd}x^2.$$

Since we want the same properties of real numbers to hold for complex numbers also, we multiply complex numbers the same way as with binomials and simplify the result by using $i^2 = -1$.

Example 5 • Multiplying Complex Numbers

The following computations illustrate the procedure for multiplying complex numbers.

a. $(3 + 7i)(2 - 5i)$

$$\begin{aligned} &= (3)(2) + (3)(-5i) + (7i)(2) + (7i)(-5i) && \text{multiplying binomials} \\ &= 6 - 15i + 14i - 35i^2 \\ &= 6 - i - 35(-1) && \text{since } i^2 = -1 \\ &= 41 - i \end{aligned}$$

b. $$\begin{aligned} (5 + 3i)(5 - 3i) &= 5^2 - (3i)^2 && \text{difference of two squares} \\ &= 25 - 9(-1) && \text{since } i^2 = -1 \\ &= 34 \end{aligned}$$

□

If the same procedure is performed with arbitrary complex numbers $a + bi$ and $c + di$, the rule given in the next definition is obtained.

Definition of Multiplication

Let $a + bi$ and $c + di$ be arbitrary complex numbers. Then

$$(a + bi)(c + di) = (ac - bd) + (ad + bc)i.$$

Most people find that they make fewer errors by performing the multiplication as done in Example 5 rather than memorizing and using the rule stated formally in the definition.

◆ Practice Problem 2 ◆

Write the result of the following multiplication in standard form.

$(5 - i)(-2 + i)$

In part b of Example 5, the product had the form $(a + bi)(a - bi)$ and turned out to be a positive real number. The following multiplication shows this was no accident.

$$\begin{aligned}(a + bi)(a - bi) &= a^2 - abi + abi - b^2i^2 \\ &= a^2 - b^2(-1) \\ &= a^2 + b^2\end{aligned}$$

If $a + bi \neq 0$, then $a^2 + b^2$ is a positive real number. In connection with this result, we make the following definition.

Definition of Conjugate

For any complex number $z = a + bi$, the **conjugate** of z is the complex number $\bar{z} = a - bi$.

Example 6 • Conjugates

Some examples of conjugates are as follows.

a. If $z = 3 + 4i$, then $\bar{z} = 3 - 4i$. That is,
$$\overline{3 + 4i} = 3 - 4i.$$

b. The conjugate of $2 - 3i$ is $2 + 3i$. That is,
$$\overline{2 - 3i} = 2 + 3i.$$

c. $\bar{4} = 4$ □

The next example illustrates how a quotient of two complex numbers can be found by multiplying the numerator and denominator by the conjugate of the denominator.

Example 7 • Dividing Complex Numbers

Perform the following divisions and express each result in standard form.

a. $\dfrac{6 - 2i}{3 + i}$ b. $\dfrac{1}{2 - 3i}$

Solution

a.
$$\begin{aligned}\frac{6 - 2i}{3 + i} &= \frac{(6 - 2i)(3 - i)}{(3 + i)(3 - i)} \\ &= \frac{16-12i}{9 + 1} \\ &= \frac{(2)(8 - 6i)}{(2)(5)} \\ &= \frac{8 - 6i}{5} \\ &= \frac{8}{5} - \frac{6}{5}i\end{aligned}$$

b.
$$\begin{aligned}\frac{1}{2 - 3i} &= \frac{(1)(2 + 3i)}{(2 - 3i)(2 + 3i)} \\ &= \frac{2 + 3i}{4 + 9} \\ &= \frac{2}{13} + \frac{3}{13}i\end{aligned}$$

□

◆ Practice Problem 3 ◆

Perform the following division and express the result in standard form.

$$\frac{3 + 5i}{4 - 2i}$$

According to the definition of a complex number, -1 has a square root in the set of complex numbers since

$$i^2 = -1.$$

We find that -1 also has $-i$ as another square root since

$$(-i)^2 = (-i)(-i) = +i^2 = -1.$$

Thus i and $-i$ are both square roots of -1. Similarly, $2i$ and $-2i$ are square roots of -4. Every negative real number is of the form $-a$, where a is a positive real number, and $-a$ has two square roots, $i\sqrt{a}$ and $-i\sqrt{a}$. The square root of $-a$ which has the positive imaginary part is designated as the **principal square root** and is denoted by $\sqrt{-a}$. Thus

$$\sqrt{-1} = i, \quad \sqrt{-4} = 2i, \quad \sqrt{-25} = 5i, \quad \sqrt{-7} = i\sqrt{7},$$

and so on. We usually write, and accept as standard form, $i\sqrt{7}$ instead of $\sqrt{7}\,i$. This is done to avoid confusion between the two different numbers $\sqrt{7i}$ and $\sqrt{7}\,i$.

The property $\sqrt{a}\sqrt{b} = \sqrt{ab}$, where $a > 0$ and $b > 0$, *does not carry over to principal square roots of negative numbers:*

$$\sqrt{-4}\sqrt{-9} = (2i)(3i) = 6i^2 = -6$$

and

$$\sqrt{(-4)(-9)} = \sqrt{36} = 6 \neq \sqrt{-4}\,\sqrt{-9}.$$

Computations involving square roots of negative numbers are best handled by writing all numbers involved in standard form $a + bi$ and then performing the computations. This procedure is followed in Example 8.

Example 8 • Square Roots of Negative Numbers

a. $-\sqrt{-36} = -6i$

b. $\sqrt{-36}\,\sqrt{-4} = (6i)(2i) = 12i^2 = -12$

c. $\dfrac{\sqrt{-36}}{\sqrt{-9}} = \dfrac{6i}{3i} = 2$

□

The integral powers of i are of some interest. Beginning with i^1, the next several powers of i are given by

$$i^1 = i$$
$$i^2 = -1$$
$$i^3 = i^2 \cdot i = (-1)i = -i$$
$$i^4 = i^2 \cdot i^2 = (-1)(-1) = 1.$$

Using the fact that $i^2 = -1$, any integral power of the complex number i can be reduced to one of the values 1, -1, i, or $-i$. The technique is demonstrated in Example 9.

Example 9 • Powers of *i*

Write i^{47} in standard form.

Solution

$i^{47} = i^{46} \cdot i = (i^2)^{23} \cdot i = (-1)^{23} \cdot i = -i$ □

◆ Practice Problem 4 ◆

Write i^{-23} in standard form.

EXERCISES 1.9

Use the definition of equality of complex numbers to solve for x and y, where x and y are real numbers. (See Example 2.)

1. $2 - yi = x + 6i$
2. $x - yi = 4 + 2i$
3. $x - 7i = yi$
4. $2x - 3yi = 6 + 9i$
5. $x - yi = 4 + i$
6. $8 - yi = x - 2i$

Evaluate each expression. (See Example 8.)

7. $\sqrt{-16}$
8. $\sqrt{-25}$
9. $-\sqrt{-49}$
10. $-\sqrt{-64}$
11. $\sqrt{-25}\sqrt{-9}$
12. $\sqrt{-16}\sqrt{-36}$
13. $\dfrac{\sqrt{-45}}{\sqrt{-5}}$
14. $\dfrac{\sqrt{-75}}{\sqrt{-3}}$

Perform the indicated operations and write the results in standard form. (See Examples 3–5, and 7 and Practice Problems 1–3.)

15. $(3 + 2i) + (6 - 3i)$
16. $(7 - i) - (3 - 6i)$
17. $(65 - 3i) - (50 - 70i)$
18. $(64 + 32i) + (-59 - 75i)$
19. $(2 - 5i) - (6 - 3i)$
20. $(6 - 19i) + (32 - 7i)$
21. $(2 - 3i) - (7 - 6i)$
22. $(4 - 7i) + (16 - 5i)$
23. $(3 - 7i)(2 - i)$
24. $(-8 - 2i)(5 - 3i)$
25. $(2 + 3i)(3 - i)$
26. $(11 - 5i)(2 - 3i)$
27. $(6 + 4i)(-7 + 3i)$
28. $(2 - 3i)(2 + 3i)$
29. $(6 - 3i)^2$
30. $(3 - 2i)^2$
31. $i(12 - 4i)(3 + i)$
32. $i(2 - 5i)(3 - 7i)$
33. $-i(-2 + 3i)(4 - i)$
34. $-2i(1 + 5i)(2 - 5i)$
35. $(6 - 7i) \div 3$
36. $(3 - 9i) \div 9$

37. $\dfrac{2}{i}$

38. $\dfrac{5}{i}$

39. $\dfrac{4}{3i}$

40. $\dfrac{7}{2i}$

41. $\dfrac{6 - i}{-3i}$

42. $\dfrac{4 + i}{-2i}$

43. $\dfrac{1}{3 + 4i}$

44. $\dfrac{1}{4 - 5i}$

45. $\dfrac{1}{5 - 12i}$

46. $\dfrac{1}{15 + 8i}$

47. $6 \div (3 - i)$

48. $4 \div (2 - i)$

49. $\dfrac{5 - 3i}{6 + i}$

50. $\dfrac{9 - i}{3 + 2i}$

51. $\dfrac{4 - 7i}{1 - 2i}$

52. $\dfrac{7 - 5i}{3 - 7i}$

53. $(4 - 3i) \div (3 + 4i)$

54. $(5 - 10i) \div (2 + i)$

55. $(-1 + 4i) \div (1 - 3i)$

56. $(-2 - 3i) \div (-3 + i)$

57. $(1 - i)^3$

58. $(3 + 2i)^3$

59. $\left(\dfrac{i}{1 - i}\right)^2$

60. $\left(\dfrac{4i}{2 + i}\right)^2$

Write each power of i in standard form. (See Example 9 and Practice Problem 4.)

61. i^{72} 62. i^{56} 63. i^{29} 64. i^{13}

65. i^{91} 66. i^{39} 67. $\dfrac{1}{i^{15}}$ 68. $\dfrac{1}{i^{95}}$

69. i^{-62} 70. i^{-78} 71. i^{915} 72. i^{231}

If P is a polynomial in the variable z, a **zero** of P is a value of z that makes P equal to 0.

73. Show that $z = 3i$ is a zero of $P = z^2 + 9$.
74. Show that $z = -2i$ is a zero of $P = z^2 + 4$.
75. Show that $z = 2 - i$ is a zero of $P = z^2 - 4z + 5$.
76. Show that $z = 1 - 2i$ is a zero of $P = z^2 - 2z + 5$.
77. Show that $z = -1 + 2\sqrt{3}i$ is a zero of $P = z^3 - 8$.
78. Show that $z = -i$ is a zero of $P = 2z^4 + z^3 - 8z^2 + z - 10$.

Critical Thinking: Exploration and Writing

79. The **absolute value** of a complex number $a + bi$ in standard form is denoted by $|a + bi|$ and is defined by the equation

$$|a + bi| = \sqrt{a^2 + b^2}.$$

Explain why the complex number 0 is the only complex number whose absolute value is equal to zero.

Critical Thinking: Exploration and Writing

80. Use the definition of absolute value in Exercise 79 and decide if either of the equations $|wz| = |w| \cdot |z|$ or $|w + z| = |w| + |z|$ is true for all complex numbers w and z. Explain the reasoning for your decision.

◆ Solutions for Practice Problems

1. a. $(6 + 3i) + (-10 + 5i) = (6 - 10) + (3 + 5)i = -4 + 8i$
   b. $(-2 - i) - (3 - 2i) = (-2 - 3) + (-1 + 2)i = -5 + i$

2. $(5 - i)(-2 + i) = -10 + 5i + 2i - i^2 = -10 + 7i + 1 = -9 + 7i$

3. $$\frac{3 + 5i}{4 - 2i} = \frac{3 + 5i}{4 - 2i} \cdot \frac{4 + 2i}{4 + 2i} = \frac{2 + 26i}{20} = \frac{(2)(1 + 13i)}{(2)(10)}$$
   $$= \frac{1 + 13i}{10} = \frac{1}{10} + \frac{13}{10}i$$

4. $$i^{-23} = \frac{1}{i^{23}} = \frac{1}{i^{22} \cdot i} = \frac{1}{(i^2)^{11}i} = \frac{1}{(-1)^{11}i} = \frac{1}{-i} = \frac{1 \cdot i}{-i \cdot i} = \frac{i}{-(-1)} = i$$

CHAPTER REVIEW

Summary of Important Concepts and Formulas

Basic Properties for the Real Numbers

1. Addition properties: The operation of addition defined in $\mathcal{R}$ has the following properties:
   a. Closure property: $a + b \in \mathcal{R}$ for all $a \in \mathcal{R}$ and $b \in \mathcal{R}$.
   b. Associative property: $a + (b + c) = (a + b) + c$ for all $a, b, c \in \mathcal{R}$.
   c. Additive identity: There is a real number 0 such that $a + 0 = 0 + a = a$ for all $a \in \mathcal{R}$.
   d. Additive inverses: For each $a \in \mathcal{R}$, there is an element $-a \in \mathcal{R}$ such that $a + (-a) = (-a) + a = 0$.
   e. Commutative property: $a + b = b + a$ for all a and $b \in \mathcal{R}$.
2. Multiplication properties: The operation of multiplication defined in $\mathcal{R}$ has the following properties:
   a. Closure property: $a \cdot b \in \mathcal{R}$ for all $a \in \mathcal{R}$ and $b \in \mathcal{R}$.
   b. Associative property: $a \cdot (b \cdot c) = (a \cdot b) \cdot c$ for all $a, b, c \in \mathcal{R}$.
   c. Multiplicative identity: There is a real number 1 such that $1 \neq 0$ and $a \cdot 1 = 1 \cdot a = a$ for all $a \in \mathcal{R}$.
   d. Multiplicative inverses: For each *nonzero* $a \in \mathcal{R}$, there is an element
   $$\frac{1}{a} \in \mathcal{R} \text{ such that } a\left(\frac{1}{a}\right) = \left(\frac{1}{a}\right)a = 1.$$
   e. Commutative property: $a \cdot b = b \cdot a$ for all $a, b \in \mathcal{R}$.
3. Distributive property: $a \cdot (b + c) = (a \cdot b) + (a \cdot c)$ for all $a, b, c \in \mathcal{R}$.

Order of Operations and Symbols of Grouping

1. In an expression involving powers and fundamental operations we proceed as follows:
   a. All powers are evaluated first.
   b. All multiplications and divisions are performed next, from left to right.
   c. All additions and subtractions are then performed, from left to right.

2. Parentheses, ( ), brackets, [ ], or braces, { }, may be used to indicate the order of operations, with the understanding that the innermost symbols of grouping are removed first. That is, the innermost operations are performed first.

Basic Formulas for Fractions

Let a, b, c, and d be real numbers with $b \neq 0$ and $d \neq 0$.

1. $\dfrac{a}{b} = \dfrac{c}{d}$ if and only if $ad = bc$ — equality
2. $\dfrac{ad}{bd} = \dfrac{a}{b}$ — fundamental principle of fractions
3. $\dfrac{a}{b} + \dfrac{c}{b} = \dfrac{a+c}{b}$ — addition
4. $\dfrac{a}{b} + \dfrac{c}{d} = \dfrac{ad+bc}{bd}$ — addition
5. $\dfrac{a}{b} \cdot \dfrac{c}{d} = \dfrac{ac}{bd}$ — multiplication
6. If b, c, and d are all nonzero, then

$$\frac{\frac{a}{b}}{\frac{c}{d}} = \frac{a}{b} \div \frac{c}{d} = \frac{a}{b} \cdot \frac{d}{c} = \frac{ad}{bc}.$$ — division

Compact Notation for Inequalities

1. Compact forms such as $a < x < b$ or $a > x > b$ are reserved for compound statements that use *and.* (Compact forms are not used for compound statements that involve *or.*)
2. The two inequality symbols involved must have the same direction.
3. In a statement of the form $a < x < b$, the smaller number goes on the left. In a statement of the form $a > x > b$, the smaller number goes on the right.

Addition and Multiplication Properties of Inequalities

For any real numbers a, b, and c:

1. If $a > b$, then $a + c > b + c$. — addition property
2. If $a > b$ and $c > 0$, then $ac > bc$. — multiplication properties
3. If $a > b$ and $c < 0$, then $ac < bc$. — multiplication properties

Definition of Absolute Value

$$|a| = \begin{cases} a, & \text{if } a \geq 0 \\ -a, & \text{if } a < 0 \end{cases}$$

Absolute Value and the Fundamental Operations

Let a and b be real numbers. Then

1. $|ab| = |a| \cdot |b|$ — multiplication property
2. $\left|\dfrac{a}{b}\right| = \dfrac{|a|}{|b|}$, if $b \neq 0$ — division property
3. $|a + b| \leq |a| + |b|$ — the triangle inequality

Absolute Value and Distance

Let x and d be real numbers with $d > 0$. Then

1. $|x| < d$ if and only if $-d < x < d$,
2. $|x| > d$ if and only if either $x > d$ or $x < -d$.

The distance between x and a is $|x - a|$.

Laws of Exponents

Let a and b be real numbers. The following properties hold for all integers m and n for which the quantities involved are defined.

1. $a^m \cdot a^n = a^{m+n}$ — product rule
2. $(a^m)^n = a^{mn}$ — power of a power rule
3. $\dfrac{a^m}{a^n} = a^{m-n}$, for $a \neq 0$ — quotient rule
4. $(ab)^n = a^n b^n$ — power of a product rule
5. $\left(\dfrac{a}{b}\right)^n = \dfrac{a^n}{b^n}$, for $b \neq 0$ — power of a quotient rule

Scientific Notation

$N = a \times 10^n$, $1 \leq a < 10$ and n is an integer.

Formulas for Factoring

Common factor:	$ax + ay = a(x + y)$
Difference of two squares:	$x^2 - y^2 = (x + y)(x - y)$
Square of a binomial:	$x^2 + 2xy + y^2 = (x + y)^2$ $x^2 - 2xy + y^2 = (x - y)^2$
Sum of two cubes:	$x^3 + y^3 = (x + y)(x^2 - xy + y^2)$
Difference of two cubes:	$x^3 - y^3 = (x - y)(x^2 + xy + y^2)$

The Fundamental Principle of Fractions

Let P, Q, and R be polynomials. Then

$$\frac{PR}{QR} = \frac{P}{Q} \quad \text{if } Q \neq 0 \text{ and } R \neq 0.$$

Multiplication and Division of Rational Expressions

Let P, Q, R, and S be polynomials. Then

$$\frac{P}{Q} \cdot \frac{R}{S} = \frac{PR}{QS} \quad \text{and} \quad \frac{P}{Q} \div \frac{R}{S} = \frac{P}{Q} \cdot \frac{S}{R} = \frac{PS}{QR}$$

where all denominators are nonzero.

Addition and Subtraction of Rational Expressions

Let P, Q, and R be polynomials with $Q \neq 0$. Then

$$\frac{P}{Q} + \frac{R}{Q} = \frac{P+R}{Q} \quad \text{and} \quad \frac{P}{Q} - \frac{R}{Q} = \frac{P-R}{Q}.$$

Finding the Least Common Denominator

1. Factor each denominator completely (that is, into prime factors).
2. Form the product of all the different prime factors, with each prime factor raised to the highest exponent with which it appears in any of the given denominators. In the prime factors, the leading coefficients are chosen to be positive.

Definition of nth root

Let n be a positive integer. The **principal nth root** of a real number a is the real number $\sqrt[n]{a}$ defined by the following statements:

1. If n is even and $a \geq 0$, then $\sqrt[n]{a}$ is the *nonnegative number* such that $(\sqrt[n]{a})^n = a$.
2. If n is odd, then $\sqrt[n]{a}$ is the real number such that $(\sqrt[n]{a})^n = a$.

The symbol $\sqrt[n]{a}$ is called a **radical,** a is called the **radicand,** and n is called the **index** or **order** of the radical.

Basic Properties of Radicals

Let n and m be positive integers, and let a and b be real numbers. Whenever each indicated root is defined, then

1. $(\sqrt[n]{a})^n = a$
2. $\sqrt[n]{a^n} = \begin{cases} a, & \text{if } n \text{ is odd} \\ |a|, & \text{if } n \text{ is even} \end{cases}$ (if $a \geq 0$, $\sqrt[n]{a^n} = a$)
3. $\sqrt[n]{ab} = \sqrt[n]{a}\sqrt[n]{b}$
4. $\sqrt[n]{\dfrac{a}{b}} = \dfrac{\sqrt[n]{a}}{\sqrt[n]{b}}$, $b \neq 0$
5. $\sqrt[m]{\sqrt[n]{a}} = \sqrt[mn]{a}$

Definition of $a^{1/n}$

Let n be a positive integer. Then

$$a^{1/n} = \sqrt[n]{a}.$$

Definition of $a^{m/n}$

If m is an integer and n is a positive integer, then we define $a^{m/n}$ by

$$a^{m/n} = \sqrt[n]{a^m} = (\sqrt[n]{a})^m,$$

with the conditions that $a \geq 0$ if n is even and $a \neq 0$ if $m \leq 0$.

Definition of Simplest Radical Form

A radical expression is in **simplest radical form** if the following conditions are satisfied:

1. The expression contains no negative or zero exponents.
2. The radicand contains no factor to a power equal to or greater than the index of the radical.
3. The denominator contains no radicals, and no fraction appears under a radical.
4. The index of the radical is as small as possible.

Definition of Complex Number

The number i is, by definition, a number such that

$$i^2 = -1.$$

That is, $i = \sqrt{-1}$. A **complex number** is a number of the form

$$a + bi$$

where a and b are real numbers. The real number a is called the **real part** of the complex number, and the real number b

is called the **imaginary part** of the complex number. The set of all complex numbers is denoted by $\mathscr{C}$.

Definition of Equality

For two complex numbers $a + bi$ and $c + di$,

$$a + bi = c + di \quad \text{if and only if} \quad a = c \text{ and } b = d.$$

Addition, Subtraction, Multiplication, and Conjugates of Complex Numbers

Let $a + bi$ and $c + di$ be arbitrary complex numbers.

$$(a + bi) + (c + di) = (a + c) + (b + d)i$$
$$(a + bi) - (c + di) = (a - c) + (b - d)i$$
$$(a + bi)(c + di) = (ac - bd) + (ad + bc)i$$
$$\overline{a + bi} = a - bi$$

Critical Thinking: Find the Errors

Each of the following nonsolutions has one or more errors. Can you find them?

Problem 1

Simplify $4[-(2 - x) - (1 + 2x)]$ by removing grouping symbols and combining like terms.

Nonsolution

$$\begin{aligned} 4[-(2 - x) - (1 + 2x)] &= 4[-2 - x - 1 + 2x] \\ &= 4[-3 + x] \\ &= 4 - 3 + x \\ &= 1 + x \end{aligned}$$

Problem 2

If $x = 1$, $y = -2$, and $z = 3$, evaluate $x - y(z + 3) - 2$.

Nonsolution

$$\begin{aligned} x - y(z + 3) - 2 &= 1 - (-2)(3 + 3) - 2 \\ &= 1 + 2(6) - 2 \\ &= 3(6) - 2 \\ &= 18 - 2 \\ &= 16 \end{aligned}$$

Problem 3

Draw the graph of $|x| \le 2$.

Nonsolution

The graph is shown in Figure 1.22.

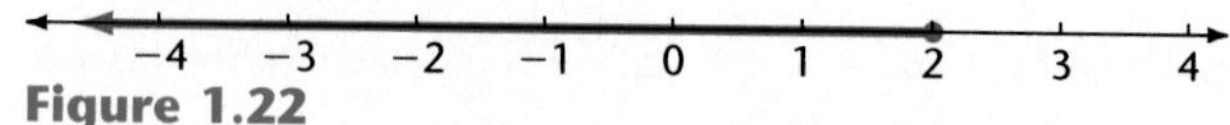

Figure 1.22

Problem 4

Simplify $(x^{-2}y)^{-3}$.

Nonsolution

$$(x^{-2}y)^{-3} = \frac{1}{(x^2y)^3} = \frac{1}{x^5y^3}$$

Problem 5

Simplify $(x^0 - x^{-1})^{-1}$.

Nonsolution

$$\begin{aligned} (x^0 - x^{-1})^{-1} &= (x^0)^{-1} - (x^{-1})^{-1} \\ &= x^0 - x = 1 - x \end{aligned}$$

Problem 6

Factor $x^3 + 8$.

Nonsolution

$$x^3 + 8 = (x + 2)(x^2 + 4x + 4)$$

Problem 7

Factor by grouping: $r^2 - s^2 - r - s$.

Nonsolution

$$\begin{aligned} r^2 - s^2 - r - s &= (r^2 - s^2) - (r - s) \\ &= (r - s)(r + s) - (r - s) \\ &= (r - s)(r + s - 1) \end{aligned}$$

Problem 8

Divide and write the results in lowest terms:
$\dfrac{x^3 - y^3}{x - y}$.

Nonsolution

$$\frac{x^3 - y^3}{x - y} = \frac{\overset{x^2}{\cancel{x^3}} - \overset{y^2}{\cancel{y^3}}}{\cancel{x} - \cancel{y}} = \frac{x^2 - y^2}{1} = x^2 - y^2$$

Problem 9

Perform the subtraction and write the result in lowest terms:
$\dfrac{p - 3}{p + 1} - \dfrac{2p - 1}{p + 2}$.

Nonsolution

$$\begin{aligned} \frac{p - 3}{p + 1} - \frac{2p - 1}{p + 2} &= \frac{p - 3 - 2p + 1}{(p + 1)(p - 2)} \\ &= \frac{-p - 2}{(p + 1)(p + 2)} \\ &= \frac{-(p + 2)}{(p + 1)(p + 2)} \\ &= \frac{-1}{p + 1} \end{aligned}$$

Problem 10

Evaluate $(-27)^{2/3}$.

Nonsolution

$$(-27)^{2/3} = [(-27)^{1/3}]^2 = (-9)^2 = 81$$

Problem 11

If possible, combine $2\sqrt{3} - 4\sqrt{2}$ by using the distributive property.

Nonsolution

$$\begin{aligned} 2\sqrt{3} - 4\sqrt{2} &= (2 - 4)(\sqrt{3} - \sqrt{2}) \\ &= (-2)(\sqrt{1}) \\ &= -2 \end{aligned}$$

Problem 12

Change $\sqrt[3]{8x^2y^3}$ to simplest radical form, where x and y represent positive real numbers.

Nonsolution

$$\sqrt[3]{8x^2y^3} = \sqrt{2 \cdot 4 \cdot x^2 \cdot y \cdot y^2} = 2xy\sqrt{2y}$$

Problem 13

Change $\dfrac{\sqrt{8x^3y}}{\sqrt{16xy^3}}$ to simplest radical form, where x and y represent positive real numbers.

Nonsolution

$$\frac{\sqrt{8x^3y}}{\sqrt{16xy^3}} = \sqrt{\frac{8x^3y}{16xy^3}} = \sqrt{\frac{x^2}{2y^2}} = \frac{x}{y\sqrt{2}}$$

Problem 14

Evaluate $\sqrt{-25}\sqrt{-4}$.

Nonsolution

$$\sqrt{-25}\sqrt{-4} = \sqrt{100} = 10$$

Problem 15

Write $\dfrac{1}{2 + i}$ in standard form.

Nonsolution

$$\frac{1}{2 + i} = \frac{1}{2 + i} \cdot \frac{2 - i}{2 - i} = \frac{2 - i}{4 - 1} = \frac{2}{3} - \frac{1}{3}i$$

Review Problems for Chapter 1

Let N be the set of natural numbers, W the set of whole numbers, Z the set of integers, Q the set of rational numbers, $\mathcal{R}$ the set of real numbers, and H the set of irrational numbers. In which of these sets does each of the following belong?

1. -4
2. 0
3. $\frac{3}{2}$
4. $-\sqrt{2}$
5. -0.7666
6. $\sqrt{25}$

Fill in each blank to make the given statement an application of the stated property.

7. $c(p+q) =$ ______ distributive property
8. $ab =$ ______ commutative property, $\cdot$
9. $a(bc) =$ ______ associative property, $\cdot$
10. $x($______$) = 1$ multiplicative inverse
11. $ax + ay =$ ______ commutative property, $+$
12. $1 \cdot x + 0 =$ ______ additive identity

Identify the basic property for the real numbers that justifies the given statement.

13. $(1)(3) + 0 = (1)(3)$
14. $(1)(3) + 0 = 3 + 0$
15. $2 \cdot 5 + 2 \cdot 3 = 2(5 + 3)$
16. $2 \cdot 5 + 2 \cdot 3 = 2 \cdot 5 + 3 \cdot 2$
17. $2 \cdot (3 \cdot 4) = (2 \cdot 3) \cdot 4$
18. $2 + (5 + 3) = (2 + 5) + 3$
19. $[6 + (-6)] + 5 = 0 + 5$
20. $3 \cdot 4 + 7 \cdot 2 = 7 \cdot 2 + 3 \cdot 4$

If $x = 2$ and $y = 3$, find the value of the following expressions.

21. $[2x(3-y)] - (7-y)(2x-3y)$
22. $[(x+10) \div y] + x^2 - 2y^2$
23. $[(x-4)(y+2) - (3x-y)(7+y)] \div (x+y)$
24. $\dfrac{(-x)(-y) - y\left(\dfrac{2x+y}{2x-y}\right)}{x - x\left(\dfrac{x+2y}{y-2x}\right)}$

Simplify the following expressions by removing symbols of grouping and combining like terms.

25. $2[x - 2(x-3) - 4]$
26. $3\{2 - [3z - 4(z+2)] + z\}$

Draw the graph of the given statement.

27. $x \le -2$
28. $x \le 0$ or $x > 4$
29. $-2 < x \le 1$ or $x < -4$
30. $-2 < x < 2$ or $x \ge 5$

Write a statement that has the given graph.

31.
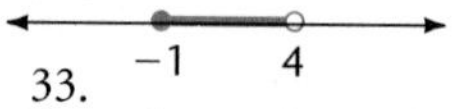

32.
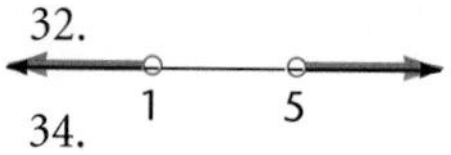

33.
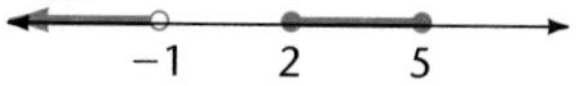

34.
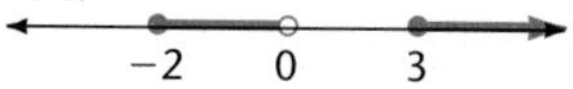

Draw the graph of each inequality.

35. $|x| \le 1$
36. $|x| > 2$
37. $|x-3| < 4$
38. $|x+2| \ge 3$

Express each of the following without using absolute value symbols.

39. $|5-7|$
40. $|\sqrt{35} - 6|$
41. $|6 - \sqrt{37}|$
42. $\left|\dfrac{x}{3}\right|$, if $x < 0$
43. $|y+3|$, if $y > -3$
44. $|14 - 3x|$, if $x < 4$

Label each of the following as true or false.

45. $|x| < 3$, if $x < 3$
46. $|x| > 4$, if $x > 5$ or $x < -5$
47. $|x-3| \le |x| - |3|$
48. $|-3x| \ge -3|x|$

Compute the value if possible; otherwise, simplify by using the laws of exponents.

49. 4^{-2}
50. $(-5x^3)^{-2}$
51. x^2y^{-2}
52. $\dfrac{(3x^4)(5x^2)}{6x^5}$
53. $(x^2y)^2(2xy^3)^2$
54. $(a+b)^0$
55. $\left(\dfrac{3a^{-1}b^2}{2ab^{-3}}\right)^{-1}$

Write the following numbers in scientific notation.

56. 29,300
57. 5,790,000
58. 0.00016
59. 0.0706

Write the following numbers in decimal notation.

60. 4.93×10^5
61. 8.07×10^{-3}
62. 7.68×10^{-1}
63. 1.19×10

Perform the indicated operations and write the results in simplest form.

64. $(5x^2 - 3x + 2) + (-7x^3 - 3x^2 + x - 1)$
65. $(9x^4 - 3x^2 + 7x) - (5x^3 + 4x^2 - 3x - 2)$
66. $(5r - 3)(7r^2 - 3)$
67. $(6x^2 - 3x + 1)(5x^3 - 3x)$
68. $(2p - 3q)(2p + 3q)$
69. $(3x - y)^2$
70. $(u - v)(u^2 + uv + v^2)$
71. $(4x - 2y)^3$
72. $(2x - y)(2x + y)^2$
73. $(3a + b)(9a^2 - 3ab + b^2)$

Factor each polynomial completely.

74. $a^2 - 16b^2$
75. $49x^2 - 14x + 1$
76. $4x^2 - 8xy + 4y^2$
77. $8pq^2 + 2p^2q - 4pq$
78. $m^2 - 3mn + 2n^2$
79. $8x^3 - 27$
80. $a^4 + 64ab^3$
81. $6t^3 - 14t^2 - 12t$
82. $z^3 - 4z^2 - 9z + 36$
83. $4x^2 - y^2 + 4yz - 4z^2$

Perform the indicated operations and reduce the results to lowest terms.

84. $(9x^2 - 16y^2) \cdot \left(\dfrac{4x}{3x - 4y}\right)$
85. $\dfrac{x^2 - y^2}{4x^2y} \cdot \dfrac{20xy^3}{(x - y)^2}$
86. $\dfrac{2x + 4}{x^2 - 4x + 4} \cdot \dfrac{x^2 - 4}{x + 2}$
87. $\dfrac{16 - 9x^2}{x^3 - 8} \cdot \dfrac{x^3 + 2x^2 + 4x}{4 + x - 3x^2}$
88. $\dfrac{w^2 - 5w + 6}{w^2 - 7w + 10} \cdot \dfrac{w^2 + 6w + 9}{w^2 - 9}$
89. $(5x^2 - 20) \div \left(1 + \dfrac{x}{2}\right)$
90. $\dfrac{(x - y)^2}{2x^2 + 3xy + y^2} \div \dfrac{x^2 - y^2}{x^2 + 2xy + y^2}$
91. $\dfrac{2a - 3}{a - 3} - \dfrac{2a^2}{a^2 - 9}$
92. $\dfrac{y}{6x^2} - \dfrac{x}{2xy}$
93. $\dfrac{3}{2x - 2y} - \dfrac{4y}{x^2 - y^2}$
94. $\dfrac{1}{2x - 2} - \dfrac{x}{x^2 - 4x + 3}$
95. $\dfrac{2}{x - 2} + \dfrac{3}{x + 4} + \dfrac{18}{x^2 + 2x - 8}$
96. $\dfrac{3x + 2}{2x^2 - x - 3} + \dfrac{4 + 3x}{5x^2 + 3x - 2}$
97. $(25x^2 - 9y^2) \div \left(1 + \dfrac{3y}{5x}\right)$
98. $\dfrac{1 - \dfrac{2}{x}}{x - \dfrac{4}{x}}$
99. $\dfrac{4x^2y}{\dfrac{x - y}{x - 2y} - 1}$
100. $\dfrac{9x^2 - 16y^2}{\dfrac{x - 2y}{2y - 2x} - 1}$
101. $\left(\dfrac{x^{-1} - y^{-1}}{x^{-2} - y^{-2}}\right)^{-1}$

Find the value of the given expression or simplify it. Assume all variables represent positive real numbers.

102. $\sqrt{\dfrac{121}{36}}$
103. $(\sqrt[4]{79})^4$
104. $\sqrt[3]{\dfrac{-64}{125}}$
105. $\sqrt[3]{-27x^3y^6}$
106. $\sqrt{\sqrt[3]{64}}$
107. $\sqrt{\sqrt[3]{729}}$
108. $\sqrt[3]{-27x^{-3}y^6}$
109. $\sqrt{\sqrt[3]{64x^6y^{12}}}$
110. $(\sqrt[3]{5ab^3})^6$
111. $\sqrt[5]{\left(-\dfrac{32x^5}{y^{10}}\right)^{-2}}$

Change from radical form to exponential form, or from exponential form to radical form, whichever is appropriate.

112. $\sqrt[7]{x^3}$
113. $(5ab^2)^{2/5}$

Evaluate each of the following expressions.

114. $(-32)^{3/5}$
115. $\left(\dfrac{1}{125}\right)^{-2/3}$
116. $(36)^{-3/2}$
117. $(-32a^5)^{1/5}$
118. $(0.04a^4)^{1/2}$
119. $(-125a^6)^{-2/3}$

In Problems 120–149, assume all variables represent positive real numbers.
In Problems 120–125, simplify by removing all possible factors from the radicand.

120. $\sqrt{27a^3}$
121. $\sqrt[6]{64x^6y^5}$
122. $\sqrt[4]{128a^5b^4}$
123. $\sqrt[5]{-x^7y^4}$
124. $\sqrt[3]{81a^4b^5}$
125. $\sqrt[3]{\dfrac{216x^6}{64y^3}}$

Combine whenever possible by using the distributive property.

126. $\sqrt{8} + \sqrt{32}$
127. $3\sqrt{18} - 5\sqrt{50}$
128. $\sqrt{8a^2b} - \sqrt{32a^4b^3}$
129. $\sqrt[3]{8ab^4} + \sqrt[3]{64a^4b}$

Perform the indicated operations and remove all possible factors from the radicand.

130. $\sqrt[3]{4} \cdot \sqrt[3]{16a^4}$

131. $\sqrt[3]{9} \cdot \sqrt[3]{9}$

132. $\sqrt[4]{16a^2} \cdot \sqrt[4]{4a^2b}$

133. $\dfrac{\sqrt[3]{40}}{\sqrt[3]{5a^3}}$

134. $(3 - 2\sqrt{5})(2 + 4\sqrt{5})$

135. $(\sqrt{5} - \sqrt{3})(\sqrt{5} - 2\sqrt{3})$

Reduce the radical index and remove any factors possible from the radicand.

136. $\sqrt[12]{125}$

137. $\sqrt[9]{125x^{12}}$

138. $\sqrt[12]{4x^4y^6}$

139. $\sqrt[15]{27x^6y^{12}}$

Rationalize the denominator.

140. $\sqrt{\dfrac{3}{5}}$

141. $\sqrt[3]{\dfrac{3}{4}}$

142. $\dfrac{\sqrt{3} - 2}{2\sqrt{3} + 1}$

Change to simplest radical form. Assume all variables represent positive real numbers.

143. $\sqrt{27x^4y^{-2}}$

144. $\dfrac{\sqrt[3]{2a^2b^4}}{\sqrt[3]{16ab^5}}$

145. $\sqrt[3]{40x^2y^{-1}}$

146. $\dfrac{3 + \sqrt{y}}{2 - \sqrt{y}}$

147. $\sqrt{\dfrac{2}{x - 2}}$

148. $\sqrt[3]{9\sqrt{36}}$

149. $\sqrt[5]{x^2y^3} \cdot \sqrt{xy}$

Perform the indicated operations and write the result in standard form.

150. $(8 + 3i) - (5 - 7i)$

151. $\sqrt{-36}\,\sqrt{-25}$

152. $(2 - i)(3 + 4i)$

153. $\dfrac{3 - i}{2 + 5i}$

154. i^{-5}

ENCORE

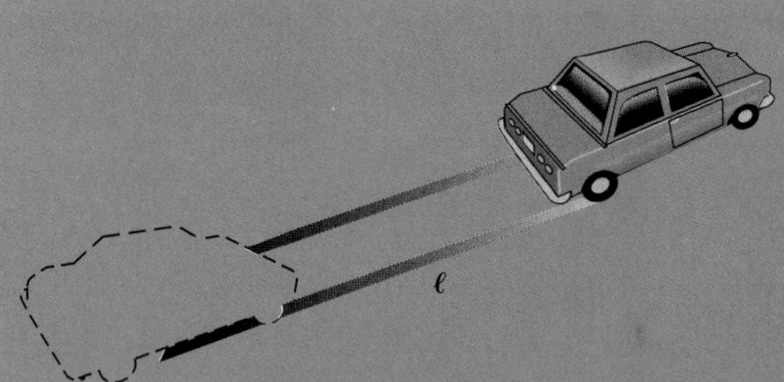

How fast was the car going?

Skid marks are often used by investigators to estimate the speed of a car involved in an accident. The length ℓ (in feet) of skid marks left on the pavement by the rear wheels of a car with its brakes locked can be used to approximate the speed s (in miles per hour) of the car at the moment the brakes were applied. For the car shown above, the formula that relates s and ℓ is $s = 5\sqrt{\ell}$. If the skid marks are measured and found to be 144 feet long, the car was traveling at about

$$s = 5\sqrt{144} = 5(12) = 60$$

miles per hour.

Equations and Inequalities in One Variable

Real-world problems usually lead to a need for the solution of an equation or an inequality. This entire chapter involves such solutions, and the content is very rich in applications.

2.1 SOLUTIONS OF EQUATIONS

In this section, we consider several types of equations that involve a single real variable,† real coefficients, and real constants. A very simple example is provided by the equation

$$3x - 6 = 0.$$

As is frequently the case, this equation is true for some values of x and false for others.

Those values of the variable that make a given equation true are called **solutions** of that equation, and the set of all solutions of the equation is called the **solution set.** To **solve** an equation is to find all the solutions of the equation.

Two equations are said to be **equivalent** if they have the same solution set. It is easy to see that any two of the equations

$$\begin{aligned} 3x - 6 &= 0 \\ 3x &= 6 \\ x &= 2 \end{aligned}$$

are equivalent since $\{2\}$ is the solution set for each of them.

One of the simplest types of equations is a linear equation. A **linear equation** in the variable x is an equation that can be written in the form

$$ax + b = 0,$$

where $a \neq 0$. Generalizing from the equation $3x - 6 = 0$, we see that if $a \neq 0$, then

$$\begin{aligned} ax + b &= 0 \\ ax &= -b \\ x &= -\frac{b}{a} \end{aligned}$$

are equivalent equations. Thus a linear equation always has exactly one solution.

The equations that we solve in this section are equations with solution sets that can be found by solving certain related linear equations. The method that we use to solve these equations is based on the fact that an equivalent equation is obtained if either

1. The same quantity is added to both sides of an equation, or
2. Both sides of an equation are multiplied by the same nonzero quantity.

We can use steps of these two types to isolate the variable on one side of the equation. Formal statements that justify our work are as follows.

†The terms *variable, coefficient,* and *constant* are defined in Section 1.1.

Addition and Multiplication Properties of Equations

Let R, S, and T be algebraic expressions. The equation

$$R = S$$

is equivalent to

a. $R + T = S + T$ — addition property

and

b. $RT = ST$, if $T \neq 0$. — multiplication property

Example 1 • Using the Addition and Multiplication Properties

Solve the following equation.

$$4x - 3 = 7x + 3$$

Solution

We use the addition and multiplication properties as follows:

$4x = 7x + 6$	adding 3 to both sides
$-3x = 6$	adding $-7x$ to both sides
$x = -2$	multiplying both sides by $-\frac{1}{3}$, the reciprocal of the coefficient of x

That is, the solution set is $\{-2\}$. □

In Example 1, adding $-7x$ to both sides of the equation has the same effect as subtracting $7x$ from both sides, and multiplying both sides by $-\frac{1}{3}$ has the same effect as dividing both sides by -3. The addition and multiplication properties have general implications of this nature: equivalent equations can be obtained by subtracting the same expression from both sides, or by dividing both sides by the same nonzero expression.

The work in Example 1 illustrates the general procedure that we use to solve an equation. Generally, we assume that the original equation is true and then perform operations that isolate one or more possible values for the variable. Sometimes the operations that are performed lead to possible values for the variable that do not satisfy the original equation.

Notice in the multiplication property of equations that the multiplier T is required to be different from zero in order to have an equivalent equation. Consider the next example, in which it happens that $T = 0$ for certain values of the variable.

Example 2 • An Equation with No Solutions

Solve $\dfrac{y+4}{y-1} = \dfrac{y-3}{y-2} + \dfrac{5y-9}{(y-1)(y-2)}$.

Solution

If we multiply both sides by the least common denominator, $(y-1)(y-2)$, we obtain

$$(y-2)(y+4) = (y-3)(y-1) + 5y - 9$$

or

$$y^2 + 2y - 8 = y^2 - 4y + 3 + 5y - 9.$$

This equation simplifies to

$$2y - 8 = y - 6$$

and then to

$$y = 2.$$

But we note that the multiplier $(y-1)(y-2)$ is equal to zero whenever $y = 2$ and that $y = 2$ yields undefined quantities in the original equation. This means $y = 2$ cannot be a solution to the original equation, hence the solution set is the empty set, $\varnothing$. □

In the preceding example, clearing the equation of fractions led to the solution $y = 2$ that did not satisfy the original equation. Such solutions are called **extraneous solutions,** or **extraneous roots.** Multiplication of an equation by a variable quantity is one way that extraneous solutions may be introduced, and other ways come up later in this book. To determine whether or not a solution is extraneous, it can be checked in the original equation.

We recall that the definition of the absolute value of a real number x is given by

$$|x| = \begin{cases} x, & \text{if } x \geq 0 \\ -x, & \text{if } x < 0 \end{cases}.$$

The next four examples illustrate the solution of certain types of equations that involve absolute values. Since the definition of absolute value involves two cases, we expect two solutions to an absolute value equation. We find there are indeed two solutions in Example 3, but Examples 4 and 5 show that other possibilities occur.

Example 3 • An Absolute Value Equation with Two Solutions

Solve $|x-5| = 4$.

Solution

We consider cases corresponding to the definition of absolute value. First, if $x - 5 \geq 0$, we have $|x-5| = x - 5$, and we must solve the linear equation

$$x - 5 = 4.$$

Clearly, the solution to this equation is $x = 9$. But if $x - 5 < 0$, then $|x - 5| = -(x-5)$ and we must solve

$$-(x-5) = 4.$$

The solution to this equation is $x = 1$. Thus the solutions to $|x - 5| = 4$ are given by $x = 1$ and $x = 9$. That is, the solution set is $\{1, 9\}$. □

Example 4 • An Absolute Value Equation with One Solution

Determine the solution set of $|t| + 2 = 3t - 1$. (1)

Solution

If $t \geq 0$, then $|t| = t$, and $t = \frac{3}{2}$ is the only solution to

$$t + 2 = 3t - 1.$$

Whenever $t < 0$, then $|t| = -t$, and $t = \frac{3}{4}$ is the only solution to

$$-t + 2 = 3t - 1.$$

But $t = \frac{3}{4}$ does not satisfy the condition that $t < 0$, so $t = \frac{3}{4}$ is not a solution to Equation (1). Thus the only solution to the original equation is $t = \frac{3}{2}$, and the solution set is $\{\frac{3}{2}\}$. □

The next example points out that not every equation involving absolute value has a solution.

Example 5 • An Absolute Value Equation with No Solutions

Solve $|3x - 7| = -4$.

Solution

Now $|3x - 7|$ is either positive or zero and cannot be negative. Hence

$$|3x - 7| \neq -4$$

for every value of x, and the equation has no solution. In other words, the solution set is $\varnothing$. □

In the last three examples, the equations involved an absolute value of only one expression. Consider now an equation of the form

$$|R| = |S|,$$

where R and S are algebraic expressions. Thinking of absolute value in terms of distance on the number line, $|R|$ is the distance between R and 0 and $|S|$ is the distance between S and 0, so the equality $|R| = |S|$ means that R and S are equally distant from 0. Therefore $R = S$ or $R = -S$, which can be expressed compactly as $R = \pm S$. The $\pm$ sign is read "plus or minus." Summarizing, we have

$|R| = |S|$ is equivalent to $R = \pm S$.

Example 6 • An Equation Involving Two Absolute Values

Solve $|w - 2| = |3w + 1|$.

Solution

The solutions to the given equation occur when the expressions within the absolute value symbols are equal or when they are opposites of each other.

$$\begin{aligned} w - 2 &= 3w + 1 \\ -2w &= 3 \\ w &= -\tfrac{3}{2} \end{aligned} \qquad \text{or} \qquad \begin{aligned} w - 2 &= -(3w + 1) \\ 4w &= 1 \\ w &= \tfrac{1}{4} \end{aligned}$$

Thus the solutions to $|w - 2| = |3w + 1|$ are $w = -\frac{3}{2}$ and $w = \frac{1}{4}$. □

◆ Practice Problem 1 ◆

Solve $|r + 3| = 7$.

The ideas we have considered in this section can be extended to equations that involve more than one variable. Such equations are called **literal equations.** An equation of this type can be solved for a specified variable in terms of the others by treating the other variables as constants.

Example 7 • A Literal Equation

Solve for r in the equation $P = A(1 - rn)$.

Solution

We must isolate r on one side of the equation. To begin, we perform the indicated multiplication on the right-hand side and get

$$\begin{aligned} P &= A - Arn \\ Arn &= A - P && \text{adding } Arn - P \text{ to both sides} \\ r &= \frac{A - P}{An}. && \text{multiplying both sides by } \frac{1}{An} \end{aligned}$$

□

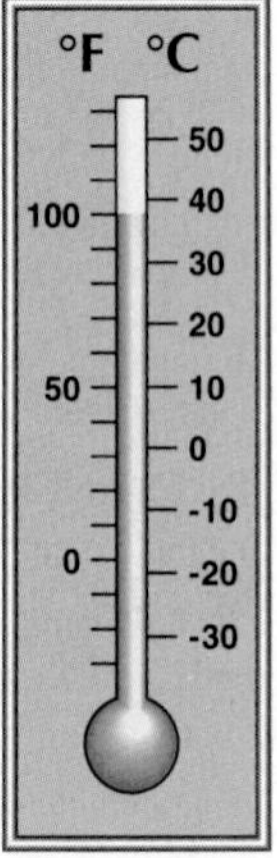

Figure 2.1 Fahrenheit and Celsius thermometer

Example 8 • Record-Breaking Temperatures

A record-breaking temperature of 103°F was recorded at Denver on July 8, 1989, and this extended the city's consecutive days of 100°F or higher to five. Records at Denver going back 118 years had never before recorded more than two consecutive days of triple-digit temperatures.

a. Celsius temperature C and Fahrenheit temperature F are related by the equation $F = \frac{9}{5}C + 32$. (See Figure 2.1.) Find the Celsius temperature to the nearest degree that corresponds to the record-breaking 103°F at Denver.

b. Solve for C in the equation $F = \frac{9}{5}C + 32$.

Solution

a. With $F = 103$, we have

$$\begin{aligned}\tfrac{9}{5}C + 32 &= 103\\ \tfrac{9}{5}C &= 71\\ C &= \tfrac{5}{9}(71) = 39,\end{aligned}$$

rounded to the nearest degree.

b. From $F = \frac{9}{5}C + 32$, we get

$$\begin{aligned}\tfrac{9}{5}C + 32 &= F\\ \tfrac{9}{5}C &= F - 32\\ C &= \tfrac{5}{9}(F - 32).\end{aligned}$$

□

EXERCISES 2.1

Determine the solution set for each of the following equations. (See Examples 1 and 2.)

1. $9x + 5 = 32$
2. $8x + 7 = 39$
3. $5y + 3(1 - y) = 3y - 3$
4. $3y + 2(1 + y) = y + 2$
5. $\frac{10}{3} - \frac{1}{2}z = \frac{3}{4}z$
6. $12z + \frac{3}{2} = \frac{3}{5}z + 3$
7. $3.6 - 0.2x = -1.4x$
8. $3.3 - 0.4x = -1.5x$
9. $-1.09 + 1.21x = 0.13x + 1.07$
10. $-1.21 + 0.27x = 0.32x + 0.79$
11. $6w - [5w - 3(w - 4)] + 1 = 2(1 - w) + 5$
12. $3w - [4w - 2(w - 5)] - 1 = 2(3 - w) - 17$
13. $(5w - 1)(4w + 2) = (10w - 1)(2w + 3) - 2(w + 13)$
14. $(3y - 1)(4y + 5) = (6y + 1)(2y + 7) - 3(y - 1)$
15. $\dfrac{t - 2}{2} = \dfrac{3t - 1}{4}$
16. $\dfrac{t + 8}{4} = \dfrac{3t + 14}{2}$
17. $\dfrac{7x - 4}{3} = x - \dfrac{2}{5}$
18. $\dfrac{3x - 7}{5} = x - \dfrac{7}{4}$
19. $\dfrac{1}{2 - t} = \dfrac{-3}{2 + t}$
20. $\dfrac{1}{3 - t} = \dfrac{5}{3 + t}$
21. $\dfrac{1}{r + 1} = -\dfrac{r}{r + 1}$
22. $\dfrac{3r + 5}{r} = 4 + \dfrac{5}{r}$
23. $\dfrac{-3}{p - 2} + 1 = \dfrac{2p}{3(p - 2)}$
24. $\dfrac{-2}{p + 1} + 3 = \dfrac{7p}{3(p + 1)}$
25. $\dfrac{y + 2}{y + 3} - 1 = \dfrac{3y + 8}{y + 3}$
26. $\dfrac{y}{y - 6} + 3 = \dfrac{6}{y - 6}$
27. $\dfrac{1}{1 - z} + \dfrac{3}{z} = \dfrac{-1}{z}$
28. $\dfrac{1}{z - 1} + \dfrac{1}{z} = \dfrac{-2}{z}$
29. $\dfrac{-1}{2x + 3} - \dfrac{2}{x - 2} = \dfrac{3x + 4}{(2x + 3)(x - 2)}$

30. $\dfrac{-1}{2x + 5} - \dfrac{3}{x - 3} = \dfrac{4x - 1}{(2x + 5)(x - 3)}$

31. $\dfrac{3}{p + 4} + \dfrac{4}{p + 3} = \dfrac{4}{p^2 + 7p + 12}$

32. $\dfrac{2}{p + 2} + \dfrac{3}{p - 2} = \dfrac{12}{p^2 - 4}$

Solve each of the following equations for the variable indicated. (See Example 7.)

33. $ax + by = c$, for x — straight line
34. $P = \frac{1}{2}A + 110$, for A — normal blood pressure
35. $A = P + Prt$, for r — amount of a simple interest loan
36. $y = mx + b$, for x — straight line
37. $A = \dfrac{h}{2}(a + b)$, for b — area of a trapezoid
38. $A = \pi(r + R)S$, for R — surface area of a frustrum of a cone
39. $V = \frac{1}{6}h(b + 4M + B)$, for B — volume of a prismatoid
40. $S = \dfrac{n}{2}[2a + (n - 1)d]$, for a — sum of the terms in an arithmetic progression
41. $p = \dfrac{S}{S + F}$, for S — probability of a success
42. $d = \dfrac{i}{1 + ni}$, for i — discount rate
43. $S = \dfrac{a - rL}{1 - r}$, for r — sum of the terms in a geometric progression
44. $m = \dfrac{C(100 - p)}{100 - d}$, for d — marked price of an item

Solve each of the following equations involving real variables. (See Examples 3–6 and Practice Problem 1.)

45. $|x - 3| = 6$
46. $|x - 2| = 4$
47. $|2r + 7| = 11$
48. $|3t + 4| = 13$
49. $|2z - 7| = 0$
50. $|4y - 9| = 0$
51. $|4w + 13| = -3$
52. $|3x + 5| = -2$
53. $|3y + 8| = |2 - y|$
54. $|8q + 5| = |q + 4|$
55. $\dfrac{|2x - 5|}{|x + 2|} = 1$
56. $\dfrac{|4p + 3|}{|p - 5|} = 1$
57. $\dfrac{|9 - 4t|}{|t + 2|} = 0$
58. $\dfrac{|3 - 2t|}{|t + 9|} = 0$
59. $|x| + 3x - 9 = 0$
60. $|x + 5| = 3x - 2$

61–64

Solve each of the following equations, rounding the solutions to the nearest hundredth.

61. $2.79 - 4.20x = 6.44x + 9.56$

62. $\dfrac{9.82 - 6.10x}{2.77} = 4.21(6.05x - 2.74)$

63. $\sqrt{2.41}x - \sqrt{7.63} = \sqrt{0.42} + 3\sqrt{5.87}x$

64. $\pi x + 2\pi^2 = \sqrt{2}x - \sqrt{3}$

Most Comfortable Temperature

65. (See Example 8.) Find the Celsius temperature that corresponds to 74°F, considered to be the most comfortable household temperature.

Fahrenheit and Celsius

66. (See Example 8.) Find the temperature at which the Fahrenheit and Celsius readings are the same.

Amount of a Loan

67. If a sum of money P (the **principal**) is loaned at an interest rate of r per time period for t time periods, the **simple interest** I on the loan is computed by the formula $I = Prt$. The **amount** A owed on the loan at the end of t time periods is given by $A = P + I$, or

$$A = P(1 + rt).$$

Find the amount owed at the end of 1 year if \$2800 is borrowed at a simple interest rate of 1.5% per month.

Principal of a Loan

68. (See Exercise 67.) A sum of money was borrowed at 11% per year, and the amount required to pay off the loan after 3 years was \$4522. How much money was borrowed?

Discount Rate

69. If a bank note with maturity value A is discounted for n years at a discount rate r, the proceeds P of the note are given by

$$P = A(1 - rn).$$

Find the discount rate if a note discounted for 3 years has a maturity value of \$3000 and proceeds of \$1470.

Maturity Value

70. (See Exercise 69.) Find the maturity value of a note discounted for 2 years if the proceeds are \$889.20 and the discount rate is 11%.

Retail Sales

71. The retail price R and the cost C of an item in a store are related by the equation

$$R = C + rC,$$

where r is the percent markup. Find the cost of an article with retail price \$245 if the markup is 25%.

Markup 72. (See Exercise 71.) Find the percent markup if an article that costs \$241.75 has a retail price of \$319.11.

Critical Thinking: Discussion 73. If R and S are algebraic expressions, explain why $|R| = |S|$ and $R^2 = S^2$ are equivalent equations.

Critical Thinking: Discussion 74. Write a discussion of the equation $|2x - 5| = d$ that describes the values of d which correspond to each of the three cases (a) two solutions, (b) one solution, and (c) no solution. Explain the reasoning behind your answer in each case.

◆ Solution for Practice Problem

1. We have $|r + 3| = 7$ if and only if either $r + 3 = 7$ or $r + 3 = -7$. Thus the solution set is $\{4, -10\}$.

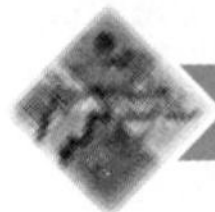

2.2 APPLICATIONS

In this section, we are concerned with problems that can be represented by a linear equation in one variable. To solve such a problem, we first decide what the unknown is in that particular situation. Next, we use the given information to write an equation involving that unknown. Finally, we solve the equation by the methods used in the last section.

Example 1 • Consecutive Even Integers

Find two consecutive positive even integers such that the difference of their squares is 52.

Solution

Let x and $x + 2$ represent the two consecutive positive even integers. Then we have the equation

$$(x + 2)^2 - x^2 = 52.$$

Expanding and simplifying, we have

$$x^2 + 4x + 4 - x^2 = 52$$

and

$$4x + 4 = 52.$$

This gives $x = 12$ and $x + 2 = 14$. That is, the two consecutive positive even integers are 12 and 14. □

Example 2 • Digits of a Number

The sum of the digits in a two-digit number is 5. If the digits are reversed, the new number is 9 more than the original number. What is the original number?

Solution

Suppose that x is the units digit in the original number. Then the tens digit is $5 - x$, and the original number expressed in terms of x has a value $10(5 - x) + x$. Similarly, in the new number, x is the tens digit and $5 - x$ is the units digit, so that its value is $10x + (5 - x)$. Thus we have

$$10x + (5 - x) = 10(5 - x) + x + 9.$$

This equation can be solved as follows.

$$\begin{aligned} 9x + 5 &= -9x + 59 \\ 18x &= 54 \\ x &= 3 \qquad \text{units digit} \end{aligned}$$

Also,

$$5 - x = 2. \qquad \text{tens digit}$$

This means that the original number is 23. □

Example 3 • An Absolute Value Application

On the foul tip shown in Figure 2.2, the baseball left the bat and traveled upward. After t seconds, its speed s in feet per second was given by $s = |80 - 32t|$ for $0 \le t \le 5$.

a. Find the speed of the ball when it left the bat and at the end of 4 seconds.
b. Find the time t when the speed of the ball was 0.
c. The catcher caught the ball at the end of 5 seconds. Find its speed when it was caught.

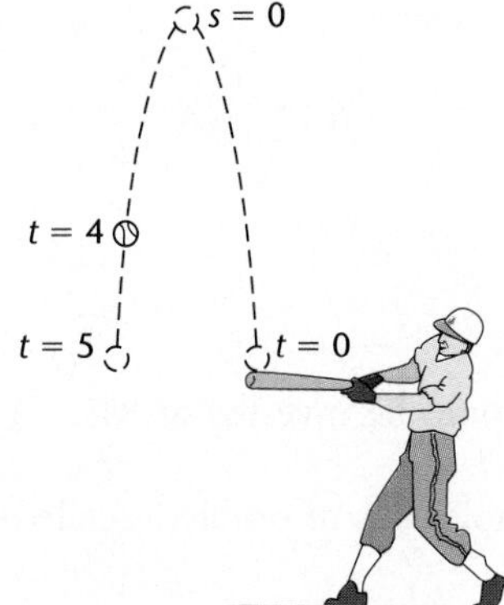

Figure 2.2 The foul tip in Example 3

Solution

a. The ball left the bat when $t = 0$, and $t = 0$ gives $s = |80 - 0| = 80$ feet per second. At the end of 4 seconds, the speed was

$$s = |80 - 32(4)| = |-48| = 48 \text{ feet per second.}$$

b. If we set the speed equal to 0, we have

$$\begin{aligned} |80 - 32t| &= 0 \\ 80 - 32t &= 0 \\ 80 &= 32t \\ t &= \frac{80}{32} = 2.5. \end{aligned}$$

Thus $s = 0$ (the ball stops) when $t = 2.5$ seconds.

c. To find its speed when the ball was caught, we put $t = 5$ and get

$$s = |80 - 32(5)| = |-80| = 80 \text{ feet per second.}$$

As the ball travels upward, its speed steadily slows down and reaches 0 at the end of 2.5 seconds when it is at its highest point. This follows from our work in part b. Also, the results in parts a and c show that the ball was caught at the same speed with which it left the bat. □

The next example illustrates the solution of an investment problem that involves simple interest.† We find in this solution that the organization of information in a table is helpful.

Example 4 • Simple Interest

Part of \$9000 is to be invested at 12%, and the remainder in a more secure investment at 8%. How much should be invested at each rate to yield an annual interest income of \$860?

Solution

Let y be the amount of money invested at 12%. Then $9000 - y$ is the amount invested at 8%, and we can describe the situation as in Table 2.1.

Table 2.1 The Two Investments in Example 4

Principal (dollars)	Rate (percent)	Time (years)	Interest (dollars)
y	0.12	1	$0.12y$
$9000 - y$	0.08	1	$0.08(9000 - y)$

The sum of the expressions in the interest column in Table 2.1 must equal \$860 since this is the required total interest. We set up this equation and solve it as follows.

$$\begin{aligned} 0.12y + 0.08(9000 - y) &= 860 \\ 0.12y + 720 - 0.08y &= 860 \\ 0.04y &= 140 \\ y &= \frac{140}{0.04} = 3500 \end{aligned}$$

Thus \$3500 should be invested at 12% and \$5500 should be invested at 8%. □

Example 5 illustrates a method of solution for a whole class of problems called **mixture problems.**

†The related formulas for simple interest are given in Exercise 67 in Section 2.1. Unless stated otherwise, the interest rate in a simple interest problem is understood to be an annual rate.

Example 5 • Mixture

A chemist has 16 liters of a mixture that is 65% acid. How much of an 85% solution should she add to make a mixture that is 70% acid?

Solution

Let x represent the number of liters of the 85% solution to be added. The key to working mixture problems is to set up an equation using amounts of a pure substance as shown in Figure 2.3.

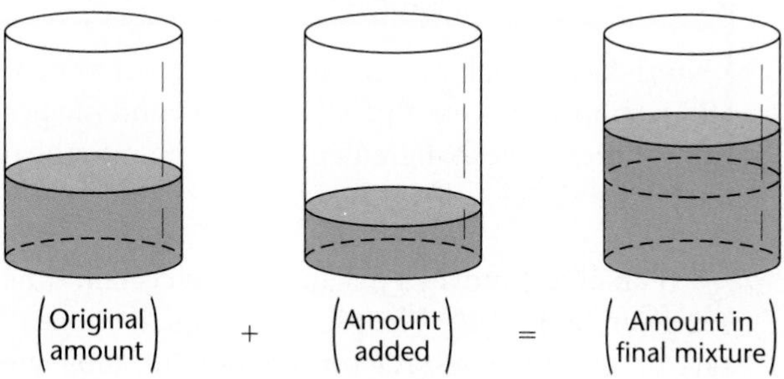

Figure 2.3 Mixing acid

In this problem, we need to set up an equation in pure acid. A table similar to Table 2.2 is helpful.

Table 2.2 The Mixtures in Example 5

Mixture	Percent Acid	Quantity of Solution (liters)	Amount of Pure Acid (liters)
Original	65% = 0.65	16	0.65(16)
Added	85% = 0.85	x	$0.85x$
Final	70% = 0.70	$16 + x$	$0.70(16 + x)$

Since the sum of the amounts of pure acid in the parts must equal the amount of pure acid in the final mixture, we have the following equation.

$$0.65(16) + 0.85x = 0.70(16 + x)$$

Solving for x, we have

$$\begin{aligned} 10.40 + 0.85x &= 11.20 + 0.70x \\ 0.15x &= 0.80 \\ x &= \frac{0.80}{0.15} = \frac{16}{3}. \end{aligned}$$

The chemist should add $\frac{16}{3}$ or $5\frac{1}{3}$ liters of the 85% solution in order to obtain a mixture that is 70% acid. □

One variation of mixture problems involves mixing together substances that have different monetary values. In this type of problem, we set up an equation of the following form.

$$\begin{pmatrix}\text{Value of}\\ \text{first ingredient}\end{pmatrix} + \begin{pmatrix}\text{Value of}\\ \text{second ingredient}\end{pmatrix} = \begin{pmatrix}\text{Value of}\\ \text{mixture}\end{pmatrix}$$

This sort of analysis extends easily to any number of ingredients.

◆ Practice Problem 1 ◆

Nutti-Korn snack mix consists of caramel popcorn and honey-roasted peanuts. If the popcorn is worth \$0.80 a pound and the peanuts are worth \$2.50 a pound, how much of each ingredient should go into the mixture for a 1-pound box that sells for \$1.82?

If an object moves a distance d in a certain time t, then its average rate of speed r is given by $r = d/t$. The movement of the object is called **uniform motion** if the rate of speed is constant throughout the movement. Uniform motion is usually described by the following distance formula.

Uniform Motion

$$\text{Distance} = \text{rate} \cdot \text{time}$$
$$d = rt$$

In using this distance formula, we must be sure that the units in which the quantities are expressed are in agreement. For instance, if d is measured in miles and t in hours, then r must be in miles per hour.

The following example shows how a table can be used as an effective tool in solving uniform motion problems.

Example 6 • Uniform Motion

Sue Kowalczyk drove her boat upstream on the Mississippi River from Moline, Illinois, to Dubuque, Iowa, in 5.5 hours, and then made the return trip downstream in 4.5 hours. (See Figure 2.4.) Throughout both trips, the boat was running at the rate of 20 miles per hour in still water. Find the rate of the current.

Solution

Let c represent the rate of the current. Then the boat travels at the rate of $(20 - c)$ miles per hour upstream and at the rate of $(20 + c)$ miles per hour downstream.

Our work is easier if we organize our information in a table with columns headed by distance, rate, and time, and with rows corresponding to travel upstream and downstream. (See Table 2.3.)

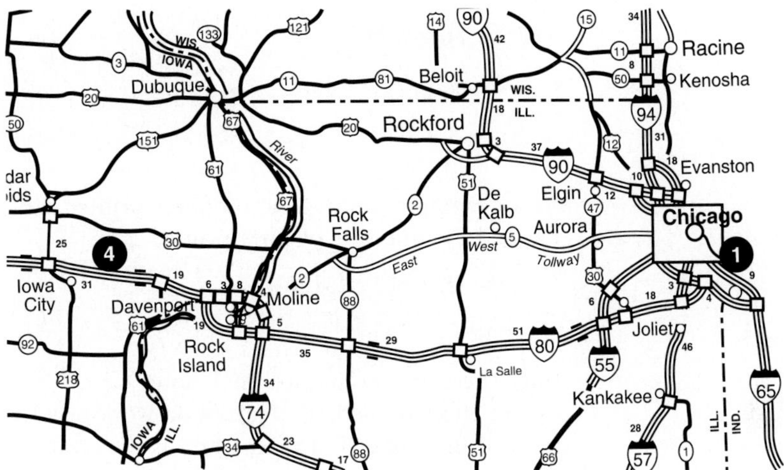

Figure 2.4 Map for Example 6

Table 2.3 Distances Upstream and Downstream

Travel	Rate (mph)	Time (hours)	Distance (miles)
Upstream	$20 - c$	5.5	$5.5(20 - c)$
Downstream	$20 + c$	4.5	$4.5(20 + c)$

Since the distance traveled upstream is equal to the distance traveled downstream, we can write the equation

$$5.5(20 - c) = 4.5(20 + c)$$

and then solve for c.

$$110 - 5.5c = 90 + 4.5c$$
$$20 = 10c$$
$$c = 2$$

The rate of the current is 2 miles per hour. □

◆ Practice Problem 2 ◆

A certain plane travels 600 kilometers per hour in still air. The plane flies north with the wind for 5 hours and makes the return trip against the wind in 5 hours and 40 minutes. What is the rate of the wind?

Some problems involving work can be analyzed in a manner similar to that used with uniform motion. If a job is performed at a constant rate r for a certain time t, the output or **work** w is given by the following work formula.

Work Formula

Work = rate · time

$w = rt$

As an example, suppose a computer printer can print 52 lines per minute. In 1 hour, the number w of lines printed is

$$w = \underset{\text{Rate}}{(52)}\underset{\text{Time}}{(60)} = 3120$$

since there are 60 minutes in 1 hour.

An effective method of attack on work problems is to write an equation in the rates of the workers. This method is illustrated in the next example.

Example 7 • Work

Herman has a small backhoe that can dig a certain ditch in 5 days, and Toby has a larger backhoe that can dig the ditch in 2 days. How long will it take for both Herman and Toby to dig the ditch if they work together?

Solution

Let x represent the number of days it takes both to dig the ditch working together. Then $1/x$ represents the portion of the ditch that they can dig together in 1 day. Similarly, since it takes Herman 5 days to dig the ditch, he can dig $\frac{1}{5}$ of the ditch in 1 day. Toby can dig $\frac{1}{2}$ of the ditch in 1 day, and, working together, they can dig $\frac{1}{5} + \frac{1}{2}$ of the ditch in 1 day. Thus we have

$$\frac{1}{5} + \frac{1}{2} = \frac{1}{x}.$$

We multiply this equation by $10x$ and then solve for x as follows.

$$\begin{aligned} 2x + 5x &= 10 \\ 7x &= 10 \\ x &= \tfrac{10}{7} = 1\tfrac{3}{7} \end{aligned}$$

Working together, Herman and Toby can dig the ditch in $1\frac{3}{7}$ days. □

EXERCISES 2.2

Number Problems 1–6

1. Find three consecutive integers whose sum is 147.
2. Find two consecutive even integers whose sum is 90.
3. The sum of the digits in a two-digit number is 11, and the tens digit is 1 more than four times the units digit. Find the number.
4. The sum of the digits in a two-digit number is 12, and the tens digit is 3 times the units digit. Find the number.

5. In a two-digit number, the units digit is 1 less than twice the tens digit. If the digits are reversed, the new number is 20 less than twice the original number. Find the original number.

6. In a three-digit number, the tens digit is twice the hundreds digit and half the units digit. If the digits are reversed, the new number is 49 more than 3 times the original number. Find the original number.

Absolute Value Applications 7–8

7. Find the times when the speed of the ball in Example 3 was 16 feet per second.

8. (See Example 3.) On a certain foul tip, the baseball traveled straight up with its speed s after t seconds given by $s = |112 - 32t|$ feet per second for $0 \leq t \leq 7$.

   a. Find the speed of the ball when it left the bat and at the end of 5 seconds.

   b. Find the time when the ball was at its highest point.

Investment Problems 9–14

9. If \$4000 is invested at 8% per year, how much additional money needs to be invested at 12% per year so that the total annual interest income from the investments is \$968?

10. If \$6000 is invested at the rate of 16%, how much additional money must be invested at 12% in order that the annual interest income from the two investments together will be \$2280?

11. Part of \$14,000 is to be invested at 9%, and the remainder at 12%. How much should be invested at each rate in order to yield an annual interest income of \$1500?

12. Suppose that Matthew Pallai plans to invest part of \$17,000 at 10% and the remainder at 8%. How much should he invest at each rate in order for the annual interest income to be \$1592?

13. Ed and Connie have twice as much money invested in bonds paying 10% as they do in stocks paying 12%. What is the total amount they have invested if the total annual interest income from both investments is \$8640?

14. Adrienne has three times as much money invested in bonds paying 8% as she does in stocks paying 14%. If her total annual interest income from these investments is \$24,700, how much does she have invested in each?

Coin Problems 15–16

15. Suppose that 12 dimes and nickels are worth 95 cents. How many dimes and how many nickels are there?

16. Suppose that 25 nickels, dimes, and quarters are worth \$2.75 and there are twice as many dimes as nickels. How many of each denomination are there?

Mixture Problems 17–18

17. Mrs. Phillips went to the grocery store to buy milk and eggs. Suppose that a carton of milk cost \$0.93 and a carton of eggs cost \$0.99. How many cartons each of milk and eggs did she buy if she spent \$7.74 (excluding tax) and bought eight items?
18. There are two types of tickets available for a concert. The reserved-seat tickets cost \$18.00 each, and the cheap-seat tickets cost \$12.50 each. How many tickets were sold if 3 times as many cheap-seat tickets were sold as reserved-seat tickets and the total proceeds were \$688,200?

Perimeter of a Rectangle

19. The perimeter of a rectangular garden is 112 meters, and the length is 4 meters less than twice the width. Find the length and width of the garden.

Angles in a Triangle

20. The sum of the three angles in a triangle is 180°. Find the measures of the angles if the largest is 4 times as large as the smallest and equal to the sum of the two smaller.

Mixture Problems 21–26

21. Suppose that we wish to mix peanuts worth \$2.10 per pound with cashews worth \$2.40 per pound to obtain 12 pounds of a mixture worth \$2.30 per pound. How many pounds of each type should we use?
22. How many pounds of coffee worth \$2.75 per pound must be mixed with 15 pounds of coffee worth \$2.00 per pound to obtain a blend worth \$2.30 per pound?
23. Suppose that we wish to produce a 40-gram bar of metal that is 15% pure silver by melting together parts of one bar of metal that is 20% pure silver and another that is 12% pure silver. How many grams of each should be used?
24. How much pure acid must be added to 10 liters of a 10% solution to obtain a solution that is 50% acid?
25. Suppose that a chemist wishes to dilute 40 liters of a solution that is 80% acid to one that is 50% acid. How much water should be added to the acidic solution?
26. If 15 liters of an acidic solution mixed with 20 liters of a 55% acidic solution produces a 40% acidic solution, what is the strength of the first solution?

Uniform Motion Problems 27–32

27. Suppose that Dan and Fran live 450 kilometers apart and at the same time they begin driving toward each other with Dan traveling an average rate of 50 kilometers per hour and with Fran's average rate 55 kilometers per hour. How long will it be before they meet?
28. In Exercise 27, suppose that Dan leaves at 12 noon and Fran leaves at 1 P.M. At what time will they meet?

29. Suppose that a boat travels 1 hour downstream and then returns in 1 hour 20 minutes. If the speed of the current is 3 kilometers per hour, find the speed of the boat and the distance it traveled downstream.
30. It takes a plane as long to fly 400 kilometers against the wind as it does to fly 450 kilometers with the same wind. If the speed of the wind is 20 kilometers per hour, find the speed of the plane in still air.
31. Two cars leave a city at the same time traveling in opposite directions. At the end of 5 hours, they are 725 kilometers apart. How fast was each car traveling if one traveled 5 kilometers per hour faster than the other?
32. One car traveling 77 kilometers per hour overtook another car traveling the same highway in 2 hours. If the faster car started at the same point 45 minutes after the slower car, find the speed of the slower car.

Work Problems 33–38

33. If it takes 8 hours for Jill to mow the grass with her push mower, and it takes Martin 5 hours to mow the grass with his riding mower, how long will it take to mow the grass if both Jill and Martin work together?
34. If it takes 12 hours for Beckie to clean the house, 6 hours for her sister Lisa to clean the same house, and 10 hours for the third sister, Donna, to clean the same house, how long will it take the three sisters to clean the house if they work together?
35. Suppose that Jim and Ruth can roof a house working together in 20 hours, and Jim can do the job alone in 36 hours. How long would it take Ruth to do the job alone?
36. Suppose that it takes Mary and Sam 3 hours to grade a set of homework papers if they work together and Sam works twice as fast as Mary. How long would it take Mary working alone?
37. A tank can normally be filled in 10 hours. But after the tank developed a leak, it took 12 hours to fill it. How long would it take the leak to empty the full tank?
38. Suppose that it takes 6 hours to fill an empty swimming pool with one hose and it takes 5 hours to fill the same pool with a different hose. Also, suppose that it takes 15 hours to completely drain the full pool. How long will it take to fill the pool if both hoses are used and the drain is left open?

Critical Thinking: Discussion

39. In Example 3, the absolute value expression $|80 - 32t|$ represented the speed of a ball. This is realistic because a number that tells how fast a ball is traveling should not be negative. Describe some other physical quantities where an absolute value expression might be appropriate because negative numbers cannot describe the quantities.

Critical Thinking: Writing

40. Let n and $n + 1$ represent two consecutive integers. Write a description in words for each of the expressions.

a. $2[n + (n + 1)]$ b. $n^2 + (n + 1)^2$

◆ Solutions for Practice Problems

1. Let x be the number of pounds of popcorn in the mixture. Then $1 - x$ is the number of pounds of peanuts in the mixture, and we can fill out Table 2.4.

Table 2.4 Values of the Ingredients

Ingredient	Value per pound (dollars)	Quantity of Ingredient (pounds)	Total Value (dollars)
Popcorn	\$0.80	x	\$0.80x
Peanuts	\$2.50	$1 - x$	\$2.50$(1 - x)$
Mixture	\$1.82	1	\$1.82(1)

Since the sum of the values of the ingredients is equal to the value of the mixture, we can set up and solve the following equation.

$$0.80x + 2.50(1 - x) = 1.82$$
$$0.80x + 2.50 - 2.50x = 1.82$$
$$0.68 = 1.70x$$
$$x = \frac{0.68}{1.70} = 0.4 \text{ pound}$$
$$1 - x = 0.6 \text{ pound}$$

Thus a 1-pound box of Nutti-Korn contains 0.4 pound of popcorn and 0.6 pound of peanuts.

2. Let r represent the rate of the wind. Then the plane's ground speed is $(60 + r)$ kilometers per hour with the wind and $(600 - r)$ kilometers per hour against the wind. Converting 5 hours and 40 minutes to $5\frac{2}{3} = \frac{17}{3}$ hours and following the same pattern as in Example 6, we obtain Table 2.5.

Table 2.5 Distances With and Against the Wind

Travel	Rate (km/hr)	Time (hours)	Distance (kilometers)
With the wind	$600 + r$	5	$(600 + r)(5)$
Against the wind	$600 - r$	$\frac{17}{3}$	$(600 - r)(\frac{17}{3})$

Since the distance traveled on both trips is the same, we set the distances in Table 2.5 equal and solve the resulting equation.

$$(600 + r)(5) = (600 - r)(\tfrac{17}{3})$$
$$3000 + 5r = 3400 - \tfrac{17}{3}r$$
$$\tfrac{32}{3}r = 400$$
$$r = 37.5$$

That is, the rate of the wind is 37.5 kilometers per hour.

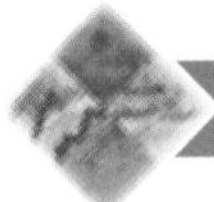

2.3 LINEAR AND ABSOLUTE VALUE INEQUALITIES

In this section we again restrict our work to real numbers and consider inequalities of the form $ax + b > 0$, $a \neq 0$, and similar inequalities where the inequality symbol may be $\geq$, $<$, or $\leq$. Any inequality of this form is called a **linear inequality.** The **solution set** for such an inequality is the set of all values of the variable which make the inequality true. Two inequalities are said to be **equivalent** if they have the same solution set.

We will determine the solution set of a given linear inequality by finding an inequality whose solution set is easily determined and which is equivalent to the original inequality. The following properties will be useful.

Addition and Multiplication Properties of Inequalities

Let R, S, and T be algebraic expressions. Then

$$R > S$$

is equivalent to

a. $R + T > S + T$, — addition property
b. $RT > ST$, if $T > 0$,
c. $RT < ST$, if $T < 0$. — (b, c) multiplication properties

Similar statements can be made using one of the symbols $\geq$, $<$, or $\leq$. The solution procedure is illustrated in the following examples.

Example 1 • Using the Properties of Inequalities

Determine the solution set for

$$x - 5 > 2x + 7.$$

Solution

We proceed as follows.

$x > 2x + 12$ — adding 5
$-x > 12$ — adding $-2x$
$x < -12$ — multiplying by -1 and *reversing* the inequality

Thus the solution set is

$$\{x \mid x < -12\}.$$

□

In working with inequalities, one frequently encounters what are called compound inequalities. A **compound inequality** is a compound statement that consists of two inequalities connected by the word **and** or the word **or.**

When two inequalities in a variable x are connected by the word **and** to form a compound inequality, the solution set of the compound inequality is the set of all values of x that satisfy **both** of the inequalities involved in the compound statement. This is illustrated in the next example.

Example 2 • Solving a Compound Inequality with *and*

Determine the solution set for

$$3x - 2 > 4 - 5x \quad \text{and} \quad 9 - 3x \geq 4x + 2.$$

Solution

We must find the values of x that make *both* of the stated inequalities true. To do this, we solve each of the two inequalities separately and then form the *intersection* of their solution sets.

$$\begin{array}{rr} 3x - 2 > 4 - 5x & 9 - 3x \geq 4x + 2 \\ 8x > 6 & -7x \geq -7 \\ x > \frac{3}{4} & x \leq 1 \end{array}$$

The solution set for the compound inequality is

$$\{x \mid x > \tfrac{3}{4}\} \cap \{x \mid x \leq 1\} = \{x \mid \tfrac{3}{4} < x \leq 1\},$$

and its graph is shown in Figure 2.5. □

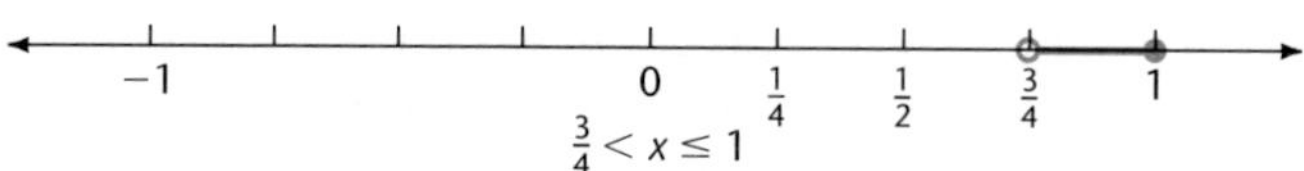

Figure 2.5 The solution set for Example 2

In Figure 2.5, the open circle at $\frac{3}{4}$ and the heavy dot at 1 have the same meaning as in Section 1.2. An open circle indicates that the endpoint is not included in the set, and a heavy dot indicates that the endpoint is included in the set. In an alternate notation, the endpoint is marked with a parenthesis if it is not included in the set, and with a bracket if it is included in the set. With this notation, the set $\{x \mid \frac{3}{4} < x \leq 1\}$ would appear as in Figure 2.6.

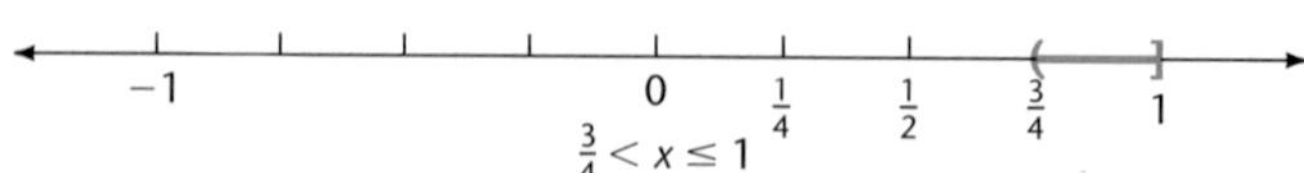

Figure 2.6 Alternate notation for the set in Figure 2.5

The use of parentheses and brackets described in the preceding paragraph leads to a very compact and useful notation for the following types of sets.

Interval Notation

The letters a and b represent real numbers with $a < b$.

$(a, b) = \{x \mid a < x < b\}$	$(a, \infty) = \{x \mid x > a\}$
$[a, b] = \{x \mid a \leq x \leq b\}$	$[a, \infty) = \{x \mid x \geq a\}$
$(a, b] = \{x \mid a < x \leq b\}$	$(-\infty, a] = \{x \mid x \leq a\}$
$[a, b) = \{x \mid a \leq x < b\}$	$(-\infty, a) = \{x \mid x < a\}$
	$(-\infty, \infty) = \mathscr{R}$

The symbols ∞ and $-\infty$ are read "infinity" and "negative infinity," respectively. They are purely notational and do not represent real numbers. Any set of the types shown above is called an **interval,** and the compact notation is referred to as **interval notation.** With interval notation, the set pictured in Figures 2.5 and 2.6 would be written simply as $(\frac{3}{4}, 1]$.

When two inequalities in a variable x are connected by the word **or** to form a compound inequality, the solution set of the compound inequality is the set of all values of x that satisfy **either** one or the other of the two inequalities involved in the compound statement. This is illustrated in Example 3.

Example 3 • Solving a Compound Inequality with *or*

Determine the solution set for

$$2x - 7 \leq 17 - x \qquad \text{or} \qquad x > 4x - 6.$$

Solution

We must find the values of x that make *either* one or the other of the stated inequalities true. To do this, we solve each of the two inequalities separately and then form the *union* of their solution sets.

$$\begin{aligned} 2x - 7 &\leq 17 - x \qquad & x &> 4x - 6 \\ 3x &\leq 24 & -3x &> -6 \\ x &\leq 8 & x &< 2 \end{aligned}$$

The solution set for the compound inequality is

$$(-\infty, 8] \cup (-\infty, 2) = (-\infty, 8],$$

and its graph is shown in Figure 2.7. □

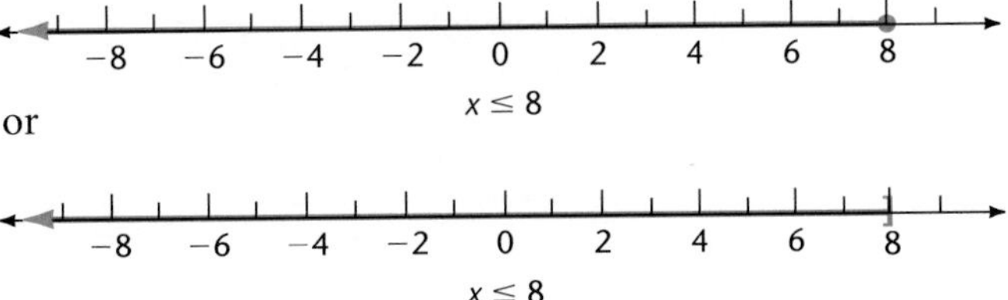

Figure 2.7 The solution set for Example 3

In Section 1.2, we considered the graphs of inequalities such as $|x - 3| < 2$ and $|x + 4| > 1$. The inequalities that we worked with there are special cases of the type that we consider now.

With the methods of this section, we can solve inequalities of the form

$$|ax + b| < d$$

or

$$|ax + b| > d,$$

where $a \neq 0$, and similar inequalities using $\geq$ or $\leq$.

Absolute Value Inequalities

Suppose d is *positive.* Then

$|ax + b| < d$ is equivalent to $-d < ax + b < d$

and

$|ax + b| > d$ is equivalent to $ax + b > d$ or $ax + b < -d$.

Thus absolute value inequalities can be solved by changing to the related compound inequalities and using the techniques in Examples 2 and 3. This is demonstrated in the following examples.

Example 4 • Solving an Absolute Value Inequality

Solve $|3x - 1| < 4$.

Solution

We must solve the following inequality.

$$-4 < 3x - 1 < 4$$

$$-3 < 3x < 5 \qquad \text{adding 1}$$

$$-1 < x < \tfrac{5}{3} \qquad \text{multiplying by } \tfrac{1}{3}$$

The solution set is $(-1, \frac{5}{3})$, as shown in Figure 2.8. □

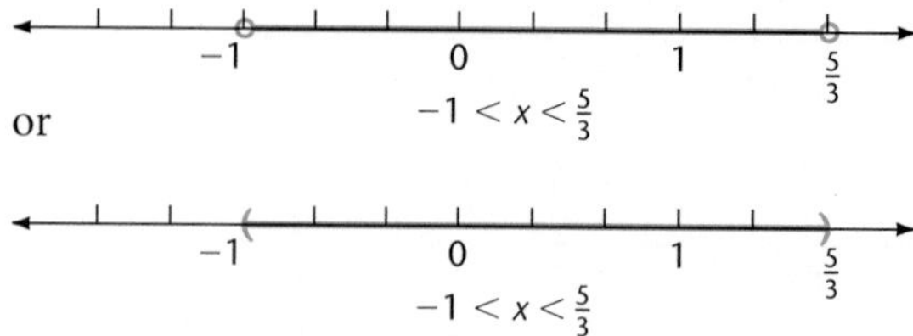

Figure 2.8 Solution set for $|3x - 1| < 4$

◆ Practice Problem 1 ◆

Solve $|9 - 2x| > 5$.

There are special cases with inequalities involving absolute value that correspond to the special cases that came up in Section 2.1. Some illustrations are given in the next example and practice problem.

Example 5 • An Absolute Value Inequality with No Solution

Solve $|x - 5| < -1$.

Solution

Since the absolute value of a real number is either zero or a positive number, $|x - 5| < -1$ is impossible for any x, and the solution set is $\varnothing$. □

◆ Practice Problem 2 ◆

Solve $|5x + 6| \geq -4$.

Example 6 • Water Pressure

The water tank in Figure 2.9 sits on a platform 40 feet above ground and has the shape of a cylinder with an altitude of 20 feet. When the tank is full of water, the pressure p in the tank at height h feet above ground level is $p = 62.5(60 - h)$ pounds per square foot for $40 \leq h \leq 60$. Find the values of h for which the pressure is greater than 750 pounds per square foot.

Solution

We set $p > 750$ and obtain

$$62.5(60 - h) > 750.$$

This inequality can be solved for h as follows.

$$\begin{aligned} 3750 - 62.5h &> 750 \\ -62.5h &> -3000 \\ h &< \frac{3000}{62.5} \\ h &< 48 \end{aligned}$$

The pressure will be greater than 750 pounds per square foot when $h < 48$ feet. Since the bottom of the tank is at a height 40 feet above ground, the answer to the problem is $40 \leq h < 48$, or $h \in [40, 48)$. □

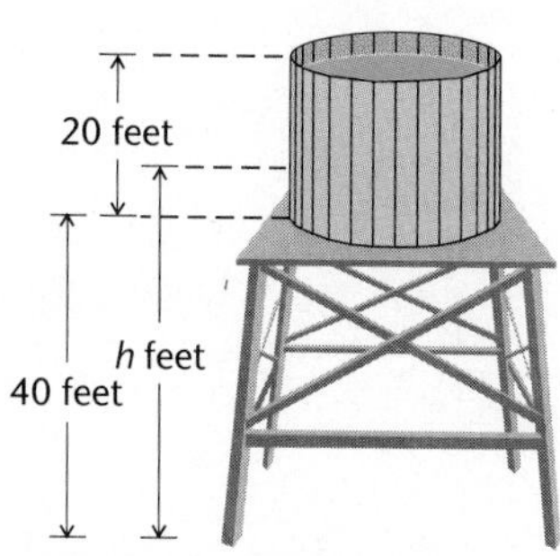

Figure 2.9 The water tank in Example 6

EXERCISES 2.3

Determine the solution sets for the following inequalities. (See Example 1.)

1. $2x < 8$
2. $3x - 5 \geq 1$
3. $25 - 5x < 0$
4. $10 \geq 3y + 4$
5. $2y > \frac{8}{3}y - 4$
6. $-4x \leq -16$
7. $17 - 4x > 2x + 5$
8. $-6x + 3 \geq -5x - 2$
9. $-3x + 2\sqrt{2} < x - 2\sqrt{2}$
10. $3\sqrt{2} - 2x \geq 7\sqrt{2} - 6x$
11. $\frac{1}{2}x \leq \frac{3}{4}x + 9$
12. $\frac{5}{6}x + \frac{1}{3} \geq \frac{2}{3}x$
13. $17 + 5(x + 2) \leq x - 3(x - 2)$
14. $3(x + 2) - 4x > 1 - 2(3x - 2)$
15. $\dfrac{x}{3} + \dfrac{x}{2} - \dfrac{x}{4} \geq \dfrac{x}{5}$
16. $\dfrac{1}{2}(x + 2) > \dfrac{x}{3} - 2$
17. $\dfrac{7 - 2x}{3} \geq 11$
18. $\dfrac{3 - x}{4} < \dfrac{2}{3}$
19. $\dfrac{2 - x}{3} > \dfrac{1 - 2x}{5}$
20. $\dfrac{x - 2}{3} \geq \dfrac{7x + 1}{2}$

Solve the following compound inequalities. (See Examples 2 and 3.)

21. $-17 < 5x - 7 \leq 13$
22. $-4 < \dfrac{x - 1}{3} < 3$
23. $-3 < 5 - 2t \leq 3$
24. $-2 \leq 10 - 3t \leq 7$
25. $x - 1 \geq 0$ and $3 \geq x - 1$
26. $x + 2 \leq 4$ and $x + 2 \geq 1$
27. $1 - 2x < 3$ and $2(x - 2) < x - 1$
28. $5 - 2x < 9$ and $3(x - 2) < 2x - 1$
29. $7 - 2x \leq 5$ and $1 - 3x > 2(2 - x)$
30. $8 - 3x \leq 2$ and $6 - 5x > 3(4 - x)$
31. $3x + 5 > 2$ and $9x + 2 \geq 4(x + 3)$
32. $2x + 5 > 1$ and $7x + 6 \geq 3(x + 2)$
33. $3x + 11 \leq 5$ or $4x - 3 > x + 6$
34. $2x + 3 \leq 9$ or $5x - 10 > 3x - 8$
35. $-2x + 7 \geq 3$ or $5x - 2 > 2(x + 5)$
36. $-3x + 2 \geq -1$ or $4x - 5 > 3(x - 1)$
37. $\dfrac{7x + 6}{6} > \dfrac{x + 2}{2}$ or $4(x + 4) > 2(2 - x)$
38. $\dfrac{x + 3}{2} < \dfrac{1 - 2x}{4}$ or $9(x - 4) \leq 3(4 - 7x)$
39. $7x - 8 \geq -43$ or $-x > 8 + x$
40. $7 - 2x < 5$ or $1 + 3x < 2x + 3$

Solve the following inequalities and graph the solution sets. (See Examples 4 and 5 and Practice Problems 1 and 2.)

41. $\lvert 2x - 5 \rvert < 3$	42. $\lvert 10x - 3 \rvert < 12$	43. $\lvert 7 - x \rvert \le 2$
44. $\lvert 2 - x \rvert \le 5$	45. $\lvert 8x + 5 \rvert < 25$	46. $\lvert 6x + 4 \rvert < 18$
47. $\lvert 2x - 4 \rvert \ge 3$	48. $\lvert x + 4 \rvert \ge 4$	49. $\lvert 3 - 4x \rvert > 2$
50. $\lvert 1 - 3x \rvert > 2$	51. $\lvert 5x + 3 \rvert > 7$	52. $\lvert 3x + 2 \rvert > 7$
53. $\lvert 4 - 5x \rvert < 0$	54. $\lvert 7x - 2 \rvert < -3$	55. $\lvert 3x - 6 \rvert \le 0$
56. $\lvert 2x + 7 \rvert \le 0$	57. $\lvert 1 - x \rvert > 0$	58. $\lvert 3x + 1 \rvert > 0$
59. $\lvert 2x + 5 \rvert > -5$	60. $\lvert 4x + 7 \rvert \ge -2$	

61–68

Solve the following inequalities and write your answers in interval notation rounded to the nearest hundredth.

61. $4.38x - 6.15 < 2.17 - 1.95x$

62. $73.9x + 42.6 \ge 85.4 + 99.2x$

63. $-7.21 < \dfrac{1.55x - 2.74}{6.01} < 10.2$

64. $\sqrt{4.13}x - \sqrt{9.82} \ge 4.61$ or $52.1 - \sqrt{47.3}x > 72.5 - \sqrt{21.3}x$

65. $\lvert x - \sqrt{2} \rvert < \sqrt{3}$

66. $\lvert x + 2.47^{2/3} \rvert \le \pi$

67. $\lvert 10.2^{0.56}x - \sqrt{2} \rvert \ge 17.3$

68. $\left\lvert \dfrac{2.31 - 6.92x}{4.33} \right\rvert > 7.68$

Appliance Repair 69. A certain appliance repairman charges \$28 for a house call plus \$40 per hour while working. Find the number of hours he can work and keep the charge for a house call at \$200 or less.

Pizza Delivery 70. The cost of operating a pizza delivery car is \$0.70 per mile after an initial investment of \$9000. What mileage on the car will keep the cost at or below \$36,400?

Trade-in Value 71. A forklift costs \$16,000 and loses value at the rate of \$3000 per year. Find the time span for which the trade-in value is \$5500 or more.

Return on an Investment 72. On an investment of \$5000, the Atlantic Klipper Investment Company gives a guaranteed return R after t years time of $R = 5000(1 + 0.08t)$ dollars, and t is not required to be a whole number. Find the time span for which the return R will be no more than \$10,000.

Water Pressure in a Cistern 73. Many homes in the West Indies have an underground cistern for storage of rainwater. The top of the cistern shown in the accompanying figure is 2 feet below ground level, and the cistern is in the shape of a cylinder with an altitude of 10 feet. When the cistern is full of water, the pressure p in the water at a depth d feet below ground level is $p = 62.5(d - 2)$ pounds per square foot. Find the values of d for which p is greater than 250 pounds per square foot.

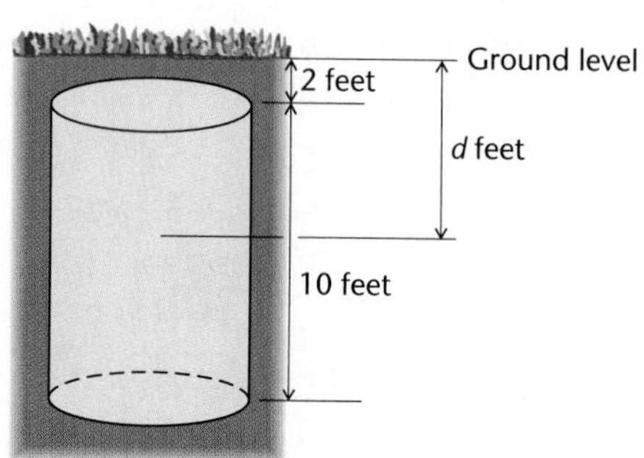

Figure for Exercise 73

Water Pressure in a Swimming Pool

74. All across the deep end of a rectangular swimming pool, the water is 12 feet deep, and the water pressure p at a point x feet above the bottom of the pool is $p = 62.5(12 - x)$ pounds per square foot. (See the accompanying figure.) Find the values of x for which p is greater than 500 pounds per square foot.

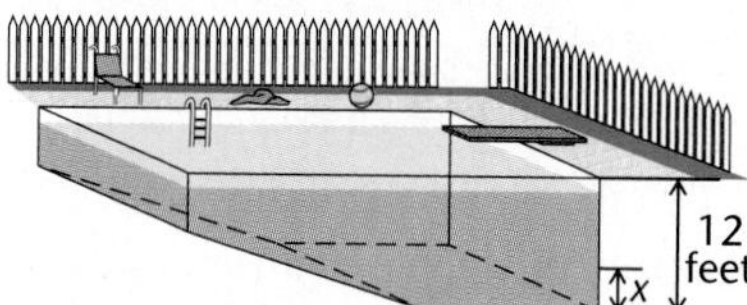

Figure for Exercise 74

Temperatures in Atlanta

75. In this exercise, let t be the number of hours that have elapsed since midnight. As shown in the accompanying figure, $t = 15$ at 3 P.M. On a certain July day in Atlanta, the Fahrenheit temperature F could be found from the formula

$$F = 94 - 3|t - 15|$$

for $t \in [7, 20]$. Find the interval of t values for which the temperature was over 82°F.

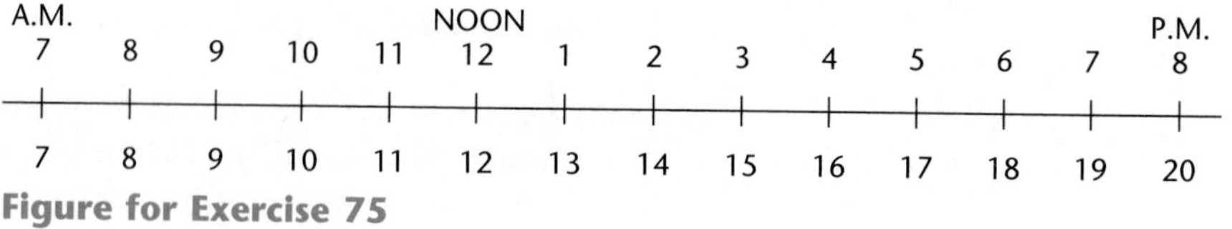

Figure for Exercise 75

Extreme Temperatures

76. Using the information in Exercise 75, find the coolest temperature and hottest temperature between 7 A.M. and 8 P.M.

Critical Thinking: Writing 77. State the addition property of inequalities in words.

Critical Thinking: Writing 78. State both of the multiplication properties of inequalities in words.

◆ Solutions for Practice Problems

1. The inequality $|9 - 2x| > 5$ is equivalent to

$$9 - 2x > 5 \quad \text{or} \quad 9 - 2x < -5.$$

Solving these inequalities separately, we get

$$\begin{array}{ll} -2x > -4 & -2x < -14 \\ x < 2 & x > 7. \end{array}$$

The solution set for the compound inequality is the union of these two solution sets.

$$(-\infty, 2) \cup (7, \infty)$$

The solution set is graphed in Figure 2.10.

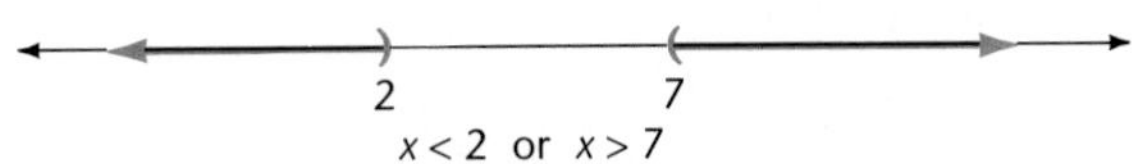

Figure 2.10 Solution set for $|9 - 2x| > 5$

2. Since the absolute value of a real number is either zero or a positive number, $|5x + 6| \geq -4$ is true for all x and the solution set is the set $\mathcal{R}$ of all real numbers.

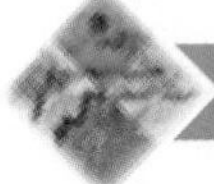

2.4 QUADRATIC EQUATIONS

Equations that can be written in the form $ax^2 + bx + c = 0$, where $a \neq 0$, are called **quadratic equations.** There are at most two distinct values of x that satisfy a quadratic equation. These two values are called the **solutions,** or **roots,** of the equation. In this section, we study two methods of solving a quadratic equation that has **real coefficients:** (1) factoring and (2) completing the square.

The method of factoring relies on this fact: *The product of two factors is zero if and only if at least one of the factors is zero.*

Zero Product Property

$xy = 0$ if and only if either $x = 0$ or $y = 0$

Example 1 • Solving by Factoring

Solve $3x^2 + 3x - 60 = 0$.

Solution

Factoring the quadratic expression on the left, we have

$$3(x^2 + x - 20) = 0$$
$$3(x + 5)(x - 4) = 0.$$

We can multiply both sides of the last equation by $\frac{1}{3}$ and obtain the equivalent equation

$$(x + 5)(x - 4) = 0.$$

The zero product property leads to the problem of solving two linear equations.

$$x + 5 = 0 \qquad \text{or} \qquad x - 4 = 0$$

Thus $x = -5$ and $x = 4$ are the two solutions to the quadratic equation $3x^2 + 3x - 60 = 0$, and the solution set is $\{-5, 4\}$. □

Example 2 • Solving by Factoring

Solve $x^2 = 9$.

Solution

The given equation is equivalent to

$$x^2 - 9 = 0.$$

Factoring the quadratic expression on the left, we have

$$(x - 3)(x + 3) = 0,$$

and the solutions to the two linear equations are $x = 3$ and $x = -3$. □

Example 2 easily generalizes to any equation of the form $x^2 = a$, where a is a real number. The general result can be stated as follows.

Square Root Property

If a is a real number, the solutions to $x^2 = a$ are $x = \sqrt{a}$ and $x = -\sqrt{a}$.

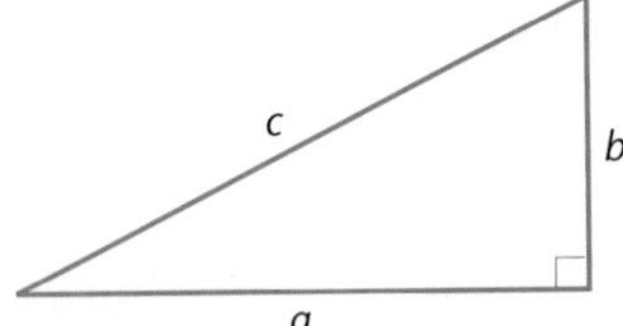

Figure 2.11 A right triangle with hypotenuse c

We write $x = \pm\sqrt{a}$ to indicate the solutions of $x^2 = a$, and we say that these solutions are obtained by "taking the squre root of both sides." If $a < 0$, the values of $\pm\sqrt{a}$ are not real numbers. For example, $\pm\sqrt{-16} = \pm 4i$.

Our next example shows how the solution of an equation of the form $x^2 = a$ can be useful in a practical situation. In this example, we need to use the **Pythagorean thorem** from geometry. Suppose a right triangle is labeled as in Figure 2.11, where a, b, and c are the lengths of the sides with c the length of the hypotenuse. Then the Pythagorean theorem states that the square of the hypotenuse is equal to the sum of the squares of the other sides.

Pythagorean Theorem

$c^2 = a^2 + b^2$ or $c = \sqrt{a^2 + b^2}$

Example 3 • Locating a Radio Tower

Ranger station R is located at a distance of 11 miles across the lake and due west of park headquarters H as shown in Figure 2.12. A short-wave radio tower with a range of 30 miles is to be located on a north-south road and due north of H. How far from H should the tower T be placed to make the distance from T to R equal the range of the radio?

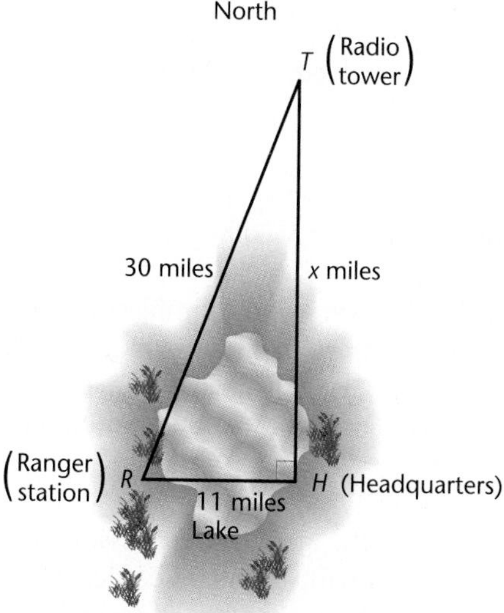

Figure 2.12 Radio tower location

Solution

From the figure, we see that triangle RHT is a right triangle with hypotenuse 30 and one side 11. If we let x be the number of miles from H to T, we can use the Pythagorean theorem and write

$$x^2 + (11)^2 = (30)^2.$$

Solving for x, we get

$$x^2 + 121 = 900$$
$$x^2 = 779$$
$$x = \sqrt{779} \approx 27.9.$$

(The solution $x = -\sqrt{779}$ is rejected since it is meaningless in this problem.) The tower should be located about 27.9 miles north of the park headquarters. □

In the remainder of this section, we deal with more complicated quadratic equations.

Example 4 • Using the Square Root Property

Solve $(2x + 3)^2 = 4$.

Solution

Equating square roots as described, we have

$$2x + 3 = \pm\sqrt{4}$$

or

$$2x + 3 = \pm 2.$$

This equation can be solved as follows.

$$2x = -3 \pm 2 \quad \text{adding } -3$$
$$x = -\tfrac{3}{2} \pm 1 \quad \text{multiplying by } \tfrac{1}{2}$$

Thus the solution set is $\{-\frac{1}{2}, -\frac{5}{2}\}$. □

The method of factoring is the simplest and most direct approach to solving a quadratic equation if the quadratic expression can be factored easily using the methods of Section 1.5. Such a factoring is not always possible. For example, suppose that we wish to solve the quadratic equation

$$2x^2 + 4x - 1 = 0.$$

If we attempt to factor the quadratic expression, we discover that it is impossible to find two linear factors with rational coefficients whose product is $2x^2 + 4x - 1$. However, another method can be used to solve the equation. It is the method of **completing the square,** which is illustrated in the following six steps.

1. Isolate the constant term on one side of the equation.

$$2x^2 + 4x = 1$$

2. Divide both sides of the equation by the leading coefficient, that is, the coefficient of x^2.

$$x^2 + 2x = \tfrac{1}{2}$$

We want the left side of the equation to become a perfect-square trinomial $x^2 + 2kx + k^2$. In other words, we wish to add some constant to both sides of the equation so the left side factors as $(x + k)^2$, where k is a constant. This is described in the next step.

3. Compute $\frac{1}{2}$ times the coefficient of x, square the result, and add this to both sides of the equation.

$$[\tfrac{1}{2}(2)]^2 = (1)^2 = 1$$

$$x^2 + 2x + 1 = \tfrac{1}{2} + 1$$

4. Write the left side of the equation as a perfect square.

$$(x + 1)^2 = \tfrac{3}{2}$$

5. Equate the square roots of both sides.

$$x + 1 = \pm\sqrt{\frac{3}{2}}$$

$$x + 1 = \pm\frac{\sqrt{6}}{2}$$

6. Solve the resulting linear equations.

$$x + 1 = \pm\frac{\sqrt{6}}{2} \text{ have the solutions } x = \frac{-2 \pm \sqrt{6}}{2}.$$

Thus the two solutions to $2x^2 + 4x - 1 = 0$ are $x = (-2 \pm \sqrt{6})/2$.

Example 5 • Solving by Completing the Square

Solve

$$2x^2 - 2x + 1 = 0$$

by completing the square.

Solution

We follow the six steps described in the preceding discussion.

1. $2x^2 - 2x = -1$
2. $x^2 - x = -\frac{1}{2}$
3. $x^2 - x + (-\frac{1}{2})^2 = -\frac{1}{2} + (-\frac{1}{2})^2$

   $x^2 - x + \frac{1}{4} = -\frac{1}{4}$
4. $(x - \frac{1}{2})^2 = -\frac{1}{4}$
5. $x - \frac{1}{2} = \pm\frac{1}{2}i$
6. $x = \frac{1}{2} \pm \frac{1}{2}i$

The solution set is $\{\frac{1}{2} \pm \frac{1}{2}i\}$. □

◆ Practice Problem 1 ◆

Solve

$$x^2 + 6x + 13 = 0$$

by completing the square.

Some equations can be rewritten so that they become quadratic in form by making an appropriate change of variable. This is illustrated in Example 6.

Example 6 • An Equation Quadratic in Form

Solve $4z^4 = 13z^2 - 9$.

Solution

With an appropriate change of variables, this equation can be expressed as a quadratic equation. If we let $x = z^2$, then $z^4 = x^2$, and the equation can be written as

$$4x^2 = 13x - 9 \quad \text{or} \quad 4x^2 - 13x + 9 = 0.$$

This factors as

$$(4x - 9)(x - 1) = 0.$$

Thus $x = \frac{9}{4}$ and $x = 1$ are the two solutions of the quadratic equation in x. However, we must solve for the variable z. When $x = \frac{9}{4}$, we have $z^2 = \frac{9}{4}$ and $z = \pm\frac{3}{2}$. When $x = 1$, then $z^2 = 1$ and $z = \pm 1$. Thus the solution set for $4z^4 = 13z^2 - 9$ is $\{\frac{3}{2}, -\frac{3}{2}, 1, -1\}$. □

EXERCISES 2.4

Solve by factoring. (See Examples 1 and 2.)

1. $4x^2 - 25 = 0$
2. $9x^2 - 16 = 0$
3. $r^2 + 3r = 0$
4. $u^2 - 5u = 0$
5. $y^2 + 3y - 10 = 0$
6. $x^2 + 2x - 15 = 0$
7. $x^2 + 6x + 9 = 0$
8. $16z^2 - 40z + 25 = 0$
9. $3t^2 - 10t + 3 = 0$
10. $2y^2 - 5y + 2 = 0$
11. $49x^3 + 7x^2 = 2x$
12. $2x^3 + 5x^2 = 3x$

Solve by completing the square. (See Example 5 and Practice Problem 1.)

13. $p^2 + 4p + 3 = 0$
14. $q^2 + 2q - 3 = 0$
15. $x^2 + x - 1 = 0$
16. $x^2 + 2x - 2 = 0$
17. $w^2 - 2w + 2 = 0$
18. $z^2 - 6z + 10 = 0$
19. $2x^2 - 8x = -16$
20. $2x^2 + 8x = -10$
21. $2y^2 + 5y + 1 = 0$
22. $4x^2 - 10x + 3 = 0$
23. $6x^2 + 13x = -6$
24. $3r^2 - 17r + 10 = 0$

Solve by either factoring or completing the square.

25. $7 - 15x + 2x^2 = 0$
26. $3z^2 = 11z + 4$
27. $x^2 + 2x - 1 = 0$
28. $y^2 + 5y + 5 = 0$
29. $4t^2 - 5t - 6 = 0$
30. $6w^2 - w - 12 = 0$
31. $-27y^2 + 3y + 2 = 0$
32. $15x^2 - x - 28 = 0$

33. $2z^2 + 2z - 5 = 0$

34. $2y^2 = 6y + 1$

35. $4x^2 - 8x + 5 = 0$

36. $4t^2 - 16t + 17 = 0$

Solve by first expressing the equation in quadratic form. (See Example 6.)

37. $y^4 - 4y^2 + 4 = 0$

38. $y^4 - 2y^2 + 1 = 0$

39. $27z^6 + 215z^3 - 8 = 0$

40. $z^6 + 16z^3 + 64 = 0$

41. $(y + 2)^2 - 5(y + 2) = 14$

42. $(w - 4)^2 + 2(w - 4) = 63$

43. $\left(\dfrac{1}{x}\right)^2 + \dfrac{6}{x} + 5 = 0$

44. $2p^{-2} + 7p^{-1} + 3 = 0$

45. $2y^{2/3} + y^{1/3} - 1 = 0$

46. $y^{2/3} - y^{1/3} - 12 = 0$

47. $x = 8\sqrt{x} - 15$

48. $(x - 3)^{1/2} - 5(x - 3)^{1/4} + 6 = 0$

Distance Across a Sinkhole

49. To find the distance across a sinkhole in Florida, a surveyor drove stakes at points A, B, and C, forming a right angle at C as shown in the accompanying figure. She measured the distance from B to A as 90 yards, and the distance from B to C as 60 yards. Find the distance across the sinkhole, rounded to the nearest yard.

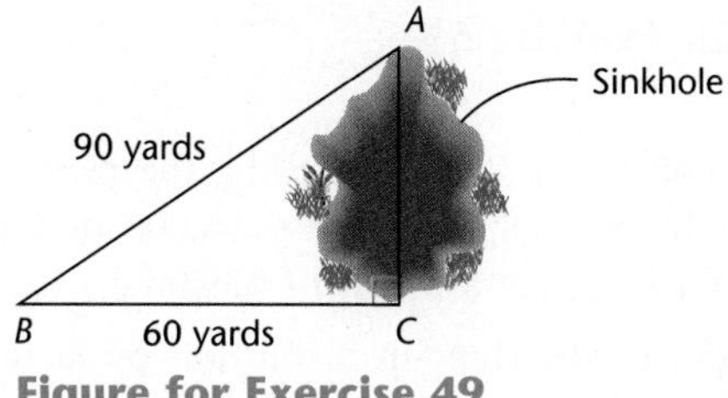

Figure for Exercise 49

Right Triangle

50. One side of a right triangle is 3 meters longer than the shortest side, and the hypotenuse is 15 meters in length. Find the length of the shortest side.

Dimensions of a Field 51–52

51. The length of a rectangular field is twice its width, and the area is 1800 square meters. Find the dimensions of the field.

52. Suppose the perimeter of a rectangular field is 480 meters and the area is 10,800 square meters. Determine the dimensions of the field.

Special Right Triangles

53. Is the right triangle with sides measuring 3, 4, and 5 units the only right triangle which has sides with lengths that are consecutive positive integers?

Height of an Object

54. If an object is thrown vertically upward with an initial velocity of v_0 feet per second, then the distance s (in feet) that the object will be above the earth at time t (in seconds) is given by the equation $s = v_0 t - 16t^2$.

a. Find t if $v_0 = 128$ feet per second and $s = 192$ feet.

b. Find t if $v_0 = 128$ feet per second and $s = 0$ feet.

Plowing a Field 55. A farmer is plowing a rectangular field that is 100 meters wide and 200 meters long, as shown in the accompanying figure. How wide a strip must she plow around the field so that 52 percent of the field is plowed?

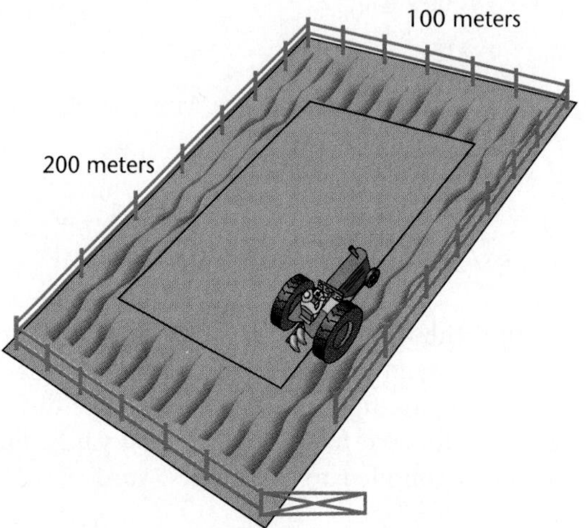

Figure for Exercise 55

Profit from Sales 56. At Jacobsohn's Boat Company, the profit P from selling x power boats is given by $P = 1800x - 6x^2 - 96{,}000$ dollars. How many power boats must be sold to make a profit of \$39,000?

Critical Thinking: Exploration and Discussion 57. a. Reverse the steps in a solution by factoring to find a quadratic equation whose solutions are $x = 3$ and $x = 7$.

b. Describe a procedure for finding a quadratic equation that has two given numbers as its solutions, say $x = r$ and $x = s$.

Critical Thinking: Writing 58. Michael Johnson solved the equation $(x - 1)(x + 4) = 0$ by setting $x - 1 = 0$ or $x + 4 = 0$. He tried the same procedure on the equation $(x - 1)(x + 4) = 6$, setting $x - 1 = 6$ or $x + 4 = 6$, and he doesn't understand why the procedure failed this time. Write an explanation for Michael.

◆ Solutions for Practice Problems

1.
$$\begin{aligned} x^2 + 6x + 13 &= 0 \\ x^2 + 6x &= -13 \\ x^2 + 6x + (3)^2 &= -13 + (3)^2 \\ (x + 3)^2 &= -4 \\ x + 3 &= \pm 2i \\ x &= -3 \pm 2i \end{aligned}$$

The solution set is $\{-3 \pm 2i\}$.

2.5 THE QUADRATIC FORMULA

Let us consider again the quadratic equation $ax^2 + bx + c = 0$ with real coefficients. As we have already seen, this equation can be solved either by factoring the quadratic expression or by completing the square on the variable x. Suppose that we wish to find, in general, the solutions of the quadratic equation $ax^2 + bx + c = 0$, $a \neq 0$. This can be done by following the steps outlined in the preceding section to complete the square on the variable x.

1. $$ax^2 + bx = -c$$

2. $$x^2 + \frac{b}{a}x = -\frac{c}{a}$$

3. $$x^2 + \frac{b}{a}x + \left[\frac{1}{2}\left(\frac{b}{a}\right)\right]^2 = -\frac{c}{a} + \left[\frac{1}{2}\left(\frac{b}{a}\right)\right]^2 = \frac{b^2 - 4ac}{4a^2}$$

4. $$\left(x + \frac{b}{2a}\right)^2 = \frac{b^2 - 4ac}{4a^2}$$

5. $$x + \frac{b}{2a} = \pm\frac{\sqrt{b^2 - 4ac}}{2a}$$

6. $$x = -\frac{b}{2a} \pm \frac{\sqrt{b^2 - 4ac}}{2a} = \frac{-b \pm \sqrt{b^2 - 4ac}}{2a}$$

Thus we have the following result.

The Quadratic Formula

The solutions to the quadratic equation $ax^2 + bx + c = 0$, $a \neq 0$, are given by

$$x = \frac{-b \pm \sqrt{b^2 - 4ac}}{2a}.$$

The **quadratic formula** can be used to solve any quadratic equation after it has been written in the form $ax^2 + bx + c = 0$. We first check to see if the equation can be simplified by removing any common factors of a, b, and c. After this is done, we then identify the coefficients of x^2, x, and the constant term and substitute these values into the quadratic formula. As mentioned in Section 2.4, the method of factoring is usually the most efficient approach whenever applicable and is usually tried first. However, use of the quadratic formula is preferred when factoring is too difficult.

Example 1 • Using the Quadratic Formula

Solve $9x^2 + 7x = 1$ by using the quadratic formula.

Solution

We first write the equation in the form $ax^2 + bx + c = 0$.

$$9x^2 + 7x - 1 = 0$$

Then we note that $a = 9$, $b = 7$, and $c = -1$. Substituting these values into the quadratic formula yields

$$\begin{aligned} x &= \frac{-7 \pm \sqrt{(7)^2 - 4(9)(-1)}}{2(9)} \\ &= \frac{-7 \pm \sqrt{85}}{18}, \end{aligned}$$

and the solution set is

$$\left\{\frac{-7 + \sqrt{85}}{18}, \frac{-7 - \sqrt{85}}{18}\right\}.$$

□

Example 2 • Consecutive Even Integers

Find two consecutive even integers whose product is 168.

Solution

Let x represent the first even integer and $x + 2$ the next even integer. Then we have

$$x(x + 2) = 168.$$

This equation can be written in the general quadratic form $ax^2 + bx + c = 0$ by performing the multiplication on the left and subtracting 168 from both sides.

$$x^2 + 2x - 168 = 0$$

Using the quadratic formula with $a = 1$, $b = 2$, and $c = -168$, we have

$$\begin{aligned} x &= \frac{-2 \pm \sqrt{(2)^2 - 4(1)(-168)}}{2(1)} \\ &= \frac{-2 \pm \sqrt{676}}{2} \\ &= \frac{-2 \pm 26}{2} \\ &= -1 \pm 13. \end{aligned}$$

This means that $x = 12$ and $x = -14$ are the two solutions to $x(x + 2) = 168$. When $x = 12$, $x + 2 = 14$, and when $x = -14$, $x + 2 = -12$. Thus there are two pairs of consecutive even integers whose product is 168. □

◆ Practice Problem 1 ◆

Solve $4x^2 + 4x + 5 = 0$ by using the quadratic formula.

Example 3 • Designing a Box

An open box is to be made from a rectangular piece of tin that is 12 centimeters wide and 14 centimeters long by cutting equal squares from the four corners and turning up the sides as shown in Figure 2.13. How large a square must be cut from each corner if the area of the base is to be 48 square centimeters?

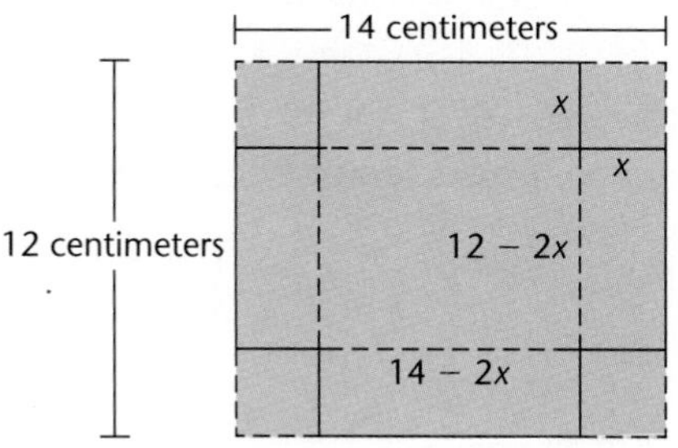

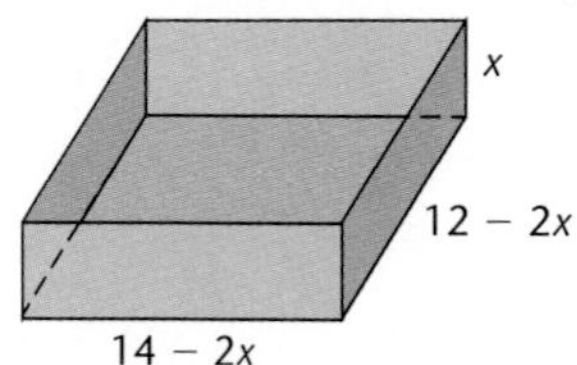

Figure 2.13 Constructing an open box

Solution

Let x centimeters be the length of a side of the squares, as shown in the accompanying figure. The length of the long side on the base of the box is $(14 - 2x)$ centimeters because x centimeters is cut off each end. Similarly, the short side on the base measures $(12 - 2x)$ centimeters in length. This means that the area of the base is $(14 - 2x)(12 - 2x)$ square centimeters. We need to set this area equal to 48 and solve for x.

$$
\begin{aligned}
(14 - 2x)(12 - 2x) &= 48 \\
(2)(7 - x)(2)(6 - x) &= 48 \qquad \text{factoring 2 from each binomial} \\
(7 - x)(6 - x) &= 12 \qquad \text{dividing by 4} \\
42 - 13x + x^2 &= 12 \\
x^2 - 13x + 30 &= 0
\end{aligned}
$$

The quadratic formula gives

$$x = \frac{13 \pm \sqrt{169 - 120}}{2} = \frac{13 \pm 7}{2}$$

and $x = 10$ or $x = 3$. The value $x = 10$ must be rejected because the piece of tin is only 12 centimeters wide. The value $x = 3$ is the only one that fits the physical situation, and it is easy to check that it actually works. □

The expression $b^2 - 4ac$, which occurs under the radical in the quadratic formula, is called the **discriminant.** The value of $b^2 - 4ac$ will be positive, negative, or zero. We can classify the roots of any quadratic equation by using the discriminant as follows.

Classification of Roots by Use of the Discriminant

1. If $b^2 - 4ac = 0$, then the two roots are equal real numbers.
2. If $b^2 - 4ac > 0$, then the two roots are distinct (unequal) real numbers.
3. If $b^2 - 4ac < 0$, then the two roots are nonreal complex numbers that are conjugates of each other.

Example 4 • Classifying the Roots

Use the discriminant to determine the type of roots of the following equations.

a. $2x^2 - x - 15 = 0$ b. $4x^2 - 20x + 25 = 0$ c. $x^2 + 4x + 13 = 0$

Solution

a. Since

$$b^2 - 4ac = (-1)^2 - 4(2)(-15) = 121 > 0,$$

the equation has two distinct real roots.

b. Since

$$b^2 - 4ac = (-20)^2 - 4(4)(25) = 0,$$

the two roots are equal real numbers.

c. The discriminant

$$b^2 - 4ac = 4^2 - 4(1)(13) = -36 < 0$$

and the two roots are complex numbers that are conjugates of each other. □

Suppose now that we have two values, r_1 and r_2, and we want to determine a quadratic equation that has these values as its roots. Now r_1 is a solution to the linear equation $x - r_1 = 0$, and r_2 is a solution to the linear equation $x - r_2 = 0$. Furthermore, the product of $x - r_1$ and $x - r_2$ is zero if and only if one of the factors is zero. Thus

$$(x - r_1)(x - r_2) = 0$$

is a quadratic equation with the desired roots. Expanding and simplifying, we have

$$x^2 - (r_1 + r_2)x + r_1r_2 = 0.$$

Since multiplying both sides of an equation by the same nonzero constant does not change the solution set, any constant multiple of this equation also has solutions r_1

and r_2. It is important to observe that whenever the leading coefficient† (the coefficient of x^2) is 1, the constant term is the product of the two roots and the coefficient of x is the negative of the sum of the two roots.

Example 5 • Finding a Quadratic Equation with Certain Properties

Determine a monic quadratic equation if the sum of its roots is -2 and the product of its roots is -8.

Solution

Suppose that r_1 and r_2 are the two roots. Then

$$r_1 + r_2 = -2 \qquad \text{and} \qquad r_1 r_2 = -8.$$

Thus a quadratic equation with the desired roots is

$$x^2 + 2x - 8 = 0.$$ □

Example 6 • Finding a Quadratic Equation with Specified Roots

Find a monic quadratic equation whose roots are 0 and -3.

Solution

We can write the desired quadratic expression as a product of its linear factors and obtain

$$(x - 0)[x - (-3)] = 0$$

or

$$x^2 + 3x = 0.$$ □

EXERCISES 2.5

Solve the following equations by using the quadratic formula. (See Examples 1 and 2 and Practice Problem 1.)

1. $x^2 + 3x - 28 = 0$
2. $x^2 - 16x + 64 = 0$
3. $4x^2 - 8x + 3 = 0$
4. $6x^2 - 11x - 35 = 0$
5. $x^2 + x = 0$
6. $4x^2 = -7x$
7. $-x^2 - 4x + 12 = 0$
8. $-x^2 - 2x + 1 = 0$
9. $16x^2 - 25 = 0$
10. $2x^2 = 1$
11. $4x^2 = -7x - 2$
12. $2x^2 + 7x = 4$
13. $x^2 + 5x + 5 = 0$
14. $x^2 + 3x + 1 = 0$
15. $3x^2 + 8x + 3 = 0$
16. $-5x^2 + 2x + 1 = 0$

†Recall that such a polynomial is called **monic.** The same terminology is used with equations arising from setting a polynomial equal to zero.

17. $x^2 - 2x + 4 = 0$
18. $x^2 + 3x + 9 = 0$
19. $28x^2 = 45 - x$
20. $6x^2 = 29x - 28$

Use the quadratic formula to approximate the solutions of the following equations, rounded to two decimal places.

21–24

21. $0.26x^2 + 1.02x + 0.82 = 0$
22. $91.31x^2 + 21.81x - 11.24 = 0$
23. $1.01x^2 + 3.72x + 1.82 = 0$
24. $6.51x^2 - 9.15x + 2.11 = 0$

Use the discriminant to determine the type of roots of each of the following equations. Do not evaluate the roots. (See Example 4.)

25. $x^2 - 4 = 0$
26. $4x^2 + 7x - 15 = 0$
27. $2x^2 - 5x + 7 = 0$
28. $-x^2 - x - 2 = 0$
29. $3x^2 + 5x - 1 = 0$
30. $2x^2 + 3x - 1 = 0$
31. $49x^2 + 14x + 1 = 0$
32. $-x^2 + 2x - 1 = 0$

33–36

33. $5.09x^2 + 9.21x + 1.82 = 0$
34. $61.28x^2 + 45.19x + 21.54 = 0$
35. $27.81x^2 - 15.25x + 11.84 = 0$
36. $8.880x^2 - 27.21x + 15.62 = 0$

Determine the value(s) of k necessary in order for the roots of the given quadratic equation to be equal real numbers.

37. $kx^2 + 6x - 2 = 0$
38. $2x^2 + 5x + k = 0$
39. $4x^2 + kx + 3 = 0$
40. $6x^2 + kx + 5 = 0$

Find a monic quadratic equation that has the given numbers as solutions. (See Example 6.)

41. $1, -2$
42. $-5, 2$
43. $\frac{3}{2}, 4$
44. $\frac{1}{2}, -\frac{1}{2}$
45. $-2, -2$
46. $0, -2$

Find a monic quadratic equation with solutions that have the sum and product given. (See Example 5.)

	Sum	*Product*		*Sum*	*Product*
47.	2	1	48.	5	6
49.	−3	−4	50.	0	4
51.	2	0	52.	−11	28

Number Problems 53–56

53. Find two consecutive integers whose product is 462.
54. Find two consecutive odd integers whose product is 1443.
55. The sum of the squares of four consecutive integers is 174. Find the integers.

56. Find three consecutive odd integers such that the sum of their squares is 371.

Height of an Object 57–58

57. If a ball is thrown straight upward with a speed of 48 feet per second when it is released, it reaches a height of h feet after t seconds given by $h = 48t - 16t^2$. How long will it take the ball to reach a height of 32 feet on the way up? On the way down?

58. If air resistance is neglected, a bullet fired straight upward with a muzzle velocity of 2000 feet per second will reach a height of h feet after t seconds given by $h = 2000t - 16t^2$. How long will it take the bullet to reach a height of 9600 feet on its way up? On its way down?

Production Cost

59. The cost of producing x lawn mowers is given in dollars by $C = 1500 - 60x + 3x^2$. How many lawn mowers can be produced at a cost of $1200?

Revenue 60–61

60. The revenue R (in dollars) that results from selling x birdhouses from Jeremy's woodworking shop can be found from $R = 40x - x^2 - 150$. How many birdhouses must be sold so that the revenue is $250?

61. The sale of n boats by the Odessa Boat Company produces revenue of $R = 15{,}000n - 150n^2$ dollars. Find the number of boats that must be sold in order to produce $240,000 in revenue.

Profit 62–64

62. If Boston Anchor Works sells n fishing boat anchors, it makes a profit of $P = 80n - 0.02n^2$ dollars. How many anchors need to be sold to make a profit of $1592?

63. The profit that results from the sale of t tons of topsoil is given by $P = 4t - 150 - 0.02t^2$, where P is in hundreds of dollars. How many tons of topsoil must be sold for a profit of $5000?

64. If Dr. Mazeres sells w loads of firewood from his wood lot, he makes a profit of $P = 54w - w^2 - 9$ dollars. How many loads of wood sold will produce a profit of $720?

Production Cost

65. The cost C of producing p ceramic pots is given in dollars by $C = 5 + 10p - p^2$. How many pots can be produced for $29?

Demand

66. The demand D for the book *Algebra in Your Own Home* is given by $D = 10 + 4p - 0.5p^2$, where D is the demand (in thousands of books) and p is the price (in dollars). What price p will produce a demand for 16,000 books?

Supply 67–68

67. The supply S of Chesapeake Flyer sailboats can be computed from the formula $S = p^2 + 4p - 31$, where S is in thousands of boats and p is the price in thousands of dollars. Find the price that produces a supply of 29,000 sailboats.

68. The supply S of $8'' \times 10''$ picture frames at the Northside Flea Market is given by $S = p^2 - 2p - 8$, where S is the supply (in thousands of frames) and p is the price of a frame (in dollars). Find the price that will result in a supply of 16,000 frames.

Field Trip Cost 69. Ms. Carlisle's third-grade class is raising funds for a field trip to a pumpkin patch. The pumpkin patch charges \$3.00 per child; however, for class sizes between 20 and 40, the ticket price is reduced by \$0.10 times the number of children in excess of 20. The class has raised \$62.50. How many children can go to the pumpkin patch for this amount?

Magazine Subscriptions 70. A publisher can sell 1000 magazine subscriptions per year at a price of \$40 per year. For each \$1 reduction in subscription price, an additional 50 subscriptions can be sold. If the revenue from subscriptions last year was \$45,000, how many subscriptions were sold?

Constructing a Box 71. An open box is to be made from a square sheet of tin that is 8 inches on a side by cutting square pieces of the same size from the four corners of the sheet and folding up the sides. What size squares should be cut from the corners if the area of the base is to be 49 square inches?

Constructing Dog Pens 72. A veterinarian is building seven dog pens by enclosing a rectangular region and constructing six cross fences in the rectangle, as shown in the accompanying figure. Find the dimensions of the rectangular region enclosed if 208 feet of fencing is used and the total area enclosed is 676 square feet.

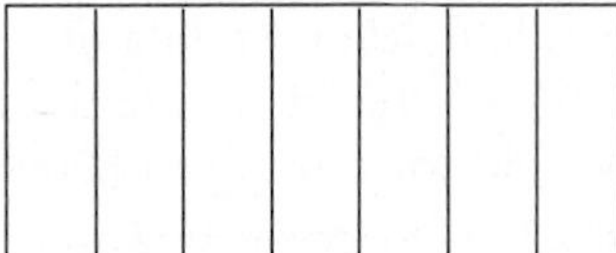

Figure for Exercise 72

Critical Thinking: Writing 73. Explain how you can find the values of two numbers if both their sum and their product are known.

Critical Thinking: Writing and Discussion 74. Write a word description for the discriminant of a quadratic equation without using any variables in your description. Then discuss the advantages of using variables in the statements of mathematical formulas.

◆ Solution for Practice Problem

1. We have $a = 4$, $b = 4$, and $c = 5$. Hence

$$x = \frac{-4 \pm \sqrt{16 - 4(4)(5)}}{2(4)} = \frac{-4 \pm \sqrt{-64}}{8} = \frac{-4 \pm 8i}{8} = \frac{-1 \pm 2i}{2}$$
$$= -\tfrac{1}{2} \pm i.$$

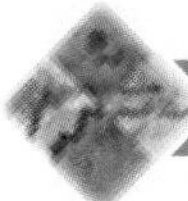

2.6 MORE ON QUADRATIC EQUATIONS

In this section we consider equations that at first glance do not appear to be quadratic, but that can be transformed into a quadratic equation and solved using the techniques of the previous section.

Example 1 • An Equation Involving a Fraction

Solve $x = 3 + \dfrac{5}{x+1}$.

Solution

As long as $x + 1 \neq 0$, we can multiply both sides of this equation by $x + 1$ to eliminate the fraction and then simplify. The resulting quadratic equation can be solved by factoring.

$$x = 3 + \frac{5}{x+1}$$

$$(x+1)x = (x+1)3 + (x+1)\frac{5}{x+1} \qquad \text{multiplying by } x+1$$

$$x^2 + x = 3x + 3 + 5 \qquad \text{simplifying}$$

$$x^2 - 2x - 8 = 0 \qquad \text{rewriting}$$

$$(x-4)(x+2) = 0 \qquad \text{factoring}$$

$$x - 4 = 0 \quad \text{or} \quad x + 2 = 0 \qquad \text{using the zero product property}$$

Thus the solutions to the quadratic equation are $x = 4$ and $x = -2$. Both of these values satisfy the original equation, so the solution set for the original equation is $\{4, -2\}$. □

We saw in Example 2 in Section 2.1 that clearing an equation of fractions may introduce extraneous solutions. This possibility occurs again in the next example, where clearing the fractions leads to an equation that is not equivalent to the original equation.

Example 2 • Extraneous Solution

Solve $\dfrac{x}{4} + \dfrac{x}{x-2} = \dfrac{2}{x-2}$.

Solution

We first clear the equation of fractions by multiplying both sides by the least common denominator, $4(x - 2)$.

$$4(x-2)\frac{x}{4} + 4(x-2)\frac{x}{x-2} = 4(x-2)\frac{2}{x-2} \qquad \text{multiplying by } 4(x-2)$$

$$x^2 - 2x + 4x = 8 \qquad \text{simplifying}$$

$$x^2 + 2x - 8 = 0 \qquad \text{rewriting}$$

$$(x+4)(x-2) = 0 \qquad \text{factoring}$$

$$x = -4 \quad \text{or} \quad x = 2 \qquad \text{using the zero product property}$$

The solutions to the quadratic equation are $x = -4$ and $x = 2$. However, the value $x = 2$ must be rejected because it makes a denominator zero in the original equation. The value $x = -4$ can be checked as follows.

Check: $x = -4$
Since

$$\frac{-4}{4} + \frac{-4}{-4-2} = -1 + \frac{2}{3} = -\frac{1}{3}$$

and

$$\frac{2}{-4-2} = -\frac{1}{3},$$

we see that $x = -4$ is a solution to the original equation, and the solution set is $\{-4\}$. □

Uniform motion problems can lead to quadratic equations, as Example 3 shows.

Example 3 • Uniform Motion

Matt rode his bike to a store 3 miles from home. If he traveled 3 miles per hour faster going to the store than he did returning and his total travel time was 27 minutes, how fast did he go each way?

Solution

Let r miles per hour represent Matt's speed returning from the store. Then $r + 3$ is his speed going to the store. From the uniform motion formula $rt = d$, we have $t = d/r$, so his travel time going is $3/(r + 3)$ and his travel time returning is $3/r$. We fill out the distance, rate, and time as shown in Table 2.6.

Table 2.6 Travel Time

Travel	Rate (mph)	Time (hours)	Distance (miles)
Going	$r + 3$	$\frac{3}{r+3}$	3
Returning	r	$\frac{3}{r}$	3

Since the total travel time is 27 minutes, or $\frac{9}{20}$ hours, we must solve the following equation.

$$\begin{aligned}
\frac{3}{r+3} + \frac{3}{r} &= \frac{9}{20} && \\
\frac{1}{r+3} + \frac{1}{r} &= \frac{3}{20} && \text{dividing by 3} \\
20r + 20(r+3) &= 3r(r+3) && \text{multiplying by } 20r(r+3) \\
20r + 20r + 60 &= 3r^2 + 9r && \text{simplifying} \\
0 &= 3r^2 - 31r - 60 && \text{making one side 0} \\
0 &= (3r+5)(r-12) && \text{factoring} \\
3r + 5 = 0 \quad &\text{or} \quad r - 12 = 0 && \text{using the zero product property} \\
r = -\tfrac{5}{3} \quad &\text{or} \quad r = 12 &&
\end{aligned}$$

The value $r = -\frac{5}{3}$ does not make sense as a solution. Thus Matt's rate returning was 12 miles per hour and his rate going was 15 miles per hour. □

We next consider equations in one real variable that contain a radical in one or both members. The standard method for solving an equation of this type is to raise both sides of the equation to a power that will eliminate one or more of the radicals, simplify the resulting equation, and solve for the variable by repeating the process, if necessary.

Example 4 • An Equation Containing a Cube Root

To solve the equation

$$\sqrt[3]{x+1} = 2$$

we cube both sides of the equation and obtain

$$x + 1 = 8.$$

Solving for x then yields $x = 7$. □

In some instances, raising both sides of an equation to a power leads to extraneous solutions (that is, solutions that do not satisfy the original equation). Consider, for example, the equation $x = -3$. Squaring both sides yields $x^2 = 9$, whose solutions are $x = 3$ and $x = -3$. But $x = 3$ does not satisfy the original equation. That is, $x = 3$ is an extraneous root. To determine whether or not a solution is extraneous, we must check the solution in the original equation (or any equivalent equation, before both sides are raised to a power).

Example 5 • An Equation Containing a Square Root

Solve $\sqrt{8x+1} - 4 = 1 - 2x$.

Solution

Squaring both sides will not eliminate the radical. To eliminate the radical, we must first isolate it on one side of the equation.

$\sqrt{8x+1} = 5 - 2x$	adding 4 to both sides
$(\sqrt{8x+1})^2 = (5-2x)^2$	squaring both sides
$8x + 1 = 25 - 20x + 4x^2$	performing the indicated operations
$0 = 24 - 28x + 4x^2$	rewriting to force zero on one side
$0 = 6 - 7x + x^2$	dividing both sides by 4
$0 = (1-x)(6-x)$	factoring

The solutions here are $x = 1$ and $x = 6$. Since one or both solutions might be extraneous, they need to be checked in the original equation.

Check: $x = 1$
Since

$$\sqrt{8(1)+1} - 4 = 1 - 2(1),$$

the value $x = 1$ is a solution to the original equation.

Check: $x = 6$
Since

$$\sqrt{8(6)+1} - 4 \neq 1 - 2(6),$$

then $x = 6$ is an extraneous solution, and the solution set to the original equation is $\{1\}$. □

◆ Practice Problem 1 ◆

Solve $\sqrt[3]{7x+1} = x + 1$.

In some equations involving radicals, it is necessary to apply the procedure of raising both sides of the equation to a power more than once in order to eliminate all the radicals. In such cases, it is important that the checking procedure be performed in the original equation (or any equivalent equation, before both sides are raised to a power). Consider the following example.

Example 6 • An Equation Containing Two Square Roots

Solve $\sqrt{3x+1} = \sqrt{x+4} + 1$.

Solution

Although squaring both sides will not eliminate both radicals, one of them will be eliminated.

$(\sqrt{3x+1})^2 = (\sqrt{x+4}+1)^2$	squaring both sides
$3x + 1 = x + 4 + 2\sqrt{x+4} + 1$	performing the indicated operations
$2x - 4 = 2\sqrt{x+4}$	isolating $2\sqrt{x+4}$
$x - 2 = \sqrt{x+4}$	dividing by 2
$x^2 - 4x + 4 = x + 4$	squaring both sides
$x^2 - 5x = 0$	simplifying
$x(x - 5) = 0$	factoring

The solutions to the quadratic equation are $x = 0$ and $x = 5$. These must be checked in the original equation.

Check: $x = 0$
Since

$$\sqrt{3(0)+1} \neq \sqrt{0+4} + 1,$$

then $x = 0$ is an extraneous solution.

Check: $x = 5$
Since

$$\sqrt{3(5)+1} = 4 \qquad \text{and} \qquad \sqrt{5+4} + 1 = 4,$$

then $x = 5$ is a solution and $\{5\}$ is the solution set for the original equation. □

The procedure illustrated in the preceding examples is summarized in the following steps.

Solving Equations Involving Radicals

Step 1 Isolate a radical on one side.
Step 2 Raise both sides to the smallest power so that the isolated radical will be removed.
Step 3 Simplify the equation obtained in step 2. If a radical remains, repeat the procedure, starting with step 1. If no radicals remain, proceed to step 4.
Step 4 Solve the equation obtained in step 3.
Step 5 Check the solutions in the original equation.

EXERCISES 2.6

Determine the solution set for each of the following. (See Examples 1 and 2.)

1. $x = \dfrac{6}{x+1}$

2. $2x + 5 = \dfrac{12}{x}$

3. $2x = \dfrac{4}{x} - 3 + x$

4. $6 = \dfrac{2(1 - 2x)}{x} - 1$

5. $2x - 1 = \dfrac{-5(3x + 2)}{2x + 1}$

6. $2(x + 1) = 13 - \dfrac{4}{x - 1}$

7. $\dfrac{1}{x} = \dfrac{x}{x + 1}$

8. $\dfrac{1}{x} = \dfrac{1}{1 - \dfrac{1}{x}}$

9. $\dfrac{6}{(y + 2)^2} + \dfrac{1}{y + 2} = 1$

10. $\dfrac{1}{z + 1} + \dfrac{12}{(z + 1)^2} = 1$

11. $\dfrac{4}{(t - 2)^2} + \dfrac{7}{t - 2} + 3 = 0$

12. $\dfrac{9}{(r - 2)^2} = \dfrac{11}{r - 2} - 2$

13. $x - \dfrac{2x}{x + 1} = \dfrac{2}{x + 1}$

14. $2p + \dfrac{p + 3}{p - 3} = \dfrac{2p}{p - 3}$

15. $\dfrac{2z}{3} - \dfrac{4 - z}{z + 4} = \dfrac{2z}{z + 4}$

16. $\dfrac{x}{4} + \dfrac{2x}{x - 5} = \dfrac{x + 5}{x - 5}$

Determine the solution set for each of the following equations. (See Examples 4–6.)

17. $\sqrt{x + 7} = \sqrt{2x + 1}$

18. $\sqrt{x + 5} = \sqrt{3x - 1}$

19. $x = \sqrt{8 - 2x}$

20. $x = \sqrt{x + 12}$

21. $\sqrt{x + 10} = x - 2$

22. $\sqrt{4x + 1} = x - 5$

23. $\sqrt{10 + x} - x = 10$

24. $\sqrt{x - 1} + 2 = 2x$

25. $\sqrt{x + 8} - 1 = 2x$

26. $4 + \sqrt{x + 2} = x$

27. $\sqrt{2 - t} - t = 10$

28. $z - \sqrt{7 - z} = 1$

29. $\sqrt{2r + 3} + 6 = r$

30. $\sqrt{y - 1} + y = 7$

31. $\sqrt{v - 3} - v = -5$

32. $5 - \sqrt{t + 1} = t$

33. $\sqrt[3]{x - 1} = 3$

34. $\sqrt[3]{5 - 20x + 2x^2} + 3 = 0$

35. $\sqrt[3]{3x^2 + 4x - 1} = \sqrt[3]{3x^2 + 7}$

36. $\sqrt[5]{x^2 + 2x} = \sqrt[5]{3}$

37. $\sqrt[3]{x^2 + 2x} = -1$

38. $\sqrt[4]{2x - 1} + \sqrt[4]{x} = 0$

39. $\sqrt[3]{3x - 1} + 1 = x$

40. $\sqrt[5]{7x - 4} + 1 = 0$

41. $\sqrt{9 - x} + \sqrt{x + 8} = 3$

42. $\sqrt{2x + 3} - \sqrt{2x} = 1$

43. $\sqrt{2x - 3} - \sqrt{x + 2} = 1$

44. $\sqrt{x - 2} = 3 + \sqrt{x + 2}$

45. $\sqrt{2x + 2 + \sqrt{2x + 6}} = 4$

46. $\sqrt{3 - x} + \sqrt{2 + x} = 3$

47. $\sqrt{11 - x} - \sqrt{x + 6} = 3$

48. $\sqrt{1 - 5x} + \sqrt{1 - x} = 2$

49. $\sqrt{2x - 3} - \sqrt{x + 7} + 2 = 0$

50. $\sqrt{x + 3} = \sqrt{2x + 7} - 1$

51. $\sqrt{5 + \sqrt{x + 1}} + 1 = \sqrt{x + 1}$

52. $\sqrt{x} + \sqrt{3} = \sqrt{x + 3}$

Uniform Motion Problems 53–56

53. A car traveled 600 miles at a uniform rate. If the rate had been 5 miles per hour more, the trip would have taken 1 hour less. Find the rate of the car.

54. A plane traveled 800 miles at a uniform rate of speed. If the rate had been 40 miles per hour more, the trip would have taken 1 hour less. Find the rate of the plane.

55. Adrienne drove her boat 12 miles upstream and returned, making the round-trip in 54 minutes. If the rate of the current was 3 miles per hour, what was the speed of her boat in still water?

56. Beth flew her airplane against the wind a one-way distance of 540 miles to Anaheim. If the speed of her plane in still air was 150 miles per hour and her round-trip travel time was 7.5 hours, what was the speed of the wind? Assume the wind was blowing uniformly with the same speed and direction during the entire flight.

Critical Thinking: Exploration and Writing

57. Suppose that a, b, and c are *all integers* in the quadratic equation $ax^2 + bx + c = 0$. If the discriminant is positive, explain why both solutions to this equation are irrational or neither solution is irrational.

Critical Thinking: Exploration and Writing

58. The result in Exercise 57 is sometimes described by saying, "If the coefficients are integers, irrational roots of quadratic equations occur in conjugate pairs." Explain why the word *conjugate* is appropriate here.

Critical Thinking: Exploration and Writing

59. Discuss why extraneous real solutions are *never* introduced by cubing both sides of an equation. Can you extend this result to powers other than the third power?

◆ Solution for Practice Problem

1. Cubing both sides eliminates the radical.

$$\begin{aligned} 7x + 1 &= (x + 1)^3 \\ 7x + 1 &= x^3 + 3x^2 + 3x + 1 && \text{expanding} \\ 0 &= x^3 + 3x^2 - 4x && \text{simplifying} \\ 0 &= x(x^2 + 3x - 4) && \text{factoring} \\ 0 &= x(x + 4)(x - 1) && \text{factoring completely} \end{aligned}$$

Thus the solutions to the cubic equation are $x = 0, -4, 1$. It is easy to check these solutions in the original equation. Hence the solution set is $\{0, -4, 1\}$.

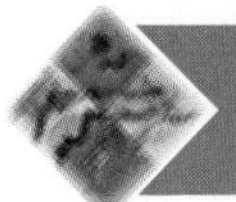

2.7 NONLINEAR INEQUALITIES IN ONE REAL VARIABLE

In this section, our work is restricted to the set of real numbers. A **quadratic inequality** in the real variable x is an inequality which can be written in the form

$$ax^2 + bx + c > 0,$$

with $a \neq 0$, or a similar statement with one of the other inequality symbols: $\geqslant$, $<$, or $\leqslant$. The **solution set** of the quadratic inequality is the set of all real numbers x for which the statement of the inequality is true.

One way to determine the solution set for a given inequality is to first write the quadratic expression as a product of linear factors and then construct a diagram picturing the signs of each factor. Such a diagram is called a **sign graph.** We illustrate this technique in the first example.

Example 1 • Solving a Quadratic Inequality

Solve the inequality and graph the solution set.

$$x^2 - 2x - 3 \leq 0$$

Solution

Factoring the quadratic expression, we have

$$(x - 3)(x + 1) \leq 0.$$

The solution set for the inequality consists of all real numbers x for which the product $(x - 3)(x + 1)$ is zero or negative. The product is zero at $x = 3$ and at $x = -1$. The sign of the product depends upon the sign of each of the factors. Considering each factor individually, we know that

$$\begin{aligned} x - 3 &> 0, && \text{when } x > 3, \\ x - 3 &= 0, && \text{when } x = 3, \\ x - 3 &< 0, && \text{when } x < 3; \end{aligned}$$

and

$$\begin{aligned} x + 1 &> 0, && \text{when } x > -1, \\ x + 1 &= 0, && \text{when } x = -1, \\ x + 1 &< 0, && \text{when } x < -1. \end{aligned}$$

We can summarize this information on a sign graph, which is constructed by first locating $x = 3$ and $x = -1$ on a number line in Figure 2.14. To indicate the sign of each factor, we place + signs in the interval where each factor is positive, − signs where each factor is negative, and 0 where each factor assumes the value zero. Finally, we consider the product of the two factors. Since quantities with *like* signs multiply to yield a positive number and quantities with *opposite* signs multiply to yield a negative number, we can determine the sign of the product. Thus the solution set for the inequality

$$x^2 - 2x - 3 \leq 0$$

is the set of real numbers in the interval where the product is negative or zero.

From the sign graph, we see that the solution set is

$$\{x \mid -1 \leq x \leq 3\} = [-1,3].$$

The graph is drawn in Figure 2.14. □

Factors	$x-3$	−	−	−	−	−	−	0	+	+
	$x+1$	−	−	0	+	+	+	+	+	+
Product		+	+	0	−	−	−	0	+	+

-1 3

$-1 \leqslant x \leqslant 3$

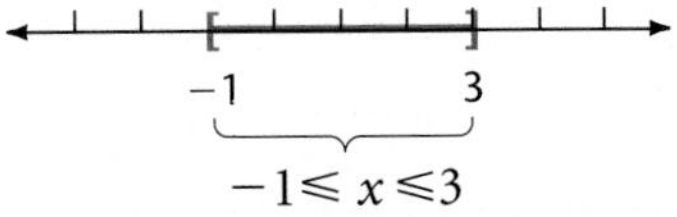

Figure 2.14 Sign graph for $x^2 - 2x - 3 \leq 0$

The sign graph method extends easily to products that contain more than two linear factors and also to quotients made up of linear factors.

Example 2 • A Nonlinear Inequality Involving a Quotient

Solve and graph the inequality

$$\frac{2}{w-1} > \frac{1}{w+2}.$$

Solution

To use the sign graph method, we must compare a product or quotient to *zero*. Thus we first obtain an inequality with one member zero.

$$\frac{2}{w-1} - \frac{1}{w+2} > 0 \qquad \text{subtracting } \frac{1}{w+2}$$

$$\frac{2(w+2) - 1(w-1)}{(w-1)(w+2)} > 0 \qquad \text{combining fractions}$$

$$\frac{w+5}{(w-1)(w+2)} > 0 \qquad \text{simplifying}$$

The sign of the quotient depends upon the sign of each linear factor used in forming the quotient. The signs of the linear factors are displayed on the first three lines of the sign graph in Figure 2.15. The last line indicates the sign of the quotient in each interval, and "U" indicates that the quotient is undefined at that point.

Factors	$w+5$	−	−	0	+	+	+	+	+	+	+	+	+
	$w-1$	−	−	−	−	−	−	−	−	0	+	+	+
	$w+2$	−	−	−	−	−	0	+	+	+	+	+	+
Quotient		−	−	0	+	+	U	−	−	U	+	+	+

-5 -2 1

$-5 < w < -2$ $1 < w$

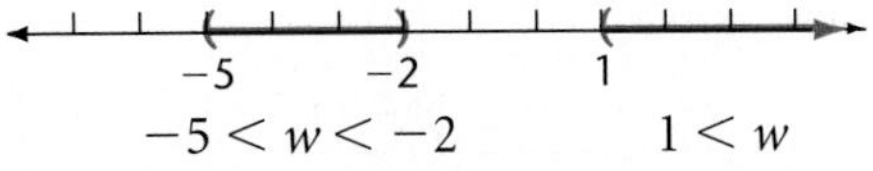

Figure 2.15 Sign graph for $\dfrac{w+5}{(w-1)(w+2)} > 0$

Therefore the solution set for the inequality

$$\frac{2}{w-1} > \frac{1}{w+2}$$

is given by $(-5, -2) \cup (1, \infty)$, and the graph is shown in Figure 2.15. □

When solving an inequality such as

$$\frac{2}{w-1} > \frac{1}{w+2}$$

in Example 2, it is very tempting to multiply both sides by $(w-1)(w+2)$ and then solve the resulting inequality

$$2(w+2) > 1(w-1).$$

Multiplication by $(w-1)(w+2)$ is *not* a valid step, since the product $(w-1)(w+2)$ is sometimes positive and sometimes negative, depending upon the value of w. (See Critical Thinking Problem 5 in the Chapter Review and its solution in the answer section of the book.)

At this point, it is easy to see that the solution of an inequality in x amounts to determining the values of x that make an expression positive, negative, or zero, depending on which symbol of inequality is involved. These values of x can be determined without drawing a sign graph as we did in the last two examples.

We must first determine the linear factors that are involved in the expression and then find the points where the linear factors are zero. These points separate the real numbers into intervals over which the expression has the same sign (positive or negative) because the expression cannot change sign unless one of its factors changes sign. We can determine the sign of expression in each interval by testing a value of x from that region. Systematically testing all the intervals leads to the solution set of the given inequality. This method is called the **algebraic method.**

Example 3 • Using the Algebraic Method

Use the algebraic method to solve the inequality

$$\frac{x+2}{x-1} \geq 2.$$

Solution

We begin by transforming the given inequality into one in which one member is zero.

$$\frac{x+2}{x-1} - 2 \geq 0 \qquad \text{subtracting 2}$$

$$\frac{x+2-2(x-1)}{x-1} \geq 0 \qquad \text{adding fractions}$$

$$\frac{-x+4}{x-1} \geq 0 \qquad \text{simplifying}$$

We reason now that the value of the quotient

$$\frac{-x+4}{x-1}$$

does not change sign without taking on the value zero or becoming undefined because of a zero denominator. The quotient is zero at $x = 4$, and it is undefined at $x = 1$. These values of x separate the real numbers into the intervals

$$(-\infty,1), \qquad (1,4), \qquad \text{and} \qquad (4,\infty).$$

Selecting a test point in each interval (any point in each interval can be chosen), we obtain the results recorded in Figure 2.16. Keeping in mind that the quotient $(-x + 4)/(x - 1)$ is zero at $x = 4$ and undefined at $x = 1$, we find that the solution set is

$$\{x \mid 1 < x \leq 4\} = (1,4]$$

as graphed in Figure 2.16. □

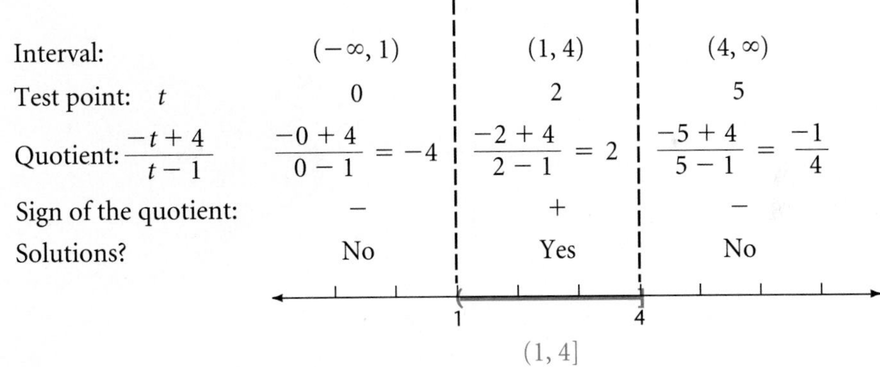

Figure 2.16 Solution set for $\frac{x+2}{x-1} \geq 2$

It is important to notice in Example 3 that *we did not multiply both sides by* $x - 1$. This is not a valid step because $x - 1$ may be positive or it may be negative. It is possible to consider the two cases (1) when $x - 1 > 0$, and (2) when $x - 1 < 0$, and solve the inequality by considering these cases separately. However, this method is somewhat tedious and inefficient, and we do not go into it here.

Solve and graph the inequality $\frac{1}{x-1} \leq \frac{1}{x+2}$.

Example 4 • Height of a Baseball

In the World Series playoffs, Slugger Sam hits a foul ball. The ball leaves his bat at a height of 3 feet and is forced vertically upward with an initial velocity of 80 feet per second. At t seconds after leaving the bat, the height h of the ball is given in feet by

$$h = 3 + 80t - 16t^2.$$

At what times will the ball be above the lights if the lights are 67 feet high? (See Figure 2.17.)

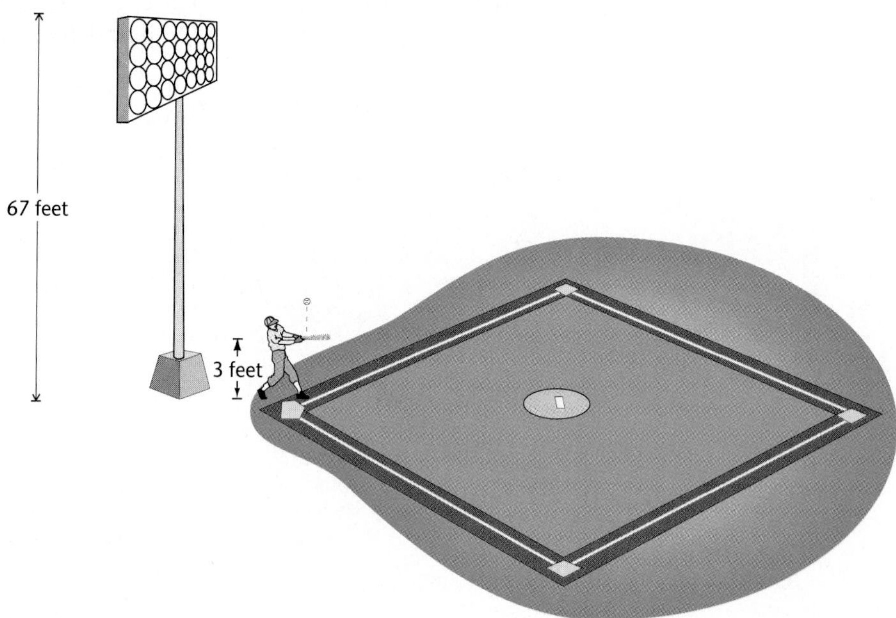

Figure 2.17 Sam hits a foul ball

Solution

The ball will be above the lights when the height h is greater than 67. Thus we need to solve the quadratic inequality

$$3 + 80t - 16t^2 > 67, \qquad t \geq 0.$$

We rewrite this inequality so that one side is 0 and factor the other side.

$-64 + 80t - 16t^2 > 0$	subtracting 67
$4 - 5t + t^2 < 0$	dividing by -16 and reversing the inequality
$(4 - t)(1 - t) < 0$	factoring

Figure 2.18 indicates that the ball will be above the lights between the first and fourth seconds after leaving the bat. Notice that only the nonnegative portion of the number line is used in Figure 2.18 because of the restriction that $t \geq 0$. □

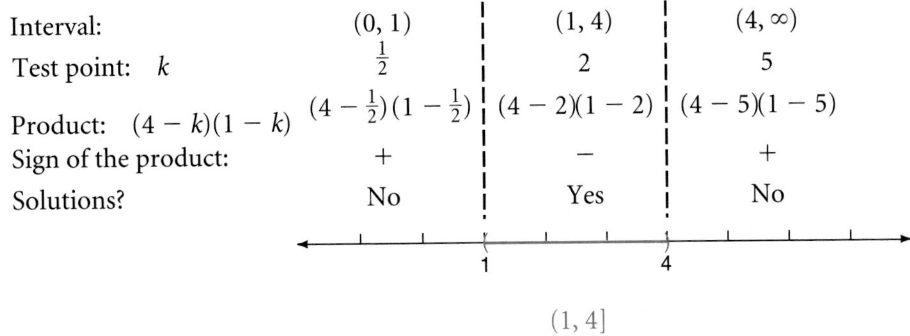

Interval:	$(0, 1)$	$(1, 4)$	$(4, \infty)$
Test point: k	$\frac{1}{2}$	2	5
Product: $(4 - k)(1 - k)$	$(4 - \frac{1}{2})(1 - \frac{1}{2})$	$(4 - 2)(1 - 2)$	$(4 - 5)(1 - 5)$
Sign of the product:	$+$	$-$	$+$
Solutions?	No	Yes	No

Figure 2.18 Solution set for $3 + 80t - 16t^2 > 67$, $t \geq 0$

EXERCISES 2.7

Solve the following quadratic inequalities and graph the solution sets. (See Example 1.)

1. $x^2 - x - 12 > 0$
2. $x^2 + 7x + 10 > 0$
3. $x^2 - 2x - 3 \leq 0$
4. $x^2 - 8x + 12 \leq 0$
5. $-x^2 + 4x - 3 \leq 0$
6. $9 - 8x - x^2 \leq 0$
7. $-2x^2 + 6 < 0$
8. $x^2 - 4 < 0$
9. $x^2 - 9x > 0$
10. $6x + x^2 > 0$
11. $8x - 2x^2 \geq 0$
12. $4x - x^2 \geq 0$
13. $-2x^2 + 4 < 0$
14. $-3x^2 + 9 < 0$
15. $2x - x^2 \leq 0$
16. $6x - 2x^2 \leq 0$
17. $x^2 + 6x + 16 < 8$
18. $x^2 + 4x + 6 \geq 3$
19. $z^2 + 2z \leq 15$
20. $w^2 + 2w > 99$
21. $3w^2 + 13w - 10 > 0$
22. $4s^2 + 3s - 1 \geq 0$
23. $-15x^2 + 28x - 12 > 0$
24. $-2z^2 + 11z - 15 > 0$
25. $3z^2 > 4z$
26. $6w^2 < w$
27. $r - 6r^2 > -35$
28. $4x^2 > 2 - 7x$
29. $21 - 4x^2 > -5x$
30. $2x^2 < 12 - 5x$
31. $-4s^2 - 4s + 1 \geqslant 0$
32. $w^2 + 8w + 13 \geq 0$

Solve and graph the following inequalities. (See Examples 2 and 3.)

33. $(x - 1)(x - 2)(x - 3) > 0$
34. $(x - 1)(x - 2)(x - 3)(x - 4) \leq 0$
35. $\dfrac{x + 1}{2x - 1} > 3$
36. $\dfrac{2x - 1}{x} < 0$

√ 37. $\dfrac{3x-1}{x+2} < 1$

38. $\dfrac{3w+2}{w} \leqslant 2$

39. $\dfrac{w+1}{w+3} \geqslant 2$

40. $\dfrac{1}{w-2} > \dfrac{1}{3}$

41. $\dfrac{7-z}{(z-2)(z-3)} < 0$

42. $\dfrac{x-5}{(3x-1)(2x-3)} < 0$

√ 43. $\dfrac{2}{w+2} \geq \dfrac{1}{2w+1}$

44. $\dfrac{1}{r-3} \leq \dfrac{3}{r+1}$

45. $\dfrac{5}{2w+3} \geqslant \dfrac{-5}{w}$

46. $\dfrac{3}{x+1} < \dfrac{2}{x+2}$

47–50

Solve the nonlinear inequalities. Write each solution set using interval notation, rounding to the nearest hundredth.

47. $\pi x^2 + 3\sqrt{2}x > \sqrt{15}$

48. $1.02x^2 < 5.89x + 4\sqrt{5.92}$

49. $3\pi x^2 + 0.0537 \leq -3.96x$

50. $4.12x^2 - \pi^3 x \geq -8\pi$

(See Example 4.)

Dimensions of a Field 51–52

51. The length of a rectangular field is twice its width. For what dimensions is the area at least 1800 square feet?

52. Suppose the perimeter of a rectangular field is 480 meters. For what dimensions is the area at least 10,800 square meters?

Height of an Object

√ 53. Suppose an object is thrown vertically upward with an initial velocity of 128 feet per second. If s is the height above ground in feet and t is the time elapsed (in seconds), then $s = 128t - 16t^2$. At what times will the height of the object be 192 feet or more?

Plowing a Field

54. A farmer is plowing a rectangular field that is 100 meters wide and 200 meters long. How wide a strip must he plow around the field so that at least 52 percent of the field is plowed?

Critical Thinking: Exploration and Discussion

55. Michael Johnson attempted to solve the inequality $(x-4)(x+3) > 0$ by setting $x - 4 > 0$ and $x + 3 > 0$. This led him to $(4,\infty)$ as the answer. When he checked, he found he had only one part of the correct solution set $(-\infty,-3) \cup (4,\infty)$. Explain where the part $(-\infty,-3)$ comes from.

Critical Thinking: Exploration and Discussion

56. Profit P, revenue R, and cost C in the operation of a business are related by the equation $P = R - C$. Explain why the solutions to $P < 0$ would be of great concern to the owners of a business.

◆ **Solution for Practice Problem**

1. We first rewrite the inequality forcing 0 on the right.

$$\frac{1}{x-1} \le \frac{1}{x+2}$$

$$\frac{1}{x-1} - \frac{1}{x+2} \le 0 \qquad \text{subtracting } \frac{1}{x+2}$$

$$\frac{(x+2)-(x-1)}{(x-1)(x+2)} \le 0 \qquad \text{combining fractions}$$

$$\frac{3}{(x-1)(x+2)} \le 0 \qquad \text{simplifying}$$

The sign graph for this inequality is given in Figure 2.19, where we let "U" signify that the quotient is undefined.

$x-1$	− − − − − 0 + + +
$x+2$	− − 0 + + + + + +
$\frac{3}{(x-1)(x+2)}$	+ + U − − U + + +

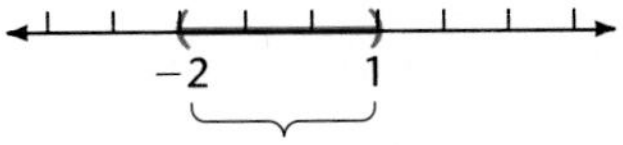

−2   1

$-2 < x < 1$

Figure 2.19 Sign graph for $\frac{1}{x-1} \le \frac{1}{x+2}$

From the sign graph, we see that the solution set is $\{x \mid -2 < x < 1\} = (-2,1)$, as shown in Figure 2.19.

CHAPTER REVIEW

Summary of Important Concepts and Formulas

Addition and Multiplication Properties of Equations

Let R, S, and T be algebraic expressions. The equation

$R = S$

is equivalent to

a. $R + T = S + T$ addition property

and

b. $RT = ST$, if $T \neq 0$. multiplication property

Addition and Multiplication Properties of Inequalities

Let R, S, and T be algebraic expressions. Then

$R > S$

is equivalent to

a. $R + T > S + T$, addition property

b. $RT > ST$, if $T > 0$,

c. $RT < ST$, if $T < 0$. } multiplication properties

Interval Notation

The letters a and b represent real numbers with $a < b$.

$(a, b) = \{x \mid a < x < b\}$ $\quad (a, \infty) = \{x \mid x > a\}$
$[a, b] = \{x \mid a \leq x \leq b\}$ $\quad [a, \infty) = \{x \mid x \geq a\}$
$(a, b] = \{x \mid a < x \leq b\}$ $\quad (-\infty, a] = \{x \mid x \leq a\}$
$[a, b) = \{x \mid a \leq x < b\}$ $\quad (-\infty, a) = \{x \mid x < a\}$
$(-\infty, \infty) = \mathcal{R}$

Absolute Value Inequalities

Suppose d is *positive.* Then

$|ax + b| < d$ is equivalent to $-d < ax + b < d$

and

$|ax + b| > d$ is equivalent to $ax + b > d$ or $ax + b < -d$.

Zero Product Property

$xy = 0$ if and only if either $x = 0$ or $y = 0$

Pythagorean Theorem

If a right triangle has sides a, b, and c, with c as the hypotenuse, then $c^2 = a^2 + b^2$.

The Quadratic Formula

The solutions to the quadratic equation $ax^2 + bx + c = 0$, $a \neq 0$, are given by

$$x = \frac{-b \pm \sqrt{b^2 - 4ac}}{2a}.$$

Classification of Roots by Use of the Discriminant

1. If $b^2 - 4ac = 0$, then the two roots are equal real numbers.
2. If $b^2 - 4ac > 0$, then the two roots are distinct (unequal) real numbers.
3. If $b^2 - 4ac < 0$, then the two roots are nonreal complex numbers that are conjugates of each other.

Critical Thinking: Find the Errors

Each of the following nonsolutions has one or more errors. Can you find them?

Problem 1

Solve the compound inequality:
$9 + 2x < 5$ or $4 + 3x < 4x + 1$.

Nonsolution

$2x < -4 \qquad 4 - 1 < x$
$x < -2 \qquad 3 < x$

The solution set is $\{x \mid 3 < x < -2\}$.

Problem 2

Solve by factoring: $6x^2 - x = 1$.

Nonsolution

$x(6x - 1) = 1$
$x = 1 \quad$ or $\quad 6x - 1 = 1$
$6x = 2$
$x = \frac{1}{3}$

The solution set is $\{1, \frac{1}{3}\}$.

Problem 3

Solve $2 - \sqrt{1 - x} = 2x$.

Nonsolution

$4 + 1 - x = 4x^2$
$0 = 4x^2 + x - 5$
$0 = (4x + 5)(x - 1)$
$4x + 5 = 0 \quad$ or $\quad x - 1 = 0$
$x = -\frac{5}{4} \qquad x = 1$

Check: $x = -\frac{5}{4}$

Since

$$2 - \sqrt{1 - (-\tfrac{5}{4})} = \tfrac{1}{2} \neq 2(-\tfrac{5}{4}),$$

$x = -\frac{5}{4}$ is not a solution.

Check: $x = 1$

Since

$$2 - \sqrt{1 - 1} = 2 = 2(1),$$

$x = 1$ is a solution, and the solution set is $\{1\}$.

Problem 4

Solve the following inequality: $x^2 + 6x + 8 < 3$.

Nonsolution

$$(x + 4)(x + 2) < 3$$

$$x + 4 < 3 \quad \text{or} \quad x + 2 < 3$$

$$x < -1 \quad \text{or} \quad x < 1$$

The solution set is $\{x \mid x < -1\} \cup \{x \mid x < 1\} = (-\infty, 1)$.

Problem 5

Solve the following inequality: $\dfrac{2}{w - 1} > \dfrac{1}{w + 2}$.

Nonsolution

$$\frac{2}{w - 1}(w - 1)(w + 2) > \frac{1}{w + 2}(w - 1)(w + 2)$$

$$2(w + 2) > 1(w - 1)$$

$$2w + 4 > w - 1$$

$$w > -5$$

The solution set is $\{w \mid w > -5\} = (-5, \infty)$.

Review Problems for Chapter 2

Solve the following equations.

1. $6x + 5 = 4$
2. $7x + 4(3 - 2x) = 5 + x$
3. $(3x - 1)(4x + 2) = (12x - 1)(x + 2)$
4. $\dfrac{3}{x + 1} + 5 = \dfrac{4x}{x + 1}$
5. $\dfrac{3}{x - 1} + \dfrac{4}{x + 2} = \dfrac{5x}{(x - 1)(x + 2)}$
6. $|2x - 9| = 0$
7. $|3x + 4| = 19$
8. $|x + 2| = -5$
9. $|7x - 1| = |4x + 3|$
10. $\dfrac{|3 - 2x|}{|2 - x|} = 1$
11. Solve for y in the equation

    $ax + by = bx + ay,\ a \neq b.$
12. Solve for z in the equation

    $$\frac{1}{x} + \frac{1}{y} + \frac{1}{z} = 1$$

Speed of a Rocket

13. A rocket in an Independence Day fireworks display traveled straight up with its speed at the end of t seconds given by $s = |176 - 32t|$ feet per second for $0 \leq t \leq 11$.
    a. Find the times when the speed of the rocket was 96 feet per second.
    b. Find the time when the rocket was at its highest point.

Student Population

14. The number of sophomores at Upstate University is 2300 less than twice the number of juniors. Find the number of sophomores and the number of juniors if the total number for the two groups is 8650 students.

Dimensions of a Rectangle

15. A long side of a certain rectangle is 20 inches longer than a short side, and the perimeter of the rectangle is 240 inches. Find the dimensions of the rectangle.

Age

16. Four years from now, Beckie will be 3 times as old as she was 6 years ago. Find her present age.

Interest Income 17–18

17. Part of \$36,000 is to be invested at 8% and the remainder at 9%. How much should be invested at each rate in order to yield an annual interest income of \$3090?
18. If \$4000 is invested at the rate of 15%, how much additional money should be invested at 17% for the total annual interest income from the investments to be \$1144?

Ticket Sales

19. The Shoestring Players sold 14 more children's tickets than adult's tickets to their summer play. How many tickets of each type were sold if children's tickets sold for \$5.80, adult's tickets sold for \$7.50, and the total revenue was \$1836.80?

Acid Mixture

20. How much pure acid must be added to 5 liters of a 2% solution to obtain a solution that is 25% acid?

Silver Alloy

21. Jennifer wishes to produce a bar of metal that is 25% silver by melting together 500 grams of an alloy containing 15% silver and a quantity of another alloy containing 30% silver. How many grams of the 30% alloy should be used?

Fertilizer

22. Sacks of lawn and garden fertilizer are labeled with numbers such as 8-8-8, 5-10-5, and 12-12-12. The numbers listed state the percentage of nitrogen, phosphoric acid, and potash, in that order. How many pounds of 8-8-8 should be mixed with 700 pounds of 5-10-5 to obtain a mixture that is 6% nitrogen?

Uniform Motion 23–24

23. It took a tugboat 5 hours to push a barge up the Hudson River to Albany, and only 2 hours to push it back downstream to where it started. If the tugboat can push the barge at 7 miles per hour in still water, find the rate of the current.
24. At 9:00 A.M., Stella Friedman started her morning walk at 4 miles per hour. At 9:30 A.M., her mother set out on her bike and caught up with Stella at 9:50 A.M. How fast was the bike traveling?

Work

25. Ron can mow a yard in 4 hours. If he and his friend Jane mow the yard together with two mowers, it takes only 1 hour. How long would it take Jane to mow the yard if she worked alone?

Solve the following inequalities. Express your answers in interval notation.

26. $7x - 3 < 4$
27. $5y + 7 < 7y + 3$
28. $5z - 11 \geq 7z - 3$
29. $4t - 21 < 9t + 14$
30. $3x + 5 > 4$ and $-5x + 6 < 4x - 1$
31. $13 > 5 - 2t \geq 7$
32. $-10 < 2 - 3y \leq 11$
33. $3x - 5 < 4$ or $7x + 2 > 5$

Solve the following inequalities and graph the solution sets. Express your answers in interval notation.

34. $3(2 - u) < 4u + 27$ or $3(u - 2) > 7(u + 2)$
35. $|5t - 6| \leq 3$
36. $\dfrac{|2x - 7|}{3} < 5$
37. $\left|\dfrac{3x - 8}{2}\right| \geq 2$
38. $|5 - 6y| \geq 5$

Solve by factoring.

39. $3x^2 + 7x = 6$
40. $2z^2 + 5z = 3$
41. $5t^2 - 7t - 6 = 0$
42. $3v(6v - 5) - 25 = 0$

Solve by completing the square.

43. $x^2 - 45 = 4x$
44. $z^2 + 4 = -2 - 4z$
45. $2t(t - 3) + 5 = 0$
46. $1 + 12w + 2w^2 = 0$
47. $0 = 1 - 4r^2 - 8r$

Solve by first expressing the equation in quadratic form.

48. $(2t - 1)^2 - 3(2t - 1) - 10 = 0$
49. $9z^4 - 10z^2 + 1 = 0$
50. $9y^4 + 4 = 13y^2$
51. $7 - 15\sqrt{x} + 2x = 0$

Solve the following equations by using the quadratic formula.

52. $y^2 - 2y - 5 = 0$
53. $2 - r - 2r^2 = 0$
54. $2t^2 = -5 - 2t$
55. $5x(x - 1) = -1 - 2x$

Use the discriminant to determine the type of roots of each of the following equations. Do not solve the equation.

56. $p^2 - 7p - 10 = 0$
57. $2t^2 + 6 = 5t$
58. $2y^2 + 3y = 4$
59. $z(z - 1) + 4 = 0$
60. $4x^2 - 12x = -9$

Profit

61. Ron Salmon has found that the profit from selling x wind chimes at the flea market is given (in dollars) by $P = 30x - x^2 - 160$. Find the number of wind chimes that he needs to sell to produce a profit of \$40.

Demand

62. The demand D for Halloween pumpkins at the flea market is given by $D = 22 + 6p - 0.5p^2$, where D is in hundreds of pumpkins and p is the price of a pumpkin in dollars. What price will result in a demand for 3800 pumpkins?

Profit

63. Ed Dixon has found that the sale of n baskets of strawberries from his home garden produces a profit P given by $P = 24n - n^2$ dollars. How many baskets must be sold to obtain a profit of \$144?

Demand

64. The demand D for shrimp in the seafood section of a supermarket is given by $D = 18 + 5p - 0.5p^2$, where D is in thousands of pounds and p is the price per pound of shrimp. What price for shrimp will produce a demand for 6000 pounds?

Poster Design

65. The Math Club is designing a recruitment poster with 252 square inches of printed area and margins of 2 inches at the sides and 3 inches at the top and bottom. What size poster board should they purchase if the height is to be one-third more than the width?

Plowing a Garden

66. A gardener is plowing a garden that is 100 feet by 80 feet by plowing around the outside edge and working inward. What will be the width of the plowed area when 40 percent of the garden has been plowed?

Find the solution set for each of the following equations. Assume all variables represent real numbers.

67. $3r + 1 - \dfrac{20}{3r + 2} = 0$
68. $y - \dfrac{6}{2y - 3} = 2$
69. $\dfrac{8}{2z - 5} + 7 = 2z$
70. $t + 2 - \dfrac{4t}{2t + 3} = \dfrac{2t + 9}{2t + 3}$
71. $\sqrt{3x + 1} = x - 1$
72. $\sqrt{2q + 4} = 2q - 8$
73. $3t = 5 - \sqrt{7 - 3t}$
74. $\sqrt{5y + 1} = y + 1$
75. $\sqrt{2z - 1} + z = 2$
76. $\sqrt{2r - 1} = 1 + \sqrt{r - 1}$
77. $\sqrt{2z - 1} - \sqrt{z + 3} = 1$
78. $\sqrt{2x + 11} - 2 - \sqrt{x + 2} = 0$

Solve the following inequalities and graph each solution set.

79. $w^2 + 6w + 8 < 0$
80. $2t^2 \geq 10 - t$
81. $5t < 2 - 12t^2$
82. $x^2 + 3x + 2 > 6$
83. $2y^3 + 5y^2 > 3y$
84. $\dfrac{2x + 5}{x + 3} \leq 1$
85. $\dfrac{2}{x + 3} \geq \dfrac{1}{2x + 3}$

ENCORE

The Photo Works

Many people dream of owning their own business. Over a period of years, Martin worked his way up to a position as store manager in the National Natural Foods Company chain of stores. Now he would like to withdraw his funds from the company's profit-sharing plan, buy a small convenience store, and go into business for himself.

After an extensive search, he finds a store priced at \$300,000 that he considers ideal for him. He has \$100,000 toward the purchase price and needs to borrow the remainder. A local lender will furnish the needed \$200,000 with a loan to be paid off in two installments from Martin's assets in the profit-sharing plan. The first payment will be \$100,000 at the end of 1 year, and the remaining payment will be \$132,000 at the end of the second year. Martin would like to know the annual interest rate (APR) that he will be paying under this plan.

If we let r be the unknown annual interest rate, the interest for 1 year on \$200,000 will be $200{,}000r$ dollars and the total debt at the end of 1 year will be

$$200{,}000 + 200{,}000r \text{ dollars.}$$

The \$100,000 payment at the end of the year will leave a balance of

$$\begin{aligned}(200{,}000 + 200{,}000r) - 100{,}000 &= 100{,}000 + 200{,}000r \\ &= 100{,}000(1 + 2r) \text{ dollars.}\end{aligned}$$

The interest that will accumulate on this balance during the second year will be

$$[100{,}000(1 + 2r)]r = 100{,}000r(1 + 2r) \text{ dollars,}$$

and the amount owed at the end of the second year will be

$$\underbrace{100{,}000(1 + 2r)}_{\text{Balance}} + \underbrace{100{,}000r(1 + 2r)}_{\text{Interest}} \text{ dollars.}$$

This amount is known to be \$132,000, so we have

$$100{,}000(1 + 2r) + 100{,}000r(1 + 2r) = 132{,}000.$$

If we divide both sides of this equation by 100,000 and simplify, we obtain

$$\begin{aligned}1 + 2r + r(1 + 2r) &= 1.32 \\ 2r^2 + 3r + 1 &= 1.32 \\ 2r^2 + 3r - 0.32 &= 0.\end{aligned}$$

This equation can be solved by the quadratic formula as follows.

$$\begin{aligned}r &= \frac{-3 \pm \sqrt{9 - 4(2)(-0.32)}}{4} \\ &= \frac{-3 \pm \sqrt{11.56}}{4} \\ &= \frac{-3 \pm 3.4}{4}\end{aligned}$$

The negative value of r makes no sense in this situation, and the positive value of

$$r = \frac{-3 + 3.4}{4} = 0.10 = 10\%$$

is easy to check. Thus Martin will be paying an APR of 10% under the lender's payment plan.

Functions

The examples and exercises in this chapter illustrate the use of functions in manufacturing design, dosages of medicine, business costs and profits, and other areas. Graphing is introduced in the first section and provides an extremely useful visual interpretation of the relative changes in physical quantities.

3.1 GRAPHING IN TWO-DIMENSIONAL COORDINATES

One of the most important ideas in college-level mathematics is the function concept. This concept is defined in Section 3.2, and the last two sections of the chapter are devoted to special types of functions that are used most frequently. In this chapter and the next, we restrict our attention to real variables.

An adequate treatment of functions must be based on a knowledge of graphing. With this purpose in mind, we briefly review the construction of the usual **rectangular coordinate system** in a plane.†

To construct a rectangular coordinate system in a plane, we begin with a horizontal line, called the ***x*-axis,** and a vertical line, called the ***y*-axis,** which intersect at a point O, called the **origin.** (See Figure 3.1.)

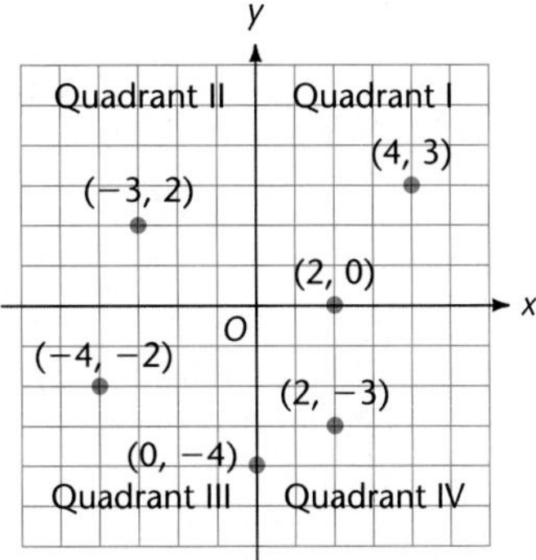

Figure 3.1 The rectangular coordinate system

On each of these lines, we set up a one-to-one correspondence between the points on the line and the real numbers, as described in Section 1.2. The correspondence on the x-axis is set up so that zero is at the origin with the positive direction to the right, indicated by an arrowhead, and the negative direction to the left. Similarly, the correspondence on the y-axis is set up so that zero is at the origin with the positive direction upward, indicated by an arrowhead, and the negative direction downward.

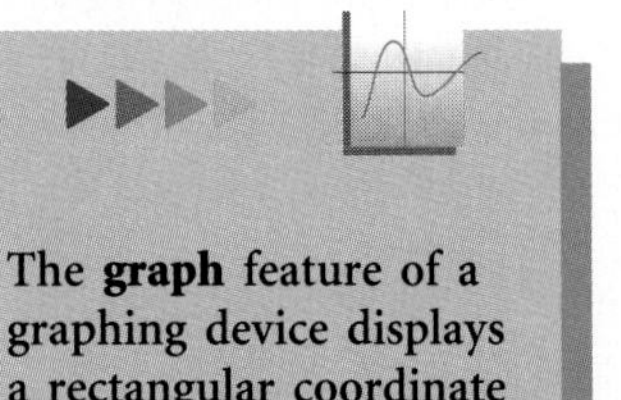

The **graph** feature of a graphing device displays a rectangular coordinate system in the window. Other features of the device can be used to change the scales on the coordinate axes.

We can now set up a one-to-one correspondence between points in the plane and ordered pairs (x, y) of real numbers. To describe this correspondence, let P be a point in the plane. We let x denote the directed horizontal distance from the y-axis to the point P, so that x is positive if P is to the right of the y-axis, x is negative if P is to the left of the y-axis, and x is zero if P is on the y-axis. Similarly, we let y denote the directed vertical distance from the x-axis to the point P, where y is positive in the upward direction, negative in the downward direction, and zero on the x-axis. The ordered pair (x, y) is then assigned to the point P. The first

†The rectangular coordinate system is also frequently referred to as the **Cartesian coordinate system.** This is in honor of René Descartes (1596–1650), the French mathematician who is credited with inventing the system.

entry, x, is called the **abscissa** or **x-coordinate,** of P, and the second entry, y, is called the **ordinate,** or **y-coordinate,** of P. The ordered pair (x, y) is referred to as the **coordinates** of P.

Conversely, each ordered pair (x, y) determines a unique point P that has the pair (x, y) as coordinates. The point P is located simply by using x as the directed horizontal distance from the y-axis, and y as the directed vertical distance from the x-axis. Several points are located by their coordinates in Figure 3.1.

The coordinate axes separate the plane into four regions, called **quadrants.** These quadrants are numbered as shown in Figure 3.1.

A rectangular coordinate system can be used to obtain a "picture" of an equation in two variables. The picture is formed by sketching the graph of the equation.

Graph of an Equation

The graph of an equation in x and y is the set of all points whose coordinates (x, y) satisfy the equation.

If a point with coordinates (x, y) is on the graph of an equation, then the coordinates (x, y) make the equation a true statement. On the other hand, if the coordinates make the equation true, then the point is on the graph of the equation.

If there is a point where a graph crosses the y-axis, the y-coordinate of that point is called a **y-intercept** of the graph. Similarly, the x-coordinate of a point where a graph crosses the x-axis is called an **x-intercept.**

A sketch of the graph of an equation is made by plotting several points on the graph of the equation and then drawing a curve through these points. This procedure is illustrated in the following examples.

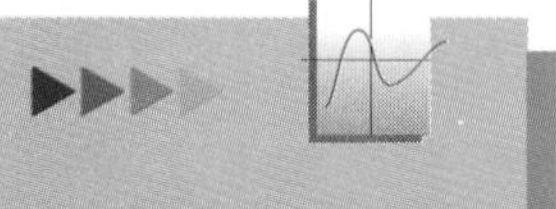

If it is possible to solve an equation for y in terms of x, a graphing device can be used to graph the equation even if the expression for y is very complicated. (See Exercises 65–70.)

Example 1 • The Graph of an Equation

Sketch the graph of the equation $y = x^2$.

Solution

We first assign several sample values to x and compute the corresponding values of y. The resulting ordered pairs are recorded in the table in Figure 3.2. We then locate the points corresponding to these coordinates and sketch the graph as well as possible from these points. □

Example 2 • An Equation Involving Absolute Value

Sketch the graph of the equation $x + 1 = |y|$.

Solution

Following the same procedure as in Example 1, we obtain the table in Figure 3.3. We soon discover that two values of y are sometimes obtained from a given value of x. Also, we find that if x is assigned a value less than -1, there is no y that satisfies the equation since an absolute value cannot be negative. When the points

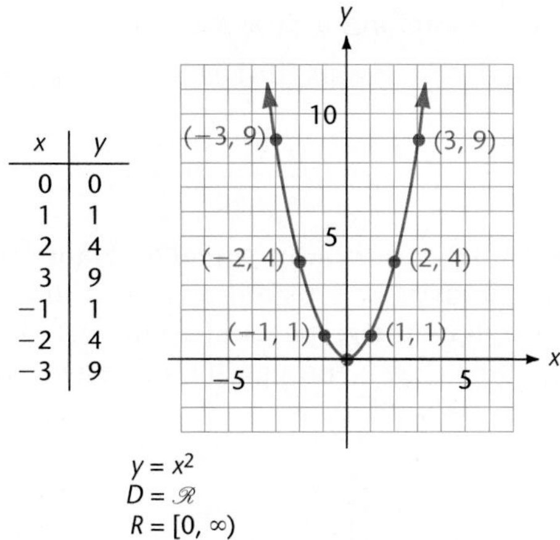

x	y
0	0
1	1
2	4
3	9
−1	1
−2	4
−3	9

Figure 3.2 Graph of $y = x^2$

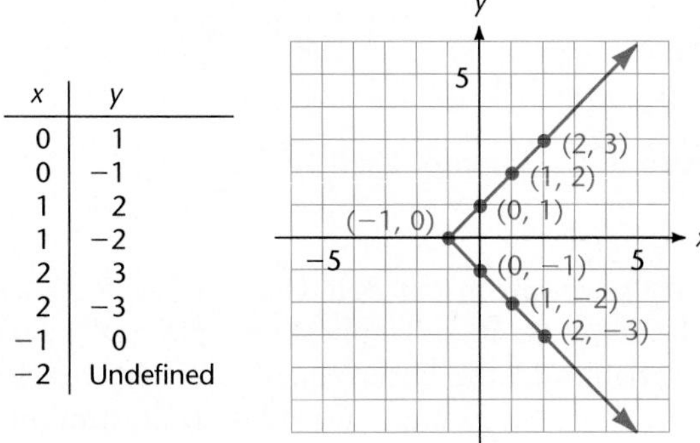

x	y
0	1
0	−1
1	2
1	−2
2	3
2	−3
−1	0
−2	Undefined

Figure 3.3 Graph of $x + 1 = |y|$

corresponding to the coordinates in the table are plotted, it appears that they lie along two half-lines that have a common endpoint at $(-1, 0)$, and the graph is drawn accordingly. □

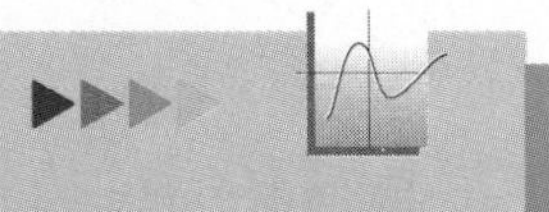

The cursor on a graphing device can be used with the **trace** feature to estimate both x- and y-intercepts with a reasonable degree of accuracy.

A **linear equation** in x and y is an equation that can be written in the form

$$Ax + By = C,$$

with at least one of A and B different from zero. Linear equations get their name from the fact that their graphs are always straight lines. That is, *the graph of any equation of the form*

$$Ax + By = C,$$

with at least one of A and B different from zero, is a **straight line.** To find a y-intercept, we could set $x = 0$ in the equation $Ax + By = C$. If there is a solution for y, this y-value will be a y-intercept. An x-intercept, if there is one, is obtained by setting $y = 0$ in $Ax + By = C$ and then solving for x.

Example 3 • Graphing a Straight Line

Find the x- and y-intercepts, if they exist, and sketch the graph of the equation $2x + 3y = 6$.

Solution

To find the y-intercept, we let $x = 0$ and obtain $3y = 6$. Thus 2 is the y-intercept. Similarly, $y = 0$ gives $2x = 6$, and 3 is the x-intercept. Plotting the points $(0, 2)$ and $(3, 0)$, we draw the graph as in Figure 3.4. The point $(2, \frac{2}{3})$ is included as a check on the line drawn through the other two points. □

x	y
0	2
3	0
2	$\frac{2}{3}$

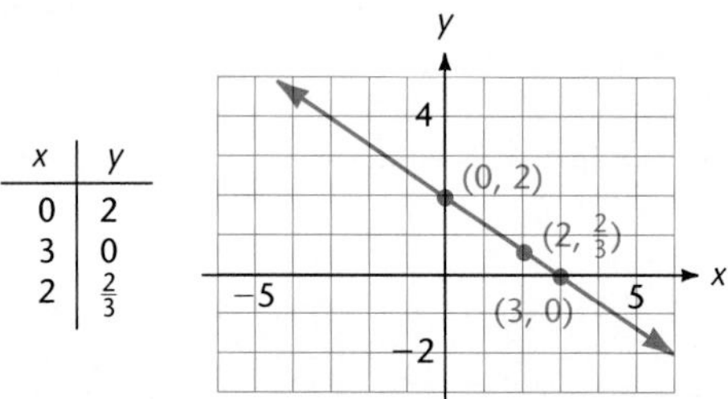

Figure 3.4 Graph of $2x + 3y = 6$

We saw some instances in Chapter 2 where linear equations were very useful in applications. In many applications, physical considerations place restrictions on the variables in the linear equation. These restrictions are then reflected in the graph. Our next example provides an illustration.

Example 4 • Appliance Repair

If a certain appliance repairman makes a house call and works for x hours on the job, his charge C (in dollars) is given by $C = 28 + 40x$ (and he refuses to work more than 8 hours). Graph the equation $C = 28 + 40x$ using appropriate restrictions on the variables to represent this situation.

Solution

It is clear that x cannot be negative in this example, and we shall interpret $x = 0$ to mean that the repairman made a house call, did no work after he got there, and charged \$28 for the call. At the other extreme, if he works 8 hours, the charge (in dollars) is given by

$$\begin{aligned} C &= 28 + 40(8) \\ &= 348. \end{aligned}$$

Thus $(0, 28)$ and $(8, 348)$ are endpoints on the graph, and we can draw the graph as in Figure 3.5. □

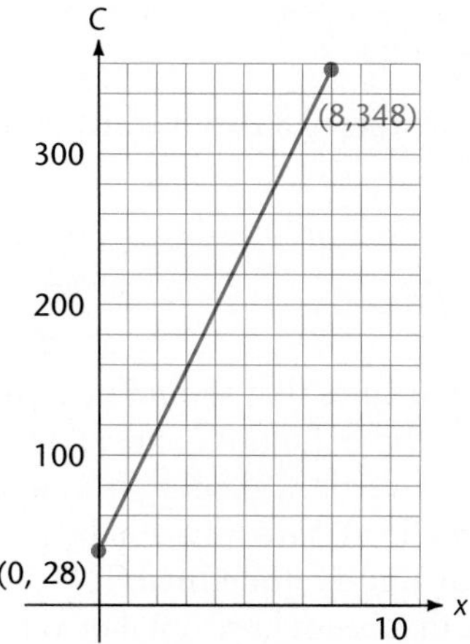

Figure 3.5 Graph of $C = 28 + 40x$, $x \in [0,8]$

There are two special cases of the equation $Ax + By = C$ that are worth noting. One of these is

$$y = C.$$

The graph of $y = C$ consists of all points with coordinates of the form (x, C), where x may be any real number. Thus the graph of $y = C$, where C is a constant, is a **horizontal straight line.** Similarly, the graph of the equation

$$x = C,$$

where C is a constant, is a **vertical straight line.** As examples, the graphs of $y = 4$ and $x = -2$ are drawn in Figure 3.6.

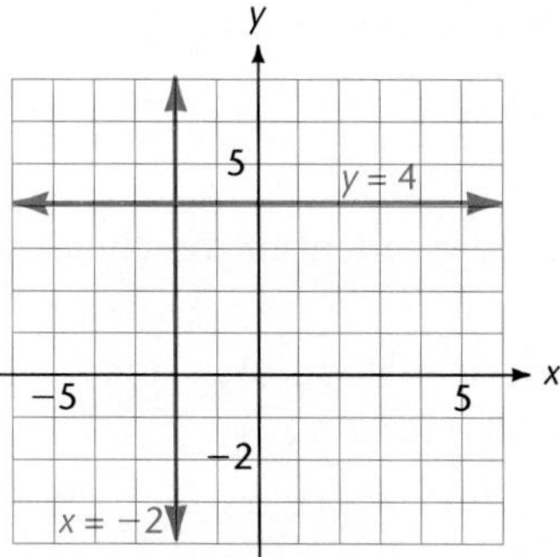

Figure 3.6 Examples of horizontal and vertical straight lines

A straight line can be determined by two of its points. Another way to determine a straight line is to specify one of its points and its direction. The direction of a line can be described by its slope.

Definition of Slope

The **slope** m of the line through two distinct points (x_1, y_1) and (x_2, y_2) is given by

$$m = \frac{y_2 - y_1}{x_2 - x_1}.$$

There are two special cases that should be noted. If the line is horizontal, then $y_1 = y_2$ and $x_1 \neq x_2$ in the definition, and $m = 0$. That is, **a horizontal line has slope zero.** If the line is vertical, then $x_1 = x_2$ and $y_1 \neq y_2$ in the definition, and $m = (y_2 - y_1)/0$ is undefined, since division by zero is impossible. This means that **the slope of a vertical line is undefined.**

The slope of a line that is not horizontal or vertical is a real number $m \neq 0$, and m is *independent of the choice of the points* (x_1, y_1) and (x_2, y_2). In Figure 3.7 another pair of points (x_1', y_1') and (x_2', y_2') on the same line is indicated, and we see that

$$\frac{y_2' - y_1'}{x_2' - x_1'} = \frac{y_2 - y_1}{x_2 - x_1}.$$

This follows from the fact that the ratios of corresponding sides of similar triangles are always equal.

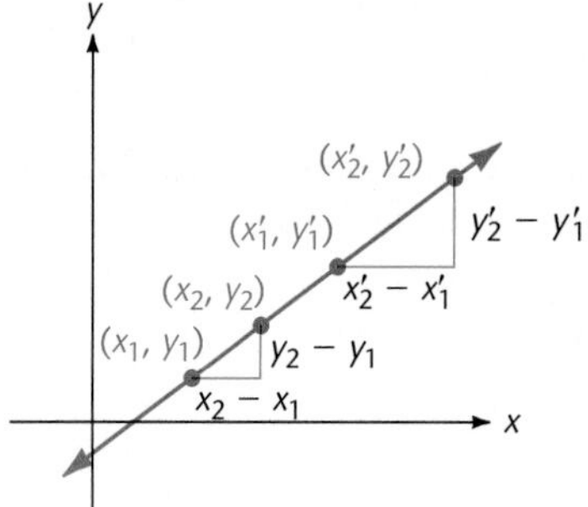

Figure 3.7 Slope is independent of the choice of points

Example 5 • Finding the Slope of a Line

Find the slope of the line $2x - y = 4$.

Solution

To find the slope, we locate two points on the line and use the slope formula. If we let $x = 1$ and solve for y in the equation of the line, we get $y = -2$. That is, $(1, -2)$ is on the graph. For $x = 3$, we get $y = 2$, and $(3, 2)$ is on the graph. Thus the slope of the line is

$$m = \frac{2 - (-2)}{3 - 1} = \frac{4}{2} = 2.$$

□

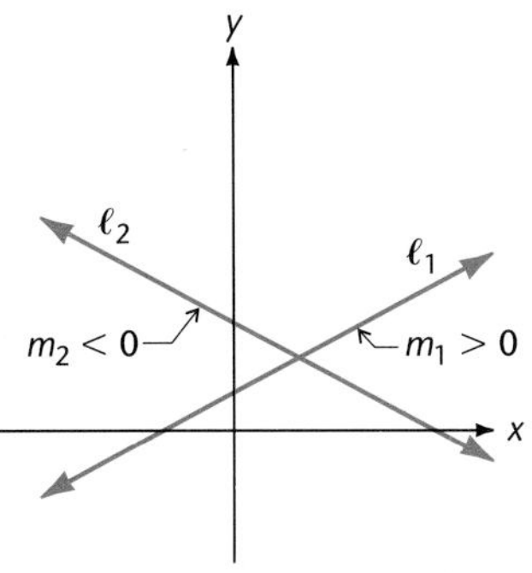

Figure 3.8 Positive and negative slopes

If the graph of a line slants upward to the right, this indicates that y increases as x increases along the line, and the slope of the line is **positive.** On the other hand, if the graph slants downward to the right, then y decreases as x increases along the line, and the slope of the line is **negative.** These situations are pictured in Figure 3.8, where line ℓ_1 has slope $m_1 > 0$ and line ℓ_2 has slope $m_2 < 0$.

◆ Practice Problem 1 ◆

Find the slope of the line with equation $5x - 6y + 12 = 0$.

It will be useful in later work to have a formula for the coordinates (x, y) of the midpoint of the line segment joining two points (x_1, y_1) and (x_2, y_2). Starting with (x_1, y_1), (x_2, y_2), and (x, y) as located in Figure 3.9, we can draw two right triangles with $90°$ angles at (x_2, y_1) and (x, y_1) as shown in the figure. These two right triangles are similar triangles, so ratios of corresponding sides are equal. Since (x, y) is located halfway up the hypotenuse in the large right triangle, the point (x, y_1) is located halfway across the base of the same triangle. This means that the change from x_1 to x is half the change from x_1 to x_2, and we have

$$x - x_1 = \tfrac{1}{2}(x_2 - x_1).$$

Solving this equation for x gives

$$\begin{aligned} x &= \tfrac{1}{2}x_2 - \tfrac{1}{2}x_1 + x_1 \\ &= \tfrac{1}{2}x_2 + \tfrac{1}{2}x_1 \\ &= \frac{x_1 + x_2}{2}. \end{aligned}$$

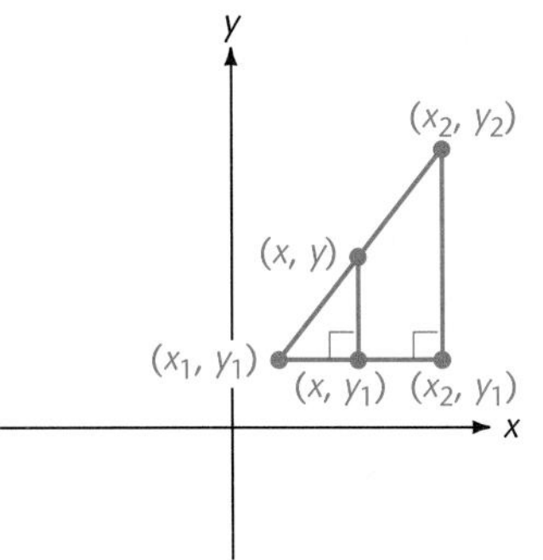

Figure 3.9 Midpoint of a line segment

As might have been expected, the x-coordinate of the midpoint is just the average of the x-coordinates at each end. A similar discussion for y-coordinates would show that

$$y = \frac{y_1 + y_2}{2},$$

and hence establish the following formula.

Midpoint Formula

The midpoint of the segment joining (x_1, y_1) and (x_2, y_2) has coordinates

$$\left(\frac{x_1 + x_2}{2}, \frac{y_1 + y_2}{2}\right).$$

Example 6 • Finding the Midpoint of a Line Segment

Find the midpoint of the line segment joining $(-7, 2)$ and $(3, 9)$.

Solution

It makes no difference which point is taken as (x_1, y_1). If we take $(x_1, y_1) = (-7, 2)$ and $(x_2, y_2) = (3, 9)$, the midpoint formula yields

$$\left(\frac{-7+3}{2}, \frac{2+9}{2}\right) = \left(-2, \frac{11}{2}\right)$$

as the midpoint of the segment. □

EXERCISES 3.1

Graph each linear equation and find the x-intercepts and y-intercepts, if they exist. (See Example 3.)

1. $y = 2x - 3$
2. $2x - 3y = 6$
3. $4x = 5$
4. $x = 2$
5. $2x = 3y$
6. $2y = -3x$
7. $2x + 5y = 7$
8. $5x + 2y = 10$
9. $y = -3$
10. $3y = 5$

Find the slope, if it exists, of the line through the given pair of points. (See Example 5.)

11. $(2, -1)$ and $(-1, 3)$
12. $(0, 4)$ and $(3, 7)$
13. $(0, 0)$ and $(3, 2)$
14. $(7, 11)$ and $(11, 7)$
15. $(3, -1)$ and $(3, 3)$
16. $(-2, 3)$ and $(-2, 5)$
17. $(-4, 5)$ and $(2, 5)$
18. $(2, 7)$ and $(-3, 7)$
19. $(-1, 1)$ and $(7, 5)$
20. $(-3, -4)$ and $(7, -2)$

Find the midpoint of the line segment joining each of the following pairs of points. (See Example 6.)

21. $(1, -2)$ and $(-3, 1)$
22. $(2, -1)$ and $(2, 7)$
23. $(-3, 2)$ and $(-5, 2)$
24. $(4, 0)$ and $(-10, 3)$
25. $(2, 7)$ and $(1, -3)$
26. $(-1, 1)$ and $(10, -4)$
27. $(-\pi, 0)$ and $(3\pi, 0)$
28. $(\pi/4, 0)$ and $(9\pi/4, 0)$
29. (a, b) and (b, a)
30. (a, b) and $(-a, b)$
31. Suppose $(1, 5)$ is the midpoint of a line segment with one endpoint $(2, 8)$. Find the other endpoint.
32. Suppose $(-2, 7)$ is the midpoint of a line segment with one endpoint $(3, -5)$. Find the other endpoint.

Find the slope, if it exists, of the line that has the given equation. (See Example 5 and Practice Problem 1.)

33. $3x - 4y = 6$
34. $4x - 2y = 20$

35. $x + 4 = 0$
36. $x - 7 = 0$
37. $y = 3$
38. $y = -2$
39. $7x - 5y = 3$
40. $3x - 7y = 0$
41. $2y = 7x - 5$
42. $3y = 5 - 2x$
43. $5x + 2y = 10$
44. $3x + 5y = 15$
45. $x = 2y + 6$
46. $x + 8 = 2y$

Suppose the two given points lie on a straight line. Find the missing coordinate if the line has the given slope m.

47. $(1, y)$, $(2, 3)$, $m = 5$
48. $(-3, y)$, $(1, -5)$, $m = -\frac{3}{5}$

Suppose the three given points lie on a straight line. Find the missing coordinate.

49. $(0, 0)$, $(1, 2)$, $(3, y)$
50. $(1, -1)$, $(-2, 8)$, $(x, -7)$

Determine whether the four given points are the vertices of a parallelogram.

51. $(-1, 3)$, $(-2, 0)$, $(2, 0)$, $(3, 3)$
52. $(-1, 1)$, $(-2, -1)$, $(1, -4)$, $(0, -6)$
53. $(-2, 8)$, $(-6, 4)$, $(-4, 1)$, $(2, 2)$
54. $(2, 4)$, $(3, 8)$, $(1, -8)$, $(4, 4)$

Sketch the graph of the following equations. (See Examples 1 and 2.)

55. $x = |y| + 1$
56. $y = -|x|$
57. $x = |y + 1|$
58. $x = |y - 1|$
59. $x = y^2 - 4$
60. $x = y^2 + 1$

Pizza Delivery

61. The cost C (in dollars) of operating a pizza delivery car is given by $C = 9000 + 0.7x$, where x is the number of miles that the car is driven. Use appropriate restrictions on the variables and graph this equation.

Trade-in Value

62. The trade-in value V of a forklift machine after t years of use is $V = 16{,}000 - 2000t$ dollars. Graph this equation with appropriate restrictions on the variables.

Water Pressure

63. In a swimming pool where the water is 12 feet deep, the water pressure p at a point x feet above the bottom of the pool is $p = 62.5(12 - x)$ pounds per square foot. Sketch the graph of this equation using appropriate restrictions on the variables.

Temperature in Atlanta

64. On a certain July day in Atlanta, the Fahrenheit temperature F could be found from the formula $F = 94 - 3|t - 15|$, where t is the number of hours that have elapsed since midnight and $7 \leq t \leq 20$. Sketch the graph of this equation with appropriate restrictions on the variables.

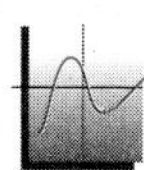

65–70

Match each equation in Exercises 65–70 with its computer-generated graph.

65. $y = \sqrt{9 - x^2}$ 66. $y = -\sqrt{3 - x}$ 67. $y = x^4 - 1$

68. $y = x^3 - 1$ ✓69. $y = -x|x|$ 70. $y = -\sqrt{9 + x^2}$

a.

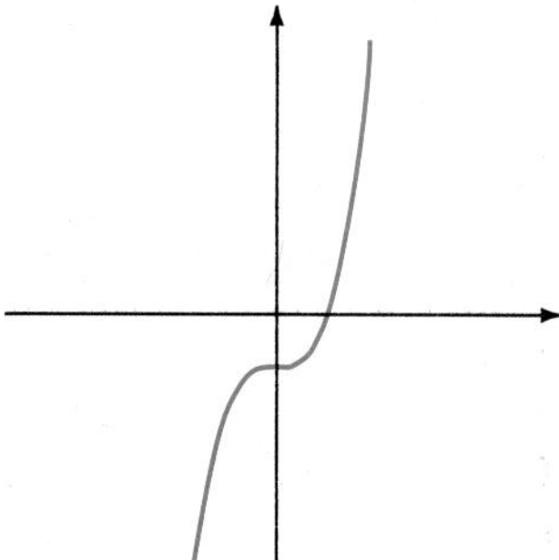

b.

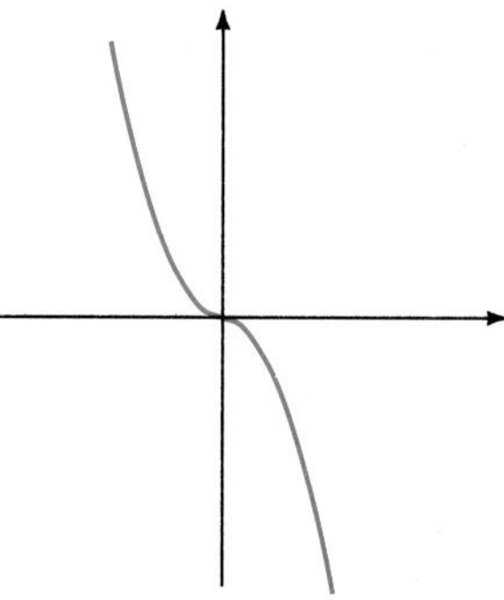

c.

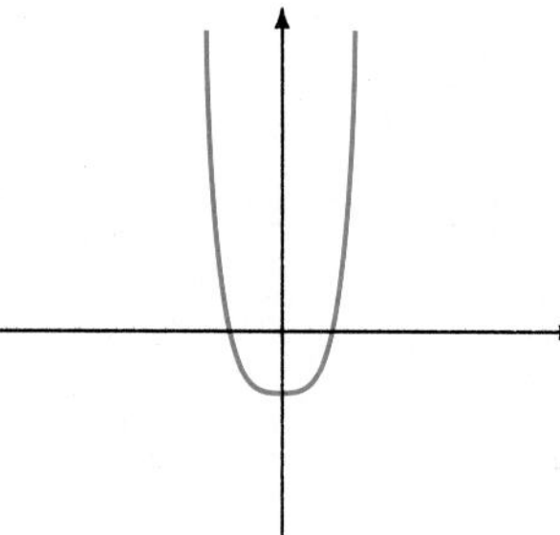

d.

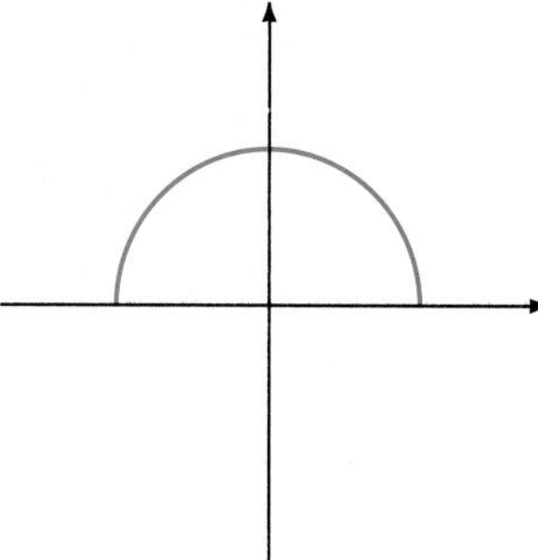

e.

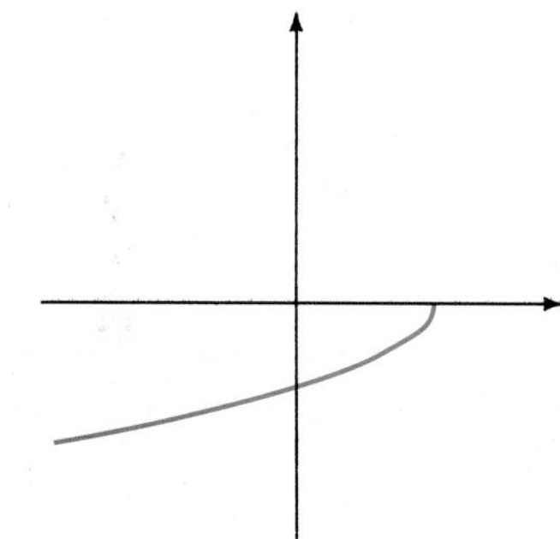

f.

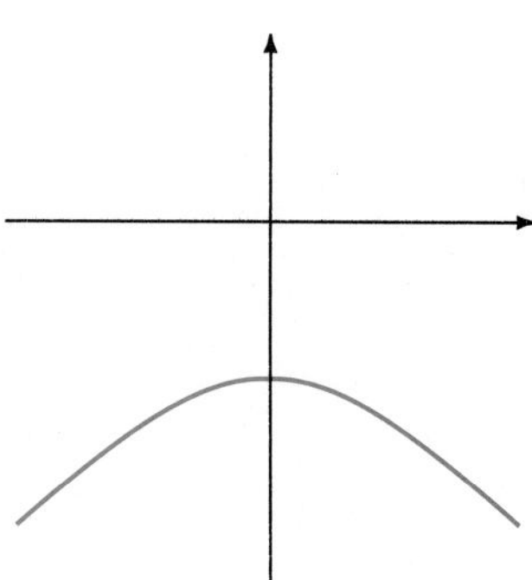

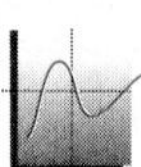

Use the cursor on a graphing device to estimate the x-intercept to the nearest hundredth.

71–74

71. $y = 3.87x - 11.64$ 72. $y = -0.487x + 13.65$

73. $y = -9.73x - 42.88$ 74. $y = 42.7x + 17.6$

Critical Thinking: Exploration and Writing

75. In a classroom discussion, the group leader asked for an equation of the line through (3, 7) with no slope. Lou said $y = 7$, and Lulu said $x = 3$. Write a justification for both answers and explain how the term *no slope* can lead to confusion.

Critical Thinking: Exploration and Writing

76. Describe a situation where the midpoint formula could be used to find the center of a circle.

Critical Thinking: Exploration and Writing

77. Explain why a straight line with slope 0 must be a horizontal line.

Critical Thinking: Exploration and Writing

78. Explain why a straight line whose slope is undefined must be a vertical line.

◆ Solution for Practice Problem

1. An easy way to locate two points on the line is to find the intercepts. If we put $y = 0$, we get $5x + 12 = 0$ and $x = -\frac{12}{5}$. Thus $(-\frac{12}{5}, 0)$ is on the graph. Similarly, $x = 0$ gives $-6y + 12 = 0$ and $y = 2$, and so $(0, 2)$ is on the graph. Using the slope formula with these two points, we get

$$m = \frac{2 - 0}{0 - (-\frac{12}{5})} = 2\left(\frac{5}{12}\right) = \frac{5}{6}.$$

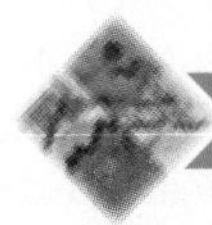

3.2 FUNCTIONS

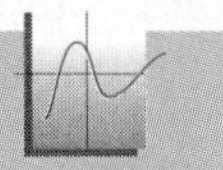

An understanding of the function concept is necessary to use a graphing device effectively because the performance of a graphing device is based exclusively on functions.

There are two quite different formulations of the function concept that are commonly used. Some mathematicians prefer one method of presentation because it is natural and intuitive, while others prefer another method because it is more precise and rigorous. In both presentations, functions are denoted by letters such as f, g, h, and r.

In one formulation, a **function** f is defined as a **correspondence** of a certain type between the elements of two sets. In this correspondence, there is a rule of association between the elements of a first set D and those of a second set. The association must be such that for each element x in D, there is *one and only one* associated element y in the second set. This discussion is summarized in the following definition.

Definition of a Function

Let D be a nonempty set of real numbers. A **function** f with **domain** D is a correspondence such that for each element x of D, there is one and only one associated element y in a second set of real numbers.

To indicate that y is associated with x by the function f, we write $y = f(x)$ and say that y is the **value of** f **at** x. The notation $f(x)$ *does not indicate multiplication.* It is read "f of x" or "f at x." The set R of all values of $f(x)$ is called the **range** of f.

With this definition of a function, it is natural to think of starting with a value of x in D and using the rule or correspondence to obtain a value $y = f(x)$ in R. In this way, the value y is assigned to x. We call x the **independent variable** and y the

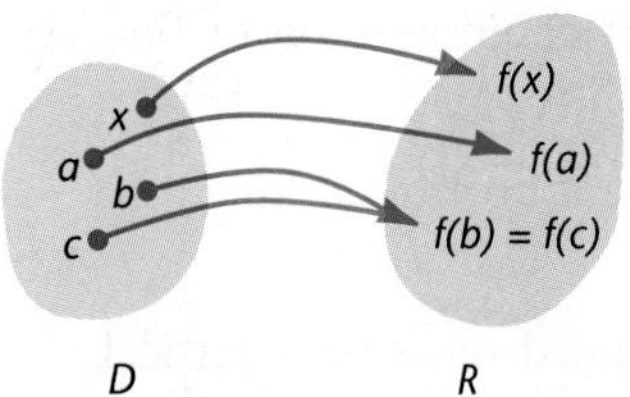

Figure 3.10 A function as a correspondence

dependent variable since the value obtained for y depends on the value selected for x. Figure 3.10 illustrates this intuition behind defining a function as a correspondence or a rule of association.

In Figure 3.10, it is indicated that $f(b) = f(c)$ with $b \neq c$ in D. This may happen with a function, and it is illustrated in Example 1. However, it is important to keep in mind that, for each x in D, there is a *unique value* $f(x)$.

Example 1 • A Simple Function

Consider the function f with domain $D = \{-2, 1, 2\}$ and $f(x) = x^2$. The values of f are given by

$$f(-2) = (-2)^2 = 4, \quad f(1) = 1^2 = 1, \quad f(2) = 2^2 = 4.$$

We note that $f(-2) = f(2)$ with $-2 \neq 2$, as indicated in Figure 3.11. □

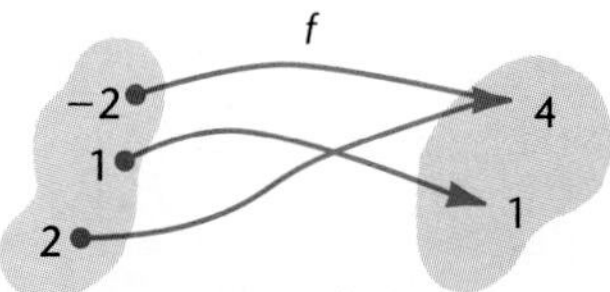

Figure 3.11 The function f in Example 1

The second formulation of the definition of a function has achieved significant popularity in the last half of the twentieth century. This formulation makes use of the idea of ordered pairs of real numbers.

By an **ordered pair of real numbers** we mean a pairing (a, b) of real numbers a and b, where there is a distinction between the pair (a, b) and the pair (b, a), if a and b are not equal. That is, there is a first position and a second position in the pair of numbers. It is traditional to use parentheses to indicate ordered pairs, where the number on the left is called the **first entry,** or **first component,** and the number on the right is called the **second entry,** or **second component.** Two ordered pairs (a, b) and (c, d) are **equal** if and only if $a = c$ and $b = d$.

Alternate Definition of a Function

A **relation** is a nonempty set of ordered pairs (x, y) of real numbers x and y. The **domain** D of a relation is the set of all first-entry elements, or the set of all x-values, that occur in the relation. The **range** of a relation is the set of all y-values that occur in the relation. A **function** is a relation in which no two distinct ordered pairs have the same x-value.

Example 2 • A Relation That Is Not a Function

The set

$$g = \{(5, -2), (5, 2), (-5, -2), (-5, 2)\}$$

is a relation with domain $D = \{-5, 5\}$ and range $R = \{-2, 2\}$. We can describe g by the rule

$$(x, y) \in g \text{ if an only if } |x| = y^2 + 1 \text{ and } x \in \{-5, 5\}.$$

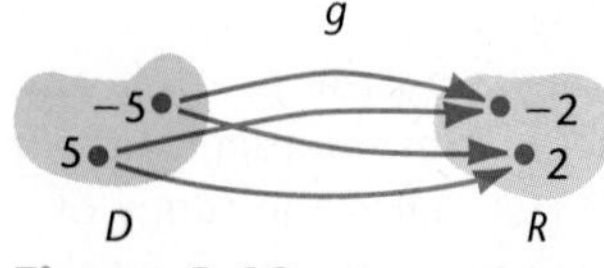

Figure 3.12 The relation g in Example 2

The relation g can be visualized as in Figure 3.12. Since both $(5, -2)$ and $(5, 2)$ are in g, there are two y-values for $x = 5$ and g is not a function. □

In our work here, we have little use for relations with domains that have only two or three elements. Our relations are usually determined by a rule relating the entries x and y of the ordered pairs (x, y), and we make the following convention regarding these rules.

> Unless otherwise specified, the domain of a relation described by a certain rule is the set of all real numbers x that yield a real number y in the rule for the relation.

This convention is illustrated in the next example.

Example 3 • Finding the Domain and Range

In each of the following relations, list three sample ordered pairs in the relation and state the domain D and range R of the relation.

a. $h = \{(x, y) \mid y = x^2\}$ b. $r = \{(x, y) \mid y = \sqrt{x}\}$

Solution

a. Three ordered pairs in the relation h are $(1, 1)$, $(2, 4)$, and $(-7, 49)$. The domain of h is the set of all real numbers since any real number x has a square that is a real number. However, the range consists only of nonnegative real numbers since the rule gives $y = x^2$, and the square of a real number is never negative. That is, h has domain†

$$D = \mathcal{R}$$

and range

$$R = \{y \mid y \geq 0\} = [0, \infty).$$

b. Three ordered pairs in r are $(1, 1)$, $(2, \sqrt{2})$, and $(4, 2)$. In this case, the only real numbers that are in the domain are nonnegative real numbers, for they are the only ones that have a square root in the real numbers. The range consists of all nonnegative real numbers since $\sqrt{x}$ denotes the nonnegative square root of x. That is, r has domain

$$D = [0, \infty)$$

and range

$$R = [0, \infty).$$

Since each of the equations $y = x^2$ and $y = \sqrt{x}$ never gives more than one y-value for the same x-value, both h and r are functions. □

†Recall that $\mathcal{R}$ denotes the set of all real numbers.

The diagrams in Figure 3.13 illustrate the difference between a function and a relation that is not a function. In Figure 3.13a we see that for each x-value there is a unique y-value. Hence the relation f is a function. But the relation f in Figure 3.13b is not a function since the x-value, x_1, is associated with two different y-values, y_1 and y_2 $(y_1 \neq y_2)$.

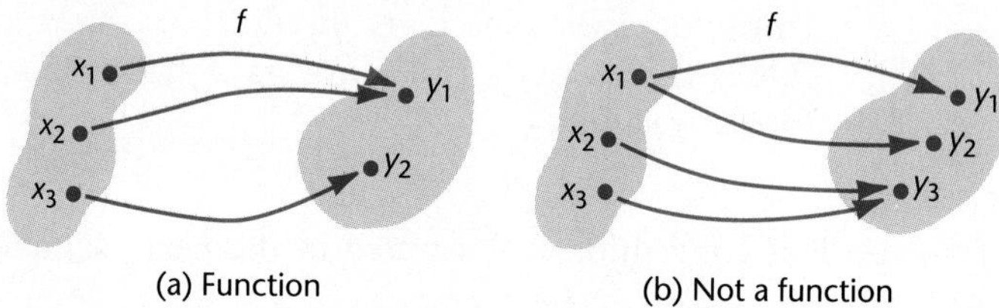

Figure 3.13 Uniqueness of function values

In Example 3, we could simply write

$$h(x) = x^2 \qquad \text{and} \qquad r(x) = \sqrt{x}$$

to define these functions. This kind of notation is emphasized in the next two examples.

Example 4 • Finding Function Values

For $F(x) = x^2 - 1$ and $G(x) = \sqrt{x - 1}$, find the values of $F(2)$, $F(-3)$, $G(1)$, and $G(14)$.

Solution

We have

$$\begin{aligned} F(2) &= (2)^2 - 1 = 3 \\ F(-3) &= (-3)^2 - 1 = 8 \\ G(1) &= \sqrt{1 - 1} = \sqrt{0} = 0 \\ G(14) &= \sqrt{14 - 1} = \sqrt{13}. \end{aligned}$$

□

Example 5 • Using Function Notation

Suppose $f(x) = 2x^2 - 3$. Evaluate the following.

a. $f(-x)$ b. $f(x + h)$

Solution

a. $f(-x) = 2(-x)^2 - 3 = 2x^2 - 3 = f(x)$

b. $f(x + h) = 2(x + h)^2 - 3 = 2(x^2 + 2xh + h^2) - 3$
$= 2x^2 + 4xh + 2h^2 - 3$

□

The last part of the next example illustrates an expression, called the **difference quotient,** used in the study of tangent lines to curves in the calculus.

Example 6 • The Difference Quotient

If $g(x) = x^2 + 3x$, find each of the following.

a. $g(x + h) - g(x)$ b. $\dfrac{g(x + h) - g(x)}{h}$ (the difference quotient for g)

Solution

a.
$$\begin{aligned} g(x + h) - g(x) &= [(x + h)^2 + 3(x + h)] - (x^2 + 3x) \\ &= x^2 + 2hx + h^2 + 3x + 3h - x^2 - 3x \\ &= 2hx + h^2 + 3h \end{aligned}$$

b. We can use the result from part a to determine the difference quotient.

$$\begin{aligned} \frac{g(x + h) - g(x)}{h} &= \frac{2hx + h^2 + 3h}{h} \\ &= \frac{h(2x + h + 3)}{h} \\ &= 2x + h + 3 \end{aligned}$$

□

◆ Practice Problem 1 ◆

If $f(x) = 4x^2 - 5$, find the difference quotient, $\dfrac{f(x + h) - f(x)}{h}$, in lowest terms.

A rectangular coordinate system can be used to draw the graph of a relation or function. The **graph** of a relation consists of all points whose coordinates (x, y) are members of the relation. In the case of a function, this is all points with coordinates (x, y) such that $y = f(x)$. An illustration is provided in Example 7.

Example 7 • Graphing an Absolute Value Function

Sketch the graph of the function $f(x) = |x - 2|$.

Solution

Since an absolute value cannot be negative, the graph of f must lie on or above the x-axis. We assign several sample values to x and obtain the pairs in the table in Figure 3.14. When the points with coordinates listed in the table are plotted, they are seen to lie along two half-lines forming a "V," and we draw the graph as shown in Figure 3.14. The graph provides a picture of the behavior of the function, and it clearly shows that the domain D of f is the set $\mathcal{R}$ of all real numbers and the range R of f is the set of all nonegative real numbers. □

It is easy to see that if a relation is defined by an equation in the manner used in Example 3, the graph of the relation is the same as the graph of the defining equation.

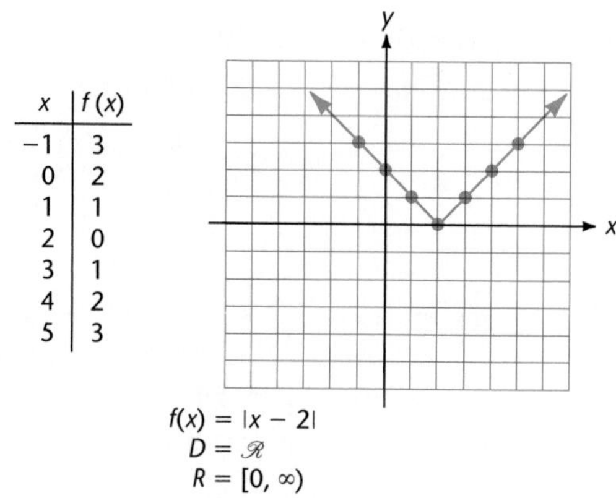

x	f(x)
−1	3
0	2
1	1
2	0
3	1
4	2
5	3

Figure 3.14 Graph of the function in Example 7

Example 8 • Graphing a Relation

Sketch the graph of the relation g defined by the equation $x = y^2 - 1$.

Solution

Since we have already graphed the equation $y = x^2$ in Example 1 in Section 3.1, we expect to see a graph similar to what we had before. Plotting the points listed in the table in Figure 3.15 leads us to draw the graph as shown in that figure. □

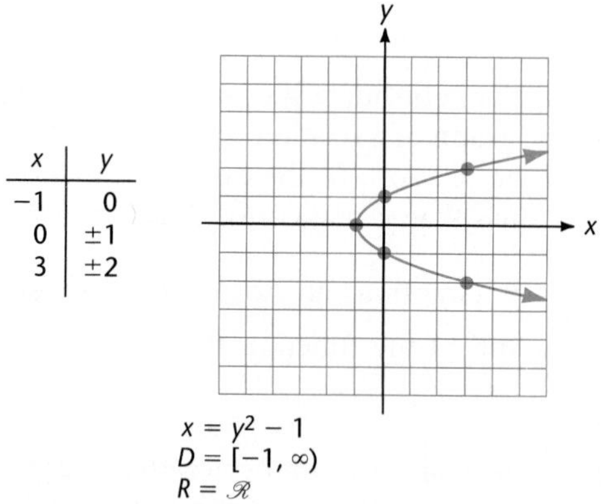

x	y
−1	0
0	±1
3	±2

Figure 3.15 Graph of the relation in Example 8

The graph of f in Figure 3.14 makes it easy to see that f is a function, since there is exactly one y-value for each x-value. Similarly, the graph of g in Figure 3.15 shows that there are two y-values for any $x > -1$, so that g is not a function. A graphical description of this situation is that some vertical lines intersect the graph of g at more than one point, and that no vertical line crosses the graph of f at more than one point. This illustrates the **vertical line test** for functions, which is stated next.

Vertical Line Test

If any vertical line intersects a graph at two or more points, the graph *does not* represent a function. On the other hand, if no vertical line intersects the graph at more than one point, the graph *does* represent a function.

For any real number x, the notation $[\![x]\!]$ denotes the *greatest integer less than or equal to x.* Some values of $[\![x]\!]$ are given by

$$[\![2]\!] = 2, \quad [\![4.12]\!] = 4, \quad [\![\sqrt{10}]\!] = 3, \quad [\![-5.9]\!] = -6, \quad [\![-\sqrt{10}]\!] = -4.$$

A more precise definition of $[\![x]\!]$ is provided by the following statement.

If n is an integer such that $n \le x < n + 1$, then $[\![x]\!] = n$.

The **greatest integer function** is the function defined by $f(x) = [\![x]\!]$.

Example 9 • Graphing a Greatest Integer Function

Sketch the graph of the greatest integer function $f(x) = [\![x]\!]$.

Solution

To sketch the graph, we assign some values to n in the boxed statement. For $n = 1$, the statement says that if $1 \le x < 2$, then $[\![x]\!] = 1$. This and some similar statements are tabulated in Figure 3.16. From the table, we see that the graph consists of horizontal line segments over intervals of length 1, as shown in Figure 3.16. □

Although a graphing device will graph a greatest integer function, the graph it generates may lack quality in detail at places where a jump occurs.

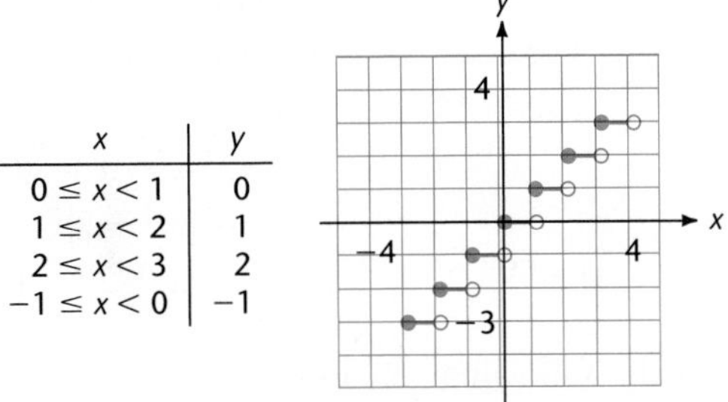

x	y
$0 \le x < 1$	0
$1 \le x < 2$	1
$2 \le x < 3$	2
$-1 \le x < 0$	−1

Figure 3.16 Graph of $f(x) = [\![x]\!]$

Because of the resemblance of its graph to stair steps, the greatest integer function is called a **step function.** The next example illustrates another type of step function.

Example 10 • Drug Dosage

Suppose the dosage of a certain pediatric drug depends upon the weight of the child in the following manner: 30 milligrams of the drug plus an additional 10 milligrams for each 5 pounds or portion of 5 pounds of body weight above 20 pounds. Sketch a graph of the dosage function for children weighing 10 through 40 pounds.

Solution

Let x represent the body weight of a child, and $D(x)$ the corresponding dosage. The domain of the dosage function is the closed interval $[10, 40]$. For a child weighing 10 through 20 pounds, the dosage is 30 milligrams. For children weighing more than 20 pounds through 25 pounds, an additional 10 milligrams of drug should be administered. Following this type of reasoning, we complete the chart and sketch the graph in Figure 3.17. □

Weight, x	Dosage, $D(x)$
$10 \leq x \leq 20$	30
$20 < x \leq 25$	$30 + 1(10) = 40$
$25 < x \leq 30$	$30 + 2(10) = 50$
$30 < x \leq 35$	$30 + 3(10) = 60$
$35 < x \leq 40$	$30 + 4(10) = 70$

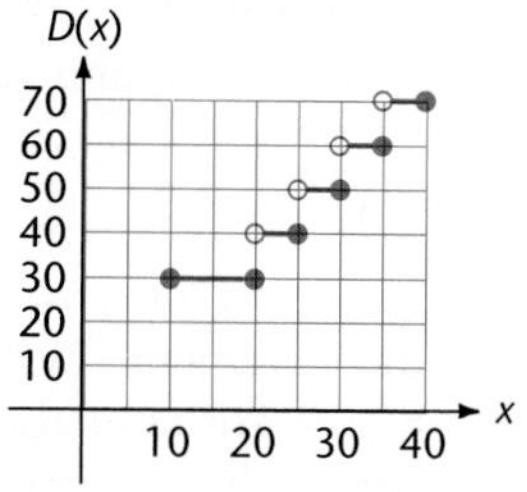

Figure 3.17 Graph of the dosage function

EXERCISES 3.2

Give the domain D and range R of each function. (See Example 3.)

1. $f(x) = 3x$
2. $g(x) = 4x - 3$
3. $h(x) = -\sqrt{x - 4}$
4. $F(x) = \sqrt{x + 2}$
5. $G(x) = x^2 - 3$
6. $p(x) = 2(x - 1)^2 + 1$
7. $y = \dfrac{1}{x - 1}$
8. $y = \dfrac{2}{x - 3}$
9. $y = |4 - x|$
10. $y = |5x - 7|$
11. $y = |x - 1| + 3$
12. $y = |x + 2| - 1$

Which of the following relations are functions? (See Examples 2 and 3.)

13. $g = \{(1, 2), (2, 1), (3, 2)\}$
14. $g = \{(-1, 0), (0, -1), (1, 0)\}$
15. $h = \{(2, 1), (1, 2), (2, 3)\}$
16. $h = \{(0, -1), (-1, 0), (0, 1)\}$

17. $p = \{(x, y) \mid x = 3\}$
18. $p = \{(x, y) \mid x = -2\}$
19. $h = \{(x, y) \mid x = y + 1\}$
20. $h = \{(x, y) \mid x = y - 3\}$
21. $r = \{(x, y) \mid y = |x|\}$
22. $r = \{(x, y) \mid y = |x + 1|\}$

If $f(x) = 4x - 2$ and $g(x) = x^2$, find each of the following. (See Examples 4–6 and Practice Problem 1.)

23. $f(3)$
24. $f(-5)$
25. $g(4)$
26. $g(-7)$
27. $f(b)$
28. $g(a)$
29. $f(-x)$
30. $g(-x)$
31. $g(x + 1)$
32. $f(x + 1)$

If $f(x) = 2x$ and $g(x) = x^2 - x$, find each of the following.

33. $f(3)$
34. $f(-2)$
35. $g(-3)$
36. $g(2)$
37. $f(0)$
38. $g(0)$
39. $f(-x)$
40. $g(-x)$
41. $f(x + h)$
42. $g(x + h)$
43. $g(x + h) - g(x)$
44. $f(x + h) - f(x)$

Find the difference quotient, $\dfrac{f(x+h) - f(x)}{h}$, in lowest terms for each of the following.

45. $f(x) = 3x - 4$
46. $f(x) = 4x + 5$
47. $f(x) = 2x^2 + 1$
48. $f(x) = 3x^2 - 1$
49. $f(x) = 3x^2 - 2x + 4$
50. $f(x) = 2x^2 - 3x + 5$
51. $f(x) = 2x^3 + 1$
52. $f(x) = x^3 + 2x$

53–56

Evaluate each of the following functions at the given value of x. Round each result to three decimal places.

53. $f(x) = 2x^5 - 3x^4 + 5x - 2,\ x = 3.6$
54. $f(x) = \dfrac{4x^3 - x + 2}{x^4 - 5x + 1},\ x = 1.3$

Evaluate the difference quotient, $\dfrac{f(x+h) - f(x)}{h}$, for the given function f and the given values of x and h. Round each result to two decimal places.

55. $f(x) = x^3 + 4x^2 - 2,\ x = -1,\ h = 0.01$
56. $f(x) = \sqrt{x^2 + 2x},\ x = 4,\ h = 0.02$

Sketch the graph of the following functions. (See Examples 7–10.)

57. $r(x) = |2x|$
58. $f(x) = -|\frac{1}{2}x|$
59. $g(x) = |x| - 1$
60. $g(x) = |x - 1|$
61. $h(x) = x^2 - 4$
62. $h(x) = x^2 + 1$

65–68

63. $F(x) = [\![x + 1]\!]$
64. $r(x) = [\![2x]\!]$
65. $f(x) = |4 - x^2|$
66. $f(x) = x\sqrt{x + 2}$
67. $h(x) = 3[\![0.5x + 2]\!]$
68. $g(x) = |(x + 3)(x - 4)(x)|$ (*Hint:* Change the viewing rectangle.)

Use the vertical line test to determine whether or not the given graph is the graph of a function.

69.

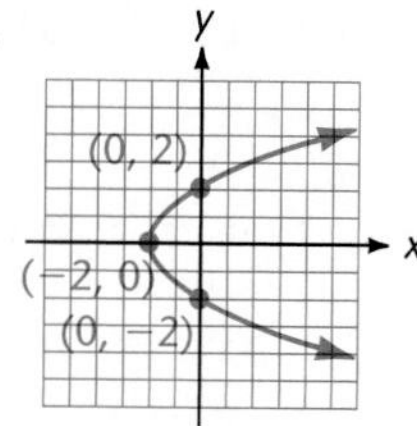

70.

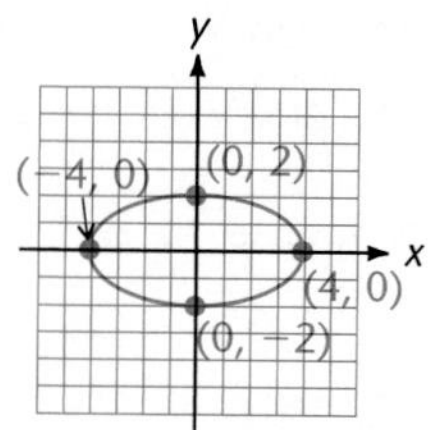

71.

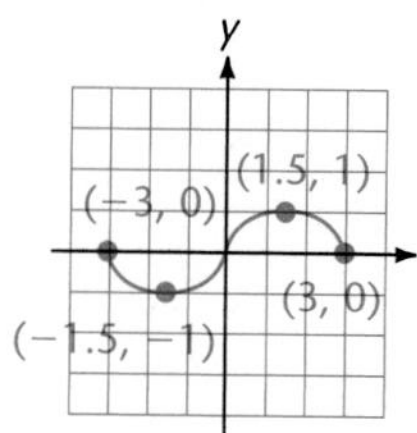

72.

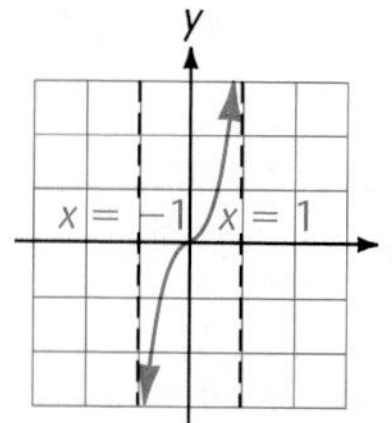

73.

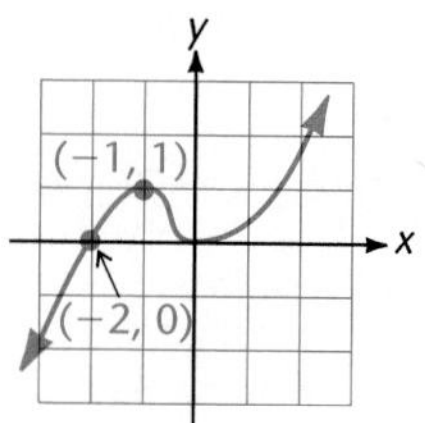

74.

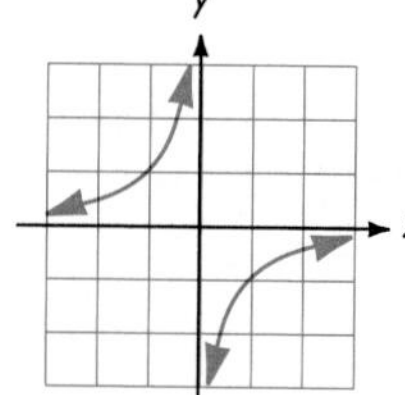

75.

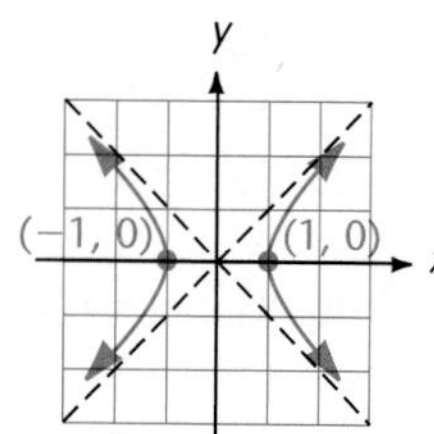

76.

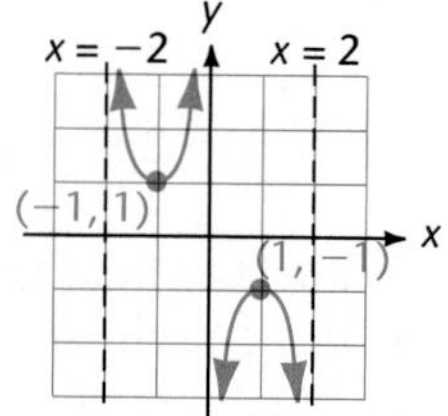

Delivery Charge Function 77. Suppose a delivery service charges for delivery of items based on mileage in the following way: \$10.00 plus an additional \$0.50 per mile or fraction of a mile. Sketch the graph of the delivery charge function for mileage of 0 through 10 miles.

Postal Rate Function 78. Postage rates for a first-class letter are determined as follows: 29 cents for the first ounce plus an additional 23 cents for each ounce or fraction of an ounce above 1 ounce but less than 12 ounces. Graph the function representing first-class postage rates for letters between 0 and 12 ounces.

Area Function 79. The members of the Big Woods Hunting Club have 120 feet of fencing to build a dog pen. They plan to build a rectangular pen with one cross fence as shown in the accompanying figure.

a. Using the fact that there is a total of 120 feet of fencing, express the length ℓ in terms of the width w.

b. Use the results of part a to express the area A of the pen as a function of the width w.

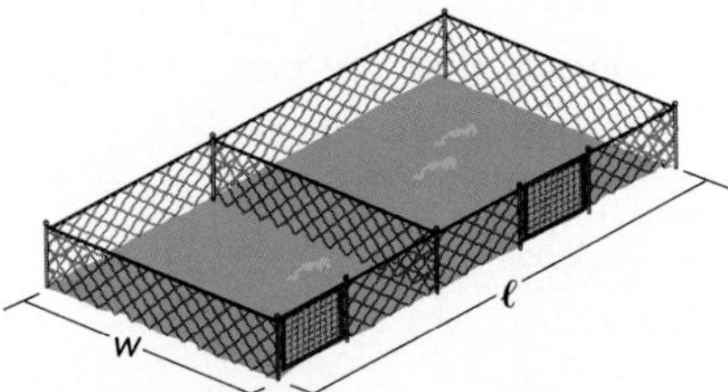

Figure for Exercise 79

Poster Area Function

80. Suppose a political campaign poster is to contain 600 square inches of printing with 6-inch margins at the top and bottom and 4-inch margins along the sides, as shown in the accompanying figure.

    a. Using the fact that the printed area totals 600 square inches, express the width w of the printed area in terms of its length ℓ.

    b. Using the results of part a, express the total area A of the poster as a function of ℓ.

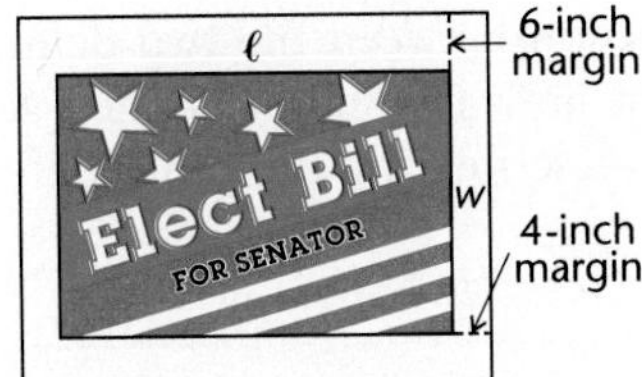

Figure for Exercise 80

Volume Function

81. An open box (without a top) is to be made from a piece of cardboard 18 inches square by cutting equal square pieces from each corner and folding up the sides as shown in the accompanying figure. Express the volume V of the box as a function of x.

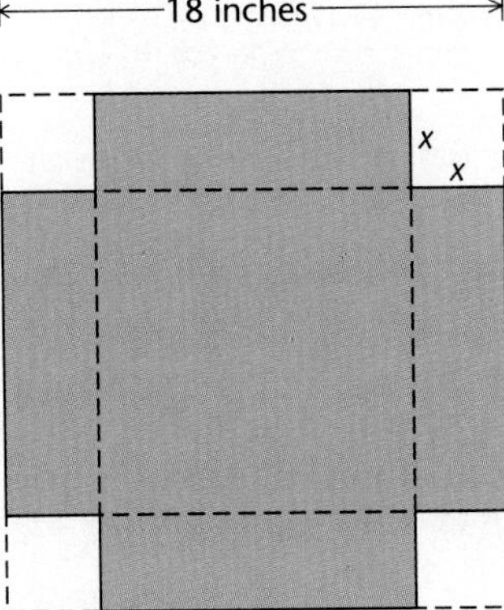

Figure for Exercise 81

Volume Function

82. The closed box with a square bottom and top in the accompanying figure is to be made using 400 square inches of plastic.
    a. Express the height h of the box in terms of the width w of the bottom.
    b. Write the volume V of the box as a function of w.

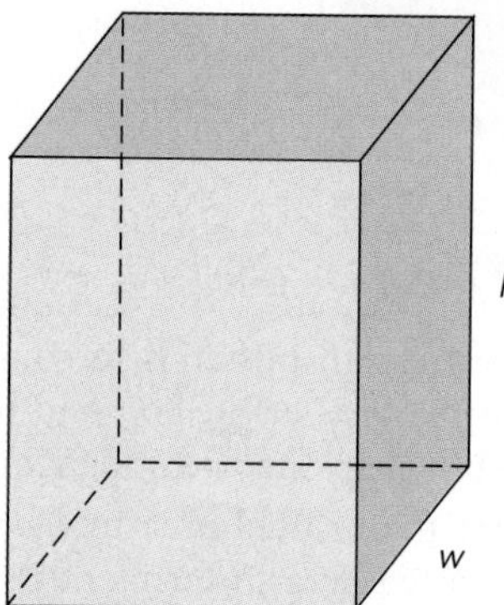

Figure for Exercise 82

Critical Thinking: Discussion

83. Discuss the connection between the two definitions of a function in this section and explain how a function defined according to either one of the definitions can be described by the other definition.

◆ Solution for Practice Problem

1. $$\begin{aligned}\frac{f(x+h)-f(x)}{h} &= \frac{4(x+h)^2-5-(4x^2-5)}{h}\\ &= \frac{4(x^2+2hx+h^2)-5-(4x^2-5)}{h}\\ &= \frac{4x^2+8hx+4h^2-5-4x^2+5}{h}\\ &= \frac{8hx+4h^2}{h} = 8x+4h\end{aligned}$$

3.3 THE ALGEBRA OF FUNCTIONS

Given two functions f and g, it is natural to consider creating new functions by using the operations of addition, subtraction, multiplication, and division with f and g. For example, if $f(x) = x^2$ and $g(x) = x - 2$, it is natural to define the sum $f + g$ by

$$(f+g)(x) = x^2 + x - 2 = f(x) + g(x),$$

the difference $f - g$ by

$$(f - g)(x) = x^2 - (x - 2) = f(x) - g(x),$$

the product $f \cdot g$ by

$$(f \cdot g)(x) = x^2(x - 2) = f(x) \cdot g(x),$$

and the quotient f/g by

$$\left(\frac{f}{g}\right)(x) = \frac{x^2}{x - 2} = \frac{f(x)}{g(x)}, \qquad \text{if } g(x) \neq 0.$$

Care must be taken when forming these new functions. Consider the next example.

Example 1 • The Sum of Two Functions

Suppose $f(x) = -x$ and $g(x) = \sqrt{x}$. Describe $f + g$ and evaluate $f + g$ at $x = -4$.

Solution

The sum $f + g$ is

$$(f + g)(x) = -x + \sqrt{x}.$$

Now suppose $x = -4$. Then

$$(f + g)(-4) = -(-4) + \sqrt{-4} = 4 + \sqrt{-4},$$

which is not defined in the real numbers. This is because -4 is not in the domain of g. Any element in the domain of $f + g$ has to be an element of both the domain of f and the domain of g. Thus a complete description of $f + g$ requires that the domain be fully described. Hence the sum $f + g$ is

$$(f + g)(x) = -x + \sqrt{x}, \qquad x \geq 0.$$

□

Example 2 • Operations on Functions

Let $f(x) = \sqrt{x - 4}$ and $g(x) = \sqrt{5 - x}$. Describe completely $f + g$, $f - g$, $f \cdot g$, and f/g.

Solution

The domain of f is $D_f = [4, \infty)$. The domain of g is $D_g = (-\infty, 5]$. Since $D_f \cap D_g = [4, 5]$, we have

$$(f + g)(x) = \sqrt{x - 4} + \sqrt{5 - x}, \qquad x \in [4, 5];$$
$$(f - g)(x) = \sqrt{x - 4} - \sqrt{5 - x}, \qquad x \in [4, 5];$$
$$(f \cdot g)(x) = \sqrt{x - 4}\sqrt{5 - x} = \sqrt{(x - 4)(5 - x)}, \qquad x \in [4, 5];$$
$$\left(\frac{f}{g}\right)(x) = \frac{\sqrt{x - 4}}{\sqrt{5 - x}} = \sqrt{\frac{x - 4}{5 - x}}, \qquad x \in [4, 5).$$

Note that 5 had to be excluded from the domain of f/g since f/g is not defined whenever $g(x) = 0$.

□

We summarize the results of the preceding discussion in the following definition.

Definition of Fundamental Operations on Functions

Suppose f and g are real-valued functions whose domains are D_f and D_g, respectively. The functions $f + g$, $f - g$, $f \cdot g$, and f/g are defined by the following equations.

$$(f + g)(x) = f(x) + g(x);$$
$$(f - g)(x) = f(x) - g(x);$$
$$(f \cdot g)(x) = f(x) \cdot g(x);$$
$$\left(\frac{f}{g}\right)(x) = \frac{f(x)}{g(x)}, \qquad g(x) \neq 0.$$

The domain of each of these functions is $D_f \cap D_g$, with the additional condition that f/g is not defined when $g(x) = 0$.

Whether a student graduates is a function of passing each required course. Whether she passes each course is a function of what grade she earns on her examinations. What grades she earns on her examinations is a function of how much she studies. How much she studies is certainly a function of some other variable, and on, and on, and on. This sort of chain of events has a mathematical counterpart called the composition of functions.

Consider the diagram in Figure 3.18 of the two functions $f(x) = x^2 - 3$ and $g(x) = 2x$. For any x in the domain of g, we can write

$$f(g(x)) \underset{\nearrow}{=} f(2x) \underset{\nwarrow}{=} (2x)^2 - 3 = 4x^2 - 3.$$

Definition of g Definition of f

Thus the functions g and f have been used to obtain yet another function. We call this new function the composition (or composite) function and denote it by $f \circ g$. For this example, we have

$$(f \circ g)(x) = f(g(x)) = 4x^2 - 3,$$

for any real x.

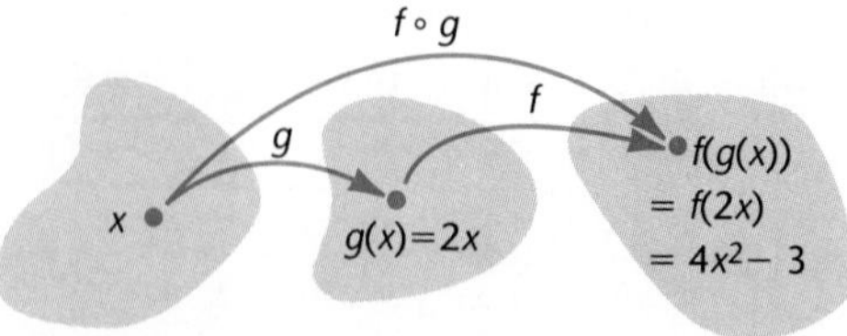

Figure 3.18 The composition $f \circ g$

Definition of Composition Function

If f and g are functions, the **composition** (or **composite**) **function** $f \circ g$ is defined by

$$(f \circ g)(x) = f(g(x)).$$

The domain of $f \circ g$ is the set of all x in the domain of g such that f is defined at $g(x)$.

Example 3 • Finding Composite Function Values

Let $f(x) = 2x - 1$ and $g(x) = x^2 + 1$. Determine $(f \circ g)(2)$ and $(g \circ f)(2)$.

Solution

Since $g(2) = 5$ and $f(2) = 3$, we have

$$(f \circ g)(2) = f(g(2)) = f(5) = 2 \cdot 5 - 1 = 9$$

and

$$(g \circ f)(2) = g(f(2)) = g(3) = 3^2 + 1 = 10.$$

□

◆ **Practice Problem 1** ◆

If $g(x) = x^2 - 1$ and $f(x) = \sqrt{x - 1}$, find each of the following.

a. $(g \circ f)(3)$ b. $(g \circ f)(x)$

Note in Example 3 that $(f \circ g)(x) \neq (g \circ f)(x)$, which is usually the case. In Section 4.6 we shall examine the situation when $f \circ g$ and $g \circ f$ are equal, and in particular when $(f \circ g)(x) = (g \circ f)(x) = x$.

Example 4 • Function Compositions

Let $f(x) = 1/x$ and $g(x) = x - 3$. Determine the following.

a. $(f \circ g)(x)$ and the domain of $f \circ g$
b. $(g \circ f)(x)$ and the domain of $g \circ f$

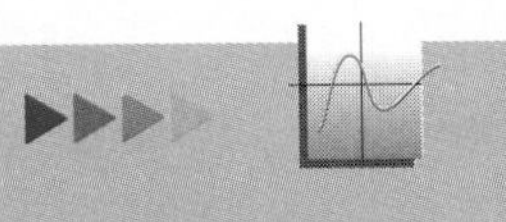

A graphing device automatically uses the correct domain when graphing a function formed using the operations defined in this section.

Solution

a. $(f \circ g)(x) = f(g(x)) = f(x - 3) = \dfrac{1}{x - 3}$

The domain of $f \circ g$ is $\{x \mid x \neq 3\} = (-\infty, 3) \cup (3, \infty)$.

b. $(g \circ f)(x) = g(f(x)) = g\left(\dfrac{1}{x}\right) = \dfrac{1}{x} - 3$

The domain of $g \circ f$ is $\{x \mid x \neq 0\} = (-\infty, 0) \cup (0, \infty)$. □

EXERCISES 3.3

Suppose $f(x) = x + 2$, $g(x) = 2x - 3$, and $h(x) = x^2$. Evaluate each of the following. (See Examples 1 and 3.)

1. $(f + g)(3)$
2. $(f + g)(2)$
3. $(h - g)(-1)$
4. $(g - h)(0)$
5. $(f \cdot h)(-3)$
6. $(h \cdot g)(-2)$
7. $(f \cdot f)(0)$
8. $(g \cdot g)(1)$
9. $(f/g)(1)$
10. $(g/h)(1)$
11. $(h/g)(0)$
12. $(f/h)(-2)$
13. $(f - g + h)(1)$
14. $(f + g - h)(1)$
15. $((f \cdot g) - h)(2)$
16. $(f - (g \cdot h))(-1)$
17. $(f \circ g)(3)$
18. $(g \circ f)(3)$
19. $(h \circ g)(0)$
20. $(h \circ f)(-3)$
21. $(f \circ f)(-1)$
22. $(g \circ g)(0)$
23. $((f \circ g) \circ h)(-2)$
24. $((f \circ g) \circ h)(1)$

For the given functions, determine $f + g$, $f - g$, $f \cdot g$, and f/g and state their domains. (See Example 2.)

25. $f(x) = x - 2$, $g(x) = 2x + 1$
26. $f(x) = 3x$, $g(x) = x - 2$
27. $f(x) = 4x - 1$, $g(x) = x(x - 1)$
28. $f(x) = x - 3$, $g(x) = (x - 1)(x - 2)$
29. $f(x) = \sqrt{x}$, $g(x) = 2x$
30. $f(x) = \sqrt{x - 1}$, $g(x) = 4x$
31. $f(x) = 2x + 1$, $g(x) = \sqrt{x + 2}$
32. $f(x) = 3x - 4$, $g(x) = \sqrt{x - 2}$
33. $f(x) = \sqrt{x - 2}$, $g(x) = \sqrt{3 - x}$
34. $f(x) = \sqrt{x + 1}$, $g(x) = \sqrt{4 - x}$
35. $f(x) = \sqrt{x + 1}$, $g(x) = \sqrt{x + 6}$
36. $f(x) = \sqrt{x - 2}$, $g(x) = \sqrt{x - 4}$
37. $f(x) = x - 4$, $g(x) = x^2 - 5x + 6$
38. $f(x) = 2x + 7$, $g(x) = x^2 - 4x - 8$
39. $f(x) = 6x^2 + 2x - 5$, $g(x) = x^2 + 3x - 4$
40. $f(x) = 5x^2 - 7$, $g(x) = x^2 - 4$

For the given functions, determine $f \circ g$ and $g \circ f$ and state their domains. (See Example 4.)

41. $f(x) = x + 1$, $g(x) = x^2$
42. $f(x) = 2x - 1$, $g(x) = x^3$
43. $f(x) = x^{10}$, $g(x) = x - 1$
44. $f(x) = x^{40}$, $g(x) = 2x + 5$
45. $f(x) = 2x^2 + x$, $g(x) = x + 3$
46. $f(x) = x - 1$, $g(x) = 3x^2 + 4x - 1$
47. $f(x) = \sqrt{x}$, $g(x) = x - 3$
48. $f(x) = \sqrt{x - 1}$, $g(x) = 2x$
49. $f(x) = 1/x$, $g(x) = 1 - x$
50. $f(x) = 1/(x + 2)$, $g(x) = x - 1$

51. $f(x) = \sqrt{x+2}$, $g(x) = x - 3$
52. $f(x) = \sqrt{x+4}$, $g(x) = x - 5$
53. $f(x) = 3x - 5$, $g(x) = (x+5)/3$
54. $f(x) = 7x - 1$, $g(x) = (x+1)/7$
55. $f(x) = 1/(x+1)$, $g(x) = (1-x)/x$
56. $f(x) = 2/(x-1)$, $g(x) = (x+2)/x$
57. $f(x) = \sqrt{x-1}$, $g(x) = x^2 + 1$
58. $f(x) = \sqrt[3]{x+2}$, $g(x) = x^3 - 2$

Determine two functions f and g such that $(f \circ g)(x)$ is the given function.

59. $(f \circ g)(x) = x^3 - 1$
60. $(f \circ g)(x) = (x-1)^3$
61. $(f \circ g)(x) = \sqrt{x+3}$
62. $(f \circ g)(x) = \sqrt{x} + 3$
63. $(f \circ g)(x) = (2x-9)^{50}$
64. $(f \circ g)(x) = (4x+7)^{23}$

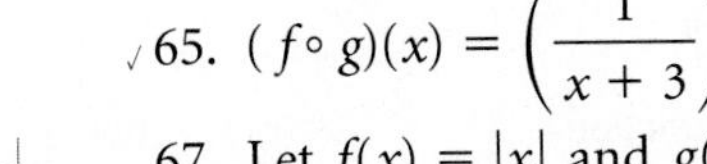

65. $(f \circ g)(x) = \left(\dfrac{1}{x+3}\right)^3$
66. $(f \circ g)(x) = \left(\dfrac{1}{1-x}\right)^4$

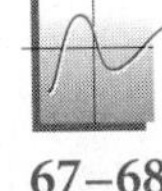

67–68

67. Let $f(x) = |x|$ and $g(x) = \sqrt{x}$.
    a. Graph $Y_1 = (f \circ g)(x-2)$ and $Y_2 = (g \circ f)(x-2)$.
    b. Compare the graphs of Y_1 and Y_2.
68. Repeat Exercise 67 with $f(x) = |x|$ and $g(x) = \sqrt[3]{x}$.

Critical Thinking: Writing

69. Describe physical situations that could be expressed as each of the following combinations of the functions f and g.
    a. $f + g$ b. $f - g$ c. $f \cdot g$ d. f/g e. $f \circ g$

Critical Thinking: Writing

70. Write a word description that tells how the value of $(f \circ g)(x)$ can be found.

◆ Solution for Practice Problem

1. a. $(g \circ f)(3) = g(f(3)) = g(\sqrt{2}) = 1$
   b. $(g \circ f)(x) = g(f(x)) = g(\sqrt{x-1}) = (\sqrt{x-1})^2 - 1 = x - 2$, where $x \geq 1$

3.4 LINEAR FUNCTIONS

One of the simplest and most useful types of functions is the linear function. A linear function is a function with a special type of defining equation.

Definition of Linear Function

A **linear function** is a function with a defining equation that is a *linear equation.*

There are several different but equivalent ways to define a linear function. Graphically, a linear function is a function that has a straight line for its graph. Any *nonvertical* straight line is the graph of a linear function, but any *vertical* straight line has many different y-values for the same x-value and therefore is *not* the graph of a function. This discussion is illustrated in Figure 3.19.

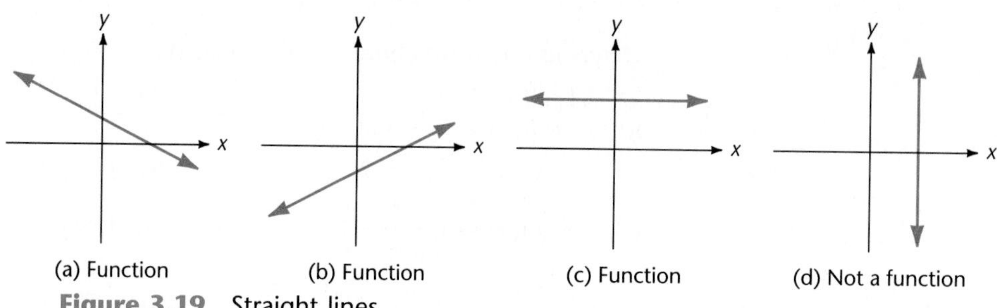

Figure 3.19 Straight lines

The defining equation for any linear function can be written in the form

$$Ax + By = C,$$

and for each x-value there must be exactly one y-value. Clearly, this will be the case if and only if it is possible to solve for y in the equation $Ax + By = C$, and this is possible if and only if $B \neq 0$.

A given straight line is the graph of many different equations because the linear equation $Ax + By = C$ can be written in many equivalent ways. The form

$$Ax + By = C$$

is called a **standard form** of the equation of a straight line, but other forms are sometimes more useful. Two of these other forms are presented in this section.

Suppose that a line has slope m and that b is the y-intercept. Then $(0,b)$ is on the line, and any other point (x, y) is on the line if and only if

$$\frac{y - b}{x - 0} = m.$$

Solving for y, we have

$$y = mx + b,$$

and this is an equation for the line. This discussion is summarized in the following statement.

Slope-Intercept Form

An equation of the straight line that has slope equal to m and y-intercept equal to b is

$$y = mx + b.$$

This form of the equation of the line is called the **slope-intercept form.**

Example 1 • Using Slope-Intercept Form

Determine the slope of the straight line with equation $2x + 3y = 6$.

Solution

We rewrite the equation in slope-intercept form as follows:

$$\begin{aligned} 2x + 3y &= 6 \\ 3y &= -2x + 6 && \text{subtracting } 2x \\ y &= -\tfrac{2}{3}x + 2 && \text{dividing by 3} \end{aligned}$$

Thus the slope m is $-\frac{2}{3}$. □

Another special form for the equation of a straight line is the point-slope form. This can be stated as follows.

Point-Slope Form

An equation of the straight line that has slope m and passes through the point (x_1, y_1) is

$$y - y_1 = m(x - x_1). \tag{1}$$

This form of the equation of the line is called the **point-slope form.**

To see why Equation (1) is valid, suppose that m is the slope of the line and (x_1, y_1) is a point on the line. If (x, y) is any other point on the line, then

$$\frac{y - y_1}{x - x_1} = m,$$

by the definition of the slope. When both sides of this equation are multiplied by $x - x_1$, we have

$$y - y_1 = m(x - x_1),$$

and this is the stated equation. We note that this last equation is satisfied when $x = x_1$ and $y = y_1$, so every point on the line, including (x_1, y_1), satisfies Equation (1).

Example 2 • Using Point-Slope Form

Find an equation of the straight line that has slope -2 and passes through the point $(-1, 4)$, and draw the graph.

Solution

Using $m = -2$ and $(x_1, y_1) = (-1, 4)$ in the point-slope form of the equation, we have

$$y - 4 = -2(x + 1)$$

as an equation of the line. The graph is drawn in Figure 3.20. □

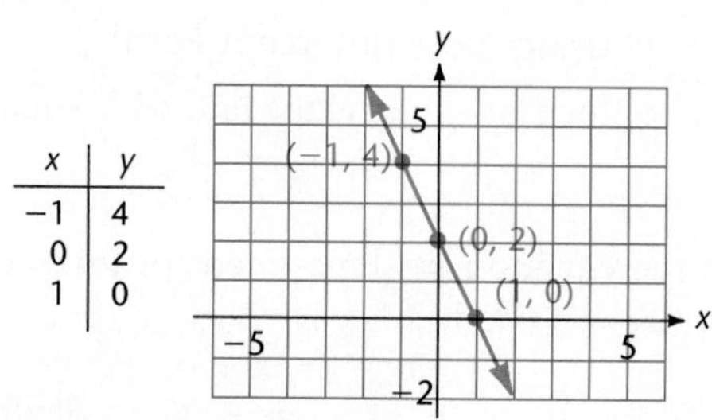

Figure 3.20 Graph of $y - 4 = -2(x + 1)$

Example 3 • Straight Line Through Two Given Points

Find an equation of the straight line passing through the points $(-1, 3)$ and $(2, -4)$.

Solution

Here we are given two distinct points, and any two distinct points determine a unique line. We can find the slope of the line by using these two points in the slope formula. We get

$$m = \frac{-4 - 3}{2 - (-1)} = -\frac{7}{3}.$$

Now we can use the point-slope form with this slope and either of the points $(-1, 3)$ or $(2, -4)$ to write an equation of the line. Using $(-1, 3)$, we have

$$y - 3 = -\tfrac{7}{3}(x + 1).$$

On the other hand, using $(2, -4)$, we have

$$y + 4 = -\tfrac{7}{3}(x - 2).$$

Both of these equations reduce to $7x + 3y = 2$. □

It is easy to see that **two nonvertical lines are parallel if and only if they have the same slope.** Slopes can also be used to determine whether or not two nonvertical lines are perpendicular to each other. If line ℓ_1 has slope m_1 and line ℓ_2 has slope m_2, then **ℓ_1 and ℓ_2 are perpendicular to each other if and only if $m_1 m_2 = -1$.** In other words, *ℓ_1 and ℓ_2 intersect at right angles if and only if the slope of one line is the negative reciprocal of the slope of the other line.*

Example 4 • Parallel Lines

Find an equation of the line that is parallel to the line $-3x + y = 7$ and contains the point $(-1, 1)$.

Solution

Solving for y in $-3x + y = 7$, we have

$$y = 3x + 7,$$

so the slope of this line is 3. Thus we are seeking an equation of the line with slope 3 that passes through $(-1, 1)$. Using the point-slope form, we have

$$y - 1 = 3(x + 1)$$

as an equation of the line. □

◆ Practice Problem 1 ◆

Find an equation of the line that is perpendicular to the line $-3x + y = 7$ and has -5 as the y-intercept.

Three special forms of an equation of a straight line have been presented in this section. These forms are listed here together with their names and key facts:

Special Form	Name of Form and Relevant Facts
$Ax + By = C$	**Standard form;** at least one of A and B is not zero.
$y = mx + b$	**Slope-intercept form;** slope is m, y-intercept is b.
$y - y_1 = m(x - x_1)$	**Point-slope form;** slope is m, passes through (x_1, y_1).

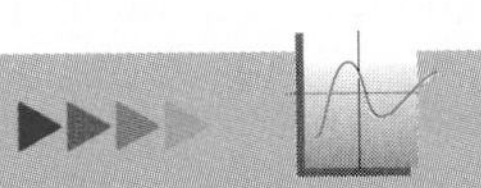

The instruction booklet for a graphing device describes the procedure for graphing a piecewise-defined function.

Some functions have graphs that are composed of parts of lines instead of one complete line. Such graphs are sometimes called **broken-line graphs,** and the functions are said to be defined **piecewise.** A function with a graph of this type is given in the next example.

Example 5 • A Broken-Line Graph

Draw the graph of the function f defined piecewise by

$$f(x) = \begin{cases} x + 1, & \text{if } x \leq 2 \\ 2 - \dfrac{x}{2}, & \text{if } x > 2 \end{cases}.$$

Solution

For $x \leq 2$, the graph of f coincides with the line $y = x + 1$, and for $x > 2$, it coincides with the line $y = 2 - x/2$. A table of values and the graph are shown in Figure 3.21. Note that the point $(2, 1)$ is plotted as an open circle to show that it is not on the graph of f. □

x	y
−1	0
2	3
3	$\frac{1}{2}$
4	0

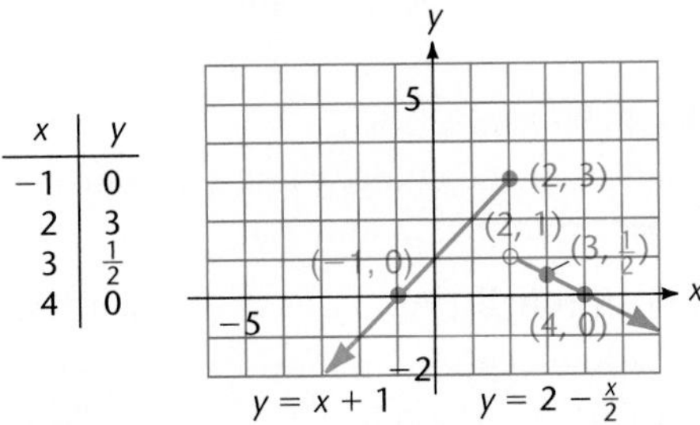

Figure 3.21 Graph of f in Example 5

Most businesses use equipment that loses its value over a period of time. Such a loss in value is called **depreciation.** We define the net cost N to be the difference between the original cost C and the scrap value S:

$$N = C - S.$$

Suppose that an item has a life span of n years. One of the simplest methods for computing depreciation is based on the assumption that an item depreciates $1/n$ of its net cost each year so that after n years it is totally depreciated. This method is called **linear depreciation** or **straight-line depreciation.**

Example 6 • Straight-Line Depreciation

Suppose a bagging machine cost \$42,000, has a life span of 5 years, and can be scrapped for \$12,000. What is the amount of the annual straight-line depreciation? Determine the linear function f that gives the undepreciated value of the machine during its life span and sketch its graph.

Solution

The net cost N is the difference between the original cost and the scrap value. Thus

$$N = \$42{,}000 - \$12{,}000 = \$30{,}000.$$

Over 5 years the bagging machine will depreciate

$$\frac{\$30{,}000}{5} = \$6000 \text{ annually.}$$

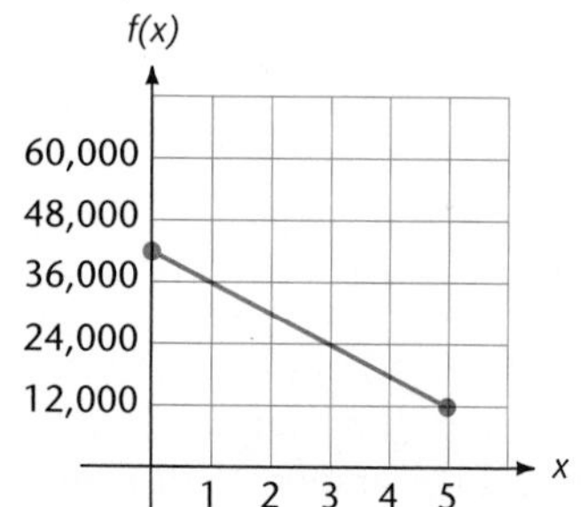

Figure 3.22 Graph of $f(x) = -6000x + 42{,}000$, $x \in [0, 5]$

To determine the linear function that gives the straight-line depreciation, we first let x represent the number of years after the purchase of the bagging machine and let $f(x)$ be the corresponding undepreciated value of the machine. Hence x is restricted to the interval $[0, 5]$. Setting $y = f(x)$, we note that when $x = 0$, $y =$ \$42,000, and when $x = 5$, $y =$ \$12,000. The straight line in Figure 3.22 drawn through these two points has the equation

$$y = -6000x + 42{,}000, \qquad \text{for } x \in [0, 5].$$

Thus the linear function f that gives the undepreciated value for year x is

$$f(x) = -6000x + 42{,}000, \qquad \text{for } x \in [0, 5].$$ □

EXERCISES 3.4

Find an equation of the straight line that satisfies the given conditions and draw the graph. (See Examples 2 and 3.)

1. Slope -1, through $(-1, 1)$
2. Through $(-1, 3)$, slope -2
3. Slope 2, y-intercept 4
4. Slope 3, x-intercept -5
5. Through $(-7, 5)$, slope 0
6. Through $(3, 4)$, slope does not exist

Find an equation, in standard form, of the straight line that satisfies the given conditions. (See Examples 1–4 and Practice Problem 1.)

7. y-intercept 3, slope $\frac{1}{2}$
8. Slope -7, through $(3, -1)$
9. Through $(3, 4)$ and $(4, 3)$
10. Through $(2, 2)$ and $(-1, 3)$
11. x-intercept 7, slope 4
12. x-intercept -3, slope -2
13. x-intercept 2, y-intercept -1
14. y-intercept 3, x-intercept -3
15. y-intercept 3, through $(-1, -1)$
16. y-intercept -2, through $(-2, -3)$
17. y-intercept 3, no x-intercept
18. y-intercept -7, slope 0
19. x-intercept 5, slope does not exist
20. x-intercept -3, no y-intercept
21. Through $(-2, 3)$, parallel to $3x - y = 6$
22. Through $(-7, 2)$, parallel to $-2x = -3y + 7$
23. Through $(-5, -3)$, perpendicular to $5x = 6y - 1$
24. Through $(-1, 3)$, perpendicular to $5x - 2y = 11$
25. Through $(0, 0)$, perpendicular to $y = x$
26. x-intercept 3, perpendicular to $3x - y = 2$
27. y-intercept -1, parallel to $5x - 7y = 5$
28. x-intercept 5, parallel to $9x - 7 = 3y$

Find an equation, in slope-intercept form, of the straight line whose graph is given.

29.

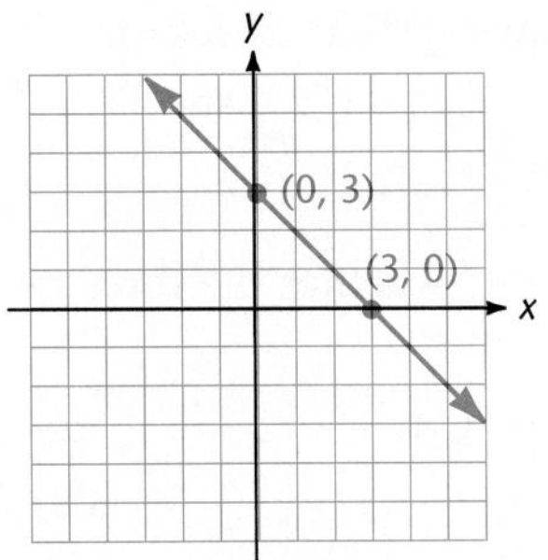

30.

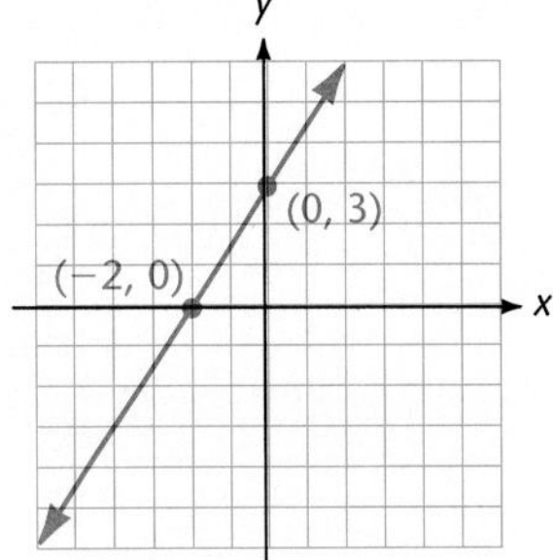

31.

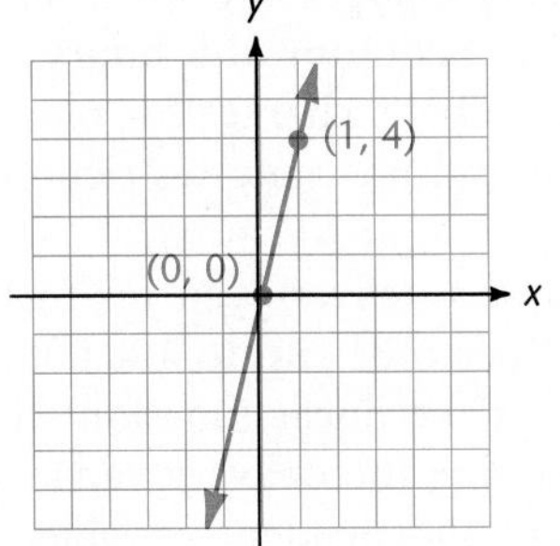

32.

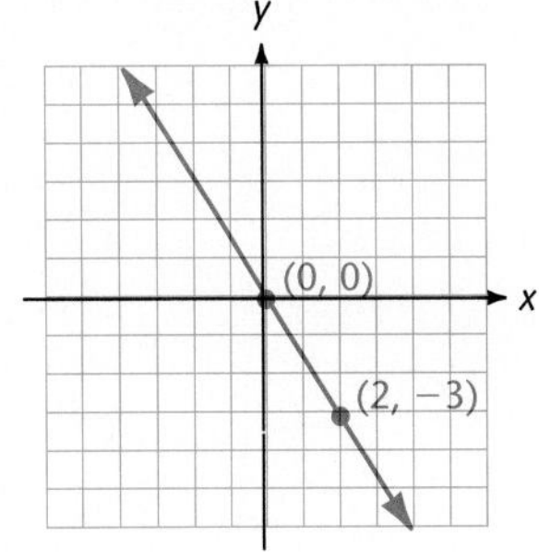

33.

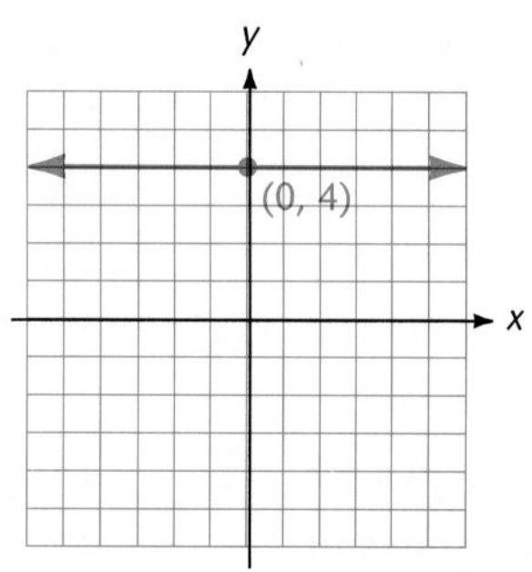

34.

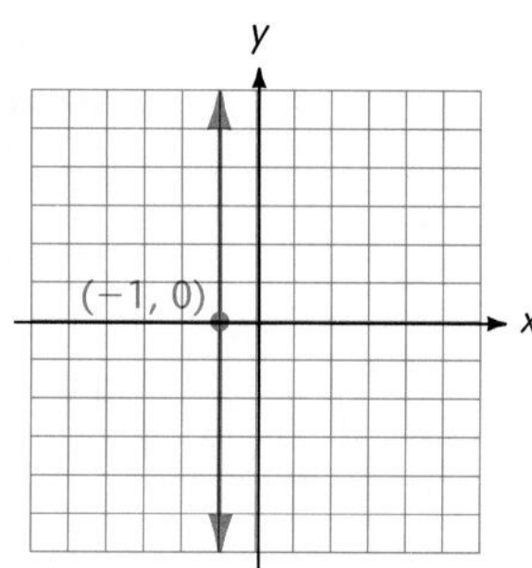

Graph each of the following functions. (See Example 5.)

35. $f(x) = \begin{cases} 1 - 2x, & \text{if } x < 0 \\ 1 + 2x, & \text{if } x \geq 0 \end{cases}$

36. $f(x) = \begin{cases} 2 - 3x, & \text{if } x < 1 \\ -1, & \text{if } x \geq 1 \end{cases}$

37. $f(x) = \begin{cases} 2 + x, & \text{if } x \leq 2 \\ 2 - x, & \text{if } x > 2 \end{cases}$

38. $f(x) = \begin{cases} 3x + 4, & \text{if } x < -1 \\ 1 - x, & \text{if } x \geq -1 \end{cases}$

39. $f(x) = \begin{cases} -5 - 3x, & \text{if } x \leq -2 \\ 1 - x, & \text{if } x > -2 \end{cases}$

40. $f(x) = \begin{cases} 4 - 2x, & \text{if } x \leq 3 \\ 2 - x, & \text{if } x > 3 \end{cases}$

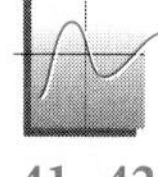

41–42

Use a graphing device to graph each of the following functions.

41. $f(x) = \begin{cases} 1 - x, & \text{if } x < -3 \\ |1 - x^2|, & \text{if } x \geq -3 \end{cases}$

42. $f(x) = \begin{cases} 2 + \sqrt{4x - 1}, & \text{if } x \geq 2 \\ -|x|, & \text{if } x < 2 \end{cases}$

43. Do the points $(-1, 3)$, $(4, 7)$, and $(6, 10)$ lie on the same straight line? Justify your answer. (*Hint:* Check the slopes between the points.)

44. Do the points $(3, 7)$, $(-1, 2)$, and $(7, 12)$ lie on the same straight line? Justify your answer.

Use slopes to determine whether the three given points are the vertices of a right triangle.

45. $(-1, 5)$, $(1, 2)$, $(-2, 0)$

46. $(-3, 3)$, $(1, 4)$, $(-1, -5)$

47. $(0, 6)$, $(-3, 0)$, $(3, 0)$

48. $(-2, 3)$, $(-7, 0)$, $(3, 1)$

Plumber's Charges

49. Suppose a plumber charges \$30 for a house call plus \$28 per hour while she is there for a maximum of 8 hours. Write out the linear function f that gives the plumber's charges for one house call in terms of the number of hours worked and note any restrictions on the independent variable.

Delivery Van Expense

50. A florist purchases a delivery van for \$20,000. Suppose it costs an additional \$1.50 per mile (up to 100,000 miles) to operate and maintain the van. Write out the linear function whose value is the total expense of buying and operating the delivery van in terms of mileage. Note any restrictions on the independent variable.

Depreciation of a Car 51. A new car is purchased for \$18,000. After 6 years of use it is traded in for a value of \$3000. Write out the linear function f for the undepreciated value of the car in terms of the number of years since its purchase. Assume the car depreciated linearly and note any restrictions on the independent variable.

Depreciation of a Copy Machine 52. A company purchases a copy machine for \$2500 and uses it for 4 years. If the trade-in value of the machine is \$1100 and the company depreciates it linearly, write out the function f that gives the undepreciated value of the machine and state the domain of the independent variable.

Slope of a Stairway 53. A typical slope for a household stairway is $\frac{7}{9}$. Referring to the accompanying figure, if the treads are 12 inches wide, how high are the risers?

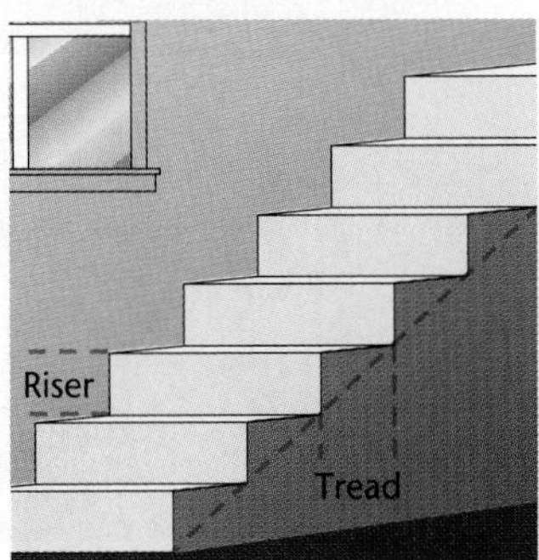

Figure for Exercises 53 and 54

Required Space for a Stairway 54. If the upper floor is 9 feet 4 inches above the lower floor in the accompanying figure, how much horizontal distance must be allowed for the stairway?

Grade of a Highway 55. The slope of a railroad or a highway is called its **grade** and is usually written as a percent. For example, a 5 percent grade means that the slope is $\frac{5}{100}$. Suppose the grade of a straight roadway is 4 percent. How far has a truck traveled along the road if its altitude has changed from 3402 feet above sea level to 3349 feet above sea level?

Incline of a Roller Coaster 56. Suppose the steepest incline on a roller coaster has a 30 percent grade. (See Exercise 55.) On this incline the passengers in the front seat are 9 feet above those in the rear seat. How long is the roller coaster train?

Critical Thinking: Exploration and Writing

57. Suppose the coordinates of two points on a straight line are given, with neither of them on a coordinate axis. Decide whether the standard form, slope-intercept form, or point-slope form is easiest to use to find an equation of the straight line and write an explanation for your choice.

Critical Thinking: Exploration and Writing

58. Some college algebra texts define a **standard form** of an equation of a straight line to be the form $ax + by + c = 0$ instead of $Ax + By = C$. Decide which of these forms is easier to use when finding the intercepts of the straight line and explain the basis for your decision.

◆ Solution for Practice Problem

1. The line $-3x + y = 7$ has slope 3. In order for a line t[illegible]e perpendicular to $-3x + y = 7$, it must have slope equal to $-\frac{1}{3}$. Using $m =$ [illegible]$\frac{1}{3}$ and $b = -5$ in the slope-intercept form, we obtain

$$y = -\tfrac{1}{3}x - 5$$

as an equation of the line.

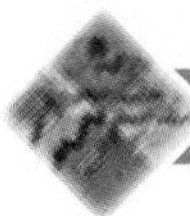

3.5 QUADRATIC FUNCTIONS

Although they are not as simple as linear functions, quadratic functions are fully as useful and important because they can often be used satisfactorily to represent a quantity in real life.

Definition of a Quadratic Function

A function f is a **quadratic function** if there are real numbers a, b, and c, with $a \neq 0$, such that the function value for f is given by

$$f(x) = ax^2 + bx + c.$$

That is, a quadratic function is a function that has a defining equation of the form

$$y = ax^2 + bx + c, \qquad a \neq 0.$$

Example 1 • Identifying Quadratic Functions

Which of the following equations define a quadratic function?

a. $y = 2x^2$ b. $2x + x^2 = y$ c. $y^2 = 4x$

d. $x^3 + 4x = y$ e. $y = 2x + 7$

Solution

The equations in parts a and b can be written in the form $y = ax^2 + bx + c$, with $a \neq 0$, so each of them defines a quadratic function. The relation defined by $y^2 = 4x$ in part c is not a function and so is not a quadratic function. The equations in parts d and e cannot be written in the required form, so they do not define quadratic functions. □

The graph of a quadratic function is called a **parabola.** We have seen the graph of the quadratic function $y = x^2$ in Figure 3.2 (Example 1 in Section 3.1). Another example is provided below.

Example 2 • Graphing a Quadratic Function

Sketch the graph of $y = 2x^2 - 12x + 17$.

Solution

One technique that can be used to great advantage in graphing a quadratic function is to complete the square on the x-terms, using the same sort of procedure that was used in solving quadratic equations in Section 2.4. Instead of dividing both sides by the coefficient of x^2, we factor the coefficient away from the x-terms in the first step.

$$\begin{aligned} y &= 2x^2 - 12x + 17 \\ &= 2(x^2 - 6x + \qquad) + 17 \qquad \text{factoring 2 out of the } x\text{-terms} \end{aligned}$$

To complete the square on $x^2 - 6x$, we add $[\frac{1}{2}(-6)]^2 = 9$ inside the parentheses. In doing this, we are actually adding $2(9) = 18$, so we must subtract 18 from 17 to have an equivalent equation.

$$y = 2(x^2 - 6x + 9) + 17 - 18 \qquad \text{Adding } 2(9) \text{ and subtracting } 18$$
$$= 2(x - 3)^2 - 1 \qquad \text{rewriting with a perfect square}$$

This form of the equation gives some important information about the graph. Since $2(x - 3)^2 \geq 0$ for all x, the smallest possible value for y is -1, and this occurs when $2(x - 3)^2 = 0$, that is, when $x = 3$. This means that *the lowest point on the graph is at* $(3, -1)$.

We observe now that $(x - 3)^2 = |x - 3|^2$ is the square of the distance from x to 3; this means we get the same value for y when we move a certain distance to the left from $x = 3$ as we get when we move the same distance to the right from $x = 3$. In other words, *the graph is symmetric, or balanced, with respect to the line* $x = 3$.

We now assign some values to x and compute the corresponding y-values. These are recorded in the table in Figure 3.23, and a sketch of the graph is shown there. The extreme (lowest) point on the graph is called the **vertex.** □

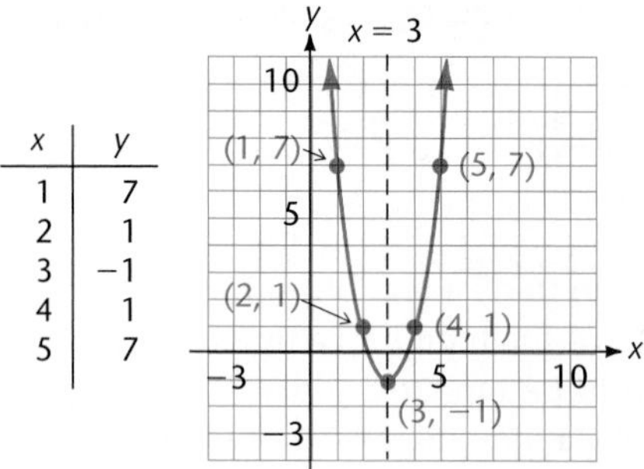

x	y
1	7
2	1
3	−1
4	1
5	7

Figure 3.23 Graph of $y = 2x^2 - 12x + 17$

As mentioned just before Example 2, the graph of a quadratic function is called a **parabola.** The parabolas in Figures 3.2 and 3.23 are typical as to shape. Parabolas always have a "bullet" shape, with an extreme point (the **vertex**), and a **line of symmetry** through the vertex. However, the vertex may be the highest point on the graph instead of the lowest point. The graph opens downward and has a highest point when $a < 0$ in $y = ax^2 + bx + c$. A parabola of this type is shown in Example 3.

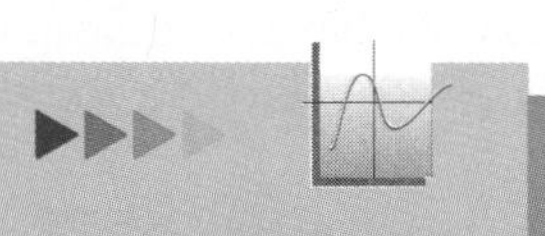

The cursor on a graphing device can be used with the **trace** feature to locate the vertex with a reasonable degree of accuracy.

The discussion given in Example 2 can be extended to the general case.

Graph of a Quadratic Function: $y = a(x - h)^2 + k$

If the equation $y = ax^2 + bx + c$ is rewritten as $y = a(x - h)^2 + k$, then

1. The vertex is at (h, k).
2. The line $x = h$ is an axis of symmetry.
3. The graph opens upward if $a > 0$ and downward if $a < 0$.

Example 3 • Using the Vertex and Axis of Symmetry

Sketch the graph of $y = -2x^2 + 8x - 7$.

Solution

We first rewrite the equation, completing the square on the x-terms to obtain the form $y = a(x - h)^2 + k$.

$$\begin{aligned} y &= -2(x^2 - 4x) - 7 \\ &= -2(x^2 - 4x + 4) - 7 + 8 \\ &= -2(x - 2)^2 + 1 \end{aligned}$$

We observe that $a = -2$, $h = 2$, and $k = 1$. The vertex is at $(2, 1)$, the line $x = 2$ is an axis of symmetry, and the parabola opens downward since $a < 0$. With this much knowledge of the graph, it is sufficient to plot five points and then sketch the graph. (See Figure 3.24.) In selecting these points, it is efficient to make use of symmetry. □

x	y
0	−7
1	−1
2	1
3	−1
4	−7

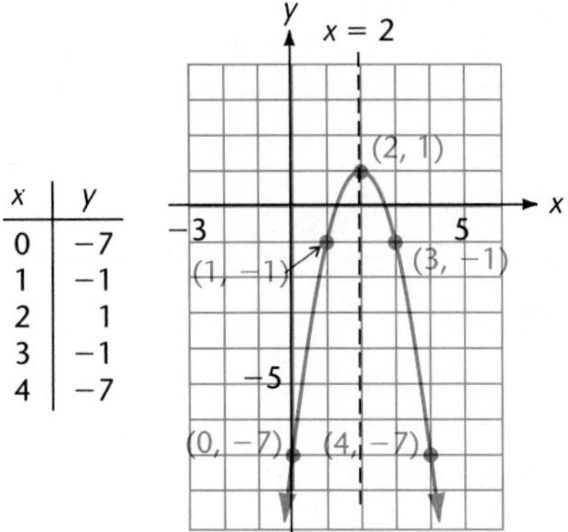

Figure 3.24 Graph of $y = -2x^2 + 8x - 7$

It is possible to obtain formulas that give the vertex and the axis of symmetry for

$$y = ax^2 + bx + c, \qquad a \neq 0,$$

in terms of a, b, and c. All we need do is complete the square in the general case. We have

$$\begin{aligned} y &= a\left(x^2 + \frac{b}{a}x\right) + c \\ &= a\left(x^2 + \frac{b}{a}x + \frac{b^2}{4a^2}\right) + c - \frac{b^2}{4a} \\ &= a\left(x + \frac{b}{2a}\right)^2 + \frac{4ac - b^2}{4a}. \end{aligned}$$

Careful examination of this equation leads to the following statements.

Graph of a Quadratic Function: $y = ax^2 + bx + c$

The graph of $y = ax^2 + bx + c$, $a \neq 0$, is a parabola that

1. has its vertex at $\left(-\frac{b}{2a}, f\left(-\frac{b}{2a}\right)\right)$, where $f(x) = ax^2 + bx + c$,
2. has the line $x = -\frac{b}{2a}$ as an axis of symmetry,
3. opens upward if $a > 0$ and downward if $a < 0$.

When the value $x = -\frac{b}{2a}$ is used to find the y-coordinate of the vertex, it turns out that

$$y = f\left(-\frac{b}{2a}\right) = \frac{4ac - b^2}{4a}.$$

It is *not* necessary to memorize this value because it can be readily obtained by substituting $x = -\frac{b}{2a}$ into the equation $y = ax^2 + bx + c$.

◆ Practice Problem 1 ◆

Sketch the graph of $y = -2x^2 - 2x + 4$.

It is often useful to find the intercepts for a parabola $y = ax^2 + bx + c$. The y-intercept is easy: when $x = 0$, $y = c$. In looking for the x-intercepts, we are at once confronted with the problem of solving

$$ax^2 + bx + c = 0.$$

Recall from Section 2.5 that this equation may have two equal real roots, two unequal real roots, or no real roots, depending on the discriminant $b^2 - 4ac$. The graphical interpretation of this situation is that the parabola may be tangent to the x-axis at the vertex, or it may cross the x-axis in two places, or it may not cross the x-axis at all. All three cases are illustrated in Figure 3.25.

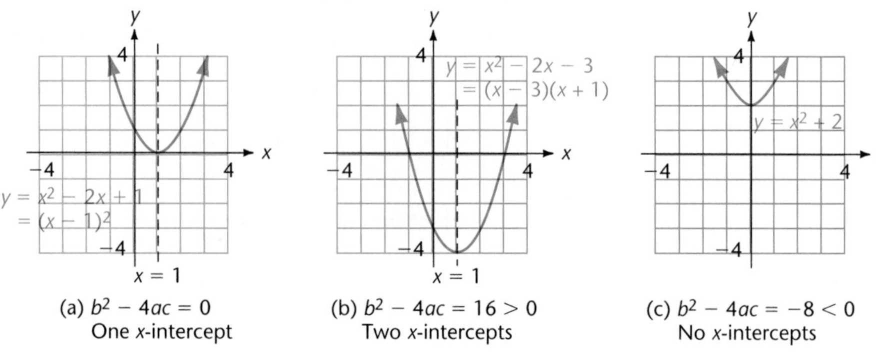

Figure 3.25 Possibilities for the x-intercepts

Since the y-coordinate of the vertex of a parabola represents the extreme value (that is, the maximum or minimum function value) of the function $y = f(x) = ax^2 + bx + c$, the parabola is often useful in solving applied problems in which it is necessary to determine the extreme value of some quadratic function. We illustrate the technique in the next example.

Example 4 • Maximizing an Area

The Vandever sisters wish to fence off part of their pasture in a rectangle for their prize-winning bull. One side is bordered by a straight river and they need no fence there. Using 120 yards of fencing, what should the dimensions of the rectangle be if they want to maximize the grazing area for their bull?

Solution

Suppose we let the dimensions of the rectangle in Figure 3.26a be length $= \ell$ and width $= w$. Since there is a total of 120 yards of fencing, we can express ℓ in terms of w as $\ell = 120 - 2w$. The area A of the rectangle is

$$\begin{aligned} A &= \ell w \\ &= (120 - 2w)w \\ &= -2w^2 + 120w. \end{aligned}$$

But this is the equation of the parabola in Figure 3.26b. Since the parabola opens downward, the second coordinate of the vertex represents the maximum area. Hence we see that the area is a maximum of 1800 square yards when the dimensions of the rectangle are $w = 30$ yards and $\ell = 120 - 2(30) = 60$ yards. □

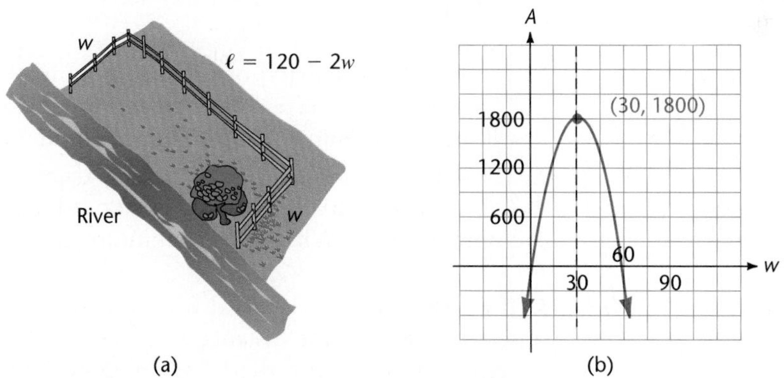

Figure 3.26 The area function and its graph

EXERCISES 3.5

Graph each of the following parabolas. Locate the vertex and the axis of symmetry, and plot at least two other points on each graph. (See Examples 2 and 3.)

1. $y = x^2 - 4$
2. $y = -2x^2 + 6$
3. $y = \frac{2}{3}(x - 2)^2 - 1$
4. $y = -\frac{4}{5}(x + 1)^2 + 2$
5. $y = 4x - x^2$
6. $y = x^2 - 9x$

7. $y = x^2 + 8x + 13$
8. $y = 5 + 6x + x^2$
9. $y = -x^2 - 4x - 1$
10. $y = -x^2 + 6x - 6$
11. $y = 3x^2 + 12x + 17$
12. $y = 2x^2 + 8x + 9$
13. $y = -2x^2 + 4x - 5$
14. $y = -4x^2 - 4x + 1$

Find the extreme value of the given function and state whether that value represents a maximum or a minimum. (See Example 4.)

15. $f(x) = x^2 - 8x + 5$
16. $f(x) = x^2 + 6x + 10$
17. $f(x) = x^2 + 10x - 20$
18. $f(x) = x^2 + 4x - 9$
19. $f(x) = -x^2 + 6x + 3$
20. $f(x) = -x^2 + 8x - 6$
21. $f(x) = -2x^2 + 4x - 3$
22. $f(x) = -2x^2 + 8x - 11$
23. $f(x) = -9 - 6x - x^2$
24. $f(x) = -16 - 8x - x^2$
25. $f(x) = 3x^2 + 9x + 14$
26. $f(x) = 2x^2 - 5x + 7$

Maximum Product 27–28

27. Find two positive numbers whose sum is 54 and whose product is a maximum.
28. Find two positive numbers whose sum is 92 and whose product is a maximum.

Maximum Area 29–30

29. Find the dimensions of a rectangle with maximum area whose perimeter is 48 feet.
30. Find the dimensions of a rectangle with maximum area whose perimeter is 88 feet.

Minimum Cost 31–32

31. The cost C (in dollars) of producing x units of a certain item is given by $C = 0.001x^2 - 0.5x + 800$. Find the production level x for which the cost will be a minimum. What is the minimum cost?
32. The cost C (in dollars) of producing x units of a certain item is given by $C = 0.002x^2 - 3x + 9000$. Find the production level x that yields a minimum cost. What is the minimum cost?

Maximum Height 33–34

33. If a rock is thrown upward from the ground with an initial velocity of 64 feet per second, the distance s in feet of the rock from the ground after t seconds is $s = -16t^2 + 64t$. When does the rock reach its maximum height, and what is that maximum height?
34. Do Exercise 33 if the initial velocity is 128 feet per second and s is given by $s = -16t^2 + 128t$.

Maximum Area 35–38

35. A rectangular dog pen is fenced off along the wall of a house, with no fence needed along the wall as indicated in the accompanying figure. If 40 feet of fence is available, find the dimensions of the dog pen with maximum area.

Figure for Exercise 35

36. A rectangular dog pen is to be made in the corner of a barn. There is 16 feet of fencing material available for the remaining two sides of the pen. Determine the dimensions of the pen with maximum area.
37. Suppose the window in the figure is made by placing a semicircle of radius r atop a rectangle with a height of h. Determine the dimensions r and h that yield the maximum area if the outside perimeter of the glass in the window is 20 feet.

Figure for Exercise 37

38. The mirror in the accompanying figure is constructed by placing two semicircular pieces of radius r against a rectangular piece of width w. Determine the value of r that yields the maximum area of the rectangular piece if the outside perimeter of the mirror is 40 feet.

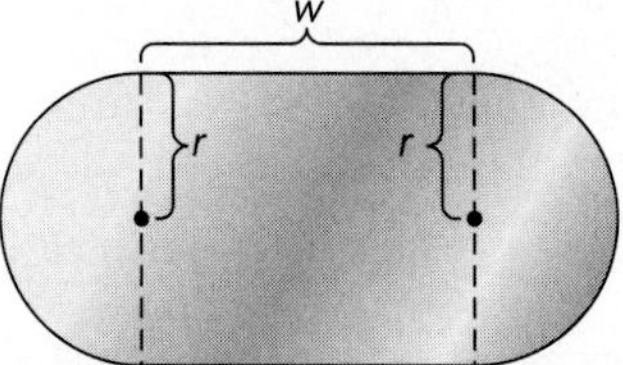

Figure for Exercise 38

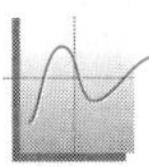

39–42

Use the graph (changing the viewing rectangle as needed) to find the x-intercepts and the vertex of the given parabola. Round the coordinates to the nearest hundredth.

39. $y = x^2 - 4x + 3.96$

40. $y = x^2 + 7x - 120$

Graph each of the following functions.

41. $f(x) = \begin{cases} 2x^2 + 1, & \text{if } x \leq 1 \\ \sqrt{8 + x}, & \text{if } x > 1 \end{cases}$

42. $f(x) = \begin{cases} -4x - x^2, & \text{if } x < 1 \\ \dfrac{x^2 - 8x - 8}{3}, & \text{if } x \geq 1 \end{cases}$

Critical Thinking: Exploration and Writing

43. Sketch the graph of each of the following parabolas on the same coordinate system. Describe the effect of k on the graph of $y = x^2 + k$.
   a. $y = x^2$ b. $y = x^2 + 2$ c. $y = x^2 - 2$ d. $y = x^2 + 4$

Critical Thinking: Exploration and Writing

44. Sketch the graph of each of the following parabolas on the same coordinate system. Describe the effect of a on the graph of $y = ax^2$, when $a > 0$.
   a. $y = x^2$ b. $y = 2x^2$ c. $y = 3x^2$ d. $y = \frac{1}{2}x^2$

Critical Thinking: Exploration and Writing

45. Sketch the graph of each of the following parabolas on the same coordinate system. Describe the effect of a on the graph of $y = ax^2$, when $a < 0$.
   a. $y = x^2$ b. $y = -x^2$ c. $y = 2x^2$ d. $y = -2x^2$

Critical Thinking: Exploration and Writing

46. Sketch the graph of each of the following parabolas on the same coordinate system. Describe the effect of h on the graph of $y = (x - h)^2$.
   a. $y = x^2$ b. $y = (x - 1)^2$ c. $y = (x + 1)^2$ d. $y = (x - 2)^2$

◆ Solution for Practice Problem

1. We have $y = ax^2 + bx + c$, with $a = -2$, $b = -2$, and $c = 4$. Using the formula for the x-coordinate of the vertex, we get

$$x = -\frac{b}{2a} = -\frac{-2}{-4} = -\frac{1}{2}.$$

The corresponding value for y is found by substituting $x = -\frac{1}{2}$ into the equation of the parabola.

$$y = -2(-\tfrac{1}{2})^2 - 2(-\tfrac{1}{2}) + 4 = -\tfrac{1}{2} + 1 + 4 = \tfrac{9}{2}.$$

Thus the vertex is located at $(-\frac{1}{2}, \frac{9}{2})$. The graph opens downward since $a = -2 < 0$. A table of values is shown with the graph in Figure 3.27.

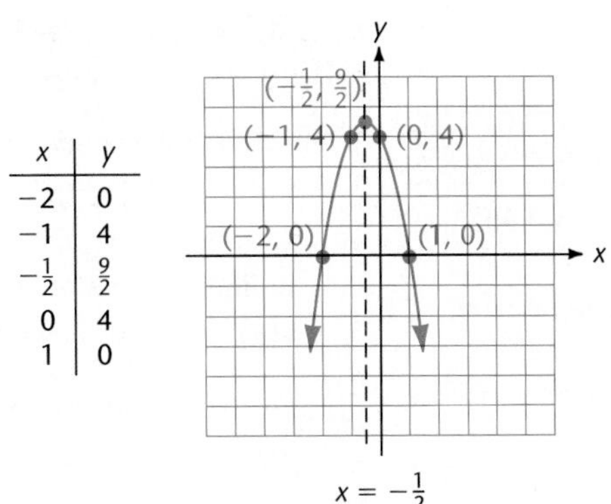

Figure 3.27 Graph of $y = -2x^2 - 2x + 4$

CHAPTER REVIEW

Summary of Important Concepts and Formulas

Definition of Slope

The **slope** m of the line through two distinct points (x_1, y_1) and (x_2, y_2) is given by

$$m = \frac{y_2 - y_1}{x_2 - x_1}.$$

Midpoint Formula

The midpoint of the segment joining (x_1, y_1) and (x_2, y_2) has coordinates

$$\left(\frac{x_1 + x_2}{2}, \frac{y_1 + y_2}{2}\right).$$

Definition of a Function

Let D be a nonempty set of real numbers. A **function** f with **domain** D is a correspondence such that for each element x of D, there is one and only one associated element y in a second set of real numbers.

Alternate Definition of a Function

A **relation** is a nonempty set of ordered paris (x, y) of real numbers x and y. The **domain** D of a relation is the set of all first-entry elements, or the set of all x-values, that occur in the relation. The **range** of a relation is the set of all y-values that occur in the relation. A **function** is a relation in which no two distinct ordered pairs have the same x-value.

Definition of a Composition Function

If f and g are functions, the **composition function** $f \circ g$ is defined by

$$(f \circ g)(x) = f(g(x)).$$

The domain of $f \circ g$ is the set of all x in the domain of g such that f is defined at $g(x)$.

Vertical Line Test

If any vertical line intersects a graph at two or more points, the graph *does not* represent a function. On the other hand, if no vertical line intersects the graph in more than one point, the graph *does* represent a function.

Definition of Linear Function

A **linear function** is a function with a defining equation that is a *linear equation.*

Equations of Straight Lines

Special Form	Name of Form and Relevant Facts
$Ax + By = C$	**Standard form;** at least one of A and B is not zero.
$y = mx + b$	**Slope-intercept form;** slope is m, y-intercept is b.
$y - y_1 = m(x - x_1)$	**Point-slope form;** slope is m, passes through (x_1, y_1).

Definition of a Quadratic Function

A function f is a **quadratic function** if there are real numbers a, b, and c, with $a \neq 0$, such that the function value for f is given by

$$f(x) = ax^2 + bx + c.$$

Graph of a Quadratic Function: $y = a(x - h)^2 + k$

If the equation $y = ax^2 + bx + c$ is rewritten as $y = a(x - h)^2 + k$, then

1. The vertex is at (h, k).
2. The line $x = h$ is an axis of symmetry.
3. The graph opens upward if $a > 0$ and downward if $a < 0$.

Graph of a Quadratic Function: $y = ax^2 + bx + c$

The graph of $y = ax^2 + bx + c$, $a \neq 0$, is a parabola that

1. has its vertex at $\left(-\frac{b}{2a}, f\left(-\frac{b}{2a}\right)\right)$, where $f(x) = ax^2 + bx + c$,
2. has the line $x = -\frac{b}{2a}$ as an axis of symmetry,
3. opens upward if $a > 0$ and downward if $a < 0$.

Critical Thinking: Find the Errors

Each of the following nonsolutions has one or more errors. Can you find them?

Problem 1

If $f(x) = x^2 - 1$ and $g(x) = \sqrt{x - 3}$, find the value of $f(g(7))$.

Nonsolution

$$\begin{aligned} f(g(7)) &= (7^2 - 1)(\sqrt{7 - 3}) \\ &= (48)(2) \\ &= 96 \end{aligned}$$

Problem 2

Write the equation, in standard form, of the line with x-intercept 3 and y-intercept -2.

Nonsolution

$$m = \frac{-2}{3} = -\frac{2}{3}$$

$$\begin{aligned} y - (-2) &= -\frac{2}{3}(x - 3) \\ 3(y + 2) &= -2(x - 3) \\ 3y + 6 &= -2x + 6 \\ 2x + 3y &= 0 \end{aligned}$$

Review Problems for Chapter 3

Graph each linear equation and find the x-intercepts and y-intercepts, if they exist.

1. $y = 4$
2. $x = -2$
3. $2x - 5y = 10$

Find the slope, if it exists, of the line through the given pair of points.

4. $(-2, 3)$ and $(5, 3)$
5. $(-2, 3)$ and $(-3, 1)$
6. $(2, -7)$ and $(-1, 5)$
7. $(5, 6)$ and $(5, -4)$

Find the midpoint of the line segment joining each of the following pairs of points.

8. $(5, 2)$ and $(3, 6)$
9. $(-3, 8)$ and $(7, 2)$
10. $(0, -3)$ and $(7, 5)$
11. $(-4, 6)$ and $(-1, -3)$

Find the slope, if it exists, of the line that has the given equation.

12. $3x - y = 10$
13. $x + 3y = 0$
14. $2x + 7y = 14$
15. $5x - 3y = 15$

Sketch the graph of the following equations.

16. $x = -y^2$
17. $x = |y - 2|$

Speed of a Baseball

18. A certain foul tip in baseball was caught 5 seconds after it left the bat. While in flight at the end of t seconds, its speed s in feet per second was given by $s = |80 - 32t|$. Graph this equation, using appropriate restrictions on the variables.

Population of the United States

19. The population P of the United States of America between 1900 and 1950 is given approximatley by the formula $P = 80 + 1.5t$, where P is in millions of people and t is the number of years since 1900. Sketch the graph of this equation with appropriate restrictions on the variables.

Give the domain D and range R of each function.

20. $f(x) = x^2 + 1$
21. $g(x) = \sqrt{x - 9}$
22. $h(x) = |3x + 2|$
23. $q(x) = \dfrac{1}{x - 4}$

If $f(x) = x^2 - 2x$ and $g(x) = \sqrt{x} + 3$, find each of the following.

24. $f(4)$
25. $(f \circ g)(9)$
26. $(f \circ g)(a)$
27. $f(x + h) - f(x)$
28. Sketch the graph of the function $G(x) = [\![x/2]\!]$.

Let $f(x) = \sqrt{x - 3}$ and $g(x) = x - 4$. Determine the following functions and state their domains.

29. a. $f + g$ b. $f - g$
30. a. $f \cdot g$ b. f/g

Decide whether or not the given graph is the graph of a function.

31.

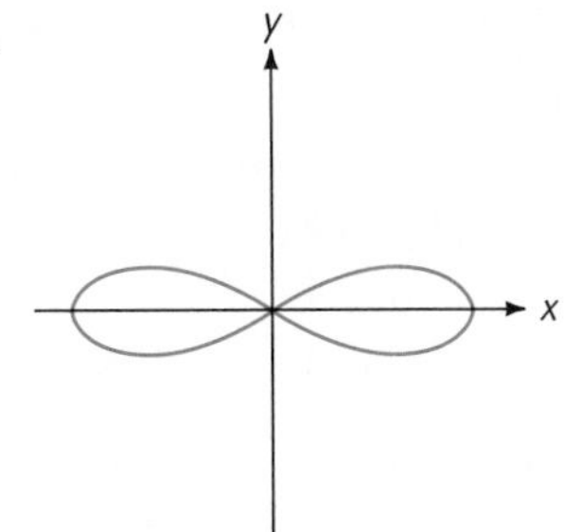

32.

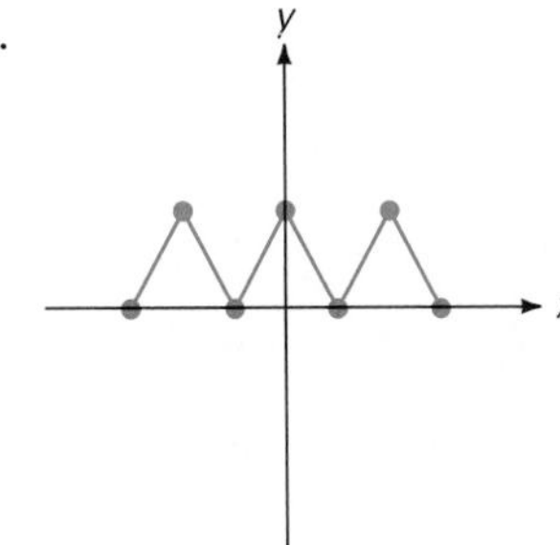

33.

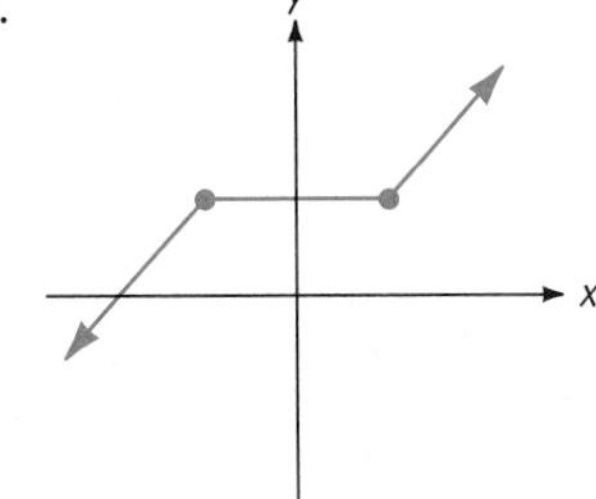

34.

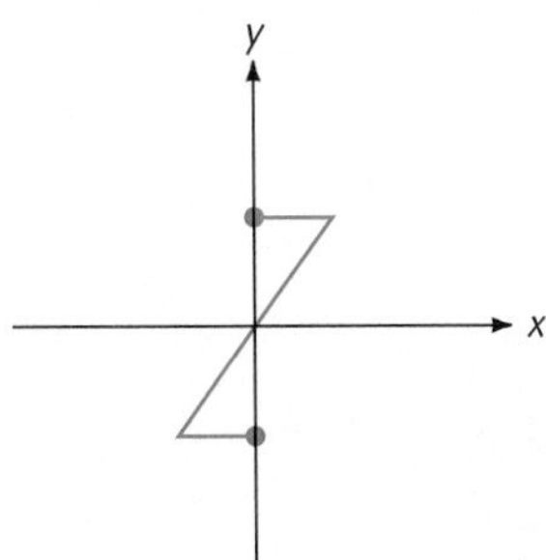

35. Find an equation of the straight line with slope -2 and x-intercept -1, and graph the line.

Find an equation, in standard form, of the straight line that satisfies the given conditions.

36. Through $(1, 1)$ and $(-3, 4)$
37. x-intercept 4, y-intercept -3
38. Through $(-2, -1)$, parallel to $5x - y = 10$
39. Through $(4, -5)$, perpendicular to $4x + y = 9$
40. Through $(-2, 6)$, slope does not exist
41. Find the value of y if the line through $(-1, 1)$ and $(3, y)$ has slope 2.

Find an equation, in standard form, of the straight line whose graph is given.

42.

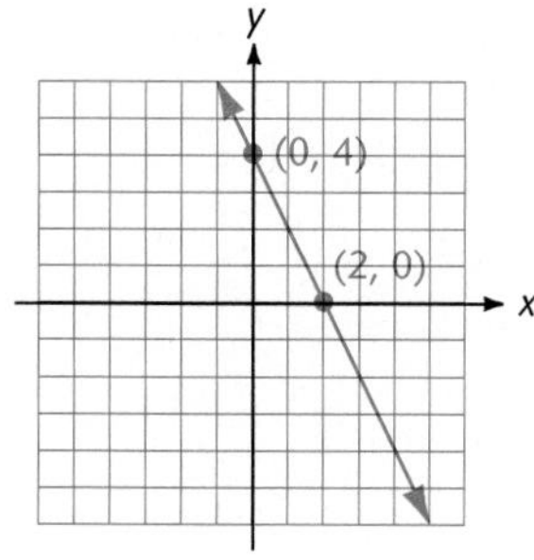

43.

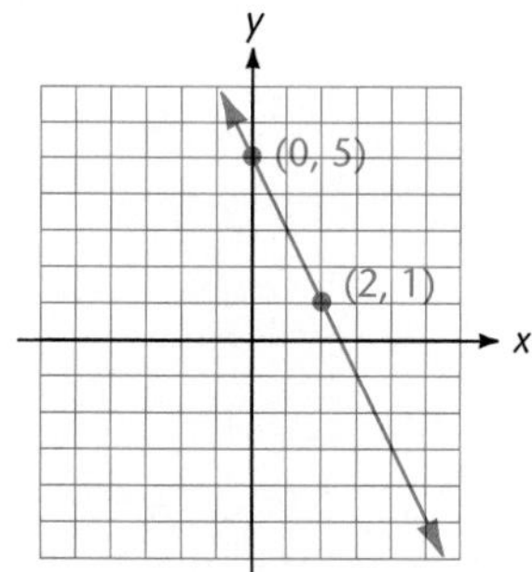

44.

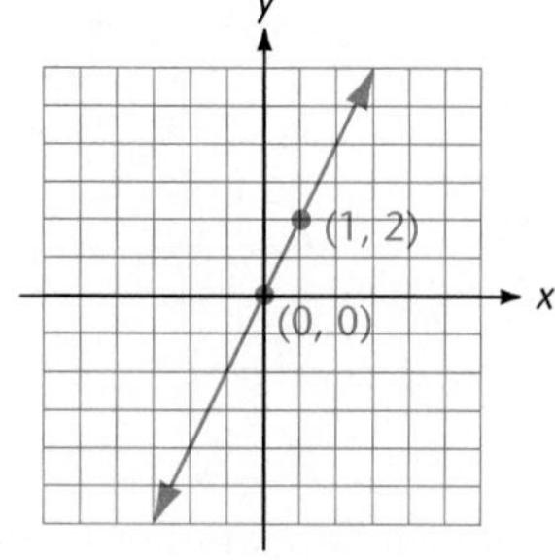

45.

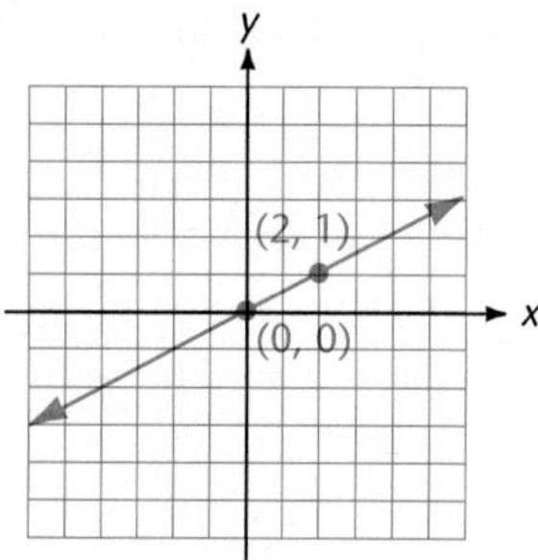

46.

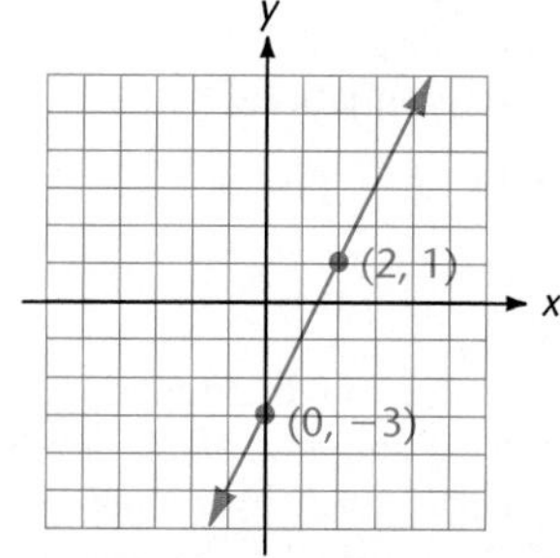

47. Graph the function $f(x) = \begin{cases} 2x + 1, & \text{if } x \leq 1 \\ 4 - x, & \text{if } x > 1 \end{cases}$.

Linear Depreciation

48. Emily Bradof bought a copier for \$1800 and sold it 15 months later for \$1200.
    a. Using linear depreciation, how much value did the copier lose each month?
    b. Write the value $V(t)$ of the copier after t months as a linear function in the form $V(t) = mt + b$.

Production Cost

49. As a hobby, Ms. Volk has started making ceramic wind chimes. She has learned that the cost C (in dollars) of producing x wind chimes is a linear function of x. Her records show that it costs \$800 to produce 32 wind chimes and \$968 to produce 44 wind chimes. Express C as a function of x.

Graph each of the following parabolas. Locate the vertex and the axis of symmetry, and plot at least two other points on each graph.

50. $y = 3(x - 1)^2 + 2$
51. $y = 4x - x^2$
52. $y = x^2 + 2x - 3$
53. $y = -2x^2 + 3x + 1$

Find the extreme value of the given function and state whether that value represents a maximum or minimum.

54. $f(x) = 2(x - 3)^2 + 1$
55. $f(x) = 1 - (x - 2)^2$
56. $f(x) = x^2 - 4x + 2$
57. $f(x) = 1 - 6x - x^2$
58. $f(x) = -2x^2 - 4x + 1$
59. $f(x) = 2x^2 - 8x + 13$

Maximum Profit

60. The profit P (in dollars) from producing x units of a certain item is given by $P = 0.8x - 0.001x^2$. Find the production level for which the profit is a maximum and find the maximum profit.

Maximum Height

61. If a ball is thrown straight up with a speed of 48 feet per second when it is released, it reaches a height of h feet after t seconds given by $h = 48t - 16t^2$. Find the length of time required for the ball to reach its maximum height and find the maximum height.

Minimum Cost

62. The cost C of producing x lawn mowers is given in dollars by $C = 1500 - 60x + 3x^2$. Find the number of lawn mowers produced that minimizes C and find the minimum cost.

Maximum Revenue

63. The revenue R (in dollars) that results from selling x birdhouses from Jeremy Pallai's woodworking shop can be found from $R = 40x - x^2 - 150$. How many birdhouses must be sold for maximum revenue, and what is the maximum value of R?

Maximum Profit

64. If Boston Anchor Works sells n fishing boat anchors, it makes a profit $P = 80n - 0.02n^2$ dollars. How many anchors need to be sold to make a maximum profit, and what is the maximum profit?

ENCORE

Veterinarians use the medication oxibendazole in the prevention of certain parasites in dogs. The dosage administered is a function of the body weight x of the dog in pounds, determined by this rule: for a body weight x from 20 to 45 pounds, the daily dosage $D(x)$ is 45 milligrams of oxibendazole plus an additional 11 milligrams for each 5 pounds or portion of 5 pounds of body weight above 20 pounds.

The domain of this dosage function is the closed interval [20,45]. For a dog weighing 20 pounds, the dosage is $D(20) = 45$ milligrams. For dogs weighing more than 20 through 25 pounds, the dosage requires an additional 11 milligrams for a total of 56 milligrams. Additional increases in body weight lead to the table and graph shown in Figure 3.28.

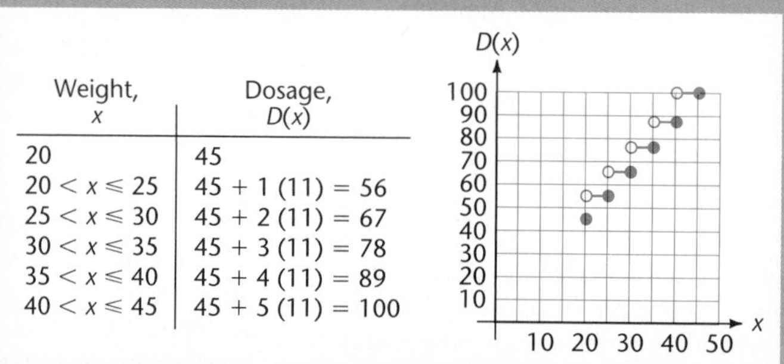

Weight, x	Dosage, $D(x)$
20	45
$20 < x \leq 25$	$45 + 1\,(11) = 56$
$25 < x \leq 30$	$45 + 2\,(11) = 67$
$30 < x \leq 35$	$45 + 3\,(11) = 78$
$35 < x \leq 40$	$45 + 4\,(11) = 89$
$40 < x \leq 45$	$45 + 5\,(11) = 100$

Figure 3.28 Dosage function

Cumulative Test Chapters 1–3

1. Simplify the following expressions by removing symbols of grouping and combining like terms.

   a. $2[x - 2(x + 3) - (x - 1)]$

   b. $-\{m - (1 - m) + 2[m - 3(m + 2)]\}$

2. Draw the graph of each inequality.

   a. $|x + 3| \leq 4$

   b. $|x - 2| > 3$

3. Compute the value if possible; otherwise, simplify by using the laws of exponents.

   a. $(x^{-2}y^{-1})^{-1}$ b. $\left(\dfrac{-2x^2}{y}\right)^3$ c. $\dfrac{3^{-2}x^2y^{-1}}{4^{-2}x^{-2}y^2}$

4. Perform the indicated operations and write the results in simplest form.

   a. $(x^2 + 5x - 2)(2x + 3)$ b. $(5a + 2b)(5a - 2b)$

   c. $(4q^2 - 5p)(3q^2 + p)$ d. $(2m - 3n)^2$

5. Perform the indicated operations and reduce the results to lowest terms.

   a. $\dfrac{x^2 + 2xy - 3y^2}{x^2 - y^2} \cdot \dfrac{x^2 + 2xy + y^2}{2x^2 + 5xy - 3y^2}$

   b. $\dfrac{2z^2 - 9z + 10}{3z^2 - 6z + 3} \div \dfrac{4z^2 - 25}{z^2 - 3z + 2}$

   c. $\dfrac{7}{3x^2 + 11x + 6} - \dfrac{11}{6x^2 + 19x + 10}$

   d. $\dfrac{\dfrac{x}{y} - \dfrac{y}{x}}{\dfrac{1}{x} - \dfrac{1}{y}}$

6. Evaluate each of the following expressions.

   a. $(216)^{-4/3}$ b. $(0.027x^6)^{-2/3}$

7. Change to simplest radical form. Assume all variables represent positive real numbers.

   a. $\sqrt[3]{-\dfrac{125}{8}}$ b. $\sqrt{\left(\dfrac{3x}{4z^2}\right)^{-2}}$

   c. $\dfrac{3}{2 - \sqrt{5}}$ d. $\sqrt{\dfrac{x^3y^{-2}}{18xy}}$

8. Perform the indicated operations and write the result in standard form.

   a. $(-2 + 7i) - i(1 + i)$ b. $\dfrac{i}{1 + 3i}$ c. i^{26}

9. Solve the equation: $\dfrac{3}{x + 1} = \dfrac{4}{x + 2}$.

10. Solve the equation: $|3x - 5| = 4$.

1. Solve for x in the equation $\dfrac{3}{x} + \dfrac{4}{y} = \dfrac{1}{xy}$.

Coffee Mixture

12. How many pounds of coffee worth \$2.88 per pound should be mixed with 9 pounds of coffee worth \$2.24 per pound to obtain a mixture worth \$2.52 per pound?

Uniform Motion

13. Walt made a 238-mile trip in 4.5 hours. He drove the first part of the trip at 50 miles per hour and then increased his speed to 60 miles per hour. How far did he travel at each speed?

14. Solve the following inequalities and graph the solution sets. Express your answers in interval notation.

    a. $3t - 2(6 - t) < 4 - 3t$ or $2(t - 3) + 4 > 8 - t$

    b. $|7 - 4x| < 5$

15. Solve the following equations by any method.

    a. $7y - 20 = -6y^2$ b. $(3x + 2)^2 + 2(3x + 2) = 8$

    c. $z^2 - 4z - 3 = 0$

16. Find the solution set for each of the following equations. Assume all variables represent real numbers.

    a. $x = 1 + \dfrac{5}{3x - 1}$ b. $\sqrt{4y + 9} - y = 1$

17. Solve the following inequalities and graph each solution set.

    a. $0 \le 12x - x^2 - x^3$ b. $\dfrac{3r - 1}{r + 1} > 2$

18. Graph $4x + 3y = 12$.

19. Find the slope and the midpoint of the line segment joining $(5, -2)$ and $(-3, -8)$.

20. Give the domain D and range R of the function $f(x) = \dfrac{1}{\sqrt{x - 2}}$.

21. Let $f(x) = \sqrt{x + 1}$ and $g(x) = x - 1$. Determine the following functions and state their domains.

    a. $f - g$ b. $f \cdot g$ c. $g \circ f$

22. Find an equation of the straight line through $(-3, 1)$ and perpendicular to $2x - y = 3$.

23. Graph the function $f(x) = \begin{cases} 2x + 4, & \text{if } x \le -1 \\ 3 - x, & \text{if } x > -1 \end{cases}$.

24. Graph the parabola $y = 2x^2 - 8x - 3$. Locate the vertex and the axis of symmetry, and plot at least two other points on the graph.

Maximum Profit

25. The profit that results from the sale of t tons of topsoil is given by $P = 4t - 150 - 0.02t^2$, where P is in hundreds of dollars. How many tons of topsoil must be sold for a maximum profit, and what is the maximum profit?

Graphing Techniques

Most of this chapter is concerned with certain types of curves called **conic sections.** These curves are important because they furnish good mathematical models for such diverse entities as the paths of planets and comets about the sun, satellite receivers, cables on suspension bridges, and some architectural designs.

4.1 THE PARABOLA AND THE CIRCLE

In this chapter, as in Chapter 3, our work is restricted to real variables. We saw in Section 3.5 that the graph of a quadratic function is a parabola. The parabola is one of three types of curves known as **conic sections.** The term *conic sections* comes from the fact that each type of curve can be obtained by intersecting a plane with a right circular cone.†

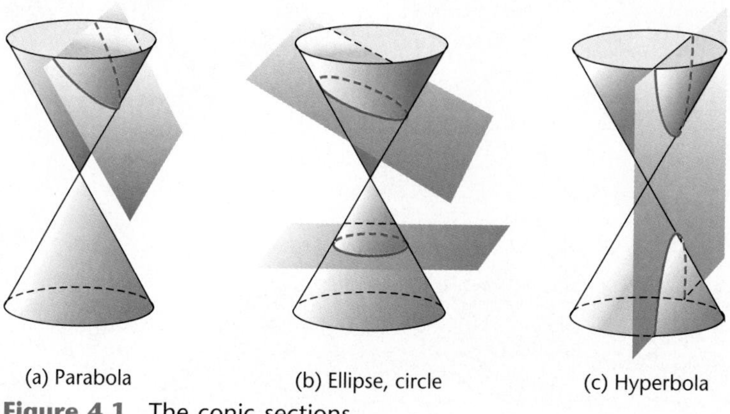

(a) Parabola (b) Ellipse, circle (c) Hyperbola

Figure 4.1 The conic sections

If the intersecting plane has a suitable inclination to the axis of the cone, the curve of intersection is a **parabola,** as shown in Figure 4.1a. With a greater inclination to the axis of the cone, the curve of intersection may be an **ellipse,** which is an oval-shaped curve, as shown in Figure 4.1b. The **circle** is a special case of the ellipse. With a lesser inclination, the curve of intersection may be a **hyperbola,** which is a curve having two branches, as shown in Figure 4.1c.

In Section 3.5 we studied the graphs of quadratic functions with equations of the form

$$y = ax^2 + bx + c, \qquad a \neq 0.$$

We can interchange the roles of x and y and obtain a complete dual to the results of that section. That is, we can study the graphs of equations of the form

$$x = ay^2 + by + c, \qquad a \neq 0.$$

We omit the details for this development and simply state the corresponding results.

†By a *cone,* we mean a complete cone, with two nappes, as shown in Figure 4.1.

The Parabola

Equation, $a \neq 0$ Alternate equation	$y = ax^2 + bx + c$ $y = a(x - h)^2 + k$	$x = ay^2 + by + c$ $x = a(y - k)^2 + h$
Vertex	$\left(-\dfrac{b}{2a}, \dfrac{4ac - b^2}{4a}\right) = (h, k)$	$\left(\dfrac{4ac - b^2}{4a}, -\dfrac{b}{2a}\right) = (h, k)$
Axis of symmetry	$x = -\dfrac{b}{2a}, x = h$	$y = -\dfrac{b}{2a}, y = k$
Direction of opening	Upward, if $a > 0$ Downward, if $a < 0$	To the right, if $a > 0$ To the left, if $a < 0$
Graphs	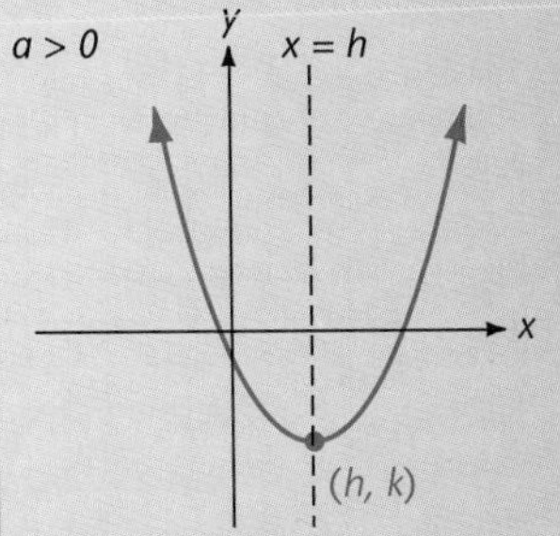 $a > 0$; $x = h$; (h, k); x; y	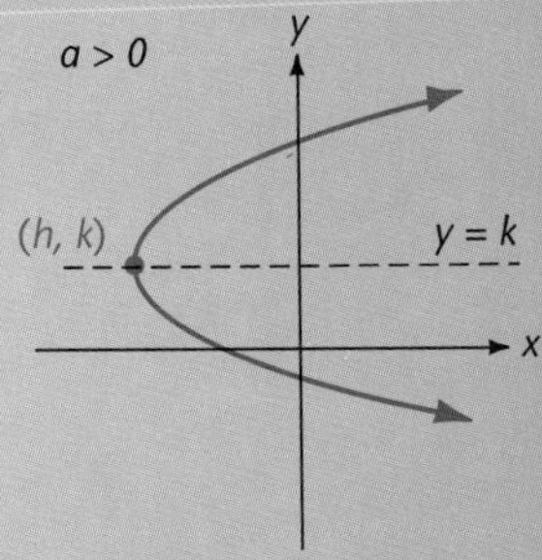 $a > 0$; $y = k$; (h, k); x; y
	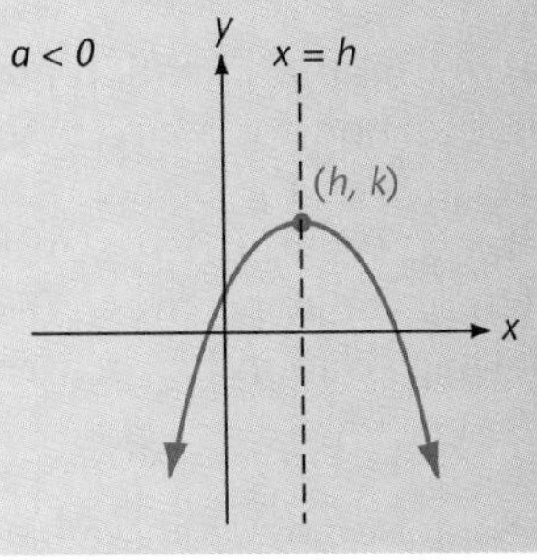 $a < 0$; $x = h$; (h, k); x; y	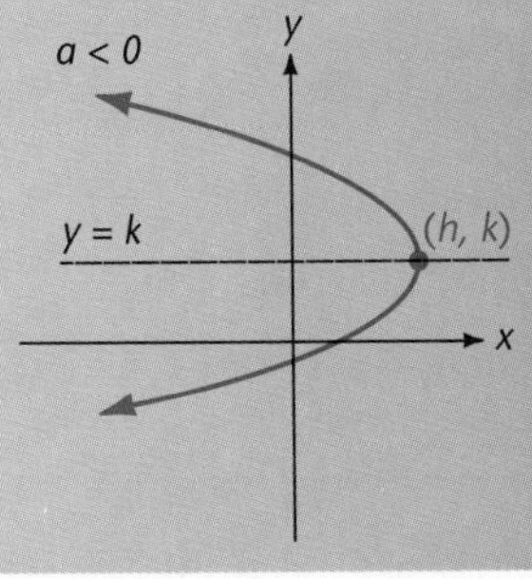 $a < 0$; $y = k$; (h, k); x; y

◆ **Practice Problem 1** ◆

Sketch the graph of $x = -y^2 + 6y - 5$.

When equations of the form $x = ay^2 + by + c$ are encountered in applications, they usually appear in a form similar to that in the following example.

Example 1 • Graphing Half a Parabola

Sketch the graph of the function $y = 1 + \sqrt{x - 2}$.

Solution

The graph is not familiar in the given form, but it can be recognized after we square both sides to eliminate the radical.

$$
\begin{aligned}
y - 1 &= \sqrt{x - 2} && \text{isolating the radical}\\
(y - 1)^2 &= x - 2 && \text{squaring both sides}\\
x &= (y - 1)^2 + 2 && \text{solving for } x
\end{aligned}
$$

We recognize this as the equation of a parabola that has vertex at (2, 1) and opens to the right. However, this parabola is not the graph of the original equation because $y = 1 + \sqrt{x - 2}$ requires that $y \geq 1$. (Recall that $\sqrt{x - 2}$ indicates the nonnegative square root of $x - 2$.) The graph of the original equation consists only of the top half of the parabola and is shown in Figure 4.2. □

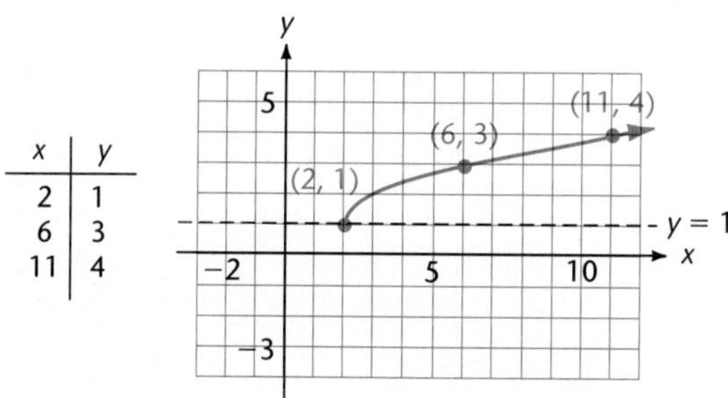

x	y
2	1
6	3
11	4

Figure 4.2 Graph of $y = 1 + \sqrt{x - 2}$

We turn our attention now to equations of circles. The derivation of the standard equation for a circle requires the use of a formula for the distance between two points in the plane.

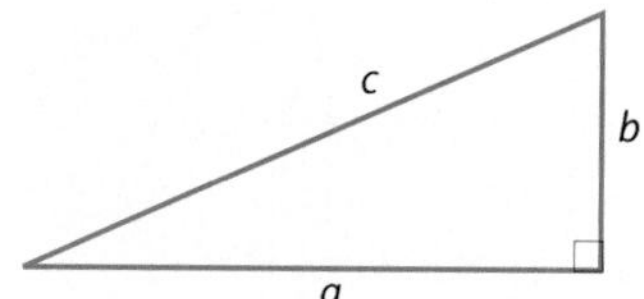

Figure 4.3 A right triangle with hypotenuse c

Let a, b, and c represent the lengths of the sides of a right triangle, with c the length of the hypotenuse. (See Figure 4.3.) The **Pythagorean theorem** states the following relationship between the lengths of the sides:

$$c^2 = a^2 + b^2 \qquad \text{or} \qquad c = \sqrt{a^2 + b^2}.$$

Suppose now that we have a right triangle drawn as indicated in Figure 4.4. The coordinates of the vertices of the right triangle are (x_1, y_1), (x_2, y_1), and (x_2, y_2), as indicated in the figure. For the figure as drawn, the length of the side parallel to the x-axis is $x_2 - x_1$, and the length of the side parallel to the y-axis is $y_2 - y_1$. The Pythagorean theorem can be used to determine the length d of the hypotenuse:

$$d = \sqrt{(x_2 - x_1)^2 + (y_2 - y_1)^2}.$$

The length d of the hypotenuse represents the distance between the points with coordinates (x_1, y_1) and (x_2, y_2). This equation holds independently of the quadrants in which the points lie and independently of the points' orientation to each other. This result is known as the **distance formula.**

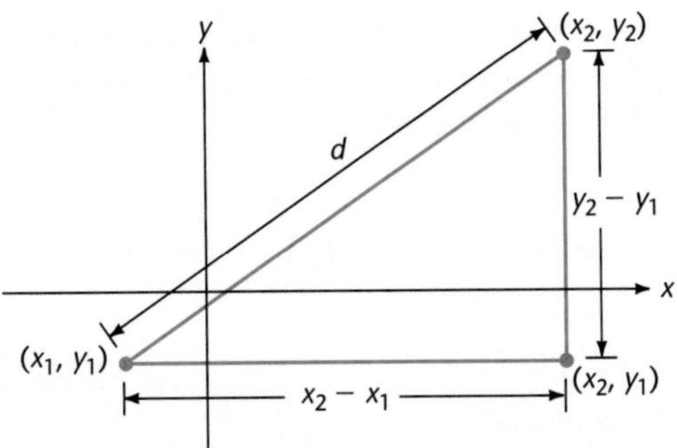

Figure 4.4 The distance between (x_1, y_1) and (x_2, y_2)

Distance Formula

The **distance** d between the points with coordinates (x_1, y_1) and (x_2, y_2) is given by

$$d = \sqrt{(x_2 - x_1)^2 + (y_2 - y_1)^2}.$$

Example 2 • Finding the Distance Between Two Points

The distance between the points $(-1, 3)$ and $(7, -3)$ is given by

$$\begin{aligned} d &= \sqrt{[7 - (-1)]^2 + (-3 - 3)^2} \\ &= \sqrt{64 + 36} \\ &= 10. \end{aligned}$$

In using the distance formula, we chose $(x_1, y_1) = (-1, 3)$, and $(x_2, y_2) = (7, -3)$. The same result would have been obtained, however, if we had chosen $(x_1, y_1) = (7, -3)$, and $(x_2, y_2) = (-1, 3)$. □

Example 3 • The Distance Between a Point and the Origin

The distance between the point (x, y) and the origin $(0, 0)$ is given by

$$\begin{aligned} d &= \sqrt{(x - 0)^2 + (y - 0)^2} \\ &= \sqrt{x^2 + y^2}. \end{aligned}$$

Squaring both sides yields

$$x^2 + y^2 = d^2.$$ □

Definition of a Circle

A **circle** is the set of all points in a plane equally distant from a given point, called the **center** of the circle. The distance from the center of the circle to any point on the circle is called the **radius** of the circle.

Suppose we wish to obtain an equation for a circle with radius r and center $(0, 0)$. Referring to Example 3, we see that the set of all points (x, y) whose distance from $(0, 0)$ is r is given by

$$\{(x, y) \mid x^2 + y^2 = r^2\},$$

hence an equation for a circle with center at $(0, 0)$ and radius r is

$$x^2 + y^2 = r^2.$$

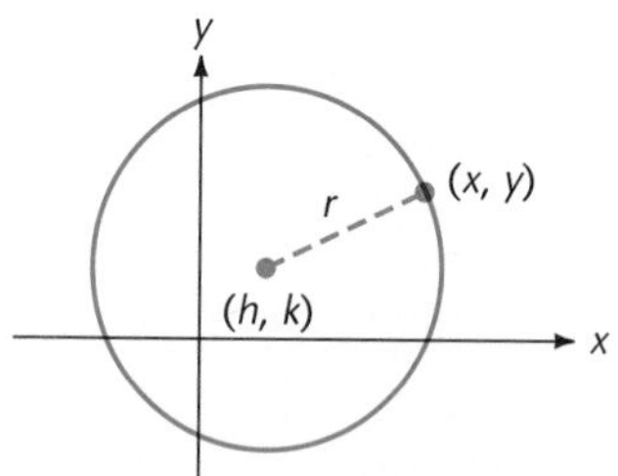

Figure 4.5 The circle with center (h, k) and radius r

Similarly, a point (x, y) is at a distance r from the fixed point (h, k) if and only if its coordinates satisfy the equation

$$\sqrt{(x - h)^2 + (y - k)^2} = r$$

or

$$(x - h)^2 + (y - k)^2 = r^2. \tag{1}$$

That is, a point (x, y) is on the circle with center (h, k) and radius r if and only if its coordinates (x, y) satisfy Equation (1). This important result is in the following statement and is represented graphically in Figure 4.5.

Standard Equation of a Circle

The **standard equation** for a circle with center (h, k) and radius r is

$$(x - h)^2 + (y - k)^2 = r^2. \tag{2}$$

Example 4 • A Point Circle

If we put $r = 0$ in Equation (2), we have

$$(x - h)^2 + (y - k)^2 = 0. \tag{3}$$

Since the square of a real number cannot be negative, this equation requires that both $x - h = 0$ and $y - k = 0$. Thus Equation (3) is satisfied only by the coordinates of the single point (h, k) and the graph of Equation (3) consists of this one point. For this reason, the graph of Equation (3) is sometimes referred to as a **point circle,** or a **degenerate circle.** □

Example 5 • Graphing Circles

Determine whether or not the graph of each equation below is a circle. If it is, find the center and radius and draw the graph.

a. $(x - 5)^2 + (y + 1)^2 = 9$ b. $x^2 + 2x + y^2 - 6y = 0$

c. $x^2 + y^2 = 10x - 8y - 41$ d. $2x^2 + 2y^2 - 4x + 10 = 0$

Solution

a. Expressing $(x - 5)^2 + (y + 1)^2 = 9$ in the form of Equation (2), we have

$$(x - 5)^2 + [y - (-1)]^2 = (3)^2.$$

Thus the graph is the circle in Figure 4.6a with center $(5, -1)$ and radius 3.

b. We must complete the squares on the variables x and y in order to change the equation into the standard form.

$(x^2 + 2x \qquad) + (y^2 - 6y \qquad) = 0$	grouping
$(x^2 + 2x + \mathbf{1}) + (y^2 - 6y + \mathbf{9}) = \mathbf{1} + \mathbf{9}$	completing the squares
$[x - (-1)]^2 + (y - 3)^2 = (\sqrt{10})^2$	writing in standard form

This is the equation of the circle in Figure 4.6b with center at $(-1, 3)$ and radius $\sqrt{10}$.

c. Completing the squares on the variables x and y and rewriting in the standard form, we obtain the following equations.

$$x^2 - 10x + \mathbf{25} + y^2 + 8y + \mathbf{16} = -41 + \mathbf{25} + \mathbf{16}$$
$$(x - 5)^2 + [y - (-4)]^2 = 0$$

This is the equation of the circle in Figure 4.6c with center at $(5, -4)$ and radius 0, that is, a point circle. The values $x = 5$, $y = -4$ are the only ones that satisfy the equation $x^2 + y^2 = 10x - 8y - 41$.

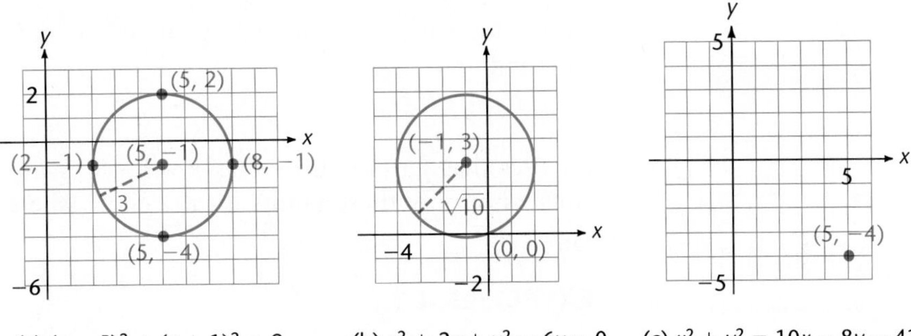

(a) $(x - 5)^2 + (y + 1)^2 = 9$ (b) $x^2 + 2x + y^2 - 6y = 0$ (c) $x^2 + y^2 = 10x - 8y - 41$

Figure 4.6 The circles in Example 5

d. The coefficients of x^2 and y^2 must be 1 before we can complete the square.

$2x^2 + 2y^2 - 4x + 10 = 0$	
$x^2 + y^2 - 2x + 5 = 0$	dividing by 2
$(x^2 - 2x \qquad) + y^2 = -5$	grouping
$(x^2 - 2x + \mathbf{1}) + y^2 = -5 + \mathbf{1}$	completing the square on x
$(x - 1)^2 + y^2 = -4$	rewriting

Since the sum of the squares of two real numbers can never be negative, there are no values of x and y that satisfy this equation. Hence the equation

$$2x^2 + 2y^2 - 4x + 10 = 0$$

does not represent a circle. The set of solutions to this equation is the empty set $\varnothing$. □

◆ Practice Problem 2 ◆

Graph the equation $x^2 + y^2 = 6y - 2x - 6$.

Although a circle is not the graph of a function, a complete circle can be graphed with a graphing device. One function must be used for the top half and another for the bottom half.

Certain equations have a graph that consists of only part of a circle. Such an equation is presented in the next example.

Example 6 • Graphing a Semicircle

Graph the equation $y - 1 = -\sqrt{4 - x^2}$.

Solution

In the given form, the equation does not much resemble the standard form for the equation of a circle. If we square both sides, however, we obtain

$$(y - 1)^2 = 4 - x^2$$

or

$$x^2 + (y - 1)^2 = 4.$$

This is the equation of a circle with center $(0, 1)$ and radius 2. But the original equation

$$y - 1 = -\sqrt{4 - x^2}$$

requires that $y - 1 \le 0$ or that $y \le 1$. Thus the graph of the original equation consists of only those points on the circle where $y \le 1$. This graph is shown in Figure 4.7. □

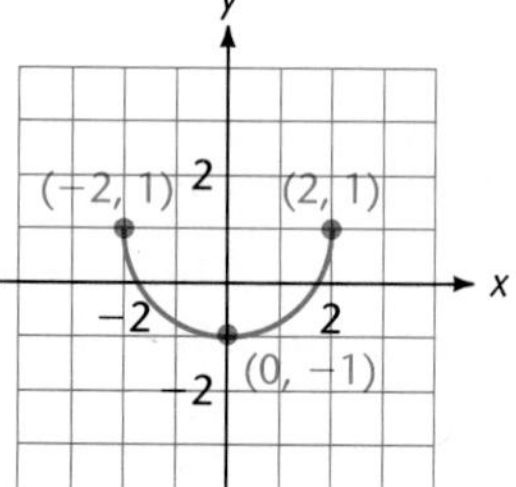

Figure 4.7 Graph of $y - 1 = -\sqrt{4 - x^2}$

EXERCISES 4.1

Determine the distance between the given points. (See Example 2.)

1. $(0, 0)$ and $(4, 3)$
2. $(-1, 0)$ and $(2, 0)$
3. $(0, -3)$ and $(3, 1)$
4. $(-4, -6)$ and $(-1, -2)$
5. $(-9, 22)$ and $(11, 1)$
6. $(-3, -12)$ and $(-4, 12)$
7. $(2, 9)$ and $(-2, -9)$
8. $(4, -8)$ and $(-1, -2)$
9. Find the shortest distance between $(3, 6)$ and the line $y = 10$.
10. Find the shortest distance between $(-2, 7)$ and the line $x = 4$.
11. If $f(x) \le 10$, find the vertical distance between $y = f(x)$ and $y = 10$.
12. If $f(x) \le g(x)$, find the vertical distance between $y = f(x)$ and $y = g(x)$.

Determine whether or not the graph of each equation is a circle. If it is, find the center and radius. (See Example 5.)

13. $x^2 + y^2 = 16$
14. $(x + 2)^2 + (y - 2)^2 = 4$
15. $x^2 + (y + 4)^2 = \frac{3}{25}$
16. $(x - 3)^2 + (y + 5)^2 = 25$
17. $x^2 + 2x + y^2 + 6y = 0$
18. $x^2 - 2x + y^2 + 4y = 4$

19. $x^2 + y^2 = 3x + 4y - 4$
20. $x^2 + y^2 = 8y$
21. $x^2 + y^2 = 10x - 6y - 36$
22. $x^2 + y^2 - 3x + 3y + 12 = 0$
23. $x^2 + y^2 - 4x + 16y + 68 = 0$
24. $x^2 + y^2 = 12x + 2y - 37$

Write an equation of the circle with the given properties.

25. Center $(0, 1)$, radius 2
26. Center $(-1, 3)$, radius 1
27. Center $(2, -2)$, through $(5, 1)$
28. Center $(-3, 0)$, through $(3, -8)$

Write an equation of the given circle with endpoints of a diameter as shown.

29.

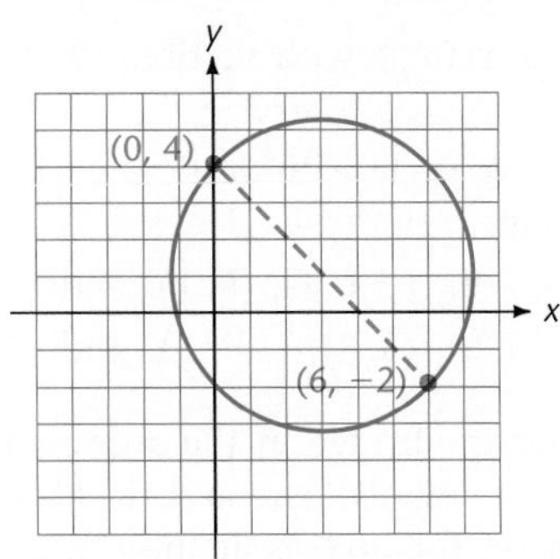

30.

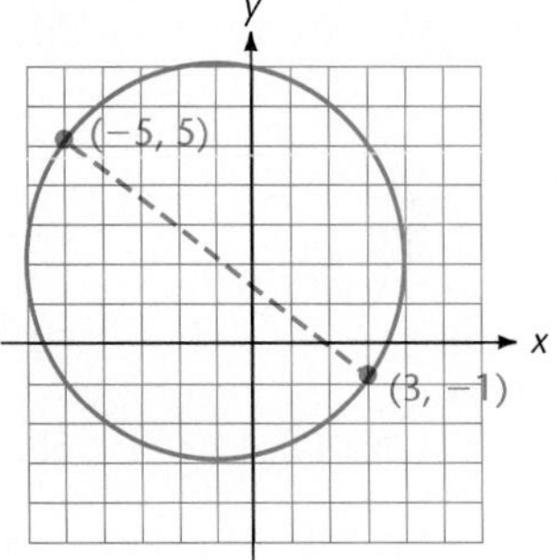

Graph the following equations. (See Examples 1, 5, and 6.)

31. $x^2 + y^2 = 4$
32. $(x - 2)^2 + (y - 1)^2 = 9$
33. $x^2 + y^2 - 4x = 21$
34. $x^2 + y^2 + 6x - 12y + 9 = 0$
35. $x^2 + y^2 = 10x$
36. $x^2 + y^2 = -4y$
37. $x = 12 - 3(y + 2)^2$
38. $x = y^2 - 4y + 4$
39. $x = 3y^2 + 6y - 20$
40. $x = -2y^2 + 8y - 11$
41. $y = \sqrt{x - 3}$
42. $y = 1 - \sqrt{x - 2}$
43. $x = \sqrt{y - 2}$
44. $y = -\sqrt{1 - x^2}$
45. $x = 2 - 2\sqrt{y - 1}$
46. $x + 2 = -\sqrt{25 - y^2}$

In Exercises 47–50, write an equation for the given parabola.

47.

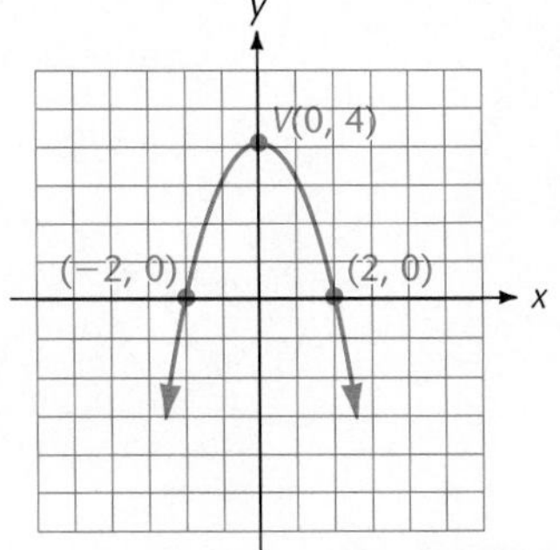

48.

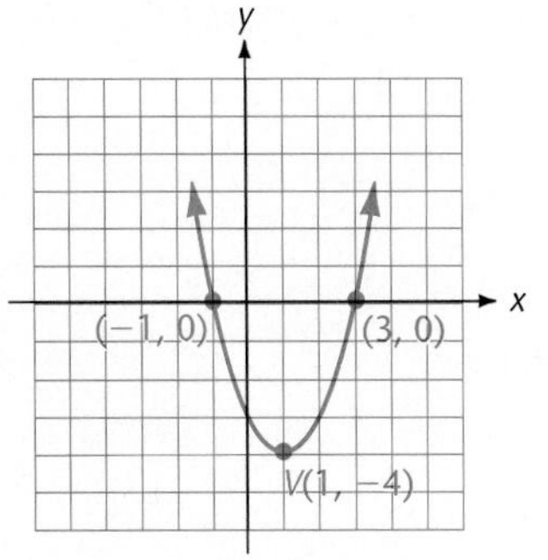

49. 50.

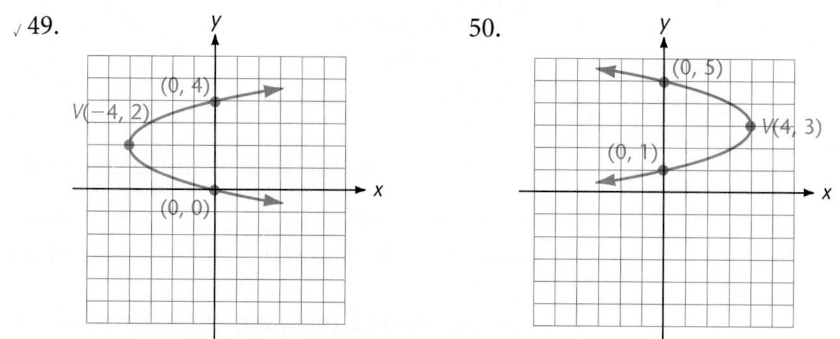

51. Show that the triangle with vertices (2, 2), (−2, −2), and (−1, 5) is a right triangle.
52. Show that the triangle with vertices (−3, 5), (4, 3), and (−3, 1) is an isosceles triangle.
53. Show that (6, −9), (−1, 4), (9, 1), and (−4, −6) are vertices of a square.
54. Show that (−3, 0), (3, 8), (1, −3), and (7, 5) are vertices of a rectangle.

Parabolic Arch 55. An arch under the bridge in the accompanying figure spans an opening in the shape of a parabola. The arch is 40 meters wide at the base, and the highest point on the arch is at the center, 12 meters above the base. Find the height of the arch at a point 10 meters from the center of the base.

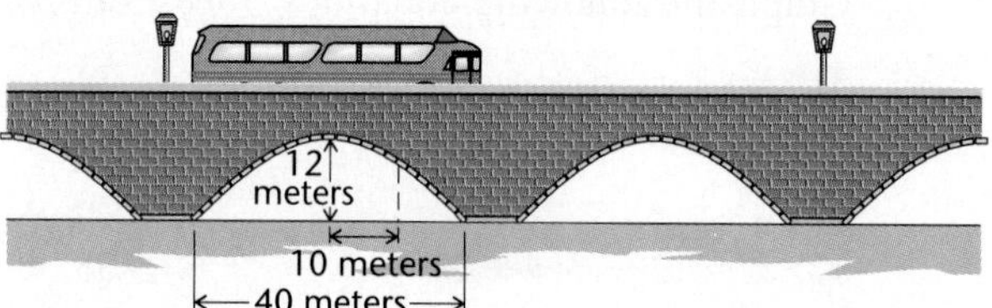

Figure for Exercise 55

Satellite Dish 56. The satellite dish in the accompanying figure has a cross section in the shape of a parabola 8 feet in diameter across the opening and 3 feet deep. Find the depth of a second dish which has the same shape but with a diameter of 10 feet.

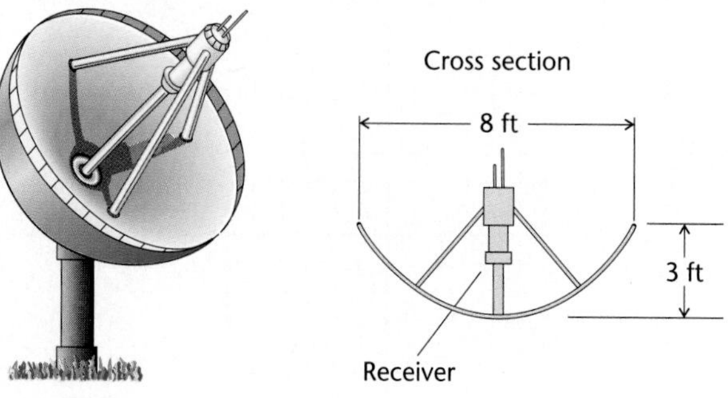

Figure for Exercise 56

Parabolic Reflector

57. A spotlight reflector has a cross section in the shape of a parabola that is 60 centimeters in diameter at the opening and 20 centimeters deep. Find the diameter of a second reflector that has the same shape and a depth of 15 centimeters.

Height of an Object

58. If a ball is thrown straight up from 6 feet above ground level with an initial velocity of 88 feet per second, its height $h(t)$ in feet above ground after t seconds is given by $h(t) = 6 + 88t - 16t^2$. Find the maximum height that the ball reaches.

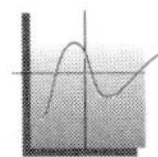

59–62

59. Use a graphing device to graph the circle $x^2 + y^2 = 25$ by graphing the two semicircles $Y_1 = \sqrt{25 - x^2}$ and $Y_2 = -\sqrt{25 - x^2}$.

60. Use the technique in Exercise 59 to graph the circle $(x - 1)^2 + (y + 2)^2 = 16$.

61. Use a graphing device to graph the parabola $4x = y^2 - 8$ by graphing the two functions $Y_1 = \sqrt{4x + 8}$ and $Y_2 = -\sqrt{4x + 8}$.

62. Use the technique in Exercise 61 to graph the parabola $x = -(y - 3)^2 + 2$.

Critical Thinking: Exploration and Writing

63. Sketch the graph of each of the following parabolas on the same coordinate system. Then describe the effect of h on the graph of $x = y^2 + h$.

a. $x = y^2$ b. $x = y^2 + 2$ c. $x = y^2 - 2$ d. $x = y^2 + 4$

Critical Thinking: Exploration and Writing

64. Sketch the graph of each of the following parabolas on the same coordinate system. Then describe the effect of a on the graph of $x = ay^2$ when $a > 0$.

a. $x = y^2$ b. $x = 2y^2$ c. $x = 3y^2$ d. $x = \frac{1}{2}y^2$

Critical Thinking: Exploration and Writing

65. Sketch the graph of each of the following parabolas on the same coordinate system. Then describe the effect of a on the graph of $x = ay^2$ when $a < 0$.

a. $x = y^2$ b. $x = -y^2$ c. $x = 2y^2$ d. $x = -2y^2$

Critical Thinking: Exploration and Writing

66. Sketch the graph of each of the following parabolas on the same coordinate system. Then describe the effect of k on the graph of $x = (y - k)^2$.

a. $x = y^2$ b. $x = (y - 1)^2$ c. $x = (y + 1)^2$ d. $x = (y - 2)^2$

◆ Solutions for Practice Problems

1. We have an equation of the form $x = ay^2 + by + c$, with $a = -1$, $b = 6$, and $c = -5$. Using the vertex formula, we set

$$y = -\frac{b}{2a} = -\frac{6}{-2} = 3$$

and compute the corresponding value

$$x = -(3)^2 + 6(3) - 5 = 4.$$

Thus the vertex is located at (4, 3). The graph opens to the left since $a = -1 < 0$. A table of values is shown with the graph in Figure 4.8.

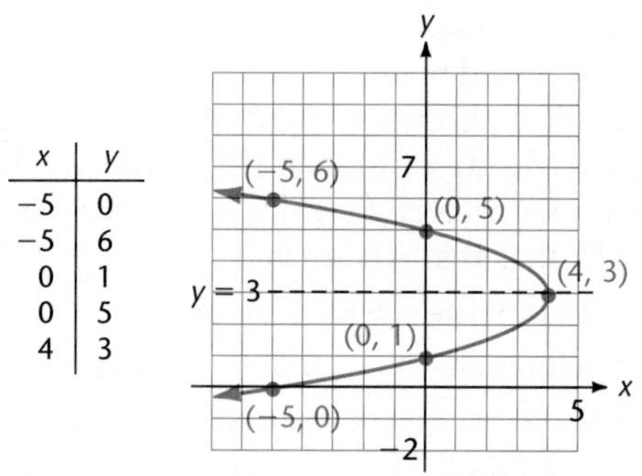

Figure 4.8 Graph of $x = -y^2 + 6y - 5$

2. We must write the equation in standard form.

$$x^2 + y^2 = 6y - 2x - 6$$
$$x^2 + 2x + \mathbf{1} + y^2 - 6y + \mathbf{9} = -6 + \mathbf{1} + \mathbf{9}$$
$$(x + 1)^2 + (y - 3)^2 = 4$$

This is the equation of a circle with center $(-1, 3)$ and radius 2. The graph is shown in Figure 4.9.

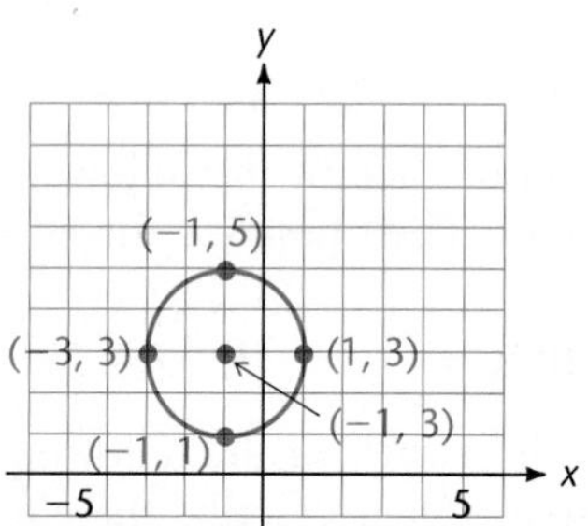

Figure 4.9 Graph of $x^2 + y^2 = 6y - 2x - 6$

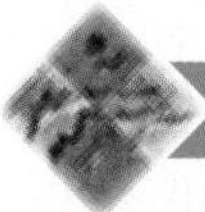

4.2 SYMMETRY AND TRANSLATIONS

In Section 3.5, we saw that a parabola is symmetric with respect to its axis of symmetry. The parabola $y = x^2$ is sketched in Figure 4.10, and we see that its axis of symmetry is the line with equation $x = 0$, the y-axis. Therefore we say that this parabola is symmetric with respect to the y-axis. Note that whenever the point (x, y) is on the graph, so is $(-x, y)$.

Other types of symmetry can be considered. For example, a graph is said to be symmetric with respect to the x-axis if the graph below the x-axis is a mirror image

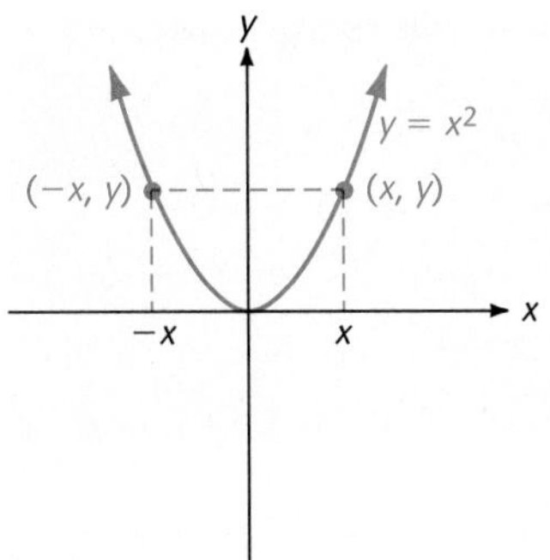

Figure 4.10 Symmetry with respect to the y-axis

of the graph above the x-axis. That is, whenever (x, y) lies on the graph, so does $(x, -y)$. This means that whenever (x, y) satisfies the equation that corresponds to the graph, so does $(x, -y)$. The graph in Figure 4.11 is symmetric with respect to the x-axis.

A graph is said to be symmetric with respect to the origin if whenever (x, y) lies on the graph of the relation, $(-x, -y)$ does also. Thus the ordered pair $(-x, -y)$ satisfies the corresponding equation whenever (x, y) does. The curve in Figure 4.12 is an example of a graph that is symmetric with respect to the origin.

Knowledge about symmetry is a very useful tool in graphing an equation, as illustrated in the following example.

Example 1 • Using Symmetry in Graphing

The graph in Figure 4.13 is a portion of a graph. Complete the graph so that it is symmetric with respect to the

a. y-axis b. x-axis c. origin.

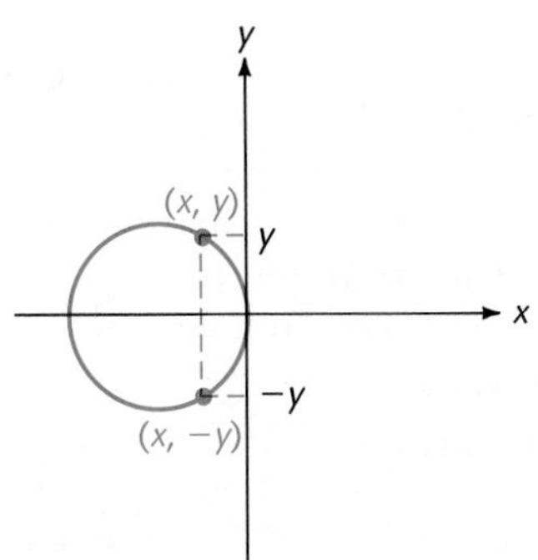

Figure 4.11 Symmetry with respect to the x-axis

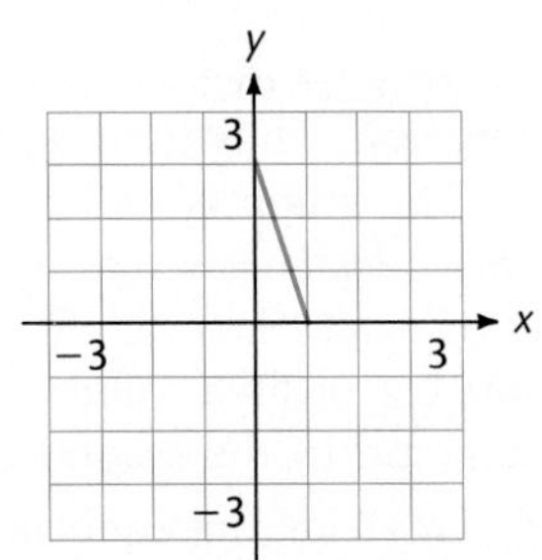

Figure 4.13 Initial graph for Example 1

Solution

The graphs are completed in the corresponding parts of Figure 4.14. □

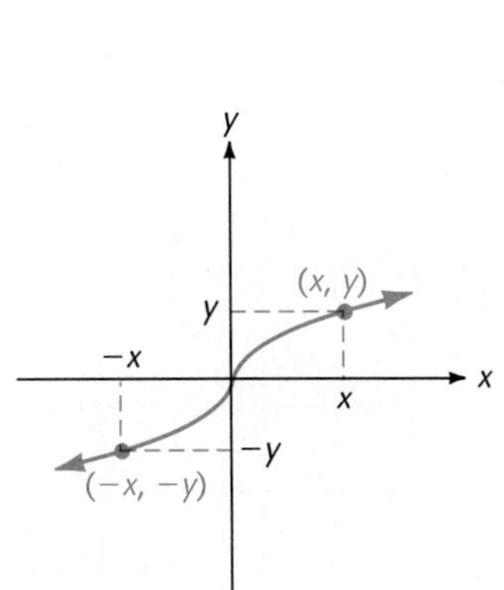

Figure 4.12 Symmetry with respect to the origin

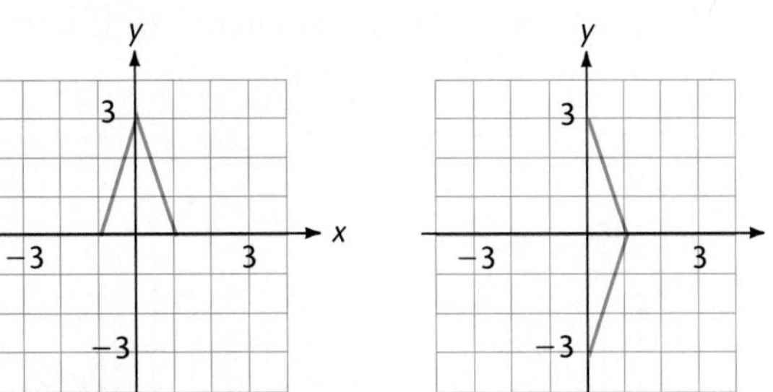

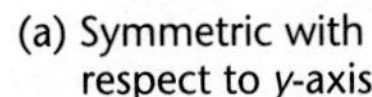

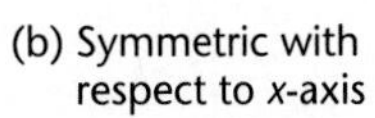

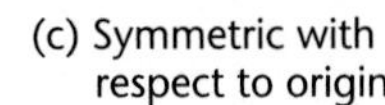

Figure 4.14 Completed graphs for Example 1

When the defining equation is given for a graph, the following tests for symmetry can be utilized.

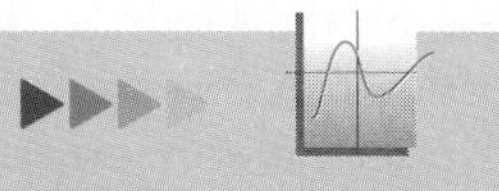

The graph drawn by a graphing device may be sufficient to reveal symmetry. This is especially useful when an equation is too complicated to apply the tests for symmetry.

Tests for Symmetry

The graph of an equation is symmetric with respect to the	if the following substitution yields an equivalent equation
y-axis	$-x$ for x
x-axis	$-y$ for y
origin	$-x$ for x and $-y$ for y

Example 2 • Using the Tests for Symmetry

Test each of the following equations for symmetry and sketch its graph.

a. $y = x^4$ b. $y^2 = x$ c. $y = x^3$

Solution

a. Substituting $-x$ for x in $y = x^4$ yields

$$y = (-x)^4 = x^4,$$

which is an equivalent equation. Thus the graph of $y = x^4$ is symmetric with respect to the y-axis. The graph is not symmetric with respect to the x-axis or to the origin since replacing y by $-y$ yields $-y = x^4$ and replacing both x by $-x$ and y by $-y$ yields $-y = x^4$. To sketch the graph in Figure 4.15a, we plot a few points for positive values of x and use symmetry to obtain the portion of the graph in the second quadrant.

b. Replacing y by $-y$ in $y^2 = x$ yields the equivalent equation

$$(-y)^2 = x \qquad \text{or} \qquad y^2 = x.$$

Thus the graph is symmetric with respect to the x-axis. The other tests for symmetry fail. The graph in Figure 4.15b is sketched by plotting a few points in the first quadrant and using symmetry to complete the graph.

c. The graph of $y = x^3$ is symmetric with respect to the origin since replacing x by $-x$ and y by $-y$ yields

$$-y = (-x)^3 = -x^3 \qquad \text{or} \qquad y = x^3.$$

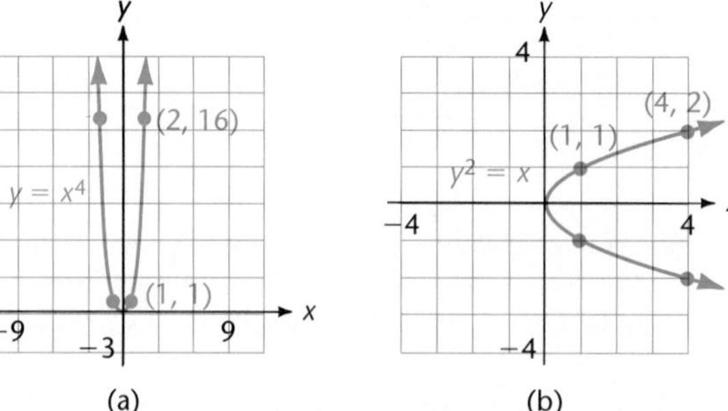

Figure 4.15 Graphs for Example 2

The graph in Figure 4.15c is sketched by plotting points in the first quadrant and using symmetry to obtain the portion in the third quadrant. □

Other aids for graphing are described in the next examples.

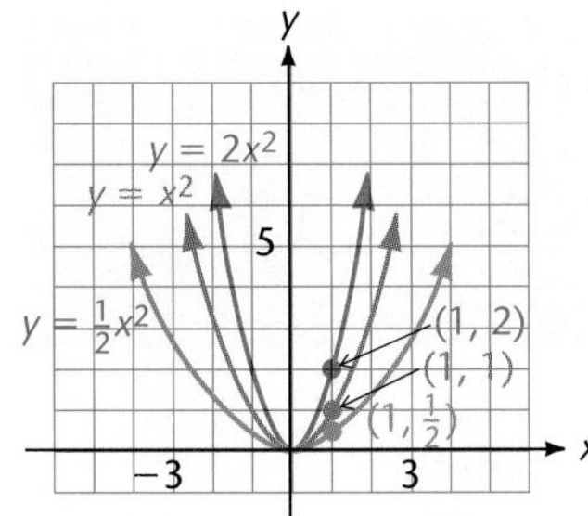

Figure 4.16 Stretchings of $y = x^2$

Example 3 • Stretching a Parabola

Sketch the graphs of $y = x^2$, $y = 2x^2$, and $y = \frac{1}{2}x^2$ on the same coordinate and compare their shapes.

Solution

Each of the graphs is sketched in Figure 4.16. The graph of $y = 2x^2$ can be obtained by multiplying each y-coordinate on the graph of $y = x^2$ by 2. Similarly, if each y-coordinate on the graph of $y = x^2$ is multiplied by $\frac{1}{2}$, the resulting point is on the graph of $y = \frac{1}{2}x^2$. A change in graphs, such as we see here, is called a **stretching** of the function $y = x^2$. □

Stretching of $y = f(x)$

To graph $y = cf(x)$, $c > 0$, multiply each y-coordinate on the graph of $y = f(x)$ by c.

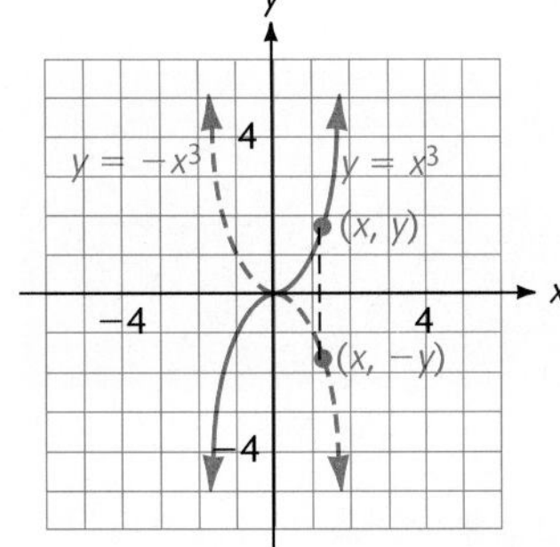

Figure 4.17 The reflection of $y = x^3$ through the x-axis

Example 4 • Reflecting a Cubic Function

Sketch the graphs of $y = x^3$ and $y = -x^3$ on the same coordinate system.

Solution

The multiplier -1 in $y = -x^3$ multiplies all the function values of $y = x^3$ by -1. Thus for every point (x, y) on the graph of $y = x^3$, there is a corresponding point $(x, -y)$ on the graph of $y = -x^3$. (See Figure 4.17.) We call the graph of $y = -x^3$ a **reflection** through the x-axis of the graph of $y = x^3$. □

Reflection of $y = f(x)$

To graph $y = -f(x)$, reflect the graph of $y = f(x)$ through the x-axis.

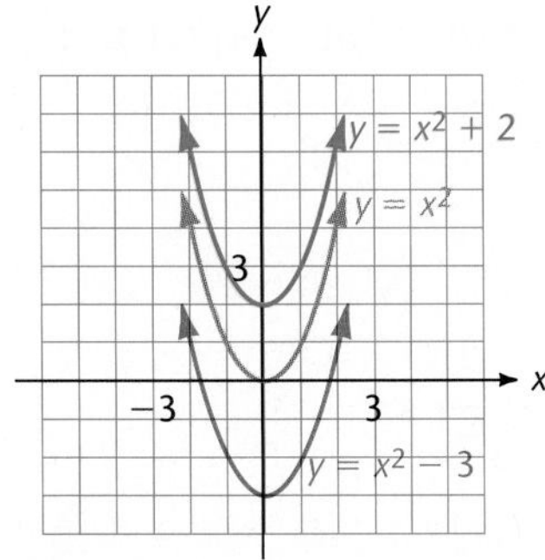

Figure 4.18 Vertical translations of $y = x^2$

Example 5 • Translating a Parabola Vertically

Sketch the graphs of $y = x^2$, $y = x^2 + 2$, and $y = x^2 - 3$ on the same coordinate system.

Solution

The graph of $y = x^2 + 2$ can be obtained by shifting the graph of $y = x^2$ upward 2 units. That is, if 2 is added to the y-coordinate of each point on the parabola $y = x^2$, the result is the y-coordinate of a point on $y = x^2 + 2$. Similarly, if the graph of $y = x^2$ is shifted downward 3 units, the result is the graph of $y = x^2 - 3$. These parabolas are sketched in Figure 4.18. □

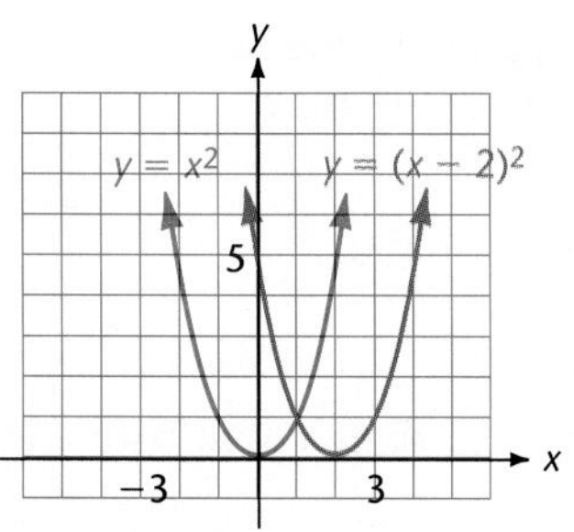

Figure 4.19 Horizontal translation of $y = x^2$

Vertical Translation of $y = f(x)$

To graph $y = f(x) + c$,

1. If $c > 0$, shift the graph of $y = f(x)$ upward c units.
2. If $c < 0$, shift the graph of $y = f(x)$ downward $|c|$ units.

Another type of translation is illustrated in the next example.

Example 6 • Translating a Parabola Horizontally

Sketch the parabolas $y = x^2$ and $y = (x - 2)^2$ on the same coordinate system.

Solution

In Figure 4.19, we see that the graph of $y = (x - 2)^2$ can be obtained by shifting (or translating horizontally) the graph of $y = x^2$ to the right, 2 units. □

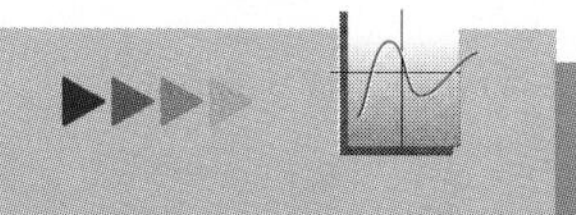

A graphing device lends itself well to observing the effects of changes in a constant that produce a stretching, reflection, or translation.

Horizontal Translation of $y = f(x)$

To graph $y = f(x - c)$,

1. If $c > 0$, shift the graph of $y = f(x)$ to the right c units.
2. If $c < 0$, shift the graph of $y = f(x)$ to the left $|c|$ units.

◆ **Practice Problem 1** ◆

Sketch the graph of $y = |x - 1| - 2$ by using either one of or a combination of a stretching, a reflection, or a translation of a familiar function.

EXERCISES 4.2

In Exercises 1–8, a portion of a graph is given. Use it to construct three additional graphs: (a) one that is symmetric with respect to the y-axis, (b) one that is symmetric with respect to the x-axis, and (c) one that is symmetric with respect to the origin. (See Example 1.)

1.

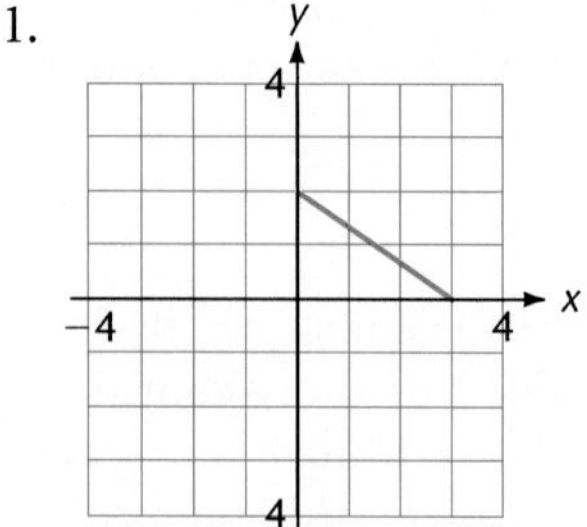

2.

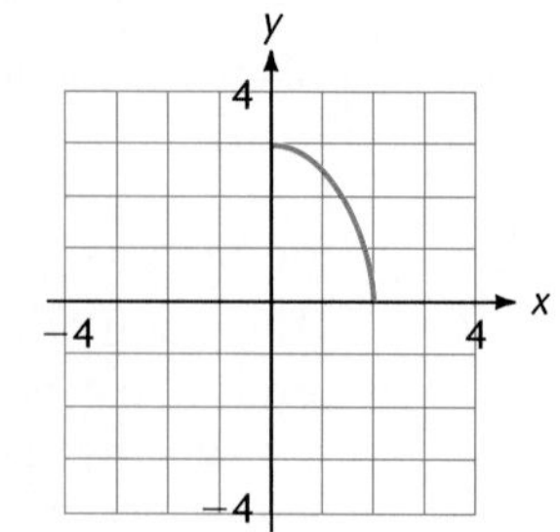

3.

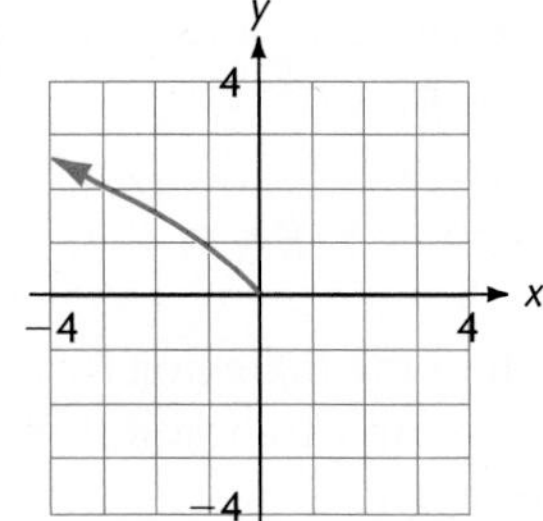

4.

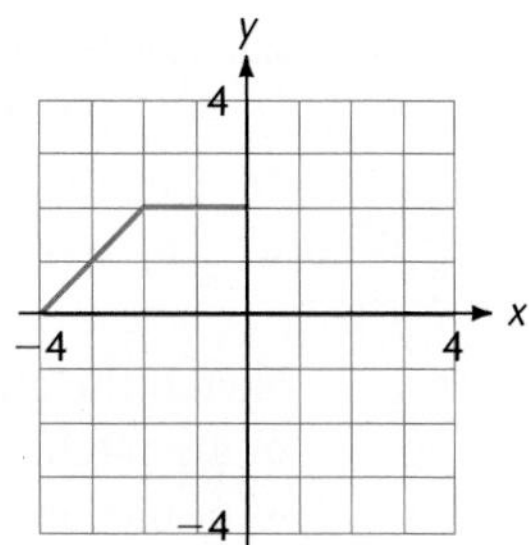

5.

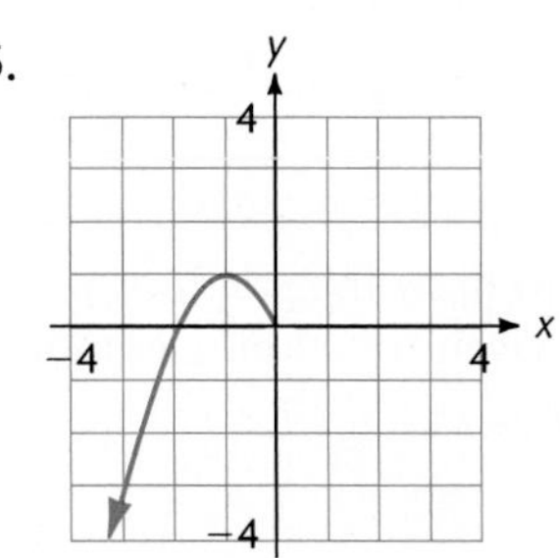

6.

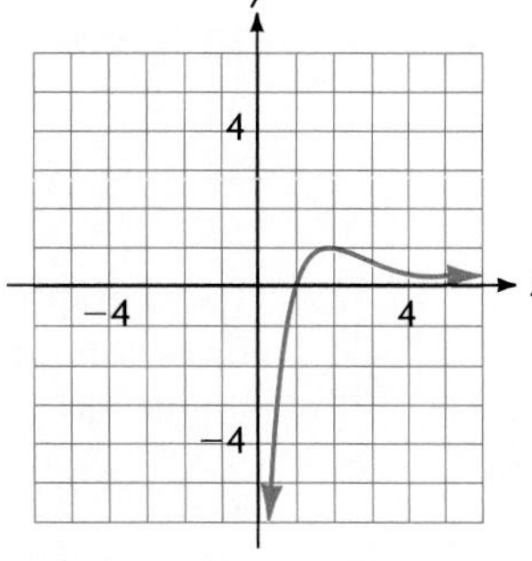

7.

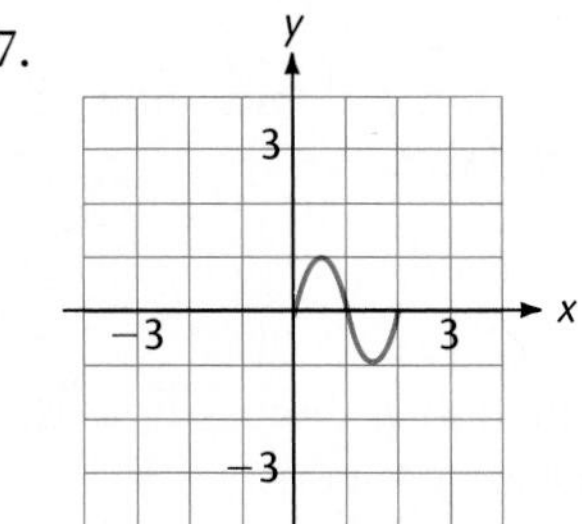

8.

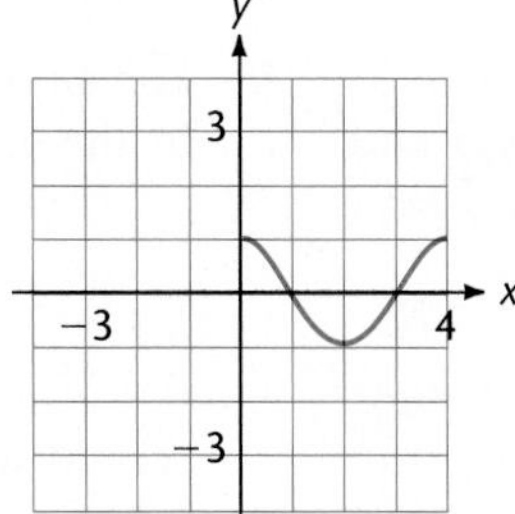

Test each of the following equations for symmetry. (See Example 2.)

9. $y = |x|$
10. $|y| = x$
11. $|y| = |x|$
12. $y = 2x^4$
13. $y^2 = x + 1$
14. $y^4 = x$
15. $y = x$
16. $y = x^5$
17. $x^2 + y^2 = 25$
18. $4x^2 + y^2 = 4$
19. $xy = 9$
20. $4 = xy$
21. $y = x^4 - 4x^2$
22. $y = x^5 - 4x^3$

(See Examples 3–6.)

23. Let the graph in Exercise 1 be the graph of $y = f(x)$. Sketch the graph of (a) $y = 2f(x)$ and (b) $y = f(x) - 2$.
24. Let the graph in Exercise 2 be the graph of $y = f(x)$. Sketch the graph of (a) $y = -f(x)$ and (b) $y = f(x) + 1$.

25. Let the graph in Exercise 3 be the graph of $y = f(x)$. Sketch the graph of (a) $y = f(x - 1)$ and (b) $y = -f(x)$.
26. Let the graph in Exercise 4 be the graph of $y = f(x)$. Sketch the graph of (a) $y = -2f(x)$ and (b) $y = f(x - 2)$.

Sketch the graphs of each of the following by using either one of or a combination of a stretching, a reflection, or a translation of a familiar function. (See Examples 3–6 and Practice Problem 1.)

27. $y = \frac{1}{2}x^3$
28. $y = -2|x|$
29. $y = |x + 1|$
30. $y = |x| + 1$
31. $y = (x - 1)^3$
32. $y = (x + 2)^3$
33. $y = x^3 - 1$
34. $y = x^3 + 2$
35. $y = (x + 1)^3 - 2$
36. $y = (x - 1)^3 + 2$

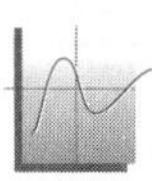
37–42

Use a graphing device to draw the graph of $y = f(x) = x^{1/3}$. Without erasing, then draw the graph of the given function. Describe the manner in which the original graph has been changed.

37. $y = f(x - 5)$
38. $y = f(x) - 5$
39. $y = 3f(x)$
40. $y = -3f(x)$
41. $y = f(-x)$
42. $y = f(-10x)$

Critical Thinking: Exploration and Discussion

43. For an arbitrary function f, discuss how the graph of $y = f(-x)$ is related to the graph of $y = f(x)$.

Critical Thinking: Exploration and Discussion

44. For an arbitrary function f, discuss how the graph of $y = -f(-x)$ is related to the graph of $y = f(x)$.

Critical Thinking: Exploration and Writing

45. Is there a function f whose graph is symmetric with respect to the origin? If so, give an example; otherwise, explain why not.

Critical Thinking: Exploration and Writing

46. Is there a function f whose graph is symmetric with respect to the x-axis? If so, give an example; otherwise, explain why not.

◆ Solution for Practice Problem

1. The familiar graph of $y = |x|$ is shown in Figure 4.20a. The graph of $y = |x - 1|$ can be obtained from this by translating to the right 1 unit, as shown in Figure 4.20b. A vertical translation downward 2 units then yields the graph of $y = |x - 1| - 2$, as shown in Figure 4.20c. As a check on our work, we determine that the coordinates $(0, -1)$ and $(2, -1)$ do indeed satisfy the equation $y = |x - 1| - 2$.

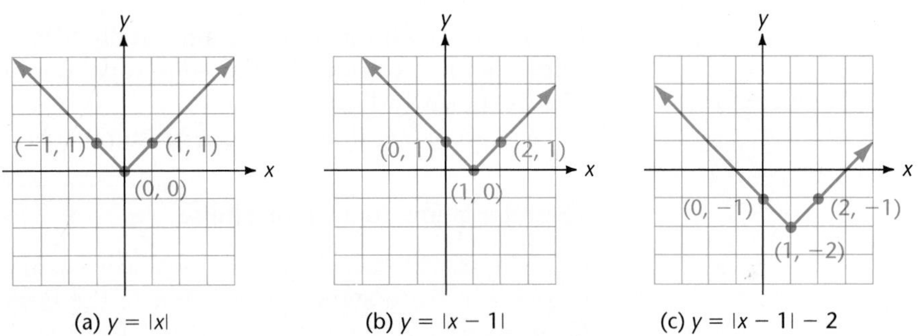

Figure 4.20 Translating $y = |x|$ horizontally and vertically

4.3 THE ELLIPSE

The sketches of the ellipse and the circle in Figure 4.1b show that their shapes are somewhat similar, with the ellipse being oval-shaped while the circle is round.

The concept of distance was used to define a circle in the last section, and distances can also be used to define an ellipse in the following way.

> **Definition of an Ellipse**
>
> Let F_1 and F_2 denote two fixed points in a plane. An **ellipse** is the set of all points P in the plane such that the sum of the distance from P to F_1 and the distance from P to F_2 is a constant. Each of the fixed points is called a **focus** (plural: **foci**) of the ellipse.

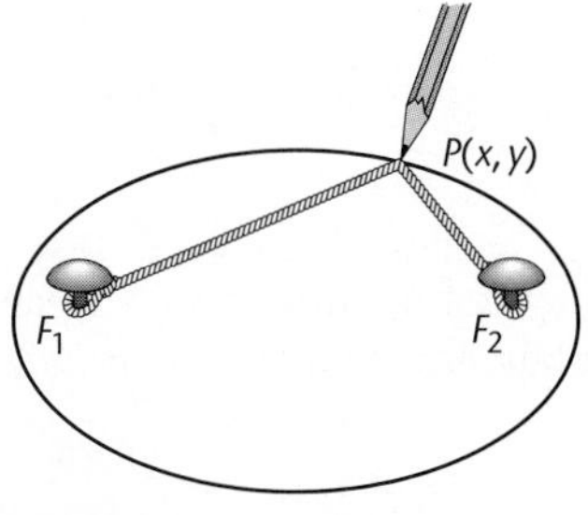

Figure 4.21 Drawing an ellipse

Figure 4.21 shows a method of drawing an ellipse that illustrates the definition very well. With this method, we fasten the two ends of a string with two thumbtacks, leaving slack in the string. When we pull the string taut, but not stretched, with a pencil and move the pencil to all possible positions, the pencil tip traces out an ellipse with foci at the thumbtacks.

We do not go through all the details here, but the distance formula can be used to obtain a standard equation for an ellipse that is analogous to the standard equation for a circle. Suppose first that the foci are located at equal distances from the origin on the x-axis as shown in Figure 4.22a. The ellipse is then symmetric with respect to the x-axis, the y-axis, and the origin. If we choose a and b to be positive numbers such that $\pm a$ denotes the x-intercepts and $\pm b$ denotes the y-intercepts, the ellipse will have an equation of the form

$$\frac{x^2}{a^2} + \frac{y^2}{b^2} = 1, \qquad \text{where } a > b.$$

If the foci are on the y-axis symmetrically placed with respect to the origin, the ellipse can be described by the same equation, but with $b > a$ instead of $a > b$. This is shown in Figure 4.22b.

The preceding discussion can be summarized in the following way.

Standard Equation of an Ellipse

An ellipse with foci on a coordinate axis symmetrically placed with respect to the origin has a **standard equation** of the form

$$\frac{x^2}{a^2} + \frac{y^2}{b^2} = 1.$$

Typical graphs are shown in Figure 4.22, and it is worth noting that the larger square occurs under the variable that corresponds to the location of the foci.

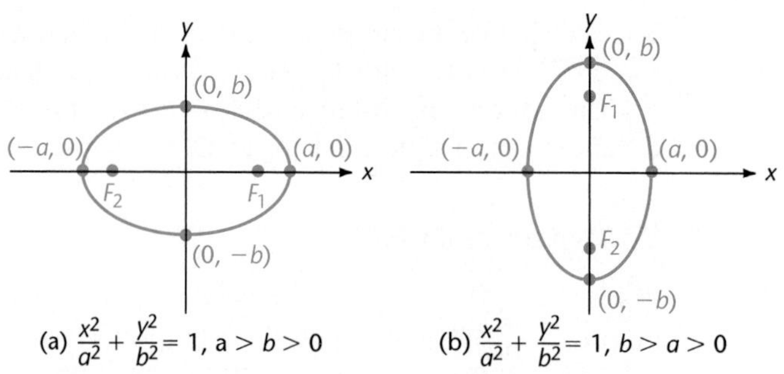

Figure 4.22 Typical ellipses

Example 1 • Graphing an Ellipse

Sketch the graph of the ellipse $4x^2 + 9y^2 = 36$.

Solution

Since the given equation is not in standard form, it may not be immediately clear that this is an equation of an ellipse. In order to obtain the standard form, we divide both members by 36. This gives

$$\frac{x^2}{9} + \frac{y^2}{4} = 1.$$

This fits the standard form with $a = 3$ and $b = 2$. Since $a > b$, we have an ellipse with foci on the x-axis. It is easy to see that the x-intercepts are ± 3 and the y-intercepts are ± 2. With this information, we can draw the graph as in Figure 4.23. □

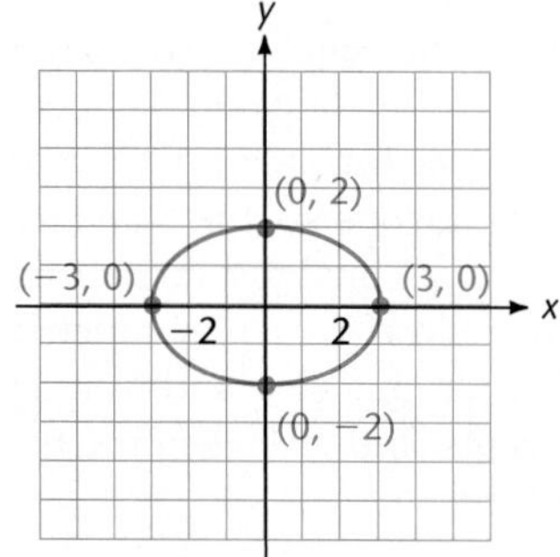

Figure 4.23 Graph of $4x^2 + 9y^2 = 36$

Example 1 illustrates how easy it is to sketch the graph of an ellipse by using the intercepts. The intercepts can be read by inspection from the standard equation. They are always given by $\pm a$ and $\pm b$, *with the square of the x-intercepts under* x^2 *and the square of the y-intercepts under* y^2. It is usually easier to find the intercepts directly from the original equation than it is to obtain the standard form. This is done in the solution to Practice Problem 1.

◆ Practice Problem 1 ◆

Sketch the graph of the ellipse $16x^2 + 9y^2 = 144$.

Consider again the ellipses in Figure 4.22. The two line segments, one joining the x-intercepts and the other joining the y-intercepts, are called the **axes** of the ellipse. The longer axis is the **major axis,** and it always contains the foci. The endpoints of the major axis are called the **vertices** of the ellipse. The shorter axis is designated as the **minor axis,** and the **center** of the ellipse is the point where the two axes cross.

There are variations in the graphs of ellipses that are analogous to the variations in the graphs of parabolas and circles that we saw earlier. One such variation is demonstrated in the next example.

Example 2 • The Top Half of an Ellipse

Sketch the graph of $y = \dfrac{\sqrt{4 - 9x^2}}{2}$.

Solution

In the given form, the equation does not fit the standard form. In order to recognize the type of graph we are dealing with, we multiply both sides by 2 and then square both sides. This gives

$$(2y)^2 = (\sqrt{4 - 9x^2})^2 \qquad \text{or} \qquad 4y^2 = 4 - 9x^2.$$

Adding $9x^2$ to both sides, we have

$$4y^2 + 9x^2 = 4.$$

We recognize this as the equation of an ellipse. Setting $x = 0$, we find that the y-intercepts are ± 1. With $y = 0$, we obtain x-intercepts $\pm\frac{2}{3}$.

The graph of this equation is given in Figure 4.24a. But this is *not* the graph of the original equation $y = \sqrt{4 - 9x^2}/2$ because the original equation requires that y not be negative. Taking only the top half of the ellipse, we obtain the graph of the original equation as shown in Figure 4.24b. □

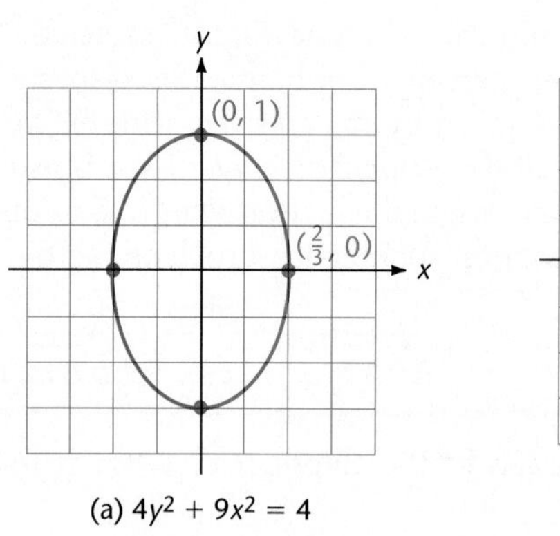

(a) $4y^2 + 9x^2 = 4$

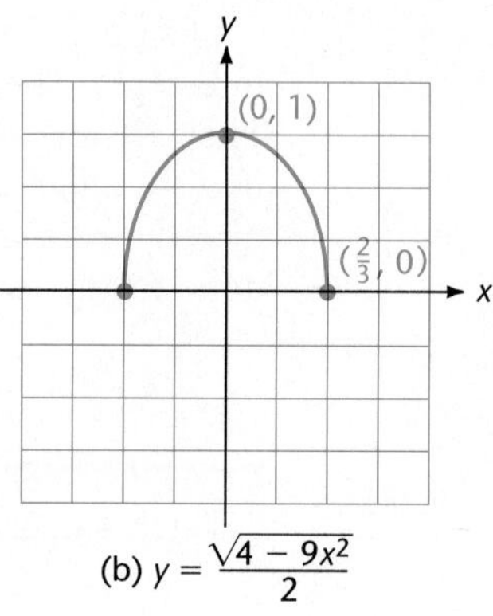

(b) $y = \frac{\sqrt{4 - 9x^2}}{2}$

Figure 4.24 Graphing half an ellipse

EXERCISES 4.3

Sketch the graphs of the following equations.

1. $\frac{x^2}{4} + \frac{y^2}{9} = 1$
2. $\frac{x^2}{25} + \frac{y^2}{16} = 1$
3. $25x^2 + 4y^2 = 100$
4. $9x^2 + 16y^2 = 144$
5. $4x^2 + 25y^2 = 25$
6. $9x^2 = 9 - 16y^2$
7. $\frac{4x^2}{9} + \frac{9y^2}{16} = 1$
8. $\frac{9x^2}{4} + \frac{16y^2}{25} = 1$
9. $64x^2 + 36y^2 = 100$
10. $36x^2 + 16y^2 = 25$
11. $2x^2 + y^2 = 8$
12. $3x^2 + y^2 = 12$
13. $x^2 + 4y^2 = 0$
14. $4x^2 + 9y^2 = 0$
15. $2y = \sqrt{16 - x^2}$
16. $3y = \sqrt{25 - 4x^2}$
17. $5x = -\sqrt{9 - 4y^2}$
18. $3x = -\sqrt{25 - 6y^2}$
19. $y = -\sqrt{16 - 9x^2}$
20. $y = -\sqrt{4 - 9x^2}$
21. $x = \frac{\sqrt{25 - 16y^2}}{2}$
22. $y = \frac{-\sqrt{100 - 9x^2}}{4}$

In Exercises 23–26, write an equation for the given ellipse.

23.

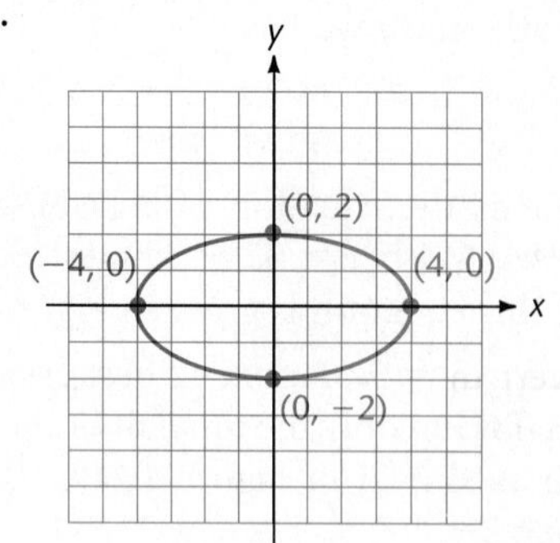

24.

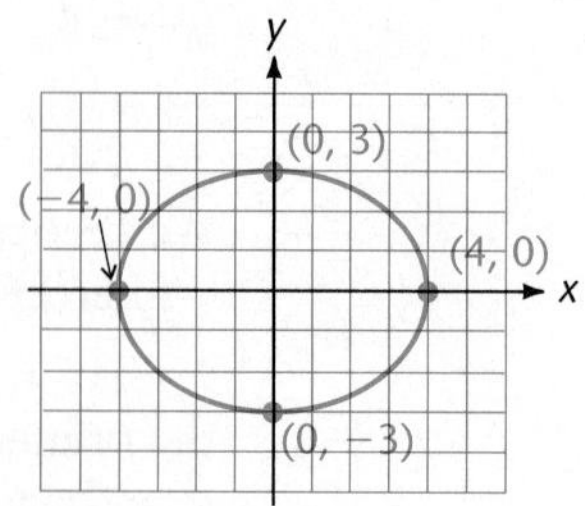

25. 26.

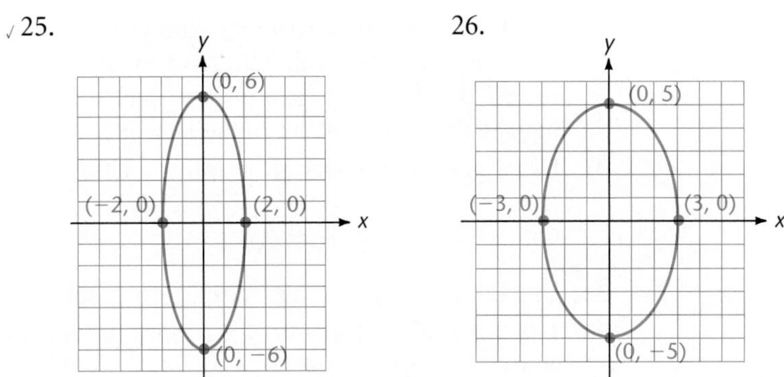

Elliptical Arch 27. The arch over the gateway in the accompanying figure is in the shape of half an ellipse, with the major axis vertical. The arch is 12 feet across at the bottom and 9 feet high at the center. How high is the arch at a point 3 feet from the center of the gateway?

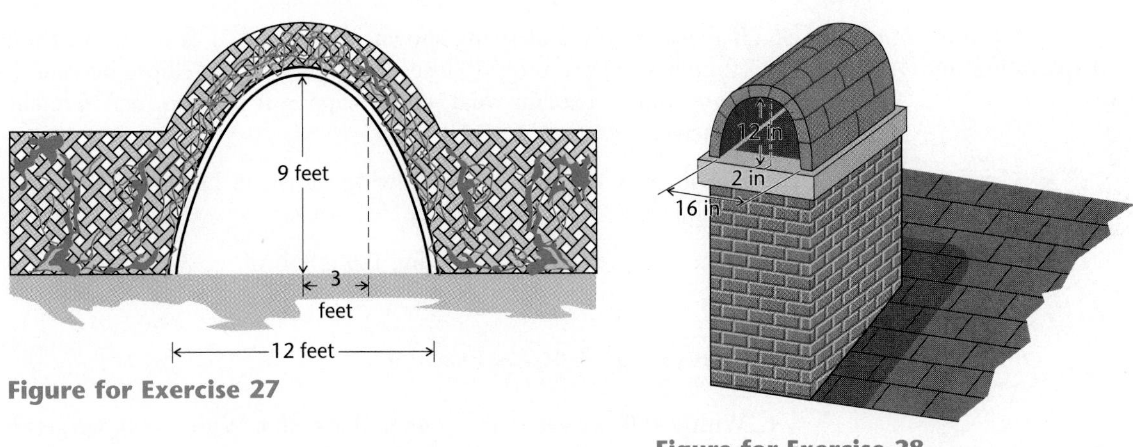

Figure for Exercise 27

Figure for Exercise 28

Chimney Cap 28. The chimney cap in the accompanying figure forms an arch in the shape of half an ellipse. The base of the arch is 16 inches across, and the highest point on the arch is 12 inches above the base. Find the height of the arch above the base at a point 2 inches from the center of the base.

Elliptical Culvert 29. A metal culvert under a street has a cross section that is an ellipse with a horizontal major axis of length 6 feet and a vertical minor axis of length 4 feet. Find the horizontal distance across the culvert 1 foot above the center of the ellipse.

Roadway Tunnel

30. A cross section of a roadway tunnel has the shape of half an ellipse with the minor axis vertical. The road in the tunnel is 24 feet wide, and the tunnel is 9 feet high at a point 6 feet from the center of the road. How high is the tunnel at its highest point?

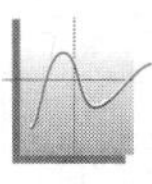

31–34

31. Use a graphing device to graph the ellipse $5x^2 + 21y^2 = 420$ by graphing the two functions $Y_1 = \sqrt{\dfrac{420 - 5x^2}{21}}$ and $Y_2 = -\sqrt{\dfrac{420 - 5x^2}{21}}$.
32. Use the technique in Exercise 31 to graph the ellipse $17x^2 + 5y^2 = 340$.
33. Use a graphing device to draw only the fourth-quadrant portion of the ellipse $y^2 + 6x^2 = 150$.
34. Graph the piecewise-defined function f on a graphing device.

$$f(x) = \begin{cases} \sqrt{50 - 2x^2}, & \text{if } x \le 0 \\ 5\sqrt{2} - x, & \text{if } x > 0 \end{cases}$$

Critical Thinking: Exploration and Writing

35. If p and q are two positive numbers, write a description of the way the graphs of $\dfrac{x^2}{p^2} + \dfrac{y^2}{q^2} = 1$ and $\dfrac{x^2}{q^2} + \dfrac{y^2}{p^2} = 1$ are related.

Critical Thinking: Exploration and Writing

36. If the same piece of string shown in Figure 4.21 is used, but the foci (thumbtacks) are moved closer together, will the ellipse become flatter or more round? Explain what would happen if the two foci were allowed to coincide.

Critical Thinking: Writing

37. Explain why the method of drawing shown in Figure 4.21 always produces an ellipse.

Critical Thinking: Writing

38. Write a procedure for sketching the graph of $\dfrac{x^2}{a^2} + \dfrac{y^2}{b^2} = 1$.

◆ Solution for Practice Problem

1. With $y = 0$, we get $16x^2 = 144$ and $x = \pm 3$. With $x = 0$, we get $9y^2 = 144$ and $y = \pm 4$. The graph is shown in Figure 4.25.

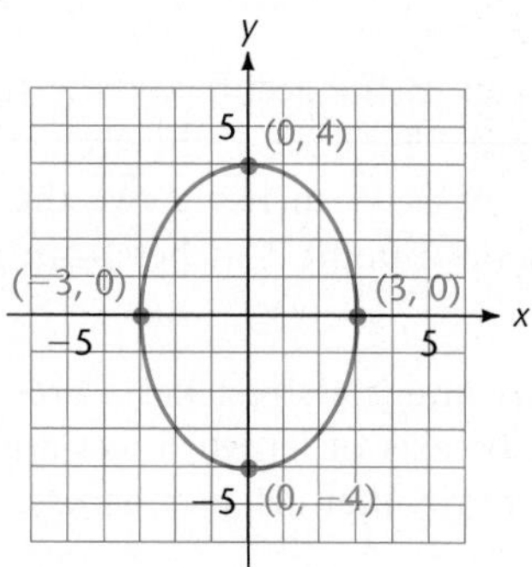

Figure 4.25 Graph of $16x^2 + 9y^2 = 144$

4.4 THE HYPERBOLA

In this section, we formulate some standard forms for an equation of a hyperbola. As with the ellipse, our development is limited to special situations. A hyperbola can be defined in a manner that is analogous to the way in which we defined an ellipse.

Definition of a Hyperbola

Let F_1 and F_2 denote two fixed points in a plane. A **hyperbola** is the set of all points P in the plane such that the absolute value of the difference of the distances from P to the fixed points is a constant. Each of the fixed points is called a **focus** (plural: **foci**) of the hyperbola.

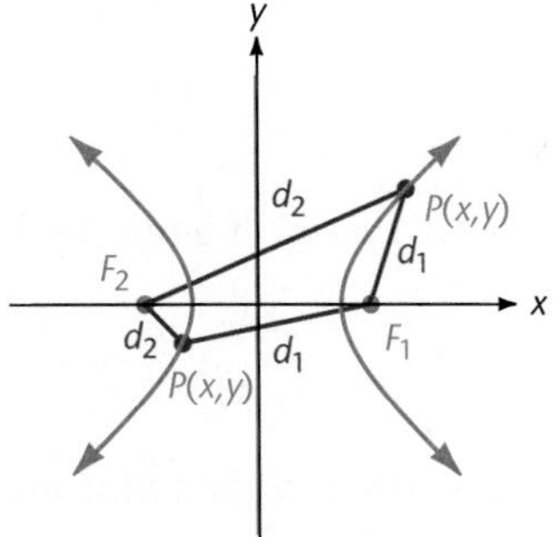

Figure 4.26 Hyperbola with foci on the x-axis

We consider first the case where the foci F_1 and F_2 are located on the x-axis at equal distances from the origin. We let d_1 be the distance from $P(x, y)$ to F_1, and d_2 be the distance from $P(x, y)$ to F_2. This is illustrated in Figure 4.26, where $P(x, y)$ is drawn at two locations where $|d_1 - d_2|$ has the constant value referred to in the definition. The distance formula can be used to derive standard equations for the hyperbola, but the derivation is not included in this text.

A hyperbola with foci on the x-axis has an equation of the form

$$\frac{x^2}{a^2} - \frac{y^2}{b^2} = 1. \tag{1}$$

Figure 4.27 shows the hyperbola that has this equation, and the points labeled in the figure indicate how the numbers a and b in the equation are related to the graph. The points on the branches of the curve that are nearest the other branch are called **vertices** and are labeled $V_1(a, 0)$ and $V_2(-a, 0)$ in the figure. The dashed lines and line segments are *not* part of the graph but serve as guides in drawing the hyperbola. The dashed lines passing through the origin and the vertices of the

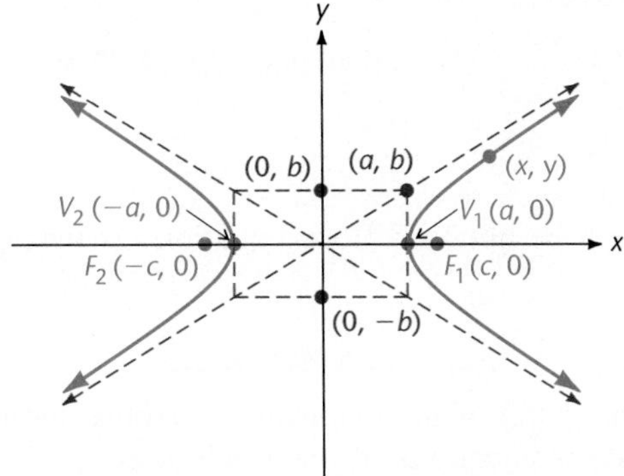

Figure 4.27 Graph of $\frac{x^2}{a^2} - \frac{y^2}{b^2} = 1$

rectangle have equations $y = \pm(b/a)x$. As the distance between x and 0 increases from a, the points (x, y) on the hyperbola are nearer these dashed lines. These dashed lines are called the **asymptotes** of the hyperbola. The rectangle is drawn to serve as a guide in sketching the asymptotes, and it emphasizes that the slope of the asymptotes are $\pm b/a$.

It is an interesting fact that if the number on the right-hand side of Equation (1) is replaced by 0, we obtain an equation of the asymptotes. This is easy to confirm, for the equation

$$\frac{x^2}{a^2} - \frac{y^2}{b^2} = 0$$

is equivalent to

$$b^2x^2 - a^2y^2 = 0$$

or

$$(bx - ay)(bx + ay) = 0.$$

This product is zero if and only if one of the factors is zero, that is, if and only if

$$y = \pm\frac{b}{a}x.$$

The main results of the foregoing discussion are summarized in the following statements.

Hyperbola with Foci on the *x*-Axis

A hyperbola with foci on the x-axis symmetrically placed with respect to the origin has a **standard equation** of the form

$$\frac{x^2}{a^2} - \frac{y^2}{b^2} = 1.$$

The equations of the asymptotes are given by

$$y = \pm\frac{b}{a}x,$$

which can be obtained from Equation (1) by replacing the 1 by 0.

Example 1 • Graphing a Hyperbola

Sketch the graph of the following hyperbola, locate the vertices and asymptotes, and write the equations of the asymptotes.

$$9x^2 - 4y^2 = 36$$

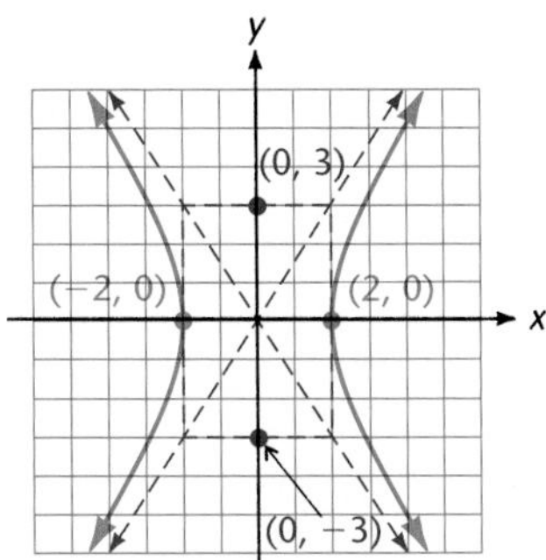

Figure 4.28 Graph of $9x^2 - 4y^2 = 36$

Solution

We first divide both sides by 36 to obtain the standard form of the equation. This gives

$$\frac{x^2}{4} - \frac{y^2}{9} = 1 \quad \text{or} \quad \frac{x^2}{(2)^2} - \frac{y^2}{(3)^2} = 1.$$

Using $a = 2$ and $b = 3$ in the same manner as in Figure 4.27, we obtain the graph in Figure 4.28. The vertices are located at $(\pm 2, 0)$, and the asymptotes are given by $y = \pm\frac{3}{2}x$. □

As one would expect, the hyperbola with foci on the x-axis has a dual where the foci are on the y-axis. This dual can be stated as follows, and it is illustrated in Figure 4.29.

Hyperbola with Foci on the y-Axis

A hyperbola with foci on the y-axis symmetrically placed with respect to the origin has a **standard equation** of the form

$$\frac{y^2}{a^2} - \frac{x^2}{b^2} = 1. \tag{2}$$

The equations of the asymptotes are given by

$$y = \pm\frac{a}{b}x,$$

which can be obtained from Equation (2) by replacing the 1 by 0.

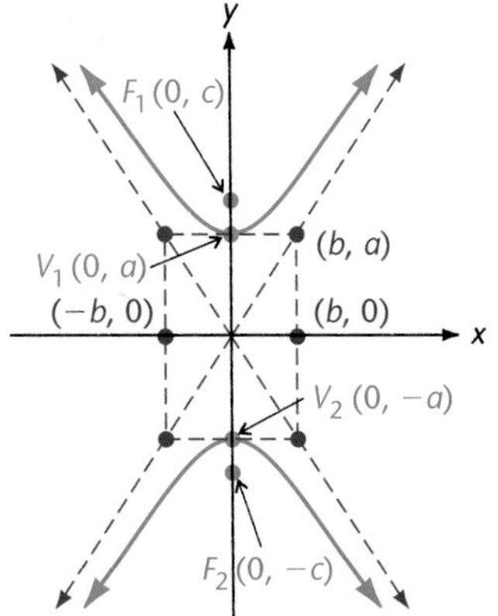

Figure 4.29 Graph of $\frac{y^2}{a^2} - \frac{x^2}{b^2} = 1$

It is easy to see that this dual statement can be obtained from the first one simply by interchanging x and y.

◆ Practice Problem 1 ◆

Sketch the graph of the following hyperbola, locate the vertices and asymptotes, and write the equations of the asymptotes.

$$\frac{y^2}{16} - \frac{x^2}{4} = 1$$

Once again, there are variations in the equations that correspond to only portions of a hyperbola.

Example 2 • The Bottom Half of a Hyperbola

Sketch the graph of $y = -\dfrac{5\sqrt{x^2 + 16}}{4}$.

Solution

If we multiply both sides of the given equation by 4 and then square both sides, we have

$$(4y)^2 = 25(x^2 + 16) \qquad \text{or} \qquad 16y^2 = 25x^2 + 400.$$

To obtain a standard equation, we subtract $25x^2$ from both sides and then divide both sides by 400. This gives

$$\frac{y^2}{25} - \frac{x^2}{16} = 1 \qquad \text{or} \qquad \frac{y^2}{(5)^2} - \frac{x^2}{(4)^2} = 1.$$

We recognize this as the standard equation of a hyperbola with $a = 5$, $b = 4$, and foci on the y-axis. Using Figure 4.29 as a guide, we obtain the graph of this last equation as given in Figure 4.30a. But the original equation

$$y = -\frac{5\sqrt{x^2 + 16}}{4}$$

requires that y be nonpositive. Thus we must take only the bottom half of the hyperbola to have the graph of the original equation. This is shown in Figure 4.30b. □

A graphing device can be used to observe the asymptotic behavior of a hyperbola by drawing both the hyperbola and its asymptotes in the same viewing rectangle.

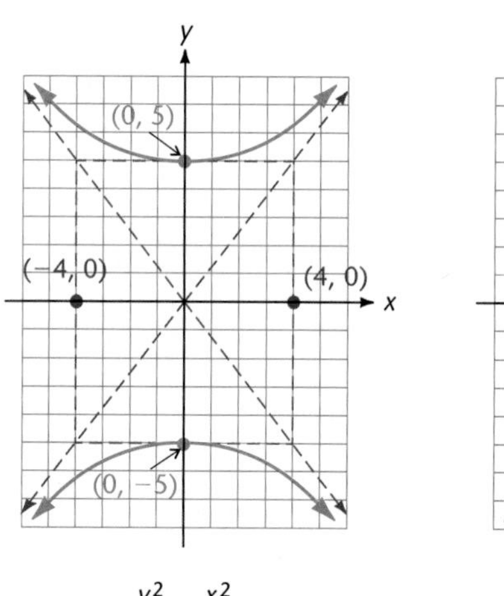

(a) $\frac{y^2}{25} - \frac{x^2}{16} = 1$

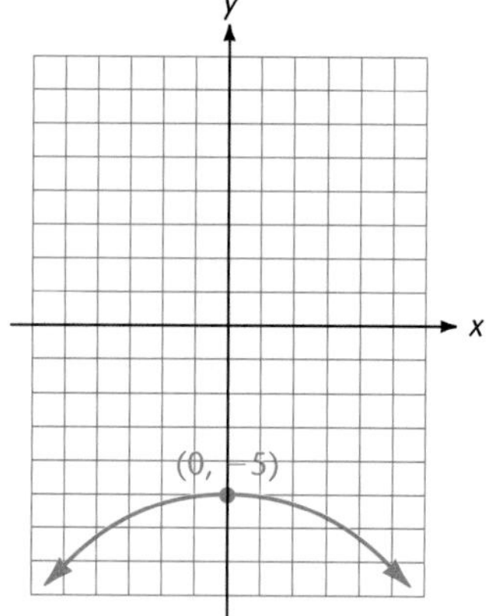

(b) $y = -\frac{5\sqrt{x^2+16}}{4}$

Figure 4.30 Graphing half a hyperbola

EXERCISES 4.4

Sketch the graphs of the following equations. Locate the vertices and asymptotes of all hyperbolas and write the equations of the asymptotes.

1. $\frac{x^2}{9} - \frac{y^2}{16} = 1$
2. $\frac{x^2}{9} - \frac{y^2}{4} = 1$
3. $\frac{y^2}{4} - \frac{x^2}{25} = 1$
4. $\frac{y^2}{36} - \frac{x^2}{25} = 1$

5. $y^2 - x^2 = 9$

6. $y^2 - x^2 = 16$

7. $4x^2 - 9y^2 = 36$

8. $36x^2 - 4y^2 = 144$

9. $4x^2 - 25y^2 = 0$

10. $16x^2 - 9y^2 = 0$

11. $x^2 - 4y^2 = 4$

12. $9x^2 - y^2 = 9$

13. $\dfrac{16y^2}{25} - \dfrac{9x^2}{4} = 1$

14. $\dfrac{25y^2}{16} - \dfrac{4x^2}{9} = 1$

15. $2x^2 - y^2 = 8$

16. $y^2 - 3x^2 = 75$

17. $5y = \sqrt{9 + 4x^2}$

18. $3y = \sqrt{25 + 16x^2}$

19. $2x = \sqrt{16 + y^2}$

20. $3x = \sqrt{25 + 4y^2}$

In Exercises 21–24, write an equation for the given hyperbola.

21.

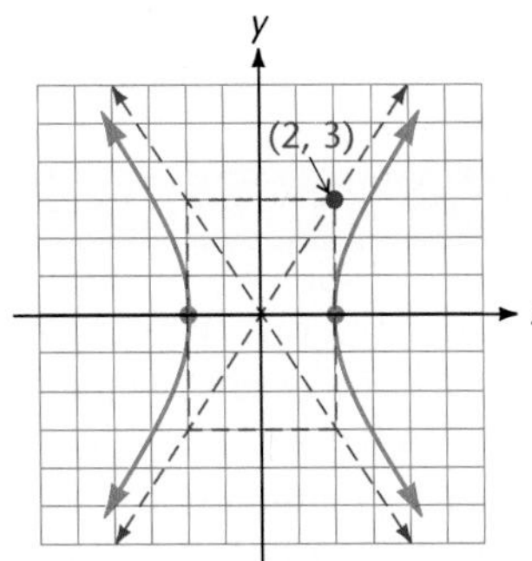

22.

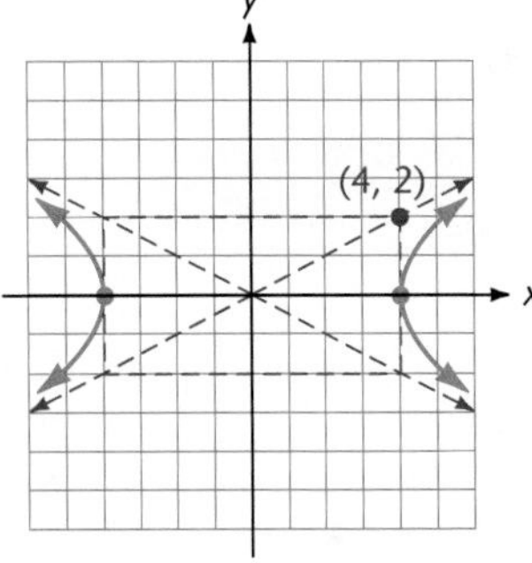

23.

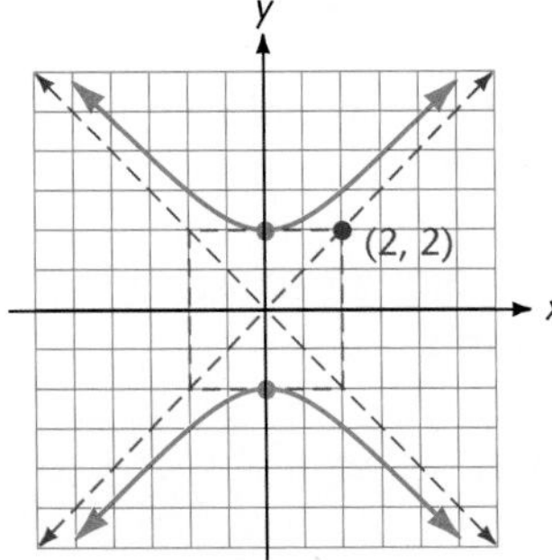

24.

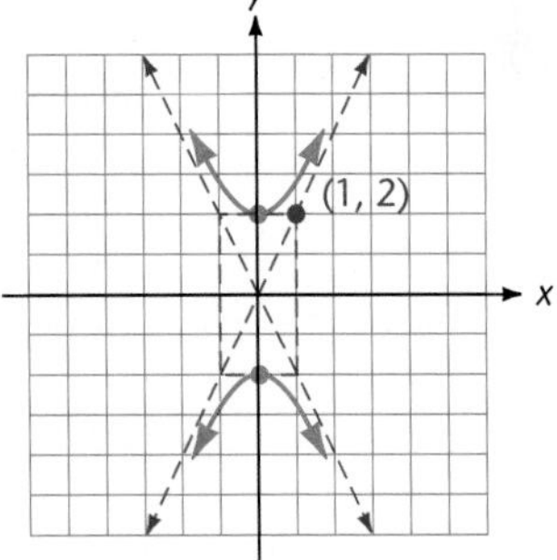

25–28

25. Use a graphing device to draw the hyperbola $y^2 - x^2 = 16$ by graphing the two functions $Y_1 = \sqrt{16 + x^2}$ and $Y_2 = -\sqrt{16 + x^2}$.

26. Use the technique in Exercise 25 to graph $x^2 - y^2 = 16$.

27. Use a graphing device to draw only the second-quadrant portion of the hyperbola $2y^2 - x^2 = 30$.

28. Use a graphing device to draw only the third-quadrant portion of the hyperbola $3x^2 - 5y^2 = 25$.

Critical Thinking: Exploration and Writing

29. If p and q are two positive numbers, write a description of the way the graphs of $\dfrac{x^2}{p^2} - \dfrac{y^2}{q^2} = 1$ and $\dfrac{x^2}{q^2} - \dfrac{y^2}{p^2} = 1$ are related.

Critical Thinking: Exploration and Writing

30. With p and q as in Exercise 29, write a description of the way the graphs of $\dfrac{x^2}{p^2} - \dfrac{y^2}{q^2} = 1$ and $\dfrac{y^2}{q^2} - \dfrac{x^2}{p^2} = 1$ are related.

◆ **Solution for Practice Problem**

1. We have $a = 4$ and $b = 2$ from the equation $y^2/16 - x^2/4 = 1$. Using Figure 4.29 as a guide, we draw the graph as shown in Figure 4.31. The vertices are at $(0, \pm 4)$, and the asymptotes have equations $y = \pm 2x$.

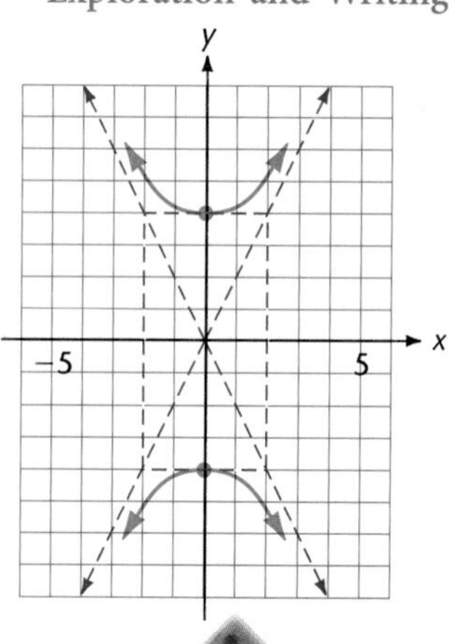

Figure 4.31 Graph of $\dfrac{y^2}{16} - \dfrac{x^2}{4} = 1$

4.5 TRANSLATIONS OF ELLIPSES AND HYPERBOLAS

Some of the earlier material in this chapter has paved the way for this section. Our experience in Section 4.1 shows that when a circle with radius r has its center translated from $(0, 0)$ to (h, k), the equation changes from $x^2 + y^2 = r^2$ to $(x - h)^2 + (y - k)^2 = r^2$. In studying translations of the graphs of functions, we found that the replacement of x in $y = f(x)$ by $x - h$ to obtain $y = f(x - h)$ causes a horizontal translation that shifts the graph $|h|$ units left or right. Similarly, replacement of y in $y = f(x)$ by $y - k$ to obtain $y - k = f(x)$ causes a vertical translation that shifts the graph $|k|$ units up or down. The translations of ellipses and hyperbolas that we do in this section fit in very naturally with this previous work.

If an ellipse with center at the origin is translated h units horizontally and k units vertically, the result is an ellipse with center at (h, k). Such an ellipse has the following **standard form.**

Ellipse with Center at (*h*, *k*)

Standard Equation: $\dfrac{(x-h)^2}{a^2} + \dfrac{(y-k)^2}{b^2} = 1$

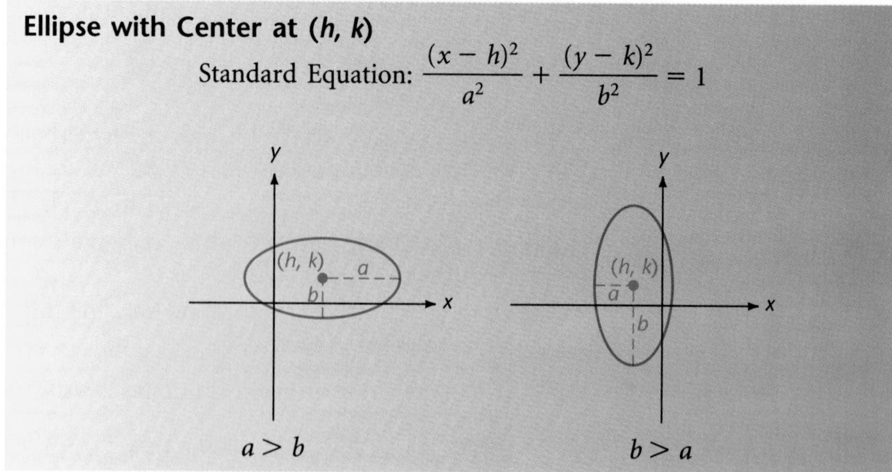

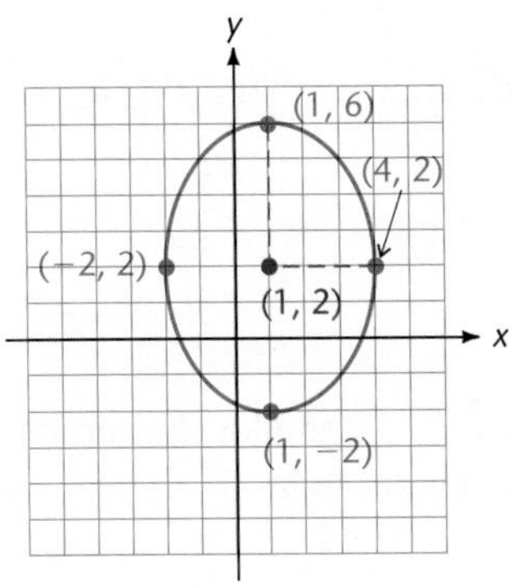

Figure 4.32 Graph of $\frac{(x-1)^2}{9} + \frac{(y-2)^2}{16} = 1$

Example 1 • Translating an Ellipse

Sketch the graph of

$$\frac{(x-1)^2}{9} + \frac{(y-2)^2}{16} = 1.$$

Solution

We see that $a^2 = 9$ and $b^2 = 16$, so $a = 3$ and $b = 4$. The center is at $(1, 2)$. We locate the ends of the minor axis by moving 3 units left or right from the center to the points

$$(1 + 3, 2) = (4, 2) \qquad \text{and} \qquad (1 - 3, 2) = (-2, 2).$$

We locate the vertices by moving 4 units up or down from the center to

$$(1, 2 + 4) = (1, 6) \qquad \text{and} \qquad (1, 2 - 4) = (1, -2).$$

With the ends of the axes located, the graph can then be drawn as in Figure 4.32. □

It may happen that an equation of an ellipse is written in a way that differs from the standard form. To gain information about the center and the axes, it is necessary to rewrite the equation in standard form. This is done in Example 2.

Example 2 • Finding the Standard Form

Rewrite $4x^2 + y^2 + 16x - 6y = -9$ in standard form, find the center, and state the values of a and b.

Solution

We begin by completing the square on the x-terms and also on the y-terms.

$$4(x^2 + 4x \qquad) + (y^2 - 6y \qquad) = -9 \qquad \text{grouping } x\text{-terms and } y\text{-terms}$$

$$4(x^2 + 4x + 4) + (y^2 - 6y + 9) = -9 + 4(4) + 9 \qquad \text{completing the square}$$

$$4(x + 2)^2 + (y - 3)^2 = 16 \qquad \text{factoring and simplifying}$$

$$\frac{(x+2)^2}{4} + \frac{(y-3)^2}{16} = 1 \qquad \text{dividing by 16}$$

The equation is now in standard form with $a^2 = 4$ and $b^2 = 16$. We can see from this that the center is at $(-2, 3)$, $a = 2$, and $b = 4$. □

Our next example also illustrates the procedure for changing an equation of an ellipse to standard form.

Example 3 • Graphing an Ellipse

Sketch the graph of $x^2 + 4x + 4y^2 - 8y = 8$.

Solution

We must complete the square on x and y to locate the center of the ellipse.

$$x^2 + 4x + 4 + 4(y^2 - 2y + 1) = 8 + 4 + 4$$
$$(x + 2)^2 + 4(y - 1)^2 = 16$$
$$\frac{(x + 2)^2}{4^2} + \frac{(y - 1)^2}{2^2} = 1$$

Thus the ellipse has center at $(-2, 1)$. We use $a = 4$ and $b = 2$ to plot the vertices and the endpoints of the minor axes in the ellipse in Figure 4.33. □

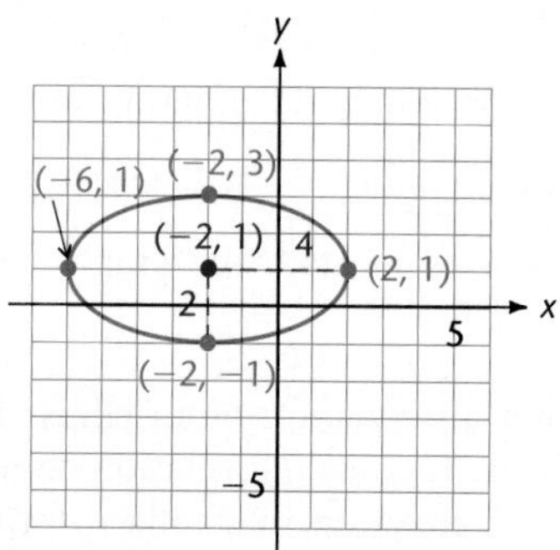

Figure 4.33 Graph of $x^2 + 4x + 4y^2 - 8y = 8$

Example 4 • Finding the Standard Equation of an Ellipse

Write the standard equation of the ellipse that has center at $(-3, 2)$, vertices at $(-3, -3)$ and $(-3, 7)$, and endpoints of the minor axis at $(-5, 2)$ and $(-1, 2)$.

Solution

The major axis extends vertically from one vertex at $(-3, -3)$ to the other at $(-3, 7)$ and therefore has length

$$\sqrt{(-3 + 3)^2 + (7 + 3)^2} = 10.$$

Thus $2b = 10$ and $b = 5$. Similarly, the minor axis extends horizontally from $(-5, 2)$ to $(-1, 2)$ and consequently has length $2a = 4$. This means that $a = 2$, and we can write the standard equation as

$$\frac{(x + 3)^2}{4} + \frac{(y - 2)^2}{25} = 1.$$

□

If a hyperbola with center at the origin is translated h units horizontally and k units vertically, the result is a hyperbola with center at (h, k). The corresponding **standard equations** and graphs are as follows.

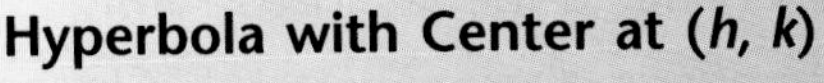

Hyperbola with Center at (h, k)

Standard Equation: $\dfrac{(x-h)^2}{a^2}-\dfrac{(y-k)^2}{b^2}=1$ $\qquad$ $\dfrac{(y-k)^2}{a^2}-\dfrac{(x-h)^2}{b^2}=1$

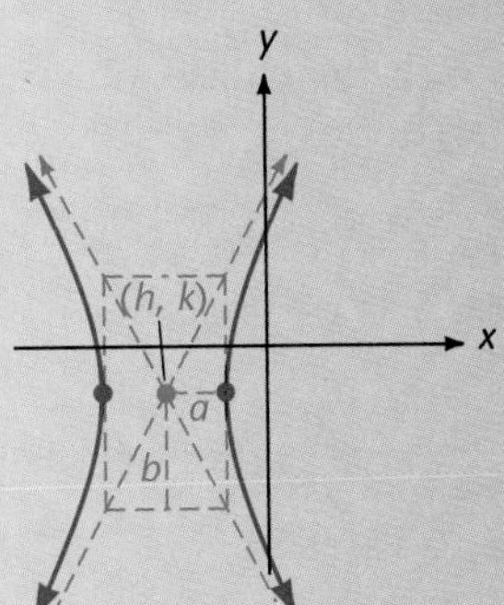

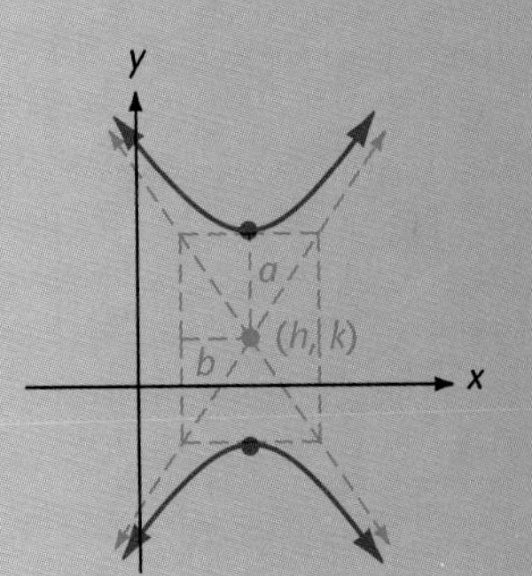

Example 5 • Translating a Hyperbola

Sketch the graph of $\dfrac{(x+1)^2}{9}-\dfrac{(y-1)^2}{4}=1.$

Solution

This is a hyperbola with center at $(-1, 1)$, $a = 3$, and $b = 2$. Since $a = 3$, we move 3 units to the right and left of the center to obtain the vertices of the hyperbola at $(-4, 1)$ and $(2, 1)$. Since $b = 2$, we move 2 units up and down from each vertex to locate the corners of the rectangle. With these points we can first sketch the rectangle, then the asymptotes, and finally the hyperbola in Figure 4.34. □

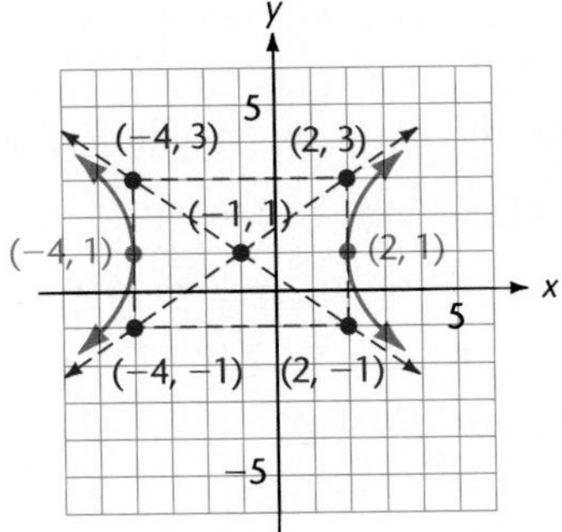

Figure 4.34 Graph of $\dfrac{(x+1)^2}{9}-\dfrac{(y-1)^2}{4}=1$

In the next example, we rewrite the equation of a hyperbola so that it fits one of the standard forms.

Example 6 • Finding the Standard Form

Rewrite $16x^2 - 9y^2 - 64x - 72y = 224$ in standard form, find the center, and state the values of a and b.

Solution

Just as with the ellipse, our first steps are to complete the square on the x-terms and the y-terms.

$$16(x^2 - 4x \qquad) - 9(y^2 + 8y \qquad) = 224 \qquad \text{factoring}$$

$$16(x^2 - 4x + 4) - 9(y^2 + 8y + 16) = 224 + 16(4) - 9(16) \qquad \text{completing the square}$$

$$16(x - 2)^2 - 9(y + 4)^2 = 144 \qquad \text{factoring and simplifying}$$

$$\frac{(x - 2)^2}{9} - \frac{(y + 4)^2}{16} = 1 \qquad \text{dividing by 144}$$

The last equation is in standard form with $a^2 = 9$ and $b^2 = 16$. Thus the center is at $(2, -4)$, $a = 3$, and $b = 4$. □

◆ Practice Problem 1 ◆

Rewrite $9x^2 - y^2 + 54x + 8y + 74 = 0$ in standard form, find the center, and state the values of a and b.

Example 7 • Graphing a Hyperbola

Sketch the graph of $y^2 - 4x^2 + 2y + 16x = 31$.

Solution

We begin by rewriting the equation in standard form.

$$(y^2 + 2y \qquad) - 4(x^2 - 4x \qquad) = 31 \qquad \text{factoring}$$

$$(y^2 + 2y + 1) - 4(x^2 - 4x + 4) = 31 + 1 - 4(4) \qquad \text{completing the square}$$

$$(y + 1)^2 - 4(x - 2)^2 = 16 \qquad \text{factoring and simplifying}$$

$$\frac{(y + 1)^2}{16} - \frac{(x - 2)^2}{4} = 1 \qquad \text{dividing by 16}$$

From the last equation, we see that the center is at $(2, -1)$, $a = 4$, and $b = 2$. Since the positive sign is on the square involving y, the branches open upward and downward with vertices at $(2, -1 \pm 4)$ or

$$(2, 3) \qquad \text{and} \qquad (2, -5).$$

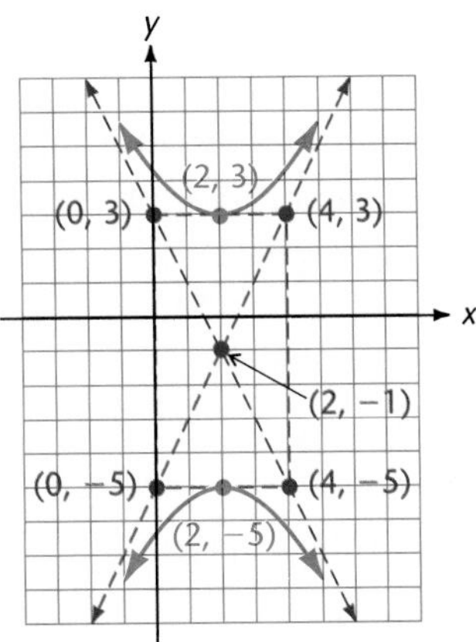

Figure 4.35 Graph of $y^2 - 4x^2 + 2y + 16x = 31$

We use $b = 2$ to locate the corners of the rectangle at

$$(0, 3),\ (4, 3),\ (0, -5),\ \text{and}\ (4, -5).$$

With these points located, we draw the rectangle, asymptotes, and hyperbola as shown in Figure 4.35. □

EXERCISES 4.5

Sketch the graphs of the following equations. Locate the center and vertices of ellipses and hyperbolas. (See Examples 1 and 5.)

1. $\dfrac{(x-2)^2}{4} + \dfrac{(y-1)^2}{9} = 1$
2. $\dfrac{(x+2)^2}{25} + \dfrac{(y-1)^2}{16} = 1$
3. $\dfrac{(x-1)^2}{9} - \dfrac{(y-2)^2}{25} = 1$
4. $\dfrac{(x+2)^2}{4} - \dfrac{(y-1)^2}{9} = 1$
5. $\dfrac{(y+3)^2}{4} - (x-2)^2 = 1$
6. $\dfrac{(y-1)^2}{16} - \dfrac{(x+4)^2}{4} = 1$
7. $4(x+2)^2 - 9(y+2)^2 = 36$
8. $4(x-3)^2 - y^2 = 4$
9. $25(x+1)^2 + 4(y-3)^2 = 100$
10. $9x^2 + 16(y+2)^2 = 144$
11. $9(x+2)^2 + (y-3)^2 = 81$
12. $4(x-1)^2 + 25(y+3)^2 = 100$

Rewrite each of the following equations in standard form. Identify the graph of the equation as an ellipse or a hyperbola, find the center, and state the values of a and b. (See Examples 2 and 6 and Practice Problem 1.)

13. $4x^2 + 9y^2 - 8x - 36y + 4 = 0$
14. $x^2 + 16y^2 - 2x + 64y + 49 = 0$
15. $9x^2 - y^2 - 36x + 2y + 44 = 0$
16. $4x^2 - 9y^2 - 8x - 36y + 4 = 0$
17. $9x^2 + 16y^2 - 54x + 64y + 1 = 0$
18. $9x^2 + 16y^2 + 18x - 32y = 119$
19. $x^2 - 4y^2 + 6x + 32y = 155$
20. $x^2 - 4y^2 - 2x - 32y = 79$

Sketch the graph of each of the following equations. (See Examples 3 and 7.)

21. $4x^2 - y^2 + 2y = 17$
22. $y^2 - x^2 - 4y - 2x + 2 = 0$
23. $4x^2 - 8x + y^2 + 6y = 3$
24. $9x^2 + 4y^2 + 36x - 8y + 4 = 0$
25. $x^2 + 9y^2 + 6x = 27$
26. $4x^2 + 25y^2 - 32x - 50y = 11$
27. $4y^2 - 9x^2 + 8y - 36x = 68$
28. $x^2 - 25y^2 + 4x + 150y = 246$

In Exercises 29–36, write the standard equation of the ellipse satisfying the given conditions. (See Example 4.)

29. Center: $(0, 0)$; x-intercepts: ± 3; y-intercepts: ± 2
30. Center: $(0, 0)$; x-intercepts: ± 6; y-intercepts: ± 3
31. Center: $(0, 0)$; x-intercepts: ± 1; y-intercepts: ± 4
32. Center: $(0, 0)$; x-intercepts: ± 2; y-intercepts: ± 7
33. Center: $(1, 0)$; vertices: $(3, 0)$, $(-1, 0)$; endpoints of minor axis: $(1, 1)$, $(1, -1)$

34. Center: (0, 2); vertices: (−4, 2), (4, 2); endpoints of minor axis: (0, 5), (0, −1)
35. Center: (2, 3); vertices: (2, 6), (2, 0); endpoints of minor axis: (0, 3), (4, 3)
36. Center: (1, −2); vertices: (1, 2), (1, −6); endpoints of minor axis: (4, −2), (−2, −2)

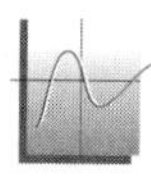
37–40

Use a graphing device to draw the graphs of the following equations.

37. $\dfrac{(x+3)^2}{16}+\dfrac{(y-1)^2}{25}=1$ 38. $\dfrac{(x-1)^2}{16}-\dfrac{(y+2)^2}{25}=1$

39. $x^2-y^2-4x-8y=21$ 40. $16x^2+9y^2-32x+18y=119$

Critical Thinking: Writing

41. Write a step-by-step procedure for sketching the graph of

$$\frac{(x-h)^2}{a^2}+\frac{(y-k)^2}{b^2}=1.$$

Critical Thinking: Writing

42. Write a step-by-step procedure for sketching the graph of the following.

a. $\dfrac{(x-h)^2}{a^2}-\dfrac{(y-k)^2}{b^2}=1$ b. $\dfrac{(y-k)^2}{a^2}-\dfrac{(x-h)^2}{b^2}=1$

◆ Solution for Practice Problem

1. $9(x^2+6x\qquad)-(y^2-8y\qquad)+74=0$ — factoring

$9(x^2+6x+9)-(y^2-8y+16)=-74+9(9)-16$ — completing the square

$9(x+3)^2-(y-4)^2=-9$ — factoring and simplifying

$\dfrac{(y-4)^2}{9}-(x+3)^2=1$ — dividing by −9

The center is at (−3, 4), $a=3$, and $b=1$.

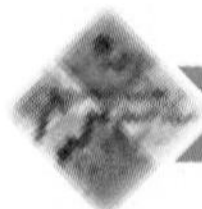

4.6 INVERSE FUNCTIONS

In the definition of a function *f*, it is required that for each *x*-value in the domain, there must be one and only one *y*-value, and this *y*-value is denoted by $y=f(x)$. That is,

for each x in the domain, there is a unique $y=f(x)$ in the range.

This requirement does allow different numbers in the domain to have the same function value. An illustration of this type of function f is shown in Figure 4.36.

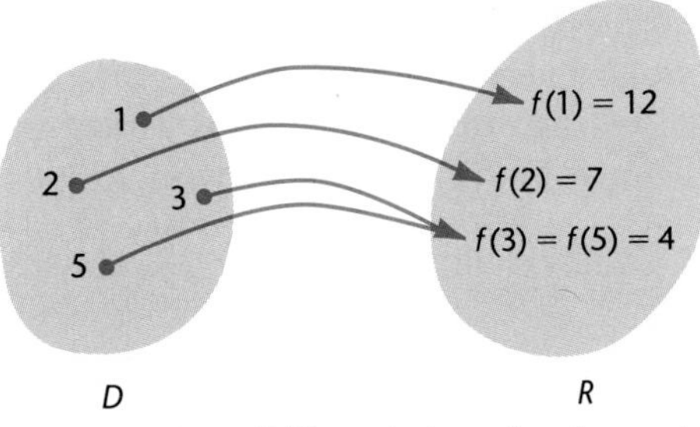

Figure 4.36 Different domain elements with the same function value

We concentrate in this section on those functions that have the following additional property:

If $a \neq b$ in the domain, then $f(a) \neq f(b)$ in the range.

A function g with this property is shown in Figure 4.37 to make a contrast with the function f in Figure 4.36. We see that $3 \neq 5$ in the domain, but $f(3) = f(5) = 4$ in Figure 4.36, while $g(3) = 4$ and $g(5) = 6$ are not equal in Figure 4.37.

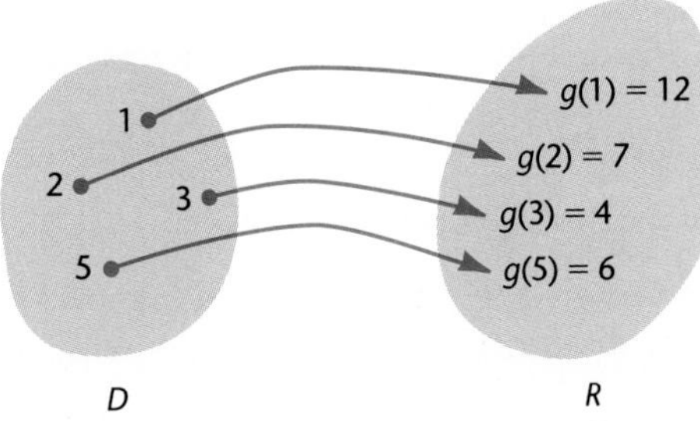

Figure 4.37 A one-to-one function

A function with the additional property that $a \neq b$ implies $f(a) \neq f(b)$ is called a **one-to-one function** because of the one-to-one pairing between x-values and y-values that is required by the additional property. Corresponding to the vertical line test for functions, we can state the following horizontal line test for one-to-one functions.

A quick way to determine if a function is one-to-one is to use the horizontal line test on the graph provided by a graphing device.

Horizontal Line Test

If any horizontal line intersects the graph of the function f at more than one point, then f *is not* a one-to-one function. If no horizontal line intersects the graph of the function f at more than one point, then f *is* a one-to-one function.

Example 1 • Using the Horizontal Line Test

The function with defining equation

$$f(x) = x^2$$

is *not* a one-to-one function since any horizontal line above the x-axis cuts the graph in more than one place. This is shown in Figure 4.38a. The graph of the function defined by $f(x) = x^3$ is shown in Figure 4.38b. The horizontal line test shows that $f(x) = x^3$ defines a one-to-one function. □

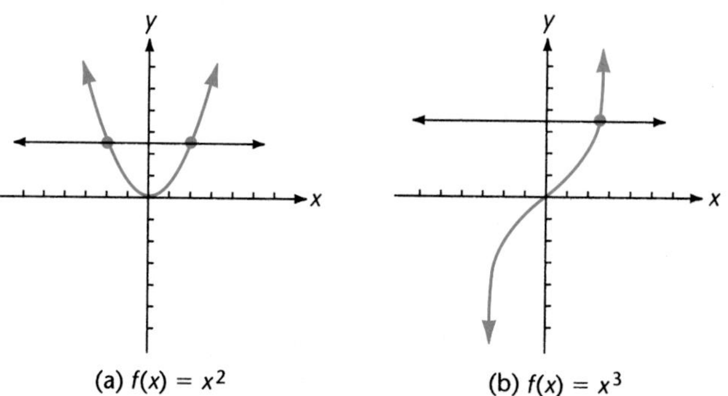

(a) $f(x) = x^2$ (b) $f(x) = x^3$

Figure 4.38 Illustrations of the horizontal line test

Every one-to-one function has an inverse function that can be defined as follows.

Definition of an Inverse Function

Let f be a one-to-one function with domain D and range R. The **inverse function** of f is the function denoted by f^{-1} and is defined by

$$f^{-1}(b) = a \qquad \text{if and only if} \qquad b = f(a).$$

The notation f^{-1} is read as "the inverse of f" or as "f-inverse." This must not be confused with the reciprocal of the function f, which is defined by $[f(x)]^{-1} = 1/[f(x)]$. That is,

$$f^{-1}(x) \neq \frac{1}{f(x)}.$$

The following facts are direct consequences of the definition of an inverse function.

1. The rule

   $f^{-1}(b) = a$ if and only if $b = f(a)$, defines a function since f is one-to-one.

2. The domain of f^{-1} is the range of f.
3. The range of f^{-1} is the domain of f.

A diagram of this situation is shown in Figure 4.39.

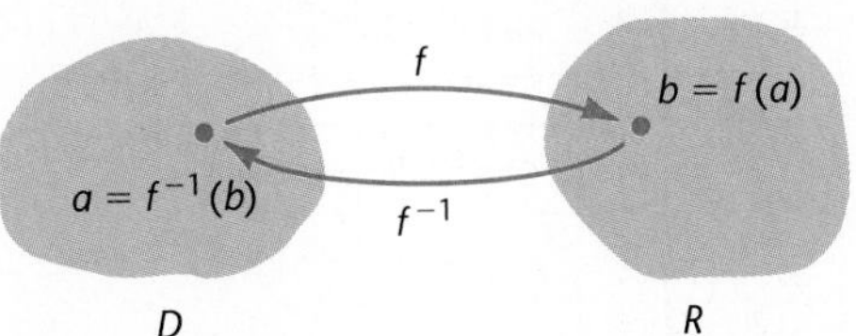

Figure 4.39 Function values of f and f^{-1}

Often a function f is described by a defining equation in x and y. In such cases, a defining equation for f^{-1} can be obtained by interchanging x and y in the defining equation for f. It is customary to solve for y in the equation that results from the interchange if this can be done with a reasonable amount of work. An illustration is given in Example 2.

Example 2 • Finding the Inverse of a One-to-One Function

Find an expression for $g^{-1}(x)$ if

$$g(x) = 3x + 6.$$

Also, draw the graphs of g and g^{-1} on the same coordinate system.

Solution

The defining equation for g is $y = 3x + 6$. Interchanging x and y in this equation, we obtain

$$x = 3y + 6$$

as a defining equation for g^{-1}. Solving for y in the last equation, we get

$$y = \tfrac{1}{3}x - 2.$$

Thus

$$g^{-1}(x) = \tfrac{1}{3}x - 2.$$

The graphs of g and g^{-1} are straight lines, as shown in Figure 4.40. The line $y = x$ is drawn in the figure as a dashed line, and the graphs of g and g^{-1} are symmetric about the line $y = x$. That is, the graph of g^{-1} is the reflection (mirror image) of the graph of g through the line $y = x$. □

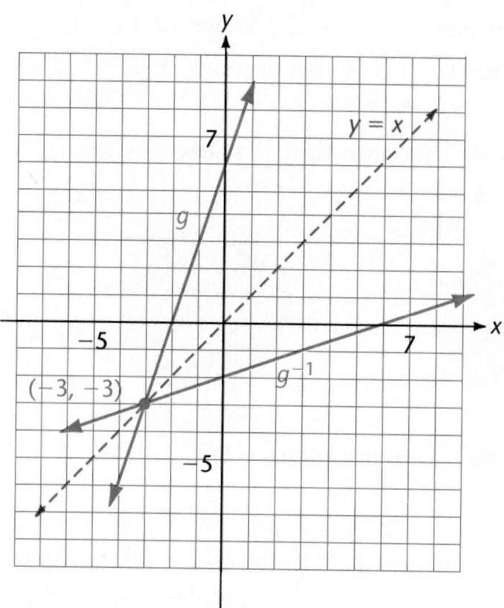

Figure 4.40 Graphs of g and g^{-1}

If a and b are any real numbers, the points (a, b) and (b, a) are located symmetrically with respect to the line $y = x$. For this reason, the graph of the inverse of a function is always the reflection of the graph of the original function through the line $y = x$.

It follows from the properties diagrammed in Figure 4.39 that

$$f^{-1}(f(x)) = x, \qquad \textbf{for all } x \textbf{ in the domain of } f,$$

and

$$f(f^{-1}(x)) = x, \qquad \textbf{for all } x \textbf{ in the domain of } f^{-1}.$$

Example 3 • Verifying Properties of f and f^{-1}

Each of the following equations defines a function f. Find an expression for $f(x)$ and also for $f^{-1}(x)$ if f is one-to-one. If f is one-to-one, verify each of the equations $f^{-1}(f(x)) = x$ and $f(f^{-1}(x)) = x$.

a. $x^2 - y + 3 = 0$ b. $3x - 2y = 6$

Solution

a. Solving for y in the defining equation, we get $y = x^2 + 3$, so

$$f(x) = x^2 + 3.$$

Since $f(1) = 4$ and $f(-1) = 4$, f is not one-to-one. (The horizontal line test on the graph of $y = x^2 + 3$ would also show that f is not one-to-one.)

b. Solving for y, we obtain $y = \frac{3}{2}x - 3$, so

$$f(x) = \tfrac{3}{2}x - 3.$$

The graph of f is a nonhorizontal straight line, so f is one-to-one. A defining equation for f^{-1} is

$$3y - 2x = 6.$$

Solving for y in this equation, we get $y = \frac{2}{3}x + 2$, so

$$f^{-1}(x) = \tfrac{2}{3}x + 2.$$

Verifying the equation $f^{-1}(f(x)) = x$, we have

$$\begin{aligned} f^{-1}(f(x)) &= f^{-1}(\tfrac{3}{2}x - 3) \\ &= \tfrac{2}{3}(\tfrac{3}{2}x - 3) + 2 \\ &= x - 2 + 2 \\ &= x. \end{aligned}$$

Also,

$$\begin{aligned} f(f^{-1}(x)) &= f(\tfrac{2}{3}x + 2) \\ &= \tfrac{3}{2}(\tfrac{2}{3}x + 2) - 3 \\ &= x + 3 - 3 \\ &= x, \end{aligned}$$

and the equation $f(f^{-1}(x)) = x$ is verified. □

The equations $f^{-1}(f(x)) = x$ and $f(f^{-1}(x)) = x$ are sometimes used to define the inverse of a function in the following way. Let f be a function with domain D and range R. If g is a function with domain R and range D such that

$$f(g(x)) = x, \qquad \text{for all } x \in R,$$

and

$$g(f(x)) = x, \qquad \text{for all } x \in D,$$

then $g = f^{-1}$ and $f = g^{-1}$.

Example 4 • Checking for Inverse Functions

Determine if the following functions f, g are inverse functions of each other.

$$f(x) = \frac{x}{x - 1}, \qquad g(x) = \frac{x - 1}{x}$$

Solution

Both the equations $f(g(x)) = x$ and $g(f(x)) = x$ must hold in order for f and g to be inverse functions of each other. However,

$$\begin{aligned} f(g(x)) &= f\left(\frac{x-1}{x}\right) \\ &= \frac{\frac{x-1}{x}}{\frac{x-1}{x} - 1} \cdot \frac{x}{x} \\ &= \frac{x-1}{x-1-x} \\ &= -x + 1 \\ &\neq x \end{aligned}$$

Since $f(g(x)) \neq x$, f and g are not inverse functions of each other. □

◆ Practice Problem 1 ◆

Which of the following pairs of functions are inverses of each other?

a. $f(x) = x^3 - 2$, $g(x) = \sqrt[3]{x + 2}$
b. $f(x) = x^2$, $g(x) = \sqrt{x}$

EXERCISES 4.6

In Exercises 1–4, use the horizontal line test to decide whether or not the given graph is the graph of a one-to-one function. (See Example 1.)

1.

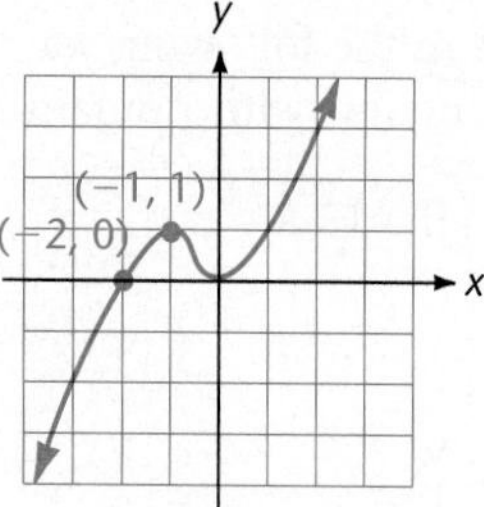

2.

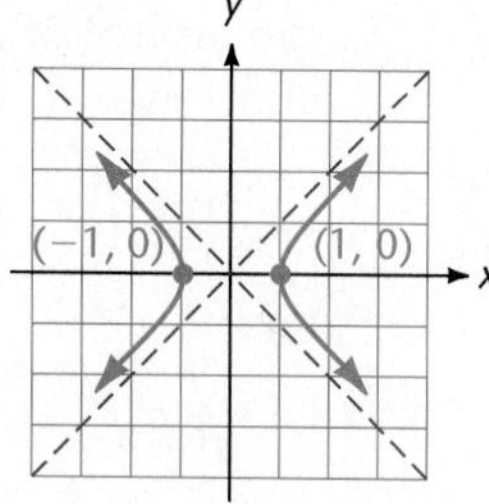

3.

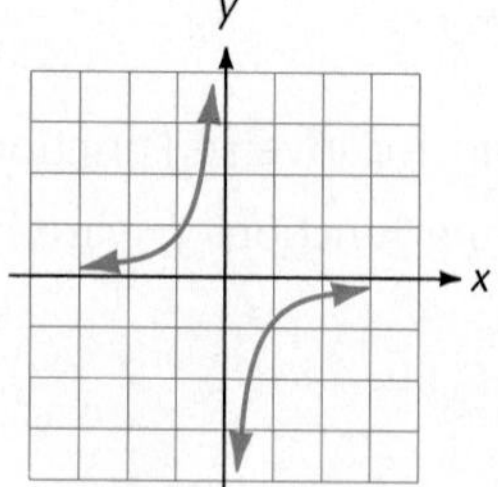

4.

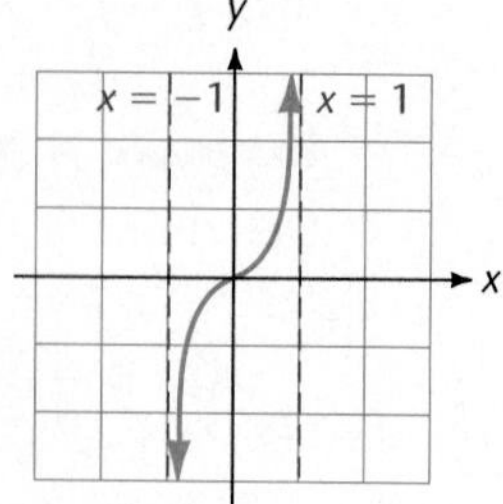

Each of the following equations defines a one-to-one function g. Find an expression for $g^{-1}(x)$, draw $y = x$ as a dashed line, and on the same coordinate system draw the graphs of g and g^{-1}. (See Example 2.)

5. $g(x) = x + 3$
6. $g(x) = 4 - x$
7. $g(x) = \frac{4}{3}x - 4$
8. $g(x) = \frac{2}{3}x - 4$
9. $g(x) = 8 - x^3$
10. $g(x) = (x - 1)^3$

Each of the following equations defines a function f. Find an expression for $f(x)$ and also for $f^{-1}(x)$ if f is one-to-one. If f is one-to-one, verify each of the equations $f^{-1}(f(x)) = x$ and $f(f^{-1}(x)) = x$. (See Example 3.)

11. $2x - y = 4$
12. $3x - y = 9$
13. $y = x^2 + 2$
14. $y = x^2 - 1$
15. $y = (x - 1)^2$
16. $y = (x + 2)^2$
17. $y = x^3 - 1$
18. $y = x^3 + 1$

Which of the following pairs of functions f and g are inverse functions of each other? (See Example 4 and Practice Problem 1.)

19. $f(x) = -\frac{3}{4}x + 3$, $g(x) = -\frac{4}{3}x + 4$
20. $f(x) = -\frac{3}{5}x + 3$, $g(x) = -\frac{5}{3}x + 5$
21. $f(x) = 3x - 9$, $g(x) = \dfrac{x + 9}{3}$
22. $f(x) = \dfrac{4x - 10}{5}$, $g(x) = \dfrac{5x + 10}{4}$
23. $f(x) = \dfrac{2x + 1}{x}$, $g(x) = \dfrac{1}{x - 2}$
24. $f(x) = \dfrac{x - 1}{2x}$, $g(x) = \dfrac{1}{1 - 2x}$
25. $f(x) = \dfrac{2x}{x + 1}$, $g(x) = \dfrac{x + 1}{2x}$
26. $f(x) = \dfrac{x}{2x - 1}$, $g(x) = \dfrac{2x - 1}{x}$
27. $f(x) = \sqrt{x - 1}$, $g(x) = x^2 + 1$ where $x \geq 0$
28. $f(x) = \sqrt{x^2 - 4}$, where $x \geq 2$, $g(x) = \sqrt{x^2 + 4}$, where $x \geq 0$
29. $f(x) = \sqrt{x + 2}$, $g(x) = x^2 - 2$
30. $f(x) = \sqrt{2x - 6}$, $g(x) = \frac{1}{2}x^2 + 3$

Critical Thinking: Writing

31. Write an explanation that justifies the use of the exponent -1 in both $f^{-1}(x)$ and $[f(x)]^{-1}$, even though $f^{-1}(x)$ and $[f(x)]^{-1}$ represent different quantities. (You may want to consult a dictionary for the meaning of *inverse.*)

Critical Thinking: Writing

32. To find the value of the function g at a number, we first multiply the number by 5 and then add 7. Write a similar description for the inverse of this g.

Critical Thinking: Exploration and Writing

33. Suppose f and g are one-to-one functions. Decide which of $f^{-1} \circ g^{-1}$ or $g^{-1} \circ f^{-1}$ is the inverse of $f \circ g$ and explain your choice with the aid of a diagram similar to those in Figures 3.18 and 4.39.

◆ Solution for Practice Problem

1. a. We have

$$f(g(x)) = f(\sqrt[3]{x+2}) = (\sqrt[3]{x+2})^3 - 2 = x$$

and

$$g(f(x)) = g(x^3 - 2) = \sqrt[3]{x^3 - 2 + 2} = x.$$

The function f has as both domain and range the set of all real numbers, and g does also. Thus f and g are inverses of each other.

b. We have

$$f(g(x)) = f(\sqrt{x}) = (\sqrt{x})^2 = x,$$

and this equation holds for all x in the domain of g. However,

$$g(f(x)) = g(x^2) = \sqrt{x^2} = |x| \neq x,$$

for any negative x. Thus one equation fails, and the functions f and g are not inverses of each other.

4.7 VARIATION

When we speak of the **ratio** of one quantity a to a second quantity b, we mean the quotient a/b. We write $a\!:\!b$ or a/b to indicate the ratio of a to b. In working with ratios, care must be taken in choosing the units of measure to be associated with each of the quantities. If the two measurements are of the same type of quantities, we should use the same units of measure. For example, the ratio of 1 yard to 18 inches could be expressed as:

$$\frac{1}{\frac{1}{2}} \quad \text{in terms of yards}$$

or

$$\frac{3}{\frac{3}{2}} \quad \text{in terms of feet}$$

or

$$\frac{36}{18} \quad \text{in terms of inches.}$$

Notice that each of these ratios is equal to the quotient 2/1, or 2.

If the quantities are not of the same type, then the units of measure must be different. For example, if we consider the ratio of pressure to surface area, our units might be pounds and square inches. That is, the ratio of 3 pounds to 400 square inches would be

$$\frac{3 \text{ pounds}}{400 \text{ square inches}}.$$

This ratio is read as "3/400 pounds per square inch."

Sometimes it is necessary to express a ratio in units that are different from the ones given. Consider the following example.

Example 1 • Converting Units of Measure

Express 60 miles per hour in terms of feet per second.

Solution

We recall that 5280 feet = 1 mile and 3600 seconds = 1 hour. Now 60 miles per hour can be expressed as the fraction

$$\frac{60 \text{ miles}}{1 \text{ hour}}.$$

Since multiplying the numerator and denominator of any fraction by equal quantities does not change the value of the fraction, we can multiply this fraction by

$$\frac{1 \text{ hour}}{3600 \text{ seconds}} \quad \text{and} \quad \frac{5280 \text{ feet}}{1 \text{ mile}} \quad \text{to obtain}$$

$$\begin{aligned}\frac{60 \text{ miles}}{1 \text{ hour}} &= \frac{60 \cancel{\text{miles}}}{1 \cancel{\text{hour}}} \cdot \frac{1 \cancel{\text{hour}}}{3600 \text{ seconds}} \cdot \frac{5280 \text{ feet}}{1 \cancel{\text{mile}}} \\ &= \frac{(60)(1)(5280) \text{ feet}}{(1)(3600)(1) \text{ seconds}} \\ &= \frac{88 \text{ feet}}{1 \text{ second}}.\end{aligned}$$

Therefore 60 miles per hour is equal to 88 feet per second. Notice that we chose the multipliers in the form $\frac{1 \text{ hour}}{3600 \text{ seconds}}$ and $\frac{5280 \text{ feet}}{1 \text{ mile}}$ $\Big($instead of $\frac{3600 \text{ seconds}}{1 \text{ hour}}$

and $\left.\dfrac{5280 \text{ feet}}{1 \text{ mile}}\right)$ so that in our product the units "hours" would occur in the numerator and denominator and the units "miles" would occur in the numerator and denominator. □

An equality of two ratios is called a **proportion.** If the ratio of a to b is equal to the ratio of c to d, we write

$$a:b = c:d \qquad \text{or} \qquad \frac{a}{b} = \frac{c}{d}.$$

This proportion is sometimes read as "a is to b as c is to d."

Example 2 • Using a Proportion

If a car moves at a uniform rate of speed and travels 400 kilometers in 8 hours, how far does it travel in 6 hours?

Solution

Let x represent the number of kilometers traveled in 6 hours. Then a proportion can be set up as follows.

$$\frac{400 \text{ kilometers}}{8 \text{ hours}} = \frac{x \text{ kilometers}}{6 \text{ hours}}$$

This gives a linear equation in one variable to solve:

$$\frac{400}{8} = \frac{x}{6},$$

and we find

$$x = 300 \text{ kilometers}.$$

□

Many of the formulas that occur in mathematics and other sciences can be described by using the terminology of **variation.** Each variation statement can be translated into a simple mathematical equation.

Direct Variation

The statement "y varies directly as x" or "y is directly proportional to x" means that $y = kx$ for some nonzero constant k. In direct variation, the word *directly* is often omitted, and the variation statement reads as "y varies as x" or "y is proportional to x."

If y varies directly as x, then y doubles as x doubles, y triples as x triples, and so on. If k is positive, then y increases as x increases and y decreases as x decreases.

On the other hand, if $y = k/x$ for a positive constant k, then y decreases as x increases, and vice versa. This leads to the following definition.

Inverse Variation

The statement "y varies inversely as x" or "y is inversely proportional to x" means that $y = k/x$ for some nonzero constant k.

The next two definitions are extensions of the concepts of direct and inverse variation.

Joint Variation

The statement "y varies jointly as x and z" or "y is jointly proportional to x and z" means that $y = kxz$ for some nonzero constant k.

Combined Variation

The statement "y varies directly as x and inversely as z" or "y is directly proportional to x and inversely proportional to z" means that $y = kx/z$, for some nonzero constant k.

In all of the statements of variation, the constant k is called the **constant of variation,** or the **proportionality constant.**

Most problems involving variation can be solved by the following steps:

Solving Variation Problems

1. Express the variation statement as an equation with a constant of variation k.
2. Use a known set of values for the variables to find the value of the constant of variation k.
3. Use the equation with the value of the constant k and another set of known values for all but one of the variables to find a corresponding value of the unknown variable.

This procedure is illustrated in the following examples.

Example 3 • Using Proportions to Determine Dosage

In the prevention of certain parasites in dogs, the medication diethylcarbamazine citrate is administered with a daily dosage that is directly proportional to body weight. If the dosage is 45 milligrams for a dog weighing 15 pounds, find the dosage for a dog that weighs 42 pounds.

Solution

1. Let D denote the dosage in milligrams and let w denote the weight of the dog in pounds. Since D is directly proportional to w, we have

$$D = kw,$$

for a constant k.

2. It is given that $D = 45$ when $w = 15$, so

$$45 = k(15) \qquad \text{and} \qquad k = 3.$$

Thus the equation for the variation is

$$D = 3w.$$

3. For a dog that weighs 42 pounds, the dosage is

$$D = 3(42) = 126 \text{ milligrams per day.}$$

□

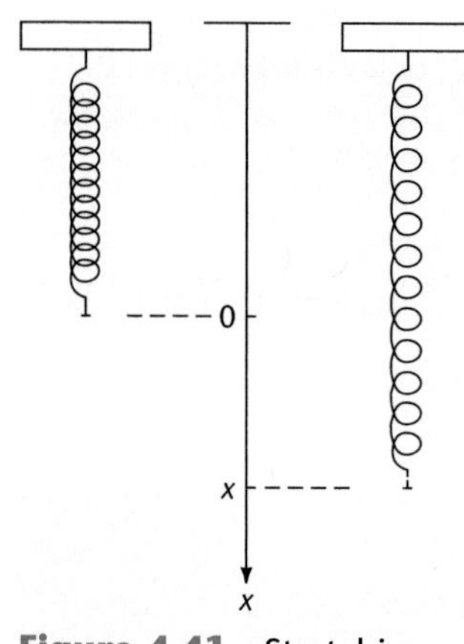

Figure 4.41 Stretching a spring

Example 4 • Work Required to Stretch a Spring

In physics we learn that work W may be measured in foot-pounds and that the work required to stretch a spring x feet beyond its natural length varies directly as the square of x. (See Figure 4.41.) If the work required to stretch a certain spring 2 feet beyond its natural length is 20 foot-pounds, find the work required to stretch the same spring from its natural length to a point 3 feet beyond its natural length.

Solution

1. The equation for the variation is

$$W = kx^2.$$

2. With the given values of $W = 20$ and $x = 2$, we have

$$20 = k(4) \qquad \text{and} \qquad k = 5.$$

Substitution for k yields

$$W = 5x^2.$$

3. When the spring is stretched 3 feet beyond its natural length, the required work is given by

$$W = 5(3)^2 = 45 \text{ foot-pounds.}$$

□

◆ Practice Problem 1 ◆

The height h of a certain type of right circular cylinder varies inversely as the square of the radius r. If the height is 9 meters when the radius is 4 meters, find the height of a right circular cylinder of this type whose radius is 2 meters.

The last example in this section deals with a more complicated situation.

Example 5 • Combined Variation

Suppose that z varies jointly as x and t^2 and inversely as $3w - 1$. If $z = 4$ when $x = -2$, $t = 1$, and $w = 5$, find z when $x = -3$, $t = 4$, and $w = -1$.

Solution

1. The variation equation is

$$z = \frac{kxt^2}{3w - 1}.$$

2. Since $z = 4$ when $x = -2$, $t = 1$, and $w = 5$, we have

$$4 = \frac{k(-2)(1)^2}{3(5) - 1} \quad \text{or} \quad 4 = \frac{-2k}{14}.$$

This yields $k = -28$, and the variation equation is

$$z = \frac{-28xt^2}{3w - 1}.$$

3. To determine the value of z when $x = -3$, $t = 4$, $w = -1$, we substitute these values into the variation equation and obtain

$$z = \frac{-28(-3)(4)^2}{3(-1) - 1} = -336.$$

□

EXERCISE 4.7

Express each of the following ratios as a quotient.

Ratios 1–6

1. 3 pounds to 4 pounds
2. 4 feet to 2 inches
3. 12 meters to 150 centimeters
4. \$20 to 2 tickets
5. 90 boys to 30 girls
6. 40 miles to $\frac{1}{3}$ hour

Converting Units of Measure 7–12

Perform the following conversions. (See Example 1.)

7. 44 feet per second to miles per hour
8. 40 pounds per second to tons per hour
9. 48 ounces per square inch to pounds per square foot
10. 55 meters per second to kilometers per hour
11. 20 miles per gallon to miles per quart
12. 90 miles per hour to yards per minute

Proportions 13–18

Solve the following exercises by using proportions. (See Example 2.)

13. If a car can travel 60 miles on 2.5 gallons of gas, how far can it travel on 9 gallons of gas?

14. If a punch recipe calls for 5 parts water for each part syrup, how much syrup must be used to make 10 gallons of punch?
15. If gas must be mixed with oil at a ratio of 50:1, how many pints of oil must be used with 6 gallons of gas?
16. Suppose that the ratio of gas to oil in Problem 15 is 40:1. How many ounces of oil must be used with 1 gallon of gas? (16 ounces = 1 pint.)
17. If two boys divide \$40 in a ratio of 3 to 5, how much money will each receive?
18. If a mixture contains 40% alcohol, how many ounces of alcohol are in 6 ounces of the mixture?

Variation Problems 19–28

Solve the following variation problems by first determining the value of the constant of variation. (See Examples 3–5 and Practice Problem 1.)

19. Suppose y varies directly as x^2. If $y = 12$ when $x = 2$, find y when $x = 5$.
20. If y varies directly as $x + 2$ and $y = 10$ when $x = 3$, determine y when $x = 7$.
21. Assume y varies inversely as t^2 and $y = 2$ when $t = 2$. Find y when $t = 4$.
22. Suppose z varies jointly as x and y. Find z when $x = 3$ and $y = 2$ if $z = 16$ when $x = -1$ and $y = 8$.
23. Given that y is directly proportional to z and inversely proportional to w, find y when $z = 1$ and $w = 4$ if $y = \frac{1}{3}$ when $w = 18$ and $z = \frac{1}{6}$.
24. Suppose that w varies directly as the product of x and y and inversely as the square of z. If $w = 9$ when $x = 6$, $y = 27$, and $z = 3$, find the value of w when $x = 4$, $y = 7$, and $z = 2$.
25. Assume that the square of t varies inversely as the cube of z and that $t = 4$ when $z = \frac{1}{2}$. Find t when $z = 2$.
26. Suppose that x varies directly as the square root of $y - 1$ and that $x = -8$ when $y = 5$. Find x when $y = 50$.
27. If x varies jointly as y and t^3, and inversely as $4z - 3$, find x when $y = 7$, $z = -1$, and $t = 4$, if it is given that $x = 2$ when $y = -4$, $t = -2$, and $z = 1$.
28. Suppose that z varies directly as w, and inversely as $x - 1$ and the square of t. Find z when $x = -1$, $w = -1$, and $t = -1$, if $z = -4$ when $w = 7$, $x = 2$, and $t = -3$.

Hooke's Law 29–30

29. According to Hooke's law, the force f required to stretch a spring x units beyond its natural length varies directly with x. If a 30-pound force is required to stretch a 12-inch spring to 15 inches, how much force is required to stretch it to 16 inches?
30. The force f required to compress a spring x units from its natural length is directly proportional to x. Suppose a force of 16 pounds is required to compress a spring of length 28 inches to 26 inches. Find the force required to compress the same spring to 20 inches.

Water Pressure 31. The water pressure p at a point on an object varies directly as the depth d. If the pressure at a point is 1250 pounds per square foot at a depth of 20 feet, find the pressure at a depth of 50 feet.

Andrew J. Martines/Photo Researchers

The weight of the water above a diver exerts a pressure on the diver that varies directly as the depth. (See Exercise 31.)

Fluid Pressure 32. The fluid pressure p on a submerged object varies directly as the depth d. Suppose the pressure 16 feet deep in an oil tank is 960 pounds per square foot. Find the pressure at a point 5 feet deep.

Production Cost 33. Suppose the total cost C of producing x units of a certain product varies directly as the square of x. Find the cost of producing 80 units of the product if 15 units cost $450 to produce.

Effect of Gravity 34. The weight w of a body above the surface of the earth varies inversely as the square of its distance d from the center of the earth. Assuming the radius of the earth is 4000 miles, how much does a person weigh 200 miles above the surface of the earth if he weighs 180 pounds on the earth's surface?

Pendulum 35. The time t required for one complete swing of a pendulum varies directly as the square root of its length x. If a pendulum 4 feet long makes one complete swing in 2.5 seconds, how long would it take for a pendulum 2.25 feet long to make a complete swing?

Equilateral Triangle 36. The area of an equilateral triangle varies as the square of its base. If the area is $\sqrt{3}$ square meters when the length of the base is 2 meters, find the area of an equilateral triangle whose base has a length of 5 meters.

Free-Falling Body 37. The distance in feet traveled by a free-falling body varies directly as the square of the time in seconds that it falls. If such a body falls 144 feet in 3 seconds, how far will it fall in 10 seconds?

Strength of a Beam 38. Suppose that the strength S of a beam with a constant length and a rectangular cross section varies directly as the width x and the square of the depth y. If a beam 2 inches wide and 4 inches in depth has strength of 40 pounds, how wide must a beam be cut to have the same strength and be 8 inches in depth?

Cone 39. The altitude h of a cone varies directly as its volume and inversely as the square of the radius of its base. If a cone with altitude 32 centimeters and radius of the base 4 centimeters has volume $512\pi/3$ cubic centimeters, find the altitude of a cone with volume 6π cubic centimeters and radius of the base 3 centimeters.

Stiffness of a Beam 40. The stiffness S of a rectangular beam varies directly as the width x and the cube of the depth d. If $S = 90$ when $x = 2$ and $d = 3$, find the stiffness S for a beam with width 3 and depth 2.

Right Circular Cone 41. The slant height S of a right circular cone varies directly as its lateral surface area and inversely as the radius of its base. (See the accompanying figure.) If the lateral surface area is 400π square inches when the radius of the base is 20 inches and the slant height is 20 inches, find the slant height of a right circular cone with a lateral surface area of 12 square inches and a radius of the base of 2 inches.

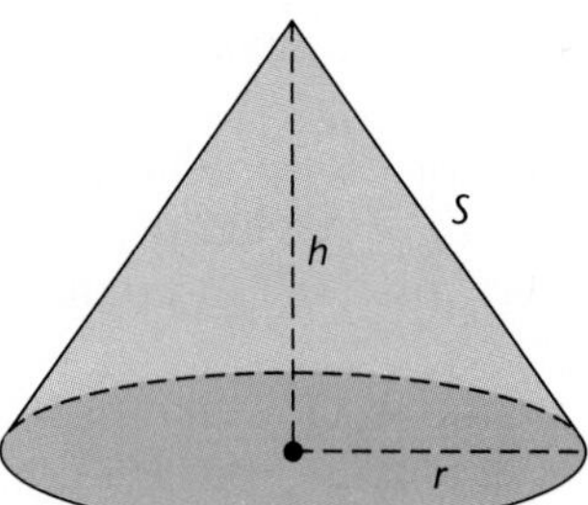

Figure for Exercise 41

Stopping a Car 42. The distance in feet required to stop a car varies directly as the square of its speed in miles per hour. If the car can stop in 49 feet from a speed of 35 miles per hour, how many feet are required to stop from a speed of 70 miles per hour?

Work ✓ 43. The time required to finish a job varies inversely as the number of persons working. If 5 workers take 12 days to finish a job, how many workers are needed to finish the same job in 10 days?

Boyle's Law 44. Boyle's law states that the pressure P of a gas varies directly as the absolute temperature T and inversely as the volume V. If the pressure is 350 pounds per square inch when the temperature is 70 K and the volume is 2 cubic feet, what is P when the temperature is 35 K and the volume is 4 cubic feet?

Critical Thinking: Exploration and Writing 45. If y varies directly as x, explain why x varies directly as y.

Critical Thinking: Exploration and Writing 46. If y varies inversely as x, explain why x varies inversely as y.

Critical Thinking: Exploration and Writing 47. If z varies jointly as x and y, is it true that z varies directly as y? Explain your decision.

Critical Thinking: Exploration and Writing 48. Suppose y varies inversely as x with a negative constant of variation. Describe the effect on y (a) as x increases, and (b) as x decreases.

◆ Solution for Practice Problem

1. We follow the procedure outlined in this section for solving variation problems.

   The equation describing the variation is

$$h = \frac{k}{r^2}.$$

Since $h = 9$ when $r = 4$, we have

$$9 = \frac{k}{4^2}$$

and $k = 144$. Thus the variation equation is

$$h = \frac{144}{r^2}.$$

To find the value of h when $r = 2$, we substitute $r = 2$ into the variation equation and obtain

$$h = \frac{144}{2^2}.$$

Thus $h = 36$ meters when $r = 2$ meters.

CHAPTER REVIEW

Summary of Important Concepts and Formulas

The Parabola

Equation, $a \neq 0$	$y = ax^2 + bx + c$	$x = ay^2 + by + c$
Alternate equation	$y = a(x - h)^2 + k$	$x = a(y - k)^2 + h$
Vertex	$\left(-\frac{b}{2a}, \frac{4ac - b^2}{4a}\right) = (h, k)$	$\left(\frac{4ac - b^2}{4a}, -\frac{b}{2a}\right) = (h, k)$
Axis of symmetry	$x = -\frac{b}{2a}, x = h$	$y = -\frac{b}{2a}, y = k$
Direction of opening	Upward, if $a > 0$ Downward, if $a < 0$	To the right, if $a > 0$ To the left, if $a < 0$

Distance between (x_1, y_1) and (x_2, y_2)

$$d = \sqrt{(x_2 - x_1)^2 + (y_2 - y_1)^2}$$

Standard Equation for a Circle

$$(x - h)^2 + (y - k)^2 = r^2$$

Tests for Symmetry

The graph of an equation is symmetric with respect to the	if the following substitution yields an equivalent equation
y-axis	$-x$ for x
x-axis	$-y$ for y
origin	$-x$ for x and $-y$ for y

Stretching of $y = f(x)$

To graph $y = cf(x)$, $c > 0$, multiply each y-coordinate on the graph of $y = f(x)$ by c.

Reflection of $y = f(x)$

To graph $y = -f(x)$, reflect the graph of $y = f(x)$ through the x-axis.

Vertical Translation of $y = f(x)$

To graph $y = f(x) + c$,

1. If $c > 0$, shift the graph of $y = f(x)$ upward c units.
2. If $c < 0$, shift the graph of $y = f(x)$ downward $|c|$ units.

Horizontal Translation of $y = f(x)$

To graph $y = f(x - c)$,

1. If $c > 0$, shift the graph of $y = f(x)$ to the right c units.
2. If $c < 0$, shift the graph of $y = f(x)$ to the left $|c|$ units.

Standard Equation of an Ellipse with Center at (0, 0)

$$\frac{x^2}{a^2} + \frac{y^2}{b^2} = 1$$

Standard Equation of a Hyperbola with Foci on the x-Axis and Center at (0, 0)

$$\frac{x^2}{a^2} - \frac{y^2}{b^2} = 1$$

Standard Equation of a Hyperbola with Foci on the y-Axis and Center at (0, 0)

$$\frac{y^2}{a^2} - \frac{x^2}{b^2} = 1$$

Standard Equation of an Ellipse with Center at (h, k)

$$\frac{(x - h)^2}{a^2} + \frac{(y - k)^2}{b^2} = 1$$

Standard Equations of Hyperbolas with Center at (h, k)

$$\frac{(x - h)^2}{a^2} - \frac{(y - k)^2}{b^2} = 1, \qquad \frac{(y - k)^2}{a^2} - \frac{(x - h)^2}{b^2} = 1$$

Horizontal Line Test

If any horizontal line intersects the graph of the function f at more than one point, then f *is not* a one-to-one function. If no horizontal line intersects the graph of the function f at more than one point, then f *is* a one-to-one function.

Inverse Function Equations

$f^{-1}(f(x)) = x,\quad$ for all x in the domain of f
$f(f^{-1}(x)) = x,\quad$ for all x in the domain of f^{-1}

Direct Variation

The statement "y varies directly as x" or "y is directly proportional to x" means that $y = kx$, for some nonzero constant k.

Inverse Variation

The statement "y varies inversely as x" or "y is inversely proportional to x" means that $y = k/x$ for some nonzero constant k.

Joint Variation

The statement "y varies jointly as x and z" or "y is jointly proportional to x and z" means that $y = kxz$ for some nonzero constant k.

Combined Variation

The statement "y varies directly as x and inversely as z" or "y is directly proportional to x and inversely proportional to z" means that $y = kx/z$ for some nonzero constant k.

Critical Thinking: Find the Errors

Each of the following nonsolutions has one or more errors. Can you find them?

Problem 1

Find the distance between the points (4, 6) and (1, 2).

Nonsolution

$$\begin{aligned} d &= \sqrt{(4-1)^2 + (6-2)^2} \\ &= \sqrt{3^2 + 4^2} \\ &= 3 + 4 \\ &= 7 \end{aligned}$$

Problem 2

If the graph of the following equation is a circle, find the center and the radius.

$$x^2 + 2x + y^2 = 11 - 4y$$

Nonsolution

$$\begin{aligned} x^2 + 2x + y^2 + 4y &= 11 \\ x^2 + 2x + 1 + y^2 + 4y + 4 &= 11 + 1 + 4 \\ (x+1)^2 + (y+2)^2 &= 16 \end{aligned}$$

The center is at (1, 2), and the radius is 16.

Problem 3

Find the inverse f^{-1} of the function defined by $y = \sqrt{x+1}$.

Nonsolution

$$f(x) = \sqrt{x+1},$$

so

$$f^{-1}(x) = \frac{1}{\sqrt{x+1}}.$$

Review Problems for Chapter 4

Find the distance between the given points.

1. $(7, 2)$ and $(3, -1)$
2. $(3, -2)$ and $(1, 0)$
3. $(-2, -3)$ and $(4, 0)$
4. $(-1, 7)$ and $(-1, -1)$

Determine whether or not the graph of the given equation is a circle. If it is, find the center and the radius.

5. $(x - 3)^2 + (y + 5)^2 = 16$
6. $x^2 + y^2 + 2x + 4y = 4$
7. $x^2 - 12x + y^2 = 6y - 50$

Write an equation of the circle with the given properties.

8. Center $(2, -3)$, radius 5
9. Center $(-3, -4)$, through $(-7, -1)$

Write an equation of the given circle with endpoints of a diameter as shown.

10.

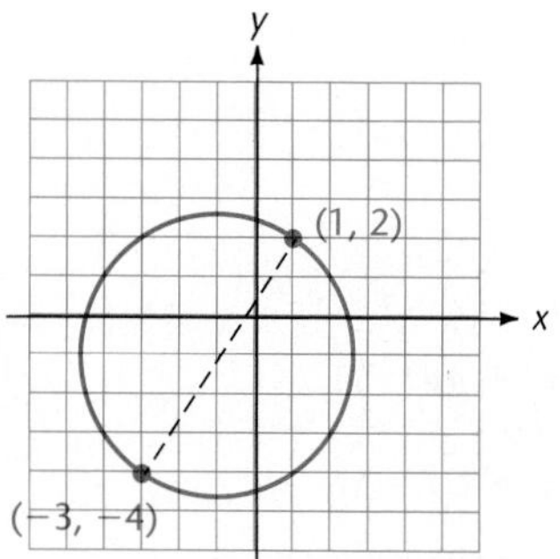

11.

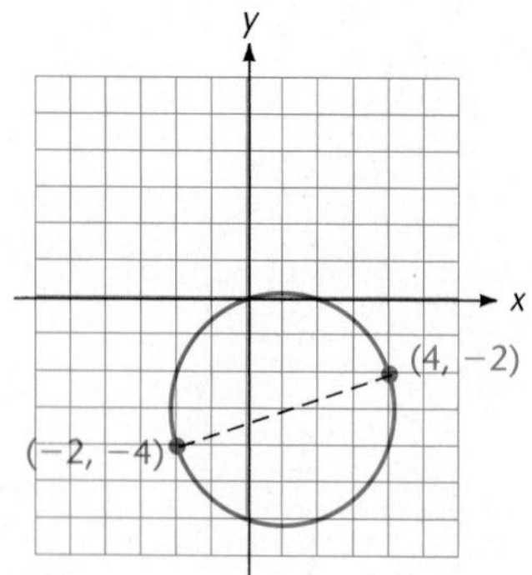

12. A portion of a graph is given. Use it to complete the graph in three ways:
   a. So that it becomes symmetric with respect to the y-axis
   b. So that it becomes symmetric with respect to the x-axis
   c. So that it becomes symmetric with respect to the origin.

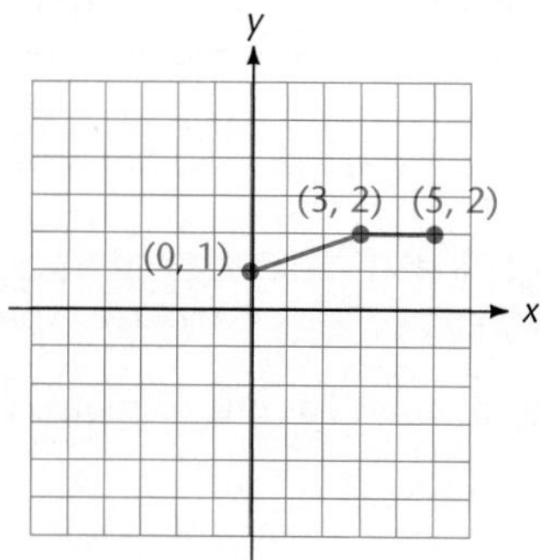

Test the given equation for symmetry with respect to the x-axis, the y-axis, and the origin, and sketch its graph.

13. $y^2 = x^2$
14. $y = |x| - 1$

The graph of $y = f(x)$ is given in the accompanying figure. Use it to sketch the graphs of the functions given in Problems 15–18.

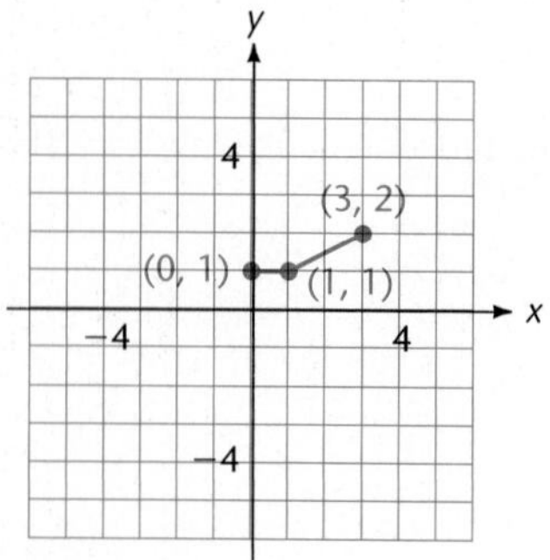

Figure for Problems 15–18

15. $y = 2f(x)$
16. $y = f(x) + 1$
17. $y = f(x + 1)$
18. $y = -f(x)$

Sketch the graphs of the following equations. Locate the vertices and asymptotes of all hyperbolas and write the equations of the asymptotes.

19. $y = 4 - (x - 2)^2$
20. $4x^2 + 9y^2 = 144$
21. $4x^2 - y^2 = 36$
22. $25y^2 - 4x^2 = 100$
23. $x = -2y^2 + 8y - 6$
24. $2y = \sqrt{1 - 4x^2}$
25. $\dfrac{4x^2}{9} + \dfrac{9y^2}{16} = 1$
26. $3x = -\sqrt{16 - 9y^2}$
27. $2y = -\sqrt{16 + x^2}$

Sketch the graphs of the following equations. Locate the center and vertices of ellipses and hyperbolas.

28. $(x + 3)^2 - 4(y - 2)^2 = 16$
29. $x^2 - 2x + y^2 = 12y - 38$

Elliptical Underpass

30. As shown in the accompanying figure, an underpass has the shape of half an ellipse with the major axis vertical. The arch is 12 feet across the bottom and 8 feet high in the center. Can a truck that is 8 feet wide and 6 feet high pass through the arch?

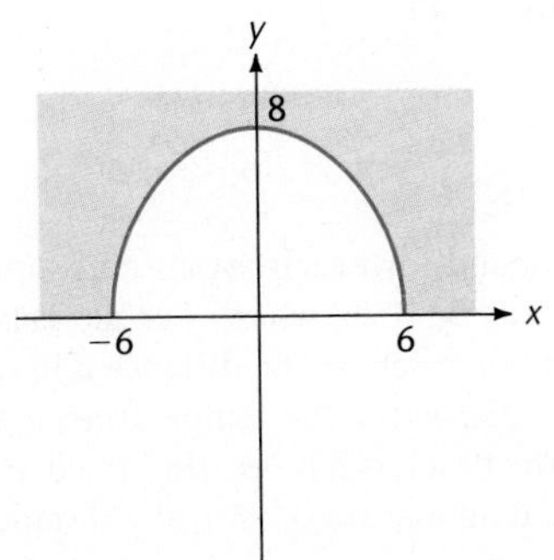

Figure for Problem 30

In each of Problems 31–34, decide whether or not the given graph is the graph of a one-to-one function.

31.

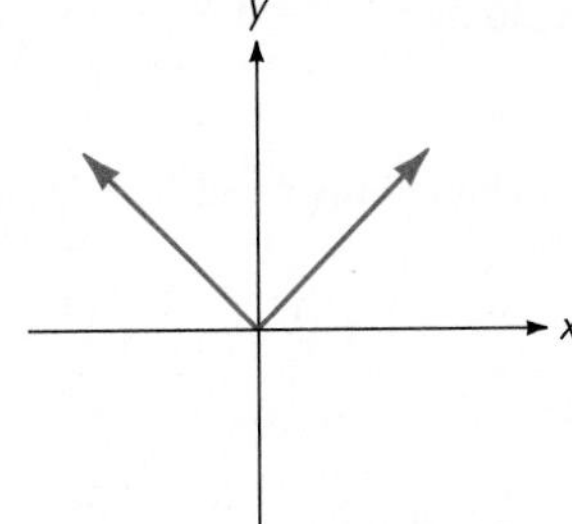

32.

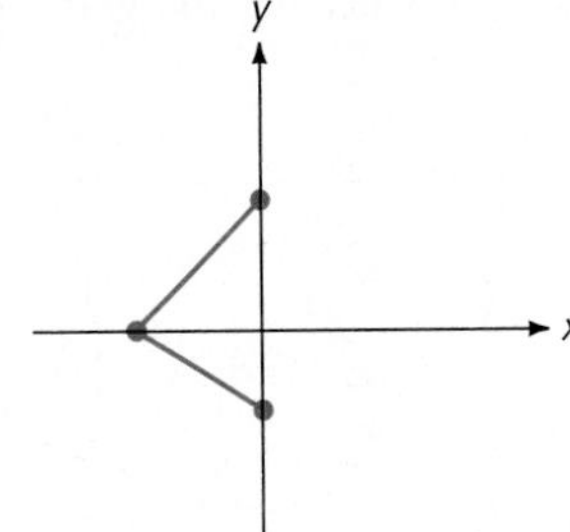

33.

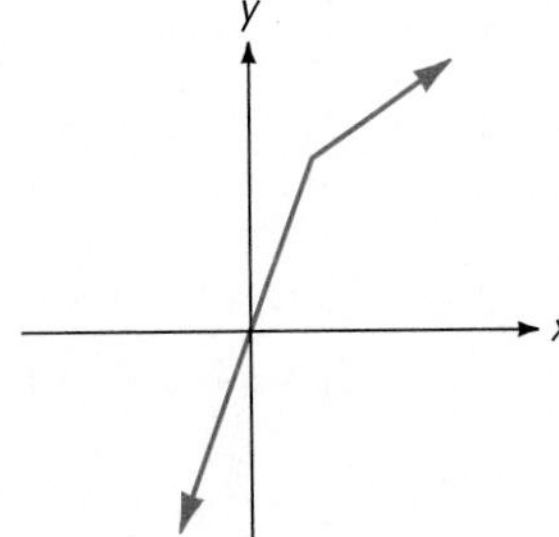

34.

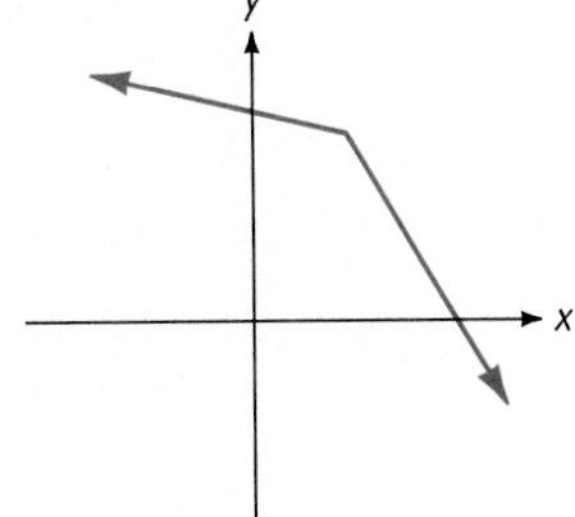

Each of the following equations defines a function f. Find an expression for $f(x)$ and also for $f^{-1}(x)$ if f is one-to-one.

35. $2x - 3y = 6$

36. $y = x^2 - 4$

37. $y - x^2 = 1$

Decide if the following pairs of functions f and g are inverse functions of each other.

38. $f(x) = 2x - 1,\ g(x) = \dfrac{1}{2x - 1}$

39. $f(x) = \dfrac{2}{3}x + 6,\ g(x) = \dfrac{3}{2}x - 9$

In Problems 40–42, the graph of a one-to-one function f is given. Sketch the graph of f^{-1} by reflecting the graph of f through the line $y = x$.

40.

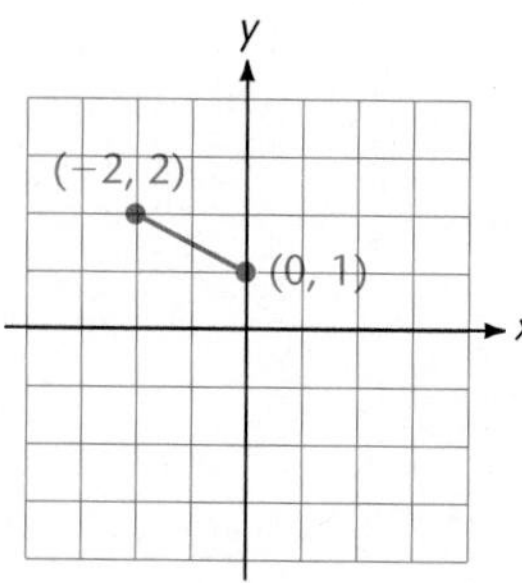

41.

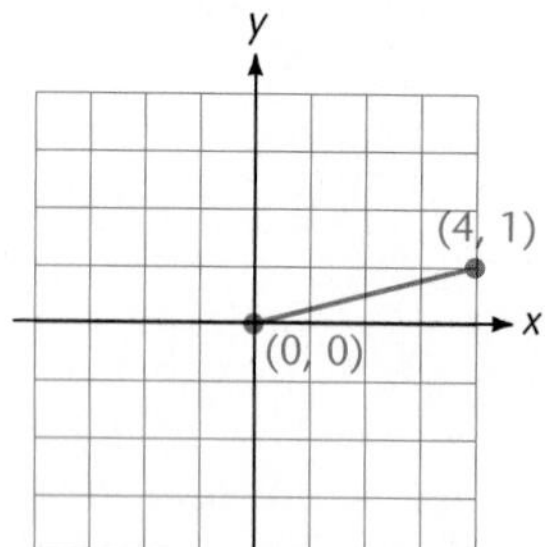

42.

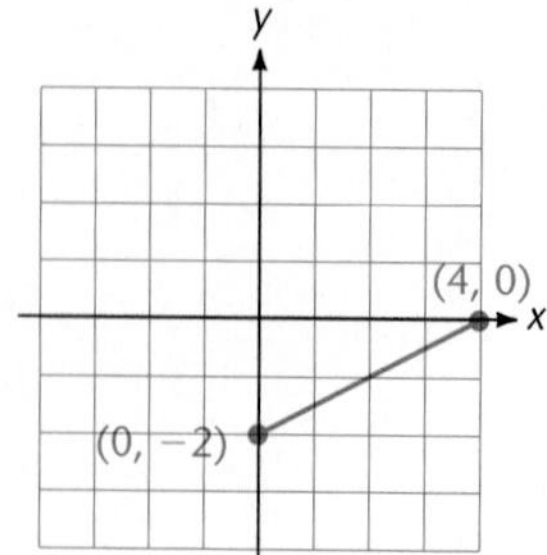

Concrete Mixture

43. A concrete mixture calls for 1 part cement, 2 parts sand, and 3 parts gravel by volume. How much concrete can be made from 4 cubic meters of cement?

44. Convert 45 miles per hour to feet per second.

45. It is known that y varies directly as t and inversely as z^2. If $y = 20$ when $t = 15$ and $z = 3$, find y when $t = 9$ and $z = 6$.

46. Suppose that x varies jointly as y and z, and inversely as w^2. If $x = 40$ when $y = 12$, $z = 6$, and $w = 3$, find x when $y = 4$, $z = 10$, and $w = 5$.

Ohm's Law

47. For a simple electric circuit as shown in the accompanying figure, Ohm's law states that the current I varies directly as the voltage E and inversely as the resistance R. If $I = 22$ amps when $E = 154$ volts and $R = 7$ ohms, find I when $E = 220$ volts and $R = 5$ ohms.

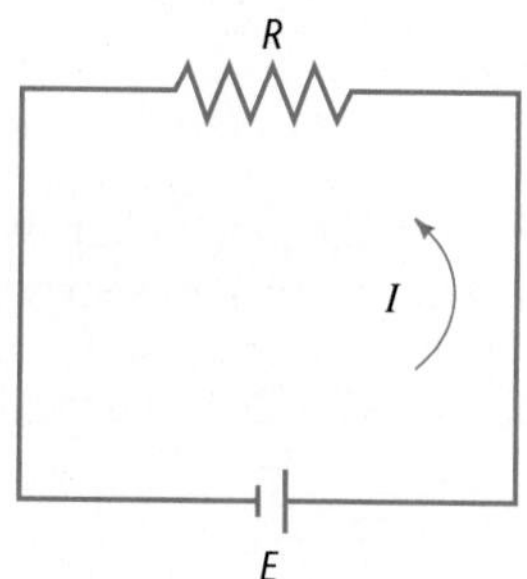

Figure for Problem 47

Demand for Beach Towels

48. Let D denote the demand for beach towels and suppose that D varies directly as $T - 75$, where T is the Fahrenheit temperature, and inversely as the distance d in miles to the beach. If $D = 350$ when the temperature is 90°F and the distance to the beach is 3 miles, find the demand for beach towels at a temperature of 95°F at a distance of 2 miles from the beach.

ENCORE

The elliptical orbits of the planets about the sun are shown in this diagram (not drawn to scale).

Eccentricity is a number used to measure roundness of ellipses. An ellipse with equation

$$\frac{x^2}{a^2} + \frac{y^2}{b^2} = 1$$

and $a > b$ has **eccentricity** e defined by

$$e = \frac{\sqrt{a^2 - b^2}}{a}.$$

It can be shown that $0 < e < 1$ for any ellipse that is not a circle. Ellipses with eccentricity close to 0 are nearly circular, and ellipses with eccentricity close to 1 are extremely elongated.

The planets in our solar system revolve about the sun in elliptical orbits as shown in the accompanying diagram. The orbit of the earth is an ellipse with eccentricity $e = 0.017$. Using a coordinate system with the sun at a focus on the positive x-axis, the equation of the path of the earth is

$$\frac{x^2}{(92.895)^2} + \frac{y^2}{(92.882)^2} = 1,$$

where $a = 92.895$ and $b = 92.882$ are in millions of miles. The eccentricities of all the planets are given in the accompanying table.

Planet	Eccentricity
Mercury	0.206
Venus	0.007
Earth	0.017
Mars	0.093
Jupiter	0.048
Saturn	0.056
Uranus	0.047
Neptune	0.008
Pluto	0.249

Polynomials and Rational Functions

The polynomial functions and rational functions studied in this chapter are used extensively in mathematical models of production costs, consumer demands, wildlife management, biological processes, and many other scientific studies. One section of the chapter deals with the approximation of real zeros of a polynomial function. Approximation techniques steadily gain in importance with the growing use of computers and calculators.

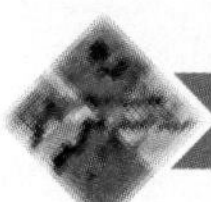

5.1 SYNTHETIC DIVISION

Polynomial division is one of the basic tools needed in the study of the theory of polynomials. In fact, division of polynomials in which the divisor is of the form $x - c$, c being a constant, is of special interest in this chapter. We begin with an example that reviews the procedure for long division of polynomials.

Example 1 • Long Division of Polynomials

Divide $8y^3 - 18y^2 - 6 + 11y$ by $-3y + 2 + 4y^2$ and write the results in both forms

$$\frac{\text{Dividend}}{\text{Divisor}} = \text{quotient} + \frac{\text{remainder}}{\text{divisor}}$$

and

$$\text{Dividend} = (\text{divisor})(\text{quotient}) + \text{remainder}.$$

Solution

We write the dividend and divisor in descending powers of y and record them in the form

$$4y^2 - 3y + 2\overline{\big)\,8y^3 - 18y^2 + 11y - 6}.$$

We first divide $4y^2$ into $8y^3$ to obtain $2y$ and proceed as follows.

$$\begin{array}{rl}
 2y & \text{writing } 2y \text{ as part of the quotient} \\
4y^2 - 3y + 2\overline{\big)\,8y^3 - 18y^2 + 11y - 6} & \\
\underline{8y^3 - 6y^2 + 4y} & \text{multiplying the divisor by } 2y \\
-12y^2 + 7y - 6 & \text{subtracting}
\end{array}$$

Next, we divide $4y^2$ into $-12y^2$ to obtain -3 and repeat the procedure.

$$\begin{array}{lrl}
 & 2y - 3 \quad \text{Quotient} & \\
\text{Divisor} & 4y^2 - 3y + 2\overline{\big)\,8y^3 - 18y^2 + 11y - 6} & \text{Dividend} \\
 & \underline{8y^3 - 6y^2 + 4y} & \\
 & -12y^2 + 7y - 6 & \\
 & \underline{-12y^2 + 9y - 6} & \\
 & -2y & \text{Remainder}
\end{array}$$

The remainder, $-2y$, has degree less than the degree of the divisor, $4y^2 - 3y + 2$. Thus the procedure terminates here, and we have

$$\frac{8y^3 - 18y^2 + 11y - 6}{4y^2 - 3y + 2} = 2y - 3 + \frac{-2y}{4y^2 - 3y + 2}$$

or

$$8y^3 - 18y^2 + 11y - 6 = (4y^2 - 3y + 2)(2y - 3) - 2y. \qquad \square$$

Many people prefer a shortened form that allows more rapid division of a polynomial by a binomial of the form $x - c$. This procedure is called **synthetic division,** or **detached coefficients.** To understand why the synthetic division procedure works, we will work an example with a divisor of the form $x - c$. Then we will rework it with the goal of streamlining the procedure in mind.

Example 2 • Development of Synthetic Division

Divide $P(x) = x^3 - 3x^2 - x + 4$ by $x - 2$ and write the results in the form

$$\frac{\text{Dividend}}{\text{Divisor}} = \text{quotient} + \frac{\text{remainder}}{\text{divisor}}.$$

Solution

We divide $P(x)$ by $x - 2$ as follows.

$$\begin{array}{rrl}
 & x^2 - x \quad - 3 & \text{Quotient} \\
\text{Divisor} \quad x - 2 & \overline{)\,x^3 - 3x^2 - \quad x + 4} & \text{Dividend} \\
 & x^3 - 2x^2 \qquad\qquad\quad & \\
 & - \quad x^2 - \quad x + 4 & \\
 & - \quad x^2 + 2x \qquad & \\
 & - 3x + 4 & \\
 & - 3x + 6 & \\
 & - 2 & \text{Remainder}
\end{array}$$

Thus

$$\frac{x^3 - 3x^2 - x + 4}{x - 2} = x^2 - x - 3 + \frac{-2}{x - 2}.$$

As a first step in explaining synthetic division, we refer back to the long division and eliminate the terms in color to obtain the following more compact form.

$$\begin{array}{rl}
 & x^2 - x - 3 \\
x - 2 & \overline{)\,x^3 - 3x^2 - \quad x + 4} \\
 & \quad - 2x^2 + 2x + 6 \\
\hline
 & x^3 - \quad x^2 - 3x - 2
\end{array}$$

If we go a step further and eliminate the variables and omit the quotient, we have

$$\begin{array}{r|rrrr}
1 - 2 & 1 & -3 & -1 & 4 \\
 & & -2 & 2 & 6 \\
\hline
 & 1 & -1 & -3 & -2
\end{array}.$$

As final refinements, we drop the coefficient of x in the divisor and change the sign of the constant in the divisor $x - 2$ to 2, so that we can **add** at each stage rather than subtract. This yields the routine known as synthetic division.

$$\begin{array}{r|rrrrl}
\text{Divisor } x - 2 \rightarrow 2 & 1 & -3 & -1 & 4 & \leftarrow \text{Dividend } x^3 - 3x^2 - x + 4 \\
 & & 2 & -2 & -6 & \\
\hline
 & 1 & -1 & -3 & -2 &
\end{array}$$

Quotient $= x^2 - x - 3$ Remainder $= -2$ □

The mechanics of synthetic division are indicated by the arrows shown in the diagram in Example 2 constructed by using the coefficients of the dividend $P(x)$ and the constant c in the divisor $x - c$. The first step is to bring down the leading coefficient of $P(x)$. Then repeat the following two steps until all the coefficients of $P(x)$ are exhausted.

1. Multiply each number in the bottom row by c to obtain the next number in the second row.
2. Add the numbers in the first and second rows to obtain the next number in the bottom row.

The last entry in the bottom row is the remainder, and the remaining entries in the last row are the coefficients of the quotient polynomial. This quotient polynomial always has degree 1 less than the degree of $P(x)$. We also note that the synthetic division process cannot be used if the divisor is not a linear monic polynomial.

We demonstrate the synthetic division procedure in the next example, where the divisor does not appear to be in the form $x - c$.

Example 3 • Using Synthetic Division

Use synthetic division to divide $P(x) = x^4 - 2x^3 - x^2 - x + 1$ by $x + 1$.

Solution

The divisor must be of the form $x - c$ to use the synthetic division process. Thus we write $x + 1$ as $x - (-1)$ and identify $c = -1$. The synthetic division appears as follows.

$$\begin{array}{l r|r r r r r l} \text{Divisor} & -1 & 1 & -2 & -1 & -1 & 1 & \text{Dividend} \\ & & & -1 & 3 & -2 & 3 & \\ \hline & & 1 & -3 & 2 & -3 & 4 & \\ & & \multicolumn{4}{c}{\underbrace{\hphantom{1 \quad -3 \quad 2 \quad -3}}_{\text{Quotient}}} & \underbrace{\hphantom{4}}_{\text{Remainder}} & \end{array}$$

Hence we can write

$$\frac{x^4 - 2x^3 - x^2 - x + 1}{x + 1} = x^3 - 3x^2 + 2x - 3 + \frac{4}{x + 1}.$$

□

The next example illustrates the fact that if any coefficients in the dividend are 0, these 0's *must* be recorded in the synthetic division procedure, and that the dividend must be written in decreasing powers of x.

Example 4 • Synthetic Division with Missing Powers

Divide $2x^4 - 5 - 12x^2$ by $x + 3$.

Solution

We note that

$$2x^4 - 5 - 12x^2 = 2x^4 + 0x^3 - 12x^2 + 0x - 5$$

and $x + 3 = x - (-3)$ before beginning the synthetic division procedure.

$$\begin{array}{r|rrrrr} -3 & 2 & 0 & -12 & 0 & -5 \\ & & -6 & 18 & -18 & 54 \\ \hline & 2 & -6 & 6 & -18 & 49 \end{array}$$

Hence

$$\frac{2x^4 - 12x^2 - 5}{x + 3} = 2x^3 - 6x^2 + 6x - 18 + \frac{49}{x + 3}.$$ □

◆ Practice Problem 1 ◆

Use synthetic division to divide

$P(x) = 3x^4 - 10x^2 - x + 2$ by $x + 2$.

The next example illustrates the use of synthetic division with polynomials containing nonreal coefficients. For the remainder of this chapter, the domain of a variable is the set $\mathscr{C}$ of all complex numbers unless otherwise stated.

Example 5 • Using Synthetic Division with Complex Numbers

Divide $P(x) = x^3 - ix^2 - x + 2$ by $x - i$, using synthetic division.

Solution

The synthetic division appears as follows.

$$\begin{array}{r|rrrr} i & 1 & -i & -1 & 2 \\ & & i & 0 & -i \\ \hline & 1 & 0 & -1 & 2 - i \end{array}$$

$$\underbrace{1 \quad 0 \quad -1}_{\text{Quotient}} \quad \underbrace{2 - i}_{\text{Remainder}}$$

The degree of the quotient polynomial is 2, and we write

$$\frac{x^3 - ix^2 - x + 2}{x - i} = x^2 - 1 + \frac{2 - i}{x - i}.$$ □

◆ Practice Problem 2 ◆

Divide $x^4 + 2x^2 + 1$ by $x + i$.

EXERCISES 5.1

In Exercises 1–10, use long division and write the results in the form

$$\frac{\text{Dividend}}{\text{Divisor}} = \text{quotient} + \frac{\text{remainder}}{\text{divisor}}.$$

(See Example 1.)

1. $\dfrac{x^3 - 4x^2 + 6x - 3}{x - 1}$

2. $\dfrac{x^3 + 6x^2 + 3x - 6}{x + 2}$

3. $\dfrac{2x^3 - 11x^2 + 19x - 10}{2x - 5}$

4. $\dfrac{3x^3 - 11x^2 + 5x + 3}{3x + 1}$

5. $\dfrac{3x^3 - 5x^2 + 14x + 3}{3x - 2}$

6. $\dfrac{22 - 33x + 5x^2 + 2x^3}{2x - 5}$

7. $\dfrac{2x^4 + 5x - 8 + 3x^3 - 3x^2}{x^2 + 2x - 2}$

8. $\dfrac{2x^4 + 9x - 9 - 3x^3 + x^2}{x^2 - 2x + 3}$

9. $\dfrac{4x^4 + 5x^2 - 7x + 3}{2x^2 + 3x - 2}$

10. $\dfrac{6x^4 - 5x^2 + 9x - 5}{2x^2 + 2x - 3}$

In Exercises 11–18, use synthetic division and write the results in the form

$$\frac{\text{Dividend}}{\text{Divisor}} = \text{quotient} + \frac{\text{remainder}}{\text{divisor}}.$$

(See Examples 3 and 4.)

11. $(2x^3 - 4x^2 + 5x - 1) \div (x - 2)$

12. $(5x^3 - 10x^2 - 11x + 8) \div (x - 3)$

13. $(-3x^3 + 2x - 75) \div (x + 3)$

14. $(x^4 - 7x^2 - 6x) \div (x + 2)$

15. $(x^5 - x^4 - 7x^3 + 24) \div (x - 3)$

16. $(4x^5 + 30x^2 + 3) \div (x + 2)$

17. $(x + 7 + 4x^3) \div (x + 1)$

18. $(-22x - x^4 - 40 + 6x^3) \div (x - 4)$

In Exercises 19–28, divide using synthetic division when appropriate and write the results in the form

$$\text{Dividend} = (\text{divisor})(\text{quotient}) + \text{remainder}.$$

19. $\dfrac{4m^3 - 3m - 2}{m + 1}$

20. $\dfrac{-5 + x^3 + 2x^2}{x + 3}$

21. $\dfrac{3x^4 - 10x^3 - 10x^2 - 8x + 3x^5 - 8}{x^2 + 3x + 2}$

22. $\dfrac{5x^2 + 2x^5 - 14x^3 - x^4 + x + 1}{x^2 + 3x + 1}$

23. $\dfrac{18k^2 - 3rk - 10r^2}{6k - 5r}$

24. $\dfrac{9k^2 - 27rk + 20r^2}{3k - 4r}$

25. $\dfrac{4x^4 - 13ax^3 + 12a^2x^2 - 5a^3x + 2a^4}{4x^2 - ax + a^2}$

26. $\dfrac{4x^4 - 3ax^3 + 12a^2x^2 + 2a^3x + 3a^4}{4x^2 + ax + a^2}$

27. $\dfrac{x^3 - 27}{x + 3}$

28. $\dfrac{x^4 + 16}{x - 2}$

In Exercises 29–40, find the quotient $Q(x)$ and the remainder r when $P(x)$ is divided by $D(x)$.

29. $P(x) = x^2 - 7x + 12,\ D(x) = x - 3$
30. $P(x) = x^3 + 8,\ D(x) = x + 2$
31. $P(x) = 2x^3 - x^2 - 2x + 2,\ D(x) = x - 2$
32. $P(x) = x^4 + 3x^3 + 2x^2 + 5x + 2,\ D(x) = x + 3$
33. $P(x) = x^5 - 4x^4 - x^2 + 2,\ D(x) = x - 2$
34. $P(x) = x^6 + 7x^3 - 2,\ D(x) = x - 1$
35. $P(x) = -x^2 - 2 + 4x^4 + 6x,\ D(x) = x + \frac{1}{2}$
36. $P(x) = 7x + 5 + 6x^2 + 9x^3,\ D(x) = x + \frac{1}{3}$
37. $P(x) = x^3 + 3x^2 + x + 3,\ D(x) = x + i$
38. $P(x) = x^4 + 6x^2 + 8,\ D(x) = x - 2i$
39. $P(x) = -ix^3 + 4x^2 + 5,\ D(x) = x - 3i$
40. $P(x) = -2ix^3 + 4x^2 + (1 + i)x + (1 + i),\ D(x) = x + i$

In Exercises 41–50, find the remainder when $P(x)$ is divided by $x - c$.

41. $P(x) = x^4 - 8x^2 - 8x + 1,\ c = -2$
42. $P(x) = -3x^4 + 8x^3 + 10x^2 + 25x - 4,\ c = 4$
43. $P(x) = x^5 - x^4 + 2x^3 + 3x^2 + 4,\ c = 1$
44. $P(x) = x^4 - 3x - 4,\ c = -1$
45. $P(x) = x^5 - ix^4 + 2x^3 + 4ix + 8,\ c = 2i$
46. $P(x) = -2x^4 - 2ix^3 + ix^2 + 2,\ c = -i$
47. $P(x) = 2x^3 - 9x^2 + 14x - 5,\ c = 2 + i$
48. $P(x) = x^3 - x^2 + 2,\ c = 1 - i$
49. $P(x) = x^4 - a^4,\ c = -a$ 50. $P(x) = x^5 - a^5,\ c = a$
51. Find k so that the remainder is zero when $P(x) = x^7 + 30x^2 + k$ is divided by $x + 2$.
52. Find k so that the remainder is zero when $P(x) = -x^4 + kx^3 + x^2 - x + 1$ is divided by $x + 1$.
53. Determine the values of k so that the remainder is positive when $P(x) = -x^3 + kx^2 - 5x - 20$ is divided by $x + 2$.
54. Determine the values of k so that the remainder is negative when $P(x) = kx^5 + x^2 + 1$ is divided by $x + 1$.

A calculator with a memory cell can be helpful in synthetic division if it is used to store c in memory. Find the remainder (rounded to two decimal places) when $P(x)$ is divided by $x - c$.

55–58

55. $P(x) = x^5 + x^3 - 11x^2 + 43,\ c = 1.01$
56. $P(x) = -2x^4 + 5x^3 + x^2 + 1,\ c = 2.13$
57. $P(x) = -3x^6 + x^4 - 5.1x^3 - x - 2.9,\ c = -0.91$
58. $P(x) = 4x^5 - x^3 + x - 9.31,\ c = -1.05$

Critical Thinking: Exploration and Discussion

59. Discuss why the degree of the dividend is equal to the sum of the degree of the quotient and the degree of the divisor in any division of polynomials.

Critical Thinking: Exploration and Discussion

60. If $a \neq 0$, explain how the division of a polynomial $P(x)$ by $ax - b$ can be accomplished by using the synthetic division of $P(x)$ by $x - b/a$.

◆ Solutions for Practice Problems

1. $$\begin{array}{r|rrrrr} -2 & 3 & 0 & -10 & -1 & 2 \\ & & -6 & 12 & -4 & 10 \\ \hline & 3 & -6 & 2 & -5 & 12 \end{array}$$

$$\frac{3x^4 - 10x^2 - x + 2}{x + 2} = 3x^3 - 6x^2 + 2x - 5 + \frac{12}{x + 2}$$

2. $$\begin{array}{r|rrrrr} -i & 1 & 0 & 2 & 0 & 1 \\ & & -i & -1 & -i & -1 \\ \hline & 1 & -i & 1 & -i & 0 \end{array}$$

$$\frac{x^4 + 2x^2 + 1}{x + i} = x^3 - ix^2 + x - i + \frac{0}{x + i} = x^3 - ix^2 + x - i$$

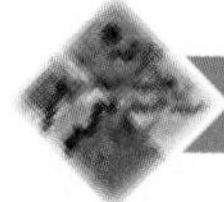

5.2 THE REMAINDER AND FACTOR THEOREMS

When we divide a polynomial $P(x)$ by a linear polynomial $x - c$, there is a quotient $Q(x)$ and a remainder r such that

$$P(x) = (x - c) \cdot Q(x) + r,$$

where r is a constant. This equality of polynomials holds for all values of x and, in particular, for $x = c$. If we assign the value c to x, we obtain

$$\begin{aligned} P(c) &= (c - c) \cdot Q(c) + r \\ &= 0 \cdot Q(c) + r \\ &= 0 + r \\ &= r. \end{aligned}$$

This means that the remainder r is the same as the value of $P(x)$ when $x = c$. This important result is known as the remainder theorem and is valid whether the coefficients of $P(x)$ and $Q(x)$ are real or nonreal numbers.

The Remainder Theorem

If a polynomial $P(x)$ is divided by $x - c$, the remainder is $P(c)$.

Example 1 • **Using the Remainder Theorem to Find a Remainder**

Use the remainder theorem to find the remainder in each of the following divisions.

a. $(x^2 + 3x + 2) \div (x - 2)$

b. $\dfrac{x^{40} - 2x^{10} + 1}{x + 1}$

c. $(x^3 - x^2 - x + 2) \div (x + i)$

Solution

a. By the remainder theorem, when $P(x) = x^2 + 3x + 2$ is divided by $x - 2$, the remainder is

$$r = P(2) = (2)^2 + 3(2) + 2 = 12.$$

b. With $P(x) = x^{40} - 2x^{10} + 1$ and $x - c = x + 1 = x - (-1)$, the remainder is

$$r = P(-1) = (-1)^{40} - 2(-1)^{10} + 1 = 1 - 2 + 1 = 0.$$

c. With $P(x) = x^3 - x^2 - x + 2$ and $x + i$ as divisor, the remainder is

$$\begin{aligned} r &= P(-i) \\ &= (-i)^3 - (-i)^2 - (-i) + 2 \\ &= i + 1 + i + 2 \\ &= 3 + 2i. \end{aligned}$$

□

The remainder theorem can also be used in the other direction: the value of $P(x)$ can be found by obtaining the remainder when $P(x)$ is divided by $x - c$.

Example 2 • **Using the Remainder Theorem and Synthetic Division to Evaluate a Polynomial**

Use the remainder theorem to find the value of $P(2)$ if

$$P(x) = x^4 - 3x^2 - x + 4.$$

Solution

We use synthetic division to divide $P(x)$ by $x - 2$.

$$\begin{array}{r|rrrrl} 2 & 1 & 0 & -3 & -1 & 4 \\ & & 2 & 4 & 2 & 2 \\ \hline & 1 & 2 & 1 & 1 & 6 = r = P(2) \end{array}$$

By the remainder theorem, $P(2)$ is the same as the remainder. That is, $P(2) = 6$.

□

For an arbitrary number c, the remainder theorem states that

$$P(x) = (x - c) \cdot Q(x) + P(c).$$

Now $x - c$ is a factor of $P(x)$ if and only if the remainder is 0 when $P(x)$ is divided by $x - c$, and therefore the equation above shows that $x - c$ is a factor of $P(x)$ if and only if $P(c) = 0$. If it happens that $P(c) = 0$, then the polynomial $P(x)$ can be factored as

$$P(x) = (x - c)Q(x).$$

This establishes the following factor theorem.

The Factor Theorem

A polynomial $P(x)$ has a factor $x - c$ if and only if $P(c) = 0$.

Example 3 • Using the Factor Theorem to Test a Possible Factor

Use the factor theorem to decide whether or not $x + 1$ is a factor of

$$P(x) = x^{99} + x^{36} - x^{15} - 1.$$

Solution

By direct computation,

$$P(-1) = (-1)^{99} + (-1)^{36} - (-1)^{15} - 1 = -1 + 1 - (-1) - 1 = 0.$$

Thus $P(-1) = 0$ and $x + 1$ is a factor of $x^{99} + x^{36} - x^{15} - 1$. □

◆ Practice Problem 1 ◆

Use the factor theorem to decide whether or not $x - 5$ is a factor of

$$P(x) = x^3 - 4x^2 - 2x + 15.$$

A **zero** of the polynomial $P(x)$ is a number c such that $P(c) = 0$. The factor theorem states that $x - c$ is a factor of $P(x)$ if and only if c is a zero of $P(x)$.

Example 4 • Using the Factor Theorem to Test a Possible Zero

Use the factor theorem to decide whether or not -2 is a zero of

$$P(x) = x^4 - x^3 - 5x^2 + 3x + 2.$$

Solution

By the factor theorem, -2 is a zero of $P(x) = x^4 - x^3 - 5x^2 + 3x + 2$ if and only if $x + 2$ is a factor of $P(x)$. Synthetic division can be used to check if the remainder is 0 when $P(x)$ is divided by $x + 2 = x - (-2)$.

$$\begin{array}{r|rrrrr} -2 & 1 & -1 & -5 & 3 & 2 \\ & & -2 & 6 & -2 & -2 \\ \hline & 1 & -3 & 1 & 1 & 0 = r = P(-2) \end{array}$$

Since $P(-2) = 0$, then $x + 2$ is a factor of $P(x)$, and -2 is a zero of $P(x)$. □

Sometimes it is easier to compute $P(c)$ than to divide $P(x)$ by $x - c$. In other cases, the synthetic division process may be more efficient. The solution of Example 5 illustrates one of the major benefits of using synthetic division in conjunction with the factor theorem.

Example 5 • Solving a Cubic Equation

Given that 1 is a zero of $P(x) = x^3 - 3x^2 + x + 1$, solve the equation

$$x^3 - 3x^2 + x + 1 = 0.$$

Solution

Since it is given that 1 is a zero, we know that $x - 1$ is a factor of $P(x)$. To find the quotient, we divide by $x - 1$.

$$\begin{array}{r|rrrr} 1 & 1 & -3 & 1 & 1 \\ & & 1 & -2 & -1 \\ \hline & \underbrace{1 \quad -2 \quad -1}_{\text{Quotient}} & & & 0 = r \end{array}$$

Thus the quotient is $x^2 - 2x - 1$, and the given equation is equivalent to

$$(x - 1)(x^2 - 2x - 1) = 0.$$

Since a product is zero if and only if one of the factors is zero, the other solutions of the given equation are the same as the solutions to

$$x^2 - 2x - 1 = 0.$$

These solutions are easily found by the quadratic formula to be

$$x = \frac{2 \pm \sqrt{4 + 4}}{2} = \frac{2 \pm 2\sqrt{2}}{2} = 1 \pm \sqrt{2}.$$

Thus the solution set for the equation $x^3 - 3x^2 + x + 1 = 0$ is

$$\{1, 1 + \sqrt{2}, 1 - \sqrt{2}\}.$$

□

◆ Practice Problem 2 ◆

Given that -2 is a zero of $P(x) = x^3 + x^2 - 3x - 2$, find all the other zeros.

The next theorem can be used to locate a real zero of a real polynomial within a certain interval without knowing the exact value of the zero.

The Intermediate Value Theorem for Polynomials

Let $P(x)$ be a polynomial with real coefficients. If $[a, b]$ is an interval such that $P(a) \neq P(b)$, then $P(x)$ takes on every value between $P(a)$ and $P(b)$ over the interval $[a, b]$.

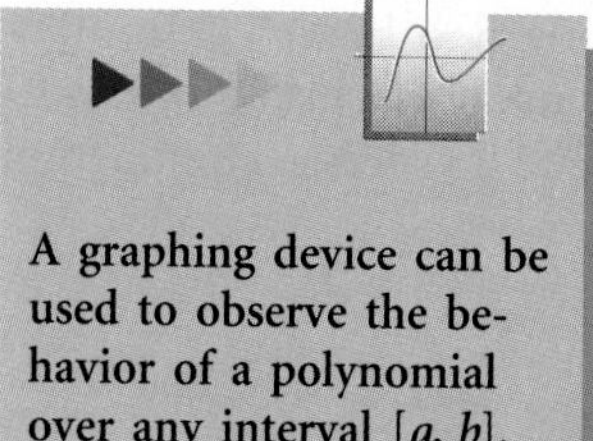

A graphing device can be used to observe the behavior of a polynomial over any interval $[a, b]$.

A special case of this theorem is that the values of $P(x)$ cannot change from positive to negative, or from negative to positive, without assuming the value 0 between. The sketches in Figure 5.1 show how the graph of $y = P(x)$ might look over the interval from $x = a$ to $x = b$. A point c where $P(c) = 0$ is indicated in each case.

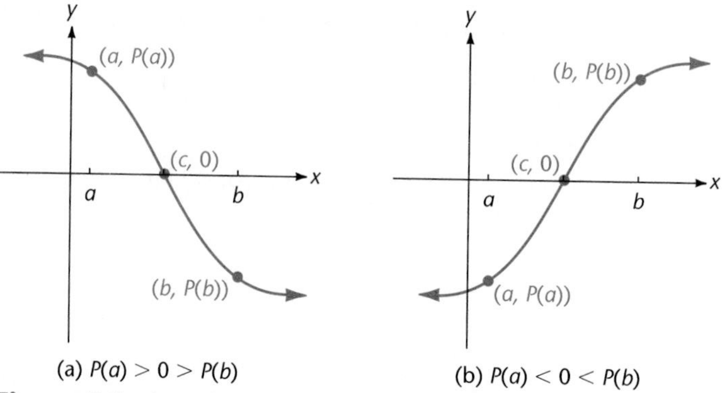

Figure 5.1 Locating a zero of $P(x)$

Example 6 • Locating a Zero Between Consecutive Integers

Use the intermediate value theorem and synthetic division to locate the positive zero of

$$P(x) = x^4 + 5x^3 - x^2 - 25x - 20$$

between two consecutive integers.

Solution

Replacing x by 0 in $P(x)$, we find $P(0) = -20$ and we note that $P(0)$ is negative. Next we use synthetic division and systematically divide with integral values until we detect a change of sign in the remainder.

$$\begin{array}{r|rrrrr} 1 & 1 & 5 & -1 & -25 & -20 \\ & & 1 & 6 & 5 & -20 \\ \hline & 1 & 6 & 5 & -20 & -40 \end{array} = P(1) < 0$$

$$\begin{array}{r|rrrrr} 2 & 1 & 5 & -1 & -25 & -20 \\ & & 2 & 14 & 26 & 2 \\ \hline & 1 & 7 & 13 & 1 & -18 \end{array} = P(2) < 0$$

$$\begin{array}{r|rrrrr} 3 & 1 & 5 & -1 & -25 & -20 \\ & & 3 & 24 & 69 & 132 \\ \hline & 1 & 8 & 23 & 44 & 112 \end{array} = P(3) > 0$$

Since $P(2)$ is negative and $P(3)$ is positive, then there must be a zero between 2 and 3.

□

Before continuing, let us observe that the three synthetic divisions in Example 5 could have been tabulated as follows, where the bottom row of each synthetic division is written to the right of the possibility that is being tested.

	1	5	−1	−25	**−20** = $P(0)$	Negative	
1	1	6	5	−20	**−40** = $P(1)$	Negative	
2	1	7	13	1	**−18** = $P(2)$	Negative	Opposite signs
3	1	8	23	44	**112** = $P(3)$	Positive	

A table such as this is efficient when several divisions are to be performed in succession. With a little practice, the necessary arithmetic can be done mentally.

EXERCISES 5.2

Use the remainder theorem to find the remainder in each of the following divisions. (See Example 1.)

1. $(x^4 - 2x^3 + 2x^2 - 1) \div (x - 2)$
2. $(2x^4 - 3x^3 + 4x - 5) \div (x - 1)$
3. $(4x^5 - 3x^2 + 5x - 1) \div (x - 1)$
4. $(-3x^4 + x^2 + 7x + 3) \div (x + 2)$
5. $(x^3 + 2x^2 - 5x + 1) \div (x + 3)$
6. $(2x^3 - 9x^2 + x - 30) \div (x - 5)$
7. $(2x^3 - 2x^2 + 4) \div (x - \sqrt{2})$
8. $(x^4 + x^3 - 6x^2 - 3x + 9) \div (x + \sqrt{3})$
9. $(x^2 - 3x + 1) \div (x - i)$
10. $(x^3 + x^2 + x + 1) \div (x - 2i)$
11. $(x^{1023} - 3x^{15} + 7) \div (x + 1)$
12. $(x^{95} - 17x^4 + 9) \div (x - 1)$

Use the remainder theorem and synthetic division to find $P(c)$. (See Example 2.)

13. $P(x) = 3x^4 - 5x^3 + x^2 - x + 7,\ c = -1$
14. $P(x) = 5x^4 + x^2 + 4x - 9,\ c = 2$
15. $P(x) = -x^6 + x^4 + 5x^3 - 2,\ c = -2$
16. $P(x) = -2x^5 + x^4 + 2x^2 - 9,\ c = -1$
17. $P(x) = 4x^3 + 6x^2 - 2,\ c = \frac{1}{2}$
18. $P(x) = 6x^4 - 10x^3 - 8x^2 - x,\ c = \frac{2}{3}$
19. $P(x) = 4x^3 - 2x + 6,\ c = \sqrt{2}$
20. $P(x) = 5x^3 + 2x^2 - \sqrt{3},\ c = \sqrt{3}$
21. $P(x) = 3x^4 - 2x^2 + x - 5,\ c = 1 - i$
22. $P(x) = 2x^4 - 4x^2 + 2x - 3,\ c = 1 + i$
23. $P(x) = x^4 + 2x^2 - 3x,\ c = 1 + 2i$
24. $P(x) = x^4 - 3x^2 + 4x,\ c = 2 - i$

Use the factor theorem to determine whether or not $D(x)$ is a factor of $P(x)$. (See Example 3 and Practice Problem 1.)

25. $P(x) = x^2 - 3x + 4,\ D(x) = x + 1$
26. $P(x) = x^3 - 4x^2 + 9,\ D(x) = x - 3$

27. $P(x) = x^{19} - x^{17} + x^2 - 1,\ D(x) = x - 1$
28. $P(x) = x^{30} - x^5 - 2,\ D(x) = x + 1$
29. $P(x) = x^4 - 3x^3 + 7x - 1,\ D(x) = x + 2$
30. $P(x) = 7x^3 - 3x^2 - 19,\ D(x) = x - 3$
31. $P(x) = x^3 + 5x^2 + x + 5,\ D(x) = x - i$
32. $P(x) = x^2 + x + 4,\ D(x) = x - (1 - 2i)$
33. $P(x) = x^2 - (3 - i)x + 8 + i,\ D(x) = x - (1 + 2i)$
34. $P(x) = x^3 - 3ix + i + 7,\ D(x) = x - (1 - i)$

Use the factor theorem to decide whether or not the given number c is a zero of $P(x)$. (See Example 4.)

35. $P(x) = x^3 - x^2 - x - 5,\ c = 2$
36. $P(x) = x^3 - 2x^2 - 2x + 7,\ c = 3$
37. $P(x) = 8x^4 + 2x^2 - 4x - 3,\ c = -\frac{1}{2}$
38. $P(x) = -3x^4 + 4x^3 + 8x - 2,\ c = \frac{1}{3}$
39. $P(x) = 4x^3 + 5x^2 - 10x + 4,\ c = \frac{3}{4}$
40. $P(x) = 3x^4 - x^3 + 4x^2 + 10x + 6,\ c = -\frac{2}{3}$
41. $P(x) = x^3 - 3x^2 + 4x - 2,\ c = 1 + i$
42. $P(x) = x^3 - 3x^2 + x - 3,\ c = i$

(See Example 5 and Practice Problem 2.)

43. Given that 3 is a zero of $P(x) = x^3 - 3x^2 + x - 3$, solve the equation
$$x^3 - 3x^2 + x - 3 = 0.$$
44. Given that -1 is a zero of $P(x) = x^3 - x^2 + 2$, solve the equation
$$x^3 - x^2 + 2 = 0.$$
45. Given that -7 is a zero of $P(x) = 2x^3 + 17x^2 + 19x - 14$, find all the other zeros.
46. Given that 9 is a zero of $P(x) = x^3 - 10x^2 + 3x + 54$, find all the other zeros.
47. Determine k so that $P(x) = x^3 - kx^2 + 3x + 7k$ is divisible by $x + 2$.
48. Determine k so that $P(x) = x^4 + kx^3 - 3kx + 9$ is divisible by $x - 3$.

In Exercises 49–54, use the intermediate value theorem to show that the given polynomial has a zero between the given numbers. (See Example 6.)

49. $P(x) = x^3 + 3x^2 - 3x - 9$, 1 and 2
50. $P(x) = x^4 + 2x^3 - 8x^2 - 20x - 20$, 3 and 4
51. $P(x) = x^4 + x^3 - 11x^2 - 12x - 12$, 3 and 4
52. $P(x) = x^4 - 2x^3 - 4x^2 + 12x - 12$, 2 and 3

53. $P(x) = 2x^4 + 6x^3 + 5x^2 - 3x - 3$, 0 and -1
54. $P(x) = x^3 + x^2 - 9x + 6$, -3 and -4
55. Locate the positive zero of $P(x) = x^4 + 2x^3 + 2x^2 + x - 1$ between two consecutive integers.
56. Locate the negative zero of $P(x) = x^4 + 2x^3 + 2x^2 + x - 1$ between two consecutive integers.
57. The polynomial $P(x) = x^4 + 2x^3 - 6x - 9$ has one positive and one negative zero. Locate each between two consecutive integers.
58. The polynomial $P(x) = x^4 + 2x^3 - 21x^2 - 4x + 40$ has two positive and two negative zeros. Locate each between two consecutive integers.

Critical Thinking: Exploration and Writing

59. Suppose $P(x)$ has a factor $x - c$, where c is real. Explain how this would show up as a feature on the graph of $y = P(x)$.

Critical Thinking: Exploration and Writing

60. Suppose $c \neq d$ and both $P(c) = 0$ and $P(d) = 0$. Explain why it must be true that $P(x) = (x - c)(x - d)G(x)$ for some polynomial $G(x)$.

◆ Solutions for Practice Problems

1. Since $P(5) = 5^3 - 4(5)^2 - 2(5) + 15 = 30 \neq 0$, then $x - 5$ is not a factor of $P(x)$.

2.
$$\begin{array}{r|rrrr} -2 & 1 & 1 & -3 & -2 \\ & & -2 & 2 & 2 \\ \hline & 1 & -1 & -1 & 0 \end{array}$$

$P(x) = (x - (-2))(x^2 - x - 1)$

We use the quadratic formula to find the zeros of $x^2 - x - 1$.

$$x = \frac{-(-1) \pm \sqrt{(-1)^2 - 4(1)(-1)}}{2(1)} = \frac{1 \pm \sqrt{1 + 4}}{2} = \frac{1 \pm \sqrt{5}}{2}$$

Thus the zeros of $P(x)$ are $x = -2$, $(1 + \sqrt{5})/2$, and $(1 - \sqrt{5})/2$.

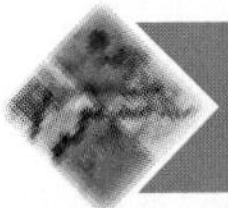

5.3 THE FUNDAMENTAL THEOREM OF ALGEBRA AND DESCARTES' RULE OF SIGNS

We recall from the last section that a **zero** of a polynomial $P(x)$ is a number c such that $P(c) = 0$. In this chapter, we are concerned primarily with the problem of finding the zeros of a given polynomial

$$P(x) = a_n x^n + a_{n-1} x^{n-1} + \cdots + a_1 x + a_0.$$

This is the same problem as that of solving the polynomial equation

$$a_n x^n + a_{n-1} x^{n-1} + \cdots + a_1 x + a_0 = 0.$$

The solutions to an equation are frequently referred to as the **roots** of the equation. We are especially interested in finding the real roots of a polynomial equation that has real coefficients.

Since the problem of finding the roots of a polynomial equation of degree 2 is completely resolved by the quadratic formula, it might be expected that similar formulas exist for the roots of polynomial equations of higher degree. This is true, up to a point. In the period 1500–1550, Italian mathematicians named Niccolò Tartaglia, Girolamo Cardano, and Ludovico Ferrari obtained formulas that could be used to solve the general equations of the third and fourth degrees. For over 200 years afterward, mathematicians searched for similar formulas for equations of degree greater than 4. It was not until 1824 that a Norwegian mathematician, N. H. Abel, proved that it is impossible to express the roots of a general equation of degree greater than 4 by a formula involving only the four fundamental operations and the extraction of roots.

Thus the problem we are dealing with in this chapter is far from simple. Our development begins with the following theorem, which was first proved in 1799 by the German mathematician C. F. Gauss (1777–1855).

The Fundamental Theorem of Algebra

Let

$$P(x) = a_n x^n + a_{n-1}x^{n-1} + \cdots + a_1 x + a_0$$

denote a polynomial of degree $n \geq 1$ with coefficients that are real or complex numbers. Then $P(x)$ has a zero in the field of complex numbers.

In other words, the conclusion of the fundamental theorem of algebra states that there is a complex number c_1 such that

$$P(c_1) = 0.$$

This complex number c_1 may be a real number; that is, c_1 may be a real zero of $P(x)$. In any case, $x - c_1$ is a factor of $P(x)$, by the factor theorem. Thus we can write

$$P(x) = (x - c_1)Q_1(x),$$

where $Q_1(x)$ has degree $n - 1$. If the quotient $Q_1(x)$ has degree ≥ 1, then by the fundamental theorem of algebra, $Q_1(x)$ has a zero c_2 and a corresponding factor $x - c_2$. That is,

$$Q_1(x) = (x - c_2)Q_2(x) \qquad \text{and} \qquad P(x) = (x - c_1)(x - c_2)Q_2(x).$$

If the quotient $Q_2(x)$ has degree ≥ 1, the procedure can be repeated again, obtaining a factor $x - c_3$ of $P(x)$. Each time another factor is obtained, the degree of the new quotient is 1 less than the degree of the previous quotient. After n applications of the fundamental theorem and the factor theorem, we arrive at the complete factorization of $P(x)$.

Factorization of $P(x) = a_nx^n + a_{n-1}x^{n-1} + \cdots + a_1x + a_0$

$P(x) = (x - c_1)(x - c_2)\cdots(x - c_n)a_n$

The last quotient must be a_n since this is the coefficient of x^n in $P(x)$. This factorization of $P(x)$ also shows that the numbers $c_1, c_2, \ldots, c_n$ are zeros of $P(x)$ and are not necessarily distinct. That is, a given factor $x - c$ may be repeated in the factorization of $P(x)$. If the factor $x - c$ is of **multiplicity *k*,** then c is a zero of multiplicity k. Thus a polynomial of degree $n \geq 1$ has exactly n zeros in the complex numbers if a zero of multiplicity k is counted as a zero k times.

Example 1 • Constructing a Polynomial with Given Zeros

Find a polynomial $P(x)$ of least degree and in simplest form that has 2, -1, and $2i$ as zeros.

Solution

We know that $P(x)$ must have degree 3 and must factor as

$$P(x) = a_3(x - 2)(x + 1)(x - 2i),$$

where a_3 is the leading coefficient in $P(x)$. The choice $a_3 = 1$ makes $P(x)$ a monic polynomial.

$$\begin{aligned} P(x) &= (x - 2)(x + 1)(x - 2i) \\ &= x^3 - (1 + 2i)x^2 - (2 - 2i)x + 4i \end{aligned}$$

□

Example 2 • Identifying Factors and Zeros of a Polynomial

State the factors $x - c$ and zeros of the polynomial

$$P(x) = 3(x - 2)^3(x + 1)^2(x - 3).$$

Solution

According to the preceding discussion, $P(x)$ has six factors of the form $x - c$ and six zeros, counting repetitions. The six factors are

$x - 2$, with multiplicity 3,

$x - (-1)$, with multiplicity 2, and

$x - 3$.

The six zeros are 2, 2, 2, -1, -1, and 3. □

◆ Practice Problem 1 ◆

Find a polynomial $P(x)$ of least degree with 1 a zero of multiplicity 2, 3 a zero of multiplicity 1, and $P(0) = 6$. Write $P(x)$ in factored form.

As indicated earlier, we are especially interested in finding the real zeros of a polynomial that has real coefficients. We know that a polynomial of degree $n \geq 1$ has exactly n zeros. However, up to this point, we have no systematic method for finding these zeros. The next theorem is often useful for this purpose when it is used in combination with results that we will present in the next section. This theorem is known as **Descartes' rule of signs.**

Descartes' rule of signs allows certain predictions to be made about the number of positive real zeros, or about the number of negative real zeros, of a polynomial that has *real coefficients.* The predictions are based on the number of "variations in sign" that occur when the terms of the polynomial are arranged in the usual order of descending powers of x:

$$P(x) = a_n x^n + a_{n-1}x^{n-1} + \cdots + a_1 x + a_0.$$

After any powers of x with zero coefficients are deleted, a *variation in sign* is said to occur when two consecutive coefficients are opposite in sign. For example,

$$P(x) = x^4 - 3x^3 + 7x + 11$$

has two variations in sign, as indicated by the arrows.

Descartes' Rule of Signs

Let

$$P(x) = a_n x^n + a_{n-1}x^{n-1} + \cdots + a_1 x + a_0$$

be a polynomial with real coefficients. The number of positive real zeros of $P(x)$ either is equal to the number of variations in sign occurring in $P(x)$, or is less than this number by an even positive integer. The number of negative real zeros of $P(x)$ either is equal to the number of variations in sign occurring in $P(-x)$, or is less than this number by an even positive integer.

The proof of Descartes' rule of signs, like the proof of the fundamental theorem of algebra, is much beyond the level of this text and so is not included. Some applications of the rule are given in Example 3, but its full usefulness is not illustrated until Section 5.4.

Example 3 • Using Descartes' Rule of Signs

Use Descartes' rule of signs to discuss the nature of the zeros of each polynomial. That is, describe the possibilities as to the number of positive real zeros, the number of negative real zeros, and the number of nonreal complex zeros.

a. $P(x) = x^3 - x^2 - 3$

b. $P(x) = x^4 - 3x^2 - 7x + 11$

Solution

a. Counting the number of variations of sign in

$$P(x) = x^3 - x^2 - 3,$$

we see there is only one variation, so $P(x)$ has one positive real zero. In

$$P(-x) = (-x)^3 - (-x)^2 - 3$$
$$= -x^3 - x^2 - 3$$

there are no variations in sign, so $P(x)$ has no negative real zeros. Thus we know that $P(x)$ has one positive real zero and therefore two nonreal complex zeros.

b. There are two variations of sign in

$$P(x) = x^4 - 3x^2 - 7x + 11.$$

Descartes' rule of signs states that there are either two or no positive real zeros of $P(x)$. Since there are two variations in sign in

$$P(-x) = x^4 - 3x^2 + 7x + 11,$$

there are either two or no negative real zeros of $P(x)$. The following table summarizes all possibilities as to the nature of the zeros of $P(x)$. Recall that $P(x)$ has exactly four zeros.

	Real		Nonreal
	Positive	Negative	
Two positive real zeros combined with negative possibilities	2	2	0
	2	0	2
Zero positive real zeros combined with negative possibilities	0	2	2
	0	0	4

Each row must total 4 [the degree of $P(x)$].

Thus any one of the following is a possibility as to the nature of the zeros of $P(x)$:

i. Two positive zeros, two negative zeros
ii. Two positive zeros, two nonreal complex zeros
iii. Two negative zeros, two nonreal complex zeros
iv. Four nonreal complex zeros. □

◆ Practice Problem 2 ◆

Use Descartes' rule of signs to discuss the nature of the zeros of

$$P(x) = x^5 - x^4 + 2x^2 - x - 1.$$

There is one more basic fact we will need concerning the zeros of a polynomial that has real coefficients. This fact, which is stated in the next theorem, involves the conjugates of the zeros of $P(x)$. We recall from Section 1.9 that if $z = a + bi$ is a complex number in standard form, the conjugate of z is the complex number $\bar{z} = a - bi$.

Conjugate Zeros Theorem

Let

$$P(x) = a_n x^n + a_{n-1}x^{n-1} + \cdots + a_1 x + a_0$$

be a polynomial with real coefficients. If $z = a + bi$ is a zero of $P(x)$, then $\bar{z} = a - bi$ is also a zero of $P(x)$.

This means that, if $P(x)$ is a polynomial with real coefficients, the complex zeros of $P(x)$ always occur in conjugate pairs.

Example 4 • Using the Conjugate Zeros Theorem to Find Zeros

Given that $1 - i$ is a zero of $P(x) = x^3 - 4x^2 + 6x - 4$, find all zeros of $P(x)$.

Solution

Since $1 - i$ is a zero of $P(x)$, $x - (1 - i)$ is a factor of $P(x)$. We use synthetic division to divide $P(x)$ by $x - (1 - i)$.

$$\begin{array}{r|rrrr} 1-i & 1 & -4 & 6 & -4 \\ & & 1-i & -4+2i & 4 \\ \hline & 1 & -3-i & 2+2i & 0 \end{array}$$

Thus $P(x)$ can be factored as

$$P(x) = [x - (1 - i)][x^2 + (-3 - i)x + (2 + 2i)].$$

By the conjugate zeros theorem, $\overline{1 - i} = 1 + i$ is also a zero of $P(x)$. Since $1 + i$ is not a zero of the factor $x - (1 - i)$, it must be a zero of the quotient. We have

$$\begin{array}{r|rrr} 1+i & 1 & -3-i & 2+2i \\ & & 1+i & -2-2i \\ \hline & 1 & -2 & 0 \end{array}$$

and the new quotient is $x - 2$. This means that the complete factorization of $P(x)$ is

$$P(x) = [x - (1 - i)][x - (1 + i)](x - 2),$$

so the zeros of $P(x)$ are $1 - i$, $1 + i$, and 2. □

Example 5 • Using the Conjugate Zeros Theorem to Find a Polynomial

Find a polynomial $Q(x)$ of least degree with *real* coefficients that has $3i$ and 4 as zeros.

Solution

Since $Q(x)$ is to have *real* coefficients and is to have $3i$ as a zero, it must also have $\overline{3i} = -3i$ as a zero. The product

$$\begin{aligned} Q(x) &= (x - 3i)(x + 3i)(x - 4) \\ &= (x^2 + 9)(x - 4) \\ &= x^3 - 4x^2 + 9x - 36 \end{aligned}$$

is a monic polynomial with the required properties. □

In connection with Example 5, we note that the monic polynomial of least degree that has $3i$ and 4 as zeros is

$$P(x) = (x - 3i)(x - 4) = x^2 - (4 + 3i)x + 12i,$$

but this polynomial does not have the real coefficients that were required in Example 5.

◆ Practice Problem 3 ◆

Find a polynomial $Q(x)$ of least degree with real coefficients that has $2 - i$ and i as zeros.

EXERCISES 5.3

For each of the following polynomials, state its degree. List the distinct zeros and their multiplicities. (See Example 2.)

1. $P(x) = (x - 1)(x + 2)^2$
2. $Q(x) = 3(x - 4)^2(x - 1)$
3. $Q(x) = -(x + 3)^2(4x - 3)^3$
4. $P(x) = x^3(2x - 1)(x + 2)^2$
5. $P(x) = x^2(x^2 + 1)(x^2 - 1)$
6. $Q(x) = 4x^2(x - 1)^2(x^2 - 1)$

Find a polynomial $P(x)$ of least degree that satisfies the given conditions. Assume each zero has multiplicity 1, unless otherwise stated. Write $P(x)$ in factored form. (See Example 1 and Practice Problem 1.)

7. Monic; the zeros are 3 (multiplicity 2) and -1.
8. Monic; the zeros are -2 and 4 (multiplicity 3).
9. Monic; the zeros are 0 (multiplicity 5) and -2 (multiplicity 3).
10. Monic; the zeros are 1, -2 (multiplicity 2), and 3 (multiplicity 2).
11. The zeros are 1, -1, and 2; $P(0) = 4$.
12. The zeros are 4, -1, and 0 (multiplicity 2); $P(1) = -12$.

Use Descartes' rule of signs to discuss the nature of the zeros of each polynomial. (See Example 3 and Practice Problem 2.)

13. $P(x) = x^3 - 4x^2 + 5x + 1$
14. $P(x) = x^3 + 3x^2 + 4x - 6$
15. $P(x) = 2x^4 + 3x^3 - 2x + 1$
16. $P(x) = 4x^4 - 7x^3 + 3x + 2$

17. $P(x) = 2x^4 - x^3 + 3x - 1$
18. $P(x) = 4x^4 - 2x^2 + 5x - 1$
19. $P(x) = x^6 + x^3 + 2x + 3$
20. $P(x) = 4x^4 + 2x^3 + 4x - 2$
21. $P(x) = 2x^5 - 5x^4 - 12x^3 + 3x^2 - 14x + 8$
22. $P(x) = 4x^5 + 5x^4 - 43x^3 - 47x^2 + 63x + 18$
23. $P(x) = x^7 + x^6 - 4x^5 - 4x^4 + x^3 + x^2 + 6x + 6$
24. $P(x) = x^7 + x^6 - 6x^5 - 6x^4 + 11x^3 + 11x^2 - 6x - 6$
25. $P(x) = x^6 + 4x^4 + x^2 + 5$
26. $P(x) = x^5 + 3x^3 + 7x$

In Exercises 27–42, some of the zeros of the polynomial are given. Find the other zeros. (See Example 4.)

27. $P(x) = x^2 + 9$; $-3i$ is a zero
28. $P(x) = x^2 + 2x + 2$; $-1 + i$ is a zero
29. $Q(x) = x^3 + x + 10$; -2 and $1 + 2i$ are zeros
30. $Q(x) = x^3 + x^2 - x + 15$; -3 and $1 + 2i$ are zeros
31. $P(x) = x^4 + 20x^2 + 64$; $-2i$ and $4i$ are zeros
32. $P(x) = x^4 + 17x^2 + 16$; $-i$ and $4i$ are zeros
33. $Q(x) = x^3 + 4x^2 + 4x + 16$; $-2i$ is a zero
34. $Q(x) = x^3 - 5x^2 + 9x - 45$; $3i$ is a zero
35. $P(x) = x^3 - 2x^2 - 3x + 10$; $2 + i$ is a zero
36. $P(x) = x^3 + x^2 - 4x + 6$; $1 - i$ is a zero
37. $Q(x) = x^4 + x^3 + 10x^2 + 9x + 9$; $-3i$ is a zero
38. $Q(x) = x^4 + 3x^3 + 6x^2 + 12x + 8$; $2i$ is a zero
39. $P(x) = x^4 - 2x^3 + x^2 + 2x - 2$; $1 + i$ is a zero
40. $P(x) = x^4 - 7x^3 + 19x^2 - 23x + 10$; $2 + i$ is a zero
41. $P(x) = x^5 - 6x^4 + 16x^3 - 24x^2 + 20x - 8$; $1 + i$ is a zero of multiplicity 2
42. $P(x) = x^5 - 9x^4 + 34x^3 - 66x^2 + 65x - 25$; $2 - i$ is a zero of multiplicity 2

Find (a) polynomial $P(x)$ of least degree and in simplest form that has the given numbers as zeros and (b) a polynomial $Q(x)$ of least degree with real coefficients and in simplest form that has the given numbers as zeros. (See Examples 1 and 5, and Practice Problem 3.)

43. $3, -5$
44. $2, -2$
45. $2i, -i$
46. $-3i, 4i$
47. $3, 2 - i$
48. $2, 1 + i$
49. $3, 1 - i, 3 + 2i$
50. $1 - 2i, 3i, -2$

Critical Thinking: Exploration and Writing

51. Explain why $P(x)$ has no positive real zeros if all the coefficients in $P(x)$ are positive.

Critical Thinking: Exploration and Writing

52. Suppose $P(x)$ has no odd powers of x and all its coefficients are of the same sign. Explain why $P(x)$ has no real zeros different from 0.

Critical Thinking: Exploration and Writing

53. Explain why every polynomial of odd degree with real coefficients must have at least one real zero.

◆ Solutions for Practice Problems

1. $P(x) = a(x-1)^2(x-3)$

$$P(0) = a(0-1)^2(0-3) = 6$$
$$-3a = 6$$
$$a = -2$$
$$P(x) = -2(x-1)^2(x-3)$$

2. There are three variations in sign in

$$P(x) = x^5 - x^4 + 2x^2 - x - 1.$$

Thus there are either exactly three or exactly one positive real zero. Since there are two variations in sign in

$$P(-x) = -x^5 - x^4 + 2x^2 + x - 1,$$

there are either exactly two or exactly zero negative real zeros. The following table summarizes all possibilities as to the nature of the zeros of $P(x)$.

Real		Nonreal
Positive	Negative	
3	2	0
3	0	2
1	2	2
1	0	4

3. Since $Q(x)$ must have real coefficients, $2+i$ and $-i$ are also zeros of $Q(x)$. Thus

$$\begin{aligned} Q(x) &= [x-(2-i)][x-(2+i)](x-i)[x-(-i)] \\ &= (x^2-4x+5)(x^2+1) \\ &= x^4-4x^3+6x^2-4x+5. \end{aligned}$$

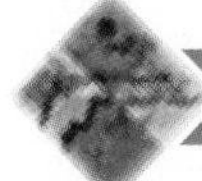

5.4 RATIONAL ZEROS

In practical applications of algebra, it is common to encounter a problem that calls for finding the rational numbers† that are zeros of a polynomial $P(x)$ with integral coefficients. In such cases, the following theorem is fundamental to the solution of the problem.

†Recall that a rational number is a quotient p/q of integers p and q, with $q \neq 0$.

Rational Zeros Theorem

Let

$$P(x) = a_n x^n + a_{n-1}x^{n-1} + \cdots + a_1 x + a_0$$

be a polynomial in which all coefficients are integers, and let p/q denote a rational number that has been reduced to lowest terms. If p/q is a zero of $P(x)$, then p is a factor of a_0 and q is a factor of a_n.

That is, if p/q is a zero of $P(x)$ and p/q is written in lowest terms, then p must be a factor of the constant term and q must be a factor of the leading coefficient of $P(x)$.

To see why the theorem is true, suppose that p/q, in lowest terms, is a zero of $P(x)$. Then

$$a_n\left(\frac{p}{q}\right)^n + a_{n-1}\left(\frac{p}{q}\right)^{n-1} + \cdots + a_1\left(\frac{p}{q}\right) + a_0 = 0.$$

Multiplying both sides by q^n, we have

$$a_n p^n + a_{n-1}p^{n-1}q + \cdots + a_1 pq^{n-1} + a_0 q^n = 0.$$

Subtracting $a_0 q^n$ from both sides yields

$$a_n p^n + a_{n-1}p^{n-1}q + \cdots + a_1 pq^{n-1} = -a_0 q^n$$

and

$$p(a_n p^{n-1} + a_{n-1}p^{n-2}q + \cdots + a_1 q^{n-1}) = -a_0 q^n.$$

This equation shows that p is a factor of $a_0 q^n$. Since p/q is in lowest terms, the greatest common divisor of p and q is 1, and this means that p is a factor of a_0. Similarly, the equation

$$a_{n-1}p^{n-1}q + \cdots + a_1 pq^{n-1} + a_0 q^n = -a_n p^n$$

can be used to show that q is a factor of a_n.

The case where $P(x)$ is a monic polynomial is important enough to state as a special result.

Rational Zeros of a Monic Polynomial

Let

$$P(x) = x^n + a_{n-1}x^{n-1} + \cdots + a_1 x + a_0$$

be a monic polynomial with integral coefficients. Then any rational zero of $P(x)$ is an integral factor of a_0.

Example 1 • Using the Rational Zeros Theorem

Find all rational zeros of $P(x) = 2x^3 - x^2 - 8x - 5$.

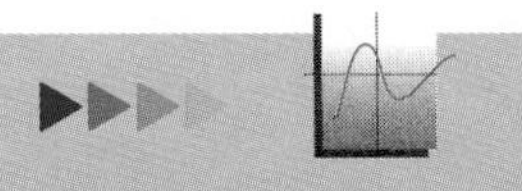

When repeated evaluation of a polynomial is desired, a graphing device can be used effectively by storing the defining polynomial expression.

Solution

By the rational zeros theorem, any rational zero of $P(x)$ has the form p/q, where p is a factor of -5 and q is a factor of 2. This means that

$$p \in \{\pm 1, \pm 5\} \qquad \text{and} \qquad q \in \{\pm 1, \pm 2\}.$$

Thus any rational zero p/q of $P(x)$ is included in the list

$$\pm 1,\ \pm 5,\ \pm\tfrac{1}{2},\ \pm\tfrac{5}{2}.$$

It is good practice to list the possible zeros in order from left to right and test them systematically. We rewrite the possibilities as

$$-5,\ -\tfrac{5}{2},\ -1,\ -\tfrac{1}{2},\ \tfrac{1}{2},\ 1,\ \tfrac{5}{2},\ 5.$$

Descartes' rule of signs indicates there is one positive zero of

$$P(x) = 2x^3 - x^2 - 8x - 5,$$

but this positive zero is not necessarily rational. Using synthetic division and testing the positive integral possibilities first, we have

	2	−1	−8	−5	⟵ $P(0) = -5 < 0$
1	2	1	−7	−12	⟵ $P(1) < 0$ } Apply the intermediate value theorem.
5	2	9	37	180	⟵ $P(5) > 0$ }
$\frac{5}{2}$	2	4	2	0	

In the table above we use the intermediate value theorem to conclude that a zero exists between 1 and 5. Since $\frac{5}{2}$ is the only rational possibility, we test it and find that $\frac{5}{2}$ is indeed a zero.

It would be straightforward to continue testing the remaining possible rational zeros, which are $\frac{1}{2}$, $-\frac{1}{2}$, -1, $-\frac{5}{2}$, -5. However, we know from the last division performed that $x - \frac{5}{2}$ is a factor of $P(x)$, and

$$P(x) = (x - \tfrac{5}{2})(2x^2 + 4x + 2).$$

Thus any remaining zeros of $P(x)$ are zeros of the quotient $2x^2 + 4x + 2 = 2(x^2 + 2x + 1)$. We need only solve

$$x^2 + 2x + 1 = 0$$

to finish the problem. We have

$$(x + 1)^2 = 0,$$

so the other zero of $P(x)$ is -1, with a multiplicity of 2. That is, the rational zeros of $P(x)$ are given by

$$\{\tfrac{5}{2}, -1, -1\}.$$ □

As Example 1 shows, the testing for rational zeros may lead to a situation where the quotient is a quadratic polynomial. In such a case, it is easier to find the zeros of the quotient than to continue checking rational zero possibilities. Sometimes this method will produce all zeros of $P(x)$, not just the rational zeros.

Example 2 • Finding All Zeros of a Polynomial

Find all zeros of $P(x) = x^3 + x^2 - x + 2$.

Solution

Since $P(x)$ is monic, any rational zero of $P(x)$ is a factor of 2. Thus the possible rational zeros are

$$-2, -1, 1, 2.$$

There are two variations of sign in

$$P(x) = x^3 + x^2 - x + 2,$$

and one in

$$P(-x) = -x^3 + x^2 + x + 2.$$

Using Descartes' rule of signs, we construct the table describing the possible numbers of zeros of each type.

Real		Nonreal
Positive	Negative	
2	1	0
0	1	2

Since there must be exactly one negative zero, we first test the rational possibilities that are negative. (If -1 and -2 both fail to be a zero, then the negative zero must be irrational.)

	1	1	−1	2
−1	1	0	−1	3
−2	1	−1	1	0

The last division shows that -2 is a zero and that

$$P(x) = (x + 2)(x^2 - x + 1).$$

To find the other zeros of $P(x)$, we set

$$x^2 - x + 1 = 0$$

and obtain

$$x = \frac{1 \pm i\sqrt{3}}{2}.$$

The zeros of $P(x)$ are given by $\left\{-2, \dfrac{1 + i\sqrt{3}}{2}, \dfrac{1 - i\sqrt{3}}{2}\right\}$. □

It is important to note that a polynomial may have a large set of possible rational zeros, and yet in fact not have any rational zeros. For example, the polynomial

$$P(x) = 4x^4 + 7x^2 + 3$$

has 12 possible rational zeros, shown in the following set:

$$\{\pm 1, \pm 3, \pm \tfrac{1}{2}, \pm \tfrac{3}{2}, \pm \tfrac{1}{4}, \pm \tfrac{3}{4}\}.$$

But Descartes' rule of signs shows that $P(x)$ has *no real* zeros, and certainly *no rational* zeros.

◆ **Practice Problem 1** ◆

List the set of possible rational zeros of the following polynomials.

a. $P(x) = x^4 + 2x^3 - 5x^2 - 4x + 6$
b. $Q(x) = 3x^4 + 5x^3 - 5x^2 - 5x + 2$

When there is a large number of possible rational zeros, the following theorem can be extremely useful. We accept the theorem without proof.

Bounds Theorem

Let

$$P(x) = a_n x^n + a_{n-1}x^{n-1} + \cdots + a_1 x + a_0$$

be a polynomial with real coefficients and $a_n > 0$, and suppose that synthetic division is used to divide $P(x)$ by $x - c$, where c is real. The last row in the synthetic division can be used in the following manner.

1. If $c > 0$ and all numbers in the last row are nonnegative, then $P(x)$ has no zero greater than c.
2. If $c < 0$ and the numbers in the last row alternate in sign (with 0 written as $+0$ or -0), then $P(x)$ has no zero less than c.

In case $P(x)$ has no zero less than the number a, we say that a is a **lower bound** for the zeros. Similarly, if $P(x)$ has no zeros greater than b, then b is called an **upper bound** for the zeros. The bounds theorem can frequently be used to obtain a positive upper bound and a negative lower bound for the zeros of $P(x)$.

Example 3 • Applying the Bounds Theorem

Let $P(x) = 18x^3 + 3x^2 - 37x - 12$.

a. Find the smallest positive integer that the bounds theorem detects as an upper bound for the zeros of $P(x)$.
b. Find the negative integer nearest zero that the bounds theorem detects as a lower bound for the zeros of $P(x)$.

Solution

a. To find an upper bound we systematically check positive integral values of c starting with 1. Our check consists of looking for *all nonnegative numbers* in the last row of the synthetic division.

		18	3	−37	−12	
	1	18	21	−16		⟵ We stop the synthetic division as soon as we detect a negative number.
2 is an upper bound. ⟶	2	18	39	41	70	⟵ All are nonnegative.

Thus 2 is the smallest positive integer that the bounds theorem detects as an upper bound. Therefore any positive real zero of $P(x)$ must be less than 2.

b. To find a lower bound, we systematically check negative integral values of c starting with −1. This time we are looking for *alternating signs* in the last row of the synthetic division.

		18	3	−37	−12	
	−1	18	−15	−22		⟵ We stop the synthetic division as soon as the signs fail to alternate.
−2 is a lower bound. ⟶	−2	18	−33	29	−70	⟵ Alternating signs.

Thus all negative real zeros of $P(x)$ must be larger than −2. □

Example 4 • Finding All the Rational Zeros of a Polynomial

Find all rational zeros of $P(x) = 2x^4 + x^3 - 8x^2 + x - 10$.

Solution

Any rational zero of $P(x)$ has the form p/q, where p is a factor of −10 and q is a factor of 2. That is,

$$p \in \{\pm 1, \pm 2, \pm 5, \pm 10\} \qquad \text{and} \qquad q \in \{\pm 1, \pm 2\}.$$

This gives the following set of possible rational zeros:

$$\{\pm 1, \pm 2, \pm 5, \pm 10, \pm \tfrac{1}{2}, \pm \tfrac{5}{2}\}.$$

Since there are three variations of sign in

$$P(x) = 2x^4 + x^3 - 8x^2 + x - 10,$$

the number of positive zeros is either three or one. We arrange the positive possibilities in order of size,

$$\tfrac{1}{2}, 1, 2, \tfrac{5}{2}, 5, 10,$$

and begin a systematic check on the integral values.

	2	1	−8	1	−10	⟵ $P(0)$
1	2	3	−5	−4	−14	⟵ $P(1)$
2	2	5	2	5	0	⟵ $P(2)$

The last division shows two things: 2 is a rational zero of $P(x)$, and 2 is an upper bound of the zeros of $P(x)$. Since

$$P(x) = (x - 2)(2x^3 + 5x^2 + 2x + 5),$$

we concentrate now on the zeros of

$$Q(x) = 2x^3 + 5x^2 + 2x + 5.$$

Since there are *no* variations in sign in $Q(x)$, there are no positive real zeros of $Q(x)$. Hence we consider the set of negative rational possibilities:

$$\{-\tfrac{1}{2}, -1, -\tfrac{5}{2}, -5\}.$$

Notice that -10 and -2 are no longer possibilities. The intermediate value theorem is used in the following synthetic division table to first isolate the zero between -1 and -5. Thus we try the only rational possibility between -1 and -5.

	2	5	2	5	
-1	2	3	-1	6	A zero exists between -1 and -5, and -5 is a lower bound for the zeros of $Q(x)$.
-5	2	-5	27	-130	
$-\frac{5}{2}$	2	-0	2	-0	⟵ $-\frac{5}{2}$ is a zero of $Q(x)$, and hence of $P(x)$.

From the last row of the table, we see that $-\frac{5}{2}$ is a rational zero, and $-\frac{5}{2}$ is also a lower bound of the zeros. We now factor $P(x)$ as

$$\begin{aligned} P(x) &= (x - 2)(x + \tfrac{5}{2})(2x^2 + 2) \\ &= 2(x - 2)(x + \tfrac{5}{2})(x^2 + 1). \end{aligned}$$

The complete set of zeros of $P(x)$ is $\{2, -\frac{5}{2}, i, -i\}$, and the rational zeros of $P(x)$ are given by $\{2, -\frac{5}{2}\}$. □

◆ Practice Problem 2 ◆

Find all the rational zeros of

$$P(x) = x^4 + x^2 + x - 2.$$

Example 5 • Profit

Kracmore Pottery Company can produce ceramic flower pots and make a profit $P(x)$ (in thousands of dollars) given by

$$P(x) = x^3 - 2x^2 - 13x - 2,$$

where x is the number of pots produced (in thousands). To stay in business, the company needs to turn a profit of \$8000 in March. How many pots should be produced in March to make \$8000?

Solution

To find the required production, we set $P(x) = 8$ and solve for x.

$$x^3 - 2x^2 - 13x - 2 = 8$$

This equation is equivalent to

$$x^3 - 2x^2 - 13x - 10 = 0.$$

Since negative values of x are not meaningful here, we consider only the possible rational solutions that are positive. They are 1, 2, 5, and 10. Testing these possibilities, we find that neither 1 nor 2 is a solution, but that 5 is a solution.

	1	−2	−13	−10
1	1	−1	−14	−24
2	1	0	−13	−36
5	1	3	2	0

We have

$$\begin{aligned} x^3 - 2x^2 - 13x - 10 &= (x - 5)(x^2 + 3x + 2) \\ &= (x - 5)(x + 1)(x + 2). \end{aligned}$$

Thus 5 is the only positive solution to the equation. Hence the company needs to produce 5000 pots in March to make \$8000 profit. □

EXERCISES 5.4

1. a. Show that 1 is an upper bound for the positive zeros of

$$P(x) = x^5 + 7x^2 - x + 3.$$

   b. Show that -3 is a lower bound for the negative zeros of $P(x)$.

2. a. Show that 2 is an upper bound for the positive zeros of

$$P(x) = 2x^4 + x^2 - 9x + 11.$$

   b. Show that -1 is a lower bound for the negative zeros of $P(x)$.

In Exercises 3–10, (a) find the smallest positive integer that the bounds theorem detects as an upper bound for the zeros of the given polynomial, and (b) find the negative integer nearest 0 that the bounds theorem detects as a lower bound for the zeros of the given polynomial. (See Example 3.)

3. $2x^3 + x^2 - 10x - 4$
4. $3x^3 - 2x^2 - 21x + 15$
5. $2x^3 - 3x^2 + 8x - 13$
6. $3x^3 - 7x^2 + 15x - 35$
7. $x^4 + x^3 + 2x^2 + 4x - 9$
8. $x^4 - x^3 - 4x^2 - 2x - 13$
9. $x^4 + x^3 - 5x^2 + x - 5$
10. $x^4 + x^3 - 11x^2 + x - 10$

For each of the following polynomials, give the set of possible rational zeros. Do not try to find the zeros. (See Practice Problem 1.)

11. $P(x) = x^4 - 5x^3 + 8x^2 - 7x + 3$
12. $P(x) = x^3 + 3x^2 - 11x - 5$
13. $P(x) = x^4 + 7x^3 + x^2 - 42x + 8$
14. $P(x) = x^4 - 5x^3 + 5x^2 + 5x - 6$
15. $P(x) = 2x^4 + 7x^3 - 17x - 12$
16. $P(x) = 3x^4 - 2x^3 + 14x^2 - 10x - 5$
17. $P(x) = 2x^4 + 2x^3 - 5x^2 - x + 2$
18. $P(x) = 9x^4 + 9x^3 + 17x^2 - x - 2$
19. $P(x) = 4x^4 + 12x^3 + 5x^2 - 9x - 6$
20. $P(x) = 6x^4 + 11x^3 + 2x^2 - 5x - 2$

Find all zeros of the given polynomial. (See Examples 1 and 2.)

21. $x^3 - 3x^2 + 4x - 12$
22. $x^3 + 2x^2 + 6x + 12$
23. $2x^3 + 7x^2 + 2x - 3$
24. $2x^3 - 3x^2 - 7x - 6$
25. $3x^3 - 5x^2 - 4$
26. $2x^3 + x^2 - 2x - 6$

Find all solutions to the given equations.

27. $x^4 - x^3 - 2x^2 + 6x - 4 = 0$
28. $x^4 + x^3 - 2x^2 - 6x - 4 = 0$
29. $2x^4 - x^3 - 13x^2 + 5x + 15 = 0$
30. $2x^4 + x^3 - 13x^2 - 5x + 15 = 0$
31. $2x^4 + 5x^3 - 7x^2 - 10x + 6 = 0$
32. $2x^4 - x^3 - x^2 - x - 3 = 0$

Find all rational zeros of the given polynomial. (See Example 4 and Practice Problem 2.)

33. $x^3 + 7x - 6$
34. $x^3 - 2x^2 + 10$
35. $x^4 + x^3 + 2x^2 + 4x - 8$
36. $x^4 + 3x^3 + 6x^2 + 12x + 8$
37. $3x^3 + 8x^2 + 3x - 2$
38. $2x^3 + 9x^2 + 7x - 6$
39. $2x^3 + 17x^2 + 38x + 15$
40. $2x^3 - 9x^2 - 8x + 15$
41. $2x^4 - x^3 - 4x^2 - x - 6$
42. $2x^4 - x^3 - x^2 - x - 3$
43. $2x^4 + 3x^3 + 2x^2 + 11x + 12$
44. $3x^4 - 2x^3 + 3x^2 + 16x - 12$
45. $2x^5 - x^4 - 8x^3 + 3x^2 + 5x - 6$
46. $3x^5 + 4x^4 - 7x^3 - x^2 + 8x - 4$
47. $3x^4 + 5x^2 + 6$
48. $5x^6 + 2x^2 + 4$
49. Show that $\sqrt{3}$ is irrational by applying the rational zeros theorem to $P(x) = x^2 - 3$.
50. Show that $\sqrt{2}$ is irrational by applying the rational zeros theorem to $P(x) = x^2 - 2$.

Constructing a Box with Maximum Volume

51. An open box is to be made from a square piece of tin that measures 3 meters on each side by cutting equal squares from the four corners and bending up the sides, as shown in the accompanying figure. By using calculus, it can be shown that the largest posible volume for such a box is

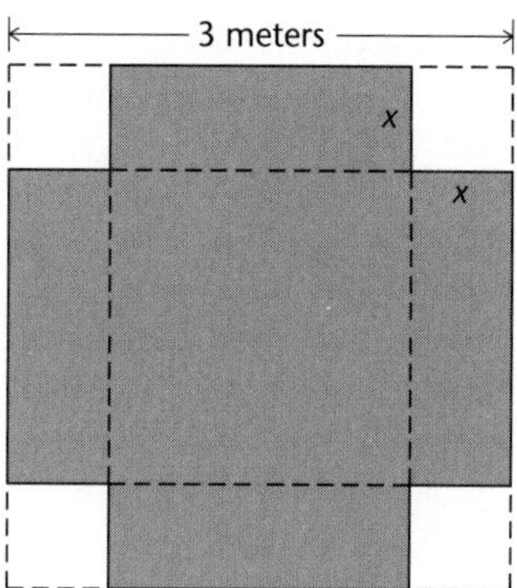

Figure for Exercise 51

2 cubic meters. How large a square must be cut from each corner to have a volume of 2 cubic meters?

Bacteria Population

52. At time $t = 0$, a bactericide is introduced into a medium in which bacteria are growing. The number $P(t)$ of thousands of bacteria present t hours later is approximated by $P(t) = 1000 + 10t - t^2$. How long will it be before $P(t) = 800$?

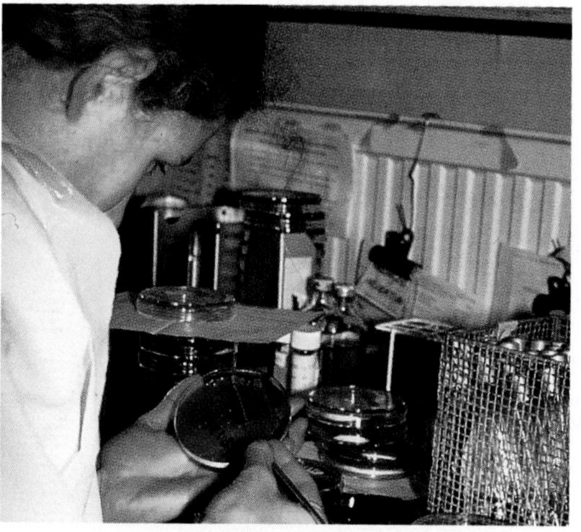

Rob Stepney/Science Photo Library/Photo Researchers

Bacterial samples being placed in serum with antibodies at a public health laboratory.

A Rectangular Box with Maximum Volume

53. A rectangular box is to be made with an open top and vertical sides. The bottom is to be a square with sides of length x meters, and only 12 square meters of material is available to make the box. By using calculus, it can be shown that the largest possible volume for such a box is 4 cubic meters. Find the value of x that yields this maximum volume.

A Package with Maximum Volume

54. A package is to be mailed that has the shape of the rectangular parallelepiped with a square base as shown in the accompanying figure. It can be mailed by parcel post only if the sum of its length and girth

(perimeter of the base) is no more than 9 feet. By using calculus, it can be shown that the largest possible volume for such a package is $\frac{27}{4}$ cubic feet. Find the dimensions that yield this maximum volume.

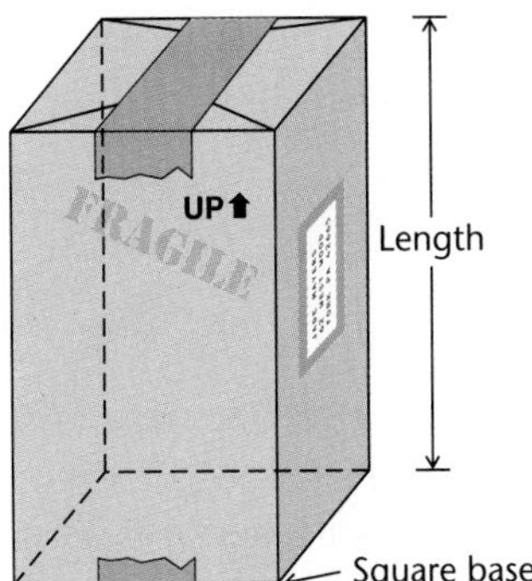

Figure for Exercise 54

A Cylinder Inscribed in a Sphere

55. If a right circular cylinder is inscribed in a sphere of radius 3, the volume of the cylinder is given by $V(h) = \left(\frac{\pi}{4}\right)(36h - h^3)$, where h is the altitude of the cylinder. Find two values of h for which $V(h) = \frac{55\pi}{4}$.

Demand

56. The demand $D(x)$ for Pedalex cars (in thousands of cars) is given by $D(x) = x(2 + x)(8 - x)$, where x is the price (in thousands of dollars). Show that there are two values of x for which the demand is 96,000.

Production Cost 57–58

57. Acme Boat Company has found that the cost $C(x)$ of producing x canoes is given (in dollars) by $C(x) = x^3 - 20x^2 + 500x$. How many canoes can be produced at a cost of \$1347?

58. Ace Gravel Company can produce t tons of gravel at a cost of $C(t) = t^3 - 5t^2 + 9t$, where $C(t)$ is in hundreds of dollars. How many tons of gravel can be produced at a cost of \$9000?

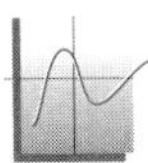

59–62

Use a graphing device to sketch the graph of the given polynomial function. State the set of possible rational zeros and use the graph to eliminate some of the possibilities. From those remaining, find all the rational zeros p/q and verify that $P(p/q) = 0$.

59. $P(x) = 8x^4 + 2x^3 - 7x^2 + 2x - 15$
60. $P(x) = 10x^4 + 7x^3 - 2x^2 + 7x - 12$
61. $P(x) = 15x^5 - 22x^4 - 5x^3 + 30x^2 - 44x - 10$
62. $P(x) = 10x^5 + 39x^4 + 14x^3 + 10x^2 + 39x + 14$

Critical Thinking: Exploration and Writing

63. Explain why 1 is a zero of a polynomial $P(x)$ if the sum of the coefficients in $P(x)$ is 0.

Critical Thinking: Exploration and Writing

64. Explain why the rational zeros theorem cannot be used directly to find the rational solutions to

$$\tfrac{1}{2}x^3 + \tfrac{1}{6}x^2 - \tfrac{4}{3}x + \tfrac{2}{3} = 0.$$

Obtain an equivalent equation to which the rational zeros theorem can be applied and solve the equation.

◆ Solutions for Practice Problems

1. a. Any rational zero of $P(x) = x^4 + 2x^3 - 5x^2 - 4x + 6$ must be a factor of 6. Thus the only possibilities are ± 6, ± 3, ± 2, and ± 1.
   b. Any rational zero of $Q(x) = 3x^4 + 5x^3 - 5x^2 - 5x + 2$ has the form p/q, where p is a factor of 2 and q is a factor of 3. Thus

$$p \in \{\pm 2, \pm 1\} \qquad \text{and} \qquad q \in \{\pm 3, \pm 1\}.$$

Forming all possible quotients using elements from the first set as the numerator and those from the second set as the denominator, we have

$$\frac{p}{q} \in \{\pm 2, \pm 1, \pm \tfrac{2}{3}, \pm \tfrac{1}{3}\}$$

as the set of possible rational zeros of $Q(x)$.

2. Since $P(x)$ is monic, any rational zero of $P(x) = x^4 + x^2 + x - 2$ is a factor of 2. Hence ± 2 and ± 1 are the only rational possibilities. Descartes' rule of signs allows us to predict the nature of the zeros.

$P(x) = x^4 + x^2 + x - 2$

$P(-x) = x^4 + x^2 - x - 2$

Real		Nonreal
Positive	Negative	
1	1	2

Thus we know that there is exactly one positive real zero and one negative real zero. This means that we will find at most two rational zeros.

	1	0	1	1	−2	
1	1	1	2	3	1	← $P(0) < 0$, $P(1) > 0$: Use the intermediate value theorem.
−1	1	−1	2	−1	−1	← $P(-1) < 0$
−2	1	−2	5	−9	16	← $P(-2) > 0$: Use the intermediate value theorem.

We conclude that the positive real zero lies between 0 and 1 and must be irrational since there are no rational possibilities between 0 and 1. Also, the negative real zero lies between −1 and −2, and hence is irrational. Thus there are no rational zeros of $P(x) = x^4 + x^2 + x - 2$.

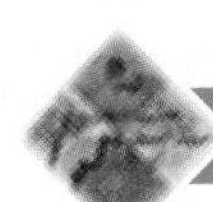

5.5 APPROXIMATION OF REAL ZEROS

In the preceding section we saw how the rational zeros of a polynomial with integral coefficients can be found. The other real zeros (that is, the irrational zeros) of such a polynomial can be found to any desired degree of accuracy by using the intermediate value theorem. The fundamental idea is to locate the zeros of $P(x)$ in intervals having length small enough to give the desired accuracy. The method is illustrated in the following example.

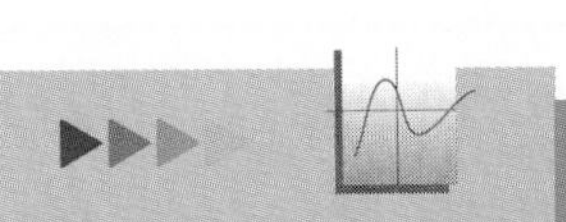

By changing the viewing rectangle, a graphing device can be used to isolate a zero of a polynomial to greater degrees of accuracy.

Example 1 • Locating Zeros

Let $P(x) = x^4 + 4x^3 + 7x^2 + 7x + 1$. Use the intermediate value theorem to show that the given interval contains one of the zeros of $P(x)$.

a. $[-1, 0]$
b. $[-0.2, -0.1]$
c. $[-0.17, -0.16]$

Solution

a. We first evaluate $P(0)$ and $P(-1)$ and note that their signs are opposite.

$$P(0) = 0^4 + 4(0)^3 + 7(0)^2 + 7(0) + 1 = 1 > 0$$
$$P(-1) = (-1)^4 + 4(-1)^3 + 7(-1)^2 + 7(-1) + 1 = -2 < 0$$

The intermediate value theorem guarantees the existence of a zero of $P(x)$ between 0 and -1. This locates a zero in an interval of length 1.

b. For the interval $[-0.2, -0.1]$ we look for a change of sign in the remainders of the synthetic division.

$$\begin{array}{r|rrrrr} -0.2 & 1 & 4 & 7 & 7 & 1 \\ & & -0.2 & -0.76 & -1.248 & -1.1504 \\ \hline & 1 & 3.8 & 6.24 & 5.752 & -0.1504 \end{array}$$

$$\begin{array}{r|rrrrr} -0.1 & 1 & 4 & 7 & 7 & 1 \\ & & -0.1 & -0.39 & -0.661 & -0.6339 \\ \hline & 1 & 3.9 & 6.61 & 6.339 & 0.3661 \end{array}$$

Since we detect a change in sign between $P(-0.2)$ and $P(-0.1)$, then there must be a zero of $P(x)$ between -0.2 and -0.1. We have located the zero in an interval of length 0.1.

c. Again we note the change in sign in the remainders of the synthetic division.

$$\begin{array}{r|rrrrr} -0.17 & 1 & 4 & 7 & 7 & 1 \\ & & -0.17 & -0.6511 & -1.079313 & -1.0065168 \\ \hline & 1 & 3.83 & 6.3489 & 5.920687 & -0.0065168 \end{array}$$

$$\begin{array}{r|rrrrr} -0.16 & 1 & 4 & 7 & 7 & 1 \\ & & -0.16 & -0.6144 & -1.021696 & -0.9565286 \\ \hline & 1 & 3.84 & 6.3856 & 5.978304 & 0.0434714 \end{array}$$

Hence the polynomial $P(x) = x^4 + 4x^3 + 7x^2 + 7x + 1$ has a zero between -0.17 and -0.16. Thus we have narrowed one zero of $P(x)$ down to an interval of length 0.01. □

◆ Practice Problem 1 ◆

Find an interval of length 0.5 containing the positive zero of

$$P(x) = x^4 + 5x^3 - x^2 - 25x - 20.$$

We end this section with an example in which we utilize as much information as possible about the polynomial in order to minimize the amount of computation needed to find its zeros. The theorems of the previous sections provide this essential information.

Example 2 • Approximating Zeros to the Nearest Tenth

Find the positive real zeros of the polynomial

$$P(x) = x^4 + 4x^2 - 6x - 5,$$

correct to the nearest tenth.

Solution

Since there is one variation in sign in $P(x)$, the polynomial has one positive real zero, by Descartes' rule of signs. The set of possible positive rational zeros is $\{1, 5\}$.

Using synthetic division, we construct the following table.

	1	0	4	−6	−5
1	1	1	5	−1	−6
5	1	5	29	139	690

There are no positive rational zeros of $P(x)$, but the work above shows that $P(1) = -6$ and $P(5) = 690$, so the positive real zero is between 1 and 5, by the intermediate value theorem. The values $P(1) = -6$ and $P(5) = 690$ indicate that the zero is probably much closer to 1 than 5, so we try 2 next in synthetic division.

2	1	0	4	−6	−5
		2	4	16	20
	1	2	8	10	15

We have $P(1) = -6$ and $P(2) = 15$, so the zero is between 1 and 2. We next evaluate $P(1.5)$ since this will locate the zero in one half of the interval or the other.

1.5	1	0	4	−6	−5
		1.5	2.25	9.375	5.0625
	1	1.5	6.25	3.375	0.0625

Since $P(1.5) = 0.0625$ is positive, the zero is between 1 and 1.5, probably nearer 1.5. We next find $P(1.4)$.

1.4	1	0	4	−6	−5
		1.4	1.96	8.344	3.2816
	1	1.4	5.96	2.344	−1.7184

Thus the zero is between 1.4 and 1.5, probably closer to 1.5. Next we compute $P(1.45)$.

1.45	1	0	4	−6	−5
		1.45	2.1025	8.8486	4.1305
	1	1.45	6.1025	2.8486	−0.8695

Since $P(1.5) = 0.0625$ and $P(1.45) = -0.8695$, the positive zero of $P(x)$ is between 1.45 and 1.5. To the nearest tenth, then, the zero is 1.5. □

The real (rational or irrational) zeros of a polynomial with real (not necessarily rational) coefficients can be approximated to the nearest hundredth, or to any needed degree of accuracy, by the method used in Example 2. The numerical calculations, of course, become more and more tedious as accuracy increases.

EXERCISES 5.5

Use the intermediate value theorem to show that the given polynomial has a zero in the given interval. (See Example 1.)

1. $P(x) = 2x^3 - 7x^2 + x + 3$; $[0, 1]$
2. $P(x) = x^3 - 5x^2 - x + 7$; $[-1, -2]$
3. $P(x) = x^4 + 2x^3 + x^2 - 5$; $[1, 1.5]$
4. $P(x) = x^4 - x^2 + 4x - 9$; $[-2.5, -2]$
5. Locate the positive zero of $P(x) = 2x^4 + 6x^3 + x^2 - 9x - 6$ in an interval of length 0.5.
6. Locate the positive zero of $P(x) = 2x^4 + x^3 + 6x^2 + x - 4$ in an interval of length 0.5.
7. Locate the negative zero of $P(x) = x^3 - 2x^2 + x + 1$ in an interval of length 0.1.
8. Locate the positive zero of $P(x) = x^3 + 4x^2 - x - 5$ in an interval of length 0.1.

Use the intermediate value theorem to approximate, to the nearest tenth, the zero of the given polynomial which is in the indicated interval. (See Example 2.)

9–14

9. $2x^3 - 11x^2 + 15x - 1$, between 2 and 3
10. $x^3 - 3x^2 + x + 1$, between 0 and -1
11. $x^3 + x^2 - 9x + 4$, between 0 and 1

12. $x^3 - 9x + 7$, between 2 and 3
13. $2x^4 - 5x^3 + 6x^2 - 22x + 7$, between 2 and 3
14. $x^4 + 2x^3 - 3x^2 + 7x - 4$, between 0 and 1

15–20

Each of the following polynomials has exactly one positive real zero. Find the value of the positive zero, correct to the nearest tenth.

15. $-x^3 - 3x^2 - x + 1$
16. $-x^3 - x^2 + 3x + 2$
17. $2x^3 + 2x^2 - 3x - 2$
18. $2x^3 - x^2 - 4x - 1$
19. $3x^4 + 6x^3 - 2x^2 - 10x - 5$
20. $4x^4 + 12x^3 + 5x^2 - 9x - 8$

21–28

Each of the following polynomials has exactly one real zero. Find the value of the zero, correct to the nearest tenth.

21. $x^3 + x^2 + x - 1$
22. $x^3 + x^2 + x - 2$
23. $4x^3 + 4x^2 + 2x + 1$
24. $x^3 + 2x^2 + 2x + 2$
25. $x^3 + 2x^2 + x - 5$
26. $x^3 - 2x^2 + 2x - 3$
27. $x^3 + 3x^2 + 3x - 10$
28. $x^3 + x^2 - x + 1$

29–32

Find all real zeros of the given polynomial, correct to the nearest tenth.

29. $x^3 - x^2 - 3x + 1$
30. $x^3 - 3x + 1$
31. $2x^3 + 2x^2 - 2x - 1$
32. $3x^3 - 7x^2 - 6x + 8$

33–36

Find all real solutions of the given equation, correct to the nearest tenth.

33. $x^3 - x^2 - 15x - 17 = 0$
34. $x^3 + 2x^2 - 14x - 32 = 0$
35. $2x^4 - 8x^3 + x^2 + 4 = 0$
36. $x^4 - 2x^3 - 2x^2 - 3 = 0$

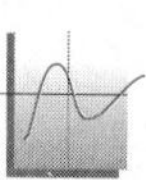

37–44

Use a graphing device to find all real zeros of the given polynomial, correct to the nearest hundredth.

37. $x^3 + x - 1$
38. $2x^3 + 3x^2 + 7$
39. $x^4 + 2x^2 + x - 4$
40. $x^4 - 3x - 5$
41. $-2x^3 + 5x^2 + 4x - 3$
42. $-2x^3 - 5x^2 + 4x + 6$
43. $0.1x^5 + 0.3x^2 - 5x - 3$
44. $x^5 - 2.6x^4 - x^3 + 9$

Critical Thinking: Exploration and Writing

45. Let $P(x)$ be a polynomial with real coefficients and suppose a and b are distinct real numbers such that $P(a)$ and $P(b)$ have the same sign. Can we conclude that $P(x)$ has no zeros between a and b? Explain.

Critical Thinking: Exploration and Writing

46. Suppose $P(x)$ is a third-degree polynomial with real coefficients and terms arranged in descending powers of x. Explain how three variations in sign could occur in $P(x)$ and yet $P(x)$ have no rational zeros.

◆ Solution for Practice Problem

1. We use synthetic division to make a systematic check on the signs of the remainders.

	1	5	−1	−25	−20	⟵ $P(0) < 0$
1	1	6	5	−20	−40	⟵ $P(1) < 0$
2	1	7	13	1	−18	⟵ $P(2) < 0$ } Use the intermediate
3	1	8	23	44	112	⟵ $P(3) > 0$ } value theorem.

We have located the zero between 2 and 3, that is, in an interval of length 1. We next check the sign of $P(x)$ at $x = 2.5$.

2.5	1	5	−1	−25	−20	
		2.5	18.75	44.375	48.4375	
	1	7.5	17.75	19.375	28.4375	⟵ $P(2.5) > 0$

Since $P(2) < 0$ and $P(2.5) > 0$, then the zero lies in the interval $[2, 2.5]$ of length 0.5.

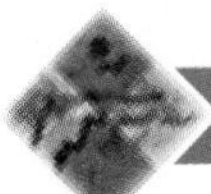

5.6 GRAPHS OF POLYNOMIAL FUNCTIONS

We restrict our attention in this section to polynomial functions defined by $y = P(x)$, where x is a real variable and $P(x)$ is a polynomial with real coefficients, and concentrate on sketching their graphs. In Section 3.4 we saw that the graph of a first-degree polynomial function is always a straight line. Then in Section 3.5 we saw that the graph of a second-degree polynomial function is always a parabola. In this section we study the graphs of polynomial functions of degree greater than 2.

The number of possible shapes of the graph increases as the degree of the polynomial function increases. Even for third-degree polynomials there are several possibilities. Some (not all) of these are indicated in Figure 5.2, where attention is called to the number of real zeros of the polynomial.

The graphs shown in Figure 5.2 represent third-degree polynomials

$$P(x) = a_3x^3 + a_2x^2 + a_1x + a_0,$$

with a leading coefficient a_3 which is positive. Similar graphs could be drawn for the case where $a_3 < 0$. These graphs would simply correspond to reflections through the x-axis, since that is the effect of multiplying a function by -1.

The point here is that there is no single characteristic shape, such as a parabola, for the graphs of higher-degree polynomials. As the degree increases, the number of possible shapes increases. However, some general remarks can be made. First, the graph of a polynomial function is a continuous, smooth curve (no breaks,

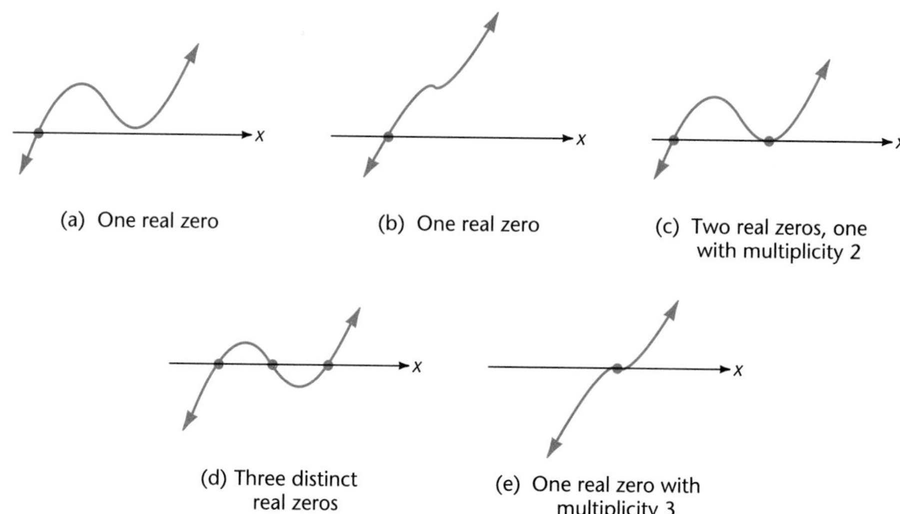

Figure 5.2 Cubic polynomials with a positive leading coefficient

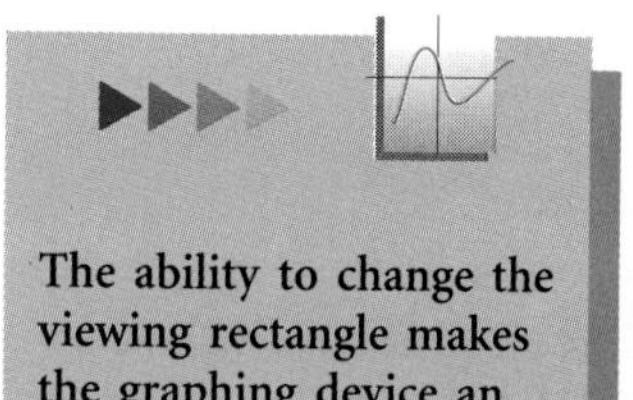

The ability to change the viewing rectangle makes the graphing device an extremely efficient tool in graphing polynomial functions.

gaps, or sharp corners). The maximum number of turning points on the graph is 1 less than the degree (note the possibilities in Figure 5.2).

For maximum efficiency in graphing a polynomial function, it is necessary to use calculus or a graphing device. For certain simpler functions, however, a reasonably good graph can be drawn by plotting suitably selected points on the curve that have integral x-coordinates and connecting these points with a smooth curve. Knowledge of the zeros of the polynomial, of symmetry, and even of the algebraic sign of the polynomial between the zeros should be utilized whenever possible. A complete analysis of the sign of the polynomial between its zeros can be carried out by using the techniques from Section 2.7 for solving nonlinear inequalities.

Example 1 • Distinct Linear Factors

Sketch the graph of P if $P(x) = (x + 1)(x - 2)(x + 3)$.

Solution

The cursor on a graphing device can be used with the **trace** feature to locate the turning points on a graph with a reasonable degree of accuracy.

The zeros of the polynomial are $x = -1$, $x = 2$, and $x = -3$, and they separate the coordinate plane into four vertical regions. We can determine the sign of $P(x)$ in each region by choosing test points. If $P(x)$ is positive for one value of x in a given interval, then it is positive for all x in that interval. Similarly, if $P(x)$ is negative for one value of x in a given interval, then it is negative for all x in that interval. The test points, along with the zeros, are used to sketch the graph of $y = P(x)$ in Figure 5.3. Additional points can be plotted as desired. Determining the exact location of the points where the curve turns requires the use of calculus. □

It is easy to see that if $x - r$ is a factor of $P(x)$ with *even* multiplicity, then $P(x)$ does not change sign as x increases through r, and the graph of $y = P(x)$ is tangent to the x-axis at $x = r$. However, if $x - r$ is a factor of $P(x)$ with *odd* multiplicity,

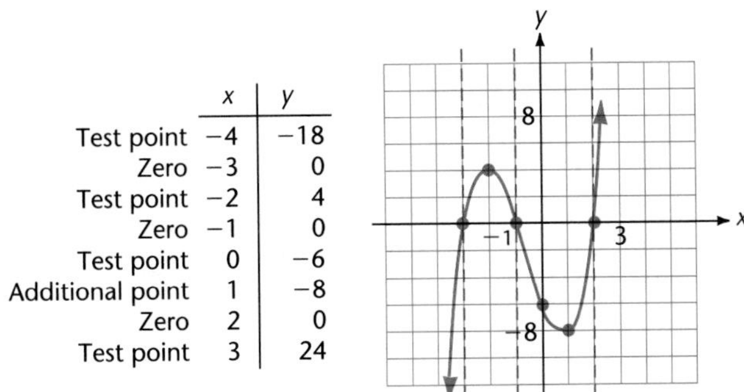

	x	y
Test point	−4	−18
Zero	−3	0
Test point	−2	4
Zero	−1	0
Test point	0	−6
Additional point	1	−8
Zero	2	0
Test point	3	24

Figure 5.3 Graph of $P(x) = (x + 1)(x - 2)(x + 3)$

then $P(x)$ does change sign as x increases through r, and the graph of $y = P(x)$ crosses the x-axis at $x = r$.

These remarks are illustrated in the following example.

Example 2 • A Repeated Linear Factor

Sketch the graph of P if $P(x) = (x - 1)(x - 2)^2$.

Solution

Since $x - 2$ is a factor with *even* multiplicity, the graph is tangent to the x-axis at $x = 2$. Also, since $x - 1$ is a factor with *odd* multiplicity, the graph crosses the x-axis at $x = 1$. A table of values and the graph are shown in Figure 5.4. □

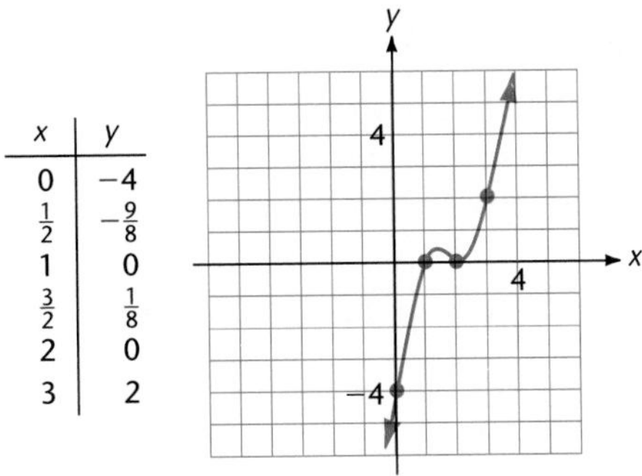

x	y
0	−4
$\frac{1}{2}$	$-\frac{9}{8}$
1	0
$\frac{3}{2}$	$\frac{1}{8}$
2	0
3	2

Figure 5.4 Graph of $y = (x - 1)(x - 2)^2$

◆ Practice Problem 1 ◆

Sketch the graph of

$$y = (x - 3)(x - 1)^2(x + 1).$$

The next example illustrates how to find a polynomial by examining its graph.

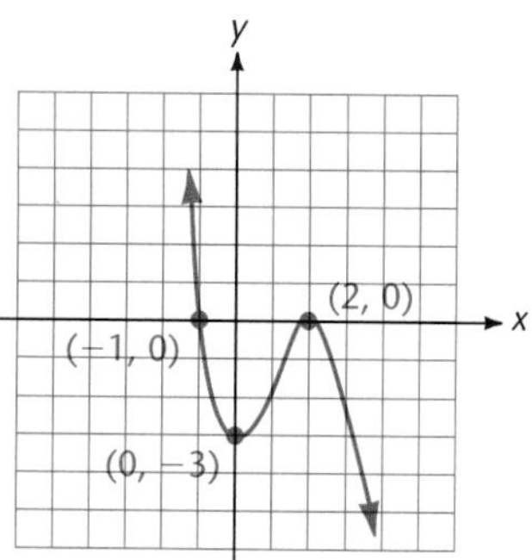

Figure 5.5 Graph for Example 3

Example 3 • Finding a Polynomial from Its Graph

Find the third-degree polynomial whose graph is given in Figure 5.5.

Solution

The polynomial has zeros at -1 (with multiplicity 1) and at 2 (with multiplicity 2). Thus the third-degree polynomial must have factored form as

$$\begin{aligned} P(x) &= a[x - (-1)]^1(x-2)^2 \\ &= a(x+1)(x-2)^2. \end{aligned}$$

We need only determine the leading coefficient a. Since $(0, -3)$ lies on the graph, then $y = -3$ when $x = 0$. In other words, if $y = P(x)$, then $-3 = P(0)$. With this information we determine a.

$$\begin{aligned} y &= P(x) \\ y &= a(x+1)(x-2)^2 \\ -3 &= a(0+1)(0-2)^2 \qquad \text{replacing } x \text{ by } 0 \text{ and } y \text{ by } -3 \\ -3 &= a(1)(4) \\ -\tfrac{3}{4} &= a \end{aligned}$$

The polynomial whose graph is given in Figure 5.5 is

$$P(x) = -\tfrac{3}{4}(x+1)(x-2)^2.$$

□

◆ **Practice Problem 2** ◆

Find the third-degree polynomial whose graph is given in Figure 5.6.

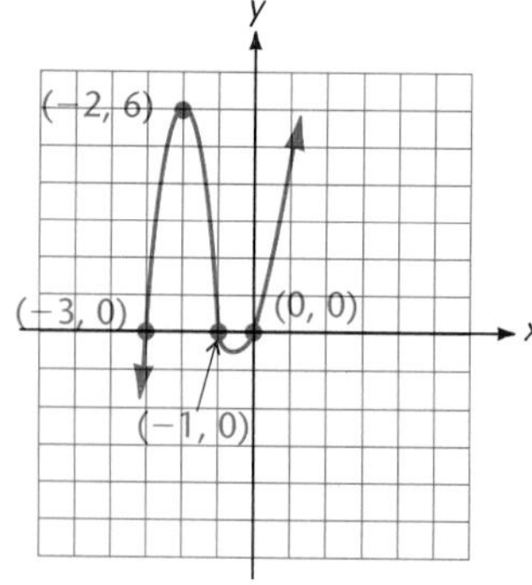

Figure 5.6 Graph for Practice Problem 2

If a polynomial is not given in factored form and cannot be easily factored, then synthetic division, Descartes' rule of signs, the intermediate value theorem, and the bounds theorem can be very useful. An example is provided here.

Example 4 • Graphing a Fourth-Degree Polynomial

Sketch the graph of the polynomial function defined by

$$P(x) = -x^4 + 24x^2 - 12x + 4.$$

Solution

Descartes' rule of signs indicates that since there are three variations of sign in

$$P(x) = -x^4 + 24x^2 - 12x + 4,$$

the graph crosses the positive x-axis in either one or three places. Also, there is one variation of sign in

$$P(-x) = -x^4 + 24x^2 + 12x + 4,$$

so the graph crosses the negative x-axis at one place. Using synthetic division, we obtain the following table, where the first and last columns contain the x- and y-coordinates, respectively, of points lying on the graph of $y = P(x)$.

x					y	
	-1	0	24	-12	4	
0	-1	0	24	-12	4	
1	-1	-1	23	11	15	
2	-1	-2	20	28	60	
3	-1	-3	15	33	103	
4	-1	-4	8	20	84	Apply the intermediate value theorem.
5	-1	-5	-1	-17	-81	↖Apply the bounds theorem.

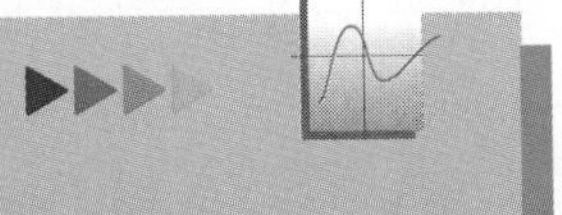

The zoom feature of a graphing device can sometimes be used to determine the number of zeros of $P(x)$ in an interval.

From the last two lines, we have $P(4) = 84$ and $P(5) = -81$, so there is a zero between 4 and 5. With a negative leading coefficient, the fact that there are all negative numbers in the last row tells us that 5 is an upper bound for the zeros. Thus it appears that there is only one positive real zero. This is not certain, however, for two more zeros of $P(x)$ may lie between consecutive integers. The methods of calculus can be used to check on possibilities such as this.

Before plotting our points, we compute the function values for some negative integers.

x					y	
	-1	0	24	-12	4	
-1	-1	1	23	-35	39	
-2	-1	2	20	-52	108	
-3	-1	3	15	-57	175	
-4	-1	4	8	-44	180	
-5	-1	5	-1	-7	39	Apply the intermediate value theorem.
-6	-1	6	-12	60	-356	↖Apply the bounds theorem.

The last two lines show that the negative zero is between -5 and -6. Because the function values vary over a very large range compared to the x-values, it is desirable to use a smaller scale on the y-axis when plotting our points. Even then, it is convenient to omit the point corresponding to $P(-6) = -356$. A plausible sketch of the graph is shown in Figure 5.7. □

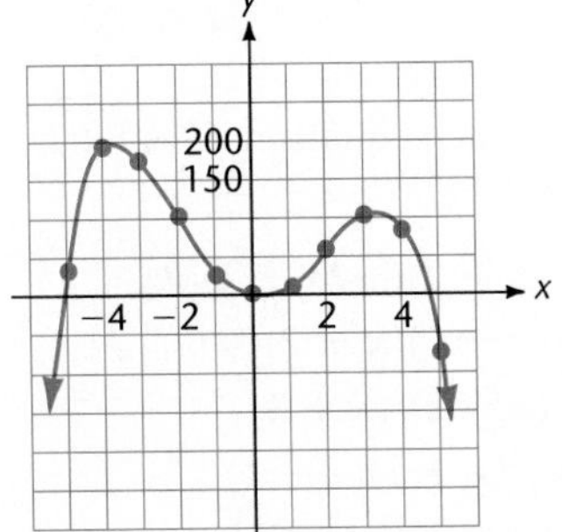

Figure 5.7 Graph of $P(x) = -x^4 + 24x^2 - 12x + 4$

Example 5 • Profit Function

Acme Anchor Company can sell up to 8000 anchors per day, but it can produce at most 4000 anchors per day with the single machine that it owns. It is considering the purchase of a second machine that could double production. An expert predicts that with two machines in operation, the company's profit P (in thousands of dollars per day) will be given by

$$P(x) = x(x - 4)(x - 6)$$

where x is the daily production (in thousands of anchors).

a. Find the nonnegative values of x for which $P(x) > 0$.
b. Sketch the graph of $P(x)$ for $0 \leq x \leq 8$.

Solution

a. We see by inspection that the zeros of P are 0, 4, and 6. Testing a point in each interval, we find that $P(2) = 16$, $P(5) = -5$, and $P(7) = 21$. Thus $P(x) > 0$ for $x \in (0, 4) \cup (6, \infty)$.

b. We use our test points and the additional function values in the table in Figure 5.8 to draw the graph shown in Figure 5.8. □

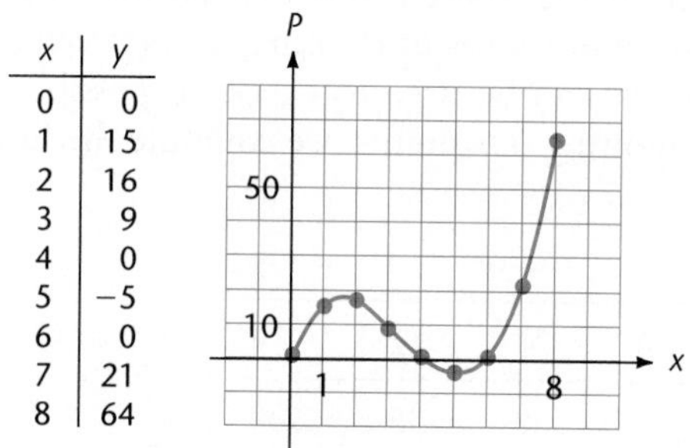

x	y
0	0
1	15
2	16
3	9
4	0
5	−5
6	0
7	21
8	64

Figure 5.8 Graph of $P(x) = x(x - 4)(x - 6)$, $x \in [0, 8]$

EXERCISES 5.6

Sketch the graph of the polynomial function defined by each equation. (See Examples 1 and 2 and Practice Problem 1.)

1. $P(x) = (x - 1)(x + 2)(x - 3)$
2. $P(x) = (x + 2)(-2x + 3)(2x + 1)$
3. $P(x) = x(x - 1)^2$
4. $P(x) = (x - 1)(x + 2)^2$
5. $P(x) = -2(x + 1)^2(1 - x)$
6. $P(x) = (x - 2)^2(x + 3)$
7. $P(x) = x^2(x + 2)(x - 2)$
8. $P(x) = x^2(x + 1)(x - 1)$
9. $P(x) = (x - 1)^2(x + 1)^2$
10. $P(x) = (x - 3)^2(x + 3)^2$
11. $P(x) = (x - 1)^2(x + 4)^2$
12. $P(x) = (x + 3)^2(x - 2)^2$

Sketch the graph of each of the following polynomials by first factoring completely.

13. $P(x) = (2 - x)(x^2 - 3x + 2)$
14. $P(x) = (1 - x)(x^2 - 3x + 2)$
15. $P(x) = x^3 - 2x^2 - 15x$
16. $P(x) = x^3 + 4x^2 + 4x$

In Exercises 17–24, find the polynomial of the specified degree whose graph is given. (See Example 3 and Practice Problem 2.)

17. Third degree

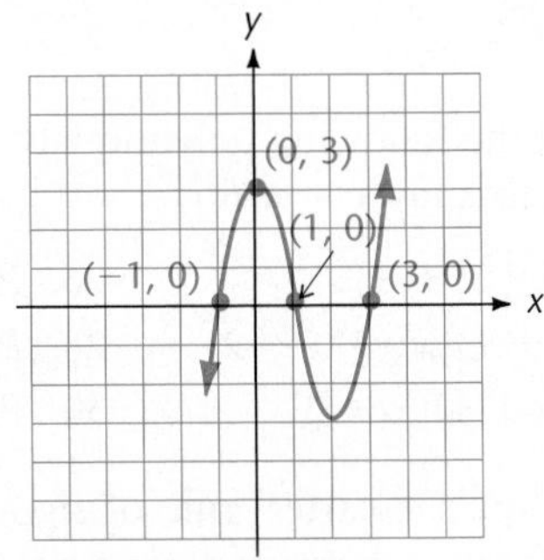

18. Third degree

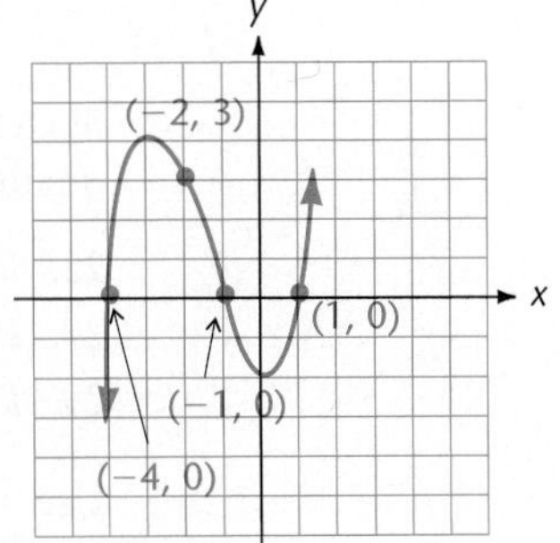

19. Third degree

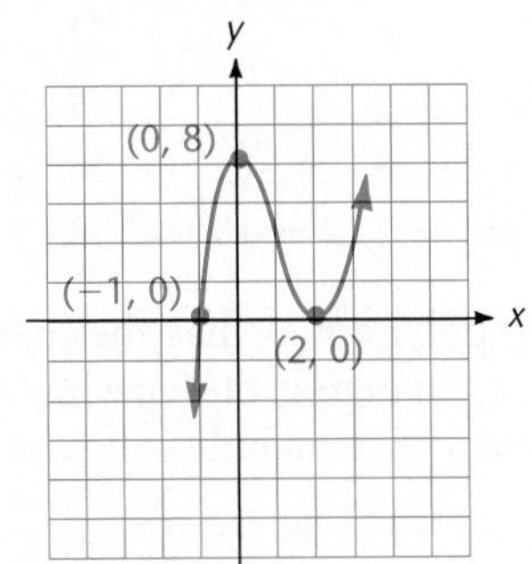

20. Third degree

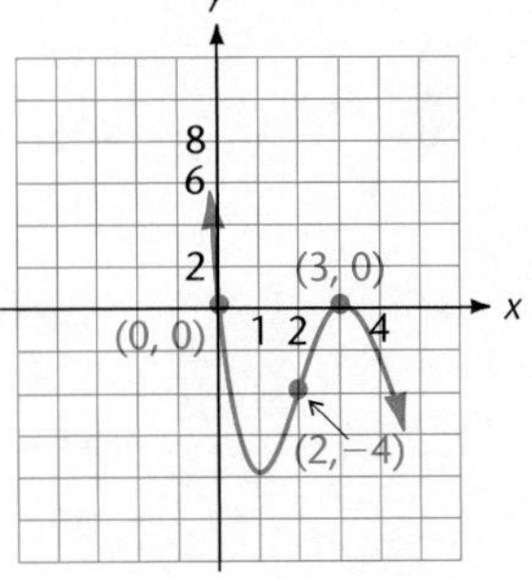

21. Fourth degree

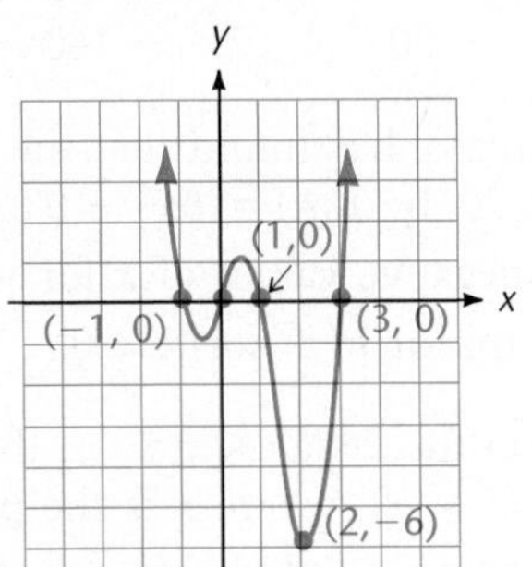

22. Fourth degree

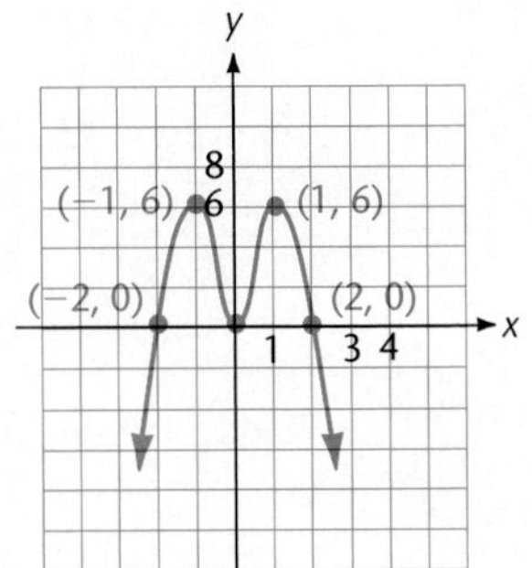

23. Fourth degree

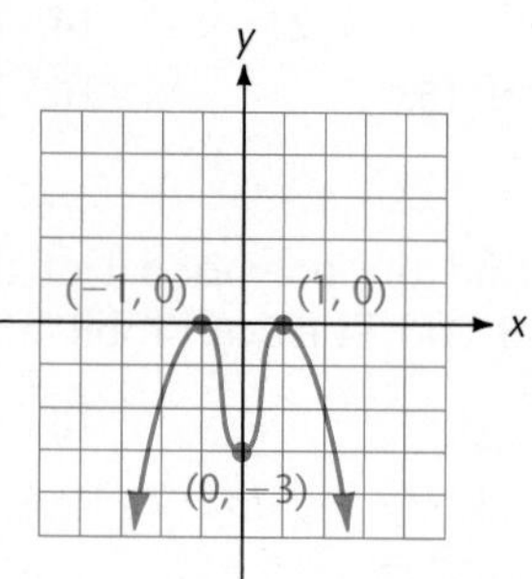

24. Fourth degree

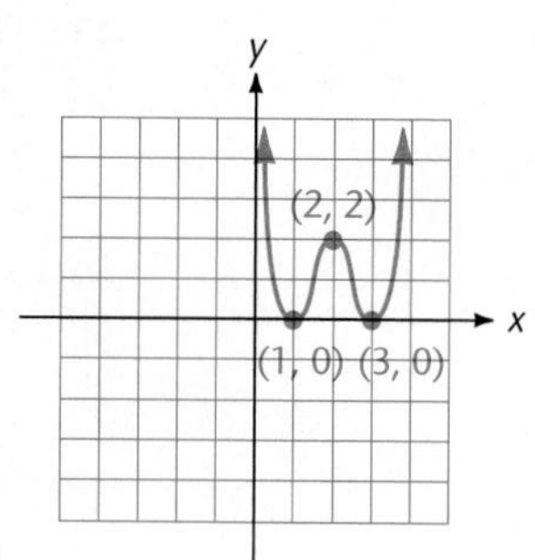

Sketch the graph of the following polynomials, noting that each has an irreducible quadratic factor that is always positive.

25. $P(x) = (x^2 + 1)(x^2 - 1)$
26. $P(x) = (x^2 + 3)x^2$
27. $P(x) = -2(x^2 + 1)(x^2 + x - 2)$
28. $P(x) = -2(x^2 + 2)(x^2 - 1)$
29. $P(x) = (x^2 + x + 1)(x - 2)$
30. $P(x) = (x^2 + x + 2)(x + 1)$

Use synthetic division, Descartes' rule of signs, the intermediate value theorem, and the bounds theorem, whenever appropriate, to sketch the graph of each of the following polynomials. (See Example 4.)

31. $y = x^3 - 3x + 4$
32. $y = x^3 - 3x - 4$
33. $y = x^3 - x^2 - 3x + 1$
34. $y = x^3 - 3x + 1$
35. $y = x^4 - 4x^3 + 4x^2 - 1$
36. $y = x^4 + 3x^3 + 3x^2 - 1$
37. $y = x^4 - 2x^3 + 3x^2 - 12x + 10$
38. $y = x^4 - 2x^3 + 2x^2 - 6x + 3$

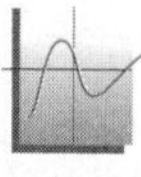
39–42

Each of the following polynomials has the maximum number of turning points. Use a graphing device, changing the viewing rectangle as necessary, to obtain a graph of the polynomial that includes all turning points and x-intercepts.

39. $P(x) = 3x^3 - 4x^2 - 9x - 28$
40. $P(x) = x^3 - 7x^2 - 38x + 140$
41. $P(x) = 0.1x^4 + 0.1x^3 - 2.1x^2 - 4.1x - 12$
42. $P(x) = -x^5 - x^4 + 30x^3 + 12x^2 - 160x + 150$

Revenue

43. Acme Boat Company has found that the revenue from producing x canoes is given (in dollars) by $R(x) = 8x^2 - 0.02x^3$.
    a. Find the nonnegative values of x for which $R(x) \geq 0$.
    b. Sketch the graph of $R(x)$ for $x \geq 0$.

Demand

44. The demand $D(x)$ for Pedalex cars (in thousands of cars) is given by $D(x) = x(2 + x)(8 - x)$, where x is the price (in thousands of dollars).
    a. Find the nonnegative values of x for which $D(x) \geq 0$.
    b. Sketch the graph of D for $x \geq 0$.

Volume of a Box

45. An open box is to be made from a rectangular piece of tin by cutting equal squares from each corner and bending up the sides as shown in the accompanying figure. If the piece of tin measures 8 inches by 12 inches, the volume of the resulting box is given by $V(x) = x(8 - 2x)(12 - 2x)$. Find the nonnegative values of x for which $V(x) \geq 0$.

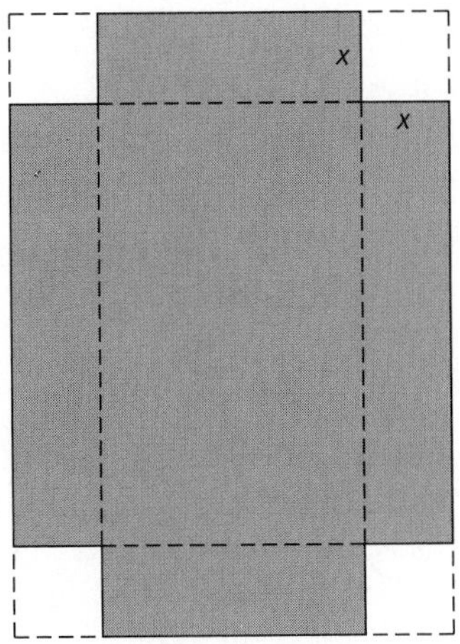

Figure for Exercise 45

Cylinder Inscribed in a Sphere

46. If a right circular cylinder is inscribed in a sphere as shown in the accompanying figure, the volume of the cylinder is $V(h) = \pi h(256 - h^2)/4$, where h is the altitude of the cylinder. Find the nonnegative values of h for which $V(h) \geq 0$.

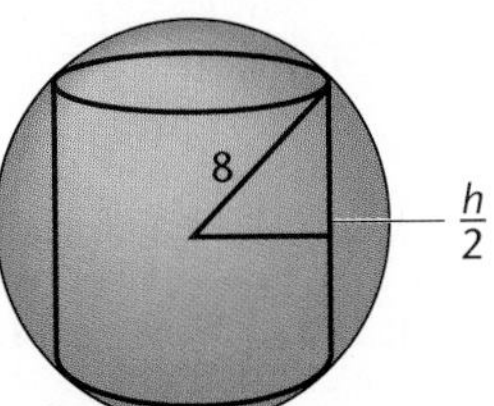

Figure for Exercise 46

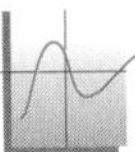

47–52

Use a graphing device to solve the given inequalities. Round the endpoints of all intervals to the nearest tenth.

47. $2x^3 + 3x^2 - 3x - 8 < 0$
48. $2x^3 + 9x^2 + 9x + 5 > 0$
49. $x^3 + 2x^2 - 6x - 5 > 0$
50. $-x^3 - x^2 + 7x + 4 > 0$
51. $-0.2x^4 + 0.4x^3 + 2x^2 - 0.8x - 4.2 > 0$
52. $x^4 - 3x^3 + x^2 - 4x - 6 < 0$

Critical Thinking: Exploration and Writing

53. Let $P(x)$ be a polynomial with real coefficients and suppose that $P(a)$ and $P(b)$ are opposite in sign. Then the intermediate value theorem assures us there is a zero between a and b. If $P(b)$ has a smaller absolute value than $P(a)$, it is reasonable to expect that the zero is nearer b than a. Explain graphically how this does not necessarily happen. Then confirm that an example is provided by $P(x) = x^4 + 4x^2 - 6x - 5$, with $a = -1$ and $b = 0$.

Critical Thinking: Exploration and Writing

54. Let $P(x)$ be a polynomial with real coefficients and suppose a and b are distinct real numbers.
    a. If $P(a)$ and $P(b)$ have opposite signs, decide what can be said about the number of zeros of $P(x)$ between a and b and explain your decision.
    b. What can be said about this number of zeros if $P(a)$ and $P(b)$ have the same sign?

◆ Solutions for Practice Problems

1. Observing that -1 and 3 are zeros of multiplicity 1 and that 1 is a zero of multiplicity 2, we know that the graph crosses the x-axis at -1 and 3 and is tangent to the x-axis at $x = 1$. A table of values and the graph are shown in Figure 5.9.

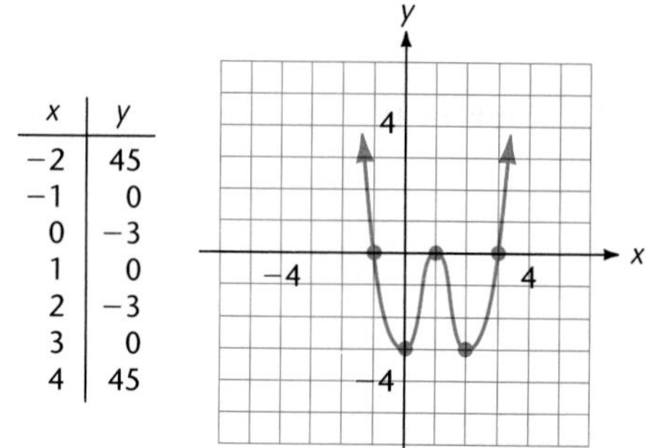

x	y
-2	45
-1	0
0	-3
1	0
2	-3
3	0
4	45

Figure 5.9 Graph of $y = (x - 3)(x - 1)^2(x + 1)$

2. The zeros of the polynomial $P(x)$ are -3, -1, and 0, each of multiplicity 1. Thus

$$\begin{aligned} P(x) &= a[x - (-3)][x - (-1)](x - 0) \\ &= a(x + 3)(x + 1)x. \end{aligned}$$

Since $(-2, 6)$ lies on the graph of $y = P(x)$, then $y = 6$ when $x = -2$. Hence

$$\begin{aligned} 6 &= P(-2) \\ 6 &= a(-2 + 3)(-2 + 1)(-2) \\ 6 &= a(1)(-1)(-2) \\ a &= 3. \end{aligned}$$

Thus $P(x) = 3(x + 3)(x + 1)x$ is the polynomial whose graph is given in the figure.

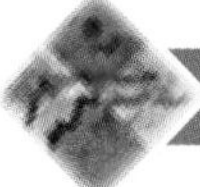

5.7 GRAPHS OF RATIONAL FUNCTIONS

In Section 1.6 a rational expression was defined as a quotient of polynomials. Similarly, a **rational function** is a function defined by an equation of the form

$$f(x) = \frac{P(x)}{Q(x)},$$

where $P(x)$ and $Q(x)$ are polynomials in the variable x, and $Q(x)$ is not the zero polynomial. In this section we consider only those rational functions that are quotients of polynomials with real coefficients, and our goal is to learn to sketch the graphs of such functions with some degree of efficiency. If such a rational function f is defined by

$$y = f(x) = \frac{P(x)}{Q(x)},$$

it is understood that the domain of f is the set of all real numbers x such that $Q(x) \neq 0$.

In very simple cases the graph can be sketched satisfactorily by plotting several points. Consider the following example.

Example 1 • The Graph of a Rational Function

Sketch the graph of the function f defined by

$$f(x) = \frac{1}{x}.$$

Solution

The domain of the function $y = f(x)$ is the set of all real numbers x such that $x \neq 0$. This means that there is no point on the graph where $x = 0$, and the graph is separated into two parts: those points (x, y) with $x > 0$ and those with $x < 0$. We first concentrate on those points where $x > 0$. We select some convenient integral values of x and also some values of x very close to zero, compute the corresponding function values, and record them in the following table:

x	1	2	3	4	⋯	0.1	0.01	0.001	⋯
$f(x)$	1	$\frac{1}{2}$	$\frac{1}{3}$	$\frac{1}{4}$	⋯	10	100	1000	⋯

We notice, first, that as the values of x get increasingly large, the function values get closer and closer to zero, and second, that as the values of x get closer to zero, the function values get increasingly large. This type of behavior is referred to as **asymptotic** behavior and the lines with equations $x = 0$ and $y = 0$ are called the **vertical asymptote** and **horizontal asymptote,** respectively, of the graph of the function. The table of values can be used to graph the portion of the graph for $x > 0$ in Figure 5.10.

Replacing x by $-x$ and y by $-y$ in

$$y = \frac{1}{x}$$

yields

$$-y = -\frac{1}{x},$$

an equivalent equation. Hence the graph of $f(x) = 1/x$ is symmetric with respect to the origin, and we use this symmetry to sketch the portion of the graph for $x < 0$. □

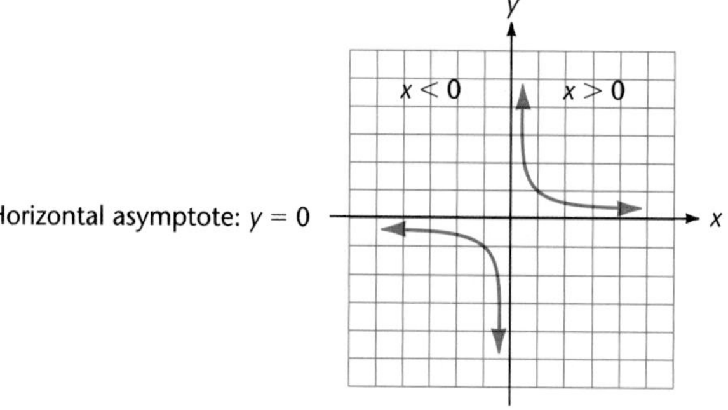

Figure 5.10 Graph of $f(x) = \dfrac{1}{x}$

The asymptotes as described in Example 1 are not part of the graph of the function but instead serve as *aids* in graphing the function. We actually consider three types of asymptotes in this section: vertical, horizontal, and oblique. Typically, they are drawn as dashed lines to distinguish them from the graph of the function. Figure 5.11 illustrates some of the common asymptotic behaviors peculiar to rational functions.

Translations, reflections, and stretchings can sometimes (although not often) be used to sketch the graph of a rational function.

Example 2 • Shifting and Stretching

Sketch the graph of the function g defined by

$$g(x) = \frac{2}{x-1}.$$

Solution

The domain of the function is the set of all real numbers x such that $x \neq 1$. We note that the graph of g can be obtained from the graph of the function $f(x) = 1/x$

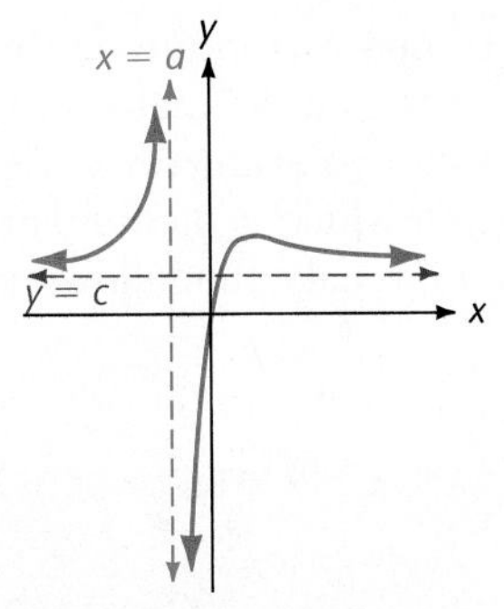

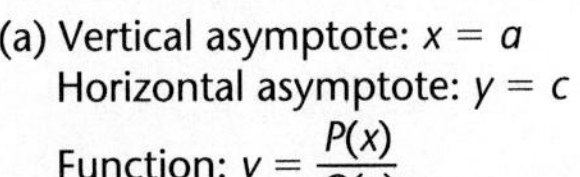

(a) Vertical asymptote: $x = a$
Horizontal asymptote: $y = c$
Function: $y = \frac{P(x)}{Q(x)}$

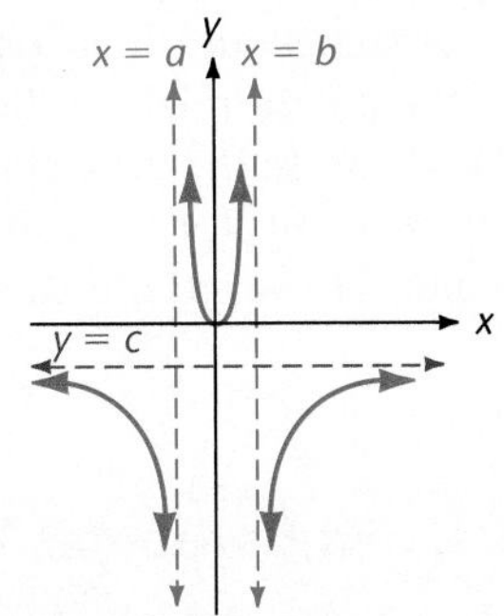

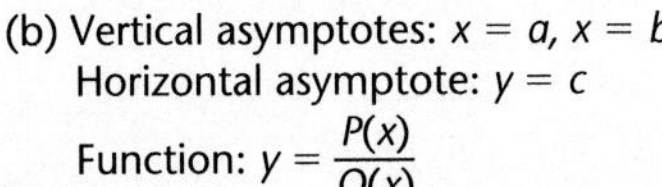

(b) Vertical asymptotes: $x = a$, $x = b$
Horizontal asymptote: $y = c$
Function: $y = \frac{P(x)}{Q(x)}$

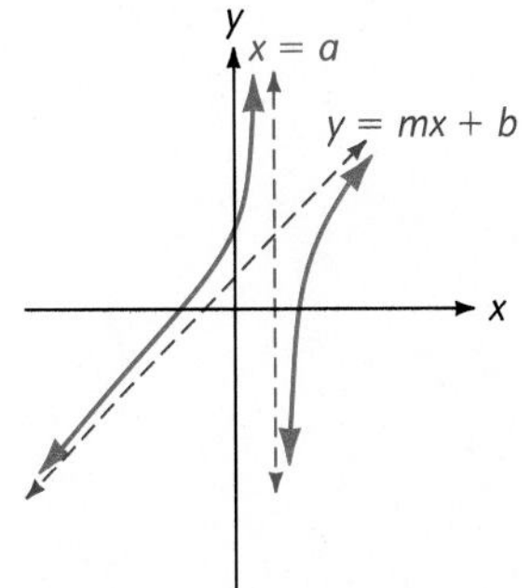

(c) Vertical asymptote: $x = a$
Oblique asymptote: $y = mx + b$
Function: $y = \frac{P(x)}{Q(x)}$

Figure 5.11 Common asymptotic behavior of rational functions

in Example 1 by a horizontal shift 1 unit to the right followed by a stretch using a factor of 2 since

$$g(x) = 2 \cdot f(x - 1) = 2 \cdot \frac{1}{x - 1} = \frac{2}{x - 1}$$

The shift followed by the stretching is illustrated in Figure 5.12 with the graph of g shown in Figure 5.12c. □

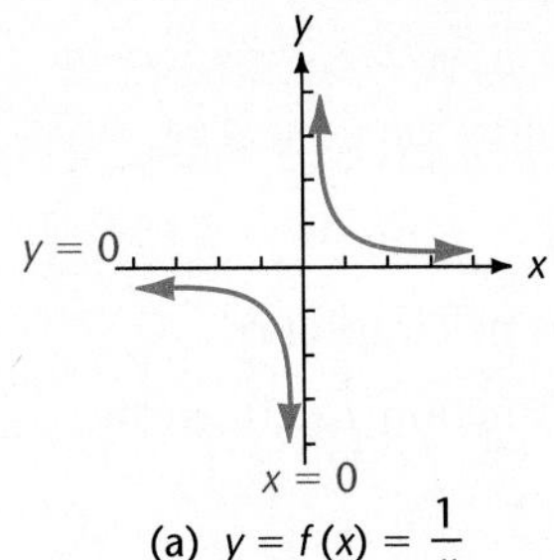

(a) $y = f(x) = \frac{1}{x}$

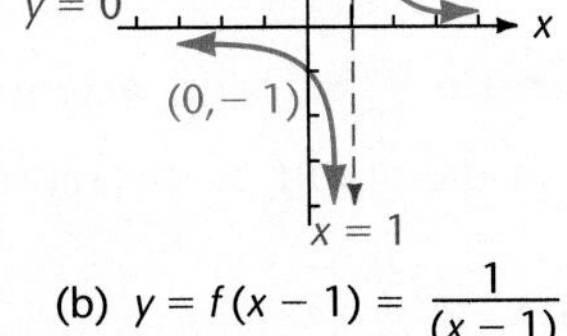

(b) $y = f(x - 1) = \frac{1}{(x - 1)}$

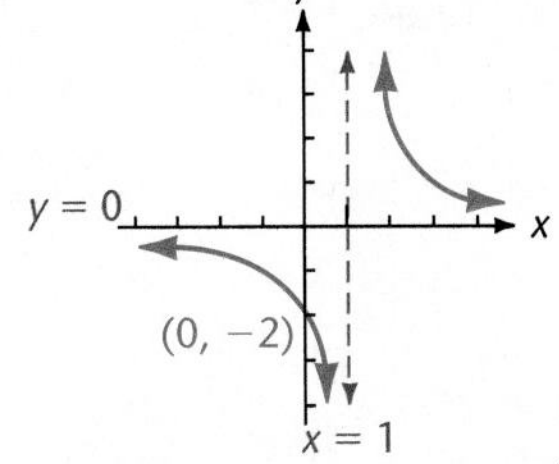

(c) $y = g(x) = 2f(x - 1) = \frac{2}{(x - 1)}$

Figure 5.12 Shifting and stretching $f(x) = 1/x$

It is not an accident that the vertical asymptotes in Examples 1 and 2 occur at a zero of the denominator of the function. In the general case, suppose that

$$y = \frac{P(x)}{Q(x)}$$

and $x = a$ is a value of x for which $Q(a) = 0$ and $P(a) \neq 0$. As x takes on values very close to a, $Q(x)$ is very close to 0 and $P(x)$ is very close to $P(a)$. This means that the quotient $P(x)/Q(x)$ grows larger numerically (that is, in absolute value) as x gets closer to a, and $|y|$ increases without bound.

The existence and location of a horizontal asymptote depends upon the degrees of the polynomials in the numerator and denominator of the rational function. We illustrate the procedures in the next two examples. A rigorous discussion of asymptotes calls for the concept of a limit, which is the fundamental concept of the calculus. However, our discussions should make the following theorem plausible.

Vertical and Horizontal Asymptotes

Let f be a rational function defined by

$$f(x) = \frac{P(x)}{Q(x)} = \frac{a_n x^n + a_{n-1}x^{n-1} + \cdots + a_1 x + a_0}{b_m x^m + b_{m-1}x^{m-1} + \cdots + b_1 x + b_0},$$

where $P(x)$ and $Q(x)$ are polynomials with real coefficients and $Q(x)$ is not the zero polynomial. The horizontal and vertical asymptotes of the graph of f may be found by the following rules.

Vertical Asymptotes

If a is a real number such that $Q(a) = 0$ and $P(a) \neq 0$, then the line $x = a$ is a vertical asymptote of the graph of f.

Horizontal Asymptotes

1. If $n < m$, then $y = 0$ is a horizontal asymptote.
2. If $n = m$, then $y = a_n/b_m$ is a horizontal asymptote.
3. If $n > m$, there are no horizontal asymptotes.

Example 3 • Using Asymptotes in Graphing

Sketch the graph of the rational function f defined by

$$f(x) = \frac{2x - 1}{x + 2}.$$

Solution

To find any vertical asymptotes we set the denominator equal to 0 and solve for x.

$$x + 2 = 0$$
$$x = -2 \qquad \text{vertical asymptote}$$

Note that the numerator is not zero for $x = -2$. Since the numerator and denominator have the same degree, the horizontal asymptote is the straight line with equation $y = \frac{2}{1} = 2$. This can be justified by the following discussion: Suppose the numerator and denominator of $f(x)$ are divided by x (the *highest* power of x that occurs in the numerator or denominator).

$$y = f(x) = \frac{2x - 1}{x + 2} = \frac{\frac{2x}{x} - \frac{1}{x}}{\frac{x}{x} + \frac{2}{x}} = \frac{2 - \frac{1}{x}}{1 + \frac{2}{x}}$$

We saw in Example 1 that as the values of x increase without bound, the values of $1/x$ decrease to zero, as does $2/x$. In fact, as $|x|$ increases, $1/x$ and $2/x$ approach zero. Thus as $|x|$ increases without bound, the quotient

$$\frac{2 - \frac{1}{x}}{1 + \frac{2}{x}} \quad \text{approaches} \quad \frac{2 - 0}{1 + 0} = 2.$$

That is, the graph of $y = f(x)$ approaches the straight line with equation $y = 2$, the horizontal asymptote.

The graph of a rational function never crosses a vertical asymptote. However, the possibility exists that the graph of a rational function may cross the horizontal asymptote. In this example, for a point (x, y) to be on the graph of the function and at the same time on the horizontal asymptote forces the y-coordinate to be 2. Thus we set $y = 2$ and solve (if possible) for x.

$$2 = \frac{2x - 1}{x + 2} \qquad \text{setting } y = 2 \text{ in } y = f(x)$$

$$2(x + 2) = 2x - 1 \qquad \text{multiplying by } x + 2$$

$$2x + 4 = 2x - 1$$

The resulting equation has *no solution.* This means that the graph of $y = (2x + 1)/(x + 2)$ does *not* cross the horizontal asymptote.

We set $x = 0$ to locate the y-intercept,

$$y = \frac{2(0) - 1}{0 + 2} = -\frac{1}{2},$$

and set $y = 0$ to locate the x-intercept,

$$0 = \frac{2x - 1}{x + 2}$$

$$0 = 2x - 1$$

$$\frac{1}{2} = x.$$

The asymptotes and a few carefully chosen points give enough information to sketch the graph in Figure 5.13. For example, when $x = -4$, $y = 9/2$. □

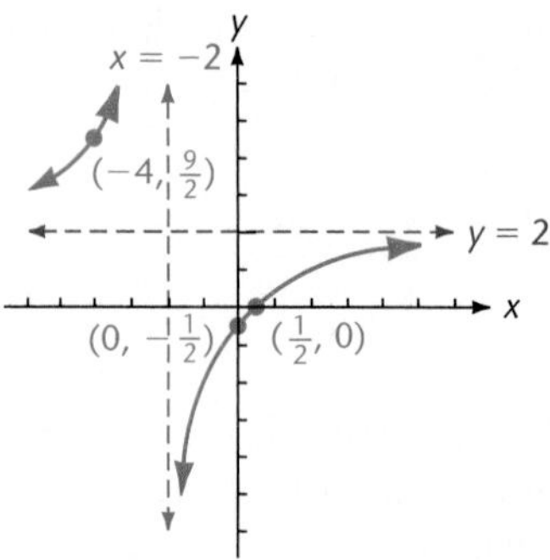

Figure 5.13 Graph of $f(x) = \dfrac{2x - 1}{x + 2}$

◆ **Practice Problem 1** ◆

Write equations for all vertical and horizontal asymptotes of the graph of the given function.

a. $f(x) = \dfrac{x - 2}{x + 3}$ b. $g(x) = \dfrac{x + 2}{(x - 1)(x - 2)}$

A systematic routine for graphing rational functions is given in the following steps.

Procedure for Graphing a Rational Function $f(x) = \dfrac{P(x)}{Q(x)}$

1. Locate all asymptotes of the graph.
2. Locate the x-intercepts and y-intercepts, if there are any.
3. Note the sign of $f(x)$ in the intervals determined by the zeros of $P(x)$ and $Q(x)$.
4. Note any symmetry detected by the tests in Section 4.2.
5. Plot a few points on either side of each vertical asymptote and check to see if the graph crosses a horizontal asymptote.
6. Sketch the graph, using the points plotted and the asymptotes as guides. The graph will be a smooth curve except for breaks at the asymptotes.

In the next example, we illustrate the use of the sign graph as another tool for graphing rational functions.

Example 4 • A Graph with Two Vertical Asymptotes

Sketch the graph of

$$f(x) = \frac{2x^2 - 1}{x^2 - 3x},$$

locating all vertical or horizontal asymptotes.

Solution

To find vertical asymptotes, we set the denominator equal to 0:

$$x^2 - 3x = 0 \qquad \text{or} \qquad x(x - 3) = 0.$$

Since $P(x) = 2x^2 - 1$ is not zero at $x = 0$ or $x = 3$, the lines $x = 0$ and $x = 3$ are vertical asymptotes. The numerator and denominator have the same degree, so $y = 2$ is a horizontal asymptote since

$$y = \frac{2x^2 - 1}{x^2 - 3x} = \frac{2 - \dfrac{1}{x^2}}{1 - \dfrac{3}{x}} \text{ approaches } \frac{2}{1} \text{ as } |x| \text{ increases without bound.}$$

Setting $y = 2$ in

$$y = \frac{2x^2 - 1}{x^2 - 3x},$$

we find that $x = \frac{1}{6} \approx 0.2$, and the graph crosses the horizontal asymptote there. Since 0 is not in the domain, there is no y-intercept. The x-intercepts are given by $x = \pm 1/\sqrt{2} \approx \pm 0.7$.

We rewrite the rational function in factored form as

$$f(x) = \frac{(\sqrt{2}x - 1)(\sqrt{2}x + 1)}{x(x - 3)}$$

and use the sign graph in Figure 5.14 to analyze the sign of $y = f(x)$ between the x-intercepts and the vertical asymptotes.

	-2	$-1/\sqrt{2}$ … 0 … $1/\sqrt{2}$	2	3	4
$\sqrt{2}x - 1$	−	− − − − 0	+	+	+
$\sqrt{2}x + 1$	−	0 + + + +	+	+	+
x	−	− − 0 + +	+	+	+
$x - 3$	−	− − − − −	−	0	+
$f(x) = \dfrac{(\sqrt{2}x - 1)(\sqrt{2}x + 1)}{x(x-3)}$	+	0 − U + 0	−	U	+

x-Intercept ($-1/\sqrt{2}$); x-Intercept ($1/\sqrt{2}$); Vertical asymptote (0); Vertical asymptote (3)

Figure 5.14 Sign graph for $f(x) = \dfrac{2x^2 - 1}{x^2 - 3x}$

No symmetry is detected by the tests in Section 4.2. The following table of values is for the graph in Figure 5.15, with coordinates rounded to the nearest tenth.

x	−3	−2	−1	−0.7	−0.5	0	0.2	0.7	1	2	3	4	5	8
$f(x)$	0.9	0.7	0.3	0	−0.3	U	1.6	0	−0.5	−3.5	U	7.8	4.9	3.2

□

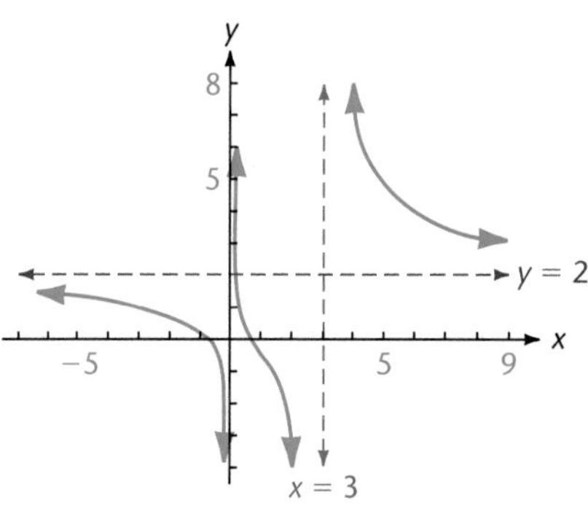

Figure 5.15 Graph of $f(x) = \dfrac{2x^2 - 1}{x^2 - 3x}$

◆ Practice Problem 2 ◆

Sketch the graph of the rational function f defined by

$$f(x) = \frac{2x^2}{(x-1)^2}.$$

Locate all vertical or horizontal asymptotes.

Under certain conditions, the graph of a rational function $f(x) = P(x)/Q(x)$ may have an **oblique asymptote.** In particular, if the degree of $P(x)$ is exactly 1 more than the degree of $Q(x)$, then we can use polynomial division to write

$$f(x) = \frac{P(x)}{Q(x)} = ax + b + \frac{R(x)}{Q(x)},$$

where the degree of $R(x)$ is less than the degree of $Q(x)$. This means that

$$y = ax + b$$

is an asymptote since $R(x)/Q(x)$ approaches 0 as $|x|$ increases.

Example 5 • An Oblique Asymptote

Sketch the graph of

$$f(x) = \frac{x^2}{x - 1}$$

and locate all asymptotes of the graph.

Solution

The graph has a vertical asymptote at $x = 1$, and it has no horizontal asymptote. The only point where the graph crosses a coordinate axis is at the origin $(0, 0)$. To find the oblique asymptote, we divide x^2 by $x - 1$ and obtain

$$f(x) = x + 1 + \frac{1}{x - 1}.$$

From this equation, it can be seen that for $|x|$ very large, the value of $1/(x - 1)$ is near zero, and points (x, y) on the graph of f are close to the line

$$y = x + 1.$$

That is, $y = x + 1$ is an oblique asymptote.

We analyze the sign of $y = f(x)$ in the following manner. Since x^2 is always nonnegative, then the sign of y depends only on the sign of $x - 1$. Thus $y > 0$ when $x > 1$, and $y \leq 0$ when $x < 1$.

The tests in Section 4.2 do not detect any symmetry. A table of values, rounded to the nearest tenth, is given below.

x	-1	-2	-3	-4	0	0.5	2	2.5	3	4
$f(x)$	-0.5	-1.3	-2.3	-3.2	0	-0.5	4	4.2	4.5	5.3

The graph is shown in Figure 5.16. □

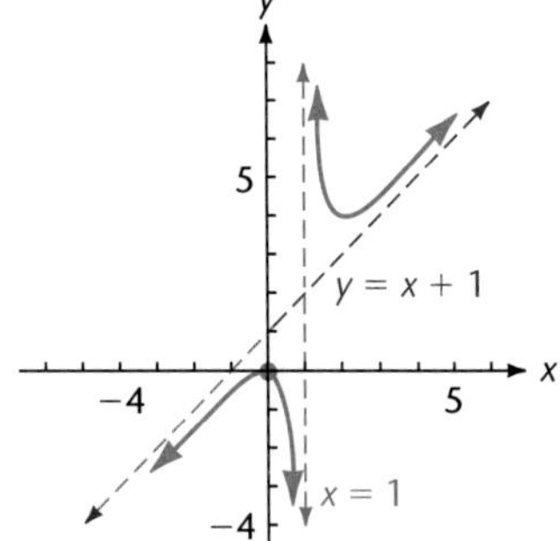

Figure 5.16 Graph of $f(x) = \dfrac{x^2}{x - 1}$

In all the examples to this point, we have considered only rational functions $f(x) = P(x)/Q(x)$, where $P(a)$ was *not* zero if $Q(a) = 0$. As our last example in this section, we consider a case where both $P(a) = 0$ and $Q(a) = 0$ for some value of a. In any case such as this, there must be a break in the graph of f at $x = a$ because

$$f(a) = \frac{P(a)}{Q(a)} = \frac{0}{0} \text{ is undefined.}$$

Our example shows that this situation is not as difficult to handle as it might seem.

Example 6 • A Graph with a "Hole"

Sketch the graph of

$$f(x) = \frac{2x - 10}{x^2 - 6x + 5}$$

and locate all asymptotes of the graph.

Solution

When $P(x) = 2x - 10$ and $Q(x) = x^2 - 6x + 5$ are written in factored form, we have

$$f(x) = \frac{2(x-5)}{(x-1)(x-5)}.$$

In this form we see that

$$f(5) = \frac{0}{0} \text{ is undefined,}$$

and that

$$f(x) = \frac{2}{x-1} \qquad \text{for all } x \neq 5.$$

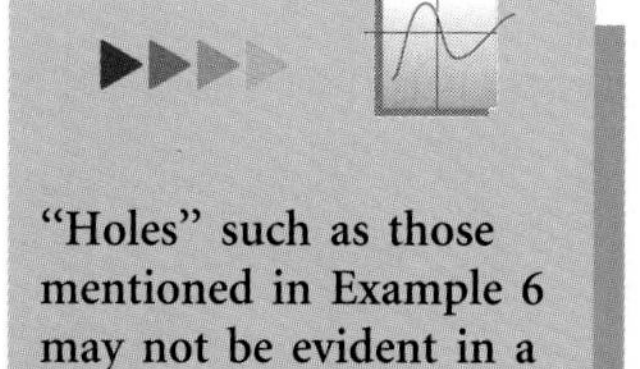

"Holes" such as those mentioned in Example 6 may not be evident in a graph provided by a graphing device.

Thus the graph of f is *almost* the same as the graph of $y = 2/(x - 1)$ in Example 2 of this section. The only difference is that the point $(5, \frac{1}{2})$ is on the graph of $y = 2/(x - 1)$, but $(5, \frac{1}{2})$ is *not* on the graph of

$$f(x) = \frac{2(x-5)}{(x-1)(x-5)}.$$

Thus the graph of f has a "hole" in it, and this is indicated in Figure 5.17 where the graph of f is drawn with an open dot at $(5, \frac{1}{2})$. □

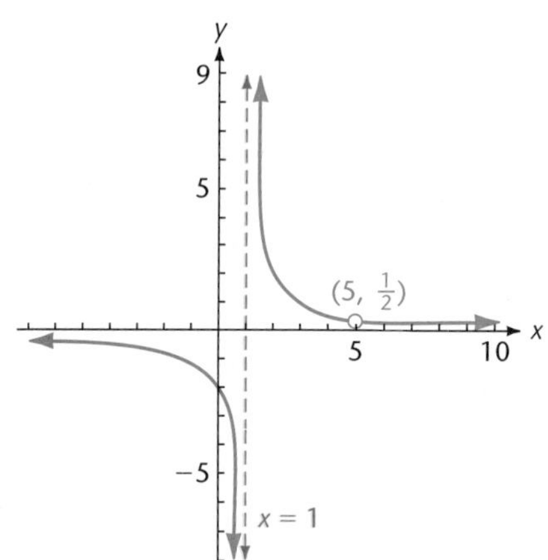

Figure 5.17 Graph of $f(x) = \dfrac{2x - 10}{x^2 - 6x + 5}$

EXERCISES 5.7

Write equations for all asymptotes of the graph of the given function. Do not sketch the graph.

1. $f(x) = \dfrac{-1}{x+3}$

2. $f(x) = \dfrac{1}{x^2}$

3. $f(x) = \dfrac{2x}{x(x+3)}$

4. $f(x) = \dfrac{x}{x^2 - 1}$

5. $f(x) = \dfrac{(3x+1)(x-2)}{(x+2)^2}$

6. $f(x) = \dfrac{4 - x^2}{(x-1)(x+3)}$

7. $f(x) = \dfrac{x^2 - 1}{(x+3)(2x-3)}$

8. $f(x) = \dfrac{5x(x+2)}{(x+1)(3x)}$

9. $f(x) = \dfrac{x^2}{x+2}$

10. $f(x) = \dfrac{x^2 - 1}{x - 2}$

11. $f(x) = \dfrac{x^3 + 2x}{x^2}$

12. $f(x) = \dfrac{3x^2}{x+3}$

13. $f(x) = \dfrac{x^2 - 1}{x^2 + 2x + 1}$

14. $f(x) = \dfrac{x+3}{x^2 - 2x - 15}$

15. $f(x) = \dfrac{x^2 - 25}{x+5}$

16. $f(x) = \dfrac{x^2 - 4}{x^2 + x - 6}$

Sketch the graphs of the rational functions defined by the following equations. Locate all vertical, horizontal, and oblique asymptotes.

17. $f(x) = \dfrac{2}{x+4}$

18. $f(x) = -\dfrac{3}{2-x}$

19. $f(x) = \dfrac{1}{x^2 - 3x}$

20. $f(x) = \dfrac{1}{x^2 - x}$

21. $f(x) = \dfrac{1}{(x+1)^2}$

22. $f(x) = \dfrac{1}{x^2}$

23. $f(x) = \dfrac{1}{x^2 + 4}$

24. $f(x) = \dfrac{1}{x^2 + 1}$

25. $f(x) = \dfrac{2x}{(x+2)^2}$

26. $f(x) = \dfrac{2x}{(x-1)(x+2)}$

27. $f(x) = \dfrac{1}{x^2(x+1)}$

28. $f(x) = \dfrac{1}{x^2(x-2)}$

29. $f(x) = \dfrac{2x-2}{x+3}$

30. $f(x) = \dfrac{2x-1}{x+2}$

31. $f(x) = \dfrac{3x}{x+2}$

32. $f(x) = \dfrac{2x}{x-1}$

33. $f(x) = \dfrac{x^2 - x - 2}{x^2 - x}$

34. $f(x) = \dfrac{2x^2 + 5x - 3}{x^2 + 2x}$

35. $f(x) = \dfrac{x^2 - 3}{2x^2 - x - 3}$

36. $f(x) = \dfrac{x^2 - 9}{2x^2 - 5x - 3}$

37. $f(x) = \dfrac{x^2}{x^2 - 4}$

38. $f(x) = \dfrac{x^2}{x^2 - 9}$

39. $f(x) = \dfrac{x^2 - 4}{x^2 - 1}$

40. $f(x) = \dfrac{x^2 - 4}{x^2 - 9}$

41. $f(x) = \dfrac{x^2 - 1}{x^2(x + 2)}$

42. $f(x) = \dfrac{x^2 - 4}{x^2(x + 3)}$

43. $f(x) = \dfrac{x^2 + 4}{x - 2}$

44. $f(x) = \dfrac{x^2 + 9}{x - 3}$

45. $f(x) = \dfrac{x^2 - 9}{x^3}$

46. $f(x) = \dfrac{x^2 - 4}{x^3}$

47. $f(x) = \dfrac{x^2 - 4}{x + 2}$

48. $f(x) = \dfrac{x^2 - 9}{x - 3}$

49. $f(x) = \dfrac{x^3 - 27}{x - 3}$

50. $f(x) = \dfrac{x^3 + 8}{x + 2}$

Medication Dosage

51. If A denotes the adult dosage of a medication in milligrams, a formula that is commonly used to find the dosage $D(x)$ for a child x years old is

$$D(x) = \frac{Ax}{x + 12}.$$

For the medication phenylpropanolamine hydrochloride, $A = 25$. Sketch the graph of D for $A = 25$ and $x > 0$.

Deer Population

52. In a restocking program, 60 deer are released in a wildlife management area. The expected population $P(x)$ of deer after x years is

$$P(x) = \frac{60 + 15x}{1 + 0.05x}.$$

Sketch the graph of P for $x \geq 0$.

Oil Spill

53. The cost $C(x)$ (in thousands of dollars) of removing x percent of a certain oil spill off the Texas coast is approximated by

$$C(x) = \frac{15x}{100 - x}.$$

Sketch the graph of C for $0 \leq x < 100$.

Behavioral Psychologist

54. A behavioral psychologist observed that a 4-year-old child memorized f lines of a certain poem, where f depended upon the number x of 5-minute time intervals that the psychologist worked with the child and

$$f(x) = \frac{20x}{x + 1}.$$

Sketch the graph of f for $x \geq 0$.

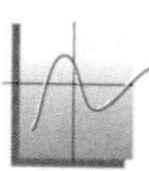
55–58

Use a graphing device to sketch the graph of the given rational functions.

55. $f(x) = \dfrac{x^4 + 4}{x^3 + 1}$ 56. $f(x) = \dfrac{x^3 - 8}{x^2 + 1}$

57. $f(x) = \dfrac{x^4 + 4}{x^2 + 1}$ 58. $f(x) = \dfrac{x^4 - 8}{x^2 + 1}$

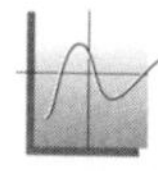
59–60

Use a graphing device to sketch the graph of the given function (a) in the viewing rectangle $[-10, 10]$ by $[-10, 10]$ and (b) in the rectangle $[-1000, 1000]$ by $[-10, 10]$. In part b, observe how closely the graph of the function resembles the graph of its horizontal asymptote.

59. $f(x) = \dfrac{2x^2 - 3}{x^2 + x + 1}$ 60. $f(x) = \dfrac{3x^2 + 4}{x^2 + 2x + 3}$

Critical Thinking: Writing

61. If a rational function contains only even powers of x, describe the symmetry present in its graph.

Critical Thinking: Exploration

62. A **parabolic asymptote** is an asymptote which has the shape of a parabola and is defined by a quadratic equation. The following function has a parabolic and a vertical asymptote. Refer to the discussion in this section on oblique asymptotes and use the same type of analysis to find the parabolic asymptote. Also, find the vertical asymptote and sketch the graph of the function.

$$f(x) = \frac{x^3 + 1}{x}$$

◆ Solutions for Practice Problems

1. a. The denominator is zero (and the numerator is nonzero) at $x = -3$. Hence the line $x = -3$ is the vertical asymptote. Since

$$f(x) = \frac{x - 2}{x + 3} = \frac{\dfrac{x}{x} - \dfrac{2}{x}}{\dfrac{x}{x} + \dfrac{3}{x}} = \frac{1 - \dfrac{2}{x}}{1 + \dfrac{3}{x}} \quad \text{approaches} \quad \frac{1 - 0}{1 + 0} = 1$$

as $|x|$ increases, then $y = 1$ is the horizontal asymptote.

b. The denominator is zero (and the numerator is nonzero) at $x = 1$ and $x = 2$. Hence there are two vertical asymptotes: $x = 1$ and $x = 2$. Dividing numerator and denominator of $g(x)$ by x^2 gives

$$g(x) = \frac{x + 2}{(x - 1)(x - 2)} = \frac{x + 2}{x^2 - 3x + 2} = \frac{\dfrac{x}{x^2} + \dfrac{2}{x^2}}{\dfrac{x^2}{x^2} - \dfrac{3x}{x^2} + \dfrac{2}{x^2}}$$

$$= \frac{\frac{1}{x} + \frac{2}{x^2}}{1 - \frac{3}{x} + \frac{2}{x^2}}.$$

This quotient approaches

$$\frac{0 + 0}{1 - 0 + 0} = 0$$

as $|x|$ increases. Hence $y = 0$ is the horizontal asymptote.

2. We note that the domain is the set of all real numbers $x \neq 1$ and that y is never negative since

$$y = 2\left(\frac{x}{x - 1}\right)^2.$$

Since $Q(x) = (x - 1)^2$ is zero at $x = 1$ and $P(x) = 2x^2$ is not zero at $x = 1$, the line $x = 1$ is a vertical asymptote. Since

$$y = \frac{2x^2}{x^2 - 2x + 1},$$

$y = 2$ is a horizontal asymptote. By setting $y = 2$ in

$$y = 2\left(\frac{x}{x - 1}\right)^2$$

and solving for x, we find that the graph crosses the horizontal asymptote at $x = 0.5$. We note that the x-intercept is 0, and the y-intercept is 0. Plotting a few points on either side of $x = 1$, we obtain the following table. The graph is shown in Figure 5.18.

x	2	3	5	10	0.5	0	−1	−2	−9
y	8	4.5	3.1	2.5	2	0	0.5	0.9	1.6

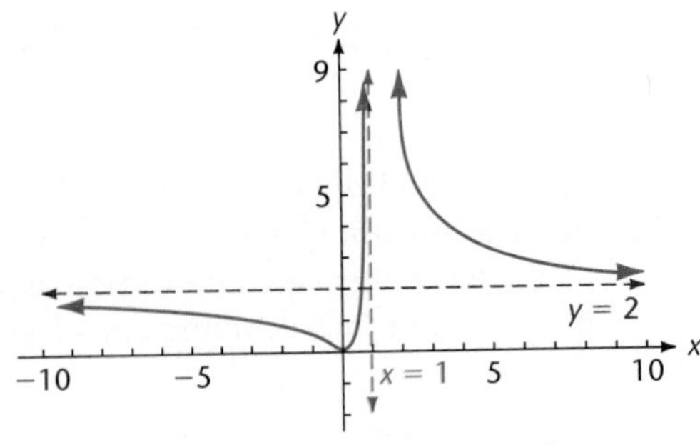

Figure 5.18 Graph of $y = \frac{2x^2}{(x - 1)^2}$

CHAPTER REVIEW

Summary of Important Theorems and Concepts

The Remainder Theorem

If a real or complex polynomial $P(x)$ is divided by $x - c$, for c real or complex, the remainder is $P(c)$.

The Factor Theorem

A polynomial $P(x)$ has a factor $x - c$ if and only if $P(c) = 0$.

The Intermediate Value Theorem for Polynomials

Let $P(x)$ be a polynomial with real coefficients. If $[a, b]$ is an interval such that $P(a) \neq P(b)$, then $P(x)$ takes on every value between $P(a)$ and $P(b)$ over the interval $[a, b]$.

The Fundamental Theorem of Algebra

Let

$$P(x) = a_n x^n + a_{n-1}x^{n-1} + \cdots + a_1 x + a_0$$

denote a polynomial of degree $n \geq 1$ with coefficients that are real or complex numbers. Then $P(x)$ has a zero in the field of complex numbers.

Factorization of $P(x)$

Let $P(x)$ be a polynomial of degree $n \geq 1$ with coefficients that are real or complex numbers. Then $P(x)$ can be factored as

$$P(x) = (x - c_1)(x - c_2) \cdot \cdots \cdot (x - c_n)a_n,$$

where $c_1, c_2, \ldots, c_n$ are n complex numbers that are zeros of $P(x)$ and a_n is the leading coefficient of $P(x)$.

Descartes' Rule of Signs

Let

$$P(x) = a_n x^n + a_{n-1}x^{n-1} + \cdots + a_1 x + a_0$$

be a polynomial with real coefficients. The number of positive real zeros of $P(x)$ either is equal to the number of variations in sign occurring in $P(x)$ or is less than this number by an even positive integer. The number of negative real zeros of $P(x)$ either is equal to the number of variations in sign occurring in $P(-x)$ or is less than this number by an even positive integer.

Conjugate Zeros Theorem

Let

$$P(x) = a_n x^n + a_{n-1}x^{n-1} + \cdots + a_1 x + a_0$$

be a polynomial with real coefficients. If $z = a + bi$ is a zero of $P(x)$, then $\bar{z} = a - bi$ is also a zero of $P(x)$.

Rational Zeros Theorem

Let

$$P(x) = a_n x^n + a_{n-1}x^{n-1} + \cdots + a_1 x + a_0$$

be a polynomial in which all coefficients are integers, and let p/q denote a rational number that has been reduced to lowest terms. If p/q is a zero of $P(x)$, then p is a factor of a_0 and q is a factor of a_n.

Rational Zeros of a Monic Polynomial

Let

$$P(x) = x^n + a_{n-1}x^{n-1} + \cdots + a_1 x + a_0$$

be a monic polynomial with integral coefficients. Then any rational zero of $P(x)$ is an integral factor of a_0.

Bounds Theorem

Let

$$P(x) = a_n x^n + a_{n-1}x^{n-1} + \cdots + a_1 x + a_0$$

be a polynomial with real coefficients and $a_n > 0$, and suppose that synthetic division is used to divide $P(x)$ by $x - r$, where r is real. The last row in the synthetic division can be used in the following manner.

1. If $r > 0$ and all numbers in the last row are nonnegative, then $P(x)$ has no zero greater than r.
2. If $r < 0$ and the numbers in the last row alternate in sign (with 0 written as $+0$ or -0), then $P(x)$ has no zero less than r.

Vertical and Horizontal Asymptotes

Let f be a rational function defined by

$$f(x) = \frac{P(x)}{Q(x)} = \frac{a_n x^n + a_{n-1}x^{n-1} + \cdots + a_1 x + a_0}{b_m x^m + b_{m-1}x^{m-1} + \cdots + b_1 x + b_0},$$

where $P(x)$ and $Q(x)$ are polynomials with real coefficients and $Q(x)$ is not the zero polynomial. The horizontal and vertical asymptotes of the graph of f may be found by the following rules.

Vertical Asymptotes

If a is a real number such that $Q(a) = 0$ and $P(a) \neq 0$, then the line $x = a$ is a vertical asymptote of the graph of f.

Horizontal Asymptotes

1. If $n < m$, then $y = 0$ is a horizontal asymptote.
2. If $n = m$, then $y = a_n/b_m$ is a horizontal asymptote.
3. If $n > m$, there are no horizontal asymptotes.

Procedure for Graphing a Rational Function $f(x) = P(x)/Q(x)$

1. Locate all asymptotes of the graph.
2. Locate the x-intercepts and y-intercepts, if there are any.
3. Note the sign of $f(x)$ in the intervals determined by the zeros of $P(x)$ and $Q(x)$.
4. Note any symmetry detected by the tests in Section 4.2.
5. Plot a few points on either side of each vertical asymptote and check to see if the graph crosses a horizontal asymptote.
6. Sketch the graph, using the points plotted and the asymptotes as guides. The graph will be a smooth curve except for breaks at the asymptotes.

Critical Thinking: Find the Errors

Each of the following nonsolutions has one or more errors. Can you find them?

Problem 1

Use the factor theorem to determine if $c = 2$ is a zero of the polynomial $P(x) = 2x^3 - 7x^2 + 6$.

Nonsolution

$$\begin{array}{r|rrr} 2 & 2 & -7 & 6 \\ & & 4 & -6 \\ \hline & 2 & -3 & 0 \end{array}$$

Since the remainder is zero, $x - 2$ is a factor of $P(x)$ and 2 is a zero of $P(x)$.

Problem 2

Find a polynomial $P(x)$ of least degree with real coefficients that has -2 and $3i$ as zeros.

Nonsolution

$$P(x) = (x + 2)(x - 3i)$$
$$= x^2 + (2 - 3i)x - 6i$$

Problem 3

Use Descartes' rule of signs to discuss the nature of the zeros of the following polynomial.

$$P(x) = x^3 + x^2 - x + 15$$

Nonsolution

There are two variations of sign in

$$P(x) = x^3 + x^2 - x + 15,$$

and one variation of sign in

$$P(-x) = -x^3 + x^2 + x + 15.$$

Therefore $P(x)$ has two positive zeros and one negative zero.

Review Problems for Chapter 5

In Problems 1–3, use synthetic division and write the results in the form

$$\frac{\text{Dividend}}{\text{Divisor}} = \text{quotient} + \frac{\text{remainder}}{\text{divisor}}.$$

1. $(2x^3 - x^2 - 4x - 30) \div (x - 3)$
2. $(3x^3 + 5x^2 + 7) \div (x + 2)$
3. $(2x^4 + 6x^2 - 2x + 1) \div (x + 1)$

In Problems 4 and 5, find the quotient $Q(x)$ and the remainder r when $P(x)$ is divided by $D(x)$.

4. $P(x) = 6x^3 + x^2 - 5x - 2,\ D(x) = x + \frac{1}{2}$
5. $P(x) = x^4 - x^3 + 3x^2 - x + 3,\ D(x) = x + i$

Use the remainder theorem and synthetic division to find $P(c)$.

6. $P(x) = 4x^3 + 5x^2 - 3x - 40,\ c = 2$
7. $P(x) = 2x^4 - 3x^2 + 4x,\ c = 1 + i$

Use the factor theorem and synthetic division to decide whether or not the given number c is a zero of $P(x)$.

8. $P(x) = x^3 - 2x^2 + 2x - 4,\ c = 2$
9. $P(x) = x^3 - 7x + 6,\ c = -3$
10. $P(x) = 3x^3 - 14x + 8,\ c = 4$

Use the factor theorem to determine whether or not $D(x)$ is a factor of $P(x)$.

11. $P(x) = 2x^3 - 5x^2 + x + 10,\ D(x) = x - 2$
12. $P(x) = 2x^3 - 4x^2 + 2x - 4,\ D(x) = x + i$
13. Given that $\frac{1}{4}$ is a zero of $P(x) = 4x^3 + 7x^2 - 62x + 15$, find all the other zeros.
14. Show that $P(x) = x^4 + x^3 - 9x^2 - 7x + 14$ has a zero between 2 and 3.
15. The polynomial $P(x) = x^4 + 2x^3 - x^2 - 6x - 6$ has one positive zero and one negative zero. Locate each between two consecutive integers.
16. Find a monic polynomial $P(x)$ of least degree that has 2 as a zero of multiplicity 2 and -1 as a zero of multiplicity 3. Write $P(x)$ in factored form.
17. Find a polynomial $P(x)$ of least degree in factored form whose zeros are 2, -1, and 3 (each with multiplicity 1) and such that $P(1) = -16$.

In Problems 18–20, use Descartes' rule of signs to discuss the nature of the zeros of the given polynomial.

18. $P(x) = 5x^3 - x^2 + 4x + 9$
19. $P(x) = 2x^4 + x^3 - 4x - 3$
20. $P(x) = 3x^4 - 5x^3 - 7x^2 - 4x + 6$

In Problems 21–23, some of the zeros of the polynomial are given. Find the other zeros.

21. $P(x) = x^3 - 5x^2 + 8x - 6$; 3 and $1 - i$ are zeros.
22. $P(x) = x^3 - 2x^2 - 3x + 10$; -2 is a zero.
23. $P(x) = x^4 - x^3 - x^2 - x - 2$; $-i$ is a zero.
24. Find a polynomial of least degree that has 4 and $2i$ as zeros.
25. Find a polynomial of least degree with real coefficients that has -1, -2, and $3i$ as zeros.
26. a. Find a positive integer, as small as possible, that the bounds theorem detects as an upper bound for the zeros of the polynomial $2x^3 + 4x^2 - 3x - 6$.
    b. Find a negative integer, as large as possible, that the bounds theorem detects as a lower bound for the zeros of the polynomial in part a.

Find all rational zeros of the given polynomial.

27. $P(x) = 2x^3 + 4x^2 - 3x - 6$
28. $P(x) = 5x^6 + 8x^4 + 6$
29. $P(x) = 3x^3 + 2x^2 - 7x + 2$

In Problems 30–32, find all solutions of the given equation.

30. $x^3 - 6x^2 + 3x + 10 = 0$
31. $x^3 - x^2 - x - 2 = 0$
32. $x^4 - x^3 - 12x^2 - 4x + 16 = 0$

In Problems 33–35, find all zeros of the given polynomial.

33. $P(x) = 3x^3 - 7x^2 + 8x - 2$
34. $P(x) = 9x^3 + 27x^2 + 8x - 20$
35. $P(x) = 3x^3 + 19x^2 + 30x + 8$
36. Find an interval of length 0.1 containing the positive zero of $P(x) = x^4 + x^3 + 3x^2 - 2x - 5$.
37. The following polynomial has one negative real zero. Find the value of the zero, correct to the nearest tenth.

$$P(x) = 2x^4 + x^2 - 4x - 3.$$

38. Find all real zeros of the following polynomial, correct to the nearest tenth.

$$P(x) = 2x^3 + 5x + 2$$

In Problems 39–41, sketch the graph of the polynomial function defined by the given equation.

39. $P(x) = x(x + 2)(x - 3)$ 40. $P(x) = -x^2(x^2 + 1)$
41. $P(x) = x^3 + x^2 - 2x - 1$

In Problems 42–45, write equations for all asymptotes of the graph of the given function. Do not sketch the graph.

42. $f(x) = \dfrac{-2}{(x - 1)^2}$ 43. $f(x) = \dfrac{x}{x + 2}$

44. $f(x) = \dfrac{-x}{(x - 1)(x + 2)}$ 45. $f(x) = \dfrac{-x^2}{x - 3}$

Sketch the graph of the rational functions defined by the following equations. Locate all vertical, horizontal, and oblique asymptotes.

46. $f(x) = \dfrac{2x - 4}{x + 1}$ 47. $f(x) = \dfrac{x^2 + 1}{x}$

48. $f(x) = \dfrac{x^2 - 1}{x + 1}$

E N C O R E

Herb Levart/Photo Researchers

Boats anchored in the harbor at Camden, Maine.

After careful consideration, Acme Anchor Company decided to buy a machine that would increase its production from 4000 to 8000 anchors per day. A consultant predicted in Example 5 of Section 5.6 that installation of this machine would result in a profit P (in thousands of dollars per day) given by

$$P(x) = x(x - 4)(x - 6),$$

where x is the daily production (in thousands of anchors).

A record was kept of the daily profits at various levels of production during the first month of operation of the new machine. This information is shown in the table in Figure 5.19. At all levels of production, profits were turning out to be larger than the predicted values.

Even though Acme's managers were happy with these profits, they asked the consultant to explain the discrepancy between predicted profits and actual profits. Her explanation was very simple: since her study was made, the price of the metal used in making anchors had decreased by \$1 for each anchor. This decrease in cost of raw materials resulted in a profit increase of \$1 for each anchor produced, and a daily production of x thousand anchors resulted in an increase of x thousand dollars in profits above her prediction. An appropriate correction to the original profit function would be to add an x term. Performing the addition, she obtained the corrected profit function

$$\begin{aligned} Q(x) &= P(x) + x \\ &= x(x-4)(x-6) + x \\ &= x^3 - 10x^2 + 25x \\ &= x(x-5)^2 \end{aligned}$$

The function values from $Q(x)$ agreed completely with the profit values in Figure 5.19, so the discrepancy was explained.

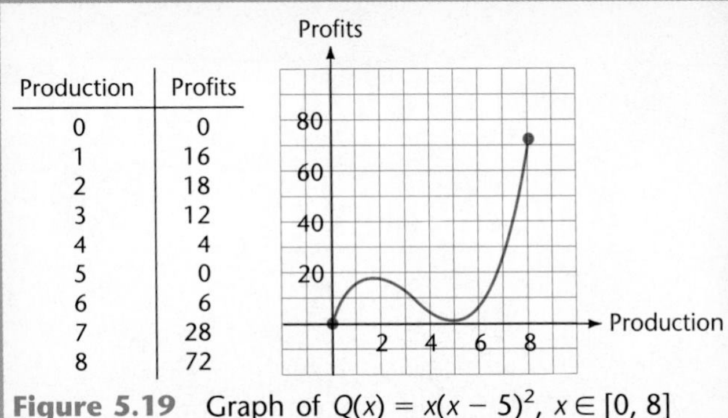

Production	Profits
0	0
1	16
2	18
3	12
4	4
5	0
6	6
7	28
8	72

Figure 5.19 Graph of $Q(x) = x(x-5)^2$, $x \in [0, 8]$

Exponential and Logarithmic Functions

The applications of exponential and logarithmic functions that appear in this chapter illustrate the importance of these functions. One example concerns one of the most famous applications in recent years, the dating of the Shroud of Turin by carbon-14 tests in 1988. Many other examples and exercises show the usefulness of these functions in a variety of fields, including engineering and business as well as the biological sciences.

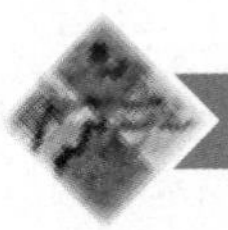

6.1 EXPONENTIAL FUNCTIONS

In this chapter we consider only real variables and two types of functions: exponential functions and logarithmic functions. These functions are indispensable in working with problems that involve population growth, decay of radioactive materials, and other processes that occur in nature.

In an exponential function, the function value is obtained by raising a fixed number, called the **base,** to a variable power. Recall that for $a \neq 0$, a^x was defined for integral values of x in Section 1.3. In Section 1.7 this definition was extended to include values $x = m/n$, which are rational numbers. However, if a is negative, there may be no real number $a^{m/n}$. [Such an example is provided by $(-4)^{1/2}$.] For this reason, *we restrict our attention in this chapter to the case where a is positive.* Once this restriction is made, a^x is defined for all rational values of x. It is possible to extend the definition of a^x to include irrational values of x, but a rigorous treatment of this extension requires a degree of mathematical sophistication that is beyond this text. A complete treatment of this sort of topic belongs to the area of mathematics known as **analysis.**

To develop some intuitive feeling for the situation involving an irrational exponent, we consider a specific example, say $2^{\sqrt{5}}$. The reasoning we use is this: if $\sqrt{5}$ is approximated by a rational number m/n, then $2^{\sqrt{5}}$ should be approximated by $2^{m/n}$. More specifically, if $\sqrt{5}$ is approximated by successively more accurate rational numbers such as

$$2.2,\ 2.23,\ 2.236,$$

then successively closer approximations to $2^{\sqrt{5}}$ should be provided by

$$2^{2.2},\ 2^{2.23},\ 2^{2.236}.$$

Each of these approximations is meaningful because the exponents are rational. For example,

$$2^{2.2} = 2^{22/10} = 2^{11/5} = (\sqrt[5]{2})^{11}.$$

This intuitive procedure is justified in more advanced courses. We accept it here without justification. In the same spirit, we accept the following generalized statement of the laws of exponents.

Laws of Exponents

If a, b, x, and y are real numbers with a and b positive, then

1. $a^x \cdot a^y = a^{x+y}$ — product rule
2. $(a^x)^y = a^{xy}$ — power of a power rule
3. $\dfrac{a^x}{a^y} = a^{x-y}$ — quotient rule

4. $(ab)^x = a^x b^x$ — power of a product rule
5. $\left(\dfrac{a}{b}\right)^x = \dfrac{a^x}{b^x}$ — power of a quotient rule

Definition of Exponential Function

If a is a positive real number and $a \neq 1$, the function f defined by

$$f(x) = a^x$$

is an **exponential function with base *a*.**

This definition excludes values of a that are not positive, so that a^x will be a real number for all real numbers x, and it excludes $a = 1$ because $1^x = 1$ defines a constant function.

Some graphs of exponential functions are shown in the next two examples.

Example 1 • An Exponential Function With $a > 1$

Sketch the graph of the exponential function

$$f(x) = 3^x.$$

Solution

After tabulating the function values displayed in Figure 6.1, we plot the corresponding points and draw a smooth curve, as shown in the figure. Notice that the x-axis is an asymptote for the graph: as x decreases without bound, 3^x approaches 0. Similarly, as x increases without bound, so does 3^x. □

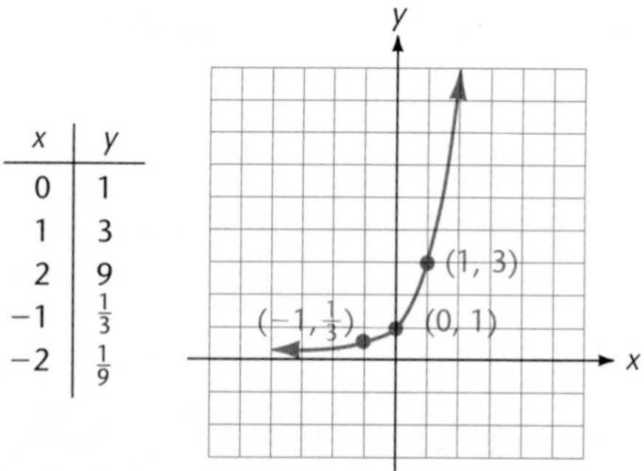

x	y
0	1
1	3
2	9
−1	$\frac{1}{3}$
−2	$\frac{1}{9}$

Figure 6.1 Graph of $f(x) = 3^x$

Example 2 • An Exponential Function with $0 < a < 1$

Sketch the graph of the exponential function

$$f(x) = \left(\frac{1}{2}\right)^x.$$

Solution

We follow the same procedure as in Example 1 and obtain the results shown in Figure 6.2. Note that the x-axis is an asymptote in this case also: as x increases without bound, $(\frac{1}{2})^x$ approaches 0. □

x	y
0	1
1	$\frac{1}{2}$
2	$\frac{1}{4}$
3	$\frac{1}{8}$
−1	2
−2	4
−3	8

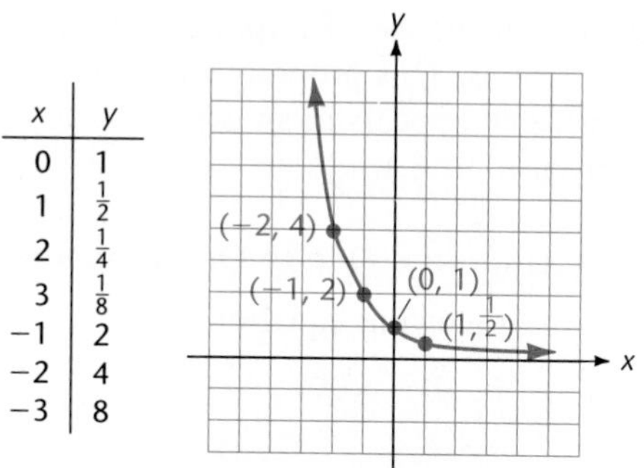

Figure 6.2 Graph of $f(x) = (\frac{1}{2})^x$

The graphs in Figures 6.1 and 6.2 illustrate the two typical graphs for exponential functions. If $a > 1$, the graph of $f(x) = a^x$ resembles the graph in Figure 6.1. If $0 < a < 1$, the graph of $f(x) = a^x$ resembles the graph in Figure 6.2. This is illustrated in Figure 6.3.

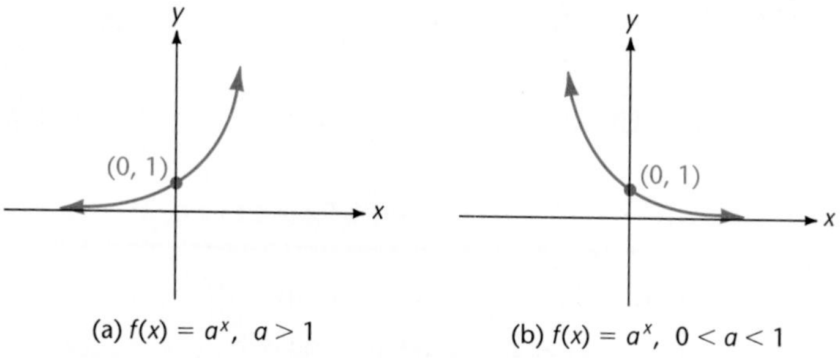

(a) $f(x) = a^x,\ a > 1$ (b) $f(x) = a^x,\ 0 < a < 1$

Figure 6.3 Typical exponential graphs

The graphs shown in Figure 6.3 are typical, and they illustrate the following observations concerning exponential functions.

Important Features of the Exponential Functions

$f(x) = a^x$, $a > 0$ and $a \neq 1$

1. The **domain** of f is the set of all real numbers.
2. The **range** of f is the set of all positive real numbers.
3. If $a > 1$, the graph of $f(x) = a^x$ resembles that of $f(x) = 3^x$. (See Figure 6.1.)

4. If $0 < a < 1$, the graph of $f(x) = a^x$ resembles that of $f(x) = (\frac{1}{2})^x$. (See Figure 6.2).
5. $a^u = a^v$ if and only if $u = v$.

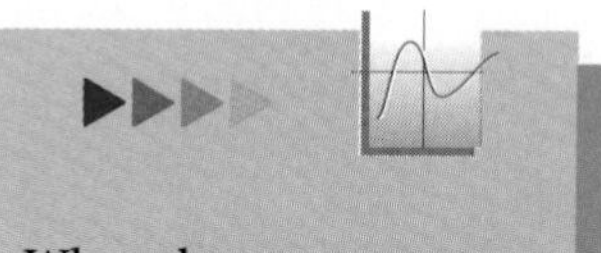

When the one-to-one property does not apply, a graphing device can be used to solve an equation involving an exponential by locating an x-intercept on the graph of an appropriate function. (See Exercises 65 and 66.)

The last item in the list of important features is the **one-to-one property** of exponential functions. This property allows us to solve certain types of equations for an unknown that appears in an exponent. A more general method of solution is presented in Section 6.4.

Example 3 • Using the One-to-One Property

Solve for x in the following equations.

a. $3^x = 81$ b. $9^{-x} = 27$

Solution

a. We recognize that $81 = 3^4$, and so the given equation can be rewritten as

$$3^x = 81 = 3^4.$$

But $3^x = 3^4$ requires that $x = 4$.

b. Each member of the given equation can be written as a power of 3.

$$\begin{aligned} 9^{-x} &= 27 \\ (3^2)^{-x} &= 3^3 \\ 3^{-2x} &= 3^3 \end{aligned}$$

Therefore $-2x = 3$, and $x = -\frac{3}{2}$ is the solution. □

◆ **Practice Problem 1** ◆

Solve for x in each equation.

a. $2^x = \frac{1}{32}$ b. $4^x = 32$

A more general type of exponential function is one defined by an equation of the form $y = a^{P(x)}$, where $P(x)$ is a polynomial in x. The graphs of two such functions are considered in Example 4 and Practice Problem 2.

Example 4 • Graphing $f(x) = a^{P(x)}$

Sketch the graph of $f(x) = 2^{-x^2}$.

Solution

We have the following table.

x	-2	-1	0	1	2
$f(x)$	$\frac{1}{16}$	$\frac{1}{2}$	1	$\frac{1}{2}$	$\frac{1}{16}$

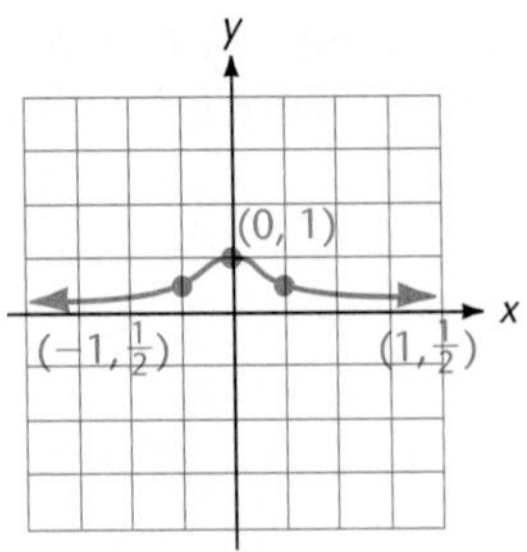

Figure 6.4 Graph of $f(x) = 2^{-x^2}$

Noting that

$$f(x) = 2^{-x^2} = \frac{1}{2^{x^2}},$$

we see that the largest possible function value of f is $f(0) = 1$ and that the function values approach zero as $|x|$ increases without bound. The graph is shown in Figure 6.4. □

◆ **Practice Problem 2** ◆

Sketch the graph of $y = 3^{2x-1}$.

An important application of exponential functions occurs in the formula for computing the value of an investment, or **original principal** P, when interest is added to the principal at the end of certain periods of time so that the accumulated interest also earns interest in the next period of time. Interest rates are usually stated at an annual rate r, and interest is converted to principal, or **compounded,** a specified number n times per year. The **rate per period** is then r/n, and the original investment P, together with accumulated interest, has the value

$$A = P\left(1 + \frac{r}{n}\right)^{nt}$$

after t years. The total amount A is called the **compound amount.**

Example 5 • Compound Interest

Suppose \$10,000 is invested at an annual rate of 12% for a period of 10 years. Find the compound amount if interest is compounded

a. annually b. daily.

Solution

From the given information, $P = \$10{,}000$, $r = 0.12$, and $t = 10$. In each part we use the formula

$$A = P\left(1 + \frac{r}{n}\right)^{nt}$$

and compute the exponential values with a calculator.

a. We have $n = 1$ and

$$A = 10{,}000(1 + 0.12)^{10} \approx \$31{,}058.48.$$

b. With interest compounded daily, we have $n = 365$ and

$$A = 10{,}000\left(1 + \frac{0.12}{365}\right)^{3650} \approx \$33{,}194.62.$$

The amounts obtained here illustrate how the frequency of compounding can affect the value of an investment. This is shown graphically in Figure 6.5. □

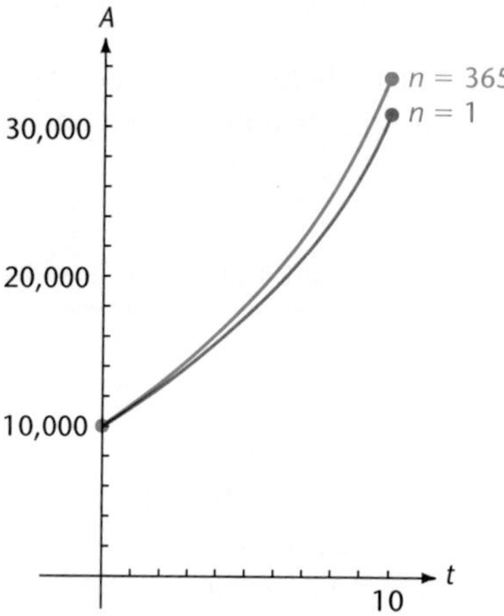

Figure 6.5 Effect of frequency on compound interest

The following example illustrates an application of exponential functions in the natural sciences. Other applications are included in the exercises for this section.

Example 6 • Carbon-14 Dating

Archaeologists are able to estimate the age of organic remains by measuring the amount of radioactive carbon, ^{14}C, in the organism's remains. If an organism contained 600 milligrams of ^{14}C at its death, the amount $A(t)$ in its remains t years after death is given by

$$A(t) = 600 \cdot 2^{-0.000175t}$$

Find the amount of ^{14}C in the remains 10,000 years after death.

Solution

We have $t = 10{,}000$ and

$$\begin{aligned} A(10{,}000) &= 600 \cdot 2^{-1.75} \\ &\approx 600(0.2973018) \\ &\approx 178. \end{aligned}$$

There will be approximately 178 milligrams of ^{14}C remaining from the original 600 milligrams. □

Peter Menzel/Stock, Boston

Archaeologists examine the remains of a prehistoric plesiosaur in Australia.

EXERCISES 6.1

If $f(x) = (\frac{3}{2})^x$, find each function value.

1. $f(0)$ 2. $f(1)$ 3. $f(2)$ 4. $f(3)$
5. $f(-1)$ 6. $f(-2)$ 7. $f(-\frac{1}{2})$ 8. $f(\frac{1}{2})$

Solve for x in each equation. (See Example 3 and Practice Problem 1.)

9. $3^x = 27$
10. $2^x = 32$
11. $2^x = \frac{1}{16}$
12. $3^x = \frac{1}{81}$
13. $2^x = 1$
14. $3^x = 1$
15. $(\frac{1}{2})^x = 8$
16. $(\frac{1}{3})^x = 9$
17. $(\frac{2}{3})^x = \frac{9}{4}$
18. $(\frac{3}{2})^x = \frac{8}{27}$
19. $8^x = \frac{1}{16}$
20. $(27)^x = \frac{1}{9}$
21. $5^{-x} = 625$
22. $4^{-x} = 128$
23. $3^{2x-1} = \frac{1}{81}$
24. $2^{3x+1} = 64$
25. $(81)^{2x+1} = \frac{1}{3}$
26. $(32)^{3x-2} = \frac{1}{4}$

Sketch the graph of each function. (See Examples 1 and 2.)

27. $f(x) = 3^x$
28. $f(x) = 10^x$
29. $f(x) = 3^{-x}$
30. $f(x) = 4^{-x}$
31. $f(x) = -2^x$
32. $f(x) = -(\frac{1}{2})^x$
33. $f(x) = -3^x$
34. $f(x) = -4^x$
35. $f(x) = -3^{-x}$
36. $f(x) = -4^{-x}$
37. $f(x) = (\frac{1}{3})^x$
38. $f(x) = (\frac{1}{4})^x$
39. $f(x) = (\frac{2}{3})^x$
40. $f(x) = (\frac{3}{2})^x$

Sketch the graph of each equation. (See Example 4 and Practice Problem 2.)

41. $y = 2^{2x-1}$
42. $y = 3^{2x+1}$
43. $y = 2 \cdot 5^x$
44. $y = 2 \cdot 3^x$
45. $y = 5^{-x} + 1$
46. $y = 2^x - 1$
47. $y = 4^{-x^2}$
48. $y = (2)^{1-x^2}$

Population Growth

49. The population of a city is now 16,000. The population t years from now is given by the formula $P = 16{,}000 \cdot 2^{t/10}$. What will the population be 40 years from now? How often does the population double?

Investment Growth

50. A woman's savings grow according to the formula $P = P_0(3)^{0.08t}$, where P_0 is the amount of her original investment and t is the time in years since the investment. If $P_0 = \$10{,}000$, how much will she have in 25 years?

51–60

Compound Amount 51–52

51. Find the compound amount if \$20,000 is deposited at 12% compounded quarterly for 8 years.
52. Find the compound amount if \$15,000 is deposited for 12 years in an account that pays interest at 8% compounded semiannually.

Compound Interest 53–54

53. Find the amount of interest earned if \$42,000 is deposited at 9% compounded monthly for 5 years.
54. Find the amount of interest earned if \$50,000 is deposited at 10% compounded quarterly for 6 years.

Undepreciated Value

55. If an item has initial cost C and depreciates at a fixed annual rate r, its value A after t years is given by $A = C(1 - r)^t$. Find the value after 4 years of a copier that cost \$5000 new and depreciates at 10% annually.

Amount of Depreciation

56. With A and C as defined in Exercise 55 the amount of depreciation is $D = C - A$. Find the amount of depreciation after 3 years on a \$10,000 computer that depreciates at 7% annually.

Bacteria Growth

57. The number of cholera bacteria in a laboratory culture is doubling every 2 hours. The culture started with 2000 bacteria at time $t = 0$, and the number $N(t)$ of bacteria after t hours is given by $N(t) = 2000 \cdot 2^{t/2}$. Find the number of bacteria present after 24 hours.

Sugar Processing

58. In processing raw sugar, the sugar's molecular structure changes in a step that is called **inversion.** If the inversion process starts with 6000 pounds of sugar, the amount of raw sugar remaining after t hours is given by $S(t) = 6000 \cdot 2^{-0.0322t}$ pounds. Find the amount of raw sugar remaining after 8 hours.

Contents of a Tank

59. When it is open, the spring-loaded valve on a water tank allows one-half of the water in the tank to flow out in 1 hour. If the tank contains 1000 gallons of water when the valve is opened, find a formula for the amount $A(t)$ of water in the tank t hours later.

Figure for Exercise 59

Bacteria Growth

60. The bacteria population in a laboratory culture doubles every day. If the culture starts with 10,000 bacteria, find a formula for the number of bacteria $B(x)$ at the end of x days.

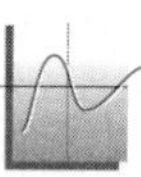

61–64

Use a graphing device to sketch the graph of the given function.

61. $y = 10(2.3^{-\sqrt{x}})$ 62. $y = 3(2^{-\sqrt[5]{x}})$

63. $y = 3^{\sqrt{x}} - x$ 64. $y = 3^x + \sqrt[3]{x}$

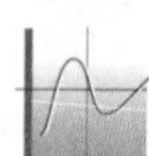

65–66

65. Use a graphing device to find the negative solution of the equation $2^x = x^2$ by locating a zero of the function $y = 2^x - x^2$. Round your answer to the nearest tenth.

66. Use the technique in Exercise 65 to find the positive solution of the equation $x^4 = 3^{-x}$, correct to the nearest tenth.

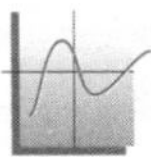

67–70

Use a graphing device to solve the following inequalities. Round the endpoints of all intervals to the nearest hundredth.

67. $2^x - 7 > 0$ 68. $5^x - 8 < 0$

69. $2^x - 4x < 0$ 70. $2^x - x^2 - 4 < 0$

Critical Thinking: Writing 71. Use the graph of $f(x) = 3^x$ shown in Figure 6.1 to explain how the equation $3^x = b$ has exactly one solution for each positive value of b.

Critical Thinking: Writing 72. The last item in the list of important features of the exponential functions states that $a^u = a^v$ if and only if $u = v$. Describe in words how this feature can be observed in the graph of $y = a^x$.

◆ Solutions for Practice Problems

1. a. Since $2^5 = 32$, we have

$$2^x = \frac{1}{2^5} = 2^{-5}.$$

Therefore $x = -5$ is the only solution.

b. In this case, we can write each side of the equation as a power of 2.

$$4^x = 32$$
$$(2^2)^x = 2^5$$
$$2^{2x} = 2^5$$

Therefore $2x = 5$, and $x = \frac{5}{2}$ is the solution.

2. A table of values is shown here, and the graph is drawn in Figure 6.6.

x	-2	-1	0	$\frac{1}{2}$	1	2
y	$\frac{1}{243}$	$\frac{1}{27}$	$\frac{1}{3}$	1	3	27

Here it is important to single out the value that makes the exponent zero, namely, $x = \frac{1}{2}$.

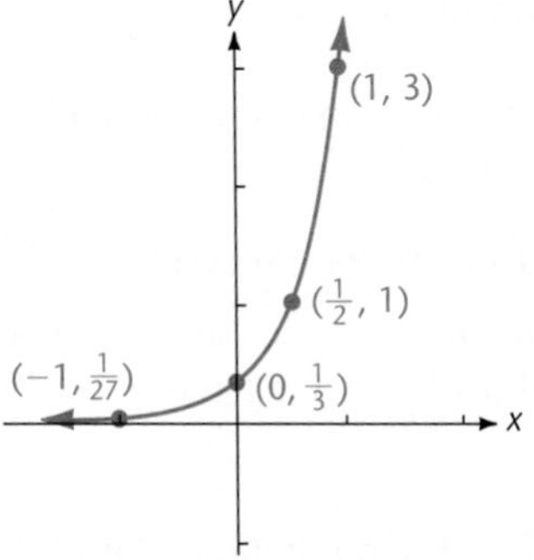

Figure 6.6 Graph of $y = 3^{2x-1}$

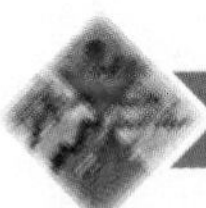

6.2 THE NATURAL EXPONENTIAL FUNCTION

Two exponential functions are especially useful in applications. One of these is the function $f(x) = 10^x$. The other is

$$f(x) = e^x,$$

where e is an irrational number with value

$$e \approx 2.71828,$$

correct to five decimal places. The number e is frequently called the **natural number** e, although it is not an element of the set of natural numbers $N = \{1, 2, 3, \ldots\}$. Many calculus formulas are simpler when the base e is used, so e is the "natural" choice in those situations.

Example 1 • Graphing the Natural Exponential Function

Sketch the graph of the exponential function f where

$$f(x) = e^x.$$

Solution

For sketching the graph, we use a calculator to obtain the function values shown in Figure 6.7. These values have been rounded to the nearest tenth. □

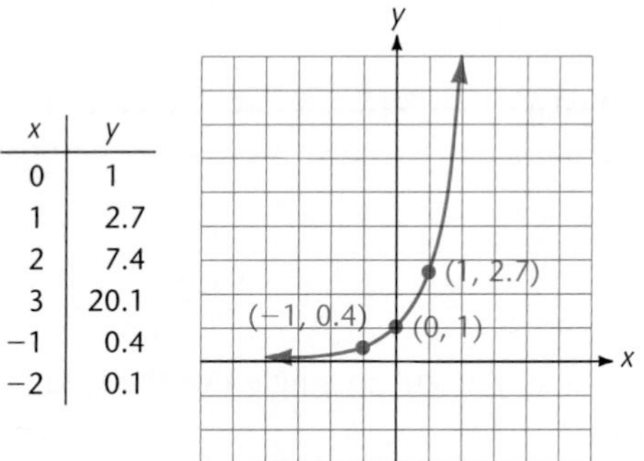

x	y
0	1
1	2.7
2	7.4
3	20.1
−1	0.4
−2	0.1

Figure 6.7 Graph of $f(x) = e^x$

The number e is useful in connection with many physical situations. One such situation is given in Example 2.

Example 2 • Atmospheric Pressure

The measure of the average atmospheric pressure P (in inches of mercury) at an altitude x miles above sea level is given approximately by

$$P(x) = 30e^{-0.21x}.$$

The swinging bridge on Grandfather Mountain in North Carolina (see the accompanying photograph) is at an altitude of 1 mile. Find the atmospheric pressure there.

William Russ/North Carolina Travel and Tourism Division

The swinging bridge on Grandfather Mountain near Linville, North Carolina.

Solution

With $x = 1$ in the given formula, we have

$$P(1) = 30e^{-0.21}.$$

Using a calculator, we compute the value of $e^{-0.21}$ as 0.8106. Therefore

$$\begin{aligned} P(1) &\approx 30(0.8106) \\ &\approx 24.32 \text{ inches of mercury.} \end{aligned}$$ □

The compound interest formula

$$A = P\left(1 + \frac{r}{n}\right)^{nt}$$

that we used in Section 6.1 has an interesting connection with the number e. If we take $r = 1$ and $t = 1$, the compound amount A is given by

$$A = P\left(1 + \frac{1}{n}\right)^{n}.$$

(This corresponds to taking 100% as an interest rate and a time of 1 year.) As the number n of periods per year increases without bound, the factor

$$\left(1 + \frac{1}{n}\right)^{n}$$

approaches $e \approx 2.718281828$ as a limiting value. This can be observed in the values shown in Table 6.1.

Table 6.1 Values of $\left(1 + \frac{1}{n}\right)^n$

n	1	10	100	10,000	1,000,000
$\left(1 + \frac{1}{n}\right)^n$	2	2.593742	2.704814	2.718146	2.718280

Thus as the number of periods per year increases without bound, an investment of 1 dollar would grow in 1 year to $e \approx 2.718281828$ dollars as a limiting value.

The result in the preceding paragraph can be generalized: for an arbitrary interest rate r and time of t years, the factor

$$\left(1 + \frac{r}{n}\right)^{nt}$$

approaches e^{rt} as the number of periods n increases without bound. The compound amount A thus approaches Pe^{rt}. The compound amount is sometimes computed from the formula

$$A = Pe^{rt}.$$

We then say that interest is being **compounded continuously.**

Example 3 • Compounding Continuously

Suppose \$10,000 is invested at an annual rate of 12% for a period of 10 years with interest compounded continuously. Find the value of the investment at the end of 10 years.

Solution

The value of the investment is given by

$$A = 10{,}000e^{(0.12)(10)} = 10{,}000e^{1.2} \approx \$33{,}201.17.$$

It is worth noting that this value differs from the one obtained by compounding annually in part a of Example 4 in the last section by \$2142.69, but it differs from that obtained by compounding daily by only \$6.55. □

◆ Practice Problem 1 ◆

Find the value of the investment in Example 3 if the rate of interest is cut in half to 6%. That is, find the value of a \$10,000 investment at the end of 10 years if interest is compounded continuously at a rate of 6%.

Example 4 • The Shroud of Turin

For centuries it was thought that the Shroud of Turin might be the burial cloth of Jesus Christ. In 1988 the archbishop of Turin announced that carbon-14 dating tests had concluded that the shroud was made between 1260 and 1390.† If A_0 was the amount of radioactive ^{14}C in the shroud when it was made, the amount $A(t)$ remaining t years later is given approximately by $A(t) = A_0e^{-0.000124t}$. Find the percentage of ^{14}C remaining in the shroud in 1988 if it was made in 1260.

Solution

The number of years from 1260 to 1988 is 728. With $t = 728$, we find that

$$A(728) = A_0e^{(-0.000124)(728)} \approx A_0e^{-0.0903} \approx A_0(0.914).$$

Hence approximately 91.4 percent of A_0 remained in the shroud in 1988. □

◆ **Practice Problem 2** ◆

What percentage of ^{14}C should remain in the Shroud of Turin if it were made in the year 33?

If f is a constant multiple of an exponential function of t and if $f(t)$ increases as t increases, then f is called an **exponential growth function.** When P and r are positive constants, the function $A(t) = Pe^{rt}$ used in Example 3 provides an illustration of exponential growth. On the other hand, if $f(t)$ *decreases* as t increases, then f is called an **exponential decay function,** in which case $r < 0$. Exponential decay functions appear in both Examples 2 and 4. Figure 6.8 shows the graph of the function $A(t) = A_0e^{-0.000124t}$ that was used in Example 4.

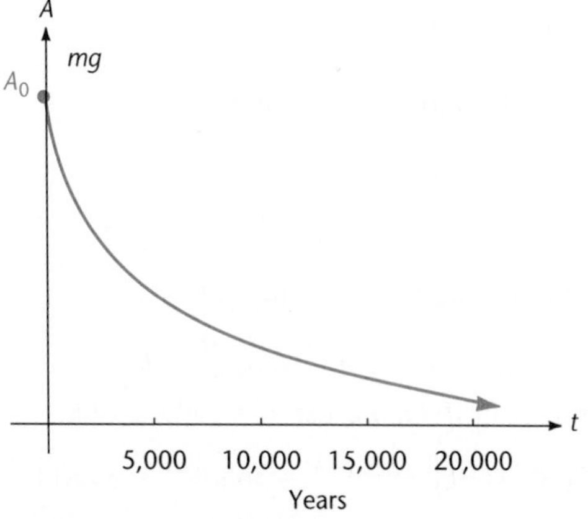

Figure 6.8 Exponential decay: $A = A_0e^{-0.000124t}$

†The accuracy of this conclusion was reported to be 95%; that is, the range of years computed in this manner includes the actual date 95% of the time.

Exponential functions can be used to approximate the number of cases of infection in a disease epidemic. Example 5 gives an illustration of this type of approximation with the number of deaths due to AIDS among U.S. women of childbearing age.

Example 5 • Deaths from AIDS

Among women in the United States between the ages of 15 and 44, the number of deaths from AIDS almost quadrupled between 1985 and 1988. Actual figures for the period are recorded in Table 6.2.

Table 6.2 Deaths from AIDS per 100,000

Year	1985	1986	1987	1988
No. of deaths per 100,000	360	631	1016	1430

In the table, let $D(t)$ denote the number of deaths that occur in year t after 1985. [$D(0) = 360$, $D(1) = 631$, and so on.]

a. Use the formula

$$D(t) = 360e^{0.46t}$$

to obtain approximate values of $D(1)$ and $D(3)$.

b. Use the same formula to predict the number of deaths in 1995.

Solution

a. With a calculator, we obtain

$$D(1) = 360e^{0.46} \approx 570$$

and

$$D(3) = 360e^{(0.46)(3)} = 360e^{1.38} \approx 1431$$

as approximations to the actual values of 631 and 1430.

b. For the year 1995, we use $t = 1995 - 1985 = 10$ and obtain

$$D(10) = 360e^{(0.46)(10)} = 360e^{4.60} \approx 35{,}814.$$

If AIDS continues to spread at the same rate, in 1995 there will be nearly 36,000 deaths due to AIDS per 100,000 U.S. women of ages 15 through 44. □

The functions in Exercises 9 and 10 are two of the **hyperbolic functions** that are included in the **advanced functions** of a graphing device. The hyperbolic functions are studied in detail in the calculus.

EXERCISES 6.2

1–22

Sketch the graph of the following functions either by using a calculator to obtain function values or by using a graphing device.

1. $f(x) = e^{0.1x}$
2. $f(x) = e^{0.2x}$
3. $f(x) = e^{-0.2x}$
4. $f(x) = e^{-0.1x}$
5. $f(x) = e^{x} + 1$
6. $f(x) = e^{-x} + 1$
7. $f(x) = e^{x-1}$
8. $f(x) = e^{-x+1}$
9. $f(x) = \dfrac{e^{x} + e^{-x}}{2}$
10. $f(x) = \dfrac{e^{x} - e^{-x}}{2}$

Continuous Compound Amount 11–12

11. Find the compound amount if \$20,000 is deposited at 12% compounded continuously for 8 years.
12. Find the compound amount if \$15,000 is deposited for 12 years in an account that pays interest at 8% compounded continuously.

Continuous Compound Interest 13–14

13. Find the interest earned if \$42,000 is deposited at 9% compounded continuously for 5 years.
14. Find the interest earned if \$50,000 is deposited at 10% compounded continuously for 6 years.

Shroud of Turin

15. (See Example 4.) Find the percentage of ^{14}C remaining in the Shroud of Turin in 1988 if it was made in 1390.

Mount Everest

16. The atmospheric pressure $P(x)$ (in inches of mercury) at an altitude x miles above sea level is approximated by $P(x) = 30e^{-0.21x}$. Find the atmospheric pressure at the top of Mount Everest, which is 5.5 miles above sea level.

Half-Life of Radium

17. We say that the **half-life** of radium is 1690 years because it takes 1690 years for half of a given amount of radium to decay into another substance. Starting with 100 milligrams of radium, the number $A(x)$ of milligrams that will be left after x years is $A(x) = 100e^{-0.000411x}$. How much radium will remain after 1000 years?

Half-Life of Carbon-14

18. The radioactive carbon atom ^{14}C is formed in the upper atmosphere and has a half-life of about 5600 years. If an organism contains 400 milligrams of ^{14}C at its death, the amount $A(x)$ of ^{14}C remaining x years later is $A(x) = 400e^{-0.000124x}$ milligrams. Find the amount of ^{14}C remaining in the organism after 1000 years.

Carbon-14 Dating 19. While Route I-85 was being moved to another location in Spartanburg County, South Carolina, a fossilized bone was discovered and estimated to be 3000 years old. Using the formula $A(t) = A_0e^{-0.000124t}$ from Example 4, find the percentage of ^{14}C left in the bone when it was discovered.

Half-Life of Polonium 20. The half-life of polonium is 140 days. Starting with A_0 milligrams of polonium, the amount $A(x)$ remaining after x days is $A(x) = A_0e^{-0.00495x}$ milligrams. Find the percentage of A_0 remaining after 100 days.

Deer Restocking 21. As part of restocking program, 60 white-tail deer were imported from Michigan and released in the Jackson–Bienville wildlife management area in Louisiana. The number $N(x)$ of these deer expected to be alive after x years is approximated by $N(x) = 60e^{-0.163x}$. Find the approximate number of deer expected to survive to the end of the second year.

Cooling a Liquid 22. According to **Newton's law of cooling,** the temperature T of an object placed in a medium with constant temperature T_0 changes at a rate proportional to $T - T_0$. If a bowl of soup with temperature 90°C is placed in a room at 20°C, its temperature $T(x)$ after x minutes is approximated by $T(x) = 20 + 70e^{-0.056x}$. Find its temperature after 10 minutes.

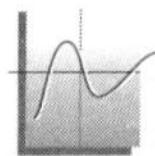

23–26

Use a graphing device to sketch the graph of the given function.

23. $y = e^x - x$

24. $y = e^{x^2}$

25. $y = 5e^{-0.1x^2}$

26. $y = \dfrac{e^x - e^{-x}}{e^x + e^{-x}}$

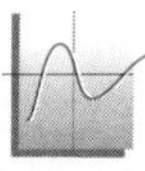

27–29

27. Use a graphing device to find the least positive solution of the equation $e^x = x^3$ by locating a zero of the function $y = e^x - x^3$. Round your answer to the nearest tenth.

28. Use the technique in Exercise 27 to find the negative solution of the equation $e^x = 4x + 7$. Round your answer to the nearest tenth.

29. Use a graphing device to sketch the graphs of both functions $y_1 = (1 + 1/x)^x$ and $y_2 = e$ at the same time in the viewing rectangle [0, 25] by [0, 5]. Observe that the graphs appear to be identical for x sufficiently large.

Critical Thinking: Exploration and Writing

30. Describe the difference in appearance between the graph of an exponential growth function and an exponential decay function.

Critical Thinking: Exploration and Writing

31. a. Describe a feature of the graph of the exponential function $f(x) = e^x$ that is not possessed by the graph of any polynomial function of degree 2 or more.

 b. Let $P(x)$ be a polynomial of odd degree. Describe a feature of the graph of $y = P(x)$ that is not possessed by the graph of $y = e^x$.

◆ Solutions for Practice Problems

1. The value of the investment is

$$A = 10{,}000e^{(0.06)(10)} = 10{,}000e^{0.6} \approx \$18{,}221.19.$$

It is interesting to observe that cutting the interest rate by one-half reduces the value of the interest earned from \$23,201.17 to \$8,221.19. The reduction is much more than one-half.

2. $A(1955) = A_0e^{(-0.000124)(1955)} \approx A_0(0.785)$

Hence approximately 78.5 percent of A_0 should remain in the shroud in 1988.

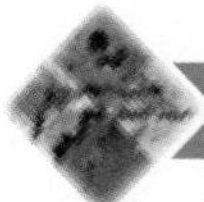

6.3 LOGARITHMIC FUNCTIONS

We noted in Section 6.1 that any exponential function $f(x) = a^x$ is a one-to-one function with domain the set of all real numbers and range the set of all positive real numbers. It follows from the results of Section 4.6 that the exponential function has an inverse function with domain the set of all positive numbers and range the set of all real numbers.

The inverse of the exponential function $f(x) = a^x$ is called the **logarithmic function with base *a*.** The value of this logarithmic function at x is called the **logarithm of *x* to the base *a*** and is abbreviated $\log_a x$.

In Section 4.6, we saw that a defining equation for f^{-1} may be obtained by interchanging x and y in the defining equation for f. The defining equation for the logarithmic function is $x = a^y$, and the defining statement is given next.

The Logarithm of *x* to the Base *a*

$y = \log_a x$ if and only if $x = a^y$

In these equations, we must keep in mind that x and a are positive and $a \neq 1$. Rewording the boxed statement, we can say that **$\log_a x$ is the power to which the base *a* must be raised to produce *x*.**

The defining statement for the logarithmic function sets up an equivalence between the logarithmic statement $L = \log_a N$ and the exponential statement $N = a^L$. A statement in either of these forms can be changed to an equivalent statement in the other form.

Example 1 • Changing to Logarithmic Form

Change the following statements to logarithmic form.

a. $16 = 2^4$ b. $\frac{1}{9} = (27)^{-2/3}$

Solution

a. $16 = 2^4$ implies that $\log_2 16 = 4$.

b. $\frac{1}{9} = (27)^{-2/3}$ implies that $\log_{27} \frac{1}{9} = -\frac{2}{3}$. □

Example 2 • Changing to Exponential Form

Change the following statements to exponential form.

a. $\log_3 81 = 4$ b. $\log_{16} \frac{1}{64} = -\frac{3}{2}$

Solution

a. $\log_3 81 = 4$ implies that $81 = 3^4$.

b. $\log_{16} \frac{1}{64} = -\frac{3}{2}$ implies that $\frac{1}{64} = (16)^{-3/2}$. □

◆ Practice Problem 1 ◆

Change the following from exponential form to logarithmic form or from logarithmic form to exponential form, whichever is appropriate.

a. $4^{-2} = \frac{1}{16}$ b. $6^0 = 1$ c. $\log_4 \frac{1}{8} = -\frac{3}{2}$

From Chapter 4, we know that $f^{-1}(f(x)) = x$ for all x in the domain of f and that $f(f^{-1}(x)) = x$ for all x in the domain of f^{-1}. For the exponential function $f(x) = a^x$, these equations read as

$$\log_a a^x = x$$

for all real numbers x, and

$$a^{\log_a x} = x$$

for all positive real numbers x. In the first equation we get

$$\log_a a = 1$$

with $x = 1$, and we get

$$\log_a 1 = 0$$

with $x = 0$.

Since the logarithmic function $y = \log_a x$ is the inverse of the exponential function $y = a^x$, the graphs of these two functions are reflections of each other through the line $y = x$. This is illustrated in Figure 6.9.

To graph a logarithmic function, keep Figure 6.9 in mind and plot a few strategic points using the equation $\log_a a^x = x$.

Example 3 • A Logarithmic Function With $a > 1$

Sketch the graph of $y = \log_2 x$.

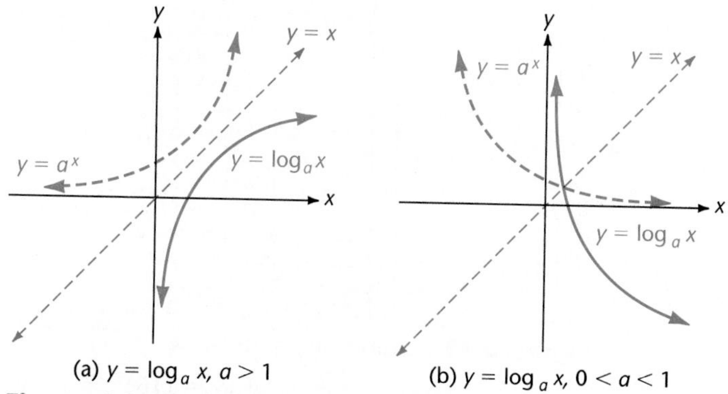

Figure 6.9 Typical logarithmic graphs

Solution

We choose some convenient powers of 2 and use the equality $y = \log_2 2^n = n$ to obtain the points tabulated in Figure 6.10. After plotting these points, we join them with a smooth curve that has the typical shape shown in Figure 6.9a. □

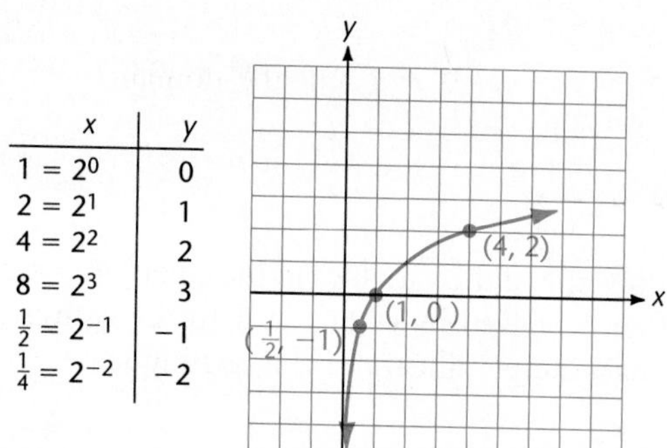

x	y
$1 = 2^0$	0
$2 = 2^1$	1
$4 = 2^2$	2
$8 = 2^3$	3
$\frac{1}{2} = 2^{-1}$	−1
$\frac{1}{4} = 2^{-2}$	−2

Figure 6.10 Graph of $y = \log_2 x$

Example 4 • A Logarithmic Function with $0 < a < 1$

Sketch the graph of $y = \log_{1/2} x$.

Solution

Following the same procedure as in Example 3, we calculate the values shown in the table in Figure 6.11, plot the corresponding points, and join them with a smooth curve that has the typical shape shown in Figure 6.9b. □

The graphs shown in Figure 6.9–6.11 illustrate the following features of the logarithmic functions. Feature 5 on the list is the one-to-one property of logarithmic functions.

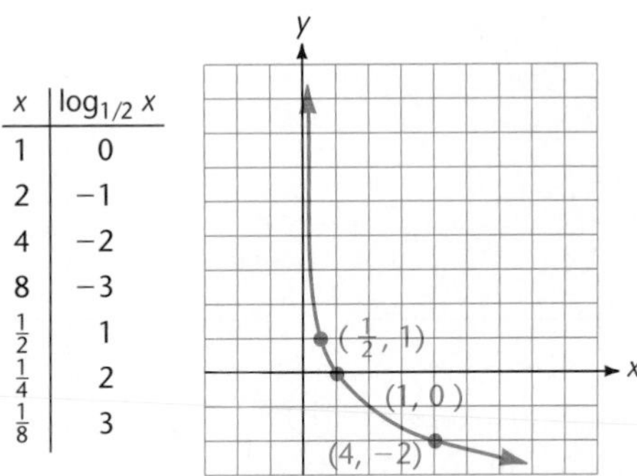

x	$\log_{1/2} x$
1	0
2	-1
4	-2
8	-3
$\frac{1}{2}$	1
$\frac{1}{4}$	2
$\frac{1}{8}$	3

Figure 6.11 Graph of $y = \log_{1/2} x$

Important Features of the Logarithmic Functions

$$f(x) = \log_a x, \qquad a > 0 \text{ and } a \neq 1$$

1. The **domain** of f is the set of all positive real numbers.
2. The **range** of f is the set of all real numbers.
3. If $a > 1$, the graph of $f(x) = \log_a x$ resembles that of $f(x) = \log_2 x$. (See Figure 6.10.)
4. If $0 < a < 1$, the graph of $f(x) = \log_a x$ resembles that of $f(x) = \log_{1/2} x$. (See Figure 6.11.)
6. $\log_a u = \log_a v$ if and only if $u = v$.

If only one of the variables in the equation $y = \log_a x$ is unknown, the equation can often be solved by changing it to the equivalent exponential statement. The following example illustrates the technique.

Example 5 • Using the Definition of a Logarithm

Solve for the unknown variable.

a. $y = \log_{3/2} \frac{9}{4}$ b. $\log_4 x = -\frac{3}{2}$ c. $\log_a 64 = -3$

Solution

a. $y = \log_{3/2} \frac{9}{4}$ is equivalent to $(\frac{3}{2})^y = \frac{9}{4}$. Since $(\frac{9}{4}) = (\frac{3}{2})^2$, we have

$$(\tfrac{3}{2})^y = (\tfrac{3}{2})^2$$

and therefore $y = 2$.

b. $\log_4 x = -\frac{3}{2}$ is equivalent to

$$x = (4)^{-3/2} = \frac{1}{4^{3/2}} = \frac{1}{(\sqrt{4})^3} = \frac{1}{8}.$$

c. $\log_a 64 = -3$ is equivalent to $a^{-3} = 64$. Therefore

$$\frac{1}{a^3} = 64$$
$$a^3 = \tfrac{1}{64}$$
$$a = \sqrt[3]{\tfrac{1}{64}} = \tfrac{1}{4}.$$

□

Since the exponential and logarithmic functions with base a are inverse functions, the laws of exponents have some important implications for logarithms. These are stated in the following properties of logarithms.

Properties of Logarithms

Let a, u, v, and r be real numbers with u, v, and a positive and $a \neq 1$. Then

1. $\log_a u = \log_a v$ if and only if $u = v$
2. $\log_a (uv) = \log_a u + \log_a v$
3. $\log_a\left(\frac{u}{v}\right) = \log_a u - \log_a v$
4. $\log_a u^r = r \log_a u$

Numerical computations can be made by using these properties with a table of logarithms, but calculators have virtually elminated this use of logarithms.

Property 1 follows from the fact that the logarithmic function is the inverse of the exponential function. More explicitly,

$$\log_a u = \log_a v \quad \text{implies that} \quad a^{\log_a u} = a^{\log_a v},$$

and therefore

$$u = v.$$

The laws of exponents can be used to derive properties 2–4. We shall prove property 2 here and leave the other proofs as exercises.

Let $x = \log_a u$ and $y = \log_a v$. Then $u = a^x$ and $v = a^y$. By the product rule in the laws of exponents,

$$uv = a^x a^y = a^{x+y},$$

and therefore

$$\log_a uv = x + y$$
$$= \log_a u + \log_a v.$$

Logarithmic functions are defined only for positive real numbers, and **we assume for the remainder of this section that all variables are appropriately restricted so that the functions involved are defined.**

The properties of logarithms can be used to expand a single logarithmic expression into terms that involve simpler logarithmic expressions, or to change an expanded expression to a single logarithm. This is illustrated in the following examples.

Example 6 • Using Properties of Logarithms

a. Express $\log_{10} 75$ in terms of $\log_{10} 3$ and $\log_{10} 5$.

b. Write the expression

$$5 \log_{10} 2 - 3 \log_{10} 2 + 1$$

as a single logarithm.

Solution

a. We have

$$\begin{aligned} \log_{10} 75 &= \log_{10} 3 \cdot 5^2 \\ &= \log_{10} 3 + \log_{10} 5^2 \\ &= \log_{10} 3 + 2 \log_{10} 5. \end{aligned}$$

b. Since $\log_{10} 10 = 1$, we can write

$$\begin{aligned} 5 \log_{10} 2 - 3 \log_{10} 2 + 1 &= \log_{10} 2^5 - \log_{10} 2^3 + \log_{10} 10 \\ &= \log_{10} \frac{2^5}{2^3} + \log_{10} 10 \\ &= \log_{10} 4 + \log_{10} 10 \\ &= \log_{10} 40. \end{aligned}$$

□

Example 7 • Expanding and Simplifying Logarithmic Expressions

a. Express $\log_a \sqrt{\dfrac{x^2y^3}{z^4}}$ in terms of logarithms of x, y, and z.

b. Write the expression

$$\log_a 6xy^3 - 2 \log_a 3xy^2z$$

as a single logarithm.

Solution

a. We have

$$\begin{aligned} \log_a \sqrt{\frac{x^2y^3}{z^4}} &= \log_a\left(\frac{x^2y^3}{z^4}\right)^{1/2} \\ &= \tfrac{1}{2} \log_a\left(\frac{x^2y^3}{z^4}\right) \\ &= \tfrac{1}{2}(\log_a x^2y^3 - \log_a z^4) \\ &= \tfrac{1}{2}(\log_a x^2 + \log_a y^3 - \log_a z^4) \\ &= \tfrac{1}{2}(2 \log_a x + 3 \log_a y - 4 \log_a z) \\ &= \log_a x + \tfrac{3}{2} \log_a y - 2 \log_a z. \end{aligned}$$

b. By reversing the direction that was followed in part a, we obtain

$$\begin{aligned}\log_a 6xy^3 - 2\log_a 3xy^2z &= \log_a 6xy^3 - \log_a(3xy^2z)^2\\ &= \log_a \frac{6xy^3}{(3xy^2z)^2}\\ &= \log_a \frac{2}{3xyz^2}.\end{aligned}$$

□

EXERCISES 6.3

Express each equation in logarithmic form. (See Example 1.)

1. $4^2 = 16$
2. $5^3 = 125$
3. $3^4 = 81$
4. $2^4 = 16$
5. $3^{-2} = \frac{1}{9}$
6. $4^{-3} = \frac{1}{64}$
7. $(10)^{-2} = \frac{1}{100}$
8. $8^{-1/3} = \frac{1}{2}$
9. $5^0 = 1$
10. $2^1 = 2$
11. $4^{-1/2} = \frac{1}{2}$
12. $(64)^{1/6} = 2$

Express each equation in exponential form. (See Example 2.)

13. $\log_2 64 = 6$
14. $\log_2 \frac{1}{8} = -3$
15. $\log_3 \frac{1}{27} = -3$
16. $\log_{64} \frac{1}{2} = -\frac{1}{6}$
17. $\log_{3/4} \frac{9}{16} = 2$
18. $\log_{1/3} 9 = -2$
19. $\log_{10} (0.01) = -2$
20. $\log_{10} 100 = 2$

Graph each function. (See Examples 3 and 4.)

21. $y = \log_3 x$
22. $y = \log_5 x$
23. $y = \log_{1/3} x$
24. $y = \log_{1/5} x$
25. $y = \log_8 x$
26. $y = \log_{10} x$
27. $y = \log_3 (x + 4)$
28. $y = \log_{10} (x - 3)$

Solve for the unknown variable. (See Example 5.)

29. $y = \log_2 1$
30. $y = \log_5 5$
31. $y = \log_2 8$
32. $y = \log_7 49$
33. $y = \log_{1/3} \frac{1}{9}$
34. $y = \log_{10} 1000$
35. $2 = \log_3 x$
36. $3 = \log_2 x$
37. $\log_3 x = -4$
38. $\log_4 x = -3$
39. $y = \log_{1/10} 100$
40. $y = \log_{1/3} 81$
41. $y = \log_{1/3} 9$
42. $y = \log_{1/2} 4$
43. $3 = \log_a 8$
44. $4 = \log_a 625$
45. $\log_a 125 = -3$
46. $\log_a 100 = -2$
47. $y = \log_5 \frac{1}{125}$
48. $y = \log_5 \frac{1}{25}$
49. $y = \log_2 4$
50. $y = \log_{1/3} 27$

Express each of the following in terms of $\log_{10} 3$ and $\log_{10} 5$. (See Example 6.)

51. $\log_{10} 15$
52. $\log_{10} 45$
53. $\log_{10} 27$
54. $\log_{10} 25$
55. $\log_{10} (5)^{1/2}$
56. $\log_{10} \sqrt{3}$
57. $\log_{10} \frac{3}{5}$
58. $\log_{10} \frac{5}{3}$
59. $\log_{10} \frac{9}{25}$
60. $\log_{10} \frac{125}{27}$

Express each of the following in terms of logarithms of x, y, and z. (See Example 7.)

61. $\log_a xy$
62. $\log_a xyz$
63. $\log_a \frac{x^2y^3}{z^4}$
64. $\log_a \frac{x^3y^4}{z^2}$
65. $\log_a \frac{x^3}{y^5z^2}$
66. $\log_a \frac{x^7}{y^3z^4}$
67. $\log_a \frac{\sqrt[5]{xz^2}}{y}$
68. $\log_a \frac{\sqrt[3]{x^4y^2}}{\sqrt{z^5}}$
69. $\log_a \sqrt{\frac{xy^4}{z^3}}$
70. $\log_a \sqrt{\frac{x^2y}{z^5}}$

Write each of the following as a single logarithm. (See Examples 6 and 7.)

71. $4 \log_{10} 5 - 2 \log_{10} 5 + 2$
72. $5 \log_{10} 4 - 3 \log_{10} 4 + 1$
73. $4 \log_{10} 2 + \log_{10} 2 - 1$
74. $3 \log_{10} 4 + \log_{10} 4 - 2$
75. $3 \log_{10} 4 - \log_{10} 4 - 2$
76. $4 \log_{10} 2 - \log_{10} 2 - 1$
77. $\log_a x - 2 \log_a y + \frac{1}{2} \log_a z$
78. $\log_a 18x^2 + 3 \log_a z - \frac{1}{3} \log_a 6y$
79. $\log_a x^2y + 2 \log_a 5xy^3 - \log_a 10x^2y^2$
80. $\log_a x^3y^4 - 3 \log_a 4y^2z + \log_a 8x^2yz$
81. $\frac{2}{3} \log_a 8x^2z^3 + \log_a 3 + \frac{1}{3} \log_a 27x^4y^6z^9$
82. $\frac{3}{4} \log_a 8x^2y^4 + \log_a 5 + \frac{1}{4} \log_a 8x^6z^4$
83. Use the laws of exponents to prove property (3) of logarithms.

$$\log_a \left(\frac{u}{v}\right) = \log_a u - \log_a v$$

84. Use the laws of exponents to prove property (4) of logarithms.

$$\log_a u^r = r \log_a u$$

Critical Thinking: Exploration and Writing

85. a. Explain why a must not be negative in the definition of $\log_a x$.
b. Explain why the condition $a \neq 1$ is imposed in the definition of $\log_a x$.

Critical Thinking: Exploration and Writing

86. In the definition of the function $\log_a x$, x and y are interchanged in the equation $y = a^x$ to obtain $x = a^y$. Write a description of the relation between the graphs of $y = P(x)$ and $x = P(y)$, where $P(x)$ is an arbitrary polynomial with real coefficients.

Critical Thinking: Exploration and Discussion

87. In Section 6.2, exponential growth and decay functions are defined by equations of the form $A(t) = Pe^{rt}$. Use the fact that $e = b^{\log_b e}$ to explain how any positive number b can be used as base and write $A(t) = P \cdot b^{kt}$ for an appropriate choice of the constant k. Discuss the advantage of choosing e as the base used.

◆ Solution for Practice Problem

1. a. $4^{-2} = \frac{1}{16}$ implies that $\log_4 \frac{1}{16} = -2$.
   b. $6^0 = 1$ implies that $\log_6 1 = 0$.
   c. $\log_4 \frac{1}{8} = -\frac{3}{2}$ implies that $4^{-3/2} = \frac{1}{8}$.

6.4 LOGARITHMIC AND EXPONENTIAL EQUATIONS

In technical work it is sometimes necessary to solve equations that contain logarithmic functions or exponential functions. Numerical values of logarithms are required to obtain satisfactory solutions to these types of equations.

Before hand-held calculators came into common use, logarithms were frequently used in performing numerical computations. Since our number system uses the base 10, logarithms to base 10 are the most practical for computational purposes, and they are referred to as **common logarithms.** For the remainder of this book we adopt the convention that

$$\log x = \log_{10} x.$$

That is, the base is understood to be 10 unless it is specified otherwise. We also use the phrase "logarithm of x" to mean "logarithm of x to the base 10."

The [log] key on most scientific calculators is used for computations with common logarithms. Tables can also be used, and a table is provided at the end of the Appendix to this book. The Appendix gives instructions on using the table and also on using logarithms to perform computations.

An equation that contains logarithmic functions is called a **logarithmic equation.** Many logarithmic equations can be solved by using the properties and the definition of a logarithm. Such an equation is solved in Example 1.

Example 1 • Solving a Logarithmic Equation

Find the solution set for the equation

$$\log(2x + 1) + \log(3x - 4) = 1.$$

Solution

Using property 2 of logarithms, the left side can be written as a single logarithm. The 1 on the right side can be replaced by log 10.

$$\log(2x + 1)(3x - 4) = \log 10$$

By the one-to-one property of a logarithmic function, this is equivalent to

$$(2x + 1)(3x - 4) = 10.$$

Therefore

$$6x^2 - 5x - 4 = 10$$
$$6x^2 - 5x - 14 = 0,$$

and

$$(x - 2)(6x + 7) = 0.$$

This leads to

$$x - 2 = 0 \quad \text{or} \quad 6x + 7 = 0,$$
$$x = 2 \quad \text{or} \quad x = -\tfrac{7}{6}.$$

We must check to see if the logarithms in the original equation are defined at these values. With $x = -\frac{7}{6}$, the left member is

$$\log\left(-\tfrac{4}{3}\right) + \log\left(-\tfrac{15}{2}\right).$$

This quantity is not defined because it involves logarithms of negative numbers. The value $x = -\frac{7}{6}$ must be rejected. With $x = 2$, all logarithms in the original equation are defined, and $x = 2$ is a solution. Thus the solution set is $\{2\}$. □

An equation that contains an exponential function is called an **exponential equation.** In many cases an exponential equation can be solved by converting it to a logarithmic equation and solving the logarithmic equation.

Example 2 • Solving an Exponential Equation

Solve the equation $2^{2x+1} = 3^{x+4}$.

Solution

Taking the logarithm of each side of the equation, we have

$$(2x + 1)\log 2 = (x + 4)\log 3$$

or

$$2x\log 2 + \log 2 = x\log 3 + 4\log 3.$$

Solving for x, we have

$$(2\log 2 - \log 3)x = 4\log 3 - \log 2$$

and

$$x = \frac{4\log 3 - \log 2}{2\log 2 - \log 3} \approx 12.87.$$

□

Solve for x if $4^x = 23$.

Example 3 • Compound Interest

If an original investment of \$1000 grows to a value of \$1500 in 5 years when interest is compounded quarterly at an annual rate r, find r.

Solution

Using the formula for compound amount from Example 5 in Section 6.1 with $A = \$1500$, $P = \$1000$, $n = 4$, and $t = 5$, we have

$$1500 = 1000\left(1 + \frac{r}{4}\right)^{20}.$$

This gives

$$1.5 = (1 + 0.25r)^{20}.$$

Taking the logarithm of each side, we have

$$\log 1.5 = 20 \log (1 + 0.25r)$$

and

$$\log (1 + 0.25r) = \frac{\log 1.5}{20} \approx 0.0088.$$

Since $10^{0.0088} \approx 1.02$, we have

$$1 + 0.25r \approx 1.02$$

and

$$r \approx \frac{0.02}{0.25} = 0.08.$$

That is, the annual interest rate is 8%. □

Example 4 • Intensity of Sound

Logarithms to the base 10 are used in the decibel scale for measuring loudness. If the intensity of the sound in watts per square meter is I, the decibel level L is given by

$$L = 10 \log (I \times 10^{12}) \text{ decibels.}$$

For ordinary conversation, L is about 65 decibels, and the threshold of pain is about 120 decibels. Find the effect of doubling the power I on an audio amplifier.

Solution

If we replace I by $2I$ in the formula for L, we obtain

$$\begin{aligned} L &= 10 \log (2I \times 10^{12}) \\ &= 10[\log 2 + \log (I \times 10^{12})] \\ &= 10 \log 2 + 10 \log (I \times 10^{12}) \\ &\approx 3 + 10 \log (I \times 10^{12}). \end{aligned}$$

Thus doubling the power increases L by only about 3 decibels. □

Although logarithms to base 10 are adequate for many purposes, other logarithms are used more extensively in scientific and technical work. These are logarithms to the base e. The notation "ln x" is commonly used to denote logarithms to the base e, and "ln x" is read "the natural logarithm of x." We write

$$\ln x = \log_e x.$$

Many of the formulas used in calculus have a particularly simple form when expressed using logarithms to the base e, and many physical situations can be described with these logarithms. Most books of mathematical tables have a table of natural logarithms, and scientific calculators have a key marked [ln x] that will yield the natural logarithm of any positive number.

When exponential equations involving the base e occur, natural logarithms are useful in solving these equations. An illustration is given in the next example.

Example 5 • Atmospheric Pressure

The measure of the atmospheric pressure P (in inches of mercury) at an altitude x miles above sea level is given approximately by

$$P = 30e^{-0.198x}.$$

If the pressure at a mountain peak is 15 inches of mercury, find the altitude of the peak.

Solution

We put $P = 15$ and solve for x.

$$\begin{aligned} 30e^{-0.198x} &= 15 \\ e^{-0.198x} &= \tfrac{1}{2} \\ \ln(e^{-0.198x}) &= \ln(2^{-1}) \\ -0.198x &= -\ln 2 \\ x &= \frac{\ln 2}{0.198} \\ x &\approx 3.50 \end{aligned}$$

The peak is approximately 3.50 miles above sea level. □

It is possible to use either common logarithms or natural logarithms to calculate the value of logarithms to any desired base. To see how this is done, we consider the general problem of changing logarithms from one base to another.

To obtain an equation relating $\log_a N$ and $\log_b N$, let

$$L = \log_b N.$$

Then

$$N = b^L$$

and

$$\log_a N = \log_a b^L$$
$$= L \log_a b.$$

Solving for L, we get

$$L = \frac{\log_a N}{\log_a b}.$$

Since $L = \log_b N$ from our first equation, we have the following change-of-base formula.

Change-of-Base Formula

$$\log_b N = \frac{\log_a N}{\log_a b}$$

To graph $y = \log_b x$ with a graphing device, enter y in one of the forms $y =$ LOG X/LOG b or $y =$ LN X/LN b.

Since scientific calculators have common and natural logarithm capabilities, the change-of-base formula is usually used in one of the forms

$$\log_b N = \frac{\log N}{\log b} \quad \text{or} \quad \log_b N = \frac{\ln N}{\ln b}.$$

Either of these formulas will yield the same results. In Example 6 we use logarithms to base 10.

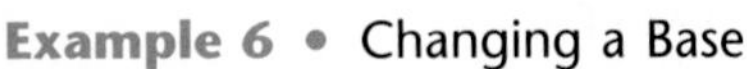

Example 6 • Changing a Base

Find a three-digit value for $\log_5 19$.

Solution

Using the change-of-base formula, we obtain

$$\log_5 19 = \frac{\log 19}{\log 5} \approx 1.83.$$

□

EXERCISES 6.4

Find the solution set for the following equations. (See Example 1.)

1. $\log(2x+1) - \log 5 = \log x$
2. $\log 2x - \log 5 = 2$
3. $\log x^2 - \log 9 = \log x$
4. $\log x^2 - \log 4 = \log x$
5. $\log_5(4x-1) - \log_5 3 = \log_5(x+2)$
6. $\log_3(9x-6) - \log_3 4 = \log_3(x+1)$
7. $\log_2 x - \log_2(x-1) = 3$
8. $\log_3 x - \log_3(x-1) = 2$
9. $\log x + \log(3x-7) = 1$
10. $\log(3x-16) + \log(x-1) = 1$
11. $\log 8x + \log(x+2) = 1$
12. $\log 5x + \log(6x+1) = 1$

13. $\log 2x + \log(13x - 1) = 2$
14. $\log(2 - 6x) + \log(8 + x) = 2$
15. $\log_4 x + \log_4(x + 6) = 2$
16. $\log_4(x - 2) + \log_4(x + 1) = 1$
17. $e^{\ln x^2} = 4$
18. $e^{\ln(6x-5)} = 7$
19. $\ln e^{4x} = 8$
20. $\ln e^{9x+8} = 10$

21–36

Find a three-digit value for x in each equation. (See Examples 2 and 3.)

21. $3^x = 17$
22. $2^x = 5$
23. $4^{1-x} = 19$
24. $7^{2-x} = 6$
25. $9^{2x^2-1} = 11$
26. $5^{x^2+2} = 39$
27. $3^x = 2^{4x+3}$
28. $5^x = 6^{x-1}$
29. $5^{x-3} = 9^{x+4}$
30. $7^{1-x} = 6^{3-x}$
31. $9^{x+2} = 7^{3x-1}$
32. $5^{2+3x} = 8^{4x-1}$
33. $(1 + x)^3 = 3.12$
34. $(1 + x)^5 = 1.34$
35. $(1 + 0.06)^x = 3.17$
36. $60(1 + 0.02)^{4x} = 130$

37–44

Find a three-digit value for each of the following logarithms. (See Example 6.)

37. $\log_3 7$
38. $\log_4 8$
39. $\log_6 43$
40. $\log_8 51$
41. $\log_2 28.6$
42. $\log_5 41.3$
43. $\log_7 132$
44. $\log_9 158$

45–52

Use natural logarithms to solve the following equations and round the answers to three digits. (See Example 5.)

45. $e^x = 4.2$
46. $e^x = 26$
47. $e^{-x} = 5.3$
48. $e^{-x} = 6.1$
49. $e^{x+1} = 67.1$
50. $e^{x+1} = 24.7$
51. $e^{2x+1} = 114$
52. $e^{2x+1} = 92.5$

53–70

Compound Amount 53. Find the compound amount A of $P = \$1000$ compounded quarterly for 5 years at an annual rate of 12%.

Compound Interest Rate 54. If $P = \$1000$ amounts to \$1250 in 2 years and interest is compounded semiannually, find the annual interest rate r.

Doubling an Investment 55. How long will it take an original principal P to double if it is invested at 14% compounded semiannually?

Tripling an Investment 56. How long will it take an original principal P to triple if it is invested at 12% compounded monthly?

Investment Principal 57–58

57. If money earns 10% compounded quarterly, how much should be invested today to accumulate to \$10,000 in 6 years?
58. If money earns 8% compounded monthly, how much should be invested today to accumulate to \$5000 in 4 years?

Compounding Continuously 59–60

59. How long will it take an original investment to double if interest is at 10% compounded continuously?
60. How long will it take an original investment to triple if interest is at 9% compounded continuously?

Population Growth 61–63

61. If a colony of bacteria starts with 2000 bacteria and increases by 15% of its population each day, the population P after t days is given approximately by $P = 2000e^{0.14t}$. How long will it take the population to reach 8000?
62. A colony of bacteria increases by 20% of its population each day. Starting with 3000 bacteria, the population P after x days is $P = 3000e^{0.18x}$. How long will it take the population to reach 6000?
63. How long will it take the population in Exercise 62 to reach 9000?

Atmospheric Pressure

64. The atmospheric pressure P at an altitude x miles above sea level is given approximately by $P = 30e^{-0.198x}$, where P is measured in inches of mercury. Find the altitude of a mountain peak if the pressure there is 16 inches of mercury.

Carbon-14 Dating 65–66

65. If an organism contains 40 milligrams of the radioactive carbon atom ^{14}C at its death, the amount A of ^{14}C remaining x years later is approximately $A = 40e^{-0.000124x}$ milligrams. How long after its death will 10 milligrams of ^{14}C remain?
66. If an organism contains 10 milligrams of the radioactive carbon atom ^{14}C at its death, the amount of ^{14}C remaining x years later is $A = 10e^{-0.000124x}$ milligrams. How long has it been since death when 4 milligrams of ^{14}C remain?

Radioactive Decay

67. Starting with 100 milligrams of radium, the amount A of radium remaining after x years of radioactive decay is $A = 100e^{-0.000411x}$ milligrams. How long will it be until half the radium remains?

pH of a Solution

68. The pH of a chemical solution is given by $\text{pH} = -\log(H_3O^+)$, where (H_3O^+) is the concentration in moles per liter of the hydronium ion. Find the hydronium ion concentration (H_3O^+) if the pH is 6.8.

Intensity of Sound 69–70

69. Using the decibel formula $L = 10 \log(I \times 10^{12})$ decibels, find the effect on L of tripling the power I on an audio amplifier.
70. Using the formula in Exercise 69, find the effect on L of increasing the power from I to $5I$.

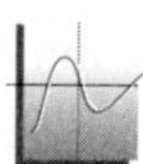

71–74 Use a graphing device to graph the given function in the viewing rectangle $[-1, 10]$ by $[-2, 2]$.

71. $y = \ln x$
72. $y = \log x$
73. $y = \log(2x - 1)$
74. $y = \ln(0.7x - 1)$

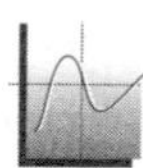

75–78 Use a graphing device to graph the given function in the given viewing rectangle.

75. $y = x \ln x$, $[0, 3]$ by $[-1, 3]$
76. $y = x + \ln x$, $[0, 3]$ by $[-1, 3]$
77. $y = \ln x^2$, $[-10, 10]$ by $[-10, 10]$
78. $y = x + \ln x^2$, $[-10, 10]$ by $[-10, 10]$

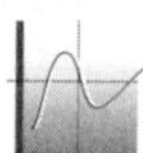

79–82 Use a graphing device and the change-of-base formula to graph the given function in the viewing rectangle $[0, 10]$ by $[-5, 5]$.

79. $y = \log_3 x$
80. $y = \log_2 x$
81. $y = \dfrac{1}{\log_2 x}$
82. $y = \log_3 \left(\dfrac{1}{x}\right)$

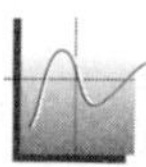

83–86 Use a graphing device to find the solution to the given equation, correct to the nearest hundredth.

83. $x + \ln x = 0$
84. $e^x + \ln x = 0$
85. $\ln(2x - 1) = \dfrac{1}{x^2}$
86. $\log x = \dfrac{1}{x^3}$

Critical Thinking: Exploration and Discussion

87. Explain why u and v must be positive in the property

$$\log_a(uv) = \log_a u + \log_a v.$$

Critical Thinking: Exploration and Discussion

88. Explain how the use of the property in Exercise 87 to solve an equation may lead to extraneous solutions.

◆ Solution for Practice Problem

1. Taking the logarithm of both sides of the equation $4^x = 23$, we get

$$\log 4^x = \log 23$$

or

$$x \log 4 = \log 23$$

by property 4 of logarithms. This readily yields

$$x = \frac{\log 23}{\log 4} \approx 2.262.$$

CHAPTER REVIEW

Summary of Important Concepts and Formulas

Laws of Exponents

1. $a^x \cdot a^y = a^{x+y}$ product rule
2. $(a^x)^y = a^{xy}$ power of a power rule
3. $\dfrac{a^x}{a^y} = a^{x-y}$ quotient rule
4. $(ab)^x = a^x \cdot b^x$ power of a product rule
5. $\left(\dfrac{a}{b}\right)^x = \dfrac{a^x}{b^x}$ power of a quotient rule

Important Features of the Exponential Functions

$f(x) = a^x$, $a > 0$ and $a \neq 1$

1. The **domain** of f is the set of all real numbers.
2. The **range** of f is the set of all positive real numbers.
3. If $a > 1$, the graph of $f(x) = a^x$ resembles the graph of $f(x) = 3^x$. (See Figure 6.1.)
4. If $0 < a < 1$, the graph of $f(x) = a^x$ resembles that of $f(x) = (\frac{1}{2})^x$. (See Figure 6.2.)
5. $a^u = a^v$ if and only if $u = v$.

Logarithmic Function

$y = \log_a x$ if and only if $x = a^y$.

Important Features of the Logarithmic Functions

$f(x) = \log_a x$, $a > 0$ and $a \neq 1$

1. The **domain** of f is the set of all positive real numbers.
2. The **range** of f is the set of all real numbers.
3. If $a > 1$, the graph of $f(x) = \log_a x$ resembles that of $f(x) = \log_2 x$. (See Figure 6.10.)
4. If $0 < a < 1$, the graph of $f(x) = \log_a x$ resembles that of $f(x) = \log_{1/2} x$. (See Figure 6.11.)
5. $\log_a u = \log_a v$ if and only if $u = v$.

Properties of Logarithms

$\log_a a^x = x$ $\quad$ $\log_a u = \log_a v$ if and only if $u = v$.

$a^{\log_a x} = x$ $\quad$ $\log_a(uv) = \log_a u + \log_a v$

$\log_a a = 1$ $\quad$ $\log_a(u/v) = \log_a u - \log_a v$

$\log_a 1 = 0$ $\quad$ $\log_a u^r = r \log_a u$

Change-of-Base Formula

$$\log_b N = \frac{\log_a N}{\log_a b}$$

Critical Thinking: Find the Errors

Each of the following nonsolutions has one or more errors. Can you find them?

Problem 1

Solve $\log_2 x - \log_2 (x + 1) = 3$.

Nonsolution

$$\log_2 \frac{x}{x+1} = 3$$

$$\frac{x}{x+1} = 8$$

$$x = 8x + 8$$

$$-7x = 8$$

$$x = -8/7$$

The solution set is $\{-8/7\}$.

Problem 2

Solve the following equation.

$\log(2x+3)+\log(3x-1)=1$

Nonsolution

$$\begin{aligned}\log[(2x+3)+(3x-1)] &= 1\\ \log(5x+2) &= 1\\ 5x+2 &= 10\\ 5x &= 8\\ x &= 8/5\end{aligned}$$

The solution is $x = 8/5$.

Problem 3

Solve the following equation.

$\log_3 x+\log_3(2x+3)=2$

Nonsolution

$$\begin{aligned}x(2x+3) &= 2\\ 2x^2+3x-2 &= 0\\ (2x-1)(x+2) &= 0\\ 2x-1=0 \quad &\text{or} \quad x+2=0\\ x=1/2 \quad &\text{or} \quad x=-2\end{aligned}$$

When $x=-2$, the left-hand side of the original equation contains logarithms of negative numbers. This value must be rejected, and the only solution is $x = 1/2$.

Review Problems for Chapter 6

1. If $f(x) = (2/3)^x$, find the following function values.
   a. $f(-2)$ b. $f(1/2)$

Solve for x.

2. $(81)^x = 1/27$
3. $8^x = 1/32$
4. $4^{x+3} = 32^{2x-5}$
5. $10^{-2x} = \dfrac{1}{1000^{x+3}}$

Sketch the graphs of the functions defined by the following equations.

6. $f(x) = 2^{x+2}$
7. $f(x) = e^x - 2$

7–8 8. $f(x) = \ln x + 1$
9. $f(x) = \log_3 x$

Find the value of the unknown variable.

10. $\log_n 64 = \frac{3}{2}$
11. $y = \log_{16}(1/32)$
12. $\log_4 x = -\frac{5}{2}$
13. $\log_3 3^{4x} = 18$
14. $e^{\ln e^3} = y$
15. $\log 10^x = -3$

Expand each of the following into terms involving the logarithms of x, y, and z. 26–32

16. $\log_a\left(\dfrac{x^2z}{y^3}\right)$
17. $\log_a \dfrac{\sqrt[3]{x^2y}}{z^4}$

Write each of the following as a single logarithm.

18. $2\log_a x^2z + \log_a 9xy^3 - 2\log_a 4yz^3$
19. $2\log_a x^2yz + 3\log_a 2xyz^2 - \log_a 4x^3y^2z^4$

Find a three-digit value for x in each equation.

20. $x = \log_3 17$
21. $e^x = 4.7$

20–23 22. $5^{x+1} = 192$
23. $5^{2x-1} = 8^{x+1}$

Find the solution set for each of the following equations.

24. $\log x + \log(3x+1) = 1$
25. $\log_8(x+5) + \log_8(3x-1) = 2$

Learning Function

26. A typing teacher has found that the number of words per minute that an average student can type after t months in typing class is given by $W(t) = 50 - 45(2)^{-0.361t}$. Find the number of words per minute that an average student can type after 4 months of typing class.

Cooling a Liquid

27. If a bowl of boiling soup (temperature 212°F) is placed in a refrigerator where the temperature is 35°F, the Fahren-

heit temperature of the soup after t minutes is given by $F(t) = 35 + 177(2)^{-0.0256t}$. Find the temperature of the soup after 1 hour.

Term of an Investment

28. How long will it take an original principal P to increase by 50% if it is invested at 10% compounded monthly?

Population Decline

29. An anthill containing 130,000 ants is sprayed with an insecticide. If 23% of the ant population dies each hour, the population P after t hours is approximately $P = 130{,}000e^{-0.26t}$. How long will it take the population to reach 10,000 ants?

Learning Function

30. After t months in typing class, the number of words per minute that an average student can type is given by $W(t) = 50 - 45(2)^{-0.361t}$. Find the length of time until the average student can type 42 words per minute.

Deer Restocking

31. When 60 deer are released in a wildlife management area, the number $N(x)$ of these deer expected to be alive at the end of x years is $N(x) = 60e^{-0.163x}$. Find the expected length of time it takes for 40 of the deer to die.

Radioactive Decay

32. Starting with A_0 milligrams of polonium, the amount $A(x)$ remaining after x days is $A(x) = A_0e^{-0.000495x}$ milligrams. Find the number of days until $0.37A_0$ milligrams of polonium remain.

ENCORE

Rescue work on the Bay Bridge after the 1989 San Francisco Earthquake.

How did the earthquake of 1989 in San Francisco compare with other major earthquakes? To answer this question, some knowledge about the measurement of earthquakes is needed. Most reports give magnitudes of earthquakes in terms of the Richter scale developed by Charles F. Richter at the California Institute of Technology.

Magnitudes on the Richter scale are computed in terms of a zero-level earthquake† by the formula

$$R = \log \frac{A}{A_0} = \log A - \log A_0,$$

where A is the amplitude measured on a seismograph of the largest wave of the earthquake and A_0 is the amplitude on the same seismograph of a zero-level earthquake with the same epicenter.

Extensive tables of values for A_0 at different distances from the epicenter have been compiled. With a reading of the amplitude A and a value for the distance from the epicenter, a value of A_0 is read from the tables and used to compute

$$R = \log A - \log A_0.$$

A value on the Richter scale is usually reported as a decimal number rounded to the nearest tenth. An earthquake of magnitude 2.0 or less would not be noticed by most people. Since the values on the scale are actually logarithms to the base 10, an increase of 1 on the Richter scale corresponds to an increase of 10 times as great in the intensity of the earthquake.

An earthquake of magnitude 5.2 would be a moderate earthquake causing slight damage, while one of magnitude 6.2 would be 10 times as severe and would be considered a strong earthquake. Similarly, an earthquake of magnitude 7.2 would be 100 times as strong as one of magnitude 5.2. The largest magnitude ever recorded on the Richter scale was 8.9 in the Pacific Ocean near the border of Colombia and Ecuador in 1906 and also in Japan in 1933.

The San Francisco earthquake of 1906 had magnitude 8.3 on the Richter scale, while the earthquake that hit there in 1989 measured 7.1 on the scale. Comparing the 1906 quake with the 1989 one, the difference $8.3 - 7.1 = 1.2$ indicates that the 1906 quake was

$$10^{1.2} \approx 16 \text{ times as strong}$$

as the one in 1989. The Mexico City earthquake of 1985 measured 8.1 on the Richter scale, so it was

$$10^{8.1-7.1} = 10 \text{ times as strong}$$

as the 1989 earthquake in San Francisco. On the other hand, the 1989 San Francisco earthquake was

$$10^{7.1-6.6} \approx 3 \text{ times as strong}$$

as the 1994 Los Angeles earthquake which measured 6.6 on the scale.

†A **zero-level earthquake** is one in which the amplitude of the ground motion of the largest wave is 1 micron (0.001 millimeter) at a distance of 100 kilometers from the epicenter.

Cumulative Test Chapters 3–6

1. Find the slope of the line with equation $2x - 7y = 4$.

2. Let $f(x) = x^2 + 3$. Find and simplify the difference quotient $\frac{f(x+h) - f(x)}{h}$.

3. Find the vertex and axis of symmetry of the parabola with equation $y = -3x^2 - 6x + 5$.

Maximum Profit

4. If Reggie sells w loads of firewood from his wood lot, he makes a profit of $P = 54w - w^2 - 9$ dollars. How many loads of wood sold will produce a maximum profit, and what will the maximum profit be?

5. Find the center and radius of the circle with equation $x^2 + y^2 + 2x + \frac{3}{4} = 0$.

6. Write an equation of the circle that goes through $(6, -6)$ and has center at $(10, -3)$.

7. A portion of a graph is given. Use it to complete the graph in three ways:

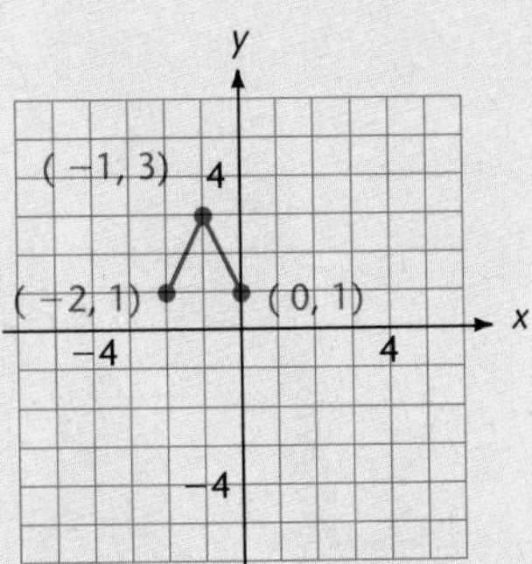

Figure for Problem 7

a. so that it becomes symmetric with respect to the y-axis

b. so that it becomes symmetric with respect to the x-axis

c. so that it becomes symmetric with respect to the origin.

Sketch the graphs of the following equations. Locate the vertices and asymptotes of all hyperbolas and write the equations of the asymptotes.

8. $x^2 - 4y^2 = 16$ 9. $3y = \sqrt{9 - x^2}$

10. The equation $3x + 2y = 12$ defines a function f. Find an expression for $f(x)$ and also for $f^{-1}(x)$ if f is one-to-one.

Stress-Related Virus

11. Suppose the incidence V of a certain stress-related virus varies directly with age A and inversely with D, where D is the income in thousands of dollars. If $V = 8$ for the 40-year-old group with income \$25,000, find V for the 60-year-old group with income \$30,000.

12. Use the remainder theorem and synthetic division to find $P(c)$ where $P(x) = 5x^4 - 2x^3 - 4x + 1$ and $c = -2$.

13. Use the factor theorem to determine whether or not $D(x) = x + 3$ is a factor of

$$P(x) = x^4 + 4x^3 + x^2 + 18.$$

14. Use Descartes' rule of signs to discuss the nature of the zeros of $P(x) = 5x^4 - 2x^3 - 7x + 4$.

15. Find all zeros of $P(x) = 6x^3 + 11x^2 + x - 4$.

16. The following polynomial has exactly one real zero. Find the value of the zero, correct to the nearest tenth.

$$P(x) = x^3 + x^2 - 1$$

17. Sketch the graph of $P(x) = (x - 1)^2(x + 2)$.

18. Sketch the graph of

$$f(x) = \frac{2x}{(x-2)^2}.$$

Locate all vertical, horizontal, and oblique asymptotes.

19. Solve for x:

$$4^{x-1} = \frac{1}{8^{2x}}.$$

Atmospheric Pressure

20. The atmospheric pressure P (in inches of mercury) at an altitude x miles above sea level is approximated by $P = 30e^{-0.198x}$. Find the atmospheric pressure at the top of Mount Mitchell, which is about 1.27 miles above sea level.

21. Sketch the graph of $f(x) = \log_2 (x - 2)$.

22. Find the value of the unknown variable.

    a. $\log_a 16 = 2/3$ b. $y = \log_9 (1/27)$

23. a. Expand $\log_a \dfrac{\sqrt{x}}{yz^3}$ into terms involving the logarithms of x, y, and z.

    b. Write $3 \log_a y^2z - 2 \log_a xy^2 + \log_a x^3yz^4$ as a single logarithm.

24. Find the solution set for the equation

    $\log (2x + 1) + \log (x + 1) = 1.$

Cooling a Liquid

25. If a bowl of soup at temperature 90°C is placed in a room at 20°C, its temperature $T(x)$ after x minutes is $T(x) = 20 + 70e^{-0.056x}$ degrees Celsius. Find how long it will take the soup to cool to 37°C (body temperature).

The Trigonometric Functions

This chapter presents many interesting and practical problems that can be solved by using the trigonometric functions. One such problem that received a great deal of publicity in 1988 is related in Section 7.6. In that year a surveyor in Louisville, Kentucky, used the tangent function to prove that the Louisville Falls Fountain on the Ohio River is the world's largest floating fountain. The rich abundance of applications of trigonometry is evident throughout the chapter.

7.1 ANGLES AND TRIANGLES

Much of trigonometry involves working with triangles. In fact, the word *trigonometry* means "triangle measurement" and comes from the Greek words *tri* ("three"), *gonia* ("angle"), and *metron* ("measure"). In order to work with triangles, some definitions and terminology concerning angles are essential.

Any point on a straight line separates the line into two parts, called **half-lines.** A half-line together with its endpoint is called a **ray.** An **angle** consists of two rays that have a common endpoint. The two rays are called the **sides** of the angle.

In trigonometry, we usually think of an angle as being formed by revolving, or rotating, a ray about its endpoint. The endpoint is called the **vertex** of the angle. We label the initial side and terminal side of the angle according to the direction of rotation (see Figure 7.1). An angle is called a **positive angle** if the direction of rotation is counterclockwise, and a **negative angle** if the direction of rotation is clockwise. The direction of rotation is usually indicated by an arrow.

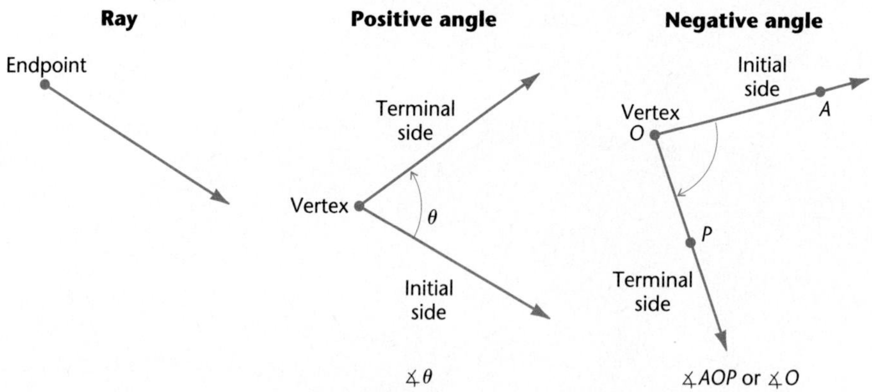

Figure 7.1 Positive and negative angles

Angles are commonly denoted by the Greek letters, α, β, γ, θ, and so on, or by capital letters A, B, C, and so on.† If an angle has its vertex at point O, its initial side passing through point A, and its terminal side passing through point P, then it might be labeled "angle at O" or "angle AOP." The symbol ∢ is used to denote an angle, but often this symbol is omitted if the context is clear.

The **measure** of an angle is given by stating the amount of rotation used to revolve from the initial position of the ray to the terminal position. We write $A = B$ to mean "the measure of angle A equals the measure of angle B." There is more than one unit for measuring angles, just as there is more than one unit of linear measure (feet, miles, meters, kilometers, and so on). The commonly used units of measure for angles are revolution, degree, and radian.

†The Greek alphabet may be found on the back endpapers of this book.

A measure of 1 **revolution** is, by definition, the amount of rotation needed for one full turn of a ray about its endpoint, so that the initial and terminal sides of the angle coincide. Historically, 1 revolution is taken to be an angle of **degree** measure 360, written 360°. Thus an angle measuring 1° is 1/360 part of 1 revolution, an angle measuring 30° is 30/360 or 1/12 of a revolution, one measuring 400° is 400/360 or 10/9 of a revolution, and so on. See Figure 7.2.

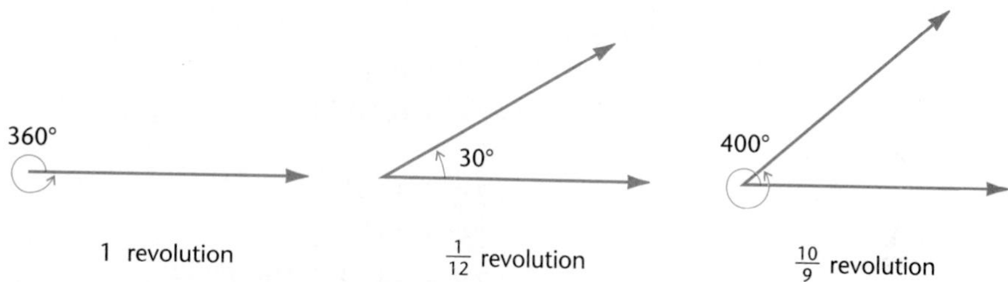

Figure 7.2 Degrees and revolutions

A **right angle** measures 90°, and a **straight angle** measures 180°. A positive angle measuring less than 90° is called an **acute angle,** and one measuring more than 90° but less than 180° is called an **obtuse angle.** If the sum of two positive angles is 90°, the angles are called **complementary angles.** If the sum of two positive angles is 180°, the angles are called **supplementary angles.**

Just as a kilometer can be subdivided into meters and further subdivided into centimeters, or a yard can be subdivided into feet and then inches, a degree can be subdivided into units called **minutes,** denoted by ′, and **seconds,** denoted by ″. There are 60 minutes in 1 degree and 60 seconds in 1 minute.

$$1° = 60' = 3600'' \qquad 1' = 60''$$

Example 1 • Finding a Complementary Angle

Find the angle A that is complementary to the angle $B = 27°20'14''$.

Solution

We know that the sum of A and B must be 90°. We rewrite 90° as 89°60′ and then as 89°59′60″, so that we can subtract B from 90°.

$$\begin{array}{rlcr} 90°: & 90° & \xrightarrow{\text{Rewriting}} & 89°59'60'' \\ B: & \underline{27°20'14''} & & \underline{-27°20'14''} \\ A: & & & 62°39'46'' \end{array}$$

Thus the angles $A = 62°39'46''$ and $B = 27°20'14''$ are complementary. □

◆ **Practice Problem 1** ◆

Find the angle A that is supplementary to the angle $B = 87°49'16''$.

Many calculators use only decimals to denote fractional parts of a degree. Thus it is desirable to become acquainted with the procedure for transforming fractional parts of a degree to minutes and seconds, and vice versa.

Example 2 • Changing Decimal Degrees to Degrees-Minutes-Seconds

Transform 23.24° to degree-minute-second format.

Solution

$$
\begin{aligned}
23.24^\circ &= 23^\circ + 0.24^\circ \\
&= 23^\circ + 0.24(60)' && \text{using } 1^\circ = 60' \text{ to transform decimal degrees to minutes} \\
&= 23^\circ + 14.4' \\
&= 23^\circ + 14' + 0.4' \\
&= 23^\circ + 14' + 0.4(60)'' && \text{using } 1' = 60'' \text{ to transform decimal minutes to seconds} \\
&= 23^\circ 14' 24''
\end{aligned}
$$

□

Example 3 • Changing Degree-Minutes-Seconds to Decimal Degrees

Convert 143°27′15″ to decimal-degree format.

Solution

$$
\begin{aligned}
143^\circ 27' 15'' &= 143^\circ + 27' + 15'' \\
&= 143^\circ + \tfrac{27}{60}^\circ + \tfrac{15}{3600}^\circ && \text{using } 1' = \tfrac{1}{60}^\circ \text{ and } 1'' = \tfrac{1}{3600}^\circ \text{ to convert minutes and seconds to degrees} \\
&= 143^\circ + 0.4500\ldots^\circ + 0.00416\ldots^\circ \\
&= 143.4542^\circ
\end{aligned}
$$

□

Technically, the last = symbol should be ≈ for "approximately equals." Here and elsewhere with our work in trigonometry, we use = for convenience in dealing with approximations.

The following terminology plays an important role in much of the subsequent material.

Definition of an Angle in Standard Position

An angle is in **standard position** if its vertex is placed at the origin of the coordinate system and its initial side lies along the positive x-axis.

An angle in standard position is called a **first, second, third,** or **fourth-quadrant angle,** depending on whether its terminal side lies in the first, second, third, or fourth quadrant, respectively.

For example, 150° is a second-quadrant angle and −20° is a fourth-quadrant angle.

Definition of Coterminal Angles

Two angles are called **coterminal angles** if, when placed in standard position, their terminal sides coincide.

For a given angle θ, in degrees, the expression

$$\theta + n \cdot 360^\circ, \qquad n \text{ any integer,}$$

characterizes all angles that are coterminal to θ.

Example 4 • Finding Coterminal Angles

Find and draw four angles, two positive and two negative, that are coterminal to the first-quadrant angle measuring 50°.

Solution

Using $n = 1$ and $n = 2$ yields positive angles coterminal to 50°. The values $n = -1$ and $n = -2$ yield negative angles coterminal to 50°.

$n = 1,$	$50^\circ + 1 \cdot 360^\circ = 410^\circ$	see Figure 7.3a
$n = 2,$	$50^\circ + 2 \cdot 360^\circ = 770^\circ$	see Figure 7.3b
$n = -1,$	$50^\circ + (-1) \cdot 360^\circ = -310^\circ$	see Figure 7.3c
$n = -2,$	$50^\circ + (-2) \cdot 360^\circ = -670^\circ$	see Figure 7.3d

□

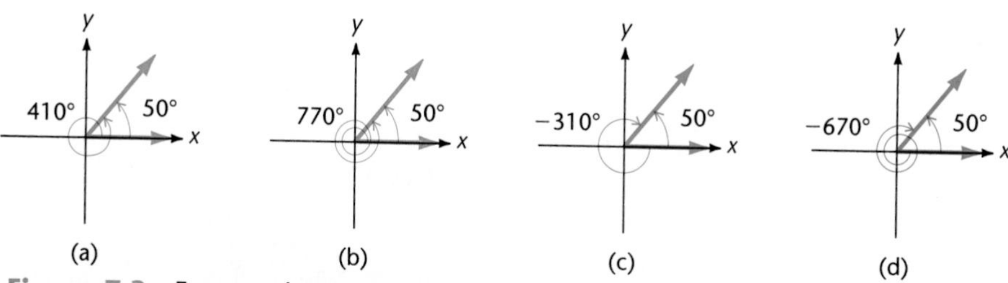

Figure 7.3 Four angles coterminal to 50°

The sum of the three angles in any triangle is 180°. Consequently, the two acute angles in a *right triangle* are complementary angles. An **oblique triangle** is a triangle that is not a right triangle. An **isosceles triangle** is one that has two equal sides (hence two equal angles). If all three sides are equal (or all three angles are equal), the triangle is called **equilateral.**

Two triangles are said to be **similar** if two angles of one triangle are equal to two angles of the other triangle. The sides of similar triangles satisfy the following property:

The corresponding sides of similar triangles are proportional.

If the sides of similar triangles are labeled as in Figure 7.4, then

$$\frac{a}{a'} = \frac{b}{b'} = \frac{c}{c'}.$$

This property is frequently useful in practical applications.

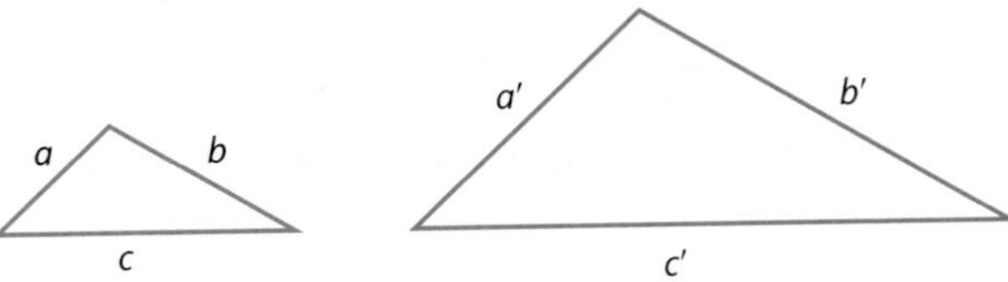

Figure 7.4 Similar triangles

Example 5 • Using Similar Triangles

Suppose 5 feet of a 31-foot tilted piling extends 2 feet vertically above the water line. If the water is uniformly deep and 6 feet of the tilted piling is embedded in the sand, how deep is the water?

Solution

In Figure 7.5, triangles *ACE* and *BCD* are similar. Hence their corresponding sides are proportional. If x is the depth of the water, we have

$$\frac{25}{5} = \frac{x+2}{2}$$

$$\frac{25(2)}{5} = x + 2$$

$$10 = x + 2$$

$$8 = x.$$

The water is 8 feet in depth. □

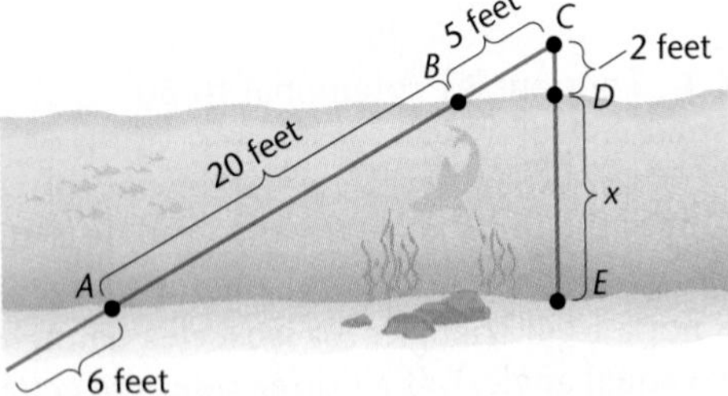

Figure 7.5 Tilted piling

◆ **Practice Problem 2** ◆

If a 6-foot man casts a 2-foot shadow at the same time that a building casts a 12-foot shadow, how tall is the building?

EXERCISES 7.1

Find the angle θ such that θ and the given angle are complementary. (See Example 1.)

1. $42°$ 2. $77°$ √3. $14°25'$ 4. $47°50'$
√5. $18°42'14''$ 6. $49°17'52''$ 7. $29°12'43''$ 8. $83°25'51''$
√9. $18.27°$ 10. $4.11°$ 11. $36.2582°$ 12. $52.0035°$

Find the angle θ such that θ and the given angle are supplementary.

√13. $44°$ 14. $137°$ 15. $138°21'$ 16. $49°18'$
17. $19°23'14''$ 18. $48°6'2''$ √19. $141°3''$ 20. $136°52''$
21. $101.37°$ 22. $88.10°$ √23. $43.5127°$ 24. $142.6228°$

Convert the given angle to degree-minute-second format. (See Example 2.)

25. $14.27°$ 26. $382.19°$ √27. $-63.72°$ 28. $-108.45°$
√29. $27.5201°$ 30. $271.9340°$ 31. $-45.0942°$ 32. $-729.5055°$

Convert the given angle to decimal-degree format. (See Example 3.)

33. $40°18'$ 34. $203°20'$ √35. $-18°50'$ 36. $-193°45'$
√37. $53°14'26''$ 38. $169°50'18''$ 39. $-421°17''$ 40. $-182°33'7''$

Suppose $r > 0$. Determine the sign of each quotient if (x, y) is in the stated quadrant.

41. y/r, quadrant I 42. y/r, quadrant III
√43. x/r, quadrant III 44. x/r, quadrant IV
√45. y/x, quadrant II 46. y/x, quadrant III
47. r/y, quadrant IV 48. r/x, quadrant III

Name the angle according to its quadrant. Find four angles (two positive and two negative) that are coterminal to the given angle. Choose your angles with measure as close to 0° as possible. (See Example 4.)

49. $42°$ 50. $37°$ 51. $150°$ 52. $175°$
√53. $214°$ 54. $263°$ 55. $320°$ 56. $295°$
57. $415°$ 58. $383°$ √59. $-28°$ 60. $-160°$

For the following problems, see Example 5.

Height of a Tree √61. If the 4-foot fence post shown in the figure on page 386 casts a 1.8-foot shadow at the same time that a pine tree casts a 36-foot shadow, how tall is the pine tree?

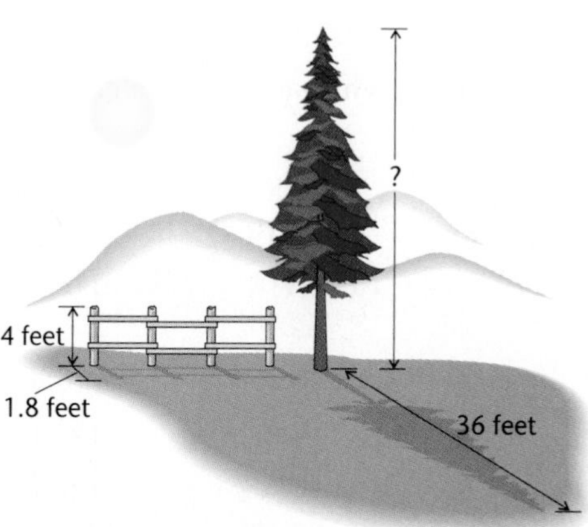

Figure for Exercise 61

Height of a Cliff

62. If a 4-foot boy casts a 7-foot shadow at the same time that a vertical cliff casts a 91-foot shadow, how high is the cliff?

Mirror Reflections 63–64

63. Sarah and Joey are positioned as indicated in the accompanying figure so that they see each other's reflections in the mirror. How far is Joey standing from the wall?

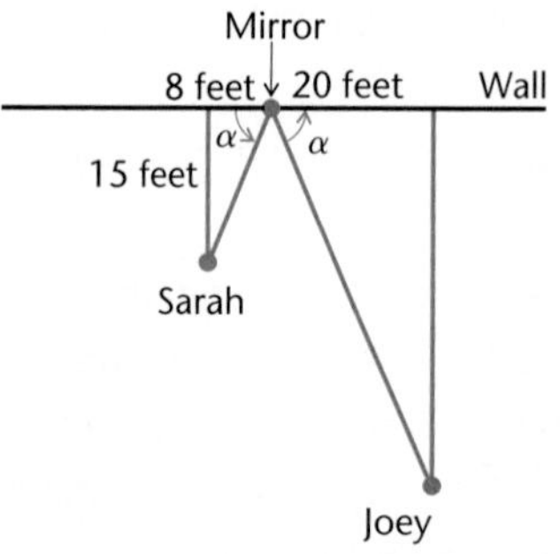

Figure for Exercise 63

64. Matt places a mirror on the pavement in such a position that he sees the top of a flagpole. The mirror is lying 3.1 feet from Matt's feet and 12.4 feet from the base of the flagpole. How tall is the flagpole if Matt's eye level is 5.2 feet high?

Slough

65. To find the distance across a slough, a surveyor drove stakes at points A, B, C, D, and E (see the accompanying figure) so that the line segment joining points A and E was parallel to the line segment joining points D and B. He measured and found the distance from A to C to be 42 yards, from B to D to be 15 yards, and from B to C to be 18 yards. Find the distance AE across the slough.

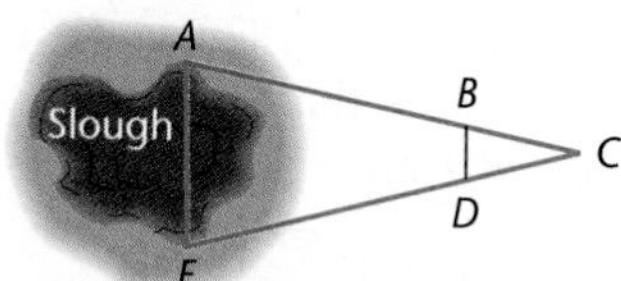

Figure for Exercise 65

Sinkhole

66. To find the distance across a sinkhole, a surveyor drove stakes at points A, B, C, D, and E (see the accompanying figure) so that the line segment joining points A and E was parallel to the line segment joining points B and D. She measured and found the distance from A to C to be 215 meters, from B to D to be 60 meters, and from B to C to be 50 meters. Find the distance AE across the sinkhole.

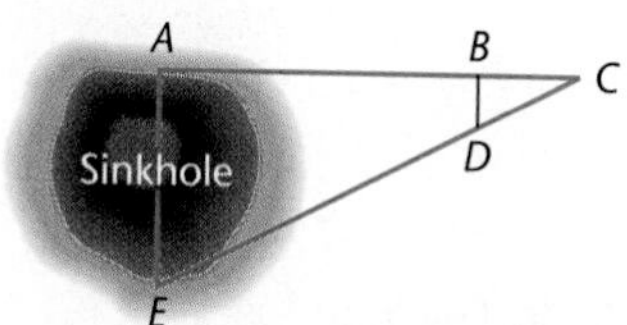

Figure for Exercise 66

Tilted Piling 67–68

67. Suppose 7 feet of a 40-foot tilted piling extends 2 feet vertically above the water line. If 12 feet of the tilted piling is embedded in the sand and the water is uniformly deep, how deep is the water?

68. Suppose 9 feet of a tilted piling extends 5 feet vertically above the water line. Assuming the water is uniformly 20 feet deep and 20 feet of the tilted piling is embedded in the sand, how long is the piling?

Critical Thinking: Exploration and Writing

69. a. If a triangle is labeled with a, b, and c as shown in Figure 7.4, there are six ratios such as a/b that can be formed using the sides of the triangle. Make a list of a/b with the other five ratios of the sides that can be formed using a, b, and c.

b. As stated just before Figure 7.4, the sides of the similar triangles in that figure satisfy

$$\frac{a}{a'} = \frac{b}{b'} = \frac{c}{c'}.$$

Use this property to show that

$$\frac{a}{b} = \frac{a'}{b'}.$$

c. Since the labeling used in Figure 7.4 is arbitrary, we can conclude from part b of this exercise that similar equalities hold for all six of the ratios listed in part a:

$$\frac{a}{c} = \frac{a'}{c'}, \text{ and so on.}$$

Write a statement that describes these equalities with words.

◆ Solutions for Practice Problems

1. We write 180° in the form $179°59'60''$. Then $A = 179°59'60'' - 87°49'16'' = 92°10'44''$.

2. The two triangles in Figure 7.6 are similar. Hence their corresponding sides are proportional. If h is the height of the building, we have

$$\frac{2}{12} = \frac{6}{h}$$
$$2h = 72$$
$$h = 36.$$

The building is 36 feet tall.

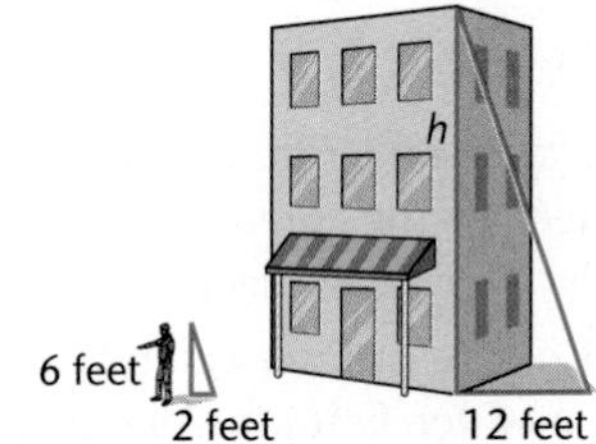

Figure 7.6 The similar triangles in Practice Problem 2

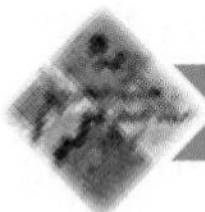

7.2 RADIAN MEASURE

Degree measure is commonly used in fields such as surveying and navigation, but radian measure is more conventional in applications that require the methods of calculus.

In order to formulate the definition of a radian, we consider an angle that has its vertex at the center of a circle. Such an angle is called a **central angle.** Since an angle measuring 1° is $\frac{1}{360}$ of a revolution, a central angle of 1° subtends an arc with length $\frac{1}{360}$ of the circumference C of the circle. The radian unit of measure is defined in terms of the length of the intercepted arc.

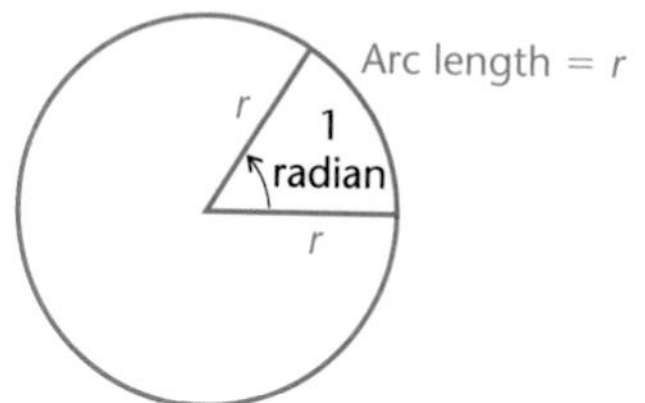

Figure 7.7 An angle measuring one radian

> **Definition of a Radian**
>
> A central angle has measure 1 **radian** if it intercepts an arc with length equal to the radius of the circle.

A central angle of measure 1 radian is drawn in Figure 7.7.

It follows from the definition of a radian than an angle has measure 2 radians if it intercepts an arc that has length $2r$ when placed with its vertex at the center of a circle having radius r. Similarly, a central angle of 3 radians intercepts an arc of length $3r$. The number of radians in 1 revolution is the number of times the radius r can be measured off along the circumference C of the circle. Since $C = 2\pi r$, we conclude that there are 2π radians in 1 revolution and that

$$2\pi \text{ radians} = 1 \text{ revolution} = 360°.$$

Dividing by 2, we obtain the **basic equation** relating radians and degrees.

$$\pi \text{ radians} = 180°$$

Dividing both sides of the basic equation by π, we get

$$1 \text{ radian} = \frac{180°}{\pi} \approx 57.2958° \approx 57°17'45''.$$

Also, dividing both sides of the basic equation by 180, we get

$$1° = \frac{\pi}{180} \text{ radians} \approx 0.0174533 \text{ radians}.$$

These equations lead to the following conversion procedures.

> 1. To change radians to degrees, multiply by $180/\pi$.
> 2. To change degrees to radians, multiply by $\pi/180$.

Example 1 • Changing Radians to Degrees

Convert each of the following to degree measure.

a. $-\dfrac{\pi}{3}$ radians b. $\dfrac{3\pi}{4}$ radians

Solution

In each part, we multiply by $\dfrac{180}{\pi}$.

a. $-\dfrac{\pi}{3} \text{ radians} = -\left(\dfrac{\pi}{3} \cdot \dfrac{180}{\pi}\right)^{\circ} = -60°$

b. $\dfrac{3\pi}{4} \text{ radians} = \left(\dfrac{3\pi}{4} \cdot \dfrac{180}{\pi}\right)^{\circ} = 135°$

□

Example 2 • Changing Degrees to Radians

Convert each of the following to radian measure in terms of π.

a. $210°$ b. $-405°$

Solution

In each part, we multiply by $\dfrac{\pi}{180}$.

a. $210° = \left(210 \cdot \dfrac{\pi}{180}\right) \text{ radians} = \dfrac{7\pi}{6} \text{ radians}$

b. $-405° = -\left(405 \cdot \dfrac{\pi}{180}\right) \text{ radians}$

$= -\dfrac{9\pi}{4} \text{ radians}$ □

◆ Practice Problem 1 ◆

a. Convert $\dfrac{5\pi}{6}$ radians to degree measure.

b. Convert $-270°$ to radian measure in terms of π.

Radian measures are frequently left in terms of π, as they were in the preceding examples. **It is customary to omit the word *radian* when radian measure is being used.**† Thus we write equations such as

$$30° = \frac{\pi}{6}, \quad 45° = \frac{\pi}{4}, \quad 60° = \frac{\pi}{3}, \quad 90° = \frac{\pi}{2}.$$

In Figure 7.8 several special angles are drawn and labeled in degrees and radians.

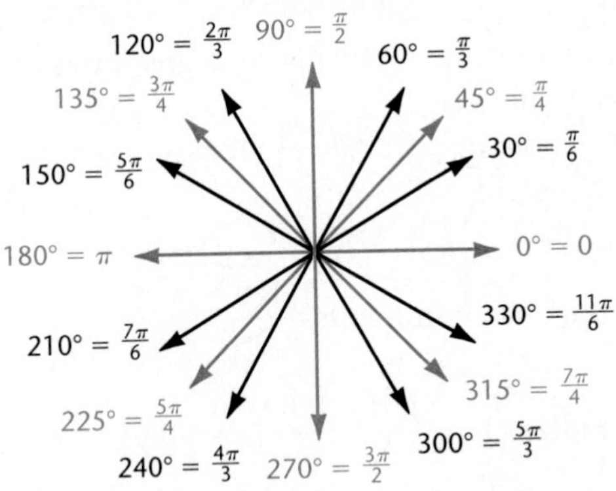

Figure 7.8 Special angles in degree and radian measure

†The word *radian* is sometimes abbreviated "rad."

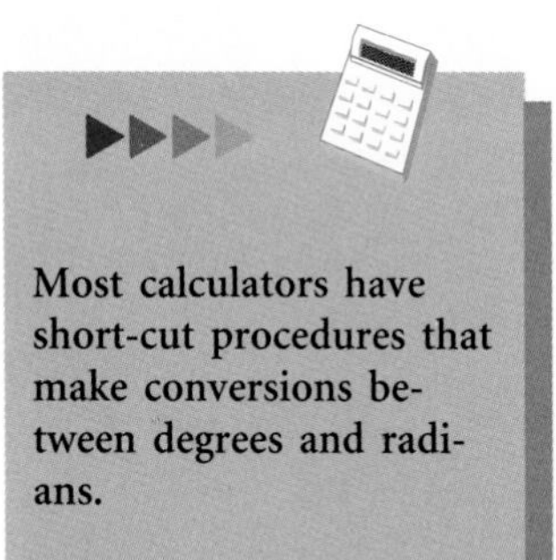

Most calculators have short-cut procedures that make conversions between degrees and radians.

It is sometimes necessary to work with radian measures that are expressed as decimal numbers instead of multiples of π. In such cases we may have to convert a decimal radian measure to decimal degrees, or vice versa. To accomplish this, we can use a calculator and multiply by the appropriate multiplier, $180/\pi$ or $\pi/180$. Some illustrations of conversions going both ways are given in the following example.

Example 3 • Changes with Decimal Measure

a. Convert 2.05 radians to decimal degrees, rounding to four decimal places.

b. Convert $14°35'22''$ to decimal radians, rounding to four decimal places.

Solution

a. 2.05 radians $= 2.05(180/\pi)° = 117.4564°$

b. We first change $14°35'22''$ to decimal degrees and then multiply by $\pi/180$.

$$14°35'22'' = 14° + \frac{35°}{60} + \frac{22°}{3600} = 14° + 0.58333° + 0.00611°$$

$$= 14.5894° = 14.5894\left(\frac{\pi}{180}\right) \text{ radians} = 0.2546 \text{ radians}$$

□

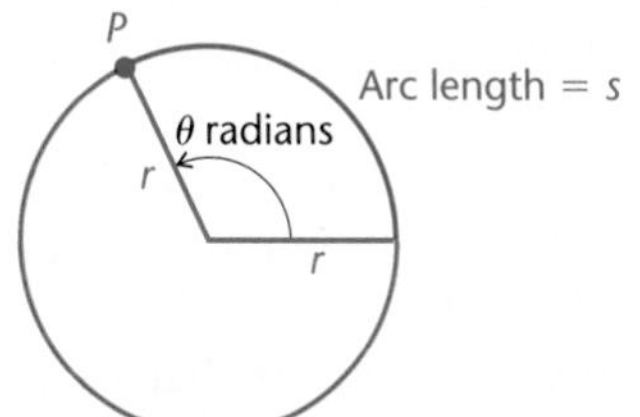

Figure 7.9 Central angle θ and arc length s

Consider a central angle of measure θ radians in a circle with radius r. As in Figure 7.9, let s denote the length of the arc intercepted by θ on the circle and assume that r and s are measured in the same linear units (for instance, both in feet or both in miles).

We have seen that $s = r$ when $\theta = 1$ radian. It is not hard to see, then, that $s = 2r$ when $\theta = 2$ radians, $s = r/2$ when $\theta = \frac{1}{2}$ radian, and generally that

$$s = r\theta.$$

There are two important facts to keep in mind when using this formula.

1. θ is the radian measure of the central angle.
2. s and r are measured in the same linear units.

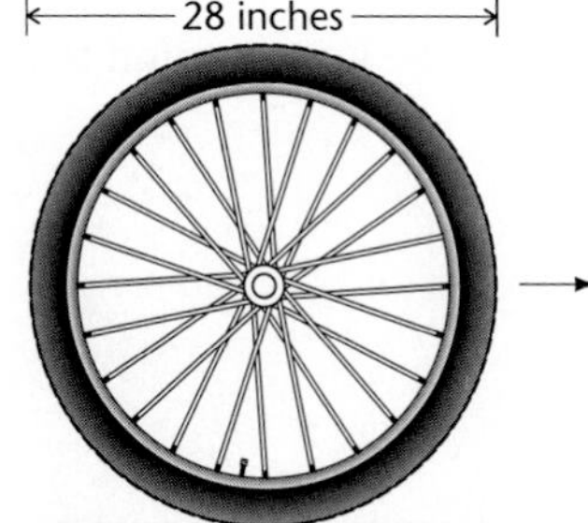

Figure 7.10 A 28-inch bicycle wheel

Example 4 • An Arc Length as a Distance

A bicycle has wheels that are 28 inches in diameter. How far does the bicycle move as the wheels roll through an angle of 15°?

Solution

The distance that the bicycle moves is the same as the arc length intercepted on a wheel by a central angle of 15° (see Figure 7.10). The radius of a wheel is

$$r = \tfrac{1}{2}(28) = 14 \text{ inches},$$

and

$$\theta = 15^\circ = 15\left(\frac{\pi}{180}\right) = \frac{\pi}{12} \text{ radians.}$$

Thus the bicycle moves

$$s = 14\left(\frac{\pi}{12}\right) = \frac{7\pi}{6} = 3.7 \text{ inches,}$$

rounded to two digits. □

◆ Practice Problem 2 ◆

A central angle θ intercepts an arc length of 3.0 feet in a circle with radius 44 inches. Find the radian measure of θ, rounded to two decimal places.

If an object moves in a circular path, two speeds are involved: the rate at which distance is traveled along the circle, and the rate at which the object revolves about the center of the circle.

In Figure 7.9, suppose that the particle P has traveled at a constant rate along the arc length s intercepted by the central angle θ, and that t is the time that has elapsed during this motion. The ratio $v = s/t$ is called the **linear speed** of the particle since it gives the rate at which distance is covered along the circular path. The ratio $\omega = \theta/t$ is called the **angular speed** of the particle since it describes the rate at which the particle is revolving about the center of the circle.

If θ is the measure of the central angle *in radians,* then $s = r\theta$, where s and r are measured *in the same units.* When both sides of this equation are divided by t, we get

$$\frac{s}{t} = r \cdot \frac{\theta}{t},$$

or

$$\mathbf{v = r\omega.}$$

This equation relates linear speed and angular speed. It must be kept in mind that

1. ω is angular speed in radians per unit time;
2. The time units used in v and ω must be the same;
3. The linear units used in v and r must be the same.

Example 5 • Finding an Angular Speed

A car has wheels 28 inches in diameter and is traveling at 45 miles per hour. Find the angular speed of the wheels in radians per minute.

Solution

The radius of a wheel is half the diameter.

$$r = \tfrac{1}{2}(28) = 14 \text{ inches}$$

We are given $v = 45$ miles per hour, and we need the same linear units in r and v. A good plan would be to change to feet in both cases and also to change time units to minutes since we want ω in radians per minute.

$$r = 14 \text{ inches} = \tfrac{14}{12} \text{ feet} = \tfrac{7}{6} \text{ feet}$$

$$\begin{aligned} v &= \frac{45 \text{ miles}}{1 \text{ hour}} \cdot \frac{5280 \text{ feet}}{1 \text{ mile}} \cdot \frac{1 \text{ hour}}{60 \text{ minutes}} \\ &= \frac{45(5280) \text{ feet}}{60 \text{ minutes}} \\ &= 3960 \text{ feet per minute} \end{aligned}$$

Solving $v = r\omega$ for ω, we obtain $\omega = v/r$ and

$$\omega = \frac{3960}{\frac{7}{6}} = 3960 \cdot \frac{6}{7} = 3394 \text{ radians per minute},$$

rounded to the nearest radian per minute. □

◆ **Practice Problem 3** ◆

A pulley with diameter 10 inches is connected by a belt to a larger pulley with diameter 14 inches. If the larger pulley is turning at 60 revolutions per minute, find the angular speed of the smaller pulley in radians per second. (*Hint:* The linear speed of a point on the belt is the same for both pulleys.)

EXERCISES 7.2

Convert each of the following radian measures to degree measure. (See Example 1.)

1. $\frac{\pi}{6}$
2. $\frac{\pi}{9}$
3. $-\frac{\pi}{12}$
4. $-\frac{\pi}{5}$
5. $-\frac{7\pi}{5}$
6. $-\frac{3\pi}{10}$
7. $\frac{11\pi}{4}$
8. $\frac{7\pi}{30}$

Convert each of the following to radian measure in terms of π. (See Example 2.)

9. $-90°$
10. $-60°$
11. $120°$
12. $135°$
13. $520°$
14. $1260°$
15. $105°$
16. $-345°$

Convert the following radian measures to decimal degrees. Round answers to three decimal places. (See Example 3.)

17–20 17. 3 radians 18. 4 radians 19. 0.513 radians 20. 1.67 radians

Convert each of the following to decimal radians. Round answers to four decimal places. (See Example 3.)

21–24 21. 18°30′ 22. 23°40′ 23. 72°14′20″ 24. 144°11′45″

Convert each of the following to radian measure.

25. 2 revolutions 26. 4 revolutions
27. $\frac{3}{2}$ revolutions 28. $\frac{9}{4}$ revolutions

Convert each of the following to revolutions.

29. $\dfrac{\pi}{36}$ radians 30. $\dfrac{\pi}{16}$ radians 31. $\dfrac{2\pi}{3}$ radians 32. $\dfrac{3\pi}{4}$ radians

In Exercises 33–38, s is the length of the arc intercepted by the central angle θ in a circle of radius r. Find the exact value of the missing variable. (See Example 4.)

33. $r = 213$ meters, $\theta = \dfrac{5\pi}{3}$ 34. $r = 246$ meters, $\theta = \dfrac{5\pi}{6}$
35. $r = 1.8$ feet, $\theta = 210°$ 36. $r = 6.3$ feet, $\theta = 120°$
37. $s = 9$ centimeters, $\theta = 2$ 38. $s = 13$ centimeters, $\theta = 4$

For the following exercises, see Example 4.

Wagon Wheel 39. A wagon has wheels that are 3.6 feet in diameter. How far does the wagon move as the wheels turn through 72°?

Truck Wheel 40. A wheel on a truck has a radius of 1.5 feet. How far will the truck move if the wheel turns through 36°?

Pendulum 41. How far does the tip of the 18-inch pendulum shown in the accompanying figure travel as it swings through an angle of 10°?

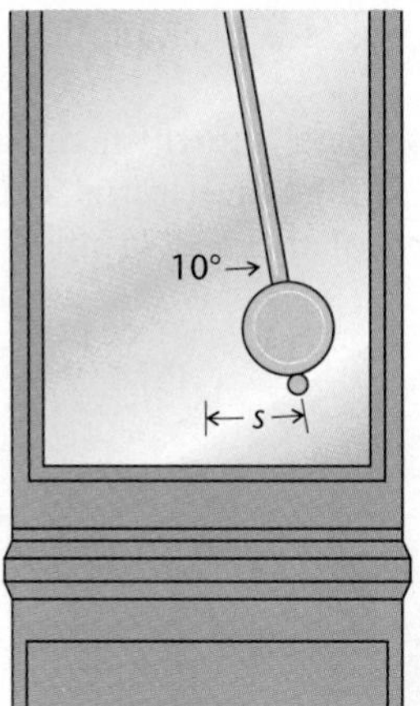

Figure for Exercise 41

Tower Clock

42. The minute hand on a tower clock is 2.0 feet in length. How far does its tip travel in 5 minutes?

Latitudes and Distance 43–44

43. Madison, Wisconsin, is almost due north of Poplarville, Mississippi. The latitude of Madison is 43°N, and that of Poplarville is 31°N (see the accompanying figure). Use 4000 miles as the radius of the earth and find the distance between the two cities, rounded to the nearest 10 miles.

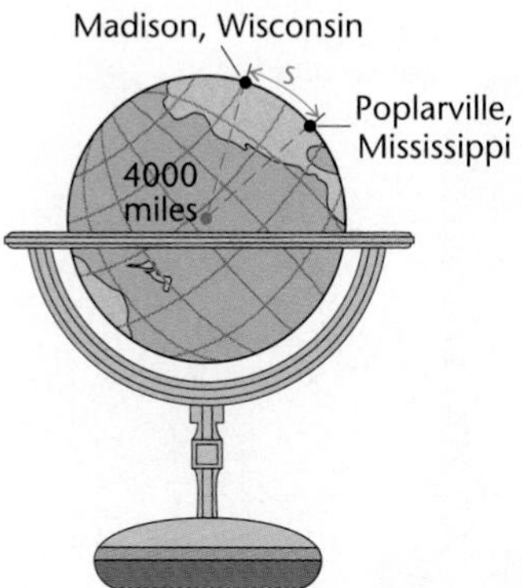

Figure for Exercise 43

44. (See Exercise 43.) Memphis, Tennessee, is almost due north of New Orleans, Louisiana. The latitude of Memphis is 35°N, and that of New Orleans is 30°N. Find the distance between the cities, rounded to the nearest 10 miles.

For the following exercises, see Example 5.

Ferris Wheel 45. A seat on the ferris wheel shown in the accompanying figure is located 20 feet from the axle. If the wheel turns at the rate of 18 degrees per second, find the linear speed of the seat in miles per hour.

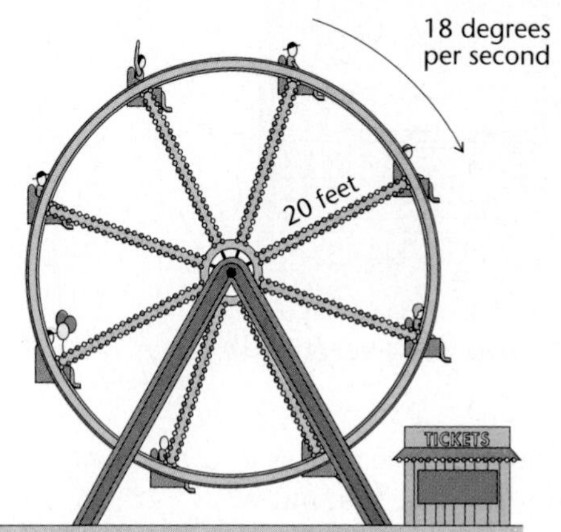

Figure for Exercise 45

Pulley 46. The weight in the accompanying figure is being raised by a rope that passes over a pulley with a diameter of 10 inches. If the pulley turns through 240 degrees per second, find the rate at which the weight is rising in feet per minute.

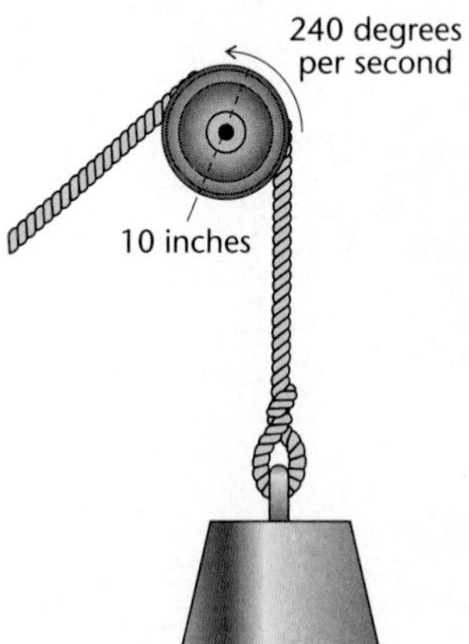

Figure for Exercise 46

Bicycle Wheels 47. The front and rear wheels of a bicycle have diameters of 20 inches and 30 inches, respectively. If the bike is moving so that the angular speed of the rear wheel is 60 revolutions per minute, find the angular speed of the front wheel in radians per second.

Meshed Gears

48. The gear having radius 3 inches in the accompanying figure is meshed with another gear having radius 7 inches. If the larger gear is turning at 84 revolutions per minute, find the angular speed of the smaller gear in radians per second.

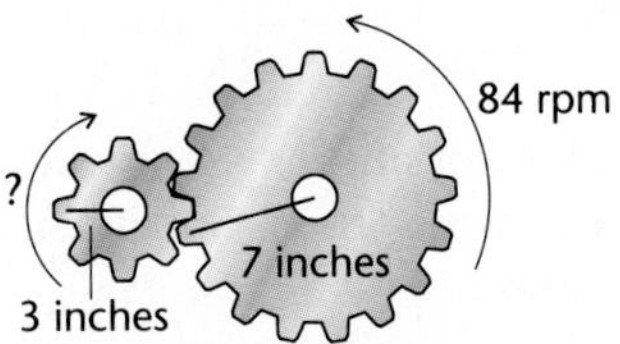

Figure for Exercise 48

Pulley Belt

49. Find the diameter d of the pulley shown in the accompanying figure that is driven at 360 revolutions per minute by a belt moving at 40 feet per second.

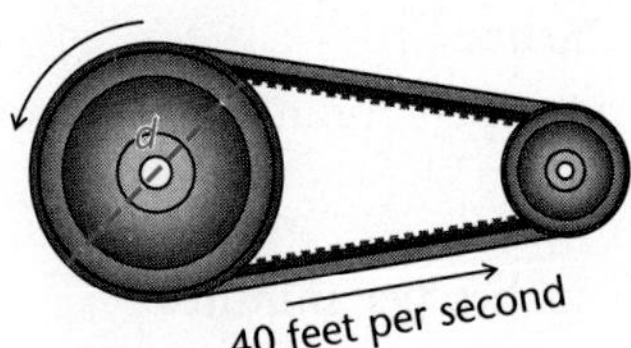

Figure for Exercise 49

Truck Wheels

50. The wheels on a truck turn at the rate of $30/\pi$ revolutions per second when the truck is traveling at 90 feet per second. Find the diameter of a wheel.

Linear Speed at the Equator

51. Quito, Ecuador, is located on the equator of the earth. Take 4000 miles as the radius of the earth and find the linear speed, to the nearest 10 miles per hour, of a person in Quito due to the rotation of the earth.

Satellite Speed

52. (See Exercise 51.) What must be the linear speed of a satellite if it is to stay in an orbit that keeps it 300 miles directly above Quito, Ecuador?

Another unit sometimes used to measure angles is the grad. If 1 revolution is divided into 400 equal parts, a single part is 1 **grad.** Thus 1 right angle is equal to 100 grads.

Critical Thinking: Exploration and Writing

53. a. Describe a procedure for changing degrees to grads.
    b. Describe a procedure for changing grads to degrees.

Critical Thinking: Exploration and Writing

54. a. Describe a procedure for changing radians to grads.
    b. Describe a procedure for changing grads to radians.

◆ Solutions for Practice Problems

1. a. $\frac{5\pi}{6}$ radians $= \left(\frac{5\pi}{6} \cdot \frac{180}{\pi}\right)^\circ = 150^\circ$

   b. $-270^\circ = -\left(270 \cdot \frac{\pi}{180}\right) = -\frac{3\pi}{2}$ radians

2. It is given that $s = 3.0$ feet and $r = 44$ inches. To use the formula $s = r\theta$, s and r must be in the same units. Solving for θ and using $s = 36$ inches, $r = 44$ inches, we obtain

$$\theta = \frac{s}{r} = \frac{36}{44} = \frac{9}{11} = 0.82 \text{ radians},$$

   rounded to two decimal places.

3. The linear speed of each pulley is the same as the linear speed of the belt that connects them. For the larger pulley,

$$r = \tfrac{1}{2}(14) = 7 \text{ inches},$$
$$\omega = 60 \text{ revolutions per minute} = 120\pi \text{ radians per minute},$$

   and

$$v = r\omega = 840\pi \text{ inches per minute}.$$

   This value of v is also the linear speed of the smaller pulley. The radius of the smaller pulley is

$$r = \tfrac{1}{2}(10) = 5 \text{ inches},$$

   and its angular speed is

$$\omega = \frac{v}{r} = \frac{840\pi}{5} = 168\pi \text{ radians per minute}$$

   because the time unit in v is minutes. As the last step, we change this value of ω to radians per second.

$$\omega = \frac{168\pi \text{ radians}}{1 \text{ minute}} \cdot \frac{1 \text{ minute}}{60 \text{ seconds}} = \frac{14\pi}{5} \text{ radians per second},$$

   which is approximately 8.8 radians per second.

7.3 TRIGONOMETRIC FUNCTIONS OF ANGLES

Suppose θ is an angle placed in standard position, as shown in Figure 7.11. Let (x, y) be a point, other than the origin, on the terminal side of θ. Then the distance r from the origin to (x, y) is always positive and is given by

$$r = \sqrt{x^2 + y^2}.$$

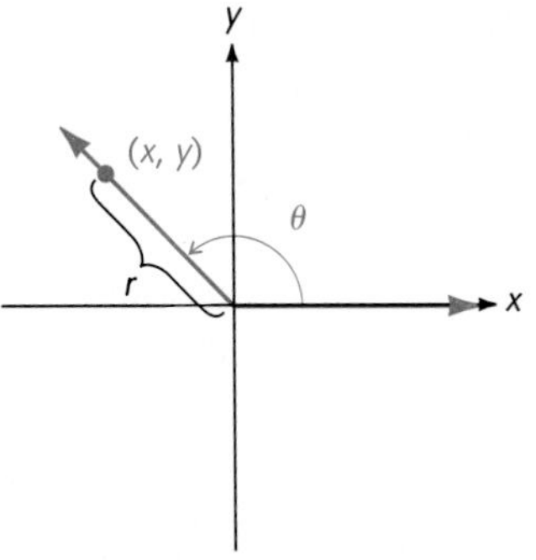

Figure 7.11 Angle in standard position

Six possible ratios can be formed using the three values x, y, and r. They are

$$\frac{y}{r}, \frac{x}{r}, \frac{y}{x}, \frac{x}{y}, \frac{r}{x}, \text{ and } \frac{r}{y}.$$

These six ratios are given the special names sine θ, cosine θ, tangent θ, cotangent θ, secant θ, and cosecant θ, with respective abbreviations $\sin\theta$, $\cos\theta$, $\tan\theta$, $\cot\theta$, $\sec\theta$, and $\csc\theta$. This is formalized in the following definition.

Definition of the Trigonometric Functions of an Angle

Let θ be an angle in standard position and let (x, y) be a point, other than the origin, on the terminal side of θ. With $r = \sqrt{x^2 + y^2}$, the **trigonometric functions** of θ are defined as follows.

$$\sin\theta = \frac{y}{r} \qquad \cot\theta = \frac{x}{y}$$

$$\cos\theta = \frac{x}{r} \qquad \sec\theta = \frac{r}{x}$$

$$\tan\theta = \frac{y}{x} \qquad \csc\theta = \frac{r}{y}$$

The values of the six trigonometric functions of an angle θ in standard position are independent of the choice of the point (x, y) on the terminal side of θ. This can be shown by using similar triangles. Hence the value of $\sin\theta = y/r$ is *unique* for a given θ, and we are justified in using the term *function.* Similar remarks can be made concerning the other trigonometric functions.

Example 1 • Using the Definitions of The Trigonometric Functions

Write out the exact values of all the trigonometric functions of the angle β in standard position whose terminal side passes through the point $(-2, 1)$ (see Figure 7.12).

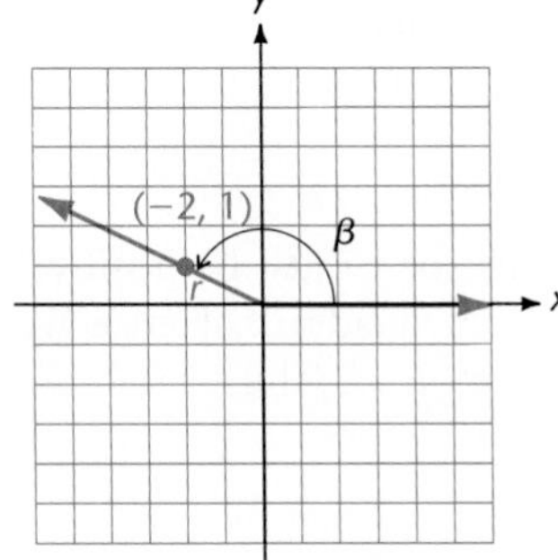

Figure 7.12 Angle β in standard position

Solution

With $x = -2$ and $y = 1$, we find $r = \sqrt{(-2)^2 + 1^2} = \sqrt{5}$. Then the exact values of the trigonometric functions of β are

$$\sin\beta = \frac{y}{r} = \frac{1}{\sqrt{5}} = \frac{\sqrt{5}}{5}, \qquad \cot\beta = \frac{x}{y} = \frac{-2}{1} = -2,$$

$$\cos\beta = \frac{x}{r} = \frac{-2}{\sqrt{5}} = \frac{-2\sqrt{5}}{5}, \qquad \sec\beta = \frac{r}{x} = \frac{\sqrt{5}}{-2} = -\frac{\sqrt{5}}{2},$$

$$\tan\beta = \frac{y}{x} = \frac{1}{-2} = -\frac{1}{2}, \qquad \csc\beta = \frac{r}{y} = \frac{\sqrt{5}}{1} = \sqrt{5}.$$

□

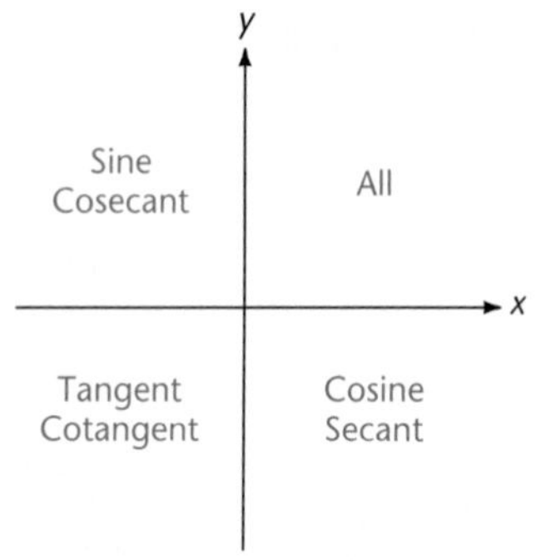

Figure 7.13 Positive functions in each quadrant

The signs of the trigonometric function values of θ depend on the quadrant in which θ terminates. Since r is *always* positive, the sign of $\sin \theta$ and $\csc \theta$ depends only on the sign of y. Hence $\sin \theta$ and $\csc \theta$ are both positive whenever y is positive—that is, whenever θ is in quadrant I or II. Sin θ and $\csc \theta$ are both negative in quadrants III and IV.

The signs of $\cos \theta$ and $\sec \theta$ depend only on the sign of x. Since x is positive in quadrants I and IV, $\cos \theta$ and $\sec \theta$ are positive there and negative in quadrants II and III.

The remaining two functions, tangent and cotangent, depend on both x and y. We see that $\tan \theta$ and $\cot \theta$ are positive whenever x and y have the same sign, and negative whenever x and y have opposite signs. Thus whenever θ is a first- or third-quadrant angle, $\tan \theta$ and $\cot \theta$ are positive; whenever θ is a second- or fourth-quadrant angle, $\tan \theta$ and $\cot \theta$ are negative. These results are summarized in Figure 7.13.

Example 2 • Using the Signs of Two Functions to Determine the Quadrant

If $\sin \theta < 0$ and $\cos \theta > 0$, in which quadrant does θ terminate?

Solution

Sin θ is negative in quadrants III and IV; $\cos \theta$ is positive in quadrants I and IV. Thus for $\sin \theta < 0$ and $\cos \theta > 0$, θ must be a fourth-quadrant angle. □

The next example illustrates how the values of all the trigonometric functions of an angle θ can be determined if one function value is known along with the quadrant in which θ terminates.

Example 3 • Using the Quadrant and a Known Function Value

Determine the values of all the trigonometric functions of the fourth-quadrant angle α if $\cos \alpha = \frac{8}{17}$.

Solution

We must determine the values x, y, and r where (x, y) is a point on the terminal side of α and r is the distance from the origin to (x, y). Since

$$\cos \alpha = \frac{x}{r} = \frac{8}{17},$$

we choose x and r so that their ratio is 8 to 17. The value of r must be positive. Also, x must be positive since α terminates in the fourth quadrant. Hence we choose $x = 8$ and $r = 17$. To determine y, we use $r^2 = x^2 + y^2$. Thus

$$\begin{aligned} 289 &= 64 + y^2 \\ 225 &= y^2 \\ \pm 15 &= y. \end{aligned}$$

The value of y is chosen to be negative since α terminates in quadrant IV. Now with $x = 8$, $y = -15$, and $r = 17$, we write out the values of all the other trigonometric functions.

$$\sin\alpha = -\tfrac{15}{17} \qquad \sec\alpha = \tfrac{17}{8}$$
$$\tan\alpha = -\tfrac{15}{8} \qquad \csc\alpha = -\tfrac{17}{15}$$
$$\cot\alpha = -\tfrac{8}{15}$$

□

Recall that the values of the trigonometric functions are independent of the choice of the point on the terminal side of the angle. Thus in Example 3 we could have used the values $x = 16$ and $r = 34$ since the ratio of x to r is

$$\frac{x}{r} = \frac{16}{34} = \frac{8}{17}.$$

◆ Practice Problem 1 ◆

Find the exact values of all the trigonometric functions of the third-quadrant angle β if $\tan\beta = 1.5$.

Later in this chapter we shall find values of trigonometric functions by using a calculator or a table of values. However, function values for certain special angles can be found by using the definitions. These special angles are used so much in future work that it is worthwhile to learn how to find their function values without using tables or a calculator.

Among the special angles are those whose terminal sides lie along one of the coordinate axes. Angles such as these, whose terminal sides do not lie in any quadrant, are called **quadrantal angles.**

Example 4 • Function Values of a Quadrantal Angle

Determine the values of the trigonometric functions of 180°.

Solution

Figure 7.14 shows 180° placed in standard position, with the point $(-1, 0)$ on its terminal side. With $x = -1$, $y = 0$, we find $r = \sqrt{(-1)^2 + 0^2} = 1$. Then, remembering that division by 0 is undefined, we have

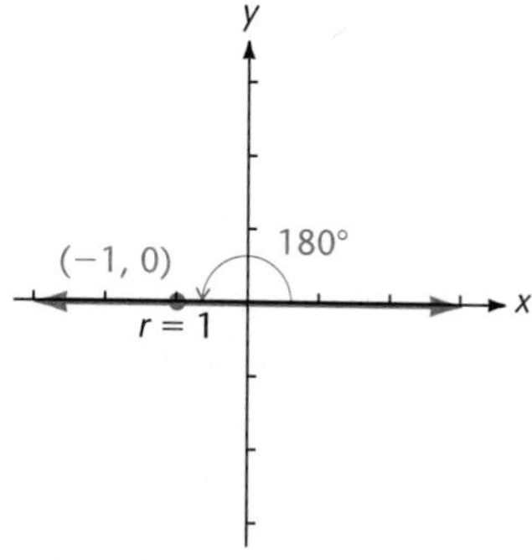

Figure 7.14 180° in standard position

$$\sin 180° = \frac{0}{1} = 0, \qquad \cot 180° = \frac{-1}{0} = \text{undefined},$$

$$\cos 180° = \frac{-1}{1} = -1, \qquad \sec 180° = \frac{1}{-1} = -1,$$

$$\tan 180° = \frac{0}{-1} = 0, \qquad \csc 180° = \frac{1}{0} = \text{undefined}.$$

□

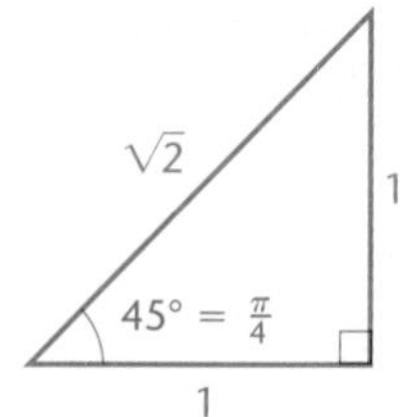

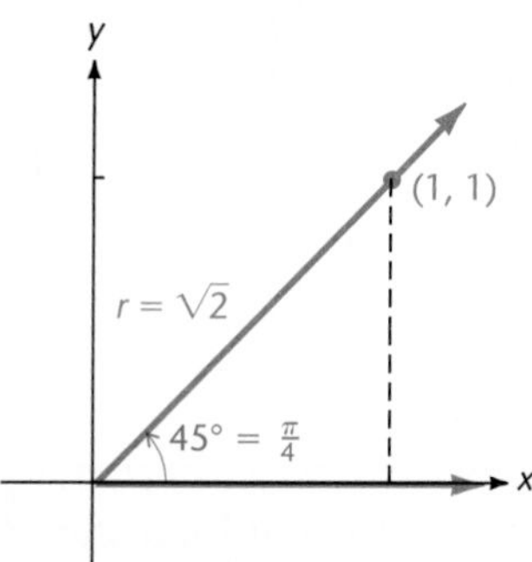

Figure 7.15 The 45° right triangle

The values of the trigonometric functions of any quadrantal angle are found by using the same procedure as in Example 4.

We use some knowledge of geometry to determine the values of the trigonometric functions of the special angles 30°, 45°, and 60°. In the right triangle in Figure 7.15 with 45° acute angles and legs of length 1, the hypotenuse has length $c = \sqrt{1^2 + 1^2} = \sqrt{2}$. If we place this triangle in a coordinate system with one of the 45° angles in standard position, then we find that the point (1, 1) lies on the terminal side of a 45° angle. Using $r = \sqrt{2}$, we write out the values of the trigonometric functions of $45° = \pi/4$.

$$\sin\frac{\pi}{4} = \sin 45° = \frac{1}{\sqrt{2}} = \frac{\sqrt{2}}{2} \qquad \cot\frac{\pi}{4} = \cot 45° = \frac{1}{1} = 1$$

$$\cos\frac{\pi}{4} = \cos 45° = \frac{1}{\sqrt{2}} = \frac{\sqrt{2}}{2} \qquad \sec\frac{\pi}{4} = \sec 45° = \frac{\sqrt{2}}{1} = \sqrt{2}$$

$$\tan\frac{\pi}{4} = \tan 45° = \frac{1}{1} = 1 \qquad \csc\frac{\pi}{4} = \csc 45° = \frac{\sqrt{2}}{1} = \sqrt{2}$$

To obtain the values of the trigonometric functions of 30°, we place a 30°-60° right triangle in the coordinate system as shown in Figure 7.16a, with the 30° angle in standard position. If the length of the hypotenuse is 2, then the length of the shorter side is 1, and the length of the remaining side is $\sqrt{3}$. With this information we can write out the values of the trigonometric functions of 30°. Similarly, in Figure 7.16b we position the 60° angle in standard position to obtain its function values.

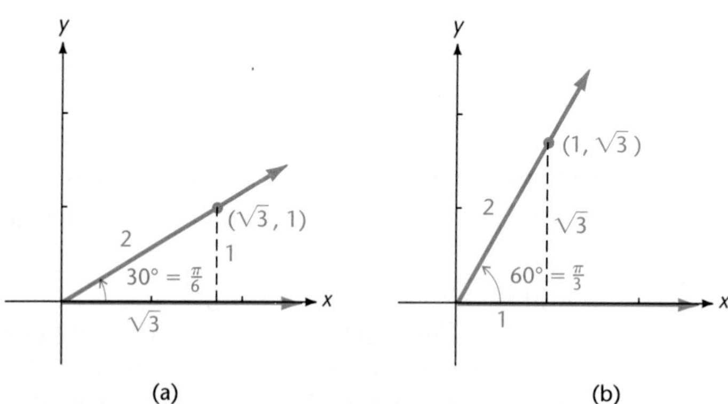

Figure 7.16 The 30°-60° right triangle

$$\sin\frac{\pi}{6} = \sin 30° = \frac{1}{2} \qquad \sin\frac{\pi}{3} = \sin 60° = \frac{\sqrt{3}}{2}$$

$$\cos\frac{\pi}{6} = \cos 30° = \frac{\sqrt{3}}{2} \qquad \cos\frac{\pi}{3} = \cos 60° = \frac{1}{2}$$

$$\tan\frac{\pi}{6} = \tan 30^\circ = \frac{1}{\sqrt{3}} = \frac{\sqrt{3}}{3} \qquad \tan\frac{\pi}{3} = \tan 60^\circ = \sqrt{3}$$

$$\cot\frac{\pi}{6} = \cot 30^\circ = \sqrt{3} \qquad \cot\frac{\pi}{3} = \cot 60^\circ = \frac{1}{\sqrt{3}} = \frac{\sqrt{3}}{3}$$

$$\sec\frac{\pi}{6} = \sec 30^\circ = \frac{2}{\sqrt{3}} = \frac{2\sqrt{3}}{3} \qquad \sec\frac{\pi}{3} = \sec 60^\circ = 2$$

$$\csc\frac{\pi}{6} = \csc 30^\circ = 2 \qquad \csc\frac{\pi}{3} = \csc 60^\circ = \frac{2}{\sqrt{3}} = \frac{2\sqrt{3}}{3}$$

The next definition is invaluable in determining the values of the trigonometric functions of many angles.

Definition of the Related Angle

Suppose θ is an angle in standard position and is not a quadrantal angle. The **related** (or **reference**) **angle** θ' for the angle θ is the positive acute angle that the terminal side of θ makes with the x-axis.

Example 5 • Finding Related Angles

Determine the related angle θ' for the given angle θ.

a. $\theta = 115^\circ$ b. $\theta = \dfrac{4\pi}{3}$ c. $\theta = 290^\circ$ d. $\theta = -\dfrac{5\pi}{3}$

Solution

First visualize the situation with a sketch such as those shown in Figure 7.17. Notice that the related angle θ' is *always* drawn between the terminal side of θ and the x-axis, *never* the y-axis. □

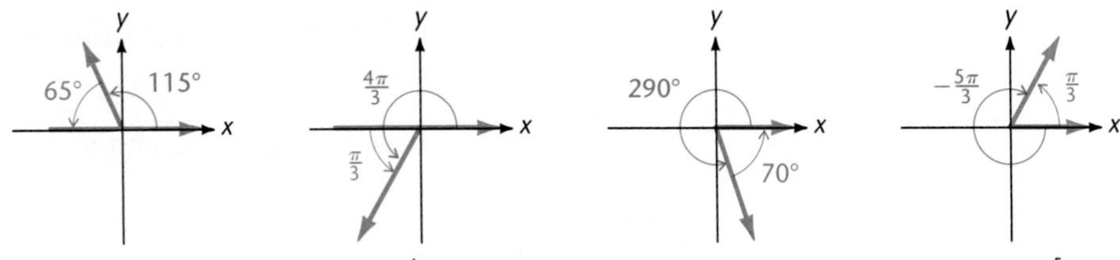

(a) $\theta' = 180^\circ - 115^\circ = 65^\circ$ (b) $\theta' = \frac{4\pi}{3} - \pi = \frac{\pi}{3}$ (c) $\theta' = 360^\circ - 290^\circ = 70^\circ$ (d) $\theta' = 2\pi - \frac{5\pi}{3} = \frac{\pi}{3}$

Figure 7.17 Related angles in Example 5

Example 6 • Using a Special Angle

Write out the exact values of the trigonometric functions of $5\pi/4$.

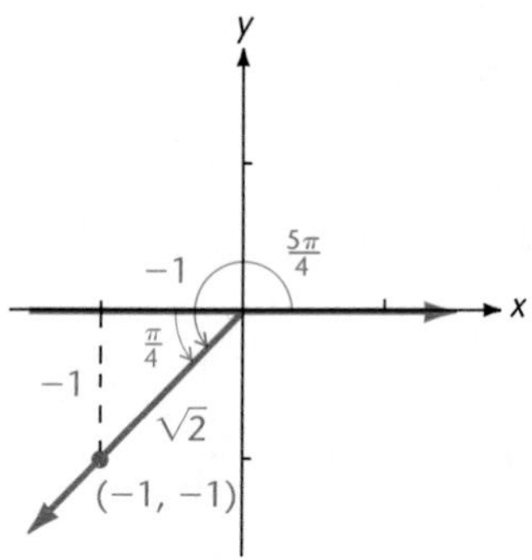

Figure 7.18 The related angle for $\frac{5\pi}{4}$

Solution

The related angle for $5\pi/4$ is $\pi/4 = 45°$, one of the special angles. When we place a 45° right triangle as shown in Figure 7.18, we see that $(-1, -1)$ is a point on the terminal side of $5\pi/4$ with $r = \sqrt{2}$. Using the definitions of the trigonometric functions, we can write out their values.

$$\sin\frac{5\pi}{4} = \frac{-1}{\sqrt{2}} = -\frac{\sqrt{2}}{2} \qquad \cot\frac{5\pi}{4} = \frac{-1}{-1} = 1$$

$$\cos\frac{5\pi}{4} = \frac{-1}{\sqrt{2}} = -\frac{\sqrt{2}}{2} \qquad \sec\frac{5\pi}{4} = \frac{\sqrt{2}}{-1} = -\sqrt{2}$$

$$\tan\frac{5\pi}{4} = \frac{-1}{-1} = 1 \qquad \csc\frac{5\pi}{4} = \frac{\sqrt{2}}{-1} = -\sqrt{2}$$

□

When one of the special triangles is positioned with hypotenuse coincident with the terminal side of θ and one side lying along the x-axis, it is called the **related** (or **reference**) **triangle.** The related angle θ' for θ is the angle in the related triangle between the hypotenuse and the side along the x-axis.

◆ Practice Problem 2 ◆

Find the exact values of the trigonometric functions of $-210°$.

The values of the trigonometric functions of an angle θ can be found by considering the values of the trigonometric functions of the related angle. Let θ_1, θ_2, θ_3, and θ_4 be angles in quadrants I, II, III, and IV, respectively, each having related angle the same measure as θ_1 (see Figure 7.19).

Let (a, b) be a point on the terminal side of θ_1, with $r = \sqrt{a^2 + b^2}$. Then,

1. The point $(-a, b)$ lies on the terminal side of θ_2, r units away from the origin;
2. The point $(-a, -b)$ lies on the terminal side of θ_3, r units away from the origin;
3. The point $(a, -b)$ lies on the terminal side of θ_4, r units away from the origin.

The sine of each of the angles θ_2, θ_3, and θ_4 can be expressed in terms of $\sin\theta_1$.

$$\sin\theta_1 = \frac{b}{r}$$

$$\sin\theta_2 = \frac{b}{r} = \sin\theta_1$$

$$\sin\theta_3 = \frac{-b}{r} = -\frac{b}{r} = -\sin\theta_1$$

$$\sin\theta_4 = \frac{-b}{r} = -\frac{b}{r} = -\sin\theta_1$$

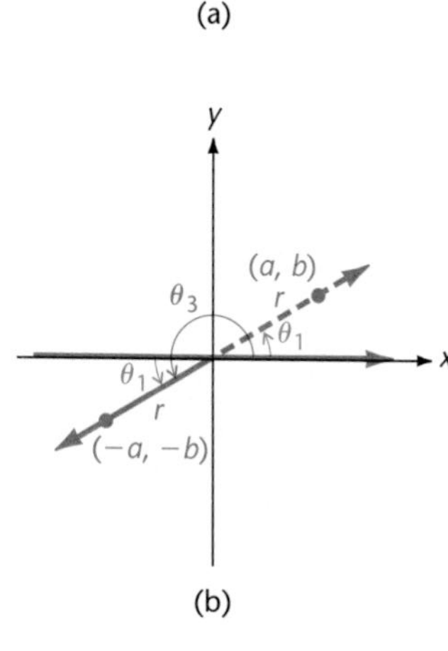

Figure 7.19 Angles with the same related angle

The values of the sine of each angle θ_1, θ_2, θ_3, and θ_4 are the same except for sign. Each sign depends on the quadrant in which the corresponding angle terminates. As before, the sine function assumes positive values for angles in quadrants I and II and negative values in quadrants III and IV.

Similar remarks can be made for each of the other trigonometric functions, and we have the following result.

> **Related Angle Theorem**
>
> The value of any trigonometric function of an angle θ is equal to the value of the corresponding trigonometric function of the related angle, except possibly for the sign. The sign depends on the quadrant θ is in and can be determined by using the diagram in Figure 7.13.

Example 7 • Using the Related Angle Theorem

Write each of the following in terms of the same trigonometric function of a related angle.

a. $\sin 261°$ b. $\cos \dfrac{35\pi}{6}$ c. $\tan(-218°30')$

Solution

The related angle for each angle is shown in the various parts of Figure 7.20.

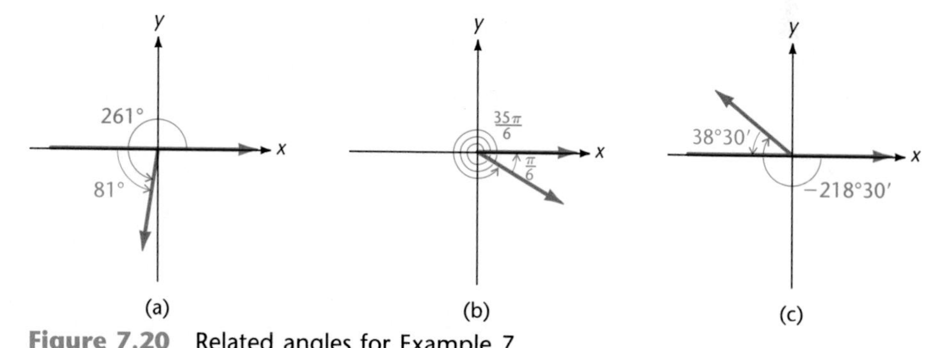

Figure 7.20 Related angles for Example 7

a. Since the sine function assumes negative values in quadrant III,

$$\sin 261° = -\sin 81°.$$

b. The cosine function values are positive in quadrant IV, so

$$\cos \frac{35\pi}{6} = \cos \frac{\pi}{6}.$$

c. The tangent function values are negative in quadrant II. Thus

$$\tan(-218°30') = -\tan 38°30'.$$

□

EXERCISES 7.3

Determine the exact values of the six trigonometric functions of an angle θ in standard position if the terminal side of θ goes through the given point. (See Example 1.)

1. 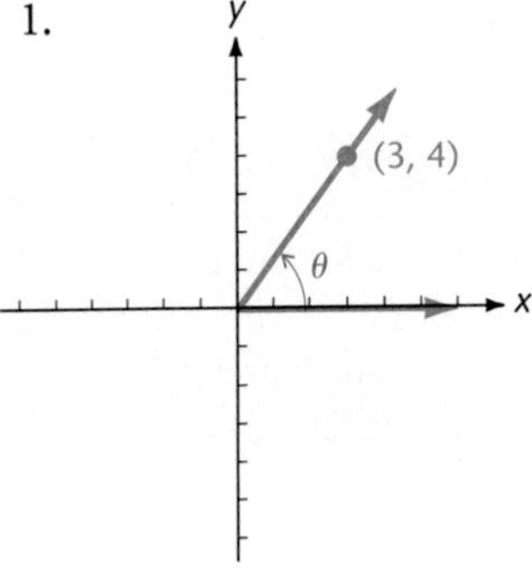

2. 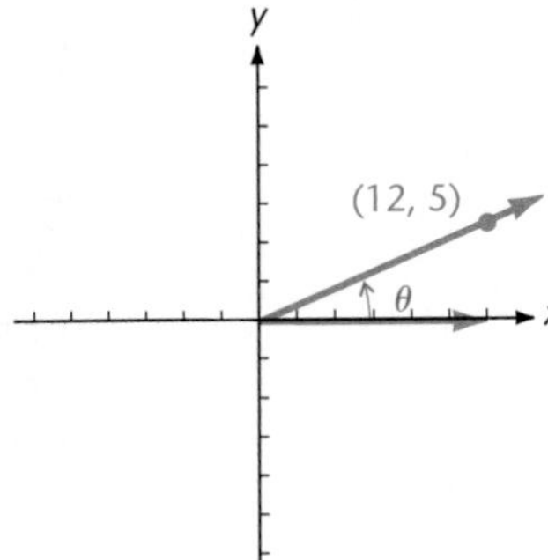

3. 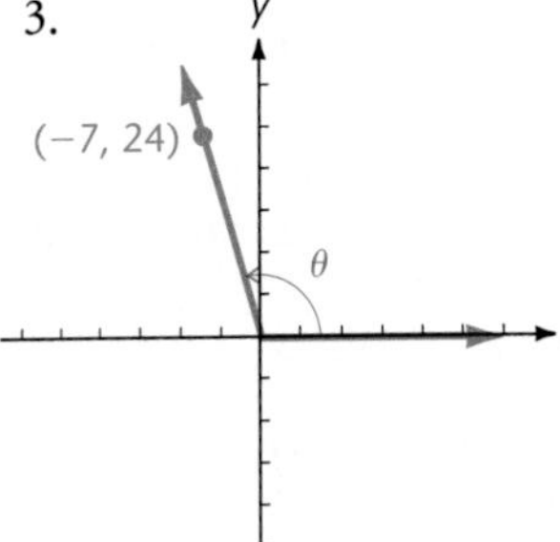

4. 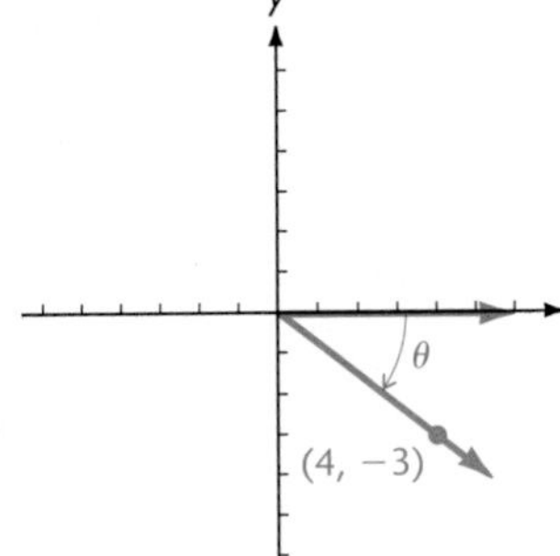

5. 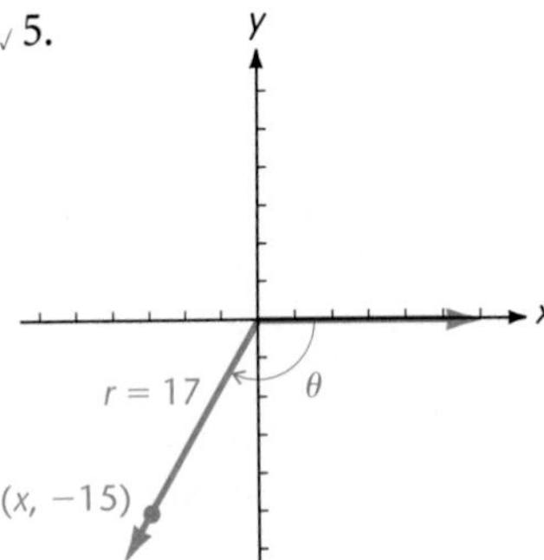

6. 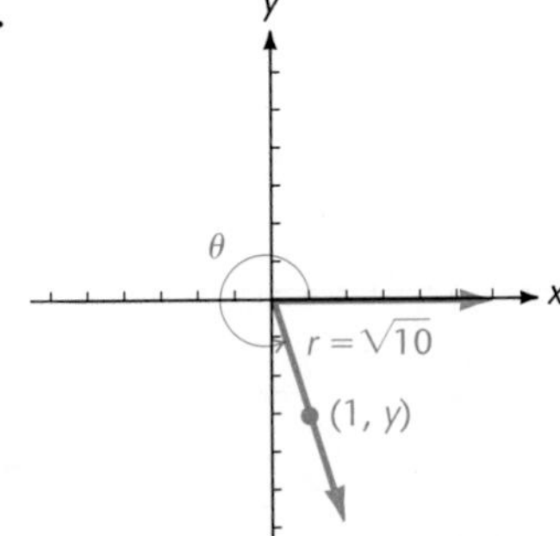

7. $(-\sqrt{2}, -7)$
8. $(\sqrt{10}, -2)$
9. $(x, 1),\ x > 0$
10. $(1, y),\ y > 0$
11. $(x, \sqrt{1 - x^2}),\ x > 0$
12. $(\sqrt{1 - y^2}, y),\ y > 0$

Use a calculator to determine the approximate values of the six trigonometric functions of an angle α in standard position if the terminal side of α passes through the given point. Round your answers to three decimal places.

13–16
13. $(-1.271, 3.122)$
14. $(-0.511, -0.698)$
15. $(10.211, -35.513)$
16. $(-44.821, 13.514)$

Determine the quadrant(s) in which the angle lies if the following conditions are satisfied. (See Example 2.)

17. $\sin \alpha > 0$ and $\sec \alpha < 0$
18. $\cos \beta < 0$ and $\tan \beta < 0$
19. $\cot \gamma > 0$ and $\sin \gamma < 0$
20. $\sin \theta < 0$ and $\tan \theta < 0$
21. $\tan \alpha < 0$ and $\csc \alpha > 0$
22. $\cot \theta > 0$ and $\sec \theta < 0$
23. $\sin \alpha < 0$ and $\csc \alpha < 0$
24. $\sec \alpha > 0$ and $\cos \alpha > 0$
25. $\tan \beta < 0$
26. $\sec \alpha < 0$

Determine the exact values of all the trigonometric functions of the angle that terminates in the given quadrant. (See Example 3.)

27. $\sin \alpha = \frac{3}{5}$, Q I
28. $\tan \beta = \frac{4}{3}$, Q I
29. $\cos A = -\frac{7}{25}$, Q II
30. $\cot B = -\frac{8}{15}$, Q II
31. $\tan \theta = \frac{1}{3}$, Q III
32. $\cos \alpha = -\frac{2}{5}$, Q III
33. $\tan \alpha = -\frac{2}{3}$, Q IV
34. $\csc \alpha = -3$, Q IV
35. $\cos \theta = -0.4$, Q III
36. $\cot \theta = -1.3$, Q IV
37. $\sin \theta = y$, Q I
38. $\cos \theta = x$, Q I
39. $\sec \theta = 1/x$, Q II
40. $\tan \theta = y$, Q IV

Determine the values of the trigonometric functions of each of the following. (See Example 4.)

41. $360°$
42. $-90°$
43. $-\dfrac{5\pi}{2}$
44. 7π

Determine the related angle of each of the given angles. (See Example 5.)

45. $147°$
46. $223°$
47. $\dfrac{9\pi}{5}$
48. $\dfrac{7\pi}{10}$
49. $-\dfrac{8\pi}{9}$
50. $-423°$

Write each of the following in terms of the same trigonometric function of a related angle. (See Example 7.)

51. $\sin \dfrac{16\pi}{9}$
52. $\sin \dfrac{13\pi}{8}$
53. $\cos(-73°)$
54. $\cos 140°$
55. $\tan \dfrac{7\pi}{9}$
56. $\tan\left(-\dfrac{13\pi}{5}\right)$
57. $\csc(-120°30')$
58. $\cot 83°20'$

Write the exact values of the trigonometric functions of each angle. (See Example 6.)

59. $150°$
60. $135°$
61. $\dfrac{4\pi}{3}$
62. $\dfrac{7\pi}{6}$

63. $-60°$ 64. $-135°$ 65. $-\frac{5\pi}{4}$ 66. $-\frac{\pi}{3}$

Find the exact value of each of the following expressions. The notation $\sin^2 \theta$ means $(\sin \theta)^2$, and $\cos^2 \theta$ means $(\cos \theta)^2$.

67. $\cos^2 \frac{\pi}{3} - \sin^2 \frac{\pi}{3}$

68. $3 \tan 180° - 5 \csc 270°$

69. $2 \sec 0 - 3 \csc \frac{\pi}{2}$

70. $\sin \frac{3\pi}{4} + \cos \frac{5\pi}{4}$

71. $\sec \frac{7\pi}{6} + 6 \cot \frac{4\pi}{3}$

72. $5 \cot 150° + 6 \tan 300°$

Determine whether each of the following is true or false.

73. $\sin (30° + 60°) = \sin 30° + \sin 60°$

74. $\cos \frac{\pi}{3} = \cos^2 \frac{\pi}{6} - \sin^2 \frac{\pi}{6}$

75. $\cos \frac{2\pi}{3} = \frac{1}{2} \cos \frac{4\pi}{3}$

76. $\sin 150° = \frac{1}{2} \sin 300°$

Critical Thinking: Exploration and Writing

77. Find two line segments in the accompanying figure that have their lengths equal to $\sin \theta$ and $\tan \theta$. Then use the figure to explain why $\sin \theta < \tan \theta$ for any first-quadrant angle θ.

Critical Thinking: Exploration and Writing

78. Find a line segment in the accompanying figure that has length equal to $\cos \theta$. Use the figure to write a description of the change in the value of $\cos \theta$ as θ increases steadily from 0° to 90°.

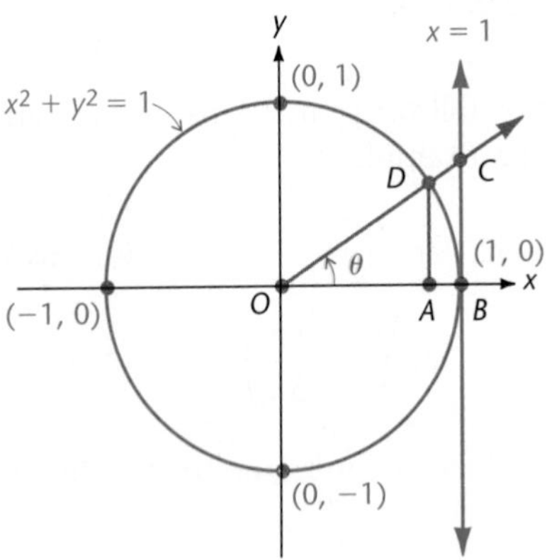

Figure for Exercises 77 and 78

◆ Solutions for Practice Problems

1. Since β is a third-quadrant angle, we write 1.5 as the quotient of -3 divided by -2:

$$\tan \beta = 1.5,$$
$$\frac{y}{x} = \frac{-3}{-2}.$$

So $y = -3$, $x = -2$, and $r = \sqrt{(-2)^2 + (-3)^2} = \sqrt{13}$. The values of the remaining five trigonometric functions of β are as follows.

$$\sin \beta = \frac{-3}{\sqrt{13}} = -\frac{3\sqrt{13}}{13} \qquad \sec \beta = \frac{\sqrt{13}}{-2} = -\frac{\sqrt{13}}{2}$$
$$\cos \beta = \frac{-2}{\sqrt{13}} = -\frac{2\sqrt{13}}{13} \qquad \csc \beta = \frac{\sqrt{13}}{-3} = -\frac{\sqrt{13}}{3}$$
$$\cot \beta = \frac{-2}{-3} = \frac{2}{3}$$

2. The related angle for $-210°$ is $30°$. We place a $30°$-$60°$ triangle as shown in Figure 7.21 and use this related triangle to locate the point $(-\sqrt{3}, 1)$ with $r = 2$ on the terminal side of $-210°$. Using the definition of each trigonometric function, we obtain the following values.

$$\sin(-210°) = \frac{1}{2} \qquad \cot(-210°) = -\sqrt{3}$$
$$\cos(-210°) = -\frac{\sqrt{3}}{2} \qquad \sec(-210°) = -\frac{2}{\sqrt{3}} = -\frac{2\sqrt{3}}{3}$$
$$\tan(-210°) = -\frac{1}{\sqrt{3}} = -\frac{\sqrt{3}}{3} \qquad \csc(-210°) = 2$$

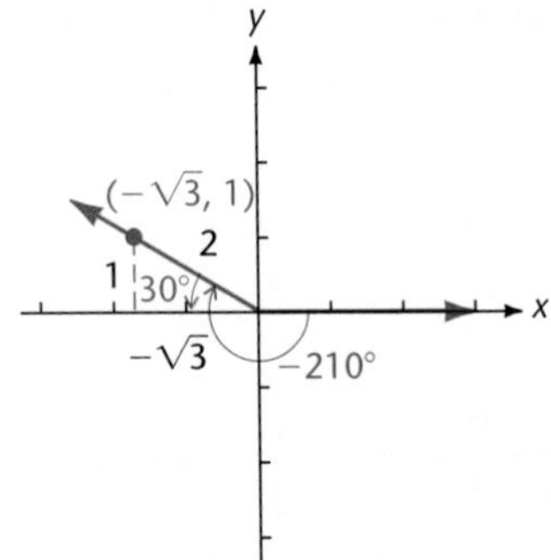

Figure 7.21 Related triangle for $-210°$

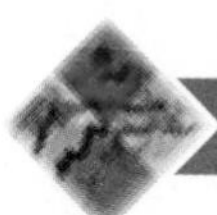

7.4 SOME FUNDAMENTAL PROPERTIES

In this section we obtain several useful results that are consequences of the definitions of the trigonometric functions. To derive the first of these results, we notice that the sine and cosecant functions are both defined in terms of y and r. When $y \neq 0$,

$$\sin\theta\csc\theta = \frac{y}{r}\cdot\frac{r}{y} = 1.$$

The equation $\sin\theta\csc\theta = 1$ is true for all values of θ for which $\sin\theta$ and $\csc\theta$ are defined. Equations that are always true whenever both sides are defined are called **identities.** We call the identity

$$\sin\theta\csc\theta = 1$$

a **reciprocal† identity,** and we say that $\sin\theta$ and $\csc\theta$ are reciprocals. Other forms of this relationship are

$$\sin\theta = \frac{1}{\csc\theta} \quad \text{and} \quad \csc\theta = \frac{1}{\sin\theta}.$$

Similarly, $\cos\theta$ and $\sec\theta$ are reciprocals, as are $\tan\theta$ and $\cot\theta$. We record these results, labeling them the reciprocal identities. It is understood that the equations hold only when each side is defined.

Reciprocal Identities

$\sin\theta\csc\theta = 1$	$\cos\theta\sec\theta = 1$	$\tan\theta\cot\theta = 1$
$\sin\theta = \dfrac{1}{\csc\theta}$	$\cos\theta = \dfrac{1}{\sec\theta}$	$\tan\theta = \dfrac{1}{\cot\theta}$
$\csc\theta = \dfrac{1}{\sin\theta}$	$\sec\theta = \dfrac{1}{\cos\theta}$	$\cot\theta = \dfrac{1}{\tan\theta}$

Two additional identities can be derived directly from the definitions of the trigonometric functions. Consider the quotients formed using $\sin\theta$ and $\cos\theta$.

$$\text{If } \cos\theta \neq 0,\ \frac{\sin\theta}{\cos\theta} = \frac{\frac{y}{r}}{\frac{x}{r}} = \frac{y}{x} = \tan\theta.$$

$$\text{If } \sin\theta \neq 0,\ \frac{\cos\theta}{\sin\theta} = \frac{\frac{x}{r}}{\frac{y}{r}} = \frac{x}{y} = \cot\theta.$$

†If $a \neq 0$, the reciprocal of a is $1/a$.

These relationships are true for all θ as long as all denominators are different from zero. We call these relationships the **quotient identities.**

Quotient Identities

$$\tan\theta = \frac{\sin\theta}{\cos\theta} \qquad \cot\theta = \frac{\cos\theta}{\sin\theta}$$

The next example illustrates how useful the reciprocal and quotient identities can be.

Example 1 • Using Identities to Find Function Values

Suppose $\sin\theta = -1/5$ and $\cos\theta = -2\sqrt{6}/5$. Use identities to determine the values of the remaining trigonometric functions of θ.

Solution

Using the quotient identities for $\tan\theta$ and $\cot\theta$, we have

$$\tan\theta = \frac{\sin\theta}{\cos\theta} = \frac{-\frac{1}{5}}{-\frac{2\sqrt{6}}{5}} = \frac{1}{2\sqrt{6}} = \frac{1\sqrt{6}}{2\sqrt{6}\sqrt{6}} = \frac{\sqrt{6}}{12},$$

$$\cot\theta = \frac{\cos\theta}{\sin\theta} = \frac{-\frac{2\sqrt{6}}{5}}{-\frac{1}{5}} = 2\sqrt{6}.$$

The reciprocal identities are useful for obtaining the values of $\sec\theta$ and $\csc\theta$.

$$\sec\theta = \frac{1}{\cos\theta} = \frac{5}{-2\sqrt{6}} = -\frac{5\sqrt{6}}{2\sqrt{6}\sqrt{6}} = -\frac{5\sqrt{6}}{12}$$

$$\csc\theta = \frac{1}{\sin\theta} = -5$$

□

◆ Practice Problem 1 ◆

Suppose $\sec\alpha = 2$ and $\sin\alpha = -\sqrt{3}/2$. Use identities to determine the values of the remaining trigonometric functions of α.

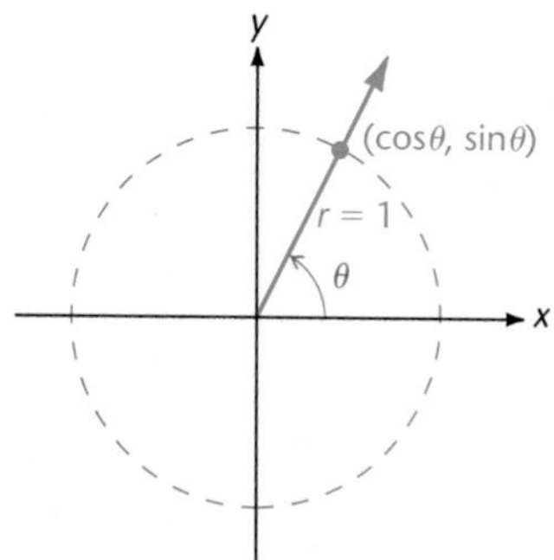

Figure 7.22 Cos θ and sin θ when $r = 1$

Next we turn our attention to the ranges of the sine and cosine functions. Let θ be an angle in standard position with (x, y) a point on the terminal side chosen at a distance one unit from the origin. Then, as indicated in Figure 7.22, $r = 1$, and

$$\sin\theta = \frac{y}{1} = y, \qquad \cos\theta = \frac{x}{1} = x.$$

Any point (x, y) on a circle of radius 1 has x- and y-coordinates that satisfy the following inequalities.

$$-1 \leq x \leq 1 \quad \text{and} \quad -1 \leq y \leq 1$$

Thus we have

$$-1 \leq \cos \theta \leq 1 \quad \text{and} \quad -1 \leq \sin \theta \leq 1.$$

We know that the values of the trigonometric functions are independent of the choice of the point on the terminal side of the angle. Hence these inequalities always hold and in fact describe the range of the cosine and sine functions.

We have seen that $\sec \theta$ and $\csc \theta$ are reciprocals of $\cos \theta$ and $\sin \theta$, respectively. It follows then that the ranges of the secant and cosecant functions can be described by the following inequalities.

$$\sec \theta \leq -1 \quad \text{or} \quad \sec \theta \geq 1$$
$$\csc \theta \leq -1 \quad \text{or} \quad \csc \theta \geq 1$$

To study the range of the tangent functions, we consider the angle in standard position, choosing the point (x, y) on the terminal side with $x = 1$, as shown in Figure 7.23a. Then

$$\tan \theta = \frac{y}{x} = \frac{y}{1} = y,$$

and we see that **the tangent function ranges over all real numbers.** In Figure 7.23b we choose (x, y) with $y = 1$. Then

$$\cot \theta = \frac{x}{y} = \frac{x}{1} = x,$$

and we see that the cotangent function also ranges over all real numbers.

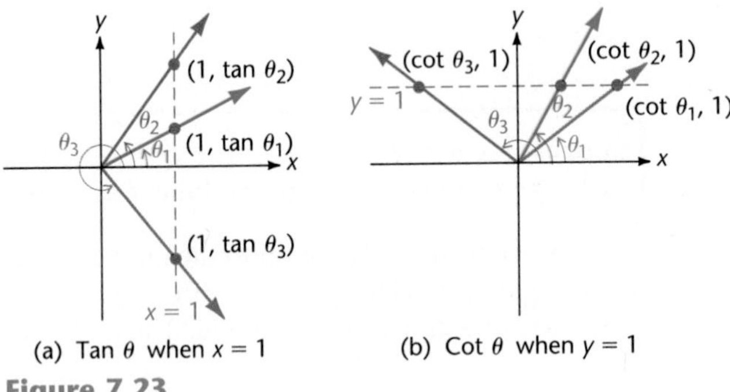

(a) Tan θ when $x = 1$ (b) Cot θ when $y = 1$

Figure 7.23

Example 2 • Using the Ranges of the Trigonometric Functions

Determine whether each of the following statements is possible or impossible.

a. $\sin \theta = 2$ b. $\cos \theta = 1$ and $\sec \theta = -1$.
c. $\sec \theta = -3994$ d. $\csc \theta = 0$

Solution

a. Impossible. The value of $\sin \theta$ cannot be larger than 1.
b. Impossible. Since $\cos \theta$ and $\sec \theta$ are reciprocals, their product must be 1, not -1.
c. Possible. The value -3994 lies in the range of the secant function.
d. Impossible. The value 0 lies outside the range of the cosecant function. □

EXERCISES 7.4

Use identities and a calculator to determine the indicated function value. Round each result to four decimal places. (See Example 1.)

1–16

1. $\sin t = 0.2193$, $\csc t = ?$
2. $\sec \theta = 4.2711$, $\cos \theta = ?$
3. $\tan u = 2.1370$, $\cot u = ?$
4. $\cot \alpha = 1.5131$, $\tan \alpha = ?$
5. $\cos t = -0.9132$, $\sec t = ?$
6. $\sin \beta = -0.3196$, $\csc \beta = ?$
7. $\csc v = -1.4576$, $\sin v = ?$
8. $\tan \theta = -0.8352$, $\cot \theta = ?$
9. $\sin \beta = 0.4722$, $\cos \beta = 0.8815$, $\tan \beta = ?$
10. $\sin t = -0.5798$, $\cos t = -0.8148$, $\tan t = ?$
11. $\sin \theta = -0.7813$, $\cos \theta = 0.6242$, $\cot \theta = ?$
12. $\sin \alpha = 0.1391$, $\cos \alpha = -0.9903$, $\cot \alpha = ?$
13. $\sin A = 0.1521$, $\tan A = 0.1539$, $\cos A = ?$
14. $\cos B = -0.6731$, $\tan B = 1.0987$, $\sin B = ?$
15. $\sin D = 0.1216$, $\cot D = -8.1627$, $\cos D = ?$
16. $\cos C = -0.4561$, $\cot C = 0.5125$, $\sin C = ?$

Use identities to determine the exact values of the remaining trigonometric functions of the angle satisfying the given conditions. (See Example 1.)

17. $\sin \theta = -\frac{3}{5}$, $\cos \theta = \frac{4}{5}$
18. $\sin s = \frac{12}{13}$, $\cos s = -\frac{5}{13}$
19. $\tan \theta = \frac{12}{5}$, $\sec \theta = \frac{13}{5}$
20. $\cot t = \frac{8}{15}$, $\sin t = \frac{15}{17}$
21. $\csc \alpha = \frac{17}{8}$, $\tan \alpha = -\frac{8}{15}$
22. $\csc \theta = -\frac{5}{4}$, $\tan \theta = -\frac{4}{3}$
23. $\sin v = -\frac{7}{25}$, $\sec v = -\frac{25}{24}$
24. $\tan \beta = \frac{24}{7}$, $\sec \beta = -\frac{25}{7}$
25. $\cos t = -\frac{1}{3}$, $\tan t = -2\sqrt{2}$
26. $\sin A = \sqrt{21}/5$, $\tan A = -\sqrt{21}/2$
27. $\csc s = -\sqrt{2}$, $\tan s = -1$

28. $\cot B = -\sqrt{3}$, $\sin B = \frac{1}{2}$
29. $\sin \alpha = -\sqrt{7}/3$, $\cos \alpha = -\sqrt{2}/3$
30. $\sin u = -\sqrt{11}/4$, $\cos u = -\sqrt{5}/4$
31. $\cot \theta = -\sqrt{10}/2$, $\csc \theta = \sqrt{14}/2$
32. $\sec \theta = \frac{3}{2}$, $\sin \theta = -\sqrt{5}/3$

Determine whether each of the following statements is possible or impossible. (See Example 2.)

33. $\cos \alpha = 1.1$
34. $\sin t = 2.3$
35. $\tan s = 0$
36. $\sec \beta = 0$
37. $\csc \gamma = -1.1$
38. $\cot u = -1$
39. $\sec \theta = -0.99$
40. $\cos \alpha = -0.33333$
41. $\sin u = \frac{5}{4}$
42. $\cos \alpha = -\frac{10}{3}$
43. $\sec t = -\frac{1}{2}$
44. $\csc v = -\frac{3}{4}$
45. $\sin \alpha > 0$ and $\csc \alpha < 0$
46. $\cos s < 0$ and $\sec s > 0$
47. $\tan \theta < 0$ and $\cot \theta > 0$
48. $\csc \alpha > 0$ and $\sin \alpha < 0$
49. $\cos t = 3$ and $\sec t = \frac{1}{3}$
50. $\tan \alpha = 2$ and $\cot \alpha = -2$
51. $\sin \beta = \frac{2}{3}$ and $\csc \beta = -\frac{3}{2}$
52. $\cos v = -\frac{1}{4}$ and $\sec v = -4$

◆ Solution for Practice Problem

1. $\cos \alpha = \dfrac{1}{\sec \alpha} = \dfrac{1}{2}$

$$\tan \alpha = \frac{\sin \alpha}{\cos \alpha} = \frac{-\sqrt{3}/2}{1/2} = -\sqrt{3}$$

$$\cot \alpha = \frac{1}{\tan \alpha} = \frac{1}{-\sqrt{3}} = -\frac{\sqrt{3}}{3}$$

$$\csc \alpha = \frac{1}{\sin \alpha} = \frac{1}{-\sqrt{3}/2} = -\frac{2}{\sqrt{3}} = \frac{-2\sqrt{3}}{3}$$

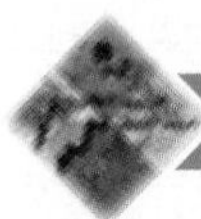

7.5 VALUES OF TRIGONOMETRIC FUNCTIONS

Values of trigonometric functions for angles other than the special ones considered in Section 7.3 must be found by using a table or a calculator. A calculator is easier to use if that option is available.

Scientific and graphing calculators have keys labeled [sin], [cos], and [tan] that will yield values of the corresponding trigonometric functions of an angle. The sequence of keystrokes varies with the type of calculator, and the correct procedure for a given calculator can be found in its accompanying instruction booklet. The following example illustrates typical procedures for many calculators in common use.

Example 1 • Finding Function Values with a Calculator

Use a calculator to find the value of each of the following and round the results to four digits.†

a. $\cos 14°23'$ b. $\csc(-214.4°)$ c. $\sin 39.7$

Solution

a. To find $\cos 14°23'$, the calculator must be set in degree mode. Some calculators require that the angle be entered first, while others require that the [cos] key be pressed first. Some calculators allow an angle to be entered in degrees and minutes, but most require that $14°23'$ be changed to decimal degrees.

$$14°23' = (14 + \tfrac{23}{60})° = 14.38333333$$

Following the appropriate keystroke procedure with the angle $14°23'$ and the [cos] key results in a displayed value such as 0.9686554611. The number of digits displayed will of course depend on the calculator in use. Rounding gives

$$\cos 14°23' = 0.9687.$$

b. To find $\csc(-214.4°)$, we use the fact that $\csc\theta = 1/\sin\theta$. We set the calculator for degrees and follow the appropriate keystroke procedure for the angle $-214.4°$ and the [sin] key to obtain

$$\sin(-214.4°) = 0.5649670034.$$

Pressing the [1/x] or other key to obtain a reciprocal then yields

$$\csc(-214.4°) = \frac{1}{\sin(-214.4°)} = 1.770,$$

where the last number has been rounded to four digits.

c. We recall from our work in previous sections of this chapter that it is customary to omit the word *radian* or the abbreviation *rad* when working with radian measure. Since no unit of measure is indicated in $\sin 39.7$, it is understood that the angle measure is 39.7 radians. Thus we set the calculator in radian mode and follow the appropriate sequence of keystrokes with 39.7 and the [sin] key to obtain the value 0.9089274645. This rounds to give

$$\sin 39.7 = 0.9089.$$

□

†By a four-digit number, we mean that there are four digits when the number is written in scientific notation. That is, zeros used only to place the decimal are not counted.

Values of trigonometric functions may also be found by using a table. Two such tables are found in the Appendix to this book. Table I gives values of all six trigonometric functions of θ at increments of 0.0029, from 0 to 1.5708. The increment 0.0029 in radian measure corresponds to 10 minutes in degree measure, and the interval from 0 to 1.5708 radians corresponds to the interval from 0° to 90°. The use of trigonometric function tables is described in Section A.2 of the Appendix.

In using a calculator to find a value of the angle from a given function value, the procedure varies from one type of calculator to another. With most calculators, it is necessary to have a value of one of the functions sine, cosine, or tangent. If a function value different from these is given, one of the reciprocal identities can be used to obtain the value of either sine, cosine, or tangent.

We consider first the case in which a positive number is given as one of the values $\sin \theta$, $\cos \theta$, or $\tan \theta$, and we want to find a first-quadrant angle θ in degrees or radians. The calculator must be set for the desired measure. Depending on the type of calculator, the correct keystroke procedure will involve the given positive function value and one or more of the following keys.

[inv], [arc], [2nd], [sin], [cos], [tan], [$\sin^{-1}$], [$\cos^{-1}$], [$\tan^{-1}$].

The instruction booklet for the calculator describes the sequence of keystrokes that will cause the value of the angle to be displayed by the calculator.

If a positive value of $\csc \theta$, $\sec \theta$, or $\tan \theta$ is given, most calculators require that the reciprocal key ([$1/x$] or its equivalent) be used to obtain a corresponding value of $\sin \theta$, $\cos \theta$, or $\tan \theta$. The reciprocal value obtained in this manner is used at the appropriate point in the routine discussed in the preceding paragraph.

When a negative number is given as a value of a trigonometric function of θ, we use the related angle θ' to find θ. The reason for this is explained after the next example.

Example 2 • Finding the Degree Measure of an Angle

If $180° \leq \theta \leq 270°$ and $\cos \theta = -0.9641$, find θ to the nearest tenth of a degree.

Solution

We first find the related angle θ' in degrees.

$$\begin{aligned} \cos \theta' &= 0.9641 \\ \theta' &= 15.4° \end{aligned}$$

Since $180° \leq \theta \leq 270°$, we have

$$\theta = 180° + \theta' = 195.4°.$$

This is the desired third-quadrant angle θ that has $\cos \theta = -0.9641$.

Consider now the result if the negative number -0.9641 is used with the same keystroke sequence that produces a first-quadrant angle for a positive cosine value. When the calculator is set in degree mode and -0.9641 is used as a value of $\cos \theta$, the usual procedure yields the displayed value 164.6009916°. This is not the desired third-quadrant angle but is another angle that has −0.9641 for its cosine. We note that both 195.4° and 164.6° have $\theta' = 15.4°$ as their related angle. □

The last part of Example 2 illustrates that a calculator will yield an angle that has a given negative function value, although the calculator's answer may not be the one desired. The reason behind the calculator's answer will become clear in the last section of Chapter 8. In the meantime, the simplest procedure that we can follow is to use the related angle.

Example 3 • Finding the Radian Measure of an Angle

If $\tan\theta = -1.419$, $0 \le \theta < 2\pi$, and θ lies in quadrant II, find the value of θ in radians, rounded to four decimal places.

Solution

This time we find the related angle θ' in radians.

$$\tan\theta' = 1.419$$
$$\theta' = 0.9569$$

Since $\pi/2 < \theta < \pi$, we have

$$\theta = \pi - \theta' = 2.1847.$$

□

Figure 7.24 Reflection and refraction

Example 4 • Snell's Law

Consider a beam of yellow sodium light that travels in a vacuum and makes an angle of incidence θ_1 with the normal to the surface of substance a, as shown in Figure 7.24. The beam is in part reflected and in part refracted, and the angle θ_a in Figure 7.24 is called the **angle of refraction** of the substance.

One form of **Snell's law** in physics states that

$$\frac{\sin\theta_1}{\sin\theta_a} = n_a,$$

where n_a is a constant, called the **index of refraction** of the substance a. Table 7.1 gives the index of refraction for several substances. We shall find the angle of refraction θ_a of water at 20°C when the angle of incidence θ_1 is 45°.

Table 7.1 Index of Refraction for Yellow Sodium Light

Substance	Index of Refraction
Ice	1.309
Water at 20°C	1.333
Ethyl alcohol	1.360
Rock salt	1.544
Diamond	2.417

Solution

We have $\theta_1 = 45°$, and Table 7.1 gives $n_a = 1.333$ for water at 20° C. According to Snell's law,

$$\frac{\sin 45°}{\sin\theta_a} = 1.333 \quad \text{and} \quad \sin\theta_a = \frac{\sin 45°}{1.333}.$$

Using a calculator or Table I or II in the Appendix, we get

$$\sin\theta_a = 0.5305 \quad \text{and} \quad \theta_a = 32°.$$

Although Snell's law applies to light traveling in a vacuum, ordinary atmospheric pressure does not have enough of an effect to change the result that $\theta_a = 32°$ for water at 20°C when $\theta_1 = 45°$. □

EXERCISES 7.5

Find the values of the following trigonometric functions by using either a calculator or Tables I and II, as instructed by the teacher. In either case, give answers correct to four digits. (See Example 1 or Section A.2 in the Appendix.)

1. sin 31.3°
2. sin 42.2°
3. cos 58.7°
4. cos 76.5°
5. tan 63.2°
6. tan 81.8°
7. cot 115.6°
8. cot 295.6°
9. cos 32°20′
10. cos 57°40′
11. tan 108°40′
12. tan 260°50′
13. sec (−112°50′)
14. sec (−278°30′)
15. sin (−401°30′)
16. sin (−526°20′)
17. sin 0.3054
18. csc 0.4422
19. sec 5.6258
20. cos 3.7816
21. tan 2.8158
22. cot 3.2667
23. cos 5.1720
24. sin 5.4745

In Exercises 25–36, θ has the given function value and lies in the given quadrant, with $0° \le \theta < 360°$. According to the teacher's instructions, use either a calculator or Table II to find θ in decimal degrees to the nearest tenth of a degree. (See Example 2.)

25. $\sin\theta = 0.5299$, Q I
26. $\sin\theta = 0.6587$, Q I
27. $\cot\theta = 1.6842$, Q I
28. $\cot\theta = 1.2482$, Q I
29. $\cos\theta = 0.5793$, Q I
30. $\cos\theta = 0.6756$, Q I
31. $\tan\theta = -0.6924$, Q II
32. $\tan\theta = -0.4813$, Q II
33. $\cos\theta = -0.8171$, Q III
34. $\cos\theta = -0.8616$, Q III
35. $\sin\theta = -0.4478$, Q IV
36. $\cot\theta = -8.028$, Q IV

In Exercises 37–48, θ has the given function value and lies in the given quadrant, with $0 \le \theta < 2\pi$. According to the teacher's instructions, use either a calculator or Table I to find θ in radians, rounded to four decimal places. (See Example 3.)

37. $\tan\theta = 1.767$, Q I
38. $\tan\theta = 0.3121$, Q I
39. $\sec\theta = 1.231$, Q I
40. $\sin\theta = 0.9051$, Q I
41. $\cos\theta = -0.8732$, Q III
42. $\tan\theta = 1.756$, Q III
43. $\sin\theta = -0.7214$, Q IV
44. $\cot\theta = -0.9827$, Q IV
45. $\cos\theta = -0.5901$, Q II
46. $\cos\theta = -0.5446$, Q II
47. $\csc\theta = -1.244$, Q III
48. $\sec\theta = -1.448$, Q II

In Exercises 49–60, θ has the given function value and lies in the given quadrant, with $0° \leq \theta < 360°$. According to the teacher's instructions, use either a calculator or Table I to find θ in degrees and minutes, correct to the nearest 10 minutes.

49. $\cos\theta = 0.9013$, Q I
50. $\cos\theta = 0.6450$, Q I
51. $\sin\theta = 0.9250$, Q I
52. $\sin\theta = 0.7698$, Q I
53. $\cot\theta = 1.473$, Q III
54. $\cot\theta = 2.066$, Q III
55. $\sec\theta = -1.111$, Q II
56. $\sec\theta = -1.023$, Q II
57. $\csc\theta = -1.485$, Q III
58. $\csc\theta = -1.951$, Q III
59. $\tan\theta = -1.626$ Q IV
60. $\tan\theta = -2.590$, Q IV

(See Example 4.)

Angle of Refraction for Rock Salt

61. Assume that a beam of yellow sodium light travels in a vacuum and makes an angle of incidence of 30° with the normal to a flat surface of rock salt. Use either a calculator or Table I or II to find the angle of refraction to the nearest degree.

Angle of Refraction for a Diamond

62. Assume that a beam of yellow sodium light travels in a vacuum and makes an angle of incidence of 60° with the normal to a flat diamond surface. Use either a calculator or Table I or II to find the angle of refraction to the nearest degree.

63–64

Critical Thinking: Exploration and Writing

63. Use a calculator to find the values of $\sin\theta$ at intervals of 10 grads from 0 to 100 grads. Then write a description of the change in the value of $\sin\theta$ as θ increases steadily from 0 grads to 100 grads.

Critical Thinking: Exploration

64. Let x be the measure of an angle in radians and assume that the intermediate value theorem for polynomials from Section 5.2 also holds for the function $f(x) = \cos x - \tan x$ on the interval $[0,1]$. Use a calculator with the methods of Section 5.5 to find a solution for the equation $\cos x = \tan x$ that lies in the interval $[0,1]$ and is correct to the nearest tenth.

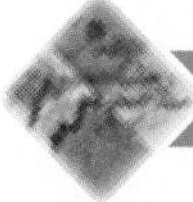

7.6 RIGHT TRIANGLES

Numerous practical problems that involve right triangles can be solved by using the trigonometric functions. One problem of this type received much publicity in 1988 when a surveyor in Louisville, Kentucky, used the tangent function to prove that the Louisville Falls Fountain on the Ohio River is the world's largest floating

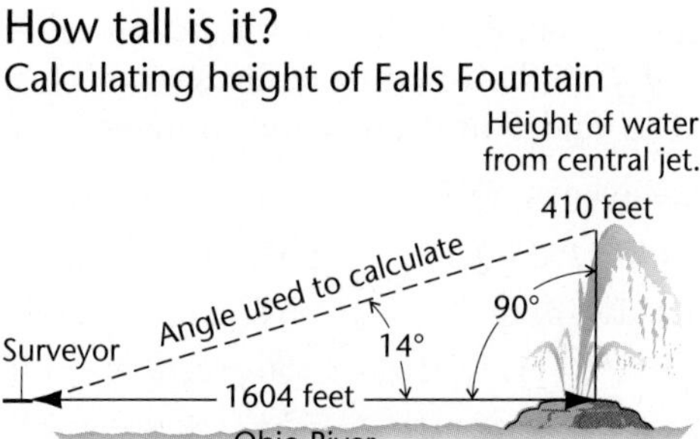

Figure 7.25 Louisville Falls Fountain. Using two angles and one side of a triangle, surveyors used trigonometry to determine the height of the Falls Fountain. Numbers are rounded off. Adapted from a chart by Steve Durbin.

fountain. A diagram of the procedure employed is shown in Figure 7.25, and the calculation of the height of the fountain is given in Example 3 of this section.

We adopt conventional notation by labeling the right angle in a right triangle C and the two acute angles A and B. The hypotenuse is always labeled c; side a is opposite angle A; and side b is opposite angle B. Thus side b is adjacent to angle A, and side a is adjacent to angle B. The six trigonometric functions of an acute angle in a right triangle can be restated in terms of the sides of the triangle. We first place the triangle ABC in Figure 7.26a in a coordinate system in Figure 7.26b so that angle A is in standard position.

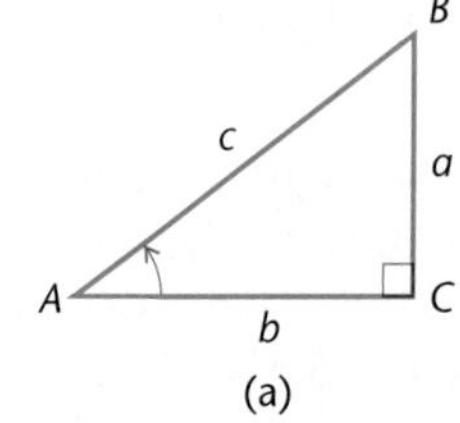

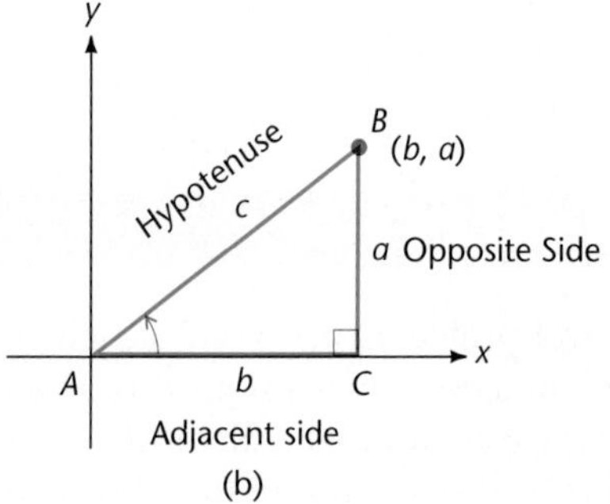

Figure 7.26 Acute angle A in standard position

The point with coordinates $x = b$, $y = a$ is located on the terminal side of A and at the vertex of B. The distance from the origin to this point is $r = c$. From the definition of the trigonometric functions of angle A, it follows that

$$\sin A = \frac{a}{c} = \frac{\text{opposite side}}{\text{hypotenuse}}, \qquad \cot A = \frac{b}{a} = \frac{\text{adjacent side}}{\text{opposite side}},$$
$$\cos A = \frac{b}{c} = \frac{\text{adjacent side}}{\text{hypotenuse}}, \qquad \sec A = \frac{c}{b} = \frac{\text{hypotenuse}}{\text{adjacent side}},$$
$$\tan A = \frac{a}{b} = \frac{\text{opposite side}}{\text{adjacent side}}, \qquad \csc A = \frac{c}{a} = \frac{\text{hypotenuse}}{\text{opposite side}}.$$

Although we have formulated the definitions for angle A, the same relationships hold for angle B:

$$\sin B = \frac{b}{c} = \frac{\text{opp}}{\text{hyp}}, \quad \cos B = \frac{a}{c} = \frac{\text{adj}}{\text{hyp}}, \quad \text{and so on.}$$

In connection with these relationships, we note that A and B are complementary angles, and we can write

$$\sin A = \frac{a}{c} = \cos(90° - A), \qquad \cos A = \frac{b}{c} = \sin(90° - A),$$
$$\tan A = \frac{a}{b} = \cot(90° - A), \qquad \cot A = \frac{b}{a} = \tan(90° - A),$$
$$\sec A = \frac{c}{b} = \csc(90° - A), \qquad \csc A = \frac{c}{a} = \sec(90° - A),$$

Because of these relations, we say that sine and cosine are **complementary functions,** or **cofunctions.** Similarly, tangent and cotangent are cofunctions, and secant and cosecant are cofunctions. In Chapter 8 we shall see that any trigonometric function of an angle equals the cofunction of the complementary angle and that this is true for all angles (positive, negative, or zero) as long as the functions involved are defined. This is the origin of the cofunctions' names. For example, *cosine* is a shortened form of *complementary sine.*

To **solve** a triangle means to find the value of all six parts of the triangle: the three sides and the three angles. With the convention that $C = 90°$ in effect, all parts of a triangle are determined when one side and one other quantity are specified.

In a calculation involving approximate numbers, the digits known to be correct are called **significant digits.** The **number of significant digits** in a number is the total when the digits are counted from left to right, starting with the first nonzero digit and ending with the last digit in the number. The placing of the decimal in a number has nothing to do with the number of significant digits, but there may be confusion when zeros are present.

Zeros between nonzero digits are always significant, and **zeros used only to place the decimal to the left are not significant.** Thus each of the numbers 405.8 and 0.004295 has four significant digits.

Zeros at the right end of a number are sometimes significant and sometimes not. **Zeros that are at the right end of a number and to the right of the decimal are significant.** For example, writing a measurement as 43.60 meters indicates that the measurement is accurate to the nearest hundredth of a meter and that the true value lies between 43.595 meters and 43.605 meters.

In a measurement such as 6400 kilometers, the zeros may or may not be significant. In the absence of additional information or some prior agreement, this could represent a measurement to the nearest kilometer, to the nearest 10 kilometers, or to the nearest 100 kilometers. Such uncertainty can be avoided by writing the number in scientific notation.

6.4×10^3 kilometers indicates accuracy to the nearest 100 kilometers.
6.40×10^3 kilometers indicates accuracy to the nearest 10 kilometers.
6.400×10^3 kilometers indicates accuracy to the nearest kilometer.

Rather than use scientific notation, we shall adopt the convention in this text that **all zeros in whole numbers are significant unless it is explicitly stated otherwise.**

In solving triangles, we need a guide as to how the accuracy in measuring sides corresponds to the accuracy in measuring angles. Such a guide is given in the accompanying table.

Number of Significant Digits in Side Measures	Angle Accuracy to the Nearest
2	1°
3	10′, or 0.1°
4	1′, or 0.01°

We follow this guide in the examples as well as in the exercises throughout this chapter.

Example 1 • Solving a Right Triangle

Solve the right triangle that has $c = 33$, $B = 22°$.

Solution

It may be helpful to sketch the triangle, as shown in Figure 7.27.

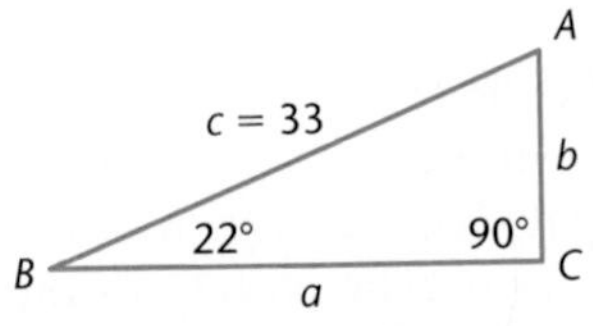

Figure 7.27 Right triangle for Example 1

As a first step, we find angle A to be 68°. To find side b, we may use either equation

$$\sin B = \frac{b}{c} \qquad \text{or} \qquad \csc B = \frac{c}{b}$$

since both B and c are known. Solving the first equation for b, we get

$$\begin{aligned} b &= c \sin B \\ &= 33 \sin 22^\circ \\ &= 33(0.3746) \\ \mathbf{b} &= \mathbf{12.} \end{aligned}$$

Solving now for a, we use the equation

$$\cos B = \frac{a}{c}.$$

Multiplying by c,

$$\begin{aligned} a &= c \cos B \\ &= 33 \cos 22^\circ \\ &= 33(0.9272) \\ \mathbf{a} &= \mathbf{31.} \end{aligned}$$

All parts of the triangle are now known.

$\mathbf{a = 31}$	$\mathbf{A = 68^\circ}$
$\mathbf{b = 12}$	$B = 22^\circ$
$c = 33$	$C = 90^\circ$

□

In most cases there is more than one correct procedure for solving a triangle. Occasionally two correct procedures will yield results that differ by 1 or 2 in the rightmost significant digit. This is nothing to worry about, however. It merely reflects the fact that we are working with approximate numbers.

◆ Practice Problem 1 ◆

Solve the right triangle that has $a = 358$, $b = 226$.

Example 2 • Computing the Length of a Guy Wire

A guy wire on a vertical utility pole is anchored at a point on level ground 18 feet from the base of the pole. If the wire makes an angle of 63° with the ground, how long is the wire?

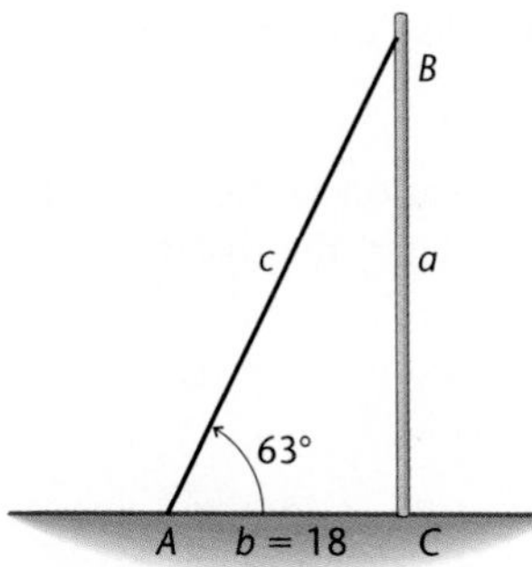

Figure 7.28 The right triangle formed using the guy wire and the pole

Solution

Let c be the length of the wire, as shown in Figure 7.28. From the figure we see that

$$\sec A = \frac{c}{b}$$

$$c = b \sec A = 18 \sec 63^\circ = 18(2.203) = 40.$$

The guy wire is 40 feet in length. □

Louisville Falls Fountain

Whenever an observer looks at an object, an acute angle is formed in a vertical plane between the horizontal direction and the line of sight. This acute angle is called an **angle of elevation** if the line of sight is above the horizontal, and an **angle of depression** if the line of sight is below the horizontal. This is illustrated in Figure 7.29.

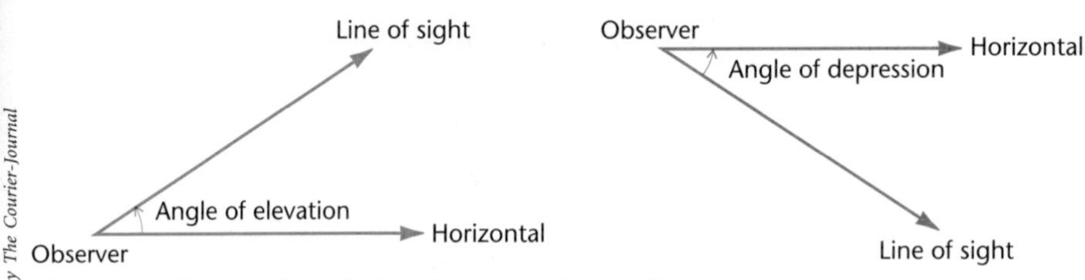

Figure 7.29 Angles of elevation and depression

Example 3 • Computing the height of the Louisville Falls Fountain

In this example we relate how the height h of the Louisville Falls Fountain was calculated.

Solution

Roger Basham, a Louisville Water Company surveyor, obtained a measurement of the distance along a horizontal line between the fountain and his instrument on River Road. This distance was approximately 1604 feet, as shown in Figure 7.30. Since the central stream of water in the fountain was vertical, it formed a 90° angle with the horizontal line.

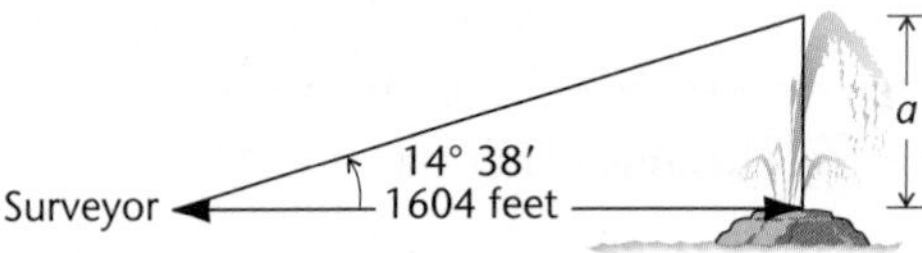

Figure 7.30 The angle used to measure the Louisville Falls Fountain

With his instrument, Basham measured the angle of elevation of the top of the fountain stream as 14°38′ (see Figure 7.30). He then calculated the distance a in Figure 7.30 as follows.

$$\tan 14°38' = \frac{a}{1604}$$

$$\begin{aligned} a &= 1604 \tan 14°38' \\ &= 1604\ (0.2611) \\ &= 418.8 \end{aligned}$$

Since the opening of the fountain's central jet was 8.8 feet above the horizontal line, he subtracted to obtain

$$h = a - 8.8 = 410.$$

Thus the height of the fountain was found to be 410 feet. □

In surveying and navigation, two methods are commonly used to describe the direction, or **bearing,** of a line. The method ordinarily used in surveying is to state the acute angle that measures the variation from one of the directions north or south and to indicate whether this variation is to the east or to the west. For example,

N 40° E indicates a bearing 40° to the east of north;

N 70° W indicates a bearing 70° to the west of north;

S 60° E indicates a bearing 60° to the east of south;

S 30° W indicates a bearing 30° to the west of south.

All these directions are pictured in Figure 7.31.

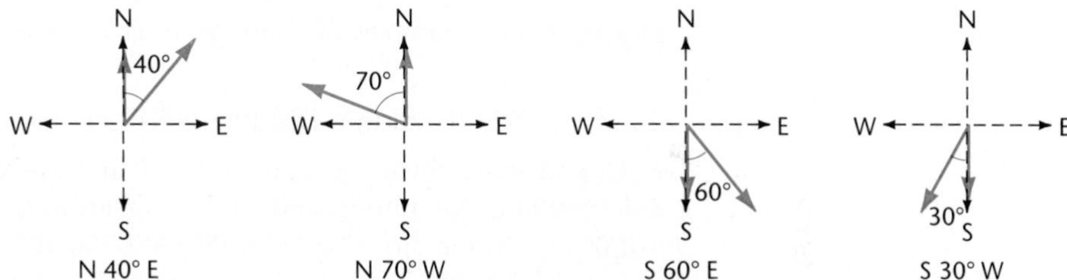

Figure 7.31 Examples of bearings

In all cases, either N or S comes first, and this tells the direction from which the acute angle is measured. Either E or W comes last, and this tells on which side of the north-south line the acute angle lies.

Example 4 • Finding a Bearing

The distance from Vicksburg to Laurel is 114 miles, on a bearing S 60.0° E. From Laurel, the distance to Meridian is 53.0 miles, on a bearing N 30.0° E. Find the bearing from Vicksburg to Meridian.

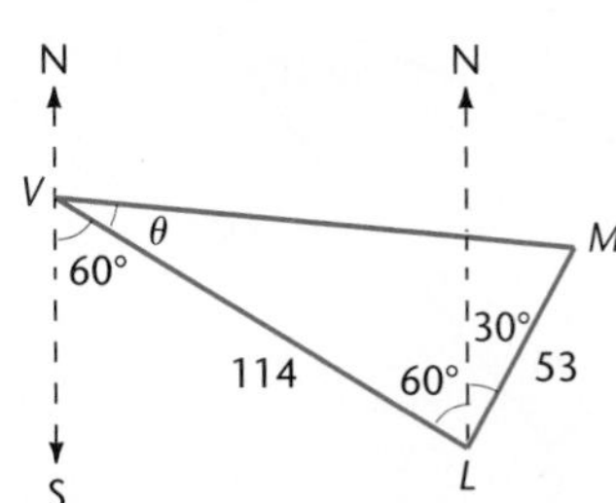

Figure 7.32 The triangle determined by Vicksburg, Laurel, and Meridian

Solution

In Figure 7.32, the cities are represented by the first letters in their names. As shown in the figure, the bearing from Laurel to Vicksburg is N 60.0° W. Thus triangle VLM is a right triangle with the 90° angle at L. To determine the bearing from Vicksburg to Meridian, we need to find the angle θ with vertex at V. We have

$$\tan \theta = \frac{53}{114} = 0.4649,$$

and therefore

$$\theta = 24.9°,$$

to the nearest tenth of a degree. Since $60° + 24.9° = 84.9°$, the direction from Vicksburg to Meridian is S 84.9° E. □

The second method referred to earlier is used to give bearings in air navigation. With this method, a bearing is given by stating the angle measured positive in a clockwise direction from due north. The bearings shown earlier in Figure 7.31 are restated in Figure 7.33 using the air navigation method. (Note that an angle representing a bearing is drawn differently than an angle in standard position.)

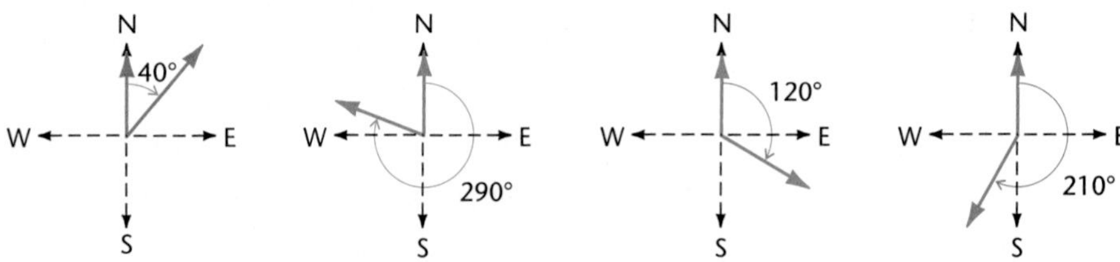

Figure 7.33 Examples of bearings in air navigation

Example 5 • Finding a Distance Between Towns

P, *Q*, and *R* are towns. A plane leaves *P* and flies in the direction 220°0′ at 215 miles per hour for 4 hours and lands at *Q*. From *Q*, the plane flies in the direction 310°0′ to *R*. If the direction from *P* to *R* is 250°10′, find the distance from *Q* to *R*.

Solution

As shown in Figure 7.34a, angle *QPS* is $220° - 180° = 40°$. Using the fact that

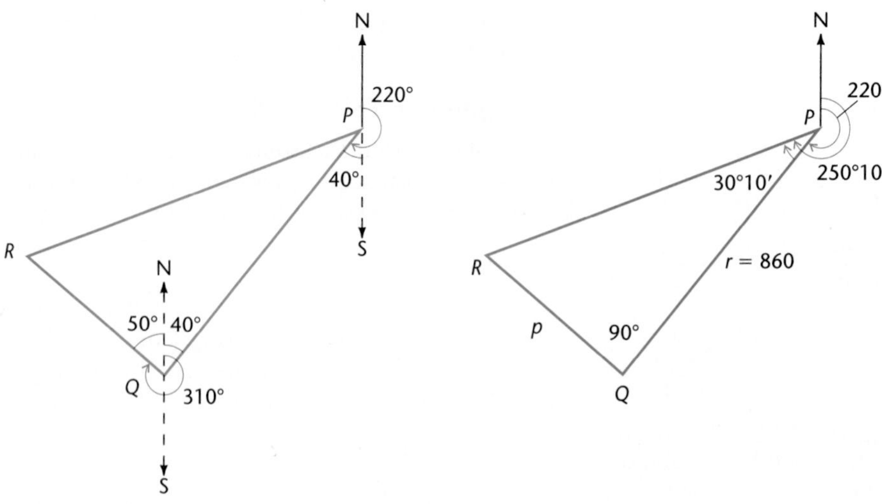

Figure 7.34 The triangle determined by towns *P*, *Q*, and *R*

alternate interior angles are equal, we see that the angle PQN is also 40°. Since angle RQN is $360° - 310° = 50°$, we have a 90° angle at Q, and triangle PQR is a right triangle.

To fly from P to Q, it took 4 hours at 215 miles per hour, so the distance from P to Q is 860 miles, as shown in Figure 7.34b. From this figure, we see that angle QPR is $250°10' - 220° = 30°10'$.

To find the distance p from Q to R, we write

$$\tan 30°10' = \frac{p}{860}$$

$$p = 860 \tan 30°10' = 860(0.5812) = 500.$$

The distance from Q to R is 500 miles. □

EXERCISES 7.6

Solve the following right triangles to the degree of accuracy consistent with the given information. (See Example 1.)

1.

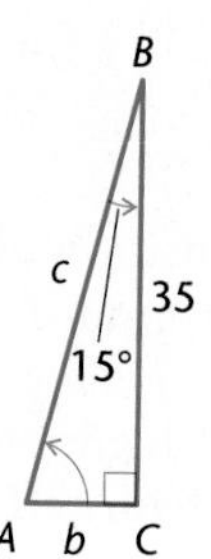

2.

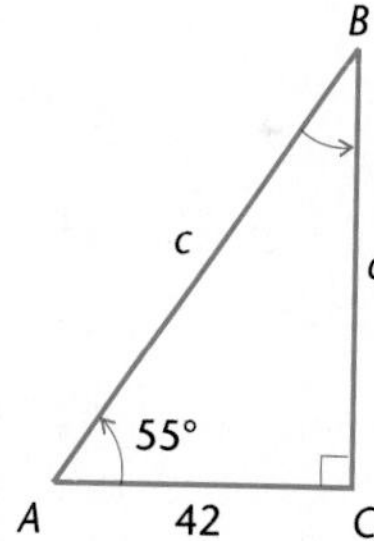

3.

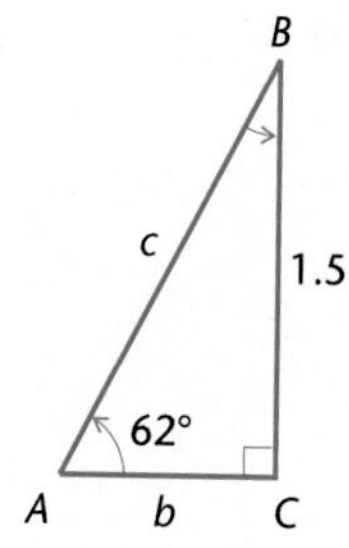

4.

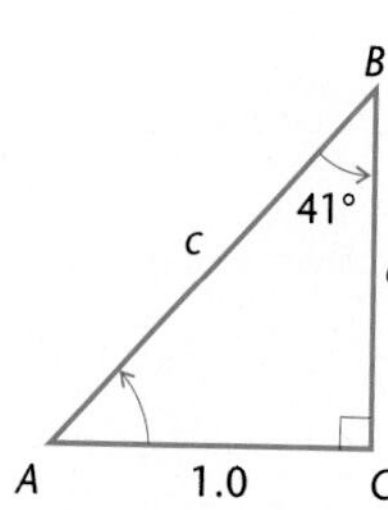

5.

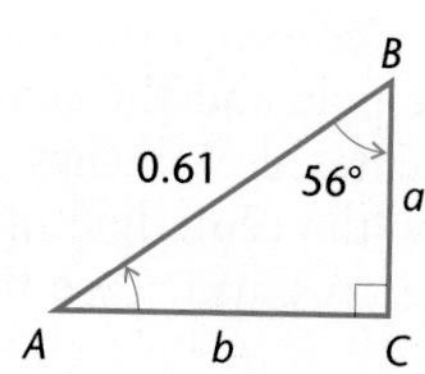

6.

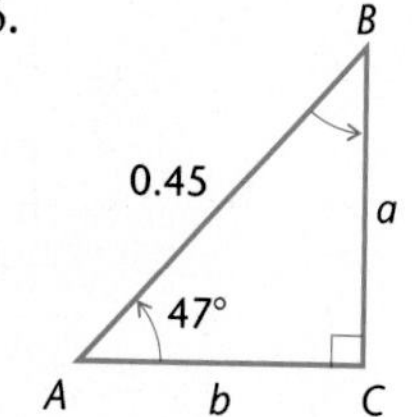

7.

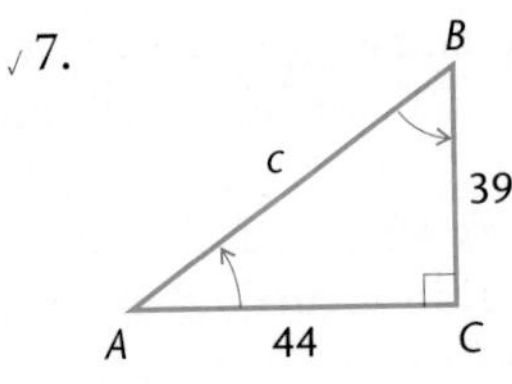

8.

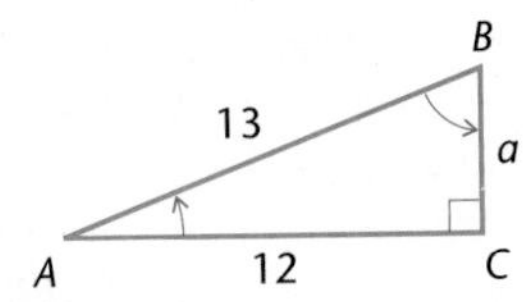

9. $c = 0.19$, $A = 36°$
10. $c = 0.79$, $B = 68°$
11. $a = 3.0$, $c = 5.0$
12. $a = 76$, $b = 51$
13. $a = 172$, $A = 39°40'$
14. $b = 537$, $B = 38°10'$
15. $a = 62.9$, $B = 71.8°$
16. $b = 1.17$, $A = 53.3°$
17. $c = 0.321$, $A = 62°30'$
18. $c = 0.413$, $A = 74°40'$
19. $c = 253$, $B = 42.2°$
20. $c = 358$, $A = 37.4°$
21. $a = 4.05$, $b = 12.8$
22. $a = 8.13$, $b = 3.91$
23. $a = 0.0625$, $c = 0.144$
24. $b = 0.0502$, $c = 0.209$
25. $a = 143.8$, $B = 63°28'$
26. $a = 342.9$, $b = 512.6$
27. $c = 91.08$, $A = 17.91°$
28. $a = 57.20$, $c = 314.5$
29. $a = 6.013$, $b = 11.54$
30. $b = 0.4772$, $A = 14°34'$
31. $b = 17.24$, $c = 98.48$
32. $c = 117.7$, $B = 66.88°$

(See Example 2.)

Leaning Ladder 33. A ladder 19 feet long leans against a wall and makes an angle of 58° with the ground. How high from the ground is the top end of the ladder?

Fence Brace 34. The posts in a fence are 5.0 feet high, with 12 feet of space between posts, as shown in the accompanying figure. If a brace is placed from the bottom of one post to the top of the next, what angle does the brace make with the ground, and how long is it?

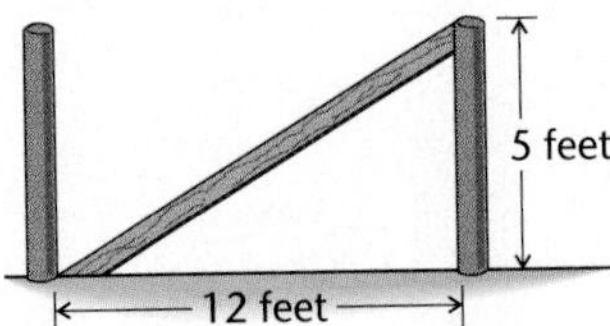

Figure for Exercise 34

Crossarm Brace 35. A metal brace supports a crossarm on a utility pole. It is attached by bolts in each end through the pole and the centerline of the crossarm. If the holes in the brace are centered 32 inches apart and one end is attached to the pole 14 inches below the centerline of the crossarm, what angle does the brace make with the crossarm? (See the accompanying figure.)

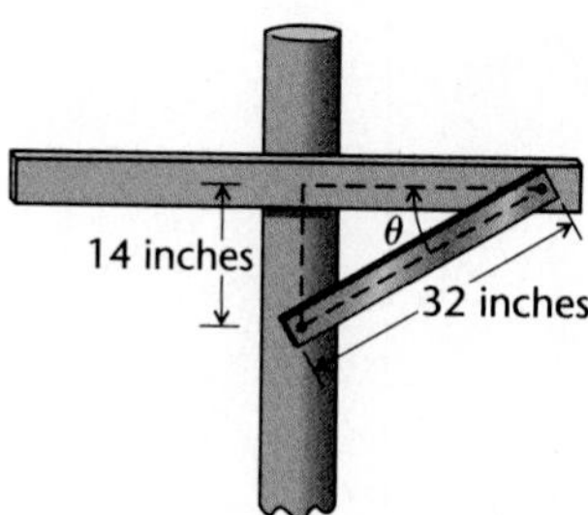

Figure for Exercise 35

TV Antenna Mast

36. A mast for a TV antenna stands on top of a flat hotel roof. A 48-foot guy wire has one end attached to the top of the mast and the other to the roof at a point 26 feet from the base of the mast. How high is the mast?

(See Example 3.)

Angles of Depression or Elevation 37–40

37. From a point on the ground 200 feet from the base of a flagpole, the angle of elevation of the top of the pole is 15°20′. Find the height of the pole.
38. From the top of a tower that is 300 feet high, the angle of depression of a rock is 35.7°. Find the distance from the base of the tower to the rock.
39. The angle of depression from the top of a pine tree to the base of an oak tree is 66°. If the trees are 30 feet apart, how tall is the pine tree?
40. From atop a vertical cliff 100 feet above the surface of the ocean, the angle of depression to a buoy is 15.0°, as shown in the accompanying figure. Find the distance from the buoy to the base of the cliff.

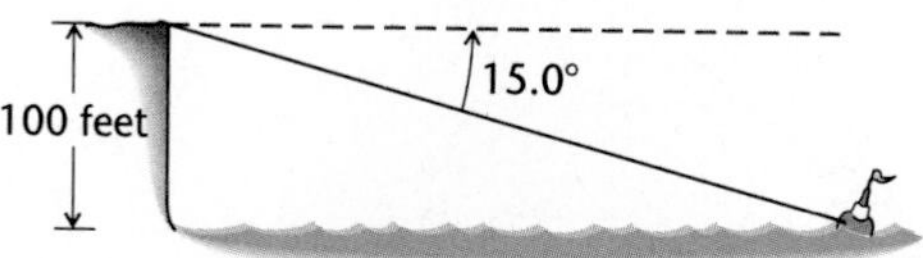

Figure for Exercise 40

World's Largest Fountain

41. When it ran at its maximum height, the giant fountain in Fountain Hills, Arizona, was known as the world's largest fountain. It has since been throttled back so that the angle of elevation of the top of the fountain is 56.0° from a point 236 feet from its base, as shown in the figure on page 430. Find the height h of the fountain since it has been throttled back.

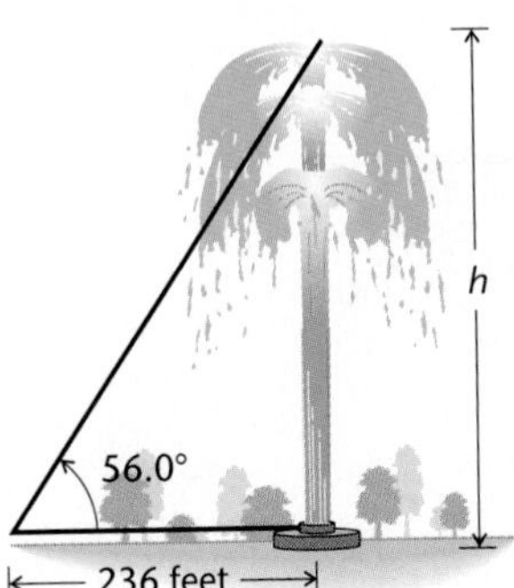

Figure for Exercise 41

Geneva Fountain

42. The famous fountain in Geneva, Switzerland, is reported to be 400 feet tall. Find the angle of elevation of the top of the fountain from a point 115 feet from its base.

(See Example 4.)

Sea Navigation 43–44

43. A ship sails 128 kilometers from its home port on a bearing S 37°20′ W to a point *A*. From *A* it sails due east to a point *B* directly south of the home port. Find the distance from *B* to the home port.

44. The cruise ship in the accompanying figure sails from its home port *H* on a bearing of N 58°10′ E for a distance of 158 kilometers to point *A*. From *A*, the ship sails 423 kilometers on a bearing S 31°50′ E to point *B*. Find the bearing of the home port from *B*.

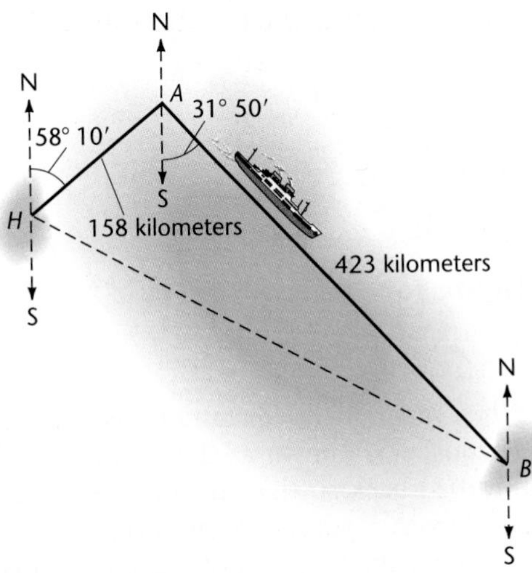

Figure for Exercise 44

Surveying 45–46

45. A corner lot in the shape of a triangle is bounded on the north and west by streets that intersect at a right angle. The boundary on the third side is a fence that is 322 feet long and makes an angle of 39°40′ with the street along the west side of the lot. Find the length of frontage on the street along the west side of the lot.

46. From one corner of a triangular tract of land, the boundaries run in the directions S 42°40′ E and S 47°20′ W. The boundary opposite this corner has length 892 yards, and the boundary in the direction S 42°40′ E has length 507 yards. Find the angles at the three corners of the triangular tract.

(See Example 5.)

Bearing Between Towns 47–50

47. Grambling is due north of Quitman at a distance of 13 miles, and Monroe is due east of Grambling at a distance of 35 miles. Find the bearing from Quitman to Monroe.

48. The distance from Lubbock to Abilene is 150 miles, on a bearing of 123.0°. From Abilene, the distance to Wichita Falls is 126 miles, on a bearing of 33.0°. Find the bearing from Lubbock to Wichita Falls.

49. The bearing from Lafayette to Alexandria is 342°, and the bearing from Lafayette to Baton Rouge is 72°. The distance from Lafayette to Alexandria is 74 miles, and the bearing from Alexandria to Baton Rouge is 128°. Find the distance from Alexandria to Baton Rouge.

50. The distance from Kingman to Las Vegas is 87.0 miles, on a bearing of 316.0°. The distance from Kingman to San Diego is 241 miles, on a bearing of 226.0°. Find the bearing from San Diego to Las Vegas.

Angles of Depression and Elevation 51–54

51. The two hotels shown in the accompanying figure face each other across a street that is 97 feet wide. From a window in one hotel, the angle of elevation of the top of the hotel across the street is 41°, and the angle of depression of the bottom at street level is 22°. Find the height of the hotel across the street.

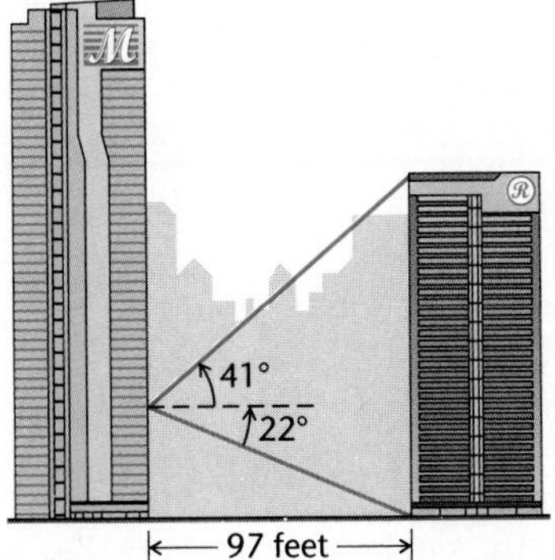

Figure for Exercise 51

52. A ship travels in a straight line toward a lighthouse that is 300 feet high. A woman on the ship observes that the angle of elevation of the top of the lighthouse is 23.0°. Later she observes that the angle of elevation of the top is 34.0°. How far has the ship traveled between the observations?

53. From the top of a building 152 feet above ground level in the accompanying figure, the angle of depression of the top of a campus street light is 18.0°, and the angle of depression of the base of the light is 32.0°. How high is the light?

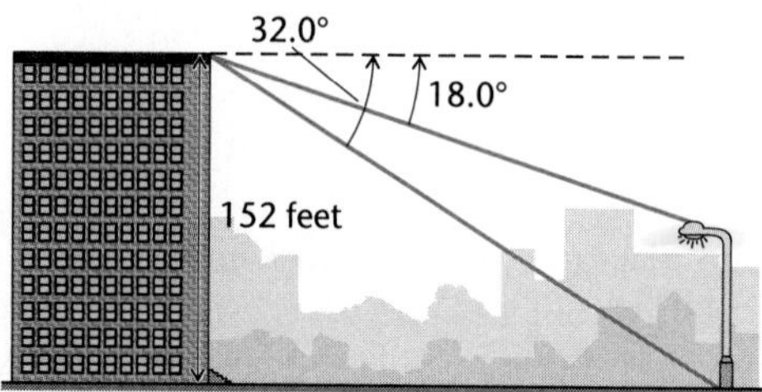

Figure for Exercise 53

54. From a certain point on the ground in the accompanying figure, a sand flea observes that the angle of elevation of the top of a vertical flagpole is 20°. He then advances 30 feet toward the pole and finds the angle of elevation to be 30°. How high is the pole?

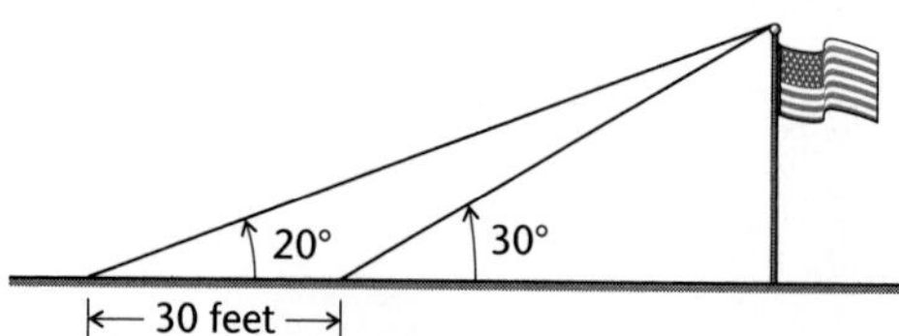

Figure for Exercise 54

Critical Thinking: Exploration and Writing

55. In the accompanying figure, C represents an arbitrary point above the x-axis and on the circle $x^2 + y^2 = r^2$. Use the distance formula and the Pythagorean theorem to show that triangle ABC is always a right triangle with the 90° angle at C.

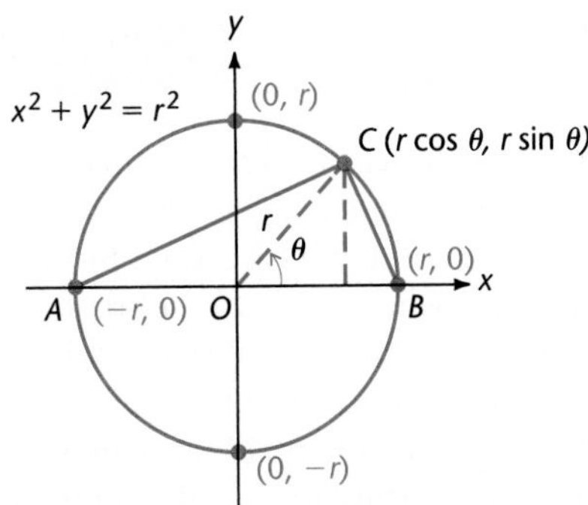

Figure for Exercise 55

◆ Solution for Practice Problem

1. The known information is shown in Figure 7.35. We can find angle A by using the tangent function.

$$\tan A = \frac{a}{b} = \frac{358}{226} = 1.584$$

Table II or a calculator then yields

$$\mathbf{A = 57.7°},$$

to the nearest tenth of a degree. To find B, we use

$$B = 90° - A \qquad \mathbf{B = 32.3°}.$$

The Pythagorean theorem yields

$$c = \sqrt{a^2 + b^2} = \sqrt{(358)^2 + (226)^2} \qquad \mathbf{c = 423}.$$

Summarizing, the parts of the triangle are as follows.

$a = 358$	$\mathbf{A = 57.7°}$
$b = 226$	$\mathbf{B = 32.3°}$
$\mathbf{c = 423}$	$C = 90°$

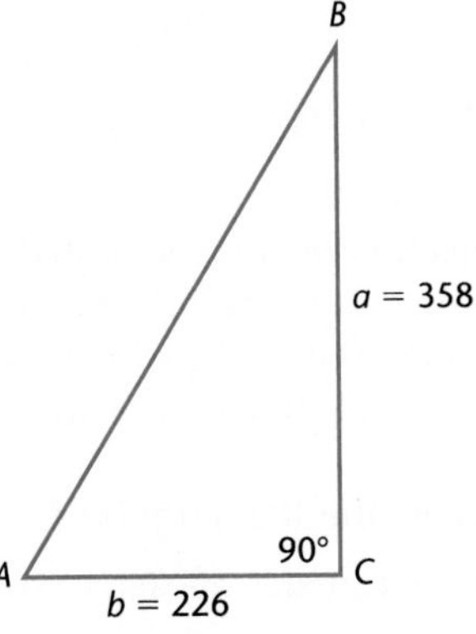

Figure 7.35 The right triangle for Practice Problem 1

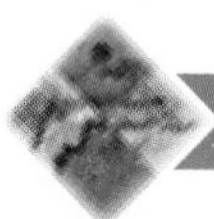

7.7 TRIGONOMETRIC FUNCTIONS OF REAL NUMBERS

When the trigonometric functions were defined in Section 7.3, we emphasized that they were *functions of an angle.* There are many practical applications for the trigonometric functions defined in this way. But this situation is in strong contrast to the study of algebra, where only functions of numbers are considered.

In order that the methods of calculus and other more advanced courses may be applied to the trigonometric functions, the definitions of the trigonometric functions must be reformulated so they are *functions of real numbers* rather than functions of angles. Such a reformulation is the main goal of this section. The method that we use to achieve this goal utilizes a **unit circle,** that is, a circle that has radius 1 unit in length.

Consider a unit circle in a coordinate plane with center at the origin, as shown in each part of Figure 7.36. With each real number t, we associate a point $P(t) = (x, y)$ *on the unit circle* that is located by the following two-part rule.

1. If $t \geq 0$, $P(t)$ is at a distance t units from $(1, 0)$ along an arc of the circle in a counterclockwise direction.
2. If $t < 0$, $P(t)$ is at a distance $|t|$ units from $(1, 0)$ along an arc of the circle in a clockwise direction.

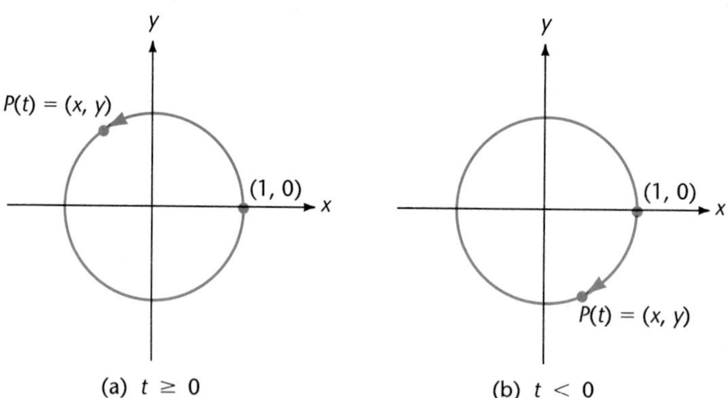

Figure 7.36 Location of $P(t)$ on a unit circle

Figure 7.36a indicates how $P(t)$ is located for $t \geq 0$, and Figure 7.36b indicates how $P(t)$ is located for $t < 0$. The rule may be summarized by saying that t is the directed arc length along the circle, with t positive in the counterclockwise direction and negative in the clockwise direction.

Example 1 • Finding the Rectangular Coordinates of $P(t)$

Find the rectangular coordinates of each of the following points on the unit circle.

a. $P(0)$ b. $P(\pi/2)$ c. $P(-\pi/2)$

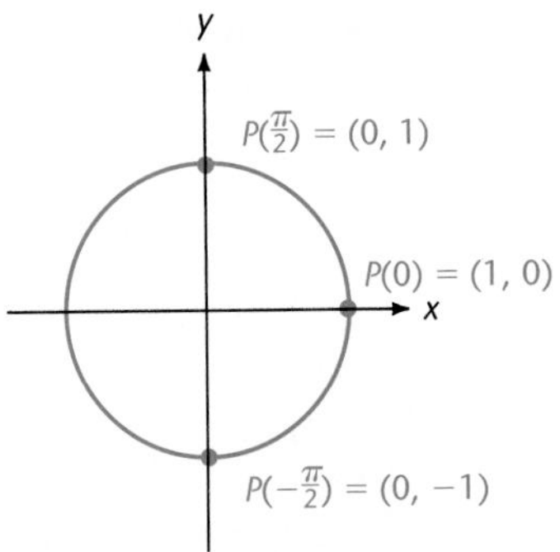

Figure 7.37 The three points in Example 1

Solution

All three points are graphed in Figure 7.37.

a. With $t = 0$ in the first part of the rule, we see that $P(0) = (1, 0)$.

b. Since the unit circle has radius $r = 1$, its circumference is $C = 2\pi r = 2\pi$. The distance $t = \pi/2$ reaches one-fourth of the way around the circle since $\pi/2 = \frac{1}{4}(2\pi)$. Thus $P(\pi/2)$ is located on the y-axis at $(0, 1)$.

c. By the same reasoning as in part b, $P(-\pi/2)$ is located one-fourth of the way around the circle in a clockwise direction, at $(0, -1)$. □

As pointed out in Example 1, the circumference of the unit circle is 2π. This means that the number $t + 2\pi$ corresponds to the same point on the unit circle as t. That is,

$$P(t + 2\pi) = P(t)$$

since adding 2π to the distance results in one additional full turn counterclockwise about the circle. Similarly,

$$P(t - 2\pi) = P(t).$$

These results generalize easily to any integral multiple of 2π, and

> $P(t) = P(t + 2\pi n)$
>
> for an arbitrary integer n (including negative n) and every real number t.

Example 2 • Finding the Quadrant for $P(t)$

Determine which quadrant contains the given point $P(t)$.

a. $P(13\pi/5)$ b. $P(12)$

Solution

In each part our first step is to locate the given value of t between multiples of 2π. After this is done, we can find a number between 0 and 2π that differs from t by a multiple of 2π and that locates the same point on the unit circle.

a. Since

$$\frac{13\pi}{5} = 2\pi + \frac{3}{5}\pi,$$

we have

$$P\left(\frac{13\pi}{5}\right) = P\left(\frac{3}{5}\pi + 2\pi\right) = P\left(\frac{3}{5}\pi\right).$$

Since $\frac{1}{2}\pi < \frac{3}{5}\pi < \pi$, $P(13\pi/5)$ lies in quadrant II.

b. Using $\pi = 3.1416$, we compute successive multiples of 2π and find that

$$2\pi = 6.2832, \qquad 4\pi = 12.5664.$$

This locates 12 between 2π and 4π. We can write 12 as

$$\begin{aligned} 12 &= 5.7168 + 6.2832 \\ &= 5.7168 + 2\pi, \end{aligned}$$

so

$$P(12) = P(5.7168).$$

Therefore $P(12)$ lies in the fourth quadrant since

$$\frac{3\pi}{2} < 5.7168 < 2\pi.$$

□

The pairing that we have made between the real numbers t and the corresponding points $P(t)$ on the unit circle is frequently called the **wrapping function.** This name comes about because the association that we have made between numbers and points on the circle is the same as would occur if a number line were placed with its origin at $(1, 0)$ and wrapped around the unit circle, as indicated in Figure 7.38. The positive part of the number line is wrapped in a counterclockwise direction, whereas the negative part is wrapped in a clockwise direction.

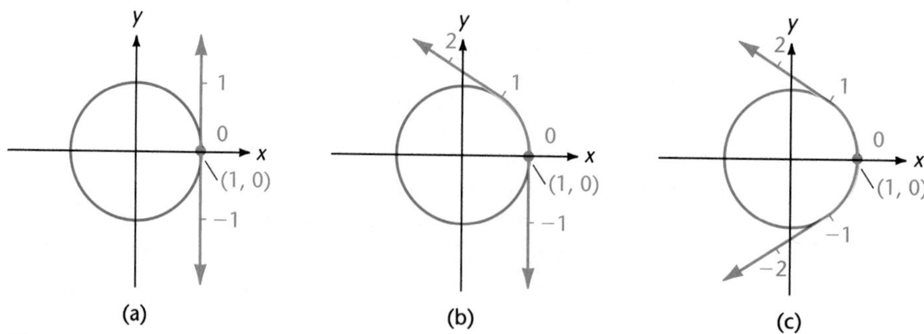

Figure 7.38 The wrapping function

The wrapping function can be used in the following manner to define trigonometric functions of real numbers. In this definition we use the same abbreviations for the six trigonometric functions that we used in the definition of the trigonometric functions of an angle in Section 7.3.

Definition of the Trigonometric Functions of a Real Number

For any real number t, let $P(t) = (x, y)$ be the point on the unit circle that corresponds to t with the wrapping function. The six trigonometric functions of t are defined as follows.

$$\sin t = y \qquad \cot t = \frac{x}{y}$$

$$\cos t = x \qquad \sec t = \frac{1}{x}$$

$$\tan t = \frac{y}{x} \qquad \csc t = \frac{1}{y}$$

The definition above reformulates the definitions of the trigonometric functions so that they appear as *functions of the real number t*, with no reference whatever to an angle.† This accomplishes the main goal of this section.

A natural question arises immediately, however. What, if any, relation exists between these new functions of the real number t and the original trigonometric functions of an angle θ? This question is answered in the following two paragraphs.

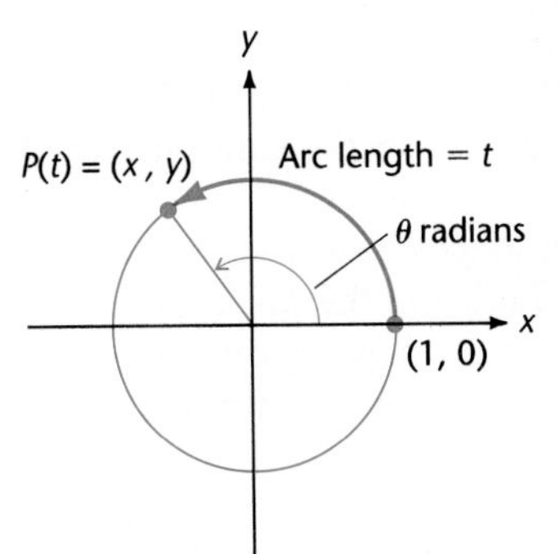

Figure 7.39 Arc length and radian measure

Consider a point $P(t) = (x, y)$ on the unit circle that is located by the arc with directed length t, and let θ be the radian measure of the central angle in the unit circle that intercepts this arc. This is shown in Figure 7.39. Since $r = 1$ in the unit circle, it follows from the arc length formula $s = r\theta$ that

$$t = \text{arc length} = r\theta = 1 \cdot \theta = \theta.$$

That is, the measure of the central angle in radians is equal to the directed length of the intercepted arc on the unit circle.

Referring to Figure 7.39 and using the definitions of the trigonometric functions of the angle θ and the real number t, we have

$$\sin \theta = \frac{y}{r} = \frac{y}{1} = \sin t,$$

$$\cos \theta = \frac{x}{r} = \frac{x}{1} = \cos t,$$

$$\tan \theta = \frac{y}{x} = \tan t.$$

In similar fashion, we find that $\csc \theta = \csc t$, $\sec \theta = \sec t$, and $\cot \theta = \cot t$. Thus we arrive at the following conclusion.

†The trigonometric functions defined in this way are sometimes called **circular functions.** However, this term has been used with different meanings in various texts.

Any trigonometric function of a real number t is equal to the same function of an angle with measure t radians.

For example, the value of $\sin t$ that the definition in this section gives for a real number t is the same as the value of $\sin t$ that the definition of the sine of an angle in Section 7.3 gives for an angle of t radians.

Example 3 • Using Trigonometric Functions to Find $P(t)$

Find the rectangular coordinates of the point $P(-5\pi/6)$ on the unit circle.

Solution

The coordinates (x, y) of $P(t)$ are given by $x = \cos t$ and $y = \sin t$, for any value of t. For $P(-5\pi/6)$, we have

$$x = \cos\left(-\frac{5\pi}{6}\right) = -\frac{\sqrt{3}}{2},$$

$$y = \sin\left(-\frac{5\pi}{6}\right) = -\frac{1}{2}.$$

□

◆ Practice Problem 1 ◆

Find the rectangular coordinates of the point $P(-7\pi/6)$ on the unit circle.

Values of trigonometric functions of numbers (or angles measured in radians) may be found by using a calculator or a table such as Table I in the Appendix. With a suitable calculator, we simply first set the calculator in radian mode and then follow the procedure described in Section 7.5. The use of Table I is described in Section A.2 of the Appendix.

EXERCISES 7.7

Find the rectangular coordinates of each of the following points on the unit circle. (See Examples 1 and 3.)

1. $P(\pi)$ 2. $P(3\pi/2)$ 3. $P(-3\pi/2)$ 4. $P(2\pi)$
5. $P(\pi/6)$ 6. $P(-\pi/3)$ 7. $P(-\pi/4)$ 8. $P(\pi/4)$
9. $P(5\pi/6)$ 10. $P(4\pi/3)$ 11. $P(5\pi/3)$ 12. $P(3\pi/4)$
13. $P(-3\pi/4)$ 14. $P(-2\pi/3)$ 15. $P(-7\pi/3)$ 16. $P(-5\pi/3)$
17. $P(13\pi/4)$ 18. $P(10\pi/3)$ 19. $P(23\pi/6)$ 20. $P(19\pi/6)$

In Exercises 21–28, determine the quadrant that contains the given point $P(t)$. (See Example 2.)

21. $P(17\pi/5)$ 22. $P(22\pi/7)$ 23. $P(-43\pi/9)$ 24. $P(-37\pi/7)$
25–28 25. $P(10)$ 26. $P(14)$ 27. $P(-20)$ 28. $P(-17)$

Find the exact value of the given trigonometric function.

29. $\tan 3\pi$
30. $\sec(-\pi)$
31. $\sin \dfrac{7\pi}{4}$
32. $\cos \dfrac{11\pi}{6}$
33. $\cot \dfrac{2\pi}{3}$
34. $\csc\left(-\dfrac{5\pi}{6}\right)$
35. $\cos \dfrac{7\pi}{6}$
36. $\tan \dfrac{11\pi}{6}$

Find the exact value of each of the following expressions. The notation $\sin^2 t$ means $(\sin t)^2$, and $\cos^2 t$ means $(\cos t)^2$.

37. $\sin^2 \dfrac{\pi}{2} + \cos^2 \dfrac{\pi}{2}$
38. $2 \sin \dfrac{\pi}{6} \cos \dfrac{\pi}{6}$
39. $2 \cos^2 \dfrac{\pi}{4} - 1$
40. $\cos \dfrac{11\pi}{6} - 4 \sin \dfrac{2\pi}{3}$

Determine whether each of the following is true or false.

41. $2 \sin \dfrac{\pi}{4} = \sin \dfrac{\pi}{2}$
42. $\cos \dfrac{2\pi}{3} = 2 \cos^2 \dfrac{\pi}{3} - 1$
43. $\tan \dfrac{2\pi}{3} = \dfrac{2 \tan \dfrac{\pi}{3}}{1 - \tan^2 \dfrac{\pi}{3}}$
44. $\tan \dfrac{\pi}{3} = \dfrac{1 - \cos \dfrac{2\pi}{3}}{\sin \dfrac{2\pi}{3}}$
45. Use the given figure to show that $\tan(-t) = -\tan t$ for $0 \le t < 2\pi$.

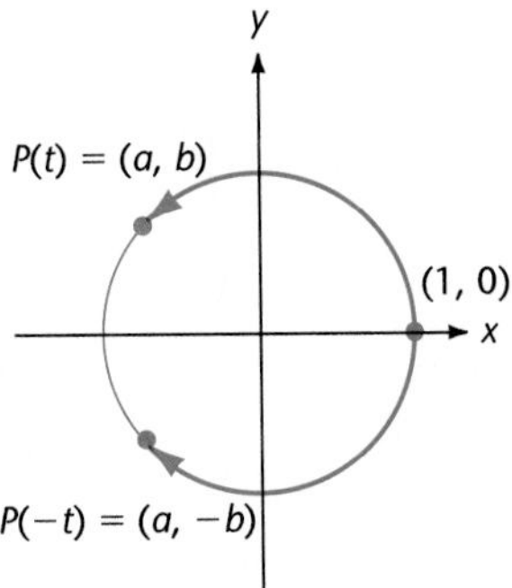

Figure for Exercise 45

46. a. Use the figure given for Exercise 45 to show that $\sin(-t) = -\sin t$ for $0 \le t < 2\pi$.
    b. Use the result in part a and the fact that $P(t) = P(t + 2\pi n)$ to show that $\sin(-t) = -\sin t$ for any real number t.
47. a. Use the figure given for Exercise 45 to show that $\cos(-t) = \cos t$ for $0 \le t < 2\pi$.

b. Use the result in part a and the fact that $P(t) = P(t + 2\pi n)$ to show that $\cos(-t) = \cos t$ for any real number t.

Critical Thinking: Exploration and Writing

48. Let A denote the area of a sector with central angle θ radians in a circle of radius r, as shown in the accompanying figure. It is clear from the figure that A varies directly as θ. Use this fact to derive the formula

$$A = \tfrac{1}{2}r^2\,\theta.$$

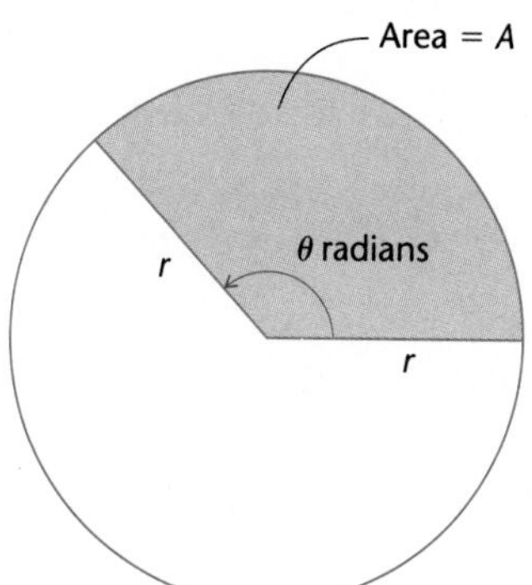

Figure for Exercise 48

Critical Thinking: Exploration and Writing

49. Using the accompanying figure, explain why

$$\text{area } \Delta OBD \le \text{area sector } OBD.$$

Then use this inequality and the formula in Exercise 48 to show that

$$\sin t \le t, \qquad \text{for } 0 \le t \le \frac{\pi}{2}.$$

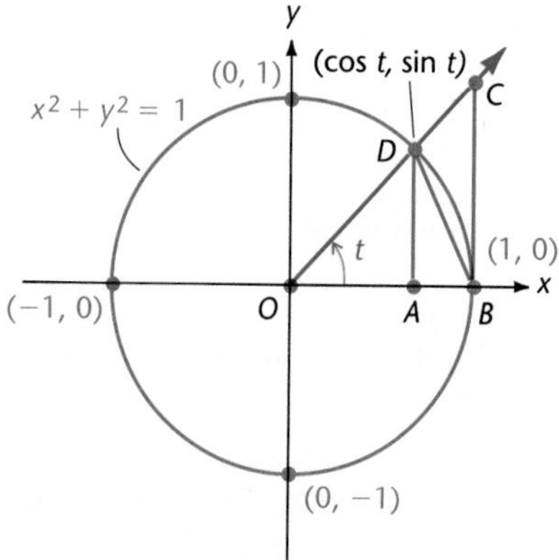

Figure for Exercise 49

◆ **Solution for Practice Problem**

1. For $P(-7\pi/6)$, we have

$$x = \cos\left(\frac{-7\pi}{6}\right) = -\frac{\sqrt{3}}{2},$$

$$y = \sin\left(\frac{-7\pi}{6}\right) = \frac{1}{2}.$$

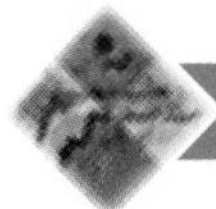

7.8 SINUSOIDAL GRAPHS

We saw in Section 7.7 that the trigonometric functions may be regarded as functions of a real number. When considered in that way, they have graphs just as do the more familiar functions studied in algebra.† In this chapter we learn to sketch the graphs of trigonometric functions.

Since we plan to draw graphs in the usual *xy*-plane, we shall write the equations that define functions in a form such as $y = \sin x$ instead of $y = \sin\theta$ or $y = \sin t$.

We start with the sine function, $y = \sin x$, and make a table of values for points with $0 \le x \le 2\pi$, spaced at intervals of $\pi/6$ on the *x*-axis. This is shown in Table 7.2.

Table 7.2 $y = \sin x$

x	0	$\frac{\pi}{6}$	$\frac{\pi}{3}$	$\frac{\pi}{2}$	$\frac{2\pi}{3}$	$\frac{5\pi}{6}$	π	$\frac{7\pi}{6}$	$\frac{4\pi}{3}$	$\frac{3\pi}{2}$	$\frac{5\pi}{3}$	$\frac{11\pi}{6}$	2π
y	0	$\frac{1}{2}$	$\frac{\sqrt{3}}{2}$	1	$\frac{\sqrt{3}}{2}$	$\frac{1}{2}$	0	$-\frac{1}{2}$	$-\frac{\sqrt{3}}{2}$	-1	$-\frac{\sqrt{3}}{2}$	$-\frac{1}{2}$	0

In Section 7.7 values of the sine function were interpreted as ordinates of points on the unit circle. That geometric interpretation, together with the values tabulated in Table 7.2, leads us to draw the smooth curve shown in Figure 7.40. This is the graph of $y = \sin x$ over the interval from 0 to 2π.

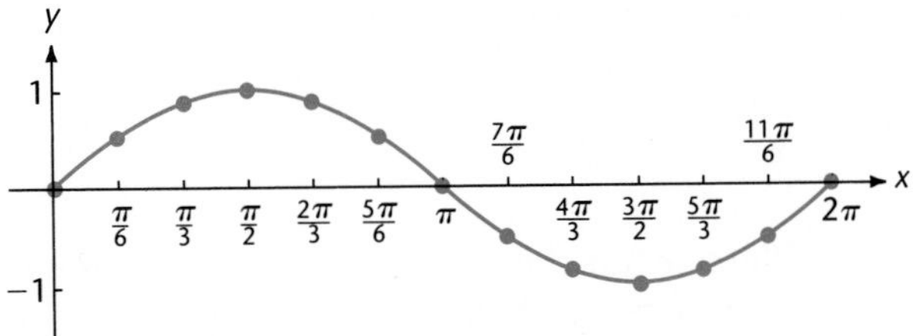

Figure 7.40 Graph of $y = \sin x$, $x \in [0, 2\pi]$

†Recall that the *graph* of a function f is the set of all points with coordinates of the form $(x, f(x))$.

We saw in Section 7.7 that $P(x) = P(x + 2\pi)$ for any point $P(x)$ on the unit circle. Since $P(x)$ has coordinates $(\cos x, \sin x)$ and $P(x + 2\pi)$ has coordinates $(\cos (x + 2\pi), \sin (x + 2\pi))$, we see that

$$\sin (x + 2\pi) = \sin x$$

for every real number x. This means that the graph of $y = \sin x$ repeats itself over intervals of width 2π along the x-axis. The complete graph of $y = \sin x$ extends indefinitely both to the left and to the right, as indicated in Figure 7.41.

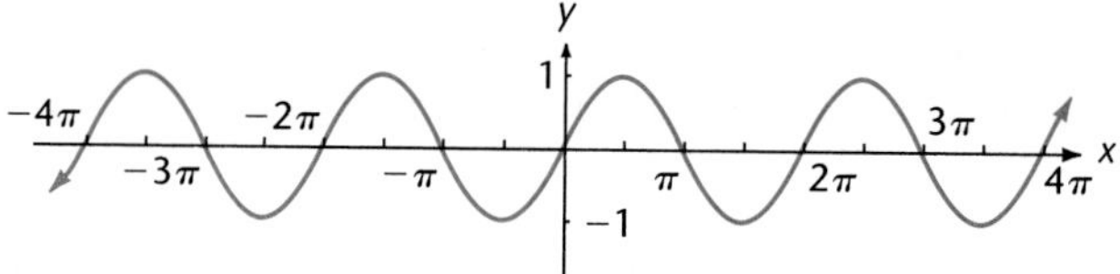

Figure 7.41 Graph of $y = \sin x$

The graphs of the other trigonometric functions repeat in a manner similar to that of the sine function. Functions that have this repeating property are called periodic.

Definition of a Periodic Function

A function is said to be **periodic** if there exists a positive number p such that $f(x + p) = f(x)$ for all x in the domain of f. If P is the smallest value of p for which this condition holds, then P is called the **period** of f.

Thus the sine function is periodic with period $P = 2\pi$. Another useful term is given in the following definition.

Definition of Amplitude

Suppose f is a periodic function that has a maximum value of M and a minimum value of m. The **amplitude** of f, abbreviated **Amp,** is defined to be one-half the difference between M and m.

$$\text{Amp} = \frac{M - m}{2}$$

The sine function is periodic with a maximum value of $M = 1$ and a minimum value of $m = -1$. According to the definition, its amplitude is given by

$$\text{Amp} = \frac{1 - (-1)}{2} = 1.$$

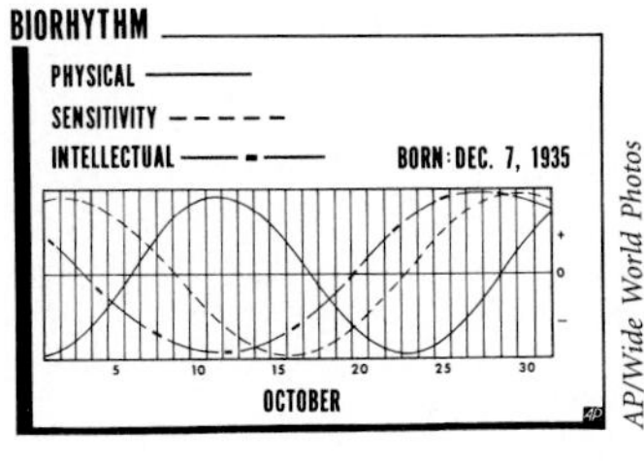

AP/Wide World Photos

The term **biorhythm** is used to refer to a periodic cycle of change in the functions of an organism. Theoretically, there are three sinusoidal biorhythms, which begin at birth and continue until death, that can be used to predict variations in a person's feelings or abilities. According to this belief, each person has a 23-day physical cycle, a 28-day sensitivity cycle, and a 33-day intellectual cycle. Conditions are supposedly favorable during the first half of each cycle and unfavorable during the last half. Critical days occur on the first day of each cycle and on days when a cycle changes from the favorable phase to the unfavorable phase. A person is thought to be more likely to have bad luck on critical days.

The objective of this section is to learn to sketch quickly the graphs of functions of the types $y = a\sin(bx + c)$ and $y = a\cos(bx + c)$. Such functions are called **sinusoidal† functions,** and $bx + c$ is referred to as the **argument** of the function.

We note that the function $f(x) = \sin x$ has amplitude 1 and period 2π. Its graph is called a **sine curve,** and a part of the graph that extends over one period is called **one cycle** of the sine curve. The cycle that extends over the interval from $x = 0$ to $x = 2\pi$ is called a **typical cycle** for $y = \sin x$, or a **basic sine wave.** The complete graph consists of basic sine waves repeated endlessly in both directions along the x-axis.

We summarize the important features of the graphs in Figures 7.40 and 7.41 as follows.

Graph of $y = \sin x$

1. Amplitude: Amp $= 1$
2. Period: $P = 2\pi$
3. The graph crosses the x-axis at the beginning, middle, and end of a typical cycle.
4. The maximum value occurs halfway between the beginning and middle of a typical cycle, at one-quarter of the period.
5. The minimum value occurs halfway between the middle and the end of a typical cycle, at three-quarters of the period.

The values of the constants a, b, and c in $y = a\sin(bx + c)$ change the graph, but they do not alter its basic shape. The graph of such a function always has **key points** corresponding to the last three items in the preceding list of features for $y = \sin x$. These key points are labeled in Figure 7.42 on a curve that we refer to as the **typical cycle for a sine curve.**

The following examples illustrate the effect that the constants a, b, and c have on the graph of $y = a\sin(bx + c)$. We consider the effect of a in $y = a\sin x$ in our first two examples.

†The word *sinusoidal* means "sinelike."

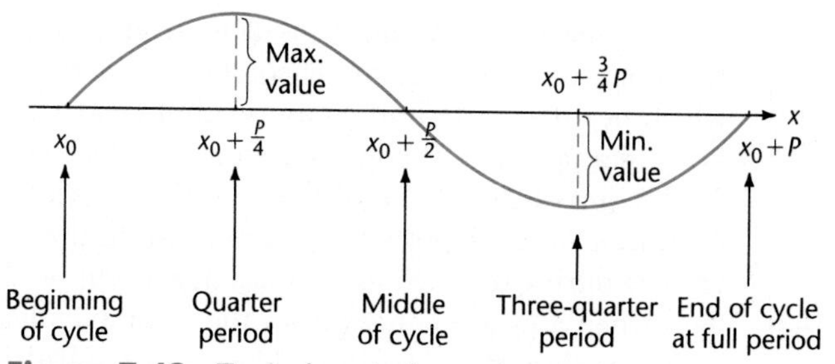

Figure 7.42 Typical cycle for a sine curve

Example 1 • Stretching a Sine Curve

Sketch the graph of the function $y = 3 \sin x$ over the interval $0 \leq x \leq 2\pi$.

Solution

The multiplier $a = 3$ in $y = 3 \sin x$ simply multiplies all the function values from $y = \sin x$ by 3. The key points on the graph are tabulated in Figure 7.43, and a sketch of the graph is shown there. We note that

1. Amp = 3.
2. $P = 2\pi$.
3. The typical cycle for a sine curve starts at $x = 0$. □

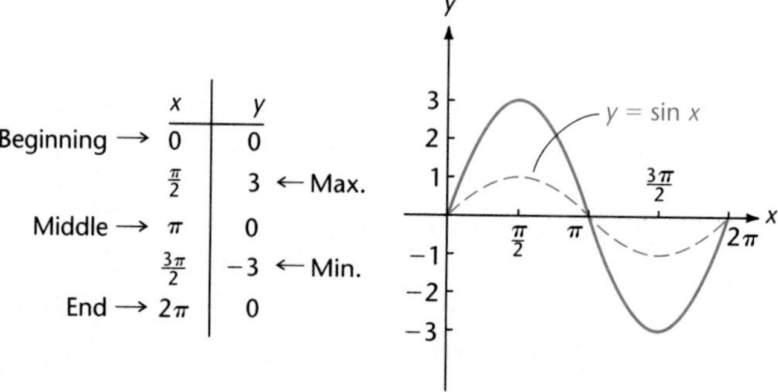

	x	y	
Beginning →	0	0	
	$\frac{\pi}{2}$	3	← Max.
Middle →	π	0	
	$\frac{3\pi}{2}$	-3	← Min.
End →	2π	0	

Figure 7.43 Graph of $y = 3 \sin x$, $x \in [0, 2\pi]$

Example 2 • Stretching and Reflecting a Sine Curve

Sketch the graph of $y = -2 \sin x$ over the interval $0 \leq x \leq 2\pi$.

Solution

The multiplier $a = -2$ in $y = -2 \sin x$ multiplies all the function values from $y = \sin x$ by -2. The key points on the graph are tabulated in Figure 7.44, and a sketch of the graph is shown there. We note that

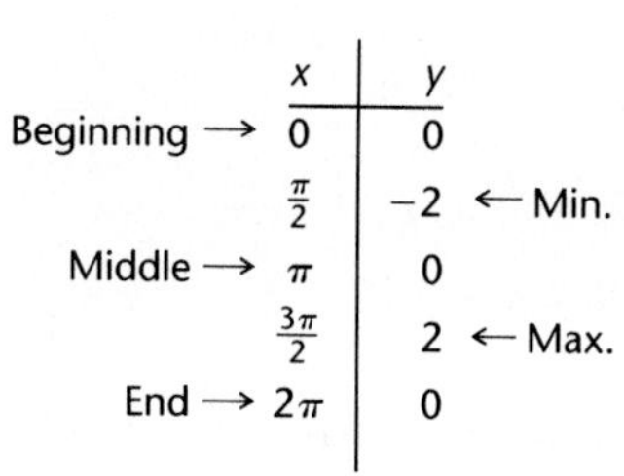

	x	y	
Beginning →	0	0	
	$\frac{\pi}{2}$	-2	← Min.
Middle →	π	0	
	$\frac{3\pi}{2}$	2	← Max.
End →	2π	0	

Figure 7.44 Graph of $y = -2 \sin x$, $x \in [0, 2\pi]$

1. Amp = $2 = |-2|$.
2. $P = 2\pi$.

3. The graph is a reflection through the x-axis of a typical sine curve, starting at $x = 0$. □

Generalizing from these two examples, we conclude that the effect of a multiplier a in $y = a \sin x$ is to change the amplitude to the value $\text{Amp} = |a|$. The usefulness of this number lies in the fact that it dictates how high the graph of $y = a \sin x$ is at the maximum value and how low it is at the minimum value.

We next consider the effect of b in $y = \sin bx$. Our attention in this chapter is confined to the case in which $b > 0$. The case in which $b < 0$ is treated in Section 8.2.

Example 3 • Changing the Period of a Sine Curve

Sketch the graph of $y = \sin 2x$ for $0 \leq x \leq 2\pi$.

Solution

Those values that correspond to key points are tabulated in Figure 7.45. From the graph in Figure 7.45, we observe that the multiplier $b = 2$ in $y = \sin 2x$ has changed the period and that

1. $\text{Amp} = 1$;
2. $P = \dfrac{2\pi}{2} = \pi$;
3. The typical sine curve shape starts at $x = 0$. □

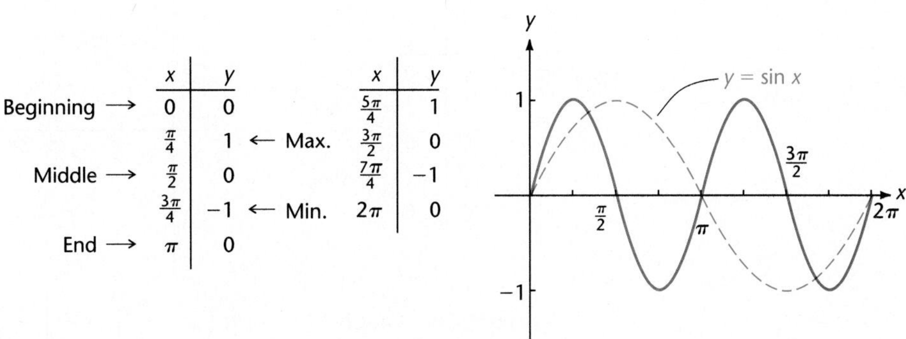

	x	y	
Beginning →	0	0	
	$\frac{\pi}{4}$	1	← Max.
Middle →	$\frac{\pi}{2}$	0	
	$\frac{3\pi}{4}$	-1	← Min.
End →	π	0	

x	y
$\frac{5\pi}{4}$	1
$\frac{3\pi}{2}$	0
$\frac{7\pi}{4}$	-1
2π	0

Figure 7.45 Graph of $y = \sin 2x$, $x \in [0, 2\pi]$

Example 3 illustrates how a positive multiplier b in $y = \sin bx$ changes the period to $2\pi/b$. This happens because an increase in x from 0 to $2\pi/b$ causes the argument bx to increase from 0 to 2π.

$$0 \leq x \leq \frac{2\pi}{b} \qquad \text{and} \qquad 0 \leq bx \leq 2\pi$$

are equivalent inequalities.

The next example illustrates the shifting effect of the constant c in the graph of $y = \sin(x + c)$.

Example 4 • Horizontal Translation of a Sine Curve

Sketch the graph of $y = \sin\left(x + \dfrac{\pi}{4}\right)$ over one complete period, locating the key points on the graph.

Solution

The table in Figure 7.46 includes only those values that correspond to key points on the graph. The value $x = -\pi/4$ was chosen to begin with because it locates the point that corresponds to the beginning of a typical cycle on a sine curve. The corresponding y-value is calculated as

$$y = \sin\left(-\frac{\pi}{4} + \frac{\pi}{4}\right) = \sin 0 = 0.$$

The key points located on the graph in Figure 7.46 indicate that the function has been graphed through one complete period over the interval $-\pi/4 \le x \le 7\pi/4$. The constant $c = \pi/4$ caused the graph to be translated horizontally $\pi/4$ units to the left. A horizontal translation such as this is called a **phase shift.** □

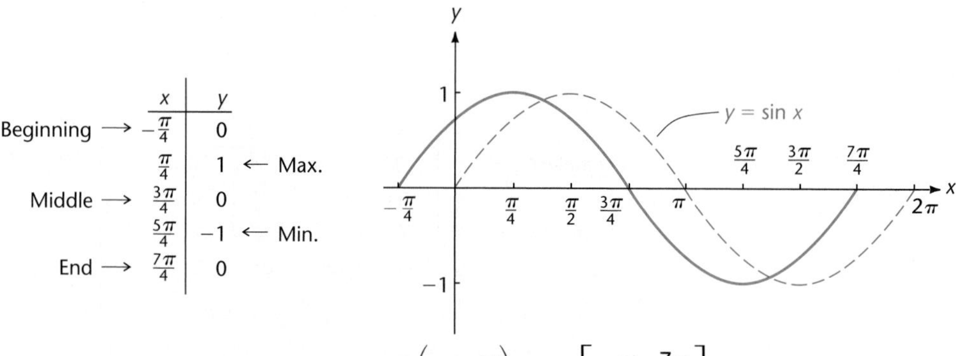

	x	y	
Beginning →	$-\frac{\pi}{4}$	0	
	$\frac{\pi}{4}$	1	← Max.
Middle →	$\frac{3\pi}{4}$	0	
	$\frac{5\pi}{4}$	−1	← Min.
End →	$\frac{7\pi}{4}$	0	

Figure 7.46 Graph of $y = \sin\left(x + \dfrac{\pi}{4}\right)$, $x \in \left[-\dfrac{\pi}{4}, \dfrac{7\pi}{4}\right]$

Example 4 indicates that the constant c in $y = a \sin(bx + c)$ causes a phase shift. The phase shift can be found by locating any two corresponding points on the graph. As standard practice, we shall use the beginning of a cycle on the typical sine curve. The phase shift can also be found by formula, and some people prefer this method. With $b > 0$, the shift is c/b units to the left if $c > 0$, and $|c/b|$ units to the right if $c < 0$.

Generalizing from our examples, we arrive at the following guide.

Guide for Graphing $y = a \sin(bx + c)$, where $b > 0$

1. Amp $= |a|$.
2. $P = \dfrac{2\pi}{b}$.
3. A cycle begins at the value of x where the argument is 0, that is, where $bx + c = 0$.
4. If $a > 0$, the graph is a typical sine curve. If $a < 0$, the graph is the reflection through the x-axis of a typical sine curve.

The next example illustrates the use of this guide.

Example 5 • Using the Guide to Graph a Sine Curve

Sketch the graph of $y = \dfrac{1}{2}\sin\left(3x - \dfrac{\pi}{2}\right)$ over one complete period, locating the key points on the graph.

Solution

With $a = 1/2$, $b = 3$, and $c = -\pi/2$, we find that the amplitude is 1/2 and the period is $2\pi/3$. A cycle begins where

$$3x - \frac{\pi}{2} = 0$$

$$x = \frac{\pi}{6}.$$

To find the value of x at the end of the cycle, we add the period to the starting point.

$$x = \frac{\pi}{6} + \frac{2\pi}{3} = \frac{5\pi}{6}$$

To find the value of x at the middle of the cycle, we simply average the x-values at the beginning and end of the cycle.

$$x = \frac{1}{2}\left(\frac{\pi}{6} + \frac{5\pi}{6}\right) = \frac{\pi}{2}$$

Similarly, we locate the other x-values tabulated in Figure 7.47 on page 448 and use the properties of the typical sine curve in the following list.

1. The graph crosses the x-axis at $x = \pi/6$, $x = \pi/2$, and $x = 5\pi/6$.
2. A maximum of $y = |a| = 1/2$ occurs at $x = \pi/3$.
3. A minimum of $y = -|a| = -1/2$ occurs at $x = 2\pi/3$.

The graph is drawn in Figure 7.47. □

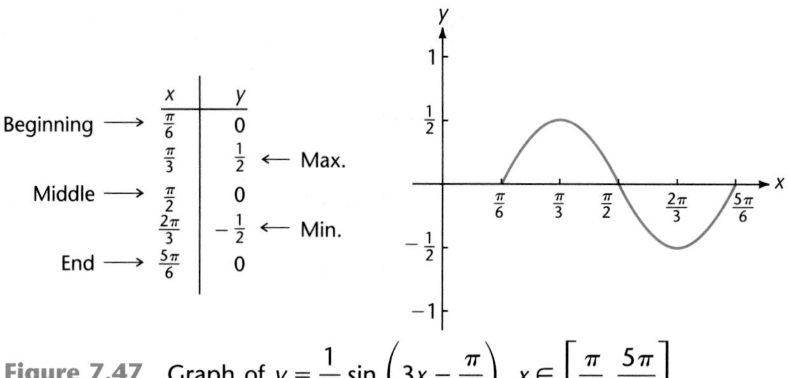

Figure 7.47 Graph of $y = \frac{1}{2}\sin\left(3x - \frac{\pi}{2}\right),\ x \in \left[\frac{\pi}{6}, \frac{5\pi}{6}\right]$

Functions of the form $y = a\cos(bx + c)$ can be graphed by using a method very similar to the one we have used for sine curves. The procedure is presented here with a minimum of details.

The graph of $y = \cos x$ over the interval $0 \le x \le 2\pi$ is shown in Figure 7.48.

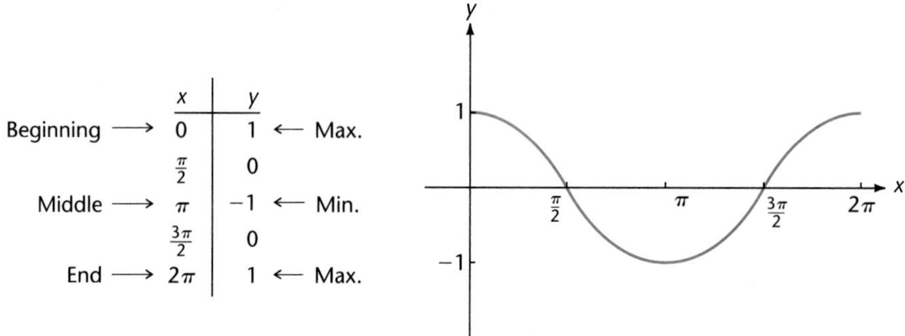

Figure 7.48 Graph of $y = \cos x,\ x \in [0, 2\pi]$

The unit circle can be used in the same way that it was with the sine function to conclude that the cosine function has period 2π. As it was with $y = \sin x$, the complete graph of $y = \cos x$ extends indefinitely in both directions along the x-axis.

We take the shape of the graph in Figure 7.48 as a pattern for the **typical cycle for a cosine curve** and select as **key points** those points where the graph crosses the x-axis, where it has a maximum value, or where it has a minimum value. These key points are labeled in Figure 7.49.

To sketch the graph of a function of the type $y = a\cos(bx + c)$, we use the typical cycle of a cosine curve and the following guide.

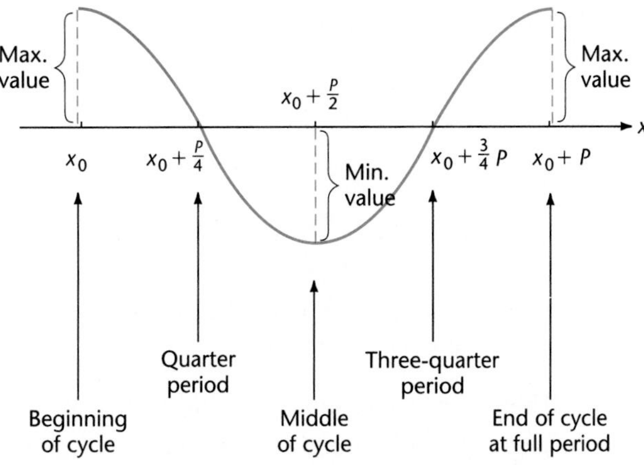

Figure 7.49 Typical cycle for a cosine curve

Guide for Graphing $y = a \cos(bx + c)$, where $b > 0$

1. Amp $= |a|$.
2. $P = \dfrac{2\pi}{b}$.
3. A cycle begins at the value of x where the argument is 0, that is, where $bx + c = 0$.
4. If $a > 0$, the graph is a typical cosine curve. If $a < 0$, it is the reflection through the x-axis of a typical cosine curve.

The following example illustrates the procedure.

Example 6 • Using the Guide to Graph a Cosine Curve

Sketch the graph of $y = 4\cos\left(\dfrac{1}{2}x + \dfrac{\pi}{4}\right)$ over one complete period, locating the key points on the graph.

Solution

With $a = 4$, $b = 1/2$, and $c = \pi/4$, we find that the amplitude is 4 and the period $2\pi/(1/2) = 4\pi$. A cycle begins where

$$\frac{1}{2}x + \frac{\pi}{4} = 0$$

$$x = -\frac{\pi}{2}.$$

The value of x at the end of the cycle is

$$x = -\frac{\pi}{2} + 4\pi = \frac{7\pi}{2},$$

and the middle of the cycle is where

$$x = \frac{1}{2}\left(-\frac{\pi}{2} + \frac{7\pi}{2}\right) = \frac{3\pi}{2}.$$

The key points are tabulated in Figure 7.50. According to the properties of the typical cycle of a cosine curve,

1. There is a maximum of $y = |a| = 4$ at $x = -\pi/2$ and $x = 7\pi/2$;
2. There is a minimum of $y = -|a| = -4$ at $x = 3\pi/2$;
3. The graph crosses the x-axis at $x = \pi/2$ and $x = 5\pi/2$.

The graph is shown in Figure 7.50. □

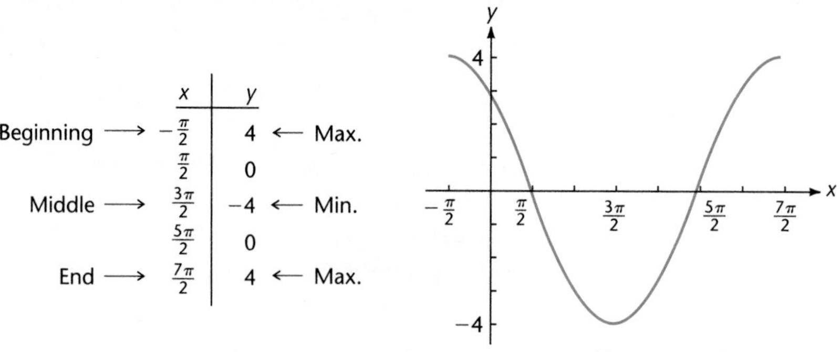

Figure 7.50 Graph of $y = 4\cos\left(\frac{1}{2}x + \frac{\pi}{4}\right)$, $x \in \left[-\frac{\pi}{2}, \frac{7\pi}{2}\right]$

The preceding examples have concentrated on sketching the graph of a function over one complete period. It is not difficult to extend such graphs to cover any desired interval.

Example 7 • Extending a Sinusoidal Graph

Sketch the graph of $y = 4\cos\left(\frac{1}{2}x + \frac{\pi}{4}\right)$ over the interval $-\frac{5\pi}{2} \le x \le \frac{9\pi}{2}$.

Solution

The function $y = 4\cos\left(\frac{1}{2}x + \frac{\pi}{4}\right)$ is graphed for $-\pi/2 \le x \le 7\pi/2$ in Example 6. All we need do is locate key points on the graph until the required interval is covered and then sketch the graph over that interval. The result is shown in Figure 7.51. □

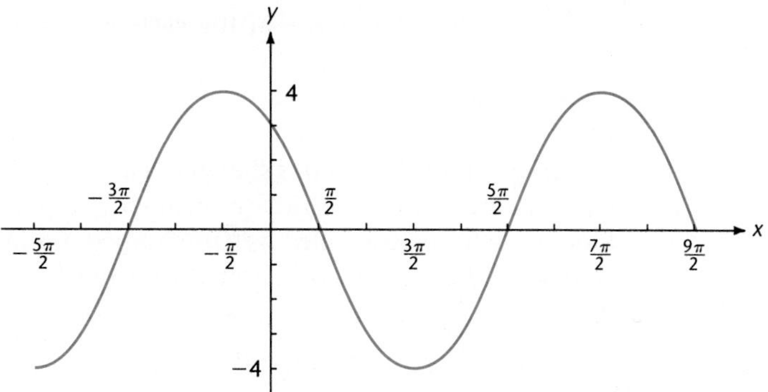

Figure 7.51 Graph of $y = 4\cos\left(\frac{1}{2}x + \frac{\pi}{4}\right),\ x \in \left[-\frac{5\pi}{2}, \frac{9\pi}{2}\right]$

Sinusoidal graphs are useful in describing many physical quantities. For instance, they are used to describe the motion of a spring under various conditions. We relate one of the simplest situations in the next example.

Example 8 • Free Undamped Motion

A certain spring is such that an object weighing 2 pounds stretches the spring $\frac{1}{2}$ foot and reaches an equilibrium position as shown in Figure 7.52b. If the spring is stretched an additional $\frac{1}{3}$ foot and released with initial velocity zero, it is possible to describe the position of the object with a cosine function. We assume that the only forces acting on the object are the force of the spring and the weight of the object.

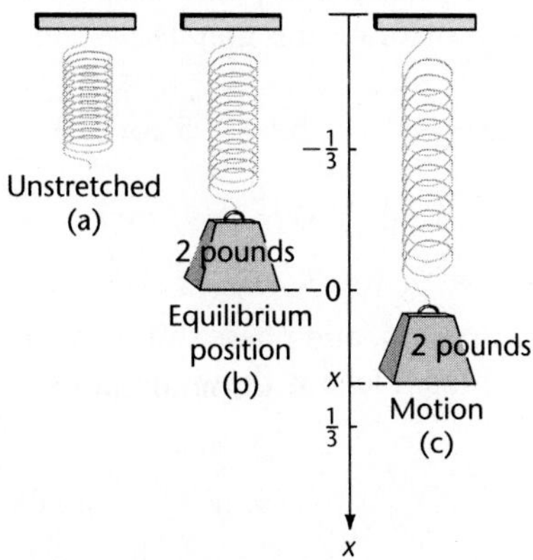

Figure 7.52 Stretching a spring

We let x denote the displacement of the object from its equilibrium position t seconds after its release, with x in feet measured positive downward and negative upward. This is illustrated in Figure 7.52c.

With these notations, the displacement x is given by the formula

$$x = \frac{1}{3} \cos 8t, \text{ for } t \geq 0.$$

The graph of this function is shown in Figure 7.53. Our formula is valid only under the assumption that no friction or other damping force is exerted on the object. Motion of this type is called **free undamped motion.** More complex situations can be described by more complicated sinusoidal functions. □

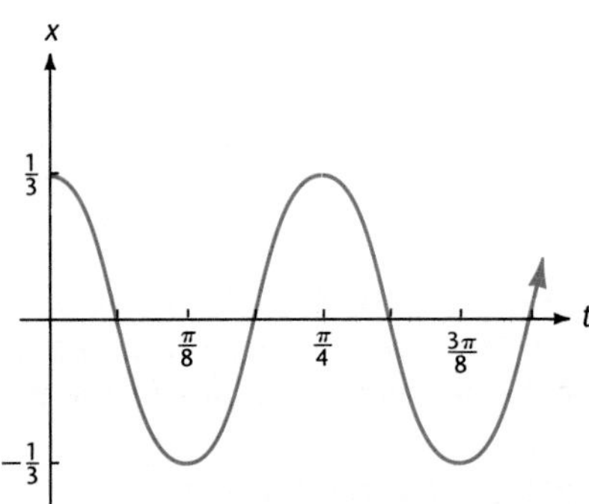

Figure 7.53 The graph of the displacement in Example 8

◆ Practice Problem 1 ◆

Find the amplitude, period, and phase shift for the graph of each of the following functions. Do not draw the graphs.

a. $y = \frac{1}{3} \sin (2x + \pi)$ b. $y = 5 \cos \left(3x - \frac{\pi}{4}\right)$

EXERCISES 7.8

Find the amplitude, period, and phase shift (if there is one) for the graph of each of the following functions. Do not draw the graphs.

√1. $y = 4 \sin 3x$

2. $y = 3 \cos 2x$

3. $y = -5 \cos 4x$

4. $y = -2 \sin 8x$

√5. $y = \frac{1}{2} \cos \left(x - \frac{\pi}{4}\right)$

6. $y = \frac{1}{3} \sin \left(x + \frac{\pi}{6}\right)$

7. $y = \frac{1}{3} \sin (3x + \pi)$

8. $y = \frac{1}{2} \cos (2x - \pi)$

9. $y = -4 \sin \left(3x - \frac{\pi}{4}\right)$

10. $y = -3 \cos \left(\frac{1}{2}x - \frac{\pi}{2}\right)$

√11. $y = -3\cos\left(2x + \frac{\pi}{4}\right)$

12. $y = -4\sin\left(2x + \frac{\pi}{3}\right)$

Sketch the graph of the given function through one complete period, locating the key points on each graph. (See Examples 1–6.)

13. $y = 2\sin x$
14. $y = 3\cos x$
15. $y = -2\cos x$
16. $y = -\sin x$
√17. $y = -\frac{1}{3}\sin 3x$
18. $y = -\frac{1}{2}\cos 4x$
19. $y = 2\cos\frac{1}{2}x$
20. $y = 6\sin\frac{1}{3}x$
√21. $y = 2\sin\left(2x - \frac{\pi}{2}\right)$
22. $y = 2\sin(3x + \pi)$
23. $y = 3\sin\left(x + \frac{\pi}{3}\right)$
24. $y = 3\sin(4x + \pi)$
25. $y = 3\cos\left(2x - \frac{\pi}{4}\right)$
26. $y = 4\cos\left(2x + \frac{\pi}{4}\right)$
27. $y = 2\cos\left(3x + \frac{\pi}{4}\right)$
28. $y = 4\cos\left(x + \frac{\pi}{3}\right)$
29. $y = 2\sin\left(\frac{1}{2}x + \frac{\pi}{4}\right)$
30. $y = 3\sin\left(\frac{1}{2}x - \frac{\pi}{4}\right)$
√31. $y = 3\sin\left(\frac{1}{3}x + \frac{\pi}{3}\right)$
32. $y = 3\sin\left(\frac{1}{2}x - 2\pi\right)$
33. $y = 3\cos\left(\frac{1}{2}x + \frac{\pi}{4}\right)$
34. $y = 2\cos\left(\frac{1}{4}x + \frac{\pi}{4}\right)$
√35. $y = 2\cos\left(\frac{1}{4}x - \frac{\pi}{4}\right)$
36. $y = 3\cos\left(\frac{1}{3}x + \frac{\pi}{9}\right)$

In Exercises 37–40, sketch the graph of the given function over the indicated interval. (See Example 7.)

37. $y = \sin\left(2x + \frac{\pi}{2}\right),\ 0 \le x \le 2\pi$

38. $y = 2\sin\left(\frac{1}{4}x - \frac{\pi}{2}\right),\ -2\pi \le x \le 6\pi$

√39. $y = 3\cos\left(\frac{1}{2}x - \frac{\pi}{2}\right),\ -\pi \le x \le 5\pi$

40. $y = \cos\left(3x + \frac{\pi}{2}\right),\ -\frac{\pi}{6} \le x \le \frac{5\pi}{6}$

In Exercises 41 and 42, find an equation of the form $y = a \sin(bx + c)$ that has the given graph.

41.

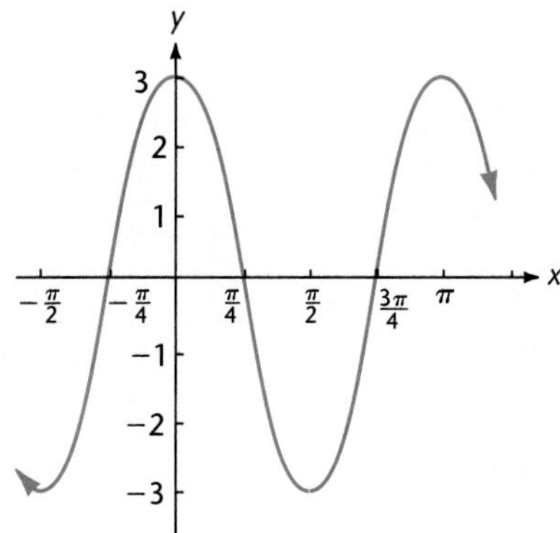

42.

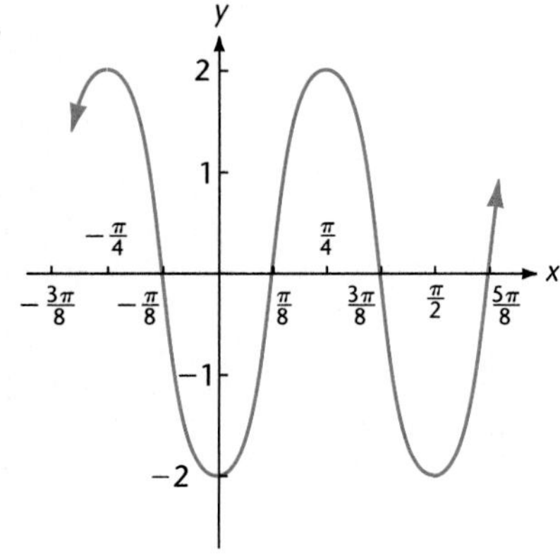

In Exercises 43 and 44, find an equation of the form $y = a \cos(bx + c)$ that has the given graph.

43.

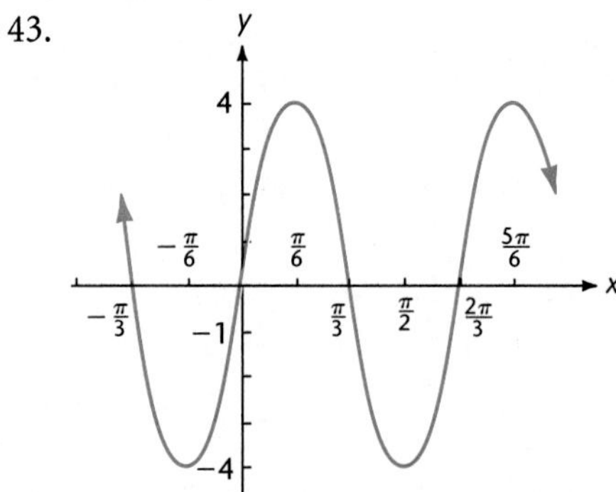

44.

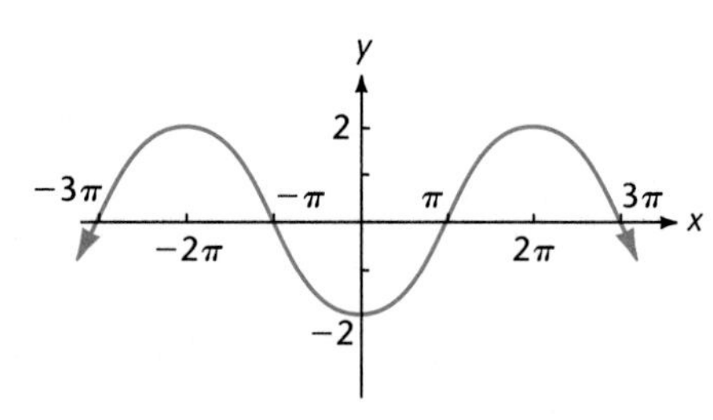

Electrical Circuit 45–46

45. The accompanying figure shows a diagram of an electrical circuit that contains an electromotive force $E(t)$ measured in volts, an inductor with inductance L measured in henrys, and a capacitor with capacitance C

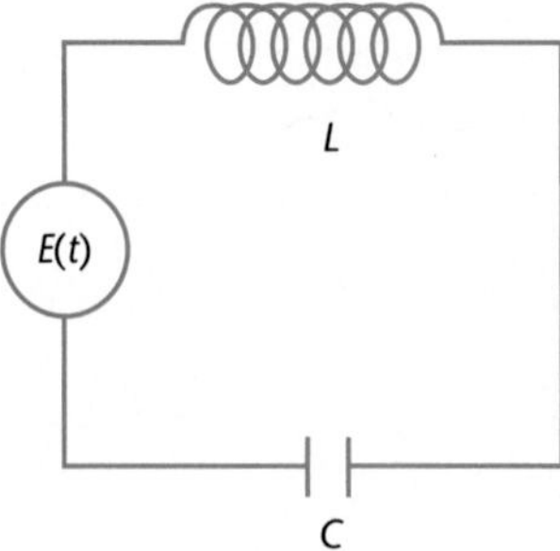

Figure for Exercise 45

measured in farads. We assume that $E(0) = 0$ and that the charge on the capacitor is 0 at $t = 0$. If $E(t) = 60$ volts for $t > 0$, $L = 1$ henry, and $C = \frac{1}{16}$ farad, the current $I(t)$ is given in amperes by $I(t) = 15 \sin 4t$. Sketch the graph of I through one complete period.

46. We assume that the electrical circuit shown in Exercise 45 has $E(0) = 0$ and that the charge on the capacitor is 0 at $t = 0$. If $E(t) = 36$ volts for $t > 0$, $L = 2$ henrys, and $C = \frac{1}{18}$ farad, the current $I(t)$ is given in amperes by $I(t) = 6 \sin 3t$. Sketch the graph of I through one complete period.

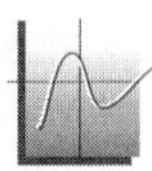
47–54

Use a graphing device to graph the following functions.

47. $y = x \sin x$
48. $y = x \cos x$
49. $y = \frac{1}{2}x^2 + 4 \cos x$
50. $y = \frac{1}{2}x^2 + 4 \sin x$
51. $y = \cos^3 x$
52. $y = \sin^2 x$
53. $y = \sqrt{\sin x}$
54. $y = |\sin x|$

◆ Solution for Practice Problem

1. a. In $y = \frac{1}{3} \sin (2x + \pi)$, we have $a = \frac{1}{3}$, $b = 2$, $c = \pi$. By the formulas,

$$\text{Amp} = |a| = \frac{1}{3}, \qquad P = \frac{2\pi}{b} = \frac{2\pi}{2} = \pi.$$

To find the phase shift, we set the argument equal to zero and thereby locate the beginning of a typical cycle.

$$2x + \pi = 0,$$
$$x = -\frac{\pi}{2}.$$

The starting point for a typical cycle of the sine curve has moved from $x = 0$ to $x = -\pi/2$, so the phase shift is $\pi/2$ units to the left.

b. In $y = 5 \cos\left(3x - \frac{\pi}{4}\right)$, we have $a = 5$, $b = 3$, $c = -\frac{\pi}{4}$. Hence

$$\text{Amp} = |a| = 5, \qquad P = \frac{2\pi}{b} = \frac{2\pi}{3}.$$

Locating the beginning of a typical cycle, we have

$$3x - \frac{\pi}{4} = 0,$$
$$x = \frac{\pi}{12}.$$

The phase shift is $\pi/12$ units to the right.

7.9 OTHER TRIGONOMETRIC GRAPHS

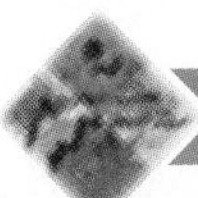

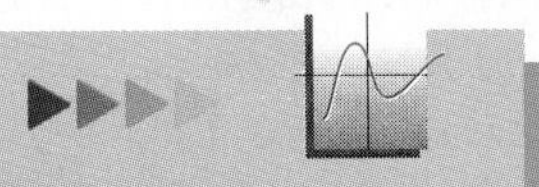

In order to view the graphs of the secant, cosecant, or cotangent functions on a graphing device, it is necessary to use the reciprocal identities.

We consider first the graph of $y = \csc x$. Since $\csc x = 1/\sin x$, values of $\csc x$ can be obtained by using the reciprocals of nonzero values of $\sin x$.

Since 1 is its own reciprocal, $\csc x = 1$ where $\sin x = 1$, and similarly $\csc x = -1$ where $\sin x = -1$. The value of $\csc x$ is undefined at each x where $\sin x = 0$. As values of $\sin x$ get closer to 0, the values of $\csc x$ increase in absolute value without bound. Thus the graph of $y = \csc x$ has a vertical asymptote at each x where $\sin x = 0$.

In Figure 7.54, the graph of $y = \sin x$ is shown as a dashed curve. By taking reciprocals of the ordinates on the dashed curve, we obtain the solid graph in the figure as the graph of $y = \csc x$.

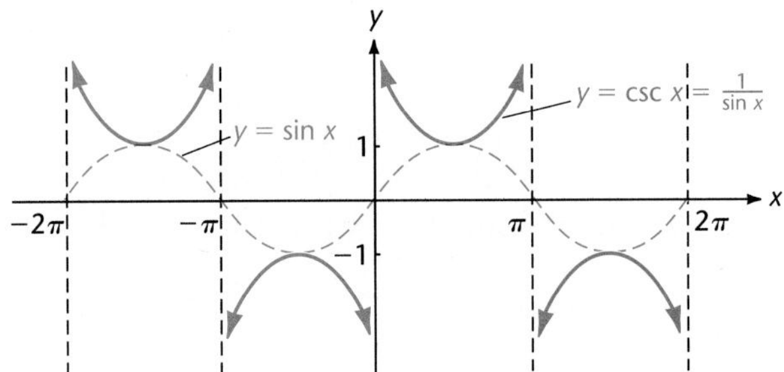

Figure 7.54 Graph of $y = \csc x$

Since $y = \sin x$ has period 2π and $\csc x = 1/\sin x$, it follows that $y = \csc x$ has period 2π. The amplitude is not defined for the cosecant function since it has neither a maximum nor a minimum.

In the same way that we obtained the graph of $y = \csc x$ from that of $y = \sin x$, the graph of a function of the form $y = a \csc (bx + c)$ can be obtained from the graph of $y = a \sin (bx + c)$.† Since $\csc (bx + c) = 1/\sin (bx + c)$, the two functions have the same period, $P = 2\pi/b$. The graph of $y = a \csc (bx + c)$ has vertical asymptotes where $a \sin (bx + c) = 0$. The graphs of $y = a \csc (bx + c)$ and $y = a \sin (bx + c)$ intersect where $\sin (bx + c) = \pm 1$, that is, at maximum and minimum points on the sine curve. Finally, the phase shift for the cosecant function is the same as for the corresponding sine function.

Example 1 • Using a Sine Curve to Graph a Cosecant Curve

Sketch the graph of $y = \dfrac{1}{4} \csc \left(3x + \dfrac{\pi}{2}\right)$ over one complete period.

†We confine our attention here to the case in which $b > 0$, just as we did in Section 7.8.

Solution

We start out as if we were going to graph $y = \frac{1}{4}\sin\left(3x + \frac{\pi}{2}\right)$. The period is $\frac{2\pi}{3}$, and a cycle begins where

$$3x + \frac{\pi}{2} = 0$$
$$x = -\frac{\pi}{6}.$$

As in the last section, we locate the x-values that correspond to key points on the graph of $y = \frac{1}{4}\sin\left(3x + \frac{\pi}{2}\right)$. These x-values are tabulated in Figure 7.55, and the graph of $y = \frac{1}{4}\sin\left(3x + \frac{\pi}{2}\right)$ is drawn as a dashed curve in the figure. The graph of $y = \frac{1}{4}\csc\left(3x + \frac{\pi}{2}\right)$ is drawn as a solid curve, and it has vertical asymptotes where the sine curve crosses the x-axis. The two function values are equal at maximum and minimum points on the sine curve. We note that the phase shift for the cosecant curve is $\frac{\pi}{6}$ units to the left, the same as it is for $y = \frac{1}{4}\sin\left(3x + \frac{\pi}{2}\right)$.

□

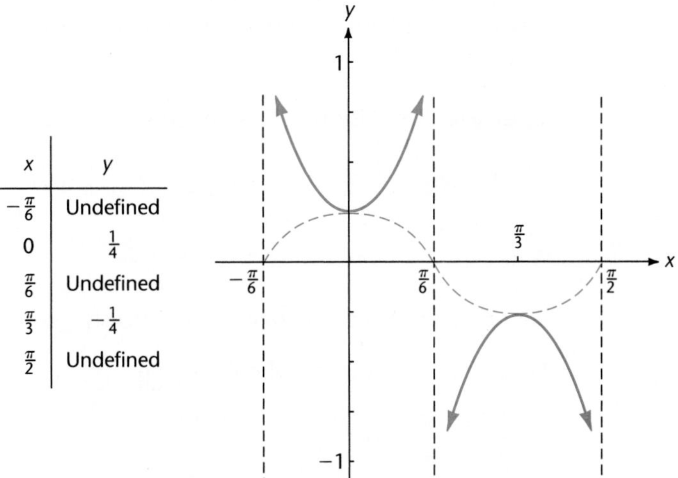

x	y
$-\frac{\pi}{6}$	Undefined
0	$\frac{1}{4}$
$\frac{\pi}{6}$	Undefined
$\frac{\pi}{3}$	$-\frac{1}{4}$
$\frac{\pi}{2}$	Undefined

Figure 7.55 Graph of $y = \frac{1}{4}\csc\left(3x + \frac{\pi}{2}\right)$, $x \in \left[-\frac{\pi}{6}, \frac{\pi}{2}\right]$

After a little practice, it is no longer necessary to draw the corresponding sine curve as we did in Example 1.

We turn our attention now to the graph of the secant function. Since $\sec x = 1/\cos x$, the graph of $y = \sec x$ is related to that of $y = \cos x$ in the same way that the graph of $y = \csc x$ is related to that of $y = \sin x$. This is shown in Figure 7.56, where the graph of $y = \cos x$ appears as a dashed curve.

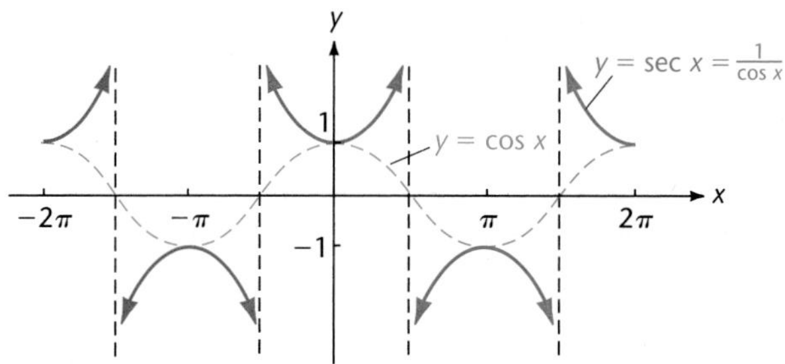

Figure 7.56 Graph of $y = \sec x$

The graph of $y = a \sec (bx + c)$ can be obtained by using the graph of $y = a \cos (bx + c)$. Since $\sec (bx + c) = 1/\cos (bx + c)$, the two functions have the same period, $P = 2\pi/b$, and the same phase shift. The amplitude is not defined for the secant function, and the graph of $y = a \sec (bx + c)$ has vertical asymptotes at each x where

$$a \cos (bx + c) = 0.$$

The graphs of the two functions intersect where $\cos (bx + c) = \pm 1$, that is, at maximum and minimum points on the cosine curve. All this is illustrated in Example 2.

Example 2 • Using a Cosine Curve to Graph a Secant Curve

Sketch the graph of $y = 2 \sec \left(\frac{1}{2}x - \frac{\pi}{6}\right)$ over one complete period.

Solution

We relate this graph to that of $y = 2 \cos \left(\frac{1}{2}x - \frac{\pi}{6}\right)$. The period is $2\pi/(1/2) = 4\pi$, and a cycle begins where

$$\frac{1}{2}x - \frac{\pi}{6} = 0$$

$$x = \frac{\pi}{3}.$$

The x-values that correspond to key points on the cosine curve are tabulated in Figure 7.57. The graph of $y = 2 \cos \left(\frac{1}{2}x - \frac{\pi}{6}\right)$ is drawn as a dashed curve, and

x	y
$\frac{\pi}{3}$	2
$\frac{4\pi}{3}$	Undefined
$\frac{7\pi}{3}$	-2
$\frac{10\pi}{3}$	Undefined
$\frac{13\pi}{3}$	2

Figure 7.57 Graph of $y = 2 \sec\left(\frac{1}{2}x - \frac{\pi}{6}\right)$, $x \in \left[\frac{\pi}{3}, \frac{13\pi}{3}\right]$

the graph of $y = 2 \sec\left(\frac{1}{2}x - \frac{\pi}{6}\right)$ is drawn as a solid curve in the figure. Note that both graphs have a phase shift of $\pi/3$ units to the right. □

All four of the functions $\sin x$, $\cos x$, $\csc x$, and $\sec x$ have period 2π. Unlike them, $\tan x$ and $\cot x$ have period π. For this reason, we restrict our attention at first to the interval $-\pi/2 < x < \pi/2$ in graphing $y = \tan x$. We start by making a table of values at convenient points, much as we did for the sine function in Table 7.2. These values are displayed in Table 7.3.

Table 7.3 $y = \tan x$

x	$-\frac{\pi}{3}$	$-\frac{\pi}{4}$	$-\frac{\pi}{6}$	0	$\frac{\pi}{6}$	$\frac{\pi}{4}$	$\frac{\pi}{3}$
y	$-\sqrt{3}$	-1	$-\frac{\sqrt{3}}{3}$	0	$\frac{\sqrt{3}}{3}$	1	$\sqrt{3}$

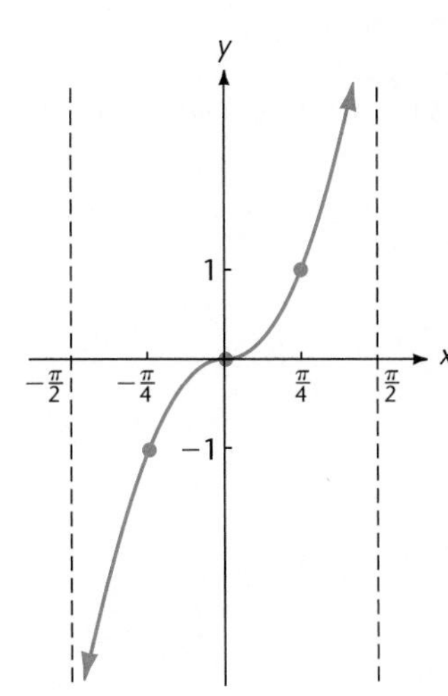

Figure 7.58 Graph of $y = \tan x$, $x \in \left(-\frac{\pi}{2}, \frac{\pi}{2}\right)$

Since $\tan x = \sin x/\cos x$ and $\cos(\pi/2) = 0$, $y = \tan x$ is undefined at $x = \pi/2$. For the moment, we concentrate on values of x between $\pi/3$ and $\pi/2$. As x gets closer to $\pi/2$, $\sin x$ gets closer to 1 and $\cos x$ gets closer to 0. As this happens, $\tan x = \sin x/\cos x$ increases without bound. (This fact can be observed to some extent in the tables or by use of a calculator.) Thus $y = \tan x$ has a vertical asymptote at $x = \pi/2$. A similar situation occurs at $x = -\pi/2$, as shown in Figure 7.58.

We could check on the fact that $y = \tan x$ has period π by plotting more points,† but we do not bother with this. Instead, we assume that the graph of $y = \tan x$ given in Figure 7.59 is correct.

†This fact can be proved by using the identities in Chapter 8.

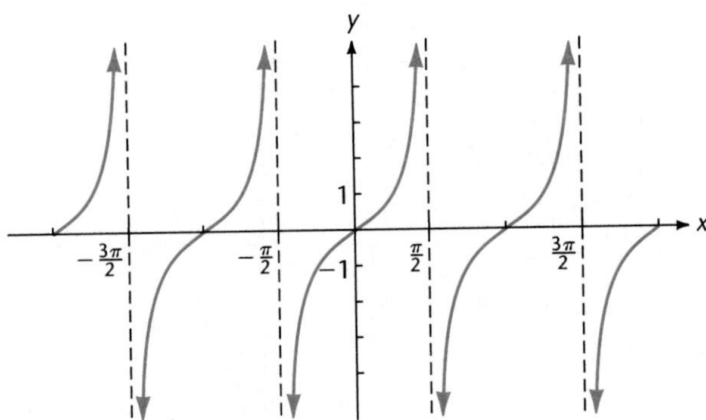

Figure 7.59 Graph of $y = \tan x$

We shall now investigate the effect of the constants a, b, and c on the graph of $y = a \tan (bx + c)$. As we have done before, we restrict ourselves to the case where b is positive.

It is easy to see that the factor a in $y = a \tan x$ simply multiplies the function values in $y = \tan x$ by a. This effect is most conspicuous at $x = \pi/4$ and $x = -\pi/4$ since these are the points where $\tan x = \pm 1$ and $y = \pm a$.

The positive multiplier b in $y = a \tan bx$ has the same sort of effect that it did in previous cases. It changes the period from π to π/b. (The difference here is that we are starting with a period of π instead of 2π.)

Finally, the constant c in $y = a \tan (bx + c)$ causes a phase shift in a similar way as it did with the previous trigonometric functions.

We take the shape of the graph in Figure 7.58 as a pattern for the **typical cycle for a tangent curve.** This typical cycle is shown in Figure 7.60. Those points marked with large dots in the figure are our **key points** on the graph.

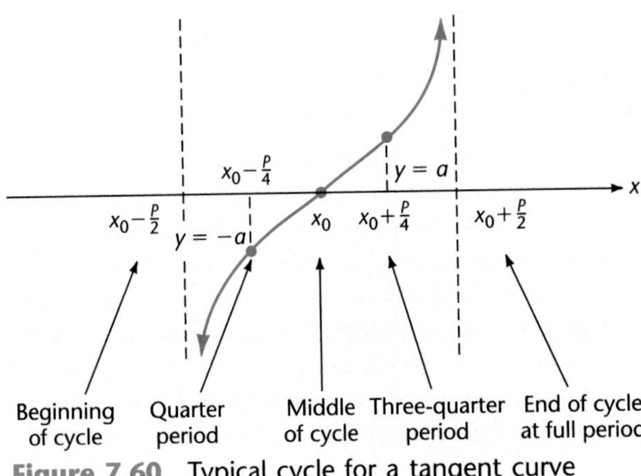

Figure 7.60 Typical cycle for a tangent curve

To sketch the graph of a function of the type $y = a \tan(bx + c)$ with $b > 0$, we use the typical cycle of a tangent curve in Figure 7.60 and the following guide.

Guide for Graphing $y = a \tan(bx + c)$, where $b > 0$

1. $P = \dfrac{\pi}{b}$.
2. The middle of a cycle is at the value of x where $bx + c = 0$.
3. Vertical asymptotes are located at the beginning and end of the cycle.
4. If $a > 0$, the graph is a typical tangent curve. If $a < 0$, it is the reflection through the x-axis of a typical tangent curve.

Example 3 • Using the Guide to Graph a Tangent Curve

Sketch the graph of $y = 2\tan\left(\dfrac{1}{3}x + \dfrac{\pi}{6}\right)$ over one complete period.

Solution

We have $a = 2$, $b = 1/3$, and $c = \pi/6$. The period is $P = \pi/b = \pi/(1/3) = 3\pi$. The middle of a cycle is where

$$\frac{1}{3}x + \frac{\pi}{6} = 0$$

$$x = -\frac{\pi}{2}.$$

To locate the beginning of the cycle, we mark off half the length of the period to the left of $x = -\pi/2$. That is, we subtract $\frac{1}{2}(3\pi)$ from $-\pi/2$.

$$x = -\frac{\pi}{2} - \frac{1}{2}(3\pi) = -2\pi$$

Similarly, we add $3\pi/2$ to $-\pi/2$ to locate the end of the cycle.

$$x = -\frac{\pi}{2} + \frac{1}{2}(3\pi) = \pi$$

We plot the value $y = a = 2$ halfway between the middle and end of the cycle, at $x = \pi/4$. Similarly, $y = -a = -2$ at $x = -5\pi/4$. Using the typical cycle for a tangent curve, we obtain the graph in Figure 7.61. □

Since $\cot x = 1/\tan x$, the function $y = \cot x$ has period π. We can use the graph of $y = \tan x$ to obtain the graph of $y = \cot x$, noting that each of the two functions has a vertical asymptote where the other is zero. To illustrate the rela-

x	y
-2π	Undefined
$-\frac{5\pi}{4}$	-2
$-\frac{\pi}{2}$	0
$\frac{\pi}{2}$	2
π	Undefined

Figure 7.61 Graph of $y = 2\tan\left(\frac{1}{3}x + \frac{\pi}{6}\right)$, $x \in (-2\pi, \pi)$

tionship between the graphs, the graph of $y = a\cot x$ is drawn as a solid curve and the graph of $y = a\tan x$ is drawn as a dashed curve in Figure 7.62.

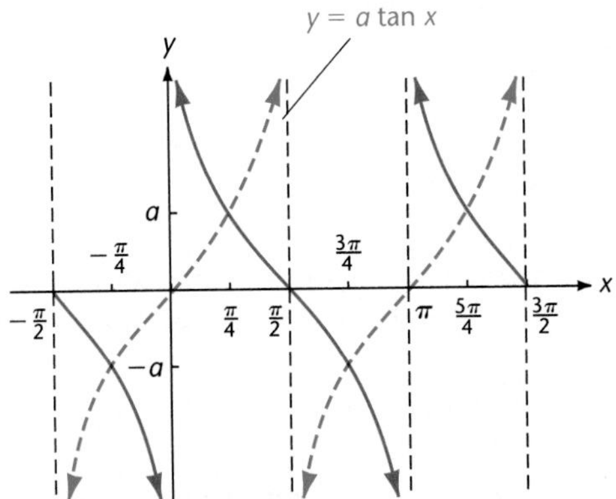

Figure 7.62 Graph of $y = a\cot x$, $a > 0$

To be consistent with our procedure for graphing tangent functions, we use the portion of the solid graph in Figure 7.62 between $-\pi/2$ and $\pi/2$ as the **typical cycle of a cotangent curve.**

Example 4 • Graphing a Cotangent Curve

Sketch the graph of $y = 2\cot\left(\frac{1}{3}x + \frac{\pi}{6}\right)$ over one complete period.

Solution

The period is $P = \pi/b = \pi/(1/3) = 3\pi$. The middle of a cycle is where

$$\frac{1}{3}x + \frac{\pi}{6} = 0$$

$$x = -\frac{\pi}{2}.$$

The beginning of the cycle is at

$$x = -\frac{\pi}{2} - \frac{1}{2}(3\pi) = -2\pi,$$

and the end of the cycle is at

$$x = -\frac{\pi}{2} + \frac{1}{2}(3\pi) = \pi.$$

We plot $y = a = 2$ at $x = \pi/4$ and $y = -a = -2$ at $x = -5\pi/4$. (The work to this point is exactly the same as for the tangent function in Example 3.) We plot zeros at each end of the period, draw a vertical asymptote in the middle of the period, and sketch the graph as shown in Figure 7.63. □

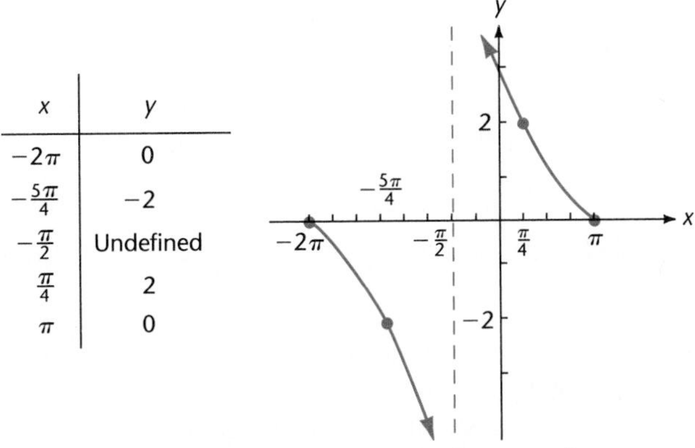

x	y
-2π	0
$-\frac{5\pi}{4}$	-2
$-\frac{\pi}{2}$	Undefined
$\frac{\pi}{4}$	2
π	0

Figure 7.63 Graph of $y = 2 \cot\left(\frac{1}{3}x + \frac{\pi}{6}\right)$, $x \in [-2\pi, \pi]$

A graphing device can be used to examine graphs of complicated functions, but the choice of a suitable viewing rectangle is required.

Suppose that a given function $y = f(x)$ can be expressed as the sum $f(x) = g(x) + h(x)$ of two simpler functions, $g(x)$ and $h(x)$. The graph of the sum $y = f(x)$ can be obtained by graphing $y_1 = g(x)$ and $y_2 = h(x)$ separately and then adding their ordinates: $y = y_1 + y_2$ for each value of x. That is, for any value of x, we add the y-value from $g(x)$ to the y-value from $h(x)$ to obtain the y-value for $f(x)$. This method, called **addition of ordinates,** is illustrated in the next two examples.

Example 5 • Using the Addition of Ordinates Technique

Sketch the graph of $y = x + \cos x$ over the interval $0 \le x \le 2\pi$.

Solution

We first graph $y_1 = x$ and $y_2 = \cos x$ separately. These graphs are shown in Figure 7.64 as dashed curves. We then select some x-values at which it is convenient to add the ordinates y_1 and y_2. The large dots in Figure 7.64 show the results of the selections that were made. Whenever possible we choose points where the value of either y_1 or y_2 is known without computation. As soon as enough points are plotted to reveal the shape of the graph, we draw a smooth curve through them. □

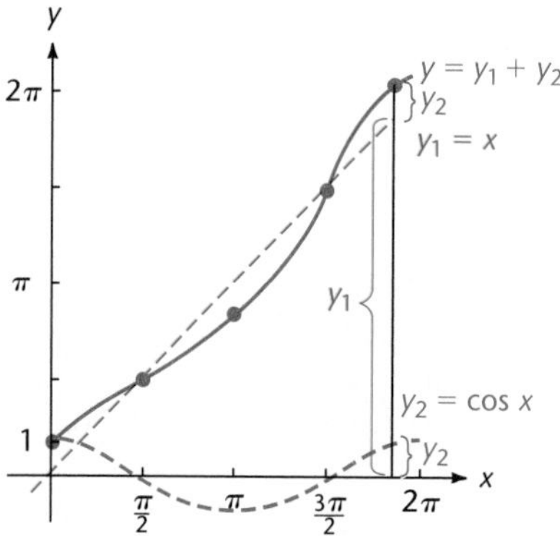

Figure 7.64 Graph of $y = x + \cos x$, $x \in [0, 2\pi]$

Our next example shows how we handle a situation that involves the *difference* of two functions rather than the sum.

Example 6 • Graphing the Difference of Two Functions

Sketch the graph of $y = \sin x - \cos 2x$ over the interval $0 \le x \le 2\pi$.

Solution

The graph of $y = \sin x - \cos 2x$ could be obtained by graphing $\sin x$ and $\cos 2x$ separately and then graphically *subtracting* ordinates of $\cos 2x$ from ordinates of $\sin x$. However, it is easier to graph $y_1 = \sin x$ and $y_2 = -\cos 2x$ and then *add* ordinates as we did in Example 5 (see Figure 7.65). □

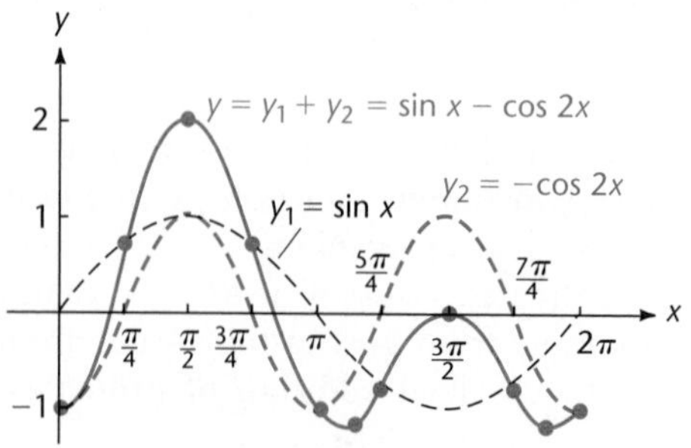

Figure 7.65 Graph of $y = \sin x - \cos 2x$, $x \in [0, 2\pi]$

◆ **Practice Problem 1** ◆

Find the period and phase shift for the graph of the following functions. Do not draw the graphs.

a. $y = 2\csc(3x + \pi)$

b. $y = 3\sec\left(4x - \dfrac{2\pi}{3}\right)$

c. $y = 2\tan\left(\dfrac{1}{2}x + \dfrac{\pi}{6}\right)$

d. $y = \dfrac{1}{3}\cot\left(2x - \dfrac{\pi}{4}\right)$

EXERCISES 7.9

Find the period, amplitude (if it is defined), and any phase shift or vertical translation for the graph of each of the following functions. Do not draw the graphs.

1. $y = 3\csc\left(2x - \dfrac{\pi}{4}\right)$
2. $y = 2\csc\left(3x + \dfrac{\pi}{2}\right)$
3. $y = 2\sec\left(3x + \dfrac{\pi}{3}\right)$
4. $y = 3\sec\left(4x - \dfrac{2\pi}{3}\right)$
5. $y = -\tan\left(2x + \dfrac{\pi}{4}\right)$
6. $y = 4\tan\left(\dfrac{1}{3}x - \dfrac{\pi}{6}\right)$
7. $y = 4\cot\left(\dfrac{1}{2}x - \dfrac{\pi}{4}\right)$
8. $y = -2\cot\left(x + \dfrac{\pi}{4}\right)$
9. $y = 2\cos x$
10. $y = \frac{1}{2}\sin 4x$
11. $y = 1 + 2\sin\left(\dfrac{1}{2}x + \dfrac{\pi}{6}\right)$
12. $y = 2 + 6\cos\left(\dfrac{1}{3}x - \dfrac{\pi}{6}\right)$
13. $y = 2 + 3\csc\left(\dfrac{1}{3}x - \dfrac{\pi}{2}\right)$
14. $y = 3 + 4\sec\left(2x + \dfrac{\pi}{2}\right)$
15. $y = 1 + 5\tan(4x - \pi)$
16. $y = 4 + 2\cot(3x - 2\pi)$

Sketch the graph of the given function through one complete period. (See Examples 1–4.)

17. $y = \csc 2x$
18. $y = \sec 2x$
19. $y = \tan 2x$
20. $y = \cot 2x$
21. $y = \frac{1}{3}\sec 2x$
22. $y = \frac{1}{4}\tan 3x$
23. $y = \frac{1}{2}\cot 3x$
24. $y = 2\csc\left(\dfrac{x}{3}\right)$
25. $y = 3\csc\left(\dfrac{1}{2}x - \dfrac{\pi}{2}\right)$
26. $y = \dfrac{1}{4}\csc\left(2x + \dfrac{\pi}{4}\right)$
27. $y = \frac{1}{3}\sec(2x - \pi)$
28. $y = \dfrac{1}{3}\sec\left(2x + \dfrac{\pi}{3}\right)$

29. $y = \tan(2x - \pi)$

30. $y = \tan\left(\frac{1}{2}x + \frac{\pi}{16}\right)$

31. $y = \frac{1}{2}\tan\left(x + \frac{\pi}{6}\right)$

32. $y = \frac{1}{3}\tan\left(x - \frac{\pi}{4}\right)$

33. $y = 3\tan\left(\frac{1}{3}x - \frac{\pi}{6}\right)$

34. $y = 2\tan\left(\frac{1}{3}x + \frac{\pi}{6}\right)$

35. $y = 2\cot\left(\frac{1}{3}x - \frac{\pi}{12}\right)$

36. $y = 3\cot\left(\frac{1}{2}x + \frac{\pi}{6}\right)$

Sketch the graph of each of the following functions over the interval $0 \le x \le 2\pi$. (See Examples 5 and 6.)

37. $y = 2 + \sin x$

38. $y = -2 + \sin x$

39. $y = x - \cos x$

40. $y = x + \sin x$

41. $y = \sin x + \cos 2x$

42. $y = \cos x + \sin 2x$

43. $y = 2\cos x + \sin 2x$

44. $y = 3\sin x + \cos 2x$

Electrical Circuit

45. The accompanying figure shows a diagram of an electrical circuit that contains an electromotive force $E(t) = 60$ volts, an inductance $L = 1$ henry, and a capacitance $C = \frac{1}{16}$ farad. The charge $q(t)$ on the capacitor in the circuit is given in coulombs by

$$q(t) = \tfrac{15}{4} - \tfrac{15}{4}\cos 4t.$$

Sketch the graph of q through one complete period.

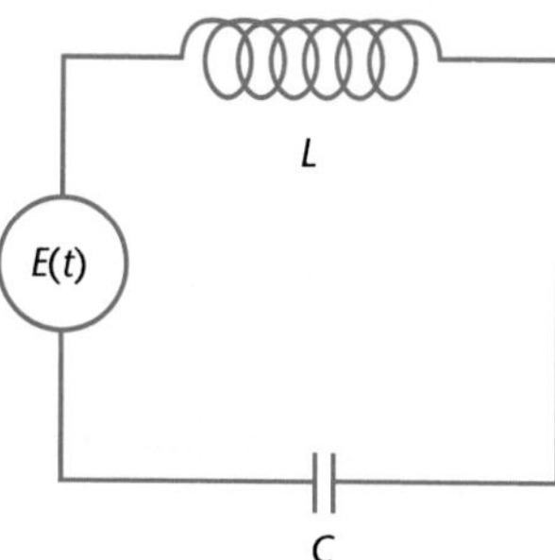

Figure for Exercise 45

Hours of Daylight

46. For a given location, the number H of hours of daylight in each day of the year can be expressed as a function of the date. If n is the number of days after January 1, the number $H(n)$ of hours of daylight in Atlanta, Georgia, is approximated by

$$H(n) = \frac{9}{4}\sin\frac{2\pi}{365}(n - 79) + \frac{49}{4}.$$

Sketch the graph of H from $n = 0$ to $n = 365$ and find the values of n that give a maximum value or a minimum value of H.

Stretching a Spring

47. An object weighing 2 pounds stretches a spring $\frac{1}{2}$ foot and reaches an equilibrium position as shown in the accompanying figure. The spring is then stretched an additional $\frac{2}{3}$ foot, and the object is released with an upward velocity of $\frac{4}{3}$ foot per second. The only forces acting on the object are its weight and the force of the spring. If x denotes the displacement of the object from its equilibrium position t seconds after its release, then

$$x = \tfrac{2}{3}\cos 8t - \tfrac{1}{6}\sin 8t,$$

where x is in feet, measured positive downward and negative upward. Sketch the graph of x through one complete period.

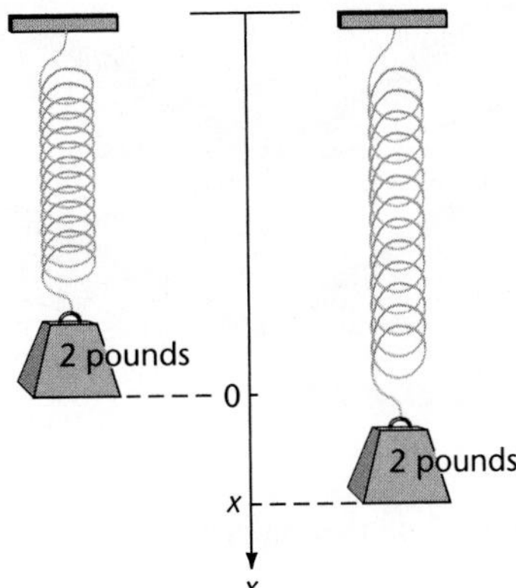

Figure for Exercise 47

Critical Thinking: Exploration

48. a. Sketch the graph of $y = \sin(x + \pi)$ over the interval $[-2\pi, 2\pi]$.
    b. Sketch the graph of $y = \cos(x + \pi)$ over the interval $[-2\pi, 2\pi]$.
    c. Observe from the graphs in parts a and b that $\sin(x + \pi) = -\sin x$ and $\cos(x + \pi) = -\cos x$.
    d. Use the observations in part c to show that $\tan(x + \pi) = \tan x$.

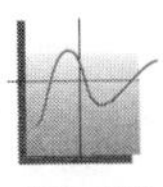

49–58

Use a graphing device to graph the following functions.

49. $y = \dfrac{\tan x}{x}$ 50. $y = x\cot x$ 51. $y = x\sec x$ 52. $y = x\csc x$

53. $y = \sqrt{\tan x}$ 54. $y = \tan^2 x$ 55. $y = |\tan x|$ 56. $y = \csc |x|$

Clearance Around a Hall Corner

57. Two halls in a building have widths 3 feet and 4 feet and intersect at an angle of 120°, as shown in the accompanying figure. The longest straight board that can be carried horizontally around the corner is the same as

the minimum value of the length ℓ shown. Show that the value of ℓ is given by

$$\ell = \frac{3}{\sin \theta} + \frac{4}{\sin\left(\frac{\pi}{3} - \theta\right)}$$

and use a graphing device to find the minimum value of ℓ, to the nearest tenth of a foot.

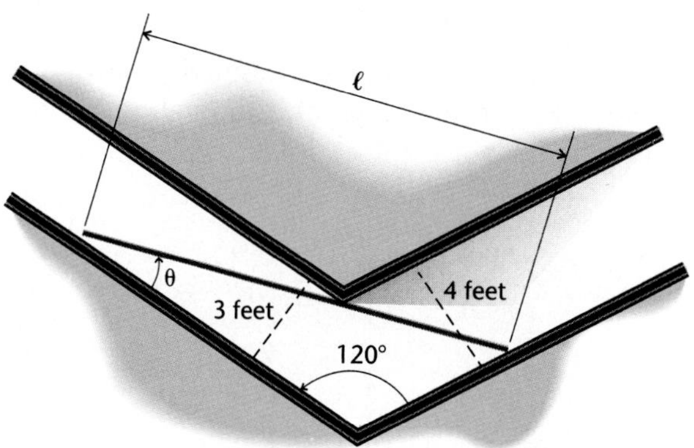

Figure for Exercise 57

58. Work Exercise 57 if the halls intersect at right angles.

◆ Solution for Practice Problem

1. a. For $y = 2 \csc(3x + \pi)$, $P = 2\pi/b = 2\pi/3$. When $3x + \pi = 0$, we obtain $x = -\pi/3$, so the phase shift is $\pi/3$ units to the left.

   b. For $y = 3 \sec\left(4x - \frac{2\pi}{3}\right)$, $P = 2\pi/b = 2\pi/4 = \pi/2$.

$$4x - \frac{2\pi}{3} = 0$$

$$x = \frac{\pi}{6}.$$

   The phase shift is $\pi/6$ units to the right.

   c. For $y = 2 \tan\left(\frac{1}{2}x + \frac{\pi}{6}\right)$, $P = \pi/b = \pi/(1/2) = 2\pi$.

$$\frac{1}{2}x + \frac{\pi}{6} = 0$$

$$x = -\frac{\pi}{3}.$$

The phase shift is $\pi/3$ units to the left.

d. For $y = \frac{1}{3}\cot\left(2x - \frac{\pi}{4}\right)$, $P = \pi/b = \pi/2$.

$$2x - \frac{\pi}{4} = 0$$

$$x = \frac{\pi}{8}.$$

The phase shift is $\pi/8$ units to the right.

CHAPTER REVIEW

Summary of Important Concepts and Formulas

Conversions in Degree Measure

$1° = 60'$, $1' = 60''$

Basic Equation Relating Radians and Degrees

$\pi = 180°$

Conversion Procedures for Radians and Degrees

To change radians to degrees, multiply by $180/\pi$.
To change degrees to radians, multiply by $\pi/180$.

Arc Length

$s = r\theta$

Linear and Angular Speed

$v = r\omega$

Trigonometric Functions

$\sin\theta = \frac{y}{r}$ $\cos\theta = \frac{x}{r}$ $\tan\theta = \frac{y}{x}$

$\cot\theta = \frac{x}{y}$ $\sec\theta = \frac{r}{x}$ $\csc\theta = \frac{r}{y}$

Reciprocal Identities

$\csc\theta = \frac{1}{\sin\theta}$ $\sec\theta = \frac{1}{\cos\theta}$ $\cot\theta = \frac{1}{\tan\theta}$

Quotient Identities

$\tan\theta = \frac{\sin\theta}{\cos\theta}$ $\cot\theta = \frac{\cos\theta}{\sin\theta}$

Ranges of the Trigonometric Functions

$-1 \le \cos\theta \le 1$ $-1 \le \sin\theta \le 1$

$\sec\theta \le -1$ or $\sec\theta \ge 1$ $\csc\theta \le -1$ or $\csc\theta \ge 1$

$\tan\theta$: all real numbers $\cot\theta$: all real numbers

Trigonometric Functions in a Right Triangle

$\sin A = \frac{\text{opp}}{\text{hyp}}$ $\cos A = \frac{\text{adj}}{\text{hyp}}$ $\tan A = \frac{\text{opp}}{\text{adj}}$

$\cot A = \frac{\text{adj}}{\text{opp}}$ $\sec A = \frac{\text{hyp}}{\text{adj}}$ $\csc A = \frac{\text{hyp}}{\text{opp}}$

Wrapping Function

$P(t) = (\cos t, \sin t)$

Trigonometric Functions of a Real Number t

Any trigonometric function of t is equal to the same function of an angle with measure t radians.

Guide for Graphing $y = a \sin(bx + c),\ b > 0$

1. Amp $= |a|$.
2. $P = \dfrac{2\pi}{b}$.
3. A cycle begins at the value of x where the argument is 0, that is, where $bx + c = 0$.
4. If $a > 0$, the graph is a typical sine curve. If $a < 0$, the graph is the reflection through the x-axis of a typical sine curve.

Guide for Graphing $y = a \cos(bx + c),\ b > 0$

1. Amp $= |a|$.
2. $P = \dfrac{2\pi}{b}$.
3. A cycle begins at the value of x where the argument is 0, that is, where $bx + c = 0$.
4. If $a > 0$, the graph is a typical cosine curve. If $a < 0$, it is the reflection through the x-axis of a typical cosine curve.

Guide for Graphing $y = a \tan(bx + c),\ b > 0$

1. $P = \dfrac{\pi}{b}$.
2. The middle of a cycle is at the value of x where $bx + c = 0$.
3. Vertical asymptotes are located at the beginning and end of the cycle.
4. If $a > 0$, the graph is a typical tangent curve. If $a < 0$, it is the reflection through the x-axis of a typical tangent curve.

Critical Thinking: Find the Errors

Each of the following nonsolutions has one or more errors. Can you find them?

Problem 1

The angle of depression from the top of a 62-foot pine tree to a point on the ground directly behind a log skidder is 38°. How far from the pine tree is the log skidder?

Nonsolution

The length x of the side of the right triangle in Figure 7.66 is the required distance. We use the tangent function to find x.

$$\tan 38° = \frac{x}{62}$$

$$x = 62 \tan 38° = 48 \text{ feet}$$

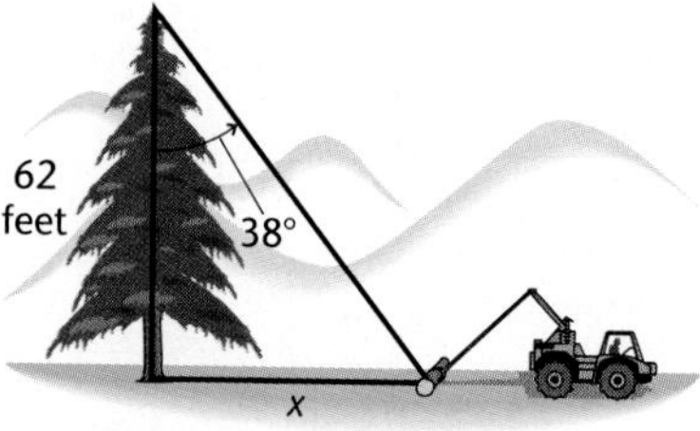

Figure 7.66

Problem 2

Convert 48°24′ to decimal-degree format.

Nonsolution

$48°24' = 48.24°$

Problem 3

Find $\cos\theta$ if $\sin\theta = \frac{4}{5}$ and θ is in quadrant II.

Nonsolution

$$\sin\theta = \frac{y}{r} = \frac{4}{5}$$

$$y = 4 \qquad \text{and} \qquad r = 5$$

$$5^2 = x^2 + 4^2$$

$$9 = x^2$$

$$x = 3$$

$$\cos\theta = \frac{x}{r} = \frac{3}{5}$$

Problem 4

Use special angles to find cos 120°.

Nonsolution

In Figure 7.67 the angle between the terminal side of 120° and the y-axis is 30°. Therefore $\cos 120° = -\cos 30° = -\sqrt{3}/2$.

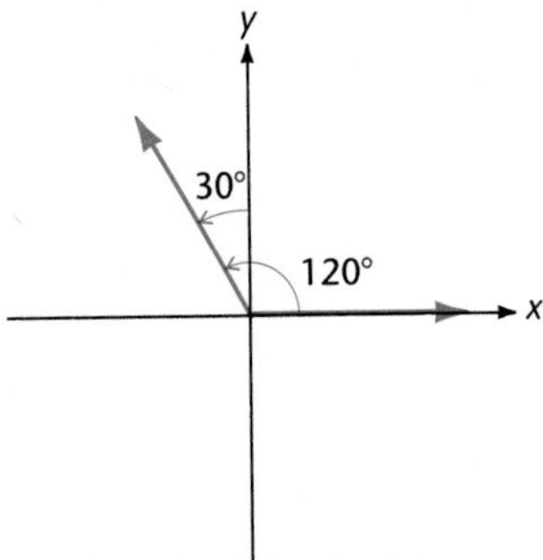

Figure 7.67

Problem 5

A wheel on a trailer has a radius of 9 inches. If the wheel is traveling at 36 feet per second, find its angular speed in radians per second.

Nonsolution

$$\omega = \frac{v}{r} = \frac{36}{9} = 4 \text{ radians per second}$$

Problem 6

Sketch the graph of $y = 2\sin\left(\frac{x}{2} - \frac{\pi}{4}\right)$ over one complete period, locating the key points on the graph.

Nonsolution

We have

$$y = 2\sin\left(\frac{x}{2} - \frac{\pi}{4}\right) = \sin 2\left(\frac{x}{2} - \frac{\pi}{4}\right) = \sin\left(x - \frac{\pi}{2}\right)$$

The amplitude is 1, and the period is 2π. A period starts at $x = \frac{\pi}{2}$. The graph is shown in Figure 7.68.

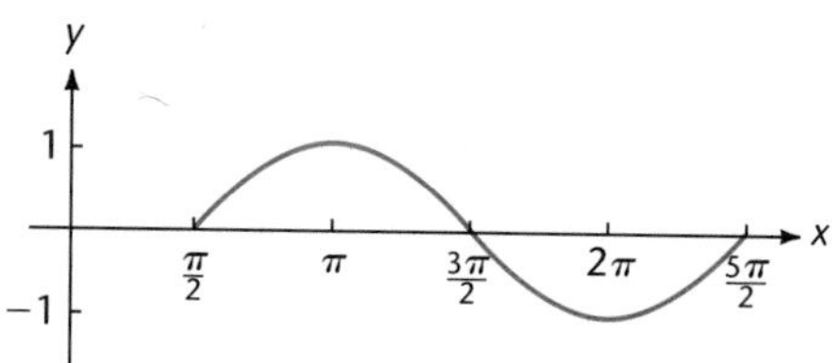

Figure 7.68

Problem 7

Sketch the graph of $y = \cos\left(x + \frac{\pi}{2}\right)$ over one complete period, locating the key points on the graph.

Nonsolution

The amplitude is 1, and the period is 2π. A period starts at $x = -\pi/2$. The graph is shown in Figure 7.69.

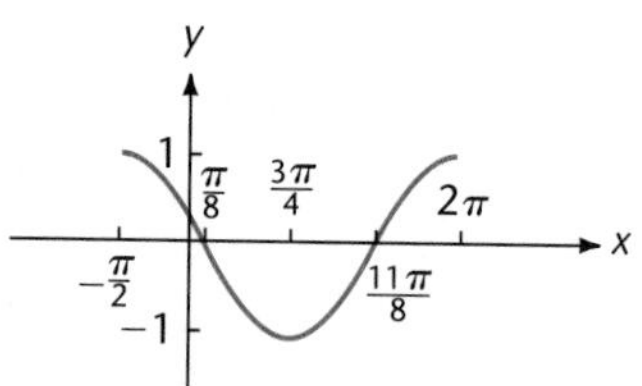

Figure 7.69

Review Problems for Chapter 7

1. Name the angle according to its quadrant. Find four angles (two positive and two negative) that are coterminal to the given angle. Choose your angles with measure as close to 0° as possible.

   a. 123° b. −44°

Swing Set Brace

2. The brace at the end of a swing set is made from three pipes in the shape of an A. The pipes for the legs are 10 feet long, and the pipe for the crossbar is 4 feet long. How far down the legs should the crossbar be bolted so that the legs spread 6 feet apart on the ground?
3. Determine the exact values of the six trigonometric functions of an angle θ in standard position if the terminal side of θ goes through $(-2\sqrt{10}, 3)$.
4. Determine the exact values of all the other trigonometric functions of the fourth quadrant angle α if $\tan\alpha = -24/7$.
5. Use identities to determine the exact values of the remaining trigonometric functions of the angle β, where

$$\csc\beta = \frac{3}{2} \quad \text{and} \quad \cot\beta = \frac{\sqrt{5}}{2}.$$

Latitudes and Distance 6–7

6. Grand Rapids, Michigan, is almost due north of Gadsden, Alabama. Their latitudes are 43°N and 34°N respectively. Use 4000 miles as the radius of the earth and find the distance between the cities to the nearest 10 miles.
7. (See Problem 6.) Walla Walla, Washington, is almost due north of Los Angeles, California; their latitudes are 46°N and 34°N, respectively. Find the distance between the two cities to the nearest 10 miles.

Angular and Linear Speed 8–10

8. A belt-driven roller with diameter 3.0 feet has an angular speed of 7π radians per minute. Find the speed at which the belt is moving in feet per second.

9. The wheels of a cart are 28 inches in diameter and are turning at the rate of 24 revolutions per minute. Find the speed of the cart in miles per hour.

10. A bicycle with wheels 28 inches in diameter is traveling at a speed of 14 miles per hour. Find the angular speed of a wheel in radians per minute.

11. Convert each of the following radian measures to degree measure.

a. $-\frac{7\pi}{6}$ b. $\frac{5\pi}{3}$

12. Convert each of the following to radian measure in terms of π.

a. $630°$ b. $42°11'15''$

13. In a circle of diameter 36 centimeters, find the length of the arc intercepted by a central angle measuring $30°$.

14. Write out the exact values of all the trigonometric functions of the given angle.

a. $-270°$ b. $510°$ c. $315°$

d. $-\frac{\pi}{2}$ e. $\frac{7\pi}{6}$ f. $\frac{4\pi}{3}$

15. Determine whether each of the following statements is possible or impossible.

a. $\cos\theta = -3/2$

b. $\tan\alpha = 5/6$

c. $\sin\beta = 5/4$ and $\csc\beta = 4/5$

d. $\cot\alpha < 0$ and $\sec\alpha > 0$

e. $\cot\alpha < 0$ and $\tan\alpha > 0$

16. Find the values of the following trigonometric functions, correct to four digits. Use a calculator or Tables I and II, as instructed by the teacher.

a. $\sec 13.7°$ b. $\tan 143°10'$

c. $\csc(-217°20')$ d. $\cos(-125.4°)$

e. $\sin 438.2°$

17. Find the values of the following trigonometric functions by using either a calculator or Table I. Give answers correct to four significant digits.

a. $\cos 1.4341$ b. $\tan 2.3969$

18. Suppose θ has the given function value and lies in the given quadrant, with $0° \le \theta < 360°$. According to the instructions of the teacher, use a calculator or Table II to find θ in decimal degrees to the nearest tenth of a degree.

a. $\sin\theta = 0.4179$, Q I

b. $\cot\theta = 0.4813$, Q I

c. $\cos\theta = -0.8829$, Q II

d. $\tan\theta = 1.1383$, Q III

e. $\cos\theta = 0.3305$, Q IV

19. Suppose θ has the given function value and lies in the given quadrant, with $0 \le \theta < 2\pi$. According to the teacher's instructions, use a calculator or Table I to find θ in radians, rounded to four decimal places.

a. $\tan\theta = -7.596$, Q II

b. $\sin\theta = -0.2812$, Q III

c. $\cos\theta = -0.3827$, Q II

d. $\cot\theta = -0.6619$, Q IV

e. $\sec\theta = -3.388$, Q III

20. Use Table I or a calculator to find a value of t, with $0 \le t \le 1.5708$, that satisfies the given equation. Give the answers to four decimal places.

a. $\tan t = 0.3314$ b. $\sec t = 2.203$

21. With θ in radian measure and in terms of π, find all values of θ for which $0 \le \theta < 2\pi$ and the given equation is true.

a. $\tan\theta = 1$ b. $\sin\theta = -1/2$

22. Solve the right triangle that has $a = 84$ feet, $A = 15°$.

Surveying

23. From one corner of a triangular field, the boundary fences run in the direction N 37°40′ E for 516 yards and in the direction S 52°20′ E. The fence opposite this corner runs in a north-south direction. Find its length.

Angle of Depression

24. The television broadcasting tower in the accompanying figure is known to be 527 meters in height. From the top of the tower, the angle of depression of an irrigation pump is 29°30′. Find the distance from the base of the tower to the pump.

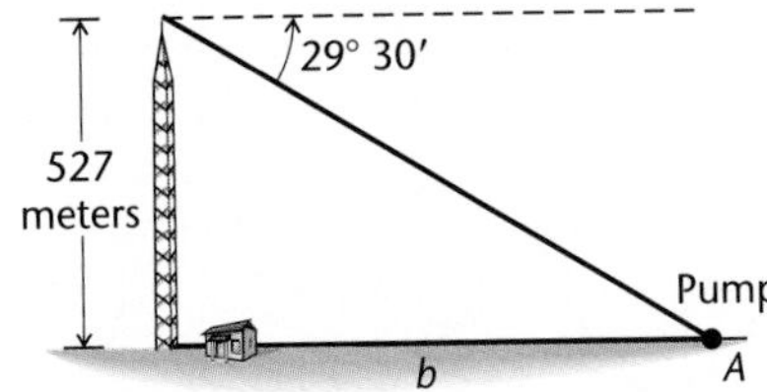

Figure for Problem 24

25. Find the rectangular coordinates of each of the following points on the unit circle.

a. $P\left(-\frac{3\pi}{2}\right)$ b. $P\left(\frac{2\pi}{3}\right)$

26. Find the period, amplitude (if it is defined), and any phase shift or vertical translation for the graph of each of the following functions. Do not draw the graphs.

    a. $y = 1 + 2\cos(2x - \pi)$

    b. $y = -2 + \dfrac{1}{2}\sin(4x + \pi)$

    c. $y = 4 + \tan 3x$

    d. $y = 3\csc\left(x + \dfrac{\pi}{2}\right)$

    e. $y = 4\sec\left(\dfrac{1}{2}x - \dfrac{\pi}{6}\right)$

In Problems 27–35, sketch the graph of the given function through one complete period, locating key points on each graph.

27. $y = \dfrac{1}{3}\cos 2x$

28. $y = \tan 3x$

29. $y = \dfrac{1}{4}\csc\left(\dfrac{3}{2}x + \dfrac{\pi}{4}\right)$

30. $y = \sec\left(\dfrac{1}{2}x - \dfrac{\pi}{6}\right)$

31. $y = \cot\left(x + \dfrac{\pi}{4}\right)$

32. $y = \dfrac{2}{5}\sin\left(3x - \dfrac{\pi}{2}\right)$

33. $y = \tan\left(x + \dfrac{\pi}{4}\right)$

34. $y = 2\sin\left(\dfrac{1}{2}x + \dfrac{\pi}{2}\right)$

35. $y = 1 + 2\cos\left(\dfrac{1}{2}x + \dfrac{\pi}{4}\right)$

36. Use the addition of ordinates method to sketch the graph of $y = 2\cos x + \sin x$ over the interval $0 \le x \le 2\pi$.

Find an equation of the form $y = a\sin(bx + c)$ that has the given graph.

37.

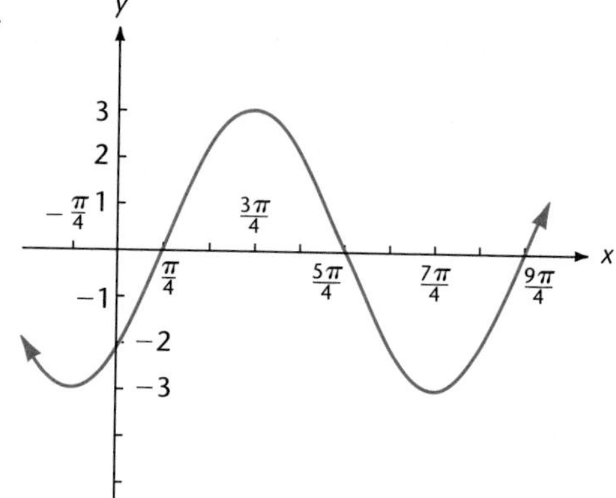

38.

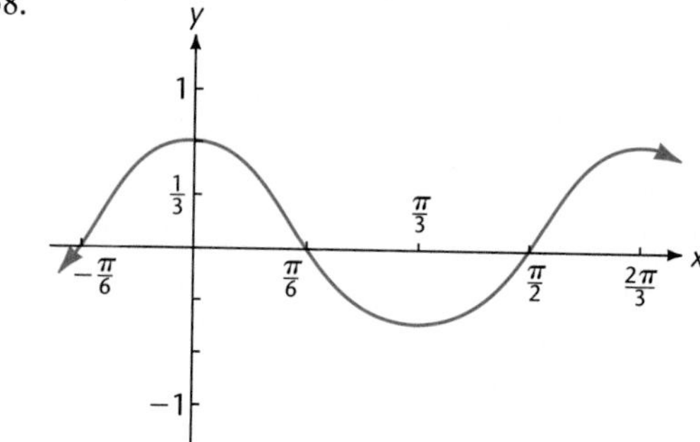

Find an equation of the form $y = a\cos(bx + c)$ that has the given graph.

39.

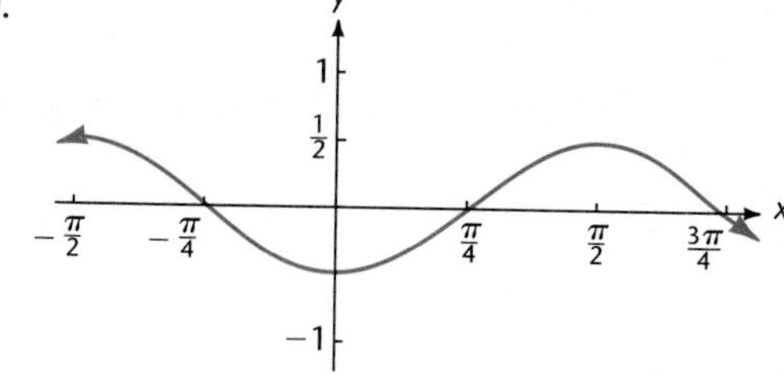

40.

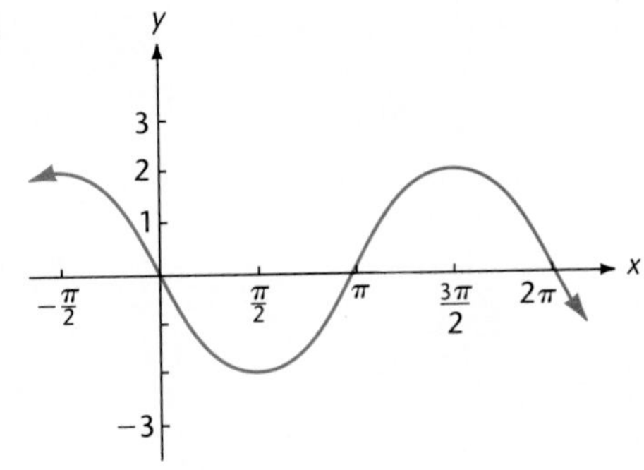

Hours of Daylight

41. If n is the number of days after January 1, the number $H(n)$ of hours of daylight in Tampa, Florida, is approximated by

$$H(n) = \frac{11}{6} \sin \frac{2\pi}{365}(n - 79) + \frac{73}{6}.$$

Sketch the graph of H from $n = 0$ to $n = 365$ and find the values of n that give a maximum value or a minimum value of H.

ENCORE

Bohdan Hrynewych/Stock, Boston

Mount Everest, on the Tibet-Nepal frontier in the Himalaya range, is the highest mountain in the world.

"Everest Shrinks 7 Feet in Eyes of Scientists" was the headline on an Associated Press news release that appeared in April 1993. According to the Associated Press story, a new measurement of Mount Everest placed its peak at 29,022 feet 7 inches. This made the famous mountain 6 feet 8 inches lower than previously believed, but still well above the second highest peak, K2 in Kashmir, at 28,250 feet.

The new measurement was made by a scientific team sponsored by Baume and Mercier, A Geneva-based watch company that also provided precision equipment for the team. The scientists made the measurement by aiming laser beams at prisms located on the peak and at reference points off the mountain. The fact that the prisms reflected the light at known angles provided enough information about a triangle to enable the scientists to calculate the height of the mountain.

Some drawings of prisms are shown in Figure 7.70. Prisms made of glass or quartz are used in many optical instruments because of their ability to reflect light in a desired direction. They can also change the direction of light by refraction, and this property makes them an important part of instruments used to measure the spectral composition of light. Angle of reflection and angle of refraction are illustrated in Figure 7.24 of this chapter, and an angle of refraction is calculated in Example 4 in Section 7.5.

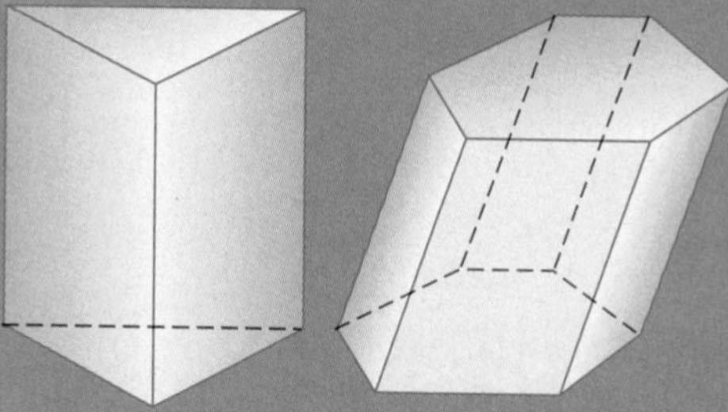

Figure 7.70 Examples of prisms

Analytic Trigonometry

Most of the material in this chapter concerns identities in trigonometric functions. A knowledge of these identities is essential in calculus, and they play a key role in the analysis of electrical circuits and other physical phenomena. An insight into this role can be obtained by considering the applications in Exercises 8.3 and 8.6.

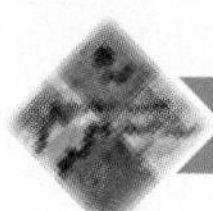

8.1 VERIFYING IDENTITIES

In Section 7.4 the reciprocal identities

$$\sin\theta\csc\theta = 1, \qquad \cos\theta\sec\theta = 1, \qquad \tan\theta\cot\theta = 1$$

and the quotient identities

$$\tan\theta = \frac{\sin\theta}{\cos\theta}, \qquad \cot\theta = \frac{\cos\theta}{\sin\theta}$$

are used to find the values of the trigonometric functions of an angle. More important, these identities can be used to rewrite or reduce expressions involving trigonometric functions (called **trigonometric expressions**).

Frequently the algebra involved in working with trigonometric expressions is more troublesome to students than the trigonometry itself. Fractions involving trigonometric functions can be combined in the same way as fractions involving algebraic expressions. The techniques of factoring extend to trigonometric expressions. Any of the operations on polynomials can be performed when the polynomials contain trigonometric functions as their variables. Once the algebraic operations are mastered, then the trigonometric identities can be used concurrently with the algebra to simplify or reduce trigonometric expressions.

Example 1 • Using a Quotient Identity

Reduce $\cos\theta\tan\theta$ to $\sin\theta$ by using one of the quotient identities.

Solution

We first replace $\tan\theta$ by $\sin\theta/\cos\theta$.

$$\cos\theta\tan\theta = \cos\theta\frac{\sin\theta}{\cos\theta} = \sin\theta$$

The last equality follows by dividing out the common factor, $\cos\theta$. □

Recall that the trigonometric functions of θ are defined in terms of x, y, and r, where (x,y) is a point on the terminal side of θ. Three identities can be derived from the Pythagorean relationship existing among x, y, and r.

$$x^2 + y^2 = r^2$$

Dividing both sides of this equation by r^2 (note that $r^2 > 0$) yields

$$\left(\frac{x}{r}\right)^2 + \left(\frac{y}{r}\right)^2 = 1.$$

Since $\cos\theta = x/r$ and $\sin\theta = y/r$, we obtain what is known as one of the **Pythagorean identities,**

$$\cos^2\theta + \sin^2\theta = 1.$$

Note that this equation is true for *all* values of θ.

Dividing both sides of $\cos^2\theta + \sin^2\theta = 1$ by $\cos^2\theta$, for $\cos\theta \neq 0$, yields

$$1 + \left(\frac{\sin\theta}{\cos\theta}\right)^2 = \left(\frac{1}{\cos\theta}\right)^2$$

and then another Pythagorean identity,

$$\mathbf{1 + \tan^2\theta = \sec^2\theta.}$$

Similarly, dividing both sides of $\cos^2\theta + \sin^2\theta = 1$ by $\sin^2\theta$, for $\sin\theta \neq 0$, yields

$$\left(\frac{\cos\theta}{\sin\theta}\right)^2 + 1 = \left(\frac{1}{\sin\theta}\right)^2,$$

which gives the last Pythagorean identity,

$$\mathbf{\cot^2\theta + 1 = \csc^2\theta.}$$

The Pythagorean identities, along with the reciprocal and quotient identities, are known as the eight **fundamental identities.** They should be memorized. The student must then be able quickly to obtain or recognize any alternative form.

Fundamental Identities

Reciprocal identities:	$\sin\theta\csc\theta = 1$ $\cos\theta\sec\theta = 1$ $\tan\theta\cot\theta = 1$
Quotient identities:	$\tan\theta = \dfrac{\sin\theta}{\cos\theta}$ $\cot\theta = \dfrac{\cos\theta}{\sin\theta}$
Pythagorean identities:	$\sin^2\theta + \cos^2\theta = 1$ $1 + \tan^2\theta = \sec^2\theta$ $1 + \cot^2\theta = \csc^2\theta$

A comparison of graphs on a graphing device will often provide an indication (not a proof) as to whether a given equation is an identity. (See Exercises 111–115.)

The fundamental identities can be used to prove other identities. The phrases "prove the identity" and "verify the identity" mean that we should prove that a given equation is true for all values of the variable such that both members are defined. The typical plan of attack is to reduce the trigonometric expression on one side of the equation to the expression on the other side of the equation.†

Usually the proof of an identity is easier if we *start with the more complicated* of the two expressions. It is also helpful to know what trigonometric expression we are trying to obtain. If we keep one eye on the trigonometric expression we are working toward, the path to that result will be easier to find. In other words, *keep the goal in mind and work toward it.*

†Although this is not the only method of proving identities, it is a method that ensures that each step is reversible.

Sometimes it is advantageous to replace a fraction by a sum (or difference) of two fractions. This technique is illustrated in Example 2.

Example 2 • Replacing a Fraction by a Sum

Prove the identity: $2\cos\theta = \dfrac{\cos\theta\tan\theta + \sin\theta}{\tan\theta}$.

Solution

We reduce the more complicated side

$$\frac{\cos\theta\tan\theta + \sin\theta}{\tan\theta}$$

to the simpler side, $2\cos\theta$.

$$\begin{aligned}
\boldsymbol{\frac{\cos\theta\tan\theta + \sin\theta}{\tan\theta}} &= \frac{\cos\theta\tan\theta}{\tan\theta} + \frac{\sin\theta}{\tan\theta} && \text{algebra}\\
&= \cos\theta + \frac{\sin\theta}{\tan\theta} && \text{algebra}\\
&= \cos\theta + \sin\theta\cot\theta && \text{reciprocal identity}\\
&= \cos\theta + \sin\theta\frac{\cos\theta}{\sin\theta} && \text{quotient identity}\\
&= \cos\theta + \cos\theta && \text{algebra}\\
&\boldsymbol{= 2\cos\theta} && \text{algebra}
\end{aligned}$$

□

Note that although we now have another identity,

$$\frac{\cos\theta\tan\theta + \sin\theta}{\tan\theta} = 2\cos\theta,$$

at our disposal, it is not necessary (or even advisable) to memorize it. We restrict our memory work only to those identities listed on the back endpapers of this text.

◆ Practice Problem 1 ◆

Verify the identity: $\dfrac{1 + \cot^2\theta}{\cot^2\theta} = \sec\ \theta.$

The conjugate of the binomial $a + b$ is $a - b$. If a binomial appears as the denominator of a fraction, the fraction can often be simplified by multiplying both numerator and denominator by the conjugate of the denominator. This algebraic technique can be applied to fractions containing trigonometric functions, as illustrated in Example 3.

Example 3 • Using Conjugates

Verify the identity: $\dfrac{\sin\beta}{\csc\beta - 1} = \dfrac{1+\sin\beta}{\cot^2\beta}$.

Solution

A binomial, $\csc\beta - 1$, appears in the denominator of the left side of the equation. Our first step is to multiply numerator and denominator by the conjugate, $\csc\beta + 1$.

$$\frac{\sin\beta}{\csc\beta - 1} = \frac{\sin\beta(\csc\beta + 1)}{(\csc\beta - 1)(\csc\beta + 1)} \quad \text{algebra (multiplying by the conjugate)}$$

$$= \frac{\sin\beta\csc\beta + \sin\beta}{\csc^2\beta - 1} \quad \text{algebra (using the distributive law)}$$

$$= \frac{1+\sin\beta}{\csc^2\beta - 1} \quad \text{reciprocal identity}$$

$$= \frac{1+\sin\beta}{\cot^2\beta} \quad \text{Pythagorean identity} \quad \square$$

Sometimes a simplification might involve multiplying or factoring polynomials, as shown in the next example.

Example 4 • Factoring

Verify the identity: $1 + 2\tan^2\theta = \sec^4\theta - \tan^4\theta$.

Solution

We choose $\sec^4\theta - \tan^4\theta$ as the more complicated side since it contains the higher powers of $\sec\theta$ and $\tan\theta$.

$$\sec^4\theta - \tan^4\theta = (\sec^2\theta - \tan^2\theta)(\sec^2\theta + \tan^2\theta) \quad \text{algebra (factoring)}$$

$$= 1(\sec^2\theta + \tan^2\theta) \quad \text{Pythagorean identity}$$

$$= 1 + \tan^2\theta + \tan^2\theta \quad \text{Pythagorean identity}$$

$$= 1 + 2\tan^2\theta \quad \text{algebra (combining like terms)} \quad \square$$

In some cases, it might be necessary to rewrite all the trigonometric functions in terms of sines and cosines.

Example 5 • Rewriting in Terms of Sines and Cosines

Prove the identity: $\dfrac{1}{\tan\alpha + \cot\alpha} = \sin\alpha\cos\alpha$.

Solution

It is tempting to multiply numerator and denominator of the left side by the conjugate, $\tan\alpha - \cot\alpha$, of the denominator. However, doing so would lead us astray. A more direct path to the goal of $\sin\alpha\cos\alpha$ is to rewrite the left side in terms of sines and cosines.

$$\frac{1}{\tan\alpha + \cot\alpha} = \frac{1}{\dfrac{\sin\alpha}{\cos\alpha} + \dfrac{\cos\alpha}{\sin\alpha}} \qquad \text{quotient identities}$$

$$= \frac{1}{\dfrac{\sin^2\alpha + \cos^2\alpha}{\sin\alpha\cos\alpha}} \qquad \text{algebra (adding the fractions)}$$

$$= \frac{1}{\dfrac{1}{\sin\alpha\cos\alpha}} \qquad \text{Pythagorean identity}$$

$$= \sin\alpha\cos\alpha \qquad \text{algebra (simplifying the fraction)} \ \square$$

The following list summarizes our suggestions for verifying identities. Note that these are only suggestions: There is no rigid step-by-step procedure that guarantees success. Success at verifying identities requires *practice and a thorough knowledge of the fundamental identities.*

Verifying Trigonometric Identities

1. Know the fundamental identities and be familiar with all of the alternate forms.
2. Choose the more complicated side and reduce it to the simpler side.
3. Keep the goal (or alternate forms of the goal) in mind and work toward it. Remember, you cannot score unless you have a goal.
4. If there are fractions, you may need to
   a. combine two or more fractions into one,
   b. separate one fraction into two or more, or
   c. multiply numerator and denominator by the conjugate of the denominator (or in some cases, the conjugate of the numerator).
5. If there are trigonometric polynomials, you may need to
   a. factor or
   b. multiply.
6. If several trigonometric functions are involved, use the fundamental identities to eliminate some of those functions.
7. In some cases, you may need to rewrite all trigonometric functions in terms of sines and cosines and rely on algebraic manipulation of the resulting expression.

◆ Practice Problem 2 ◆

Prove that $\dfrac{1 - \tan^2 A}{1 + \tan^2 A} = 2\cos^2 A - 1.$

In our next example we prove that an equation is not an identity. It is sufficient to find one value of the variable for which both members of the equation are defined but for which the equation is false. Such equations are called **conditional equations.**

Example 6 • Providing a Counterexample

Show that $\sin\theta + \cos\theta = 1$ is not an identity.

Solution

For $\theta = 180°$, we have

$$\sin 180° + \cos 180° = 0 + (-1) = -1 \neq 1.$$

Thus $\sin\theta + \cos\theta = 1$ is not an identity. In Section 8.6 we learn how to find the values of θ that do satisfy this equation. □

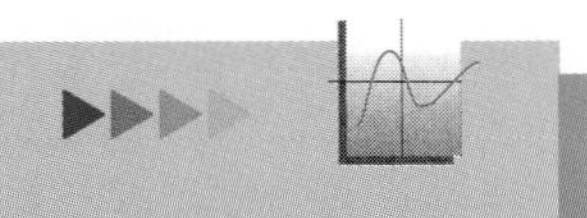

By graphing the functions that represent the two members of an equation, a graphing device can often be used to easily find a counterexample.

In the solution of Example 6, it was necessary only to demonstrate that for one value of the variable both members of the equation were defined but had different values. A demonstration of this type is called a **counterexample.**

◆ Practice Problem 3 ◆

Show that $\cos A = \sqrt{\cos^2 A}$ is not an identity.

Substitutions that involve a trigonometric function can sometimes be used with a Pythagorean identity to simplify an expression that contains a radical. Our last example in this section illustrates the technique.

Example 7 • Making a Trigonometric Substitution

Simplify the radical $\sqrt{x^2 - a^2}$ by making the substitution $x = a \sec u$, $a > 0$ and $0 \leq u < \pi/2$.

Solution

With the given substitution, we have

$$\begin{aligned}\sqrt{x^2 - a^2} &= \sqrt{a^2\sec^2 u - a^2}\\ &= \sqrt{a^2(\sec^2 u - 1)}\\ &= \sqrt{a^2\tan^2 u}.\end{aligned}$$

Now $\sqrt{a^2} = a$ since $a > 0$. Similarly, $\sqrt{\tan^2 u} = \tan\ u$ since $\tan u \geq 0$ for $0 \leq u < \pi/2$. Thus we can write

$$\sqrt{x^2 - a^2} = a \tan u$$

as our final simplification. □

EXERCISES 8.1

Reduce the first expression to the second. (See Example 1.)

1. $\sin \alpha \cot \alpha$; $\cos \alpha$
2. $\sec \beta \cot \beta$; $\csc \beta$
3. $\dfrac{\cos^2 A + \sin^2 A}{\cos^2 A}$; $\sec^2 A$
4. $\dfrac{\cos^2 A + \sin^2 A}{\sin^2 A}$; $\csc^2 A$
5. $\dfrac{\sec^2 \alpha}{\csc^2 \alpha}$; $\tan^2 \alpha$
6. $\dfrac{\csc^2 \theta}{\sec^2 \theta}$; $\cot^2 \theta$
7. $\dfrac{1 - \sin^2 \alpha}{1 - \cos^2 \alpha}$; $\cot^2 \alpha$
8. $\dfrac{1 - \cos^2 \beta}{1 - \sin^2 \beta}$; $\tan^2 \beta$
9. $(1 + \tan^2 \theta) \sin^2 \theta$; $\tan^2 \theta$
10. $(1 + \cot^2 \theta) \cos^2 \theta$; $\cot^2 \theta$
11. $\dfrac{\cos \theta + \sin \theta \tan \theta}{\cos \theta}$; $\sec^2 \theta$
12. $\dfrac{\sin \theta + \cos \theta \cot \theta}{\sin \theta}$; $\csc^2 \theta$
13. $\tan \beta + \cot \beta$; $\csc \beta \sec \beta$
14. $\sec^2 \beta + \csc^2 \beta$; $\sec^2 \beta \csc^2 \beta$

In Exercises 15–18, write all the trigonometric functions in terms of the given function.

15. $\cos \theta$
16. $\sin \theta$
17. $\tan \theta$
18. $\sec \theta$

In each of the following radicals, replace x by the indicated trigonometric expression and simplify by using the fundamental identities. (See Example 7.)

19. $\sqrt{1 + x^2}$; $x = \tan \theta$, $0° \leq \theta < 90°$
20. $\sqrt{25 - x^2}$; $x = 5 \cos \theta$, $0° \leq \theta < 90°$
21. $\sqrt{4 - x^2}$; $x = 2 \sin \theta$, $0° \leq \theta < 90°$
22. $\sqrt{x^2 - 9}$; $x = 3 \sec \theta$, $0° \leq \theta < 90°$
23. $\dfrac{1}{\sqrt{16 - x^2}}$; $x = 4 \sin u$, $-\dfrac{\pi}{2} < u < \dfrac{\pi}{2}$
24. $\dfrac{1}{\sqrt{4 + x^2}}$; $x = 2 \tan u$, $-\dfrac{\pi}{2} < u < \dfrac{\pi}{2}$
25. $(x^2 + 9)^{3/2}$; $x = 3 \tan u$, $-\dfrac{\pi}{2} < u < \dfrac{\pi}{2}$
26. $(x^2 - 25)^{3/2}$; $x = 5 \sec u$, $0 \leq u < \dfrac{\pi}{2}$
27. $\dfrac{\sqrt{x^2 + a^2}}{x}$; $x = a \tan t$, $a > 0$ and $0 < t < \dfrac{\pi}{2}$

28. $\dfrac{x}{\sqrt{a^2 - x^2}}$; $x = a \sin t$, $a > 0$ and $-\dfrac{\pi}{2} < t < \dfrac{\pi}{2}$

29. $\dfrac{1}{x\sqrt{x^2 - a^2}}$; $x = a \sec t$, $a > 0$ and $0 < t < \dfrac{\pi}{2}$

30. $\dfrac{1}{x^2 + a^2}$; $x = a \tan t$, $a > 0$ and $-\dfrac{\pi}{2} < t < \dfrac{\pi}{2}$

In Exercises 31–100, verify that each equation is an identity. (See Examples 2–5.)

31. $(\sec^2 \theta - 1)(\cot^2 \theta + 1) = \sec^2 \theta$
32. $(1 - \sin^2 \theta)(\sec^2 \theta - 1) = \sin^2 \theta$
33. $\dfrac{1 - \sin^2 \theta}{\cot \theta} = \sin \theta \cos \theta$
34. $\dfrac{1 - \cos^2 \theta}{\tan \theta} = \sin \theta \cos \theta$
35. $\cot^2 \beta - \cos^2 \beta = \cos^2 \beta \cot^2 \beta$
36. $\tan^2 \beta - \sin^2 \beta = \sin^2 \beta \tan^2 \beta$
37. $\sin^2 \alpha - \cos^2 \alpha \tan^2 \alpha = 0$
38. $\cos^2 \alpha - \sin^2 \alpha \cot^2 \alpha = 0$
39. $\sec^2 \alpha = \sin^2 \alpha + \tan^2 \alpha + \cos^2 \alpha$
40. $\cos^2 \alpha = \csc^2 \alpha - \sin^2 \alpha - \cot^2 \alpha$
41. $(1 - \sin \gamma)(1 + \sin \gamma) = \cos^2 \gamma$
42. $(\sec \gamma + 1)(\sec \gamma - 1) = \tan^2 \gamma$
43. $(\csc \gamma - \cot \gamma)(\csc \gamma + \cot \gamma) = 1$
44. $(\sec \gamma + \tan \gamma)(\sec \gamma - \tan \gamma) = 1$
45. $\dfrac{1}{\cos^2 A} = \tan A\,(\cot A + \tan A)$
46. $\dfrac{1}{\sin^2 A} = \cot A\,(\tan A + \cot A)$
47. $\dfrac{\cos A}{\sec A} + \dfrac{\sin A}{\csc A} = 1$
48. $\dfrac{\sec A}{\cos A} - \dfrac{\tan A}{\cot A} = 1$
49. $\dfrac{\csc B - \sin B}{\sin B} = \cot^2 B$
50. $\dfrac{\sec B - \cos B}{\cos B} = \tan^2 B$
51. $\dfrac{\cos B \tan B + \sin B}{1 - \cos^2 B} = 2 \csc B$
52. $\dfrac{\sin B \cot B + \cos B}{1 - \sin^2 B} = 2 \sec B$
53. $\cos t = \dfrac{1}{\sin t(\tan t + \cot t)}$
54. $-\sec^2 t = \dfrac{1}{\sin t(\sin t - \csc t)}$
55. $(\tan t + 1)^2 = \sec t(\sec t + 2 \sin t)$
56. $(\cot t + 1)^2 = \csc t(\csc t + 2 \cos t)$
57. $(\cos s + \sin s)^2 + (\cos s - \sin s)^2 = 2$
58. $(\tan s + \sec s)^2 - (\tan s - \sec s)^2 = 4 \sin s \sec^2 s$
59. $(\csc s - \sec s)(\sin s + \cos s) = \cot s - \tan s$

60. $(\cot s - \csc s)(\tan s + \sin s) = \cos s - \sec s$

61. $\dfrac{\cos\theta}{1-\sin\theta} = \dfrac{1+\sin\theta}{\cos\theta}$

62. $\dfrac{\sin\theta}{1-\cos\theta} = \dfrac{1+\cos\theta}{\sin\theta}$

63. $\dfrac{\tan\theta}{\sec\theta+1} = \dfrac{\sec\theta-1}{\tan\theta}$

64. $\dfrac{\cot\theta}{\csc\theta-1} = \dfrac{\csc\theta+1}{\cot\theta}$

65. $\dfrac{\sec u-1}{\tan^2 u} = \dfrac{\cos u}{\cos u+1}$

66. $\dfrac{\csc u+1}{\cot^2 u} = \dfrac{\sin u}{1-\sin u}$

67. $\dfrac{\cos u}{\sec u-\tan u} = 1+\sin u$

68. $\dfrac{\sin u}{\csc u+\cot u} = 1-\cos u$

69. $\dfrac{\tan\alpha-\cot\alpha}{\cot\alpha} - \dfrac{\cot\alpha-\tan\alpha}{\tan\alpha} = \sec^2\alpha - \csc^2\alpha$

70. $\dfrac{\sec\alpha-\cos\alpha}{\sec\alpha} - \dfrac{\sin\alpha-\csc\alpha}{\csc\alpha} = 1$

71. $\dfrac{1+\cos\alpha}{1-\cos\alpha} = \dfrac{\sec\alpha+1}{\sec\alpha-1}$

72. $\dfrac{\tan\alpha-1}{\tan\alpha+1} = \dfrac{1-\cot\alpha}{1+\cot\alpha}$

73. $\dfrac{\cot\gamma-\tan\gamma}{\cot\gamma+\tan\gamma} = \cos^2\gamma - \sin^2\gamma$

74. $\dfrac{\tan^2\gamma-1}{1-\cot^2\gamma} = \tan^2\gamma$

75. $\dfrac{\csc\gamma+\sin\gamma}{\csc\gamma-\sin\gamma} = 1+2\tan^2\gamma$

76. $\dfrac{\sec\gamma+\cos\gamma}{\sec\gamma-\cos\gamma} = 1+2\cot^2\gamma$

77. $\dfrac{\cos x}{1-\sin x} - \dfrac{\cos x}{1+\sin x} = 2\tan x$

78. $\dfrac{\tan x}{1-\cos x} - \dfrac{\tan x}{1+\cos x} = 2\csc x$

79. $\dfrac{\sin x}{\sec x-1} + \dfrac{\sin x}{\sec x+1} = 2\cot x$

80. $\dfrac{\cot x}{\csc x-1} - \dfrac{\cot x}{\csc x+1} = 2\tan x$

81. $\cos^4 t - \sin^4 t = 2\cos^2 t - 1$

82. $\sec^4 t - \tan^4 t = 2\sec^2 t - 1$

83. $\dfrac{\cos^3 t - \sin^3 t}{\cos t - \sin t} = 1 + \cos t \sin t$

84. $\dfrac{1 - \tan^3 t}{1 - \tan t} = \sec^2 t + \tan t$

85. $\dfrac{\sin \theta}{1 - \cos \theta} + \dfrac{1 - \cos \theta}{\sin \theta} = 2 \csc \theta$

86. $\dfrac{\sec \theta}{1 - \tan \theta} - \dfrac{1 - \tan \theta}{\sec \theta} = \dfrac{2 \sin \theta}{1 - \tan \theta}$

87. $(\cot \theta - \csc \theta)^2 = \dfrac{1 - \cos \theta}{1 + \cos \theta}$

88. $(\tan \theta + \sec \theta)^2 = \dfrac{\sin \theta + 1}{1 - \sin \theta}$

89. $\dfrac{\sin^2 x}{1 - \cos x} = \dfrac{\sin x}{\csc x - \cot x}$

90. $\dfrac{\tan^2 x}{1 - \sec x} = \dfrac{\tan x}{\cot x - \csc x}$

91. $\dfrac{\tan x + 1}{\sin x + \cos x} = \sec x$

92. $\dfrac{\csc x + 1}{\sin x + 1} = \csc x$

93. $\dfrac{\sin A \cos B + \cos A \sin B}{\cos A \cos B - \sin A \sin B} = \dfrac{\tan A + \tan B}{1 - \tan A \tan B}$

94. $\dfrac{\sin A \cos B - \cos A \sin B}{\cos A \cos B + \sin A \sin B} = \dfrac{\cot B - \cot A}{\cot A \cot B + 1}$

95. $\dfrac{\sin A + \sin B}{\csc A + \csc B} = \sin A \sin B$

96. $\dfrac{\tan A + \tan B}{\cot A + \cot B} = \tan A \tan B$

97. $2 - \cos^2 \theta \csc \theta = \dfrac{\cos \theta}{\sec \theta - \tan \theta} - \dfrac{\cot \theta}{\sec \theta + \tan \theta}$

98. $2 + \sin^2 \theta \sec \theta = \dfrac{\sin \theta}{\csc \theta + \cot \theta} + \dfrac{\tan \theta}{\csc \theta - \cot \theta}$

99. $\dfrac{\cot \theta + \cos \theta}{\cot \theta \cos \theta} = \dfrac{\cot \theta \cos \theta}{\cot \theta - \cos \theta}$

100. $\dfrac{\tan \theta \sin \theta}{\tan \theta + \sin \theta} = \dfrac{\tan \theta - \sin \theta}{\tan \theta \sin \theta}$

Show that each equation in Exercises 101–108 is not an identity by providing a counterexample. (See Example 6.)

101. $(\sin \theta + \cos \theta)^2 = \sin^2 \theta + \cos^2 \theta$

102. $\tan \alpha = \sec \alpha - 1$

103. $\cot^2 \beta = (\csc \beta - 1)^2$

104. $\sin 2\alpha = 2 \sin \alpha$

105. $\tan 3A = 3 \tan A$

106. $\cot \frac{1}{2}\beta = \frac{1}{2} \cot \beta$

107. $\cos (\alpha + \beta) = \cos \alpha + \cos \beta$

108. $\sec A + \csc A = \dfrac{1}{\cos A + \sin A}$

109–110

Use a calculator to verify that each equation is true for the given value of θ.

109. $\dfrac{1}{\csc \theta + \cot \theta} = \csc \theta - \cot \theta,\ \theta = 247°$

110. $\cos^2 \theta - \sin^2 \theta = \cos 2\theta,\ \theta = 123°$

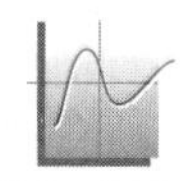

111–115

Use a graphing device to graph the given functions Y_1 and Y_2 on the same coordinate system and make a guess as to whether the equation $Y_1 = Y_2$ is an identity. Then explain why your decision is correct.

Critical Thinking: Exploration and Writing

111. $Y_1 = \cos 3x - \cos x,\ Y_2 = \cos 2x$

Critical Thinking: Exploration and Writing

112. $Y_1 = 2 \sin x \cos x,\ Y_2 = (\sin x + \cos x)^2 - 1$

Critical Thinking: Exploration and Writing

113. $Y_1 = \sec x + \tan x,\ Y_2 = \dfrac{1}{\sec x - \tan x}$

Critical Thinking: Exploration and Writing

114. $Y_1 = \sin x + \cos x,\ Y_2 = \dfrac{1}{\sin x - \cos x}$

Critical Thinking: Exploration and Writing

115. Explain why a graphing device cannot be used with complete certainty to prove that two functions are identical.

◆ Solutions for Practice Problems

1.
$$\begin{aligned} \frac{1 + \cot^2 \theta}{\cot^2 \theta} &= \frac{1}{\cot^2 \theta} + \frac{\cot^2 \theta}{\cot^2 \theta} && \text{algebra} \\ &= \tan^2 \theta + 1 && \text{reciprocal identity and algebra} \\ &= \sec^2 \theta && \text{Pythagorean identity} \end{aligned}$$

2.
$$\begin{aligned} \frac{1 - \tan^2 A}{1 + \tan^2 A} &= \frac{1 - \tan^2 A}{\sec^2 A} && \text{Pythagorean identity} \\ &= \cos^2 A(1 - \tan^2 A) && \text{reciprocal identity} \\ &= \cos^2 A - \cos^2 A \tan^2 A && \text{algebra} \\ &= \cos^2 A - \cos^2 A \frac{\sin^2 A}{\cos^2 A} && \text{quotient identity} \\ &= \cos^2 A - \sin^2 A && \text{algebra} \\ &= \cos^2 A - (1 - \cos^2 A) && \text{Pythagorean identity} \\ &= 2 \cos^2 A - 1 && \text{algebra} \end{aligned}$$

3. Replacing A by 120° in $\cos A = \sqrt{\cos^2 A}$, we have a false statement because

$$\cos 120^\circ = -\frac{1}{2},$$

whereas

$$\sqrt{\cos^2 120^\circ} = \sqrt{\left(\frac{-1}{2}\right)^2} = \sqrt{\frac{1}{4}} = \frac{1}{2}.$$

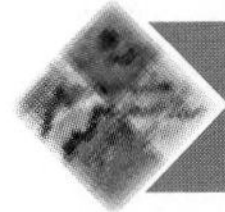

8.2 COSINE OF THE SUM OR DIFFERENCE OF TWO ANGLES

The fundamental identities involve trigonometric functions of a single angle, say θ, α, or β. The sum, $\alpha + \beta$, the difference, $\alpha - \beta$, and multiples, 2θ and $\frac{1}{2}\theta$, of these angles are called **composite angles.** In this chapter we study the interesting and sometimes surprising relationships that exist between the trigonometric functions of composite angles and the trigonometric functions of a single angle. These relationships are known as the **composite angle identities.**

It would be nice if $\sin 2\theta$ were the same as $2 \sin \theta$, for all θ. But unfortunately $\sin 2\theta = 2 \sin \theta$ is true only for integral multiples of 180°. Similarly, it would be nice if $\cos (\alpha - \beta)$ were equal to $\cos \alpha - \cos \beta$ for all α and β. Again, this is not true. In fact, we now prove the intriguing result that

$$\cos (\alpha - \beta) = \cos \alpha \cos \beta + \sin \alpha \sin \beta.$$

In Figure 8.1a, angles α and β are drawn in standard position with $\alpha > \beta$. Points A and B are located where the unit circle intersects the terminal sides of α and β, respectively. The angle formed between the terminal sides of α and β is $\alpha - \beta$, and the points A and B have coordinates $(\cos \alpha, \sin \alpha)$ and $(\cos \beta, \sin \beta)$, respectively.

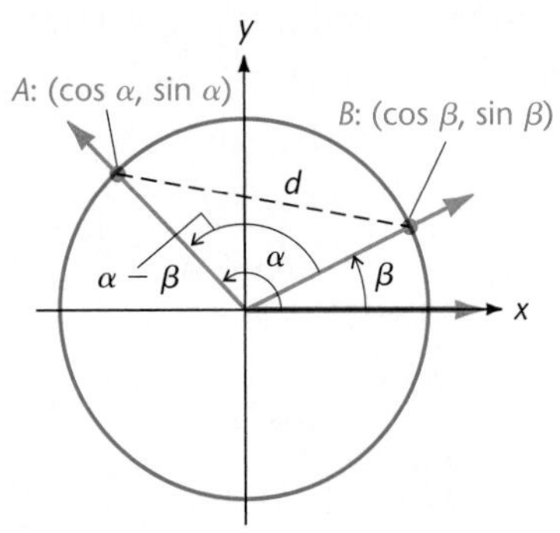

(a) Angles α and β in standard position

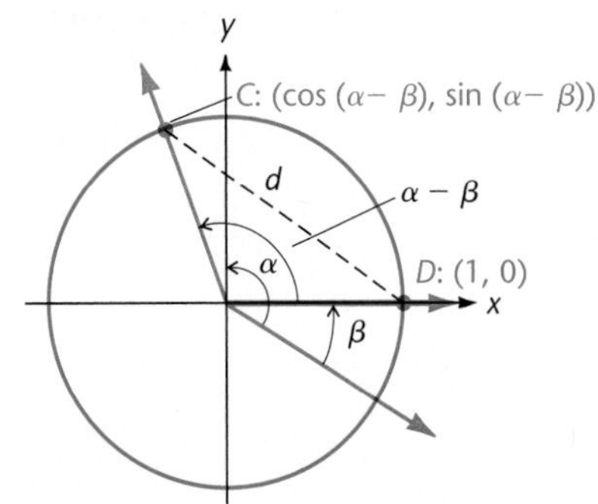

(b) Angle $\alpha - \beta$ in standard position

Figure 8.1

In Figure 8.1a the length of the chord subtended by the angle $\alpha - \beta$ is the distance d between the points A and B. The distance formula gives

$$\begin{aligned} d^2 &= (\cos\alpha - \cos\beta)^2 + (\sin\alpha - \sin\beta)^2 \\ &= \cos^2\alpha - 2\cos\alpha\cos\beta + \cos^2\beta + \sin^2\alpha - 2\sin\alpha\sin\beta + \sin^2\beta \\ &= (\cos^2\alpha + \sin^2\alpha) + (\cos^2\beta + \sin^2\beta) - 2(\cos\alpha\cos\beta + \sin\alpha\sin\beta) \\ &= 2 - 2(\cos\alpha\cos\beta + \sin\alpha\sin\beta). \end{aligned}$$

Next we rotate the angles in Figure 8.1a so that $\alpha - \beta$ is in standard position, as in Figure 8.1b. Then d, the length of the chord subtended by $\alpha - \beta$, is the distance between the points C with coordinates $(\cos(\alpha - \beta), \sin(\alpha - \beta))$ and D with coordinates $(1, 0)$. Then

$$\begin{aligned} d^2 &= [\cos(\alpha - \beta) - 1]^2 + [\sin(\alpha - \beta) - 0]^2 \\ &= \cos^2(\alpha - \beta) - 2\cos(\alpha - \beta) + 1 + \sin^2(\alpha - \beta) \\ &= [\cos^2(\alpha - \beta) + \sin^2(\alpha - \beta)] + 1 - 2\cos(\alpha - \beta) \\ &= 2 - 2\cos(\alpha - \beta). \end{aligned}$$

Thus we have

$$2 - 2\cos(\alpha - \beta) = 2 - 2(\cos\alpha\cos\beta + \sin\alpha\sin\beta),$$

and this leads to the following identity.

Cosine of the Difference of Two Angles

$$\cos(\alpha - \beta) = \cos\alpha\cos\beta + \sin\alpha\sin\beta$$

We illustrate some uses of this identity in the next examples.

Example 1 • Evaluating the Cosine of a Difference

Without determining α or β, find the exact value of $\cos(\alpha - \beta)$ if $\cos\alpha = -\frac{4}{5}$, $\sin\beta = -\frac{7}{25}$, α is in quadrant III, and β is in quadrant IV.

Solution

Knowing that $\cos\alpha = -\frac{4}{5}$ and α is in quadrant III, we find $\sin\alpha = -\frac{3}{5}$. Also since β is in quadrant IV with $\sin\beta = -\frac{7}{25}$, then $\cos\beta = \frac{24}{25}$. Then

$$\begin{aligned} \cos(\alpha - \beta) &= \cos\alpha\cos\beta + \sin\alpha\sin\beta \\ &= \left(-\frac{4}{5}\right)\left(\frac{24}{25}\right) + \left(-\frac{3}{5}\right)\left(-\frac{7}{25}\right) \\ &= \frac{-96 + 21}{125} \\ &= -\frac{75}{125} \\ &= -\frac{3}{5}. \end{aligned}$$

□

◆ Practice Problem 1 ◆

Write $\cos 4x \cos x + \sin 4x \sin x$ as a single term.

In the next example we prove that the equation $\cos (90° - \theta) = \sin \theta$ is an identity. This generalizes to arbitrary angles a cofunction relation that was noted for right triangles in Section 7.6.

Example 2 • Verifying a Cofunction Identity

Verify that the equation $\cos (90° - \theta) = \sin \theta$ is an identity.

Solution

For any θ,

$$\begin{aligned} \mathbf{\cos (90° - \boldsymbol{\theta})} &= \cos 90° \cos \theta + \sin 90° \sin \theta \\ &= 0 \cdot \cos \theta + 1 \cdot \sin \theta \\ &\mathbf{= \sin \boldsymbol{\theta},} \end{aligned}$$

and the identity is proved. □

If we replace θ by $90° - \theta$ in the identity in Example 2, we obtain another identity,

$$\begin{aligned} \mathbf{\sin (90° - \boldsymbol{\theta})} &= \cos [90° - (90° - \theta)] \\ &\mathbf{= \cos \boldsymbol{\theta}.} \end{aligned}$$

Proofs are requested in the exercises to show that **any function of a given angle equals the cofunction of the complementary angle** whenever both functions are defined. Stated as equations, these relationships are called the **cofunction identities.**

Cofunction Identities

$$\cos (90° - \theta) = \sin \theta \qquad \sin (90° - \theta) = \cos \theta$$
$$\cot (90° - \theta) = \tan \theta \qquad \tan (90° - \theta) = \cot \theta$$
$$\csc (90° - \theta) = \sec \theta \qquad \sec (90° - \theta) = \csc \theta$$

Corresponding statements can be made for radian measure or functions of real numbers: $\cos (\pi/2 - \theta) = \sin \theta$, $\sin (\pi/2 - \theta) = \cos \theta$, and so on.

We write $-\theta = 0 - \theta$ as a first step in expressing the cosine of $-\theta$ in terms of the cosine of θ. The next step is to apply the identity for the cosine of the difference of two angles.

$$\begin{aligned} \cos (-\theta) &= \cos (0 - \theta) \\ &= \cos 0 \cos \theta + \sin 0 \sin \theta && \text{cosine of the difference} \\ &= 1 \cdot \cos \theta + 0 \cdot \sin \theta && \text{special angles} \\ &= \cos \theta. \end{aligned}$$

This identity,

$$\cos(-\theta) = \cos\theta,$$

is one of the **negative-angle identities.**

To derive the negative-angle identity for the sine function, we write

$$
\begin{aligned}
\sin(-\theta) &= \cos\left[\frac{\pi}{2} - (-\theta)\right] && \text{cofunction identity}\\
&= \cos\left(\frac{\pi}{2} + \theta\right) && \text{algebra}\\
&= \cos\left[\theta - \left(-\frac{\pi}{2}\right)\right] && \text{algebra}\\
&= \cos\theta\cos\left(-\frac{\pi}{2}\right) + \sin\theta\sin\left(-\frac{\pi}{2}\right) && \text{cosine of the difference}\\
&= \cos\theta\cdot 0 + \sin\theta\cdot(-1) && \text{special angles}\\
&= -\sin\theta.
\end{aligned}
$$

The corresponding negative-angle identities for the remaining trigonometric functions can easily be established by using the fundamental identities.

Negative-Angle Identities

$\sin(-\theta) = -\sin\theta$	$\cos(-\theta) = \cos\theta$	$\tan(-\theta) = -\tan\theta$
$\csc(-\theta) = -\csc\theta$	$\sec(-\theta) = \sec\theta$	$\cot(-\theta) = -\cot\theta$

The application of negative-angle identities is unnecessary when using a graphing device.

The negative-angle identities are sometimes useful in graphing. This is illustrated in the next example.

Example 3 • Using a Negative-Angle Identity in Graphing

Sketch the graph of $y = 2\sin(-x)$ over one complete period.

Solution

We use the negative-angle identity to write

$$
\begin{aligned}
y &= 2\sin(-x)\\
&= -2\sin x.
\end{aligned}
$$

As shown in Figure 8.2, the graph is a reflection through the x-axis of a typical sine curve with Amp = 2, $P = 2\pi$, and no phase shift. □

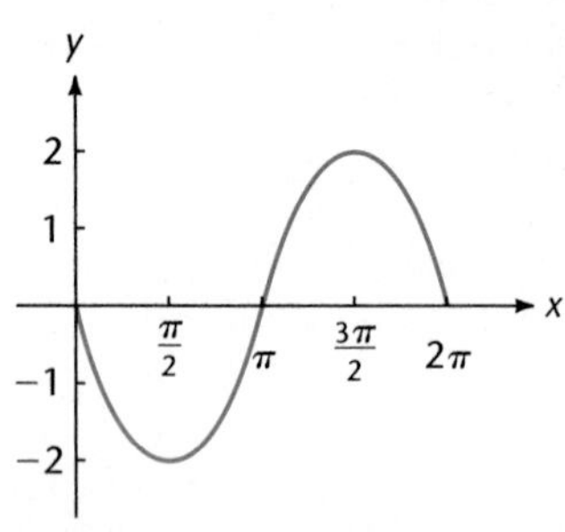

Figure 8.2 Graph of $y = 2\sin(-x)$, $x \in [0, 2\pi]$

We can use the identities

$$
\begin{aligned}
\cos(\alpha - \beta) &= \cos\alpha\cos\beta + \sin\alpha\sin\beta,\\
\cos(-\theta) &= \cos\theta,\\
\sin(-\theta) &= -\sin\theta,
\end{aligned}
$$

to derive an identity for the cosine of the sum, $\alpha + \beta$. We first write $\alpha + \beta$ as $\alpha - (-\beta)$.

$$\begin{aligned}
\boldsymbol{\cos(\alpha + \beta)} &= \cos[\alpha - (-\beta)] && \text{algebra}\\
&= \cos\alpha\cos(-\beta) + \sin\alpha\sin(-\beta) && \text{cosine of the difference}\\
&= \cos\alpha\cos\beta + \sin\alpha(-\sin\beta) && \text{negative-angle identities}\\
&= \boldsymbol{\cos\alpha\cos\beta - \sin\alpha\sin\beta} && \text{algebra}
\end{aligned}$$

This establishes the desired identity.

Cosine of the Sum of Two Angles

$\cos(\alpha + \beta) = \cos\alpha\cos\beta - \sin\alpha\sin\beta$

The identities for the cosine of the sum and difference of two angles can be used with the function values of the special angles to find the *exact* values of the cosine of certain angles.

Example 4 • Evaluating the Cosine of a Sum

Find the exact value of $\cos 285°$.

Solution

We can write 285° as the sum of two special angles, 225° and 60°. Thus

$$\begin{aligned}
\cos 285° &= \cos(225° + 60°)\\
&= \cos 225° \cos 60° - \sin 225° \sin 60°\\
&= \left(-\frac{\sqrt{2}}{2}\right)\left(\frac{1}{2}\right) - \left(-\frac{\sqrt{2}}{2}\right)\left(\frac{\sqrt{3}}{2}\right)\\
&= \frac{\sqrt{2}}{4}(\sqrt{3} - 1).
\end{aligned}$$

□

◆ **Practice Problem 2** ◆

Find the exact value of $\cos(\pi/12)$.

As a final example, we verify an identity.

Example 5 • Proving an Identity

Verify the identity: $\dfrac{\cos(\alpha + \beta)}{\cos(\alpha - \beta)} = \dfrac{\cot\alpha\cot\beta - 1}{\cot\alpha\cot\beta + 1}$.

Solution

$$\frac{\cos(\alpha+\beta)}{\cos(\alpha-\beta)} = \frac{\cos\alpha\cos\beta - \sin\alpha\sin\beta}{\cos\alpha\cos\beta + \sin\alpha\sin\beta}$$ cosine of the sum / cosine of the difference

$$= \frac{\dfrac{\cos\alpha\cos\beta}{\sin\alpha\sin\beta} - \dfrac{\sin\alpha\sin\beta}{\sin\alpha\sin\beta}}{\dfrac{\cos\alpha\cos\beta}{\sin\alpha\sin\beta} + \dfrac{\sin\alpha\sin\beta}{\sin\alpha\sin\beta}}$$ algebra (dividing numerator and denominator by $\sin\alpha\sin\beta$)

$$= \frac{\cot\alpha\cot\beta - 1}{\cot\alpha\cot\beta + 1}$$ quotient identities

□

EXERCISES 8.2

Use an identity to find the exact value of each of the following.

1. $\cos 127° \cos 37° + \sin 127° \sin 37°$
2. $\cos 200° \cos 20° + \sin 200° \sin 20°$
3. $\cos 1° \cos 361° + \sin 1° \sin 361°$
4. $\cos 257° \cos 527° + \sin 257° \sin 527°$
5. $\cos 42° \cos 138° - \sin 42° \sin 138°$
6. $\cos 23° \cos 67° - \sin 23° \sin 67°$

Express each of the following in terms of trigonometric functions of θ.

7. $\cos(\pi + \theta)$
8. $\cos(\pi - \theta)$
9. $\cos\left(\theta - \frac{3\pi}{2}\right)$
10. $\cos\left(\theta + \frac{\pi}{3}\right)$
11. $\cos\left(\theta - \frac{5\pi}{3}\right)$
12. $\cos\left(\theta + \frac{7\pi}{6}\right)$

Without determining α or β, find the exact values of $\cos(\alpha - \beta)$ and $\cos(\alpha + \beta)$ if α and β satisfy the given conditions. (See Example 1.)

13. $\cos\alpha = \frac{3}{5}$, $\sin\beta = \frac{3}{5}$; α and β in Q I
14. $\cos\alpha = \frac{7}{25}$, $\cos\beta = \frac{5}{13}$; α and β in Q IV
15. $\cos\alpha = -\frac{4}{5}$, $\sin\beta = -\frac{12}{13}$; α in Q II, β in Q III
16. $\sin\alpha = \frac{15}{17}$, $\sin\beta = \frac{5}{13}$; α in Q II, β in Q I
17. $\cos\alpha = -\frac{24}{25}$, $\sin\beta = -1$; α in Q III
18. $\sin\alpha = 1$, $\sin\beta = -\frac{3}{5}$; β in Q III
19. $\sin\alpha = -\frac{1}{2}$, $\cos\beta = -\frac{5}{13}$; α in Q III, β in Q II
20. $\sin\alpha = \frac{8}{17}$, $\sin\beta = \frac{2}{3}$; α in Q II, β in Q II
21. $\cos\alpha = \sqrt{7}/3$, $\sin\beta = \sqrt{5}/4$; α in Q IV, β in Q II
22. $\cos\alpha = -\frac{3}{7}$, $\cos\beta = -\sqrt{2}/3$; α in Q II, β in Q III

Write each of the following as a single term.

23. $\cos 2x \cos x - \sin 2x \sin x$

24. $\cos 3y \cos 2y + \sin 3y \sin 2y$

25. $\cos \frac{x}{4} \cos \frac{x}{6} + \sin \frac{x}{4} \sin \frac{x}{6}$

26. $\cos \frac{x}{2} \cos \frac{x}{3} - \sin \frac{x}{2} \sin \frac{x}{3}$

Use the negative-angle identities to simplify and then sketch the graph through one complete period. (See Example 3.)

27. $y = \cos(-x)$

28. $y = \tan(-x)$

29. $y = 3 \sin(-x)$

30. $y = 2 \cos(-x)$

31. $y = -3 \cos(-2x)$

32. $y = -3 \sin(-2x)$

Find the exact value of each of the following by using special angles. (See Example 4.)

33. $\cos 75°$

34. $\cos 195°$

35. $\cos 15°$

36. $\cos 165°$

37. $\cos \frac{7\pi}{12}$

38. $\cos \frac{5\pi}{12}$

39. $\cos\left(-\frac{\pi}{12}\right)$

40. $\cos\left(-\frac{7\pi}{12}\right)$

Verify the identities. (See Example 5.)

41. $\cos(t + s) + \cos(t - s) = 2 \cos t \cos s$

42. $\cos(\alpha + \beta) - \cos(\alpha - \beta) = -2 \sin \alpha \sin \beta$

43. $\frac{\cos(\alpha + \beta)}{\cos \alpha \cos \beta} = 1 - \tan \alpha \tan \beta$

44. $\frac{\cos(x - y)}{\sin x \sin y} = \cot x \cot y + 1$

45. $\cos(\alpha + \beta) \cos(\alpha - \beta) + \sin(\alpha + \beta) \sin(\alpha - \beta) = \cos 2\beta$

46. $\cos(\alpha + \beta) \cos(\alpha - \beta) - \sin(\alpha + \beta) \sin(\alpha - \beta) = \cos 2\alpha$

47. $\frac{\cos(\alpha + \beta) + \cos(\alpha - \beta)}{\cos(\alpha + \beta) - \cos(\alpha - \beta)} = -\cot \alpha \cot \beta$

48. $\frac{\cos(\alpha - \beta)}{\cos(\alpha + \beta)} = \frac{1 + \tan \alpha \tan \beta}{1 - \tan \alpha \tan \beta}$

49. $\frac{\cos \alpha}{\sin \beta} - \frac{\sin \alpha}{\cos \beta} = \frac{\cos(\alpha + \beta)}{\sin \beta \cos \beta}$

50. $\frac{\cos 2\theta}{\sin 3\theta} + \frac{\sin 2\theta}{\cos 3\theta} = \frac{\cos \theta}{\sin 3\theta \cos 3\theta}$

51. $\cos(\alpha + \beta) \cos(\alpha - \beta) = \cos^2 \alpha - \sin^2 \beta$

52. $\dfrac{\cos(\theta + h) - \cos\theta}{h} = \cos\theta\left(\dfrac{\cos h - 1}{h}\right) - \sin\theta\left(\dfrac{\sin h}{h}\right)$

Prove the following cofunction identities and negative-angle identities.

53. $\tan(90° - \theta) = \cot\theta$
54. $\cot(90° - \theta) = \tan\theta$
55. $\sec(90° - \theta) = \csc\theta$
56. $\csc(90° - \theta) = \sec\theta$
57. $\tan(-\theta) = -\tan\theta$
58. $\cot(-\theta) = -\cot\theta$
59. $\sec(-\theta) = \sec\theta$
60. $\csc(-\theta) = -\csc\theta$

Show that each equation is not an identity by providing a counterexample.

61. $\cos(x - y) = \cos x - \cos y$
62. $\cos\left(\dfrac{\pi}{2} - \theta\right) = \cos\theta$
63. $\tan\left(t - \dfrac{\pi}{2}\right) = \cot t$
64. $\sin\left(v - \dfrac{\pi}{2}\right) = \cos v$

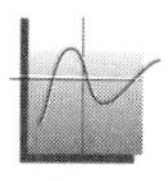

65–70 In Exercises 65–68, use a graphing device to graph the function on each side of the given equation on the same coordinate system. Basing your decision on the graphical results, decide if the given equation is an identity. Then prove your decision is correct.

Critical Thinking: Exploration and Writing

65. $\cos\left(\dfrac{\pi}{4} + x\right) = \sin\left(\dfrac{\pi}{4} - x\right)$

Critical Thinking: Exploration and Writing

66. $\cos(\pi + x) = -\cos x$

Critical Thinking: Exploration and Writing

67. $\cos\left(\dfrac{\pi}{2} + x\right) = \sin x$

Critical Thinking: Exploration and Writing

68. $\cos(\pi - x) = \cos x$

Critical Thinking: Exploration and Writing

69. Graph the following pair of functions on the same coordinate system, using first the viewing rectangle $[-1, 1]$ by $[-2, 2]$ and then the rectangle $[-3, 3]$ by $[-2, 2]$. Explain the connection between these graphs and Exercise 115 in Exercises 8.1.

$$Y_1 = \cos 2x, \qquad Y_2 = 1 - 2x^2 + \frac{2}{3}x^4 - \frac{4}{45}x^6$$

Critical Thinking: Exploration and Writing

70. Follow the instructions in Exercise 69 with the given pair of functions.

a. $Y_1 = \cos 2x$

$Y_2 = 1 - 2x^2 + \dfrac{2}{3}x^4$

b. $Y_1 = \cos 2x$

$Y_2 = 1 - 2x^2$

◆ **Solutions for Practice Problems**

1. $\cos 4x \cos x + \sin 4x \sin x = \cos(4x - x) = \cos 3x$

2. $$\begin{aligned} \cos\frac{\pi}{12} &= \cos\left(\frac{4\pi}{12} - \frac{3\pi}{12}\right) \\ &= \cos\left(\frac{\pi}{3} - \frac{\pi}{4}\right) \\ &= \cos\frac{\pi}{3}\cos\frac{\pi}{4} + \sin\frac{\pi}{3}\sin\frac{\pi}{4} \\ &= \frac{1}{2}\frac{\sqrt{2}}{2} + \frac{\sqrt{3}}{2}\frac{\sqrt{2}}{2} \\ &= \frac{\sqrt{2}}{4}(1 + \sqrt{3}) \end{aligned}$$

8.3 IDENTITIES FOR THE SINE AND TANGENT

The major goal of this section is to extend our list of identities to include identities for the sine and tangent of the sum and difference of two angles.

Sine and Tangent of the Sum and Difference of Two Angles

$$\sin(\alpha + \beta) = \sin\alpha\cos\beta + \cos\alpha\sin\beta$$
$$\sin(\alpha - \beta) = \sin\alpha\cos\beta - \cos\alpha\sin\beta$$
$$\tan(\alpha + \beta) = \frac{\tan\alpha + \tan\beta}{1 - \tan\alpha\tan\beta}$$
$$\tan(\alpha - \beta) = \frac{\tan\alpha - \tan\beta}{1 + \tan\alpha\tan\beta}$$

The first identity is established by using three identities derived in the preceding section.

$$\begin{aligned} \boldsymbol{\sin(\alpha + \beta)} &= \cos[90^\circ - (\alpha + \beta)] && \text{cofunction identity} \\ &= \cos[(90^\circ - \alpha) - \beta] && \text{algebra} \\ &= \cos(90^\circ - \alpha)\cos\beta \\ &\quad + \sin(90^\circ - \alpha)\sin\beta && \text{cosine of the difference} \\ &= \boldsymbol{\sin\alpha\cos\beta + \cos\alpha\sin\beta} && \text{cofunction identities} \end{aligned}$$

The identity for the sine of the difference of two angles follows from the identity for the sine of the sum; we first write $\alpha - \beta$ as $\alpha + (-\beta)$.

$$\begin{aligned} \boldsymbol{\sin(\alpha - \beta)} &= \sin[\alpha + (-\beta)] && \text{algebra} \\ &= \sin\alpha\cos(-\beta) + \cos\alpha\sin(-\beta) && \text{sine of the sum} \\ &= \boldsymbol{\sin\alpha\cos\beta - \cos\alpha\sin\beta} && \text{negative-angle identities} \end{aligned}$$

Next we derive the identity for the tangent of the sum of two angles by expressing the tangent function as the quotient of the sine and cosine functions.

$$\begin{aligned} \boldsymbol{\tan(\alpha + \beta)} &= \frac{\sin(\alpha + \beta)}{\cos(\alpha + \beta)} && \text{quotient identity} \\ &= \frac{\sin\alpha\cos\beta + \cos\alpha\sin\beta}{\cos\alpha\cos\beta - \sin\alpha\sin\beta} && \begin{matrix}\text{sine of the sum} \\ \text{cosine of the sum}\end{matrix} \\ &= \frac{\dfrac{\sin\alpha\cos\beta}{\cos\alpha\cos\beta} + \dfrac{\cos\alpha\sin\beta}{\cos\alpha\cos\beta}}{\dfrac{\cos\alpha\cos\beta}{\cos\alpha\cos\beta} - \dfrac{\sin\alpha\sin\beta}{\cos\alpha\cos\beta}} && \text{algebra} \\ &= \boldsymbol{\frac{\tan\alpha + \tan\beta}{1 - \tan\alpha\tan\beta}} && \text{quotient identities} \end{aligned}$$

Thus the identity for the tangent of the sum of two angles is established. We leave the proof of the identity for the tangent of the difference of two angles to the student. (See Exercise 39 at the end of this section.)

Some uses of these identities are demonstrated in the next examples.

Example 1 • Using Identities to Find Exact Values

Use identities to find the exact value of each of the following.

a. $\sin 20° \cos 110° - \cos 20° \sin 110°$

b. $\dfrac{\tan 43° + \tan 137°}{1 - \tan 43° \tan 137°}$

c. $\tan(\alpha - \beta)$, if the exact values of $\tan\alpha$ and $\tan\beta$ are 1.1 and 3.7, respectively

Solution

In each part we apply one of the identities listed at the beginning of this section.

a. $\sin 20° \cos 110° - \cos 20° \sin 110° = \sin(20° - 110°) = \sin(-90°) = -1$

b. $\dfrac{\tan 43° + \tan 137°}{1 - \tan 43° \tan 137°} = \tan(43° + 137°) = \tan 180° = 0$

c. $\tan(\alpha - \beta) = \dfrac{\tan\alpha - \tan\beta}{1 + \tan\alpha\tan\beta} = \dfrac{1.1 - 3.7}{1 + 1.1(3.7)} = \dfrac{-2.6}{5.07} = -\dfrac{20}{39}$ □

Example 2 • Evaluating the Sine of a Sum

Use the special angles to find the exact value of sin 165°.

Solution

Since we can write 165° as 135° + 30°, we have

$$\begin{aligned}\sin 165° &= \sin(135° + 30°)\\ &= \sin 135° \cos 30° + \cos 135° \sin 30°\\ &= \frac{\sqrt{2}}{2}\frac{\sqrt{3}}{2} + \left(-\frac{\sqrt{2}}{2}\right)\left(\frac{1}{2}\right) = \frac{\sqrt{2}}{4}(\sqrt{3} - 1).\end{aligned}$$

□

Example 3 • Proving an Identity

Verify the identity: $\sin(\alpha + \beta)\cos(\alpha - \beta) = \sin\alpha\cos\alpha + \sin\beta\cos\beta$.

Solution

$$\begin{aligned}\boldsymbol{\sin(\alpha + \beta)\cos(\alpha - \beta)} &= (\sin\alpha\cos\beta + \cos\alpha\sin\beta)\\ &\quad\times(\cos\alpha\cos\beta + \sin\alpha\sin\beta)\\ &= \sin\alpha\cos\alpha\cos^2\beta + \cos^2\alpha\sin\beta\cos\beta\\ &\quad + \sin^2\alpha\sin\beta\cos\beta + \cos\alpha\sin\alpha\sin^2\beta\\ &= \sin\alpha\cos\alpha(\cos^2\beta + \sin^2\beta)\\ &\quad + \sin\beta\cos\beta(\cos^2\alpha + \sin^2\alpha)\\ &= \sin\alpha\cos\alpha\cdot 1 + \sin\beta\cos\beta\cdot 1\\ &= \boldsymbol{\sin\alpha\cos\alpha + \sin\beta\cos\beta}\end{aligned}$$

□

◆ **Practice Problem 1** ◆

Verify the identity: $\cos(\alpha + \beta)\cos(\alpha - \beta) = \cos^2\beta - \sin^2\alpha$.

In the area of mathematics known as differential equations, expressions of the form $a\sin\theta + b\cos\theta$ are often encountered. The identity for the sine of the sum can be used to rewrite this type of expression in a more convenient form. Given the real numbers a and b, we can draw, as in Figure 8.3, an angle α in standard position whose terminal side passes through the point with coordinates (a, b). Then

$$\cos\alpha = \frac{a}{\sqrt{a^2 + b^2}} \quad\text{and}\quad \sin\alpha = \frac{b}{\sqrt{a^2 + b^2}}.$$

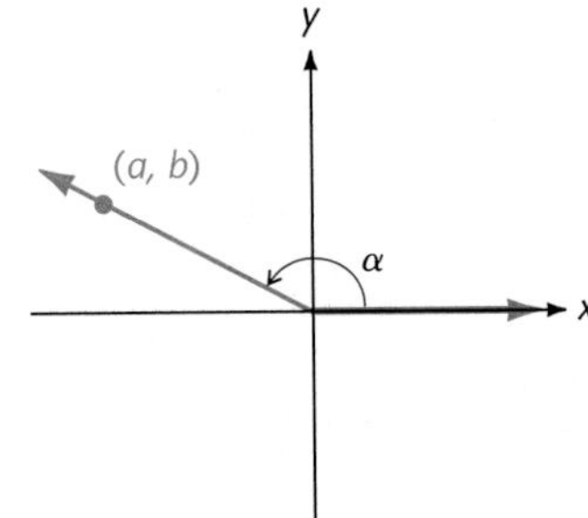

Figure 8.3 The angle α in the reduction identity

Now

$$\begin{aligned}&\boldsymbol{\sqrt{a^2 + b^2}\sin(\theta + \alpha)}\\ &\quad = \sqrt{a^2 + b^2}(\sin\theta\cos\alpha + \cos\theta\sin\alpha) \qquad \text{sine of the sum}\\ &\quad = \sqrt{a^2 + b^2}\left(\sin\theta\cdot\frac{a}{\sqrt{a^2 + b^2}} + \cos\theta\cdot\frac{b}{\sqrt{a^2 + b^2}}\right)\\ &\quad = \boldsymbol{a\sin\theta + b\cos\theta}.\end{aligned}$$

This result is known as the **reduction identity.** We use it in this section for graphing and in Section 8.6 to solve certain trigonometric equations.

Reduction Identity

$$a \sin \theta + b \cos \theta = \sqrt{a^2 + b^2} \sin (\theta + \alpha),$$

where α is determined by the equations

$$\cos \alpha = \frac{a}{\sqrt{a^2 + b^2}} \quad \text{and} \quad \sin \alpha = \frac{b}{\sqrt{a^2 + b^2}}.$$

Example 4 • Using the Reduction Identity in Graphing

Sketch the graph of $y = \sin x - \cos x$ over the interval $0 \le x \le 2\pi$.

Solution

It is possible to sketch the graph of $y = \sin x - \cos x$ by using the method of addition of ordinates, as described in Section 7.9. However, it can be sketched more easily if we rewrite this equation using the reduction identity with $a = 1$, $b = -1$. Then

$$\sqrt{a^2 + b^2} = \sqrt{2} \quad \text{and} \quad \alpha = -\frac{\pi}{4}$$

since

$$\cos \alpha = \frac{1}{\sqrt{2}} \quad \text{and} \quad \sin \alpha = \frac{-1}{\sqrt{2}}.$$

Thus the graph of

$$y = \sin x - \cos x$$

is the same as the graph of

$$y = \sqrt{2} \sin \left(x - \frac{\pi}{4} \right).$$

This is a sine curve with Amp $= \sqrt{2}$, $P = 2\pi$, and phase shift $\pi/4$ units to the right. The curve is sketched in Figure 8.4. □

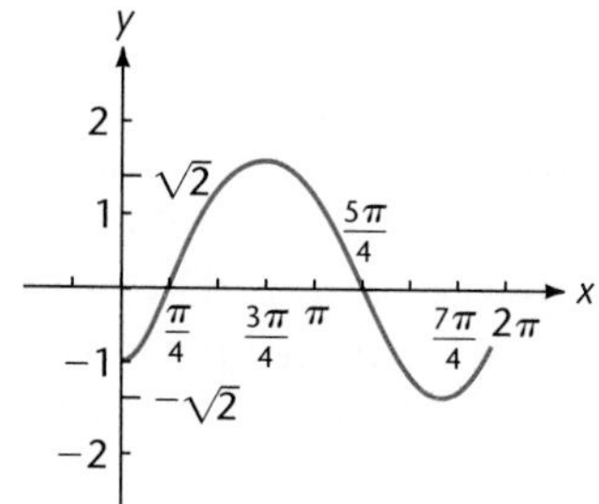

Figure 8.4 Graph of $y = \sin x - \cos x$, $x \in [0, 2\pi]$

EXERCISES 8.3

Use an identity to find the exact value of each of the following. (See Example 1.)

1. $\sin 29^\circ \cos 31^\circ + \cos 29^\circ \sin 31^\circ$
2. $\sin 115^\circ \cos 20^\circ + \cos 115^\circ \sin 20^\circ$
3. $\sin 198^\circ \cos 18^\circ - \cos 198^\circ \sin 18^\circ$

4. $\sin 625° \cos 310° - \cos 625° \sin 310°$

5. $\dfrac{\tan 66° + \tan 54°}{1 - \tan 66° \tan 54°}$

6. $\dfrac{\tan 100° + \tan 80°}{1 - \tan 100° \tan 80°}$

7. $\dfrac{\tan 17° - \tan 47°}{1 + \tan 17° \tan 47°}$

8. $\dfrac{\tan 74° - \tan 314°}{1 + \tan 74° \tan 314°}$

9. $\tan(\alpha + \beta)$ if $\tan \alpha = 2.1$ (exact) and $\tan \beta = -0.3$ (exact)

10. $\tan(\alpha - \beta)$ if $\tan \alpha = -1.2$ (exact) and $\tan \beta = 4$ (exact)

11. $\tan(\alpha - \beta)$ if $\cot \alpha = 15$ (exact) and $\cot \beta = 20$ (exact)

12. $\tan(\alpha + \beta)$ if $\cot \alpha = 0.7$ (exact) and $\cot \beta = 1.4$ (exact)

Without determining α or β, find the exact values of (a) $\sin(\alpha + \beta)$, (b) $\sin(\alpha - \beta)$, (c) $\tan(\alpha + \beta)$, and (d) $\tan(\alpha - \beta)$, if α and β satisfy the given conditions.

13. $\cos \alpha = \frac{3}{5}$, $\sin \beta = \frac{3}{5}$; α and β in Q I

14. $\cos \alpha = \frac{7}{25}$, $\cos \beta = \frac{5}{13}$; α and β in Q IV

15. $\cos \alpha = -\frac{4}{5}$, $\sin \beta = -\frac{12}{13}$; α in Q II, β in Q III

16. $\sin \alpha = \frac{15}{17}$, $\sin \beta = \frac{5}{13}$; α in Q II, β in Q I

Find the exact value of each of the following by using special angles. (See Example 2.)

17. $\sin 75°$

18. $\sin 105°$

19. $\sin \dfrac{\pi}{12}$

20. $\sin\left(-\dfrac{7\pi}{12}\right)$

21. $\tan \dfrac{7\pi}{12}$

22. $\tan \dfrac{5\pi}{12}$

23. $\tan 165°$

24. $\tan 285°$

Express each of the following in terms of trigonometric functions of θ.

25. $\sin(\pi - \theta)$

26. $\sin\left(\theta - \dfrac{3\pi}{2}\right)$

27. $\sin\left(\theta + \dfrac{5\pi}{6}\right)$

28. $\sin\left(\theta + \dfrac{3\pi}{4}\right)$

29. $\tan(\theta + \pi)$

30. $\tan\left(\theta + \dfrac{5\pi}{3}\right)$

31. $\tan\left(\theta - \dfrac{3\pi}{4}\right)$

32. $\tan(\theta - 3\pi)$

Sketch the graph over the interval $0 \le x \le 2\pi$. (See Example 4.)

33. $y = \sin x + \cos x$

34. $y = -2 \sin x + 2 \cos x$

35. $y = \sqrt{3} \sin x - \cos x$

36. $y = -\sin x - \sqrt{3} \cos x$

Verify the identities. (See Example 3.)

37. $\sin(\alpha+\beta)+\sin(\alpha-\beta)=2\sin\alpha\cos\beta$

38. $\sin(\alpha+\beta)-\sin(\alpha-\beta)=2\cos\alpha\sin\beta$

39. $\tan(\alpha-\beta)=\dfrac{\tan\alpha-\tan\beta}{1+\tan\alpha\tan\beta}$

40. $\cot\left(\dfrac{\pi}{2}-\theta\right)=\tan\theta$

41. $\sin(\alpha+\beta)\sin(\alpha-\beta)=\sin^2\alpha-\sin^2\beta$

42. $\tan(\alpha-\beta)\tan(\alpha+\beta)=\dfrac{\tan^2\alpha-\tan^2\beta}{1-\tan^2\alpha\tan^2\beta}$

43. $\dfrac{\sin(\alpha+\beta)}{\sin\alpha\sin\beta}=\cot\beta+\cot\alpha$

44. $\dfrac{\sin(\alpha+\beta)}{\cos\alpha\cos\beta}=\tan\alpha+\tan\beta$

45. $\dfrac{\sin(\alpha+\beta)}{\sin(\alpha-\beta)}=\dfrac{\tan\alpha+\tan\beta}{\tan\alpha-\tan\beta}$

46. $\dfrac{\cos(\alpha+\beta)}{\sin(\alpha-\beta)}=\dfrac{1-\tan\alpha+\tan\beta}{\tan\alpha-\tan\beta}$

47. $\cot(\alpha+\beta)=\dfrac{\cot\alpha\cot\beta-1}{\cot\alpha+\cot\beta}$

48. $\cot(\alpha-\beta)=\dfrac{\cot\alpha\cot\beta+1}{\cot\beta-\cot\alpha}$

49. $\dfrac{\tan(\alpha+\beta)+\tan(\alpha-\beta)}{1-\tan(\alpha+\beta)\tan(\alpha-\beta)}=\tan 2\alpha$

50. $\tan\left(\theta+\dfrac{\pi}{4}\right)=\dfrac{1+\tan\theta}{1-\tan\theta}$

51. $\sin(\alpha+\beta+\gamma)=\sin\alpha\cos\beta\cos\gamma-\sin\alpha\sin\beta\sin\gamma+\cos\alpha\sin\beta\cos\gamma+\cos\alpha\cos\beta\sin\gamma$

52. $\sin(\alpha-\beta-\gamma)=\sin\alpha\cos\beta\cos\gamma-\sin\alpha\sin\beta\sin\gamma-\cos\alpha\sin\beta\cos\gamma-\cos\alpha\cos\beta\sin\gamma$

53. $\cos(\alpha+\beta+\gamma)=\cos\alpha\cos\beta\cos\gamma-\sin\alpha\sin\beta\cos\gamma-\sin\alpha\cos\beta\sin\gamma-\cos\alpha\sin\beta\sin\gamma$

54. $\tan(\alpha+\beta+\gamma)=\dfrac{\tan\alpha+\tan\beta+\tan\gamma-\tan\alpha\tan\beta\tan\gamma}{1-\tan\alpha\tan\beta-\tan\gamma\tan\alpha-\tan\beta\tan\gamma}$

55. $\dfrac{\sin(x+h)-\sin x}{h}=\sin x\left(\dfrac{\cos h-1}{h}\right)+\cos x\left(\dfrac{\sin h}{h}\right)$

56. $\dfrac{\tan(x+h)-\tan x}{h}=\left(\dfrac{\sin h}{h}\right)\dfrac{\sec^2 x}{\cos h-\tan x\sin h}$

Pendulum 57. The pendulum shown in the accompanying figure has a length of 2 feet. If it is released at $t=0$ with a displacement of $\frac{1}{4}$ radian to the right of the vertical and an angular velocity of $\sqrt{3}$ radians per second, the resulting motion is described by the equation

$$\theta(t)=\frac{1}{4}\cos 4t+\frac{\sqrt{3}}{4}\sin 4t,$$

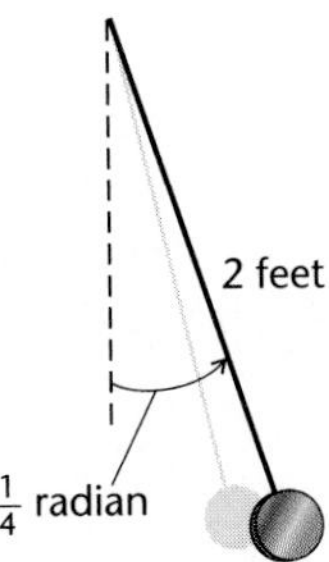

Figure for Exercise 57

where θ is the displacement angle in radians and t is the time (in seconds). Use the reduction identity to write this equation in the form

$$\theta(t) = A \sin (Bt + C)$$

and then state the amplitude and period for the motion.

Stretching a Spring

58. In Exercise 47 in Exercises 7.9 the displacement $x(t)$ of an object attached to a spring is described by the equation

$$x(t) = \tfrac{2}{3} \cos 8t - \tfrac{1}{6} \sin 8t.$$

Use the reduction identity to write this equation in the form

$$x(t) = A \sin (Bt + C)$$

and then state the amplitude and period for the motion of the object.

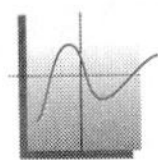

59–64

Use a graphing device to compare the graphs of the functions on each side of the given equation and decide if the equation is an identity.

59. $\sin x + \sin x = \sin 2x$

60. $\tan x + \tan 3x = \tan 4x$

61. $\sin x + \sin 3x = 2 \sin 2x \cos x$

62. $\sin 3x - \sin x = 2 \sin x \cos 2x$

63. (See Exercise 57.) Use a graphing device to graph the function

$$Y = \frac{1}{4} \cos 4x + \frac{\sqrt{3}}{4} \sin 4x.$$

Use the cursor to locate a maximum value of Y and compare this result with the amplitude found in Exercise 57.

64. (See Exercise 58.) Follow the instructions in Exercise 63 to find a maximum value of

$$Y = \tfrac{2}{3} \cos 8x - \tfrac{1}{6} \sin 8x$$

and compare your result with the amplitude found in Exercise 58.

◆ **Solution for Practice Problem**

1. $$\begin{aligned}\cos(\alpha+\beta)\cos(\alpha-\beta) &= (\cos\alpha\cos\beta - \sin\alpha\sin\beta)\\ &\quad \times (\cos\alpha\cos\beta + \sin\alpha\sin\beta)\\ &= \cos^2\alpha\cos^2\beta - \sin^2\alpha\sin^2\beta\\ &= \cos^2\alpha\cos^2\beta + \sin^2\alpha\cos^2\beta - \sin^2\alpha\cos^2\beta\\ &\quad - \sin^2\alpha\sin^2\beta\\ &= \cos^2\beta(\cos^2\alpha + \sin^2\alpha) - \sin^2\alpha\\ &\quad \times (\cos^2\beta + \sin^2\beta)\\ &= \cos^2\beta - \sin^2\alpha\end{aligned}$$

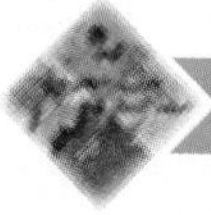

8.4 DOUBLE-ANGLE AND HALF-ANGLE IDENTITIES

We devote the first part of this section to special cases of the identities for the sum of two angles, which are called the **double-angle identities.**

Double-Angle Identities

$$\begin{aligned}\sin 2\theta &= 2\sin\theta\cos\theta\\ \cos 2\theta &= \cos^2\theta - \sin^2\theta\\ &= 2\cos^2\theta - 1\\ &= 1 - 2\sin^2\theta\\ \tan 2\theta &= \frac{2\tan\theta}{1-\tan^2\theta}\end{aligned}$$

Each of these identities follows directly from an identity for the sum of two angles. To determine the sine of twice an angle, we replace both α and β by θ in

$$\sin(\alpha+\beta) = \sin\alpha\cos\beta + \cos\alpha\sin\beta$$

to yield

$$\sin(\theta+\theta) = \sin\theta\cos\theta + \cos\theta\sin\theta$$

Thus we have the double-angle identity for the sine function

$$\boldsymbol{\sin 2\theta = 2\sin\theta\cos\theta.}$$

Similarly, we replace both α and β by θ in

$$\cos(\alpha+\beta) = \cos\alpha\cos\beta - \sin\alpha\sin\beta$$

to yield

$$\cos(\theta + \theta) = \cos\theta\cos\theta - \sin\theta\sin\theta.$$

The resulting double-angle identity for cosine is

$$\mathbf{\cos 2\theta = \cos^2\theta - \sin^2\theta.}$$

A Pythagorean identity is used to write this identity in two alternative forms. First, we can express $\cos 2\theta$ strictly in terms of $\cos\theta$.

$$\begin{aligned} \cos 2\theta &= \cos^2\theta - \sin^2\theta && \text{double-angle identity} \\ &= \cos^2\theta - (1 - \cos^2\theta) && \text{Pythagorean identity} \\ &= 2\cos^2\theta - 1 && \text{algebra} \end{aligned}$$

Second, we can express $\cos 2\theta$ strictly in terms of $\sin\theta$.

$$\begin{aligned} \cos 2\theta &= \cos^2\theta - \sin^2\theta && \text{double-angle identity} \\ &= (1 - \sin^2\theta) - \sin^2\theta && \text{Pythagorean identity} \\ &= 1 - 2\sin^2\theta && \text{algebra} \end{aligned}$$

The double-angle identity for the tangent can be derived by replacing both α and β by θ in

$$\tan(\alpha + \beta) = \frac{\tan\alpha + \tan\beta}{1 - \tan\alpha\tan\beta}.$$

Thus

$$\tan(\theta + \theta) = \frac{\tan\theta + \tan\theta}{1 - \tan\theta\tan\theta}$$

gives

$$\tan 2\theta = \frac{2\tan\theta}{1 - \tan^2\theta}.$$

The double-angle identities can be used to find the values of the trigonometric functions of 2θ if certain information is known about θ.

Example 1 • Using Double-Angle Identities

Suppose $\sin\theta = -(\sqrt{5}/3)$ and $\cos\theta < 0$. Find the exact values of all the trigonometric functions of 2θ.

Solution

The angle θ is in quadrant III since both $\sin\theta$ and $\cos\theta$ are negative. We find

$$\cos\theta = -\frac{2}{3} \qquad \text{and} \qquad \tan\theta = \frac{\sqrt{5}}{2}$$

by using the fundamental identities. Now we turn our attention to the angle 2θ and apply the double-angle identities.

$$\sin 2\theta = 2 \sin \theta \cos \theta = 2\left(-\frac{\sqrt{5}}{3}\right)\left(-\frac{2}{3}\right) = \frac{4\sqrt{5}}{9}$$

$$\cos 2\theta = \cos^2 \theta - \sin^2 \theta = \left(-\frac{2}{3}\right)^2 - \left(-\frac{\sqrt{5}}{3}\right)^2 = \frac{4}{9} - \frac{5}{9} = -\frac{1}{9}$$

$$\tan 2\theta = \frac{2 \tan \theta}{1 - \tan^2 \theta} = \frac{2\left(\frac{\sqrt{5}}{2}\right)}{1 - \left(\frac{\sqrt{5}}{2}\right)^2} = \frac{\sqrt{5}}{1 - \frac{5}{4}} = \frac{\sqrt{5}}{-\frac{1}{4}} = -4\sqrt{5}$$

Finally, we use the fundamental identities to evaluate the remaining trigonometric functions.

$$\cot 2\theta = \frac{1}{\tan 2\theta} = \frac{1}{-4\sqrt{5}} = -\frac{\sqrt{5}}{20}$$

$$\sec 2\theta = \frac{1}{\cos 2\theta} = -9$$

$$\csc 2\theta = \frac{1}{\sin 2\theta} = \frac{9}{4\sqrt{5}} = \frac{9\sqrt{5}}{20}$$

□

In Example 1 we chose to illustrate the double-angle identities by evaluating all three of $\sin 2\theta$, $\cos 2\theta$, and $\tan 2\theta$. Note that once any one of $\sin 2\theta$, $\cos 2\theta$, or $\tan 2\theta$ is evaluated, then the fundamental identities can be used to evaluate the others.

◆ Practice Problem 1 ◆

Express each of the following as a trigonometric function of twice the given angle.

a. $\cos^2 110° - \sin^2 110°$

b. $2 \sin \frac{\pi}{9} \cos \frac{\pi}{9}$

c. $1 - 2 \sin^2 \frac{7\pi}{8}$

d. $\frac{2 \tan 47°}{1 - \tan^2 47°}$

e. $\frac{1}{2 \sin 17° \cos 17°}$

f. $\frac{1 - \tan^2 3\gamma}{2 \tan 3\gamma}$

Example 2 • Proving an Identity

Verify the identity: $2 \cos^3 \theta - \cos \theta = \frac{\sin 4\theta}{4 \sin \theta}$.

Solution

We use the double-angle identities on the right-hand side.

$$\frac{\sin 4\theta}{4 \sin \theta} = \frac{\sin 2(2\theta)}{4 \sin \theta} \qquad \text{algebra}$$

$$= \frac{2 \sin 2\theta \cos 2\theta}{4 \sin \theta} \qquad \text{double-angle identity}$$

$$= \frac{2\,(2 \sin \theta \cos \theta)(2 \cos^2 \theta - 1)}{4 \sin \theta} \qquad \text{double-angle identities}$$

$$= \cos \theta\,(2 \cos^2 \theta - 1) \qquad \text{algebra}$$

$$= 2 \cos^3 \theta - \cos \theta \qquad \text{algebra}$$

□

The trigonometric functions of half an angle, $\theta/2$, can also be expressed in terms of trigonometric functions of the angle θ. The resulting identities are called the **half-angle identities,** and we list them next.

Half-Angle Identities

$$\sin \frac{\theta}{2} = \pm\sqrt{\frac{1 - \cos \theta}{2}} \qquad \tan \frac{\theta}{2} = \frac{\sin \theta}{1 + \cos \theta}$$

$$\cos \frac{\theta}{2} = \pm\sqrt{\frac{1 + \cos \theta}{2}} \qquad = \frac{1 - \cos \theta}{\sin \theta}$$

We choose either + or − on each radical, depending on the quadrant in which $\theta/2$ terminates.

We establish the half-angle identity for sine by first solving the double-angle identity

$$\cos 2A = 1 - 2 \sin^2 A$$

for $\sin A$.

$$2 \sin^2 A = 1 - \cos 2A$$

$$\sin^2 A = \frac{1 - \cos 2A}{2}$$

$$\sin A = \pm\sqrt{\frac{1 - \cos 2A}{2}}$$

Next, if we replace A by $\theta/2$, we have the desired form,

$$\sin \frac{\theta}{2} = \pm\sqrt{\frac{1 - \cos \theta}{2}}.$$

Since sine is positive in the first and second quadrants, we use + when $\theta/2$ terminates in quadrant I or II and − when $\theta/2$ terminates in quadrant III or IV.

The half-angle identity for cosine can be derived in a similar way using $\cos 2A = 2\cos^2 A - 1$.

The first form of the half-angle identity for tangent can be verified as follows.

$$\begin{aligned}
\frac{\sin\theta}{1+\cos\theta} &= \frac{\sin\left(2\cdot\frac{\theta}{2}\right)}{1+\cos\left(2\cdot\frac{\theta}{2}\right)} && \text{rewriting } \theta \\
&= \frac{2\sin\frac{\theta}{2}\cos\frac{\theta}{2}}{1+\left(2\cos^2\frac{\theta}{2}-1\right)} && \text{double-angle identities} \\
&= \frac{2\sin\frac{\theta}{2}\cos\frac{\theta}{2}}{2\cos^2\frac{\theta}{2}} && \text{algebra} \\
&= \frac{\sin\frac{\theta}{2}}{\cos\frac{\theta}{2}} && \text{algebra} \\
&= \tan\frac{\theta}{2} && \text{quotient identity}
\end{aligned}$$

The derivation of the other form of the half-angle identity for tangent is left as an exercise.

Example 3 • Using Half-Angle Identities

Write the exact values of all the trigonometric functions of $\theta/2$ if $0 \le \theta < 2\pi$, $\cos\theta = \frac{7}{25}$, and $\cot\theta < 0$.

Solution

Since $\cos\theta > 0$ and $\cot\theta < 0$, then θ lies in quadrant IV, and we write

$$\frac{3\pi}{2} < \theta < 2\pi.$$

Dividing all three members of this inequality by 2 locates the angle $\theta/2$ in quadrant II.

$$\frac{3\pi}{4} < \frac{\theta}{2} < \pi$$

We use this information to determine the signs to be used in the half-angle identities.

$$\sin\frac{\theta}{2} = +\sqrt{\frac{1-\cos\theta}{2}} = \sqrt{\frac{1-\frac{7}{25}}{2}} = \sqrt{\frac{\frac{18}{25}}{2}} = \sqrt{\frac{9}{25}} = \frac{3}{5}$$

$$\cos\frac{\theta}{2} = -\sqrt{\frac{1+\cos\theta}{2}} = \sqrt{\frac{1+\frac{7}{25}}{2}} = -\sqrt{\frac{\frac{32}{25}}{2}} = -\sqrt{\frac{16}{25}} = -\frac{4}{5}$$

We can use the quotient identity to obtain

$$\tan\frac{\theta}{2} = \frac{\sin\frac{\theta}{2}}{\cos\frac{\theta}{2}} = \frac{\frac{3}{5}}{-\frac{4}{5}} = -\frac{3}{4}.$$

The reciprocal identities then give

$$\csc\frac{\theta}{2} = \frac{5}{3}, \qquad \sec\frac{\theta}{2} = -\frac{5}{4}, \qquad \cot\frac{\theta}{2} = -\frac{4}{3}.$$ □

Again we note that once one trigonometric function for $\theta/2$ is evaluated, then the fundamental identities can be used to evaluate all others.

◆ **Practice Problem 2** ◆

Use the half-angle identities to find the exact values of the sine, cosine, and tangent of $3\pi/8$.

EXERCISES 8.4

Express each of the following in terms of a trigonometric function of twice or half the given angle. Assume all variables are such that all expressions are defined.

1. $2 \sin 22° \cos 22°$
2. $8 \sin 17° \cos 17°$
3. $\dfrac{1}{\sin 105° \cos 105°}$
4. $\dfrac{1}{\sin 43° \cos 43°}$
5. $\cos^2\dfrac{\pi}{18} - \sin^2\dfrac{\pi}{18}$
6. $1 - 2\sin^2\dfrac{\pi}{10}$
7. $\dfrac{1}{2\cos^2\gamma - 1}$
8. $\dfrac{1}{\cos^2 2\alpha - \sin^2 2\alpha}$
9. $\dfrac{2\tan\frac{\pi}{7}}{1-\tan^2\frac{\pi}{7}}$
10. $\dfrac{2\tan 115°}{1-\tan^2 115°}$

11. $\dfrac{1 - \tan^2 \alpha}{2 \tan \alpha}$

12. $\dfrac{1 - \tan^2 4\beta}{2 \tan 4\beta}$

13. $\sqrt{\dfrac{1 - \cos 242°}{2}}$

14. $\sqrt{\dfrac{1 - \cos 48°}{2}}$

15. $-\sqrt{\dfrac{1 - \cos 198°}{1 + \cos 198°}}$

16. $-\sqrt{\dfrac{1 - \cos \dfrac{16\pi}{9}}{1 + \cos \dfrac{16\pi}{9}}}$

17. $-\sqrt{\dfrac{1 + \cos 378°}{2}}$

18. $\sqrt{\dfrac{1 + \cos \dfrac{4\pi}{7}}{2}}$

19. $\dfrac{\sin 408°}{1 + \cos 408°}$

20. $\dfrac{1 - \cos 130°}{\sin 130°}$

Determine the exact value of all the trigonometric functions of 2θ if θ satisfies the given conditions. (See Example 1.)

21. $\sin \theta = \frac{4}{5}$; θ in Q II

22. $\cos \theta = \frac{5}{13}$; $\tan \theta < 0$

23. $\sec \theta = -\frac{3}{2}$; $\sin \theta < 0$

24. $\tan \theta = -2\sqrt{2}$; $\sec \theta > 0$

Use the composite-angle identities to express each of the following trigonometric functions of the given multiple angle in terms of the same trigonometric function of θ.

25. $\cos 4\theta$

26. $\cos 3\theta$

27. $\tan 3\theta$

28. $\tan 4\theta$

Determine the exact values of all trigonometric functions of $\theta/2$ if θ satisfies the given conditions and $0 \leq \theta < 2\pi$. (See Example 3.)

29. $\sin \theta = \frac{4}{5}$; θ in Q II

30. $\cos \theta = \frac{5}{13}$; $\tan \theta < 0$

31. $\sec \theta = -\frac{3}{2}$; $\sin \theta < 0$

32. $\tan \theta = -2\sqrt{2}$; $\sec \theta > 0$

Determine the exact values of the sine, cosine, and tangent of each of the following.

33. $\dfrac{\pi}{8}$

34. $\dfrac{5\pi}{8}$

35. $\dfrac{7\pi}{12}$

36. $\dfrac{11\pi}{12}$

37. 157.5°

38. 105°

39. 195°

40. 202.5°

Verify each of the following identities. (See Example 2.)

41. $(\sin \theta - \cos \theta)^2 = 1 - \sin 2\theta$

42. $\cos^4 \theta - \sin^4 \theta = \cos 2\theta$

√ 43. $\dfrac{2}{1 + \cos 2A} = \sec^2 A$

44. $\dfrac{2 + 2\cos 2A}{\sin^2 2A} = \csc^2 A$

45. $\dfrac{1 - \cos 2\alpha}{\sin 2\alpha} = \tan \alpha$

46. $\dfrac{1 + \cos 2\alpha}{\sin 2\alpha} = \cot \alpha$

47. $\dfrac{\sin 2B \cos B}{1 - \sin^2 B} = 2 \sin B$

48. $\dfrac{\cos B \sin 2B}{1 + \cos 2B} = \sin B$

√ 49. $\dfrac{\cos 2\beta - \sin \beta}{\cos^2 \beta} = \dfrac{1 - 2\sin \beta}{1 - \sin \beta}$

50. $\dfrac{\sin^2 \beta}{\cos \beta - \cos 2\beta} = \dfrac{1 + \cos \beta}{1 + 2\cos \beta}$

√ 51. $\dfrac{1 + \sin 2x - \cos 2x}{1 + \cos 2x + \sin 2x} = \tan x$

52. $\dfrac{4 \sin 2x}{(1 + \cos 2x)^2} = 2 \tan x \sec^2 x$

53. $\cot 2\theta = \dfrac{\cot^2 \theta - 1}{2 \cot \theta}$

54. $\sec 2\theta = \dfrac{\sec^2 \theta}{2 - \sec^2 \theta}$

√ 55. $\cot 2\theta + \tan \theta = \csc 2\theta$

56. $\tan 2\alpha - \tan \alpha = \tan \alpha \sec 2\alpha$

57. $\csc 2\alpha = \frac{1}{2} \csc \alpha \sec \alpha$

58. $\csc 6\alpha \tan 3\alpha = \dfrac{\sec^2 3\alpha}{2}$

59. $\dfrac{\sin 4\alpha}{\cos \alpha} + \dfrac{\cos 4\alpha}{\sin \alpha} = 2 \csc 2\alpha \cos 3\alpha$

60. $\dfrac{\sin 2\alpha}{\sin 3\alpha} + \dfrac{\cos 2\alpha}{\cos 3\alpha} = 2 \sin 5\alpha \csc 6\alpha$

61. $\tan 3y - \tan y = 2 \sin y \sec 3y$

62. $\cot y - \cot 3y = 2 \cos y \csc 3y$

√ 63. $\dfrac{(1 + \tan t)^2}{1 + \tan^2 t} = 1 + \sin 2t$

64. $\dfrac{1 - \tan^2 t}{1 + \tan^2 t} = \cos 2t$

65. $\dfrac{1 + \sin 2\gamma}{\cos 2\gamma} = \dfrac{1 + \tan \gamma}{1 - \tan \gamma}$

66. $1 + \tan \gamma \tan 2\gamma = \sec 2\gamma$

67. $\tan \dfrac{\theta}{2}(1 + \sec \theta) = \tan \theta$

68. $\tan \dfrac{\theta}{2} + 2 \sin^2 \dfrac{\theta}{2} \cot \theta = \sin \theta$

√ 69. $\tan \dfrac{\theta}{2} + \cot \dfrac{\theta}{2} = 2 \csc \theta$

70. $\tan^2 \dfrac{\theta}{2} + 1 = 2 \csc \theta \tan \dfrac{\theta}{2}$

71. Derive this alternate form of the half-angle identity for tangent.

$$\tan\frac{\theta}{2} = \frac{1-\cos\theta}{\sin\theta}$$

72. Use identities stated in this section and prove that

$$\tan\frac{\theta}{2} = \pm\sqrt{\frac{1-\cos\theta}{1+\cos\theta}}.$$

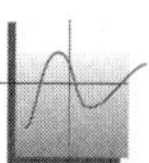

Use a graphing device to compare the graphs of the functions on each side of the given equation and decide if the equation is an identity.

73–76

73. $(\sin x + \cos x)^2 = 1 + \sin 2x$

74. $\cot x - \tan x = 2\cot 2x$

75. $\cos^4 x + \sin^4 x = 1$

76. $\dfrac{\cos^2 x + \sin^2 x}{(\cos x + \sin x)^2} = \dfrac{1}{\sin 2x}$

◆ Solutions for Practice Problems

1. a. $\cos^2 110° - \sin^2 110° = \cos 2(110°) = \cos 220°$

   b. $2\sin\frac{\pi}{9}\cos\frac{\pi}{9} = \sin 2\left(\frac{\pi}{9}\right) = \sin\frac{2\pi}{9}$

   c. $1 - 2\sin^2\frac{7\pi}{8} = \cos 2\left(\frac{7\pi}{8}\right) = \cos\frac{7\pi}{4}$

   d. $\dfrac{2\tan 47°}{1-\tan^2 47°} = \tan 2(47°) = \tan 94°$

   e. $\dfrac{1}{2\sin 17°\cos 17°} = \dfrac{1}{\sin 2(17°)} = \dfrac{1}{\sin 34°} = \csc 34°$

   f. $\dfrac{1-\tan^2 3\gamma}{2\tan 3\gamma} = \dfrac{1}{\tan 2(3\gamma)} = \dfrac{1}{\tan 6\gamma} = \cot 6\gamma$

2. We can write $\frac{3\pi}{8}$ as $\frac{1}{2}\left(\frac{3\pi}{4}\right)$. Then we use the half-angle identities with $\theta = \frac{3\pi}{4}$.

$$\sin\frac{3\pi}{8} = +\sqrt{\frac{1-\cos\frac{3\pi}{4}}{2}} = \sqrt{\frac{1-\left(\frac{-\sqrt{2}}{2}\right)}{2}}$$

$$= \sqrt{\frac{2+\sqrt{2}}{4}} = \frac{\sqrt{2+\sqrt{2}}}{2}$$

$$\cos\frac{3\pi}{8} = +\sqrt{\frac{1+\cos\frac{3\pi}{4}}{2}} = \sqrt{\frac{1+\left(\frac{-\sqrt{2}}{2}\right)}{2}}$$

$$= \sqrt{\frac{2-\sqrt{2}}{4}} = \frac{\sqrt{2-\sqrt{2}}}{2}$$

$$\tan\frac{3\pi}{8} = \frac{1-\cos\frac{3\pi}{4}}{\sin\frac{3\pi}{4}} = \frac{1-\left(\frac{-\sqrt{2}}{2}\right)}{\frac{\sqrt{2}}{2}} = \frac{2+\sqrt{2}}{\sqrt{2}}$$

$$= \frac{2\sqrt{2}+2}{2} = \sqrt{2}+1$$

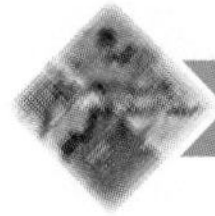

8.5 PRODUCT AND SUM IDENTITIES

In some instances it is advantageous to rewrite a product involving sines and/or cosines as a sum of sines and/or cosines, and vice versa. The identities of this section can be applied to those situations.

We begin by first rewriting the four identities (from Sections 8.2 and 8.3) for the sine and cosine of the sum and difference of two angles.

$$\sin(\alpha+\beta) = \sin\alpha\cos\beta + \cos\alpha\sin\beta$$
$$\sin(\alpha-\beta) = \sin\alpha\cos\beta - \cos\alpha\sin\beta$$
$$\cos(\alpha+\beta) = \cos\alpha\cos\beta - \sin\alpha\sin\beta$$
$$\cos(\alpha-\beta) = \cos\alpha\cos\beta + \sin\alpha\sin\beta$$

Adding and subtracting the first two of these equations yield the first two of the following four identities labeled **product identities.** Similarly, adding and subtracting the last two of these equations yield the last two of the product identities.

Product Identities

$$2\sin\alpha\cos\beta = \sin(\alpha+\beta) + \sin(\alpha-\beta)$$
$$2\cos\alpha\sin\beta = \sin(\alpha+\beta) - \sin(\alpha-\beta)$$
$$2\cos\alpha\cos\beta = \cos(\alpha+\beta) + \cos(\alpha-\beta)$$
$$2\sin\alpha\sin\beta = \cos(\alpha-\beta) - \cos(\alpha+\beta)$$

Example 1 • Using the Product Identities

Write each of the following products as a sum. Use the negative-angle identities if necessary so that no argument contains a minus sign.

a. $2 \sin 3\alpha \cos 2\alpha$
b. $\cos 4\gamma \cos \gamma$
c. $6 \sin 2\theta \sin 6\theta$

Solution

a. Using the first identity for the product of a sine and a cosine, we have

$$\begin{aligned} 2 \sin 3\alpha \cos 2\alpha &= \sin (3\alpha + 2\alpha) + \sin (3\alpha - 2\alpha) \\ &= \sin 5\alpha + \sin \alpha. \end{aligned}$$

b. Since we have a product of two cosines, we use the third of the product identities.

$$\begin{aligned} \cos 4\gamma \cos \gamma &= \tfrac{1}{2} [\cos (4\gamma + \gamma) + \cos (4\gamma - \gamma)] \\ &= \tfrac{1}{2} \cos 5\gamma + \tfrac{1}{2} \cos 3\gamma \end{aligned}$$

c. For the product of two sines we use the last of the product identities.

$$\begin{aligned} 6 \sin 2\theta \sin 6\theta &= 3[\cos (2\theta - 6\theta) - \cos (2\theta + 6\theta)] \\ &= 3[\cos (-4\theta) - \cos 8\theta] \\ &= 3 \cos 4\theta - 3 \cos 8\theta \end{aligned}$$

The last equality results from a negative-angle identity. □

Next we derive four additional identities, called **sum identities.** To do this, in each of the four product identities we let

$$x = \alpha + \beta \qquad \text{and} \qquad y = \alpha - \beta.$$

The corresponding expressions for α and β are

$$\alpha = \frac{x + y}{2} \qquad \text{and} \qquad \beta = \frac{x - y}{2}.$$

Rewriting the four product identities in terms of x and y results in the following sum identities.

Sum Identities

$$\sin x + \sin y = 2 \sin \left(\frac{x + y}{2}\right) \cos \left(\frac{x - y}{2}\right)$$

$$\sin x - \sin y = 2 \cos \left(\frac{x + y}{2}\right) \sin \left(\frac{x - y}{2}\right)$$

$$\cos x + \cos y = 2\cos\left(\frac{x+y}{2}\right)\cos\left(\frac{x-y}{2}\right)$$
$$\cos x - \cos y = -2\sin\left(\frac{x+y}{2}\right)\sin\left(\frac{x-y}{2}\right)$$

Example 2 • Using the Sum Identities

Write each of the following sums as a product. Use the negative-angle identities if necessary so that no argument contains a minus sign.

a. $\sin 135° + \sin 45°$ b. $\cos 3\theta - \cos 9\theta$

Solution

In each part we apply the appropriate sum identity.

$$\text{a. } \sin 135° + \sin 45° = 2\sin\left(\frac{135° + 45°}{2}\right)\cos\left(\frac{135° - 45°}{2}\right)$$
$$= 2\sin 90° \cos 45°$$

$$\text{b. } \cos 3\theta - \cos 9\theta = -2\sin\left(\frac{3\theta + 9\theta}{2}\right)\sin\left(\frac{3\theta - 9\theta}{2}\right)$$
$$= -2\sin 6\theta \sin(-3\theta)$$
$$= 2\sin 6\theta \sin 3\theta$$

□

Example 3 • Proving an Identity

Verify the identity: $\dfrac{\sin 4\theta + \sin 2\theta}{\cos 4\theta + \cos 2\theta} = \tan 3\theta.$

Solution

$$\frac{\sin 4\theta + \sin 2\theta}{\cos 4\theta + \cos 2\theta} = \frac{2\sin\left(\frac{4\theta + 2\theta}{2}\right)\cos\left(\frac{4\theta - 2\theta}{2}\right)}{2\cos\left(\frac{4\theta + 2\theta}{2}\right)\cos\left(\frac{4\theta - 2\theta}{2}\right)} \quad \text{sum identities}$$
$$= \frac{2\sin 3\theta \cos\theta}{2\cos 3\theta \cos\theta} \quad \text{algebra}$$
$$= \tan 3\theta \quad \text{quotient identity}$$

□

◆ Practice Problem 1 ◆

Verify the identity: $\cos(A + B)\cos(A - B) = \cos^2 A - \sin^2 B$.

EXERCISES 8.5

Express each of the following products as a sum. Use the negative-angle identities if necessary so that no argument contains a minus sign. (See Example 1.)

1. $2 \sin 84° \cos 27°$
2. $4 \cos 93° \sin 16°$
3. $8 \cos 12° \cos 13°$
4. $6 \sin 93° \sin 116°$
5. $\cos 3\alpha \sin 8\alpha$
6. $\cos 4\theta \cos 5\theta$
7. $\sin x \sin 2x$
8. $\sin 2y \cos 3y$

Express each of the following sums as a product. Use the negative-angle identities if necessary so that no argument contains a minus sign. (See Example 2.)

9. $\sin 42° + \sin 56°$
10. $\sin 112° - \sin 48°$
11. $\cos 17° - \cos 93°$
12. $\cos 80° + \cos 20°$
13. $\sin 9\alpha - \sin 11\alpha$
14. $\sin 4\gamma + \sin 6\gamma$
15. $\cos x + \cos 3x$
16. $\cos 2t - \cos 4t$

Verify each of the following identities. (See Example 3.)

17. $\dfrac{\sin 3\theta - \sin \theta}{\cos 3\theta + \cos \theta} = \tan \theta$
18. $\dfrac{\cos 5x + \cos 7x}{\sin 5x - \sin 7x} = -\cot x$
19. $\dfrac{\sin 6x - \sin 2x}{\sin 6x + \sin 2x} = \dfrac{\tan 2x}{\tan 4x}$
20. $\dfrac{\cos 3\theta - \cos \theta}{\cos 3\theta + \cos \theta} = \dfrac{2 \tan^2 \theta}{\tan^2 \theta - 1}$
21. $\cos (A + B) \sin (A - B) = \sin A \cos A - \sin B \cos B$
22. $\sin (A + B) \sin (A - B) = \cos^2 B - \cos^2 A$
23. $\sin 2x - \sin 4x - \sin 6x = -4 \cos 3x \sin 2x \cos x$
24. $\sin 2x - \sin 4x + \sin 6x = 4 \cos 3x \cos 2x \sin x$
25. $\sin 6t \sin 4t + \cos 8t \cos 2t = \cos 4t \cos 2t$
26. $\sin 12t \sin 4t + \cos 6t \cos 10t = \cos 6t \cos 2t$
27. $\sin 6A \cos 4A - \cos 8A \sin 2A = \sin 4A \cos 2A$
28. $\cos 7A \sin A - \sin 2A \cos 6A = -\cos 5A \sin A$
29. $\sin 2\theta + \sin 4\theta + 2 \cos 3\theta \sin \theta = 2 \sin 4\theta$
30. $\sin 3\theta - \sin \theta + 2 \sin 2\theta \cos \theta = 2 \sin 3\theta$
31. $\cos 4y + \cos 8y - 2 \sin 6y \sin 2y = 2 \cos 8y$
32. $\cos 9y - \cos 7y + 2 \cos 8y \cos y = 2 \cos 9y$

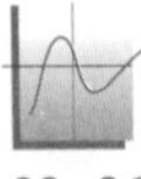

33–36

Use a graphing device to graph the function on each side of the given equation. By locating the point of intersection with a cursor, find the least positive value of x that satisfies the equation. Round your answer to the nearest tenth.

33. $\cos x + \cos 2x = \cos 3x$
34. $\sin 3x - \sin x = \sin 2x$
35. $\sin 2x \sin 3x = \sin 6x$
36. $\cos 2x \cos 3x = \cos 6x$

◆ **Solution for Practice Problem**

1. $\cos(A + B)\cos(A - B)$
$$= \tfrac{1}{2}\{\cos(A + B + A - B) + \cos[(A + B) - (A - B)]\}$$
$$= \tfrac{1}{2}(\cos 2A + \cos 2B)$$
$$= \tfrac{1}{2}(2\cos^2 A - 1 + 1 - 2\sin^2 B)$$
$$= \cos^2 A - \sin^2 B$$

8.6 TRIGONOMETRIC EQUATIONS

A **trigonometric equation** is an equation that involves at least one trigonometric function. To this point we have concentrated mainly on trigonometric equations that are identities. We turn our attention now to conditional trigonometric equations.†

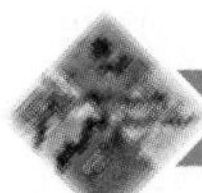

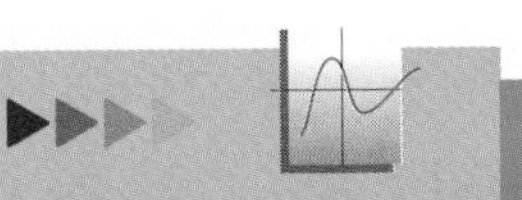

The techniques used with a graphing device to solve algebraic equations also apply to trigonometric equations, as illustrated in Exercises 33–36 of Section 8.5.

The techniques used to solve equations in algebra can be applied to trigonometric equations. We illustrate the parallel procedures in the next three examples.

Example 1 • Solving a Linear Equation

Solve $4\sin\theta - \sqrt{3} = 2\sin\theta$.

Solution

If you can do this. . . . . .then you can do this!

$4x - \sqrt{3} = 2x$. . . . .$4\sin\theta - \sqrt{3} = 2\sin\theta$

$2x = \sqrt{3}$ . . . algebra . . . . $2\sin\theta = \sqrt{3}$

$x = \dfrac{\sqrt{3}}{2}$. . . algebra . . . .$\sin\theta = \dfrac{\sqrt{3}}{2}$

related angle . . . . $\theta' = 60°$

Since $\sin\theta > 0$ only in quadrants I and II, then $\theta = 60°$ and $\theta = 120°$ are solutions, as well as any angle coterminal to 60° and 120°. Thus all solutions are given by

$$\theta = \begin{cases} 60° + n \cdot 360°, & n \text{ any integer;} \\ 120° + n \cdot 360°, & n \text{ any integer.} \end{cases}$$ □

In most cases we restrict our attention to solutions that are nonnegative angles measuring less than 1 revolution.

†A conditional equation is an equation that is a false statement for at least one value in its domain.

Example 2 • Solving a Quadratic Equation

Solve for the radian measure of x, where $0 \le x < 2\pi$.

$$\sin^2 x + 2\sin x - 3 = 0$$

Solution

If you can do this . . . . then you can do this!

If you can do this		then you can do this!
$y^2 + 2y - 3 = 0$		$\sin^2 x + 2\sin x - 3 = 0$
$(y+3)(y-1) = 0$	algebra	$(\sin x + 3)(\sin x - 1) = 0$
$y + 3 = 0$ or $y - 1 = 0$	algebra	$\sin x + 3 = 0$ or $\sin x - 1 = 0$
$y = -3$ or $y = 1$	algebra	$\sin x = -3$ or $\sin x = 1$
	trigonometry	no solution $\quad x = \frac{\pi}{2}$
	solution	$x = \frac{\pi}{2}$

□

If a quadratic equation cannot be easily factored, the quadratic formula can be used.

◆ Practice Problem 1 ◆

Use the quadratic formula to solve

$$2\sin^2\theta - 2\sqrt{2}\sin\theta + 1 = 0, \qquad 0° \le \theta < 360°.$$

Example 3 • Using Factoring

Solve $2\sin\theta\cos\theta = \cos\theta$, for $0° \le \theta < 360°$.

Solution

If you can do this . . . . then you can do this!

If you can do this		then you can do this!
$2xy = y$		$2\sin\theta\cos\theta = \cos\theta$
$2xy - y = 0$	algebra	$2\sin\theta\cos\theta - \cos\theta = 0$
$y(2x-1) = 0$	algebra	$\cos\theta(2\sin\theta - 1) = 0$
$y = 0$ or $2x - 1 = 0$	algebra	$\cos\theta = 0$ or $2\sin\theta - 1 = 0$
$x = \frac{1}{2}$	algebra	$\cos\theta = 0 \quad \sin\theta = \frac{1}{2}$
	trigonometry	quadrantal angles: 90°, 270° $\quad \theta' = 30°$, $\theta = 30°, 150°$
	solutions	$\theta = 90°, 270°, 30°, 150°$

□

In Example 3 we must resist the temptation to divide out the common factor $\cos\theta$. Doing so would lose the solutions $\theta = 90°, 270°$. Dividing both sides of an equation by 0 ($\cos\theta = 0$ for $\theta = 90°, 270°$) is *not* a valid operation. As a general procedure, we rewrite the equation so that one side is 0 and the other side factors.

The fundamental identities are often useful and sometimes necessary in solving trigonometric equations involving more than one trigonometric function.

Example 4 • Using a Fundamental Identity

Solve for the radian measure of θ in

$$\sec\theta + 1 = \tan^2\theta,\ 0 \le \theta < 2\pi.$$

Solution

We rewrite the equation $\sec\theta + 1 = \tan^2\theta$ strictly in terms of $\sec\theta$ by using one of the Pythagorean identities.

$\sec\theta + 1 = \sec\ \theta - 1$		Pythagorean identity
$\sec^2\theta - \sec\theta - 2 = 0$		algebra
$(\sec\theta - 2)(\sec\theta + 1) = 0$		algebra
$\sec\theta - 2 = 0$ or $\sec\theta + 1 = 0$		algebra
$\sec\theta = 2$ or	$\sec\theta = -1$	algebra
Related angle: $\theta' = \dfrac{\pi}{3}$	quadrantal angle	trigonometry
$\theta = \dfrac{\pi}{3}, \dfrac{5\pi}{3}$	$\theta = \pi$	solutions

□

The preceding examples illustrate some suggestions for solving trigonometric equations. In the following list, we summarize these suggestions, emphasizing that they are only suggestions and not a rigid step-by-step procedure. Rely on your intuition and do not hesitate to use a trigonometric identity or to perform some algebraic operation that might simplify the equation.

Solving Trigonometric Equations

1. If the equation is linear in one trigonometric function,
   a. Solve for that trigonometric function;
   b. Then solve for the angle by recognizing the function values of the quadrantal or special angles or by using a calculator or the trigonometric tables.
2. If more than one trigonometric function occurs, you might use the fundamental identities to rewrite the equation in terms of one trigonometric function.

3. If the equation is not linear, rewrite it so that one side is identically zero and try to factor the other side. If a quadratic expression will not factor, use the quadratic formula.
4. Rely heavily on algebraic techniques for solving equations.

Trigonometric equations involving composite angles can be separated into two categories: those that require simplification by using composite angle identities and those that do not. We begin with the simpler situation in Example 5, keeping in mind the preceding suggestions for solving trigonometric equations.

Example 5 • An Equation with a Multiple Angle

Solve for the values of x in

$$2\cos 3x = 1, \qquad 0 \leq x < 2\pi.$$

Solution

We have

$$\cos 3x = \tfrac{1}{2},$$

so the related angle for $3x$ is $\pi/3$. The problem requires solutions for x to be such that

$$0 \leq x < 2\pi.$$

Multiplying all three members of this inequality by 3 gives the required range on the variable $3x$.

$$0 \leq 3x < 6\pi$$

Thus $\pi/3$ is the related angle (or number) for nonnegative angles (or numbers) measuring up to 3 revolutions (up to 6π). Hence the solutions for $3x$ in

$$\cos 3x = \tfrac{1}{2}, \qquad 0 \leq 3x < 6\pi$$

are

$$3x = \frac{\pi}{3},\ \frac{5\pi}{3},\ \frac{7\pi}{3},\ \frac{11\pi}{3},\ \frac{13\pi}{3},\ \text{and}\ \frac{17\pi}{3}.$$

Finally, solving for x, we have

$$x = \frac{\pi}{9},\ \frac{5\pi}{9},\ \frac{7\pi}{9},\ \frac{11\pi}{9},\ \frac{13\pi}{9},\ \text{and}\ \frac{17\pi}{9}.$$

These are all the solutions for $0 \leq x < 2\pi$. □

◆ Practice Problem 2 ◆

Solve for all x in

$$2\sin 2x = \csc 2x, \qquad 0 \leq x < 2\pi.$$

The composite angle identities play key roles in the solutions of the remaining examples.

Example 6 • **Using a Double-Angle Identity**

Solve $\cos 2x - \sin x = 0$ for all values of x such that $0 \le x < 2\pi$.

Solution

Replacing $\cos 2x$ by $1 - 2\sin^2 x$ in the equation

$$\cos 2x - \sin x = 0, \qquad 0 \le x < 2\pi,$$

results in a quadratic equation in $\sin x$.

$$1 - 2\sin^2 x - \sin x = 0, \qquad 0 \le x < 2\pi$$

We solve this equation by the method of factoring.

$$
\begin{array}{rclrl}
 & & 2\sin^2 x + \sin x - 1 = 0 & & \text{algebra} \\
 & & (2\sin x - 1)(\sin x + 1) = 0 & & \text{algebra} \\
2\sin x - 1 = 0 & \text{or} & \sin x + 1 = 0 & & \text{algebra} \\
\sin x = \frac{1}{2} & \text{or} & \sin x = -1 & & \text{algebra} \\
x = \frac{\pi}{6}, \frac{5\pi}{6} & & x = \frac{3\pi}{2} & & \text{special angles}
\end{array}
$$

The solutions are $x = \dfrac{\pi}{6}$, $\dfrac{5\pi}{6}$, and $\dfrac{3\pi}{2}$. □

Example 7 • **Using a Composite Angle Identity**

Solve for all x in

$$\sin 4x \cos x - \cos 4x \sin x = 1, \ 0 \le x < 2\pi.$$

Solution

Recognizing the left side of

$$\sin 4x \cos x - \cos 4x \sin x = 1$$

as part of the identity for the sine of the difference of two angles, we can rewrite the equation in the form

$$\sin(4x - x) = 1, \qquad 0 \le x < 2\pi,$$

or

$$\sin 3x = 1, \qquad 0 \le 3x < 6\pi.$$

Solving for nonnegative values of $3x$ up to 3 revolutions gives

$$3x = \frac{\pi}{2}, \quad \frac{5\pi}{2}, \quad \text{and} \quad \frac{9\pi}{2}.$$

Finally, solving for x yields the solutions in the range $0 \le x < 2\pi$:

$$x = \frac{\pi}{6}, \quad \frac{5\pi}{6}, \quad \text{and} \quad \frac{3\pi}{2}.$$

□

◆ Practice Problem 3 ◆

Solve for all x in $\cos 3x + \cos x = 0$, $0 \le x < 2\pi$.

Example 8 • Using the Reduction Identity

Solve

$$\sin x + \cos x = 1$$

for all values of x such that $0 \le x < 2\pi$.

Solution

The reduction identity can be used to write

$$\sin x + \cos x = \sqrt{2} \sin \left(x + \frac{\pi}{4} \right).$$

Then the equation

$$\sin x + \cos x = 1, \qquad 0 \le x < 2\pi,$$

becomes

$$\sqrt{2} \sin \left(x + \frac{\pi}{4} \right) = 1, \qquad \frac{\pi}{4} \le x + \frac{\pi}{4} < 2\pi + \frac{\pi}{4},$$

or

$$\sin \left(x + \frac{\pi}{4} \right) = \frac{1}{\sqrt{2}}.$$

Solving for $x + (\pi/4)$, we have

$$x + \frac{\pi}{4} = \frac{\pi}{4} \qquad \text{and} \qquad x + \frac{\pi}{4} = \frac{3\pi}{4}.$$

Subtracting $\pi/4$ finally yields the solutions

$$x = 0 \quad \text{and} \quad x = \frac{\pi}{2}.$$

□

EXERCISES 8.6

Solve the equation in Exercises 1–12 for all θ. (See Examples 1 and 2.)

1. $3 \tan \theta = -\sqrt{3}$
2. $\sec \theta + \sqrt{2} = 0$
3. $2 \csc \theta - 2 = \csc \theta$
4. $5 \cot \theta + 3\sqrt{3} = 2\sqrt{3} + 4 \cot \theta$

5. $2\cos^2\theta + \cos\theta - 1 = 0$
6. $\sin^2\theta + 3\sin\theta + 2 = 0$
7. $2\cos^2\theta = 5\cos\theta + 3$
8. $2\sin^2\theta - 3 = \sin\theta$
9. $\sec^2\theta - 3\sec\theta = -2$
10. $\tan^2\theta = 2\tan\theta - 1$
11. $\cot^2\theta - 3 = 0$
12. $3\csc^2\theta - 4 = 0$

In Exercises 13–20, use the quadratic formula and solve for the degree measure of θ, where $0° \leq \theta < 360°$. In Exercises 17–20, express θ to the nearest tenth of a degree.

13. $3\tan^2\theta - 2\sqrt{3}\tan\theta + 1 = 0$
14. $4\sin^2\theta - 4\sqrt{3}\sin\theta + 3 = 0$
15. $\sec^2\theta + 2\sqrt{2}\sec\theta + 2 = 0$
16. $2\cos^2\theta + 2\sqrt{2}\cos\theta + 1 = 0$
17. $\sin^2\theta + 3\sin\theta - 1 = 0$
18. $\cos^2\theta - 2\cos\theta - 2 = 0$

17–20

19. $\tan^2\theta - \tan\theta - 3 = 0$
20. $\sin^2\theta + 4\sin\theta + 2 = 0$

Solve the equations in Exercises 21–54 for $0 \leq x < 2\pi$. (See Examples 3–5.)

21. $\sin x \tan x = \sin x$
22. $\cos x \cot x = \cos x$
23. $2\cos^2 x \sin x = \sin x$
24. $4\sin^2 x \cos x = \cos x$
25. $3\sin x = 2\cos^2 x$
26. $2\cos^2 x = 3\sin x + 3$
27. $1 + \cos x = 2\sin^2 x$
28. $5\cos x = 1 + 2\sin^2 x$
29. $\cot^2 x + 3\csc x + 3 = 0$
30. $\csc^2 x = 2\sqrt{3}\cot x - 2$
31. $\tan^2 x = 3(\sec x - 1)$
32. $\sec^2 x = 2\tan x$
33. $\cos x - \sec x = 0$
34. $\sin x - \csc x = 0$
35. $\tan x - \cot x = 0$
36. $2\sin x - 1 - \csc x = 0$
37. $2\sin x\cos x + 2\cos x - \sin x - 1 = 0$
38. $4\sin x\cos x - 2\cos x + 2\sin x - 1 = 0$
39. $\sin 2x = 1$
40. $\cos 2x = 1$
41. $\tan\dfrac{x}{2} = -1$
42. $2\sin\dfrac{x}{2} = 1$
43. $2\cos\dfrac{x}{2} = -1$
44. $\cot\dfrac{x}{2} = \sqrt{3}$
45. $2\tan 3x - 2\sqrt{3} = 0$
46. $2\cot 3x = \sqrt{3} - \cot 3x$
47. $\sin^2 3x - 1 = 0$
48. $\cos^2 3x - 1 = 0$
49. $\sin 2x = \cos 2x$
50. $4\cos 2x = \sec 2x$
51. $2\sin^2 2x + \sin 2x = 1$
52. $2\cos^2 2x - \cos 2x = 1$
53. $2\cos^2 3x + 3\cos 3x - 2 = 0$
54. $2\sin^2 3x - 5\sin 3x - 3 = 0$

Use the composite angle identities to solve each of the equations in Exercises 55–68 for all values of x such that $0 \leq x < 2\pi$. (See Examples 6 and 7.)

55. $\sin 2x = \sin x$
56. $\sin 2x = \cos x$
57. $2 - \cos x = \cos 2x$
58. $-2 - \sin x = \cos 2x$

59. $\sin 2x = \cos 2x + 1$

60. $\sin 2x = \cos 2x - 1$

61. $2\cos^2 x + \cos 2x = 0$

62. $\cos 2x - 2\sin^2 x = 0$

63. $\cos 2x \cos x + \sin 2x \sin x = \frac{1}{2}$

64. $\sin 3x \cos x + \cos 3x \sin x = -1$

65. $2\cos^2 \dfrac{x}{2} - 3\cos x = 2$

66. $\tan \dfrac{x}{2} + 1 = -\cot x$

67. $\cos 4x - \cos 2x = \sin 3x$

68. $\sin 3x - \sin x = \cos 2x$

Use the reduction identity to solve the following equations for all x such that $0 \le x < 2\pi$. (See Example 8.)

69. $\sin x + \cos x = \sqrt{2}$

70. $\sin x - \cos x = -1$

71. $\sqrt{3}\sin x + \cos x = 2$

72. $\sin x + \sqrt{3}\cos x = -1$

Electrical Circuit 73. The current $I(t)$ in an electrical circuit is given by

$$I(t) = 15 \sin 4t$$

amperes. Find the nonnegative values of t for which the current is zero.

Charge on a Capacitor 74. The charge $q(t)$ on a capacitor in an electrical circuit is given by

$$q(t) = \tfrac{15}{4} - \tfrac{15}{4}\cos 4t$$

coulombs. Find the positive values of t for which the charge on the capacitor is zero.

75. Exercise 42 of Exercises 7.9 calls for a sketch of the graph of the function

$$f(x) = \cos x + \sin 2x$$

over the interval $[0, 2\pi]$. Find the x-intercepts of the graph in this interval.

76. Exercise 44 of Exercises 7.9 calls for a sketch of the graph of the function

$$f(x) = 3\sin x + \cos 2x$$

over the interval $[0, 2\pi]$. Find the x-intercepts of the graph in this interval.

Pendulum 77. In Exercise 57 of Exercises 8.3, the angular displacement $\theta(t)$ of a pendulum is given by

$$\theta(t) = \frac{1}{4}\cos 4t + \frac{\sqrt{3}}{4}\sin 4t.$$

Find the positive values of t for which the pendulum is in the equilibrium position.

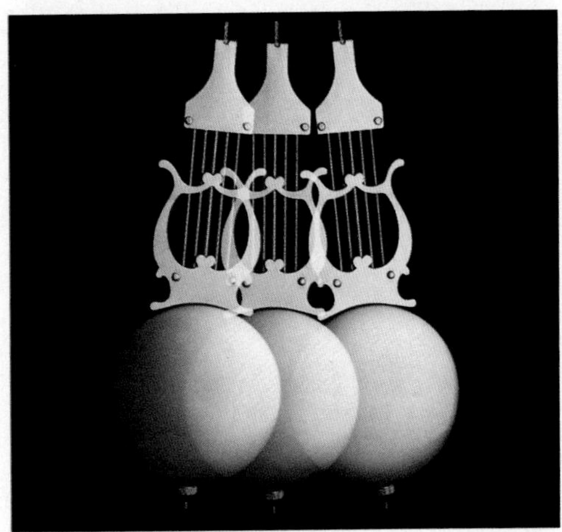

Motion of a pendulum

Stretching a Spring 78. In Exercise 47 of Exercises 7.9, the displacement $x(t)$ of an object attached to a spring is given by

$$x(t) = \tfrac{2}{3}\cos 8t - \tfrac{1}{6}\sin 8t.$$

Find the smallest positive value of t for which the object is in the equilibrium position.

Turning Points on a Graph 79–80

79. The graph of $y = 2\sin x - \cos 2x$ over the interval $[0, 2\pi]$ is drawn in the accompanying figure. By using the methods of calculus, it can be shown that the turning points A, B, C, D occur at x values where

$$\cos x + \sin 2x = 0.$$

Find the coordinates of these four points.

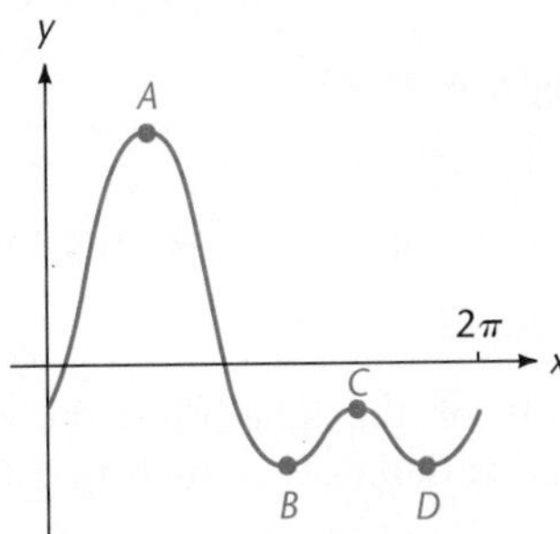

Figure for Exercise 79

80. The figure on page 526 shows the graph of $y = \sqrt{3}\sin x + \frac{1}{2}\cos 2x$ over the interval $[0, 2\pi]$. By using calculus, the turning points A, B, C, D can be located at values of x for which

$$\sqrt{3}\cos x - \sin 2x = 0.$$

Find the coordinates of these four points.

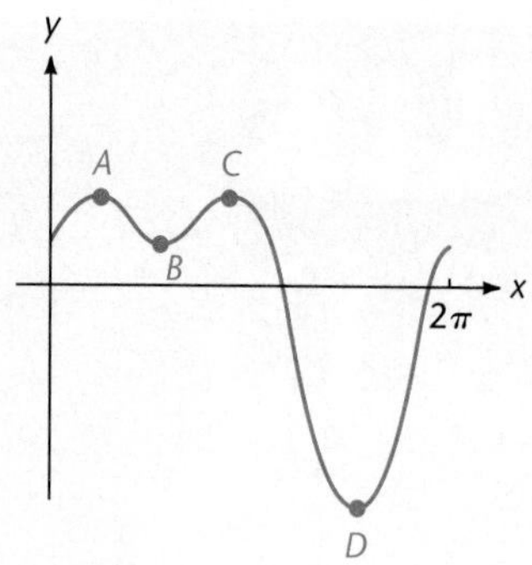

Figure for Exercise 80

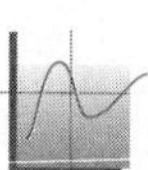

81–86 Use a graphing device to find all solutions to the given equation that lie in the indicated interval. Round your answers to the nearest tenth.

81. $2 \sin x = x$, $(-\infty, \infty)$
82. $2 \cos x = x$, $(-\infty, \infty)$
83. $2 \cos x = e^x$, $(-\pi, \pi)$
84. $\tan x = 2 - x^2$, $(0, \pi/2)$
85. $\cot x = x^3$, $(0, \pi)$
86. $\csc x = 3 - x^2$, $(0, \pi)$

◆ Solutions for Practice Problems

1. $2 \sin^2 \theta - 2\sqrt{2} \sin \theta + 1 = 0$, $0° \leq \theta < 360°$

$$\sin \theta = \frac{2\sqrt{2} \pm \sqrt{(-2\sqrt{2})^2 - 4(2)(1)}}{2(2)} = \frac{\sqrt{2}}{2}$$

Related angle: $\theta' = 45°$

Solutions: $\theta = 45°, 135°$

2. We use a reciprocal identity to rewrite the equation

$$2 \sin 2x = \csc 2x$$

strictly in terms of the sine function. We also observe that whenever $0 \leq x < 2\pi$, the variable $2x$ is such that $0 \leq 2x < 4\pi$.

$2 \sin 2x = \dfrac{1}{\sin 2x}$	reciprocal identity
$2 \sin^2 2x = 1$	algebra
$\sin^2 2x = \dfrac{1}{2}$	algebra
$\sin 2x = \pm\dfrac{1}{\sqrt{2}}$	algebra

Solving for nonnegative values of $2x$ up to 2 revolutions gives

$$2x = \frac{\pi}{4}, \frac{3\pi}{4}, \frac{5\pi}{4}, \frac{7\pi}{4}, \frac{9\pi}{4}, \frac{11\pi}{4}, \frac{13\pi}{4}, \text{ and } \frac{15\pi}{4}.$$

Then dividing each of these by 2 yields

$$x = \frac{\pi}{8}, \frac{3\pi}{8}, \frac{5\pi}{8}, \frac{7\pi}{8}, \frac{9\pi}{8}, \frac{11\pi}{8}, \frac{13\pi}{8}, \text{ and } \frac{15\pi}{8},$$

which are the solutions in the required range $0 \le x < 2\pi$.

3. The sum

$$\cos 3x + \cos x$$

can be written as a product,

$$2 \cos\left(\frac{3x + x}{2}\right) \cos\left(\frac{3x - x}{2}\right),$$

using one of the sum identities. Then the equation

$$\cos 3x + \cos x = 0, \qquad 0 \le x < 2\pi,$$

becomes

$$2 \cos 2x \cos x = 0.$$

Setting each factor equal to zero, we have two linear equations to solve.

$$\cos 2x = 0, \qquad 0 \le 2x < 4\pi \qquad \text{and} \qquad \cos x = 0, \qquad 0 \le x < 2\pi.$$

$$2x = \frac{\pi}{2}, \frac{3\pi}{2}, \frac{5\pi}{2}, \text{ and } \frac{7\pi}{2} \qquad\qquad x = \frac{\pi}{2} \text{ and } \frac{3\pi}{2}$$

$$x = \frac{\pi}{4}, \frac{3\pi}{4}, \frac{5\pi}{4}, \text{ and } \frac{7\pi}{4}$$

The solutions are $x = \frac{\pi}{4}, \frac{\pi}{2}, \frac{3\pi}{4}, \frac{5\pi}{4}, \frac{3\pi}{2}$, and $\frac{7\pi}{4}$.

8.7 INVERSE TRIGONOMETRIC FUNCTIONS

Our main objective in this section is to define an inverse function for each of the six trigonometric functions. We first consider the sine function, $y = \sin x$, where x is a real number.

It is clear from the graph of $y = \sin x$ in Figure 7.41 that any horizontal line between $y = -1$ and $y = 1$ intersects the graph of $y = \sin x$ at more than one point, so the sine function is *not* one-to-one. As we learned in Section 4.6, this means that the sine function does not have an inverse function when its domain is the set of all real numbers.

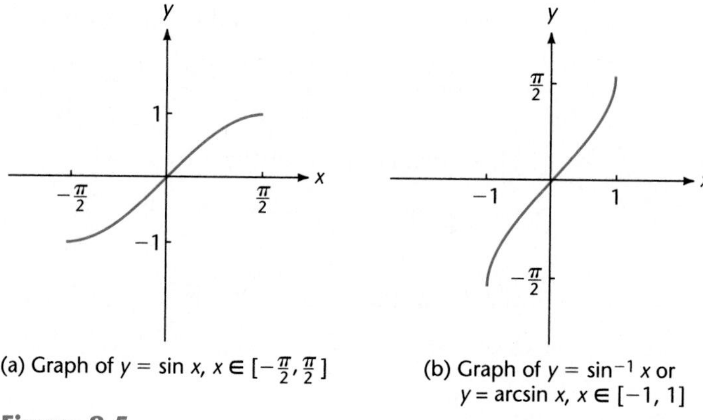

(a) Graph of $y = \sin x$, $x \in [-\frac{\pi}{2}, \frac{\pi}{2}]$

(b) Graph of $y = \sin^{-1} x$ or $y = \arcsin x$, $x \in [-1, 1]$

Figure 8.5

In order to obtain an inverse function, we restrict the domain of the sine function to the interval $-\pi/2 \leq x \leq \pi/2$. (See Figure 8.5a.) As x varies from $-\pi/2$ to $\pi/2$, $\sin x$ takes on each value from -1 to 1 exactly once.† With this restricted domain, an inverse function exists. This inverse function is called the **inverse sine function,** or the **arcsine function.**‡ Both of the following notations are commonly used to denote this function.

Either $y = \sin^{-1} x$ or $y = \arcsin x$

is equivalent to

$$x = \sin y \quad \text{and} \quad -\frac{\pi}{2} \leq y \leq \frac{\pi}{2}.$$

The graph of the inverse sine function is shown in Figure 8.5b. Note that it has domain $-1 \leq x \leq 1$ and range $-\pi/2 \leq y \leq \pi/2$.

Before considering an example, we note that the variables in the equation $x = \sin y$ have been *real numbers* throughout our discussion. We may think of y as radian measure of an angle if we wish, but we shall avoid any conversions to degrees because of the potential confusion.

Example 1 • Evaluating an Inverse Sine Function

Find the exact value of $\arcsin\left(-\frac{1}{2}\right)$ without using a calculator or tables.

†The interval $-\pi/2 \leq x \leq \pi/2$ is sometimes called the **interval of principal values** of the sine function.

‡The term *arcsine* relates to the fact that y is an arc length on the unit circle whose sine is x.

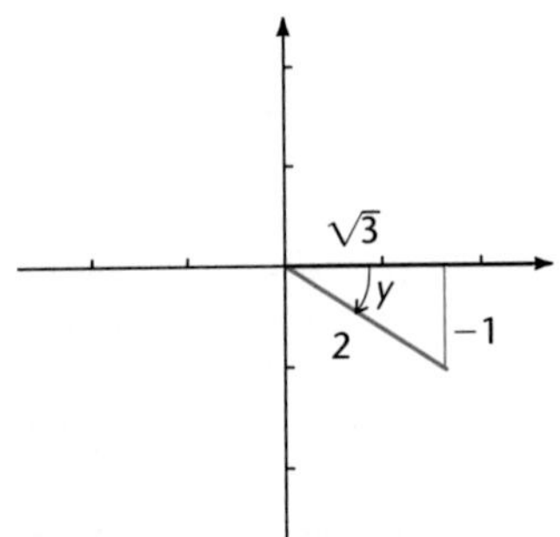

Figure 8.6 Reference triangle for arcsin $(-\frac{1}{2})$

Solution

Let $y = \arcsin(-\frac{1}{2})$. Then $\sin y = -\frac{1}{2}$ and $-\pi/2 \le y \le \pi/2$. Because of the restricted range for y and the fact that $\sin y$ is negative, we draw the reference triangle in quadrant IV, as shown in Figure 8.6. From the figure we see that

$$\arcsin\left(-\frac{1}{2}\right) = y = -\frac{\pi}{6}.$$

□

For each of the other trigonometric functions, we restrict the domain so that the new function is one-to-one and has an inverse function. The resulting **inverse trigonometric functions** have these restricted domains as their ranges. They are listed in Table 8.1 so that their ranges may be easily compared and memorized.

Table 8.1 Inverse Trigonometric Functions

Function	Defining equation	Range
$y = \sin^{-1} x = \arcsin x$	$x = \sin y$	$-\frac{\pi}{2} \le y \le \frac{\pi}{2}$
$y = \csc^{-1} x = \operatorname{arccsc} x$	$x = \csc y$	$-\frac{\pi}{2} \le y \le \frac{\pi}{2},\ y \ne 0$
$y = \tan^{-1} x = \arctan x$	$x = \tan y$	$-\frac{\pi}{2} < y < \frac{\pi}{2}$
$y = \cot^{-1} x = \operatorname{arccot} x$	$x = \cot y$	$0 < y < \pi$
$y = \cos^{-1} x = \arccos x$	$x = \cos y$	$0 \le y \le \pi$
$y = \sec^{-1} x = \operatorname{arcsec} x$	$x = \sec y$	$0 \le y \le \pi,\ y \ne \frac{\pi}{2}$

The graphs of the inverse trigonometric functions other than the inverse sine are shown in Figure 8.7 on page 530.

Example 2 • Evaluating Inverse Trigonometric Functions

Evaluate without using a calculator or tables.

a. $\cos^{-1}(-\frac{1}{2})$ b. $\sin^{-1} 2$

Solution

a. $y = \cos^{-1}(-\frac{1}{2})$ is equivalent to $\cos y = -\frac{1}{2}$ and $0 \le y \le \pi$. Since $\cos y$ is negative and $0 \le y \le \pi$, we find an angle $y = 2\pi/3$ in quadrant II with $\cos y = -\frac{1}{2}$. Thus

$$\cos^{-1}\left(-\frac{1}{2}\right) = y = \frac{2\pi}{3}.$$

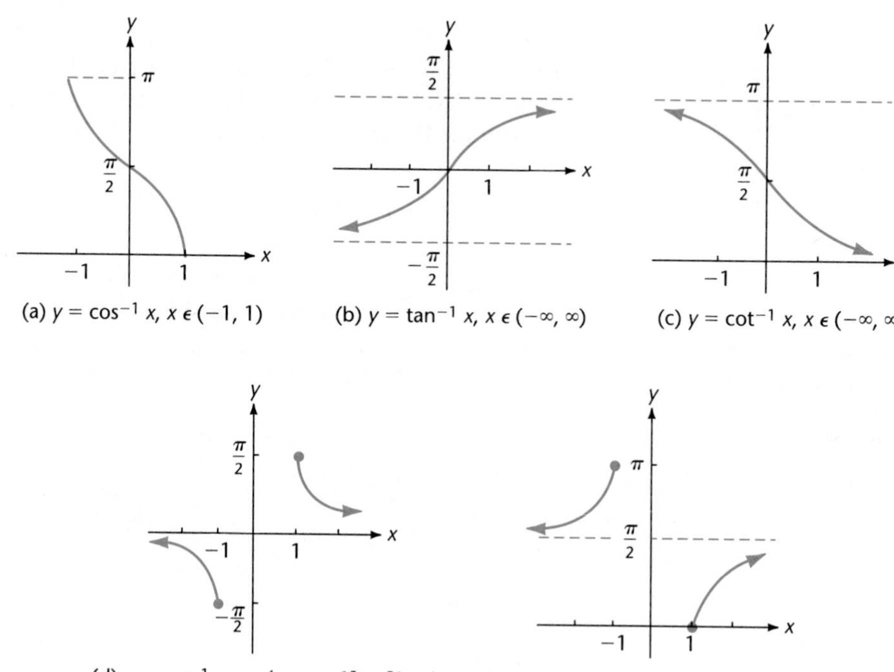

Figure 8.7 Graphs of five inverse trigonometric functions

It is worth noting that $\cos^{-1}(-\frac{1}{2}) \neq -4\pi/3$ because $-4\pi/3$ is *not in the range* of the inverse cosine function.

b. $y = \sin^{-1} 2$ is equivalent to $\sin y = 2$ and $-\pi/2 \leq y \leq \pi/2$. This is impossible since $-1 \leq \sin y \leq 1$ for all y. In other words, 2 is not in the domain of $y = \sin^{-1} x$. Thus $\sin^{-1} 2$ is not defined. □

◆ Practice Problem 1 ◆

Evaluate each of the following without using a calculator or tables.

a. $\tan^{-1}(-1)$ b. $\operatorname{arccsc}(-2)$

The equations $f^{-1}(f(x)) = x$ and $f(f^{-1}(x)) = x$ that were obtained in Section 4.6 for inverse functions in general have specific applications here.

$$\sin^{-1}(\sin x) = x \quad \text{or} \quad \arcsin(\sin x) = x, \quad \text{if } -\frac{\pi}{2} \leq x \leq \frac{\pi}{2}$$

$$\sin(\sin^{-1} x) = x \quad \text{or} \quad \sin(\arcsin x) = x, \quad \text{if } -1 \leq x \leq 1$$

Of course, corresponding equations hold for the other inverse trigonometric functions.

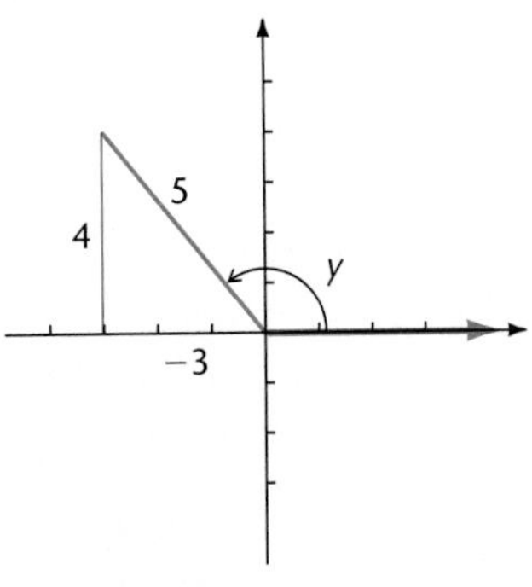

Figure 8.8 Reference triangle for $\cos^{-1}(-\frac{3}{5})$

Example 3 • Finding Values of Composite Functions

Evaluate each of the following without using a calculator or tables.

a. csc (arccsc 3) b. $\sin(\cos^{-1}(-\frac{3}{5}))$

Solution

a. It follows from the discussion in the preceding paragraph that

$$\csc(\text{arccsc } 3) = 3.$$

b. Let $y = \cos^{-1}(-\frac{3}{5})$. Then $\cos y = -\frac{3}{5}$ and $0 \le y \le \pi$. We draw y in quadrant II and complete the reference triangle as shown in Figure 8.8. From the figure,

$$\sin(\cos^{-1}(-\tfrac{3}{5})) = \sin y = \tfrac{4}{5}.$$ □

Some remarks are in order about the use of calculators. The values that a calculator yields are values of $\sin^{-1} x$, $\cos^{-1} x$, and $\tan^{-1} x$, given in either degrees or radians, depending on the mode setting of the calculator. In either case, they are values in the ranges specified in Table 8.1. Values of the other trigonometric functions can be obtained by using the following equations.

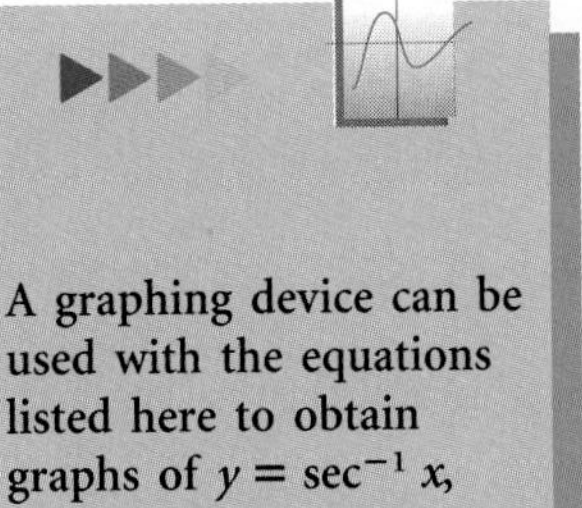

A graphing device can be used with the equations listed here to obtain graphs of $y = \sec^{-1} x$, $y = \csc^{-1} x$, and $y = \cot^{-1} x$.

$$\sec^{-1} x = \cos^{-1}\frac{1}{x}, \quad \text{if } x \le -1 \text{ or } x \ge 1.$$

$$\csc^{-1} x = \sin^{-1}\frac{1}{x}, \quad \text{if } x \le -1 \text{ or } x \ge 1.$$

$$\cot^{-1} x = \begin{cases} \tan^{-1}\dfrac{1}{x}, & \text{if } x > 0; \\[2ex] \dfrac{\pi}{2}, & \text{if } x = 0; \\[2ex] \pi + \tan^{-1}\dfrac{1}{x}, & \text{if } x < 0. \end{cases}$$

In order to evaluate more complicated expressions involving inverse trigonometric functions, it is frequently necessary to use the identities from the first part of this chapter. The following examples illustrate how substitutions can be used to simplify the work.

Example 4 • Using Composite Angle Identities

Evaluate $\sin(\tan^{-1}\frac{4}{3} + \cos^{-1}(-\frac{12}{13}))$ without using a calculator or tables.

Solution

Let $\alpha = \tan^{-1}\frac{4}{3}$ and $\beta = \cos^{-1}(-\frac{12}{13})$. We can find $\sin(\alpha + \beta)$ by using the identity $\sin(\alpha + \beta) = \sin\alpha\cos\beta + \cos\alpha\sin\beta$. Both α and β are shown in Figure 8.9 on page 532. From the figure, we get

$$\sin\alpha = \tfrac{4}{5}, \ \cos\alpha = \tfrac{3}{5}, \ \sin\beta = \tfrac{5}{13}, \ \cos\beta = -\tfrac{12}{13}.$$

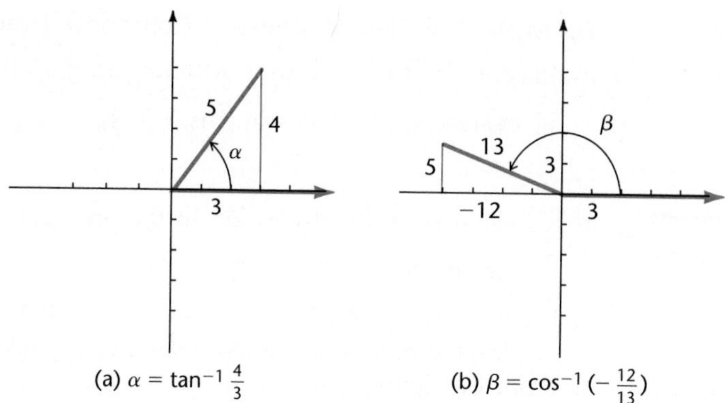

Figure 8.9 Reference triangles for α and β

Substitution of these values into the identity yields

$$\sin\left(\tan^{-1}\tfrac{4}{3} + \cos^{-1}\left(-\tfrac{12}{13}\right)\right) = \sin(\alpha + \beta)$$
$$= \left(\tfrac{4}{5}\right)\left(-\tfrac{12}{13}\right) + \left(\tfrac{3}{5}\right)\left(\tfrac{5}{13}\right) = -\tfrac{33}{65}.$$

□

◆ Practice Problem 2 ◆

Without using a calculator or tables, evaluate

$$\sin\left(2\cos^{-1}\left(-\tfrac{3}{5}\right)\right).$$

Example 5 • Finding a Composite Function

Assume that x is in the domain of the inverse sine function and find an algebraic expression for $\cos(\sin^{-1} x)$.

Solution

Let $\theta = \sin^{-1} x$. Then $\sin\theta = x$ and θ is one of the angles drawn in Figure 8.10. In either case,

$$\cos(\sin^{-1} x) = \cos\theta = \sqrt{1 - x^2}.$$

□

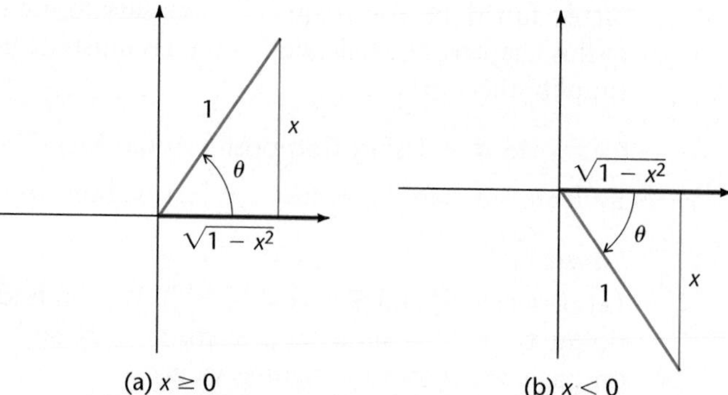

Figure 8.10 Reference triangles for $\sin^{-1} x$

We can now solve equations in which the variable occurs in an inverse trigonometric function. The basic technique for solving this type of equation is to isolate an inverse trigonometric function on one side of the equation and then use a property of the type $\sin(\sin^{-1} x) = x$. The technique is illustrated in the following example.

Example 6 • Solving an Equation

Solve the equation $\arccos x - \arcsin x = 5\pi/6$.

Solution

We add $\arcsin x$ to both sides and then take the cosine of both sides.

$$\arccos x = \frac{5\pi}{6} + \arcsin x$$

$$\cos(\arccos x) = \cos\left(\frac{5\pi}{6} + \arcsin x\right)$$

$$x = \cos\left(\frac{5\pi}{6} + \arcsin x\right)$$

If we let $\theta = \arcsin x$, then θ is one of the angles shown in Figure 8.10. In both of the cases shown there, $\sin\theta = x$ and $\cos\theta = \sqrt{1 - x^2}$. Using the identity for $\cos(\alpha + \beta)$, we have

$$x = \cos\left(\frac{5\pi}{6} + \theta\right)$$

$$x = \cos\frac{5\pi}{6}\cos\theta - \sin\frac{5\pi}{6}\sin\theta$$

$$x = \left(-\frac{\sqrt{3}}{2}\right)\sqrt{1 - x^2} - \frac{1}{2}x$$

$$2x = -\sqrt{3}\sqrt{1 - x^2} - x$$

$$3x = -\sqrt{3}\sqrt{1 - x^2}$$

$$9x^2 = 3 - 3x^2$$

$$12x^2 = 3$$

$$x = \pm\frac{1}{2}.$$

We must check these proposed solutions since we squared both sides of an equation.

Check: $x = \frac{1}{2}$

$$\arccos\frac{1}{2} - \arcsin\frac{1}{2} = \frac{\pi}{3} - \frac{\pi}{6} \neq \frac{5\pi}{6}$$

$x = \frac{1}{2}$ is *not* a solution.

Check: $x = -\frac{1}{2}$

$$\arccos\left(-\frac{1}{2}\right) - \arcsin\left(-\frac{1}{2}\right) = \frac{2\pi}{3} - \left(-\frac{\pi}{6}\right) = \frac{5\pi}{6}$$

$x = -\frac{1}{2}$ is a solution, and the solution set is $\left\{-\frac{1}{2}\right\}$. □

EXERCISES 8.7

Evaluate each of the following. (See Examples 1 and 2.)

√ 1. $\sin^{-1} 1$
2. $\cos^{-1} 1$
3. $\sec^{-1} \sqrt{2}$
4. $\csc^{-1} 2$
√ 5. $\arccos\left(-\frac{1}{\sqrt{2}}\right)$
6. $\text{arcsec}\,(-2)$
7. $\text{arccsc}\,(-\sqrt{2})$
8. $\sin^{-1}(-1)$
√ 9. $\arcsin\left(-\frac{\sqrt{3}}{2}\right)$
10. $\cos^{-1}\left(-\frac{\sqrt{3}}{2}\right)$
11. $\cot^{-1}(-\sqrt{3})$
12. $\arctan(-\sqrt{3})$
√ 13. $\cos^{-1} 2$
14. $\sin^{-1} 3$

Find the exact value of each of the following. (See Example 3.)

√ 15. $\tan(\sin^{-1} \frac{1}{2})$
16. $\cos(\tan^{-1} \sqrt{3})$
17. $\csc(\csc^{-1}(-1.3))$
18. $\cot(\cot^{-1}(-3))$
√ 19. $\cos(\sec^{-1} \frac{6}{5})$
20. $\sin(\csc^{-1} \frac{7}{6})$
√ 21. $\cos(\tan^{-1}(-\frac{3}{4}))$
22. $\csc(\text{arccot}\,(-\frac{5}{12}))$
23. $\cot(\sin^{-1}(-\frac{5}{13}))$
24. $\tan(\cos^{-1}(-\frac{12}{13}))$
√ 25. $\sin^{-1}(\sin 0.99)$
26. $\cos^{-1}(\cos 1.3)$
√ 27. $\cos^{-1}\left(\cos \frac{7\pi}{6}\right)$
28. $\sin^{-1}\left(\sin \frac{4\pi}{3}\right)$

Find the values of the following functions by using either a calculator or Table I, as instructed by the teacher. In either case, give answers to four decimal places.

29. $\arccos 0.9805$
30. $\tan^{-1} 1.829$
31. $\cot^{-1}(-0.7445)$
32. $\sin^{-1}(-0.4514)$

Find the values of the following functions by using either a calculator or Table I, as instructed by the teacher. Give answers to four digits.

33. $\sin(\cos^{-1} 0.8511)$
34. $\cos(\tan^{-1} 1.621)$
35. $\cot(\cos^{-1}(-0.5760))$
36. $\csc(\arctan(-3.108))$

Use identities to find the exact value of each of the following. (See Example 4.)

37. $\cos\left(2\cos^{-1}\frac{4}{7}\right)$
38. $\cos\left(2\sin^{-1}\frac{5}{6}\right)$
39. $\sin\left(2\arccos\left(-\frac{4}{5}\right)\right)$
40. $\sin\left(2\operatorname{arccot}\left(-\frac{4}{3}\right)\right)$
41. $\cos\left(2\arctan\left(-\frac{5}{12}\right)\right)$
42. $\cos\left(2\operatorname{arccsc}\left(-\frac{13}{12}\right)\right)$
43. $\sin\left(\cot^{-1}\frac{5}{12} + \tan^{-1}\left(-\frac{3}{4}\right)\right)$
44. $\tan\left(\sec^{-1}\frac{17}{15} - \cot^{-1}\left(-\frac{3}{4}\right)\right)$
45. $\cos\left(\arcsin\left(-\frac{4}{5}\right) - \arctan\left(-\frac{5}{12}\right)\right)$
46. $\cos\left(\arcsin\left(-\frac{3}{5}\right) + \arccos\left(-\frac{5}{13}\right)\right)$

In Exercises 47–54, assume that x and y are such that each function is defined, and find an algebraic expression for the function value. (See Example 5.)

47. $\tan\left(\cos^{-1} x\right)$
48. $\sin\left(\cos^{-1} x\right)$
49. $\sec\left(\cot^{-1} x\right)$
50. $\csc\left(\cot^{-1} x\right)$
51. $\cos\left(2\tan^{-1} x\right)$
52. $\cos\left(2\cot^{-1} x\right)$
53. $\sin\left(\sin^{-1} x + \cos^{-1} y\right)$
54. $\cos\left(\cos^{-1} x + \sin^{-1} y\right)$

Solve the following equations. (See Example 6.)

55. $\tan^{-1} x = -\dfrac{\pi}{3}$
56. $\sin^{-1} x = -\dfrac{\pi}{6}$
57. $2\cos^{-1} x - \pi = \dfrac{\pi}{2}$
58. $3\cot^{-1} x - 2\pi = \dfrac{\pi}{2}$
59. $\arcsin x = \arctan(-1)$
60. $\tan^{-1} x = \sin^{-1}\left(-\frac{1}{2}\right)$
61. $\sin^{-1} x = \cos^{-1} x$
62. $\tan^{-1} x = \cot^{-1} x$
63. $\arccos x + \arctan\dfrac{5}{12} = \dfrac{\pi}{2}$
64. $\cos^{-1} x + \tan^{-1}\dfrac{3}{4} = \dfrac{\pi}{2}$
65. $\cos^{-1} x + \sin^{-1}\left(-\frac{3}{5}\right) = \sin^{-1}\frac{4}{5}$
66. $\sin^{-1} x + \sin^{-1}\left(-\frac{5}{13}\right) = \sin^{-1}\frac{12}{13}$
67. $\arccos x + \arcsin(3 - 5x) = \dfrac{\pi}{2}$
68. $\sin^{-1}(-x) + \cos^{-1}(2 + 3x) = \dfrac{\pi}{2}$
69. $\arccos 2x - \arcsin x = \dfrac{7\pi}{6}$
70. $\arccos x - \arcsin x = \dfrac{7\pi}{6}$
71. $\cos^{-1} x + \sin^{-1} x = \dfrac{\pi}{6}$
72. $\cos^{-1} x + \sin^{-1} x = \dfrac{5\pi}{6}$
73. $\arcsin x + \arccos x = \dfrac{\pi}{2}$
74. $\tan^{-1} x + \cot^{-1} x = \dfrac{\pi}{2}$

Sketch the graph of the given equation.

75. $y = \dfrac{\pi}{2} + \sin^{-1} x$
76. $y = 2\sin^{-1} x$
77. $y = \sin^{-1} 2x$
78. $y = \sin^{-1}(x - 2)$

79. $y = -\cos^{-1} x$

80. $y = \cos^{-1} 2x$

81. $y = \tan^{-1} 2x$

82. $y = \tan^{-1}(-x)$

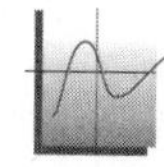
83–86

83. Verify that the graphs provided by a graphing device agree with those in Figures 8.5 and 8.7 for $\sin^{-1} x$, $\cos^{-1} x$, and $\tan^{-1} x$.

84. Verify that the graphs provided by a graphing device for $\sin^{-1}(1/x)$ and $\cos^{-1}(1/x)$ agree, respectively, with those for $\csc^{-1} x$ and $\sec^{-1} x$ in Figure 8.7.

With the help of a graphing device, determine the values of x for which $Y_1 = Y_2$.

85. $Y_1 = \sin^{-1} x$, $Y_2 = \cos^{-1}\sqrt{1 - x^2}$

86. $Y_1 = \sec^{-1} x$, $Y_2 = \tan^{-1}\sqrt{x^2 - 1}$

Critical Thinking: Exploration and Writing

87. Suppose a is negative and $\cos\theta = a$, where $\pi < \theta < 3\pi/2$. Express θ in terms of $\cos^{-1} a$ and explain why your formula works.

Critical Thinking: Exploration and Writing

88. Suppose a is negative and $\tan\theta = a$, where $\pi/2 < \theta < \pi$. Express θ in terms of $\tan^{-1} a$ and explain why your formula works.

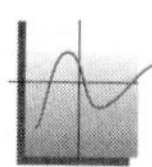
89–92

Critical Thinking: Exploration and Writing

89. Explain why the graph provided by a graphing device for $y = \tan^{-1}(1/x)$ does not agree with the graph of $y = \cot^{-1} x$ in Figure 8.7.

Critical Thinking: Exploration and Writing

In Exercises 90–92, explain why the graphs provided by a graphing device for the given pair of functions are not the same.

90. $Y_1 = \sin(\sin^{-1} x)$, $Y_2 = \sin^{-1}(\sin x)$

91. $Y_1 = \cos(\cos^{-1} x)$, $Y_2 = \cos^{-1}(\cos x)$

92. $Y_1 = \tan(\tan^{-1} x)$, $Y_2 = \tan^{-1}(\tan x)$

◆ Solutions for Practice Problems

1. a. $y = \tan^{-1}(-1)$ is equivalent to $\tan y = -1$ and $-\pi/2 < y < \pi/2$, so we draw y in quadrant IV and label the reference triangle as shown in Figure 8.11. We see that $\tan(-\pi/4) = -1$ and

$$\tan^{-1}(-1) = y = -\frac{\pi}{4}.$$

Note that $\tan^{-1}(-1) \neq 7\pi/4$.

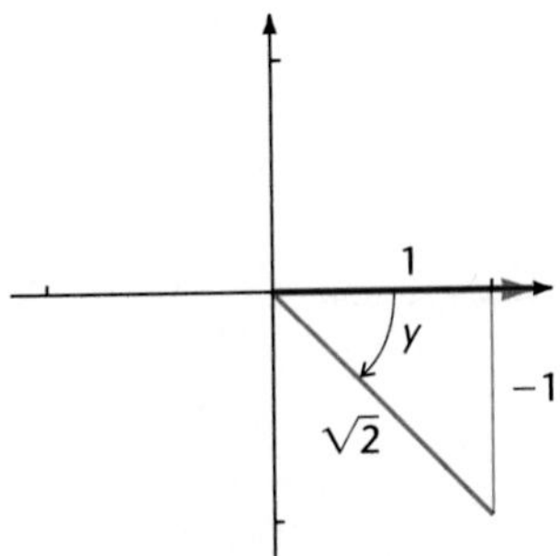

Figure 8.11 Reference triangle for $\tan^{-1}(-1)$

b. $y = \text{arccsc}\,(-2)$ is equivalent to $\csc y = -2$ and $-\pi/2 \le y \le \pi/2,\ y \ne 0$. We draw and label a reference triangle as shown in Figure 8.12. From the figure, we see that

$$\text{arccsc}\,(-2) = y = -\frac{\pi}{6}.$$

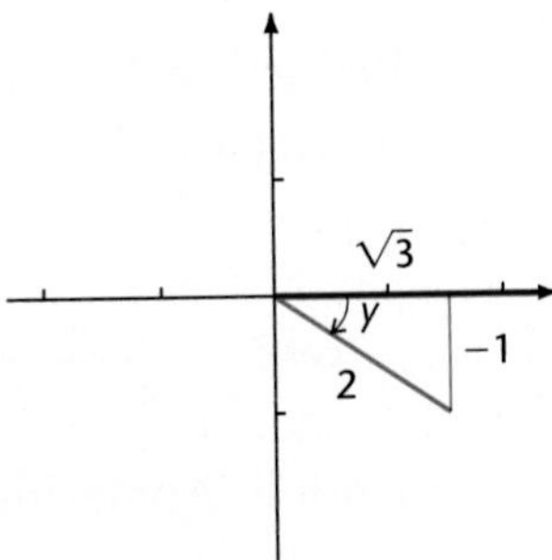

Figure 8.12 Reference triangle for arccsc (-2)

2. Let $\theta = \cos^{-1}(-\frac{3}{5})$. To use the identity

$$\sin 2\theta = 2 \sin \theta \cos \theta,$$

we need the values of both $\cos \theta$ and $\sin \theta$. The angle θ is drawn in Figure 8.13 on page 538. From the figure, $\sin \theta = \frac{4}{5}$ and $\cos \theta = -\frac{3}{5}$. Thus

$$\sin(2 \cos^{-1}(-\tfrac{3}{5})) = \sin 2\theta = 2(\tfrac{4}{5})(-\tfrac{3}{5}) = -\tfrac{24}{25}.$$

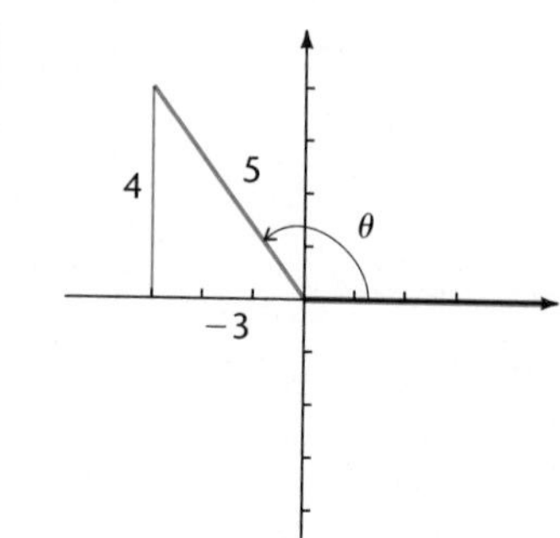

Figure 8.13 Reference triangle for $\cos^{-1}(-\frac{3}{5})$

CHAPTER REVIEW

Summary of Important Concepts and Formulas

Fundamental Identities

$\sin\theta\csc\theta = 1 \qquad \sin^2\theta + \cos^2\theta = 1$

$\cos\theta\sec\theta = 1 \qquad 1 + \tan^2\theta = \sec^2\theta$

$\tan\theta\cot\theta = 1 \qquad 1 + \cot^2\theta = \csc^2\theta$

$\tan\theta = \dfrac{\sin\theta}{\cos\theta} \qquad \cot\theta = \dfrac{\cos\theta}{\sin\theta}$

Cofunction Identities

$\cos\left(\dfrac{\pi}{2} - \theta\right) = \sin\theta \qquad \sin\left(\dfrac{\pi}{2} - \theta\right) = \cos\theta$

$\cot\left(\dfrac{\pi}{2} - \theta\right) = \tan\theta \qquad \tan\left(\dfrac{\pi}{2} - \theta\right) = \cot\theta$

$\csc\left(\dfrac{\pi}{2} - \theta\right) = \sec\theta \qquad \sec\left(\dfrac{\pi}{2} - \theta\right) = \csc\theta$

Negative-Angle Identities

$\sin(-\theta) = -\sin\theta \qquad \csc(-\theta) = -\csc\theta$

$\cos(-\theta) = \cos\theta \qquad \sec(-\theta) = \sec\theta$

$\tan(-\theta) = -\tan\theta \qquad \cot(-\theta) = -\cot\theta$

Identities for the Sum and Difference of Two Angles

$\cos(\alpha - \beta) = \cos\alpha\cos\beta + \sin\alpha\sin\beta$

$\cos(\alpha + \beta) = \cos\alpha\cos\beta - \sin\alpha\sin\beta$

$\sin(\alpha + \beta) = \sin\alpha\cos\beta + \cos\alpha\sin\beta$

$\sin(\alpha - \beta) = \sin\alpha\cos\beta - \cos\alpha\sin\beta$

$\tan(\alpha + \beta) = \dfrac{\tan\alpha + \tan\beta}{1 - \tan\alpha\tan\beta}$

$\tan(\alpha - \beta) = \dfrac{\tan\alpha - \tan\beta}{1 + \tan\alpha\tan\beta}$

Reduction Identity

$$a\sin\theta + b\cos\theta = \sqrt{a^2 + b^2}\sin(\theta + \alpha),$$

where

$$\cos\alpha = \frac{a}{\sqrt{a^2 + b^2}} \quad \text{and} \quad \sin\alpha = \frac{b}{\sqrt{a^2 + b^2}}$$

Double-Angle Identities

$$\sin 2\theta = 2\sin\theta\cos\theta$$

$$\cos 2\theta = \cos^2\theta - \sin^2\theta = 2\cos^2\theta - 1 = 1 - 2\sin^2\theta$$

$$\tan 2\theta = \frac{2\tan\theta}{1 - \tan^2\theta}$$

Half-Angle Identities

$\sin\dfrac{\theta}{2} = \pm\sqrt{\dfrac{1 - \cos\theta}{2}} \qquad \tan\dfrac{\theta}{2} = \dfrac{\sin\theta}{1 + \cos\theta}$

$\cos\dfrac{\theta}{2} = \pm\sqrt{\dfrac{1 + \cos\theta}{2}} \qquad \phantom{\tan\dfrac{\theta}{2}} = \dfrac{1 - \cos\theta}{\sin\theta}$

Product Identities

$$2\sin\alpha\cos\beta = \sin(\alpha+\beta) + \sin(\alpha-\beta)$$
$$2\cos\alpha\sin\beta = \sin(\alpha+\beta) - \sin(\alpha-\beta)$$
$$2\cos\alpha\cos\beta = \cos(\alpha+\beta) + \cos(\alpha-\beta)$$
$$2\sin\alpha\sin\beta = \cos(\alpha-\beta) - \cos(\alpha+\beta)$$

Sum Identities

$$\sin x + \sin y = 2\sin\left(\frac{x+y}{2}\right)\cos\left(\frac{x-y}{2}\right)$$
$$\sin x - \sin y = 2\cos\left(\frac{x+y}{2}\right)\sin\left(\frac{x-y}{2}\right)$$
$$\cos x + \cos y = 2\cos\left(\frac{x+y}{2}\right)\cos\left(\frac{x-y}{2}\right)$$
$$\cos x - \cos y = -2\sin\left(\frac{x+y}{2}\right)\sin\left(\frac{x-y}{2}\right)$$

Inverse Trigonometric Functions

Function	Defining equation	Range
$y = \sin^{-1} x = \arcsin x$	$x = \sin y$	$-\frac{\pi}{2} \le y \le \frac{\pi}{2}$
$y = \csc^{-1} x = \operatorname{arccsc} x$	$x = \csc y$	$-\frac{\pi}{2} \le y \le \frac{\pi}{2},\ y \ne 0$
$y = \tan^{-1} x = \arctan x$	$x = \tan y$	$-\frac{\pi}{2} < y < \frac{\pi}{2}$
$y = \cot^{-1} x = \operatorname{arccot} x$	$x = \cot y$	$0 < y < \pi$
$y = \cos^{-1} x = \arccos x$	$x = \cos y$	$0 \le y \le \pi$
$y = \sec^{-1} x = \operatorname{arcsec} x$	$x = \sec y$	$0 \le y \le \pi,\ y \ne \frac{\pi}{2}$

Values of the Inverse Trigonometric Functions

$$\sec^{-1} x = \cos^{-1}\frac{1}{x}, \text{ if } x \le -1 \text{ or } x \ge 1$$

$$\csc^{-1} x = \sin^{-1}\frac{1}{x}, \text{ if } x \le -1 \text{ or } x \ge 1$$

$$\cot^{-1} x = \begin{cases} \tan^{-1}\frac{1}{x}, & \text{if } x > 0 \\ \frac{\pi}{2}, & \text{if } x = 0 \\ \pi + \tan^{-1}\frac{1}{x}, & \text{if } x < 0 \end{cases}$$

Critical Thinking: Find the Errors

Each of the following nonsolutions has one or more errors. Can you find them?

Problem 1

Verify the identity: $\tan\theta + \cot\theta = \sec\theta\csc\theta$.

Nonsolution

$$\tan\theta + \cot\theta = \frac{\sin\theta}{\cos\theta} + \frac{\cos\theta}{\sin\theta}$$
$$= \frac{\sin\theta + \cos\theta}{\cos\theta\sin\theta}$$
$$= \frac{1}{\cos\theta} + \frac{1}{\sin\theta}$$

$$= \frac{1}{\cos\theta \sin\theta}$$

$$= \frac{1}{\cos\theta} \cdot \frac{1}{\sin\theta}$$

$$= \sec\theta \csc\theta$$

Problem 2

Solve the following equation for θ, where $0° \le \theta < 360°$.

$$2\cos^2\theta - \cos\theta = 1$$

Nonsolution

$$\cos\theta\,(2\cos\theta - 1) = 1$$

$$\cos\theta = 1 \quad \text{or} \quad 2\cos\theta - 1 = 1$$

$$\theta = 0° \qquad 2\cos\theta = 2$$

$$\cos\theta = 1$$

$$\theta = 0°$$

The solution is $\theta = 0°$.

Problem 3

Verify the identity: $\dfrac{\cos 2\theta}{\cos^2\theta} = 2 - \sec^2\theta$.

Nonsolution

$$\frac{\cos 2\theta}{\cos^2\theta} = \frac{\cos^2\theta + \sin^2\theta}{\cos^2\theta}$$

$$= \frac{\cos^2\theta}{\cos^2\theta} + \frac{\sin^2\theta}{\cos^2\theta}$$

$$= 1 + \tan^2\theta$$

$$= 1 + (1 - \sec^2\theta)$$

$$= 2 - \sec^2\theta$$

Problem 4

Verify the identity: $(\cos\theta - \sin\theta)^2 = 1 - \sin 2\theta$.

Nonsolution

$$(\cos\theta - \sin\theta)^2 = \cos^2\theta - \sin^2\theta$$

$$= 1 - \sin^2\theta - \sin^2\theta$$

$$= 1 - 2\sin^2\theta$$

$$= \cos 2\theta$$

$$= 1 - \sin 2\theta$$

Problem 5

Solve the following equation for all x such that $0 \le x < 2\pi$.

$$\sin 2x = \cos x$$

Nonsolution

$$2\sin x\cos x = \cos x$$

$$2\sin x = 1$$

$$\sin x = \frac{1}{2}$$

$$x = \frac{\pi}{6} \text{ or } x = \frac{5\pi}{6}$$

Problem 6

Find the exact value of $\operatorname{arccot}(-\sqrt{3})$.

Nonsolution

From Figure 8.14, we get

$$\operatorname{arccot}(-\sqrt{3}) = \arctan\left(-\frac{1}{\sqrt{3}}\right) = -\frac{\pi}{6}.$$

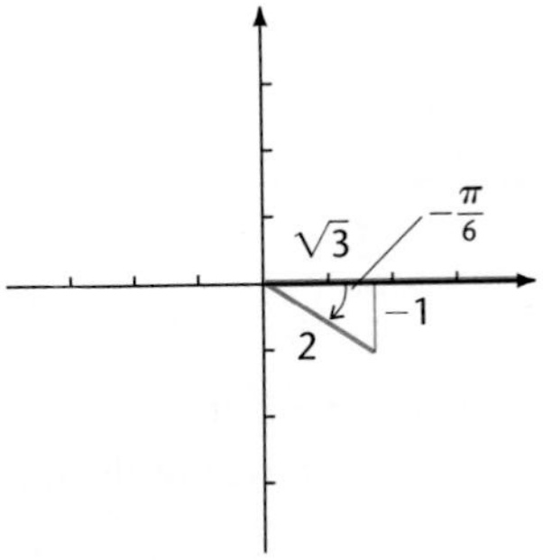

Figure 8.14

Problem 7

Find the exact value of $\sin\left(\tan^{-1}\left(-\frac{3}{4}\right)\right)$.

Nonsolution

Let $y = \tan^{-1}\left(-\frac{3}{4}\right)$, as shown in Figure 8.15. Then $r = \sqrt{(-4)^2 + (3)^2} = 5$, and $\sin(\tan^{-1}\left(-\frac{3}{4}\right)) = \sin y = \frac{3}{5}$.

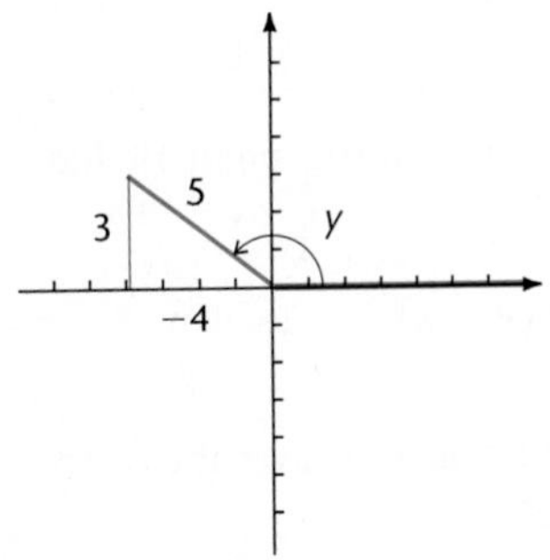

Figure 8.15

Problem 8

Find the exact value of $\sin^{-1}\left(\sin\frac{5\pi}{6}\right)$.

Nonsolution

$$\sin^{-1}\left(\sin\frac{5\pi}{6}\right) = \frac{5\pi}{6}$$

Review Problems for Chapter 8

Reduce the first expression to the second.

1. $(\sec^2\alpha - 1)\cos^2\alpha$; $\sin^2\alpha$
2. $\dfrac{1+\cot^2\gamma}{1+\tan^2\gamma}$; $\cot^2\gamma$
3. $\dfrac{\sec\theta - \sin\theta\tan\theta}{\cos\theta}$; 1
4. Write all the trigonometric functions in terms of $\csc\theta$.

In Problems 5–10, verify each identity.

5. $(\sin\theta - \csc\theta)^2 = \cot^2\theta - \cos^2\theta$
6. $\dfrac{1+\cot^2\theta}{\cot^2\theta} = \sec^2\theta$
7. $\dfrac{1-\tan^2 A}{1+\tan^2 A} = 2\cos^2 A - 1$

8. $\dfrac{1 - \sin \gamma}{1 + \sin \gamma} = \dfrac{\csc \gamma - 1}{\csc \gamma + 1}$

9. $\dfrac{\cos \alpha - \sec \alpha}{\sec \alpha} + \dfrac{\sin \alpha - \csc \alpha}{\csc \alpha} = -1$

10. $\dfrac{\tan x}{1 - \cos x} = (\sec x + 1) \csc x$

11. Prove that $\tan \theta + \sec \theta = 1$ is not an identity.

12. Use identities to find the exact value of each of the following expressions.

a. $\dfrac{1}{2 \sin 15° \cos 15°}$

b. $\cos 93° \cos 3° + \sin 93° \sin 3°$

c. $\dfrac{2 \tan 22.5°}{1 - \tan^2 22.5°}$

d. $\sin 17° \cos 43° + \cos 17° \sin 43°$

e. $\tan \dfrac{3\pi}{8}$

Write each of the following in terms of trigonometric functions of θ.

13. $\cos \left(\theta + \dfrac{\pi}{4}\right)$ 14. $\sin (\theta + \pi)$ 15. $\tan (\theta - \pi)$

Sketch the graph of each of the following over the interval $0 \leq x \leq 2\pi$.

16. $y = -2 \sin (-x)$ 17. $y = \sin x + \sqrt{3} \cos x$

18. Without determining α or β, find the exact values of all trigonometric functions of $\alpha - \beta$ if $\cos \alpha = -5/13$, $\sin \beta = 4/5$, α is in quadrant III, and β is in quadrant II.

If $0 \leq \theta < 2\pi$, $\tan \theta = -3/4$, and $\cos \theta > 0$, find the exact value of each of the following.

19. $\sin 2\theta$ 20. $\cos 2\theta$ 21. $\tan 2\theta$

22. $\sin \dfrac{\theta}{2}$ 23. $\cos \dfrac{\theta}{2}$ 24. $\tan \dfrac{\theta}{2}$

25. Express $\cos 4\theta$ in terms of $\sin \theta$.

In Problems 26–30, verify the identities.

26. $\dfrac{\sin (\alpha - \beta)}{\cos \alpha \cos \beta} = \tan \alpha - \tan \beta$

27. $\dfrac{\cot \beta + \cot \alpha}{\cot \beta \cot \alpha - 1} = \tan (\alpha + \beta)$

28. $\dfrac{\cos 2x + \sin 2x + 1}{\cos 2x - \sin 2x - 1} = -\cot x$

29. $\cos 4\theta + \cos 2\theta + 2 \sin 3\theta \sin \theta = 2 \cos^2 \theta - 2 \sin^2 \theta$

30. $\sin (\alpha + \beta) \cos (\alpha - \beta) = \sin \alpha \cos \alpha + \sin \beta \cos \beta$

31. Write each of the following products as a sum in which no argument contains a minus sign.

a. $2 \cos 5\theta \cos 3\theta$

b. $2 \cos 4x \sin 6x$

c. $\cos (A + B) \cos (A - B)$

32. Write each of the following sums as a product in which no argument contains a minus sign.

a. $\sin 4t + \sin 2t$

b. $\cos 5\alpha + \cos 7\alpha$

c. $\cos 3t - \cos 9t$

In Problems 33–36, solve for θ, where $0° \leq \theta < 360°$.

33. $\sin \theta + 3 = 2 - \sin \theta$

34. $2 \cos^2 \theta = 1 + \cos \theta$

35. $2 \sin^2 \theta - 2\sqrt{2} \sin \theta + 1 = 0$

36. $\sec \theta \sin \theta - \sqrt{2} \sin \theta + \sec \theta - \sqrt{2} = 0$

In Problems 37–42, solve the equations for all x such that $0 \leq x < 2\pi$.

37. $\sec x \tan x = \tan x$ 38. $2 \sin x + 3 \csc x = 7$

39. $\cos 2x + \sin x = 0$ 40. $\sin x - \sqrt{3} \cos x = 1$

41. $2 \tan \dfrac{x}{2} + \sqrt{3} = \tan \dfrac{x}{2}$ 42. $\cos 3x + \cos x = 0$

Find the exact value of each of the following.

43. $\arcsin 0$ 44. $\arccos \left(-\dfrac{1}{\sqrt{2}}\right)$

45. $\tan^{-1} (-1)$ 46. $\csc^{-1} \sqrt{2}$

47. $\sin^{-1} 1.2$ 48. $\cos (\cos^{-1} 0.2)$

49. $\csc \left(\cos^{-1} \dfrac{1}{\sqrt{2}}\right)$ 50. $\tan (\arccos (-1))$

51. $\sin (\operatorname{arccsc} 4)$ 52. $\sin (\tan^{-1} (-4/3))$

53. $\cos (2 \sin^{-1} (3/4))$ 54. $\sin (2 \arccos (-3/5))$

55. $\cos (\sin^{-1} (-4/5) + \tan^{-1} (-5/12))$

Solve the following equations.

56. $\arctan x = \arcsin \dfrac{1}{\sqrt{1 + x^2}}$

57. $\cos^{-1} x + \sin^{-1} (4/5) = \cos^{-1} (-12/13)$

58. $\cos^{-1} 2x + \sin^{-1} x = \dfrac{\pi}{6}$

Sketch the graph of the given equation.

59. $y = \sin^{-1}(-x)$

60. $y = \dfrac{\pi}{2} + \cos^{-1} x$

61. $y = 2 \tan^{-1} x$

62. $y = \tan^{-1}(x - 1)$

Prove that the given equation is not an identity by providing a counterexample.

63. $\sin^{-1} x = \dfrac{1}{\sin x}$

64. $\cos^{-1}(-x) = -\cos^{-1} x$

65. $\tan^{-1}(x - 1) = \tan^{-1} x - \tan^{-1} 1$

66. $\cos^{-1}(x + 1) = \cos^{-1} x + \cos^{-1} 1$

E N C O R E

Culver

Rockets were used in battle by the Chinese in the thirteenth century.

Ron Sherman/Stock, Boston

TOW missile launcher.

As the above illustrations indicate, projectiles have been in use for centuries, and modern technology has enhanced their capabilities. In physics and multivariable calculus, the motion of a projectile is modeled by a particle that moves in a vertical coordinate plane where the only force acting on it is the constant force of gravity straight downward. Calculus can be used to show that the projectile's path of flight under these ideal conditions will form a parabola.

The drawing in Figure 8.16b represents the flight of an ideal projectile launched from ground level at the origin with initial speed v_0 at an angle θ with the positive x-axis. The coordinates (x, y) in the figure represent the location of the projectile in flight. These coordinates are given by the formulas

$$x = v_0 t \cos \theta, \qquad y = v_0 t \sin \theta - 16t^2$$

when distance is measured in feet and time is measured in seconds.

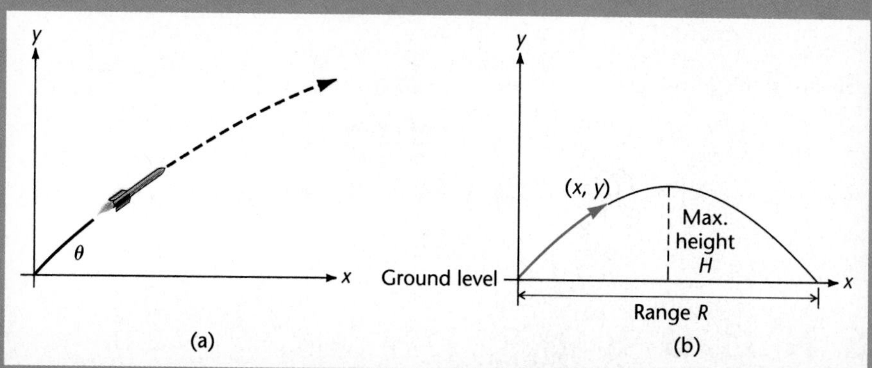

Figure 8.16 Flight of an ideal projectile

The calculus can be used to derive formulas for the range R and maximum height H that are diagrammed in Figure 8.16b. These formulas also involve trigonometric functions. They are given by

$$R = \frac{v_0^2}{32}\sin 2\theta, \qquad H = \frac{v_0^2}{32}\sin^2\theta.$$

If a projectile is launched under ideal conditions with $v_0 = 800$ feet per second, its range is given in feet by

$$\begin{aligned} R &= \frac{(800)^2}{32}\sin 2\theta \\ &= 20{,}000 \sin 2\theta. \end{aligned}$$

In order to hit a target 10,000 feet away, the elevation θ would have to be a solution to the equation

$$\begin{aligned} 20{,}000 \sin 2\theta &= 10{,}000 \\ \sin 2\theta &= \tfrac{1}{2}. \end{aligned}$$

The solutions with $0 \leq 2\theta \leq 180^\circ$ are determined by

$$2\theta = 30^\circ \qquad \text{and} \qquad 2\theta = 150^\circ.$$

Thus $\theta = 15^\circ$ and $\theta = 75^\circ$ are both elevations that would cause the projectile to hit the target.

Additional Topics in Trigonometry

The classical applications of trigonometry to surveying and navigation are treated in this chapter. Vectors are introduced and used where applicable. Some modern settings that relate to illegal drug trafficking and offshore oil exploration are included to make these applications more interesting.

9.1 THE LAW OF SINES

We now turn our attention to **oblique triangles** (triangles that are not right triangles). In this and the following section we develop some formulas that are used to solve oblique triangles. The first is called the **law of sines.**

Consider the triangles in Figure 9.1, in which the sides and angles are labeled according to the convention described in Section 7.6, with the exception that angle C is not required to be a right angle. In each triangle in the figure, we draw the perpendicular h from the vertex of angle C to side c (extended if necessary). Then we have

$$\sin B = \frac{h}{a} \quad \text{and} \quad \sin A = \frac{h}{b},$$

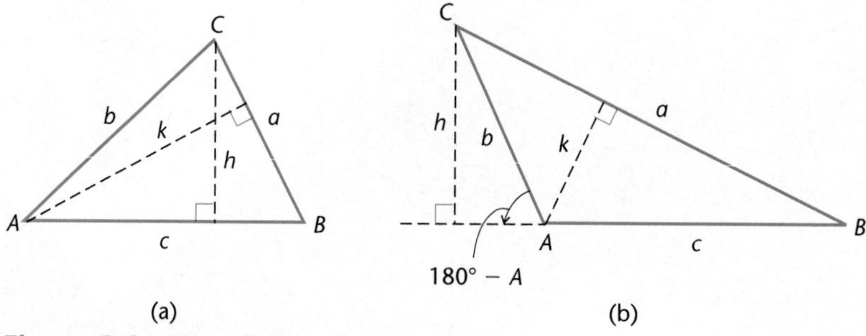

Figure 9.1 Triangles for the law of sines

where we use the fact that $\sin(180° - A) = \sin A$ for the obtuse triangle. Solving for h gives

$$a \sin B = h \quad \text{and} \quad b \sin A = h,$$

and therefore

$$a \sin B = b \sin A.$$

Dividing the equation by $\sin A \sin B$ results in one part of the law of sines.

$$\frac{a}{\sin A} = \frac{b}{\sin B}$$

Similarly, if the perpendicular k is drawn from the vertex of angle A to side a, we can obtain

$$\frac{b}{\sin B} = \frac{c}{\sin C}.$$

Combining the two parts results in the following general statement.

Law of Sines

$$\frac{a}{\sin A} = \frac{b}{\sin B} = \frac{c}{\sin C}$$

There are actually three parts to the law of sines.

$$\frac{a}{\sin A} = \frac{b}{\sin B}, \qquad \frac{a}{\sin A} = \frac{c}{\sin C}, \qquad \frac{b}{\sin B} = \frac{c}{\sin C}$$

The law of sines is especially useful in solving triangles in two particular cases when certain information is known about a triangle:

Case 1 Two angles and one side are known.

Case 2 Two sides and one angle opposite one of those sides are known.

The next example illustrates the solution of a case 1 triangle.

Example 1 • Solving a Case 1 Triangle

Solve the triangle in Figure 9.2, where $A = 37°$, $B = 82°$, and $a = 23$.

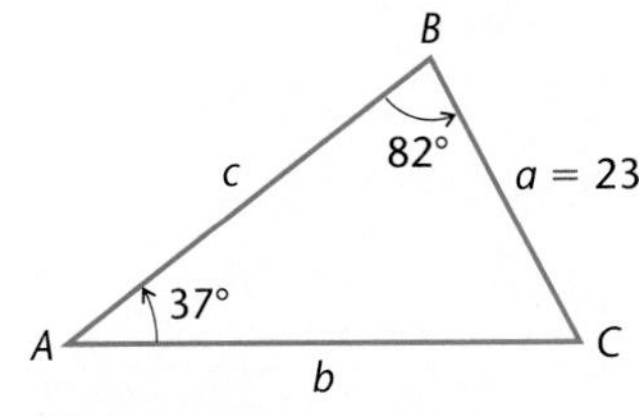

Figure 9.2 Triangle for Example 1

Solution

Since the sum of the three angles A, B, and C must be 180°, we find C to measure 61°. Sides b and c are found by using two parts of the law of sines.

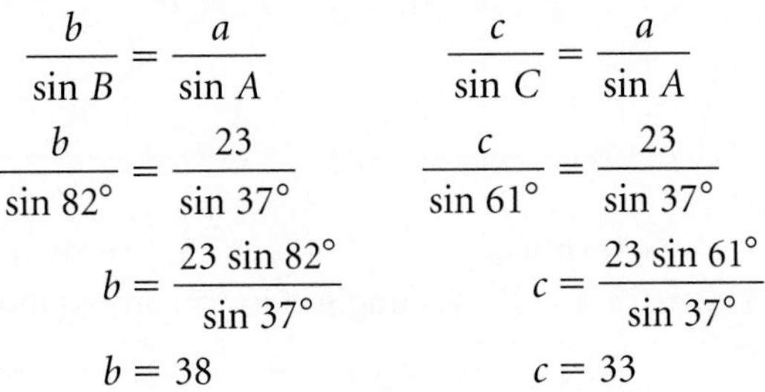
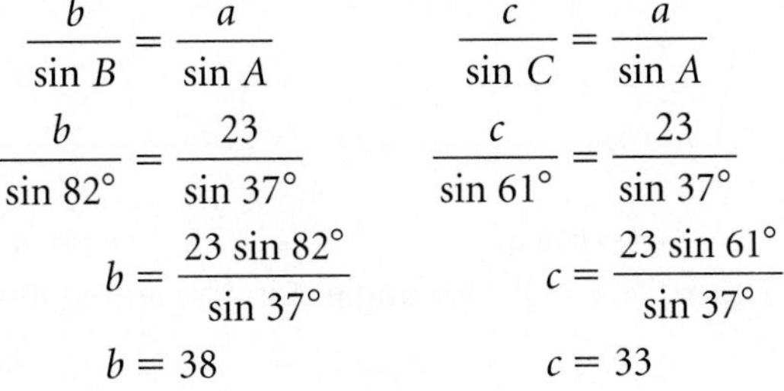

$$\frac{b}{\sin B} = \frac{a}{\sin A} \qquad \frac{c}{\sin C} = \frac{a}{\sin A}$$

$$\frac{b}{\sin 82°} = \frac{23}{\sin 37°} \qquad \frac{c}{\sin 61°} = \frac{23}{\sin 37°}$$

$$b = \frac{23 \sin 82°}{\sin 37°} \qquad c = \frac{23 \sin 61°}{\sin 37°}$$

$$b = 38 \qquad c = 33$$

All parts of the triangle are now known.

$a = 23$ $A = 37°$

$\mathbf{b = 38}$ $B = 82°$

$\mathbf{c = 33}$ $\mathbf{C = 61°}$

□

◆ **Practice Problem 1** ◆

Farmer Fred is fencing his pasture. To incorporate a pond within the fence he must detour around the pond as shown in Figure 9.3. How much extra fencing is needed over what Fred would have used if the pond had not been there?

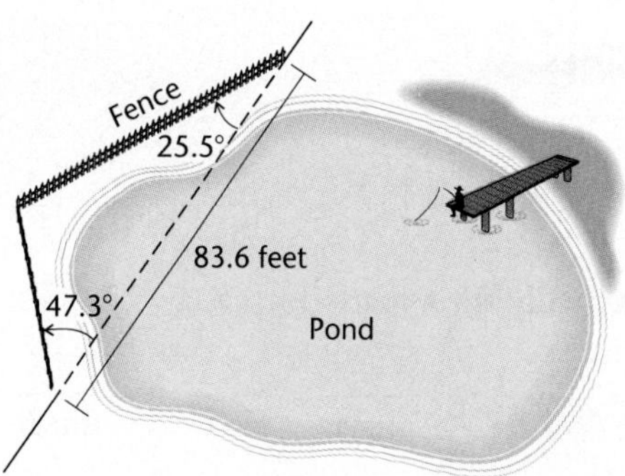

Figure 9.3 The detour around the pond

We now investigate the case 2 triangles: those in which the known parts are two sides and one angle opposite one of those sides. This case is often called the **ambiguous case** since several possible solutions may occur, depending on the sizes of the known parts.

In a triangle labeled conventionally, suppose angle A and sides a and b are the known parts. We first consider the situation when the known angle is obtuse. By examining Figure 9.4a, we see that no triangle exists if side a is too short, that is, if $a \leq b$. However, if $a > b$, as in Figure 9.4b, a unique triangle is determined.

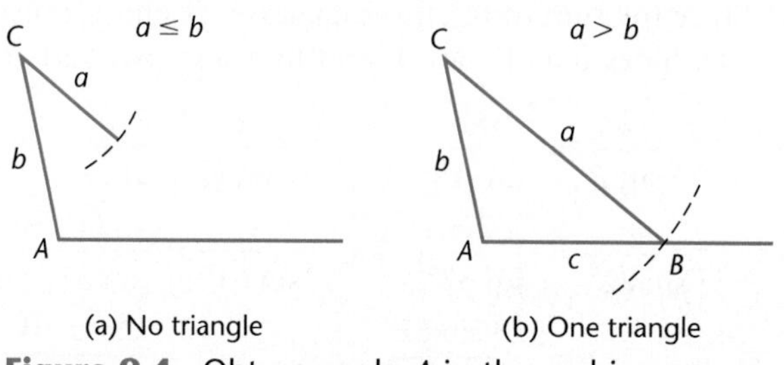

(a) No triangle (b) One triangle

Figure 9.4 Obtuse angle A in the ambiguous case

In case 2 triangles we must first solve for an angle. Therefore it is more convenient to use the law of sines in an alternative form.

$$\frac{\sin A}{a} = \frac{\sin B}{b} = \frac{\sin C}{c}$$

With an obtuse angle A, if the computations result in a value of $\sin B$ between 0 and 1, then there will be a unique triangle satisfying the given conditions. Otherwise, no such triangle exists.

Example 2 • The Ambiguous Case with No Solution

Solve the triangle, if one exists, where $A = 123°$, $a = 14$, and $b = 21$.

Solution

We use the law of sines to determine angle B.

$$\frac{\sin B}{b} = \frac{\sin A}{a}$$

$$\frac{\sin B}{21} = \frac{\sin 123°}{14}$$

$$\sin B = \frac{21 \sin 123°}{14}$$

$$\sin B = 1.2580 \qquad \text{impossible}$$

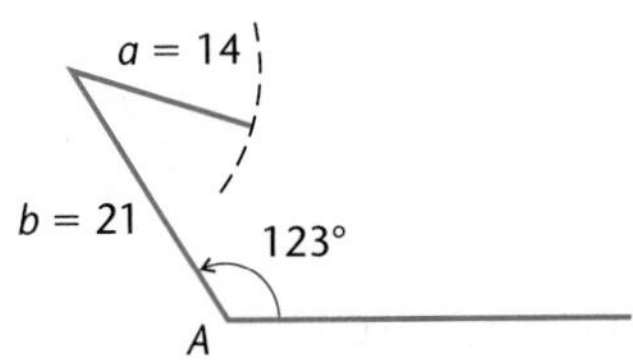

Figure 9.5 No triangle possible

Since it is true that $-1 \leq \sin B \leq 1$ for any angle B, there is no angle B such that $\sin B = 1.2580 > 1$. Hence *no triangle* satisfies the given conditions. A quick sketch, as in Figure 9.5, where we can see the consequences of the fact that $a \leq b$, would have led us directly to the same conclusion. □

Next we consider the case 2 triangles in which the known angle is acute. Again we choose to use angle A and sides a and b as the known parts. There are four possibilities to consider, as diagrammed in Figure 9.6.

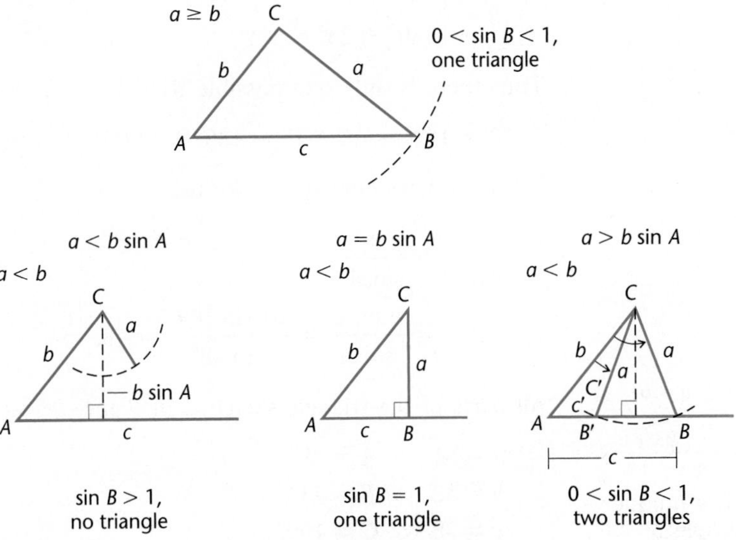

Figure 9.6 Acute angle A in the ambiguous case

Rather than memorizing the conditions stated in the figure, we can again determine which possibility occurs according to the calculations resulting from using the law of sines. If it turns out that $\sin B > 1$, then there is no triangle. If $\sin B = 1$, we have a right triangle. If $0 < \sin B < 1$, then there will be either one triangle or two triangles. We use the phrase "solve the triangle" to mean "solve the triangle(s) if one (or two) exists; otherwise show that there is no triangle satisfying the given conditions."

Example 3 • The Ambiguous Case with a Unique Solution

Solve the triangle where $A = 48°$, $b = 32$, and $a = 61$.

Solution

Solving for angle B, we have

$$\frac{\sin B}{b} = \frac{\sin A}{a}$$

$$\frac{\sin B}{32} = \frac{\sin 48°}{61}$$

$$\sin B = \frac{32 \sin 48°}{61} = 0.3898.$$

Since the sine function is positive in the first and second quadrants (B can be either acute or obtuse), we consider two possible values for B.

$$B = \begin{cases} 23° \\ 180° - 23° = 157° & \text{impossible since } A + B = 48° + 157° > 180° \end{cases}$$

Thus there is only one possible triangle. With $B = 23°$, we find C to be

$$C = 180° - (A + B) = 180° - (48° + 23°) = 109°.$$

Then, to determine side c, we use

$$\frac{c}{\sin C} = \frac{a}{\sin A}$$

$$c = \frac{a \sin C}{\sin A} = \frac{61 \sin 109°}{\sin 48°} = 78.$$

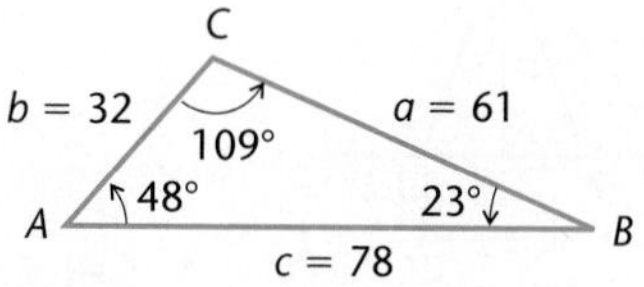

Figure 9.7 The unique triangle in Example 3

All parts of the triangle sketched in Figure 9.7 are known.

$a = 61$ $\quad A = 48°$
$b = 32$ $\quad \mathbf{B = 23°}$
$\mathbf{c = 78}$ $\quad \mathbf{C = 109°}$ □

The next example illustrates how two triangles satisfying the same given conditions can be determined.

Example 4 • The Ambiguous Case with Two Solutions

Solve the triangle where $A = 40.2°$, $a = 128$, and $b = 179$.

Solution

First we use the law of sines to determine angle B.

$$\frac{\sin B}{b} = \frac{\sin A}{a}$$

$$\frac{\sin B}{179} = \frac{\sin 40.2^\circ}{128}$$

$$\sin B = \frac{179 \sin 40.2^\circ}{128} = 0.9026$$

$$B = \begin{cases} 64.5^\circ \\ 180^\circ - 64.5^\circ = 115.5^\circ \end{cases}$$

We have two values, 64.5° and 115.5°, for angle B since for either value we have $A + B < 180^\circ$. Thus there exist *two triangles,* one acute and one obtuse. They are sketched in Figure 9.8.

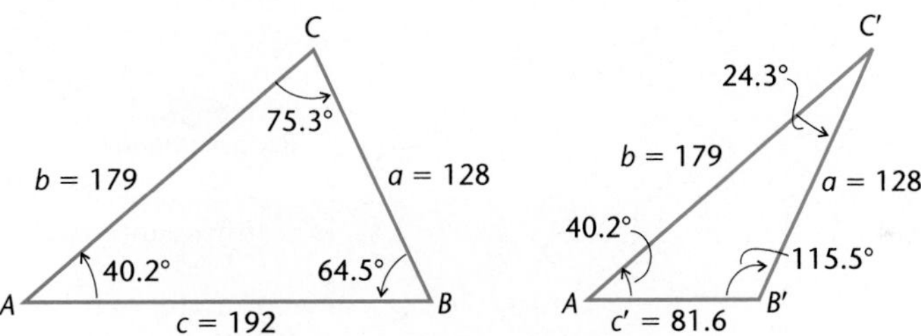

Figure 9.8 The two triangles in Example 4

Acute Angle B

$$\begin{aligned} C &= 180^\circ - (A + B) \\ &= 180^\circ - (40.2^\circ + 64.5^\circ) \\ &= 75.3^\circ \end{aligned}$$

$$\frac{c}{\sin C} = \frac{a}{\sin A}$$

$$c = \frac{128 \sin 75.3^\circ}{\sin 40.2^\circ}$$

$$c = 192$$

The parts of the acute triangle are

$a = 128$, $A = 40.2^\circ$,
$b = 179$, $\mathbf{B = 64.5^\circ}$,
$\mathbf{c = 192}$, $\mathbf{C = 75.3^\circ}$.

Obtuse Angle B′

$$\begin{aligned} C' &= 180^\circ - (A + B') \\ &= 180^\circ - (40.2^\circ + 115.5^\circ) \\ &= 24.3^\circ \end{aligned}$$

$$\frac{c'}{\sin C'} = \frac{a}{\sin A}$$

$$c' = \frac{128 \sin 24.3^\circ}{\sin 40.2^\circ}$$

$$c' = 81.6$$

The parts of the obtuse triangle are

$a = 128$, $A = 40.2^\circ$,
$b = 179$, $\mathbf{B' = 115.5^\circ}$,
$\mathbf{c' = 81.6}$, $\mathbf{C' = 24.3^\circ}$. □

To be consistent with the labeling in Figure 9.6, the known parts of the triangles in each example have been angle A and sides a and b. However, the method of solution described in each example applies no matter which angle and sides are the known parts of the case 2 triangles.

◆ Practice Problem 2 ◆

Solve the triangle for which $B = 73°$, $b = 81$, and $c = 55$.

Example 5 • A Surveying Application

Point C in Figure 9.9 is on the opposite bank of a river from park headquarters at A. The park director wishes to establish a ranger station at point C and needs to know the distance from A to C. She measures the distance from A along a straight road to a point B where the road ends and finds this distance is 11 miles. She then measures the angles at A and B, as indicated in the figure. How far is it from A to C?

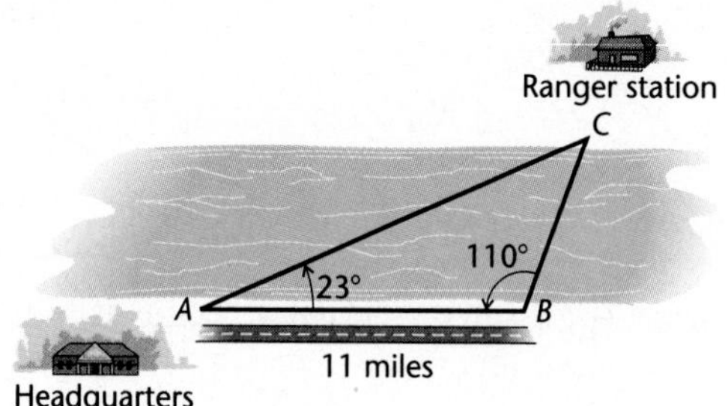

Figure 9.9 Location of the ranger station

Solution

Using conventional notation in Figure 9.9, we have $A = 23°$, $B = 110°$, $c = 11$, and we need to find b. Our plan is to find angle C and then use the law of sines in the form

$$\frac{b}{\sin B} = \frac{c}{\sin C}.$$

We have

$$C = 180° - (A + B) = 180° - 133° = 47°.$$

Thus

$$\frac{b}{\sin 110°} = \frac{11}{\sin 47°} \quad \text{and hence} \quad b = \frac{11 \sin 110°}{\sin 47°} = 14.$$

The distance from A to C is 14 miles. □

EXERCISES 9.1

Solve for the indicated part of the triangle to the degree of accuracy consistent with the given information. (See Examples 1–4.)

1. $A = 42°$, $B = 19°$, $a = 1.3$, $c = ?$
2. $B = 39°$, $C = 58°$, $a = 2.5$, $b = ?$
3. $A = 66.6°$, $B = 33.3°$, $c = 392$, $b = ?$
4. $A = 108.2°$, $C = 24.7°$, $c = 423$, $b = ?$
5. $A = 42.13°$, $B = 91.46°$, $a = 12.86$, $b = ?$
6. $B = 49.61°$, $C = 88.56°$, $b = 41.39$, $a = ?$
7. $A = 150°$, $a = 40$, $b = 10$, $B = ?$
8. $B = 100°$, $b = 9.0$, $c = 3.0$, $C = ?$
9. $B = 30°$, $a = 14$, $b = 7.0$, $A = ?$
10. $C = 30°$, $b = 2.4$, $c = 1.2$, $B = ?$
11. $B = 60°$, $b = 18$, $c = 15$, $C = ?$
12. $C = 70°$, $a = 12$, $c = 20$, $A = ?$
13. $C = 140°$, $a = 20$, $c = 15$, $A = ?$
14. $B = 72°$, $a = 20$, $b = 17$, $A = ?$
15. $A = 54°$, $a = 80$, $b = 90$, $B = ?$
16. $C = 30°$, $b = 3.4$, $c = 2.0$, $B = ?$

In Exercises 17–32, solve the triangle to the degree of accuracy consistent with the given information. (See Examples 1–4.)

17. $A = 68°$, $B = 62°$, $c = 48$
18. $B = 102°$, $C = 48°$, $a = 52$
19. $B = 79.2°$, $C = 35.1°$, $a = 11.3$
20. $A = 80.8°$, $C = 43.0°$, $c = 423$
21. $A = 120°$, $a = 20$, $b = 40$
22. $C = 121°$, $b = 3.4$, $c = 1.8$
23. $A = 127°$, $a = 40$, $c = 30$
24. $C = 105°$, $b = 22$, $c = 43$
25. $A = 47°$, $a = 80$, $b = 70$
26. $C = 51.3°$, $a = 115$, $c = 127$
27. $A = 17.8°$, $a = 1.21$, $b = 4.89$
28. $A = 68°20'$, $a = 143$, $c = 308$
29. $B = 47°$, $a = 20$, $b = 18$
30. $A = 22.8°$, $a = 1.83$, $b = 4.29$
31. $A = 62.3°$, $a = 178$, $c = 187$
32. $C = 38°40'$, $a = 4.28$, $c = 2.99$

Guy Wires 33. Two guy wires are attached 8.0 feet apart on a TV antenna mast and anchored at the same point on the ground, as shown in the accompanying figure. The shorter wire makes an angle of 45° with the horizontal, and the longer an angle of 53°. How long is each of the wires?

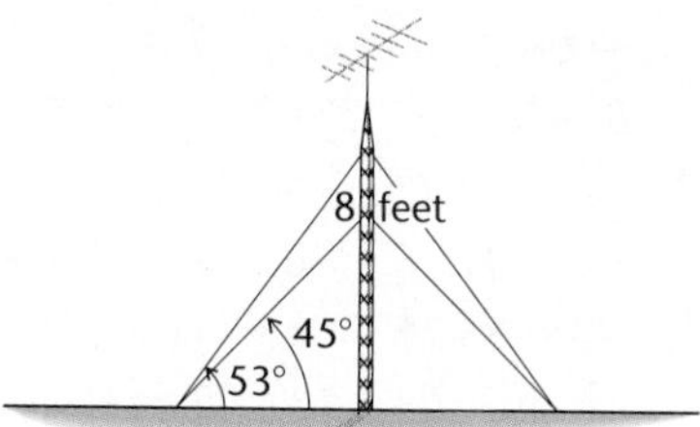

Figure for Exercise 33

Angle of Elevation 34. A hill in the accompanying figure slopes 12° with the horizontal. The angle of elevation to the top of a pine tree standing 52 feet from the base of the hill is 63°. How tall is the pine tree?

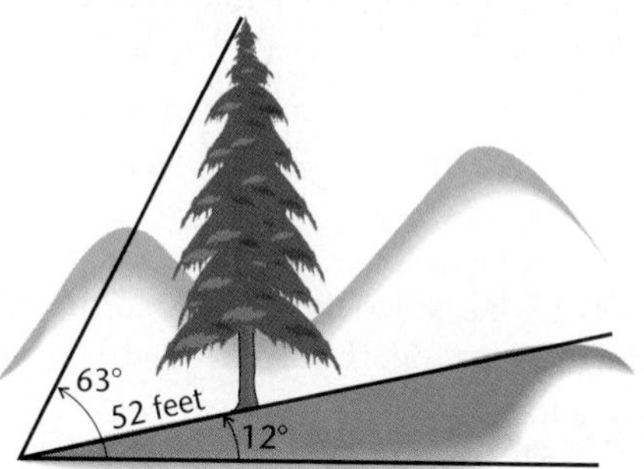

Figure for Exercise 34

A-Frame Cottage 35. The wall of the A-frame cottage in the accompanying figure makes an angle of 68° with the floor. Richard places the top of his 15-foot ladder against the base of a window in which he needs to replace a broken pane. If his ladder forms an angle of 53° with the horizontal, how far from the lower edge of the wall is the base of the ladder?

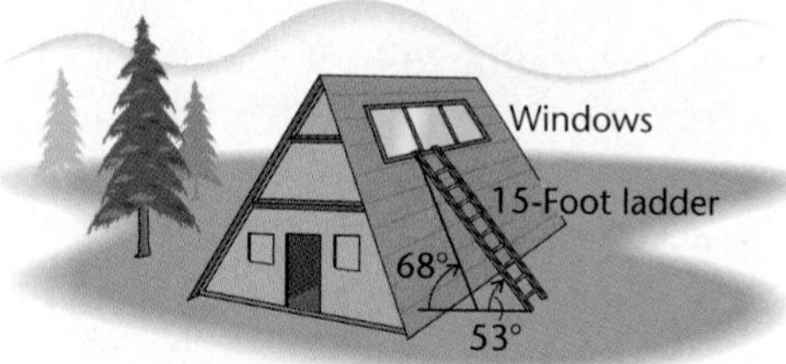

Figure for Exercise 35

Picnic Table

36. The legs on a picnic table are made from 2×4's bolted together to form an X, as shown in the accompanying figure. Find the length of 2×4 needed for each leg.

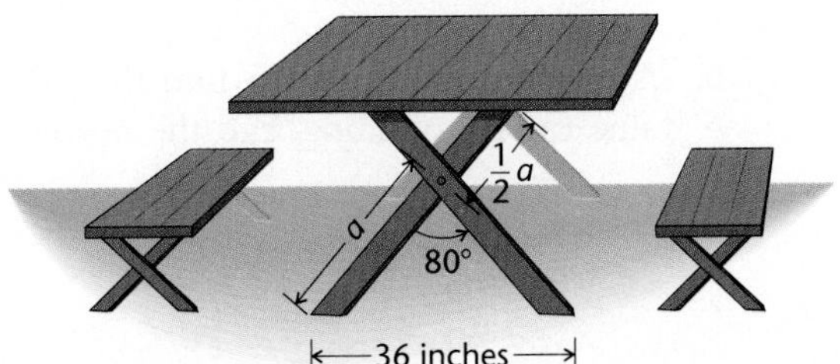

Figure for Exercise 36

Angle of Elevation 37–38

37. A vertical post stands on a hill that is inclined at 16° with the horizontal. The angle of elevation of the top of the post from a point 81 feet down the hill from the base of the post is 26°. Find the height of the post.

38. A flagpole leaning away from the sun forms a 10° angle with the vertical and casts a 22-foot shadow on level ground. From the end of the shadow the angle of elevation to the top of the pole is 47°. Find the length of the pole.

Sea Navigation

39. The harbor master receives a call for a pilot boat from a freighter in the direction N 58° E from the harbor. The pilot boat is located 60 miles due west of the harbor. The direction from the pilot boat to the freighter is N 75° E. What is the distance between the pilot boat and the freighter?

Surveying

40. Two angles of a triangular pasture are 36.1° and 45.2°. The side between the angles measures 326 yards. How many yards of fencing are required to enclose the pasture?

Angle of Elevation

41. A vertical tower that is 123 feet high stands on a hill. The angle of elevation of the top of the tower from a point 296 feet down the hill from the base is 26.2°. Find the angle of inclination of the hill with the horizontal.

Air Navigation

42. Memphis, Tennessee, is due north of New Orleans, Louisiana, at a distance of 350 miles. If a plane leaves New Orleans and flies in the direction 58.2° until it is 540 miles from Memphis, find the bearing from the plane to Memphis.

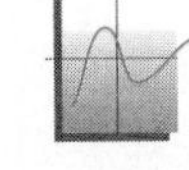

Surveying 43–44

43. A triangular parking lot on a street corner has an angle of $150° = 5\pi/6$ radians at the street corner, and the length of the back boundary is 60 meters, as shown in the figure on page 556.

a. Use the law of sines to show that the perimeter P of the lot is given in meters by

$$P = 60 + 120 \sin x + 120 \sin\left(\frac{\pi}{6} - x\right).$$

b. Use a graphing device to find the value of x that gives a maximum value for P, rounding x to the nearest hundredth of a radian.

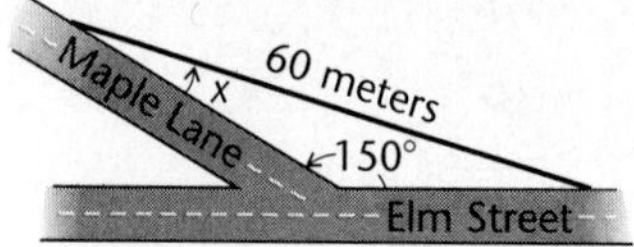

Figure for Exercise 43

44. Suppose the angle at the street corner in Exercise 43 is 120° and the length of the back boundary is 80 meters. Use a graphing device to find the value of the angle x adjacent to Maple Lane that gives a maximum value for P. Round x to the nearest hundredth of a radian.

◆ Solutions for Practice Problems

1. The triangle to be solved is shown in Figure 9.10. We first determine angle A.

$$\begin{aligned} A &= 180° - (47.3° + 25.5°) \\ &= 107.2° \end{aligned}$$

Then we apply the law of sines.

$$\frac{b}{\sin 25.5°} = \frac{83.6}{\sin 107.2°}$$

$$b = \frac{83.6 \sin 25.5°}{\sin 107.2°} = 37.7$$

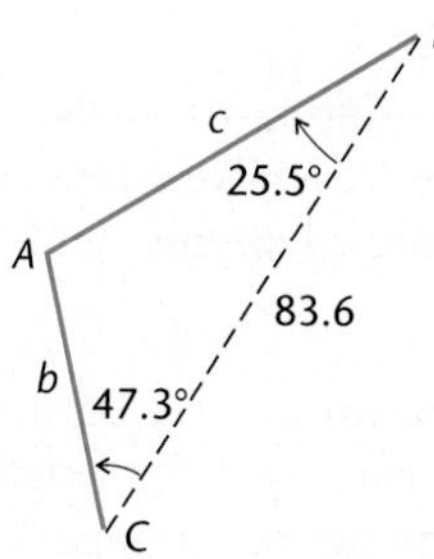

Figure 9.10 Triangle for Practice Problem 1

$$\frac{c}{\sin 47.3°} = \frac{83.6}{\sin 107.2°}$$

$$c = \frac{83.6 \sin 47.3°}{\sin 107.2°} = 64.3$$

Farmer Fred needs 37.7 + 64.3 = 102.0 feet of fencing to make the detour. This amounts to 102.0 − 83.6 = 18.4 extra feet of fencing needed.

2. To solve for angle C we write the following.

$$\frac{\sin C}{c} = \frac{\sin B}{b}$$

$$\sin C = \frac{55 \sin 73°}{81} = 0.6493$$

$$C = \begin{cases} 40° \\ 180° - 40° = 140° \qquad \text{impossible} \end{cases}$$

The value 140° for C is impossible because 140° + 73° > 180°. To find angle A we write

$$A = 180° - (B + C) = 180° - (73° + 40°) = 67°.$$

We use the law of sines again to solve for side a.

$$\frac{a}{\sin A} = \frac{b}{\sin B}$$

$$a = \frac{81 \sin 67°}{\sin 73°} = 78$$

All parts of the triangle in Figure 9.11 are now known.

$\boldsymbol{a = 78}$	$\mathbf{A = 67°}$
$b = 81$	$B = 73°$
$c = 55$	$\mathbf{C = 40°}$

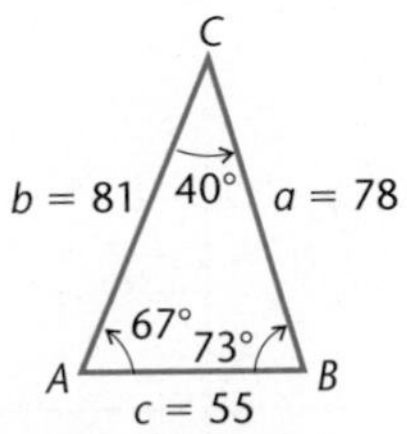

Figure 9.11 Triangle for Practice Problem 2

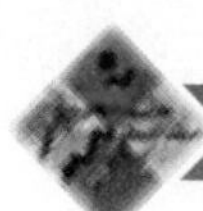

9.2 THE LAW OF COSINES

There are two more cases to consider when certain information is known about a triangle:

Case 3 Two sides and the included angle are known.
Case 4 All three sides are known.

In either of these two cases, no ambiguity arises: If such a triangle exists, it will be unique. One problem does arise, however. If we attempt to use the law of sines, we shall discover that it is impossible to find an equation containing only one unknown. This is illustrated in the first example of a case 3 triangle.

Example 1 • Case 3 Triangle

Solve the triangle where $A = 52°$, $b = 88$, and $c = 67$.

Solution

In an attempt to solve for angle B by using the law of sines, we write one of the two equations:

$$\frac{\sin B}{b} = \frac{\sin A}{a} \qquad \text{or} \qquad \frac{\sin B}{b} = \frac{\sin C}{c}$$

$$\frac{\sin B}{88} = \frac{\sin 52°}{a} \qquad\qquad \frac{\sin B}{88} = \frac{\sin C}{67}.$$

We cannot solve for angle B in either equation since we have no value for side a in the first and no value of angle C in the second. □

Example 1 illustrates the need for an additional tool for solving triangles. The **law of cosines** fills this need. We derive it next.

Suppose angle A and sides b and c are the known parts of a triangle labeled conventionally. We place angle A in standard position with side c lying along the positive x-axis. Angle A may be an acute, obtuse, or right angle. Figure 9.12 illustrates all three possibilities.

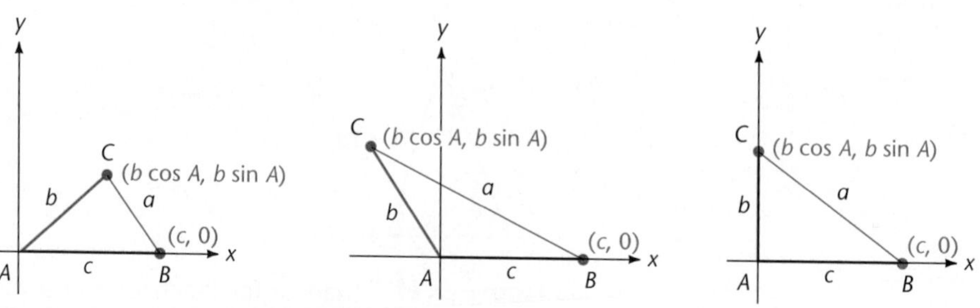

Figure 9.12 Three possibilities for angle A

In each of the triangles in Figure 9.12, the length of side a is the distance between the vertices of angles C and B. The coordinates of the vertex of angle C are $(b\cos A, b\sin A)$, and those of the vertex of angle B are $(c, 0)$. Applying the distance formula, we have

$$\begin{aligned} \mathbf{a^2} &= (b\cos A - c)^2 + (b\sin A - 0)^2 \\ &= b^2\cos^2 A - 2bc\cos A + c^2 + b^2\sin^2 A \\ &= b^2(\cos^2 A + \sin^2 A) + c^2 - 2bc\cos A \\ &\mathbf{= b^2 + c^2 - 2bc\cos A,} \end{aligned}$$

which is one of the forms of the law of cosines.

When angle A is a right angle, side a is the hypotenuse of a right triangle, and the law of cosines reduces to the Pythagorean theorem, $a^2 = b^2 + c^2$.

Two other forms of the law of cosines arise when (1) sides a and c and the included angle B are known and (2) sides a and b and the included angle C are known.

Law of Cosines

In any triangle labeled conventionally,

$$\begin{aligned} a^2 &= b^2 + c^2 - 2bc\cos A, \\ b^2 &= a^2 + c^2 - 2ac\cos B, \\ c^2 &= a^2 + b^2 - 2ab\cos C. \end{aligned}$$

We are now in a position to solve the case 3 triangle described in Example 1.

Example 2 • Solving a Case 3 Triangle

Solve the triangle where $A = 52°$, $b = 88$, and $c = 67$.

Solution

We use the law of cosines to find the length of side a.

$$\begin{aligned} a^2 &= b^2 + c^2 - 2bc\cos A \\ &= 88^2 + 67^2 - 2(88)(67)\cos 52° \\ &= 7744 + 4489 - 11792(0.6157) \\ &= 7744 + 4489 - 7260 \\ &= 4973 \end{aligned}$$

Taking square roots,

$$a = \sqrt{4973} = 71.$$

We now use the law of sines to find angle C. We choose angle C (rather than angle B) since side c is the shortest side. This guarantees that angle C will be acute.

$$\frac{\sin C}{c} = \frac{\sin A}{a}$$

$$\sin C = \frac{c \sin A}{a} = \frac{67 \sin 52^\circ}{71} = 0.7436$$

Since angle C must be acute, we use the first-quadrant value for angle C.

$$C = 48^\circ$$

Then we determine angle B by

$$B = 180^\circ - (A + C) = 180^\circ - (52^\circ + 48^\circ) = 80^\circ.$$

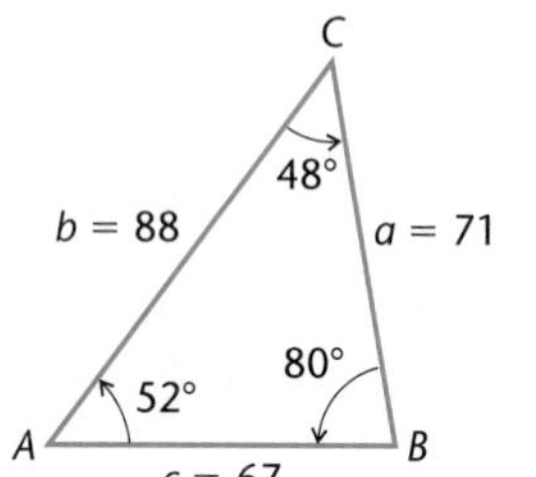

Figure 9.13 Triangle in Example 2

All parts of the triangle sketched in Figure 9.13 are now known.

$\boldsymbol{a = 71}$	$A = 52^\circ$
$b = 88$	$\boldsymbol{B = 80^\circ}$
$c = 67$	$\boldsymbol{C = 48^\circ}$

□

◆ Practice Problem 1 ◆

Solve the triangle for which $C = 108^\circ$, $a = 83$, and $b = 32$.

The law of cosines can also be used to solve case 4 triangles: those in which all three sides are known. For such a triangle to exist, the length of the longest side must be less than the sum of the lengths of the shorter two sides. That is, one of the following three inequalities must be satisfied:

$$a + b > c \qquad \text{or} \qquad a + c > b \qquad \text{or} \qquad b + c > a$$

for sides a, b, and c of any triangle. Then each of the three forms of the law of cosines can be solved for the cosine of one of the angles A, B, or C in terms of the three sides a, b, and c.

Law of Cosines

In any triangle labeled conventionally,

$$\cos A = \frac{b^2 + c^2 - a^2}{2bc},$$

$$\cos B = \frac{a^2 + c^2 - b^2}{2ac},$$

$$\cos C = \frac{a^2 + b^2 - c^2}{2ab}.$$

Example 3 • Solving a Case 4 Triangle

Solve the triangle where $a = 5.3$, $b = 4.3$, and $c = 7.2$.

Solution

Since $a + b > c$, a triangle exists satisfying the given information. We use the law of cosines to solve for angle C, the largest angle in the triangle. (This assures us that angles A and B will be acute.)

$$\cos C = \frac{a^2 + b^2 - c^2}{2ab}$$
$$= \frac{(5.3)^2 + (4.3)^2 - (7.2)^2}{2(5.3)(4.3)}$$
$$= \frac{28.09 + 18.49 - 51.84}{45.58}$$
$$= -0.1154$$

The cosine is negative in the second and third quadrants. But a third quadrant angle cannot be an angle in a triangle. Thus C must be the second-quadrant angle:

$$C = 97°.$$

Next we use the law of sines to find angle B, the smaller of the two remaining angles.

$$\frac{\sin B}{b} = \frac{\sin C}{c}$$
$$\sin B = \frac{b \sin C}{c}$$
$$= \frac{4.3 \sin 97°}{7.2}$$
$$= 0.5928$$

Since B must be acute, we have

$$B = 36°.$$

Then we determine angle A by

$$A = 180° - (B + C) = 180° - (36° + 97°) = 47°.$$

All parts of the case 4 triangle shown in Figure 9.14 on page 562 are now known.

$a = 5.3$	$\mathbf{A = 47°}$
$b = 4.3$	$\mathbf{B = 36°}$
$c = 7.2$	$\mathbf{C = 97°}$

□

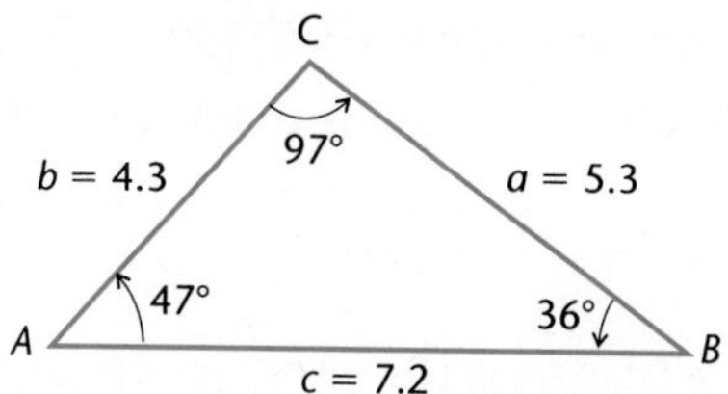

Figure 9.14 Triangle in Example 3

Example 4 • Glass Cutting

A triangular piece of glass is to be cut to fit in a side window of a small fishing boat. The dimensions of the glass should be 14 inches, 30 inches, and 36 inches. In order to mark off the triangle, the glass cutter needs to know the angle opposite the longest side. Find this angle.

Solution

A sketch of the triangular piece of glass is shown in Figure 9.15.

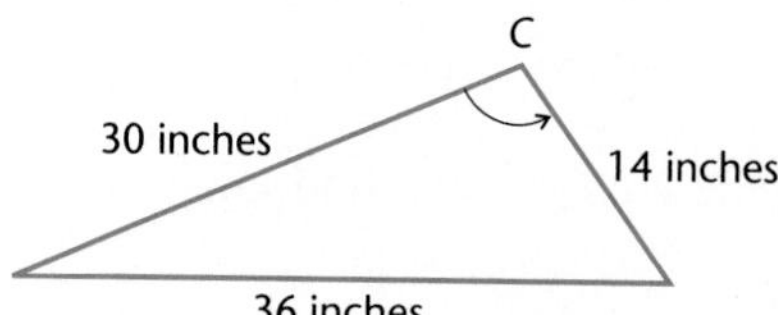

Figure 9.15 The triangular boat window

If we use C to denote the angle opposite the longest side, then $c = 36$ and

$$\begin{aligned}\cos C &= \frac{a^2 + b^2 - c^2}{2ab}\\ &= \frac{(14)^2 + (30)^2 - (36)^2}{2(14)(30)}\\ &= -0.2381.\end{aligned}$$

Thus $C = \cos^{-1}(-0.2381) = 104°$. □

EXERCISES 9.2

In Exercises 1–16, solve for the indicated part of the triangle to the degree of accuracy consistent with the given information. (See Examples 2 and 3.)

1. $A = 60°, \quad b = 20, \quad c = 10, \quad a = ?$
2. $B = 60°, \quad a = 10, \quad c = 40, \quad b = ?$

3. $C = 120°$, $a = 4.0$, $b = 6.0$, $c = ?$
4. $A = 120°$, $b = 5.0$, $c = 2.0$, $a = ?$
5. $B = 23°$, $a = 10$, $c = 17$, $b = ?$
6. $C = 48°$, $a = 41$, $b = 36$, $c = ?$
7. $A = 115°$, $b = 3.5$, $c = 2.4$, $a = ?$
8. $B = 108°$, $a = 17$, $c = 22$, $b = ?$
9. $a = 3.0$, $b = 5.0$, $c = 7.0$, $C = ?$
10. $a = 42$, $b = 18$, $c = 30$, $A = ?$
11. $a = 60$, $b = 28$, $c = 45$, $C = ?$
12. $a = 28$, $b = 84$, $c = 70$, $A = ?$
13. $a = 14$, $b = 20$, $c = 80$, $C = ?$
14. $a = 1.8$, $b = 4.5$, $c = 1.2$, $A = ?$
15. $a = 8.23$, $b = 6.81$, $c = 1.01$, $B = ?$
16. $a = 12.4$, $b = 49.8$, $c = 11.3$, $C = ?$

In Exercises 17–28, solve the triangle to the degree of accuracy consistent with the given information. (See Examples 2 and 3.)

17. $A = 56°$, $b = 20$, $c = 30$
18. $C = 35°$, $a = 8.0$, $b = 12$
19. $B = 112°$, $a = 1.5$, $c = 7.6$
20. $A = 128°$, $b = 29$, $c = 42$
21. $a = 3.0$, $b = 4.0$, $c = 6.0$
22. $a = 8.0$, $b = 5.0$, $c = 4.0$
23. $a = 20$, $b = 30$, $c = 11$
24. $a = 18$, $b = 10$, $c = 20$
25. $A = 27°10'$, $b = 308$, $c = 544$
26. $B = 46°50'$, $a = 177$, $c = 293$
27. $B = 122.6°$, $a = 14.3$, $c = 27.9$
28. $C = 128.7°$, $a = 459$, $b = 627$

Hot-Air Balloon

29. Matt and Beckie are standing on level ground and holding ropes attached to a hot-air balloon. Matt's rope is 125 yards long, and Beckie's is 105 yards long. If the ropes form an angle of 132.1° where they are attached to the balloon, how far apart are Matt and Beckie standing?

Sea Navigation 30–31

30. If one boat left a harbor and traveled N 47.1° E for 25.2 miles, and a second boat left the harbor and traveled S 21.5° E for 18.7 miles, how far apart are the boats?
31. A crewboat leaves harbor H in the figure on page 564 and goes 46 miles in the direction S 32° W to deliver the mail to a drilling rig D. From D the boat travels 68 miles in a westerly direction to carry supplies to a pipe barge B anchored 97 miles from H. What course should the boat set to go from B back to the harbor?

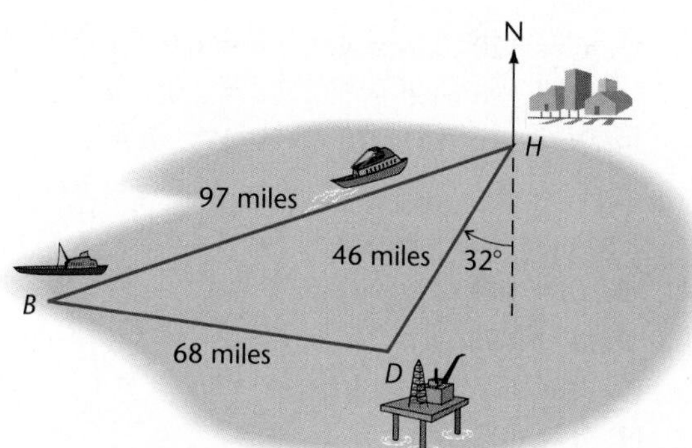

Figure for Exercise 31

Triangular Flower Bed 32. A triangular flower bed has sides measuring 3.2 feet, 4.7 feet, and 5.9 feet. Find the measure of the smallest angle.

Surveying 33–34 33. The village of Oak Hill O has built a water reservoir with a pumping station at point A in the accompanying figure and installed a pipeline that is 6.2 miles from A to the village. The reservoir furnishes enough water for Oak Hill and the neighboring village of Bell Camp as well, and plans are under consideration to build a pipeline from A to Bell Camp, as indicated in the accompanying figure. Bell Camp is 12 miles directly west of Oak Hill, and the direction from Oak Hill to A is S 55° W. Find the length of the proposed pipeline from A to B.

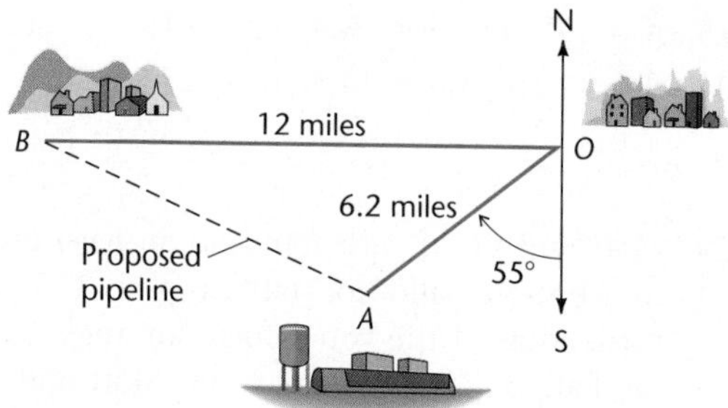

Figure for Exercise 33

34. Two streets intersect at an angle 137°. A triangular parking lot on the corner has 92-foot frontage on one street and 78-foot frontage on the other. Find the length of the back boundary of the lot.

Sonar Detection 35. The Coast Guard detects a sunken bargelike vessel and suspects that it was used for smuggling illegal drugs to Florida. With sonar devices they determine that it is located directly in front of the Coast Guard cutter at a distance of 80 feet and an angle of 37° with the horizontal. The Coast Guard then moves forward 100 feet as shown in the accompanying figure, passing over the bargelike object, and stops. How far must a diver descend in order to reach the object?

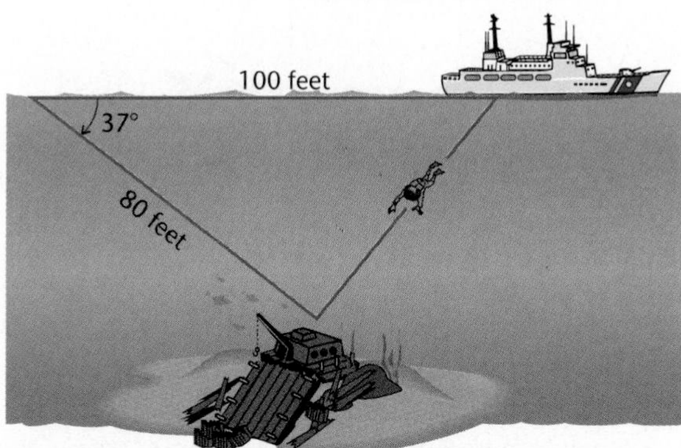

Figure for Exercise 35

Appliance Design

36. A 60-inch hose attaches to a 47-inch wand on the vacuum cleaner in the accompanying figure. When the hose is completely outstretched (with no sags), the most comfortable position for use by a person of average height occurs when a 130° angle is formed between the hose and the wand. In this position, how far from the vacuum cleaner is the carpet attachment?

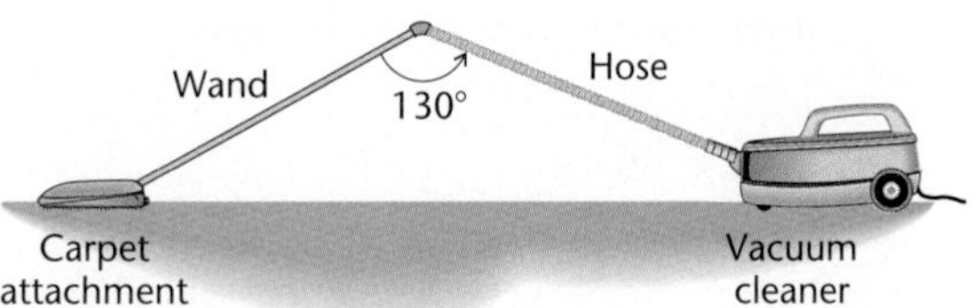

Figure for Exercise 36

Air Navigation 37–38

37. Hideaki is at Charleston, West Virginia, and he would like to fly his small plane to Norfolk, Virginia, to see the women's basketball national championship game at Old Dominion University. He knows that the direction from Charleston to Washington, D.C., is 81.0° and the distance is 342 miles. He also knows that the direction from Washington to Norfolk is 166.0° and the distance is 189 miles. Find the distance from Charleston to Norfolk.

38. A plane flies 175 miles in the direction of 150.0° from Toledo Express Airport T to point A and then turns and flies 88.0 miles in the direction of 70.0° before making an emergency landing at point B. (See the figure on page 566.) How far is point B from Toledo Express Airport?

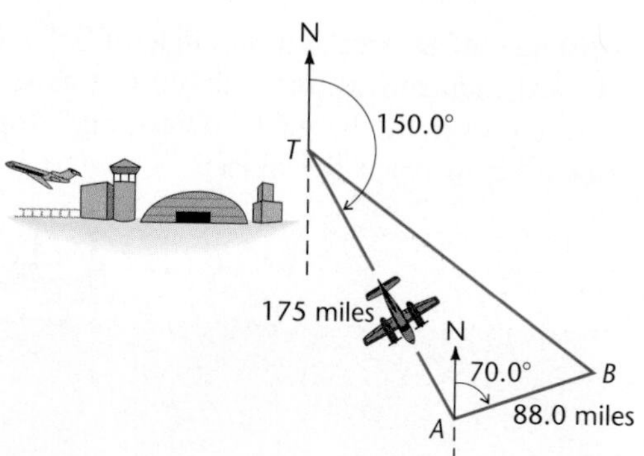

Figure for Exercise 38

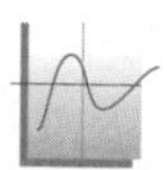

Sea Navigation
39–40

39. At 10 A.M. a freighter left Ft. Lauderdale traveling due east at 10.0 kilometers per hour. At noon a cruise ship left the same port sailing in the direction N 60.0° E at 25.0 kilometers per hour.

a. Use the law of cosines to show that the distance d between the ships t hours after noon is given in kilometers by

$$d = 5[(29 - 10\sqrt{3})t^2 + (16 - 20\sqrt{3})t + 16]^{1/2}, \qquad \text{for } t \geq 0.$$

b. Use a graphing device to find the shortest distance between the ships after noon.

40. Work Exercise 39 if the cruise ship sails in the direction N 30.0° E and all other information is unchanged.

◆ Solution for Practice Problem

1. We use the law of cosines to solve for side c.

$$\begin{aligned} c^2 &= a^2 + b^2 - 2ab\cos C \\ &= (83)^2 + (32)^2 - 2(83)(32)\cos 108° \\ &= 9554 \\ c &= 98 \end{aligned}$$

Next we solve for the smaller of angles A and B.

$$\frac{\sin B}{b} = \frac{\sin C}{c}$$

$$\sin B = \frac{32 \sin 108°}{98} = 0.3105$$

$$B = 18°$$

Then angle A is found by

$$A = 180° - (B + C) = 180° - (18° + 108°) = 54°.$$

All parts of the triangle in Figure 9.16 are now known.

$a = 83$ $\mathbf{A = 54°}$
$b = 32$ $\mathbf{B = 18°}$
$\mathbf{c = 98}$ $C = 108°$

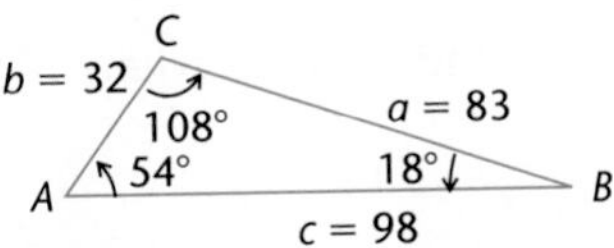

Figure 9.16 Triangle in Practice Problem 1

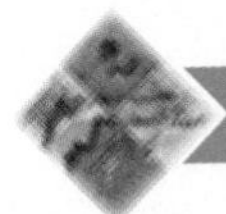

9.3 THE AREA OF A TRIANGLE

Formulas exist for finding the area of a triangle in terms of the known parts for the following three cases.

Case 1 Two angles (hence all three angles) and one side are known.
Case 3 Two sides and the included angle are known.
Case 4 All three sides are known.

First we derive the area formula for the case 3 triangle. Since the area of a triangle is one-half the product of the base and the altitude, we see in Figure 9.17 that the area K can be written in three different ways.

1. $\mathbf{K = \frac{1}{2}ch = \frac{1}{2}bc \sin A}$
2. $\mathbf{K = \frac{1}{2}ch = \frac{1}{2}ac \sin B}$
3. $\mathbf{K = \frac{1}{2}ak = \frac{1}{2}ab \sin C}$

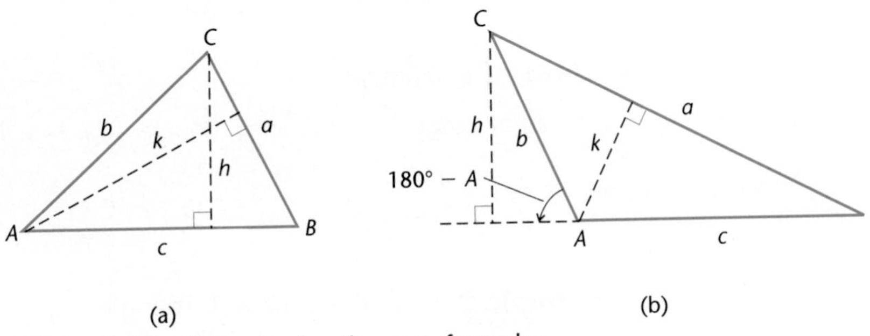

Figure 9.17 Triangles for the area formulas

Thus we have established the formulas for the area of case 3 triangles.

Area of a Triangle

In any triangle labeled conventionally, the area K is

$$K = \tfrac{1}{2}bc \sin A = \tfrac{1}{2}ac \sin B = \tfrac{1}{2}ab \sin C.$$

Example 1 • Area of a Case 3 Triangle

Determine the area of the triangle where $A = 42°$, $b = 12$ inches, and $c = 16$ inches.

Solution

Applying the formula for the area, we have

$$\begin{aligned} K &= \tfrac{1}{2}bc \sin A \\ &= \tfrac{1}{2}(12)(16) \sin 42° \\ &= 64 \text{ square inches.} \end{aligned}$$

□

We use the formulas just derived and the law of sines to obtain formulas for the area of case 1 triangles.

In the formula for area,

$$K = \tfrac{1}{2}bc \sin A,$$

we replace c by

$$c = \frac{b \sin C}{\sin B} \qquad \text{law of sines}$$

to yield a formula for area in terms of three angles and one side.

$$\mathbf{K = \frac{b^2 \sin A \sin C}{2 \sin B}}$$

Similar replacements yield two additional formulas for the area of case 1 triangles.

Area of a Triangle

In any triangle labeled conventionally, the area K is

$$K = \frac{a^2 \sin B \sin C}{2 \sin A} = \frac{b^2 \sin A \sin C}{2 \sin B} = \frac{c^2 \sin A \sin B}{2 \sin C}.$$

Example 2 • Area of a Case 1 Triangle

Compute the area of the triangle where $A = 92°$, $B = 28°$, and $b = 10$ feet.

Solution

We first determine angle C by

$$C = 180^\circ - (A + B) = 180^\circ - (92^\circ + 28^\circ) = 60^\circ.$$

Then

$$\begin{aligned} K &= \frac{b^2 \sin A \sin C}{2 \sin B} \\ &= \frac{10^2 \sin 92^\circ \sin 60^\circ}{2 \sin 28^\circ} \\ &= 92 \text{ square feet.} \end{aligned}$$

□

Finally, we derive Heron's formula for the area of case 4 triangles.

Heron's Formula

The area K of a triangle with sides a, b, and c is

$$K = \sqrt{s(s-a)(s-b)(s-c)},$$

where s is the semiperimeter of the triangle,

$$s = \tfrac{1}{2}(a + b + c).$$

To derive Heron's formula, we write the square of the area of a triangle as

$$K^2 = \tfrac{1}{4}b^2c^2 \sin^2 A \qquad \text{area of a triangle}$$

$$= \frac{b^2c^2}{4}(1 - \cos^2 A) \qquad \text{Pythagorean identity}$$

$$= \frac{b^2c^2}{4}(1 + \cos A)(1 - \cos A) \qquad \text{algebra}$$

$$= \frac{b^2c^2}{4}\left(1 + \frac{b^2 + c^2 - a^2}{2bc}\right)\left(1 - \frac{b^2 + c^2 - a^2}{2bc}\right) \qquad \text{law of cosines}$$

$$= \frac{b^2c^2}{4}\left[\frac{(b + c)^2 - a^2}{2bc}\right]\left[\frac{a^2 - (b - c)^2}{2bc}\right] \qquad \text{algebra}$$

$$= \frac{(b + c + a)(b + c - a)(a - b + c)(a + b - c)}{16} \qquad \text{algebra}$$

$$= \left(\frac{a + b + c}{2}\right)\left(\frac{b + c - a}{2}\right)\left(\frac{a + c - b}{2}\right)\left(\frac{a + b - c}{2}\right) \qquad \text{algebra}$$

If we let $s = \frac{1}{2}(a + b + c)$, then

$$s - a = \tfrac{1}{2}(a + b + c) - a = \frac{b + c - a}{2},$$

$$s - b = \tfrac{1}{2}(a + b + c) - b = \frac{a + c - b}{2},$$

$$s - c = \tfrac{1}{2}(a + b + c) - c = \frac{a + b - c}{2}.$$

Thus

$$K^2 = s(s - a)(s - b)(s - c),$$

and the area K is

$$\boldsymbol{K = \sqrt{s(s - a)(s - b)(s - c)}.}$$

Example 3 • Using Heron's Formula

Find the area of the triangle with sides measuring $a = 42$ meters, $b = 17$ meters, and $c = 35$ meters.

Solution

We first compute the semiperimeter s.

$$s = \tfrac{1}{2}(a + b + c) = \tfrac{1}{2}(42 + 17 + 35) = 47$$

Then

$$\begin{aligned} s - a &= 47 - 42 = 5, \\ s - b &= 47 - 17 = 30, \\ s - c &= 47 - 35 = 12. \end{aligned}$$

Finally, the area K is

$$\begin{aligned} K &= \sqrt{s(s - a)(s - b)(s - c)} \\ &= \sqrt{47(5)(30)(12)} \\ &= \sqrt{84{,}600} \\ &= 290 \text{ square meters,} \end{aligned}$$

rounded to two significant digits. □

We note that we can also find the area of case 2 triangles (the ambiguous case). However, we must use the law of sines first to determine some additional information about the triangle. Then we can compute the area, remembering that there may be two possible triangles to consider.

◆ Practice Problem 1 ◆

Find the area of the triangle for which $A = 128°$, $a = 32$, and $c = 18$.

EXERCISES 9.3

Find the area of each of the following triangles. (See Examples 1–3.)

1. $A = 56°, \quad b = 20, \quad c = 30$
2. $C = 35°, \quad a = 8.0, \quad b = 12$
3. $B = 112°, \quad a = 1.5, \quad c = 7.6$
4. $A = 128°, \quad b = 29, \quad c = 42$
5. $A = 48°, \quad C = 92°, \quad a = 1.3$
6. $B = 82°, \quad C = 59°, \quad c = 12$
7. $A = 68°, \quad B = 62°, \quad c = 48$
8. $B = 102°, \quad C = 48°, \quad a = 52$
9. $a = 3.0, \quad b = 5.0, \quad c = 7.0$
10. $a = 42, \quad b = 18, \quad c = 30$
11. $a = 60, \quad b = 28, \quad c = 45$
12. $a = 28, \quad b = 84, \quad c = 70$
13. $A = 47°, \quad a = 80, \quad b = 70$
14. $B = 22°, \quad b = 0.70, \quad c = 0.40$
15. $A = 127°, \quad a = 40, \quad c = 32$
16. $C = 140°, \quad a = 5.0, \quad c = 7.0$
17. $A = 120°, \quad a = 20, \quad b = 40$
18. $A = 150°, \quad a = 50, \quad b = 30$
19. $A = 17.8°, \quad a = 1.21, \quad b = 4.89$
20. $B = 48.2°, \quad b = 204, \quad c = 591$
21. $B = 47°, \quad a = 20, \quad b = 30$
22. $C = 53°, \quad a = 80, \quad c = 90$
23. $A = 22.8°, \quad a = 1.83, \quad b = 4.29$
24. $B = 40.2°, \quad b = 14.1, \quad c = 21.3$

Carpeting

25. How many square feet (to the nearest square foot) of carpet are needed to cover the floor of a triangular conference room with sides of length 20 feet, 22 feet, and 30 feet?

Hang Glider

26. How many square feet (to the nearest square foot) of cloth are needed for a triangular hang glider of dimensions 20 feet, 12 feet, and 12 feet?

Surveying 27–30

27. Compute the area of the pasture shown in the accompanying figure. (*Hint:* Divide the pasture into two triangles, as indicated in the figure.)

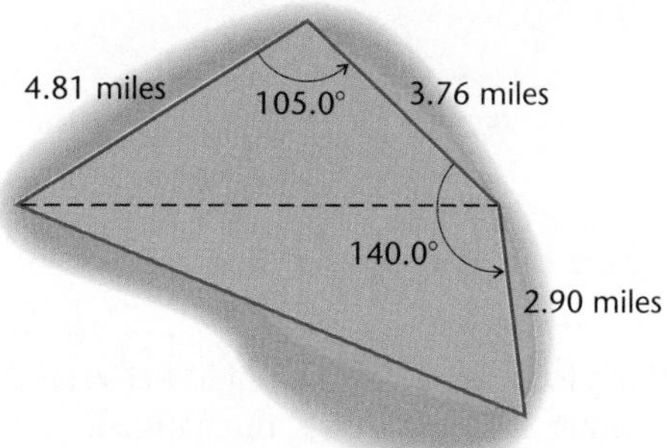

Figure for Exercise 27

28. Compute the area of the tract of land pictured in the figure on page 572. (*Hint:* Divide the land into three triangles, as indicated in the figure.)

Figure for Exercise 28

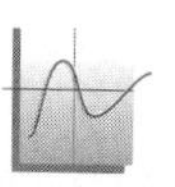

29–30

29. The angle between the fences at one corner of a pasture is 120°. A triangle is to be cut off from the corner with 100 meters of fencing as shown in the accompanying figure. Show that the area of the triangle is given in square meters by

$$A = \frac{10{,}000\sqrt{3}}{3}\sin\theta\sin(60^\circ - \theta)$$

and use a graphing device to find the value of θ that gives a maximum value for A. Round your answer to the nearest degree.

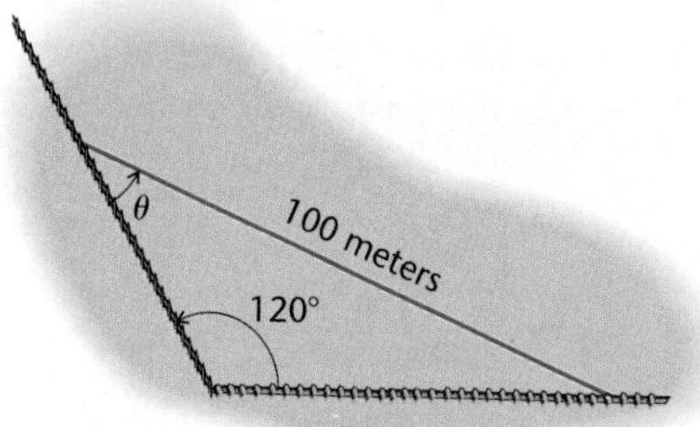

Figure for Exercise 29

30. Work Exercise 29 if the angle between the fences at the corner is 100° and all other information is unchanged.

◆ Solution for Practice Problem

1. We use the law of sines to determine angle C.

$$\sin C = \frac{c \sin A}{a} = \frac{18 \sin 128^\circ}{32} = 0.4433$$

$$C = \begin{cases} 26^\circ \\ 180^\circ - 26^\circ = 154^\circ & \text{impossible since } A + C > 180^\circ. \end{cases}$$

Next we find angle B.

$$B = 180^\circ - (A + C) = 180^\circ - (128^\circ + 26^\circ) = 26^\circ$$

The area of the triangle is

$$\begin{aligned} K &= \frac{a^2 \sin B \sin C}{2 \sin A} \\ &= \frac{32^2 \sin 26^\circ \sin 26^\circ}{2 \sin 128^\circ} \\ &= 120, \text{ rounded to two significant digits.} \end{aligned}$$

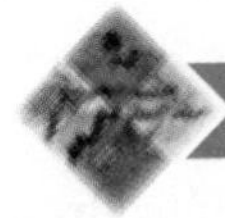

9.4 VECTORS

In order to describe many physical quantities, it is necessary to specify both a direction and a magnitude. For example, a statement that the wind is blowing at 30 miles per hour is an incomplete description. For this information to be useful, the direction also needs to be given.

Quantities that possess both magnitude and direction are called **vector quantities.** Displacements, forces, velocities, and accelerations are some of the simpler vector quantities. In contrast, a quantity that can be described by magnitude alone is called a **scalar quantity.** Pressure and temperature are examples of scalar quantities.

A vector quantity can be represented by a directed line segment, which is called a vector. That is, a **vector** is a line segment that has been given a direction from one endpoint to the other. An arrowhead is placed at one end to indicate the direction of the segment. The vector from point O to point P may be denoted by $\overrightarrow{OP}$, as indicated in Figure 9.18. The point O is called the **initial point,** or **tail,** of $\overrightarrow{OP}$, and P is called the **terminal point,** or **head,** of $\overrightarrow{OP}$.

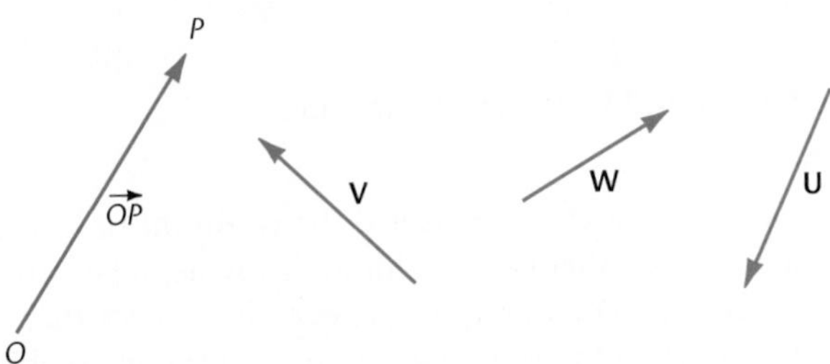

Figure 9.18 Examples of vectors

It is frequently convenient to use a single letter to name a vector. When this is done, the letter is printed in boldface, such as **V**, or an arrow is written over it, such as $\vec{V}$. Some illustrations are shown in Figure 9.18.

When a vector quantity is represented by a vector, the **magnitude,** or **length,** of the vector corresponds to the magnitude of the quantity, and the direction of the vector represents the direction of the quantity. The magnitude of the vector **V** is denoted by $\|\mathbf{V}\|$.

Vectors **U** and **V** are defined to be **equal** if they have the same direction and equal magnitudes. They need not have the same initial points. (See Figure 9.19.)

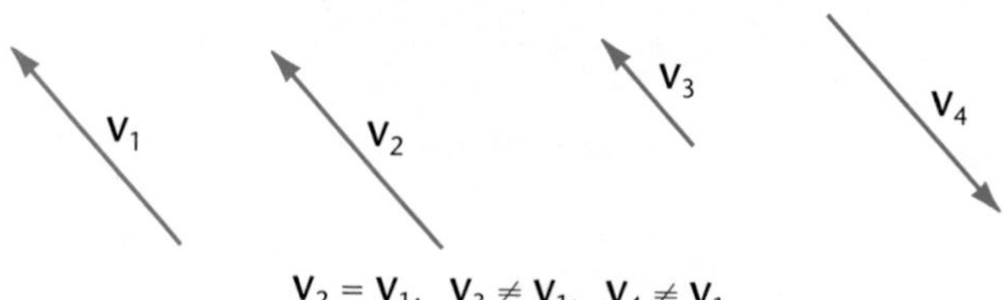

Figure 9.19 Equality of vectors

The vector from O to O is called the **zero vector** and is denoted by **0** or $\vec{0}$. The zero vector has length zero and arbitrary direction.

Many applications of vectors involve **vector addition.** When the initial point of **V** is placed at the terminal point of **U**, the vector from the initial point of **U** to the terminal point of **V** is, by definition, the sum **U** + **V**. This is shown in Figure 9.20.

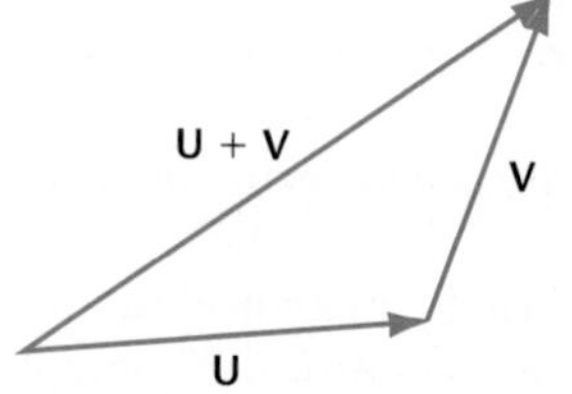

Figure 9.20 Vector addition

To construct the sum **V** + **U**, the initial point of **U** is placed at the terminal point of **V**, and **V** + **U** is drawn from the initial point of **V** to the terminal point of **U**, as shown in Figure 9.21a. In Figure 9.21b, we see that **U** + **V** = **V** + **U** is the diagonal of a parallelogram that has **U** and **V** as sides. Because of this, the rule for vector addition is frequently referred to as the **parallelogram rule.**

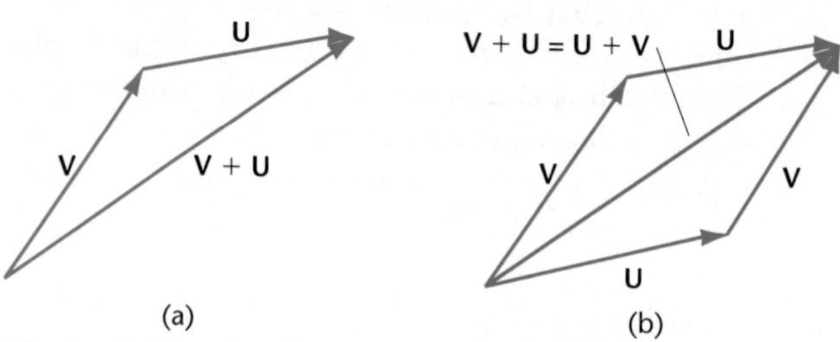

Figure 9.21 Parallelogram rule

The sum **U** + **V** is called the **resultant** of **U** and **V**. One of the reasons vectors are useful is that vector addition provides a faithful representation for many physical results. For example, we learn in physics that when forces are represented by vectors, two forces acting simultaneously on an object combine to produce a resultant force in a manner consistent with addition of vectors.

Example 1 • The Sum of Two Forces

Two forces of magnitudes 20 pounds and 30 pounds act on a point in the plane. If the angle between the directions of the two forces is 27°, find the magnitude of the resultant force.

Figure 9.22 Resultant of two forces

Solution

The vectors representing the two forces are shown in Figure 9.22a. The magnitude of the resultant force is the length a of the diagonal in the parallelogram in Figure 9.22b. We see that angle A measures 153°.

We use the law of cosines to find a.

$$a^2 = 30^2 + 20^2 - 2(30)(20)\cos 153° = 2369$$

Thus the magnitude of the resultant force is

$$a = \sqrt{2369} = 49 \text{ pounds.}$$

□

As the preceding example suggests, the representation of vector quantities by directed line segments has many useful applications. When calculus is applied to vector quantities, however, another representation is required. To obtain this representation, we restrict our attention to the set of all vectors in an xy-coordinate plane.

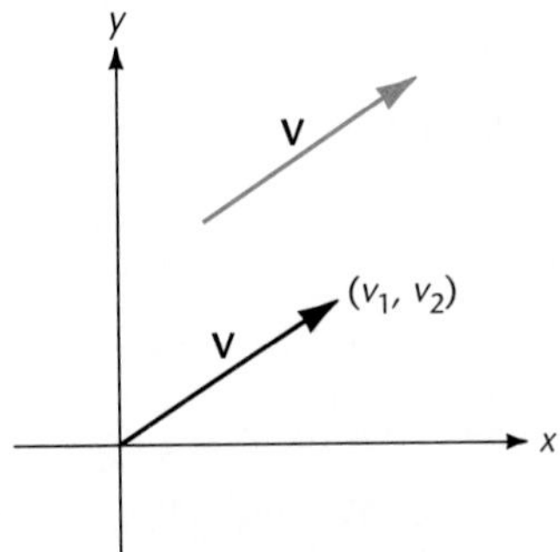

Figure 9.23 Ordered pair associated with a vector

Since vectors are equal if they have the same direction and magnitude, any vector **V** in the plane is equal to a vector with its initial point at the origin and its terminal point at (v_1, v_2), as shown in Figure 9.23. This fact associates each vector **V** in the xy plane with an ordered pair (v_1, v_2), and the association is actually a one-to-one correspondence since either the vector or the ordered pair uniquely determines the other. We use this one-to-one correspondence to represent **V** by the ordered pair $\langle v_1, v_2 \rangle$. The corner brackets $\langle$ and $\rangle$ are used instead of parentheses to avoid confusion with the coordinates of points. We write

$$\mathbf{V} = \langle v_1, v_2 \rangle,$$

and we call v_1 and v_2 the ***x*-component** and the ***y*-component** of **V**, respectively, It is clear from Figure 9.23 that

$$\|\mathbf{V}\| = \sqrt{v_1^2 + v_2^2}$$

and

$$\langle u_1, u_2 \rangle = \langle v_1, v_2 \rangle \quad \text{if and only if} \quad u_1 = v_1 \text{ and } u_2 = v_2.$$

The parallelogram rule for vectors in component form appears as

$$\langle u_1, u_2 \rangle + \langle v_1, v_2 \rangle = \langle u_1 + v_1, u_2 + v_2 \rangle.$$

This rule is diagrammed in Figure 9.24 on page 576.

The product of a number a and a vector $\mathbf{V} = \langle v_1, v_2 \rangle$ is given by

$$a\mathbf{V} = \langle av_1, av_2 \rangle.$$

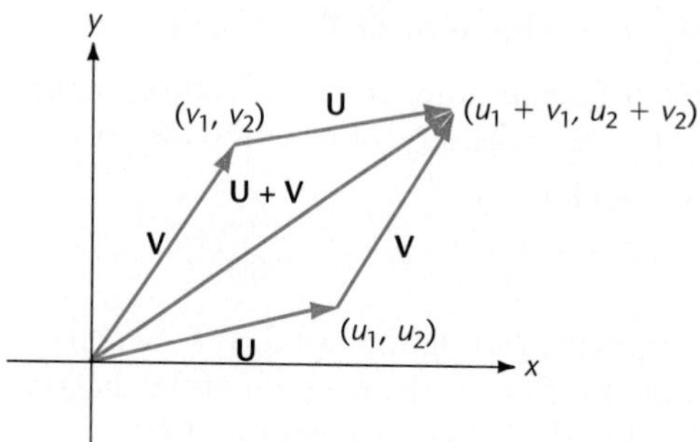

Figure 9.24 Parallelogram rule in component form

Direct computation can be used to show that

$$\|a\mathbf{V}\| = |a| \cdot \|\mathbf{V}\|.$$

In connection with the product $a\mathbf{V}$, a real number a is called a **scalar,** and the operation of computing $a\mathbf{V}$ is called **scalar multiplication.**

Considered as directed line segments, $a\mathbf{V}$ has the same direction as $\mathbf{V}$ if a is positive, and the opposite direction to $\mathbf{V}$ if a is negative. Thus $(-1)\mathbf{V}$ is the same as the vector $-\mathbf{V} = \langle -v_1, -v_2 \rangle$, which has the property that

$$\mathbf{V} + (-\mathbf{V}) = (-\mathbf{V}) + \mathbf{V} = \mathbf{0}.$$

Subtraction of vectors is defined by $\mathbf{U} - \mathbf{V} = \mathbf{U} + (-\mathbf{V})$, or

$$\langle u_1, u_2 \rangle - \langle v_1, v_2 \rangle = \langle u_1 - v_1, u_2 - v_2 \rangle.$$

Example 2 • Operations on Vectors

Let $\mathbf{U} = \langle 2, -1 \rangle$ and $\mathbf{V} = \langle 3, 4 \rangle$, as shown in Figure 9.25. Find

a. $\mathbf{U} + \mathbf{V}$ b. $2\mathbf{V}$, c. $-\mathbf{U}$, d. $5\mathbf{U} - 4\mathbf{V}$, e. $\|\mathbf{V}\|$ and $\|2\mathbf{V}\|$.

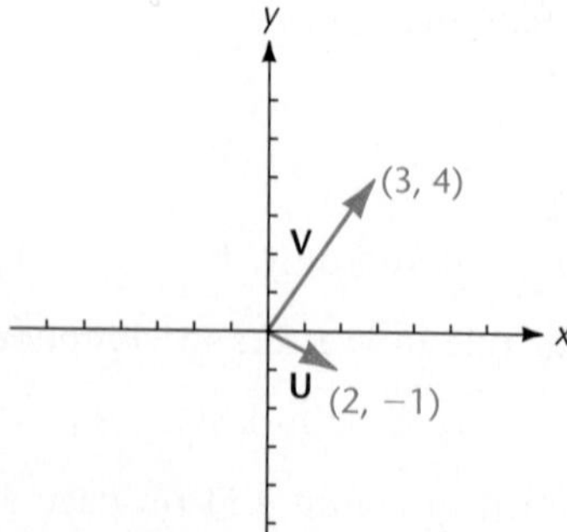

Figure 9.25 Vectors $\mathbf{U} = \langle 2, -1 \rangle$ and $\mathbf{V} = \langle 3, 4 \rangle$

Solution

a. $\mathbf{U} + \mathbf{V} = \langle 2, -1\rangle + \langle 3, 4\rangle = \langle 2 + 3, -1 + 4\rangle = \langle 5, 3\rangle$

b. $2\mathbf{V} = 2\langle 3, 4\rangle = \langle (2)(3), (2)(4)\rangle = \langle 6, 8\rangle$

c. $-\mathbf{U} = (-1)\langle 2, -1\rangle = \langle -2, 1\rangle$

d. $5\mathbf{U} - 4\mathbf{V} = 5\langle 2, -1\rangle - 4\langle 3, 4\rangle = \langle 10, -5\rangle - \langle 12, 16\rangle = \langle -2, -21\rangle$

e. We have

$$\|\mathbf{V}\| = \sqrt{3^2 + 4^2} = \sqrt{25} = 5$$

and

$$\|2\mathbf{V}\| = \sqrt{6^2 + 8^2} = \sqrt{100} = 10,$$

illustrating the fact that

$$\|2\mathbf{V}\| = 2\|\mathbf{V}\|.$$

□

The following list represents the basic properties of vector addition and scalar multiplication.

Basic Properties of Vector Operations

1. $\mathbf{U} + \mathbf{V} = \mathbf{V} + \mathbf{U}$
2. $\mathbf{U} + (\mathbf{V} + \mathbf{W}) = (\mathbf{U} + \mathbf{V}) + \mathbf{W}$
3. $\mathbf{V} + \mathbf{0} = \mathbf{V}$
4. $\mathbf{V} + (-\mathbf{V}) = \mathbf{0}$
5. $a(b\mathbf{V}) = (ab)\mathbf{V}$
6. $a(\mathbf{U} + \mathbf{V}) = a\mathbf{U} + a\mathbf{V}$
7. $(a + b)\mathbf{V} = a\mathbf{V} + b\mathbf{V}$
8. $1\,\mathbf{V} = \mathbf{V}$
9. $\|a\mathbf{V}\| = |a| \cdot \|\mathbf{V}\|$

The basic properties of vector operations are consequences of familiar properties of operations with real numbers, such as the commutative and associative properties of addition. We illustrate these with the proof of property 7 and leave the proofs of the remaining properties as exercises. For arbitrary $\mathbf{V} = \langle v_1, v_2\rangle$ and scalars a, b, we have the following.

$$\begin{aligned}
(a + b)\mathbf{V} &= (a + b)\langle v_1, v_2\rangle \\
&= \langle (a + b)v_1, (a + b)v_2\rangle && \text{scalar multiplication} \\
&= \langle av_1 + bv_1, av_2 + bv_2\rangle && \text{distributive property of real numbers} \\
&= \langle av_1, av_2\rangle + \langle bv_1, bv_2\rangle && \text{vector addition} \\
&= a\langle v_1, v_2\rangle + b\langle v_1, v_2\rangle && \text{scalar multiplication} \\
&= a\mathbf{V} + b\mathbf{V}
\end{aligned}$$

We saw in Figure 9.23 that any directed line segment in the xy plane is equal to a vector $\mathbf{V}$ with initial point at the origin and terminal point (v_1, v_2). A vector with its initial point at the origin is called a **position vector.** If the original directed line segment extended from $P(x_1, y_1)$ to $Q(x_2, y_2)$, the position vector $\mathbf{V} = \overrightarrow{PQ}$ extends from $(0, 0)$ to $(x_2 - x_1, y_2 - y_1)$, as shown in Figure 9.26.

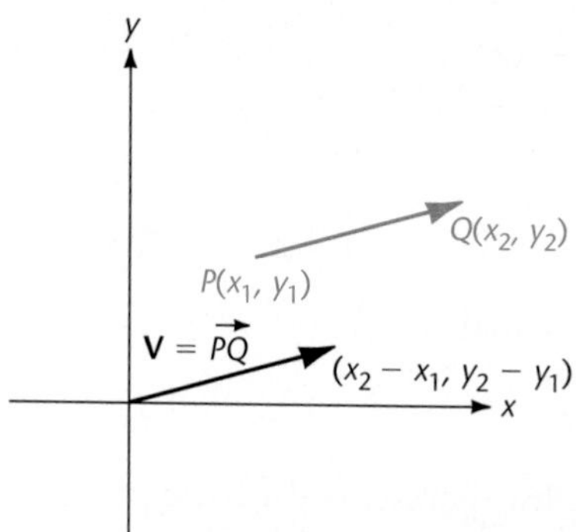

Figure 9.26 The position vector **V**

Example 3 • Finding a Position Vector

The vector $\overrightarrow{PQ}$ from $P(-4, 7)$ to $Q(2, 4)$ is given by

$$\begin{aligned}\overrightarrow{PQ} &= \langle 2 - (-4), 4 - 7\rangle \\ &= \langle 6, -3\rangle.\end{aligned}$$

It is worth noting that $\overrightarrow{QP} = \langle -6, 3\rangle$, and that $\overrightarrow{QP} = -\overrightarrow{PQ}$ is always true. □

The simple equation

$$\langle v_1, v_2\rangle = v_1\langle 1, 0\rangle + v_2\langle 0, 1\rangle$$

shows that every vector in the plane can be written as the sum of a multiple of $\langle 1, 0\rangle$ and a multiple of $\langle 0, 1\rangle$. This fact leads to an alternative notation for vectors in the plane. If we let $\mathbf{i} = \langle 1, 0\rangle$ and $\mathbf{j} = \langle 0, 1\rangle$, we have

$$\mathbf{V} = \langle v_1, v_2\rangle = v_1\mathbf{i} + v_2\mathbf{j}$$

for arbitrary $\mathbf{V}$. This notation is used in calculus and is very common in many areas of application. The vectors $\mathbf{i}$ and $\mathbf{j}$ are called **base vectors.**

In future courses it will be helpful to be familiar with the $\mathbf{i}, \mathbf{j}$ notation. We have

$$\begin{aligned}(u_1\mathbf{i} + u_2\mathbf{j}) + (v_1\mathbf{i} + v_2\mathbf{j}) &= (u_1 + v_1)\mathbf{i} + (u_2 + v_2)\mathbf{j}, \\ a(v_1\mathbf{i} + v_2\mathbf{j}) &= (av_1)\mathbf{i} + (av_2)\mathbf{j},\end{aligned}$$

and

$$\|v_1\mathbf{i} + v_2\mathbf{j}\| = \sqrt{v_1^2 + v_2^2}.$$

Example 4 • Vector Operations with the **i, j** Notation

The use of the formulas in the preceding paragraph is illustrated by the following computations.

a. $(3\mathbf{i} - 7\mathbf{j}) + (2\mathbf{i} + 5\mathbf{j}) = (3 + 2)\mathbf{i} + (-7 + 5)\mathbf{j}$
$= 5\mathbf{i} - 2\mathbf{j}$

b. $2(4\mathbf{i} - 6\mathbf{j}) = (2)(4)\mathbf{i} + (2)(-6)\mathbf{j}$
$= 8\mathbf{i} - 12\mathbf{j}$

c. $\|5\mathbf{i} - 2\mathbf{j}\| = \sqrt{5^2 + (-2)^2}$
$= \sqrt{29}$ □

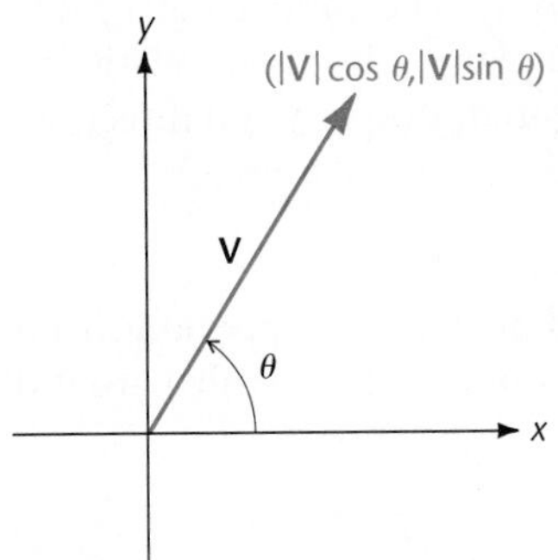

Figure 9.27 Direction angle for **V**

We note that each of $\mathbf{i} = \langle 1, 0\rangle$ and $\mathbf{j} = \langle 0, 1\rangle$ has magnitude 1. Any vector with magnitude 1 is called a **unit vector.**

There is yet another formulation that can be made for vectors. For an arbitrary nonzero position vector $\mathbf{V} = \langle v_1, v_2\rangle = v_1\mathbf{i} + v_2\mathbf{j}$, let θ be the angle in standard position with $\mathbf{V}$ as its terminal side and $0 \le \theta < 2\pi$. (See Figure 9.27.) The angle θ is called the **direction angle** for **V**. It follows at once from the definitions of $\sin\theta$ and $\cos\theta$ that $\cos\theta = v_1/\|\mathbf{V}\|$ and $\sin\theta = v_2/\|\mathbf{V}\|$. Thus $v_1 = \|\mathbf{V}\|\cos\theta$, $v_2 = \|\mathbf{V}\|\sin\theta$, and

$$\mathbf{V} = \langle \|\mathbf{V}\|\cos\theta, \|\mathbf{V}\|\sin\theta\rangle$$
$$= \|\mathbf{V}\|\cos\theta\,\mathbf{i} + \|\mathbf{V}\|\sin\theta\,\mathbf{j},$$

where $\|\mathbf{V}\| = \sqrt{v_1^2 + v_2^2}$ and $\tan\theta = v_2/v_1$. This formula expresses **V** in terms of its magnitude $\|\mathbf{V}\|$ and its direction angle θ.

Example 5 • Working with Magnitude and Direction Angle

a. Find the component form of a vector that has magnitude 6 and direction angle $5\pi/6$.

b. Find the magnitude and direction angle for the vector $\mathbf{V} = -\mathbf{i} + \sqrt{3}\mathbf{j}$.

Solution

a. Since we have $\|\mathbf{V}\| = 6$ and $\theta = 5\pi/6$, the vector **V** is given by

$$\mathbf{V} = \|\mathbf{V}\|\cos\theta\,\mathbf{i} + \|\mathbf{V}\|\sin\theta\,\mathbf{j}$$
$$= 6\cos\frac{5\pi}{6}\mathbf{i} + 6\sin\frac{5\pi}{6}\mathbf{j}$$
$$= 6\left(-\frac{\sqrt{3}}{2}\right)\mathbf{i} + 6\left(\frac{1}{2}\right)\mathbf{j}$$
$$= -3\sqrt{3}\mathbf{i} + 3\mathbf{j}.$$

b. For $\mathbf{V} = -\mathbf{i} + \sqrt{3}\mathbf{j}$, we have

$$\|\mathbf{V}\| = \sqrt{(-1)^2 + (\sqrt{3})^2} = \sqrt{4} = 2.$$

We also have $\tan\theta = \sqrt{3}/(-1) = -\sqrt{3}$, but $\theta \neq \tan^{-1}(-\sqrt{3})$ since θ is a second-quadrant angle and $\tan^{-1}(-\sqrt{3}) = -\pi/3$. We must choose the second-quadrant angle

$$\theta = \pi - \frac{\pi}{3} = \frac{2\pi}{3}$$

as the direction angle for $-\mathbf{i} + \sqrt{3}\mathbf{j}$. □

In applications involving navigation, the words *course* and *heading* have technical meanings. The **course** of a ship or a plane is the direction of the path in which the craft is moving, and the **heading** is the direction in which the craft is pointed. The word **speed** is used when speaking of magnitude only. The **airspeed** of a plane is its speed in still air, whereas the **ground speed** of a plane is its speed relative to the ground. The word **velocity** is used when referring to both speed and direction.

Example 6 • Using Vectors in Air Navigation

Find the airspeed and heading of an airplane if a wind of 30.0 miles per hour from the direction of 343.0° results in the airplane flying a course of 43.0° with a ground speed of 250 miles per hour.

Solution

In Figure 9.28, the length of side x represents the airspeed, and the angle β represents the heading of the airplane. We first find x by using the law of cosines.

$$\begin{aligned} x^2 &= (30.0)^2 + (250)^2 - 2(30.0)(250)\cos 120.0° \\ &= 900 + 62{,}500 - 15{,}000(-0.5) \\ &= 900 + 62{,}500 + 7500 \\ &= 70{,}900 \end{aligned}$$

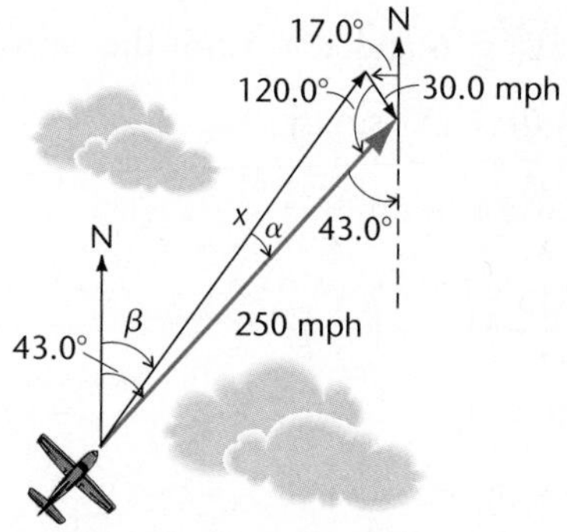

Figure 9.28 Vectors for the flight of the airplane

Thus the airspeed is

$$x = \sqrt{70{,}900}$$
$$= 266 \text{ miles per hour.}$$

rounded to the nearest mile per hour.

To find β, we first determine the angle labeled α in Figure 9.28.

$$\frac{\sin \alpha}{30.0} = \frac{\sin 120.0°}{266}$$
$$\sin \alpha = \frac{30.0 \sin 120.0°}{266}$$
$$\sin \alpha = 0.0977$$

Since α is an acute angle, we have $\alpha = 5.6°$. Then the angle β is $43.0° - 5.6° = 37.4°$, and the heading of the airplane is 37.4°. □

Example 6 illustrates the fact that velocities acting simultaneously on an object combine according to the parallelogram rule for addition of vectors.

◆ Practice Problem 1 ◆

If the airspeed of a small plane is 210 miles per hour and the wind is blowing from the north at 48 miles per hour, what heading should the pilot give his plane in order to fly on a course of 250°?

Our next example illustrates another situation in which forces combine to produce a resultant force in a manner consistent with addition of vectors.

Example 7 • Force That Holds a Barrel in Place

A barrel of oil weighing 350 pounds lies on its side on a loading ramp that makes an angle of 8.4° with the horizontal. Neglecting friction, what force parallel to the ramp is required to keep the barrel from rolling down the ramp?

Solution

As shown in Figure 9.29 on page 582, the weight of the barrel is represented by a vector $\overrightarrow{AB}$ with length 350 units and direction vertically downward. This vector can be written as the sum of a vector $\overrightarrow{AC}$ parallel to the ramp and a vector $\overrightarrow{CB}$ perpendicular to the ramp.

The triangles ABC and AOP are similar, and the angle at vertex B is 8.4°. To find the length of $\overrightarrow{AC}$, we note that

$$\sin 8.4° = \frac{\|\overrightarrow{AC}\|}{350} \quad \text{and hence} \quad \|\overrightarrow{AC}\| = 350 \sin 8.4° = 51.1,$$

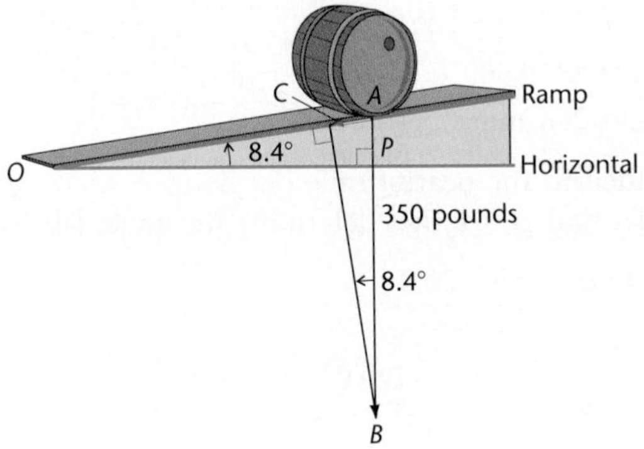

Figure 9.29 Barrel on a ramp

where the result is rounded to three digits to agree with the given data. A force of 51.1 pounds parallel to the ramp is required to hold the barrel in place. □

EXERCISES 9.4

In Exercises 1–4, sketch the vectors **U**, **V**, **U** + **V**, and **U** − **V** on the same coordinate system.

1. $\mathbf{U} = \langle 1, 2 \rangle$, $\mathbf{V} = \langle 3, -4 \rangle$
2. $\mathbf{U} = \langle -3, 2 \rangle$, $\mathbf{V} = \langle -2, -4 \rangle$
3. $\mathbf{U} = -2\mathbf{i} + 3\mathbf{j}$, $\mathbf{V} = 3\mathbf{i} - \mathbf{j}$
4. $\mathbf{U} = 2\mathbf{i} - \mathbf{j}$, $\mathbf{V} = -\mathbf{i} + 3\mathbf{j}$

In Exercises 5–8, find $2\mathbf{U} - 3\mathbf{V}$ and write your answer in both the ordered-pair notation and the **i**, **j** notation. (See Examples 2 and 4.)

5. $\mathbf{U} = \langle 1, -2 \rangle$, $\mathbf{V} = \langle -3, 5 \rangle$
6. $\mathbf{U} = \langle 0, -6 \rangle$, $\mathbf{V} = \langle 2, 7 \rangle$
7. $\mathbf{U} = 7\mathbf{i} + \mathbf{j}$, $\mathbf{V} = -5\mathbf{i} + 3\mathbf{j}$
8. $\mathbf{U} = -8\mathbf{i} + 5\mathbf{j}$, $\mathbf{V} = 7\mathbf{j}$

Find $\overrightarrow{PQ}$ and $\|\overrightarrow{PQ}\|$ for each of the following pairs P, Q. (See Examples 2 and 3.)

9. $P(-4, -1)$, $Q(-5, 3)$
10. $P(6, -3)$, $Q(1, 4)$
11. $P(-3, 0)$, $Q(-2, 4)$
12. $P(-8, 5)$, $Q(0, 7)$

In Exercises 13–20, find the magnitude and direction angle for the given vector (See Example 5.)

13. $-4\mathbf{j}$
14. $5\mathbf{j}$
15. $8\mathbf{i}$
16. $-9\mathbf{i}$
17. $\mathbf{i} - \mathbf{j}$
18. $\sqrt{3}\mathbf{i} - \mathbf{j}$
19. $-2\sqrt{2}\mathbf{i} + 2\sqrt{2}\mathbf{j}$
20. $-3\mathbf{i} - 3\mathbf{j}$

Resultant Force 21–26

21. Two forces of 25 pounds and 45 pounds act on a point in the plane. If the angle between the directions of the two forces is 120°, find the magnitude of the resultant force.

22. Two forces of 37 pounds and 13 pounds act on a point in the plane. If the angle between the directions of the two forces is 78°, find the magnitude of the resultant force.
23. A 13-pound force and a 12-pound force produce a 20-pound resultant force. Find the angle between the directions of the 13-pound force and the resultant force in the accompanying figure.

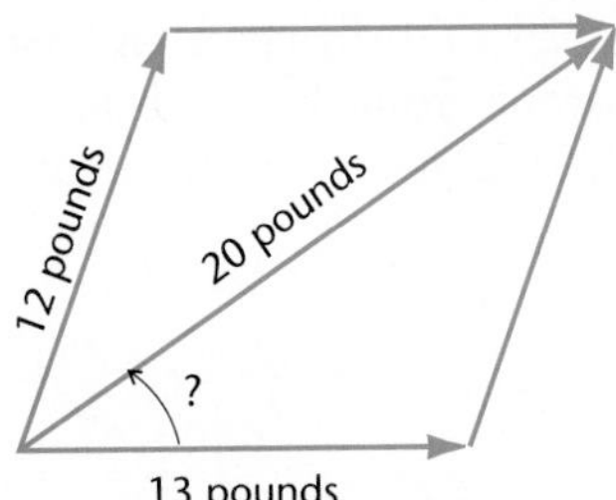

Figure for Exercise 23

24. The angle between the directions of two forces is 113.2°. One of the forces has magnitude 219 pounds, and the resultant has magnitude 242 pounds. Find the angle between the 219-pound force and the resultant force.
25. Two forces act on a point in the plane, producing a resultant force with magnitude 487 pounds. The magnitude of one of the forces is 232 pounds. Find the magnitude of the other force if the angle between the 232-pound force and the resultant force is 27.5°.
26. Two forces act on a point in the plane, producing a resultant force with magnitude 31.3 pounds. The magnitude of one of the forces is 77.1 pounds. Find the magnitude of the other force if the angle between the 77.1-pound force and the resultant force is 63.2°.

Resultant Velocity 27–29

27. Alex Jones's canoe is headed due west across a stream that is flowing south at the rate of 2.1 miles per hour. If Alex is rowing at the rate of 3.4 miles per hour in still water, find the speed and course at which the canoe is traveling. (See the accompanying figure.)

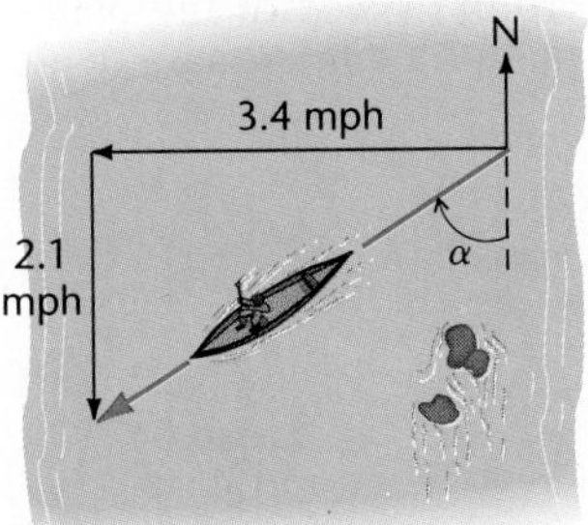

Figure for Exercise 27

28. A boat that travels at 16 miles per hour in still water wishes to travel due east across a stream that is flowing south at the rate of 2.8 miles per hour. On what heading should the boat be set?

29. A train is traveling in the direction N 41° E at 95 miles per hour, and a ball is thrown from the train at 52 miles per hour in the direction N 49° W. Find the speed and direction of the path of the ball.

Air Navigation 30–36

30. A duck heads S 20°40′ E flying at an airspeed of 63.1 miles per hour. If the wind is blowing from the north at 18.2 miles per hour, find the ground speed of the duck. (See the accompanying figure.)

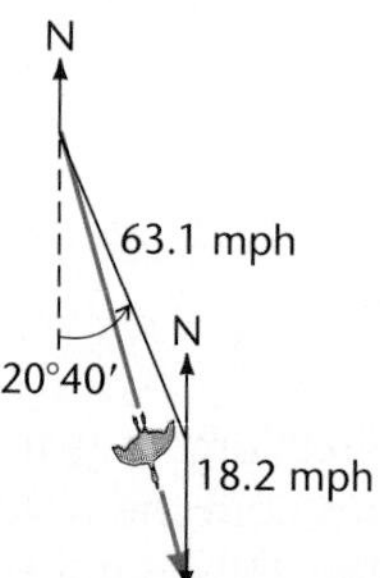

Figure for Exercise 30

31. A duck heads in the direction 159°50′ with an airspeed of 63.2 miles per hour. If the wind is blowing from the direction 49°50′ at 18.0 miles per hour, find the ground speed of the duck and the course it is traveling.

32. A plane is flying with an airspeed of 300 miles per hour and a heading of 285.0°. If the wind is blowing from the direction 230.0° at 45.0 miles per hour, find the ground speed and the course of the plane.

33. A plane is headed in the direction 128.0° with an airspeed of 350 miles per hour, and the wind is blowing from the direction 18.0°. If the course of the plane is in the direction 133.0°, find the speed of the wind.

34. The airplane in the accompanying figure is headed due east with an airspeed of 324 miles per hour in a wind blowing from the direction 200.0°. If the ground speed of the plane is 331 miles per hour, find the course of the flight and the speed of the wind.

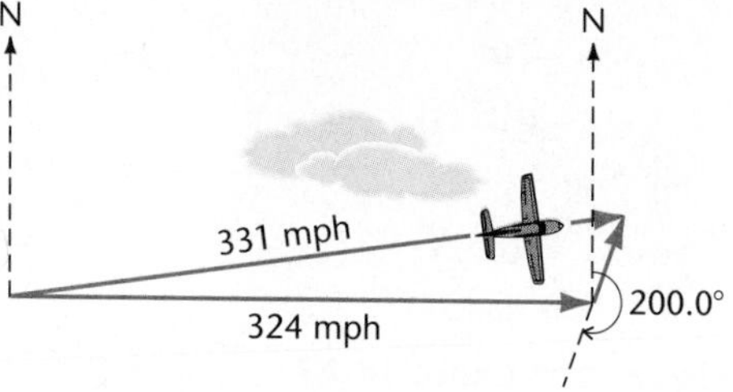

Figure for Exercise 34

35. An airplane flies with a heading of 221.0° and an airspeed of 267 miles per hour. The ground speed is 275 miles per hour, and the wind speed is 29.2 miles per hour. If the course of the plane is north of the heading, find the angle between the heading and the path of the flight.

36. An airplane flies with a heading of 167.0° and an airspeed of 318 miles per hour. The ground speed is 288 miles per hour, and the wind speed is 42.1 miles per hour. If the course of the plane is north of the heading, find the angle between the heading and the path of the flight.

Sea Navigation 37. A boat is headed N 42° E across a current flowing due south at the rate of 11 miles per hour. If the motor is driving the boat at the rate of 15 miles per hour, find the speed of the boat and the direction of its course.

Air Navigation 38. A plane is flying with an airspeed of 312 miles per hour and a heading of 287.2°. The wind is blowing from the west at 57.3 miles per hour. Find the ground speed and the course of the flight.

Gravitational Force 39–40 ✓ 39. In the accompanying figure, a car weighing 1.4 tons sits on a driveway that makes an angle of 16° with the horizontal. If friction is neglected, what force parallel to the driveway is required to hold the car in place?

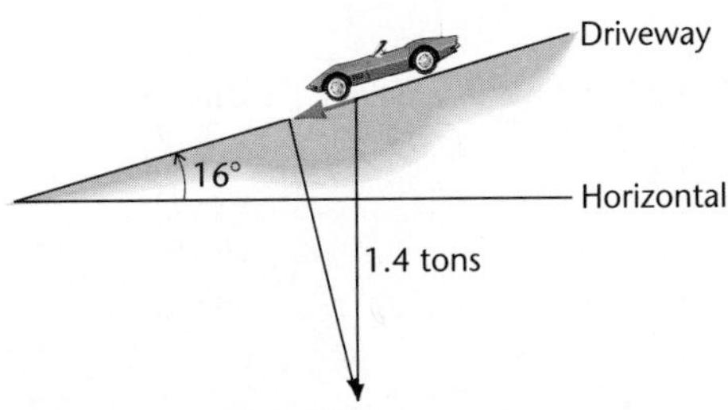

Figure for Exercise 39

40. In Exercise 39, find the magnitude of the force that the car's weight exerts perpendicular to the driveway.

41. Prove the following properties of vector addition.
   a. $\mathbf{U} + \mathbf{V} = \mathbf{V} + \mathbf{U}$
   b. $\mathbf{U} + (\mathbf{V} + \mathbf{W}) = (\mathbf{U} + \mathbf{V}) + \mathbf{W}$
   c. $\mathbf{V} + \mathbf{0} = \mathbf{V}$
   d. $\mathbf{V} + (-\mathbf{V}) = \mathbf{0}$

42. Prove the following properties of scalar multiplication.
   a. $a(b\mathbf{V}) = (ab)\mathbf{V}$
   b. $a(\mathbf{U} + \mathbf{V}) = a\mathbf{U} + a\mathbf{V}$
   c. $1\,\mathbf{V} = \mathbf{V}$
   d. $\|a\mathbf{V}\| = |a| \cdot \|\mathbf{V}\|$

43. a. Prove that $0 \cdot \mathbf{V} = \mathbf{0}$ for any $\mathbf{V}$.
    b. Prove that $a \cdot \mathbf{0} = \mathbf{0}$ for any a.
44. a. Prove that if $a\mathbf{V} = \mathbf{0}$ and $\mathbf{V} \neq \mathbf{0}$, then $a = 0$.
    b. Prove that if $a\mathbf{V} = \mathbf{0}$ and $a \neq 0$, then $\mathbf{V} = \mathbf{0}$.

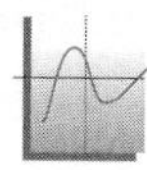

45–46

Range of a Projectile 45–46

45. In the Encore example in Chapter 8, we considered the flight of a projectile fired from the origin with an initial speed v_0 feet per second at an angle of elevation θ. If the projectile is fired uphill on an incline that makes an angle α with the horizontal, the projectile will strike the ground at a range r as shown in the figure. Using the equations for x and y from the Encore example it can be shown that

$$r = \frac{v_0^2}{16 \cos^2 \alpha} \cos \theta \sin (\theta - \alpha).$$

With $v_0 = 800$ feet per second and $\alpha = 30°$, this equation simplifies to

$$r = \frac{160{,}000}{3} \cos \theta \sin (\theta - 30°).$$

Use a graphing device to find the value of θ that yields the maximum range r up the hill. Round your answer to the nearest degree.

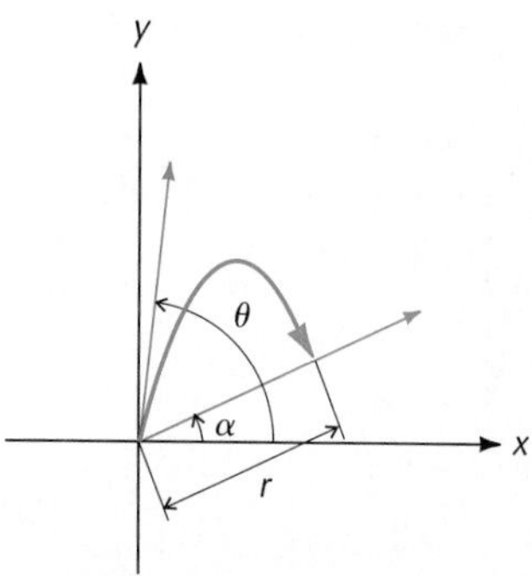

Figure for Exercise 45

46. Work Exercise 45 with $\alpha = 10°$ and all other information unchanged.

Critical Thinking: Writing

47. The **dot product** or **inner product** $\mathbf{U} \cdot \mathbf{V}$ of two vectors $\mathbf{U} = u_1\mathbf{i} + u_2\mathbf{j}$ and $\mathbf{V} = v_1\mathbf{i} + v_2\mathbf{j}$ is defined by $\mathbf{U} \cdot \mathbf{V} = u_1v_1 + u_2v_2$. Prove the following properties of the dot product.
    a. $\mathbf{U} \cdot \mathbf{V} = \mathbf{V} \cdot \mathbf{U}$
    b. $\mathbf{V} \cdot \mathbf{V} = \|\mathbf{V}\|^2$

Critical Thinking: Writing

48. (See Exercise 47.) Use the law of cosines to prove that

$$\mathbf{U} \cdot \mathbf{V} = \|\mathbf{U}\| \, \|\mathbf{V}\| \cos \theta,$$

where θ is the angle between the position vectors $\mathbf{U}$ and $\mathbf{V}$.

◆ **Solution for Practice Problem**

1. The problem is diagrammed in Figure 9.30. Using the law of sines, we have

$$\frac{\sin\theta}{48} = \frac{\sin 70^\circ}{210}$$

$$\sin\theta = \frac{48\sin 70^\circ}{210} = 0.2148.$$

Thus $\theta = 12^\circ$, to the nearest degree. The plane should be put on a heading of 262°.

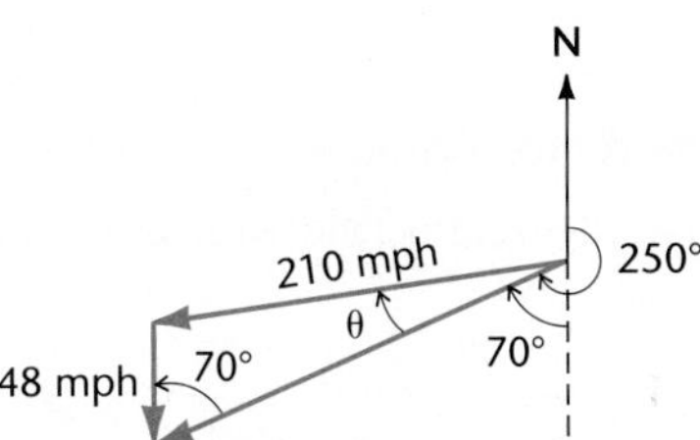

Figure 9.30 Flight of the plane

9.5 TRIGONOMETRIC FORM OF COMPLEX NUMBERS

We have seen that real numbers may be represented geometrically by the points on the real number line. It is also possible to represent complex numbers geometrically. To do this, we begin with a conventional Cartesian coordinate system in the plane. With each complex number $a + bi$ in standard form, we associate the point that has coordinates (a, b). We label this point $a + bi$ to emphasize that the point (a, b) corresponds to the complex number $a + bi$. Several complex numbers are located in Figure 9.31.

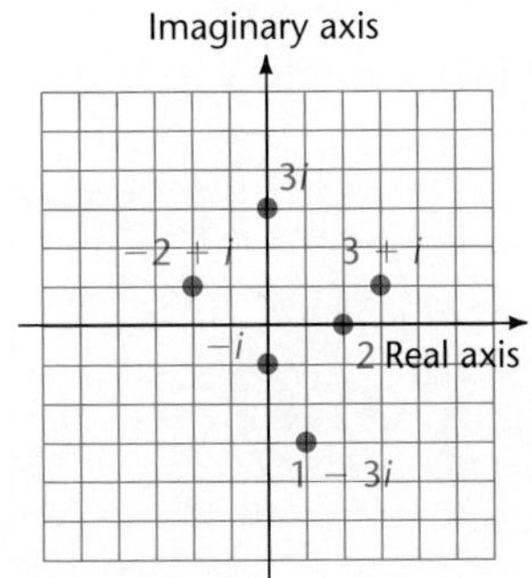

Figure 9.31 Geometric representation of complex numbers

Points on the horizontal axis correspond to the real numbers $a + 0i$, and consequently the horizontal axis is referred to as the **real axis.** Points on the vertical axis correspond to the imaginary numbers $0 + bi$, so the vertical axis is called the **imaginary axis.**

Complex numbers are sometimes represented geometrically by vectors. In this approach, the complex number $a + bi$ is represented by the vector with initial point at the origin and terminal point with coordinates (a, b), or by any other vector with the same length and direction. This is shown in Figure 9.32 on page 588.

Note that, in this interpretation, addition of complex numbers corresponds to the usual **parallelogram rule** for adding vectors. This is illustrated in Example 1.

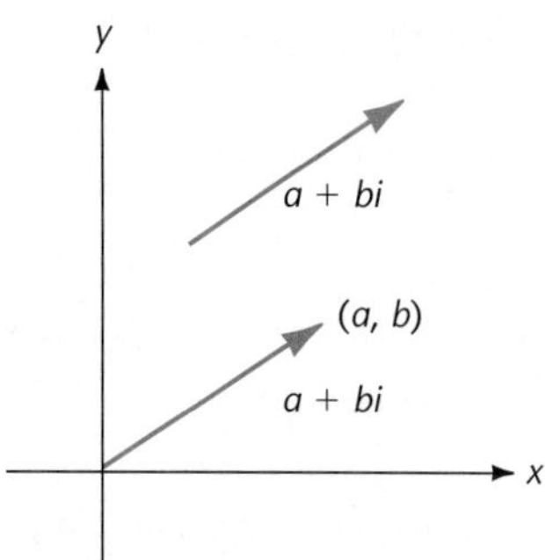

Figure 9.32 Vector representation of $a + bi$

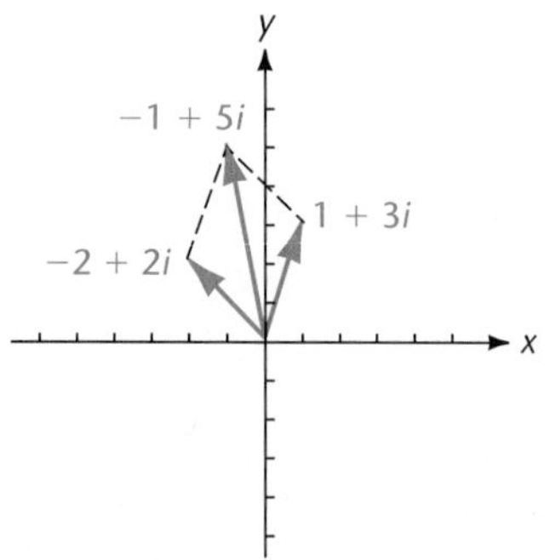

Figure 9.33 The parallelogram rule

Example 1 • Vector Addition of Complex Numbers

Illustrate the parallelogram rule with the complex numbers $1 + 3i$ and $-2 + 2i$.

Solution

In Figure 9.33, the diagonal of the parallelogram is the vector representing the sum $-1 + 5i$ of $1 + 3i$ and $-2 + 2i$. □

Any vector with initial point at the origin can be described by designating its length r and its direction θ, where we do *not* restrict θ to be between 0° and 360°. Figure 9.34 shows r and θ for a complex number $a + bi$ in standard form.

From Figure 9.34 we see that r and θ are related to a and b by the equations

$$a = r\cos\theta, \quad b = r\sin\theta, \quad r = \sqrt{a^2 + b^2}.$$

The complex number $a + bi$ can thus be written in the form

$$a + bi = r(\cos\theta + i\sin\theta).$$

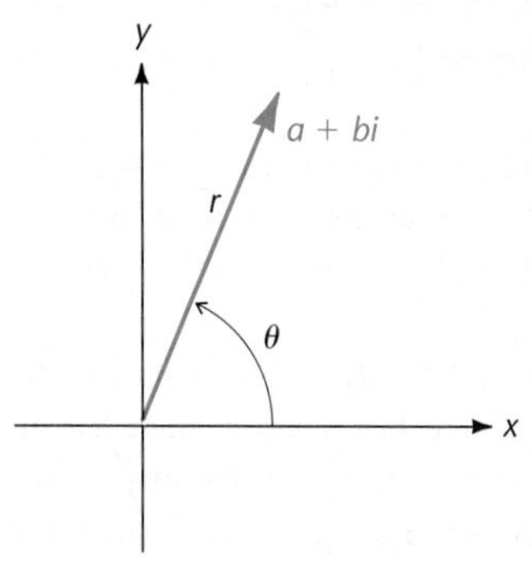

Figure 9.34 Length and direction of $a + bi$

Definition of Trigonometric Form of a Complex Number

The **trigonometric form** (or **polar form**) of the complex number $a + bi$ is†

$$r(\cos\theta + i\sin\theta),$$

where r and θ are determined by the equations

$$r = \sqrt{a^2 + b^2}, \quad a = r\cos\theta, \quad b = r\sin\theta.$$

The number r is called the **absolute value** (or **modulus**) of $a + bi$, and the angle θ is called the **argument** (or **amplitude**) of $a + bi$.

†The expression $\cos\theta + i\sin\theta$ is sometimes abbreviated as $\operatorname{cis}\theta$.

The usual absolute value notation is used for the absolute value of a complex number.

$$|a + bi| = r = \sqrt{a^2 + b^2}.$$

The absolute value r is unique, but the angle θ is not, since there are many angles in standard position that determine the same vector. We usually choose to use the smallest positive value for θ. We note that any equation of the form

$$r_1(\cos \theta_1 + i \sin \theta_1) = r_2(\cos \theta_2 + i \sin \theta_2)$$

requires that $r_1 = r_2$ and that θ_1 and θ_2 be coterminal. Hence‡

$$\theta_2 = \theta_1 + k(360°)$$

for some integer k.

Example 2 • Converting from Standard Form to Trigonometric Form

Express each complex number in trigonometric form.

a. $-1 - i$ b. $\sqrt{3} - i$ c. $-3i$

Solution

In each part of Figure 9.35, we sketch the vector representing the corresponding complex number. We then determine the values of r and θ by recognizing the special angles.

a. $-1 - i = \sqrt{2}(\cos 225° + i \sin 225°)$
b. $\sqrt{3} - i = 2(\cos 330° + i \sin 330°)$
c. $-3i = 3(\cos 270° + i \sin 270°)$ □

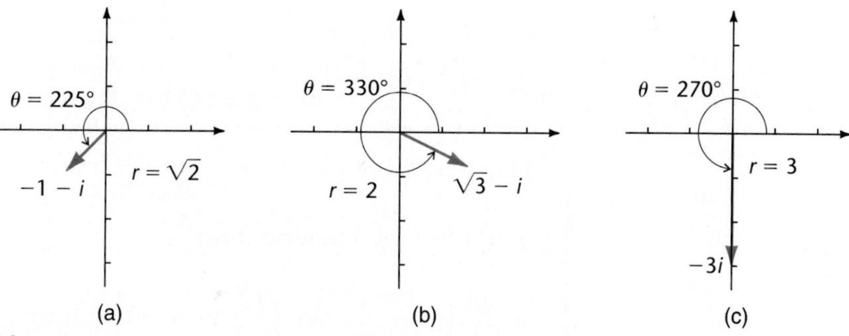

Figure 9.35 The complex numbers in Example 2

‡We use degree measure for the angle. Radian measure can be used just as well.

Example 3 • Converting from Trigonometric Form to Standard Form

Convert each of the following complex numbers from trigonometric form to standard form. Also, draw the vector representing each complex number.

a. $4(\cos 210° + i \sin 210°)$
b. $6[\cos(-45°) + i \sin(-45°)]$
c. $3(\cos \pi + i \sin \pi)$

Solution

In each part we simply evaluate the cosine and sine of the given angle and then simplify. The vectors are drawn in Figure 9.36.

a. $4(\cos 210° + i \sin 210°) = 4\left[-\frac{\sqrt{3}}{2} + i\left(-\frac{1}{2}\right)\right] = -2\sqrt{3} - 2i$

b. $6[\cos(-45°) + i \sin(-45°)] = 6\left[\frac{\sqrt{2}}{2} + i\left(-\frac{\sqrt{2}}{2}\right)\right] = 3\sqrt{2} - 3i\sqrt{2}$

c. $3(\cos \pi + i \sin \pi) = 3[(-1) + i(0)] = -3$ □

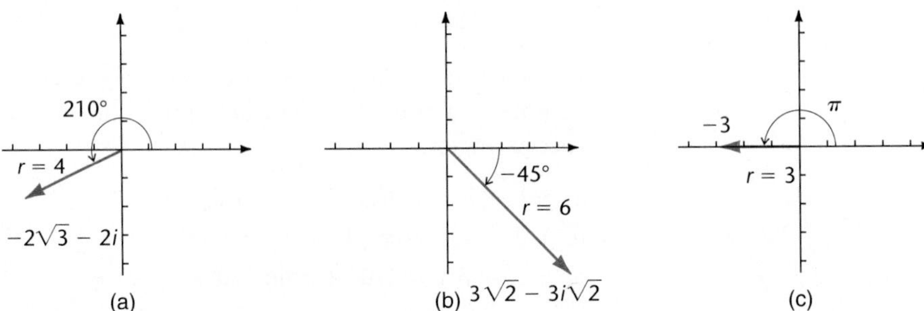

Figure 9.36 The complex numbers in Example 3

◆ **Practice Problem 1** ◆

Convert each complex number to the indicated form.

a. $\frac{1}{2}i$ to trigonometric form

b. $2\left(\cos \frac{2\pi}{3} + i \sin \frac{2\pi}{3}\right)$ to standard form

The trigonometric form for complex numbers can be used to compute products and quotients of complex numbers. Suppose z_1 and z_2 are complex numbers with trigonometric forms r_1 (cos $\theta_1 + i \sin \theta_1$) and r_2 (cos $\theta_2 + i$ sin θ_2), respectively. First, we consider the product.

$$\begin{aligned} z_1 z_2 &= r_1(\cos \theta_1 + i \sin \theta_1) \cdot r_2(\cos \theta_2 + i \sin \theta_2) \\ &= r_1 r_2[(\cos \theta_1 \cos \theta_2 - \sin \theta_1 \sin \theta_2) + i(\sin \theta_1 \cos \theta_2 + \cos \theta_1 \sin \theta_2)] \\ &= r_1 r_2[\cos(\theta_1 + \theta_2) + i \sin (\theta_1 + \theta_2)] \end{aligned}$$

The last equality follows from the identities for the sum of two angles.

Next we consider the quotient of z_1 and z_2, where $z_2 \neq 0$. This time we use the identities for the difference of two angles.

$$\begin{aligned} \frac{z_1}{z_2} &= \frac{r_1(\cos \theta_1 + i \sin \theta_1)}{r_2(\cos \theta_2 + i \sin \theta_2)} \\ &= \left(\frac{r_1}{r_2}\right) \frac{(\cos \theta_1 + i \sin \theta_1)(\cos \theta_2 - i \sin \theta_2)}{(\cos \theta_2 + i \sin \theta_2)(\cos \theta_2 - i \sin \theta_2)} \\ &= \left(\frac{r_1}{r_2}\right) \frac{(\cos \theta_1 \cos \theta_2 + \sin \theta_1 \sin \theta_2) + i(\sin \theta_1 \cos \theta_2 - \cos \theta_1 \sin \theta_2)}{\cos^2 \theta_2 - i^2 \sin^2 \theta_2} \\ &= \left(\frac{r_1}{r_2}\right) \frac{\cos (\theta_1 - \theta_2) + i \sin (\theta_1 - \theta_2)}{\cos^2 \theta_2 + \sin^2 \theta_2} \\ &= \left(\frac{r_1}{r_2}\right) [\cos (\theta_1 - \theta_2) + i \sin (\theta_1 - \theta_2)] \end{aligned}$$

We summarize these results as follows.

Product and Quotient Rules

If two complex numbers z_1 and z_2 have trigonometric forms

$$z_1 = r_1(\cos \theta_1 + i \sin \theta_1) \quad \text{and} \quad z_2 = r_2(\cos \theta_2 + i \sin \theta_2),$$

then the product is given by

$$z_1 z_2 = r_1 r_2[\cos (\theta_1 + \theta_2) + i \sin (\theta_1 + \theta_2)].$$

That is, to multiply two complex numbers, multiply their absolute values and add their arguments.

Also, if $z_2 \neq 0$, the quotient is given by

$$\frac{z_1}{z_2} = \frac{r_1}{r_2} [\cos (\theta_1 - \theta_2) + i \sin (\theta_1 - \theta_2)].$$

That is, to divide two complex numbers, divide their absolute values and subtract their arguments.

Example 4 • Using the Product and Quotient Rules

Suppose $z_1 = 12(\cos 135° + i \sin 135°)$ and $z_2 = 8(\cos 15° + i \sin 15°)$. Write the product $z_1 z_2$ and the quotient z_1/z_2 in both trigonometric form and standard form.

Solution

To multiply, we multiply the absolute values and add the arguments.

$$z_1z_2 = 12 \cdot 8[\cos(135° + 15°) + i\sin(135° + 15°)]$$
$$= \mathbf{96(\cos 150° + i\sin 150°)} \qquad \text{trigonometric form}$$
$$= 96\left(-\frac{\sqrt{3}}{2} + \frac{1}{2}i\right)$$
$$= \mathbf{-48\sqrt{3} + 48i} \qquad \text{standard form}$$

To divide, we divide absolute values and subtract the arguments.

$$\frac{z_1}{z_2} = \frac{12}{8}[\cos(135° - 15°) + i\sin(135° - 15°)]$$
$$= \mathbf{\tfrac{3}{2}(\cos 120° + i\sin 120°)} \qquad \text{trigonometric form}$$
$$= \frac{3}{2}\left(-\frac{1}{2} + \frac{\sqrt{3}}{2}i\right)$$
$$= \mathbf{-\frac{3}{4} + \frac{3\sqrt{3}}{4}i} \qquad \text{standard form}$$

We have stated each trigonometric form using the smallest positive value for the argument. □

◆ Practice Problem 2 ◆

Use the trigonometric form to find z_1z_2 and z_1/z_2, if $z_1 = -3i$ and $z_2 = 2 - 2i$. Express the results in both trigonometric form and standard form.

EXERCISES 9.5

Illustrate the parallelogram rule with each pair of complex numbers. (See Example 1.)

1. $2 + i, \quad -4 + 3i$
2. $3 - 2i, \quad 1 + 6i$
3. $-1 + 2i, \quad 3 - i$
4. $2 - 3i, \quad -4 - 2i$
5. $-2, \quad 1 + 3i$
6. $4i, \quad -3 + i$
7. $-3i, \quad 5$
8. $-4, \quad 2i$

Express each complex number in trigonometric form. In Exercises 25–28, take $r = 1$ and round the angles to the nearest tenth of a degree. (See Example 2.)

9. $-1 + i$
10. $1 + i\sqrt{3}$
11. $3 - 3i$
12. $2\sqrt{2} - 2i\sqrt{2}$
13. -4
14. $-5i$
15. $7i$
16. 3

17. $-4\sqrt{3} - 4i$

18. $-\frac{\sqrt{3}}{2} + \frac{1}{2}i$

19. $\frac{1}{4} + \frac{1}{4}i$

20. $0.8 - 0.8i$

21. $1 - i\sqrt{3}$

22. $-\sqrt{3} - i$

23. $-3\sqrt{3} + 3i$

24. $-10 + 10i$

25. $0.9033 + 0.4289i$

26. $0.6743 + 0.7385i$

27. $-0.7206 + 0.6934i$

28. $-0.6018 - 0.7986i$

Express each complex number in standard form. In Exercises 37–40, round to four decimal places. (See Example 3.)

29. $4(\cos 45° + i \sin 45°)$

30. $2(\cos 60° + i \sin 60°)$

31. $3[\cos(-120°) + i \sin(-120°)]$

32. $12(\cos 315° + i \sin 315°)$

33. $9\left(\cos \frac{3\pi}{2} + i \sin \frac{3\pi}{2}\right)$

34. $2\left[\cos\left(-\frac{\pi}{2}\right) + i \sin\left(-\frac{\pi}{2}\right)\right]$

35. $\sqrt{2}\left(\cos \frac{5\pi}{4} + i \sin \frac{5\pi}{4}\right)$

36. $\sqrt{3}\left(\cos \frac{11\pi}{6} + i \sin \frac{11\pi}{6}\right)$

37. $3(\cos 27° + i \sin 27°)$

38. $5(\cos 114° + i \sin 114°)$

39. $7(\cos 312° + i \sin 312°)$

40. $2(\cos 200° + i \sin 200°)$

Write the exact results of the indicated operations in standard form. (See Example 4.)

41. $12(\cos 93° + i \sin 93°) \cdot 2(\cos 27° + i \sin 27°)$

42. $2(\cos 47° + i \sin 47°) \cdot 8(\cos 43° + i \sin 43°)$

43. $(\cos 142° + i \sin 142°) \cdot (\cos 38° + i \sin 38°)$

44. $(\cos 182° + i \sin 182°) \cdot (\cos 133° + i \sin 133°)$

45. $3(\cos 324° + i \sin 324°) \cdot 2(\cos 396° + i \sin 396°)$

46. $10(\cos 421° + i \sin 421°) \cdot 8(\cos 119° + i \sin 119°)$

47. $8[\cos(-18°) + i \sin(-18°)] \cdot 2(\cos 138° + i \sin 138°)$

48. $(\cos 93° + i \sin 93°) \cdot 4[\cos(-153°) + i \sin(-153°)]$

49. $\frac{12(\cos 47° + i \sin 47°)}{6(\cos 2° + i \sin 2°)}$

50. $\frac{3(\cos 278° + i \sin 278°)}{2(\cos 38° + i \sin 38°)}$

51. $\frac{8(\cos 192° + i \sin 192°)}{4(\cos 102° + i \sin 102°)}$

52. $\frac{5(\cos 560° + i \sin 560°)}{4(\cos 260° + i \sin 260°)}$

53. $\frac{\cos 62° + i \sin 62°}{\cos 107° + i \sin 107°}$

54. $\frac{\cos 48° + i \sin 48°}{\cos 138° + i \sin 138°}$

55. $\frac{\cos(-20°) + i \sin(-20°)}{\cos 130° + i \sin 130°}$

56. $\frac{\cos(-142°) + i \sin(-142°)}{\cos 218° + i \sin 218°}$

Use the trigonometric forms to find the exact values of z_1z_2 and z_1/z_2. Write the results in both trigonometric form and standard form. (See Example 4.)

57. $z_1 = 4\left(\cos\frac{\pi}{8} + i\sin\frac{\pi}{8}\right), \quad z_2 = \cos\frac{5\pi}{8} + i\sin\frac{5\pi}{8}$

58. $z_1 = 3\left(\cos\frac{7\pi}{6} + i\sin\frac{7\pi}{6}\right), \quad z_2 = \cos\frac{2\pi}{3} + i\sin\frac{2\pi}{3}$

✓59. $z_1 = 2\left(\cos\frac{5\pi}{6} + i\sin\frac{5\pi}{6}\right), \quad z_2 = 3\left[\cos\left(-\frac{\pi}{6}\right) + i\sin\left(-\frac{\pi}{6}\right)\right]$

60. $z_1 = 6\left(\cos\frac{5\pi}{3} + i\sin\frac{5\pi}{3}\right), \quad z_2 = 5\left[\cos\left(-\frac{\pi}{3}\right) + i\sin\left(-\frac{\pi}{3}\right)\right]$

✓61. $z_1 = \sqrt{3} + i, \quad z_2 = -1 + i\sqrt{3}$

62. $z_1 = 1 - i\sqrt{3}, \quad z_2 = -2\sqrt{3} + 2i$

63. $z_1 = -1 - i, \quad z_2 = 2 - 2i$

64. $z_1 = -2 + 2i, \quad z_2 = 4 + 4i$

65. $z_1 = 2, \quad z_2 = 5 + 5i\sqrt{3}$

66. $z_1 = 5i, \quad z_2 = -\sqrt{3} + i$

67. $z_1 = -3i, \quad z_2 = -2 - 2i$

68. $z_1 = -8, \quad z_2 = 4 - 4i$

Use the trigonometric form to prove each of the following statements, where z, z_1, and z_2 are arbitrary complex numbers.

69. $|\bar{z}| = |z|$

70. $z\bar{z} = |z|^2$

71. $|z_1z_2| = |z_1||z_2|$

72. If $z_2 \neq 0$, $\left|\frac{z_1}{z_2}\right| = \frac{|z_1|}{|z_2|}$

Critical Thinking: Exploration and Writing

73. Let α be a constant angle and describe the geometric effect on an arbitrary complex number z when z is multiplied by $\cos\alpha + i\sin\alpha$.

Critical Thinking: Exploration and Writing

74. (See Exercise 73.) Describe the geometric effect when z is multiplied by the conjugate $\cos\alpha - i\sin\alpha$.

Critical Thinking: Exploration and Writing

75. In the nineteenth century, an Irish mathematician named William Rowan Hamilton was able to extend the complex numbers to a larger set called the **quaternions.** Use a library resource and write a short article that describes the quaternions.

◆ Solutions for Practice Problems

1. a. From the sketch in Figure 9.37 we see that $\theta = 90°$ and $r = \frac{1}{2}$. Thus

$$\frac{1}{2}i = \frac{1}{2}(\cos 90° + i\sin 90°).$$

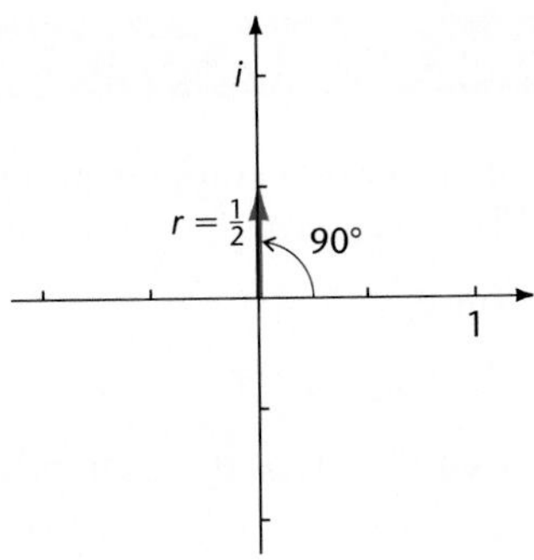

Figure 9.37 The complex number $\frac{1}{2}i$

b. Evaluating the trigonometric functions and simplifying, we have

$$2\left(\cos\frac{2\pi}{3} + i\sin\frac{2\pi}{3}\right) = 2\left(-\frac{1}{2} + i\frac{\sqrt{3}}{2}\right) = -1 + i\sqrt{3}.$$

2. We first write z_1 and z_2 in trigonometric form.

$$z_1 = 3(\cos 270° + i\sin 270°), \qquad z_2 = 2\sqrt{2}(\cos 315° + i\sin 315°)$$

Then, using the rule for the product, we obtain

$$\begin{aligned} z_1z_2 &= 3 \cdot 2\sqrt{2}[\cos(270° + 315°) + i\sin(270° + 315°)] \\ &= 6\sqrt{2}(\cos 585° + i\sin 585°) \\ &= 6\sqrt{2}(\cos 225° + i\sin 225°) && \text{trigonometric form} \\ &= 6\sqrt{2}\left(-\frac{1}{\sqrt{2}} - i\frac{1}{\sqrt{2}}\right) \\ &= -6 - 6i. && \text{standard form} \end{aligned}$$

Using the quotient rule, we obtain

$$\begin{aligned} \frac{z_1}{z_2} &= \frac{3}{2\sqrt{2}}[\cos(270° - 315°) + i\sin(270° - 315°)] \\ &= \frac{3}{2\sqrt{2}}[\cos(-45°) + i\sin(-45°)] \\ &= \frac{3}{2\sqrt{2}}(\cos 315° + i\sin 315°) && \text{trigonometric form} \\ &= \frac{3}{2\sqrt{2}}\left(\frac{\sqrt{2}}{2} - i\frac{\sqrt{2}}{2}\right) \\ &= \frac{3}{4} - \frac{3}{4}i. && \text{standard form} \end{aligned}$$

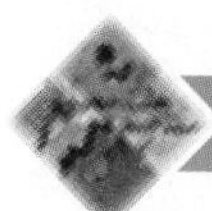

9.6 POWERS AND ROOTS OF COMPLEX NUMBERS

Any positive integral power of a complex number z written in trigonometric form can be determined by repeated application of the product rule. If

$$z = r(\cos\theta + i\sin\theta),$$

we have

$$\begin{aligned} z^2 &= r(\cos\theta + i\sin\theta)\cdot r(\cos\theta + i\sin\theta) \\ &= r^2(\cos 2\theta + i\sin 2\theta). \end{aligned}$$

Also,

$$\begin{aligned} z^3 &= z^2\cdot z \\ &= r^2(\cos 2\theta + i\sin 2\theta)\cdot r(\cos\theta + i\sin\theta) \\ &= r^3(\cos 3\theta + i\sin 3\theta). \end{aligned}$$

Similarly,

$$\begin{aligned} z^4 &= r^4(\cos 4\theta + i\sin 4\theta) \\ z^5 &= r^5(\cos 5\theta + i\sin 5\theta), \end{aligned}$$

and so on. In general, we have the next result, which begins to reveal the true usefulness of the trigonometric form.

DeMoivre's Theorem

If the complex number z has the trigonometric form

$$z = r(\cos\theta + i\sin\theta)$$

and n is any integer, then

$$z^n = r^n(\cos n\theta + i\sin n\theta).$$

Example 1 • Using DeMoivre's Theorem

Use DeMoivre's theorem to write $(-1 - i)^{10}$ in standard form.

Solution

We first write $-1 - i$ in trigonometric form (see Example 2 in Section 9.5) and then apply DeMoivre's theorem.

$$\begin{aligned} (-1 - i)^{10} &= [\sqrt{2}(\cos 225^\circ + i\sin 225^\circ)]^{10} \\ &= (\sqrt{2})^{10}[\cos(10\cdot 225^\circ) + i\sin(10\cdot 225^\circ)] \\ &= 32(\cos 2250^\circ + i\sin 2250^\circ) \\ &= 32[\cos(90^\circ + 6\cdot 360^\circ) + i\sin(90^\circ + 6\cdot 360^\circ)] \\ &= 32(\cos 90^\circ + i\sin 90^\circ) && \text{trigonometric form} \\ &= 32(0 + i) \\ &= 32i && \text{standard form} \end{aligned}$$

□

◆ Practice Problem 1 ◆

Use DeMoivre's theorem to write $(-\sqrt{3} + i)^6$ in standard form.

If n is a positive integer greater than 1 and if $u^n = z$ for complex numbers u and z, then u is called an nth root of z. There are in fact exactly n nth roots of any nonzero complex number z.

Suppose that the trigonometric forms of z and u are

$$z = r(\cos\theta + i\sin\theta) \qquad \text{and} \qquad u = s(\cos\omega + i\sin\omega).$$

Whenever $u^n = z$, we have

$$[s(\cos\omega + i\sin\omega)]^n = r(\cos\theta + i\sin\theta).$$

Applying DeMoivre's theorem to the left-hand side yields

$$s^n(\cos n\omega + i\sin n\omega) = r(\cos\theta + i\sin\theta).$$

This statement of equality of complex numbers requires that the absolute values be equal and that the arguments be coterminal. That is,

$$s^n = r \qquad \text{and} \qquad n\omega = \theta + k\cdot 360°,$$

for some integer k. Hence u can be written as

$$u = \sqrt[n]{r}\left[\cos\left(\frac{\theta + k\cdot 360°}{n}\right) + i\sin\left(\frac{\theta + k\cdot 360°}{n}\right)\right].$$

If we use any n consecutive values of k, we obtain n distinct nth roots of z. For convenience we use $k = 0, 1, 2, \ldots, n-1$.

Roots of Complex Numbers

If z is a nonzero complex number with trigonometric form

$$z = r(\cos\theta + i\sin\theta).$$

then the nth roots of z are given by

$$\sqrt[n]{r}\left[\cos\left(\frac{\theta + k\cdot 360°}{n}\right) + i\sin\left(\frac{\theta + k\cdot 360°}{n}\right)\right],$$

for $k = 0, 1, 2, \ldots, n-1$.

When the n angles

$$\frac{\theta}{n}, \quad \frac{\theta + 360°}{n}, \quad \frac{\theta + 2\cdot 360°}{n}, \quad \ldots, \quad \frac{\theta + (n-1)360°}{n}$$

are placed in standard position, their terminal sides are equally spaced $360°/n$ apart around a circle. Each nth root of z has absolute value $\sqrt[n]{r}$. Therefore the n vectors representing the n nth roots of z divide a circle of radius $\sqrt[n]{r}$ into n equal parts. This is illustrated in the next example.

Example 2 • Finding Cube Roots

Find the three cube roots of $64i$ and draw the vectors representing them.

Solution

We first write $64i$ in trigonometric form.

$$64i = 64(\cos 90° + i \sin 90°)$$

According to our formula, the three cube roots of $64i$ are

$$\sqrt[3]{64}\left[\cos\left(\frac{90° + k \cdot 360°}{3}\right) + i \sin\left(\frac{90° + k \cdot 360°}{3}\right)\right]$$
$$= 4[\cos(30° + k \cdot 120°) + i \sin(30° + k \cdot 120°)], \qquad k = 0, 1, 2.$$

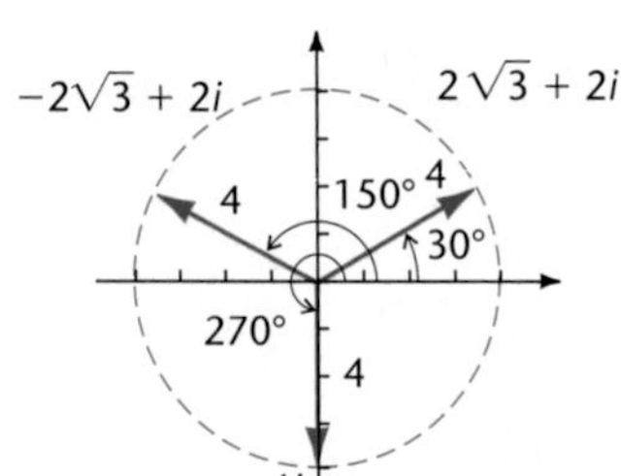

Figure 9.38 The three cube roots of $64i$

Using the three values of k, we obtain the following three cube roots of $64i$, each written in both trigonometric form and standard form.

	Trigonometric Form		Standard Form
$k = 0,$	$4(\cos 30° + i \sin 30°)$	$=$	$2\sqrt{3} + 2i$
$k = 1,$	$4(\cos 150° + i \sin 150°)$	$=$	$-2\sqrt{3} + 2i$
$k = 2,$	$4(\cos 270° + i \sin 270°)$	$=$	$-4i$

In Figure 9.38, the vectors representing the three cube roots of $64i$ are spaced 120° apart in the circle of radius 4 units. □

◆ Practice Problem 2 ◆

Write the four fourth roots of $-\frac{1}{2} - \frac{\sqrt{3}}{2}i$ in standard form and draw the vectors representing them.

EXERCISES 9.6

Use DeMoivre's theorem to find the exact value of each of the following. Leave your answers in trigonometric form. (See Example 1.)

1. $(\cos 42° + i \sin 42°)^3$
2. $(\cos 68° + i \sin 68°)^4$
3. $[2(\cos 12° + i \sin 12°)]^9$
4. $[\sqrt{2}(\cos 3° + i \sin 3°)]^{16}$
5. $\left(\cos \frac{\pi}{8} + i \sin \frac{\pi}{8}\right)^{15}$
6. $\left(\cos \frac{\pi}{10} + i \sin \frac{\pi}{10}\right)^{16}$
7. $\left[\sqrt{3}\left(\cos \frac{2\pi}{5} + i \sin \frac{2\pi}{5}\right)\right]^4$
8. $\left[\sqrt{7}\left(\cos \frac{2\pi}{7} + i \sin \frac{2\pi}{7}\right)\right]^6$

Use DeMoivre's theorem to evaluate and write the result in standard form. (See Example 1.)

9. $\left(\frac{\sqrt{3}}{2} + \frac{1}{2}i\right)^7$

10. $\left(\frac{1}{2} + \frac{\sqrt{3}}{2}i\right)^5$

√ 11. $\left(-\frac{1}{2} + \frac{\sqrt{3}}{2}i\right)^{18}$

12. $\left(\frac{\sqrt{3}}{2} - \frac{1}{2}i\right)^{21}$

13. $\left(-\frac{\sqrt{2}}{2} - \frac{\sqrt{2}}{2}i\right)^6$

14. $\left(\frac{\sqrt{2}}{2} - \frac{\sqrt{2}}{2}i\right)^7$

15. $(1 - i\sqrt{3})^8$

16. $(-\sqrt{3} + i)^9$

17. $(\sqrt{2} + i\sqrt{2})^{10}$

18. $(-\sqrt{2} + i\sqrt{2})^{12}$

19. $(2 - 2i\sqrt{3})^5$

20. $(-2\sqrt{3} - 2i)^4$

√ 21. $(-1 + i)^8$

22. $(\sqrt{3} - i)^6$

23. $(5 + 5i)^3$

24. $(3 - 3i)^4$

Find the indicated roots and write the results in standard form. Also, draw the vectors representing them. (See Example 2.)

25. Cube roots of 1

26. Fourth roots of 1

27. Fourth roots of -1

28. Sixth roots of -1

√ 29. Cube roots of $-i$

30. Square roots of i

31. Eighth roots of 256

32. Cube roots of -8

33. Square roots of $-64i$

34. Sixth roots of 64

√ 35. Fourth roots of $-8 + 8i\sqrt{3}$

36. Fourth roots of $-8 - 8i\sqrt{3}$

Find the indicated roots and write the results in trigonometric form. (See Example 2.)

37. Sixth roots of $-64i$

38. Fifth roots of $-32i$

39. Cube roots of $\frac{\sqrt{3}}{2} + \frac{1}{2}i$

40. Cube roots of $-\frac{\sqrt{3}}{2} + \frac{1}{2}i$

41. Fourth roots of $\frac{1}{2} - \frac{\sqrt{3}}{2}i$

42. Sixth roots of $-\frac{1}{2} - \frac{\sqrt{3}}{2}i$

√ 43. Fifth roots of $-\frac{\sqrt{2}}{2} - \frac{\sqrt{2}}{2}i$

44. Cube roots of $-\frac{\sqrt{2}}{2} + \frac{\sqrt{2}}{2}i$

45. Fifth roots of $16\sqrt{2} - 16i\sqrt{2}$

46. Sixth roots of $32\sqrt{3} - 32i$

47. Cube roots of $4\sqrt{2} - 4i\sqrt{2}$

48. Fifth roots of $16\sqrt{2} + 16i\sqrt{2}$

Solve each equation and write the solutions in standard form.

49. $x^3 + 27 = 0$

50. $x^8 - 1 = 0$

51. $x^3 - i = 0$

52. $x^3 + 8i = 0$

53. $x^6 + 64 = 0$

54. $x^3 + 64i = 0$

55. $x^4 + \frac{1}{2} - \frac{\sqrt{3}}{2}i = 0$

56. $x^4 + \frac{1}{2} + \frac{\sqrt{3}}{2}i = 0$

Solve the equations and write the solutions in trigonometric form.

57. $x^5 - 32i = 0$

58. $x^5 - i = 0$

59. $x^3 - \left(\frac{1}{2} + \frac{\sqrt{3}}{2}i\right) = 0$

60. $x^4 + \frac{\sqrt{3}}{2} + \frac{1}{2}i = 0$

Critical Thinking: Exploration and Writing

61. In some advanced mathematics courses, the familiar functions in this course are extended so that they become functions of a complex variable instead of a real variable. As an example, the exponential function e^x is extended by defining

$$e^{a+bi} = e^a(\cos b + i \sin b)$$

for any complex number $a + bi$ in standard form. Verify that

$$e^{z_1} \cdot e^{z_2} = e^{z_1+z_2}$$

for arbitrary complex numbers $z_1 = a_1 + b_1 i$ and $z_2 = a_2 + b_2 i$.

Critical Thinking: Exploration and Writing

62. Using the notation in Exercise 61, verify that

$$\frac{e^{z_1}}{e^{z_2}} = e^{z_1 - z_2}.$$

Critical Thinking: Exploration and Writing

63. (See Exercise 61.) Verify that

$$e^{\pi i} = -1.$$

This equation is famous in mathematics because it involves four of the most important numbers of mathematics (e, π, i, and 1) all in one equation.

◆ Solutions for Practice Problems

1. $$\begin{aligned}(-\sqrt{3} + i)^6 &= [2(\cos 150° + i \sin 150°)]^6 \\ &= 2^6[\cos(6 \cdot 150°) + i \sin (6 \cdot 150°)] \\ &= 64(\cos 900° + i \sin 900°) \\ &= 64(\cos 180° + i \sin 180°) \\ &= -64\end{aligned}$$

2. $-\frac{1}{2} - \frac{\sqrt{3}}{2}i = \cos 240° + i \sin 240°$

The four fourth roots are given by

$$\cos\left(\frac{240° + k \cdot 360°}{4}\right) + i \sin\left(\frac{240° + k \cdot 360°}{4}\right)$$
$$= \cos (60° + k \cdot 90°) + i \sin (60° + k \cdot 90°), \qquad k = 0, 1, 2, 3.$$

$$k = 0, \qquad \cos 60° + i \sin 60° = \frac{1}{2} + \frac{\sqrt{3}}{2} i$$

$$k = 1, \qquad \cos 150° + i \sin 150° = -\frac{\sqrt{3}}{2} + \frac{1}{2} i$$

$$k = 2, \qquad \cos 240° + i \sin 240° = -\frac{1}{2} - \frac{\sqrt{3}}{2} i$$

$$k = 3, \qquad \cos 330° + i \sin 330° = \frac{\sqrt{3}}{2} - \frac{1}{2} i$$

The vectors representing the four fourth roots of $-\frac{1}{2} - \frac{\sqrt{3}}{2} i$ are drawn in Figure 9.39.

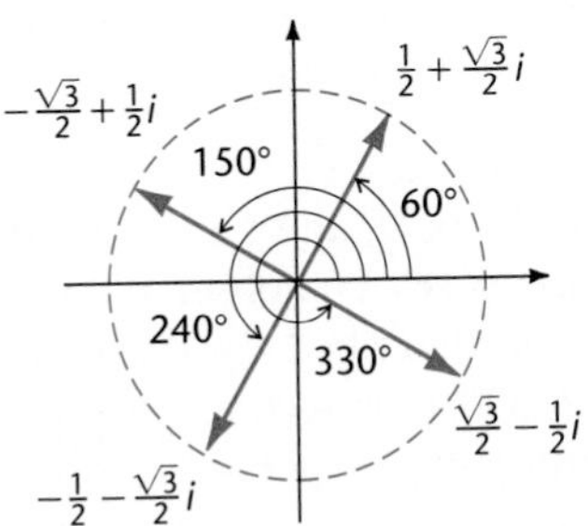

Figure 9.39 The four fourth roots of $-\frac{1}{2} - \frac{\sqrt{3}}{2} i$

9.7 POLAR COORDINATES

We use the Cartesian (rectangular) coordinate system to locate points that correspond to ordered pairs of the form (x, y). However, another coordinate system is often more convenient to use. It is called the **polar coordinate system.** Points in this system are located by specifying a distance r from a fixed point, called the **pole,** and the amount of rotation θ from a fixed ray, called the **polar axis.** These two quantities are specified in an ordered-pair format (r, θ), where r is the distance from the pole and θ is the amount of rotation from the polar axis. The ordered pair (r, θ) is called the **polar coordinates** of the point. We note that θ may be specified in either degree or radian measure.

If a Cartesian coordinate system were superimposed on a polar coordinate system, the positive x-axis would coincide with the polar axis. Hence θ can be thought of as an angle in standard position. As before, positive angles are rotated counterclockwise, and negative angles clockwise. In the polar coordinate system in Figure 9.40 on page 602, we locate several points using polar coordinates.

We also allow the possibility that r might be negative. For example, to plot a point with polar coordinates $(-2, 60°)$, we rotate 60° from the polar axis, extend

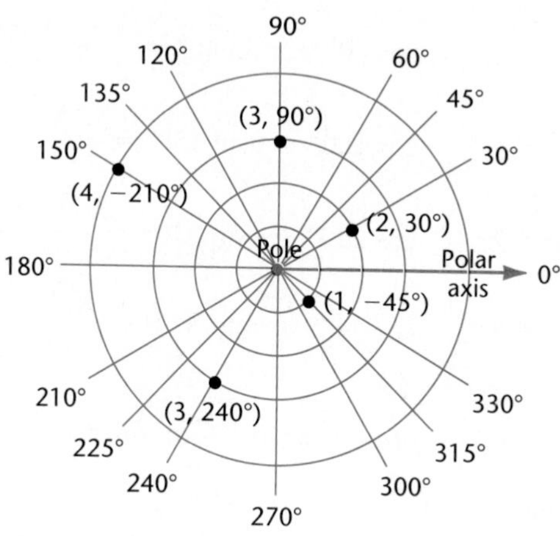

Figure 9.40 Points in a polar coordinate system

the 60° ray through the pole, and fix the point 2 units along the extension. This is illustrated in Figure 9.41. We observe that the point with polar coordinates $(-2, 60°)$ can also be described by the polar coordinates $(2, 240°)$, $(2, -120°)$, $(-2, -300°)$, or $(2, 600°)$, as well as by many others. Hence a given point may correspond to many pairs of polar coordinates; however, a given pair of polar coordinates determines one and only one point.

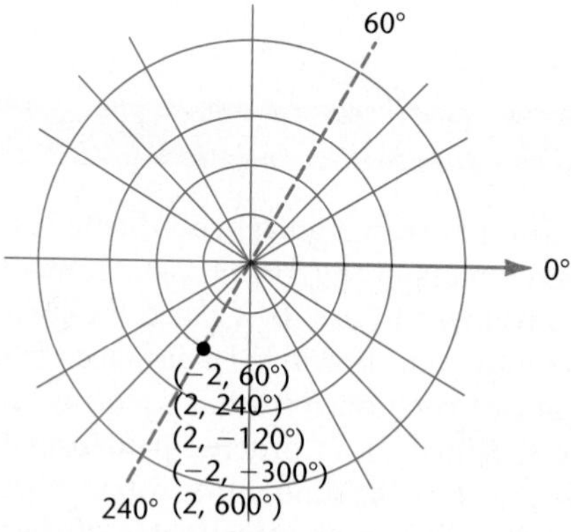

Figure 9.41 The polar coordinates of a point

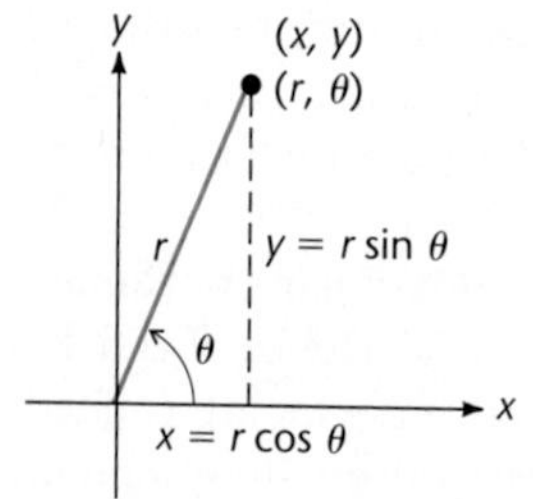

Figure 9.42 Relationship between (x, y) and (r, θ)

The relationship between the polar coordinates (r, θ) and the rectangular coordinates (x, y) of a point agrees with the relationship between the r, θ of the trigonometric form and the a, b of the standard form of a complex number. This is illustrated in Figure 9.42.

Relationship between Rectangular and Polar Coordinates

If (r, θ) are the polar coordinates of a point with rectangular coordinates (x, y), then

$$r^2 = x^2 + y^2, \quad x = r\cos\theta, \quad y = r\sin\theta.$$

Example 1 • Converting from Polar to Rectangular Coordinates

Determine the rectangular coordinates for the point with polar coordinates $(3, 150°)$.

Solution

We have $r = 3$ and $\theta = 150°$. Thus x and y are given by

$$x = 3\cos 150° = 3\left(-\frac{\sqrt{3}}{2}\right) = \frac{-3\sqrt{3}}{2} \qquad \text{and} \qquad y = 3\sin 150° = 3\left(\frac{1}{2}\right) = \frac{3}{2}.$$

The rectangular coordinates are $\left(-\frac{3\sqrt{3}}{2}, \frac{3}{2}\right)$. □

Example 2 • Converting from Rectangular to Polar Coordinates

Suppose a point has $(-2, -2)$ as its rectangular coordinates. Determine the cooresponding polar coordinates, where $0° \le \theta < 360°$.

Solution

We are given $x = -2$ and $y = -2$. Then $r = \sqrt{(-2)^2 + (-2)^2} = \sqrt{8} = 2\sqrt{2}$. The angle θ satisfies the equations

$$-2 = 2\sqrt{2}\cos\theta \qquad \text{and} \qquad -2 = 2\sqrt{2}\sin\theta$$

$$-\frac{1}{\sqrt{2}} = \cos\theta \qquad\qquad -\frac{1}{\sqrt{2}} = \sin\theta.$$

Thus $\theta = 225°$, and the polar coordinates are $(2\sqrt{2}, 225°)$. □

An equation expressed in terms of the rectangular coordinates (x, y) can be transformed into an equation in terms of the polar coordinates (r, θ), and vice versa. An equation in polar coordinates is called a **polar equation.**

Example 3 • Converting a Cartesian Equation to a Polar Equation

Write the equation $x^2 + y^2 = 4y$ in terms of polar coordinates (r, θ).

Solution

Since $x^2 + y^2 = r^2$ and $y = r\sin\theta$, the equation

$$x^2 + y^2 = 4y$$

becomes

$$r^2 = 4r\sin\theta.$$

The factor r can be divided out and still have an equivalent equation because the pole is on the graph of

$$r = 4\sin\theta.$$

□

Suppose we consider the graphs of the equation $x^2 + y^2 = 4$ and the corresponding polar coordinate equation $r^2 = 4$. The equation $x^2 + y^2 = 4$ represents the circle in Figure 9.43a, with center $(0, 0)$ and radius 2. The equation $r^2 = 4$ is equivalent to $r = \pm 2$. The graph of $r = \pm 2$ is the set of all points $(\pm 2, \theta)$, where θ is arbitrary. These points lie on the circle in Figure 9.43b, with center at the pole and radius 2. Hence we obtain the same graph whether we use rectangular or polar coordinates.

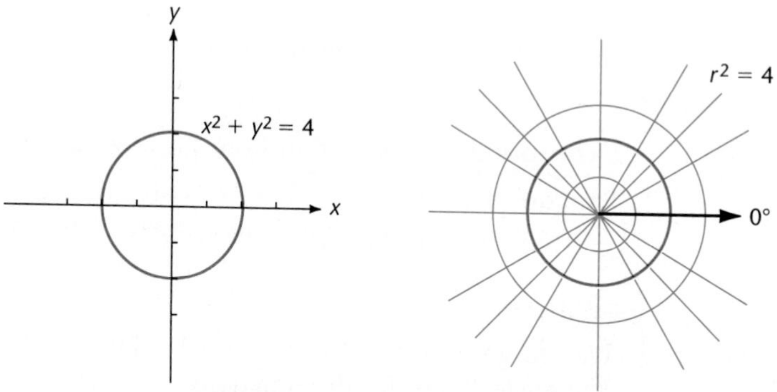

Figure 9.43 The circle centered at the origin with radius 2

Example 4 • Converting a Polar Equation to a Cartesian Equation

Convert $r\cos\theta + r\sin\theta = 2$ to an equation in rectangular coordinates.

Solution

Since $x = r\cos\theta$ and $y = r\sin\theta$, the equation

$$r\cos\theta + r\sin\theta = 2$$

becomes

$$x + y = 2.$$

□

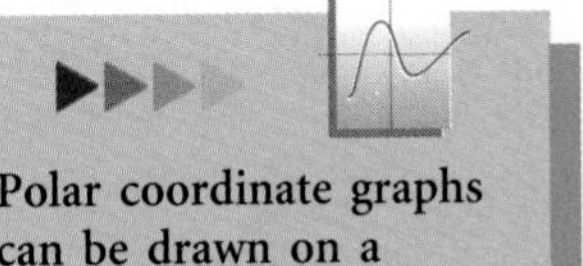

Polar coordinate graphs can be drawn on a graphing device by using the appropriate mode and features.

Many polar equations have special types of curves as their graphs. We examine a few of these next.

Example 5 • Graphing a Cardioid

Sketch the graph of the equation $r = 2 + 2\cos\theta$.

Solution

The points with coordinates $(r, \theta) = (2 + 2\cos\theta, \theta)$ lie on the graph of the equation. We can locate some of these points by choosing values of θ and computing corresponding values of $r = 2 + 2\cos\theta$. Such a tabulation is shown in Figure 9.44 for several special angles between 0° and 180°. These points all lie on the upper portion of the graph. Since $\cos(-\theta) = \cos\theta$, whenever (r, θ) lies on the graph, so does $(r, -\theta)$. Using $-\theta$ in place of θ yields points on the lower portion of the graph. The resulting curve is called a **cardioid.**

θ	0°	30°	45°	60°	90°	120°	135°	150°	180°
$2 + 2\cos\theta$	4	3.7	3.4	3	2	1	0.6	0.3	0
(r, θ)	(4, 0°)	(3.7, 30°)	(3.4, 45°)	(3, 60°)	(2, 90°)	(1, 120°)	(0.6, 135°)	(0.3, 150°)	(0, 180°)

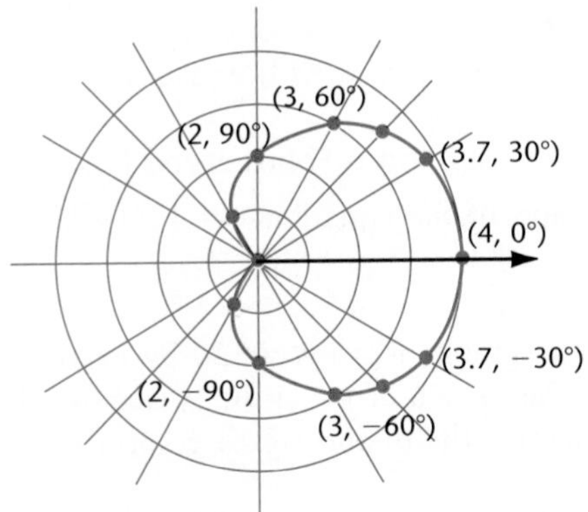

Figure 9.44 The cardioid $r = 2 + 2\cos\theta$

Alternate Solution

Graphs of this type might be analyzed more easily by examining the graphs of their rectangular counterparts; in this case $y = 2 + 2\cos x$. A typical analysis might resemble the following.

First we sketch the graph of $y = 2 + 2\cos x$ and then, as in Figure 9.45a, we rename the x-axis as the θ-axis† and the y-axis as the r-axis. We separate this graph into four parts determined by the key points of the cosine curve.

†We are switching to radian measure of θ since in Chapter 7 the x-coordinates represented real numbers.

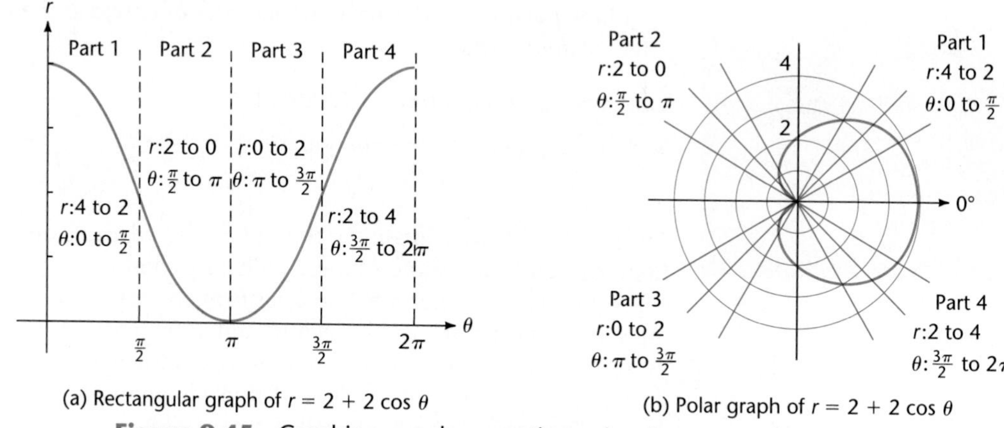

(a) Rectangular graph of $r = 2 + 2\cos\theta$ (b) Polar graph of $r = 2 + 2\cos\theta$

Figure 9.45 Graphing a polar equation using its rectangular counterpart

In Part 1 of the graph we observe that as θ ranges from 0 to $\pi/2$, the corresponding r-values decrease from 4 to 2. We transfer this information to part 1 of the polar graph in Figure 9.45b. Similarly, in part 2, as θ ranges from $\pi/2$ to π, the r-values decrease from 2 to 0. This information is transferred to part 2 of the polar graph. Continuing in part 3, as θ ranges from π to $3\pi/2$, r increases from 0 to 2. This provides sufficient information to sketch part 3 of the polar graph. In part 4, r increases from 2 to 4 as θ moves from $3\pi/2$ to 2π. This completes the graph in Figure 9.45b of the cardioid with equation $r = 2 + 2\cos\theta$. □

Example 6 • Graphing a Four-Leaved Rose

Sketch the graph of the equation $r = 4\sin 2\theta$.

Solution

We analyze the graph of $r = 4\sin 2\theta$ by examining its rectangular counterpart $y = 4\sin 2x$. The sketch of $y = 4\sin 2x$ is shown in Figure 9.46, where the x- and y-axes are renamed the θ- and r-axes, respectively. The graph is divided into eight

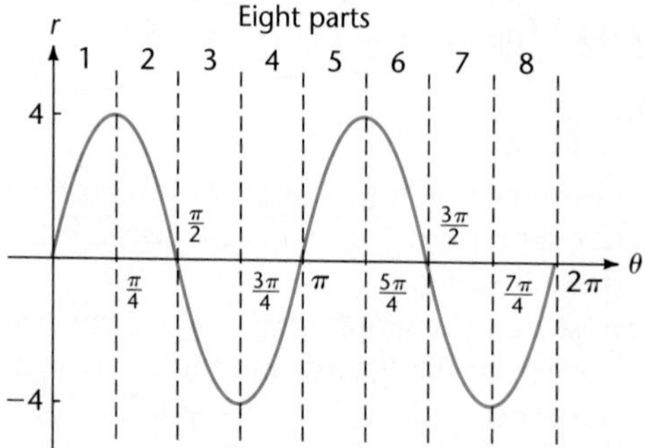

Figure 9.46 Rectangular graph of $r = 4\sin 2\theta$

parts determined by the key points of the sine curve. The following table describes how the points "act" in each of the eight parts. Notice that parts 3, 4, 7, and 8 result from points with negative r-values.

	Part 1	Part 2	Part 3	Part 4	Part 5	Part 6	Part 7	Part 8
θ	0 to $\frac{\pi}{4}$	$\frac{\pi}{4}$ to $\frac{\pi}{2}$	$\frac{\pi}{2}$ to $\frac{3\pi}{4}$	$\frac{3\pi}{4}$ to π	π to $\frac{5\pi}{4}$	$\frac{5\pi}{4}$ to $\frac{3\pi}{2}$	$\frac{3\pi}{2}$ to $\frac{7\pi}{4}$	$\frac{7\pi}{4}$ to 2π
r	0 to 4	4 to 0	0 to −4	−4 to 0	0 to 4	4 to 0	0 to −4	−4 to 0

The corresponding polar graph is sketched in Figure 9.47. This type of curve is called a **four-leaved rose.** □

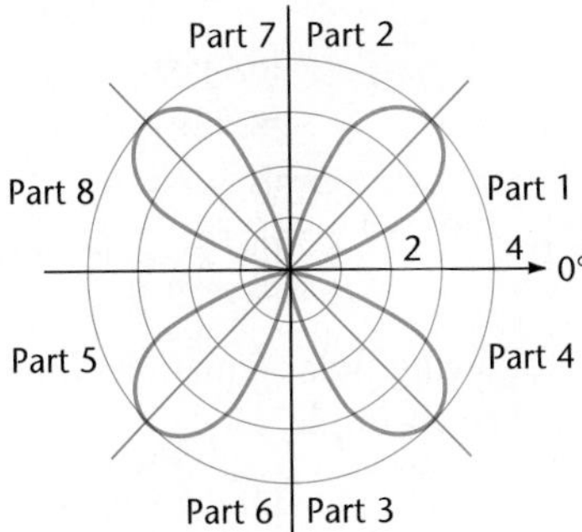

Figure 9.47 The four-leaved rose $r = 4 \sin 2\theta$

◆ Practice Problem 1 ◆

Sketch the graph of $r = 3 - \cos\theta$ in a polar coordinate system.

EXERCISES 9.7

Find the rectangular coordinates of the points whose polar coordinates are given. (See Example 1.)

1. $(4, 30°)$
2. $(3, 60°)$
3. $(1, 135°)$
4. $(2, 330°)$
5. $\left(2, \frac{2\pi}{3}\right)$
6. $\left(5, \frac{\pi}{2}\right)$
7. $\left(3, \frac{7\pi}{6}\right)$
8. $\left(1, \frac{3\pi}{4}\right)$
9. $(2, -135°)$
10. $(8, -30°)$
11. $(-5, 270°)$
12. $(-1, 300°)$

√ 13. $\left(-2, -\frac{\pi}{3}\right)$

14. $\left(-4, -\frac{5\pi}{4}\right)$

15. $(-3, -\pi)$

16. $(-5, 0)$

Find the polar coordinates for the points whose rectangular coordinates are given. Choose θ such that $0 \leq \theta < 360°$ and $r > 0$. (See Example 2.)

17. $(-\sqrt{2}, \sqrt{2})$

18. $\left(\frac{1}{2}, \frac{\sqrt{3}}{2}\right)$

√ 19. $(-\sqrt{3}, -1)$

20. $(-1, 0)$

√ 21. $(0, -1)$

22. $(3, 0)$

23. $(4, -4)$

24. $(3, 3)$

Convert each of the following equations into a polar equation. (See Example 3.)

25. $x = 3$

26. $y = -2$

√ 27. $x = y$

28. $2x + 3y = 1$

29. $x^2 + y^2 = 1$

30. $x^2 - y^2 = 4$

√ 31. $x^2 + y^2 - 4x = 0$

32. $x^2 + y^2 - 6y = 0$

Convert each of the following polar equations into an equation in the rectangular coordinates x and y. (See Example 4.)

33. $r \sin \theta = 2$

34. $r \cos \theta = -3$

35. $r = \sin \theta - \cos \theta$

36. $r = 2 \sin \theta + 2 \cos \theta$

√ 37. $r = \dfrac{1}{\sin \theta + \cos \theta}$

38. $r = \dfrac{-3}{\sin \theta + \cos \theta}$

39. $r = \dfrac{1}{r - \cos \theta}$

40. $r = \dfrac{1}{r + 4 \sin \theta}$

√ 41. $r = \dfrac{2}{1 + \sin \theta}$

42. $r = \dfrac{6}{2 + \sin \theta}$

43. $r = \dfrac{6}{2 - \cos \theta}$

44. $r = \dfrac{4}{2 + 3 \cos \theta}$

Sketch the graph of each of the following equations in a polar coordinate system. (See Examples 5 and 6.)

√ 45. $r = 2$

46. $r = 3$

√ 47. $\theta = \dfrac{\pi}{4}$

48. $\theta = \dfrac{2\pi}{3}$

49. $r = 2 \sin \theta$

50. $r = 3 \cos \theta$

√ 51. $r = -2 \cos \theta$

52. $r = -\sin \theta$

53. $r = 1 + \cos \theta$ (cardioid)

54. $r = 2 + 2 \sin \theta$ (cardioid)

55. $r = 1 - \sin\theta$ (cardioid)
56. $r = 2 - 2\cos\theta$ (cardioid)
57. $r = 2 + \cos\theta$ (This type of curve is called a limaçon.)
58. $r = 3 + \sin\theta$ (limaçon)
59. $r = 1 + 3\cos\theta$ (limaçon)
60. $r = 1 + 3\sin\theta$ (limaçon)
61. $r = 3\sin 2\theta$ (four-leaved rose)
62. $r = 3\cos 2\theta$ (four-leaved rose)
63. $r = 4\cos 3\theta$ (three-leaved rose)
64. $r = 2\sin 3\theta$ (three-leaved rose)
65. $r = 2\theta, \quad 0 \le \theta \le 2\pi$ (spiral)
66. $r = 3\theta, \quad 0 \le \theta \le 2\pi$ (spiral)
67. $r^2 = \cos 2\theta$ (This type of curve is called a lemniscate.)
68. $r^2 = \sin 2\theta$ (lemniscate)

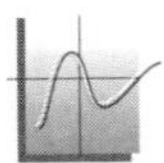

69–78

Most graphing devices can be used to draw graphs in polar coordinates by choosing the proper mode and features. Graph the following equations.

69. $r = 2 + 2\cos\theta$ (See Example 5.)
70. $r = 4\sin 2\theta$ (See Example 6.)
71. $r = 0.3\theta$
72. $r = 4\csc\theta$
73. $r = 6\cos\left(\theta - \dfrac{\pi}{4}\right)$
74. $r = 4\cos(2\theta - \pi)$
75. $r = 4\sin\left(3\theta + \dfrac{\pi}{2}\right)$
76. $r = 5\sin\left(2\theta + \dfrac{\pi}{3}\right)$
77. $r = \dfrac{6}{1 + 2\cos\theta}$
78. $r = \dfrac{2}{1 - \sin\theta}$

◆ Solution for Practice Problem

1. Choosing quadrantal angles for θ yields key values for r in the following table. The graph is sketched in Figure 9.48.

r	θ
2	0
3	$\dfrac{\pi}{2}$
4	π
3	$\dfrac{3\pi}{2}$
2	2π

Figure 9.48 The limaçon $r = 3 - \cos\theta$

CHAPTER REVIEW

Summary of Important Concepts and Formulas

Law of Sines

$$\frac{a}{\sin A} = \frac{b}{\sin B} = \frac{c}{\sin C}$$

Law of Cosines

$$a^2 = b^2 + c^2 - 2bc\cos A \qquad \cos A = \frac{b^2 + c^2 - a^2}{2bc}$$

$$b^2 = a^2 + c^2 - 2ac\cos B \qquad \cos B = \frac{a^2 + c^2 - b^2}{2ac}$$

$$c^2 = a^2 + b^2 - 2ab\cos C \qquad \cos C = \frac{a^2 + b^2 - c^2}{2ab}$$

Area of a Triangle

$$K = \tfrac{1}{2}bc\sin A = \tfrac{1}{2}ac\sin B = \tfrac{1}{2}ab\sin C$$

$$K = \frac{a^2 \sin B \sin C}{2\sin A} = \frac{b^2 \sin A \sin C}{2\sin B} = \frac{c^2 \sin A \sin B}{2\sin C}$$

$$K = \sqrt{s(s-a)(s-b)(s-c)}, \text{ where } s = \tfrac{1}{2}(a+b+c)$$

Vectors

Vector addition:

$$(u_1\mathbf{i} + u_2\mathbf{j}) + (v_1\mathbf{i} + v_2\mathbf{j}) = (u_1 + v_1)\mathbf{i} + (u_2 + v_2)\mathbf{j}$$

Scalar multiplication: $a(v_1\mathbf{i} + v_2\mathbf{j}) = (av_1)\mathbf{i} + (av_2)\mathbf{j}$

Magnitude: $\|v_1\mathbf{i} + v_2\mathbf{j}\| = \sqrt{v_1^2 + v_2^2}$

Product and Quotient Rules

If z_1 and z_2 have trigonometric forms

$$z_1 = r_1(\cos\theta_1 + i\sin\theta_1)$$

and

$$z_2 = r_2(\cos\theta_2 + i\sin\theta_2),$$

then

$$z_1z_2 = r_1r_2[\cos(\theta_1 + \theta_2) + i\sin(\theta_1 + \theta_2)]$$

$$\frac{z_1}{z_2} = \frac{r_1}{r_2}[\cos(\theta_1 - \theta_2) + i\sin(\theta_1 - \theta_2)].$$

De Moivre's Theorem

If z has the trigonometric form $z = r(\cos\theta + i\sin\theta)$, then

$$z^n = r^n(\cos n\theta + i\sin n\theta).$$

Roots of Complex Numbers

If z is a nonzero complex number with trigonometric form

$$z = r(\cos\theta + i\sin\theta),$$

then the n nth roots of z are given by

$$\sqrt[n]{r}\left[\cos\left(\frac{\theta + k \cdot 360^\circ}{n}\right) + i\sin\left(\frac{\theta + k \cdot 360^\circ}{n}\right)\right],$$

for $k = 0, 1, 2, \ldots, n - 1$.

Relationship Between Rectangular (x, y) and Polar (r, θ) Coordinates

$$r^2 = x^2 + y^2, \quad x = r\cos\theta, \quad y = r\sin\theta$$

Critical Thinking: Find the Errors

Each of the following nonsolutions has one or more errors. Can you find them?

Problem 1

Find side c in a triangle that has $a = 7.0$, $b = 3.0$, $C = 60^\circ$.

Nonsolution

$$\begin{aligned} c^2 &= a^2 + b^2 - 2ab\cos C \\ &= (7.0)^2 + (3.0)^2 - 2(7.0)(3.0)\cos 60^\circ \\ &= 49 + 9 - 42(\tfrac{1}{2}) = 16(\tfrac{1}{2}) = 8 \end{aligned}$$

$$c = \sqrt{8} = 2\sqrt{2}$$

Problem 2

A plane is flying with a heading of 120.0° and an airspeed of 380 miles per hour. If the wind is blowing from the south at 40.0 miles per hour, find the ground speed of the plane.

Nonsolution

The ground speed is represented by c in Figure 9.49. We use the law of cosines to find c.

$$\begin{aligned} c^2 &= (380)^2 + (40)^2 - 2(380)(40)\cos 150° \\ &= 144{,}400 + 1600 - 30{,}400(-0.8660) \\ &= 172{,}326 \\ c &= 415 \text{ miles per hour} \end{aligned}$$

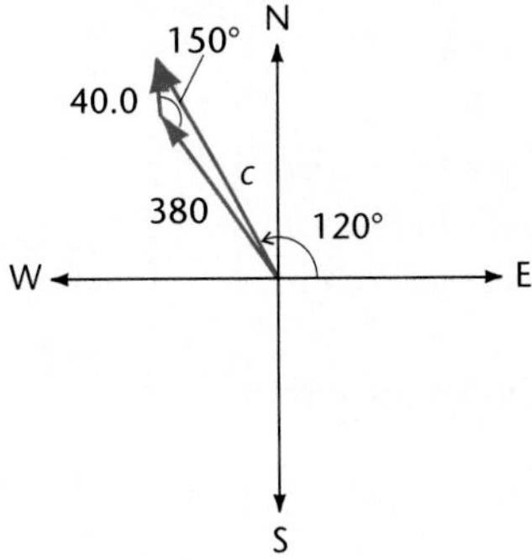

Figure 9.49

Problem 3

Express $-\sqrt{3} - i$ in trigonometric form.

Nonsolution

The vector representing $-\sqrt{3} - i$ is shown in Figure 9.50. From the figure, we get $-\sqrt{3} - i = 2(\cos 30° + i \sin 30°)$.

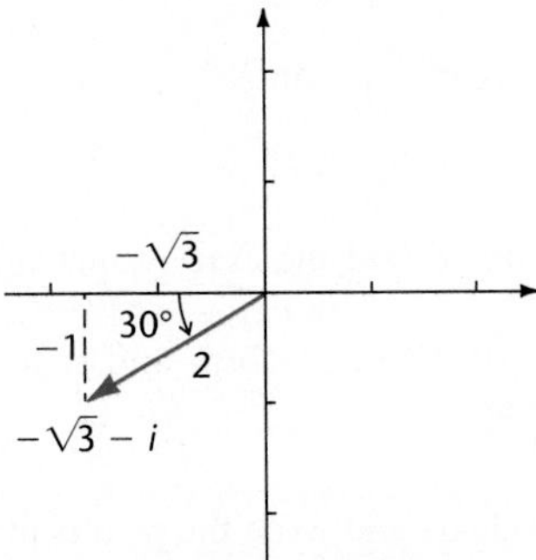

Figure 9.50

Problem 4

Use DeMoivre's theorem to evaluate $(1 + i\sqrt{3})^5$.

Nonsolution

$$\begin{aligned} (1 + i\sqrt{3})^5 &= 2(\cos 60° + i \sin 60°)^5 \\ &= 2(\cos 300° + i \sin 300°) \\ &= 2\left(\frac{1}{2} - \frac{\sqrt{3}}{2}i\right) \\ &= 1 - i\sqrt{3} \end{aligned}$$

Review Problems for Chapter 9

In Problems 1–6, solve the triangles to the degree of accuracy consistent with the given information.

1. $B = 87°$, $C = 43°$, $a = 17$
2. $A = 138°$, $a = 2.1$, $b = 8.8$
3. $C = 37.2°$, $b = 10.1$, $c = 14.2$
4. $B = 37.2°$, $c = 14.2$, $b = 10.1$
5. $a = 47.2$, $b = 41.3$, $c = 16.1$
6. $A = 120°$, $b = 14$, $c = 22$
7. Find the area of the triangle in Problem 1.
8. Find the area of the triangle in Problem 5.
9. Find the area of the triangle in Problem 6.

Write each complex number in standard form.

10. $2\left(\cos\frac{2\pi}{3} + i\sin\frac{2\pi}{3}\right)$
11. $4[\cos(-150°) + i\sin(-150°)]$

Write each complex number in trigonometric form.

12. $-3i$ 13. $2 - 2i$

Write the exact results of the indicated operations in standard form.

14. $3(\cos 17° + i\sin 17°) \cdot 4(\cos 253° + i\sin 253°)$
15. $\dfrac{4(\cos 198° + i\sin 198°)}{3(\cos 48° + i\sin 48°)}$
16. Use the trigonometric forms to find the exact values of z_1z_2 and z_1/z_2, where $z_1 = 4 - 4i\sqrt{3}$ and $z_2 = -\sqrt{3} - i$. Write the results in both trigonometric form and standard form.

Use DeMoivre's theorem to evaluate and write the results in standard form.

17. $\left[\sqrt{3}\left(\cos\frac{3\pi}{8} + i\sin\frac{3\pi}{8}\right)\right]^4$
18. $(1 - i)^{14}$ 19. $(-\sqrt{3} + i)^6$
20. Find the four fourth roots of $-\frac{1}{2} - \frac{\sqrt{3}}{2}i$ and write the results in standard form.
21. Solve the equation $x^6 - 64 = 0$ and write the solutions in standard form.

Angle of Elevation

22. Matt and Beckie are standing on a level beach where Matt is flying a kite on a 63-foot string. The angle of elevation from Matt to the kite is 47°, and the angle of elevation from Beckie to the kite is 48°. How far apart are Matt and Beckie if the kite is in a vertical plane between them?

Triangular Truss

23. Suppose three steel pipes of lengths 6.0 feet, 14 feet, and 18 feet are to be welded together to form the truss for the roof of the produce stand shown in the accompanying figure. Find the measure of the angle α of inclination of the roof to the horizontal.

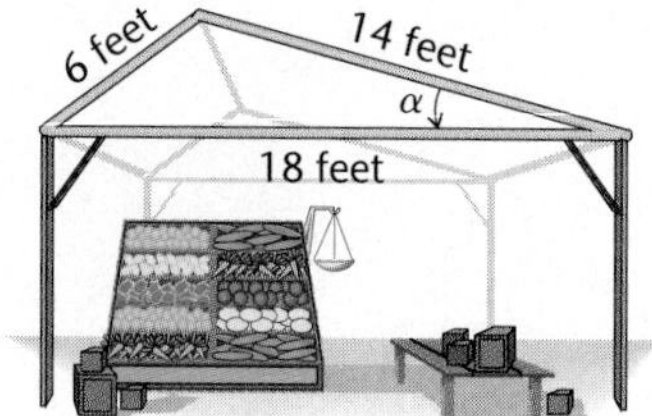

Figure for Problem 23

Billiards

24. In a billiards game a ball, struck by the cue ball, travels 12 inches, hitting the end bumper at an angle of 58°. It then bounces off the end and travels 42 inches, landing in a side pocket. What is the shortest distance from the ball's original position in the accompanying figure to the point where it dropped off into the side pocket?

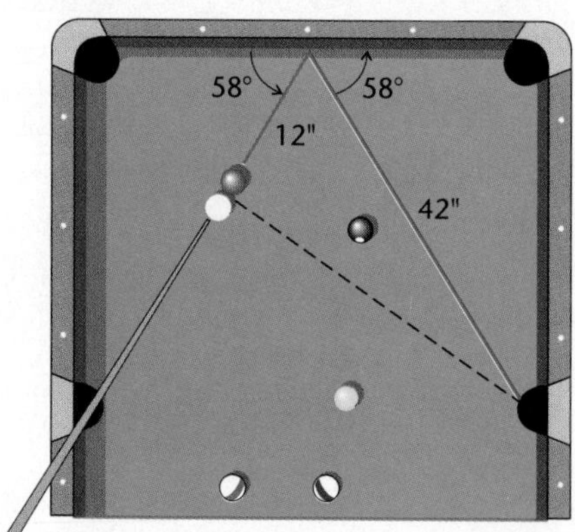

Figure for Problem 24

Air Navigation

25. An airplane is headed due north with an airspeed of 252 miles per hour, and the wind is blowing from the east at 31.8 miles per hour. Find the ground speed and the course of the plane.

Sea Navigation

26. A freighter is running at the rate of 19 miles per hour in still water and is headed in the direction N 53° E. If the ship is traveling in an ocean current flowing at 4.8 miles per hour in the direction S 37° E, what is its course? At what speed is it traveling?

Resultant Force 27–28

27. Two forces of magnitudes 12.8 pounds and 21.2 pounds act on a point in the plane. If the angle between the directions of the two forces is 108.2°, find the magnitude of the resultant force.

28. The angle between the directions of two forces is 54.0°. One of the forces has magnitude 197 pounds, and the resultant has magnitude 402 pounds. Find the angle between the 197-pound force and the resultant force.

Gravitational Force 29–30

29. A steel ball weighing 500 pounds is to be rolled up a ramp that is inclined at an angle of 7.6° with the horizontal. What force must the ramp withstand in the direction perpendicular to the ramp?

30. In Problem 29, what force parallel to the ramp is required to hold the ball in place on the ramp if friction is neglected?

Air Navigation

31. The airplane in the accompanying figure is flying with a heading of 317.0°, an airspeed of 280 miles per hour, and a ground speed of 312 miles per hour. If the wind is blowing from the south, find its speed.

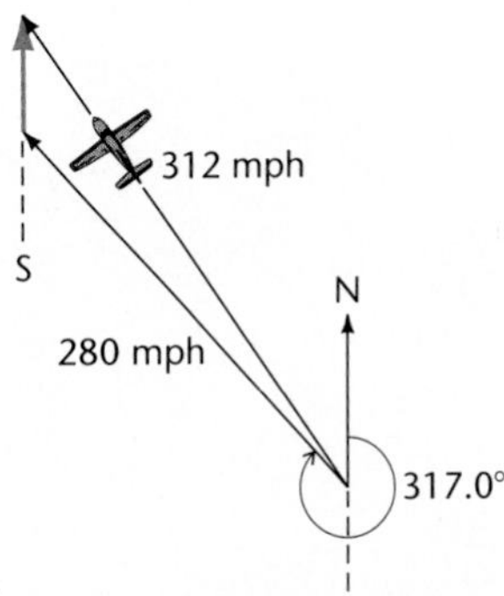

Figure for Problem 31

Convert each of the following equations into a polar equation.

32. $x^2 + y^2 = 2x$ 33. $3x + 4y = 12$

Convert each of the following into an equation in the rectangular coordinates x and y.

34. $r(\sin \theta - 2 \cos \theta) = 4$

35. $r^2 \sin \theta \cos \theta = 1$

Sketch the graph of each of the following equations in a polar coordinate system.

36. $r = 2 - \sin \theta$ (limaçon)

37. $r = 1 + \sin \theta$ (cardioid)

ENCORE

There are in fact two kinds of trigonometry: **plane trigonometry,** as presented in this book, and **spherical trigonometry** which is based on the surface of a sphere such as the earth. Spherical trigonometry is concerned with the study of spherical triangles, triangles formed from parts of great circles.

A **great circle** is the intersection of a sphere with a plane that passes through the center of the sphere. On the earth, circles that pass through both the North Pole and the South Pole are great circles and are called circles of longitude. The

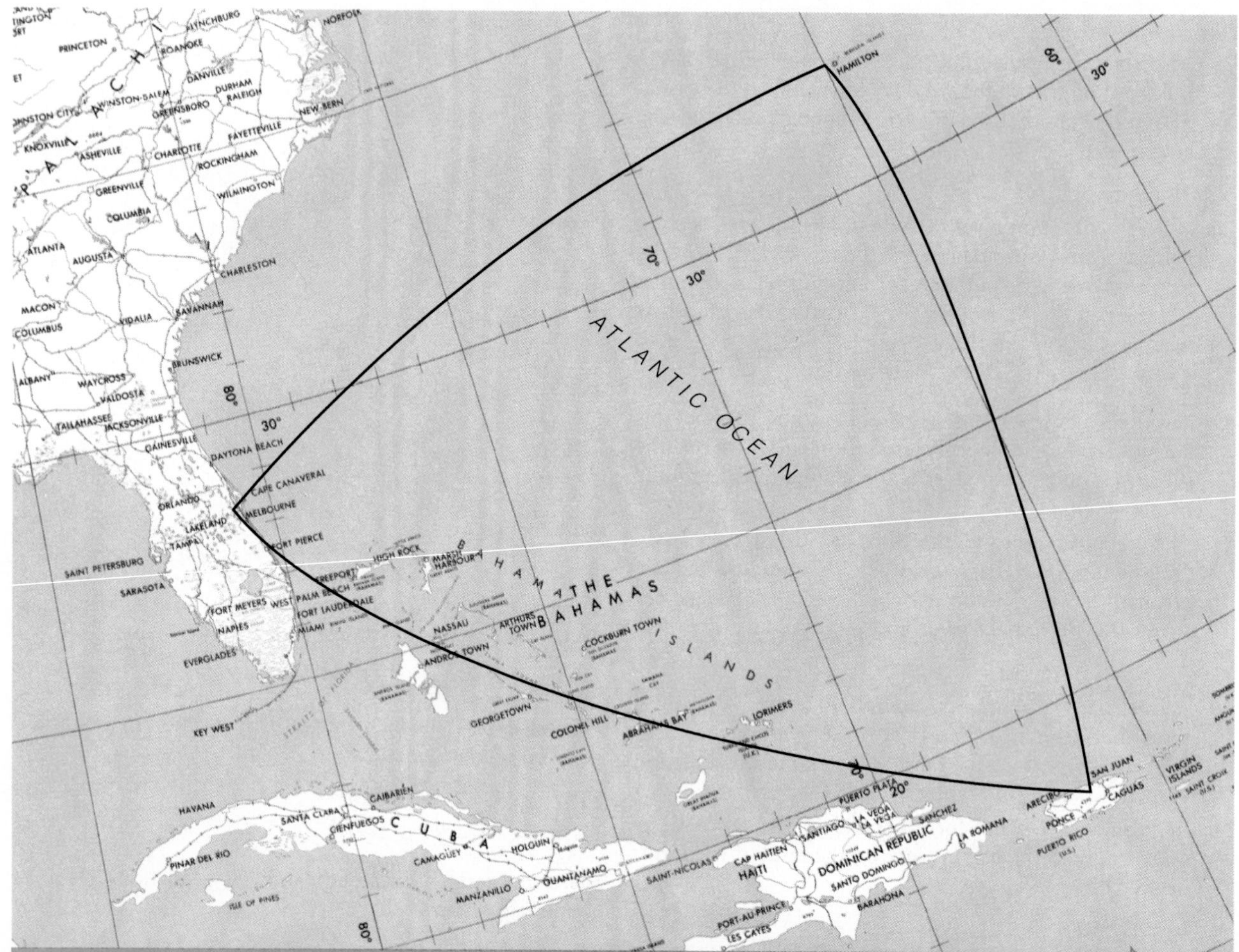

The Bermuda Triangle

equator is also a great circle. A spherical triangle is a figure formed on a sphere by three arcs of great circles. A spherical triangle is shown in Figure 9.51 with three angles labeled A, B, C having opposite sides labeled a, b, c, respectively. In the figure, O is the center of the sphere. The angle A is by definition the angle between the tangents at A to the arcs AC and AB, and the other angles are defined similarly.

There are formulas in spherical trigonometry for solving spherical triangles that are analogous to the formulas presented in this chapter. For example, there is a **law of sines** that states

$$\frac{\sin a}{\sin A} = \frac{\sin b}{\sin B} = \frac{\sin c}{\sin C}$$

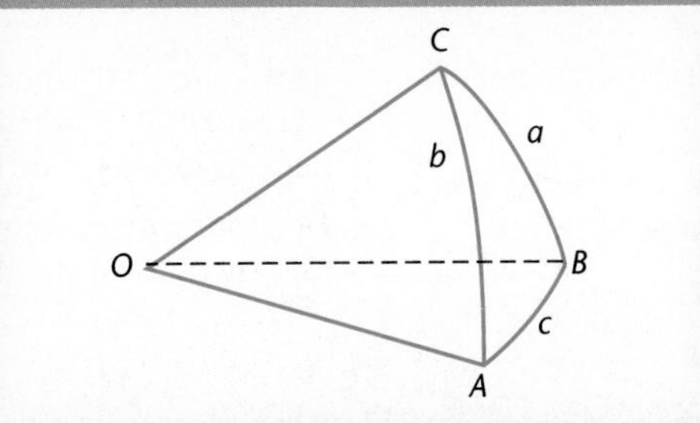

Figure 9.51 A spherical triangle *ABC*

and a **law of cosines** that states

$$\cos a = \cos b \cos c + \sin b \sin c \cos A$$

or

$$\cos A = \sin B \sin C \cos a - \cos B \cos C.$$

The best-known spherical triangle is without a doubt the famous **Bermuda Triangle,** shown on page 614. The Bermuda Triangle, also known as the **Devil's Triangle,** has vertices at Bermuda, Puerto Rico, and near Melbourne, Florida. This triangle forms a region of about 440,000 square miles, and it has become famous because many ships and planes have been lost in the region. Since 1854, over 50 planes and ships have disappeared there, many of them under mysterious circumstances with no survivors or wreckage found.

Cumulative Test Chapters 3–9

1. Write an equation, in standard form, of the straight line with x-intercept 7 and y-intercept -2.

2. Let $f(x) = \sqrt{x+2}$ and $g(x) = 1 - x$. Determine the following functions and state their domains.

   a. f/g b. $f \circ g$

Maximum Demand

3. The demand D for the book *Algebra in Your Own Home* is given by $D = 10 + 4p - 0.5p^2$, where D is the demand (in thousands of books) and p is the price (in dollars). What price will produce a maximum demand, and what is the value of the maximum demand?

4. Write an equation of the circle with center $(-3, 5)$ and radius 4.

5. Graph the equation $4(x-2)^2 + 9(y+1)^2 = 36$. Locate the center and vertices.

6. The graph of a one-to-one function f is given. Sketch the graph of f^{-1} by reflecting the graph of f through the line $y = x$.

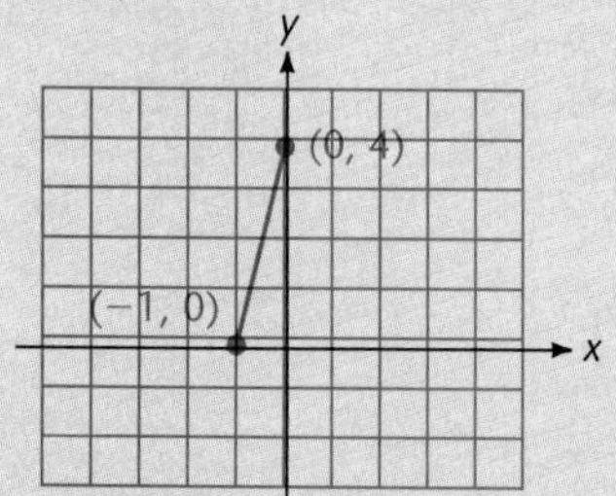

Figure for Problem 6

7. Given that $1 - i$ is a zero of $P(x) = x^3 + x^2 - 4x + 6$, find all the other zeros.

8. Find all solutions of $x^3 + 3x^2 - 10x - 24 = 0$.

9. Write equations for all asymptotes of the graph of $f(x) = \dfrac{2x^2+1}{x^2-1}$. Do not sketch the graph.

10. Sketch the graph of $f(x) = 3^{-x}$.

11. Find the solution set: $\log(6 + 3x) + \log(1 - 3x) = 1$.

Depreciation

12. The loss of value of equipment or buildings in business is called **depreciation.** If an item has initial cost C and depreciates at 8% per year, its value after t years is $A = C(1 - 0.08)^t$. Find the length of time it takes for the item to lose 65% of its original value.

13. Determine the exact values of all the other trigonometric functions of the third-quadrant angle α if $\sec\alpha = -17/8$.

Angular and Linear Speed

14. A wheel on a cart has a radius of length 2 feet and is rolling at a speed of 2 miles per hour. Find the angular speed of the wheel in radians per minute.

Angle of Elevation

15. At a point on level ground 87 feet from the base of a vertical flagpole, the angle of elevation of the top of the pole is 37°. Find the height of the pole.

16. Sketch the graph of the given function through one complete cycle, locating key points on each graph.

    a. $y = \dfrac{3}{2}\sin(2x + \pi)$ b. $y = \dfrac{1}{2}\cos\left(2\pi - \dfrac{\pi}{2}\right)$

17. Verify the identity: $\dfrac{1}{\csc\theta - \sin\theta} = \sec\theta\tan\theta$.

18. Without determining α or β, find the exact values of all trigonometric functions of $\alpha + \beta$ if $\cos\alpha = -\frac{5}{13}$, $\sin\beta = \frac{4}{5}$, α is in quadrant III, and β is in quadrant II.

19. Solve $\sin 3x = \cos 3x$ for all x such that $0 \le x < 2\pi$.

20. Solve for x: $\sin^{-1} x - \cos^{-1}\frac{12}{13} = \cos^{-1}\frac{4}{5}$.

21. Solve the triangle to the degree of accuracy consistent with $A = 33.7°$, $b = 15.3$, and $a = 11.1$.

22. Find the six sixth roots of $64i$ and write the results in trigonometric form.

Air Navigation

23. If the airspeed of a small plane is 210 miles per hour and the wind is blowing from the north at 18.0 miles per hour, what heading should the pilot give his plane in order to fly on a course due west from Alba to Whitehall?

Resultant Force

24. A 28-pound force and a 42-pound force produce a 30-pound resultant force. Find the angle between the directions of the 42-pound force and the resultant force.

25. Sketch the graph of the limaçon $r = 1 + 2\cos\theta$ in a polar coordinate system.

Systems of Equations and Inequalities

Business and industry utilize systems of equations and inequalities in making plans for efficient operation, growth, and future profits. In connection with the use of calculus, engineers and scientists also frequently apply the knowledge and methods presented in this chapter.

10.1 SYSTEMS OF LINEAR EQUATIONS IN TWO VARIABLES

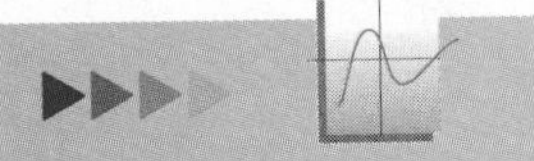

The cursor or **trace** feature of a graphing device can be used to find points of intersection whenever the equations involved can be expressed in functional form.

We consider now an important problem that frequently arises in connection with the graphs of equations. The problem is to find the coordinates (x, y) of the points that are common to the graphs of two equations in x and y, that is, the **points of intersection** of the two graphs. When we are working with this type of problem, the pair of equations involved is referred to as a **system of equations.** A set of values for x and y that satisfies both equations is called a **simultaneous solution** of the system, or simply a **solution** of the system. Since real numbers are the only ones that we can use as coordinates on graphs, **our work in this chapter is restricted to real numbers, except where stated otherwise.**

We restrict our attention in this section to systems of equations that have two linear equations, such as

$$a_1x + b_1y = c_1$$
$$a_2x + b_2y = c_2.$$

Such systems are called **linear systems.** We assume that at least one coefficient of a variable is nonzero in each equation, and this means that the graphs of the two equations are two straight lines. If a linear system has a solution, it is called **consistent.** If it has no solution, it is **inconsistent.**

There are three possibilities for the solution sets of a linear system.

1. If the graphs of the two equations are nonparallel lines, there is one point of intersection of the graphs and consequently one ordered pair in the solution set. A system such as this is called **independent,** and a typical sketch is shown in Figure 10.1a.
2. If the graphs of the two equations are different parallel lines, there are no points of intersection and consequently the solution set is empty. A system such as this is *inconsistent,* and a typical sketch is shown in Figure 10.1b.

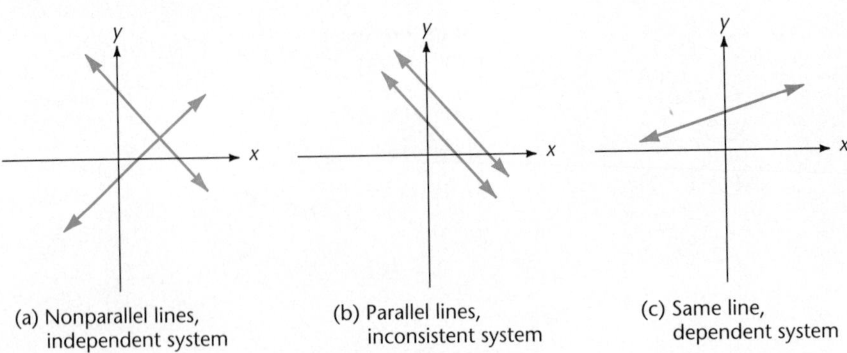

Figure 10.1 Linear systems with two equations

3. If the graphs of the two equations are the same line, then any point on that line is in the intersection and has coordinates that satisfy both equations. Thus there is an infinite number of solutions in this case, and the system is consistent. A system of this type is also called **dependent,** and a typical sketch is shown in Figure 10.1c.

We shall consider two methods in this section that can be used to solve linear systems. Each method is based on the idea that a point which is on both graphs must have coordinates that satisfy both equations. The simpler of these two methods is known as the **substitution method.** The idea with this method is to use one of the given equations, solve it for one of the variables in terms of the other, and then *substitute* this value in the other equation, thereby obtaining an equation that involves only one variable. This procedure, which sounds more complicated than it really is, is illustrated in the following example.

Example 1 • The Substitution Method

Find the point of intersection of the lines $3y + x = 7$ and $3x - 2y + 12 = 0$. That is, solve the system of equations

$$\begin{aligned} 3y + x &= 7 \\ 3x - 2y + 12 &= 0. \end{aligned}$$

Solution

As indicated in the preceding discussion, we will employ the substitution method. The first step is to solve for one of the variables in one of the given equations. The simplest possibility here is to solve for x in the first equation, obtaining

$$x = 7 - 3y.$$

When this value for x is substituted into the other equation, we have

$$3(7 - 3y) - 2y + 12 = 0.$$

This simplifies to

$$\begin{aligned} 33 - 11y &= 0 \\ 33 &= 11y \end{aligned}$$

and

$$y = 3.$$

Substituting this value for y in $x = 7 - 3y$, we get $x = -2$. Thus the point of intersection is $(-2, 3)$, and it is easy to check that the coordinates of this point do indeed satisfy both of the original equations. The situation is shown geometrically in Figure 10.2. □

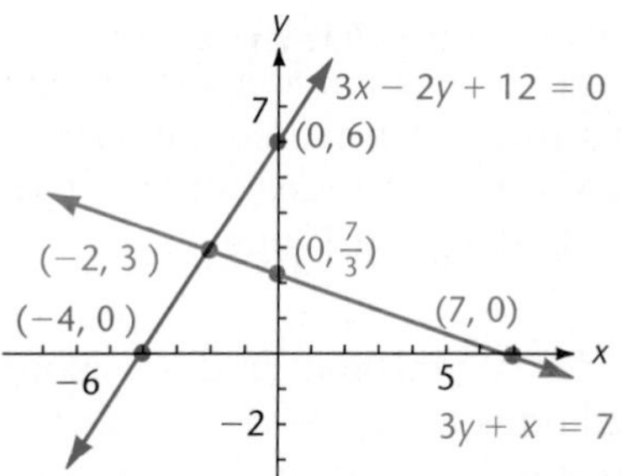

Figure 10.2 The lines in Example 1

◆ **Practice Problem 1** ◆

Solve by the substitution method.

$$x + 2y = 6$$
$$5x + 3y = 2$$

The word *equivalent* is used with systems of equations in the same way as it is with a single equation: systems of equations that have the same solution set are called **equivalent systems.**

When the substitution method leads to laborious work involving fractions, the method known as the **elimination method** is usually more efficient to use. The elimination method is based on the facts that an equivalent system of equations is obtained if one of the equations in the system is replaced by

1. an equivalent equation or
2. the sum of two equations of the system.

To solve a system by the elimination method, we multiply one or both of the original equations by numbers chosen so that one of the variables will be eliminated when the sum of the new equations is formed. This method is used in Example 2.

Example 2 • The Elimination Method

Solve the following system by using the elimination method.

$$5x + 3y = 7$$
$$3x - 2y = 8$$

Solution

We decide to eliminate y. To accomplish this, we multiply the first equation by 2, multiply the second equation by 3, and then add the new equations.

$$\begin{array}{lcr} 5x + 3y = 7 & \xrightarrow{\text{Multiply by 2}} & 10x + 6y = 14 \\ 3x - 2y = 8 & \xrightarrow{\text{Multiply by 3}} & 9x - 6y = 24 \\ & \text{Add} & \overline{19x \qquad = 38} \\ & & x = 2 \end{array}$$

We can find the value of y by substituting 2 for x in any of the equations that involve y. Using the top original equation, we get

$$\begin{aligned} 5(2) + 3y &= 7 \\ 10 + 3y &= 7 \\ 3y &= -3 \\ y &= -1. \end{aligned}$$

Thus the solution set for the system is $\{(2, -1)\}$. □

◆ Practice Problem 2 ◆

Solve the following system by the elimination method.

$$\begin{aligned} 3x - 2y &= 12 \\ 6x - 5y &= -9 \end{aligned}$$

An inconsistent system is considered in the next example.

Example 3 • An Inconsistent System

Solve by using the elimination method.

$$\begin{aligned} 12x + 8y &= 1 \\ 9x + 6y &= 4 \end{aligned}$$

Solution

We decide to eliminate y. This can be done by multiplying the first equation by 3 and the second equation by -4 before adding.

$$\begin{array}{lcr} 12x + 8y = 1 & \xrightarrow{\text{Multiply by 3}} & 36x + 24y = \quad 3 \\ 9x + 6y = 4 & \xrightarrow{\text{Multiply by } -4} & -36x - 24y = -16 \\ & \text{Add} & \overline{0 = -13} \end{array}$$

The variable x was automatically eliminated along with y, and the impossible equation $0 = -13$ has resulted. This tells us the original system is inconsistent, and the solution set is $\varnothing$. □

Our next example illustrates the elimination method with a dependent system.

Example 4 • A Dependent System

Solve by the elimination method.

$$\begin{aligned} x + 3y &= 5 \\ 2x + 6y &= 10 \end{aligned}$$

Solution

We decide to eliminate x by multiplying the first equation by -2 and adding.

$$\begin{array}{rcr} x + 3y = 5 & \xrightarrow{\text{Multiply by } -2} & -2x - 6y = -10 \\ 2x + 6y = 10 & \xrightarrow{\hspace{2cm}} & \underline{2x + 6y = 10} \\ & \text{Add} & 0 = 0 \end{array}$$

This time we find that y is automatically eliminated along with x, and the resulting equation $0 = 0$ is always true. This indicates that the original system is dependent, and the solution set is

$$\{(x, y) | x + 3y = 5\}.$$

We note that either of the original equations is a multiple of the other, and their graphs are actually the same straight line. □

Some practical problems lead to the solution of a system of linear equations. The last example in this section provides an illustration of such a problem.

Example 5 • Mixing Alloys

Bronze is an alloy of copper and tin that gave its name to the Bronze Age. Robo-Casting has one bronze alloy in stock that is 90% copper and another bronze alloy that is 70% copper. How many pounds of each alloy should be melted together to obtain 320 pounds of an alloy that is 75% copper?

Solution

Let x be the number of pounds of the 90% alloy that should be used, and let y be the number of pounds of the 70% alloy that should be used. Since there are 320 pounds of the final alloy,

$$x + y = 320.$$

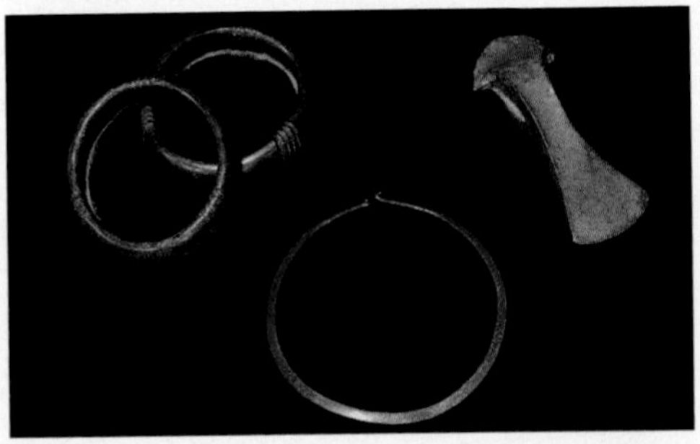

Ancient bronze ax and arm rings. Modern bronze alloys contain about 75% copper and 25% tin.

Frederick Schultz Ancient Art, photo by Maggie Nimkin.

The sum of the amounts of copper in the parts must equal the amount of copper in the final alloy, so we also have the equation

$$0.90x + 0.70y = 0.75(320).$$

Thus we need to solve the system of equations

$$\begin{aligned} x + y &= 320 \\ 0.9x + 0.7y &= 240. \end{aligned}$$

To eliminate y, we can multiply the first equation by -7 and the second equation by 10.

$$\begin{array}{rcrcl} x + y = 320 & \xrightarrow{\text{Multiply by } -7} & -7x - 7y &=& -2240 \\ 0.9x + 0.7y = 240 & \xrightarrow{\text{Multiply by } 10} & 9x + 7y &=& 2400 \\ & \text{Add} & 2x &=& 160 \\ & & x &=& 80 \end{array}$$

Substitution of $x = 80$ in $x + y = 320$ yields $y = 240$. Thus 80 pounds of 90% alloy should be melted with 240 pounds of 70% alloy to obtain 320 pounds of 75% alloy. □

EXERCISES 10.1

Solve the following systems. Sketch the graphs of the two equations on the same coordinate system and label the points of intersection. (See Example 1.)

1. $\begin{aligned} x + 2y &= 4 \\ 3x - 2y &= -12 \end{aligned}$
2. $\begin{aligned} 2y &= x + 4 \\ 3x + 2y + 12 &= 0 \end{aligned}$
3. $\begin{aligned} x - 2y &= 7 \\ 2x &= 4y - 14 \end{aligned}$
4. $\begin{aligned} 2x &= 3y - 5 \\ 9y &= 6x - 15 \end{aligned}$

Solve by the substitution method. (See Example 1 and Practice Problem 1.)

5. $\begin{aligned} 2x - 5y &= 11 \\ 5x + y &= 14 \end{aligned}$
6. $\begin{aligned} 3x - 2y &= 18 \\ 2x - y &= 11 \end{aligned}$
7. $\begin{aligned} x + 2y &= -5 \\ 2x + y &= 2 \end{aligned}$
8. $\begin{aligned} 3x - 5y &= 6 \\ 2x - y &= 11 \end{aligned}$
9. $\begin{aligned} 7x + 3y &= 5 \\ 5x - y &= 13 \end{aligned}$
10. $\begin{aligned} 4x + 3y &= 9 \\ 2x - y &= 12 \end{aligned}$
11. $\begin{aligned} 9x + 2y &= 3 \\ 2x - 3y &= 11 \end{aligned}$
12. $\begin{aligned} 7x + 4y &= 2 \\ 3x - 2y &= -3 \end{aligned}$

Solve by the elimination method. (See Example 2 and Practice Problem 2.)

13. $\begin{aligned} 2x - 5y &= -13 \\ 4x - y &= 1 \end{aligned}$
14. $\begin{aligned} x + 3y &= 1 \\ 2x + 5y &= 5 \end{aligned}$

15. $2x + 5y = 5$
$3x + 4y = 18$

16. $2x + 3y = 11$
$3x - 5y = -31$

17. $2x - 3y = -2$
$-3x + 4y = 0$

18. $4x + 3y = -5$
$5x - 4y = 17$

19. $7x + 5y = 2$
$4x + 3y = -8$

20. $2x - 9y = -1$
$3x - 4y = -2$

21. $3x + 4y = 1$
$5x - 2y = 19$

22. $2x + 9y = 3$
$5x + 7y = -8$

Solve the following systems by any method.

23. $3x - 2y = 7$
$6x - 4y = 4$

24. $4x + 3y = -5$
$5x - 4y = 17$

25. $2x - 3y = -12$
$x + 2y = 8$

26. $x + 4y = -11$
$2x + y = 6$

27. $2x = 3y - 1$
$6y = 4x + 5$

28. $5x = 2y + 1$
$4y = 10x + 1$

29. $2x - 4 = 3y$
$6x = 9y + 12$

30. $8x - 6y = 16$
$12x - 9y = 24$

31. $3x = 12 - 6y$
$4y = 8 - 2x$

32. $2x = 2 - 4y$
$6y = 3 - 3x$

Solve the following systems by first letting $u = 1/x$ and $v = 1/y$.

33. $\frac{1}{x} + \frac{1}{y} = 5$
$\frac{2}{x} + \frac{3}{y} = 14$

34. $\frac{3}{x} - \frac{2}{y} = -6$
$\frac{1}{x} + \frac{4}{y} = 40$

35. $\frac{2}{x} + \frac{3}{y} = 66$
$\frac{3}{x} + \frac{4}{y} = 91$

36. $\frac{5}{x} + \frac{2}{y} = 55$
$\frac{4}{x} + \frac{3}{y} = 79$

Investment Income

37. Georgia has \$15,000 invested, part in stocks paying a simple interest rate of 9%, and the rest in bonds paying a simple interest rate of 5%. If her total annual interest income is \$1130, find the amount invested at each rate.

The Photo Works

Interest rates on investments vary with changing economic conditions.

Investment Income

38. Dan had invested \$12,000, part at 6% annual interest and the rest at 12% annual interest. If his total annual income from both investments is \$1140 per year, how much is invested at 6%?

Uniform Motion 39–40

39. Mary drove her boat 8 miles upstream in 30 minutes and then made the return trip in 20 minutes. Find the rate of the current and the speed of her boat in still water.

40. Suppose it takes 1 hour 15 minutes for a plane to fly from Natchez, Mississippi, to Cross Anchor, South Carolina, an air distance of 600 miles, and 1 hour 30 minutes for the return flight. If the wind is blowing at a constant rate and direction, as indicated in the accompanying figure, find the rate of the wind and the rate of the plane in still air.

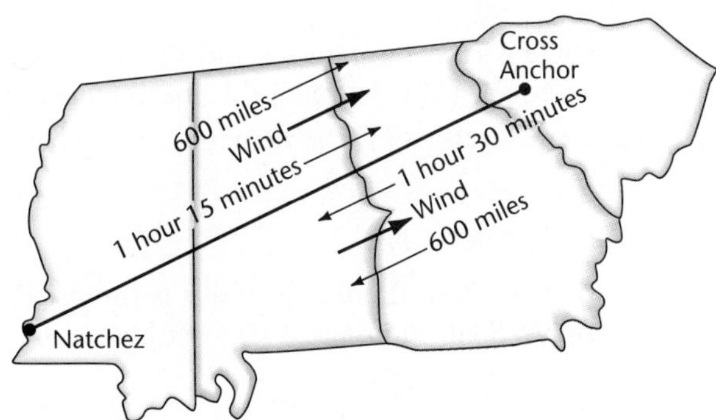

Figure for Exercise 40

Dietary Mixture 41. A dietician is doing research in which she needs 28 grams of a substance that is 30% protein. How many grams of each of two ingredients, one that is 50% protein and the other 25% protein, should she mix together?

Punch Mixture 42. A hostess needs 10 gallons of punch for her party. How much pineapple juice should be mixed with a drink that is 10% fruit juice to obtain a punch that is 50% fruit juice?

Nut Mixture 43. How many pounds of peanuts at \$2.50 per pound and cashews at \$3.60 per pound should be used to obtain 100 pounds of mixed nuts valued at \$316.00?

Coffee Blend 44. A coffee blend worth \$1.60 per pound is to be mixed with a second coffee blend worth \$3.00 per pound to obtain a mixture worth \$2.40 per pound. How many pounds of each blend should be used in order to obtain 105 pounds of the \$2.40 mixture?

Ticket Pricing 45–46

45. A circus performance is scheduled in Memorial Auditorium, and the reserved-seat tickets are to cost twice as much as the regular admission seats. There are 1000 reserved seats and a total seating capacity of 12,000 in the auditorium. What price should be charged for each ticket so that a capacity crowd will yield \$91,000 in ticket sales?
46. Work Exercise 45 if the reserved-seat tickets are to cost 3 times as much as the regular admission tickets and all other information is unchanged.

Equilibrium Price 47–48

47. The demand for David's handcrafted lamps is given by

$$D = 100 - \tfrac{5}{2}p,$$

where D is the number of people (in thousands) who want to buy one of David's lamps and p is the price of a lamp (in dollars). The supply of lamps is given by

$$S = \tfrac{5}{4}p - \tfrac{25}{2},$$

where S is in thousands. Find the price p for which supply and demand are equal. (This is called the **equilibrium price.**)

48. (See Exercise 47.) The demand for the new book *Sensible Algebra* is given by

$$D = 25 - p,$$

where D is in thousands and p is the price of the book (in dollars). The supply S (in thousands) is given by

$$S = \tfrac{3}{2}p - 20.$$

Find the equilibrium price for *Sensible Algebra*.

Production Planning 49–50

49. A furniture company manufactures two types of desks using oak and cedar lumber. The first type requires 10 board feet of oak and 5 board feet of cedar, and the second requires 6 board feet of oak and 4 board feet of cedar. From a supply of 1000 board feet of oak and 600 board feet of cedar, how many desks of each type should be made in order to use all the material on hand?

50. A small manufacturing company can sell all the chairs and bar stools that it can produce. Each chair requires $1\frac{3}{5}$ hours in the assembly room and $\frac{2}{3}$ hour in the finishing room. Each bar stool requires $\frac{4}{5}$ hour in the assembly room and $\frac{4}{3}$ hours in the finishing room. If each of the two rooms can work on only one item at a time, find the number of chairs and stools that should be manufactured in order to keep both the assembly room and the finishing room operating 24 hours a day.

Printing Work

51. Upstate Printing has two machines that do screen printing on T-shirts. On Monday, both machines were operated together and filled an order in 2 hours. On another identical order, machine A was operated alone for 4 hours and then machine B was operated alone and finished the order in 1 hour. How long would it have taken machine B alone to fill the order?

Roofing Work

52. Dave and Chuck decided to earn some extra money by putting new roofs on houses. They signed a contract with a rental agency to reroof three identical rental houses owned by the agency. On the first house, Dave worked 8 hours and Chuck worked 10 hours to complete the job. To reroof the second house, Dave worked 12 hours and Chuck worked 5 hours. Just after the second job was completed, Dave fell from a ladder and broke his leg. How long will it take Chuck to do the third job alone?

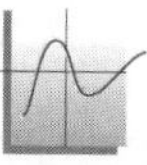

53–56

Use a graphing device to approximate the solution to the system by finding a point on each line with the same x-coordinate and y-coordinates that agree when rounded to the nearest hundredth. (Because of rounding, these approximations may not be unique.)

53. $2.37x - 4.81y = 16.5$
$6.03x + 3.22y = 11.9$

54. $54.3x + 25.7y = -510$
$49.1x - 32.5y = -911$

55. $y = \pi x - 5$
$y = \sqrt{3} - ex$

56. $y = \ln 4 - \dfrac{1}{\pi}x$
$y = \sqrt{2}x - \pi + 1$

Critical Thinking: **Exploration and Writing**

57. Consider a system of two linear equations in x and y. Describe a test for independence that could be applied without solving the system.

Critical Thinking: **Exploration and Discussion**

58. a. Suppose that

$$a_1x + b_1y = c_1$$
$$a_2x + b_2y = c_2$$

is a linear system in which the ratios a_1/a_2, b_1/b_2, and c_1/c_2 are all defined and

$$\frac{a_1}{a_2} = \frac{b_1}{b_2} = \frac{c_1}{c_2}.$$

What can be said about the number of solutions to the system?

b. Suppose the system described in part a is such that $a_1/a_2 = b_1/b_2$ but $a_1/a_2 \neq c_1/c_2$. What then can be said about the number of solutions to the system?

◆ Solutions for Practice Problems

1. It is easy to solve for x in the first equation.

$$x = 6 - 2y$$

Substituting this expression for x in the second equation, we get

$$\begin{aligned} 5(6 - 2y) + 3y &= 2 \\ 30 - 10y + 3y &= 2 \\ -7y &= -28 \\ y &= 4. \end{aligned}$$

Putting $y = 4$ in the expression for x yields

$$\begin{aligned} x &= 6 - 2(4) \\ &= -2. \end{aligned}$$

Thus $\{(-2, 4)\}$ is the solution set for the system.

2. We decide to eliminate y by multiplying the first equation by 5 and the second equation by -2 before adding.

$$\begin{array}{lcr} 3x - 2y = 12 & \xrightarrow{\text{Multiply by } 5} & 15x - 10y = 60 \\ 6x - 5y = -9 & \xrightarrow{\text{Multiply by } -2} & \underline{-12x + 10y = 18} \\ & \text{Add} & 3x \qquad = 78 \\ & & x = 26 \end{array}$$

Substituting 26 for x in the top original equation gives

$$\begin{aligned} 3(26) - 2y &= 12 \\ 78 - 2y &= 12 \\ -2y &= -66 \\ y &= 33. \end{aligned}$$

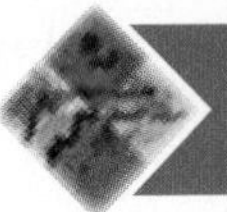

10.2 SYSTEMS OF LINEAR EQUATIONS IN THREE VARIABLES

The material in the last section extends in a natural way to linear systems of equations in three variables. A **linear system** of three equations in the variables x, y, and z has the form

$$\begin{aligned} a_1x + b_1y + c_1z &= d_1 \\ a_2x + b_2y + c_2z &= d_2 \\ a_3x + b_3y + c_3z &= d_3. \end{aligned}$$

The **solution set** of such a system is the set of all ordered triples (x, y, z) of real numbers with values of the variables that make all three equations true. To **solve** a system is to find its solution set.

Linear systems with three variables have a geometric interpretation in three-dimensional space that is analogous to the geometric interpretation of two-variable systems in terms of straight lines. A plane represents an equation $ax + by + cz = d$, and the solution set to a system is represented by a point, a straight line, or a plane. Although they are very valuable in certain advanced mathematics courses, these geometric representations are not useful enough in college algebra to justify the inclusion of their details in our presentation.

We concentrate in this section on a variation of the elimination method for solving a linear system in three variables. Other methods are explained in Chapter 11.

In the method that we shall use, the goal is to replace a given system by an equivalent system in which one of the three variables has been eliminated from two of the equations, to solve the subsystem that contains only two variables, and then to find the value of the third variable by substitution. The following example illustrates this method.

Example 1 • Solving a Linear System in Three Variables

Solve the following system.

$$\begin{aligned} x - y + z &= 9 \\ 3x + 2y - z &= 4 \\ 7x + 3y - 2z &= 5 \end{aligned}$$

Solution

Any one of the variables can be chosen as the one to be eliminated, and we decide to eliminate x. We use the first two equations as follows to obtain an equation without an x-term.

$$\begin{array}{lcr} x - y + z = 9 & \xrightarrow{\text{Multiply by } -3} & -3x + 3y - 3z = -27 \\ 3x + 2y - z = 4 & \xrightarrow{\hspace{2cm}} & 3x + 2y - z = 4 \\ & \text{Add} & \overline{5y - 4z = -23} \end{array}$$

This gives one of the needed equations in y and z. To obtain another equation without an x-term, we use the first and last equations in the original system as follows.

$$\begin{array}{rcr} x - y + z = 9 & \xrightarrow{\text{Multiply by } -7} & -7x + 7y - 7z = -63 \\ 7x + 3y - 2z = 5 & \longrightarrow & 7x + 3y - 2z = 5 \\ \hline & \text{Add} & 10y - 9z = -58 \end{array}$$

The original system is equivalent to the following system.

$$\begin{aligned} x - y + z &= 9 \\ 5y - 4z &= -23 \\ 10y - 9z &= -58 \end{aligned}$$

We can now use the elimination method on the subsystem in y and z as follows.

$$\begin{array}{rcr} 5y - 4z = -23 & \xrightarrow{\text{Multiply by } 2} & 10y - 8z = -46 \\ 10y - 9z = -58 & \xrightarrow{\text{Multiply by } -1} & -10y + 9z = 58 \\ \hline & \text{Add} & z = 12 \end{array}$$

We can find the value of y by substituting $z = 12$ in any of the y, z equations. Using $5y - 4z = -23$, we get

$$\begin{aligned} 5y - 4(12) &= -23 \\ 5y &= 25 \\ y &= 5. \end{aligned}$$

The value of x can now be found from any equation in which x appears. Using the top equation in the original system gives

$$\begin{aligned} x - 5 + 12 &= 9 \\ x &= 2. \end{aligned}$$

Thus the solution set to the system is $\{(2, 5, 12)\}$. □

The method used in Example 1 can be outlined as follows.

Elimination Method for the Solution of Linear Systems in Three Variables

1. Select a variable and use suitable multipliers to eliminate it from a pair of equations.
2. Eliminate the *same* variable from a *different* pair of equations by using appropriate multipliers.
3. Solve the resulting subsystem in two variables.
4. Substitute the values of these two variables into one of the original equations and solve for the third variable.

◆ Practice Problem 1 ◆

Solve the given system.

$$\begin{aligned} x - y + 2z &= 6 \\ 3x + y + 3z &= 7 \\ 2x - y + 3z &= 8 \end{aligned}$$

The possibilities for the solution set of a linear system in three variables are similar to those for two variables, and similar terminology is used. If there is a solution, the system is called **consistent.** Otherwise, it is **inconsistent.** If there is a unique (one and only one) solution, the system is called **independent.** If an infinite number of solutions exist, the system is **dependent** (and *consistent*).

If a system is inconsistent, the elimination method will lead to an impossible equation such as $0 = -13$. If a system is dependent, the elimination method will lead to a true equation of the form $0 = 0$. The elimination method is applied to an inconsistent system in Example 2. However, the solution of dependent systems in three variables is so much easier by the methods of Section 11.3 that we postpone their treatment until that section.

Example 2 • An Inconsistent System

Solve the system.

$$\begin{aligned} 3x - 2y + 2z &= -1 \\ 2x + y + 4z &= 2 \\ 9x - 6y + 6z &= -2 \end{aligned}$$

Solution

It is easier to eliminate y in this system because it is the only variable that has a coefficient equal to 1. We use the first and second equations as follows.

$$\begin{array}{lcl} 3x - 2y + 2z = -1 & \longrightarrow & 3x - 2y + 2z = -1 \\ 2x + y + 4z = 2 & \xrightarrow{\text{Multiply by 2}} & 4x + 2y + 8z = 4 \\ & \text{Add} & 7x \quad + 10z = 3 \end{array}$$

We obtain another equation in x and z by using the second and third equations in the following manner.

$$\begin{array}{lcl} 2x + y + 4z = 2 & \xrightarrow{\text{Multiply by 6}} & 12x + 6y + 24z = 12 \\ 9x - 6y + 6z = -2 & \longrightarrow & 9x - 6y + 6z = -2 \\ & \text{Add} & 21x \quad + 30z = 10 \end{array}$$

Working with the subsystem in x and z, we obtain the following results.

$$\begin{array}{lcl} 7x + 10z = 3 & \xrightarrow{\text{Multiply by } -3} & -21x - 30z = -9 \\ 21x + 30z = 10 & \longrightarrow & 21x + 30z = 10 \\ & \text{Add} & 0 = 1 \end{array}$$

The impossible equation $0 = 1$ indicates that the original system has no solution. □

Example 3 • Fish Kill

When a small lake was hit by a heat-related fish kill, 4500 pounds of dead fish had to be removed from the lake by three crews. Crew A removed fish by hand with buckets, crew B used wheelbarrows, and crew C used a front-end loader. Crew B removed twice as much fish as crew A, and crew C removed twice as much fish as crews A and B combined. Find the number of pounds of fish removed by each crew.

Reuters/Bettman

Workers scoop up fish killed by pollution and high temperatures in a lagoon at Rio de Janeiro, Brazil.

Solution

Let x, y, and z denote the number of pounds of fish removed by crews A, B, and C, respectively. Then $x + y + z = 4500$ since a total of 4500 pounds of dead fish was removed. We also have $y = 2x$ since crew B removed twice as much fish as crew A, and $z = 2(x + y)$ since crew C removed twice as much as crews A and B combined. This leads us to the system

$$\begin{aligned} x + y + z &= 4500 \\ y &= 2x \\ z &= 2(x + y), \end{aligned}$$

which can be rewritten as

$$\begin{aligned} x + y + z &= 4500 \\ 2x - y &= 0 \\ 2x + 2y - z &= 0. \end{aligned}$$

Adding the first and last equations, we get

$$3x + 3y = 4500 \quad \text{or} \quad x + y = 1500.$$

The last equation, together with the second equation, gives a subsystem in x and y that we can solve for x as follows.

$$\begin{aligned} x + y &= 1500 \\ 2x - y &= 0 \\ \hline 3x &= 1500 \\ x &= 500 \end{aligned}$$

The value $x = 500$ gives

$$\begin{aligned} 500 + y &= 1500 \\ y &= 1000. \end{aligned}$$

Finally, we substitute $x = 500$ and $y = 1000$ in the top original equation and obtain

$$\begin{aligned} 500 + 1000 + z &= 4500 \\ z &= 3000. \end{aligned}$$

Thus crew A removed 500 pounds of fish, crew B removed 1000 pounds of fish, and crew C removed 3000 pounds of fish. □

EXERCISES 10.2

Solve the following systems.

1. $\begin{aligned} x + y &= 1 \\ y - z &= 1 \\ x + y + z &= 2 \end{aligned}$

2. $\begin{aligned} x + 2z &= 3 \\ -2x + y &= 5 \\ y + 3z &= 9 \end{aligned}$

3. $\begin{aligned} x + y - 2z &= -3 \\ 2y + 4z &= 4 \\ x + z &= 5 \end{aligned}$

4. $\begin{aligned} x + y - z &= -1 \\ 2x + 3z &= 2 \\ 3y - 2z &= -6 \end{aligned}$

5. $\begin{aligned} x - 2y + 3z &= 3 \\ x - y + z &= 1 \\ x - y + 3z &= 5 \end{aligned}$

6. $\begin{aligned} 2x + 2y + z &= 3 \\ x + y + z &= 2 \\ 2x + y + z &= 5 \end{aligned}$

7. $\begin{aligned} 7x - 8y - z &= 3 \\ x - y &= -1 \\ 3x - 3y - z &= 4 \end{aligned}$

8. $\begin{aligned} 3x - 4y - 4z &= 5 \\ x + y + z &= 4 \\ x + 2y + z &= 0 \end{aligned}$

9. $\begin{aligned} x - 2y + z &= 1 \\ 3y - z &= -5 \\ x + y + 2z &= 0 \end{aligned}$

10. $\begin{aligned} x + 3y - z &= 6 \\ -x + 4y + 2z &= 16 \\ 2y - 3z &= 3 \end{aligned}$

11. $\begin{aligned} x - 2y - z &= 11 \\ 2x - y - z &= -5 \\ x - z &= -7 \end{aligned}$

12. $\begin{aligned} 3x + 2y + 4z &= 1 \\ -2x + y - 2z &= 1 \\ x + z &= 0 \end{aligned}$

13. $\begin{aligned} x + y - z &= 6 \\ -2x + 3z &= -14 \\ x - y - z &= 6 \end{aligned}$

14. $\begin{aligned} 2x + 3y - z &= 4 \\ -x + y + 2z &= 2 \\ 3x + 2y - 3z &= 5 \end{aligned}$

15. $\begin{aligned} x - y + z &= 0 \\ 4x + z &= 5 \\ 7x + y + z &= 11 \end{aligned}$

16. $\begin{aligned} x + 2y + 3z &= -2 \\ x - y - 4z &= 0 \\ -x + 7y + 18z &= 5 \end{aligned}$

17. $\begin{aligned} 4x + y + 2z &= -2 \\ x - y - z &= -9 \\ -5x - 5y - 7z &= 100 \end{aligned}$

18. $\begin{aligned} 2x - y + 2z &= -9 \\ x + 2y + 3z &= 0 \\ 5x + 5y + 11z &= 9 \end{aligned}$

19. $\begin{aligned} 2x - 3y + z &= 1 \\ x - y + 2z &= 0 \\ x + y + 3z &= 3 \end{aligned}$

20. $\begin{aligned} 3x - 2y + z &= 0 \\ -2x + 2y - z &= 3 \\ 4x - y - z &= 9 \end{aligned}$

21. $\begin{aligned} \frac{3}{x} + \frac{1}{y} + \frac{1}{z} &= 5 \\ \frac{1}{x} + \frac{2}{y} - \frac{1}{z} &= -3 \\ \frac{1}{x} + \frac{1}{y} + \frac{3}{z} &= 3 \end{aligned}$

22. $\begin{aligned} \frac{1}{x} - \frac{2}{y} - \frac{3}{z} &= 3 \\ \frac{2}{x} - \frac{3}{y} - \frac{12}{z} &= 20 \\ \frac{1}{x} - \frac{1}{y} - \frac{5}{z} &= 9 \end{aligned}$

23. $\begin{aligned} \frac{1}{x} - \frac{2}{y} + \frac{3}{z} &= -5 \\ \frac{2}{x} - \frac{1}{y} + \frac{1}{z} &= 2 \\ \frac{8}{x} + \frac{3}{y} + \frac{2}{z} &= 7 \end{aligned}$

24. $\begin{aligned} \frac{3}{x} - \frac{2}{y} + \frac{9}{z} &= 1 \\ \frac{2}{x} + \frac{1}{y} + \frac{6}{z} &= 3 \\ \frac{2}{x} - \frac{4}{y} + \frac{6}{z} &= 2 \end{aligned}$

25. Find a, b, and c so that the parabola with equation $y = ax^2 + bx + c$ contains the points $(-1, 1)$, $(0, 3)$, and $(1, 9)$.

26. Find a, b, and c so that the parabola with equation $x = ay^2 + by + c$ contains the points $(-1, -1)$, $(2, 0)$, and $(9, 1)$.

27. Find the equation $x^2 + y^2 + ax + by + c = 0$ of a circle that contains the points $(-1, -1)$, $(2, 2)$, and $(5, -1)$.

28. Find the equation $x^2 + y^2 + ax + by + c = 0$ of a circle that contains the points $(1, -5)$, $(-1, -3)$, and $(1, -1)$.

Number Problems 29–30

29. Thirty girls played on three basketball teams. The number of girls on the third-grade team was 8 less than the total on the fourth- and fifth-grade teams. The number of girls on the fifth-grade team was one-half the total on the third- and fourth-grade teams. How many girls played on the third-grade team?

30. The sum of three numbers is 27. The largest is 8 more than the smallest, and twice the middle number equals the sum of the largest and smallest. Find the three numbers.

Work Problems 31–32

31. An oil tanker can be filled by three pumps working together in 2 days. The smallest and largest pumps together can fill the tanker in 3 days, while the smallest and middle-sized pumps require 4 days to fill the tanker. Find the time that it would take for the smallest pump to fill the tanker.

AP/Wide World Photos

Oil is pumped from the tanker *Exxon Valdez* onto a smaller tanker during the disastrous oil spill in Prince Williams Sound during March, 1989.

32. A swimming pool has pipes A and B that can be used to fill the pool. With the drain to the pool left open, pipe A can fill the pool in 6 hours, and pipe B can fill the pool with the drain open in 4 hours. Both pumps together can fill the pool with the drain open in 2 hours. How long would it take pipe A to fill the pool with the drain closed?

Angles in a Triangle

33. The sum of the angles in any triangle is 180°. In a certain triangle, the largest angle is 20° less than the sum of the other two. Also, the largest angle is 10° less than twice the smallest angle. Find the three angles in the triangle.

Mixing Acid

34. A chemist has acid solutions A, B, and C with strengths 15%, 20%, and 30%, respectively. She needs to mix these solutions and prepare 350 liters of a 25% solution. Because a large quantity of solution C is on hand, she decides to use twice as much solution C as solution A. Find the amount of each solution that she uses.

Mixing Fertilizers 35. Lawn and garden fertilizers are labeled with numbers such as 8-8-8 and 5-10-5. The numbers listed state the percentage of nitrogen, phosphoric acid, and potash, in that order. The manager of a garden center has three types on hand that are labeled 8-4-8, 15-30-15, and 12-6-12. How many pounds of each type should be used to make 200 pounds of a 13-20-13 mixture?

Mixing Nuts 36. Matt is supplying a toasted nut mixture to a chain of convenience stores. The store management requires that the mixture contain half as many cashews as it does peanuts. To be competitive, he must sell his mix for $3.75 per pound. To stay in business, he needs to get $3.00 per pound for peanuts, $4.00 per pound for cashews, and $6.00 per pound for pecans. How many pounds of each nut should he put in to fill an order for 64 pounds?

Critical Thinking: Exploration and Discussion 37. Discuss a procedure for applying the substitution method to a linear system with three equations in three variables. Then write out steps that outline the substitution method.

Critical Thinking: Exploration and Writing 38. Give a reason why we might normally expect the elimination method to be easier to use than the substitution method for a linear system with three equations in three variables.

◆ Solution for Practice Problem

1. We decide to eliminate y and obtain a subsystem in x and z. Adding the first two equations, we have

$$\begin{array}{rcl} x - y + 2z &=& 6 \\ 3x + y + 3z &=& 7 \\ \hline 4x \qquad + 5z &=& 13. \end{array}$$

Adding the second and third equations, we have

$$\begin{array}{rcl} 3x + y + 3z &=& 7 \\ 2x - y + 3z &=& 8 \\ \hline 5x \qquad + 6z &=& 15. \end{array}$$

We solve for x in the x, z subsystem as follows.

$$\begin{array}{lcrcr} 4x + 5z = 13 & \xrightarrow{\text{Multiply by } 6} & 24x + 30z &=& 78 \\ 5x + 6z = 15 & \xrightarrow{\text{Multiply by } -5} & -25x - 30z &=& -75 \\ \hline & \text{Add} & -x &=& 3 \\ & & x &=& -3 \end{array}$$

Substituting $x = -3$ in $4x + 5z = 13$ yields

$$\begin{aligned} -12 + 5z &= 13 \\ 5z &= 25 \\ z &= 5. \end{aligned}$$

Putting $x = -3$ and $z = 5$ in the second equation of the original system yields

$$-9 + y + 15 = 7$$
$$y = 1.$$

Thus the solution set is $\{(-3, 1, 5)\}$.

10.3 SYSTEMS INVOLVING NONLINEAR EQUATIONS

The substitution method used with linear systems in Section 10.1 can also be applied to systems that involve nonlinear equations. An illustration is provided in the following example.

Example 1 • The Substitution Method

Solve the following system. Graph the two equations on the same coordinate system and label the points of intersection.

$$y^2 - 5y - 4x - 28 = 0$$
$$y - 4x = 1$$

Solution

It is convenient here to solve for y in the second equation, obtaining

$$y = 4x + 1.$$

Substituting into the first equation, we have

$$(4x + 1)^2 - 5(4x + 1) - 4x - 28 = 0,$$

which simplifies to

$$16x^2 - 16x - 32 = 0$$

and

$$x^2 - x - 2 = 0.$$

This factors as

$$(x - 2)(x + 1) = 0.$$

Thus the x-coordinates of the points of intersection are $x = 2$ and $x = -1$. When these values are substituted into $y = 4x + 1$, we find the points of intersection are given by $(2, 9)$, and $(-1, -3)$.

To draw the graph of the first equation, we solve for x, obtaining

$$x = \tfrac{1}{4}y^2 - \tfrac{5}{4}y - 7.$$

We recognize this as the equation of a parabola that opens to the right with vertex at $y = -b/(2a) = \frac{5}{2}$, $x = -\frac{137}{16}$. The graph of $y - 4x = 1$ is a straight line with slope 4 and y-intercept 1. The graphs and the points of intersection are shown in Figure 10.3.

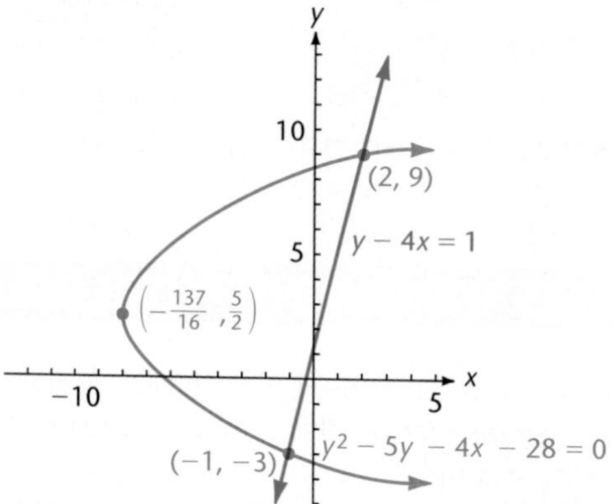

Figure 10.3 The graphs in Example 1

One more point should be made before leaving this example. It is important to notice that when the values $x = 2$ and $x = -1$ were obtained, they were substituted into the equation for the line and not into the equation for the parabola. If they had been used in the equation for the parabola, *two* values of y would have been obtained for each x, and only one of these values would have satisfied the linear equation. For example, if $x = -1$ is substituted in $y^2 - 5y - 4x - 28 = 0$, we obtain $y = -3$ and $y = 8$. The value $y = 8$ is an extraneous solution since $(-1, 8)$ is not on the line. Similarly, $x = 2$ yields $y = 9$ and $y = -4$, with $y = -4$ an extraneous solution. □

The elimination method introduced in Section 10.1 can also be extended to nonlinear systems of equations. The key to success with the elimination method is to choose multipliers for each equation in such a way as to have one of the variables eliminated when the sum is formed. This method is used in the next example.

Example 2 • The Elimination Method

Solve the following system of equations by the elimination method. Sketch the graphs of the two equations on the same coordinate system and label the points of intersection.

$$\begin{aligned} 9x^2 + y^2 &= 36 \\ x^2 + 3y^2 &= 56 \end{aligned}$$

Solution

If we multiply the first equation by 3, the second equation by -1, and then add, we have

$$\begin{aligned} 27x^2 + 3y^2 &= 108 \\ -x^2 - 3y^2 &= -56 \\ \hline 26x^2 \qquad &= 52. \end{aligned}$$

Thus y is eliminated, and this equation simplifies to

$$x^2 = 2$$

and

$$x = \pm\sqrt{2}.$$

In order to complete the solution, we substitute $x^2 = 2$ into the first equation and obtain

$$\begin{aligned} 9(2) + y^2 &= 36 \\ y^2 &= 18 \end{aligned}$$

and

$$y = \pm 3\sqrt{2}.$$

This gives the four pairs

$$(\sqrt{2}, 3\sqrt{2}), (\sqrt{2}, -3\sqrt{2}), (-\sqrt{2}, 3\sqrt{2}), (-\sqrt{2}, -3\sqrt{2}),$$

and each of these checks in the original system. It is good practice to *always check solutions in each of the original equations.* This is especially true when extraction of roots has been employed in obtaining the solutions. The graphs and points of intersection are shown in Figure 10.4. □

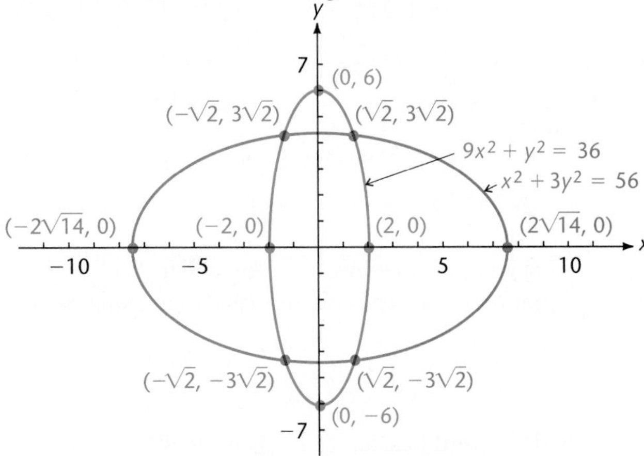

Figure 10.4 The graphs in Example 2

Graphs that picture the solutions of the system have been provided in each of the preceding examples in this section. These are valuable for understanding, and they provide a rough check on the solutions. They are not indispensable, however, and we omit them in the remainder of the section.

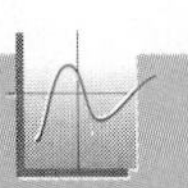

When graphing a system of equations on a graphing device, it may be necessary to change the viewing rectangle in order to exhibit all points of intersection.

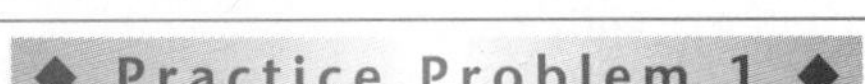

Solve the following system.

$$\begin{aligned} x^2 + y^2 &= 5 \\ y^2 - 4x &= 9 \end{aligned}$$

The two methods of solution that have been presented in this section are not adequate for the solution of all types of systems that might be encountered. The examples presented here were chosen because they lend themselves to solution by these methods. Some more difficult problems can be solved by a combination of the two methods. This is illustrated in the next example.

Example 3 • Using a Combination of the Methods

Solve the following system.

$$\begin{aligned} x^2 + 2xy + 2y^2 &= 5 \\ x^2 + xy + 2y^2 &= 4 \end{aligned}$$

Solution

It is impossible to employ the elimination method directly. Further, the substitution method is not an attractive prospect since the quadratic formula would have to be used to obtain one variable in terms of the other. However, all terms involving a square can be eliminated by subtracting the second equation from the first. When this is done, we have

$$\begin{aligned} x^2 + 2xy + 2y^2 &= 5 \\ \underline{x^2 + xy + 2y^2} &\underline{= 4} \\ xy \quad\quad &= 1 \end{aligned}$$

and

$$y = \frac{1}{x}.$$

This gives us an expression for y that can be substituted into either of the original equations. Substitution into the first equation yields

$$x^2 + 2x\left(\frac{1}{x}\right) + 2\left(\frac{1}{x}\right)^2 = 5,$$

which simplifies to

$$x^2 + \frac{2}{x^2} = 3.$$

Clearing the equation of fractions, we have

$$x^4 + 2 = 3x^2$$

and

$$x^4 - 3x^2 + 2 = 0.$$

This factors as

$$(x^2 - 2)(x^2 - 1) = 0.$$

Setting $x^2 - 2 = 0$, we obtain $x = \pm\sqrt{2}$, and $x^2 - 1 = 0$ yields $x = \pm 1$. Using each of these values for x in the equation $y = 1/x$, we obtain the solution set

$$\{(\sqrt{2}, \sqrt{2}/2), (-\sqrt{2}, -\sqrt{2}/2), (1, 1), (-1, -1)\}.$$

It is easy to confirm that all these solutions check in the original system. The interested reader may wish to verify that substitution of the values for x in either of the original equations yields extraneous solutions for y. □

As stated in the first paragraph of this chapter, we are restricting our work in this chapter to real numbers. It may well happen that a given system of equations does not have any real solutions. This is illustrated below.

Example 4 • An Inconsistent System

Solve the following system.

$$\begin{aligned} x^2 + y^2 &= 1 \\ x^2 - y \quad &= 4 \end{aligned}$$

Solution

Subtracting the second equation from the first, we have

$$\begin{array}{l} x^2 + y^2 = 1 \\ \underline{x^2 - y \quad = 4} \\ y^2 + y \quad = -3 \end{array}$$

or

$$y^2 + y + 3 = 0.$$

The discriminant's value for this equation is $b^2 - 4ac = (1)^2 - 4(1)(3) = -11$, so there is no real solution for y. This in turn indicates that the original system has no real solution. That is, the circle $(x^2 + y^2 = 1)$ and the parabola $(x^2 - y = 4)$ do not intersect. □

Example 5 • Constructing a Can

Suppose an aluminum can with volume 64π cubic inches is to be constructed so that the amount of aluminum used to form the two ends is the same as the amount used in the side of the can. Find the dimensions of such a can if it is in the shape of a right circular cylinder.

Solution

We draw the right circular cylinder in Figure 10.5a with height h and radius of the ends r. A flat sheet of aluminum as seen in Figure 10.5b with width h and length $2\pi r$ (the circumference of the circle) is used to form the side of the can. Its area is $2\pi rh$. The sum of the area of the ends is $2\pi r^2$. We obtain a nonlinear system of

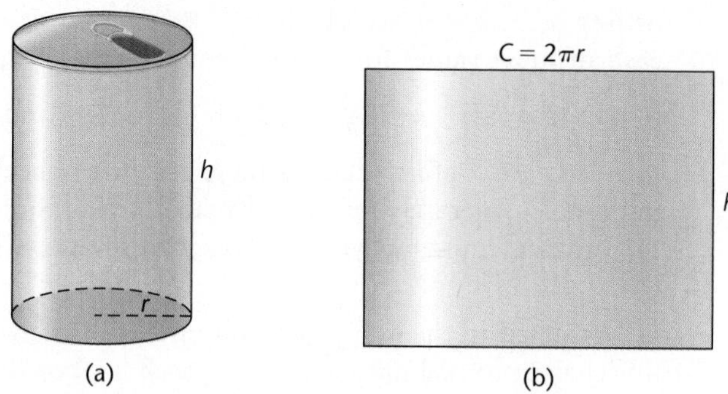

(a) (b)

Figure 10.5 The areas involved in Example 5

equations by setting the area of the ends equal to the area of the side and by using the formula for the volume of the cylinder.

Areas: $2\pi r^2 = 2\pi rh$

Volume: $\pi r^2 h = 64\pi$

Solving for h in terms of r in the second equation yields

$$h = \frac{64}{r^2}.$$

Then, dividing both sides of the first equation by 2π and substituting $h = 64/r^2$ yields an equation in r.

$$r^2 = rh \qquad \text{dividing by } 2\pi$$

$$r^2 = r\frac{64}{r^2} \qquad \text{substituting}$$

$$r^3 = 64 \qquad \text{simplifying}$$

The only real solution to this equation is $r = 4$. Then $h = 64/r^2 = 64/16 = 4$. Thus the can will have ends of radius 4 inches and will be 4 inches tall. □

EXERCISES 10.3

Solve each of the following systems for solutions in the real numbers only. Sketch the graphs of the two equations on the same coordinate system and label the points of intersection. (See Examples 1, 2, and 4.)

1. $y = 2x + 6$
   $x^2 = 2y$

2. $x^2 - y = 0$
   $2x - y + 3 = 0$

3. $y = 3x^2 + 12x$
   $2x - y = 16$

4. $y = 3 - 2x - x^2$
   $4x + y = 5$

5. $x^2 + y^2 = 4$
   $2x - y = 2$

6. $x^2 + y^2 = 25$
   $y - x = 7$

7. $x^2 + y^2 = 10x$
$4y = 3x - 8$

8. $x^2 - 2y^2 = -1$
$2x - y = -1$

9. $3x^2 + 2y^2 = 5$
$3x + 2y = -1$

10. $4x^2 + 25y^2 = 100$
$x + 2y = 8$

11. $36x^2 + 16y^2 = 25$
$x - 2y = 6$

12. $9x^2 + 16y^2 = 144$
$3x^2 + 4y^2 = 36$

13. $x^2 + y^2 = 25$
$(3x - 4y)(3x + 4y) = 0$

14. $4x^2 + 36y^2 = 100$
$(x - 4y)(x + 4y) = 0$

15. $x^2 + y^2 = 9$
$y^2 + 2x = 10$

16. $x^2 + y^2 = 4$
$x^2 - 2y = 1$

Solve the following systems of equations. It is not necessary to graph the equations. (See Examples 3–5.)

17. $x^2 + y^2 = 1$
$x^2 - y^2 = 1$

18. $y^2 + x^2 = 25$
$3y^2 - x^2 = 11$

19. $\frac{x^2}{9} + \frac{y^2}{16} = 1$
$x^2 + y^2 = 1$

20. $\frac{x^2}{25} + \frac{y^2}{4} = 1$
$x^2 + y^2 = 1$

21. $2x + 7y = 15$
$xy = 1$

22. $6x + y + 7 = 0$
$xy = 2$

23. $3x + 2y = -5$
$xy = -1$

24. $4x + 3y = 16$
$xy = -3$

25. $9y^2 - 4x^2 = 7$
$xy = -2$

26. $2x^2 + y^2 = 19$
$xy = 3$

27. $2x^2 + 3xy + y^2 = 12$
$2x^2 - xy + y^2 = 4$

28. $2x^2 - 3xy - 3y^2 = 11$
$2x^2 - xy - 3y^2 = 7$

29. $5x^2 - 4xy - 3y^2 = -8$
$5x^2 + 5xy - 3y^2 = 28$

30. $8x^2 + 6xy - 9y^2 = 8$
$8x^2 - 3xy - 9y^2 = 17$

31. $x^2 - 3y^2 = -2$
$xy + 2y^2 = 3$

32. $3x^2 - 4xy = 25$
$2x^2 - 4y^2 = 9$

33. $x^2 - y = 2$
$x = |y|$

34. $y - |x| = 2$
$y - x^2 = 0$

35. $x^2 + y^2 = 25$
$y - |x| = 1$

36. $x^2 + y^2 = \frac{25}{9}$
$3y - 4|x| = 0$

Break-Even Point 37. Suppose the cost $C(x)$ and the revenue $R(x)$ of producing and selling x items are given by $C(x) = 2000 - 50x$ and $R(x) = 50x$, where $0 \leq x \leq 30$. Find the break-even point, that is, the point where cost and revenue are equal.

Digits Problem

38. The sum of the digits of a two-digit number is 8. If the digits are reversed, the new number is 71 less than twice the original number. What is the original number?

Number Problems 39–41

39. Find two numbers whose difference is 15 and whose product is 76.

40. Find two numbers whose sum is 29 if the difference of their squares is 377.

41. Find two numbers such that the sum of their squares is 65 and the difference of their squares is 33.

Dimensions of a Rectangle

42. Find the dimensions of a rectangle if its perimeter is 42 feet and its area is 108 square feet.

Line Tangent to a Circle 43–44

43. Find a value of the real number m so that the line $y = mx - 5$ is tangent to the circle $x^2 + y^2 = 9$.

44. Find a value of the real number a so that the line $ax - 12y + 65 = 0$ is tangent to the circle $x^2 + y^2 = 25$ in the accompanying figure.

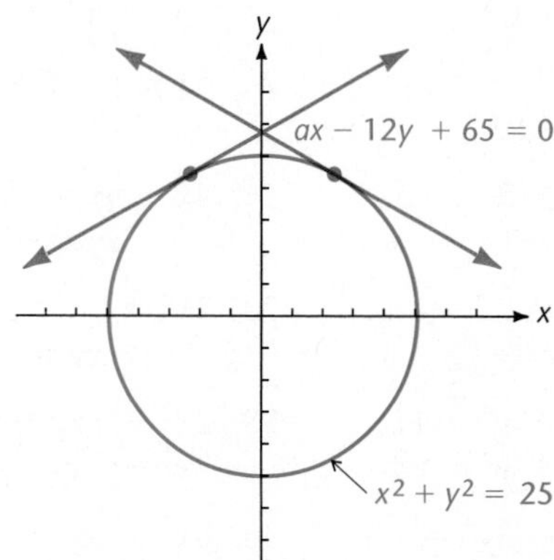

Figure for Exercise 44

Break-Even Point

45. Suppose it has been determined that the cost $C(x)$ and the revenue $R(x)$ (both in thousands of dollars) of producing and selling x hundred items are given by $C(x) = 24 + 20x$ and $R(x) = x(10 + x)$, $x \geq 0$. Find the break-even point, that is, the point where cost and revenue are equal.

Fencing Costs

46. The city council of Midcity plans to fence off a 1200-square-meter area next to an existing building for a playground. No fence is needed for the side against the building. Fencing material for the side parallel to the building costs \$4 per meter, whereas fencing for the other two sides costs

\$3 per meter. Find the dimensions of the playground that can be so constructed if the council spends \$340 on fencing.

Custom Construction

47. A wooden freight carrier with a volume of 24 cubic feet is to be specially constructed with ends in the shape of equilateral triangles, as shown in the accompanying figure. Find the dimensions s and ℓ if 3 times more wood is used for the three sides than for the two ends.

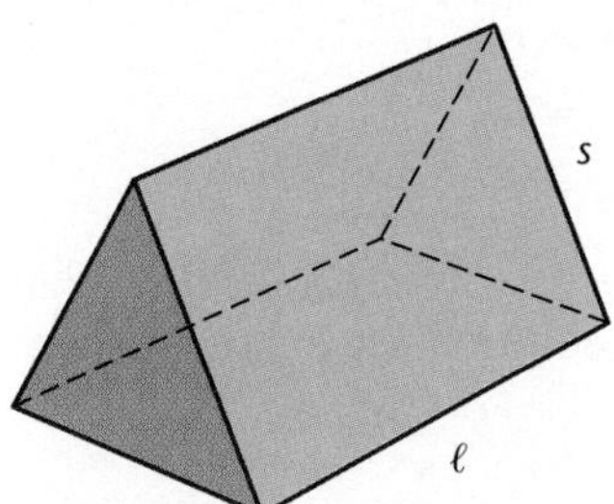

Figure for Exercise 47

Dimensions of a Box

48. Find the dimensions of a 12-cubic-foot box with square bottom and top that can be constructed from 32 square feet of cardboard (assuming no waste in construction).

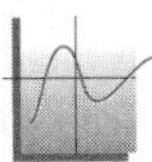

49–52

Use a graphing device to approximate the solutions to the system by finding a point on each graph with the same x-coordinate and y-coordinates that agree when rounded to the nearest hundredth. (Because of rounding, these approximations may not be unique.)

49. $y = \ln x$
 $y = e^x - e$

50. $y = x^2 + x - 2$
 $y = \sqrt{x + 2}$

51. $y = \sqrt{4 - x^2}$
 $y = \log(x + 1) + 1$

52. $y = \sqrt{1 + x^2}$
 $y = \dfrac{2}{x}$

Critical Thinking: Exploration and Writing

53. Although our work in this chapter is normally restricted to real numbers, any of the systems can be considered in the set of all complex numbers. Find the solutions in the complex numbers for the system in Example 4 in this section. Explain how these solutions can exist even though the circle $x^2 + y^2 = 1$ and the parabola $x^2 - y = 4$ do not intersect.

Critical Thinking: Exploration and Discussion

54. (See Exercise 53.) If a system consists of an equation of a circle and an equation of a straight line, discuss why the system will always have a solution in the complex numbers.

◆ Solution for Practice Problem

1. We can eliminate y by subtracting the second equation from the first.

$$\begin{aligned} x^2 + y^2 &= 5 \\ y^2 - 4x &= 9 \\ \hline x^2 + 4x &= -4 \end{aligned}$$

Solving for x in this equation, we get

$$\begin{aligned} x^2 + 4x + 4 &= 0 \\ (x+2)^2 &= 0 \\ x &= -2. \end{aligned}$$

Substitution of $x = -2$ in either of the original equations yields $y^2 = 1$ and $y = \pm 1$. Thus the solution set for the system is $\{(-2, 1), (-2, -1)\}$.

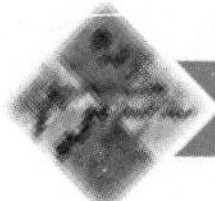

10.4 SYSTEMS OF INEQUALITIES

We have seen that a solution to a given equation in x and y is a set of values for x and y that satisfies the equation, and that the graph of the equation consists of the points with coordinates (x, y) that correspond to solutions of the equation. The situation is much the same for inequalities involving x and y. A **solution** to a given inequality in x and y is a set of values for x and y that makes the given inequality a true statement. The **graph** of the inequality consists of the points with coordinates that correspond to solutions of the inequality.

The simplest type of inequality in x and y is one that is obtained from a linear equality $ax + by = c$ by replacing the equality sign with one of the inequality symbols, $>$, $<$, $\geq$, or $\leq$. Such inequalities are called **linear inequalities.**

Suppose that a certain line in the plane has an equation given by $ax + by = c$. The line separates the points of the plane into three distinct subsets:

1. The points (x, y), where $ax + by = c$,
2. The points (x, y), where $ax + by > c$,
3. The points (x, y), where $ax + by < c$.

Geometrically, the points described in (1) are the points on the line, the points described in (2) are the points on one side of the line, and the points described in (3) are the points on the other side of the line. Each of the regions described in (2) and (3) is called a **half-plane,** and the line is the **boundary** of the half-planes.

Example 1 • Graphing a Half-Plane

Sketch the graph of $3x + 4y > 24$.

Solution

We first locate the points on the line $3x + 4y = 24$. It is easy to see that the x-intercept of the line is 8, and the y-intercept is 6. The line is drawn in Figure 10.6 as a dashed line to indicate that the points on the boundary are not part of the solution set. To see how the solutions to $3x + 4y > 24$ consist of all points on one side of the line, let us start with a particular point (x_0, y_0) on the line. If the x-coordinate increases to a new value $x > x_0$, the value of $3x$ increases and

$$3x + 4y_0 > 3x_0 + 4y_0 = 24.$$

That is, $3x + 4y_0 > 24$ for a point (x, y_0) to the right of (x_0, y_0) on the line. Any point on the right-hand side of the line is located to the right of some point on the line, so we see that all points on the right-hand side of the line satisfy $3x + 4y > 24$. Similar reasoning shows that the inequality $3x + 4y < 24$ holds for all points on the left side of the line. The solution set for $3x + 4y > 24$ is shaded in Figure 10.6, and the side of the line that yields solutions is indicated by arrows based on the line. □

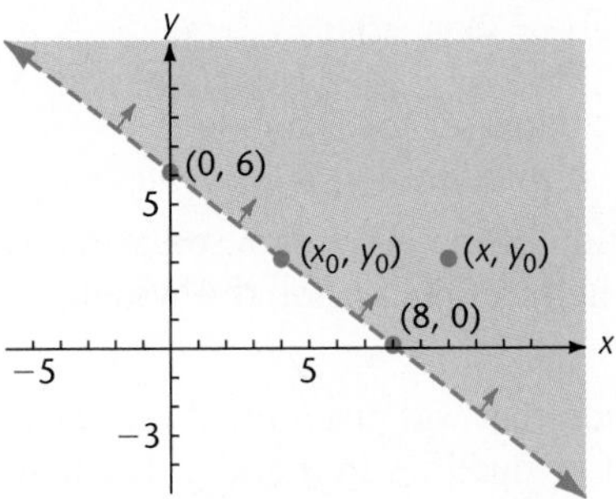

Figure 10.6 The half-plane $3x + 4y > 24$

The **shade** feature of a graphing device can be used to shade a region that lies between the graphs of two functions.

The **solution set** for a system of inequalities is the intersection of the solution sets for the individual inequalities in the system. An example of this is furnished below.

Example 2 • A Linear System of Inequalities

Graph the solution set of the following system.

$$3x + 4y > 24$$
$$x - 2y \geq -2$$

Solution

We use the solution of the first inequality from Example 1. This solution set is indicated in Figure 10.7 with shading and by blue arrows based on the boundary.

To graph the solution set of the second inequality, we first sketch the line $x - 2y = -2$. The x-intercept is -2 and the y-intercept is 1. To determine the half-plane which is the solution to $x - 2y > -2$, we simply test a point on one side of the line. The origin $(0, 0)$ is the simplest choice to use, and $0 > -2$ indicates that the solutions to $x - 2y > -2$ are on the same side of the line as the origin. This region is indicated in Figure 10.7 with shading and by pink arrows based on the boundary. The boundary is included in the solution set since equality is included in $x - 2y \geq -2$.

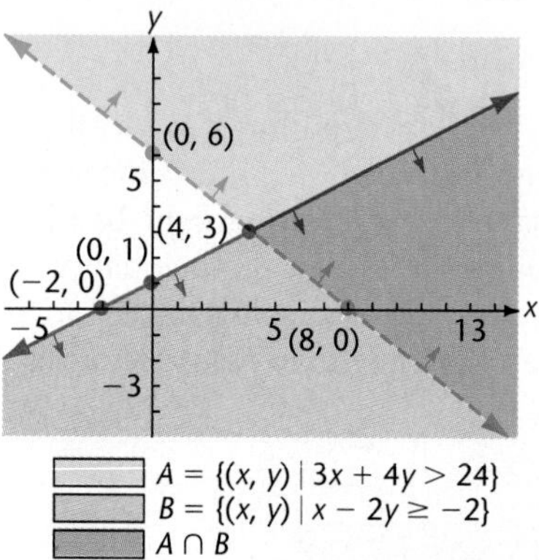

Figure 10.7 Solution set for Example 2

The intersection of the individual solution sets is indicated by the darker blue shading and is the solution set for the system. □

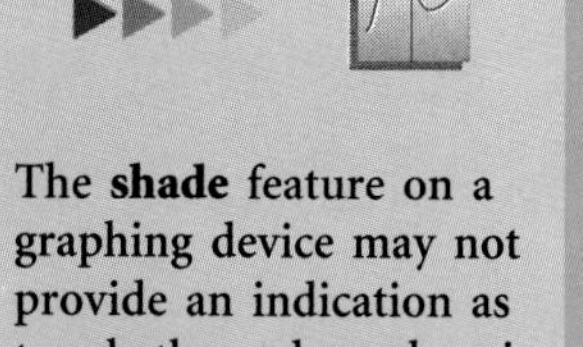

The **shade** feature on a graphing device may not provide an indication as to whether a boundary is included or omitted.

The description that was given for the graphs of linear inequalities generalizes to other types of inequalities in x and y. For a given inequality, a corresponding equation may be obtained by replacing the inequality symbol by an equality symbol. The graph of the equation separates the plane into regions, and it forms the boundary of these regions. Each region either consists entirely of solutions or contains no solutions at all. The solution set can be determined by simply testing one point from each region. The boundary is included or omitted, according to whether or not equality is permitted in the original statement of inequality.

Example 3 • A Nonlinear System of Inequalities

Graph the solution set for the following system.

$$\begin{aligned} x^2 + y^2 &\leq \tfrac{25}{9} \\ 16x - 9y^2 &> 0 \end{aligned}$$

Solution

We begin by graphing the two equations that correspond to the given inequalities.

The equation

$$x^2 + y^2 = \tfrac{25}{9}$$

is an equation of a circle with center $(0, 0)$ and radius $\frac{5}{3}$. If we solve for x in $16x - 9y^2 = 0$, we have

$$x = \tfrac{9}{16}y^2,$$

which is an equation of a parabola that has its vertex at the origin and opens to the right. To find the points of intersection, we can substitute $y^2 = \frac{16}{9}x$ into the equation of the circle. This gives

$$x^2 + \tfrac{16}{9}x = \tfrac{25}{9}$$

or

$$9x^2 + 16x - 25 = 0.$$

This can be factored as

$$(9x + 25)(x - 1) = 0,$$

so we obtain $x = -\frac{25}{9}$ and $x = 1$ as the solutions to this equation. Substituting $x = -\frac{25}{9}$ into $y^2 = \frac{16}{9}x$ yields

$$\begin{aligned} y^2 &= \tfrac{16}{9}\left(-\tfrac{25}{9}\right) \\ &= -\tfrac{400}{81}, \end{aligned}$$

which is impossible for a real number y. This indicates that there is no point of intersection at $x = -\frac{25}{9}$. Substituting $x = 1$ in $y^2 = \frac{16}{9}x$ yields

$$y^2 = \tfrac{16}{9}$$

and

$$y = \pm\tfrac{4}{3}.$$

The coordinates $(1, \frac{4}{3})$ and $(1, -\frac{4}{3})$ check in both original equations and give the points of intersection.

It is clear that the solutions to $x^2 + y^2 \leq \frac{25}{9}$ are the points interior to, or on, the circle $x^2 + y^2 = \frac{25}{9}$. To determine which points satisfy $16x - 9y^2 > 0$, we choose $(3, 0)$ as a test point. Since

$$16(3) - 9(0)^2 > 0,$$

$(3, 0)$ is a solution. This indicates that the solution set for $16x - 9y^2 > 0$ is the region to the right of the parabola $16x - 9y^2 = 0$. The solution set for the system, then, is the set of points that lie to the right of the parabola and on, or interior to, the circle. This is shown in Figure 10.8. □

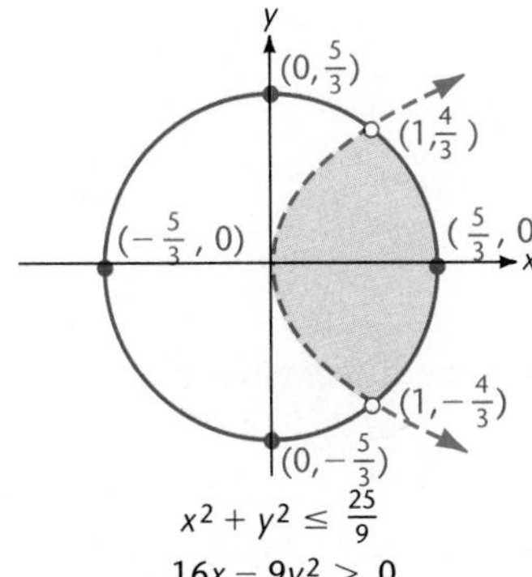

Figure 10.8 Solution set for Example 3

◆ Practice Problem 1 ◆

Graph the solution set for the following system.

$$\begin{aligned} x^2 - y^2 &\geq 16 \\ x^2 + 16y^2 &\leq 169 \end{aligned}$$

EXERCISES 10.4

Graph the solution set of the given inequality.

1. $3x + 2y \le 12$
2. $3x - 2y \ge 6$
3. $y > (x - 1)^2$
4. $y < x^2 - 1$
5. $y \le 4 - x^2$
6. $y \ge x^2 - 4x$
7. $x^2 + y^2 \le 16$
8. $(x - 1)^2 + (y + 2)^2 \le 9$

In Exercises 9–24, graph the solution set of the given system of inequalities, showing all points of intersection of the boundaries. These correspond to the problems numbered 1–16 in Exercises 10.3.

9. $y \le 2x + 6$
   $x^2 \le 2y$
10. $x^2 - y \le 0$
    $2x - y + 3 \ge 0$
11. $y > 3x^2 + 12x$
    $2x - y \ge 16$
12. $y < 3 - 2x - x^2$
    $4x + y \ge 5$
13. $x^2 + y^2 \le 4$
    $2x - y < 2$
14. $x^2 + y^2 \le 25$
    $y - x < 7$
15. $x^2 + y^2 < 10x$
    $4y > 3x - 8$
16. $x^2 - 2y^2 < -1$
    $2x - y < -1$
17. $3x^2 + 2y \le 5$
    $3x + 2y < -1$
18. $4x^2 + 25y^2 \le 100$
    $x + 2y > 8$
19. $36x^2 + 16y^2 \le 25$
    $x - 2y > 6$
20. $9x^2 + 16y^2 < 144$
    $3x^2 + 4y^2 \ge 36$
21. $x^2 + y^2 \le 25$
    $(3x - 4y)(3x + 4y) > 0$
22. $4x^2 + 36y^2 \le 100$
    $(x - 4y)(x + 4y) > 0$
23. $x^2 + y^2 \ge 9$
    $y^2 + 2x \le 10$
24. $x^2 + y^2 \le 4$
    $x^2 - 2y \le 1$

In Exercises 25–36, graph the solution set of the given system. Label all points of intersection of the boundaries.

25. $x - y > 3$
    $x + 2y > 4$
26. $3x + 2y < 6$
    $x - y > 2$
27. $y \le 2x + 2$
    $y + x + 1 \ge 0$
    $2y + 5x \le 13$
28. $y + 6 > 3x$
    $x + y < 6$
    $x \ge 1$
29. $x^2 + y^2 > 4$
    $4x^2 + 9y^2 \le 36$
30. $x^2 + y^2 \ge 9$
    $9x^2 + 16y^2 < 144$
31. $x^2 + 2y^2 \ge 18$
    $2x^2 + y^2 \le 33$
32. $9x^2 + 16y^2 \le 25$
    $16x^2 + 9y^2 \ge 25$
33. $x^2 - y^2 > 1$
    $4x^2 + 9y^2 < 36$
34. $x^2 + 2y^2 \ge 64$
    $x^2 - y^2 \le 16$

35. $y \geq 2^x$
$y \geq 2^{-x}$

36. $y < \dfrac{1}{x}$
$y < x$
$x \geq 0$
$y \geq 0$

In Exercises 37–44, write a system of inequalities that has the shaded region for its solution set.

37.

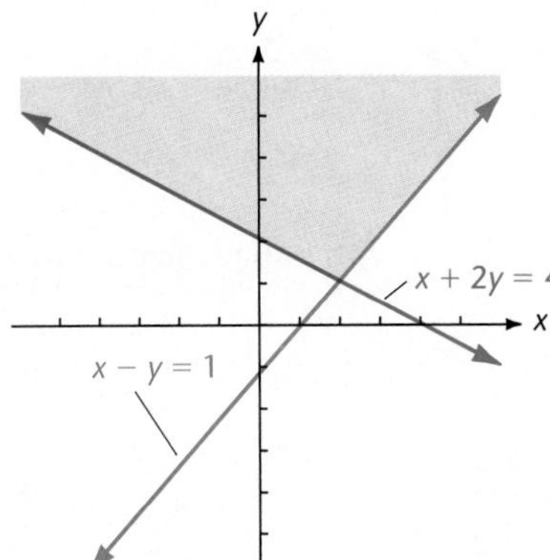

38.

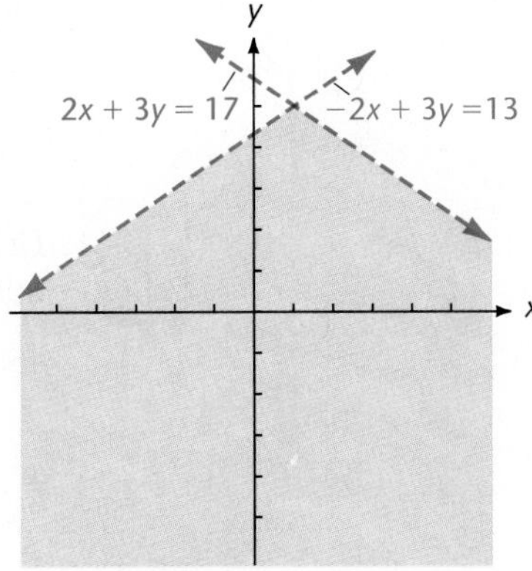

39.

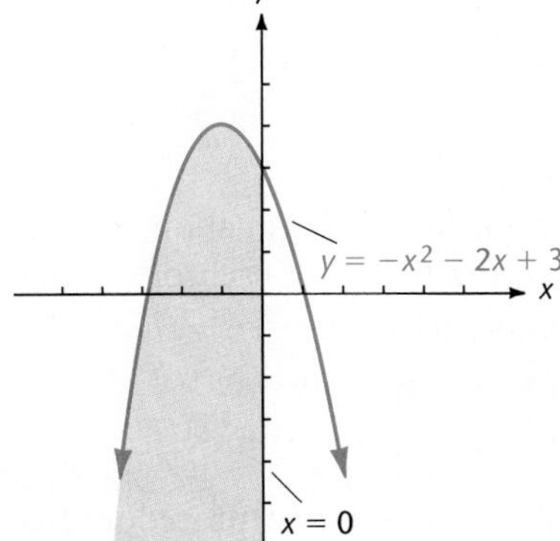

40.

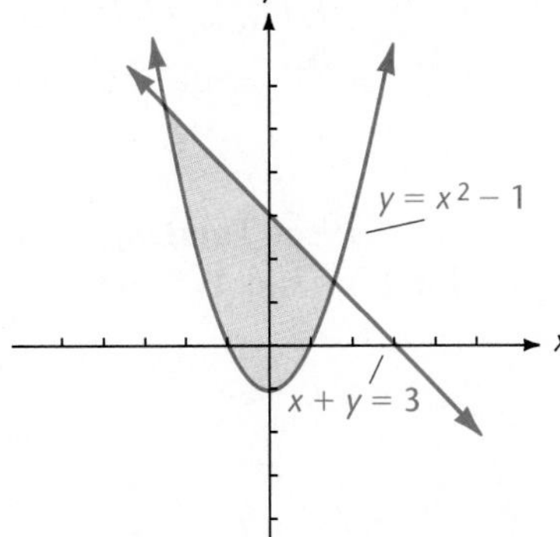

41.

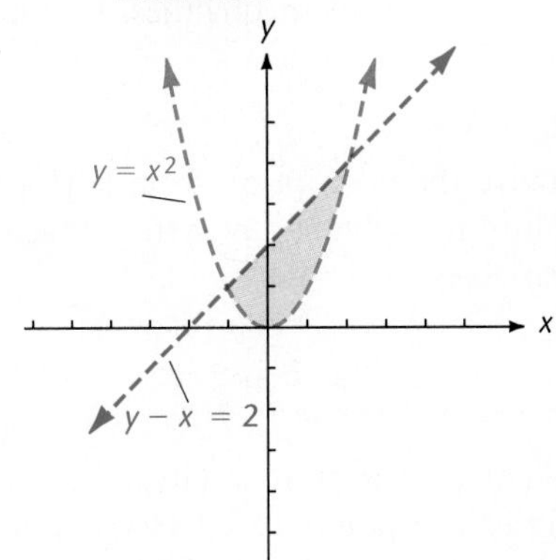

42.

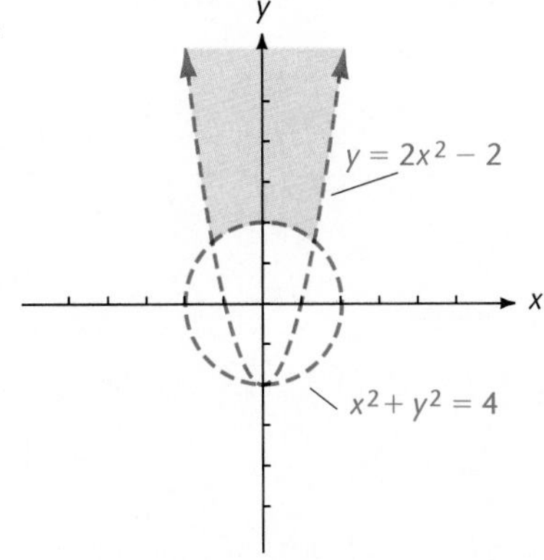

43.

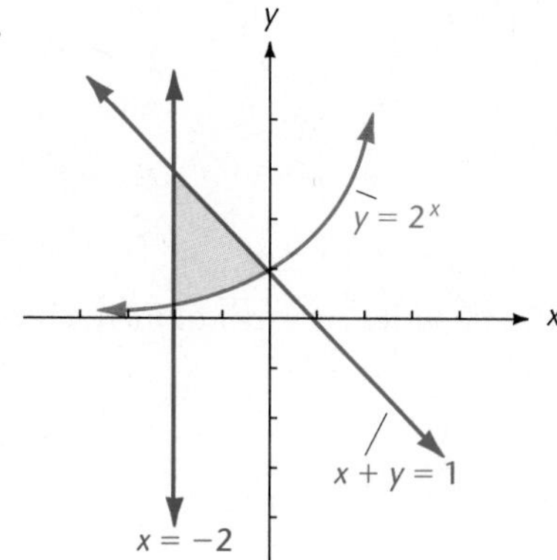

44.

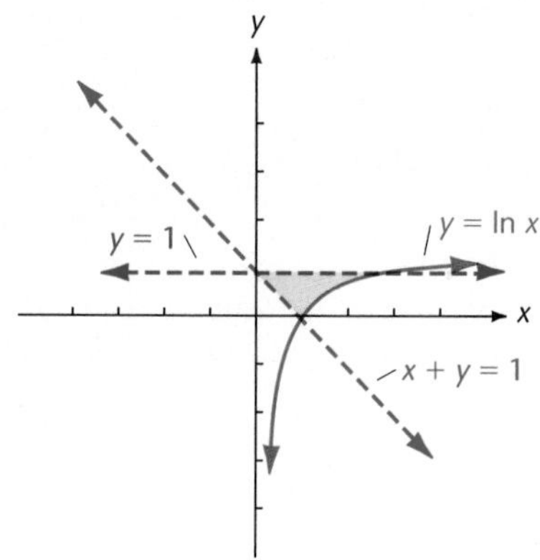

Use a graphing device with a shade feature to graph the solution set for each of the following.

45–50

45. $y \le \ln x$
$x - y \le 5$

46. $y \le x^3 + 5x^2 + 2x - 8$
$y \ge x + 2$

47. $y \ge -2\sqrt{x + 4}$
$y \le 2\sqrt{x + 4}$

48. $y \le \sqrt{25 - x^2}$
$y \ge -\sqrt{25 - x^2}$

49. $y \ge e^{0.5x} - 7$
$y \le 8 - x^2$
$-2 \le x \le 2$

50. $y \le \dfrac{5}{x^2 + 1}$
$y \ge -\sqrt{x^2 + 4}$
$-3 \le x \le 2$

Critical Thinking: **Exploration and Writing**

51. For a straight line with equation $y = mx + b$, describe the geometric location of the points in each of the following sets:
   a. The points where $y - mx = b$,
   b. The points where $y - mx > b$,
   c. The points where $y - mx < b$.

Critical Thinking: **Exploration and Writing**

52. a. Describe a step-by-step procedure that could be used to solve the following system of inequalities. It is not necessary to solve the system.

$$x^2 + y^2 \le 4$$
$$2y - x < 2$$

   b. Generalize the description from part a to describe a systematic procedure for solving any system that consists of two inequalities in two variables.

◆ Solution for Practice Problem

1. The graph of $x^2 - y^2 = 16$ is a hyperbola, and the graph of $x^2 + 16y^2 = 169$ is an ellipse. The points of intersection are easily found to be $(5, \pm 3)$ and $(-5, \pm 3)$, as shown in Figure 10.9. Using $(0, 0)$ as a test point, we find that

the origin is not a solution to the first inequality but is a solution to the second. This indicates that the solutions are on the opposite side of the hyperbola from the origin and on the same side of the ellipse as the origin. The solution set is shaded in Figure 10.9.

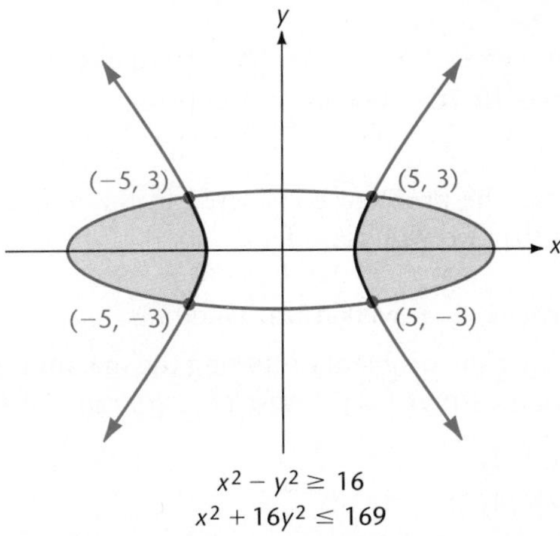

$x^2 - y^2 \geq 16$
$x^2 + 16y^2 \leq 169$

Figure 10.9 Solution set for Practice Problem 1

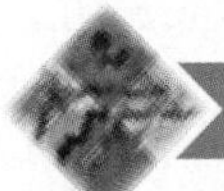

10.5 LINEAR PROGRAMMING

If the values of a variable f are given by an equation of the form

$$f = ax + by + c,$$

where a, b, and c are constants, then f is said to be a **linear function** of the two variables x and y.

In this section, we consider problems that call for a maximum or a minimum value of a linear function of two variables with a certain type of domain. Many problems in business, science, and engineering have a mathematical formulation of this type. These problems involve such things as profit and cost, diet and nutrition, or production scheduling, and they usually involve a linear function of several variables. We restrict our attention here to the two-variable case.

The linear functions that we consider have domains that are **convex regions** in the plane. A region in the plane is **convex** if for every pair of points P and Q of the region, the line segment from P to Q lies entirely in the region. This is illustrated in Figure 10.10.

A **linear programming problem** is a problem in which it is desired to find the maximum (or minimum) value of a linear function that has its domain restricted to a convex solution set of a system of linear inequalities. The linear inequalities are called **constraints,** and the solution set of the system is called the **region of feasible solutions.** A point of the region where two boundaries intersect is called a

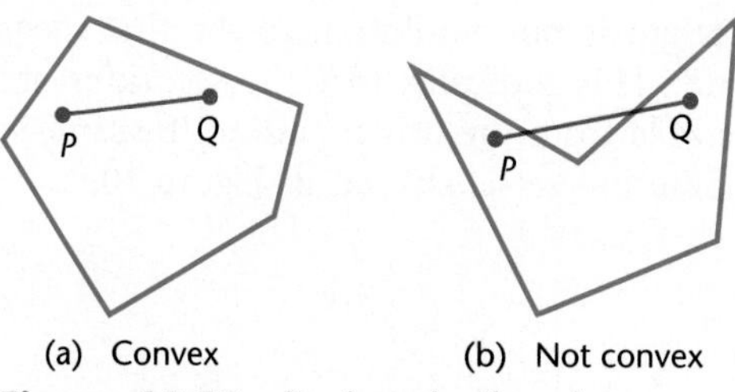

Figure 10.10 Regions in the plane

vertex. The problem considered in the following example is typical of the problems that we will consider.

Example 1 • Maximum Value

Consider the problem of finding the maximum value of $f = 2x + y$, subject to the constraints that (x, y) must be a solution of the system of inequalities

$$\begin{aligned} x + 2y &\geq 8 \\ x - 4y &\geq -10 \\ x - y &\leq 2. \end{aligned}$$

Solution

The solution set of the system of inequalities is shown in Figure 10.11. For each fixed value of d, the points (x, y) satisfying $f = d$ lie on the straight line $2x + y = d$. Different values of d give different straight lines with slope -2 and y-intercept d. Geometrically, the problem is to find the largest possible value of the y-intercept for a line that intersects the solution set. The dashed lines corresponding to $d = 2$, 7, 12, and 16 are shown in the figure. From these graphs, it is easily seen that the maximum d is for the line through the vertex $C(6, 4)$. Substituting the values $x = 6$ and $y = 4$ yields the maximum value 16 for f on the solution set. □

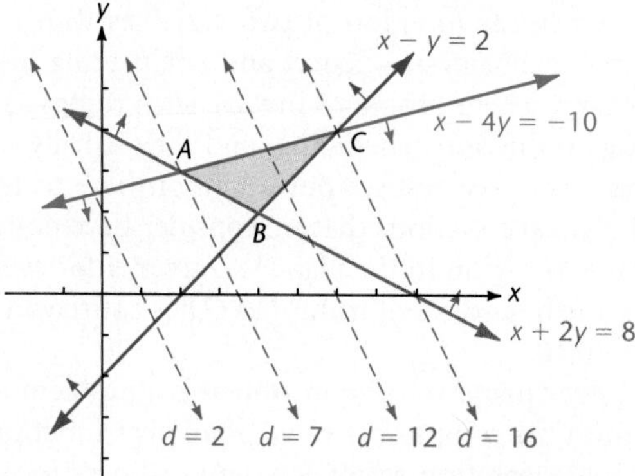

Figure 10.11 Convex set for Example 1

Following the same geometric procedure as diagrammed in Figure 10.11, a graphing device can be used to observe the lines $f = d$ at different positions across the convex region.

The following result assures us that it was no accident that the maximum value of f in Example 1 was attained at a vertex. The result states the facts that are needed in our work with linear programming problems, and we accept them without proof.

Extreme Values of a Linear Function

Suppose the linear function $f = ax + by + c$ has its domain restricted to points (x, y) in a convex solution set of a system of linear inequalities.

1. If f has a maximum or minimum value on this domain, it will occur at a vertex of the region.
2. If the domain is a convex polygon together with its interior, then f will have a maximum value and a minimum value on this domain.

The words maximize and minimize are frequently used in stating linear programming problems. To **maximize** f is to find its maximum value, and to **minimize** f is to find its minimum value.

Example 2 • Minimum Value

Minimize $f = 5x - 4y$, subject to the following constraints.

$$\begin{aligned} x - y &\leq 1 \\ x + 2y &\geq 4 \\ 2x + 3y &\leq 17 \\ 2x - 3y &\geq -13 \end{aligned}$$

Solution

We first graph the system of inequalities and locate the vertices by the methods of Section 10.4. This is shown in Figure 10.12. Since the minimum value of f occurs at

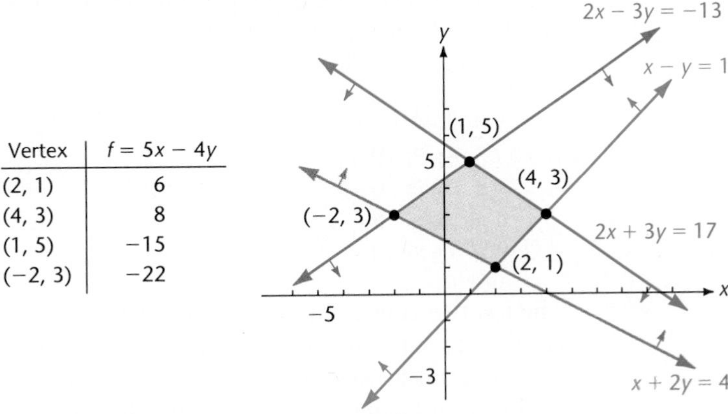

Vertex	$f = 5x - 4y$
(2, 1)	6
(4, 3)	8
(1, 5)	−15
(−2, 3)	−22

Figure 10.12 Convex set for Example 2

a vertex, all we need do is calculate the values of f at each vertex and select the smallest of them. The table in Figure 10.12 lists the value of f at each vertex. From the table, we see that f has a minimum value of -22 at the vertex $(-2, 3)$. □

◆ Practice Problem 1 ◆

Maximize $f = x - 3y + 14$ subject to the constraints in Example 2.

The next example provides some insights into the types of situations that lead to linear programming problems.

Example 3 • Fuel Refinery Processes

A fuel refinery has two processes for manufacturing three grades of gasoline: G_1, G_2, and G_3. The first process will produce 1 tank-car load of G_1, 3 loads of G_2, and 2 loads of G_3 in 10 hours. The second process will produce 3 loads of G_1, 4 loads of G_2, and 1 load of G_3 in 8 hours. The refinery has received an order for 8 loads of G_1, 19 loads of G_2, and 6 loads of G_3 that must be filled within 80 hours. How many times should each process be employed to fill the order as soon as possible, and what is the minimum time necessary?

Solution

If the first process is employed x times and the second process is employed y times, the time required is given by

$$T = 10x + 8y.$$

We need to minimize T, subject to the production requirements.

Employing the first process x times will produce x loads of G_1, $3x$ loads of G_2, and $2x$ loads of G_3. Employing the second process y times will produce $3y$ loads of G_1, $4y$ loads of G_2, and y loads of G_3. Filling the order imposes the following constraints.

$$\begin{aligned} x + 3y &\geq 8 \\ 3x + 4y &\geq 19 \\ 2x + y &\geq 6 \\ 0 \leq x &\leq 8 \\ 0 \leq y &\leq 10 \end{aligned}$$

The last two constraints are due to the facts that the number of times a process is employed cannot be negative and that the order must be filled within 80 hours.

We graph the system of inequalities and tabulate the values of T at the vertices, as shown in Figure 10.13. From the table, we see that T has a minimum value of 42 at the vertex $(1, 4)$. The first process should be employed one time, and the second should be employed four times, for a minimum time of 42 hours. □

Vertex	$T = 10x + 8y$
(0, 10)	80
(0, 6)	48
(1, 4)	42
(5, 1)	58
(8, 0)	80
(8, 10)	160

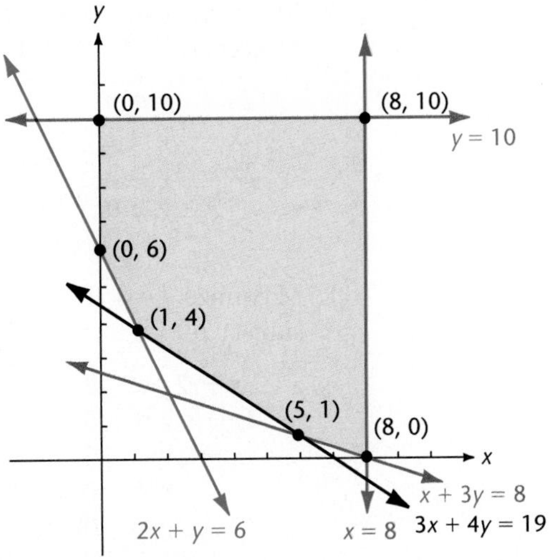

Figure 10.13 Convex set for Example 3

EXERCISES 10.5

1. Maximize $f = 2x + y$, subject to

$$\begin{aligned} 2x - y &\leq 10 \\ x + y &\leq 8 \\ x &\geq 0 \\ y &\geq 0. \end{aligned}$$

2. Maximize $f = 5x + 2y$, subject to

$$\begin{aligned} x + y &\leq 7 \\ x + 2y &\leq 10 \\ x &\geq 0 \\ y &\geq 0. \end{aligned}$$

3. Maximize $f = x + 4y + 1$, subject to

$$\begin{aligned} x + y &\leq 2 \\ x - y &\leq 1 \\ x &\geq 0 \\ y &\geq 0. \end{aligned}$$

4. Maximize $f = 2x + 5y - 3$, subject to

$$\begin{aligned} x + y &\leq 2 \\ x - 2y &\geq -1 \\ x &\geq 0 \\ y &\geq 0. \end{aligned}$$

5. Minimize $f = 3x + y + 4$, subject to

$$\begin{aligned} x - y &\leq 3 \\ x + 5y &\leq 21 \\ x + y &\geq 5. \end{aligned}$$

6. Minimize $f = 2x + y - 1$, subject to

$$\begin{aligned} x + 2y &\geq 4 \\ 2x - y &\leq 4 \\ 3x - 2y &\geq -6. \end{aligned}$$

7. Maximize $f = 2x + y$, subject to

$$\begin{aligned} 2x - y &\leq 10 \\ x + y &\leq 8 \\ x - y &\geq -5 \\ x &\geq 0 \\ y &\geq 0. \end{aligned}$$

8. Maximize $f = 14x + 3y$, subject to

$$\begin{aligned} x + y &\geq 2 \\ x - y &\leq 2 \\ x + y &\leq 4 \\ x &\geq 0. \end{aligned}$$

9. Minimize $f = y - x + 3$, subject to

$$\begin{aligned} 2x + y &\leq 9 \\ x - 2y &\leq 2 \\ 2x - 3y &\geq 3 \\ y &\geq 0. \end{aligned}$$

10. Minimize $f = y - 3x + 5$, subject to

$$\begin{aligned} x - y &\leq 2 \\ x + y &\leq 6 \\ x + 2y &\leq 10 \\ x &\geq 0 \\ y &\geq 0. \end{aligned}$$

11. Maximize $f = x + y - 2$, subject to

$$\begin{aligned} x - 2y &\leq 2 \\ 2x + y &\leq 9 \\ x - 2y &\geq -3 \\ x &\geq 0 \\ y &\geq 0. \end{aligned}$$

12. Maximize $f = 4x + y - 1$, subject to

$$\begin{aligned} 2x - 5y &\leq 10 \\ 3x - 5y &\leq 20 \\ x - 5y &\geq -30 \\ x &\geq 0 \\ y &\geq 0. \end{aligned}$$

Nut Mixtures

13. Matt supplies to a wholesaler two mixtures of nuts packed in 8-pound cans. The first mixture contains 6 pounds of peanuts and 2 pounds of cashews in each can. The second mixture contains 5 pounds of peanuts and 3 pounds of cashews. The profit on the first mixture is \$4 per can, and the profit on the second is \$5 per can. From a supply of 240 pounds of peanuts and 96 pounds of cashews, how many cans of each mixture should be made for a maximum profit?

Container Production

14. A company owns two factories that produce barrels and pressure tanks. During each day of operation, factory A produces 3000 barrels and 1000 pressure tanks; factory B produces 2000 barrels and 2000 pressure tanks. The cost at factory A is \$10 per barrel and \$20 per pressure tank, whereas the cost at factory B is \$20 per barrel and \$10 per pressure tank. The company has an order for 16,000 barrels and 8000 pressure tanks. How many days should each factory operate if the cost of filling the order is to be minimized?

Furniture Manufacture

15. A furniture company manufactures two types of desks using oak and mahogany lumber. The first type requires 10 board feet of oak and 5 board feet of mahogany, whereas the second requires 6 board feet of oak and 4 board feet of mahogany. A profit of \$45 is made on each desk of the first type, and \$30 on each desk of the second type. From a supply of 1000 board feet of oak and 600 board feet of mahogany, how many desks of each type should be made in order to yield a maximum profit?

Recreational Equipment

16. A certain company owns two small factories that produce iceboxes, ski vests, and minnow buckets. During each day of operation, factory A produces 4000 iceboxes, 1000 ski vests, and 2000 minnow buckets. Factory B produces 1000 iceboxes, 1000 ski vests, and 7000 minnow buckets each

day. The company has an order to supply 8000 iceboxes, 5000 ski vests, and 20,000 minnow buckets. It costs $6000 per day to run factory A and $2000 per day to run factory B. Find the number of days each factory must operate in order for the cost of filling the order to be a minimum.

Furniture Manufacture 17. A small manufacturing company can sell all the chairs and bar stools that it can produce. Each chair requires $1\frac{3}{5}$ hours in the assembly room and $\frac{2}{3}$ hour in the finishing room. Each bar stool requires $\frac{4}{5}$ hour in the assembly room and $\frac{4}{3}$ hours in the finishing room. Both the assembly room and the finishing room operate 24 hours a day, and each of them can work on only one item at a time. If the company makes a profit of $30 on each chair and $20 on each stool, find the number of chairs and stools that should be manufactured daily to make a maximum profit.

Vendor Sales 18. During the noon hour each day at the county fair in Rhinebeck, New York, Joe sells hero sandwiches and hot dogs from his trailer. He sells only these two items, and he makes a profit of 30 cents on each sandwich and 20 cents on each hot dog. Experience has shown him that the total number of items sold never exceeds 150 and that the number of hot dogs is always at least twice the number of sandwiches. Find the amount of sales of each item that yields a maximum profit.

Mark Antman/The Image Works

County fair, Rhinebeck, New York.

Preparation of Deli Trays 19. A supermarket prepares fresh daily two types of deli trays, a cold cut tray and a seafood tray. The meat department can supply no more than 6 hours 5 minutes of labor daily to prepare the meat, and the delicatessen can supply no more than 5 hours of labor daily to assemble the trays. Each cold cut tray requires 10 minutes in the meat department and 15 minutes in the delicatessen, whereas each seafood tray requires 25 minutes in the meat department and 10 minutes in the delicatessen. Because of demand, the supermarket wants to be sure to supply at least 4 cold cut trays daily. If there is a \$2.00 profit on each cold cut tray sold and \$3.50 profit on each seafood tray sold, how many trays of each type should the supermarket produce daily to maximize profit, assuming that all can be sold.

Clock Production 20. A small clock company produces two types of clocks, a grandfather clock and a brass mantle clock. Each grandfather clock sold nets a profit of \$150, and each mantle clock, \$30. In order to stay in business, the company must make a profit of at least \$4500 per month. Each grandfather clock requires 1 pound of brass, and each mantle clock requires 3 pounds of brass. Each month the company has available at most 100 pounds of brass. Because of space limitations in the factory, no more than 79 grandfather clocks can be produced per month. Shipping costs are \$75 per grandfather clock and \$20 per mantle clock. Determine the number of each type of clock to be produced monthly that will minimize the shipping cost.

Critical Thinking: Writing 21. Write out a procedure that will lead to the extreme values of a linear function whose domain is restricted to a convex solution set of a system of linear inequalities in x and y.

Critical Thinking: Exploration and Discussion 22. It is crucial to note that the results in this section apply only to linear functions. To illustrate other possibilities, consider the problem of minimizing $g = x^2 + y^2$, subject to the following constraints.

$$\begin{aligned} x &\leq 1 \\ y &\leq 1 \\ x + y &\geq 1 \end{aligned}$$

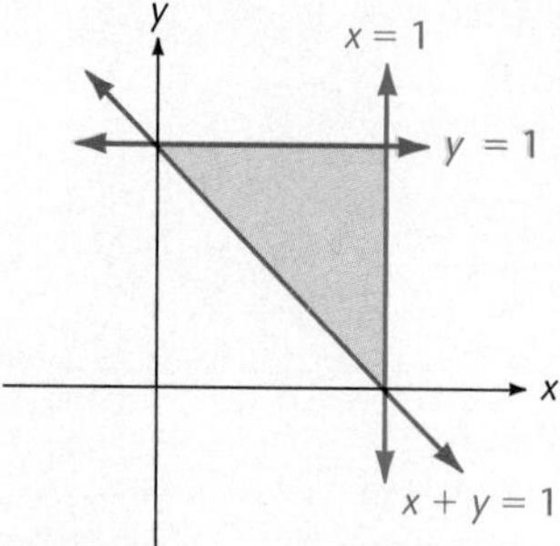

Figure for Exercise 22

The region of feasible solutions is shown in the accompanying figure. Explain why the minimum value of g does not occur at a vertex of the region.

◆ **Solution for Practice Problem**

1. Using the vertices shown in Figure 10.12, we calculate the following table of values.

Vertex	$f = x - 3y + 14$
(2, 1)	13
(4, 3)	9
(1, 5)	0
(−2, 3)	3

From the table we see that f has a maximum value of 13 at the vertex (2, 1).

CHAPTER REVIEW

Summary of Important Concepts and Formulas

Elimination Method for the Solution of Linear Systems in Three Variables

1. Select a variable and use suitable multipliers to eliminate it from a pair of equations.
2. Eliminate the *same* variable from a *different* pair of equations by using appropriate multipliers.
3. Solve the resulting subsystem in two variables.
4. Substitute the values of these two variables into one of the original equations and solve for the third variable.

Extreme Values of a Linear Function

Suppose the linear function $f = ax + by + c$ has its domain restricted to points (x, y) in a convex solution set of a system of linear inequalities.

1. If f has a maximum or minimum value on this domain, it will occur at a vertex of the region.
2. If the domain is a convex polygon together with its interior, then f will have a maximum value and a minimum value on this domain.

Critical Thinking: Find the Errors

Each of the following nonsolutions has one or more errors. Can you find them?

Problem 1

Solve the following system.

$$\begin{aligned} x - y + 2z &= 3 \\ 4x - 2y + 6z &= 8 \\ 3x + 4y - 5z &= -1 \end{aligned}$$

Nonsolution

To obtain an equation without a y-term, we add −2 times the first equation to the second equation.

$$\begin{array}{lcr} x - y + 2z = 3 & \xrightarrow{\text{Multiply by } -2} & -2x + 2y - 4z = -6 \\ 4x - 2y + 6z = 8 & \longrightarrow & 4x - 2y + 6z = 8 \\ & \text{Add} & \overline{2x \qquad + 2z = 2} \end{array}$$

To obtain another equation without a y-term, we add $-\frac{1}{2}$ times the second equation to the first equation.

$$\begin{array}{rcl} x - y + 2z = 3 & \longrightarrow & x - y + 2z = 3 \\ 4x - 2y + 6z = 8 & \xrightarrow{\text{Multiply by } -\frac{1}{2}} & -2x + y - 3z = -4 \\ & \text{Add} & \overline{-x \qquad - z = -1} \end{array}$$

We solve the x, z subsystem as follows.

$$\begin{array}{rcl} 2x + 2z = 2 & \xrightarrow{\text{Multiply by } \frac{1}{2}} & x + z = 1 \\ -x - z = -1 & \longrightarrow & -x - z = -1 \\ & \text{Add} & \overline{0 = 0} \end{array}$$

The equation $0 = 0$ is always true, so the original system is dependent, and there is an infinite number of solutions.

Problem 2

Graph the solution set of the following system of inequalities.

$$\begin{aligned} x + y^2 &< 1 \\ x + y &< -1 \end{aligned}$$

Nonsolution

The graph of $x = 1 - y^2$ is a parabola with vertex at $(1, 0)$ that opens to the left, and the graph of $x + y = -1$ is a straight line with both intercepts equal to -1. This is shown in Figure 10.14.

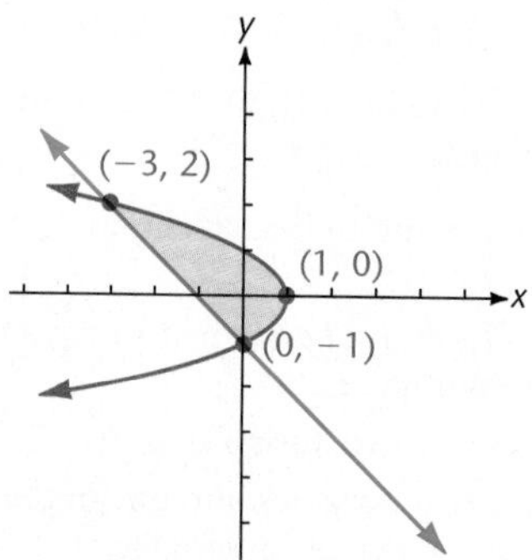

Figure 10.14

To find the points of intersection, we can substitute $x = 1 - y^2$ into the equation $x + y = -1$. This gives

$$\begin{aligned} (1 - y^2) + y &= -1 \\ 0 &= y^2 - y - 2 \\ 0 &= (y - 2)(y + 1). \end{aligned}$$

Thus $y = 2$ or $y = -1$ at the points of intersection. Substituting these values into $x + y = -1$ gives $(-3, 2)$ and $(0, -1)$ as the points of intersection.

Testing $(0, 0)$, we find that the origin is in the solution set of each inequality, and so the solution set lies to the left of the parabola and to the right of the line as shown in Figure 10.14.

Problem 3

Maximize $f = 7x + 2y$, subject to

$$\begin{aligned} 2x + y &\leq 8 \\ x + y &\leq 5 \\ x &\geq 0 \\ y &\geq 0 \end{aligned}$$

Nonsolution

The solution set for the system of inequalities is shaded in Figure 10.15, and the vertices of the solution set are labeled in the figure. Since the maximum value of f occurs at a vertex, we calculate the values of f at each vertex and select the largest of them.

Vertex	$f = 7x + 2y$
(0, 0)	0
(4, 0)	28
(5, 0)	35
(3, 2)	25
(0, 5)	10
(0, 8)	16

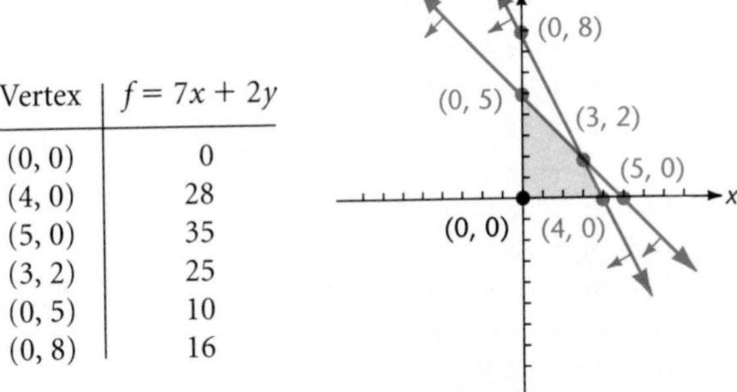

Figure 10.15

At the vertex (5, 0), f has a maximum value of 35.

Review Problems for Chapter 10

Solve by the substitution method.

1. $3x + y = 1$
   $x + 3y = -5$

2. $4x - 5y = 13$
   $x + 2y = 0$

Solve by the elimination method.

3. $3x + 2y = 1$
   $2x + 3y = 9$

4. $9x + 2y = 3$
   $7x + 5y = -8$

Solve the following systems by any method.

5. $2x - 5y = 3$
   $3x - y = -2$

6. $4x - 3y = 5$
   $3x + y = 7$

7. $3x + 5y = 7$
   $2x - 3y = 11$

8. $2x - 3y = 5$
   $4x - 6y = -10$

9. $2x + 4y = 4$
   $3x + 6y = 6$

Uniform Motion 10–11

10. Jerome Lewis's motorboat takes 3 hours to travel 36 miles upstream to Waterville and makes the return trip in 2 hours. Find the rate of the current and the rate of the boat in still water.

Figure for Problem 10

11. It took 9 hours for Chuck to fly his small plane against the wind a distance of 900 miles to New Bedford, and he made the return trip traveling with the same wind in only 6 hours. Find the speed of the wind and the speed of Chuck's plane in still air.

Aluminum Alloys

12. Accu-Casting has two aluminum alloys in stock. The first is 70% aluminum, while the second is 20% aluminum. How many pounds of each alloy should be combined to obtain 300 pounds of an alloy that is 40% aluminum?

13. Find a and b so that the points (1, 1) and (5/7, 1/7) are on the hyperbola that has equation

$$\frac{x^2}{a^2} - \frac{y^2}{b^2} = 1.$$

Solve the following systems.

14. $\begin{aligned} x + y - z &= -5 \\ 3y + 5z &= 3 \\ 2x - 3z &= -5 \end{aligned}$

15. $\begin{aligned} x - 3y - 4z &= -10 \\ 2x + y + 3z &= 0 \\ x - 3z &= 6 \end{aligned}$

16. $\begin{aligned} 2x - y - 3z &= 5 \\ x - 2y - z &= -2 \\ 3x + 8y + 2z &= 7 \end{aligned}$

17. $\begin{aligned} x - 2y - 2z &= 9 \\ 2x + y + 3z &= 0 \\ 5x + 5y + 11z &= 9 \end{aligned}$

18. $\begin{aligned} 2x - 3y - 9z &= -2 \\ x + 2y + 6z &= 6 \\ 4x - 2y - 6z &= -4 \end{aligned}$

19. Find a, b, and c so that each of (1, 3, 2), (3, −2, 1), and (2, −4, −3) is a solution to $ax + by + cz = 3$.

Coin Problem

20. Beckie has \$5.05 in nickels, dimes, and quarters. She has twice as many dimes as quarters, and a total of 41 coins. How many coins of each type does she have?

Work Problem

21. Martin Goldsworth has three backhoes that can dig an excavation for a certain model of swimming pool in 2 hours if they all work together. The same excavation can be done if backhoe A works 2 hours and backhoe C works 6 hours, and it can also be done if backhoes B and C work together for 4 hours. How long would it take each backhoe working alone to dig an excavation for one of the pools?

In Problem 22–28, solve the given system of equations for real solutions only. It is not necessary to graph the equations.

22. $\begin{aligned} x &= y^2 + 1 \\ x - y &= 3 \end{aligned}$

23. $\begin{aligned} x &= y^2 - 4y + 3 \\ x - y &= 3 \end{aligned}$

24. $\begin{aligned} x^2 - 3y &= 1 \\ 2x - y &= 3 \end{aligned}$

25. $\begin{aligned} x^2 + 3xy - 6y^2 &= 8 \\ x^2 - xy - 6y^2 &= 4 \end{aligned}$

26. $\begin{aligned} 4y^2 - x &= 0 \\ x + 2y &= 6 \end{aligned}$

27. $\begin{aligned} x^2 - 2x + 3y - 3 &= 0 \\ x + 2y &= 4 \end{aligned}$

28. $\begin{aligned} 3x^2 + y^2 &= 15 \\ x^2 + 2y^2 &= 10 \end{aligned}$

In Problems 29–33, graph the solution set of the given system of inequalities, showing all points of intersection of the boundaries.

29. $\begin{aligned} 3x + y &> 7 \\ x - y &\le 1 \\ y &< 4 \end{aligned}$

30. $\begin{aligned} 2x + y &< -2 \\ 4x - 3y &\ge -24 \\ y &\ge -4 \end{aligned}$

31. $\begin{aligned} x^2 + 2x + 3 &\le y \\ 3x + y + 1 &\le 0 \end{aligned}$

32. $\begin{aligned} x^2 + y^2 &\ge 9 \\ x^2 + 2y &\le 10 \end{aligned}$

33. $\begin{aligned} x^2 + y^2 &\le 4 \\ 3y + x^2 &\ge 0 \end{aligned}$

34. Maximize $f = 2x + 7y$, subject to

$$\begin{aligned} 3x + 4y &\le 24 \\ x - 2y &\ge -2 \\ x &\ge 0 \\ y &\ge 0. \end{aligned}$$

35. Minimize $f = 4x + 5y$, subject to

$$\begin{aligned} 3x + y &\ge 7 \\ x - y &\le 1 \\ y &\le 4. \end{aligned}$$

36. Maximize $f = 15x - 4y$, subject to

$$\begin{aligned} 3x + 2y &\le 12 \\ 3x - 2y &\le 6 \\ x - 2y &\ge -4 \\ x &\ge 0 \\ y &\ge 0. \end{aligned}$$

Maximizing Revenue

37. A nut company has 400 pounds of peanuts and 200 pounds of cashews to sell as two different mixes. One mix will contain half peanuts and half cashews and will sell for \$6 per pound. The other mix will contain three-fourths peanuts and one-fourth cashews and will sell for \$5 per pound. How many pounds of each mix should the company prepare for maximum revenue?

Maximizing Profit

38. A company manufactures two types of pocket calculators, a business calculator and a scientific calculator. It makes a profit of \$4 on each business calculator and \$3 on each scientific calculator. The company can produce at most 500 calculators each day, but because of the way production is set up it cannot produce more than 400 business calculators. Assuming that the company is able to sell all of the calculators it manufactures, how many scientific calculators should the company produce each day for the greatest profit?

ENCORE

Courtesy Betras Plastics, Inc.

Molded plastic mugs

Blasto Plastics has two molding machines that make plastic mugs. On an order from a local pizzeria, both machines were operated together and filled the order in 3 hours. To refill the same order, the faster machine was operated for 2 hours and the slower machine was operated for 6 hours. This morning a refill for the same order came in, and the slower machine is broken down. How long will it take the faster machine alone to fill the order?

This question can be answered by employing the same kind of analysis that was used on work problems in Section 2.2. If x is the number of hours it will take the faster machine to fill the order, then $1/x$ is the fractional part of the order that this machine would fill in 1 hour. Similarly, if y is the number of hours it would take the slower machine to fill the order, then $1/y$ is the fractional part of the order that the slower machine would fill in 1 hour.

Working together, the fractional part of the order the machines would fill in 1 hour is

$$\frac{1}{x} + \frac{1}{y}.$$

Since they filled the order in 3 hours working together, they would fill one-third of the order in 1 hour, so

$$\frac{1}{x} + \frac{1}{y} = \frac{1}{3}.$$

The same order was filled when the faster machine worked for 2 hours and the slower machine worked for 6 hours. Therefore

$$2\left(\frac{1}{x}\right) + 6\left(\frac{1}{y}\right) = 1,$$

and we need only solve for x in the following system of equations.

$$\begin{aligned} \frac{1}{x} + \frac{1}{y} &= \frac{1}{3} \\ \frac{2}{x} + \frac{6}{y} &= 1 \end{aligned}$$

This can be done by using the elimination method.

$$\begin{array}{lcr} \dfrac{1}{x} + \dfrac{1}{y} = \dfrac{1}{3} & \xrightarrow{\text{Multiply by } -6} & -\dfrac{6}{x} - \dfrac{6}{y} = -2 \\ \dfrac{2}{x} + \dfrac{6}{y} = 1 & \xrightarrow{\hspace{2cm}} & \dfrac{2}{x} + \dfrac{6}{y} = \quad 1 \\ \hline & \text{Add} & -\dfrac{4}{x} \qquad = -1 \end{array}$$

The last equation gives $x = 4$, so it will take the faster machine 4 hours to fill the order working alone.

Matrices and Determinants

In the last half of the twentieth century, the use of calculators and computers with the matrix methods presented in this chapter has made a revolutionary change in the kinds of applications that are possible for linear systems. Solutions of systems are routinely accomplished in business and industry today that were completely impossible at the middle of the century. As the material in this chapter becomes more familiar, the power of the methods will become clearer.

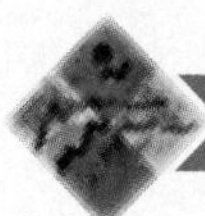

11.1 NOTATION AND DEFINITIONS

While watching a football game on television, a football fan often sees flashed on the screen a tabulation similar to the following.

	Falcons	Rams
Number of offensive plays	8	14
Yards rushing	63	84
Yards passing	102	83
Total yards	165	167
Time of possession	5:20	9:40
Number of turnovers	2	0

AP/Wide World Photos

If the fan is interested in the number of "yards passing" by the Rams, he looks at the entry in the third row and second column and finds 83.

A traveler who needs to know the distance from Los Angeles to Chicago looks at a mileage chart similar to the following.

	Chicago	New York	Los Angeles	Miami
Chicago	0	840	2090	1375
New York	840	0	2805	1332
Los Angeles	2090	2805	0	2743
Miami	1375	1332	2743	0

The entry in the third row and first column gives the desired mileage, 2090.

Suppose that we consider only the array of numbers in each of these examples, omitting the row and column headings and placing square brackets around the arrays. We have

$$\begin{bmatrix} 8 & 14 \\ 63 & 84 \\ 102 & 83 \\ 165 & 167 \\ 5{:}20 & 9{:}40 \\ 2 & 0 \end{bmatrix} \qquad \begin{bmatrix} 0 & 840 & 2090 & 1375 \\ 840 & 0 & 2805 & 1332 \\ 2090 & 2805 & 0 & 2743 \\ 1375 & 1332 & 2743 & 0 \end{bmatrix}.$$

These arrays of entries are called **matrices,** and each number in the array is called an **element** of the **matrix.** The formal definition is as follows.

Definition of a Matrix

An ***m* by *n* matrix** is a rectangular array of elements arranged in m rows and n columns. Such a matrix can be written in the form

$$A = \begin{bmatrix} a_{11} & a_{12} & \cdots & a_{1n} \\ a_{21} & a_{22} & \cdots & a_{2n} \\ a_{31} & a_{32} & \cdots & a_{3n} \\ \vdots & \vdots & & \vdots \\ a_{m1} & a_{m2} & \cdots & a_{mn} \end{bmatrix},$$

where a_{ij} denotes the element in row i and column j of the matrix A. The matrix A is referred to as a matrix of **dimension $m \times n$** (read "m by n").

It is customary to denote matrices by capital letters. An $m \times n$ matrix A can be denoted compactly by

$$A = [a_{ij}]_{(m,n)} \quad \text{or} \quad A_{(m,\, n)}.$$

The elements of the form $a_{11}, a_{22}, a_{33}, \ldots$ are called the **main diagonal elements.**

In the second example given above, the mileage from Los Angeles to Chicago is represented by the element $a_{31} = 2090$, and the matrix representing the mileage has dimension 4×4. The matrix of football statistics is of dimension 6×2.

Example 1 • Using a Formula for the Elements of a Matrix

Write the 2×2 matrix whose elements are defined by $a_{ij} = i - j$.

Solution

The element in the first row ($i = 1$) and the first column ($j = 1$) is a a_{11}, and its value is $a_{11} = 1 - 1 = 0$ by the formula. The element in the second row ($i = 2$) and the first column ($j = 1$) is a_{21}. The formula gives its values as $a_{21} = 2 - 1 = 1$. Similarly, $a_{12} = 1 - 2 = -1$ and $a_{22} = 2 - 2 = 0$. Hence the required matrix is

$$\begin{bmatrix} 0 & -1 \\ 1 & 0 \end{bmatrix}.$$ □

A matrix that has n rows and n columns is called a **square matrix of order *n*.** A matrix that has only one column is called a **column matrix,** and a matrix that has only one row is called a **row matrix.** A square matrix that has zeros for all its elements off the main diagonal is called a **diagonal matrix.**

Example 2 • Identify Types of Matrices

State the dimensions of each matrix given below. If a matrix is a column matrix, a row matrix, a square matrix, or a diagonal matrix, identify it as such.

a. $A = \begin{bmatrix} 2 & 1 \\ -3 & 4 \\ -2 & 0 \end{bmatrix}$ b. $B = [-1 \quad 0 \quad 4 \quad 7]$ c. $C = \begin{bmatrix} -\sqrt{3} \\ \pi \end{bmatrix}$

d. $D = [0]$ e. $E = \begin{bmatrix} 0 & 0 \\ 0 & -1 \end{bmatrix}$

Solution

a. The matrix A has a dimension 3×2. None of the special terms apply.

b. B is a row matrix of dimension 1×4.

c. C is a 2×1 column matrix.

d. D is a 1×1 matrix. It is a column matrix, a row matrix, a square matrix, and a diagonal matrix.

e. E is a diagonal matrix (and hence a square matrix) of order 2. □

Definition of Matrix Equality

Two matrices are **equal** if they have the same dimension and the elements placed in corresponding positions are equal.

Example 3 • Equality of Matrices

The meaning of the definition of equality is illustrated by the following pairs of matrices.

a. $\begin{bmatrix} 1 \\ 2 \\ -5 \end{bmatrix}$ and $[1 \quad 2 \quad -5]$ are not equal since they have different dimensions.

b. $\begin{bmatrix} x & 2y \\ 1 & z \\ 0 & 2 \end{bmatrix}$ and $\begin{bmatrix} -3 & 4 \\ 1 & z \\ 0 & \frac{4}{2} \end{bmatrix}$ are equal only if $x = -3$ and $y = 2$.

c. $[1 \quad 3]$ and $[3 \quad 1]$ are not equal since elements in corresponding positions are not equal. □

Suppose that we have two matrices, each representing the statistics from a quarter in a football game, as in the example at the beginning of the section.

$$\overset{\text{First quarter}}{\begin{bmatrix} 8 & 14 \\ 63 & 84 \\ 102 & 83 \\ 165 & 167 \\ 5{:}20 & 9{:}40 \\ 2 & 0 \end{bmatrix}} \qquad \overset{\text{Second quarter}}{\begin{bmatrix} 10 & 13 \\ 33 & 95 \\ 80 & 47 \\ 113 & 142 \\ 6{:}50 & 8{:}10 \\ 1 & 1 \end{bmatrix}}$$

If corresponding elements in these two matrices are added, the result is a matrix whose elements represent the statistics for the two quarters together. Addition of matrices is defined in this manner.

Definition of Matrix Addition

The **sum** of two $m \times n$ matrices A and B is the $m \times n$ matrix $A + B$ formed by adding the elements that are placed in corresponding positions of A and B.

Notice that the sum of two matrices of different dimensions is *not* defined.

Example 4 • Matrix Addition

a. $\begin{bmatrix} 2 & -2 & 1 \\ 1 & 0 & -3 \end{bmatrix} + \begin{bmatrix} 1 & -5 & 7 \\ 2 & 1 & 0 \end{bmatrix} = \begin{bmatrix} 3 & -7 & 8 \\ 3 & 1 & -3 \end{bmatrix}.$

b. $\begin{bmatrix} 1 & 4 & 0 \\ -1 & 2 & 1 \end{bmatrix} + \begin{bmatrix} 0 & 1 \\ 2 & -2 \end{bmatrix}$ is not defined since the dimensions of the two matrices are not equal. □

It can be shown that addition of matrices is associative and commutative.

Associative and Commutative Properties of Matrix Addition

Let A, B, and C be $m \times n$ matrices. Then

$$A + (B + C) = (A + B) + C \qquad \text{and} \qquad A + B = B + A$$

The **additive identity** for the set of all $m \times n$ matrices is an $m \times n$ matrix with all elements equal to 0. This additive identity matrix is also called the **zero matrix** of dimension $m \times n$.

Thus $\begin{bmatrix} 0 & 0 \\ 0 & 0 \end{bmatrix}$ is the zero matrix of order 2, and $\begin{bmatrix} 0 & 0 \\ 0 & 0 \\ 0 & 0 \end{bmatrix}$ is the 3×2 zero matrix.

The **additive inverse** of the $m \times n$ matrix A is the $m \times n$ matrix $-A$, the matrix obtained by multiplying each element of A by -1. For example, the additive inverse of

$$A = \begin{bmatrix} -2 & 4 & 1 & -1 \\ -1 & 0 & 3 & 2 \\ 0 & -2 & 0 & 1 \end{bmatrix} \quad \text{is} \quad -A = \begin{bmatrix} 2 & -4 & -1 & 1 \\ 1 & 0 & -3 & -2 \\ 0 & 2 & 0 & -1 \end{bmatrix}.$$

Subtraction of matrices is defined using additive inverses.

$$A - B = A + (-B)$$

Example 5 • Matrix Subtraction

Compute $A - B$ if $A = \begin{bmatrix} -2 & 1 \\ 3 & -2 \end{bmatrix}$ and $B = \begin{bmatrix} 1 & 7 \\ 2 & -4 \end{bmatrix}$.

Solution

$$A - B = A + (-B) = \begin{bmatrix} -2 & 1 \\ 3 & -2 \end{bmatrix} + \begin{bmatrix} -1 & -7 \\ -2 & 4 \end{bmatrix} = \begin{bmatrix} -3 & -6 \\ 1 & 2 \end{bmatrix}$$ □

There are two types of multiplications to be considered when working with matrices. The simplest is the product of a real number and a matrix. In this sort of product the real number is often called a **scalar.**

Definition of the Product of a Scalar and a Matrix

To multiply a matrix A by a scalar c, multiply each element of A by c. This **product** is denoted by cA.

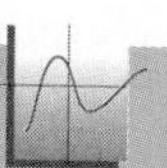

The matrix operations defined in this section can be performed by graphing devices. However, the maximum dimensions of the matrices involved are limited by the capacity of the device.

Example 6 • A Scalar Multiple of a Matrix

Compute $3A$, where $A = \begin{bmatrix} 2 & -1 & -2 \\ 1 & 4 & 0 \end{bmatrix}$.

Solution

$$3A = 3\begin{bmatrix} 2 & -1 & -2 \\ 1 & 4 & 0 \end{bmatrix} = \begin{bmatrix} 3(2) & 3(-1) & 3(-2) \\ 3(1) & 3(4) & 3(0) \end{bmatrix} = \begin{bmatrix} 6 & -3 & -6 \\ 3 & 12 & 0 \end{bmatrix} \qquad \square$$

The multiplication of two matrices is much more involved, and this is considered in the next section.

◆ Practice Problem 1 ◆

Compute $2A - 3B$ if

$$A = \begin{bmatrix} -2 & 1 & 3 \\ 1 & 5 & 1 \end{bmatrix} \quad \text{and} \quad B = \begin{bmatrix} 1 & 7 & -2 \\ 0 & 1 & -5 \end{bmatrix}.$$

EXERCISES 11.1

State the dimension of each matrix. If a matrix is a column matrix, a row matrix, a square matrix, or a diagonal matrix, identify it as such. (See Example 2.)

1. $\begin{bmatrix} 1 & 0 & -3 \\ 2 & 1 & 4 \end{bmatrix}$

2. $\begin{bmatrix} 5 & -3 & -10 & -1 \\ \frac{1}{2} & 0 & -2 & 1 \\ 4 & 0 & 0 & 5 \end{bmatrix}$

3. $\begin{bmatrix} 1 \\ 3 \\ 7 \end{bmatrix}$

4. $\begin{bmatrix} 1 \\ 1 \end{bmatrix}$

5. $\begin{bmatrix} -1 & 0.3 & 2.7 & 1.2 \\ 0.1 & 0 & -1 & 0 \end{bmatrix}$

6. $\begin{bmatrix} \sqrt{3} & \sqrt{2} \\ \pi & -\sqrt{5} \end{bmatrix}$

7. $[0]$

8. $\begin{bmatrix} -1 & 0 \\ 0 & -1 \end{bmatrix}$

9. $[1 \quad 0]$

10. $[0 \quad -1 \quad -2]$

11. $\begin{bmatrix} 1 & 0 & 0 \\ 0 & 1 & 0 \\ 0 & 0 & 1 \end{bmatrix}$

12. $\begin{bmatrix} 3 & -2 \\ 1 & 0 \\ 0 & -1 \end{bmatrix}$

Write the matrix whose elements are defined by each equation. (See Example 1.)

13. $a_{ij} = 2i + j;\ i = 1, 2;\ j = 1, 2, 3, 4$
14. $a_{ij} = i \cdot j;\ i = 1, 2, 3;\ j = 1, 2, 3, 4$

15. $a_{ij} = (-1)^i \cdot j;\ i = 1, 2, 3, 4;\ j = 1, 2$
16. $a_{ij} = (-1)^{i+j};\ i = 1, 2, 3;\ j = 1, 2, 3$
17. Write a 2×4 matrix such that $a_{ij} = 1$ if $i < j$, 0 otherwise.
18. Write a 3×4 matrix such that $a_{ij} = i + j$ if $i \geq j$, 0 otherwise.
19. Write a 4×3 matrix such that $a_{ij} = 2i + 3j$ if $i < j$, 1 otherwise.
20. Write a 4×2 matrix such that $a_{ij} = 1$ if $i \neq j$, -1 otherwise.

Determine whether or not the following pairs of matrices are equal. (See Example 3.)

21. $\begin{bmatrix} 1 & 2 \\ -1 & 3 \\ 1 & 7 \end{bmatrix}, \begin{bmatrix} 2 & 1 \\ 3 & -1 \\ 7 & 1 \end{bmatrix}$

22. $\begin{bmatrix} 0 & 0 \\ 0 & 0 \end{bmatrix}, \begin{bmatrix} 0 & 0 & 0 \\ 0 & 0 & 0 \\ 0 & 0 & 0 \end{bmatrix}$

23. $\begin{bmatrix} 1 & -7 \\ 5 & 3.2 \end{bmatrix}, \begin{bmatrix} 3-2 & -7 \\ 5 & \frac{32}{10} \end{bmatrix}$

24. $[1 \quad 3 \quad -1], \begin{bmatrix} 1 \\ 3 \\ -1 \end{bmatrix}$

25. $\begin{bmatrix} -2 & -1 \\ 3 & 4 \\ -2 & 0 \end{bmatrix}, \begin{bmatrix} -2 & -1 & 0 \\ 3 & 4 & 0 \\ -2 & 0 & 0 \end{bmatrix}$

26. $\begin{bmatrix} 10 & -2 \\ 1 & 3 \\ 0 & 0 \end{bmatrix}, \begin{bmatrix} 10 & -2 \\ 1 & 3 \end{bmatrix}$

27. $\begin{bmatrix} -1 & 4 & 3 \\ 2 & -1 & 1 \end{bmatrix}, \begin{bmatrix} -1 & 2 \\ 4 & -1 \\ 3 & 1 \end{bmatrix}$

28. $\begin{bmatrix} 1 & 0 \\ 0 & 1 \end{bmatrix}, \begin{bmatrix} 0 & 1 \\ 1 & 0 \end{bmatrix}$

29. $\begin{bmatrix} x & 1 \\ 0 & -y \end{bmatrix}, \begin{bmatrix} 2 & 1 \\ 0 & -3 \end{bmatrix}$

30. $\begin{bmatrix} x^2 \\ -y \end{bmatrix}, \begin{bmatrix} 4 \\ 2 \end{bmatrix}$

31. $[2x - 5], [-1]$

32. $[2x \quad x], [6 \quad 3]$

Perform the indicated operations, if possible. (See Examples 4–6 and Practice Problem 1.)

If a calculator with matrix features is available, work a sampling of Exercises 33–48 both by hand and with a calculator.

33. $\begin{bmatrix} 3 & -1 & 1 \\ 2 & 7 & -4 \end{bmatrix} + \begin{bmatrix} 2 & 1 & 0 \\ 1 & 3 & -1 \end{bmatrix}$

34. $\begin{bmatrix} 1 & 0 \\ 0 & 1 \end{bmatrix} + \begin{bmatrix} 2 & 1 \\ -3 & 2 \end{bmatrix}$

35. $\begin{bmatrix} -2 \\ 3 \\ -6 \end{bmatrix} + [1 \quad 1 \quad -2]$

36. $\begin{bmatrix} -3 & 1 \\ 0 & -2 \\ 1 & -1 \end{bmatrix} + \begin{bmatrix} -3 & 1 & 0 \\ -2 & 1 & -1 \end{bmatrix}$

37. $\begin{bmatrix} 4 & -2 \\ 1 & 7 \end{bmatrix} - \begin{bmatrix} 2 & 6 \\ -3 & 0 \end{bmatrix}$

38. $\begin{bmatrix} 3 & 7 & 11 \\ -2 & 4 & -1 \end{bmatrix} - \begin{bmatrix} 1 & -1 & 2 \\ 1 & -3 & 1 \end{bmatrix}$

39. $3\begin{bmatrix} -1 & 3 \\ 2 & 0 \\ -1 & 1 \end{bmatrix} - 2\begin{bmatrix} 4 & -2 \\ -5 & 4 \\ 0 & -3 \end{bmatrix}$

40. $-6\begin{bmatrix} 2 & 0 \\ -5 & 4 \end{bmatrix} - \begin{bmatrix} -1 & -3 \\ 1 & 7 \end{bmatrix}$

√ 41. $2\begin{bmatrix} -11 & 2 \\ 0 & -3 \end{bmatrix} + 3\begin{bmatrix} -11 \\ 0 \end{bmatrix}$

42. $2\begin{bmatrix} -11 & 0 \\ 0 & 0 \end{bmatrix} - 3\begin{bmatrix} -11 \\ 0 \end{bmatrix}$

43. $-3\begin{bmatrix} -1 \\ 6 \end{bmatrix} + 2\begin{bmatrix} -2 \\ 5 \end{bmatrix}$

44. $9[-1 \quad 4] - 3[-6 \quad 0]$

√ 45. $7\begin{bmatrix} 2 & 1 \\ 3 & 5 \\ 0 & 2 \end{bmatrix} - \begin{bmatrix} 6 & 1 \\ 2 & 3 \\ -2 & 1 \end{bmatrix}$

46. $-\begin{bmatrix} 1 & 1 \\ 3 & 8 \\ 0 & -2 \end{bmatrix} - 2\begin{bmatrix} 9 & 2 \\ 1 & 1 \\ 4 & 3 \end{bmatrix}$

47. $2\begin{bmatrix} 4 & -2 & 5 & 1 \\ 7 & 1 & 8 & 2 \end{bmatrix} + 3\begin{bmatrix} 6 & 2 & 1 & -3 \\ 8 & -1 & 4 & -2 \end{bmatrix} - 4\begin{bmatrix} -6 & -1 & 0 & 4 \\ 1 & 5 & 1 & 3 \end{bmatrix}$

48. $5\begin{bmatrix} 1 & 5 & -3 \\ 2 & 1 & 1 \\ 4 & 0 & -2 \end{bmatrix} - 2\begin{bmatrix} -1 & 3 & 5 \\ 4 & 8 & 6 \\ 0 & 1 & 3 \end{bmatrix} - 3\begin{bmatrix} 0 & 8 & 3 \\ 1 & 0 & 5 \\ -3 & 4 & 0 \end{bmatrix}$

49. Prove that matrix addition is associative: if A, B, and C are $m \times n$ matrices, then

$$A + (B + C) = (A + B) + C.$$

50. Prove that matrix addition is commutative: if A and B are two $m \times n$ matrices, then

$$A + B = B + A.$$

Critical Thinking: Writing

51. a. Explain why $1 \cdot A = A$ for any matrix A.
    b. Suppose c is a scalar and A is a matrix such that $cA = O$, where O is a zero matrix. Explain why one of the pair c, A must be a zero quantity.

Critical Thinking: Writing

52. Although we do not go into the proofs in this section, two distributive properties are valid that involve multiplication of scalars and matrices. They may be stated as $a(B + C) = aB + aC$ and $(a + b)A = aA + bA$. With the first one of these, appropriate conditions must be placed on B and C. Describe these appropriate conditions.

◆ Solution for Practice Problem

1. $2A - 3B = 2\begin{bmatrix} -2 & 1 & 3 \\ 1 & 5 & 1 \end{bmatrix} - 3\begin{bmatrix} 1 & 7 & -2 \\ 0 & 1 & -5 \end{bmatrix}$

$$= \begin{bmatrix} -4 & 2 & 6 \\ 2 & 10 & 2 \end{bmatrix} + \begin{bmatrix} -3 & -21 & 6 \\ 0 & -3 & 15 \end{bmatrix} = \begin{bmatrix} -7 & -19 & 12 \\ 2 & 7 & 17 \end{bmatrix}$$

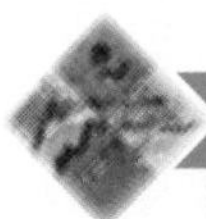

11.2 MATRIX MULTIPLICATION

The definition of matrix multiplication is much more involved than the definition of addition. We begin with a formal statement of the definition and then illustrate the definition with some examples.

Definition of Matrix Multiplication

The **product** of an $m \times n$ matrix A and an $n \times p$ matrix B is an $m \times p$ matrix $C = AB$, where the element c_{ij} in row i and column j of AB is found by using the elements in row i of A and the elements in column j of B in the following manner.

$$\begin{matrix}\text{Row } i\\ \text{of } A\end{matrix}\left\}\begin{bmatrix} \vdots & \vdots & \vdots & & \vdots \\ a_{i1} & a_{i2} & a_{i3} & \cdots & a_{in} \\ \vdots & \vdots & \vdots & & \vdots \end{bmatrix}\right. \cdot \underbrace{}_{\text{Column } j \text{ of } B}\begin{bmatrix} \cdots & b_{1j} & \cdots \\ \cdots & b_{2j} & \cdots \\ \cdots & b_{3j} & \cdots \\ & \vdots & \\ \cdots & b_{nj} & \cdots \end{bmatrix} = \underbrace{}_{\text{Column } j \text{ of } C}\begin{bmatrix} & \vdots & \\ \cdots & c_{ij} & \cdots \\ & \vdots & \end{bmatrix}\left\{\begin{matrix}\text{Row } i\\ \text{of } C\end{matrix}\right.,$$

where

$$c_{ij} = a_{i1}b_{1j} + a_{i2}b_{2j} + a_{i3}b_{3j} + \cdots + a_{in}b_{nj}.$$

That is, the element c_{ij} in row i and column j of AB is found by adding the products formed from corresponding elements of row i in A and column j in B (first times first, second times second, and so on).

Graphing devices will perform matrix multiplication, subject to the appropriate dimensional restraints.

Notice that the number of columns in A *must* equal the number of rows in B in order to form the product AB. If this is the case, then A and B are said to be **conformable for multiplication.** A simple diagram illustrates this fact.

$$A_{(m,\, n)} \cdot B_{(n,\, p)} = C_{(m,\, p)}$$

Some examples are helpful in understanding the definition of matrix multiplication.

Example 1 • Matrix Multiplication

Form the products (a) AB, (b) BA, (c) AC, and (d) CA, if possible, where

$$A = \begin{bmatrix} 3 & 1 \\ -2 & -1 \\ 4 & -2 \end{bmatrix}, \quad B = \begin{bmatrix} 2 & 1 & 0 \\ -1 & 3 & 5 \end{bmatrix}, \quad C = \begin{bmatrix} 2 & 7 \\ 0 & -4 \end{bmatrix}.$$

Solution

a. The product AB exists since A has two columns and B has two rows.

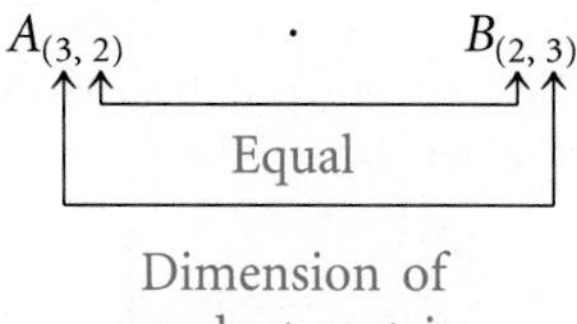

Performing the multiplication yields the following result. (The shading indicates how the element in row 1, column 3 is computed.)

$$\begin{bmatrix} 3 & 1 \\ -2 & -1 \\ 4 & -2 \end{bmatrix}\begin{bmatrix} 2 & 1 & 0 \\ -1 & 3 & 5 \end{bmatrix}$$

$$= \begin{bmatrix} 3(2) + 1(-1) & 3(1) + 1(3) & 3(0) + 1(5) \\ -2(2) + (-1)(-1) & -2(1) + (-1)(3) & -2(0) + (-1)(5) \\ 4(2) + (-2)(-1) & 4(1) + (-2)(3) & 4(0) + (-2)(5) \end{bmatrix}$$

and

$$AB = \begin{bmatrix} 5 & 6 & 5 \\ -3 & -5 & -5 \\ 10 & -2 & -10 \end{bmatrix}.$$

b. The product BA exists since B has three columns and A has three rows.

$B_{(2,3)} \cdot A_{(3,2)}$

Equal

Dimension of product matrix

Performing the multiplication, we have

$$\begin{bmatrix} 2 & 1 & 0 \\ -1 & 3 & 5 \end{bmatrix}\begin{bmatrix} 3 & 1 \\ -2 & -1 \\ 4 & -2 \end{bmatrix}$$

$$= \begin{bmatrix} 2(3) + 1(-2) + 0(4) & 2(1) + 1(-1) + 0(-2) \\ -1(3) + 3(-2) + 5(4) & -1(1) + 3(-1) + 5(-2) \end{bmatrix}$$

and

$$BA = \begin{bmatrix} 4 & 1 \\ 11 & -14 \end{bmatrix}.$$

Notice that the dimensions of the products AB and BA are not the same. Hence we have $AB \neq BA$. (The commutative law of multiplication does *not* hold for matrices.)

c. The product AC exists since A has two columns and C has two rows. The multiplication is given by

$$AC = \begin{bmatrix} 3 & 1 \\ -2 & -1 \\ 4 & -2 \end{bmatrix} \begin{bmatrix} 2 & 7 \\ 0 & -4 \end{bmatrix} = \begin{bmatrix} 6 & 17 \\ -4 & -10 \\ 8 & 36 \end{bmatrix}.$$

d. The product CA does not exist since C has two columns and A has three rows.

$C_{(2,\,2)} \cdot A_{(3,\,2)}$ does not exist.

Not equal

□

◆ Practice Problem 1 ◆

For the matrices B and C given in Example 1, compute the products (a) BC and (b) CB, if possible.

Even when the two products have the same dimension, AB and BA are usually different. This situation is illustrated in the next example. (See also the Problems 43–47 in Exercises 11.2.)

Example 2 • Failure of the Commutative Property of Multiplication

Compare the products AB and BA, where

$$A = \begin{bmatrix} 1 & 0 \\ -3 & -1 \end{bmatrix} \quad \text{and} \quad B = \begin{bmatrix} -2 & -1 \\ 1 & -3 \end{bmatrix}.$$

Solution

Multiplying, we have

$$AB = \begin{bmatrix} 1 & 0 \\ -3 & -1 \end{bmatrix} \begin{bmatrix} -2 & -1 \\ 1 & -3 \end{bmatrix} = \begin{bmatrix} -2 & -1 \\ 5 & 6 \end{bmatrix}$$

and

$$BA = \begin{bmatrix} -2 & -1 \\ 1 & -3 \end{bmatrix} \begin{bmatrix} 1 & 0 \\ -3 & -1 \end{bmatrix} = \begin{bmatrix} 1 & 1 \\ 10 & 3 \end{bmatrix}.$$

Thus $AB \neq BA$.

□

Although multiplication of matrices does not have the commutative property, it can be shown that matrix multiplication is associative, provided that the products involved are defined.

Associative Property of Matrix Multiplication

Let A be an $m \times n$ matrix, B be an $n \times p$ matrix, and C be a $p \times q$ matrix. Then

$$A(BC) = (AB)C.$$

It can also be shown that the following distributive properties hold.

Distributive Properties

Let A be an $m \times n$ matrix, B and C be $n \times p$ matrices, and D be a $p \times q$ matrix. Let a and b be real numbers. Then

1. $A(B + C) = AB + AC$
2. $(B + C)D = BD + CD$
3. $a(B + C) = aB + aC$
4. $(a + b)A = aA + bA$

One type of product is worthy of special note. For every square matrix A of order n, there is a square matrix I_n of order n such that $AI_n = I_nA = A$. The matrix I_n is called the **identity matrix of order n.** It is a diagonal matrix with 1's on the main diagonal.

$$I_n = \begin{bmatrix} 1 & 0 & 0 & \cdots & 0 \\ 0 & 1 & 0 & \cdots & 0 \\ 0 & 0 & 1 & \cdots & 0 \\ \vdots & \vdots & \vdots & & \vdots \\ 0 & 0 & 0 & \cdots & 1 \end{bmatrix}$$

If the order is understood from the context, then we write I for I_n.

Example 3 • Multiplication with an Identity Matrix

$$I_2\begin{bmatrix} 1 & 4 \\ 2 & 1 \end{bmatrix} = \begin{bmatrix} 1 & 0 \\ 0 & 1 \end{bmatrix}\begin{bmatrix} 1 & 4 \\ 2 & 1 \end{bmatrix} = \begin{bmatrix} 1 & 4 \\ 2 & 1 \end{bmatrix}$$

$$\begin{bmatrix} 1 & 4 \\ 2 & 1 \end{bmatrix}I_2 = \begin{bmatrix} 1 & 4 \\ 2 & 1 \end{bmatrix}\begin{bmatrix} 1 & 0 \\ 0 & 1 \end{bmatrix} = \begin{bmatrix} 1 & 4 \\ 2 & 1 \end{bmatrix}$$

□

Many practical problems can be formulated in terms of matrices so that matrix multiplication is instrumental in the solution of the problems.

Example 4 • An Application of Matrix Multiplication

Consider the following problem of determining the cost of 10 pounds of four chicken feed mixtures. The number of pounds of oats, barley, and corn contained in 10 pounds of each of the four mixtures A, B, C, and D are given by the following matrix.

		Ingredients		
		Oats	Barley	Corn
Mixture	A	2	4	4
	B	3	2	5
	C	2	2	6
	D	5	1	4

For example, there are 2 pounds of oats, 2 pounds of barley, and 6 pounds of corn in 10 pounds of mixture *C*.

There are two suppliers of ingredients, and the cost per pound of each ingredient from each supplier is given in the following matrix.

		Supplier	
		1	2
Ingredient	Oats	$0.17	$0.18
	Barley	$0.14	$0.16
	Corn	$0.15	$0.13

The product of these two matrices,

$$\begin{bmatrix} 2 & 4 & 4 \\ 3 & 2 & 5 \\ 2 & 2 & 6 \\ 5 & 1 & 4 \end{bmatrix} \begin{bmatrix} \$0.17 & \$0.18 \\ \$0.14 & \$0.16 \\ \$0.15 & \$0.13 \end{bmatrix},$$

will yield the matrix whose elements represent the cost of 10 pounds of each mixture with ingredients supplied by each supplier.

		Supplier	
		1	2
Mixture	A	$1.50	$1.52
	B	$1.54	$1.51
	C	$1.52	$1.46
	D	$1.59	$1.58

For example, 10 pounds of mixture B costs $1.54 when ingredients are furnished by supplier 1 and $1.51 when furnished by supplier 2. □

EXERCISES 11.2

State whether the following matrices are conformable for multiplication in the order given. If the multiplication is possible, give the dimension of the product matrix.

1. $A_{(2,3)} \cdot B_{(3,7)}$
2. $C_{(4,1)} \cdot D_{(1,4)}$
3. $B_{(3,2)} \cdot C_{(3,3)}$
4. $D_{(1,1)} \cdot A_{(2,1)}$
5. $A_{(1,3)} \cdot C_{(3,1)}$
6. $C_{(3,1)} \cdot A_{(1,3)}$
7. $I_2 \cdot A_{(2,7)}$
8. $X_{(2,7)} \cdot I_7$
9. $X_{(8,2)} \cdot Y_{(2,1)}$
10. $Z_{(10,2)} \cdot Z_{(10,2)}$
11. $A_{(3,3)} \cdot A_{(3,3)}$
12. $B_{(4,1)} \cdot A_{(2,2)}$
13. $X_{(2,2)} \cdot Z_{(3,3)}$
14. $Z_{(3,3)} \cdot X_{(2,2)}$
15. $D_{(1,1)} \cdot C_{(1,12)}$
16. $B_{(2,12)} \cdot E_{(2,2)}$

If a calculator with matrix features is available, work a sampling of Exercises 17–54 both by hand and with a calculator.

Perform the following matrix multiplications, if possible.

17. $\begin{bmatrix} 1 & 3 & 0 \\ -2 & 4 & 1 \end{bmatrix}\begin{bmatrix} 1 & 2 \\ 2 & -1 \\ 1 & 3 \end{bmatrix}$

18. $\begin{bmatrix} 1 & 2 \\ 2 & -1 \\ 1 & 3 \end{bmatrix}\begin{bmatrix} 1 & 3 & 0 \\ -2 & 4 & 1 \end{bmatrix}$

19. $\begin{bmatrix} -2 & 1 & 3 \\ 4 & 7 & 0 \\ -1 & 4 & -2 \end{bmatrix}\begin{bmatrix} 3 & 5 \\ -1 & 0 \\ 2 & 4 \end{bmatrix}$

20. $\begin{bmatrix} 3 & 5 \\ 1 & 0 \\ 2 & 4 \end{bmatrix}\begin{bmatrix} -2 & 1 & 3 \\ 4 & 7 & 0 \\ -1 & 4 & -2 \end{bmatrix}$

21. $\begin{bmatrix} 1 & 0 \\ 5 & -1 \end{bmatrix} I_3$

22. $I_2\begin{bmatrix} 1 & 7 & 0 \\ 0 & 0 & 1 \end{bmatrix}$

23. $[-3 \quad -2 \quad 0 \quad 1]\,[1 \quad 0 \quad 3 \quad -1]$

24. $\begin{bmatrix} 1 \\ 5 \\ 9 \end{bmatrix}\begin{bmatrix} -9 \\ -5 \\ -1 \end{bmatrix}$

25. $[0 \quad -3 \quad 1]\begin{bmatrix} 10 \\ -4 \\ 10 \end{bmatrix}$

26. $\begin{bmatrix} 10 \\ -4 \\ 10 \end{bmatrix}[0 \quad -3 \quad 1]$

27. $\begin{bmatrix} -2 \\ 1 \\ 1 \end{bmatrix}[4 \quad 11 \quad -2]$

28. $[4 \quad 11 \quad -2]\begin{bmatrix} -2 \\ 1 \\ 1 \end{bmatrix}$

29. $\begin{bmatrix} 1 & 0 \\ 0 & 1 \\ 0 & 0 \end{bmatrix}\begin{bmatrix} 1 & 3 \\ 2 & -2 \\ 1 & 0 \end{bmatrix}$

30. $\begin{bmatrix} 1 & 3 \\ 2 & -2 \\ 1 & 0 \end{bmatrix}\begin{bmatrix} 1 & 0 \\ 0 & 1 \\ 0 & 0 \end{bmatrix}$

31. $\begin{bmatrix} 0 & 0 & 0 & 0 \\ 0 & 0 & 0 & 0 \\ 0 & 0 & 0 & 0 \end{bmatrix} \begin{bmatrix} 1 & 0 \\ 5 & 1 \\ -3 & 4 \\ 2 & -1 \end{bmatrix}$

32. $[0]\ [9 \quad 2]$

33. $\begin{bmatrix} 3 & -1 \\ 8 & 1 \end{bmatrix} [0 \quad 0]$

34. $\begin{bmatrix} 1 & 0 \\ 0 & -1 \end{bmatrix} \begin{bmatrix} 1 & 1 & 1 \\ 1 & 1 & 1 \end{bmatrix}$

35. $\begin{bmatrix} 11 & 3 \\ 4 & 1 \end{bmatrix} \begin{bmatrix} -1 & 3 \\ 4 & -11 \end{bmatrix}$

36. $\begin{bmatrix} 1 & 1 & 2 \\ 1 & 1 & 0 \\ 0 & -1 & 1 \end{bmatrix} \begin{bmatrix} -\frac{1}{2} & \frac{3}{2} & 1 \\ \frac{1}{2} & -\frac{1}{2} & -1 \\ \frac{1}{2} & -\frac{1}{2} & 0 \end{bmatrix}$

37. $\begin{bmatrix} 10 & 1 & -4 \\ -8 & 1 & 5 \\ -1 & -1 & 4 \end{bmatrix} \begin{bmatrix} \frac{1}{9} & 0 & \frac{1}{9} \\ \frac{3}{9} & \frac{4}{9} & -\frac{2}{9} \\ \frac{1}{9} & -\frac{1}{9} & \frac{2}{9} \end{bmatrix}$

38. $\begin{bmatrix} -1 & 0 \\ 0 & -1 \end{bmatrix} \begin{bmatrix} -1 & 0 \\ 0 & -1 \end{bmatrix}$

Form all possible products using two of the three given matrices.

39. $A = \begin{bmatrix} -1 & 3 \\ -2 & 1 \end{bmatrix}, \quad B = \begin{bmatrix} -5 & -2 \\ 1 & 0 \\ 4 & -2 \end{bmatrix}, \quad C = \begin{bmatrix} 1 & 1 & 0 \\ 0 & 1 & 1 \\ -1 & 0 & -1 \end{bmatrix}$

40. $A = [1 \quad -3 \quad 3], \quad B = \begin{bmatrix} 2 & -2 \\ 3 & -4 \\ 1 & 1 \end{bmatrix}, \quad C = \begin{bmatrix} 2 & 4 & 0 \\ 1 & -1 & 1 \end{bmatrix}$

41. $A = \begin{bmatrix} 1 & -1 & 0 & 2 \\ 1 & 1 & 0 & 4 \\ 0 & 0 & 1 & 5 \\ 0 & 1 & -1 & 0 \end{bmatrix}, \quad B = \begin{bmatrix} 1 & -1 & 2 & 0 \\ 0 & 1 & 5 & 3 \end{bmatrix}, \quad C = \begin{bmatrix} 1 & 4 \\ 4 & 0 \\ 1 & 2 \\ 0 & -1 \end{bmatrix}$

42. $A = \begin{bmatrix} -2 & 3 \\ 1 & 1 \\ -1 & 0 \end{bmatrix}, \quad B = \begin{bmatrix} 3 & 1 & 0 \\ -3 & 5 & -5 \end{bmatrix}, \quad C = \begin{bmatrix} 1 & 1 \\ -1 & 2 \end{bmatrix}$

43. Find two matrices A and B such that $AB \neq BA$.
44. Find two square matrices A and B, of order 2, such that $AB \neq BA$.
45. Find two square matrices A and B, of order 3, such that $AB \neq BA$.
46. Find two square matrices A and B, of order 4, such that $AB \neq BA$.
47. Find nonzero square matrices A and B, of order 2, such that $AB = BA$.
48. Find two nonzero square matrices A and B such that $AB = O$, where O is the zero matrix.
49. Evaluate $AB + AC$ and $A(B + C)$ and compare the results for

$$A = \begin{bmatrix} -1 & 3 \\ -2 & 0 \\ -1 & 1 \end{bmatrix}, \quad B = \begin{bmatrix} 2 & 1 \\ -3 & 0 \end{bmatrix}, \quad C = \begin{bmatrix} 5 & 3 \\ 0 & -1 \end{bmatrix}.$$

50. Evaluate $A(BC)$ and $(AB)C$ and compare the results for

$$A = \begin{bmatrix} 4 & 1 \\ 0 & -1 \\ -1 & 1 \end{bmatrix}, \quad B = \begin{bmatrix} 1 & 6 & 0 & 1 \\ 5 & -3 & 4 & 0 \end{bmatrix}, \quad C = \begin{bmatrix} 2 \\ 3 \\ -1 \\ 1 \end{bmatrix}.$$

51. Evaluate $(A - B)(A + B)$ and $A^2 - B^2$ and compare the results for

$$A = \begin{bmatrix} -6 & 4 \\ 1 & 3 \end{bmatrix}, \quad B = \begin{bmatrix} 0 & 1 \\ 1 & 2 \end{bmatrix}.$$

(*Note:* $A^2 = A \cdot A$.)

52. For the matrices in Exercise 51, evaluate $(A + B)^2$ and $A^2 + 2AB + B^2$ and compare results.

Computing Grocery Bills

53. Adrienne found that she needed to purchase 2 cans of peaches, 1 sack of flour, $\frac{1}{2}$ dozen eggs, and 1 sack of sugar in order to have the ingredients for a new recipe. From advertisements in the newspaper, she found the prices of each item at three supermarkets.

	Item			
	1 Can Peaches	1 Sack Flour	1 Dozen Eggs	1 Sack Sugar
Store A	\$0.65	\$0.69	\$0.84	\$1.03
Store B	\$0.63	\$0.89	\$0.62	\$0.78
Store C	\$0.72	\$0.90	\$0.78	\$0.82

Use matrix multiplication to determine the grocery bill at each of the three stores.

Computing the Costs of Nut Mixtures

54. Suppose that a nut vendor wants to determine the cost per pound of three mixtures, each containing peanuts, cashews, and pecans, as given in the accompanying table.

	Nut		
	Peanuts	Cashews	Pecans
Mixture A	$\frac{1}{2}$	$\frac{1}{6}$	$\frac{1}{3}$
Mixture B	$\frac{1}{3}$	$\frac{1}{3}$	$\frac{1}{3}$
Mixture C	$\frac{3}{10}$	$\frac{3}{10}$	$\frac{2}{5}$

If peanuts cost \$1.90 per pound, cashews cost \$2.20 per pound, and pecans cost \$3.80 per pound, use matrix multiplication to determine the cost per pound of each mixture.

55. Prove that matrix multiplication is associative for square matrices of order 2: if A, B and C are 2×2 matrices, then $A(BC) = (AB)C$.
56. Prove the distributive property: $A(B + C) = AB + AC$ for square matrices A, B, and C of order 2.
57. Prove the distributive property: $(B + C)D = BD + CD$ for square matrices B, C, and D of order 2.
58. Prove the distributive property: $a(B + C) = aB + aC$ for a real number a and 3×2 matrices B and C.
59. If a and b are real numbers, prove that $(a + b)A = aA + bA$ for an $m \times n$ matrix A.

Critical Thinking: Exploration and Discussion

60. (See Exercise 51.) Use the distributive properties to explain why $(A - B)(A + B)$ and $A^2 - B^2$ may be unequal for some matrices A and B.

Critical Thinking: Exploration and Discussion

61. (See Exercise 52.) Use the distributive properties to explain why $(A + B)^2$ and $A^2 + 2AB + B^2$ are sometimes different when working with matrices.

◆ Solution for Practice Problem

1. a. The product BC is not defined since B has three columns and C has two rows.

   b. The product CB is defined since C has two columns and B has two rows. We have

$$CB = \begin{bmatrix} 2 & 7 \\ 0 & -4 \end{bmatrix} \begin{bmatrix} 2 & 1 & 0 \\ -1 & 3 & 5 \end{bmatrix} = \begin{bmatrix} -3 & 23 & 35 \\ 4 & -12 & -20 \end{bmatrix}.$$

11.3 SOLUTION OF LINEAR SYSTEMS BY MATRIX METHODS

Matrices are useful tools in solving systems of linear equations. In this section we consider two of the best-known matrix methods, the **Gaussian elimination method** and the **Gauss-Jordan elimination method.** Some new terminology is needed to describe these methods.

Although the methods can be presented in a more general setting, we restrict our attention to systems of n linear equations in n unknowns. A system of n linear equations in n unknowns $x_1, x_2, \ldots, x_n$ is a system of the following form.

$$\begin{aligned} a_{11}x_1 + a_{12}x_2 + a_{13}x_3 + \cdots + a_{1n}x_n &= b_1 \\ a_{21}x_1 + a_{22}x_2 + a_{23}x_3 + \cdots + a_{2n}x_n &= b_2 \\ \vdots \qquad\qquad\qquad\qquad & \quad \vdots \\ a_{n1}x_1 + a_{n2}x_2 + a_{n3}x_3 + \cdots + a_{nn}x_n &= b_n \end{aligned}$$

For this system, the matrices

$$A = \begin{bmatrix} a_{11} & a_{12} & a_{13} & \cdots & a_{1n} \\ a_{21} & a_{22} & a_{23} & \cdots & a_{2n} \\ \vdots & \vdots & \vdots & & \vdots \\ a_{n1} & a_{n2} & a_{n3} & \cdots & a_{nn} \end{bmatrix} \quad \text{and} \quad B = \begin{bmatrix} b_1 \\ b_2 \\ \vdots \\ b_n \end{bmatrix}$$

are called the **coefficient matrix** and the **constant matrix,** respectively. The matrix

$$[A \mid B] = \left[\begin{array}{ccccc|c} a_{11} & a_{12} & a_{13} & \cdots & a_{1n} & b_1 \\ a_{21} & a_{22} & a_{23} & \cdots & a_{2n} & b_2 \\ \vdots & \vdots & \vdots & & \vdots & \vdots \\ a_{n1} & a_{n2} & a_{n3} & \cdots & a_{nn} & b_n \end{array}\right]$$

is called an **augmented matrix** for the system.

Example 1 • Systems and Their Augmented Matrices

The augmented matrix for

$$\begin{aligned} x - y + 2z &= 9 \\ -2x + 3y - z &= -11 \\ 3x + y + z &= 4 \end{aligned} \quad \text{is the matrix} \quad \left[\begin{array}{ccc|c} 1 & -1 & 2 & 9 \\ -2 & 3 & -1 & -11 \\ 3 & 1 & 1 & 4 \end{array}\right].$$

Conversely, a system associated with the augmented matrix

$$\left[\begin{array}{cccc|c} 1 & 3 & 0 & 4 & 1 \\ 0 & 1 & -1 & 2 & -1 \\ 1 & -1 & 5 & 3 & 1 \\ 2 & 1 & 1 & 0 & -2 \end{array}\right] \quad \text{is} \quad \begin{aligned} x_1 + 3x_2 \phantom{{}+5x_3} + 4x_4 &= 1 \\ x_2 - x_3 + 2x_4 &= -1 \\ x_1 - x_2 + 5x_3 + 3x_4 &= 1 \\ 2x_1 + x_2 + x_3 \phantom{{}+3x_4} &= -2 \end{aligned}.$$

A system associated with the augmented matrix

$$\left[\begin{array}{ccc|c} 1 & 0 & 0 & a \\ 0 & 1 & 0 & b \\ 0 & 0 & 1 & c \end{array}\right] \quad \text{is} \quad \begin{aligned} x &= a \\ y &= b \\ z &= c \end{aligned}$$

which has the obvious solution $x = a$, $y = b$, $z = c$. □

The elimination method used in Section 10.1 can be modified to obtain the method known as *Gaussian elimination.* This modification is motivated by the following example.

Example 2 • Equivalent Systems and Their Augmented Matrices

Suppose that we wish to solve the system

$$\begin{aligned} 2x - y + z &= 6 \\ y + 2z &= 2 \\ x + y + z &= 1 \end{aligned} \quad \text{with augmented matrix} \quad \left[\begin{array}{ccc|c} 2 & -1 & 1 & 6 \\ 0 & 1 & 2 & 2 \\ 1 & 1 & 1 & 1 \end{array}\right].$$

Solution

We can interchange the first and third equations and obtain the equivalent system

$$\begin{aligned} x + y + z &= 1 \\ y + 2z &= 2 \\ 2x - y + z &= 6 \end{aligned} \quad \text{with augmented matrix} \quad \left[\begin{array}{ccc|c} 1 & 1 & 1 & 1 \\ 0 & 1 & 2 & 2 \\ 2 & -1 & 1 & 6 \end{array}\right].$$

Adding -2 times the first equation to the third equation yields the equivalent system

$$\begin{aligned} x + y + z &= 1 \\ y + 2z &= 2 \\ -3y - z &= 4 \end{aligned} \quad \text{with augmented matrix} \quad \left[\begin{array}{ccc|c} 1 & 1 & 1 & 1 \\ 0 & 1 & 2 & 2 \\ 0 & -3 & -1 & 4 \end{array}\right].$$

Adding 3 times the second equation to the third equation yields the equivalent system

$$\begin{aligned} x + y + z &= 1 \\ y + 2z &= 2 \\ 5z &= 10 \end{aligned} \quad \text{with augmented matrix} \quad \left[\begin{array}{ccc|c} 1 & 1 & 1 & 1 \\ 0 & 1 & 2 & 2 \\ 0 & 0 & 5 & 10 \end{array}\right].$$

Multiplying the third equation by $\frac{1}{5}$ yields the equivalent system

$$\begin{aligned} x + y + z &= 1 \\ y + 2z &= 2 \\ z &= 2 \end{aligned} \quad \text{with augmented matrix} \quad \left[\begin{array}{ccc|c} 1 & 1 & 1 & 1 \\ 0 & 1 & 2 & 2 \\ 0 & 0 & 1 & 2 \end{array}\right].$$

We can now find the solution by these substitutions.

Substitute $z = 2$ in $y + 2z = 2$ to obtain $y = -2$

and then

substitute $y = -2$, $z = 2$ in $x + y + z = 1$ to obtain $x = 1$.

This procedure of substitution into prior equations is called **back substitution.**

□

Focusing our attention on the matrices in Example 2, we can solve the system by making the following changes in the augmented matrix and then using back substitution.

$$\left[\begin{array}{ccc|c} 2 & -1 & 1 & 6 \\ 0 & 1 & 2 & 2 \\ 1 & 1 & 1 & 1 \end{array}\right] \xrightarrow{R_1 \leftrightarrow R_3} \left[\begin{array}{ccc|c} 1 & 1 & 1 & 1 \\ 0 & 1 & 2 & 2 \\ 2 & -1 & 1 & 6 \end{array}\right] \xrightarrow{-2R_1 + R_3} \left[\begin{array}{ccc|c} 1 & 1 & 1 & 1 \\ 0 & 1 & 2 & 2 \\ 0 & -3 & -1 & 4 \end{array}\right]$$

$$\xrightarrow{3R_2 + R_3} \left[\begin{array}{ccc|c} 1 & 1 & 1 & 1 \\ 0 & 1 & 2 & 2 \\ 0 & 0 & 5 & 10 \end{array}\right] \xrightarrow{\frac{1}{5}R_3} \left[\begin{array}{ccc|c} 1 & 1 & 1 & 1 \\ 0 & 1 & 2 & 2 \\ 0 & 0 & 1 & 2 \end{array}\right]$$

The notation $R_1 \leftrightarrow R_3$ indicates that rows 1 and 3 are interchanged; $-2R_1 + R_3$ indicates that row 3 is replaced by -2 times row 1 plus row 3; and $\frac{1}{5}R_3$ indicates that row 3 was multiplied by $\frac{1}{5}$.

In the ordinary elimination method that was used in Section 10.1, equivalent systems are obtained by using the following operations.

1. Interchange two equations.
2. Multiply (or divide) both members of an equation by the same nonzero number.
3. Add (or subtract) a multiple of one equation to (or from) another equation.

Example 2 shows how these operations on systems correspond to performing the following row operations on the augmented matrix.

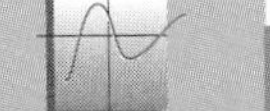

Graphing devices will perform all the row operations listed here.

Row Operations

1. Interchange two rows.
2. Multiply (or divide) every element in a row by the same nonzero number.
3. Add (or subtract) a multiple of one row to (or from) another row.

The **Gaussian elimination** method uses row operations on the augmented matrix to solve a system of linear equations. The procedure in this method follows.

Gaussian Elimination Method

1. Write the augmented matrix $[A \mid B]$ for the system, where A is the coefficient matrix and B is the constant matrix.
2. Use any of the row operations 1, 2, and 3 to change $[A \mid B]$ to a matrix in which all elements on the main diagonal are 1's and all elements below the main diagonal are 0's.
3. Use back substitution on the system that has the augmented matrix obtained in step 2.

This method is illustrated in the following example.

Example 3 • Solving a System by Gaussian Elimination

Use Gaussian elimination to solve the following system.

$$\begin{aligned} 3x + 2y - 3z &= -1 \\ x + 3y + 2z &= 1 \\ x + y - 2z &= -3 \end{aligned}$$

Solution

A straightforward approach is to change one column of the augmented matrix at a time into the desired form, working from left to right. In the first column we use appropriate row operations to obtain a 1 in the first row. We then use the third type of row operation to obtain 0's below that 1.

$$\left[\begin{array}{ccc|c} 3 & 2 & -3 & -1 \\ 1 & 3 & 2 & 1 \\ 1 & 1 & -2 & -3 \end{array}\right] \xrightarrow{R_1 \leftrightarrow R_3} \left[\begin{array}{ccc|c} 1 & 1 & -2 & -3 \\ 1 & 3 & 2 & 1 \\ 3 & 2 & -3 & -1 \end{array}\right]$$

$$\xrightarrow{-R_1 + R_2} \left[\begin{array}{ccc|c} 1 & 1 & -2 & -3 \\ 0 & 2 & 4 & 4 \\ 3 & 2 & -3 & -1 \end{array}\right] \xrightarrow{-3R_1 + R_2} \left[\begin{array}{ccc|c} 1 & 1 & -2 & -3 \\ 0 & 2 & 4 & 4 \\ 0 & -1 & 3 & 8 \end{array}\right]$$

Moving now to the second column, we obtain a 1 in the second position on the main diagonal and then use row operations of the third type to obtain 0's below that 1.

$$\xrightarrow{\frac{1}{2}R_2} \left[\begin{array}{ccc|c} 1 & 1 & -2 & -3 \\ 0 & 1 & 2 & 2 \\ 0 & -1 & 3 & 8 \end{array}\right] \xrightarrow{R_2 + R_3} \left[\begin{array}{ccc|c} 1 & 1 & -2 & -3 \\ 0 & 1 & 2 & 2 \\ 0 & 0 & 5 & 10 \end{array}\right]$$

As a final step, we get a 1 in the third position on the main diagonal.

$$\xrightarrow{\frac{1}{5}R_3} \left[\begin{array}{ccc|c} 1 & 1 & -2 & -3 \\ 0 & 1 & 2 & 2 \\ 0 & 0 & 1 & 2 \end{array}\right]$$

This augmented matrix corresponds to the following system.

$$\begin{aligned} x + y - 2z &= -3 \\ y + 2z &= 2 \\ z &= 2 \end{aligned}$$

Back substitution yields the solution $x = 3$, $y = -2$, $z = 2$. □

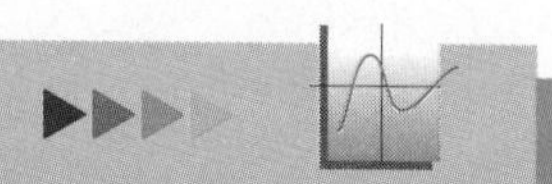

The row operations feature of a graphing device can be used to implement the steps required in either Gaussian or Gauss–Jordan elimination.

The back substitution step in Gaussian elimination can be avoided by performing additional row operations on the augmented matrix. This procedure is described in the following method, known as **Gauss–Jordan elimination.**

Gauss–Jordan Elimination

1. Write the augmented matrix $[A \mid B]$ for the system.
2. Use any of the row operations 1, 2, 3 to change $[A \mid B]$ to $[I \mid C]$, where I is an identity matrix of the same size as A, and C is a column matrix.
3. Read the solutions from the column matrix C.

This method is illustrated in the following example.

Example 4 • Solving a System by Gauss–Jordan Elimination

Use Gauss–Jordan elimination to solve the following system.

$$\begin{aligned} x + \ \ y + 2z &= \ \ 3 \\ 2x - \ \ y + \ \ z &= \ \ 6 \\ -x + 3y \qquad &= -5 \end{aligned}$$

Solution

The augmented matrix is

$$[A \mid B] = \left[\begin{array}{rrr|r} 1 & 1 & 2 & 3 \\ 2 & -1 & 1 & 6 \\ -1 & 3 & 0 & -5 \end{array}\right].$$

We shall use row operations to transform $[A \mid B]$ into

$$[I \mid C] = \left[\begin{array}{ccc|c} 1 & 0 & 0 & a \\ 0 & 1 & 0 & b \\ 0 & 0 & 1 & c \end{array}\right],$$

which is the augmented matrix for the system whose solution is

$$x = a, \qquad y = b, \qquad z = c.$$

The most straightforward approach to use to transform $[A \mid B]$ into $[I \mid C]$ is to change one column of A at a time into a column of the identity matrix I, working from left to right. In the first column we use appropriate row operations to obtain a 1 in the first row. Then, with that 1, we use the third type of row operation to obtain 0's in the remaining positions of column 1.

$$\left[\begin{array}{rrr|r} 1 & 1 & 2 & 3 \\ 2 & -1 & 1 & 6 \\ -1 & 3 & 0 & -5 \end{array}\right] \xrightarrow{-2R_1 + R_2} \left[\begin{array}{rrr|r} 1 & 1 & 2 & 3 \\ 0 & -3 & -3 & 0 \\ -1 & 3 & 0 & -5 \end{array}\right]$$

$$\xrightarrow{R_1 + R_3} \left[\begin{array}{rrr|r} 1 & 1 & 2 & 3 \\ 0 & -3 & -3 & 0 \\ 0 & 4 & 2 & -2 \end{array}\right]$$

Next, we use row operations to obtain a 1 in the second-row, second-column position. Then, with that 1, we use row operations of the third type to obtain 0's in the remaining positions of column 2.

$$\xrightarrow{-\frac{1}{3}R_2} \left[\begin{array}{rrr|r} 1 & 1 & 2 & 3 \\ 0 & 1 & 1 & 0 \\ 0 & 4 & 2 & -2 \end{array}\right] \xrightarrow{-R_2 + R_1} \left[\begin{array}{rrr|r} 1 & 0 & 1 & 3 \\ 0 & 1 & 1 & 0 \\ 0 & 4 & 2 & -2 \end{array}\right]$$

$$\xrightarrow{-4R_2 + R_3} \left[\begin{array}{rrr|r} 1 & 0 & 1 & 3 \\ 0 & 1 & 1 & 0 \\ 0 & 0 & -2 & -2 \end{array}\right]$$

Proceeding to column 3, we use row operations to obtain a 1 in the third row. Then with that 1 we use row operations of type 3 to obtain 0's in the remaining positions of column 3.

$$\xrightarrow{-\frac{1}{2}R_3} \left[\begin{array}{ccc|c} 1 & 0 & 1 & 3 \\ 0 & 1 & 1 & 0 \\ 0 & 0 & 1 & 1 \end{array}\right] \xrightarrow{-R_3 + R_2} \left[\begin{array}{ccc|c} 1 & 0 & 1 & 3 \\ 0 & 1 & 0 & -1 \\ 0 & 0 & 1 & 1 \end{array}\right]$$

$$\xrightarrow{-R_3 + R_1} \left[\begin{array}{ccc|c} 1 & 0 & 0 & 2 \\ 0 & 1 & 0 & -1 \\ 0 & 0 & 1 & 1 \end{array}\right]$$

Thus the solution is

$$x = 2, \qquad y = -1, \qquad z = 1.$$

□

◆ Practice Problem 1 ◆

Solve the following system by Gauss-Jordan elimination:

$$\begin{aligned} 2x - y &= 2 \\ -x + 3y &= 14 \end{aligned}$$

Sometimes it is impossible to transform the augmented matrix $[A \mid B]$ into $[I \mid C]$ by row operations. If any of the row operations on $[A \mid B]$ yields a row with all zero elements except possibly the last element, then either there is no solution to the system or the solution is not unique. This is illustrated in the next two examples.

Example 5 • An Inconsistent System

Solve the following system.

$$\begin{aligned} x - 2y - 2z &= -1 \\ x + y + z &= 2 \\ x + 2y + 2z &= 1 \end{aligned}$$

Solution

Using the procedure described in Example 3 or 4 yields the following result.

$$\left[\begin{array}{ccc|c} 1 & -2 & -2 & -1 \\ 1 & 1 & 1 & 2 \\ 1 & 2 & 2 & 1 \end{array}\right] \xrightarrow{-R_1 + R_2} \left[\begin{array}{ccc|c} 1 & -2 & -2 & -1 \\ 0 & 3 & 3 & 3 \\ 1 & 2 & 2 & 1 \end{array}\right]$$

$$\xrightarrow{-R_1 + R_3} \left[\begin{array}{ccc|c} 1 & -2 & -2 & -1 \\ 0 & 3 & 3 & 3 \\ 0 & 4 & 4 & 2 \end{array}\right] \xrightarrow{\frac{1}{3}R_2} \left[\begin{array}{ccc|c} 1 & -2 & -2 & -1 \\ 0 & 1 & 1 & 1 \\ 0 & 4 & 4 & 2 \end{array}\right]$$

$$\xrightarrow{-4R_2 + R_3} \left[\begin{array}{ccc|c} 1 & -2 & -2 & -1 \\ 0 & 1 & 1 & 1 \\ 0 & 0 & 0 & -2 \end{array}\right]$$

Since the third row in the last matrix contains all zero elements except the last element, it is impossible to obtain $[I \mid C]$ by row operations. The last matrix obtained is the augmented matrix for a system with a last equation that reads $0 = -2$. This indicates there is no solution to the original system. □

Example 6 • A Dependent System

Solve the following system.

$$\begin{aligned} x + 2y \quad\quad &= 1 \\ x + 3y + z &= 4 \\ 2y + 2z &= 6 \end{aligned}$$

Solution

$$\left[\begin{array}{ccc|c} 1 & 2 & 0 & 1 \\ 1 & 3 & 1 & 4 \\ 0 & 2 & 2 & 6 \end{array}\right] \xrightarrow{-R_1 + R_2} \left[\begin{array}{ccc|c} 1 & 2 & 0 & 1 \\ 0 & 1 & 1 & 3 \\ 0 & 2 & 2 & 6 \end{array}\right] \xrightarrow{-2R_2 + R_3} \left[\begin{array}{ccc|c} 1 & 2 & 0 & 1 \\ 0 & 1 & 1 & 3 \\ 0 & 0 & 0 & 0 \end{array}\right]$$

Since the last row contains only zeros, it is impossible to obtain $[I \mid C]$ by row operations. The following system corresponds to the last matrix we obtained.

$$\begin{aligned} x + 2y \quad\quad &= 1 \\ y + z &= 3 \\ 0 &= 0 \end{aligned}$$

This system has many solutions. To display them, we can let r represent an arbitrary real number, set $z = r$, and then solve for x and y in terms of r. We get

$$x = 2r - 5, \qquad y = 3 - r, \qquad z = r.$$ □

EXERCISES 11.3

Write the augmented matrix for each of the following systems. (See Example 1.)

1. $\begin{aligned} 3x - y &= 0 \\ -x + y &= 1 \end{aligned}$

2. $\begin{aligned} r + s &= 10 \\ r - s &= -4 \end{aligned}$

3. $\begin{aligned} 3x - 2y + 5z &= 0 \\ 4x + 7y - z &= 0 \\ x \quad\quad + z &= 2 \end{aligned}$

4. $\begin{aligned} x_1 + x_3 &= 4 \\ x_2 + x_3 &= 0 \\ x_1 + x_2 &= 2 \end{aligned}$

5. $\begin{aligned} p - q &= 0 \\ r + s &= 0 \\ 3p + 2s &= 0 \\ 5q - r &= 0 \end{aligned}$

6. $\begin{aligned} a + 3b + c + d &= 1 \\ a - 3b - c + 2d &= 5 \\ 2a \quad\quad + c + 3d &= 7 \\ b + c \quad\quad &= 0 \end{aligned}$

Write a system of linear equations that has the given augmented matrix. (See Example 1.)

7. $\left[\begin{array}{cc|c} 1 & 2 & 5 \\ 3 & 4 & 6 \end{array}\right]$

8. $\left[\begin{array}{cc|c} -2 & 1 & -1 \\ 4 & 3 & 0 \end{array}\right]$

9. $\left[\begin{array}{ccc|c} 1 & 0 & 0 & a \\ 0 & 0 & 1 & b \\ 0 & 1 & 0 & c \end{array}\right]$

10. $\left[\begin{array}{ccc|c} 1 & 1 & 0 & 0 \\ 3 & 0 & -3 & 0 \\ -1 & 1 & -1 & -6 \end{array}\right]$

11. $\left[\begin{array}{cccc|c} 1 & 0 & 1 & 0 & 0 \\ 0 & 2 & 1 & 3 & 7 \\ 3 & 1 & 1 & 1 & 1 \\ -3 & 1 & -1 & 2 & 5 \end{array}\right]$

12. $\left[\begin{array}{cccc|c} 0 & 0 & 0 & 1 & 1 \\ 0 & -1 & 0 & 0 & 2 \\ 1 & 0 & 0 & 0 & 3 \\ 0 & 0 & -1 & 0 & 4 \end{array}\right]$

If a calculator with matrix features is available, work a sampling of Exercises 13–36 both by hand and with a calculator.

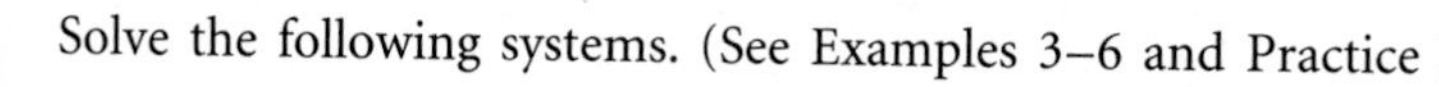

Solve the following systems. (See Examples 3–6 and Practice Problem 1.)

13. $\begin{aligned} x + y &= 1 \\ 2x + 3y &= -2 \end{aligned}$

14. $\begin{aligned} 3x + 2y &= 1 \\ x + 2y &= 7 \end{aligned}$

15. $\begin{aligned} 4a + 5b &= -22 \\ 3a - 4b &= -1 \end{aligned}$

16. $\begin{aligned} 7x_1 - x_2 &= -2 \\ -3x_1 + x_2 &= 6 \end{aligned}$

17. $\begin{aligned} x - 2y &= 0 \\ 2x - 4y &= 1 \end{aligned}$

18. $\begin{aligned} -3g + h &= 2 \\ 6g - 2h &= 0 \end{aligned}$

19. $\begin{aligned} 2x - 3y &= -1 \\ -5x + 8y &= 0 \end{aligned}$

20. $\begin{aligned} 9x - y &= 0 \\ 3x + 3y &= -10 \end{aligned}$

21. $\begin{aligned} 2x + 7y &= 0 \\ x - 2y &= 0 \end{aligned}$

22. $\begin{aligned} 3x &= 0 \\ x + 5y &= -15 \end{aligned}$

23. $\begin{aligned} x - 2y &= 1 \\ y + z &= 0 \\ 2x + 3z &= 3 \end{aligned}$

24. $\begin{aligned} x - 2y + 3z &= 0 \\ 3y - 2z &= 0 \\ -3x + 4y - z &= 0 \end{aligned}$

25. $\begin{aligned} x - y - 4z &= -4 \\ -3x + 4y + 2z &= 6 \\ -x + 3y + 2z &= 10 \end{aligned}$

26. $\begin{aligned} 2a + 2b - c &= -2 \\ -b + 4c &= 11 \\ a - b &= 6 \end{aligned}$

27. $\begin{aligned} -2y - 2z &= -2 \\ 2x - y + z &= -3 \\ x + y + 3z &= -2 \end{aligned}$

28. $\begin{aligned} 3x_1 + 2x_2 + 5x_3 &= 2 \\ 2x_1 + 4x_3 &= 2 \\ x_1 + 3x_2 - x_3 &= -2 \end{aligned}$

29. $\begin{aligned} r - t &= 1 \\ 3r + s &= 6 \\ 5s + 6t &= -12 \end{aligned}$

30. $\begin{aligned} -3x_2 + 4x_3 &= -3 \\ 3x_1 - x_3 &= 6 \\ -x_1 + 3x_2 &= 1 \end{aligned}$

31. $\begin{aligned} 2x - 2y - 2z &= -2 \\ x - 4y + z &= -2 \\ 5x - 8y - 3z &= -2 \end{aligned}$

32. $\begin{aligned} x + 4y - z &= 0 \\ 3x - 5y + z &= 1 \\ 5x - 14y + 3z &= 1 \end{aligned}$

33. $\begin{aligned} x + y - t &= 2 \\ y + z + 2t &= -3 \\ -2x + t &= -4 \\ x + y + z + t &= 0 \end{aligned}$

34. $\begin{aligned} 2x_1 - 4x_2 + x_3 + 10x_4 &= -2 \\ 3x_1 - x_2 + 3x_3 &= -3 \\ x_1 + 2x_3 &= 0 \\ x_2 - x_3 + x_4 &= 4 \end{aligned}$

35.
$$\begin{aligned} -a + b - c \quad\quad &= 5 \\ b - c + 3d &= 10 \\ a \quad - c - d &= 5 \\ a \quad\quad + 3d &= 5 \end{aligned}$$

36.
$$\begin{aligned} x + y \quad\quad - 2t &= -1 \\ 2y - z \quad\quad &= -7 \\ z - 2t &= 1 \\ x + y + z - 4t &= 0 \end{aligned}$$

Some of the coefficients and constant terms in the following systems involve an unspecified constant c. For each system, find the values of c for which the system has

a. No solution

b. Exactly one solution

c. Many solutions.

37.
$$\begin{aligned} x + 2y &= 1 \\ 2x + c^2y &= c \end{aligned}$$

38.
$$\begin{aligned} x + y &= 1 \\ x + cy &= c^2 \end{aligned}$$

39.
$$\begin{aligned} x + 2y \quad\quad - z &= 1 \\ x + y \quad\quad &= 0 \\ x + 2y + (c^2 - 1)z &= c + 1 \end{aligned}$$

40.
$$\begin{aligned} x + 2y \quad\quad - z &= 1 \\ 2x + 4y \quad\quad - 2z &= 0 \\ x + 2y + (c^2 - 1)z &= c + 1 \end{aligned}$$

Critical Thinking: Discussion

41. Compare the Gaussian elimination method with the Gauss–Jordan method and discuss the advantages of each method over the other.

Critical Thinking: Writing

42. Consider the final matrix when Gauss–Jordan elimination is used to solve a system. Write a description of
   a. The last row for a dependent system
   b. The last nonzero row for an inconsistent system.

◆ Solution for Practice Problem

1.
$$\left[\begin{array}{rr|r} 2 & -1 & 2 \\ -1 & 3 & 14 \end{array}\right] \xrightarrow{R_2 + R_1} \left[\begin{array}{rr|r} 1 & 2 & 16 \\ -1 & 3 & 14 \end{array}\right] \xrightarrow{R_1 + R_2} \left[\begin{array}{rr|r} 1 & 2 & 16 \\ 0 & 5 & 30 \end{array}\right]$$

$$\xrightarrow{\frac{1}{5}R_2} \left[\begin{array}{rr|r} 1 & 2 & 16 \\ 0 & 1 & 6 \end{array}\right] \xrightarrow{-2R_2 + R_1} \left[\begin{array}{rr|r} 1 & 0 & 4 \\ 0 & 1 & 6 \end{array}\right]$$

The solution is $x = 4$, $y = 6$.

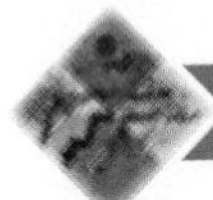

11.4 CALCULATION OF INVERSES

In Sections 11.1 and 11.2 we saw that

1. A matrix with all elements zero acts as an additive identity for matrix addition.
2. Every matrix has an additive inverse, $-A$.

3. An identity matrix I_n acts as a multiplicative identity for matrix multiplication.

A question that arises in connection with statement 3 is this: does every nonzero matrix A have a multiplicative inverse? In other words, for any nonzero matrix A, is there a matrix B such that $AB = BA = I_n$? The answer to this question might be somewhat disappointing, since *not all nonzero matrices have multiplicative inverses.*

We first notice that multiplicative inverses can exist only for square matrices since the only way that AB and BA can be equal is for A and B to be square and of the same order. But we shall see shortly that some nonzero square matrices do not have multiplicative inverses.

From now on, we shall use the term **inverse** to mean "multiplicative inverse." If the matrix A has an inverse, we shall denote it by A^{-1}.

For square matrices of order 2, we have the following formula, whose proof is requested in Exercises 40 and 41 at the end of this section.

Inverse Formula for 2 × 2 Matrices

For a given 2×2 matrix

$$A = \begin{bmatrix} a & b \\ c & d \end{bmatrix},$$

let $\delta(A)$ denote the number $\delta(A) = ad - bc$. The inverse of A exists if $\delta(A) \neq 0$ and is given by

$$A^{-1} = \frac{1}{\delta(A)} \begin{bmatrix} d & -b \\ -c & a \end{bmatrix}.$$

If $\delta(A) = 0$, then A does not have an inverse.

The matrix inversion function of a graphing device will compute an inverse directly with a single command.

Formulas for A^{-1} that are similar to the one for 2×2 matrices exist for matrices of order greater than 2, but they are much more complicated and they more properly belong in a course in linear algebra. Another method for finding A^{-1} is presented after the following example. This method is easier to use than the formulas for matrices of order greater than 2.

Example 1 • Using the Inverse Formula

Use the preceding formula to find the inverse of each of the following matrices, if it exists.

a. $\begin{bmatrix} 1 & 4 \\ 3 & -2 \end{bmatrix}$ b. $\begin{bmatrix} 3 & -2 \\ 4 & -3 \end{bmatrix}$ c. $\begin{bmatrix} 1 & -3 \\ -2 & 6 \end{bmatrix}$

Solution

a. For $A = \begin{bmatrix} 1 & 4 \\ 3 & -2 \end{bmatrix}$, we have $\delta(A) = (1)(-2) - (3)(4) = -14$. Thus A^{-1} exists and is given by

$$A^{-1} = \frac{1}{-14}\begin{bmatrix} -2 & -4 \\ -3 & 1 \end{bmatrix} = \begin{bmatrix} \frac{1}{7} & \frac{2}{7} \\ \frac{3}{14} & -\frac{1}{14} \end{bmatrix}.$$

b. For $A = \begin{bmatrix} 3 & -2 \\ 4 & -3 \end{bmatrix}$, we have $\delta(A) = (3)(-3) - (4)(-2) = -1$ and

$$A^{-1} = \frac{1}{-1}\begin{bmatrix} -3 & 2 \\ -4 & 3 \end{bmatrix} = \begin{bmatrix} 3 & -2 \\ 4 & -3 \end{bmatrix}.$$

Thus we have the unexpected result that $A^{-1} = A$.

c. For $A = \begin{bmatrix} 1 & -3 \\ -2 & 6 \end{bmatrix}$, we have $\delta(A) = (1)(6) - (-2)(-3) = 0$. Thus A^{-1} does not exist for this matrix A. □

The easier method for finding A^{-1} is the **Gauss–Jordan elimination method.** It can be used to calculate the inverse of a square matrix A of any order n, provided that A^{-1} exists. The procedure is given by the following three steps.

Gauss–Jordan Elimination Method for Finding A^{-1}

1. Augment A with an identity matrix of the same size: $[A \mid I_n]$.
2. Use row operations to transform $[A \mid I_n]$ into the form$[I_n \mid B]$. If this is not possible, then A^{-1} does not exist.
3. If $[A \mid I_n]$ is transformed into $[I_n \mid B]$ by row operations, the inverse can be read from the last n columns of $[I_n \mid B]$. That is, $B = A^{-1}$.

Example 2 • Gauss–Jordan Elimination

Use the Gauss–Jordan elimination method to find the inverse of each matrix below, if it exists.

a. $\begin{bmatrix} -2 & 3 \\ 1 & 2 \end{bmatrix}$ b. $\begin{bmatrix} 1 & 0 & 2 \\ 0 & -1 & 3 \\ 2 & 1 & 3 \end{bmatrix}$ c. $\begin{bmatrix} 1 & 1 & 1 \\ 1 & -1 & 0 \\ 2 & 0 & 1 \end{bmatrix}$

Solution

a. We augment $A = \begin{bmatrix} -2 & 3 \\ 1 & 2 \end{bmatrix}$ with a 2×2 identity and use row operations to transform $[A \mid I]$ into $[I \mid A^{-1}]$.

$$[A \mid I] = \left[\begin{array}{cc|cc} -2 & 3 & 1 & 0 \\ 1 & 2 & 0 & 1 \end{array}\right] \xrightarrow{R_1 \leftrightarrow R_2} \left[\begin{array}{cc|cc} 1 & 2 & 0 & 1 \\ -2 & 3 & 1 & 0 \end{array}\right]$$

$$\xrightarrow{2R_1 + R_2} \left[\begin{array}{cc|cc} 1 & 2 & 0 & 1 \\ 0 & 7 & 1 & 2 \end{array}\right] \xrightarrow{\frac{1}{7}R_2} \left[\begin{array}{cc|cc} 1 & 2 & 0 & 1 \\ 0 & 1 & \frac{1}{7} & \frac{2}{7} \end{array}\right]$$

$$\xrightarrow{-2R_2 + R_1} \left[\begin{array}{cc|cc} 1 & 0 & -\frac{2}{7} & \frac{3}{7} \\ 0 & 1 & \frac{1}{7} & \frac{2}{7} \end{array}\right]$$

Thus $A^{-1} = \begin{bmatrix} -\frac{2}{7} & \frac{3}{7} \\ \frac{1}{7} & \frac{2}{7} \end{bmatrix}$. This result should be checked by computing AA^{-1} and $A^{-1}A$.

b. For $A = \begin{bmatrix} 1 & 0 & 2 \\ 0 & -1 & 3 \\ 2 & 1 & 3 \end{bmatrix}$, we have

$$[A \mid I] = \left[\begin{array}{ccc|ccc} 1 & 0 & 2 & 1 & 0 & 0 \\ 0 & -1 & 3 & 0 & 1 & 0 \\ 2 & 1 & 3 & 0 & 0 & 1 \end{array}\right]$$

$$\xrightarrow{-2R_1 + R_3} \left[\begin{array}{ccc|ccc} 1 & 0 & 2 & 1 & 0 & 0 \\ 0 & -1 & 3 & 0 & 1 & 0 \\ 0 & 1 & -1 & -2 & 0 & 1 \end{array}\right]$$

$$\xrightarrow{-R_2} \left[\begin{array}{ccc|ccc} 1 & 0 & 2 & 1 & 0 & 0 \\ 0 & 1 & -3 & 0 & -1 & 0 \\ 0 & 1 & -1 & -2 & 0 & 1 \end{array}\right]$$

$$\xrightarrow{-R_2 + R_3} \left[\begin{array}{ccc|ccc} 1 & 0 & 2 & 1 & 0 & 0 \\ 0 & 1 & -3 & 0 & -1 & 0 \\ 0 & 0 & 2 & -2 & 1 & 1 \end{array}\right]$$

$$\xrightarrow{\frac{1}{2}R_3} \left[\begin{array}{ccc|ccc} 1 & 0 & 2 & 1 & 0 & 0 \\ 0 & 1 & -3 & 0 & -1 & 0 \\ 0 & 0 & 1 & -1 & \frac{1}{2} & \frac{1}{2} \end{array}\right]$$

$$\xrightarrow{3R_3 + R_2} \left[\begin{array}{ccc|ccc} 1 & 0 & 2 & 1 & 0 & 0 \\ 0 & 1 & 0 & -3 & \frac{1}{2} & \frac{3}{2} \\ 0 & 0 & 1 & -1 & \frac{1}{2} & \frac{1}{2} \end{array}\right]$$

$$\xrightarrow{-2R_3 + R_1} \left[\begin{array}{ccc|ccc} 1 & 0 & 0 & 3 & -1 & -1 \\ 0 & 1 & 0 & -3 & \frac{1}{2} & \frac{3}{2} \\ 0 & 0 & 1 & -1 & \frac{1}{2} & \frac{1}{2} \end{array}\right]$$

This gives $A^{-1} = \begin{bmatrix} 3 & -1 & -1 \\ -3 & \frac{1}{2} & \frac{3}{2} \\ -1 & \frac{1}{2} & \frac{1}{2} \end{bmatrix}$.

c. Following the same procedure with $A = \begin{bmatrix} 1 & 1 & 1 \\ 1 & -1 & 0 \\ 2 & 0 & 1 \end{bmatrix}$, we have

$$[A \,|\, I] = \left[\begin{array}{rrr|rrr} 1 & 1 & 1 & 1 & 0 & 0 \\ 1 & -1 & 0 & 0 & 1 & 0 \\ 2 & 0 & 1 & 0 & 0 & 1 \end{array}\right]$$

$$\xrightarrow{-R_1 + R_2} \left[\begin{array}{rrr|rrr} 1 & 1 & 1 & 1 & 0 & 0 \\ 0 & -2 & -1 & -1 & 1 & 0 \\ 2 & 0 & 1 & 0 & 0 & 1 \end{array}\right]$$

$$\xrightarrow{-2R_1 + R_3} \left[\begin{array}{rrr|rrr} 1 & 1 & 1 & 1 & 0 & 0 \\ 0 & -2 & -1 & -1 & 1 & 0 \\ 0 & -2 & -1 & -2 & 0 & 1 \end{array}\right]$$

$$\xrightarrow{-R_2 + R_3} \left[\begin{array}{rrr|rrr} 1 & 1 & 1 & 1 & 0 & 0 \\ 0 & -2 & -1 & -1 & 1 & 0 \\ 0 & 0 & 0 & -1 & -1 & 1 \end{array}\right].$$

The zeros in the last row of the last matrix indicate that it is impossible to transform $[A \,|\, I]$ into $[I \,|\, B]$ by row operations, and A^{-1} does not exist. □

◆ Practice Problem 1 ◆

Find the inverse of

$$A = \begin{bmatrix} 1 & 0 & 0 \\ -1 & 1 & 1 \\ 0 & 1 & 2 \end{bmatrix}.$$

Inverses of matrices can be used in solving certain types of systems of n linear equations in n unknowns. The use of inverses depends on the fact that any system of linear equations can be represented by a single matrix equation. For example, the system

$$\begin{aligned} ax + by &= e \\ cx + dy &= f \end{aligned} \tag{1}$$

is equivalent to the single matrix equation

$$\begin{bmatrix} ax + by \\ cx + dy \end{bmatrix} = \begin{bmatrix} e \\ f \end{bmatrix},$$

and this equation can be written in factored form as

$$\begin{bmatrix} a & b \\ c & d \end{bmatrix} \begin{bmatrix} x \\ y \end{bmatrix} = \begin{bmatrix} e \\ f \end{bmatrix}.$$

Thus the system of linear equations given in Equation (1) is equivalent to the matrix equation $AX = B$, where

$$A = \begin{bmatrix} a & b \\ c & d \end{bmatrix}, \quad X = \begin{bmatrix} x \\ y \end{bmatrix}, \quad B = \begin{bmatrix} e \\ f \end{bmatrix}.$$

The matrix X is called the **unknown matrix.** As in Section 11.3, the matrix A is the **coefficient matrix,** and B is the **constant matrix.**

To find the values of x and y that satisfy the system of Equation (1), we need only solve for X in the matrix equation $AX = B$. This can always be done when A^{-1} exists.

$$\begin{aligned} AX &= B & & \\ A^{-1}(AX) &= A^{-1}B & & \text{multiplying by } A^{-1} \\ (A^{-1}A)X &= A^{-1}B & & \text{associative property} \\ IX &= A^{-1}B & & \text{since } A^{-1}A = I \\ X &= A^{-1}B & & \text{since } IX = X \end{aligned}$$

To check that $X = A^{-1}B$ does indeed satisfy the equation, we can substitute $X = A^{-1}B$ into the left member of $AX = B$.

$$\begin{aligned} AX &= A(A^{-1}B) & & \text{substitution} \\ &= (AA^{-1})B & & \text{associative property} \\ &= IB & & \text{since } AA^{-1} = I \\ &= B & & \text{since } IB = B \end{aligned}$$

These results extend easily to larger systems and are recorded in the following statement.

Solution of Systems by Matrix Inverses

If A^{-1} exists, then $X = A^{-1}B$ satisfies the equation $AX = B$, and this is the only value of X that satisfies the equation.

Example 3 • Solving a Linear System

Use the inverse of the coefficient matrix to solve

$$\begin{aligned} x \quad\quad + 2z &= 1 \\ -y + 3z &= -3 \\ 2x + y + 3z &= 3. \end{aligned}$$

Solution

The matrices involved in the matrix form are given by

$$A = \begin{bmatrix} 1 & 0 & 2 \\ 0 & -1 & 3 \\ 2 & 1 & 3 \end{bmatrix}, \quad X = \begin{bmatrix} x \\ y \\ z \end{bmatrix}, \quad B = \begin{bmatrix} 1 \\ -3 \\ 3 \end{bmatrix}.$$

The inverse of the coefficient matrix was found to be

$$A^{-1} = \begin{bmatrix} 3 & -1 & -1 \\ -3 & \frac{1}{2} & \frac{3}{2} \\ -1 & \frac{1}{2} & \frac{1}{2} \end{bmatrix}$$

in Example 2b. Thus

$$X = A^{-1}B = \begin{bmatrix} 3 & -1 & -1 \\ -3 & \frac{1}{2} & \frac{3}{2} \\ -1 & \frac{1}{2} & \frac{1}{2} \end{bmatrix}\begin{bmatrix} 1 \\ -3 \\ 3 \end{bmatrix} = \begin{bmatrix} 3 \\ 0 \\ -1 \end{bmatrix},$$

and the solution to the system is

$$x = 3, \qquad y = 0, \qquad z = -1.$$

□

◆ Practice Problem 2 ◆

Use the inverse of the coefficient matrix to solve the system.

$$\begin{aligned} -2r + 3s &= 3 \\ r + 2s &= -19 \end{aligned}$$

If a calculator with matrix features is available, find the inverse in a sampling of Exercises 1–36 by hand and with the calculator using $[A]^{-1}$.

EXERCISES 11.4

Find the inverse of each matrix, if it exists. (See Example 1 and 2 and Practice Problem 1.)

√ 1. $\begin{bmatrix} 0 & 3 \\ -2 & 4 \end{bmatrix}$

2. $\begin{bmatrix} 2 & 1 \\ 1 & 1 \end{bmatrix}$

3. $\begin{bmatrix} -1 & 3 \\ 2 & 2 \end{bmatrix}$

4. $\begin{bmatrix} 4 & -2 \\ 1 & -1 \end{bmatrix}$

√ 5. $\begin{bmatrix} -3 & 5 \\ 12 & -20 \end{bmatrix}$

6. $\begin{bmatrix} 5 & -15 \\ -1 & 3 \end{bmatrix}$

7. $\begin{bmatrix} -2 & 3 \\ 5 & 7 \end{bmatrix}$

8. $\begin{bmatrix} -4 & 6 \\ 2 & 1 \end{bmatrix}$

9. $\begin{bmatrix} 0 & 1 & 0 \\ 1 & 4 & 1 \\ 0 & 3 & -1 \end{bmatrix}$

10. $\begin{bmatrix} 1 & 3 & -2 \\ 2 & 5 & -7 \\ 1 & 4 & 0 \end{bmatrix}$

√ 11. $\begin{bmatrix} 1 & -4 & 2 \\ 2 & -9 & 5 \\ 1 & -5 & 4 \end{bmatrix}$

12. $\begin{bmatrix} 2 & -3 & 3 \\ -3 & 1 & 0 \\ 1 & -1 & 1 \end{bmatrix}$

13. $\begin{bmatrix} 1 & 0 & 1 \\ -1 & 2 & 1 \\ 0 & 1 & 3 \end{bmatrix}$

14. $\begin{bmatrix} 0 & -1 & -2 \\ 2 & 4 & 8 \\ -1 & 1 & 0 \end{bmatrix}$

√ 15. $\begin{bmatrix} 5 & 3 & -2 \\ -1 & 2 & 5 \\ 11 & 4 & -9 \end{bmatrix}$

16. $\begin{bmatrix} -5 & -3 & 1 \\ 6 & 3 & 0 \\ 1 & 2 & -3 \end{bmatrix}$

17. $\begin{bmatrix} 1 & 0 & -1 & 2 \\ 3 & -1 & -1 & 6 \\ 2 & 0 & -3 & 8 \\ 1 & 2 & -2 & -9 \end{bmatrix}$

18. $\begin{bmatrix} 1 & -2 & 1 & 0 \\ 0 & 1 & 0 & -2 \\ 1 & -1 & 0 & 1 \\ 2 & -4 & 1 & 5 \end{bmatrix}$

Solve each system by using the inverse of the coefficient matrix. (See Example 3 and Practice Problem 2.)

19. $$\begin{aligned} 2x + 3y &= 1 \\ 2x + y &= 7 \end{aligned}$$

20. $$\begin{aligned} 8a - 5b &= 2 \\ -3a + 2b &= 1 \end{aligned}$$

21. $$\begin{aligned} x_1 - 3x_2 &= -6 \\ 6x_1 - 3x_2 &= 9 \end{aligned}$$

22. $$\begin{aligned} x + y &= 1 \\ x + 2y &= -3 \end{aligned}$$

23. $$\begin{aligned} 3x + 2y &= 22 \\ x + 2y &= 10 \end{aligned}$$

24. $$\begin{aligned} 2x + 8y &= 3 \\ 3x - 2y &= 1 \end{aligned}$$

25. $$\begin{aligned} x + y &= 1 \\ 7x + 3y &= 0 \end{aligned}$$

26. $$\begin{aligned} 10x + 5y &= 3 \\ 4x - y &= 0 \end{aligned}$$

27. $$\begin{aligned} x_1 + 5x_2 &= -4 \\ 2x_1 - 2x_3 &= 0 \\ 4x_2 - x_3 &= 4 \end{aligned}$$

28. $$\begin{aligned} x + 3y - 2z &= 5 \\ x + 2y - 5z &= -2 \\ x + 4y &= 5 \end{aligned}$$

29. $$\begin{aligned} x - y + z &= -2 \\ y - 3z &= 10 \\ 3x - 3y + 2z &= -3 \end{aligned}$$

30. $$\begin{aligned} a + b + 3c &= -3 \\ 2b + c &= -6 \\ a - b &= -1 \end{aligned}$$

31. $$\begin{aligned} -y + 3z &= 0 \\ -x + 5z &= 0 \\ x - y - z &= 1 \end{aligned}$$

32. $$\begin{aligned} x - 3z &= -2 \\ 2y - 3z &= -16 \\ -x - 2y + 7z &= 20 \end{aligned}$$

33. $$\begin{aligned} x + 2y + z &= 4 \\ y + z &= 0 \\ x + y + z &= 3 \end{aligned}$$

34. $$\begin{aligned} x + 2y + z &= 0 \\ x + y - 2z &= -1 \\ 2x + 4y + 3z &= \tfrac{1}{2} \end{aligned}$$

35. $$\begin{aligned} x + y + 2z - w &= 0 \\ -2x - y - 2z + 2w &= -1 \\ 4x - 2y + z &= -4 \\ y + z - w &= 1 \end{aligned}$$

36. $$\begin{aligned} a - 2b + c &= -1 \\ b - 2d &= 0 \\ b - c + d &= -2 \\ 2a - 4b + c + 5d &= -4 \end{aligned}$$

37. Find D^{-1} if $D = \begin{bmatrix} a & 0 & 0 \\ 0 & b & 0 \\ 0 & 0 & c \end{bmatrix}$,
where a, b, and c are nonzero real numbers.

38. Assuming that A^{-1} exists, solve for X in the matrix equation $XA = B$. (X and B are not column matrices here.)

39. Given that A^{-1} and C^{-1} exist, solve for X in the matrix equation $AXC = B$.

40. Let $A = \begin{bmatrix} a & b \\ c & d \end{bmatrix}$. Assume that $\delta(A) = ad - bc \neq 0$ and prove that

$$A^{-1} = \frac{1}{\delta(A)} \begin{bmatrix} d & -b \\ -c & a \end{bmatrix}.$$

41. Let $A = \begin{bmatrix} a & b \\ c & d \end{bmatrix}$. Prove that if $ad - bc = 0$, then A does not have an inverse.

Critical Thinking: Exploration and Discussion

42. Discuss and compare the effort required to solve a certain system by use of an inverse matrix or by Gauss–Jordan elimination. Make a similar comparison if there are 20 systems to be solved and they all have the same coefficient matrix.

Critical Thinking: Exploration and Discussion

43. It has been traditional to use A^{-1} instead of $1/A$ to denote the inverse of a matrix A. The use of A^{-1} is preferred because it avoids the possibility of fractional notations. Discuss how confusion might arise from the use of the following notations.

$$(B)\left(\frac{1}{A}\right), \qquad \left(\frac{1}{A}\right)(B), \qquad \frac{B}{A}$$

◆ Solutions for Practice Problems

1. $\left[\begin{array}{rrr|rrr} 1 & 0 & 0 & 1 & 0 & 0 \\ -1 & 1 & 1 & 0 & 1 & 0 \\ 0 & 1 & 2 & 0 & 0 & 1 \end{array}\right] \xrightarrow{R_1 + R_2} \left[\begin{array}{rrr|rrr} 1 & 0 & 0 & 1 & 0 & 0 \\ 0 & 1 & 1 & 1 & 1 & 0 \\ 0 & 1 & 2 & 0 & 0 & 1 \end{array}\right]$

$\xrightarrow{-R_2 + R_3} \left[\begin{array}{rrr|rrr} 1 & 0 & 0 & 1 & 0 & 0 \\ 0 & 1 & 1 & 1 & 1 & 0 \\ 0 & 0 & 1 & -1 & -1 & 1 \end{array}\right] \xrightarrow{-R_3 + R_2} \left[\begin{array}{rrr|rrr} 1 & 0 & 0 & 1 & 0 & 0 \\ 0 & 1 & 0 & 2 & 2 & -1 \\ 0 & 0 & 1 & -1 & -1 & 1 \end{array}\right]$

The inverse of $\begin{bmatrix} 1 & 0 & 0 \\ -1 & 1 & 1 \\ 0 & 1 & 2 \end{bmatrix}$ is $\begin{bmatrix} 1 & 0 & 0 \\ 2 & 2 & -1 \\ -1 & -1 & 1 \end{bmatrix}$.

2. $A = \begin{bmatrix} -2 & 3 \\ 1 & 2 \end{bmatrix}$, $\quad B = \begin{bmatrix} 3 \\ -19 \end{bmatrix}$, $\quad X = \begin{bmatrix} r \\ s \end{bmatrix}$

$A^{-1} = \begin{bmatrix} -\frac{2}{7} & \frac{3}{7} \\ \frac{1}{7} & \frac{2}{7} \end{bmatrix}$ (See Example 2a.)

$X = A^{-1}B = \begin{bmatrix} -\frac{2}{7} & \frac{3}{7} \\ \frac{1}{7} & \frac{2}{7} \end{bmatrix} \begin{bmatrix} 3 \\ -19 \end{bmatrix} = \begin{bmatrix} -9 \\ -5 \end{bmatrix}$

The solution is $r = -9$, $s = -5$.

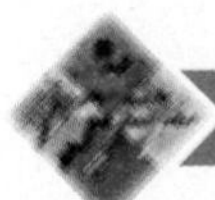

11.5 DETERMINANTS

In the formula for the inverse of a 2×2 matrix, we let $\delta(A)$ denote the number $ad - bc$ that is associated with the 2×2 matrix

$$A = \begin{bmatrix} a & b \\ c & d \end{bmatrix}.$$

This number, $\delta(A)$, is called the **determinant** of the 2×2 matrix A. Every square matrix A, of any order, has such a number associated with it. This number is called the **determinant** of A. The **order** of the determinant of A is the same as the order of the matrix A. The determinant of A is denoted by $\delta(A)$, or by $|A|$, or by simply replacing the brackets around the elements of A by straight lines. Thus if

$$A = \begin{bmatrix} a & b \\ c & d \end{bmatrix},$$

then

$$\delta(A) = |A| = \begin{vmatrix} a & b \\ c & d \end{vmatrix} = ad - bc.$$

The material in this section leads to a method of evaluating the determinant of any square matrix. It is important to note that the determinant of a matrix is a number.

Example 1 • Evaluating 2 × 2 Determinants

Evaluate the determinant of each of the following matrices.

a. $A = \begin{bmatrix} -1 & 7 \\ 2 & -3 \end{bmatrix}$ b. $B = \begin{bmatrix} 1 & 0 \\ 0 & 1 \end{bmatrix}$ c. $C = \begin{bmatrix} 2 & 0 \\ -1 & 0 \end{bmatrix}$

Solution

a. $\delta(A) = (-1)(-3) - (7)(2) = -11$, or $\begin{vmatrix} -1 & 7 \\ 2 & -3 \end{vmatrix} = -11$

b. $\delta(B) = |B| = (1)(1) - (0)(0) = 1$, or $\begin{vmatrix} 1 & 0 \\ 0 & 1 \end{vmatrix} = 1$

c. $\delta(C) = |C| = (2)(0) - (0)(-1) = 0$, or $\begin{vmatrix} 2 & 0 \\ -1 & 0 \end{vmatrix} = 0$ □

The determinant of a square matrix of order 3 is given next.

Definition of a Determinant of Order 3

For a given 3×3 matrix

$$A = \begin{bmatrix} a_{11} & a_{12} & a_{13} \\ a_{21} & a_{22} & a_{23} \\ a_{31} & a_{32} & a_{33} \end{bmatrix}$$

the **determinant** of A is the number

$$|A| = a_{11}a_{22}a_{33} + a_{12}a_{23}a_{31} + a_{13}a_{21}a_{32} - a_{11}a_{23}a_{32} - a_{12}a_{21}a_{33} - a_{13}a_{22}a_{31}.$$

Example 2 • Evaluating a 3 × 3 Determinant

Evaluate $|A|$ for

$$A = \begin{bmatrix} 3 & 2 & 1 \\ -4 & -1 & 5 \\ -3 & -4 & -2 \end{bmatrix}.$$

Solution

Using the definition, we have

$$\begin{aligned} |A| &= (3)(-1)(-2) + (2)(5)(-3) + (1)(-4)(-4) \\ &\quad - (3)(5)(-4) - (2)(-4)(-2) - (1)(-1)(-3) \\ &= 6 - 30 + 16 + 60 - 16 - 3 \\ &= 33. \end{aligned}$$

□

The determinant evaluation function of a graphing device will compute a determinant directly with a single command.

Similar and even more unwieldy definitions can be made for determinants of higher-order matrices. Although we could use these definitions to evaluate a determinant, they are very inefficient. A much more efficient way to proceed is to evaluate determinants by a method called the **cofactor expansion.** The cofactor expansion applies to determinants of any order. In order to describe the method, some preliminary definitions are needed.

Definition of Minors and Cofactors

Let A be a square matrix of order $n \geq 2$, and let a_{ij} be the element in row i and column j of A. Then the **minor,** denoted by M_{ij}, of the element a_{ij} is the determinant of the matrix formed by deleting row i and column j from the matrix A. The **cofactor,** denoted by C_{ij}, of the element a_{ij} is the product of $(-1)^{i+j}$ and the minor M_{ij}. That is,

$$C_{ij} = (-1)^{i+j}M_{ij}$$

Note that if $i + j$ is even, then $C_{ij} = M_{ij}$, and if $i + j$ is odd, then $C_{ij} = -M_{ij}$.

Example 3 • Identifying a Minor and a Cofactor

Consider the 3 × 3 matrix

$$A = \begin{bmatrix} 6 & -3 & 2 \\ 0 & 1 & -2 \\ 2 & 4 & 5 \end{bmatrix}.$$

a. The element in row 2 and column 1 is 0; that is, $a_{21} = 0$.

b. The minor of a_{21} is $\begin{vmatrix} -3 & 2 \\ 4 & 5 \end{vmatrix}$; hence $M_{21} = -23$.

c. The cofactor of a_{21} is $(-1)^{2+1}\begin{vmatrix} -3 & 2 \\ 4 & 5 \end{vmatrix}$; hence $C_{21} = 23$.

□

With the definitions of minor and cofactor in mind, we consider again the definition of a determinant of order 3. In the equation

$$|A| = a_{11}a_{22}a_{33} + a_{12}a_{23}a_{31} + a_{13}a_{21}a_{32} - a_{11}a_{23}a_{32} - a_{12}a_{21}a_{33} - a_{13}a_{22}a_{31}$$

there are two terms involving a_{11}, two terms involving a_{12}, and two terms involving a_{13}. Regrouping the terms and factoring, we have

$$|A| = a_{11}(a_{22}a_{33} - a_{23}a_{32}) - a_{12}(a_{21}a_{33} - a_{23}a_{31}) + a_{13}(a_{21}a_{32} - a_{22}a_{31}).$$

Since each expression in parentheses can be expressed as a second-order determinant, we can write

$$|A| = a_{11}\begin{vmatrix} a_{22} & a_{23} \\ a_{32} & a_{33} \end{vmatrix} - a_{12}\begin{vmatrix} a_{21} & a_{23} \\ a_{31} & a_{33} \end{vmatrix} + a_{13}\begin{vmatrix} a_{21} & a_{22} \\ a_{31} & a_{32} \end{vmatrix}.$$

Expressing each second-order determinant as a minor yields

$$|A| = a_{11}M_{11} - a_{12}M_{12} + a_{13}M_{13}$$

or

$$|A| = a_{11}C_{11} + a_{12}C_{12} + a_{13}C_{13}.$$

The last expression for $|A|$ is called the **cofactor expansion** of the determinant of A about the first row. The cofactor expansion of $|A|$ about the first row is a sum of terms formed by multiplying each element in the first row by its cofactor.

The terms in the equation defining $|A|$ for A of order 3 could also have been regrouped and factored in such a way as to have a cofactor expansion of $|A|$ about any certain row or any certain column. In general, the **cofactor expansion** of $|A|$ about row $i (i = 1, 2, \text{ or } 3)$ is

$$|A| = a_{i1}C_{i1} + a_{i2}C_{i2} + a_{i3}C_{i3},$$

and the **cofactor expansion** of $|A|$ about column j $(j = 1, 2, \text{ or } 3)$ is

$$|A| = a_{1j}C_{1j} + a_{2j}C_{2j} + a_{3j}C_{3j}.$$

Example 4 • Cofactor Expansion of a 3 × 3 Determinant

Evaluate $|A|$ by expanding about the second column.

$$|A| = \begin{vmatrix} 3 & 2 & 1 \\ -4 & -1 & 5 \\ -3 & -4 & -2 \end{vmatrix}.$$

Solution

Using

$$|A| = a_{12}C_{12} + a_{22}C_{22} + a_{32}C_{32}$$

we have

$$
\begin{aligned}
|A| &= (2)(-1)^3\begin{vmatrix} -4 & 5 \\ -3 & -2 \end{vmatrix} + (-1)(-1)^4\begin{vmatrix} 3 & 1 \\ -3 & -2 \end{vmatrix} + (-4)(-1)^5\begin{vmatrix} 3 & 1 \\ -4 & 5 \end{vmatrix} \\
&= (-2)[8 - (-15)] + (-1)[-6 - (-3)] + (4)[15 - (-4)] \\
&= (-2)(23) + (-1)(-3) + (4)(19) \\
&= 33.
\end{aligned}
$$

This result agrees with the value obtained in Example 2 for the same determinant. □

◆ Practice Problem 1 ◆

Use a cofactor expansion to evaluate

$$
\begin{vmatrix} -1 & 2 & 3 \\ 4 & -2 & 1 \\ 1 & 5 & 0 \end{vmatrix}.
$$

Determinants of order $n > 3$ can also be evaluated by the method of cofactor expansion. The cofactor expansion of an nth-order determinant is given in the following statement.

Cofactor Expansion

Let $A = [a_{ij}]_{(n,\, n)}$ be a square matrix of order $n \geq 2$, and let C_{ij} be the factor of element a_{ij} for $i = 1, 2, ..., n$ and $j = 1, 2, ..., n$. Then the cofactor expansion of $|A|$ about row i is

$$|A| = a_{i1}C_{i1} + a_{i2}C_{i2} + \cdots + a_{in}C_{in},$$

and the cofactor expansion of $|A|$ about column j is

$$|A| = a_{1j}C_{1j} + a_{2j}C_{2j} + \cdots + a_{nj}C_{nj}.$$

The cofactor expansion is proved in the study of linear algebra. That is, it is shown that $|A|$ does not depend on the choice of row number i or column number j. This proof is very complicated and does not belong in this text.

Example 5 • Cofactor Expansion of a 4 × 4 Determinant

Evaluate the following determinant.

$$
|A| = \begin{vmatrix} -2 & 1 & 0 & 1 \\ 0 & -2 & 1 & 4 \\ -3 & 1 & 0 & -2 \\ 1 & 0 & -2 & 1 \end{vmatrix}
$$

Solution

We expand $|A|$ about the third column since it contains more zeros than any other column or row, and consequently this cofactor expansion will have the fewest nonzero terms. We obtain

$$\begin{aligned}|A| &= a_{13}C_{13} + a_{23}C_{23} + a_{33}C_{33} + a_{43}C_{43}\\ &= 0 \cdot C_{13} + (1)(-1)^5\begin{vmatrix}-2 & 1 & 1\\ -3 & 1 & -2\\ 1 & 0 & 1\end{vmatrix} + 0 \cdot C_{33}\\ &\quad + (-2)(-1)^7\begin{vmatrix}-2 & 1 & 1\\ 0 & -2 & 4\\ -3 & 1 & -2\end{vmatrix}\\ &= -\begin{vmatrix}-2 & 1 & 1\\ -3 & 1 & -2\\ 1 & 0 & 1\end{vmatrix} + 2\begin{vmatrix}-2 & 1 & 1\\ 0 & -2 & 4\\ -3 & 1 & -2\end{vmatrix}.\end{aligned}$$

Expanding the first third-order determinant about its third row, and the second third-order determinant about its first column, we have

$$\begin{aligned}|A| &= -\left[(1)(-1)^4\begin{vmatrix}1 & 1\\ 1 & -2\end{vmatrix} + 0 + (1)(-1)^6\begin{vmatrix}-2 & 1\\ -3 & 1\end{vmatrix}\right]\\ &\quad + 2\left[(-2)(-1)^2\begin{vmatrix}-2 & 4\\ 1 & -2\end{vmatrix} + 0 + (-3)(-1)^4\begin{vmatrix}1 & 1\\ -2 & 4\end{vmatrix}\right]\\ &= -[(1)(-2-1) + 0 + (1)(-2-(-3))]\\ &\quad + 2[(-2)(4-4) + 0 + (-3)(4-(-2))]\\ &= -[-3+0+1] + 2[0+0-18]\\ &= 2 - 36\\ &= -34.\end{aligned}$$

□

The work in Example 5 that involved the zero elements shows why the following property is true.

Zero Expansion Property

If every element in one row (or one column) of a square matrix A is zero, then $|A| = 0$.

If a calculator with matrix features is available, work a sampling of Exercises 1–36 by hand and with the calculator using **det**.

EXERCISES 11.5

Evaluate the following determinants. Variables represent real numbers. (See Example 1.)

1. $\begin{vmatrix}-3 & 2\\ 1 & 1\end{vmatrix}$
2. $\begin{vmatrix}0 & -3\\ -1 & 3\end{vmatrix}$
3. $\begin{vmatrix}5 & 8\\ 2 & 3\end{vmatrix}$
4. $\begin{vmatrix}4 & -3\\ 3 & -2\end{vmatrix}$

5. $\begin{vmatrix} 7 & 0 \\ 0 & -3 \end{vmatrix}$ 6. $\begin{vmatrix} 0 & 0 \\ -1 & 1 \end{vmatrix}$ 7. $\begin{vmatrix} -1 & -1 \\ 1 & 1 \end{vmatrix}$ 8. $\begin{vmatrix} 2 & 9 \\ -4 & 10 \end{vmatrix}$

9. $\begin{vmatrix} -\frac{1}{2} & \frac{1}{3} \\ \frac{1}{6} & \frac{1}{9} \end{vmatrix}$ 10. $\begin{vmatrix} \frac{1}{4} & -1 \\ 0 & -4 \end{vmatrix}$ 11. $\begin{vmatrix} 0 & 3 \\ -2 & 1 \end{vmatrix}$ 12. $\begin{vmatrix} 8 & 0 \\ 1 & -5 \end{vmatrix}$

13. $\begin{vmatrix} 1 & 11 \\ -2 & 4 \end{vmatrix}$ 14. $\begin{vmatrix} 1 & 4 \\ -2 & 1 \end{vmatrix}$ 15. $\begin{vmatrix} x & -x \\ 2 & 3 \end{vmatrix}$ 16. $\begin{vmatrix} x & y \\ 2x & 3y \end{vmatrix}$

Use a cofactor expansion to evaluate the following determinants. (See Examples 4 and 5 and Practice Problem 1.)

17. $\begin{vmatrix} 2 & -1 & 2 \\ 0 & 2 & 0 \\ -3 & 1 & 0 \end{vmatrix}$ 18. $\begin{vmatrix} -2 & 6 & 1 \\ 0 & 0 & 4 \\ 1 & -4 & 0 \end{vmatrix}$

19. $\begin{vmatrix} 2 & 0 & -1 \\ -1 & 2 & 0 \\ 0 & -1 & 2 \end{vmatrix}$ 20. $\begin{vmatrix} 2 & 0 & 1 \\ 0 & 1 & 1 \\ 4 & -1 & 1 \end{vmatrix}$

21. $\begin{vmatrix} 1 & 4 & 0 \\ -1 & 2 & -1 \\ 0 & 3 & -1 \end{vmatrix}$ 22. $\begin{vmatrix} 1 & 0 & 1 \\ 0 & 1 & 3 \\ -1 & 2 & 1 \end{vmatrix}$

23. $\begin{vmatrix} 1 & 3 & -2 \\ 1 & 4 & 0 \\ 2 & 5 & -7 \end{vmatrix}$ 24. $\begin{vmatrix} 1 & 0 & 3 \\ -1 & 1 & -3 \\ 1 & -3 & 2 \end{vmatrix}$

25. $\begin{vmatrix} -1 & 1 & 1 \\ 1 & -1 & 1 \\ 1 & 1 & 1 \end{vmatrix}$ 26. $\begin{vmatrix} -2 & 3 & 3 \\ 3 & 1 & -2 \\ 1 & 1 & -1 \end{vmatrix}$

27. $\begin{vmatrix} 1 & -2 & 3 \\ 2 & 1 & 2 \\ 3 & -3 & 6 \end{vmatrix}$ 28. $\begin{vmatrix} 2 & 1 & 1 \\ 9 & 4 & 3 \\ 6 & 3 & 4 \end{vmatrix}$

29. $\begin{vmatrix} 4 & -5 & 1 \\ 5 & -9 & 2 \\ 2 & -4 & 1 \end{vmatrix}$ 30. $\begin{vmatrix} 4 & -3 & -8 \\ 1 & 3 & -2 \\ 2 & 1 & -3 \end{vmatrix}$

31. $\begin{vmatrix} 2 & 4 & 3 \\ 5 & -9 & -2 \\ -1 & 11 & 5 \end{vmatrix}$ 32. $\begin{vmatrix} -3 & -1 & 2 \\ 1 & 6 & -3 \\ -5 & 4 & 1 \end{vmatrix}$

33. $\begin{vmatrix} 1 & 0 & 1 & 2 \\ -2 & 1 & -1 & -4 \\ 1 & 0 & 0 & 1 \\ 0 & -2 & 1 & 5 \end{vmatrix}$ 34. $\begin{vmatrix} 6 & -1 & -1 & 3 \\ 8 & 0 & -3 & 2 \\ -9 & 2 & -2 & 1 \\ 2 & 0 & -1 & 1 \end{vmatrix}$

35. $\begin{vmatrix} 0 & 4 & -2 & 1 \\ -1 & -2 & -1 & 1 \\ -2 & -1 & -3 & 3 \\ 1 & 0 & 2 & -1 \end{vmatrix}$

36. $\begin{vmatrix} 1 & 3 & -1 & 1 \\ 2 & 9 & -2 & 3 \\ 2 & 10 & -3 & 2 \\ 3 & -8 & 0 & 1 \end{vmatrix}$

Solve for x in each of the following equations.

37. $\begin{vmatrix} 3 & x \\ -2 & 1 \end{vmatrix} = 1$

38. $\begin{vmatrix} -x & 3 \\ -4 & -1 \end{vmatrix} = -3$

√ 39. $\begin{vmatrix} 2x-1 & 2 \\ 1 & 4 \end{vmatrix} = 10$

40. $\begin{vmatrix} 1 & 3 \\ 2-x & -1 \end{vmatrix} = -5$

41. $\begin{vmatrix} 2 & 0 & 1 \\ 4 & x & 5 \\ 1 & 4 & -3 \end{vmatrix} = -10$

42. $\begin{vmatrix} -2 & 0 & 1 \\ 1 & -3 & -2 \\ x & 1 & 1 \end{vmatrix} = 0$

√ 43. $\begin{vmatrix} 3 & x & x \\ -2 & 1 & 0 \\ 4 & 1 & -1 \end{vmatrix} = 21$

44. $\begin{vmatrix} 1 & 3 & 2 \\ 2x & 1 & x \\ 5 & 1 & -4 \end{vmatrix} = 0$

The values of x that satisfy the equation $|A - xI| = 0$ are called the **eigenvalues,** or the **characteristic values,** of the matrix A. Find the eigenvalues of the following matrices.

45. $\begin{bmatrix} 2 & 0 \\ 0 & -3 \end{bmatrix}$

46. $\begin{bmatrix} 5 & 6 \\ 4 & 0 \end{bmatrix}$

47. $\begin{bmatrix} 1 & -2 \\ 3 & -4 \end{bmatrix}$

48. $\begin{bmatrix} 1 & 3 \\ 9 & 7 \end{bmatrix}$

49. $\begin{bmatrix} 5 & 1 & -1 \\ 0 & -3 & 2 \\ 0 & 0 & 2 \end{bmatrix}$

50. $\begin{bmatrix} 1 & 3 & 15 \\ -2 & 0 & -2 \\ 1 & 0 & 1 \end{bmatrix}$

51. $\begin{bmatrix} 1 & -2 & 0 \\ 0 & 0 & -3 \\ 2 & -4 & 0 \end{bmatrix}$

52. $\begin{bmatrix} 1 & -1 & 1 \\ 1 & -1 & -1 \\ -1 & 1 & -3 \end{bmatrix}$

53. Use the definition to show that a determinant $|A|$ of order 3 can be evaluated by any one of the following equations.
    a. $|A| = a_{11}C_{11} + a_{21}C_{21} + a_{31}C_{31}$
    b. $|A| = a_{21}C_{21} + a_{22}C_{22} + a_{23}C_{23}$
    c. $|A| = a_{12}C_{12} + a_{22}C_{22} + a_{32}C_{32}$
    d. $|A| = a_{31}C_{31} + a_{32}C_{32} + a_{33}C_{33}$
    e. $|A| = a_{13}C_{13} + a_{23}C_{23} + a_{33}C_{33}$

Critical Thinking: Writing 54. Explain the difference between a matrix and a determinant.

Critical Thinking: Writing 55. Explain how the determinant $|A|$ can be regarded as a function of the matrix A. In particular, describe the domain and the range of this function.

◆ Solution for Practice Problem

1. Expanding about row 3 we have

$$\begin{vmatrix} -1 & 2 & 3 \\ 4 & -2 & 1 \\ 1 & 5 & 0 \end{vmatrix} = 1(-1)^4\begin{vmatrix} 2 & 3 \\ -2 & 1 \end{vmatrix} + 5(-1)^5\begin{vmatrix} -1 & 3 \\ 4 & 1 \end{vmatrix} + 0 \cdot C_{33}$$

$$= 1[2 - (-6)] + 5(-1)(-1 - 12)$$

$$= 73.$$

11.6 EVALUATION OF DETERMINANTS

In Section 11.3, we used the following three types of row operations on augmented matrices to solve linear systems by Gauss-Jordan elimination.

Row Operations

1. Interchange any two rows.
2. Multiply (or divide) every element in a row by the same nonzero number.
3. Add (or subtract) a multiple of one row to (or from) another row.

Corresponding to each of these row operations, there is a similar **column operation.**

Column Operations

1. Interchange any two columns.
2. Multiply (or divide) every element in a column by the same nonzero number.
3. Add (or subtract) a multiple of one column to (or from) another column.

When a row or column operation is performed on a square matrix A, the value of $|A|$ is sometimes (but not always) changed. The effect of performing an operation of each type is described in this section by the statements of three **properties of determinants.** Each property can be stated in terms of rows or in terms of columns. We indicate these dual statements by wording each property in terms of rows and by inserting the word *column* in parentheses. Each dual is obtained simply by replacing the word *row* by the word *column.* As with the cofactor expansion, the proofs of these properties for the general case are more appropriate in a linear algebra course and are not presented here.

Interchange Property

If the matrix B is obtained from a matrix A by interchanging any two rows (columns) of A, then $|B| = -|A|$.

Example 1 • The Interchange Property

Some uses of the interchange property are presented here.

a. If $|A| = \begin{vmatrix} a & b & c \\ d & e & f \\ g & h & i \end{vmatrix}$, then $\begin{vmatrix} a & b & c \\ g & h & i \\ d & e & f \end{vmatrix} = -|A|$.

b. Every time two rows or two columns are interchanged, the sign of the determinant changes. Thus

$$\begin{vmatrix} 1 & -2 & 3 \\ 4 & -1 & 0 \\ -1 & 1 & 2 \end{vmatrix} = -\begin{vmatrix} -1 & 1 & 2 \\ 4 & -1 & 0 \\ 1 & -2 & 3 \end{vmatrix} \quad \text{since rows 1 and 3 have been interchanged}$$

$$= \begin{vmatrix} -1 & 2 & 1 \\ 4 & 0 & -1 \\ 1 & 3 & -2 \end{vmatrix}. \quad \text{since columns 2 and 3 have been interchanged}$$

□

Multiplication Property

If the matrix B is obtained from a matrix A by multiplying each element of a row (column) of A by the same number c, then $|B| = c|A|$.

Example 2 • The Multiplication Property

The determinant $\begin{vmatrix} 12 & -4 \\ 9 & 2 \end{vmatrix}$ can be expressed as $12\begin{vmatrix} 1 & -1 \\ 3 & 2 \end{vmatrix}$ by applying the multiplication property twice.

$$\begin{vmatrix} 12 & -4 \\ 9 & 2 \end{vmatrix} = 4\begin{vmatrix} 3 & -1 \\ 9 & 2 \end{vmatrix} = (4)(3)\begin{vmatrix} 1 & -1 \\ 3 & 2 \end{vmatrix}$$

□

Multiplication of a determinant by a number c has the same effect as multiplying one column or one row of the determinant by c. By contrast, in multiplication of a matrix by the number c, every element in the matrix is multiplied by c. This difference between multiplication of a matrix by a number and multiplication of a determinant by a number is illustrated in the next example.

Example 3 • The Determinant of a Scalar Multiple

If $A = \begin{bmatrix} -7 & -4 \\ 2 & 5 \end{bmatrix}$, then $|A| = -27$ and $3|A| = -81$.

Also $3A = \begin{bmatrix} -21 & -12 \\ 6 & 15 \end{bmatrix}$, and $|3A| = -243$.

Thus $|3A| \neq 3|A|$. □

Addition Property

If the matrix B is obtained from a matrix A by adding a multiple of one row (column) to another row (column), then $|B| = |A|$.

Example 4 • The Addition Property

Adding 2 times the first row to the third row in

$$\begin{vmatrix} 1 & 3 & -1 \\ 0 & 1 & 5 \\ -2 & 1 & 4 \end{vmatrix} \text{ yields } \begin{vmatrix} 1 & 3 & -1 \\ 0 & 1 & 5 \\ 0 & 7 & 2 \end{vmatrix},$$

and these determinants are equal:

$$\begin{vmatrix} 1 & 3 & -1 \\ 0 & 1 & 5 \\ -2 & 1 & 4 \end{vmatrix} = \begin{vmatrix} 1 & 3 & -1 \\ 0 & 1 & 5 \\ 0 & 7 & 2 \end{vmatrix}.$$

The value of each determinant is -33. □

The addition property proves to be most useful in evaluating determinants. By introducing zeros into a row (or column), the cofactor expansion can be reduced to only one term. This is illustrated in Example 5. In the example, we use a notation for row and column operations that is similar to that used in Section 11.3.

Notation for Row and Column Operations

1. $R_i \leftrightarrow R_j$ indicates that row i and row j are interchanged.
2. $C_i \leftrightarrow C_j$ indicates that column i and column j are interchanged.
3. $cR_i + R_j$ indicates that row j is replaced by the sum of row j and c times row i.
4. $cC_i + C_j$ indicates that column j is replaced by the sum of column j and c times column i.

Example 5 • Using Row Operations

Evaluate the following determinant by introducing zeros into column 2.

$$|A| = \begin{vmatrix} 3 & 2 & -2 \\ -1 & -1 & 4 \\ 2 & 4 & -1 \end{vmatrix}$$

Solution

We use row operations to introduce two zeros into column 2 as follows.

$$|A| = \begin{vmatrix} 3 & 2 & -2 \\ -1 & -1 & 4 \\ 2 & 4 & -1 \end{vmatrix} \underset{=}{2R_2 + R_1} \begin{vmatrix} 1 & 0 & 6 \\ -1 & -1 & 4 \\ 2 & 4 & -1 \end{vmatrix} \underset{=}{4R_2 + R_3} \begin{vmatrix} 1 & 0 & 6 \\ -1 & -1 & 4 \\ -2 & 0 & 15 \end{vmatrix}$$

The cofactor expansion of this determinant about the second column reduces to one term.

$$\begin{aligned} |A| &= 0 \cdot C_{12} + (-1)C_{22} + 0 \cdot C_{32} \\ &= (-1)\begin{vmatrix} 1 & 6 \\ -2 & 15 \end{vmatrix} \\ &= (-1)(15 + 12) \\ &= -27 \end{aligned}$$

□

The use of column operations is illustrated in the next example.

Example 6 • Using Column Operations and Row Operations

Evaluate the following determinant by introducing zeros before expanding about a row or column.

$$|A| = \begin{vmatrix} 3 & 1 & -5 & 0 \\ 2 & 1 & 4 & 1 \\ -1 & 2 & -4 & 1 \\ 0 & -3 & 1 & 5 \end{vmatrix}$$

Solution

Since there is already one zero in the first row, we choose to introduce two more zeros in this row by using column operations.

$$|A| = \begin{vmatrix} 3 & 1 & -5 & 0 \\ 2 & 1 & 4 & 1 \\ -1 & 2 & -4 & 1 \\ 0 & -3 & 1 & 5 \end{vmatrix} \underset{=}{-3C_2 + C_1} \begin{vmatrix} 0 & 1 & -5 & 0 \\ -1 & 1 & 4 & 1 \\ -7 & 2 & -4 & 1 \\ 9 & -3 & 1 & 5 \end{vmatrix}$$

$$\underset{=}{5C_2 + C_3} \begin{vmatrix} 0 & 1 & 0 & 0 \\ -1 & 1 & 9 & 1 \\ -7 & 2 & 6 & 1 \\ 9 & -3 & -14 & 5 \end{vmatrix}$$

The cofactor expansion about the first row is

$$|A| = 1(-1)\begin{vmatrix} -1 & 9 & 1 \\ -7 & 6 & 1 \\ 9 & -14 & 5 \end{vmatrix}.$$

Next we choose to introduce zeros into the third column. (Any column or row can be used.) Using row operations, we have

$$|A| = -\begin{vmatrix} -1 & 9 & 1 \\ -7 & 6 & 1 \\ 9 & -14 & 5 \end{vmatrix} \overset{-R_1 + R_2}{=} -\begin{vmatrix} -1 & 9 & 1 \\ -6 & -3 & 0 \\ 9 & -14 & 5 \end{vmatrix}$$

$$\overset{-5R_1 + R_3}{=} -\begin{vmatrix} -1 & 9 & 1 \\ -6 & -3 & 0 \\ 14 & -59 & 0 \end{vmatrix}.$$

The cofactor expansion about the third column is

$$|A| = -(1)\begin{vmatrix} -6 & -3 \\ 14 & -59 \end{vmatrix}.$$

Factoring -3 out of the first row, we have

$$|A| = -(1)(-3)\begin{vmatrix} 2 & 1 \\ 14 & -59 \end{vmatrix} = (3)(-118 - 14) = -396.$$

□

◆ Practice Problem 1 ◆

Evaluate the determinant by introducing zeros before expanding about a row or column.

$$\begin{vmatrix} 2 & 1 & -2 \\ 1 & 5 & 0 \\ 1 & -1 & 3 \end{vmatrix}$$

EXERCISES 11.6

Without evaluating the determinants, determine the value of the variable that makes each of the following statements true. (See Examples 1, 2, and 4.)

✓ 1. $\begin{vmatrix} 3 & -4 \\ 1 & 5 \end{vmatrix} = -\begin{vmatrix} 1 & 5 \\ 3 & x \end{vmatrix}$

2. $\begin{vmatrix} 11 & -2 \\ 7 & 3 \end{vmatrix} = -\begin{vmatrix} -2 & x \\ 3 & 7 \end{vmatrix}$

3. $\begin{vmatrix} 5 & -1 \\ 2 & -3 \end{vmatrix} = a\begin{vmatrix} 5 & 1 \\ 2 & 3 \end{vmatrix}$

4. $\begin{vmatrix} 3 & -2 \\ 7 & 0 \end{vmatrix} = y\begin{vmatrix} -3 & 2 \\ -7 & 0 \end{vmatrix}$

5. $\begin{vmatrix} 1 & 0 & 2 \\ -2 & 1 & 4 \\ 0 & 1 & 5 \end{vmatrix} = x\begin{vmatrix} 0 & 2 & 1 \\ 1 & 4 & -2 \\ 1 & 5 & 0 \end{vmatrix}$

6. $\begin{vmatrix} 2 & 1 & 11 \\ 3 & 1 & 4 \\ 0 & 1 & 0 \end{vmatrix} = t\begin{vmatrix} 3 & 1 & 4 \\ 0 & 1 & 0 \\ 2 & 1 & 11 \end{vmatrix}$

√7. $\begin{vmatrix} -3 & 4 \\ 9 & -2 \end{vmatrix} = x \begin{vmatrix} -1 & -2 \\ 3 & 1 \end{vmatrix}$

8. $\begin{vmatrix} 1 & 1 & 3 \\ 4 & 20 & 12 \\ 5 & 25 & 30 \end{vmatrix} = y \begin{vmatrix} 1 & 1 & 1 \\ 1 & 5 & 1 \\ 1 & 5 & 2 \end{vmatrix}$

√9. $\begin{vmatrix} 3 & 2 & 1 \\ 5 & -4 & -3 \\ 1 & 1 & 2 \end{vmatrix} = \begin{vmatrix} 3 & 2 & 1 \\ 14 & x & 0 \\ 1 & 1 & 2 \end{vmatrix}$

10. $\begin{vmatrix} -1 & -3 & 1 \\ 1 & -2 & 1 \\ 4 & -2 & 2 \end{vmatrix} = \begin{vmatrix} x & -5 & 3 \\ 1 & -2 & 1 \\ 4 & -2 & 2 \end{vmatrix}$

√11. $\begin{vmatrix} -1 & 11 & -1 & 4 \\ 0 & 3 & 0 & 1 \\ 1 & 1 & -2 & 1 \\ 4 & 1 & 5 & -2 \end{vmatrix} = \begin{vmatrix} -1 & -1 & -1 & 4 \\ 0 & 0 & 0 & 1 \\ 1 & x & -2 & 1 \\ 4 & 7 & 5 & -2 \end{vmatrix}$

12. $\begin{vmatrix} 1 & -1 & 2 & 3 \\ -2 & 1 & 1 & 4 \\ 1 & 5 & 0 & 2 \\ 1 & 1 & -2 & 3 \end{vmatrix} = \begin{vmatrix} 1 & -1 & 2 & 3 \\ -1 & 0 & x & 7 \\ 1 & 5 & 0 & 2 \\ 1 & 1 & -2 & 3 \end{vmatrix}$

Evaluate the following determinants by introducing zeros before expanding about a row or column. (See Examples 5 and 6 and Practice Problem 1.)

√13. $\begin{vmatrix} 2 & -1 & 3 \\ 1 & 2 & -1 \\ 0 & 2 & 1 \end{vmatrix}$

14. $\begin{vmatrix} 1 & 3 & -2 \\ 0 & 1 & 3 \\ 2 & 0 & 5 \end{vmatrix}$

15. $\begin{vmatrix} 7 & 4 & -2 \\ 1 & 2 & -1 \\ 4 & 2 & 0 \end{vmatrix}$

16. $\begin{vmatrix} 4 & -2 & 1 \\ -2 & 2 & -1 \\ 5 & 0 & 2 \end{vmatrix}$

17. $\begin{vmatrix} 1 & 3 & -1 \\ 3 & -2 & 4 \\ 2 & 1 & 3 \end{vmatrix}$

18. $\begin{vmatrix} 13 & 3 & 1 \\ 11 & 4 & -2 \\ 4 & -1 & 3 \end{vmatrix}$

19. $\begin{vmatrix} 1 & 2 & 3 \\ 1 & 1 & 2 \\ 1 & 1 & 3 \end{vmatrix}$

20. $\begin{vmatrix} 2 & -1 & -2 \\ 1 & 1 & -5 \\ -1 & 3 & 4 \end{vmatrix}$

√21. $\begin{vmatrix} 2 & -1 & 2 \\ 3 & -5 & 1 \\ 2 & 1 & 3 \end{vmatrix}$

22. $\begin{vmatrix} 8 & 2 & -3 \\ 5 & 3 & -1 \\ 2 & -7 & 3 \end{vmatrix}$

23. $\begin{vmatrix} -2 & 1 & -2 \\ 2 & -1 & 3 \\ -4 & 2 & 1 \end{vmatrix}$

24. $\begin{vmatrix} -1 & 1 & 7 \\ 3 & -3 & -1 \\ 1 & 3 & -3 \end{vmatrix}$

25. $\begin{vmatrix} 5 & 0 & 0 & 3 \\ 2 & 4 & -3 & 1 \\ 0 & 1 & 0 & -1 \\ 1 & -1 & 2 & -1 \end{vmatrix}$

26. $\begin{vmatrix} 4 & -3 & 1 & 0 \\ 2 & 1 & -1 & 0 \\ -1 & 0 & 1 & 2 \\ 2 & 1 & 1 & 1 \end{vmatrix}$

27. $\begin{vmatrix} 1 & 0 & 1 & 0 \\ 0 & 1 & 2 & 1 \\ 4 & 0 & 0 & -2 \\ 1 & 2 & 3 & 0 \end{vmatrix}$

28. $\begin{vmatrix} -3 & 1 & 2 & 0 \\ 1 & 0 & 1 & 2 \\ 0 & 1 & 0 & 1 \\ 4 & 5 & -2 & 2 \end{vmatrix}$

29. $\begin{vmatrix} 5 & -2 & 2 & 3 \\ 0 & 1 & 1 & 3 \\ 1 & 0 & 1 & 1 \\ -2 & -1 & 0 & 6 \end{vmatrix}$

30. $\begin{vmatrix} 24 & 10 & 4 & 1 \\ 21 & 3 & 0 & -2 \\ 3 & 2 & 3 & 0 \\ 1 & 1 & 1 & -4 \end{vmatrix}$

31. $\begin{vmatrix} 2 & 2 & -1 & 3 \\ 1 & 1 & -1 & 1 \\ 1 & -1 & 1 & 1 \\ 4 & 2 & 1 & 5 \end{vmatrix}$

32. $\begin{vmatrix} 3 & -2 & -1 & 2 \\ 4 & 1 & 2 & -3 \\ -9 & -5 & 7 & -8 \\ 1 & 5 & 3 & -2 \end{vmatrix}$

33. Show that

$$\begin{vmatrix} 1 & 1 & 1 \\ x & y & z \\ x^2 & y^2 & z^2 \end{vmatrix} = (x - y)(y - z)(z - x).$$

34. Show that

$$\begin{vmatrix} -x & 1 & 0 & 0 \\ 0 & -x & 1 & 0 \\ 0 & 0 & -x & 1 \\ -c_0 & -c_1 & -c_2 & -c_3 - x \end{vmatrix} = x^4 + c_3x^3 + c_2x^2 + c_1x + c_0.$$

35. If the rows of a 3×3 matrix A are the columns of B in the same order, prove that $|A| = |B|$. (The matrix B is called the **transpose** of A, and we write $B = A^T$.)

36. Show that if A is a square matrix of order 3 and c is a number, then $|cA| = c^3|A|$.

37. Prove the interchange property for square matrices of order 3.

38. Prove the multiplication property for square matrices of order 3.

Critical Thinking: Exploration and Writing

39. In more advanced courses, it is proved that $|AB| = |A| \cdot |B|$ for all square matrices A and B of the same order.

a. Verify this equation for arbitrary matrices A and B of order 2.

b. Explain how this implies that $|A^{-1}| = |A|^{-1}$ whenever A has an inverse.

Critical Thinking: Exploration and Writing

40. Use a cofactor expansion and explain why the following equation is true.

$$\begin{vmatrix} a & b & c \\ d & e & f \\ h & i & j \end{vmatrix} + \begin{vmatrix} a & b & c \\ k & m & n \\ h & i & j \end{vmatrix} = \begin{vmatrix} a & b & c \\ d+k & e+m & f+n \\ h & i & j \end{vmatrix}$$

◆ **Solution for Practice Problem**

1. $$\begin{vmatrix} 2 & 1 & -2 \\ 1 & 5 & 0 \\ 1 & -1 & 3 \end{vmatrix} \underset{=}{-5C_1 + C_2} \begin{vmatrix} 2 & -9 & -2 \\ 1 & 0 & 0 \\ 1 & -6 & 3 \end{vmatrix} = 1(-1)^3 \begin{vmatrix} -9 & -2 \\ -6 & 3 \end{vmatrix} = 39$$

11.7 CRAMER'S RULE

In Sections 11.3 and 11.4, we studied two matrix methods for solving systems of linear equations. The first method was that of Gauss-Jordan elimination, and the second was the method using inverses. Determinants can also be used to solve linear systems. The method that uses determinants is called **Cramer's rule.** This method is presented in this section.

We begin with a linear system of two equations in two variables.

$$\begin{aligned} a_1x + b_1y &= c_1 \\ a_2x + b_2y &= c_2 \end{aligned}$$

Suppose first that $a_1b_2 - a_2b_1 \neq 0$. We will use the elimination method to solve for x and then formulate the solution in terms of determinants. To eliminate y, we multiply the first equation by b_2, the second equation by b_1, and then subtract.

$$\begin{aligned} a_1b_2x + b_1b_2y &= c_1b_2 \\ a_2b_1x + b_1b_2y &= c_2b_1 \\ \hline (a_1b_2 - a_2b_1)x \qquad &= c_1b_2 - c_2b_1 \end{aligned}$$

Since $a_1b_2 - a_2b_1 \neq 0$, this gives

$$x = \frac{c_1b_2 - c_2b_1}{a_1b_2 - a_2b_1} = \frac{\begin{vmatrix} c_1 & b_1 \\ c_2 & b_2 \end{vmatrix}}{\begin{vmatrix} a_1 & b_1 \\ a_2 & b_2 \end{vmatrix}}.$$

Similarly, x can be eliminated by multiplying the first equation by a_2 and the second equation by a_1. The solution for y is found to be

$$y = \frac{a_1c_2 - a_2c_1}{a_1b_2 - a_2b_1} = \frac{\begin{vmatrix} a_1 & c_1 \\ a_2 & c_2 \end{vmatrix}}{\begin{vmatrix} a_1 & b_1 \\ a_2 & b_2 \end{vmatrix}}.$$

The two fractions for x and y have the same denominator, which is the determinant of the coefficients. We denote this determinant by D.

$$D = \begin{vmatrix} a_1 & b_1 \\ a_2 & b_2 \end{vmatrix}$$

The expressions we have obtained for x and y constitute the first part of Cramer's rule, as given below. The second and third parts describe the possibilities when $D = 0$, and we omit the proofs of these statements.

Cramer's Rule

1. If $D = \begin{vmatrix} a_1 & b_1 \\ a_2 & b_2 \end{vmatrix} \neq 0$ in the system $\begin{aligned} a_1x + b_1y &= c_1 \\ a_2x + b_2y &= c_2 \end{aligned}$,

   the solution is given by

$$x = \frac{\begin{vmatrix} c_1 & b_1 \\ c_2 & b_2 \end{vmatrix}}{\begin{vmatrix} a_1 & b_1 \\ a_2 & b_2 \end{vmatrix}}, \qquad y = \frac{\begin{vmatrix} a_1 & c_1 \\ a_2 & c_2 \end{vmatrix}}{\begin{vmatrix} a_1 & b_1 \\ a_2 & b_2 \end{vmatrix}}.$$

2. If $D = 0$ and either $\begin{vmatrix} c_1 & b_1 \\ c_2 & b_2 \end{vmatrix}$ or $\begin{vmatrix} a_1 & c_1 \\ a_2 & c_2 \end{vmatrix}$ is not zero, there is no solution. In this case, the system is called **inconsistent.**
3. If $D = 0$ and both $\begin{vmatrix} c_1 & b_1 \\ c_2 & b_2 \end{vmatrix}$ and $\begin{vmatrix} a_1 & c_1 \\ a_2 & c_2 \end{vmatrix}$ are zero, there are infinitely many solutions. In this case, the system is called **dependent.**

Before considering an example, we make some observations about the determinants in part 1 of Cramer's rule. If the first column of the coefficient matrix is replaced by the column of constants, the determinant of the resulting matrix is the numerator in the expression for x (the first variable in the system). If the second column of the coefficient matrix is replaced by the column of constants, the determinant of the resulting matrix is the numerator in the expression for y (the second variable in the system). The standard notations for these determinants are

$$D_x = \begin{vmatrix} c_1 & b_1 \\ c_2 & b_2 \end{vmatrix} \quad \text{and} \quad D_y = \begin{vmatrix} a_1 & c_1 \\ a_2 & c_2 \end{vmatrix}.$$

With this notation, the solutions are given by

$$x = \frac{D_x}{D}, \qquad y = \frac{D_y}{D}.$$

Example 1 • Using Cramer's Rule

Use Cramer's rule to solve the following system.

$$\begin{aligned} 3x + y &= -1 \\ 4x - y &= -13 \end{aligned}$$

Solution

We first calculate D.

$$D = \begin{vmatrix} 3 & 1 \\ 4 & -1 \end{vmatrix} = -3 - 4 = -7$$

Since $D \neq 0$, the solution is given by the formulas in part 1 of Cramer's rule. The determinants D_x and D_y are given by

$$D_x = \begin{vmatrix} -1 & 1 \\ -13 & -1 \end{vmatrix} = 1 + 13 = 14,$$

$$D_y = \begin{vmatrix} 3 & -1 \\ 4 & -13 \end{vmatrix} = -39 + 4 = -35.$$

Thus

$$x = \frac{D_x}{D} = \frac{14}{-7} = -2 \text{ and } y = \frac{D_y}{D} = \frac{-35}{-7} = 5.$$

These values can easily be checked in the original system. □

Example 2 • An Inconsistent System

Apply Cramer's rule to the following system.

$$\begin{aligned} x - 4y &= 2 \\ 2x - 8y &= 2 \end{aligned}$$

Solution

Since

$$D = \begin{vmatrix} 1 & -4 \\ 2 & -8 \end{vmatrix} = -8 + 8 = 0$$

and

$$D_y = \begin{vmatrix} 1 & 2 \\ 2 & 2 \end{vmatrix} = 2 - 4 \neq 0,$$

the system has no solution. In other words, the system is inconsistent. □

Let us consider now a general system of n linear equations in n unknowns.

$$\begin{aligned} a_{11}x_1 + a_{12}x_2 + \cdots + a_{1n}x_n &= b_1 \\ a_{21}x_1 + a_{22}x_2 + \cdots + a_{2n}x_n &= b_2 \\ \vdots \qquad\qquad\quad &\ \vdots \\ a_{n1}x_1 + a_{n2}x_2 + \cdots + a_{nn}x_n &= b_n \end{aligned}$$

We let D denote the determinant of the coefficient matrix and let D_{x_i} denote the determinant of the matrix formed by replacing the elements in the ith column of the coefficient matrix by the column of constants. Using these notations, Cramer's rule may be stated as follows.

Cramer's Rule

1. If $D \neq 0$, the solution is given by

$$x_1 = \frac{D_{x_1}}{D}, \qquad x_2 = \frac{D_{x_2}}{D}, \qquad \ldots, \qquad x_n = \frac{D_{x_n}}{D}.$$

2. If $D = 0$ and one or more D_{x_i} is not zero, there is no solution. In this case, the system is called **inconsistent.**
3. If $D = 0$ and all $D_{x_i} = 0$, the system is **dependent** and there are infinitely many solutions.

Example 3 • Using Cramer's Rule on a System of Three Equations

Use Cramer's rule to solve the following system.

$$\begin{aligned} 2x + 3y - z &= 4 \\ -x + y + 2z &= -2 \\ 3x - y - 2z &= 1 \end{aligned}$$

Solution

Evaluation of the four determinants yields

$$D = \begin{vmatrix} 2 & 3 & -1 \\ -1 & 1 & 2 \\ 3 & -1 & -2 \end{vmatrix} = 14, \qquad D_x = \begin{vmatrix} 4 & 3 & -1 \\ -2 & 1 & 2 \\ 1 & -1 & -2 \end{vmatrix} = -7,$$

$$D_y = \begin{vmatrix} 2 & 4 & -1 \\ -1 & -2 & 2 \\ 3 & 1 & -2 \end{vmatrix} = 15, \qquad D_z = \begin{vmatrix} 2 & 3 & 4 \\ -1 & 1 & -2 \\ 3 & -1 & 1 \end{vmatrix} = -25.$$

Thus the solution is given by

$$x = \frac{D_x}{D} = \frac{-7}{14} = -\frac{1}{2}, \qquad y = \frac{D_y}{D} = \frac{15}{14}, \qquad z = \frac{D_z}{D} = -\frac{25}{14}. \quad \square$$

◆ Practice Problem 1 ◆

Apply Cramer's rule to the following system.

$$\begin{aligned} 2x - y \phantom{{}-z} &= 3 \\ x + y - z &= 1 \\ 3x - 3y + z &= 5 \end{aligned}$$

If a calculator with matrix features is available, work a sampling of Exercises 1–42 both by hand and with the calculator.

EXERCISES 11.7

Use Cramer's rule to find the solution if the determinant of the coefficients is not zero. If the system is dependent or inconsistent, state so.

1. $2x + 9y = 3$, $3x - 2y = -11$

2. $x + 2y = 4$, $3x - 2y = -12$

3. $7x - y = 9$, $3x + 5y = -7$

4. $4x + 3y = -6$, $5x - y = 27$

5. $4a + 3b = 1$, $2a - 5b = -19$

6. $a + 3b = -2$, $3a + 2b = 8$

7. $3x - 12y = 9$, $-x + 4y = -3$

8. $5x + 2y = 8$, $15x + 6y = 24$

9. $7x - y = 1$, $-14x + 2y = -1$

10. $-5x + 4y = 0$, $10x - 8y = 2$

11. $x + y = 3$, $x + 2y - z = 5$, $2y + z = 1$

12. $4x + z = -6$, $y + 2z = -3$, $2x + 5y + z = 1$

13. $3x + 2y + z = -1$, $x - 2z = -3$, $y + 2z = -2$

14. $-2x - y + 3z = 1$, $2x + y - z = -2$, $-x + 3y + 2z = 4$

15. $3x + y + 2z = 3$, $x - 5y - z = 0$, $2x + 3y + 2z = 0$

16. $x + 3y - z = 4$, $2x + y + 3z = 11$, $3x - 2y + 4z = 11$

17. $a + 3b - c = 4$, $3a - 2b + 4c = 11$, $2a + b + 3c = 13$

18. $2r - s + 2t = 2$, $2r - s + t = -5$, $r - 2s + 3t = 4$

19. $2x - y + 3z = 17$, $5x - 2y + 4z = 28$, $3x + 3y - z = -1$

20. $x + 3y - 2z = 1$, $-4x + 2y - 2z = 5$, $2x - y + z = 3$

21. $2x + y - z = 5$, $x - y + 2z = -3$, $-3y + 5z = -11$

22. $4x - y + 2z = -7$, $x + y + z = 4$, $5x - 5y + z = -26$

23. $\begin{aligned} 2x_1 - x_2 + 3x_3 &= 17 \\ 5x_1 - 2x_2 + 4x_3 &= 28 \\ 3x_1 + 3x_2 - x_3 &= 1 \end{aligned}$

24. $\begin{aligned} 5x_1 + 3x_2 - x_3 &= 4 \\ 2x_1 - 7x_2 + 3x_3 &= -36 \\ 3x_1 - x_2 - 2x_3 &= -13 \end{aligned}$

25. $\begin{aligned} 6x - 4y + 3z &= 1 \\ 12x - 8y - 9z &= -3 \\ 12x - 4y + 9z &= 4 \end{aligned}$

26. $\begin{aligned} x - 2y + 3z &= 4 \\ 2x + y - 4z &= 3 \\ -x + 5y - 5z &= 1 \end{aligned}$

27. $\begin{aligned} 3x - y + 2z &= 0 \\ x + 2y + 4z &= -1 \\ 7x - 7y - 2z &= 1 \end{aligned}$

28. $\begin{aligned} 2x + 2y - 3z &= 9 \\ x - y + 2z &= -1 \\ x - 5y + 9z &= 1 \end{aligned}$

29. $\begin{aligned} x - y - 2z &= 2 \\ 2x + 3y - z &= 4 \\ 3x - y - 2z &= 1 \end{aligned}$

30. $\begin{aligned} x + y + z &= 1 \\ x - 3y - 5z &= -1 \\ 4x - 5y + 2z &= -35 \end{aligned}$

31. $\begin{aligned} x + 2y + 3z &= 0 \\ y + z &= 0 \\ x + y + 3z &= 1 \end{aligned}$

32. $\begin{aligned} 3x - y + 2z &= 4 \\ x + 2y - z &= -2 \\ x + 2y - 3z &= -2 \end{aligned}$

33. $\begin{aligned} 3x - y - 2z &= -13 \\ 5x + 3y - z &= 4 \\ 2x - 7y + 3z &= -36 \end{aligned}$

34. $\begin{aligned} x + y - 2z &= -5 \\ 3x + y + z &= 3 \\ 4x + y - 2z &= -2 \end{aligned}$

35. $\begin{aligned} x - y - z - w &= 2 \\ x + y - w &= 0 \\ 3y + 2z + w &= -2 \\ y + w &= 0 \end{aligned}$

36. $\begin{aligned} x - 2y + 3z &= -3 \\ x + 2z + w &= 0 \\ -2x + 4y + 5w &= 10 \\ 2x - 4z - 3w &= -2 \end{aligned}$

37. $\begin{aligned} x + 3y - 2z + w &= 2 \\ 2x + 4y + 2z - 3w &= -1 \\ x - y - z - 2w &= -1 \\ 2x + 5y - z - w &= 1 \end{aligned}$

38. $\begin{aligned} 4x - 2y - 5z - 6w &= 1 \\ 2x + y + 3w &= 3 \\ 7x + 2y - 3z + 2w &= 4 \\ 4x - y + 7z + 5w &= 2 \end{aligned}$

39. Find m and b so that the line with equation $y = mx + b$ passes through the points $(1, -2)$ and $(3, 2)$.

40. Find a and b so that the line with equation $ax + by = 2$ passes through the points $(-3, -4)$ and $(2, 6)$.

41. Find a, b, and c so that the parabola with equation $y = ax^2 + bx + c$ passes through the points $(2, 10)$, $(-1, -5)$, and $(-3, 5)$.

42. Find a, b, and c so that the circle with equation $x^2 + ax + y^2 + by = c$ passes through the points $(1, 1)$, $(3, 3)$, and $(5, 1)$.

Critical Thinking: Exploration and Writing

43. Let x_1 be one of the variables in a system of n linear equations in n unknowns and suppose the column of x_1-coefficients is exactly the same as the column of constants. Describe the effect of this condition on the solutions of the system.

Critical Thinking: Exploration and Writing

44. Let x_1 and x_2 be two of the variables in a system of n linear equations in n unknowns and suppose that x_1 and x_2 have equal coefficients in every equation of the system. Describe the implication that this condition has for the solutions of the system.

◆ Solution for Practice Problem

1. The system is dependent since

$$D = \begin{vmatrix} 2 & -1 & 0 \\ 1 & 1 & -1 \\ 3 & -3 & 1 \end{vmatrix} = 0, \qquad D_x = \begin{vmatrix} 3 & -1 & 0 \\ 1 & 1 & -1 \\ 5 & -3 & 1 \end{vmatrix} = 0,$$

$$D_y = \begin{vmatrix} 2 & 3 & 0 \\ 1 & 1 & -1 \\ 3 & 5 & 1 \end{vmatrix} = 0, \qquad D_z = \begin{vmatrix} 2 & -1 & 3 \\ 1 & 1 & 1 \\ 3 & -3 & 5 \end{vmatrix} = 0.$$

CHAPTER REVIEW

Summary of Important Concepts and Formulas

Properties of Matrix Addition and Multiplication

1. $A + (B + C) = (A + B) + C$
2. $A + B = B + A$
3. $A(BC) = (AB)C$
4. $A(B + C) = AB + AC$
5. $(B + C)D = BD + CD$
6. $a(B + C) = aB + aC$
7. $(a + b)A = aA + bA$
8. $AI = A = IA$
9. $AA^{-1} = I = A^{-1}A$

Row Operations on a Matrix

1. Interchange two rows.
2. Multiply (or divide) every element in a row by the same nonzero number.
3. Add (or subtract) a multiple of one row to (or from) another row.

Gaussian Elimination Method

1. Write the augmented matrix $[A \mid B]$ for the system, where A is the coefficient matrix and B is the constant matrix.
2. Use any of the row operations to change $[A \mid B]$ to a matrix in which all elements on the main diagonal are 1's and all elements below the main diagonal are 0's.
3. Use back substitution on the system that has the augmented matrix obtained in step 2.

Gauss–Jordan Elimination Method for Solving a Linear System

1. Write the augmented matrix $[A \mid B]$ for the system.
2. Use any of the row operations to change $[A \mid B]$ to $[I \mid C]$, where I is an identity matrix of the same size as A, and C is a column matrix.
3. Read the solutions from the column matrix C.

Inverse of a 2 × 2 Matrix

If $A = \begin{bmatrix} a & b \\ c & d \end{bmatrix}$ and $\delta(A) = ad - bc \neq 0$, then

$$A^{-1} = \frac{1}{\delta(A)} \begin{bmatrix} d & -b \\ -c & a \end{bmatrix}.$$

Gauss–Jordan Elimination Method for Finding A^{-1}.

1. Augment A with an identity matrix of the same size: $[A \mid I_n]$.
2. Use row operations to transform $[A \mid I_n]$ into the form $[I_n \mid B]$. If this is not possible, then A^{-1} does not exist.
3. If $[A \mid I_n]$ is transformed into $[I_n \mid B]$ by row operations, the inverse can be read from the last n columns of $[I_n \mid B]$. That is, $B = A^{-1}$.

Solution of Linear Systems by Matrix Inverses

If A^{-1} exists, $X = A^{-1}B$ is the unique solution to $AX = B$.

Cofactor Expansion of a Determinant

$$|A| = a_{i1}C_{i1} + a_{i2}C_{i2} + \cdots + a_{in}C_{in}$$
$$|A| = a_{1j}C_{1j} + a_{2j}C_{2j} + \cdots + a_{nj}C_{nj}$$

Row Operations on a Determinant

1. Interchange any two rows.
2. Multiply (or divide) every element in a row by the same nonzero number.
3. Add (or subtract) a multiple of one row to (or from) another row.

Column Operations on a Determinant

1. Interchange any two columns.
2. Multiply (or divide) every element in a column by the same nonzero number.
3. Add (or subtract) a multiple of one column to (or from) another column.

Evaluation of Determinants

Interchange property: If the matrix B is obtained from a matrix A by interchanging any two rows (columns) of A, then $|B| = -|A|$.

Multiplication property: If the matrix B is obtained from a matrix A by multiplying each element of a row (column) of A by the same number c, then $|B| = c|A|$.

Addition property: If the matrix B is obtained from a matrix A by adding a multiple of one row (column) to another row (column), then $|B| = |A|$.

Cramer's Rule

1. If $D \neq 0$, the solution is given by
$$x_1 = \frac{D_{x_1}}{D}, \quad x_2 = \frac{D_{x_2}}{D}, \quad \ldots, \quad x_n = \frac{D_{x_n}}{D}.$$
2. If $D = 0$ and one or more D_{x_i} are not zero, there is no solution. In this case the system is called **inconsistent.**
3. If $D = 0$ and all $D_{x_i} = 0$, the system is **dependent,** and there are infinitely many solutions.

Critical Thinking: Find the Errors

Each of the following nonsolutions has one or more errors. Can you find them?

Problem 1

Perform the indicated operations, if possible.

a. $[2 \quad -5 \quad 0] + [1 \quad 3]$

b. $\begin{bmatrix} 0 & 1 \\ 2 & 3 \end{bmatrix}\begin{bmatrix} 4 & 5 \\ -1 & -2 \end{bmatrix}$

Nonsolution

a. $[2 \quad -5 \quad 0] + [1 \quad 3] = [3 \quad -2 \quad 0]$

b. $\begin{bmatrix} 0 & 1 \\ 2 & 3 \end{bmatrix}\begin{bmatrix} 4 & 5 \\ -1 & -2 \end{bmatrix} = \begin{bmatrix} 0 & 5 \\ -2 & -6 \end{bmatrix}$

Problem 2

Use Gauss–Jordan elimination to find a solution, if one exists.

$$\begin{aligned} x + y + 2z &= 2 \\ y + 3z &= 1 \\ x \qquad - 2z &= 1 \end{aligned}$$

Nonsolution

$$\left[\begin{array}{rrr|r} 1 & 1 & 2 & 2 \\ 0 & 1 & 3 & 1 \\ 1 & 0 & -2 & 1 \end{array}\right] \to \left[\begin{array}{rrr|r} 1 & 1 & 2 & 2 \\ 0 & 1 & 3 & 1 \\ 0 & -1 & -4 & 1 \end{array}\right]$$

$$\to \left[\begin{array}{rrr|r} 1 & 1 & 2 & 2 \\ 0 & 1 & 3 & 1 \\ 0 & 0 & -1 & 1 \end{array}\right] \to \left[\begin{array}{rrr|r} 1 & 0 & -1 & 2 \\ 0 & 1 & 3 & 1 \\ 0 & 0 & 1 & 1 \end{array}\right]$$

$$\to \left[\begin{array}{rrr|r} 1 & 0 & 0 & 2 \\ 0 & 1 & 0 & 1 \\ 0 & 0 & 1 & 1 \end{array}\right]$$

The solution is $x = 2$, $y = 1$, $z = 1$.

Problem 3

Evaluate the following determinant by expanding about the first column.

$$\begin{vmatrix} 3 & 1 & -1 \\ -1 & 1 & 2 \\ 1 & 2 & 1 \end{vmatrix}$$

Nonsolution

$$\begin{vmatrix} 3 & 1 & -1 \\ -1 & 1 & 2 \\ 1 & 2 & 1 \end{vmatrix} = 3\begin{vmatrix} 1 & 2 \\ 2 & 1 \end{vmatrix} - 1\begin{vmatrix} 1 & -1 \\ 2 & 1 \end{vmatrix} + 1\begin{vmatrix} 1 & -1 \\ 1 & 2 \end{vmatrix}$$

$$\begin{aligned} &= 3(1 - 4) - 1(1 + 2) + 1(2 + 1) \\ &= 3(-3) - 1(3) + 1(3) \\ &= -9 \end{aligned}$$

Problem 4

Evaluate the following determinant by introducing zeros before expanding.

$$\begin{vmatrix} 0 & 1 & 4 \\ -1 & 2 & 3 \\ 1 & -2 & 0 \end{vmatrix}$$

Nonsolution

$$\begin{vmatrix} 0 & 1 & 4 \\ -1 & 2 & 3 \\ 1 & -2 & 0 \end{vmatrix} \overset{R_2 + 1}{=} \begin{vmatrix} 0 & 1 & 4 \\ 0 & 3 & 4 \\ 1 & -2 & 0 \end{vmatrix} = 1\begin{vmatrix} 1 & 4 \\ 3 & 4 \end{vmatrix} = 4 - 12 = -8$$

Problem 5

Use Cramer's rule to determine the value of y in the solution of the system

$$\begin{aligned} 2x - 4y + 2z &= -8 \\ 3y + 2z &= 4 \\ 3x \qquad + z &= 2 \end{aligned}$$

Nonsolution

$$D_y = \begin{vmatrix} 2 & -8 & 2 \\ 0 & 4 & 2 \\ 3 & 2 & 1 \end{vmatrix} \overset{\frac{1}{2}R_2}{=} \begin{vmatrix} 2 & -8 & 2 \\ 0 & 2 & 1 \\ 3 & 2 & 1 \end{vmatrix} \overset{-2C_3 + C_2}{=} \begin{vmatrix} 2 & -12 & 2 \\ 0 & 0 & 1 \\ 3 & 0 & 1 \end{vmatrix}$$

$$= -1\begin{vmatrix} 2 & -12 \\ 3 & 0 \end{vmatrix} = -36$$

Thus $y = -36$.

Review Problems for Chapter 11

Perform the indicated operations, if possible.

1. $\begin{bmatrix} 2 & -3 \\ -4 & -5 \\ 0 & 2 \end{bmatrix} - 2\begin{bmatrix} -1 & 0 \\ 3 & -11 \\ 1 & 0 \end{bmatrix}$

2. $[1 \quad 3 \quad 0] + [2 \quad -7]$

3. $2\begin{bmatrix} -2 & 1 & 3 \\ 1 & 5 & 1 \end{bmatrix} - 3\begin{bmatrix} 1 & 7 & -2 \\ 0 & 1 & -5 \end{bmatrix}$

Perform the matrix multiplications, if possible.

4. $\begin{bmatrix} 1 & 4 & 2 \\ -2 & 1 & 0 \end{bmatrix}\begin{bmatrix} -3 & 1 & 2 \\ 0 & 5 & -2 \end{bmatrix}$

5. $\begin{bmatrix} 1 & -2 & 0 \\ 0 & 3 & 2 \\ 5 & 0 & 1 \end{bmatrix}\begin{bmatrix} -1 & 0 \\ 3 & 4 \\ 0 & -1 \end{bmatrix}$

6. $\begin{bmatrix} 2 & 1 & 0 \\ -1 & 3 & 5 \end{bmatrix}\begin{bmatrix} 2 & 7 \\ 0 & -4 \end{bmatrix}$

7. $\begin{bmatrix} 2 & 7 \\ 0 & -4 \end{bmatrix}\begin{bmatrix} 2 & 1 & 0 \\ -1 & 3 & 5 \end{bmatrix}$

Use Gaussian or Gauss-Jordan elimination to find a solution, if one exists.

8. $\begin{aligned} 2x - y &= 2 \\ -x + 3y &= 14 \end{aligned}$

9. $\begin{aligned} x - 2y - 4z &= -1 \\ 3x \quad\quad - z &= 4 \\ x + 4y + 7z &= 2 \end{aligned}$

10. $\begin{aligned} -x + 2y + z &= -1 \\ 3x + y - z &= 7 \\ y + z &= -3 \end{aligned}$

11. $\begin{aligned} x + 2y + z &= 1 \\ 2x + 5y + 3z &= 2 \\ x \quad\quad - 2z &= 3 \end{aligned}$

12. $\begin{aligned} x + 2y + z &= 1 \\ x + y - z &= 1 \\ y + 3z &= 1 \end{aligned}$

13. $\begin{aligned} x \quad + z &= 3 \\ x + y + z &= 1 \\ 2x - y + 3z &= 12 \end{aligned}$

Find the inverse of the given matrix, if it exists.

14. $\begin{bmatrix} 2 & -4 \\ -1 & 3 \end{bmatrix}$

15. $\begin{bmatrix} 8 & 12 \\ 6 & 9 \end{bmatrix}$

16. $\begin{bmatrix} 1 & 0 & 0 \\ -1 & 1 & 1 \\ 0 & 1 & 2 \end{bmatrix}$

17. $\begin{bmatrix} 1 & 0 & 2 \\ 2 & 1 & 5 \\ 0 & 1 & 2 \end{bmatrix}$

18. $\begin{bmatrix} 1 & 0 & 2 \\ 2 & -2 & 5 \\ 0 & 1 & -1 \end{bmatrix}$

19. $\begin{bmatrix} 1 & 1 & -3 \\ 1 & 0 & 3 \\ -2 & 1 & -12 \end{bmatrix}$

Solve each system by using the inverse of the coefficient matrix.

20. $\begin{aligned} -2r + 3s &= 3 \\ r + 2s &= -19 \end{aligned}$

21. $\begin{aligned} 4x - y &= 2 \\ 7x - 3y &= 1 \end{aligned}$

22. $\begin{aligned} 2x + y - z &= 1 \\ y - 2z &= 9 \\ x + 3y \quad\quad &= 1 \end{aligned}$

23. $\begin{aligned} 4x - 5y - 3z &= 8 \\ 3x - 3y - 2z &= 7 \\ x - y - z &= 1 \end{aligned}$

24. Evaluate the determinant $\begin{vmatrix} 4 & -2 \\ -1 & -3 \end{vmatrix}$.

Use a cofactor expansion to evaluate the determinant.

25. $\begin{vmatrix} -1 & 2 & 3 \\ 4 & -2 & 1 \\ 1 & 5 & 0 \end{vmatrix}$

26. $\begin{vmatrix} 2 & 1 & -2 \\ 1 & 5 & 0 \\ 1 & -1 & 3 \end{vmatrix}$

27. $\begin{vmatrix} 3 & -2 & 1 \\ 4 & 0 & 1 \\ 2 & 5 & -2 \end{vmatrix}$

28. Solve for x in the equation $\begin{vmatrix} -2 & x & -2 \\ 3 & 1 & 0 \\ 1 & 2x & 1 \end{vmatrix} = 0.$

Without evaluating the determinants, determine the value of the variable that makes each statement true.

29. $\begin{vmatrix} 2 & 1 & 1 \\ -3 & 5 & -3 \\ 4 & 2 & 1 \end{vmatrix} = \begin{vmatrix} 2 & 1 & 1 \\ -3 & 5 & -3 \\ 0 & 0 & x \end{vmatrix}$

30. $\begin{vmatrix} 1 & -2 & -1 \\ 3 & 0 & 2 \\ 2 & 1 & 1 \end{vmatrix} = x\begin{vmatrix} -1 & -2 & 1 \\ 2 & 0 & 3 \\ 1 & 1 & 2 \end{vmatrix}$

31. $\begin{vmatrix} -2 & 4 & 8 \\ -4 & 2 & -6 \\ 2 & 2 & -4 \end{vmatrix} = x\begin{vmatrix} -1 & 2 & 4 \\ -2 & 1 & -3 \\ 1 & 1 & -2 \end{vmatrix}$

32. Evaluate the determinant by introducing zeros before expanding about a row or column.

$$\begin{vmatrix} -1 & 0 & 2 & 0 \\ 2 & 3 & 1 & 2 \\ 1 & 1 & 1 & 0 \\ -1 & 2 & -1 & 3 \end{vmatrix}$$

Use Cramer's rule to find the solution if the determinant of the coefficients is not zero. If the system is dependent or inconsistent, state so.

33. $\begin{aligned} 3x - y &= -2 \\ 6x - 2y &= -4 \end{aligned}$

34. $\begin{aligned} 5x + 2y &= 4 \\ -x + 2y &= -8 \end{aligned}$

35. $\begin{aligned} x + y + 2z &= 3 \\ -x + 2y + z &= 9 \\ 3x + y + 3z &= 0 \end{aligned}$

36. $\begin{aligned} 2x - y \phantom{{}+z} &= 3 \\ x + y - z &= 1 \\ 3x - 3y + z &= 5 \end{aligned}$

E N C O R E

Oldest Graeco-Latin square

Matrices have many uses that are less computational than those considered in the examples and exercises in this chapter. Some such uses involve Latin squares.

A **Latin square** is a square matrix of symbols arranged so that each symbol appears exactly once in each row and exactly once in each column. Latin squares are sometimes used in designing statistical experiments. In an industrial experiment, for instance, the rows might correspond to machines, the columns might correspond to work shifts, and symbols could be entered in the matrix positions to identify workers.

Two Latin squares are shown in Figure 11.1. In the first matrix, each of the letters *J*, *Q*, *K*, and *A* occurs exactly once in each row and once in each column. The second matrix has the same properties with respect to the symbols ♥, ♣, ♦, and ♠.

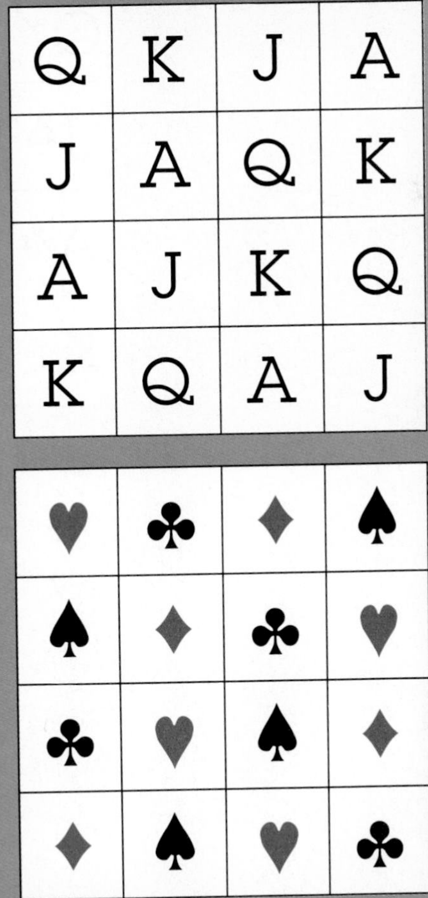

Figure 11.1 Latin squares

If the two Latin squares in Figure 11.1 are superimposed to form a combined square, each of the symbols in one of the matrices pairs up exactly once with each symbol from the other matrix. This is shown in Figure 11.2. Two Latin squares with this property are said to be **mutually orthogonal,** and the combined square is called a **Graeco-Latin square.** Replacement of the letters *J*, *Q*, *K*, and *A* by jack, queen, king, and ace, respectively, in Figure 11.2 leads to the playing card matrix shown at the beginning of this Encore. This matrix of cards is the oldest known example of a Graeco-Latin square. It comes from a 1623 puzzle by Claude-Gaspar Bachet de Méziriac (1581–1638), and it has an additional feature that each of the cards also occurs exactly once on both diagonals of the matrix.

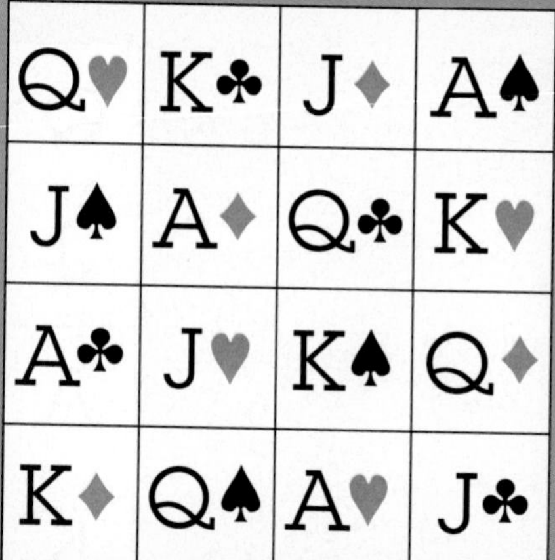

Figure 11.2 A Graeco-Latin square

Cumulative Test Chapters 3–11

1. a. Write the equation, in standard form, of the line through $(-1, 5)$ and $(4, -3)$.

   b. Find the slope and both intercepts of the line with equation $5x - 3y = 30$.

2. Find the extreme value of $f(x) = 2x^2 - 12x + 7$ and state whether that value represents a maximum or a minimum.

3. a. Find the center and radius of the circle with equation $x^2 + y^2 + 4x - 8y = 29$.

   b. Write the standard equation of the ellipse with center $(-2, 3)$, vertices $(-6, 3)$, $(2, 3)$, and endpoints of minor axis $(-2, 6)$, $(-2, 0)$.

4. The equation $3x - 4y = 24$ defines a function f. Find an expression for $f(x)$ and also for $f^{-1}(x)$ if f is one-to-one.

5. Find all the zeros of $P(x) = 2x^3 - 9x^2 + 14x - 5$, given that $2 - i$ is a zero.

6. Find all zeros of $P(x) = 2x^4 + x^3 - 4x^2 + x - 6$.

7. Solve $8^{x-3} = 4^{x+2}$.

8. Find the solution set for $\log(4x - 3) + \log(3x - 4) = 1$.

9. Write out the exact values of all the trigonometric functions of the given angle.

   a. $-240°$ b. $13\pi/4$

Angular Speed of a Roller

10. A heavy roller with a radius of 2 feet is being used to pack a roadbed for a highway. If the roller is pulled by a tractor going 10 miles per hour, find the angular speed of the roller in radians per minute.

11. Sketch the graph of the given function through one complete cycle, locating key points on each graph.

    a. $y = 3\sin\left(x - \dfrac{\pi}{6}\right)$ b. $y = \cos(3x + \pi)$

12. Verify the identity: $\dfrac{1 + \cos 2x}{\cos x \sin 2x} = \csc x$.

13. Solve $2\sin^2\theta = 1 + \sin\theta$ for θ, where $0° \leq \theta < 360°$.

14. Solve for x: $\sin^{-1}(1 + x) + \cos^{-1} x = \dfrac{5\pi}{6}$.

15. Solve the triangle to the degree of accuracy consistent with the information that $C = 108°$, $a = 83$, and $b = 32$.

Air Navigation

16. A plane is flying with a heading of 285.0°, an airspeed of 300 miles per hour, and a ground speed of 247 miles per hour. If the wind is blowing from the west and the course of the plane is in a northwesterly direction, find the speed of the wind.

17. Sketch the graph of the cardioid $r = 1 - \cos\theta$ in a polar coordinate system.

18. Solve the following systems.

    a. $\begin{aligned} 2x - 5y &= 20 \\ 3x + 4y &= 7 \end{aligned}$

    b. $\begin{aligned} 12x + 14y &= 2 \\ 6x + 7y &= -1 \end{aligned}$

    c. $\begin{aligned} 2x - 2y + z &= -3 \\ 3x - y + 5z &= 9 \\ 5x - y - 2z &= 8 \end{aligned}$

19. Solve the following system of equations. It is not necessary to graph the equations.

    $$\begin{aligned} x^2 + y^2 &= 16 \\ x^2 + y &= 4 \end{aligned}$$

20. Graph the solution set of the given system of inequalities, showing all points of intersection of the boundaries.

    $$\begin{aligned} 2x^2 - y &\leq 2 \\ 2x - y &\geq -2 \end{aligned}$$

21. Minimize $f = 4y - 5x$, subject to

    $$\begin{aligned} x - 2y &\leq 2 \\ 3x + 4y &\leq 16 \\ x &\geq 0 \\ y &\geq 0. \end{aligned}$$

22. Perform the indicated operations, if possible.

a. $3\begin{bmatrix} -1 & 0 \\ 5 & -2 \\ -2 & 4 \end{bmatrix} - \begin{bmatrix} -4 & 10 \\ 3 & 2 \\ -1 & -5 \end{bmatrix}$

b. $[1 \quad 2 \quad 3] + \begin{bmatrix} 4 \\ 5 \\ 6 \end{bmatrix}$

c. $\begin{bmatrix} 1 & 0 \\ 4 & -1 \\ 0 & 2 \end{bmatrix}\begin{bmatrix} 5 & 0 \\ -3 & 1 \\ 0 & 0 \end{bmatrix}$

d. $\begin{bmatrix} 1 & 2 & 0 \\ 5 & -3 & 1 \end{bmatrix}\begin{bmatrix} 3 & 4 & -1 \\ 0 & -1 & -1 \\ 0 & 2 & 3 \end{bmatrix}$

23. Use Gaussian or Gauss–Jordan elimination to find a solution, if one exists.

$$\begin{aligned} x - 2y - 4z &= 0 \\ 3x \qquad - z &= 1 \\ x + 4y + 8z &= 0 \end{aligned}$$

24. Solve the following system by using the inverse of the coefficient matrix.

$$\begin{aligned} x + 2y \qquad &= 0 \\ y + z &= 2 \\ 2x + 5y + 2z &= 5 \end{aligned}$$

25. Use Cramer's rule to find the solution if the determinant of the coefficients is not zero. If the system is dependent or inconsistent, state so.

$$\begin{aligned} x + 2y + z &= 1 \\ 2x + y - z &= 5 \\ x + 3y + 3z &= 1 \end{aligned}$$

Further Topics

Selections from this chapter provide excellent preparation for future courses in two directions. The contents of the first five sections are included in a typical calculus course. For the calculus-bound student, the study of these sections furnishes an advantageous prior exposure and adds depth to understanding. The last three sections form the basis for a starting point in the study of statistics.

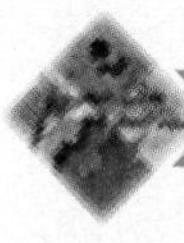

12.1 SEQUENCES AND SERIES

Higher-level mathematics makes extensive use of certain types of functions called *sequences.* Informally, a **sequence** is an ordered list in which there is a first element, a second element, a third element, and so on. Some familiar sequences are the sequence of positive integers

$$1, 2, 3, 4, \ldots$$

and the sequence of even positive integers

$$2, 4, 6, 8, \ldots.$$

These are examples of infinite sequences. Sequences that have a last, or terminal, element are called finite sequences. An example of a finite sequence is the ordered listing

$$2, 5, 8, 11, 14, 17.$$

These ideas are formalized in the following definition.

Definition of a Sequence

A **finite sequence** is a function that has a domain of the form $\{1, 2, 3, \ldots, k\}$, where k is a fixed integer. An **infinite sequence** is a function that has the set of all positive integers as its domain.

According to this definition, a finite sequence has function values of the form

$$f(1), f(2), f(3), \ldots, f(k)$$

where f is the function that has the set $\{1, 2, 3, \ldots, k\}$ as its domain. Similarly, an infinite sequence with f as its defining function has function values of the form

$$f(1), f(2), f(3), \ldots$$

where the dots at the end indicate that there is no terminal element. In working with sequences, it is traditional to use a subscripted letter, such as a_n, instead of $f(n)$ to indicate the function value at the positive integer n. A finite sequence is written as

$$a_1, a_2, a_3, \ldots, a_k$$

and an infinite sequence is written as

$$a_1, a_2, a_3, \ldots.$$

With this notation, a_n stands for the ***n*th term,** or the **general term,** in the sequence. Thus a_1 denotes the first term, a_2 denotes the second term, a_{18} denotes the eighteenth term, and so on.

Example 1 • Examples of Sequences

Write the first four terms of the sequence that has the given general term.

a. $a_n = \dfrac{1}{n}$ b. $a_n = 2n - 1$ c. $b_n = (-1)^n \dfrac{n}{n+1}$

Solution

To find the first four terms of each sequence, we assign each of the values 1, 2, 3, 4 to n in succession. This gives

a. $a_1 = 1,\quad a_2 = \frac{1}{2},\quad a_3 = \frac{1}{3},\quad a_4 = \frac{1}{4};$
b. $a_1 = 1,\quad a_2 = 3,\quad a_3 = 5,\quad a_4 = 7;$
c. $b_1 = -\frac{1}{2},\quad b_2 = \frac{2}{3},\quad b_3 = -\frac{3}{4},\quad b_4 = \frac{4}{5}.$ □

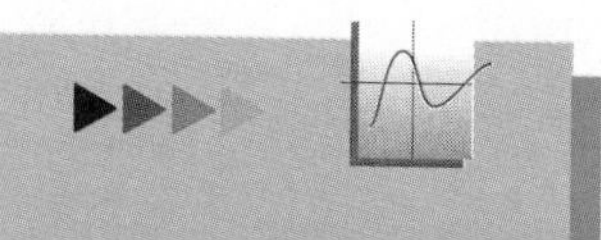

The point-plotting feature of a graphing device can be used to graph a sequence.

A graph can be drawn for a sequence, just as for any other function. The next example shows how the graph of a sequence has a distinct appearance.

Example 2 • The Graph of a Sequence

Draw the graph of the sequence that has general term $a_n = 2n - 3$ and domain $\{1, 2, 3, 4, 5\}$.

Solution

When we assign each of the values 1, 2, 3, 4, 5 to n in succession, we obtain the function values

$$a_1 = -1,\quad a_2 = 1,\quad a_3 = 3,\quad a_4 = 5,\quad a_5 = 7.$$

The graph of the sequence consists of the five points with coordinates

$$(1, -1),\quad (2, 1),\quad (3, 3),\quad (4, 5),\quad (5, 7).$$

This is shown in Figure 12.1. Note that the points on the graph are *isolated,* with a single point above each integer in the domain. This set of dots is a strong contrast to the straight line that is the graph of the equation $y = 2x - 3$. However, if the straight line were drawn, it would contain all the dots in the graph of the sequence. □

y
(5, 7)
(4, 5)
(3, 3)
(2, 1)
x
(1, −1)

$a_n = 2_n - 3$

Figure 12.1 The graph of the sequence in Example 2

Occasionally, a sequence is defined recursively by specifying the first few terms and a rule for finding the nth term from the preceding terms. Consider the following example.

Example 3 • A Sequence Defined Recursively

Find the first six terms of the following sequence.

$$a_1 = 1,\quad a_n = a_{n-1} + 2.$$

Solution

We have $a_1 = 1$ and

$$a_2 = a_1 + 2 = 1 + 2 = 3,$$
$$a_3 = a_2 + 2 = 3 + 2 = 5,$$
$$a_4 = a_3 + 2 = 5 + 2 = 7,$$
$$a_5 = a_4 + 2 = 7 + 2 = 9,$$
$$a_6 = a_5 + 2 = 9 + 2 = 11.$$

This gives the first six terms of the sequence as $1, 3, 5, 7, 9, 11$. □

◆ Practice Problem 1 ◆

Find the first six terms of the following sequence, known as the Fibonacci sequence.

$$a_1 = 1, \qquad a_2 = 1, \qquad a_{n+1} = a_n + a_{n-1}.$$

Definition of a Series

A **series** is the sum of the terms of a sequence.

In this section, we consider only series that are formed from a finite sequence. Such a series has the form

$$a_1 + a_2 + a_3 + \cdots + a_k.$$

It is possible in some cases for the sum of an infinite series to have meaning. Some series of this type are considered in Section 12.4.

If a formula for the general term is known, it is common practice to write a series in a compact form called the **sigma notation.** Using this notation, the series

$$a_1 + a_2 + a_3 + \cdots + a_k$$

is written $\sum_{n=1}^{k} a_n$. That is,

$$\sum_{n=1}^{k} a_n = a_1 + a_2 + a_3 + \cdots + a_k.$$

The capital Greek letter Σ (sigma) is used to indicate a sum, and the notations at the bottom and top of the sigma give the initial and terminal values of n. The letter n is called the **index of summation.**

Example 4 • Using Sigma Notation

Write the following series in expanded form and find the value of the sum.

$$\sum_{n=1}^{5} (2n - 1)$$

Solution

The given expression represents the series which has terms that are obtained by substituting, in succession, the values 1, 2, 3, 4, and 5 for n in $2n - 1$. Thus

$$\sum_{n=1}^{5} (2n - 1) = 1 + 3 + 5 + 7 + 9 = 25.$$

□

◆ Practice Problem 2 ◆

Write the following series in expanded form and find the value of the sum.

$$\sum_{n=1}^{6} \frac{1}{2^{n-1}}$$

There are two points that need to be made in connection with the sigma notation. First, we note that the index of summation is an arbitrary symbol, or a **dummy variable.** That is,

$$\sum_{n=1}^{k} a_n = \sum_{i=1}^{k} a_i = \sum_{j=1}^{k} a_j$$

since each of these notations represents the sum

$$a_1 + a_2 + a_3 + \cdots + a_k.$$

The second point is that the initial value of the index is not necessarily 1. For example,

$$\begin{aligned}\sum_{j=3}^{5} \frac{1}{2^j} &= \frac{1}{2^3} + \frac{1}{2^4} + \frac{1}{2^5}\\ &= \tfrac{1}{8} + \tfrac{1}{16} + \tfrac{1}{32}\\ &= \tfrac{7}{32}.\end{aligned}$$

We point out that the value $j = 3$ does not give the third term.

EXERCISES 12.1

In Exercises 1–18, write the first five terms of the sequence that has the given general term. (See Example 1.)

1. $a_n = \dfrac{1}{n+1}$

2. $a_n = 2n^2 - 1$

3. $a_n = \dfrac{n-1}{n}$

4. $a_n = \dfrac{n^2 - 1}{2n}$

5. $a_j = (-2)^j$

6. $a_j = (-1)^j$

7. $a_n = 2$

8. $a_n = 3$

√ 9. $a_i = \dfrac{(-1)^i}{i^2}$

10. $a_i = (-1)^i i^2$

11. $a_j = \dfrac{j}{2j - 1}$

12. $a_j = \dfrac{j}{j + 1}$

13. $a_i = 3i - 2$

14. $a_i = 5i + 3$

15. $a_n = (x - 1)^n$

16. $a_n = x^{n-1}, \quad x \neq 0$

√ 17. $a_j = (-1)^j x^{2j}$

18. $a_j = (-1)^j x^{2j-1}$

Draw the graph of the sequence with the given general term and domain. (See Example 2.)

√ 19. $a_n = 2n + 1$, $\{1, 2, 3, 4\}$

20. $a_n = 5 - 3n$, $\{1, 2, 3, 4, 5\}$

21. $a_n = n^2 - 6$, $\{1, 2, 3, 4\}$

22. $a_n = 2^n - 6$, $\{1, 2, 3, 4\}$

Write the first six terms of each sequence. (See Example 3 and Practice Problem 1.)

23. $a_1 = 1, \quad a_{n+1} = 2a_n$

24. $a_1 = 2, \quad a_{n+1} = a_n + 3$

√ 25. $a_1 = -5, \quad a_{n+1} = (-1)^n a_n$

26. $a_1 = 1, \quad a_{n+1} = x \cdot a_n$

√ 27. $a_1 = 2, \quad a_2 = 3, \quad a_{n+1} = 2a_n + a_{n-1}$

28. $a_1 = -1, \quad a_2 = 1, \quad a_{n+1} = 2a_n - a_{n-1}$

Write each series in expanded form and find the value of the sum. (See Example 4 and Practice Problem 2.)

29. $\sum_{n=1}^{5} 2^n$

30. $\sum_{n=1}^{7} (-2)^n$

√ 31. $\sum_{n=3}^{9} \dfrac{n - 2}{n + 1}$

32. $\sum_{n=2}^{9} \dfrac{n - 1}{n}$

33. $\sum_{j=1}^{7} \dfrac{j^2 - j}{2}$

34. $\sum_{j=1}^{11} \dfrac{j + 1}{2}$

35. $\sum_{i=1}^{7} (-1)^i$

36. $\sum_{i=4}^{10} 10^i$

√ 37. $\sum_{n=2}^{7} 2$

38. $\sum_{n=2}^{6} 3$

39. $\sum_{j=2}^{7} j^2(-1)^j$

40. $\sum_{j=-1}^{5} j$

41. $\sum_{i=-1}^{5} i^3$

42. $\sum_{i=2}^{5} 6i(-1)^i$

√ 43. $\sum_{n=1}^{6} \left(\dfrac{1}{n} - \dfrac{1}{n + 1}\right)$

44. $\sum_{n=1}^{5} \left(\frac{1}{2^n} - \frac{1}{2^{n+1}}\right)$ 45. $\sum_{n=2}^{5} (-1)^n \cdot 2^{-n}$ 46. $\sum_{n=2}^{7} 3^{-n}$

Write each of the following sums in sigma notation.

47. $1 + 2 + 3 + \cdots + 17$
48. $1^2 + 2^2 + 3^2 + \cdots + 9^2$
49. $2 + 4 + 8 + \cdots + 128$
50. $3 + 9 + 27 + \cdots + 729$
51. $1 - 3 + 5 - 7 + 9 - 11 + 13$
52. $2 - 4 + 6 - 8 + 10 - 12 + 14$
53. $\frac{1}{2} + \frac{2}{3} + \frac{3}{4} + \frac{4}{5} + \frac{5}{6}$
54. $\frac{1}{2} + \frac{3}{4} + \frac{5}{8} + \frac{7}{16} + \frac{9}{32}$
55. $\frac{1}{2} - \frac{3}{4} + \frac{5}{8} - \frac{7}{16} + \frac{9}{32} - \frac{11}{64}$
56. $\frac{1}{3} + \frac{3}{5} + \frac{5}{7} + \frac{7}{9} + \frac{9}{11} + \frac{11}{13}$
57. $\frac{1}{1 \cdot 3} + \frac{1}{3 \cdot 5} + \frac{1}{5 \cdot 7} + \cdots + \frac{1}{85 \cdot 87}$
58. $\frac{1}{1 \cdot 2} + \frac{1}{2 \cdot 3} + \frac{1}{3 \cdot 4} + \cdots + \frac{1}{49 \cdot 50}$
59. $1 + \frac{x}{2} + \frac{x^2}{3} + \cdots + \frac{x^n}{n+1}$
60. $\frac{x}{2} + \frac{x^2}{4} + \frac{x^3}{8} + \cdots + \frac{x^n}{2^n}$

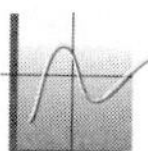

Use a graphing device with a point-plotting feature to graph the following sequences.

61–64

61. $a_n = \sqrt{e^n}$; $n = 1, 2, 3, 4, 5$
62. $a_n = e^{\sqrt{n}}$; $n = 1, 2, 3, 4, 5$
63. $a_n = n + \ln n$; $n = 1, 2, 3, 4$
64. $a_n = 1 + n \ln n$; $n = 1, 2, 3, 4, 5, 6$

Critical Thinking: Discussion

65. In Example 2 in this section, a comparison is made between the graph of a sequence with terms $a_n = 2n - 3$ and the graph of the straight line with $y = 2x - 3$. In an arbitrary sequence with general term $a_n = f(n)$, the integral variable n can be replaced by a positive real variable x to obtain a new function $y = f(x)$ with x in the domain $(0, \infty)$. Explain how the graph of the original sequence and the graph of the new function are related.

Critical Thinking: Exploration and Writing

66. The sequence of Fibonacci numbers 1, 1, 2, 3, 5, 8, 13, . . . is described in Practice Problem 1 in this section. These numbers occur frequently in nature. For instance, the needles on most pine trees always occur in bundles of 2, 3, or 5 each, and the number of scales (seeds) in a spiral around a pine cone is usually a Fibonacci number such as 8. Use a library resource to find two other instances where a Fibonacci number occurs in nature and write a short description of each.

The longleaf pine has needles that grow in bundles of three.

The needles of the bristlecone pine occur in bundles of five.

◆ Solutions for Practice Problems

1. We have $a_1 = 1$ and $a_2 = 1$. Thus

$$\begin{aligned} a_3 &= a_2 + a_1 = 1 + 1 = 2, \\ a_4 &= a_3 + a_2 = 2 + 1 = 3, \\ a_5 &= a_4 + a_3 = 3 + 2 = 5, \\ a_6 &= a_5 + a_4 = 5 + 3 = 8, \end{aligned}$$

and the first six terms are 1, 1, 2, 3, 5, 8.

2. We have

$$\begin{aligned} \sum_{n=1}^{6} \frac{1}{2^{n-1}} &= \frac{1}{2^0} + \frac{1}{2^1} + \frac{1}{2^2} + \frac{1}{2^3} + \frac{1}{2^4} + \frac{1}{2^5} \\ &= 1 + \frac{1}{2} + \frac{1}{4} + \frac{1}{8} + \frac{1}{16} + \frac{1}{32} \\ &= \frac{63}{32}. \end{aligned}$$

12.2 ARITHMETIC SEQUENCES

An arithmetic sequence is a special type of sequence defined as follows.

Definition of an Arithmetic Sequence

An **arithmetic sequence**† is a sequence in which each term after the first is obtained by adding the same number, d, to the preceding term. The constant d is called the **common difference.**

†An arithmetic sequence is also called an **arithmetic progression.**

Thus an arithmetic sequence with first term a_1 and common difference d is given by

$$a_1, a_1 + d, a_1 + 2d, a_1 + 3d, \ldots.$$

Example 1 • Writing Terms of an Arithmetic Sequence

Find the first five terms of the arithmetic sequence with first term $a_1 = -2$ and common difference $d = 3$.

Solution

The first five terms are computed as

$$\begin{aligned} a_1 &= -2, \\ a_2 &= a_1 + d = -2 + 3 = 1, \\ a_3 &= a_2 + d = 1 + 3 = 4, \\ a_4 &= a_3 + d = 4 + 3 = 7, \\ a_5 &= a_4 + d = 7 + 3 = 10. \end{aligned}$$

□

◆ **Practice Problem 1** ◆

Find the first five terms of the arithmetic sequence that has $a_1 = 5$ and $d = -2$.

It is easy to arrive at a formula for the nth term in an arithmetic sequence. Observing the pattern in

$$\begin{aligned} a_1 &= a_1, \\ a_2 &= a_1 + d, \\ a_3 &= a_1 + 2d, \\ a_4 &= a_1 + 3d, \\ a_5 &= a_1 + 4d, \end{aligned}$$

we see that the coefficient of d is always 1 less than the number of the term and

$$\mathbf{a_n = a_1 + (n - 1)d.} \tag{1}$$

Example 2 • Finding a Formula for the nth Term

Find the nineteenth term and a formula for the nth term of the following arithmetic sequence.

$$-1, 2, 5, 8, \ldots$$

Solution

The nineteenth term of the sequence is $a_{19} = a_1 + 18d$. To find the common difference, we need only subtract one of the terms from the succeeding term. Using the first two terms, we get

$$d = 2 - (-1) = 3.$$

Since $a_1 = -1$, we have

$$a_{19} = -1 + (18)(3) = 53$$

and

$$a_n = -1 + (n - 1)(3) = 3n - 4.$$

□

◆ Practice Problem 2 ◆

Find the nineteenth term and a formula for the nth term of the following arithmetic sequence.

$$3, 1, -1, -3, \ldots$$

Example 3 • Find a_1 and d

A certain arithmetic sequence has $a_5 = 5$ and $a_{13} = 21$. Find the first term a_1, the common difference d, and a_{25}.

Solution

Using the formula $a_n = a_1 + (n - 1)d$ with $a_5 = 5$ and $a_{13} = 21$, we have the system of equations:

$$\begin{aligned} a_1 + 4d &= 5 \quad \text{fifth term} \\ a_1 + 12d &= 21 \quad \text{thirteenth term} \end{aligned}$$

Subtracting the top equation from the bottom one, we get

$$\begin{aligned} 8d &= 16 \\ d &= 2. \end{aligned}$$

Using $d = 2$ in $a_1 + 4d = 5$, we have

$$\begin{aligned} a_1 + 8 &= 5 \\ a_1 &= -3. \end{aligned}$$

Then

$$a_{25} = -3 + (24)(2) = 45,$$

and this completes the solution. □

The series formed by adding the first n terms of a general arithmetic sequence is denoted by S_n.

$$S_n = a_1 + a_2 + a_3 + \cdots + a_n$$

A formula for S_n in terms of a_1 and a_n may be obtained as follows. First we write the sum S_n in the form

$$S_n = a_1 + (a_1 + d) + (a_1 + 2d) + \cdots + [a_1 + (n - 1)d]. \tag{2}$$

We then write the same sum in reverse order, subtracting d from each term to get the next one.

$$S_n = a_n + (a_n - d) + (a_n - 2d) + \cdots + [a_n - (n-1)d] \qquad (3)$$

Adding Equations (2) and (3), we have

$$2S_n = (a_1 + a_n) + (a_1 + a_n) + (a_1 + a_n) + \cdots + (a_1 + a_n). \qquad (4)$$

The right-hand side of Equation (4) has n of the terms $a_1 + a_n$, so

$$2S_n = n(a_1 + a_n)$$

and

$$\mathbf{S_n = \frac{n}{2}(a_1 + a_n).} \qquad (5)$$

We can obtain another formula for S_n in terms of a_1 and d by substituting $a_n = a_1 + (n-1)d$ in this formula. This substitution gives

$$S_n = \frac{n}{2}[a_1 + a_1 + (n-1)d]$$

or

$$\mathbf{S_n = \frac{n}{2}[2a_1 + (n-1)d].} \qquad (6)$$

For convenient reference, we record the formulas relating to arithmetic sequences in the following statement.

Arithmetic Sequence Formulas

In an arithmetic sequence with first term a_1 and common difference d,

1. The nth term is given by

$$a_n = a_1 + (n-1)d;$$

2. The sum of the first n terms is given by

$$S_n = \frac{n}{2}(a_1 + a_n) \quad \text{or} \quad S_n = \frac{n}{2}[2a_1 + (n-1)d].$$

Example 4 • Finding a_{20} and S_{20}

Use the information given for each arithmetic sequence to find the twentieth term and the sum of the first 20 terms.

a. $-3, -6, -9, -12, \ldots$

b. $a_1 = 5, \quad a_{19} = 59$

c. $S_{10} = 145, \quad a_7 = 19$

Solution

a. It is clear that $a_1 = -3$ and $d = -3$. By the formula for the nth term, we have

$$\begin{aligned} a_{20} &= a_1 + 19d \\ &= -3 + (19)(-3) \\ &= -60. \end{aligned}$$

By the second formula for the sum of an arithmetic sequence,

$$\begin{aligned} S_{20} &= \tfrac{20}{2}(2a_1 + 19d) \\ &= 10(-6 - 57) \\ &= -630. \end{aligned}$$

b. Substituting $a_1 = 5$ and $a_{19} = 59$ in the formula for the nth term, we have

$$59 = 5 + 18d \qquad \text{and then} \qquad d = 3.$$

Thus

$$a_{20} = a_{19} + d = 59 + 3 = 62$$

and

$$\begin{aligned} S_{20} &= \tfrac{20}{2}(a_1 + a_{20}) \\ &= 10(5 + 62) \\ &= 670. \end{aligned}$$

c. Using $S_{10} = 145$ in the second formula for the sum, we have

$$145 = \tfrac{10}{2}(2a_1 + 9d),$$

and this gives

$$145 = 10a_1 + 45d. \tag{7}$$

Using $a_7 = 19$ in the nth-term formula, we have

$$19 = a_1 + 6d. \tag{8}$$

To find a_1 and d, we solve Equations (7) and (8) simultaneously. Multiplying both sides of Equation (8) by 10 and rewriting Equation (7), we have the following system of equations.

$$\begin{aligned} 190 &= 10a_1 + 60d \\ 145 &= 10a_1 + 45d \end{aligned}$$

Subtraction and division yield

$$d = 3.$$

Substituting $d = 3$ in Equation (8) yields

$$a_1 = 1.$$

Thus we can compute a_{20} and S_{20} as follows.

$$\begin{aligned} a_{20} &= a_1 + 19d \\ &= 1 + (19)(3) \\ &= 58 \end{aligned} \qquad \begin{aligned} S_{20} &= \tfrac{20}{2}(2a_1 + 19d) \\ &= 10[2 + (19)(3)] \\ &= 590. \end{aligned}$$

Note that this computation of S_{20} is a lot shorter than actually performing the addition $1 + 4 + 7 + 10 + \cdots + 58$, and is less prone to errors in arithmetic. □

Example 5 • Landscaping

A landscaper is beginning work on a giant bed of begonias at the entrance to Space Age Theme Park. Plans call for a bed in the shape of a trapezoid, as shown in Figure 12.2.

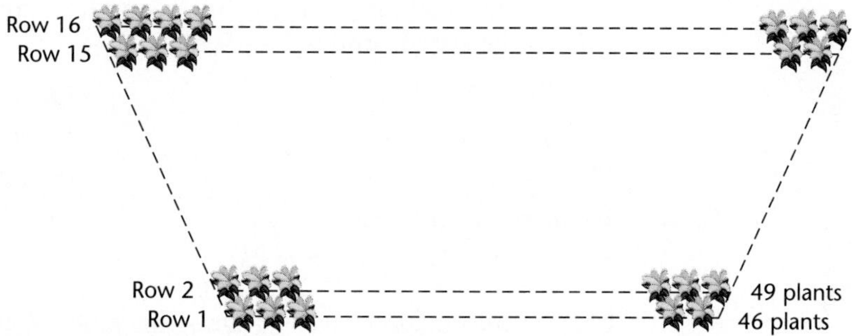

Figure 12.2

There are to be 16 rows of plants with 46 plants in the first row, 49 plants in the second row, and each row from there on with 3 more plants than the preceding row. How many begonia plants are needed for the project?

Solution

The numbers of plants in the rows form an arithmetic sequence with 16 terms, $a_1 = 46$, and $d = 3$. The total number of plants needed is the sum of the 16 terms, given by

$$\begin{aligned} S_{16} &= \tfrac{16}{2}[2(46) + 15(3)] \\ &= 8[92 + 45] \\ &= 1096. \end{aligned}$$

□

EXERCISES 12.2

Use the information given to find the first six terms of an arithmetic sequence that satisfies the stated conditions. (See Example 1 and Practice Problem 1.)

1. $a_1 = 1, \quad d = 1$
2. $a_1 = -3, \quad d = -2$
3. $a_4 = 7, \quad d = -2$
4. $a_4 = -7, \quad d = -5$

5. $a_3 = -5, \quad a_6 = -17$
6. $a_3 = 6, \quad a_4 = 5.9$
7. $a_3 - a_2 = 4, \quad a_4 = 14$
8. $a_5 = -7, \quad a_2 - a_1 = 2$

Determine whether or not each sequence is an arithmetic sequence. If it is, find d and a_n.

9. $-17, -12, -7, -2, \ldots$
10. $5, 2, -1, -4, \ldots$
11. $-6, -9, -12, -15, \ldots$
12. $\frac{5}{2}, 2, \frac{3}{2}, 1, \ldots$
13. $8, 4, 2, 1, \ldots$
14. $2, 4, 8, 16, \ldots$
15. $7, -7, -21, -35, \ldots$
16. $3, 3.04, 3.08, 3.12, \ldots$
17. $x, x + y, x + 2y, x + 3y, \ldots$
18. $-k, -k + x, -k + 2x, -k + 3x, \ldots$
19. $x, x^2, x^3, x^4, \ldots$
20. $x, x/2, x/4, x/8, \ldots$
21. $x, 2x, 3x, 4x, \ldots$
22. $2x, 4x, 6x, 8x, \ldots$

Use the information given to find a_{11} and S_{11} for an arithmetic sequence that satisfies the stated conditions. (See Example 4.)

23. $a_1 = -5, \quad d = -2$
24. $a_1 = 7, \quad a_5 = 15$
25. $a_7 = -5, \quad d = 3$
26. $a_5 = -3, \quad d = -3$
27. $a_5 = 5, \quad a_9 = -7$
28. $a_3 = 10, \quad a_7 = -2$
29. $a_1 = m, \quad d = -x$
30. $a_4 = m, \quad d = -x$
31. $a_5 = x + 2k, \quad a_7 = x - 2k$
32. $a_5 = 3 - 4p, \quad d = p$

Use the information given to find the value of the requested quantities in an arithmetic sequence. (See Examples 2–4 and Practice Problem 2.)

33. $a_1 = -7$, $d = 3$; find a_{21}.
34. $a_1 = 17$, $d = -5$; find a_{17}.
35. $a_3 = -7$, $d = 16$; find S_4
36. $a_{10} = -\frac{5}{2}$, $d = \frac{1}{2}$; find S_{17}.
37. $-4, -1, 2, 5, \ldots$; find a_{10}.
38. $\frac{5}{2}, \frac{3}{2}, \frac{1}{2}, -\frac{1}{2}, \ldots$; find a_{15}.
39. $a_1 = 16$, $a_{17} = -3$; find S_{57}.
40. $a_1 = -3$, $a_{11} = 16$; find S_{200}.
41. $S_5 = -30$, $a_7 = -4$; find a_1 and d.
42. $S_5 = \frac{5}{2}$, $a_7 = -\frac{7}{2}$; find a_1 and d.
43. $S_7 = 14$, $a_{10} = -28$; find a_1 and d.
44. $S_8 = 52$, $a_4 = 5$; find a_1 and d.

Find the sum of each of the following series, using the fact that each is formed from an arithmetic sequence. (See Exercises 53 and 54.)

45. $\sum_{i=5}^{11} (2i - 3)$
46. $\sum_{i=3}^{16} (3i - 2)$
47. $\sum_{k=0}^{10} \left(5 - \frac{k}{3}\right)$
48. $\sum_{k=0}^{20} \left(-3 - \frac{k}{2}\right)$
49. $\sum_{n=1}^{50} 2n$
50. $\sum_{n=1}^{100} n$

51. $\sum_{n=1}^{13} (5 - 2n)$

52. $\sum_{n=1}^{19} (2n - \frac{11}{2})$

53. Show that if $a_1, a_2, a_3, \ldots$ is a sequence that has a defining function f that is linear [that is, $a_n = f(n) = an + b$], then the sequence is an arithmetic sequence.

54. Show that every arithmetic sequence has a defining function that is linear (that is, $a_n = an + b$ for some fixed a and b).

55. Insert numbers in the blanks below so that the sequence forms an arithmetic sequence.

$$-15, \_\_\_, \_\_\_, \_\_\_, \_\_\_, -40$$

56. Insert seven numbers between -5 and -26 so that the sequence thus formed is an arithmetic sequence.

57. Find the sum of the multiples of 5 between 7 and 213.

58. Find the sum of the integers from -13 to -71, inclusive.

59. Find the number n of positive integers between 17 and 199 that are multiples of 5.

60. Find the number n of positive integers between 17 and 199 that are multiples of 3.

Saving Pennies 61. If a child puts 1 penny in her piggy bank one week, 3 pennies the second week, 5 pennies the third week, 7 pennies the fourth week, and so on, how many pennies will she put in the bank the twentieth week?

Total Savings 62. In Exercise 61, how much money will she have in her bank after she puts in her money the twentieth week?

Stock Display 63–64 63. A stock boy is to display cans of tomato soup in a triangular stack like that in the accompanying photograph, with 1 can in the top row, 2 cans in the second row, 3 cans in the third row, and so on. Since he has to

The Photo Works

begin stacking from the bottom, how many cans should he place in the bottom row if he must display a total of 45 cans?

64. In Exercise 63, how many cans should the stock boy place in the bottom row if he must display a total of 66 cans?

Stacking Logs 65. A pile of logs is in the shape of a trapezoid, with 50 logs on the bottom, 49 logs next to the bottom, 48 logs on the next level, and so on, with 21 logs on the top. Find the total number of logs in the pile.

Candy Tree 66. In constructing a candy Christmas tree on a flat plate, Leslie places one candy at the top, Patti places 2 candies just below, then Leslie places 3 candies below these two, and they continue, increasing by one candy at each level. If the bottom row has 27 candies in it, how many candies were used to construct the tree?

Amphitheater Seats 67. A section of seats in an amphitheater has 20 arc-shaped rows of seats as shown in the accompanying figure. The numbers of seats in the rows form an arithmetic sequence with 34 seats in the bottom row and 110 seats in the top row. How many seats are in the section?

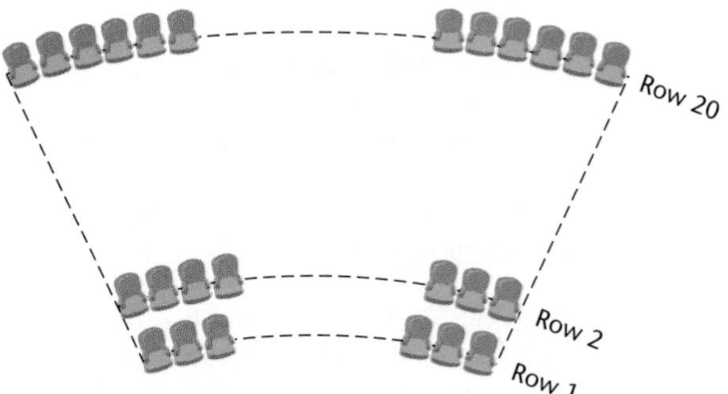

Figure for Exercises 67 and 68

Speedway Seats 68. The seats in a certain section of a NASCAR speedway are in arc-shaped rows, and there are 20 rows as indicated in the accompanying figure. In this section, the bottom row has 52 seats, and each of the other rows has 3 more seats than the next lower row. How many seats are in this section?

Critical Thinking: Discussion 69. Example 5 and Exercises 61–68 involve some realistic situations where objects are arranged in patterns with numbers that form arithmetic sequences. Describe a real-world situation, different from any of these, in which a pattern of this type occurs.

Critical Thinking: Writing 70. Use a meaning from the dictionary for the word *common* and explain why it is appropriate to label the constant d in an arithmetic sequence as the *common* difference.

◆ Solutions for Practice Problems

1. We have

$$\begin{aligned} a_1 &= 5, \\ a_2 &= a_1 + d = 5 + (-2) = 3, \\ a_3 &= a_2 + d = 3 + (-2) = 1, \\ a_4 &= a_3 + d = 1 + (-2) = -1, \\ a_5 &= a_4 + d = -1 + (-2) = -3. \end{aligned}$$

Thus the first five terms of the arithmetic sequence are 5, 3, 1, −1, −3.

2. The first term here is $a_1 = 3$, and the common difference is $d = 1 - 3 = -2$. Thus a_{19} and a_n can be computed as follows.

$$\begin{aligned} a_{19} &= a_1 + 18d \\ &= 3 + (18)(-2) \\ &= -33 \end{aligned} \qquad \begin{aligned} a_n &= a_1 + (n-1)d \\ &= 3 + (n-1)(-2) \\ &= 5 - 2n. \end{aligned}$$

12.3 GEOMETRIC SEQUENCES

In this section, we consider another special type of sequence called a geometric sequence.

Definition of a Geometric Sequence

A **geometric sequence**† is a sequence in which each term after the first is obtained by multiplying the preceding term by a fixed nonzero number r. The constant r is called the **common ratio.**

A geometric sequence with first term a_1 and common ratio r is given by

$$a_1, \quad a_1r, \quad a_1r^2, \quad a_1r^3, \quad \ldots .$$

Example 1 • Writing Terms of a Geometric Sequence

Find the first four terms of the geometric sequence that has the first term and common ratio as specified below.

a. $a_1 = -3, \quad r = -2$ b. $a_1 = 16, \quad r = \frac{1}{4}$

†A geometric sequence is also called a **geometric progression.**

Solution

a. The first four terms of the geometric sequence are computed as follows.

$$a_1 = -3$$
$$a_2 = a_1 r = (-3)(-2) = 6$$
$$a_3 = a_1 r^2 = (-3)(-2)^2 = -12$$
$$a_4 = a_1 r^3 = (-3)(-2)^3 = 24$$

Thus the first four terms of the geometric sequence are $-3, 6, -12, 24$.

b. This time we illustrate another method of computation. We obtain the terms after the first one by multiplying by r.

$$a_1 = 16$$
$$a_2 = a_1 r = (16)(\tfrac{1}{4}) = 4$$
$$a_3 = a_2 r = (4)(\tfrac{1}{4}) = 1$$
$$a_4 = a_3 r = (1)(\tfrac{1}{4}) = \tfrac{1}{4}$$

Thus the first four terms are given by $16, 4, 1, \frac{1}{4}$. □

From the pattern in the terms

$$a_1, \quad a_1 r, \quad a_1 r^2, \quad a_1 r^3, \quad a_1 r^4, \ldots,$$

of a geometric sequence, it is easy to see that the nth term in a geometric sequence is given by the formula

$$\boldsymbol{a_n = a_1 r^{n-1}}. \tag{1}$$

Example 2 • Finding a Formula for the nth Term

Find the sixth term and a formula for the nth term of the following geometric sequence.

$$1, 2, 4, 8, \ldots$$

Solution

To find r, we need only choose one of the terms, other than the first, and divide it by the preceding term. Using the third and second terms, we obtain $r = \frac{4}{2} = 2$. Since $a_1 = 1$, we have

$$a_6 = a_1 r^5 = (1)(2)^5 = 32$$

and

$$a_n = a_1 r^{n-1} = (1)(2)^{n-1} = 2^{n-1}.$$ □

◆ Practice Problem 1 ◆

Find the sixth term and a formula for the nth term of the geometric sequence with $a_1 = 4$, $r = -\frac{1}{2}$.

Example 3 • Finding a_1 and r

A geometric sequence has $a_4 = -3$ and $a_7 = -24$. Find the first term a_1 and the common ratio r.

Solution

Using the formula $a_n = a_1 r^{n-1}$ with $a_4 = -3$ and $a_7 = -24$, we have the system of equations:

$$\begin{aligned} -3 &= a_1 r^3 \quad \text{fourth term} \\ -24 &= a_1 r^6 \quad \text{seventh term} \end{aligned}$$

We can solve for r by forming quotients of corresponding sides in these two equations.

$$\frac{a_1 r^6}{a_1 r^3} = \frac{-24}{-3}$$

This gives

$$r^3 = 8 \qquad \text{and hence} \qquad r = 2.$$

Substituting $r = 2$ and $a_4 = -3$ in $a_4 = a_1 r^3$, we have

$$-3 = a_1(2)^3 \qquad \text{and then} \qquad a_1 = -\tfrac{3}{8}.$$ □

As was the case with arithmetic sequences, we can find a formula for the series formed by adding the first n terms of a general geometric sequence. To obtain the formula, let

$$S_n = a_1 + a_1 r + a_1 r^2 + \cdots + a_1 r^{n-3} + a_1 r^{n-2} + a_1 r^{n-1}. \tag{2}$$

When both sides of this equation are multiplied by r, we have

$$rS_n = a_1 r + a_1 r^2 + a_1 r^3 + \cdots + a_1 r^{n-2} + a_1 r^{n-1} + a_1 r^n. \tag{3}$$

Subtracting Equation (3) from Equation (2) yields

$$S_n - rS_n = a_1 - a_1 r^n$$

or

$$(1 - r)S_n = a_1(1 - r^n).$$

If $r \neq 1$, both sides of this equation can be divided by $1 - r$ to obtain

$$\mathbf{S_n = a_1 \frac{1 - r^n}{1 - r}, \qquad \text{if } r \neq 1.} \tag{4}$$

This formula cannot be used if $r = 1$, but Equation (2) easily gives $S_n = na_1$ for $r = 1$. For easy reference, the formulas in Equations (1) and (4) are stated in the following theorem.

Geometric Sequence Formulas

In a geometric sequence with first term a_1 and common ratio r,

1. The nth term is

$$a_n = a_1 r^{n-1}.$$

2. The sum of the first n terms is

$$S_n = a_1 \frac{1 - r^n}{1 - r} \qquad \text{if } r \neq 1.$$

Example 4 • Finding S_6

Find the sum of the first six terms in the geometric sequence

$$-\tfrac{1}{3}, -\tfrac{1}{9}, -\tfrac{1}{27}, -\tfrac{1}{81}, \ldots .$$

Solution

Using $a_1 = -\frac{1}{3}$ and $r = \frac{1}{3}$ in the formula for the sum, we have

$$\begin{aligned} S_6 &= -\frac{1}{3} \cdot \frac{1 - (\frac{1}{3})^6}{1 - \frac{1}{3}} \\ &= -\frac{1}{3} \cdot \frac{1 - \frac{1}{729}}{1 - \frac{1}{3}} \\ &= -\frac{1}{3} \cdot \frac{\frac{728}{729}}{\frac{2}{3}} \\ &= -\frac{364}{729}. \end{aligned}$$

□

Example 5 • Using Sigma Notation

Use the geometric sequence formulas to find the value of

$$\sum_{i=1}^{5} 3 \cdot 2^i.$$

Solution

Expanding the sum, we have

$$\sum_{i=1}^{5} 3 \cdot 2^i = 3 \cdot 2 + 3 \cdot 2^2 + 3 \cdot 2^3 + 3 \cdot 2^4 + 3 \cdot 2^5.$$

From this expansion, it is clear that the terms in the sum form a geometric sequence with $a_1 = 3 \cdot 2 = 6$ and $r = 2$. By the formula for the sum of a geometric sequence,

$$\sum_{i=1}^{5} 3 \cdot 2^i = 6 \cdot \frac{1 - 2^5}{1 - 2}$$
$$= 6 \cdot \frac{-31}{-1}$$
$$= 186.$$

This use of the formula is a lot shorter than doing the addition. □

EXERCISES 12.3

For the given value of n, write out the first n terms of the geometric sequence that satisfies the stated conditions. (See Example 1.)

1. $a_1 = -1, \quad r = -2, \quad n = 5$
2. $a_1 = 3, \quad r = \frac{1}{2}, \quad n = 4$
3. $a_1 = 4, \quad r = \frac{1}{4}, \quad n = 4$
4. $a_1 = -6, \quad r = 2, \quad n = 5$
5. $a_1 = \frac{3}{4}, \quad r = 4, \quad n = 4$
6. $a_1 = -\frac{1}{8}, \quad r = -2, \quad n = 5$
7. $a_2 = -\frac{1}{2}, \quad a_3 = 1, \quad n = 5$
8. $a_3 = \frac{2}{3}, \quad a_4 = 1, \quad n = 4$
9. $a_1 = -3, \quad a_3 = -12, \quad n = 3$
10. $a_1 = 4, \quad a_3 = \frac{1}{16}, \quad n = 4$

Find the fifth term, the nth term, and the sum of the first five terms of a geometric sequence that satisfies the stated conditions. (See Examples 2 and 4 and Practice Problem 1.)

11. $\frac{1}{4}, \frac{1}{2}, 1, 2, \ldots$
12. $-1, -\sqrt{3}, -3, -3\sqrt{3}, \ldots$
13. $a_1 = \frac{1}{3}, \quad r = -3$
14. $a_1 = 4, \quad r = \frac{1}{4}$
15. $a_3 = -2, \quad a_4 = 4$
16. $a_3 = -3, \quad a_4 = \frac{3}{2}$

Determine whether or not each of the following sequences is a geometric sequence. If it is, find r and a_n.

17. $7, \frac{7}{2}, \frac{7}{4}, \frac{7}{8}, \ldots$
18. $\sqrt{3}, 3, 3\sqrt{3}, 9, \ldots$
19. $4, 2\sqrt{2}, 2, \sqrt{2}, \ldots$
20. $\frac{5}{6}, \frac{5}{3}, \frac{10}{3}, \frac{20}{3}, \ldots$
21. $2, -4, 6, -8, \ldots$
22. $1, 3, 7, 15, \ldots$
23. $1, 4, 8, 12, \ldots$
24. $3, 6, 18, 108, \ldots$
25. $-4, 2, -1, \frac{1}{2}, \ldots$
26. $-5, \frac{5}{3}, -\frac{5}{9}, \frac{5}{27}, \ldots$
27. $343, -49, 7, -1, \ldots$
28. $\frac{1}{9}, -\frac{1}{3}, 1, -3, \ldots$

Find the sums of each of the following series, using the fact that each is the sum of a geometric sequence. (See Example 5.)

29. $\sum_{i=1}^{5} 2^{i-1}$
30. $\sum_{n=0}^{5} 4^n$
31. $\sum_{i=1}^{5} (\frac{3}{5})^i$
32. $\sum_{n=4}^{9} \frac{1}{3^n}$

33. $\sum_{n=0}^{4} 5(-\frac{2}{3})^n$

34. $\sum_{j=1}^{6} 128(-\frac{3}{2})^j$

35. $\sum_{j=0}^{4} 64(\frac{5}{4})^j$

36. $\sum_{n=2}^{7} 9(\frac{5}{3})^{n-2}$

37. $\sum_{j=1}^{6} 8 \cdot 2^{1-j}$

38. $\sum_{n=2}^{6} 32(\frac{1}{2})^{n-1}$

Find r and a_1 for the geometric sequence that satisfies the given conditions. (See Example 3.)

39. $a_7 = -9, \quad a_{11} = -81$

40. $a_4 = -\frac{2}{3}, \quad a_7 = -\frac{9}{4}$

41. $a_6 = 4(1.01)^4, \quad a_8 = 4(1.01)^6$

42. $a_4 = 2, \quad a_8 = \frac{2}{81}$

Saving Money

43. Adrienne decided to save money to buy a new television set. She saved one dollar the first day, and on each day following she saved twice as much as she had the day before. To her surprise, she had saved exactly enough to buy the set at the end of the ninth day. How much did the television set cost?

Population Growth

44. It is known that the population of a country in 1920 was 10,000,000 people. If the population doubles every 20 years, what will the population be in the year 2000?

Keeping Pace with Inflation

45. At a certain rate of inflation, the cost of living doubles every 6 years. If a person earns $42,000 now, how much must she earn 36 years from now to keep her salary in pace with inflation?

Bacteria Growth

46. The number of bacteria in a culture is observed to triple every 6 hours. If initially there are 1000 bacteria, approximately how long will it take for 1,000,000 bacteria to be present?

Critical Thinking: Discussion

47. Each of the situations in Exercises 43–46 involves a geometric sequence. Describe a real-world situation, different from any of these, that also involves a geometric sequence.

Critical Thinking: Writing

48. Use a meaning from the dictionary for the word *ratio* and explain why it is appropriate to label the constant r in a geometric sequence as the common *ratio.*

◆ **Solution for Practice Problem**

1. Substituting $a_1 = 4$ and $r = -\frac{1}{2}$ in the formula for a_n, we have

$$a_6 = (4)\left(-\frac{1}{2}\right)^5 = -\frac{1}{8}$$

and

$$a_n = (4)\left(-\frac{1}{2}\right)^{n-1} = \frac{4}{(-2)^{n-1}} = \frac{(-2)^2}{(-2)^{n-1}} = (-2)^{3-n}.$$

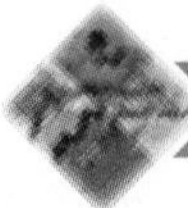

12.4 INFINITE GEOMETRIC SEQUENCES

Up to this point, we have considered only series that are formed from a finite sequence. In this section, we examine some cases where the sum of an infinite series has meaning.

Consider an infinite geometric sequence with first term a_1 and common ratio r.

$$a_1, \quad a_1r, \quad a_1r^2, \quad a_1r^3, \quad \ldots$$

We have seen in Section 12.3 that the sum S_n of the first n terms of this sequence is given by

$$\begin{aligned} S_n &= a_1 + a_1r + a_1r^2 + \cdots + a_1r^{n-1} \\ &= a_1\frac{1 - r^n}{1 - r}, \qquad \text{if } r \neq 1. \end{aligned}$$

If $|r| < 1$, that is, if $-1 < r < 1$, the term r^n steadily decreases in absolute value as n increases, getting closer and closer to zero. The fact that r^n gets nearer and nearer to zero as n takes on larger and larger values suggests that the sums S_n should themselves be getting closer and closer to some certain value S. This is actually what happens, and we write

$$\lim_{n\to\infty} S_n = S$$

to indicate that the sums S_n get closer and closer to S as n increases without bound.† The symbol $\lim_{n\to\infty} S_n$ is read "limit of S_n as n approaches infinity."

To illustrate how the sums S_n behave when $|r| < 1$, consider the particular case where $a_1 = 1$ and $r = \frac{1}{2}$. The geometric sequence is given by

$$1, \tfrac{1}{2}, \tfrac{1}{4}, \tfrac{1}{8}, \tfrac{1}{16}, \ldots,$$

†The concept of a limit is the fundamental concept of the calculus. Limits are treated there with a great deal of care and rigor.

and the sum of the first n terms is

$$S_n = 1 + \frac{1}{2} + \frac{1}{4} + \cdots + \frac{1}{2^{n-1}}$$
$$= 1 \cdot \frac{1 - (\frac{1}{2})^n}{1 - \frac{1}{2}}$$
$$= 2[1 - (\tfrac{1}{2})^n].$$

As n takes on the values 1, 2, 3, 4, 5, 6, . . . in succession, the corresponding values of S_n are given by

$$S_1 = 1 = 2(1 - \tfrac{1}{2}) = 2 - 1,$$
$$S_2 = 1 + \tfrac{1}{2} = 2(1 - \tfrac{1}{4}) = 2 - \tfrac{1}{2},$$
$$S_3 = 1 + \tfrac{1}{2} + \tfrac{1}{4} = 2(1 - \tfrac{1}{8}) = 2 - \tfrac{1}{4},$$
$$S_4 = 1 + \tfrac{1}{2} + \tfrac{1}{4} + \tfrac{1}{8} = 2(1 - \tfrac{1}{16}) = 2 - \tfrac{1}{8},$$
$$S_5 = 1 + \tfrac{1}{2} + \tfrac{1}{4} + \tfrac{1}{8} + \tfrac{1}{16} = 2(1 - \tfrac{1}{32}) = 2 - \tfrac{1}{16},$$
$$S_6 = 1 + \tfrac{1}{2} + \tfrac{1}{4} + \tfrac{1}{8} + \tfrac{1}{16} + \tfrac{1}{32} = 2(1 - \tfrac{1}{64}) = 2 - \tfrac{1}{32}.$$

From these computations it can be seen that as n increases without bound, the sums S_n get closer and closer to the value 2. Thus

$$\lim_{n \to \infty} S_n = 2$$

for this geometric sequence. We say that 2 is the **sum** of the infinite geometric sequence, and we write

$$1 + \tfrac{1}{2} + \tfrac{1}{4} + \tfrac{1}{8} + \cdots = 2,$$

even though it is impossible to find the sum by actually performing the indicated additions. Another notation that is commonly used is

$$\sum_{n=1}^{\infty} \frac{1}{2^{n-1}} = 2.$$

The development in the preceding case can be carried out for any geometric sequence with $|r| < 1$. Since r^n gets closer and closer to zero as n increases without bound, we write

$$\lim_{n \to \infty} r^n = 0.$$

With this fact in mind, it is clear that

$$\lim_{n \to \infty} S_n = \lim_{n \to \infty} a_1 \frac{1 - r^n}{1 - r}$$
$$= a_1 \frac{1 - 0}{1 - r}$$
$$= \frac{a_1}{1 - r}.$$

If $|r| > 1$ and $a_1 \neq 0$, the terms $a_1 r^n$ increase in absolute value without bound as n increases without bound, and the sums S_n do also. If $|r| = 1$ and $a_1 \neq 0$, it can be shown that $\lim_{n\to\infty} S_n$ does not exist. (See Exercise 53.) This discussion is summarized in the following statements.

Infinite Geometric Series

For an infinite geometric sequence $a_1, a_1 r, a_1 r^2, \ldots,$ let

$$S_n = a_1 + a_1 r + a_1 r^2 + \cdots + a_1 r^{n-1}.$$

1. If $|r| < 1$, the sum $a_1 + a_1 r + a_1 r^2 + \cdots$ of the infinite geometric series is given by

$$S = \lim_{n\to\infty} S_n = \frac{a_1}{1-r}.$$

2. If $|r| \geq 1$ and $a_1 \neq 0$, the sum of the series $\sum_{n=1}^{\infty} a_1 r^{n-1}$ does not exist.

In the case where $|r| < 1$, we also write

$$a_1 + a_1 r + a_1 r^2 + \cdots = \frac{a_1}{1-r} \quad \text{or} \quad \sum_{n=1}^{\infty} a_1 r^{n-1} = \frac{a_1}{1-r}.$$

Example 1 • Finding the Sum

For each of the following geometric sequences, determine whether or not the sum exists. If the sum exists, find its value.

a. $\frac{3}{5}, \frac{1}{5}, \frac{1}{15}, \frac{1}{45}, \ldots$ b. $\frac{2}{3}, 1, \frac{3}{2}, \frac{9}{4}, \ldots$

Solution

a. The common ratio is given by

$$r = \frac{\frac{1}{5}}{\frac{3}{5}} = \frac{1}{3}.$$

Since $|r| < 1$, the sum exists, and the formula gives

$$\begin{aligned} \frac{3}{5} + \frac{1}{5} + \frac{1}{15} + \frac{1}{45} + \cdots &= \frac{a_1}{1 - \mathrm{r}} \\ &= \frac{\frac{3}{5}}{1 - \frac{1}{3}} \\ &= \frac{9}{10}. \end{aligned}$$

b. It is easy to see that $r = \frac{3}{2}$. Since $|r| > 1$, the sum does not exist, by statement 2 for infinite geometric series. □

◆ Practice Problem 1 ◆

For each of the following geometric sequences, determine whether or not the sum exists. If the sum exists, find its value.

a. $\frac{9}{16}, \frac{3}{4}, 1, \frac{4}{3}, \ldots$ b. $9, -6, 4, -\frac{8}{3}, \ldots$

The formula for the sum of an infinite geometric series can be used to write a number with a repeating decimal as a quotient of integers.

Example 2 • An Infinite Repeating Decimal

Write the rational number $0.131313\ldots$ as a quotient of integers.

Solution

The given number can be written as an infinite geometric series in the following manner.

$$0.131313\ldots = 0.13 + 0.0013 + 0.000013 + \cdots$$

The terms in the sum on the right are from the geometric sequence with $a_1 = 0.13$ and $r = 0.01$. The formula for the sum gives

$$\begin{aligned} 0.131313\ldots &= \frac{0.13}{1 - 0.01} \\ &= \frac{0.13}{0.99} \\ &= \frac{13}{99}. \end{aligned}$$

□

Example 3 • An Infinite Repeating Decimal

Write the rational number $3.4173173173\ldots$ as a quotient of integers.

Solution

We first write the number as follows.

$$3.4173173173\ldots = 3.4 + 0.0173 + 0.0000173 + \cdots$$

The terms on the right after 3.4 are from the geometric sequence with $a_1 = 0.0173$ and $r = 0.001$. Thus

$$\begin{aligned} 0.0173 + 0.0000173 + \cdots &= \frac{0.0173}{1 - 0.001} \\ &= \frac{0.0173}{0.999} \\ &= \frac{173}{9990}. \end{aligned}$$

This gives

$$\begin{aligned} 3.4173173173\ldots &= 3.4 + \frac{173}{9990} \\ &= \frac{34}{10} + \frac{173}{9990} \\ &= \frac{33{,}966 + 173}{9990} \\ &= \frac{34{,}139}{9990}. \end{aligned}$$

This result may be checked by division. □

EXERCISES 12.4

For each of the following geometric sequences, determine whether or not the sum exists. If the sum exists, find its value. (See Example 1 and Practice Problem 1.)

1. $-3, \frac{3}{2}, -\frac{3}{4}, \frac{3}{8}, \ldots$
2. $-4, \frac{4}{3}, -\frac{4}{9}, \frac{4}{27}, \ldots$
3. $\frac{9}{2}, -\frac{3}{2}, \frac{1}{2}, -\frac{1}{6}, \ldots$
4. $\frac{5}{9}, \frac{1}{9}, \frac{1}{45}, \frac{1}{225}, \ldots$
5. $-\frac{4}{3}, 4, -12, 36, \ldots$
6. $8, -4, 2, -1, \ldots$
7. $1, 1.01, (1.01)^2, (1.01)^3, \ldots$
8. $1, 1.2, 1.44, 1.728, \ldots$
9. $2, \sqrt{2}, 1, 1/\sqrt{2}, \ldots$
10. $-0.9, 0.81, -0.729, 0.6561, \ldots$
11. $5, 0.5, 0.05, 0.005, \ldots$
12. $5, 2.5, 1.25, 0.625, \ldots$
13. $\frac{5}{3}, \frac{10}{9}, \frac{20}{27}, \frac{40}{81}, \ldots$
14. $\frac{4}{9}, \frac{2}{3}, 1, \frac{3}{2}, \ldots$

Find the value of each of the following sums, if it exists. If it does not exist, give a reason. (See Example 1 and Practice Problem 1.)

15. $15 + \frac{15}{2} + \frac{15}{4} + \frac{15}{8} + \cdots$
16. $52 + 4 + \frac{4}{13} + \frac{4}{169} + \cdots$
17. $4 - \frac{4}{3} + \frac{4}{9} - \frac{4}{27} + \cdots$
18. $-17 + \frac{17}{3} - \frac{17}{9} + \frac{17}{27} - \cdots$
19. $1 - 1.02 + (1.02)^2 - (1.02)^3 + \cdots$
20. $5 + 10 + 20 + 40 + \cdots$
21. $3 - \dfrac{3}{0.99} + \dfrac{3}{(0.99)^2} - \dfrac{3}{(0.99)^3} + \cdots$
22. $0.25 + 0.50 + 1 + 2 + \cdots$
23. $512 + 64 + 8 + 1 + \cdots$
24. $\frac{4}{3} + \frac{1}{3} + \frac{1}{12} + \frac{1}{48} + \cdots$
25. $\frac{1}{36} + \frac{1}{6} + 1 + 6 + \cdots$
26. $\frac{1}{16} + \frac{1}{4} + 1 + 4 + \cdots$
27. $\sum_{i=0}^{\infty} 5(\frac{1}{2})^i$
28. $\sum_{i=2}^{\infty} (-15)(\frac{2}{3})^i$
29. $\sum_{i=3}^{\infty} 17(-3)^i$
30. $\sum_{n=4}^{\infty} 7(2)^n$
31. $\sum_{i=1}^{\infty} \dfrac{13}{4^i}$
32. $\sum_{n=1}^{\infty} \dfrac{8}{5^n}$

33. $\sum_{n=5}^{\infty} 8(-\frac{1}{5})^{n-4}$

34. $\sum_{n=3}^{\infty} (-7)(-\frac{4}{3})^{n-2}$

35. $\sum_{n=2}^{\infty} 6(\frac{1}{3})^{n}$

36. $\sum_{n=2}^{\infty} 4(\frac{3}{4})^{n}$

Express each rational number as a quotient of integers. (See Examples 2 and 3.)

37. 0.9999 . . .
38. 0.1111 . . .
39. 0.010101 . . .
40. 0.313131 . . .
41. 0.013101310131 . . .
42. 0.171171171 . . .
43. 3.111111 . . .
44. 3.8787878 . . .
45. −2.2917917 . . .
46. −9.01727272 . . .
47. 10.1343434 . . .
48. 0.79191919 . . .

Distance Traveled by a Ball 49–50

49. Mary has a ball that will bounce back to three-sevenths of the height from which it is dropped. If she drops the ball from a height of 4 feet, find the approximate vertical distance it will travel before coming to rest. (See the accompanying figure.)

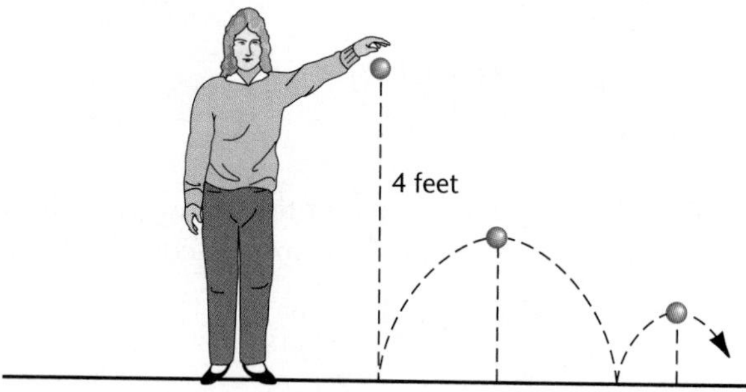

Figure for Exercise 49

50. Work Exercise 49 for a ball that bounces back each time to one-third of the height from which it falls.

Bacteria Growth

51. Suppose that a culture of bacteria doubles every 20 hours. If the culture initially contains 4000 bacteria, how many are present after 120 hours?

Doubling Interest

52. A loan shark offers to compute interest by doubling the amount of interest owed every week. If a customer owes him $50 in interest today, how much interest would he owe at the end of 10 weeks under this scheme?

Critical Thinking: Discussion

53. Discuss the values of the sums S_n in a geometric sequence with $a_1 \neq 0$ and $|r| = 1$. Explain why $\lim_{n\to\infty} S_n$ does not exist for a geometric sequence of this type.

Critical Thinking: Exploration, Discussion and Writing

54. Determine the largest possible number of digits that might occur in the repeating part of the decimal expansion of a rational number m/n with positive integers m and n in lowest terms and give an explanation for your answer.

◆ Solution for Practice Problem

1. a. The ratio is $r = \frac{4}{3}$. Since $|r| > 1$, the sum does not exist.

   b. The common ratio is

$$r = \frac{-6}{9} = -\frac{2}{3},$$

so the sum exists and

$$S = \frac{a_1}{1 - r} = \frac{9}{1 - (-\frac{2}{3})} = \frac{27}{5}.$$

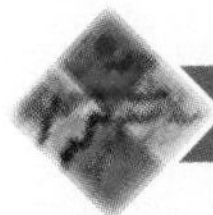

12.5 THE BINOMIAL THEOREM

In this section, a formula is presented for expanding a positive integral power $(a + b)^n$ of a binomial $a + b$ into a sum of terms. The statement of the formula is called the **binomial theorem.**

The **factorial** notation described in the next definition is needed in the statement of the binomial theorem.

Definition of *n* factorial

If n is a positive integer, then ***n* factorial,** denoted by $n!$, is the product of the n integers $n, n - 1, n - 2, \ldots, 2, 1$; that is,

$$n! = n \cdot (n - 1) \cdot (n - 2) \cdots 2 \cdot 1.$$

Also,

$$0! = 1.$$

The reason for defining $0! = 1$ will be clear after the binomial theorem is stated.

Example 1 • Using the Factorial Notation

Some illustrations of the use of the factorial notation are given below.

a. $3! = 3 \cdot 2 \cdot 1 = 6$
b. $7! = 7 \cdot 6 \cdot 5 \cdot 4 \cdot 3 \cdot 2 \cdot 1 = 5040$
c. $1! = 1$
d. $(n - 1)! = (n - 1)(n - 2)(n - 3) \cdots (2)(1)$
e. $n(n - 1)! = n(n - 1)(n - 2)(n - 3) \cdots (2)(1) = n!$ □

Many calculators have a feature that will evaluate a factorial directly. This is especially convenient when large numbers are involved.

When n is a positive integer, $(a + b)^n$ represents a product of n factors, with each factor being $(a + b)$. By direct multiplication, we have the following.

$$(a + b)^1 = a + b$$
$$(a + b)^2 = a^2 + 2ab + b^2$$
$$(a + b)^3 = a^3 + 3a^2b + 3ab^2 + b^3$$
$$(a + b)^4 = a^4 + 4a^3b + 6a^2b^2 + 4ab^3 + b^4$$
$$(a + b)^5 = a^5 + 5a^4b + 10a^3b^2 + 10a^2b^3 + 5ab^4 + b^5$$

By examination of these expansions, we can make the following observations about $(a + b)^n$.

1. There are $n + 1$ terms.
2. The highest power of a and b is n.
3. In each term, the sum of the powers of a and b is n.
4. As the power of a decreases in each successive term, the power of b increases.

The preceding expansions illustrate the fact that the terms in the expansion of $(a + b)^n$ are terms involving products of the form

$$a^n, a^{n-1}b, a^{n-2}b^2, \ldots, a^2b^{n-2}, ab^{n-1}, b^n.$$

In general, the coefficient of $a^{n-r}b^r$ in the expansion of $(a + b)^n$ is given by $\binom{n}{r}$, where

$$\binom{n}{r} = \frac{n!}{(n - r)!r!},$$

The numbers $\binom{n}{r}$, $r = 0, 1, 2, \ldots, n$, are called the **binomial coefficients.** We can now state the binomial theorem.

The Binomial Theorem

Let n be a positive integer, and let a and b be real or complex numbers. Then $(a + b)^n$ is a sum of $n + 1$ terms of the form

$$\binom{n}{r}a^{n-r}b^r, \qquad r = 0, 1, 2, \ldots, n.$$

More specifically,

$$(a + b)^n = a^n + \binom{n}{1}a^{n-1}b + \binom{n}{2}a^{n-2}b^2 + \cdots + \binom{n}{r}a^{n-r}b^r + \cdots + \binom{n}{n-2}a^2b^{n-2} + \binom{n}{n-1}ab^{n-1} + b^n.$$

We omit the proof of the binomial theorem here and concentrate on some examples that illustrate its use.

We note that with the definition $0! = 1$, the formula $\binom{n}{r}a^{n-r}b^r$ works for *all* the terms, including those with $r = 0$ and $r = n$.

Example 2 • A Binomial Expansion

Expand $(x^2 - 2)^6$.

Solution

Using the binomial theorem, we first write $(x^2 - 2)^6$ as $[x^2 + (-2)]^6$ and

$$\begin{aligned}[x^2 + (-2)]^6 &= (x^2)^6 + \binom{6}{1}(x^2)^5(-2) + \binom{6}{2}(x^2)^4(-2)^2 + \binom{6}{3}(x^2)^3(-2)^3 \\ &\quad + \binom{6}{4}(x^2)^2(-2)^4 + \binom{6}{5}(x^2)(-2)^5 + (-2)^6 \\ &= x^{12} + 6 \cdot x^{10} \cdot (-2) + 15 \cdot x^8 \cdot (4) + 20 \cdot x^6 \cdot (-8) \\ &\quad + 15 \cdot x^4 \cdot (16) + 6 \cdot x^2 \cdot (-32) + 64.\end{aligned}$$

Thus

$$(x^2 - 2)^6 = x^{12} - 12x^{10} + 60x^8 - 160x^6 + 240x^4 - 192x^2 + 64. \qquad \square$$

The binomial theorem can also be used to determine the coefficient of a particular term in the expansion of $(a + b)^n$.

Example 3 • Finding the Coefficient of a Term

Find the coefficient of t^8s^9 in the expansion of $(t^2 - s)^{13}$.

Solution

The terms in the expansion of $(t^2 - s)^{13}$ are of the form

$$\binom{n}{r}(t^2)^{n-r}(-s)^r.$$

The coefficient required is in the term with $r = 9$ and $n - r = 13 - 9 = 4$. Therefore, the term in the expansion is

$$\binom{13}{9}(t^2)^4(-s)^9,$$

and the required coefficient is

$$-\binom{13}{9} = -\frac{13!}{4!9!} = -715.$$

□

Notice in the binomial expansion that the power of b is always 1 less than the term number. For example, b^1 occurs in term number 2, b^2 occurs in term number $3, \ldots, b^n$ occurs in term number $n + 1$. Thus b^{r-1} occurs in the rth term. Since the sum of the powers of a and b must be equal to n, then the corresponding power of a is $n - (r - 1)$. Knowing the power of b leads to the correct selection of the binomial coefficient, $\binom{n}{r-1}$. Thus we have the following formula.

Formula for the rth term

The rth term $(1 \leq r \leq n + 1)$ in the expansion of $(a + b)^n$ is

$$\binom{n}{r-1}a^{n-r+1}b^{r-1}.$$

Example 4 • Finding a Particular Term

Find the twelfth term in the expansion of $(3x^2 - y^3)^{15}$.

Solution

The value of n is 15, and $r - 1$ is 11 in the twelfth term. Thus the twelfth term is given by

$$\begin{aligned}\binom{15}{11}(3x^2)^4(-y^3)^{11} &= -\frac{15!}{4!11!}3^4x^8y^{33}\\ &= -\frac{15 \cdot 14 \cdot 13 \cdot 12 \cdot 11!}{4 \cdot 3 \cdot 2 \cdot 1 \cdot 11!} \cdot 81x^8y^{33}\\ &= -110{,}565x^8y^{33}.\end{aligned}$$

□

It is shown in more advanced mathematics courses that the binomial theorem is valid under certain conditions when n is any positive or negative rational number. Before we formulate these conditions, we make the following observation. The quantity $\binom{n}{r}$ can be rewritten as

$$\begin{aligned}\binom{n}{r} &= \frac{n!}{(n-r)!r!} \\ &= \frac{n(n-1)(n-2)\cdots(n-r+1)(n-r)!}{(n-r)!r!} \\ &= \frac{n(n-1)(n-2)\cdots(n-r+1)}{r!}.\end{aligned}$$

The binomial expansion can now be restated as

$$\begin{aligned}(a+b)^n = a^n + na^{n-1}b + \frac{n(n-1)}{2!}a^{n-2}b^2 + \frac{n(n-1)(n-2)}{3!}a^{n-3}b^3 \\ + \cdots + \frac{n(n-1)(n-2)\cdots(n-r+1)}{r!}a^{n-r}b^r + \cdots.\end{aligned}$$

If n is not a positive integer, the binomial expansion never terminates, but continues indefinitely since $n - r$ is never zero. In the calculus, it is shown that if $a = 1$ and $b = x$, then the binomial expansion for

$$(a+b)^n = (1+x)^n$$

is valid for all x such that $|x| < 1$.

Example 5 • A Binomial with a Fractional Exponent

Find the first five terms of the expansion of $(1 + x)^{1/2}$.

Solution

The expansion is valid for all x such that $|x| < 1$. We have

$$\begin{aligned}(1+x)^{1/2} &= 1 + \frac{1}{2}\cdot x + \frac{(\frac{1}{2})(\frac{1}{2}-1)}{2!}x^2 + \frac{\frac{1}{2}(\frac{1}{2}-1)(\frac{1}{2}-2)}{3!}x^3 \\ &\quad + \frac{(\frac{1}{2})(\frac{1}{2}-1)(\frac{1}{2}-2)(\frac{1}{2}-3)}{4!}x^4 + \cdots \\ &= 1 + \frac{1}{2}x - \frac{1}{8}x^2 + \frac{1}{16}x^3 - \frac{5}{128}x^4 + \cdots.\end{aligned}$$

□

◆ Practice Problem 1 ◆

Approximate $(1.02)^7$ by using the first four terms in the binomial expansion.

Courtesy of the Bettmann Archive

Blaise Pascal

We close this section with an interesting pattern that emerges when only the coefficients of $(a + b)^n$ for $n = 0, 1, 2, \ldots$ are examined.

$(a + b)^n$	*Coefficients of* $(a + b)^n$
$(a + b)^0$	1
$(a + b)^1$	1 1
$(a + b)^2$	1 2 1
$(a + b)^3$	1 3 3 1
$(a + b)^4$	1 **4** **6** 4 1
$(a + b)^5$	1 5 **10** 10 5 1

This triangular pattern is called **Pascal's triangle;** it is named for the French mathematician Blaise Pascal (1623–1662). Notice that once the pattern is begun, it can be continued by adding two consecutive coefficients together in one row to obtain a coefficient in the next row.

Example 6 • Using Pascal's Triangle

Use Pascal's triangle to write out the expansion of $(x - 2y)^4$.

Solution

The coefficients for the expansion of $(x - 2y)^4$ come from row 5 of Pascal's triangle.

$$\begin{aligned}(x - 2y)^4 &= \mathbf{1}x^4 + \mathbf{4}x^3(-2y)^1 + \mathbf{6}x^2(-2y)^2 + \mathbf{4}x(-2y)^3 + \mathbf{1}(-2y)^4 \\ &= x^4 + 4(-2)x^3y + 6(4)x^2y^2 + 4(-8)xy^3 + 1(16)y^4 \\ &= x^4 - 8x^3y + 24x^2y^2 - 32xy^3 + 16y^4\end{aligned}$$

□

EXERCISES 12.5

Expand by using the binomial theorem. (See Examples 2 and 6.)

1. $(x + y)^7$
2. $(a + b)^8$
3. $(2x + y)^4$
4. $(x + 3y)^6$
5. $(x^2 + 2y)^5$
6. $(3x + y^2)^3$
7. $(x^2 - y^2)^4$
8. $(a^3 - b^2)^5$
9. $\left(2x - \dfrac{1}{y}\right)^4$
10. $\left(x + \dfrac{1}{x}\right)^3$
11. $\left(3x + \dfrac{y}{3}\right)^4$
12. $\left(\dfrac{x}{4} - 2y\right)^6$
13. $\left(x^3 - \dfrac{x^2}{2}\right)^6$
14. $(3x^2 - x)^4$
15. $(\frac{1}{2}a - 3b^2)^5$
16. $(\frac{1}{3}a^2 + 4b^3)^4$

Find the coefficient of the indicated term in the given expansion. (See Example 3.)

17. x^8y^2 in $(x - y)^{10}$
18. a^3b^6 in $(a - b)^9$
19. x^4y^9 in $(x^2 + 2y^3)^5$
20. a^2b^{10} in $(3a^2 + b^2)^6$

Find the indicated term in the given expansion. (See Example 4.)

21. 10th term of $(2x + y)^{12}$
22. 9th term of $\left(\dfrac{x}{2} + y\right)^{13}$
23. 4th term of $(x - 4y^2)^{15}$
24. 5th term of $(x^2 - 2)^9$
25. 9th term of $(p^2 + q^3)^{14}$
26. 8th term of $(x^3 + z)^{10}$
27. 14th term of $(x + 2y)^{13}$
28. 21st term of $\left(\dfrac{x}{3} - y^2\right)^{20}$

Determine the first five terms in the expansions of the following. Also determine the values of the variable for which each expansion is valid. (See Example 5.)

29. $(1 + x)^{1/4}$
30. $(1 + x)^{1/3}$
31. $(1 - y)^{2/3}$
32. $(1 - y)^{3/4}$
33. $(1 + 2x)^{-1/2}$
34. $(1 + 3x)^{-1/4}$
35. $(1 + x^2)^{-2}$
36. $(1 + x^2)^{-3}$
37. $\left(1 - \dfrac{a}{2}\right)^{-2}$
38. $(1 - 3a)^{-3}$

Expand by using the binomial theorem.

39. $(x + y + 1)^4$
40. $(x - 1 - z)^3$
41. $(x - y + z - w)^3$
42. $(x + y + z + w)^4$

Approximate the following by using the first four terms in the binomial expansion.

43. $(1.01)^6$ *Hint:* $(1.01)^6 = (1 + 0.01)^6$
44. $(1.03)^4$
45. $(0.99)^8$ *Hint:* $(0.99)^8 = (1 - 0.01)^8$
46. $(2.99)^3$
47. $(3.01)^3$
48. $(1.03)^8$
49. $\sqrt{1.02}$
50. $\sqrt[3]{1.1}$

51–54

Many calculators have a feature that will evaluate a binomial coefficient directly. In the calculator manual, this feature may be called a **combinations function,** with the notation **nCr** used in place of $\binom{n}{r}$. (See the combinations formula in Section 12.8 of this chapter.) Use a calculator to evaluate the following binomial coefficients.

51. $\binom{42}{11} = 42\mathbf{C}11$
52. $\binom{36}{18} = 36\mathbf{C}18$

53. $\binom{105}{99} = 105\mathbf{C}99$ 54. $\binom{58}{8} = 58\mathbf{C}8$

Critical Thinking: Exploration and Writing

55. Find and describe as many cases as you can where m and n are integers such that m/n is an integer and $(m/n)! = m!/n!$. Is it necessary to exclude $n = 0$?

Critical Thinking: Exploration and Writing

56. The sigma notation can be used to write the binomial theorem in the following form.

$$(a + b)^n = \sum_{r=0}^{n} \binom{n}{r} a^{n-r} b^r$$

Use the binomial theorem twice to explain why the following **trinomial theorem** is true.

$$(a + b + c)^n = \sum_{r=0}^{n} \left[\sum_{k=0}^{r} \binom{n}{r} \binom{n-r}{k} a^{n-r} b^{r-k} c^k \right]$$

◆ **Solution for Practice Problem**

1. $(1.02)^7 = (1 + 0.02)^7$

$$\approx (1)^7 + 7(1)^6(0.02) + \frac{7(6)}{2!}(1)^5(0.02)^2 + \frac{7(6)(5)}{3!}(1)^4(0.02)^3$$

$$= 1 + 0.14 + 0.0084 + 0.00028$$

$$= 1.14868$$

12.6 MATHEMATICAL INDUCTION

Mathematical induction is a method of proof used mainly to prove theorems which assert that a certain statement holds true for all positive integers.

The method of proof by mathematical induction is based on the following property of the positive integers: if T is a set such that

1. 1 is in T,
2. $k \in T$ always implies $k + 1 \in T$,

then T contains all the positive integers.

To get an intuitive feeling for the method, assume that T is a set that satisfies the two conditions above, and let us apply the conditions a few times. We have

$1 \in T$, by condition 1;

$1 \in T$ implies $1 + 1 = 2 \in T$, by condition 2;

$2 \in T$ implies $2 + 1 = 3 \in T$, by condition 2;
$3 \in T$ implies $3 + 1 = 4 \in T$, by condition 2;
$4 \in T$ implies $4 + 1 = 5 \in T$, by condition 2;

and condition 2 can be applied repeatedly, as long as we wish. By applying condition 2 enough times, we can arrive at the statement that any given positive integer is in T. Thus T must contain all the positive integers.

A proof by mathematical induction is often compared to climbing an endless ladder. To know that we can climb to any desired step, we need only know that (1) we can climb onto the first step, and that (2) we can climb from any step to the next higher step.

Mathematical induction is usually employed in connection with a certain statement $S(n)$ about the positive integer n. As an illustration, $S(n)$ might be the statement that

$$1 + 2 + 3 + \cdots + n = \frac{n(n+1)}{2},$$

for an arbitrary positive integer n. As another example, $S(n)$ might be the statement that

$$1 + 2n \leq 3^n.$$

Such statements as these can be proved using the following principle of mathematical induction.

Principle of Mathematical Induction

For each positive integer n, suppose that $S(n)$ represents a statement about n that is either true or false. If

1. $S(1)$ is true, and if
2. The truth of $S(k)$ always implies the truth of $S(k+1)$,

then $S(n)$ is true for all positive integers n.

A proof based on the principle of mathematical induction consists of three parts.

1. The statement is verified for $n = 1$.
2. The statement is *assumed*† true for $n = k$, and with this assumption made,
3. The statement is then proved to be true for $n = k + 1$.

It then follows that the statement is true for all positive integers n.

This type of proof is illustrated in the following examples.

†This assumption is frequently referred to as "the induction hypothesis."

Example 1 • Proving a Statement for any $n \geq 1$

Use mathematical induction to prove that

$$1 + 2 + 3 + \cdots + n = \frac{n(n+1)}{2}, \tag{1}$$

for any positive integer n.

Solution

Let $S(n)$ be the statement that Equation (1) holds true.

1. We first verify that $S(1)$ is true. (In a formula such as this, it is understood that when $n = 1$, there is only one term on the left side, and no addition is performed in this case.) The value of the left side is 1 when $n = 1$, and the value of the right side is

   $$\frac{1(1+1)}{2} = \frac{(1)(2)}{2} = 1.$$

   Thus $S(1)$ is true.
2. Assume that $S(k)$ is true. That is, assume that

   $$1 + 2 + 3 + \cdots + k = \frac{k(k+1)}{2}. \tag{2}$$

3. We must now prove that $S(k+1)$ is true. By adding $k + 1$ to both sides of Equation (2), we obtain

   $$\begin{aligned} 1 + 2 + 3 + \cdots + k + (k+1) &= \frac{k(k+1)}{2} + k + 1 \\ &= \frac{k(k+1) + 2(k+1)}{2} \\ &= \frac{(k+1)(k+2)}{2} \\ &= \frac{(k+1)[(k+1)+1]}{2}. \end{aligned}$$

This last expression is exactly the right member of Equation (1) with n replaced by $k + 1$. We have shown that the truth of $S(k)$ implies the truth of $S(k+1)$. Therefore, the statement $S(n)$ is true for all positive integers n. □

Example 2 • Proving a Statement for Any $n \geq 1$

Use mathematical induction to prove that

$$\frac{1}{1 \cdot 2} + \frac{1}{2 \cdot 3} + \frac{1}{3 \cdot 4} + \cdots + \frac{1}{n(n+1)} = \frac{n}{n+1}, \tag{3}$$

for any positive integer n.

Solution

1. For $n = 1$, the left member of Equation (3) is

$$\frac{1}{1 \cdot 2} = \frac{1}{2},$$

and the right member is

$$\frac{1}{1+1} = \frac{1}{2}.$$

Thus the statement is true for $n = 1$.

2. Assume that

$$\frac{1}{1 \cdot 2} + \frac{1}{2 \cdot 3} + \frac{1}{3 \cdot 4} + \cdots + \frac{1}{k(k+1)} = \frac{k}{k+1}. \tag{4}$$

3. To change the left member of Equation (4) to the left member of Equation (3) when $n = k + 1$, we need to add

$$\frac{1}{(k+1)[(k+1)+1]} = \frac{1}{(k+1)(k+2)}$$

to both sides of Equation (4). Doing this, we have

$$\begin{aligned}
\frac{1}{1 \cdot 2} + \frac{1}{2 \cdot 3} + \frac{1}{3 \cdot 4} + \cdots + \frac{1}{k(k+1)} + \frac{1}{(k+1)(k+2)} \\
&= \frac{k}{k+1} + \frac{1}{(k+1)(k+2)} \\
&= \frac{k(k+2)+1}{(k+1)(k+2)} \\
&= \frac{k^2+2k+1}{(k+1)(k+2)} \\
&= \frac{(k+1)^2}{(k+1)(k+2)} \\
&= \frac{k+1}{k+2} \\
&= \frac{k+1}{(k+1)+1}.
\end{aligned}$$

The last expression matches the right member of Equation (3) with n replaced by $k + 1$. Thus the truth of the statement for $n = k$ implies the truth of the statement for $n = k + 1$. By the principle of mathematical induction, Equation (3) holds for all positive integers n. □

Statements $S(n)$ are sometimes found that are false for a few values of the positive integer n but are true for all positive integers n that are sufficiently large. Statements of this form can be proved by a modified form of mathematical induction. Let m be a positive integer. To prove that $S(n)$ is true for all $n \geq m$, we alter our previous steps so that we

1. Verify that the statement is true for $n = m$.
2. Assume that the statement is true for $n = k$, where $k \geq m$.
3. Prove the statement is true for $n = k + 1$.

The next example demonstrates this method of proof.

Example 3 • Starting with a Value Greater Than 1

Prove that

$$1 + 2n < 2^n \tag{5}$$

for every positive integer $n \geq 3$.

Solution

For $n = 3$,

$$1 + 2n = 1 + 6 = 7 \qquad \text{and} \qquad 2^n = 2^3 = 8.$$

Since $7 < 8$, the statement is true for $n = 3$.

Assume now that the statement is true for k, where $k \geq 3$.

$$1 + 2k < 2^k$$

When $n = k + 1$, the left member of Equation (5) is $1 + 2(k + 1)$, and

$$\begin{aligned} 1 + 2(k + 1) &= 1 + 2k + 2 \\ &< 2^k + 2, \qquad \text{since } 1 + 2k < 2^k \\ &< 2^k + 2^k, \qquad \text{since } 2 < 2^k \text{ for } k \geq 3 \end{aligned}$$

But $2^k + 2^k = 2^k(1 + 1) = 2^k(2) = 2^{k+1}$ is the right member of Equation (5) when $n = k + 1$. Thus we have proved that

$$1 + 2n < 2^n$$

is true when $n = k + 1$. Therefore the statement is true for all positive integers $n \geq 3$. □

EXERCISES 12.6

Use mathematical induction to prove that each of the following statements is true for all positive integers n.

1. $1 + 3 + 5 + \cdots + (2n - 1) = n^2$
2. $2 + 4 + 6 + \cdots + (2n) = n(n + 1)$

√ 3. $1^2 + 2^2 + 3^2 + \cdots + n^2 = \dfrac{n(n+1)(2n+1)}{6}$

4. $1 \cdot 2 + 2 \cdot 3 + 3 \cdot 4 + \cdots + n(n+1) = \dfrac{n(n+1)(n+2)}{3}$

5. $1 \cdot 2 + 2 \cdot 2^2 + 3 \cdot 2^3 + \cdots + n \cdot 2^n = (n-1)2^{n+1} + 2$

6. $2 + 4 + 8 + \cdots + 2^n = 2^{n+1} - 2$

√ 7. $4 + 4^2 + 4^3 + \cdots + 4^n = \dfrac{4(4^n - 1)}{3}$

8. $4 + 8 + 12 + 16 + \cdots + 4n = 2n(n+1)$

9. $3 + 6 + 9 + 12 + \cdots + 3n = \dfrac{3n(n+1)}{2}$

√ 10. $\dfrac{1}{1 \cdot 4} + \dfrac{1}{4 \cdot 7} + \dfrac{1}{7 \cdot 10} + \cdots + \dfrac{1}{(3n-2)(3n+1)} = \dfrac{n}{3n+1}$

11. $\dfrac{1}{1 \cdot 2 \cdot 3} + \dfrac{1}{2 \cdot 3 \cdot 4} + \cdots + \dfrac{1}{n(n+1)(n+2)} = \dfrac{n(n+3)}{4(n+1)(n+2)}$

12. $1^3 + 2^3 + 3^3 + \cdots + n^3 = \dfrac{n^2(n+1)^2}{4}$

13. $\dfrac{1}{3} + \dfrac{1}{3^2} + \dfrac{1}{3^3} + \cdots + \dfrac{1}{3^n} = \dfrac{1}{2}\left[1 - \left(\dfrac{1}{3}\right)^n\right]$

14. $a + (a+d) + (a+2d) + \cdots + [a + (n-1)d] = \dfrac{n}{2}[2a + (n-1)d]$

15. $a + ar + ar^2 + \cdots + ar^{n-1} = a\dfrac{1-r^n}{1-r}$ if $r \neq 1$

√ 16. 3 is a factor of $4^n - 1$.

17. 3 is a factor of $n^3 + 2n$.

18. $1 - x$ is a factor of $1 - x^n$.

19. $a - b$ is a factor of $a^n - b^n$.
[*Hint:* $a^{k+1} - b^{k+1} = a^k(a-b) + (a^k - b^k)b$]

20. $a + b$ is a factor of $a^{2n} - b^{2n}$.

21. $x^{2n} > 0$, if $x \neq 0$.

22. $\left(\dfrac{a}{b}\right)^n < 1$, if $0 < a < b$.

√ 23. $1 + 2n \leq 3^n$

24. Use mathematical induction and the result in Example 3 to prove that $n^2 < 2^n$, if $n \geq 5$.

25. Assume that $|a + b| \leq |a| + |b|$ for all real numbers a and b. Then prove that $|a_1 + a_2 + \cdots + a_n| \leq |a_1| + |a_2| + \cdots + |a_n|$ for any real numbers $a_1, a_2, \ldots, a_n$.

26. Use mathematical induction to prove that $\overline{z^n} = (\overline{z})^n$ for any complex number z and any positive integer n.

In Exercises 27–29, use mathematical induction to prove the stated property of the sigma notation.

27. $\sum_{i=1}^{n} ca_i = c \sum_{i=1}^{n} a_i$

28. $\sum_{i=1}^{n} (a_i + b_i) = \sum_{i=1}^{n} a_i + \sum_{i=1}^{n} b_i$

29. $\sum_{i=1}^{n} (a_i - b_i) = \sum_{i=1}^{n} a_i - \sum_{i=1}^{n} b_i$

Critical Thinking: Writing

30. Show that if the statement

$$1 + 2 + 3 + \cdots + n = \frac{n(n+1)}{2} + 2$$

is assumed to be true for $n = k$, the same equation can be proved to hold for $n = k + 1$. Explain why the demonstration you have made does not prove that the statement is true for all positive integers. Is the statement true for all positive integers? Why?

Critical Thinking: Writing

31. Show that $n^2 - n + 5$ is a prime integer when $n = 1, 2, 3, 4$, but that it is not true that $n^2 - n + 5$ is always a prime integer. Write out a similar set of statements for the polynomial $n^2 - n + 11$.

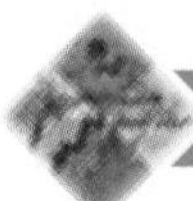

12.7 PERMUTATIONS

In the study of certain branches of mathematics, including probability, we are often faced with the problem of determining the number of ways a certain act can be performed or a certain selection can be made. As a simple example, suppose that a man has a wardrobe containing a white, a tan, and a brown shirt and a green and a gold tie. From these, he wishes to make a selection of a shirt and a tie. Since he can choose any of the 3 shirts and either of the 2 ties, there are $3 \cdot 2$, or 6, different ways he can make the selection of a shirt and tie. This is an example of the use of the **counting principle** stated in the next paragraph. These possible selections are presented in Figure 12.3. A diagram of this form is called a **tree.** The number of smaller "branches" is a number of different selections that can be made, and each possible selection is apparent in the tree. For example, the third from the top of the smaller branches indicates a selection of a tan shirt and a green tie.

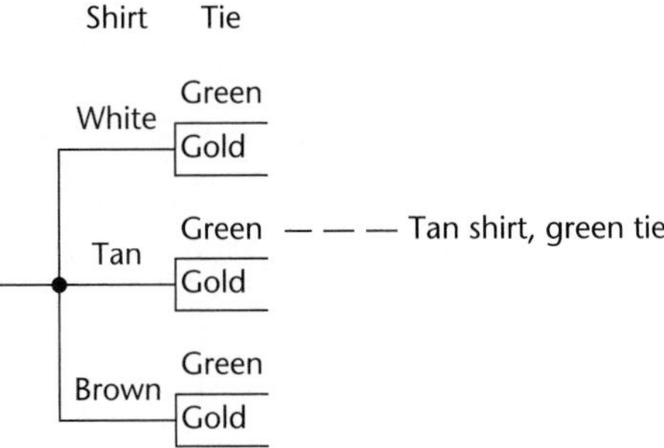

Figure 12.3 Tree diagram for "shirt-tie" selection

> **The Counting Principle**
>
> Suppose that one selection can be made in m ways, and after that selection is made, a second selection can be made in n ways, where the result of the first selection does not influence the result of the second selection. Then the two selections can be made in that order in $m \cdot n$ ways.

The use of another tree diagram is shown in Example 1, and other tree diagrams are utilized in the last section of this chapter.

Example 1 • Class Scheduling

In planning the class schedule for next semester, the mathematics chairperson finds that six evening sections of college algebra are needed. These sections must be scheduled on Monday, Wednesday, and Friday at one of the times 6:00 P.M. or 7:00 P.M. The rooms available are rooms 101, 102, 103 in the science building and rooms 205, 206, 207 in the mathematics building. Draw a tree diagram and determine the number of ways that a classroom can be assigned to Ms. Schanck's section of college algebra.

Solution

The tree diagram in Figure 12.4 shows the choices that can be made. The first set of branches corresponds to the choice between the buildings, the second set of branches shows the choice of rooms in each building, and the last set shows the choice of time. We see that there are 12 ends of branches in the tree, diagraming 12 different ways that a classroom can be assigned to Ms. Schanck's class. □

The counting principle can be extended for any finite number of selections. For instance, the work done in Example 1 shows that with 2 choices of a building, 3 choices of a room in each building, and 2 choices of time in each room, the number of ways a classroom can be assigned is

$$(2)(3)(2) = 12.$$

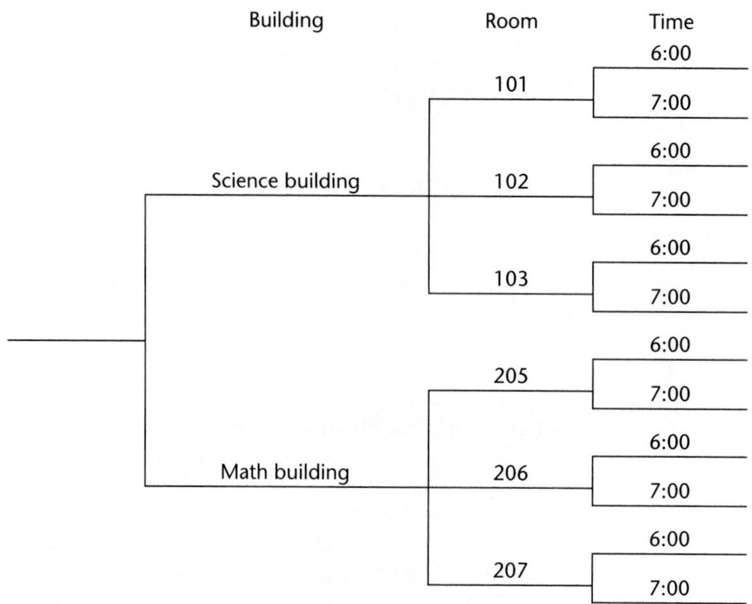

Figure 12.4 Tree diagram for classroom assignments

Example 2 • Identification Tags

Suppose that identification tags are to be made by using 2 letters from among the letters *a*, *b*, *c*, *d*, *e* followed by one of the numbers 1, 2, 3, 4, 5, 6, 7, 8, 9. How many distinct tags can be made if (a) any letter can be repeated and (b) no letter can be repeated?

Solution

The tag has the form given in Figure 12.5.

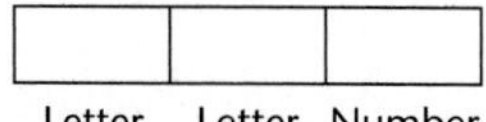

Figure 12.5 Form of identification tags

a. For the first position to be filled by one of the 5 letters, there are 5 choices. There are also 5 choices of letters to be put in the second position. For the last position, there are 9 choices of numbers. Using the counting principle, there are

$$5 \cdot 5 \cdot 9 = 225$$

different tags of the form described.

b. Any one of the 5 letters can be used in the first position. But no letter can be repeated, so once a letter is chosen for the first position, the same letter cannot be used in the second position. Hence there are only 4 choices of a letter for the second position. Any one of the 9 numbers can be used for the last position. Using the counting principle, we see that there are

$$5 \cdot 4 \cdot 9 = 180$$

different tags where no letter is repeated. □

Example 3 • A Two-Part Contest

Suppose that a contest consists of two parts: First an ordinary cubical die is rolled, and next a coin is tossed. Find the total number of possible results in the contest.

Solution

Since a cubical die has 6 distinct faces, and any one may face up when it is rolled, there are 6 possible results for the first part of the contest. Similarly, there are 2 possible results when the coin is tossed in the second part of the contest. Thus there is a total of

$$6 \cdot 2 = 12$$

possible results in the contest. □

◆ Practice Problem 1 ◆

For dessert, Grandma Davidson has a choice of 3 flavors of ice cream and 2 kinds of fruit topping. How many different fruit sundaes can she make?

A permutation is an arrangement of objects. For example,

$$\begin{array}{lllll} x & xy & yz & xyz & yzx \\ y & xz & zx & xzy & zxy \\ z & yx & zy & yxz & zyx \end{array}$$

are arrangements and hence permutations of the three letters x, y, and z. The three arrangements

$$x \quad y \quad z$$

are the permutations of the three letters taken one at a time; the six arrangements

$$\begin{array}{lll} xy & yx & zx \\ xz & yz & zy \end{array}$$

are the permutations of the three letters taken two at a time; and the six arrangements

$$\begin{array}{lll} xyz & yxz & zxy \\ xzy & yzx & zyx \end{array}$$

are the permutations of the three letters taken three at a time. In general, a **permutation** of n objects taken r at a time is an arrangement using r of the n objects.

Our interest focuses now on the determination of the *number* of permutations of n objects taken r at a time. We shall denote the number of permutations of n objects taken r at a time by the symbol

$$P(n, r), \qquad \text{where } 0 \leq r \leq n.$$

Other commonly used symbols are ${}_n\mathbf{P}_r$ and **nPr**. To determine a formula for $P(n, r)$, first consider $P(n, n)$. We use the counting principle to determine the number of distinct arrangements of n objects taken n at a time. In other words, we have n positions to fill with a choice of n objects, where repetition of any object is not allowed. For the first position there are n choices, for the second position $n - 1$ choices, for the third position $n - 2$ choices, and so on, as indicated in the chart below. Thus we have

$$P(n, n) = n \cdot (n - 1) \cdot (n - 2) \cdots 2 \cdot 1$$

or

$$\mathbf{P(n, n) = n!.} \tag{1}$$

Position	1	2	3	$\cdots$	$n-1$	n
Number of choices	n	$n-1$	$n-2$	$\cdots$	2	1

In a similar manner, suppose that we have r positions to fill with a choice of n objects ($0 \leq r \leq n$), where repetition of any object is not allowed. For the first position there are n choices, for the second position $n - 1$ choices, for the third position $n - 2$ choices, and so on. For the rth position there are $n - (r - 1)$, or $n - r + 1$, choices. This tabulation appears in the chart below. The counting principle gives

$$\mathbf{P(n, r) = n(n - 1)(n - 2) \cdots (n - r + 1).} \tag{2}$$

Position	1	2	3	$\cdots$	r
Number of choices	n	$n-1$	$n-2$	$\cdots$	$n-r+1$

To express $P(n, r)$ using factorial notation, we note that

$$\begin{aligned} P(n, r) &= n(n - 1)(n - 2) \cdots (n - r + 1) \\ &= \frac{n(n - 1)(n - 2) \cdots (n - r + 1)(n - r)(n - r - 1) \cdots (2)(1)}{(n - r)(n - r - 1) \cdots (2)(1)} \\ &= \frac{n!}{(n - r)!}. \end{aligned}$$

These results are summarized in the following formulas.

Permutations Formulas

If $0 \le r \le n$, then the number of distinct permutations of n objects taken r at a time is given by

$$P(n, r) = \frac{n!}{(n - r)!}.$$

Notice that if $r = n$, we have

$$P(n, n) = \frac{n!}{(n - n)!} = \frac{n!}{0!} = \frac{n!}{1} = n!$$

This result agrees with Equation (1) and provides another instance where the definition $0! = 1$ gives consistent results.

Example 4 • Using the Permutations Formula

Some evaluations of the number of permutations are given here.

a. $P(5, 3) = \dfrac{5!}{(5 - 3)!} = \dfrac{5!}{2!} = \dfrac{5 \cdot 4 \cdot 3 \cdot 2!}{2!} = 60$

b. $P(6, 6) = \dfrac{6!}{0!} = \dfrac{6 \cdot 5 \cdot 4 \cdot 3 \cdot 2 \cdot 1}{1} = 720$

c. $P(10, 3) = \dfrac{10!}{(10 - 3)!} = \dfrac{10!}{7!} = \dfrac{10 \cdot 9 \cdot 8 \cdot 7!}{7!} = 10 \cdot 9 \cdot 8 = 720$

d. $P(4, 0) = \dfrac{4!}{(4 - 0)!} = \dfrac{4!}{4!} = 1$

□

Example 5 • Counting Sets of Officers

From a group of 10 people, a president, vice president, secretary, and treasurer are to be chosen. How many different sets of officers can be chosen if no person can hold more than one office?

Solution

Using the counting principle, we have 10 choices for president. Once a president has been chosen, there are 9 choices for vice president, then 8 choices for secretary, and finally 7 choices for treasurer. Thus there are

$$10 \cdot 9 \cdot 8 \cdot 7 = 5040$$

different sets of officers.

Alternatively, permutations can be used to solve the problem. Each set of officers is a permutation of 10 people taken 4 at a time. Thus the number of different sets of officers is the number of distinct permutations of 10 things taken 4 at a time, or

$$P(10, 4) = \frac{10!}{6!} = 10 \cdot 9 \cdot 8 \cdot 7 = 5040.$$

□

Example 6 • Counting Seating Arrangements

In how many ways can 5 boys and 3 girls be seated in a row of 8 chairs if the girls must sit together?

Solution

Since the girls must sit together, we must choose seats for 5 boys and the group of girls. This can be done in $P(6, 6) = 6!$ ways. The number of different ways the girls can be arranged within the group is $P(3, 3) = 3!$. The counting principle gives

$$6! \cdot 3! = 4320$$

as the number of different seating arrangements of the desired type. □

Suppose that we consider next permutations of n objects that are not all distinct. For example, consider the permutations of the letters in the word *BETTER*. The "word"

BETRET

is one such permutation. If we interchange the two letters E in the arrangement, we obtain the same permutation. To count the number of distinguishable permutations, we use the counting principle. Let D represent the number of distinct permutations of the letters in the word *BETTER*. There are $P(2, 2) = 2!$ arrangements of the two E's that will not change a permutation and $P(2, 2) = 2!$ arrangements of the two T's that will not change a permutation. If all letters were distinguishable, there would be

$$D \cdot 2! \cdot 2!$$

arrangements of the six letters taken six at a time. But $P(6, 6) = 6!$ is the number of permutations of six distinguishable letters taken six at a time. Thus

$$D \cdot 2! \cdot 2! = 6! \qquad \text{and} \qquad D = \frac{6!}{2!2!} = 180.$$

The general situation is described in the next formula.

General Permutation Formula

Let D represent the number of distinct permutations of n objects taken n at a time, where there are r $(0 < r \leq n)$ distinguishable types: n_1 are of one type, n_2 are of a second type, . . . , n_r are of the rth type. Then

$$D = \frac{n!}{n_1!n_2! \cdots n_r!}, \qquad \text{where } n_1 + n_2 + \cdots + n_r = n.$$

Example 7 • Counting Permutations Made with Repeated Letters

How many distinct permutations are possible for the letters in *TENNESSEE*?

Solution

The word *TENNESSEE* has nine letters, consisting of 1 *T*, 4 *E*'s, 2 *N*'s, and 2 *S*'s. Thus there are

$$D = \frac{9!}{1!4!2!2!} = 3780$$

distinct permutations of the letters in *TENNESSEE*. □

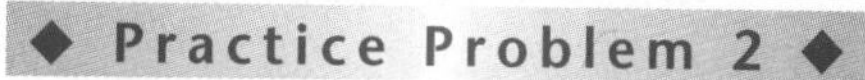

How many distinct "words" can be made using the letters in *INDIANA*?

EXERCISES 12.7

Tree Diagram for T-shirts

1. T-Shirt World has Independence Day T-shirts for sale in sizes S, M, L, and XL, with all sizes available in solid colors red or white or blue. Draw a tree diagram that shows all size and color possibilities. How many different Independence Day T-shirts are available?

Tree Diagram for Baby Names

2. A couple has narrowed down the choice of names for their baby girl so that the first name will be Whitney, Shannon, or Terry, and the second name will be Sue, Jane, or Lee. Draw a tree diagram that shows all first- and second-name possibilities. How many possibilities are there for the baby's first and second names?
3. List all the permutations of 0 and 1.
4. List all the permutations of a, b, c, and d.
5. List all the permutations of 1, 2, 3, 4, and 5, taken 2 at a time.
6. List all the permutations of 1, 2, 3, 4, and 5, taken 3 at a time.

Evaluate each of the following. (See Example 4.)

7. $P(4, 4)$
8. $P(6, 6)$
9. $P(15, 2)$
10. $P(8, 6)$
11. $P(15, 0)$
12. $P(20, 0)$
13. $P(n, 2)$
14. $P(n - 1, 3)$
15. $P(n, n - 2)$
16. $P(n - 1, n - 3)$

Simplify each of the following. (See Example 4.)

17. $\frac{n!}{(n-2)!}$
18. $\frac{(n-1)!}{(n-2)!}$
19. $\frac{n!}{(n+2)!}$
20. $\frac{(n+3)!}{(n-3)!}$

Routes Between Towns 21. To travel from town A to town B a traveler can take any of 4 routes. To travel from town B to town C, he can take any of 3 routes. In how many ways can he travel from town A to town C if he must pass through town B?

Course Schedules 22. A college offers a freshman 4 history courses, 3 math courses, 5 English courses, and 2 geography courses. If a student takes one course in each area, how many different schedules are possible?

Decorating a Room 23. A decorator has a choice of 4 carpets, 6 wallpapers, and 5 paints. How many different selections of 1 carpet, 1 wallpaper, and 1 paint can be made in decorating a room?

Menu Selection 24. A cafeteria serves 2 meats, 4 vegetables, 3 salads, and 2 desserts. A meal consists of 1 meat, 1 vegetable, 1 salad, and 1 dessert. How many different meals are possible?

True-False Tests 25. A test consists of 10 true-false questions. How many different answer sheets are possible if all questions are answered?

Multiple-Choice Tests 26. A test consists of 10 multiple-choice questions, each with 5 possible answers. How many different answer sheets are possible if all questions are answered?

Tossing Coins 27. If 4 coins are tossed on the floor and the sides facing up are observed, how many different possible results can occur?

Rolling Dice 28. If two cubical dice are rolled and the sides facing up are observed, how many different possible results can occur?

Winning Positions in a Race 29. If 6 runners enter a race, how many ways can the first 3 winning positions be awarded?

Class Officers 30. A class of 10 boys and 15 girls elects a president and a secretary. In how many ways can a president and a secretary be selected if no student can hold both offices and if

a. There are no other restrictions?

b. The president must be a girl?

c. The secretary must be a girl?

Social Security Numbers 31. A U.S. social security number consists of 3 digits followed by 2 digits, then followed by a final 4 digits. If any digit may fill any position, how many social security numbers are possible?

License Tags

32. License tags in a certain state consist of 3 letters followed by 3 digits. How many tags are possible if
    a. Repeats are not allowed on the letters?
    b. The final digit may not be zero?
    c. Repeats are not allowed on the letters or digits?
    d. There are no restrictions?

Number problems 33–36

33. How many numbers greater than 2000 and consisting of 4 digits can be formed?
34. In Exercise 33, suppose that only the digits 0, 1, 2, 3, 4, and 5 can be used.
35. How many 3-digit odd numbers can be made from the digits 1, 2, 3, 4, 5, and 6?
36. How many 3-digit even numbers can be made using the digits 1, 2, 3, 4, 5, and 6?

Seating Arrangements 37–38

37. In how many ways can 6 first graders be seated in a row of (a) 6 chairs and (b) 7 chairs?
38. In how many ways can 5 Girl Scouts be seated in a row of 10 chairs?

Arranging Books and Pictures 39–42

39. In how many ways can a set of 11 different books be arranged on a shelf?
40. In how many ways can 5 different pictures be arranged in a row on a wall?
41. In how many ways can 6 different books be arranged on a shelf if 3 of the 6 are math books and they must be together?
42. In how many ways can 8 different books be arranged on a shelf if 2 are trigonometry books that must stay together and 3 are algebra books that must stay together?

Seating Arrangements 43–44

43. In how many ways can 7 high school students be seated in a row of 7 seats if two of the students must be seated side by side?
44. In how many ways can a group of 4 college freshmen and 5 college sophomores be seated in a row of 9 seats if the 4 freshmen are seated together?

Counting "Words" 45–48

45. How many distinct "words" can be made using the letters in the word *COUNT*?
46. How many distinct "words" can be made using the letters in the word *MATH*?
47. How many distinct "words" can be made using the letters in the word (a) *ALGEBRA*, (b) *CALCULUS*, (c) *ADD*, (d) *DIVIDE*?

48. How many distinct "words" can be made using the letters in the word (a) *MISSISSIPPI*, (b) *MASSACHUSETTS*, (c) *LOUISIANA*, (d) *ILLINOIS*?

Distributing Fruit

49. In how many ways can 6 apples, 4 oranges, and 5 bananas be distributed among 15 children if each child is to receive one piece of fruit?

Awarding Prizes

50. If 5 identical television sets and 4 identical ovens are to be given to 9 prizewinners, how many ways can the prizes be awarded?

Flag Signals 51–52

51. If signals are to be made using 2 green flags, 4 black flags, and 4 white flags, how many distinct signals can be made by lining up the flags?
52. If signals are to be made using 3 red flags, 5 white flags, and 2 blue flags, how many distinct signals can be made by lining up the flags?

Alternating Seating Arrangements 53–54

53. In how many ways can 4 girls and 3 boys be seated in 7 seats if the girls and boys must alternate?
54. In how many ways can 4 girls and 5 boys be seated in 9 seats if the girls and boys must alternate?

Circular Seating Arrangements 55–58

55. In how many relative orders can 5 people be seated in 5 chairs at a round table? (*Hint:* Fix the position of one person at the table and then arrange the remaining people in the remaining positions.)
56. In how many relative orders can 7 people be seated in 7 chairs at a round table?
57. At a round table, 4 men and 4 women are to be seated in 8 chairs. How many distinct relative orders are possible if the men and women must alternate? If the men are to sit together?
58. How many distinct relative orders of seating 5 women and 5 men in 10 chairs are there around a circular table if the women and men must alternate? If the women must sit together?

Many calculators have a feature that will evaluate $P(n, r) =$ **nPr** directly. Use a calculator to evaluate each of the following.

59–62

59. 29**P**7 60. 77**P**5 61. 42**P**11 62. 36**P**18

Critical Thinking: Exploration and Writing

63. Suppose n people are to be seated in n chairs at a round table. Find a formula for the number of different relative orders in which they can be seated and explain why your formula works.

Critical Thinking: Exploration and Writing

64. Conrad Larrieu plans to make a vegetable soup by choosing 7 vegetables from this list: potatoes, carrots, tomatoes, onions, peas, corn, green beans, lima beans, celery, and okra. Decide if finding the number of different soups he can make is a permutation problem or not and explain why your decision is correct.

◆ Solutions for Practice Problems

1. The selection of ice cream can be made in 3 ways, and the selection of fruit topping for each of these can be made in 2 ways, so the sundaes can be made in $3 \cdot 2 = 6$ different ways by the counting principle.
2. The word *INDIANA* has 7 letters, consisting of two *I*'s, two *N*'s, one *D*, and two *A*'s. Hence there are

$$D = \frac{7!}{2!2!1!2!} = 630$$

distinct "words" that can be made using the letters in *INDIANA*.

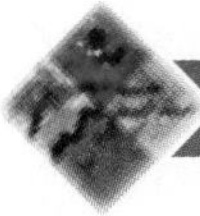

12.8 COMBINATIONS

A combination is a grouping of objects. For example,

$$\{x\} \quad \{y\} \quad \{z\} \quad \{x, y\} \quad \{x, z\} \quad \{y, z\} \quad \{x, y, z\}$$

are groupings, and hence combinations of the three letters x, y, and z. The three groupings

$$\{x\} \quad \{y\} \quad \{z\}$$

are the combinations of the three letters taken one at a time; the three groupings

$$\{x, y\} \quad \{x, z\} \quad \{y, z\}$$

are the combinations of the three letters taken two at a time; and the one grouping

$$\{x, y, z\}$$

is the only combination of the three letters taken three at a time. In general, a **combination** of n objects taken r at a time is a group of r of the n objects, where the order in which the objects appear is of no consequence.

Notice that the distinct permutations

$$xy \quad \text{and} \quad yx$$

are in fact the same combination. This important difference between combinations and permutations cannot be overemphasized. A reordering of the objects changes a permutation, but a reordering of the objects does not change a combination.

Let $C(n, r)$ denote the number of combinations of n objects taken r at a time. A formula for evaluating $C(n, r)$ can be obtained by observing the following. We can obtain $P(n, r)$, the number of permutations of n objects taken r at a time, by

1. First choosing r of the n objects, which can be done in $C(n, r)$ ways.
2. Then forming all possible distinct arrangements of the r chosen objects, which can be done in $P(r, r) = r!$ ways.

Thus we have

$$P(n, r) = C(n, r)r!$$

and

$$C(n, r) = \frac{P(n, r)}{r!}.$$

But since $P(n, r) = n!/(n - r)!$, this can be rewritten as

$$C(n, r) = \frac{n!}{(n - r)!r!}.$$

We summarize our results in the following statement.

Combinations Formula

If $0 \leq r \leq n$, then the number of combinations of n objects taken r at a time is given by

$$C(n, r) = \frac{n!}{(n - r)!r!}.$$

Thus the number $C(n, r)$ of combinations of n objects taken r at a time is the same as the binomial coefficient $\binom{n}{r}$. Other symbols commonly used in place of $C(n, r)$ are ${}_{\mathbf{n}}\mathbf{C}_{\mathbf{r}}$ and **nCr**. We will use $C(n, r)$ and $\binom{n}{r}$ interchangeably throughout the rest of this chapter.

The use of the combinations formula is demonstrated in the following examples.

Example 1 • Counting Committees

How many committees of 4 people can be chosen from a group of 7 people?

Solution

Since the order in which the committee members are chosen is of no consequence, the number of committees consisting of 4 of the 7 people is given by

$$\binom{7}{4} = \frac{7!}{4!3!} = 35.$$

□

Example 2 • Choosing a Basketball Team

A basketball team consists of 2 guards, 2 forwards, and 1 center. If a coach has 4 guards, 6 forwards, and 3 centers to choose from, how many ways can he choose a team?

Solution

The combinations formula is used three times: once to determine the number of ways he can choose the guards, again for the forwards, and again for the center. From the 4 available guards he can choose 2 in

$$C(4, 2) = \frac{4!}{2!2!} = 6$$

ways. From the 6 forwards he can choose 2 in

$$C(6, 2) = \frac{6!}{4!2!} = 15$$

ways. From the 3 centers he can choose 1 in

$$C(3, 1) = \frac{3!}{2!1!} = 3$$

ways. By the counting principle, there are

$$6 \cdot 15 \cdot 3 = 270$$

ways the coach can choose a team. □

Example 3 • Counting Poker Hands

A poker hand consists of 5 cards dealt from an ordinary deck of 52 cards. How many different poker hands are possible?

Solution

Since the order in which the cards appear in the hand is of no significance, there are

$$\binom{52}{5} = \frac{52!}{47!5!} = 2{,}598{,}960$$

different poker hands. □

◆ Practice Problem 1 ◆

A bridge hand consists of 13 cards dealt from an ordinary deck of 52 cards. How many different bridge hands are possible consisting of 7 clubs, 5 diamonds, and 1 heart?

EXERCISES 12.8

1. List all the combinations of 0 and 1.
2. List all the combinations of a, b, c, and d.
3. List all the combinations of 1, 2, 3, 4, and 5 taken 2 at a time.
4. List all the combinations of 1, 2, 3, 4, and 5 taken 3 at a time.

Evaluate each of the following.

5. $C(6, 5)$ 6. $C(8, 4)$ ✓7. $C(10, 10)$ 8. $C(4, 4)$

9. $\binom{5}{0}$ 10. $\binom{9}{0}$ ✓11. $\binom{6}{4}$ 12. $\binom{6}{2}$

Simplify each of the following.

✓13. $\binom{n}{n-1}$ 14. $\binom{n}{n-2}$ 15. $C(n, n)$ 16. $C(n, 0)$

Decorating a House 17. A decorator chooses 4 colors in decorating a house. If she has 12 colors to choose from, how many different combinations are possible?

Choosing Players 18. From a list of 9 players, 8 are to be chosen. In how many ways can this be done?

Counting Committees ✓19. From a group of 15 people, how many committees of 6 can be chosen? How many committees of 10 can be chosen?

Choosing Ushers 20. From a group of 10 students, 5 are to be chosen to serve as ushers at graduation. In how many ways can this be done?

Forming Committees 21–24 21. From a club of 8 members, an entertainment committee of 4 is to be selected. In how many ways can this be done if 1 must serve as chairperson?

22. A grievance committee is to be selected out of an algebra class of 20 students. How many different committees can be selected if there are to be 4 people on the committee, with 1 of those designated as spokesperson?

✓23. How many committees of 2 men and 3 women can be chosen from a group of 6 men and 12 women?

24. From a group of 8 freshmen and 9 sophomores, a committee consisting of 3 freshmen and 5 sophomores is to be selected. How many different committees can be chosen?

Choosing Books 25. In how many ways can a student choose 2 math books from among 6 math books, 4 novels from among 7 novels, and 3 psychology books from 9 psychology books?

Baseball Team 26. A baseball team consists of 3 outfielders, 4 infielders, 1 pitcher, and 1 catcher. From a group of 7 outfielders, 9 infielders, 6 pitchers, and 3 catchers, how many different groups of people can be chosen to form a team?

Concert Programs 27. An orchestra can play 10 marches, 8 overtures, and 15 pop songs. How many different concerts consisting of 3 marches, 2 overtures, and 5 pop songs can it give if the order of the pieces is of no consequence?

Menus 28. A restaurant can prepare 9 main dishes, 7 vegetables, 6 salads, 4 breads, and 12 desserts. If it plans a buffet consisting of 2 main dishes, 3 vegetables, 4 salads, 2 breads, and 5 desserts, how many possible menus are there?

Selecting Balls 29–30 ✓29. A bag contains 7 green, 5 white, and 2 purple balls. Three balls are to be chosen. In how many ways can the following selections be made: (a) 1 green and 2 white, (b) all white balls, (c) each of different color, (d) all purple balls?

30. From a bag containing 6 blue, 4 green, and 5 red balls, 4 are chosen. Find the number of ways each of the following selections can be made: (a) all green, (b) each of different color, (c) none green, (d) all the same color, (e) 2 blue, 1 red, and 1 green.

Drawing Lines 31. How many distinct straight lines can be drawn through 9 points, where no 3 are on the same line? (*Hint:* Two points determine a straight line.)

Drawing Triangles 32. Determine the number of triangles that can be drawn using the points in Exercise 31 as vertices.

Bridge Hands ✓33. A bridge hand consists of 13 cards dealt from an ordinary deck of 52. How many hands are there consisting of the following: (a) 5 spades, 3 hearts, 2 clubs, and 3 diamonds; (b) 12 spades and 1 heart; (c) 7 clubs and 6 spades?

Poker Hands 34. If a poker hand of 5 cards is dealt, determine the number of ways each of the following hands can occur: (a) a royal flush: ace, king, queen, jack and 10, all of the same suit; (b) a straight: five cards in consecutive numerical order, not all of the same suit (consider a jack as 11, queen as 12, king as 13, and ace as either 1 or 14); (c) four of a kind (for example, four 2's or four aces).

35. Show that $C(n, r) = C(n, n - r)$ for all $r \leq n$.

36. Show that $C(n, r - 1) + C(n, r) = C(n + 1, r)$.

37–40

Senate Committees ✓37. How many 5-member committees can be formed from the 100 senators in the U.S. Senate?

House Committees 38. How many 4-member committees can be chosen from the 435 voting representatives in the U.S. House of Representatives?

Bridge Hands

39. How many different bridge hands of 13 cards can be dealt from an ordinary deck of 52 cards?

Poker Hands

40. How many different 7-card poker hands can be dealt from an ordinary deck of 52 cards?

Critical Thinking: Writing

41. Classify each of the following as either a permutation or a combination, and explain the reasons for your classifications.
    a. Zip code
    b. Telephone number
    c. Social security number
    d. The "combination" that opens a safe

Critical Thinking: Writing

42. The flags of France, Italy, and several other countries consist of three vertical bands of different colors, as shown in the accompanying figure. Using the six colors in the rainbow (violet, blue, green, yellow, orange, and red) along with black and white, there would be 8 possible colors to select from in designing a flag of this type. Explain how the problem of finding the number of such flags can be analyzed as a permutation problem. Also, describe the problem in two steps, one of which involves combinations.

Mali

Figure for Exercise 42

◆ Solution for Practice Problem

1. There are 13 cards in each suit, and so there are

$$\binom{13}{7} = \frac{13!}{6!7!} = 1716 \text{ ways to choose 7 clubs,}$$

$$\binom{13}{5} = \frac{13!}{8!5!} = 1287 \text{ ways to choose 5 diamonds,}$$

$$\binom{13}{1} = \frac{13!}{12!1!} = 13 \text{ ways to choose 1 heart.}$$

By the counting principle, there are $(1716)(1287)(13) = 28{,}710{,}396$ ways to get a bridge hand that consists of 7 clubs, 5 diamonds, and 1 heart.

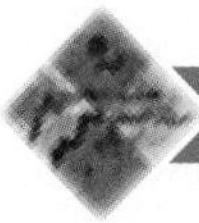

12.9 PROBABILITY

Historically, the theory of probability developed from the study of games of chance. The use of probability is significant in all aspects of modern life, from the testing of new products and the study of traffic flow to the analysis of opinion polls and the theory of the spread of disease.

One of the simplest games of chance that can be played is that of tossing a coin. We will use this game to illustrate some of the terms basic to the study of probability such as experiment, trial, outcome, sample space, event, and simple event. The **experiment** can be described as follows: toss a coin and observe which side faces up. A **trial** of the experiment is the actual performance of the experiment. The **outcomes** of the experiment are the results of the trials. In tossing a coin, all possible outcomes of the experiment can be listed: H or T, where H indicates that the side of the coin facing up is the heads side and T indicates that the tails side is up. The **sample space** is the set of all possible outcomes. Hence $\{H, T\}$ is the sample space for the experiment of tossing a coin. An **event** is any subset of the sample space, and **simple events** are those events (subsets) that contain only one element. Thus $\varnothing$, $\{H\}$, $\{T\}$, and $\{H, T\}$ are all possible events and $\{H\}$ and $\{T\}$ are the simple events.

Example 1 • Tossing Coins

Suppose two coins are tossed and the upper faces are observed. Find (a) the sample space and (b) the event of at least one head.

Solution

a. The sample space is $\{HH, HT, TH, TT\}$, where HH indicates heads on the upper faces of both coins, HT indicates heads on the first coin and tails on the second, and so on.

b. The event of at least one head is given by

$$\{HH, HT, TH\}.$$

□

We next define the concept of the probability of an event. We first note that any event can be expressed as a union of simple events. At this point, we restrict our attention to sample spaces containing a finite number of equally likely simple events. For example, the result of tossing a coin is just as likely to be heads as it is tails. Hence we say that the two simple events $\{H\}$ and $\{T\}$ are equally likely.

Definition of the Probability of an Event

Suppose S is a sample space consisting of a finite number of equally likely simple events, and suppose E is an event of S. Let $n(E)$ and $n(S)$ represent the number of simple events in E and the number of simple events in S, respectively. Then the probability that E will occur is

$$P(E) = \frac{n(E)}{n(S)}.$$

Example 2 • Rolling a Die

Suppose a die is rolled and the number on the upper face is observed. Determine the probability of obtaining (a) an even number and (b) not a 1.

Solution

The sample space $S = \{1, 2, 3, 4, 5, 6\}$ contains 6 equally likely simple events.

a. Let E be the event of obtaining an even number. Then $E = \{2, 4, 6\}$. Since there are 3 equally likely simple events in E, then

$$P(E) = \tfrac{3}{6} = \tfrac{1}{2}.$$

b. Let E be the event of not obtaining a 1. Then $E = \{2, 3, 4, 5, 6\}$ and

$$P(E) = \tfrac{5}{6}.$$ □

Tree diagrams can be very helpful in determining sample spaces and events. This is illustrated in the next example.

Example 3 • World Series

Suppose the Baltimore Orioles are playing the Atlanta Braves in the World Series and that the teams are evenly matched, with either team equally likely to win any game. Given that the Braves have won the first two games of the series, what is the probability that they will win the championship?

Solution

In the World Series, the champion is the first team to win four games. We shall use a tree diagram to analyze the possibilities. The one shown in Figure 12.6 is constructed starting with game 3 and using a W or an L for a Braves win or loss, respectively.

From the diagram, we see that the Braves will win the World Series in 10 of the 15 possible outcomes. Thus the probability that the Braves will win the championship (given that they win the first two games) is $\tfrac{10}{15} = \tfrac{2}{3}$. □

It follows from the definition of probability that when $E = \varnothing$, we have $n(E) = n(\varnothing) = 0$ and

$$\mathbf{P(\varnothing)} = \frac{n(\varnothing)}{n(S)} = \mathbf{0.}$$

Likewise if $E = S$, then $n(E) = n(S)$ and

$$\mathbf{P(S)} = \frac{n(S)}{n(S)} = \mathbf{1.}$$

For any event $E \subseteq S$, we also have

$$0 \le n(E) \le n(S)$$

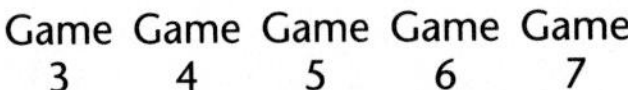

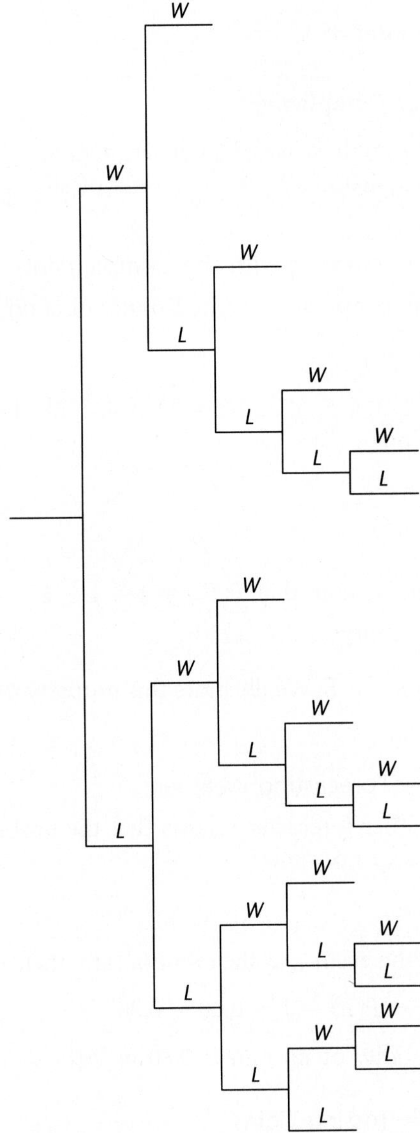

Figure 12.6 Tree diagram for the last five World Series games

and

$$0 \leq \frac{n(E)}{n(S)} \leq \frac{n(S)}{n(S)}$$

or

$$0 \le P(E) \le 1$$

since E is a subset of S.

Definition of Complement

Let S be the sample space of an experiment and E be an event. Then the **complement** of E, denoted by E', is the set of all simple events in S but not in E.

Example 4 • Working with the Complement

In Example 2b, consider the event E of not obtaining a 1. Determine E' and $P(E')$.

Solution

Since $S = \{1, 2, 3, 4, 5, 6\}$ and $E = \{2, 3, 4, 5, 6\}$, then $E' = \{1\}$ is the event of obtaining a 1. Thus

$$P(E') = \frac{n(E')}{n(S)} = \frac{1}{6}.$$

□

In Example 4, note that $P(E') = \frac{1}{6} = 1 - \frac{5}{6} = 1 - P(E)$. The property

$$P(E') = 1 - P(E)$$

is true for any event E. We illustrate this property of probability in the next example.

Example 5 • Forecasting Weather

Suppose a weather forecaster reports that the probability of rain is 20%. What is the probability of no rain?

Solution

If we describe the event E as the event of rain, then E' is the event of no rain. Thus

$$P(E') = 1 - P(E) = 1 - 0.20 = 0.80,$$

and the probability of no rain is 0.80 or 80%. □

Example 6 • Tossing Coins

Suppose two coins are tossed. Let E be the event of obtaining a head on the first coin and F be the event of obtaining a tail on the second coin. Compute $P(E \cup F)$.

Solution

The sample space S is given by $\{HH, HT, TH, TT\}$; the events are $E = \{HH, HT\}$ and $F = \{HT, TT\}$. Then $E \cup F$ is made up of the 3 simple events

$$E \cup F = \{HH, HT, TT\} \quad \text{and} \quad P(E \cup F) = \tfrac{3}{4}.$$

□

There is another way to compute the probability requested in Example 6. It would be tempting to compute

$$\begin{aligned} P(E) + P(F) &= P(\{HH, HT\}) + P(\{HT, TT\}) \\ &= \tfrac{2}{4} + \tfrac{2}{4} \\ &= 1 \end{aligned}$$

but in doing so, the simple event $\{HT\}$ has been counted twice. To compensate for this, we must subtract a quantity that is equal to the probability of the event common to the two events E and F. Any event common to the two events E and F is included in $E \cap F$. Thus we have

$$\begin{aligned} P(E \cup F) &= P(E) + P(F) - P(E \cap F) \\ &= \tfrac{2}{4} + \tfrac{2}{4} - \tfrac{1}{4} \\ &= \tfrac{3}{4}. \end{aligned}$$

We now summarize the important properties of probability, where S is the sample space and $P(E)$ represents the probability of an event E.

Properties of Probability

1. $0 \leq P(E) \leq 1$, for any event $E \subseteq S$
2. $P(S) = 1$
3. $P(\varnothing) = 0$
4. $P(E') = 1 - P(E)$, for any event $E \subseteq S$
5. $P(E \cup F) = P(E) + P(F) - P(E \cap F)$, for any events $E \subseteq S$ and $F \subseteq S$

Example 7 • Playing Cards

Consider an ordinary deck of 52 playing cards. If one card is drawn, determine the probability of its being (a) an ace, (b) a heart or a club, (c) an ace or a heart.

Solution

There are 52 simple events, one corresponding to each playing card. We assume that each of the simple events is equally likely; that is, no one card is more likely to be chosen than any other card.

a. Let A be the event of drawing an ace. Since there are 4 aces in the deck of cards,

$$P(A) = \tfrac{4}{52} = \tfrac{1}{13}.$$

b. Let H be the event of drawing a heart and C the event of drawing a club. Note that a card drawn cannot be both a heart and a club. Thus $H \cap C = \varnothing$, and the probability of drawing either a heart or a club is

$$\begin{aligned} P(H \cup C) &= P(H) + P(C) - P(H \cap C) \\ &= \tfrac{13}{52} + \tfrac{13}{52} - \tfrac{0}{52} \\ &= \tfrac{26}{52} \\ &= \tfrac{1}{2}. \end{aligned}$$

c. The event of drawing an ace or a heart is given by $A \cup H$. Also, $A \cap H$ is the event of drawing an ace and a heart, that is, the ace of hearts. Thus

$$\begin{aligned} P(A \cup H) &= P(A) + P(H) - P(A \cap H) \\ &= \tfrac{4}{52} + \tfrac{13}{52} - \tfrac{1}{52} \\ &= \tfrac{16}{52} \\ &= \tfrac{4}{13}. \end{aligned}$$

□

Example 8 • Using Combinations

Suppose a bag contains 10 marbles, of which 3 are red and 7 are blue. If two marbles are drawn from the bag, determine the probability that: (a) both are red, (b) one is red and one is blue, (c) at least one is red.

Solution

The sample space consists of all possible combinations of the 10 marbles taken 2 at a time. Hence there are $\binom{10}{2}$ elements in the sample space.

a. Let A be the event that both marbles drawn are red. The event A consists of all possible combinations of the 3 red marbles taken 2 at a time. Thus

$$P(A) = \frac{\binom{3}{2}}{\binom{10}{2}} = \frac{3}{45} = \frac{1}{15}.$$

b. Let B be the event of drawing 1 red and 1 blue marble. There are $\binom{3}{1}\binom{7}{1}$ ways that event B can occur, so

$$P(B) = \frac{\binom{3}{1}\binom{7}{1}}{\binom{10}{2}} = \frac{21}{45} = \frac{7}{15}.$$

c. Let C be the event that at least one red marble is drawn. Notice that C can be expressed as $A \cup B$. Since $A \cap B = \varnothing$, we have

$$\begin{aligned} P(C) = P(A \cup B) &= P(A) + P(B) - P(A \cap B) \\ &= \tfrac{1}{15} + \tfrac{7}{15} - 0 \\ &= \tfrac{8}{15}. \end{aligned}$$

□

◆ Practice Problem 1 ◆

Suppose a committee of 5 students is to be chosen from a class of 10 freshmen and 5 sophomores. What is the probability that the committee contains 2 freshmen and 3 sophomores?

EXERCISES 12.9

Coin and Die

1. A coin is tossed and a die is rolled. The upper face on both is noted.
   a. Determine the sample space of this experiment.
   b. Determine the event of obtaining a head on the coin and an even number on the die.

Coin Toss

2. Three coins are tossed simultaneously on the floor. The upper face on each of the 3 coins is noted. Determine (a) the sample space of the experiment, (b) the event of obtaining exactly 2 heads, and (c) the event of obtaining at least 2 heads.

Drawing Tags

3. Two bags contain tags numbered 1, 2, 3, 4, and 5. A tag is drawn from each bag and the number observed. Determine the probability of (a) choosing even-numbered tags from both bags, (b) choosing tags whose sum is 6, and (c) choosing tags whose sum is less than 6.

Rolling Dice

4. Two dice are tossed on the floor, and the numbers on their upper faces are observed. Determine the probability that the sum is (a) 7, (b) 11, (c) 7 or 11, and (d) 2 or 3 or 9.

The Photo Works

Coin Toss 5–6

5. If two coins are flipped, determine the probability of obtaining (a) 2 heads, (b) no heads, and (c) at least one tail.
6. If three coins are flipped, determine the probability of obtaining (a) 2 heads, (b) no heads, and (c) at least one tail.

World Series

7. Use a tree diagram to find the probability that a team will win the World Series, given that this team has won the first game of the series. (Assume the teams are evenly matched.)

"Three-Out-of-Five" Series

8. Suppose a team has won the first game of a "three-out-of-five" series. Use a tree diagram to find the probability that this team will win the series. (Assume the teams are evenly matched.)

Weighted Coin ✓9. Suppose a coin is weighted so that the heads side faces up more often than the tails side. If the probability of obtaining a heads is $\frac{3}{4}$, what is the probability of obtaining a tails?

"Loaded" Die 10. A die is "loaded" so that the numbers appear on the upper face with the following probabilities.

Number	1	2	3	4	5	6
Probability	$\frac{1}{9}$	$\frac{1}{9}$	$\frac{2}{9}$	$\frac{1}{6}$	$\frac{1}{6}$	?

Determine the probability of obtaining (a) a 6, (b) a 2 or a 6, (c) not a 6, (d) an even number, and (e) an odd number less than 5.

Drawing a Card 11. Suppose 1 card is drawn from an ordinary deck of 52 cards. Determine the probability of obtaining (a) a 2, (b) a black face card (jack, queen, or king), (c) a spade or a club, and (d) a face card or a club.

Balls and Blocks 12. Suppose a bag contains 10 balls and 12 blocks, where 3 balls and 4 blocks are red, 2 balls and 5 blocks are green, and 5 balls and 3 blocks are blue. If one item (either a ball or a block) is drawn from the bag, determine the probability that it is (a) red, (b) a ball, (c) not a blue block, and (d) green or a block.

Colored Marbles ✓13. If a bag contains 2 red, 5 black, and 6 blue marbles and if 3 marbles are drawn simultaneously, determine the probability that (a) all 3 are black, (b) none are red, (c) 1 is red, 1 is black, and 1 is blue, (d) at least 1 is red, and (e) all 3 are red.

Colored Tags 14. A box contains tags, of which 3 are green, 1 is white, and 6 are yellow. If 4 tags are drawn simultaneously, determine the probability that (a) all 4 are yellow, (b) none are green, (c) 1 is green, 1 is white, and 2 are yellow, (d) 2 are green and 2 are white, and (e) at least 1 is green.

Dealing Two Cards ✓15. If two cards are dealt from an ordinary deck of 52 cards, find the probability that (a) both are clubs, (b) both are kings, and (c) both are face cards.

Dealing Three Cards 16. From an ordinary deck of 52 cards, 3 cards are dealt. Find the probability that they are (a) 3 queens, (b) 2 kings and 1 jack, and (c) 1 queen, 1 jack, and 1 ace.

Poker Hand 17. A poker hand consists of 5 cards from an ordinary deck of 52 cards. What is the probability of being dealt a poker hand with (a) all 5 cards being in the same suit and (b) 3 aces and 2 kings?

Bridge Hand

18. A bridge hand consists of 13 cards dealt from an ordinary deck of 52 cards. Find the probability of being dealt a bridge hand with (a) all 13 cards being red and (b) only cards numbered 2 through 10. Leave your answers in terms of $\binom{n}{r}$.

Committee 19–20

19. From a group of 6 juniors and 5 seniors, a committee of 4 is chosen by lot. Find the probability that the committee will consist of (a) 4 juniors, (b) 2 juniors and 2 seniors, and (c) 1 junior and 3 seniors.

20. A committee is formed from a group of 4 faculty members, 5 students, and 3 administrators. If the committee has 3 members who are chosen strictly by lot, find the probability that (a) all are students, (b) none are students, (c) 2 are students and 1 is a faculty member, and (d) none are administrators.

Drawing Tags 21–22

21. Tags numbered 1 through 10 are placed in a hat. If two tags are drawn simultaneously, find the probability that (a) their sum is even and (b) their product is even.

22. Work Exercise 21, but this time suppose that the first tag is drawn, the number is recorded, and then it is placed back in the hat before the second tag is drawn.

Using Given Probabilities 23–24

23. Suppose the sample space is $S = A \cup B$, where A and B are events with $P(A) = 0.4$ and $P(B) = 0.9$. Find $P(A \cap B)$.

24. Suppose the sample space is $S = A \cup B$, where A and B are events with $P(A) = 0.6$ and $P(A \cap B) = 0.2$. Find $P(B)$.

Critical Thinking: Exploration and Writing

25. Several states have instituted state lotteries in recent years. Use library or other resources to obtain the procedure that is used to determine a winner in a state of your choice. Write a description of the procedure in terms of permutations and/or combinations, and calculate the probability of a single entry being a winner.

Critical Thinking: Discussion

26. In the casino game blackjack, a winning player receives an amount equal to the bet placed. An old scheme for winning calls for an initial bet of \$3. (This is sometimes the minimum bet.) If the first bet is lost, \$6 is bet on the next hand. This pattern continues, doubling the bet each time until the player wins. Discuss a player's chance of winning by use of this scheme. With a \$3 initial bet, what would be the amount of the sixth bet after 5 successive losses? What sort of rule might a casino institute to prohibit the use of this scheme?

◆ **Solution for Practice Problem**

1. The sample space S consists of all the 5 student committees formed from a class of 15. Thus

$$n(S) = \binom{15}{5} = 3003.$$

The event E consists of all the 5 student committees containing 2 freshmen and 3 sophomores. There are

$$\binom{10}{2}\binom{5}{3} = 450$$

such committees. Thus

$$P(E) = \frac{450}{3003} = \frac{150}{1001}.$$

CHAPTER REVIEW

Summary of Important Concepts and Formulas

Formulas for Arithmetic Sequences

$$a_n = a_1 + (n-1)d$$

$$S_n = \frac{n}{2}(a_1 + a_n)$$

$$S_n = \frac{n}{2}[2a_1 + (n-1)d]$$

Formulas for Geometric Sequences

$$a_n = a_1 r^{n-1}$$

$$S_n = a_1\frac{1-r^n}{1-r}, \text{ if } r \neq 1.$$

$$\text{If } |r| < 1, \sum_{n=1}^{\infty} a_1 r^{n-1} = \frac{a_1}{1-r}.$$

If $|r| \geq 1$ and $a_1 \neq 0$, $\sum_{n=1}^{\infty} a_1 r^{n-1}$ does not exist.

n Factorial

$n! = n \cdot (n-1) \cdots 2 \cdot 1$, if n is a positive integer.

$0! = 1$

Binomial Coefficients

$$\binom{n}{r} = \frac{n!}{(n-r)!r!}$$

Binomial Theorem

$$(a+b)^n = a^n + \binom{n}{1}a^{n-1}b + \binom{n}{2}a^{n-2}b^2 + \cdots + \binom{n}{r}a^{n-r}b^r + \cdots + \binom{n}{n-2}a^2b^{n-2} + \binom{n}{n-1}ab^{n-1} + b^n$$

rth Term in the Expansion of $(a+b)^n$

$$\binom{n}{r-1}a^{n-r+1}b^{r-1}$$

Principle of Mathematical Induction

Let $S(n)$ represent a statement about each positive integer n that is either true or false. If

$S(1)$ is true,

and if

the truth of $S(k)$ always implies the truth of $S(k+1)$,

then $S(n)$ is true for all positive integers n.

Permutations Formulas

$$P(n, r) = \frac{n!}{(n-r)!}$$

$$D = \frac{n!}{n_1!n_2!\cdots n_r!}, \text{ where } n_1 + n_2 + \cdots + n_r = n$$

Combinations Formula

$$C(n, r) = \frac{n!}{(n-r)!r!}$$

Properties of Probability

1. $0 \le P(E) \le 1$, for any event $E \subseteq S$
2. $P(S) = 1$
3. $P(\varnothing) = 0$
4. $P(E') = 1 - P(E)$, for any event $E \subseteq S$
5. $P(E \cup F) = P(E) + P(F) - P(E \cap F)$, for any events $E \subseteq S$ and $F \subseteq S$

Critical Thinking: Find the Errors

Each of the following nonsolutions has one or more errors. Can you find them?

Problem 1

Write $\sum_{n=2}^{7} (2n-1)$ in expanded form and find the value of the sum.

Nonsolution

$$\sum_{n=2}^{7} (2n-1) = 1 + 3 + 5 + 7 = 16$$

Problem 2

Find the seventh term in the arithmetic sequence whose first three terms are 2, −1, −4.

Nonsolution

$$a_7 = a_1r^6 = 2(-\tfrac{1}{2})^6 = 2(\tfrac{1}{64}) = \tfrac{1}{32}$$

Problem 3

If the following sum exists, find its value.

$$\tfrac{2}{3} - 1 + \tfrac{3}{2} - \tfrac{9}{4} + \cdots$$

Nonsolution

Since $a = \frac{2}{3}$ and $r = -\frac{3}{2}$, the value of the sum is

$$\frac{a}{1-r} = \frac{\frac{2}{3}}{1-(-\frac{3}{2})} = \frac{\frac{2}{3}}{\frac{5}{2}} = \frac{4}{15}.$$

Problem 4

A class of 10 seniors and 8 juniors elects a president and a secretary. In how many ways can this be done if no student can hold both offices and the president must be a senior?

Nonsolution

Since the president must be a senior there are 10 choices for the president and 8 choices of a junior for the secretary. Hence there are $10 \cdot 8 = 80$ ways of electing a president and secretary.

Problem 5

How many possible outcomes of win, place, and show are there in a horse race of 8 horses?

Nonsolution

We count the number of ways that 3 of the 8 horses will finish the race. There are

$$\binom{8}{3} = \frac{8}{3!5!} = \frac{8 \cdot 7 \cdot 6 \cdot 5!}{3 \cdot 2 \cdot 1 \cdot 5!} = 56$$

possible outcomes.

Problem 6

Determine the third term in the expansion of $(x - 2y)^9$.

Nonsolution

The third term is

$$\binom{9}{3}x^6(2y)^3 = \frac{9 \cdot 8 \cdot 7 \cdot 6!}{3 \cdot 2 \cdot 1 \cdot 6!}x^6 8y^3 = 672x^6y^3.$$

Review Problems for Chapter 12

Write the first five terms of the sequence that has the given general term.

1. $a_j = \dfrac{j-1}{j+1}$ 2. $a_n = \dfrac{(-2)^n}{n!}$

Write the first six terms of each sequence.

3. $a_1 = 1, \quad a_n = a_{n-1} + 2$
4. $a_1 = 2, \quad a_n = 2a_{n-1} - 3$

Write the given series in expanded form and find the value of the sum.

5. $\sum_{n=1}^{5} (-2)^n$ 6. $\sum_{n=1}^{6} \dfrac{1}{2^{n-1}}$ 7. $\sum_{n=1}^{7} 5$

8. $\sum_{n=1}^{4} \dfrac{(2n)!}{2^n}$ 9. $\sum_{n=1}^{5} \dfrac{n!}{(-1)^n}$

Write each of the following sums in sigma notation.

10. $2 + 4 + 6 + 8 + 10 + 12 + 14$
11. $(1)(2) + (2)(4) + (4)(6) + (8)(8) + (16)(10)$
12. $1 - 3 + 5 - 7 + 9 - 11$
13. $\frac{1}{2} + \frac{4}{3} + \frac{9}{4} + \frac{16}{5} + \frac{25}{6} + \frac{36}{7} + \frac{49}{8}$
14. $1 + \dfrac{1 \cdot 2}{2} + \dfrac{1 \cdot 2 \cdot 3}{2^2} + \cdots + \dfrac{1 \cdot 2 \cdot 3 \cdot 4 \cdot 5 \cdot 6}{2^5}$

In Problems 15 and 16, use the information given to find the value of the requested quantities of an arithmetic sequence.

15. $a_4 = 6$, $a_9 = -4$; find a_{11} and S_{11}.
16. $a_6 = 8$, $S_8 = 40$; find a_1 and d.
17. Find the first five terms of the arithmetic sequence that has $a_1 = 5$ and $d = -2$.
18. Find the nineteenth term and a formula for the nth term of the following arithmetic sequence.

    $3, 1, -1, -3, \ldots$

19. Determine whether or not each sequence is a geometric sequence. If it is, find r and a_n.
    a. $24, -36, 54, -81, \ldots$ b. $3, 4, 12, 20, \ldots$
20. Find r and a_1 for the geometric sequence that has $a_3 = 81$ and $a_6 = -24$.
21. Find the sixth term and a formula for the nth term of the geometric sequence with $a_1 = 4$, $r = -\frac{1}{2}$.
22. For each of the following infinite geometric sequences, determine whether or not the sum exists. If the sum exists, find its value.
    a. $\frac{9}{16}, \frac{3}{4}, 1, \frac{4}{3}, \ldots$ b. $9, -6, 4, -\frac{8}{3}, \ldots$
23. Find the value of $\sum_{n=1}^{\infty} 25(\frac{2}{3})^n$, if it exists. If it does not exist, give a reason.
24. Express $3.212121\ldots$ as a quotient of integers.

Saving Money

25. Leslie was determined to save some money for a vacation. So beginning November 1, she saved 1 cent, and on each succeeding day she saved twice as much as the day before. How much did Leslie save on November 15? What was the total amount saved if she stopped saving after November 15?

Distance Traveled by a Ball

26. Matthew had a ball that, when dropped from any height h to a concrete floor, bounced back to two-fifths of the height h. Approximately how far would the ball travel before coming to rest if he dropped it from a height of 2 meters?
27. Expand $(x - 2y)^6$ by using the binomial theorem.
28. Find the sixth term of $(2x - y^2)^9$.
29. Approximate $(1.02)^7$ by using the first four terms in a binomial expansion.

Use mathematical induction to prove that the following statements are true for all positive integers n.

30. $\frac{1}{1\cdot 3}+\frac{1}{3\cdot 5}+\cdots+\frac{1}{(2n-1)(2n+1)}=\frac{n}{2n+1}$

31. $3+6+11+\cdots+(n^2+2)=\frac{n(2n^2+3n+13)}{6}$

32. $n! > n^2$, if $n \geq 4$.

33. $a-b$ is a factor of $a^{2n}-b^{2n}$.

License Tags

34. License tags in a certain state consist of 3 digits, followed by one of the letters *A, B, C, D, E, F, G, H, X*, then followed by another 3 digits. How many different tags are possible?

Class Officers

35. A class of 12 girls and 14 boys elects a president, vice president, and secretary. In how many ways can this be done if no student can hold more than one office and
    a. There are no other restrictions?
    b. The vice president must be a boy?

License Plates

36. Suppose motorcycle license plates consist of a single letter followed by 3 digits. How many such tags are possible if
    a. No repetitions are allowed?
    b. Repetitions are allowed, and the last digit must be even?

Horse Race

37. If 8 horses enter a race, how many ways can the first 3 winning positions be awarded? (Assume that there are no ties.)

Arranging Books

38. In how many ways can 5 books with different titles be arranged between two bookends?

Seating Arrangement

39. In how many ways can 6 people be seated in row of 8 chairs?

Number Problem

40. How many 3-digit even numbers can be made from the digits 1, 2, 3, 4, 5, 6, and 7?

41. Evaluate the following.
    a. $0!$ b. $\frac{6!}{2!}$ c. $\frac{(n+1)!}{n!}$ d. $\frac{(n+1)!}{(n-1)!}$

42. Evaluate the following.
    a. $P(4, 2)$ b. $P(5, 3)$ c. $C(8, 3)$
    d. $\binom{7}{3}$ e. $C(n, n-1)$ f. $\binom{n+1}{n-1}$

Counting "Words" 43–46

How many "words" can be made from the letters in the given word?

43. *COLLEGE* 44. *DIVISION*
45. *ARKANSAS* 46. *TEXTBOOK*

Committees 47–48

47. From a group of 12 people, how many committees of 4 can be chosen?

48. From a group of 8 freshmen and 12 sophomores, a committee of 4 is to be chosen. How many different committees can be formed if exactly one sophomore serves on the committee?

Dog Show

49. If Roger has 12 dogs in his kennel, how many different sets of 5 dogs can he select to go to a dog show?

Using Given Probabilities

50. Suppose a sample space is $S = A \cup B$, where A and B are events with $P(A) = 0.8$ and $P(B) = 0.6$. Find $P(A \cap B)$.

Colored Balls

51. A bag contains 3 red and 5 blue balls. Three balls are to be chosen. Find the probability that each of the following selections is made: (a) 2 blue and 1 red; (b) 3 blue.

Committee

52. From a group of 6 girls and 5 boys a committee of 3 is chosen randomly. What is the probability that the committee contains (a) exactly 1 girl? (b) no boys?

ENCORE

Reuters/Bettmann

U.S. Air flight 5050 lies in New York's East River after skidding off the end of a La Guardia airport runway.

You are more likely to be killed by an asteroid striking the earth than by a plane crash, according to an Associated Press news release by Lee Siegel in July 1991. The release cited a 1990 report by the American Institute of Aeronautics and Astronautics which said that Asteroid 1989FC missed a collision with earth by 6 hours when it crossed the earth's orbit on March 23, 1989. The release said that Asteroid 1989FC is about 1/5 to 1/2 mile wide and would have killed millions of people if it had hit the northeastern United States, Los Angeles, or Tokyo.

David Morrison, a space scientist at NASA's Ames Research Center, estimated that the probability of an individual dying in a collision with an asteroid is 1 in 6000 to 1 in 20,000 during the next 50 years. He also said this probability is somewhat greater than the probability that an individual will be killed in an airplane crash.

Appendix

A.1 TABLE EVALUATION OF LOGARITHMS

In this section, we see how values of logarithms to base 10 can be found by using Table III on pages A-20 and A-21. As a starting point, we recall that any positive number N can be written as the product of a power of 10 and a number between 1 and 10: $N = a \times 10^n$, where $1 \le a < 10$. For example,

$$5520 = 5.52 \times 10^3 \qquad \text{and} \qquad 0.0436 = 4.36 \times 10^{-2}.$$

The expression $a \times 10^n$, where $1 \le a < 10$, is called the **scientific notation** for the number N. The equality $N = a \times 10^n$ means that $\log N$ differs from $\log a$ by an integer since

$$\begin{aligned} \log N &= \log (a \times 10^n) \\ &= \log a + \log 10^n \\ &= \log a + n \\ &= n + \log a. \end{aligned}$$

With the same numbers as above, we would have

$$\begin{aligned} \log 5520 &= \log (5.52 \times 10^3) \\ &= \log 5.52 + \log 10^3 \\ &= 3 + \log 5.52 \end{aligned}$$

and

$$\log 0.0436 = -2 + \log 4.36.$$

The significance of this is that the logarithms of numbers between 1 and 10 can be used to find the logarithms of all positive numbers. The table of common logarithms (Table III) gives the values of logarithms of numbers from 1 to 9.99, at intervals of 0.01. A portion of this table is reproduced in Figure A.1. The logarithm of a three-digit number† between 1 and 10 can be found in the table by locating the first two digits of the number in the column under N, and the last digit in the column headings at the top of the table. The logarithm of the number is located in

†By a three-digit number, we mean that there are three digits when the number is written in scientific notation. Zeros used only to place the decimal are not counted.

the row of the table that starts with the first two digits, and in the column that has the last digit at the top.

N	0	1	2	3	4	5	6	7	8	9
5.5	.7404	.7412	.7419	.7427	.7435	.7443	.7451	.7459	.7466	.7474
5.6	.7482	.7490	.7497	.7505	.7513	.7520	.7528	.7536	.7543	.7551
5.7	.7559	.7566	.7574	.7582	.7589	.7597	.7604	.7612	.7619	.7627
5.8	.7634	.7642	.7649	.7657	.7664	.7672	.7679	.7686	.7694	.7701
5.9	.7709	.7716	.7723	.7731	.7738	.7745	.7752	.7760	.7767	.7774

Figure A.1

Example 1

Use the log table to find the value of log 5.84.

Solution

In the log table, or in the portion of the table reproduced in Figure A.1, log 5.84 is located by matching up the row that has 5.8 under N and the column that has 4 at the top. The appropriate row and column are shaded in Figure A.1. The value of log 5.84 is found at their intersection.

$$\log 5.84 = 0.7664$$

Actually, the value 0.7665 is an approximation to four decimal places of log 5.84, as are most of the values in the log table. It would be more precise to write $\log 5.84 \approx 0.7664$, but we choose to write = instead of $\approx$ as a matter of convenience. □

We have noted before that any positive number N can be written in scientific notation as $N = a \times 10^n$, where $1 \leq a < 10$, and consequently,

$$\log N = n + \log a,$$

where $0 \leq \log a < 1$. Thus $\log N$ can be expressed as the sum of an integer and nonnegative decimal fraction less than 1. The integral part, n, of the logarithm is called the **characteristic,** and the fractional part, $\log a$, is called the **mantissa.** The integral part can be found by writing the number in scientific notation, and the mantissa for three-digit numbers can be read from the log table, with accuracy to four decimal places.

Example 2

Use the log table to find the value of log 584.

Solution

Writing 584 in scientific notation, we have

$$584 = 5.84 \times 10^2,$$

so the characteristic is 2 and the mantissa is

$$\log 5.84 = 0.7664$$

from the log table. Thus

$$\begin{aligned}\log 584 &= 2 + 0.7664 \\ &= 2.7664.\end{aligned}$$

□

It may happen, of course, that the characteristic is a negative number.

Example 3

Use the log table to find the value of log 0.00584.

Solution

Writing the number in scientific notation, we have

$$0.00584 = 5.84 \times 10^{-3},$$

and the characteristic is -3. As in Example 1, the mantissa is 0.7664, so

$$\log 0.00584 = -3 + 0.7664.$$

The convention in a situation like this is to avoid combining the negative integer and the positive fraction and write the logarithm in the form

$$\log 0.00584 = 7.7664 - 10.$$

□

In order to perform calculations by use of logarithms, we must be able to use the table to find a number N when $\log N$ is known.

Example 4

Find the number N, given that $\log N = 8.7657 - 10$.

Solution

The digits in N are determined by the mantissa 0.7657, and the position of the decimal in N is determined by the characteristic, which is $8 - 10 = -2$. To find the digits in N, we search through the body of the log table until we find the entry 0.7657 (note that the mantissa increases as N increases in the table). This entry is found in the row that starts with 5.8, and in the column headed by 3. Thus

$$\begin{aligned}N &= 5.83 \times 10^{-2} \\ &= 0.0583.\end{aligned}$$

□

When a number N is found by using $\log N$, the number N is referred to as the **antilogarithm** of $\log N$ (abbreviated **antilog**). Thus, in Example 4, we would say that

$$\text{antilog}\,(8.7657 - 10) = 0.0583.$$

As mentioned earlier, the log table furnishes logarithms only for numbers with three digits. Tables exist that give logarithms for numbers with more than three digits. However, the log table can be used to approximate the logarithm of a four-digit number with very good accuracy, employing a procedure called **linear interpolation.**

Example 5

To illustrate the procedure of linear interpolation, let us consider the problem of finding the value of log 10.37. The two numbers nearest 10.37 that have logarithms in the log table are 10.30 and 10.40. Their logarithms are given by

$$\log 10.30 = 1.0128 \qquad \text{and} \qquad \log 10.40 = 1.0170.$$

Thinking geometrically, this means that the points $P(10.30, 1.0128)$ and $Q(10.40, 1.0170)$ are on the graph of $y = \log x$. A portion of the graph that contains these points is shown in Figure A.2. The coordinates at R on the curve are (10.37, log 10.37). The idea behind linear interpolation is this: Use the straight-line seg-

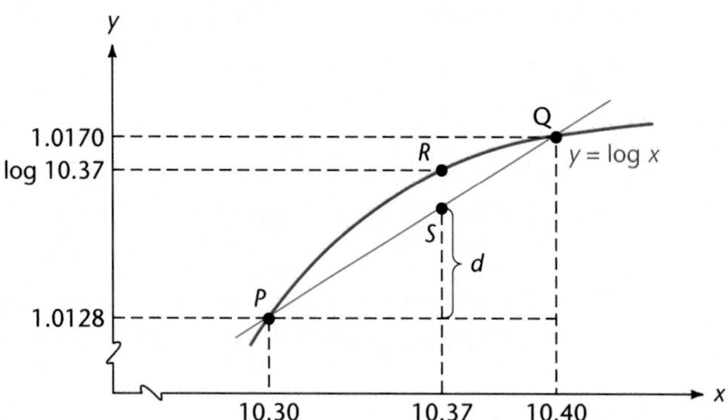

Figure A.2

ment PQ as an approximation to the curve $y = \log x$ and use the ordinate at point S as an approximation to log 10.37. The ordinate at S can be found by adding the difference d to the ordinate at P. Since 10.37 is $\frac{7}{10}$ of the distance from 10.30 to 10.40, the difference d is $\frac{7}{10}$ of the difference between the ordinates at P and Q. These differences can be set up as follows.

$$
10\left[\begin{array}{l}
7\left[\begin{array}{l}\log 10.30 = 1.0128\\ \log 10.37 = \underline{\qquad\quad}\end{array}\right]d\\
\log 10.40 = 1.0170
\end{array}\right]0.0042
$$

To find d, we use the proportion

$$\frac{d}{0.0042} = \frac{7}{10}$$

and obtain

$$\begin{aligned} d &= (\tfrac{7}{10})(0.0042) \\ &= 0.00294 \\ &= 0.0029, \end{aligned}$$

where d is rounded to the number of decimal places in the table. Adding d to the ordinate at P, we have

$$\begin{aligned} \log 10.37 &= 1.0128 + 0.0029 \\ &= 1.0157. \end{aligned}$$

(A calculator gives log 10.37 = 1.015778756.) □

The next example shows how linear interpolation can be used in finding antilogarithms.

Example 6

If $\log N = 8.4089 - 10$, use interpolation to approximate N to four digits.

Solution

The mantissa 0.4089 is located between 0.4082 and 0.4099 in the tables, and these mantissas correspond to the digits 256 and 257. Thus we have the following arrangement.

$$10\left[\begin{array}{l} x\left[\begin{array}{l}\log 0.02560 = 8.4082 - 10 \\ \log N \qquad\;\; = 8.4089 - 10\end{array}\right] 0.0007 \\ \log 0.02570 = 8.4099 - 10 \end{array}\right] 0.0017$$

$$\frac{x}{10} = \frac{0.0007}{0.0017}$$

$$x = \tfrac{70}{17} = 4$$

The number x represents the last digit in N and is rounded to the nearest whole number. Thus we have

$$N = 0.02564.$$

(A calculator gives $N = 0.0256389361$.) □

EXERCISES A.1

Find the common logarithms of each number, using the log table as necessary. (See Examples 1–3.)

1. 4.16×10^{-9}
2. 1.73×10^{3}
3. 30.7
4. 17.3
5. 4.51
6. 3.01
7. 10.0
8. 10^{-4}
9. 1,070,000
10. 10,800
11. 0.00107
12. 0.000132

Find N in each equation, using the log table as necessary. (See Example 4.)

13. $\log N = 0.8561$
14. $\log N = 0.9186$
15. $\log N = 8.4518$
16. $\log N = 4.6385$
17. $\log N = 7.6776 - 10$
18. $\log N = 6.9943 - 10$
19. $\log N = -3$
20. $\log N = 4$

Use the log table and linear interpolation to find the logarithm of each number. Check the accuracy with a calculator if one is available. (See Example 5.)

21. 10.11
22. 243.6
23. 4.171
24. 1.017
25. 417,800
26. 7103
27. 0.007717
28. 0.08354

Use the log table and linear interpolation to find N in each equation. Check the accuracy with a calculator if one is available. (See Example 6.)

29. $\log N = 0.1113$
30. $\log N = 0.7549$
31. $\log N = 4.9433$
32. $\log N = 5.4686$
33. $\log N = 7.6950 - 10$
34. $\log N = 8.9963 - 10$
35. $\log N = 9.3881 - 10$
36. $\log N = 9.5240 - 10$

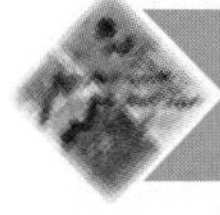

A.2 TABLE EVALUATION OF TRIGONOMETRIC FUNCTIONS

In this section, we see how values of trigonometric functions can be found by the use of tables. There are tables of trigonometric functions beginning on page A-11.

Table I gives values of all six trigonometric functions for angles from 0° to 90°, measured in degrees and minutes, at intervals of 10 minutes. A portion of this table is reproduced in Figure A.3. The use of the columns labeled *Radians* in Table I is described later in this section. Angles from 0° to 45° are listed in the leftmost column of the table, and function values for these angles are read by using the function labels at the *top* of the columns. Angles from 45° to 90° are listed in the rightmost column, and function values for these angles are read by using the function labels at the *bottom* of the column. Some illustrations of the use of Table I are given in Example 1.

Example 1

Use Table I to find the following function values.

a. $\tan 35°40'$ b. $\sec 54°50'$ c. $\csc 35°10'$

Angle θ									
Degrees	Radians	sin θ	csc θ	tan θ	cot θ	sec θ	cos θ		
27° 00′	.4712	.4540	2.203	.5095	1.963	1.122	.8910	1.0996	63° 00′
10	.4741	.4566	2.190	.5132	1.949	1.124	.8897	1.0966	50
20	.4771	.4592	2.178	.5169	1.935	1.126	.8884		40
30	.4801	.4626	2.166			.127			
40									
50			2.751	.6959			.8208		
35° 00′	.6109	.5736	1.743	.7002	1.428	1.221	.8192	.9599	55° 00′
10	.6138	.5760	1.736	.7046	1.419	1.223	.8175	.9570	50
20	.6167	.5783	1.729	.7089	1.411	1.226	.8158	.9541	40
30	.6196	.5807	1.722	.7133	1.402	1.228	.8141	.9512	30
40	.6225	.5831	1.715	.7177	1.393	1.231	.8124	.9483	20
50	.6254	.5854	1.708	.7221	1.385	1.233	.8107	.9454	10
36° 00	.6283	.5878	1.701	.7265	1.376	1.236	.8090	.9425	54° 00′
		cos θ	sec θ	cot θ	tan θ	csc θ	sin θ	Radians	Degrees
								Angle θ	

Figure A.3

Solution

a. tan 35°40′ is found in the row of the table with 35°40′ at the left end, and in the column with tan θ at the top.

$$\tan 35°40' = 0.7177$$

b. sec 54°50′ is in the row that has 54°50′ at the right end, and in the column that has sec θ at the bottom.

$$\sec 54°50' = 1.736$$

c. csc 35°10′ is in the row with 35°10′ at the left end, and in the column with csc θ at the top.

$$\csc 35°10' = 1.736$$

□

The fact that sec 54°50′ = csc 35°10′ in Example 1 is no accident. It illustrates some general facts about Table I that are worth noting. First of all, we observe that the angles at opposite ends of each row are *complementary:* their sum is 90°. Next we see that the functions paired together at the top and bottom of each column have related names:

$$\sin\theta \text{ and } \cos\theta, \qquad \tan\theta \text{ and } \cot\theta, \qquad \sec\theta \text{ and } \csc\theta.$$

Each function in one of these pairs is called the **complementary function,** or **cofunction,** of the other. The structure that we have observed in the table reflects the fact that **any function of a given angle equals the cofunction of the complementary angle.** The identities in Chapter 7 show why this is true.

For angles outside the range $0 \le \theta \le 90°$, the related angle is used as it was in Section 7.5. Some illustrations are given in Example 2.

Example 2

Use Table I to find the following function values.

a. $\cos 217°40'$ b. $\sin 476°10'$

Solution

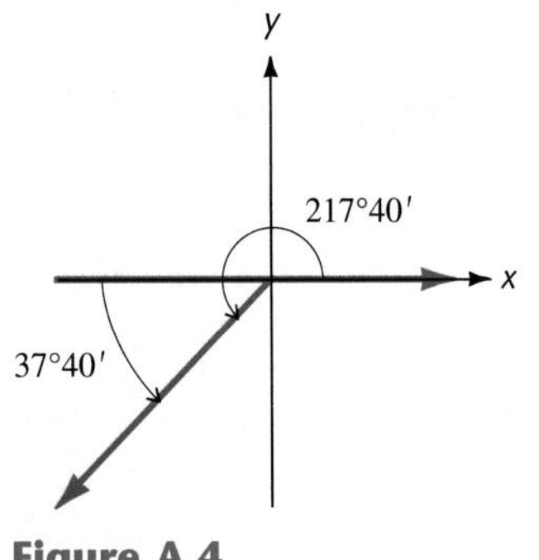

Figure A.4

a. As shown in Figure A.4, the related angle is

$$217°40' - 180° = 37°40'.$$

Since $217°40'$ is a third-quadrant angle, its cosine is a negative number. Reading $\cos 37°40'$ from Table I, we obtain

$$\begin{aligned}\cos 217°40' &= -\cos 37°40' \\ &= -0.7916.\end{aligned}$$

b. We first find an angle between 0° and 360° that is coterminal with $476°10'$.

$$476°10' - 360° = 116°10'$$

This is shown in Figure A.5, along with the related angle, which is

$$180° - 116°10' = 63°50'.$$

Since $\sin\theta$ is positive for a second quadrant angle,

$$\begin{aligned}\sin 476°10' &= \sin 116°10' \\ &= \sin 63°50' \\ &= 0.8975.\end{aligned}$$

□

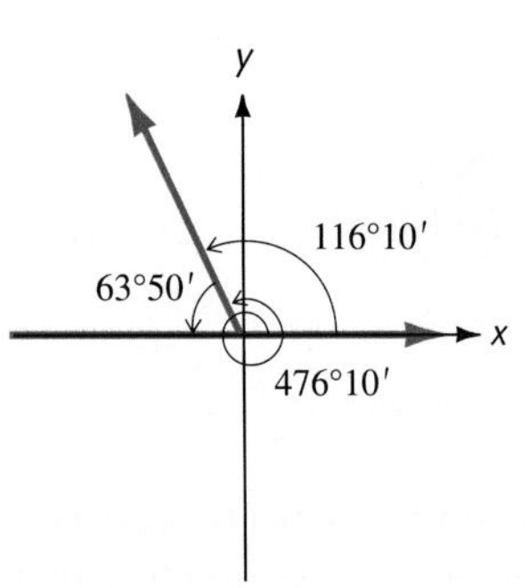

Figure A.5

If a trigonometric function value of θ is known, then Table I can be used to find the value of θ, correct to the nearest 10 minutes.† If the given function value does not appear in the table, we choose the angle whose function value is nearest the given value.

†More accuracy can be obtained by use of interpolation.

Example 3

Use Table I to find a value of θ, correct to the nearest 10 minutes, that satisfies the given equation.

a. $\cos\theta = 0.5398$ b. $\tan\theta = 0.4597$

Solution

a. The value 0.5398 is found in Table I in the column with $\cos\theta$ at the bottom and in the row with $57°20'$ at the right end. Thus

$$\cos\theta = 0.5398$$

has the solution

$$\theta = 57°20'.$$

b. The value $\tan\theta = 0.4597$ does not occur in Table I. The two tangent values nearest 0.4597 are

$$\tan 24°40' = 0.4592 \qquad \text{and} \qquad \tan 24°50' = 0.4628$$

Since 0.4597 is nearer 0.4592 than 0.4628, we take the value of θ to the nearest 10 minutes to be

$$\theta = 24°40'.$$

□

Table II is read in much the same way as Table I. The essential differences are that angles are given in decimal degrees at intervals of 0.1°, and the only functions included in the table are sine, cosine, tangent, and cotangent.

Table I gives values of all six trigonometric functions for angles in radian measure at increments of 0.0029 from 0 to 1.5708 radians. For angles outside the range $0 \le \theta \le 1.5708$, we use the radian measure of the related angle.† This is illustrated in the next example.

Example 4

Use Table I to find $\cos 2.5598$.

Solution

We think of $t = 2.5598$ as radian measure of an angle. Since

$$\frac{\pi}{2} < 2.5598 < \pi,$$

the angle terminates in quadrant II and has a related angle with radian measure

$$\pi - 2.5598 \approx 0.5818.$$

†Radian measure of the related angle is sometimes called the *related number.*

Since the cosine function is negative in quadrant II,

$$\cos 2.5598 = -\cos 0.5818 = -0.8355.$$

The value $\cos 0.5818 = 0.8355$ is read from Table I. □

As this example shows, there is little that is new in finding values of trigonometric functions of numbers.

The problems in Exercises 7.5 may be worked at this point.

Table I Trigonometric Functions—Degrees and Minutes or Radians

Angle θ									
Degrees	Radians	$\sin\theta$	$\csc\theta$	$\tan\theta$	$\cot\theta$	$\sec\theta$	$\cos\theta$		
0° 00′	.0000	.0000	No value	.0000	No value	1.000	1.0000	1.5708	90° 00′
10	.0029	.0029	343.8	.0029	343.8	1.000	1.0000	1.5679	50
20	.0058	.0058	171.9	.0058	171.9	1.000	1.0000	1.5650	40
30	.0087	.0087	114.6	.0087	114.6	1.000	1.0000	1.5621	30
40	.0116	.0116	85.95	.0116	85.94	1.000	.9999	1.5592	20
50	.0145	.0145	68.76	.0145	68.75	1.000	.9999	1.5563	10
1° 00′	.0175	.0175	57.30	.0175	57.29	1.000	.9998	1.5533	89° 00′
10	.0204	.0204	49.11	.0204	49.10	1.000	.9998	1.5504	50
20	.0233	.0233	42.98	.0233	42.96	1.000	.9997	1.5475	40
30	.0262	.0262	38.20	.0262	38.19	1.000	.9997	1.5446	30
40	.0291	.0291	34.38	.0291	34.37	1.000	.9996	1.5417	20
50	.0320	.0320	31.26	.0320	31.24	1.001	.9995	1.5388	10
2° 00′	.0349	.0349	28.65	.0349	28.64	1.001	.9994	1.5359	88° 00′
10	.0378	.0378	26.45	.0378	26.43	1.001	.9993	1.5330	50
20	.0407	.0407	24.56	.0407	24.54	1.001	.9992	1.5301	40
30	.0436	.0436	22.93	.0437	22.90	1.001	.9990	1.5272	30
40	.0465	.0465	21.49	.0466	21.47	1.001	.9989	1.5243	20
50	.0495	.0494	20.23	.0495	20.21	1.001	.9988	1.5213	10
3° 00′	.0524	.0523	19.11	.0524	19.08	1.001	.9986	1.5184	87° 00′
10	.0553	.0552	18.10	.0553	18.07	1.002	.9985	1.5155	50
20	.0582	.0581	17.20	.0582	17.17	1.002	.9983	1.5126	40
30	.0611	.0610	16.38	.0612	16.35	1.002	.9981	1.5097	30
40	.0640	.0640	15.64	.0641	15.60	1.002	.9980	1.5068	20
50	.0669	.0669	14.96	.0670	14.92	1.002	.9978	1.5039	10
4° 00′	.0698	.0698	14.34	.0699	14.30	1.002	.9976	1.5010	86° 00′
10	.0727	.0727	13.76	.0729	13.73	1.003	.9974	1.5981	50
20	.0756	.0756	13.23	.0758	13.20	1.003	.9971	1.5952	40
30	.0785	.0785	12.75	.0787	12.71	1.003	.9969	1.5923	30
40	.0814	.0814	12.29	.0816	12.25	1.003	.9967	1.5893	20
50	.0844	.0843	11.87	.0846	11.83	1.004	.9964	1.5864	10
5° 00′	.0873	.0872	11.47	.0875	11.43	1.004	.9962	1.4835	85° 00′
10	.0902	.0901	11.10	.0904	11.06	1.004	.9959	1.4806	50
20	.0931	.0929	10.76	.0934	10.71	1.004	.9957	1.4777	40
30	.0960	.0958	10.43	.0963	10.39	1.005	.9954	1.4748	30
40	.0989	.0987	10.13	.0992	10.08	1.005	.9951	1.4719	20
50	.1018	.1016	9.839	.1022	9.788	1.005	.9948	1.4690	10
6° 00′	.1047	.1045	9.567	.1051	9.514	1.006	.9945	1.4661	84° 00′
10	.1076	.1074	9.309	.1080	9.255	1.006	.9942	1.4632	50
20	.1105	.1103	9.065	.1110	9.010	1.006	.9939	1.4603	40
30	.1134	.1132	8.834	.1139	8.777	1.006	.9936	1.4573	30
40	.1164	.1161	8.614	.1169	8.556	1.007	.9932	1.4544	20
50	.1193	.1190	8.405	.1198	8.345	1.007	.9929	1.4515	10
7° 00′	.1222	.1219	8.206	.1228	8.144	1.008	.9925	1.4486	83° 00′
10	.1251	.1248	8.016	.1257	7.953	1.008	.9922	1.4457	50
20	.1280	.1276	7.834	.1287	7.770	1.008	.9918	1.4428	40
30	.1309	.1305	7.661	.1317	7.596	1.009	.9914	1.4399	30
40	.1338	.1334	7.496	.1346	7.429	1.009	.9911	1.4370	20
50	.1367	.1363	7.337	.1376	7.269	1.009	.9907	1.4341	10
8° 00′	.1396	.1392	7.185	.1405	7.115	1.010	.9903	1.4312	82° 00′
		$\cos\theta$	$\sec\theta$	$\cot\theta$	$\tan\theta$	$\csc\theta$	$\sin\theta$	Radians	Degrees
								Angle θ	

Angle θ									
Degrees	Radians	$\sin\theta$	$\csc\theta$	$\tan\theta$	$\cot\theta$	$\sec\theta$	$\cos\theta$		
8° 00′	.1396	.1392	7.185	.1405	7.115	1.010	.9903	1.4312	82° 00′
10	.1425	.1421	7.040	.1435	6.968	1.010	.9899	1.4283	50
20	.1454	.1449	6.900	.1465	6.827	1.011	.9894	1.4254	40
30	.1484	.1478	6.765	.1495	6.691	1.011	.9890	1.4224	30
40	.1513	.1507	6.636	.1524	6.561	1.012	.9886	1.4195	20
50	.1542	.1536	6.512	.1554	6.435	1.012	.9881	1.4166	10
9° 00′	.1571	.1564	6.392	.1584	6.314	1.012	.9877	1.4137	81° 00′
10	.1600	.1593	6.277	.1614	6.197	1.013	.9872	1.4108	50
20	.1629	.1622	6.166	.1644	6.084	1.013	.9868	1.4079	40
30	.1658	.1650	6.059	.1673	5.976	1.014	.9863	1.4050	30
40	.1687	.1679	5.955	.1703	5.871	1.014	.9858	1.4021	20
50	.1716	.1708	5.855	.1733	5.769	1.015	.9853	1.3992	10
10° 00′	.1745	.1736	5.759	.1763	5.671	1.015	.9848	1.3963	80° 00′
10	.1774	.1765	5.665	.1793	5.576	1.016	.9843	1.3934	50
20	.1804	.1794	5.575	.1823	5.485	1.016	.9838	1.3904	40
30	.1833	.1822	5.487	.1853	5.396	1.017	.9833	1.3875	30
40	.1862	.1851	5.403	.1883	5.309	1.018	.9827	1.3846	20
50	.1891	.1880	5.320	.1914	5.226	1.018	.9822	1.3817	10
11° 00′	.1920	.1908	5.241	.1944	5.145	1.019	.9816	1.3788	79° 00′
10	.1949	.1937	5.164	.1974	5.066	1.019	.9811	1.3759	50
20	.1978	.1965	5.089	.2004	4.989	1.020	.9805	1.3730	40
30	.2007	.1994	5.016	.2035	4.915	1.020	.9799	1.3701	30
40	.2036	.2022	4.945	.2065	4.843	1.021	.9793	1.3672	20
50	.2065	.2051	4.876	.2095	4.773	1.022	.9787	1.3643	10
12° 00′	.2094	.2079	4.810	.2126	4.705	1.022	.9781	1.3614	78° 00′
10	.2123	.2108	4.745	.2156	4.638	1.023	.9775	1.3584	50
20	.2153	.2136	4.682	.2186	4.574	1.024	.9769	1.3555	40
30	.2182	.2164	4.620	.2217	4.511	1.024	.9763	1.3526	30
40	.2211	.2193	4.560	.2247	4.449	1.025	.9757	1.3497	20
50	.2240	.2221	4.502	.2278	4.390	1.026	.9750	1.3468	10
13° 00′	.2269	.2250	4.445	.2309	4.331	1.026	.9744	1.3439	77° 00′
10	.2298	.2278	4.390	.2339	4.275	1.027	.9737	1.3410	50
20	.2327	.2306	4.336	.2370	4.219	1.028	.9730	1.3381	40
30	.2356	.2334	4.284	.2401	4.165	1.028	.9724	1.3352	30
40	.2385	.2363	4.232	.2432	4.113	1.029	.9717	1.3323	20
50	.2414	.2391	4.182	.2462	4.061	1.030	.9710	1.3294	10
14° 00′	.2443	.2419	4.134	.2493	4.011	1.031	.9703	1.3265	76° 00′
10	.2473	.2447	4.086	.2524	3.962	1.031	.9696	1.3235	50
20	.2502	.2476	4.039	.2555	3.914	1.032	.9689	1.3206	40
30	.2531	.2504	3.994	.2586	3.867	1.033	.9681	1.3177	30
40	.2560	.2532	3.950	.2617	3.821	1.034	.9674	1.3148	20
50	.2589	.2560	3.906	.2648	3.776	1.034	.9667	1.3119	10
15° 00′	.2618	.2588	3.864	.2679	3.732	1.035	.9659	1.3090	75° 00′
10	.2647	.2616	3.822	.2711	3.689	1.036	.9652	1.3061	50
20	.2676	.2644	3.782	.2742	3.647	1.037	.9644	1.3032	40
30	.2705	.2672	3.742	.2773	3.606	1.038	.9636	1.3003	30
40	.2734	.2700	3.703	.2805	3.566	1.039	.9628	1.2974	20
50	.2763	.2728	3.665	.2836	3.526	1.039	.9621	1.2945	10
16° 00′	.2793	.2756	3.628	.2867	3.487	1.040	.9613	1.2915	74° 00′
		$\cos\theta$	$\sec\theta$	$\cot\theta$	$\tan\theta$	$\csc\theta$	$\sin\theta$	Radians	Degrees
								Angle θ	

Table I Trigonometric Functions—Degrees and Minutes or Radians (*continued*)

Angle θ									
Degrees	Radians	sin θ	csc θ	tan θ	cot θ	sec θ	cos θ		
16° 00′	.2793	.2756	3.628	.2867	3.487	1.040	.9613	1.2915	74° 00′
10	.2822	.2784	3.592	.2899	3.450	1.041	.9605	1.2886	50
20	.2851	.2812	3.556	.2931	3.412	1.042	.9596	1.2857	40
30	.2880	.2840	3.521	.2962	3.376	1.043	.9588	1.2828	30
40	.2909	.2868	3.487	.2944	3.340	1.044	.9580	1.2799	20
50	.2938	.2896	3.453	.3026	3.305	1.045	.9572	1.2770	10
17° 00′	.2967	.2924	3.420	.3057	3.271	1.046	.9563	1.2741	73° 00′
10	.2996	.2952	3.388	.3089	3.237	1.047	.9555	1.2712	50
20	.3025	.2979	3.357	.3121	3.204	1.048	.9546	1.2683	40
30	.3054	.3007	3.326	.3153	3.172	1.048	.9537	1.2654	30
40	.3083	.3035	3.295	.3185	3.140	1.049	.9528	1.2625	20
50	.3113	.3062	3.265	.3217	3.108	1.050	.9520	1.2595	10
18° 00′	.3142	.3090	3.236	.3249	3.078	1.051	.9511	1.2566	72° 00′
10	.3171	.3118	3.207	.3281	3.047	1.052	.9502	1.2537	50
20	.3200	.3145	3.179	.3314	3.018	1.053	.9492	1.2508	40
30	.3229	.3173	3.152	.3346	2.989	1.054	.9483	1.2479	30
40	.3258	.3201	3.124	.3378	2.960	1.056	.9474	1.2450	20
50	.3287	.3228	3.098	.3411	2.932	1.057	.9465	1.2421	10
19° 00′	.3316	.3256	3.072	.3443	2.904	1.058	.9455	1.2392	71° 00′
10	.3345	.3283	3.046	.3476	2.877	1.059	.9446	1.2363	50
20	.3374	.3311	3.021	.3508	2.850	1.060	.9436	1.2334	40
30	.3403	.3338	2.996	.3541	2.824	1.061	.9426	1.2305	30
40	.3432	.3365	2.971	.3574	2.798	1.062	.9417	1.2275	20
50	.3462	.3393	2.947	.3607	2.773	1.063	.9407	1.2246	10
20° 00′	.3491	.3420	2.924	.3640	2.747	1.064	.9397	1.2217	70° 00′
10	.3520	.3448	2.901	.3673	2.723	1.065	.9387	1.2188	50
20	.3549	.3475	2.878	.3706	2.699	1.066	.9377	1.2159	40
30	.3578	.3502	2.855	.3739	2.675	1.068	.9367	1.2130	30
40	.3607	.3529	2.833	.3772	2.651	1.069	.9356	1.2101	20
50	.3636	.3557	2.812	.3805	2.628	1.070	.9346	1.2072	10
21° 00′	.3665	.3584	2.790	.3839	2.605	1.071	.9336	1.2043	69° 00′
10	.3694	.3611	2.769	.3872	2.583	1.072	.9325	1.2014	50
20	.3723	.3638	2.749	.3906	2.560	1.074	.9315	1.1985	40
30	.3752	.3665	2.729	.3939	2.539	1.075	.9304	1.1956	30
40	.3782	.3692	2.709	.3973	2.517	1.076	.9293	1.1926	20
50	.3811	.3719	2.689	.4006	2.496	1.077	.9283	1.1897	10
22° 00′	.3840	.3746	2.669	.4040	2.475	1.079	.9272	1.1868	68° 00′
10	.3869	.3773	2.650	.4074	2.455	1.080	.9261	1.1839	50
20	.3898	.3800	2.632	.4108	2.434	1.081	.9250	1.1810	40
30	.3927	.3827	2.613	.4142	2.414	1.082	.9239	1.1781	30
40	.3956	.3854	2.595	.4176	2.394	1.084	.9228	1.1752	20
50	.3985	.3881	2.577	.4210	2.375	1.085	.9216	1.1723	10
23° 00′	.4014	.3907	2.559	.4245	2.356	1.086	.9205	1.1694	67° 00′
10	.4043	.3934	2.542	.4279	2.337	1.088	.9194	1.1665	50
20	.4072	.3961	2.525	.4314	2.318	1.089	.9182	1.1636	40
30	.4102	.3987	2.508	.4348	2.300	1.090	.9171	1.1606	30
40	.4131	.4014	2.491	.4383	2.282	1.092	.9159	1.1577	20
50	.4160	.4041	2.475	.4417	2.264	1.093	.9147	1.1548	10
24° 00′	.4189	.4067	2.459	.4452	2.246	1.095	.9135	1.1519	66° 00′
		cos θ	sec θ	cot θ	tan θ	csc θ	sin θ	Radians	Degrees
								Angle θ	

Angle θ									
Degrees	Radians	sin θ	csc θ	tan θ	cot θ	sec θ	cos θ		
24° 00′	.4189	.4067	2.459	.4452	2.246	1.095	.9135	1.1519	66° 00′
10	.4218	.4094	2.443	.4487	2.229	1.096	.9124	1.1490	50
20	.4247	.4120	2.427	.4522	2.211	1.097	.9112	1.1461	40
30	.4276	.4147	2.411	.4557	2.194	1.099	.9100	1.1432	30
40	.4305	.4173	2.396	.4592	2.177	1.100	.9088	1.1403	20
50	.4334	.4200	2.381	.4628	2.161	1.102	.9075	1.1374	10
25° 00′	.4363	.4226	2.366	.4663	2.145	1.103	.9063	1.1345	65° 00′
10	.4392	.4253	2.352	.4699	2.128	1.105	.9051	1.1316	50
20	.4422	.4279	2.337	.4734	2.112	1.106	.9038	1.1286	40
30	.4451	.4305	2.323	.4770	2.097	1.108	.9026	1.1257	30
40	.4480	.4331	2.309	.4806	2.081	1.109	.9013	1.1228	20
50	.4509	.4358	2.295	.4841	2.066	1.111	.9001	1.1199	10
26° 00′	.4538	.4384	2.281	.4877	2.050	1.113	.8988	1.1170	64° 00′
10	.4567	.4410	2.268	.4913	2.035	1.114	.8975	1.1141	50
20	.4596	.4436	2.254	.4950	2.020	1.116	.8962	1.1112	40
30	.4625	.4462	2.241	.4986	2.006	1.117	.8949	1.1083	30
40	.4654	.4488	2.228	.5022	1.991	1.119	.8936	1.1054	20
50	.4683	.4514	2.215	.5059	1.977	1.121	.8923	1.1025	10
27° 00′	.4712	.4540	2.203	.5095	1.963	1.122	.8910	1.0996	63° 00′
10	.4741	.4566	2.190	.5132	1.949	1.124	.8897	1.0966	50
20	.4771	.4592	2.178	.5169	1.935	1.126	.8884	1.0937	40
30	.4800	.4617	2.166	.5206	1.921	1.127	.8870	1.0908	30
40	.4829	.4643	2.154	.5243	1.907	1.129	.8857	1.0879	20
50	.4858	.4669	2.142	.5280	1.894	1.131	.8843	1.0850	10
28° 00′	.4887	.4695	2.130	.5317	1.881	1.133	.8829	1.0821	62° 00′
10	.4916	.4720	2.118	.5354	1.868	1.134	.8816	1.0792	50
20	.4945	.4746	2.107	.5392	1.855	1.136	.8802	1.0763	40
30	.4974	.4772	2.096	.5430	1.842	1.138	.8788	1.0734	30
40	.5003	.4797	2.085	.5467	1.829	1.140	.8774	1.0705	20
50	.5032	.4823	2.074	.5505	1.816	1.142	.8760	1.0676	10
29° 00′	.5061	.4848	2.063	.5543	1.804	1.143	.8746	1.0647	61° 00′
10	.5091	.4874	2.052	.5581	1.792	1.145	.8732	1.0617	50
20	.5120	.4899	2.041	.5619	1.780	1.147	.8718	1.0588	40
30	.5149	.4924	2.031	.5658	1.767	1.149	.8704	1.0559	30
40	.5178	.4950	2.020	.5696	1.756	1.151	.8689	1.0530	20
50	.5207	.4975	2.010	.5735	1.744	1.153	.8675	1.0501	10
30° 00′	.5236	.5000	2.000	.5774	1.732	1.155	.8660	1.0472	60° 00′
10	.5265	.5025	1.990	.5812	1.720	1.157	.8646	1.0443	50
20	.5294	.5050	1.980	.5851	1.709	1.159	.8631	1.0414	40
30	.5323	.5075	1.970	.5890	1.698	1.161	.8616	1.0385	30
40	.5352	.5100	1.961	.5930	1.686	1.163	.8601	1.0356	20
50	.5381	.5125	1.951	.5969	1.675	1.165	.8587	1.0327	10
31° 00′	.5411	.5150	1.942	.6009	1.664	1.167	.8572	1.0297	59° 00′
10	.5440	.5175	1.932	.6048	1.653	1.169	.8557	1.0268	50
20	.5469	.5200	1.923	.6088	1.643	1.171	.8542	1.0239	40
30	.5498	.5225	1.914	.6128	1.632	1.173	.8526	1.0210	30
40	.5527	.5250	1.905	.6168	1.621	1.175	.8511	1.0181	20
50	.5556	.5275	1.896	.6208	1.611	1.177	.8496	1.0152	10
32° 00′	.5585	.5299	1.887	.6249	1.600	1.179	.8480	1.0123	58° 00′
		cos θ	sec θ	cot θ	tan θ	csc θ	sin θ	Radians	Degrees
								Angle θ	

Table I Trigonometric Functions—Degrees and Minutes or Radians *(continued)*

Angle θ Degrees	Angle θ Radians	sin θ	csc θ	tan θ	cot θ	sec θ	cos θ		
32° 00′	.5585	.5299	1.887	.6249	1.600	1.179	.8480	1.0123	58° 00′
10	.5614	.5324	1.878	.6289	1.590	1.181	.8465	1.0094	50
20	.5643	.5348	1.870	.6330	1.580	1.184	.8450	1.0065	40
30	.5672	.5373	1.861	.6371	1.570	1.186	.8434	1.0036	30
40	.5701	.5398	1.853	.6412	1.560	1.188	.8418	1.0007	20
50	.5730	.5422	1.844	.6453	1.550	1.190	.8403	.9977	10
33° 00′	.5760	.5446	1.836	.6494	1.540	1.192	.8387	.9948	57° 00′
10	.5789	.5471	1.828	.6536	1.530	1.195	.8371	.9919	50
20	.5818	.5495	1.820	.6577	1.520	1.198	.8355	.9890	40
30	.5847	.5519	1.812	.6619	1.511	1.199	.8339	.9861	30
40	.5876	.5544	1.804	.6661	1.501	1.202	.8323	.9832	20
50	.5905	.5568	1.796	.6703	1.492	1.204	.8307	.9803	10
34° 00′	.5934	.5592	1.788	.6745	1.483	1.206	.8290	.9774	56° 00′
10	.5963	.5616	1.781	.6787	1.473	1.209	.8274	.9745	50
20	.5992	.5640	1.773	.6830	1.464	1.211	.8258	.9716	40
30	.6021	.5664	1.766	.6873	1.455	1.213	.8241	.9687	30
40	.6050	.5688	1.758	.6916	1.446	1.216	.8225	.9657	20
50	.6080	.5712	1.751	.6959	1.437	1.218	.8208	.9628	10
35° 00′	.6109	.5736	1.743	.7002	1.428	1.221	.8192	.9599	55° 00′
10	.6138	.5760	1.736	.7046	1.419	1.223	.8175	.9570	50
20	.6167	.5783	1.729	.7089	1.411	1.226	.8158	.9541	40
30	.6196	.5807	1.722	.7133	1.402	1.228	.8141	.9512	30
40	.6225	.5831	1.715	.7177	1.393	1.231	.8124	.9483	20
50	.6254	.5854	1.708	.7221	1.385	1.233	.8107	.9454	10
36° 00′	.6283	.5878	1.701	.7265	1.376	1.236	.8090	.9425	54° 00′
10	.6312	.5901	1.695	.7310	1.368	1.239	.8073	.9396	50
20	.6341	.5925	1.688	.7355	1.360	1.241	.8056	.9367	40
30	.6370	.5948	1.681	.7400	1.351	1.244	.8039	.9338	30
40	.6400	.5972	1.675	.7445	1.343	1.247	.8021	.9308	20
50	.6429	.5995	1.668	.7490	1.335	1.249	.8004	.9279	10
37° 00′	.6458	.6018	1.662	.7536	1.327	1.252	.7986	.9250	53° 00′
10	.6487	.6041	1.655	.7581	1.319	1.255	.7969	.9221	50
20	.6516	.6065	1.649	.7627	1.311	1.258	.7951	.9192	40
30	.6545	.6088	1.643	.7673	1.303	1.260	.7934	.9163	30
40	.6574	.6111	1.636	.7720	1.295	1.263	.7916	.9134	20
50	.6603	.6134	1.630	.7766	1.288	1.266	.7898	.9105	10
38° 00′	.6632	.6157	1.624	.7813	1.280	1.269	.7880	.9076	52° 00′
10	.6661	.6180	1.618	.7860	1.272	1.272	.7862	.9047	50
20	.6690	.6202	1.612	.7907	1.265	1.275	.7844	.9018	40
30	.6720	.6225	1.606	.7954	1.257	1.278	.7826	.8988	30
40	.6749	.6248	1.601	.8002	1.250	1.281	.7808	.8959	20
50	.6778	.6271	1.595	.8050	1.242	1.284	.7790	.8930	10
39° 00′	.6807	.6293	1.589	.8098	1.235	1.287	.7771	.8901	51° 00′
		cos θ	sec θ	cot θ	tan θ	csc θ	sin θ	Angle θ Radians	Angle θ Degrees

Angle θ Degrees	Angle θ Radians	sin θ	csc θ	tan θ	cot θ	sec θ	cos θ		
39° 00′	.6807	.6293	1.589	.8098	1.235	1.287	.7771	.8901	51° 00′
10	.6836	.6316	1.583	.8146	1.228	1.290	.7753	.8872	50
20	.6865	.6338	1.578	.8195	1.220	1.293	.7735	.8843	40
30	.6894	.6361	1.572	.8243	1.213	1.296	.7716	.8814	30
40	.6923	.6383	1.567	.8292	1.206	1.299	.7698	.8785	20
50	.6952	.6406	1.561	.8342	1.199	1.302	.7679	.8756	10
40° 00′	.6981	.6428	1.556	.8391	1.192	1.305	.7660	.8727	50° 00′
10	.7010	.6450	1.550	.8441	1.185	1.309	.7642	.8698	50
20	.7039	.6472	1.545	.8491	1.178	1.312	.7623	.8668	40
30	.7069	.6494	1.540	.8541	1.171	1.315	.7604	.8639	30
40	.7098	.6517	1.535	.8591	1.164	1.318	.7585	.8610	20
50	.7127	.6539	1.529	.8642	1.157	1.322	.7566	.8581	10
41° 00′	.7156	.6561	1.524	.8693	1.150	1.325	.7547	.8552	49° 00′
10	.7185	.6583	1.519	.8744	1.144	1.328	.7528	.8523	50
20	.7214	.6604	1.514	.8796	1.137	1.332	.7509	.8494	40
30	.7243	.6626	1.509	.8847	1.130	1.335	.7490	.8465	30
40	.7272	.6648	1.504	.8899	1.124	1.339	.7470	.8436	20
50	.7301	.6670	1.499	.8952	1.117	1.342	.7451	.8407	10
42° 00′	.7330	.6691	1.494	.9004	1.111	1.346	.7432	.8378	48° 00′
10	.7359	.6713	1.490	.9057	1.104	1.349	.7412	.8348	50
20	.7389	.6734	1.485	.9110	1.098	1.353	.7392	.8319	40
30	.7418	.6756	1.480	.9163	1.091	1.356	.7373	.8290	30
40	.7447	.6777	1.476	.9217	1.085	1.360	.7353	.8261	20
50	.7476	.6799	1.471	.9271	1.079	1.364	.7333	.8232	10
43° 00′	.7505	.6820	1.466	.9325	1.072	1.367	.7314	.8203	47° 00′
10	.7534	.6841	1.462	.9380	1.066	1.371	.7294	.8174	50
20	.7563	.6862	1.457	.9435	1.060	1.375	.7274	.8145	40
30	.7592	.6884	1.453	.9490	1.054	1.379	.7254	.8116	30
40	.7621	.6905	1.448	.9545	1.048	1.382	.7234	.8087	20
50	.7650	.6926	1.444	.9601	1.042	1.386	.7214	.8058	10
44° 00′	.7679	.6947	1.440	.9657	1.036	1.390	.7193	.8029	46° 00′
10	.7709	.6967	1.435	.9713	1.030	1.394	.7173	.7999	50
20	.7738	.6988	1.431	.9770	1.024	1.398	.7153	.7970	40
30	.7767	.7009	1.427	.9827	1.018	1.402	.7133	.7941	30
40	.7796	.7030	1.423	.9884	1.012	1.406	.7112	.7912	20
50	.7825	.7050	1.418	.9942	1.006	1.410	.7092	.7883	10
45° 00′	.7854	.7071	1.414	1.000	1.000	1.414	.7071	.7854	45° 00′
		cos θ	sec θ	cot θ	tan θ	csc θ	sin θ	Angle θ Radians	Angle θ Degrees

Table II Trigonometric Functions—Decimal Degrees (csc $\theta = 1/\sin\theta$; sec $\theta = 1/\cos\theta$)

↓→	sin	cos	tan	cot	
0.0	0.00000	1.0000	0.00000	∞	90.0
.1	.00175	1.0000	.00175	573.0	89.9
.2	.00349	1.0000	.00349	286.5	.8
.3	.00524	1.0000	.00524	191.0	.7
.4	.00698	1.0000	.00698	143.24	.6
.5	.00873	1.0000	.00873	114.59	.5
.6	.01047	0.9999	.01047	95.49	.4
.7	.01222	.9999	.01222	81.85	.3
.8	.01396	.9999	.01396	71.62	.2
.9	.01571	.9999	.01571	63.66	89.1
1.0	0.01745	0.9998	0.01746	57.29	89.0
.1	.01920	.9998	.01920	52.08	88.9
.2	.02094	.9998	.02095	47.74	.8
.3	.02269	.9997	.02269	44.07	.7
.4	.02443	.9997	.02444	40.92	.6
.5	.02618	.9997	.02619	38.19	.5
.6	.02792	.9996	.02793	35.80	.4
.7	.02967	.9996	.02968	33.69	.3
.8	.03141	.9995	.03143	31.82	.2
.9	.03316	.9995	.03317	30.14	88.1
2.0	0.03490	0.9994	0.03492	28.64	88.0
.1	.03664	.9993	.03667	27.27	87.9
.2	.03839	.9993	.03842	26.03	.8
.3	.04013	.9992	.04016	24.90	.7
.4	.04188	.9991	.04191	23.86	.6
.5	.04362	.9990	.04366	22.90	.5
.6	.04536	.9990	.04541	22.02	.4
.7	.04711	.9989	.04716	21.20	.3
.8	.04885	.9988	.04891	20.45	.2
.9	.05059	.9987	.05066	19.74	87.1
3.0	0.05234	0.9986	0.05241	19.081	87.0
.1	.05408	.9985	.05416	18.464	86.9
.2	.05582	.9984	.05591	17.886	.8
.3	.05756	.9983	.05766	17.343	.7
.4	.05931	.9982	.05941	16.832	.6
.5	.06105	.9981	.06116	16.350	.5
.6	.06279	.9980	.06291	15.895	.4
.7	.06453	.9979	.06467	15.464	.3
.8	.06627	.9978	.06642	15.056	.2
.9	.06802	.9977	.06817	14.669	86.1
4.0	0.06976	0.9976	0.06993	14.301	86.0
	cos	sin	cot	tan	←↑

↓→	sin	cos	tan	cot	
4.0	0.06976	0.9976	0.06993	14.301	86.0
.1	.07150	.9974	.07168	13.951	85.9
.2	.07324	.9973	.07344	13.617	.8
.3	.07498	.9972	.07519	13.300	.7
.4	.07672	.9971	.07695	12.996	.6
.5	.07846	.9969	.07870	12.706	.5
.6	.08020	.9968	.08046	12.429	.4
.7	.08194	.9966	.08221	12.163	.3
.8	.08368	.9965	.08397	11.909	.2
.9	.08542	.9963	.08573	11.664	85.1
5.0	0.08716	0.9962	0.08749	11.430	85.0
.1	.08889	.9960	.08925	11.205	84.9
.2	.09063	.9959	.09101	10.988	.8
.3	.09237	.9957	.09277	10.780	.7
.4	.09411	.9956	.09453	10.579	.6
.5	.09585	.9954	.09629	10.385	.5
.6	.09758	.9952	.09805	10.199	.4
.7	.09932	.9951	.09981	10.019	.3
.8	.10106	.9949	.10158	9.845	.2
.9	.10279	.9947	.10334	9.677	84.1
6.0	0.10453	0.9945	0.10510	9.514	84.0
.1	.10626	.9943	.10687	9.357	83.9
.2	.10800	.9942	.10863	9.205	.8
.3	.10973	.9940	.11040	9.058	.7
.4	.11147	.9938	.11217	8.915	.6
.5	.11320	.9936	.11394	8.777	.5
.6	.11494	.9934	.11570	8.643	.4
.7	.11667	.9932	.11747	8.513	.3
.8	.11840	.9930	.11924	8.386	.2
.9	.12014	.9928	.12101	8.264	83.1
7.0	0.12187	0.9925	0.12278	8.144	83.0
.1	.12360	.9923	.12456	8.028	82.9
.2	.12533	.9921	.12633	7.916	.8
.3	.12706	.9919	.12810	7.806	.7
.4	.12880	.9917	.12988	7.700	.6
.5	.13053	.9914	.13165	7.596	.5
.6	.13226	.9912	.13343	7.495	.4
.7	.13399	.9910	.13521	7.396	.3
.8	.13572	.9907	.13698	7.300	.2
.9	.13744	.9905	.13876	7.207	82.1
8.0	0.13917	0.9903	0.14054	7.115	82.0
	cos	sin	cot	tan	←↑

Table II Trigonometric Functions—Decimal Degrees (csc $\theta = 1/\sin \theta$; sec $\theta = 1/\cos \theta$) (continued)

↓→	sin	cos	tan	cot		↓→	sin	cos	tan	cot	
8.0	0.13917	0.9903	0.14054	7.115	82.0	12.0	0.2079	0.9781	0.2126	4.705	78.0
.1	.14090	.9900	.14232	7.026	81.9	.1	.2096	.9778	.2144	4.665	77.9
.2	.14263	.9898	.14410	6.940	.8	.2	.2113	.9774	.2162	4.625	.8
.3	.14436	.9895	.14588	6.855	.7	.3	.2130	.9770	.2180	4.586	.7
.4	.14608	.9893	.14767	6.772	.6	.4	.2147	.9767	.2199	4.548	.6
.5	.14781	.9890	.14945	6.691	.5	.5	.2164	.9763	.2217	4.511	.5
.6	.14954	.9888	.15124	6.612	.4	.6	.2181	.9759	.2235	4.474	.4
.7	.15126	.9885	.15302	6.535	.3	.7	.2198	.9755	.2254	4.437	.3
.8	.15299	.9882	.15481	6.460	.2	.8	.2215	.9751	.2272	4.402	.2
.9	.15471	.9880	.15660	6.386	81.1	.9	.2233	.9748	.2290	4.366	77.1
9.0	0.15643	0.9877	0.15838	6.314	81.0	13.0	0.2250	0.9744	0.2309	4.331	77.0
.1	.15816	.9874	.16017	6.243	80.9	.1	.2267	.9740	.2327	4.297	76.9
.2	.15988	.9871	.16196	6.174	.8	.2	.2284	.9736	.2345	4.264	.8
.3	.16160	.9869	.16376	6.107	.7	.3	.2300	.9732	.2364	4.230	.7
.4	.16333	.9866	.16555	6.041	.6	.4	.2317	.9728	.2382	4.198	.6
.5	.16505	.9863	.16734	5.976	.5	.5	.2334	.9724	.2401	4.165	.5
.6	.16677	.9860	.16914	5.912	.4	.6	.2351	.9720	.2419	4.134	.4
.7	.16849	.9857	.17093	5.850	.3	.7	.2368	.9715	.2438	4.102	.3
.8	.17021	.9854	.17273	5.789	.2	.8	.2385	.9711	.2456	4.071	.2
.9	.17193	.9851	.17453	5.730	80.1	.9	.2402	.9707	.2475	4.041	76.1
10.0	0.1736	0.9848	0.1763	5.671	80.0	14.0	0.2419	0.9703	0.2493	4.011	76.0
.1	.1754	.9845	.1781	5.614	79.9	.1	.2436	.9699	.2512	3.981	75.9
.2	.1771	.9842	.1799	5.558	.8	.2	.2453	.9694	.2530	3.952	.8
.3	.1788	.9839	.1817	5.503	.7	.3	.2470	.9690	.2549	3.923	.7
.4	.1805	.9836	.1835	5.449	.6	.4	.2487	.9686	.2568	3.895	.6
.5	.1822	.9833	.1853	5.396	.5	.5	.2504	.9681	.2586	3.867	.5
.6	.1840	.9829	.1871	5.343	.4	.6	.2521	.9677	.2605	3.839	.4
.7	.1857	.9826	.1890	5.292	.3	.7	.2538	.9673	.2623	3.812	.3
.8	.1874	.9823	.1908	5.242	.2	.8	.2554	.9668	.2642	3.785	.2
.9	.1891	.9820	.1926	5.193	79.1	.9	.2571	.9664	.2661	3.758	75.1
11.0	0.1908	0.9816	0.1944	5.145	79.0	15.0	0.2588	0.9659	0.2679	3.732	75.0
.1	.1925	.9813	.1962	5.097	78.9	.1	.2605	.9655	.2698	3.706	74.9
.2	.1942	.9810	.1980	5.050	.8	.2	.2622	.9650	.2717	3.681	.8
.3	.1959	.9806	.1998	5.005	.7	.3	.2639	.9646	.2736	3.655	.7
.4	.1977	.9803	.2016	4.959	.6	.4	.2656	.9641	.2754	3.630	.6
.5	.1994	.9799	.2035	4.915	.5	.5	.2672	.9636	.2773	3.606	.5
.6	.2011	.9796	.2053	4.872	.4	.6	.2689	.9632	.2792	3.582	.4
.7	.2028	.9792	.2071	4.829	.3	.7	.2706	.9627	.2811	3.558	.3
.8	.2045	.9789	.2089	4.787	.2	.8	.2723	.9622	.2830	3.534	.2
.9	.2062	.9785	.2107	4.745	78.1	.9	.2740	.9617	.2849	3.511	74.1
12.0	0.2079	0.9781	0.2126	4.705	78.0	16.0	0.2756	0.9613	0.2867	3.487	74.0
	cos	sin	cot	tan	←↑		cos	sin	cot	tan	←↑

Table II Trigonometric Functions—Decimal Degrees (csc $\theta = 1/\sin\theta$; sec $\theta = 1/\cos\theta$) (*continued*)

↓→	sin	cos	tan	cot	
16.0	0.2756	0.9613	0.2867	3.487	74.0
.1	.2773	.9608	.2886	3.465	73.9
.2	.2790	.9603	.2905	3.442	.8
.3	.2807	.9598	.2924	3.420	.7
.4	.2823	.9593	.2943	3.398	.6
.5	.2840	.9588	.2962	3.376	.5
.6	.2857	.9583	.2981	3.354	.4
.7	.2874	.9578	.3000	3.333	.3
.8	.2890	.9573	.3019	3.312	.2
.9	.2907	.9568	.3038	3.291	73.1
17.0	0.2924	0.9563	0.3057	3.271	73.0
.1	.2940	.9558	.3076	3.251	72.9
.2	.2957	.9553	.3096	3.230	.8
.3	.2974	.9548	.3115	3.211	.7
.4	.2990	.9542	.3134	3.191	.6
.5	.3007	.9537	.3153	3.172	.5
.6	.3024	.9532	.3172	3.152	.4
.7	.3040	.9527	.3191	3.133	.3
.8	.3057	.9521	.3211	3.115	.2
.9	.3074	.9516	.3230	3.096	72.1
18.0	0.3090	0.9511	0.3249	3.078	72.0
.1	.3107	.9505	.3269	3.060	71.9
.2	.3123	.9500	.3288	3.042	.8
.3	.3140	.9494	.3307	3.024	.7
.4	.3156	.9489	.3327	3.006	.6
.5	.3173	.9483	.3346	2.989	.5
.6	.3190	.9478	.3365	2.971	.4
.7	.3206	.9472	.3385	2.954	.3
.8	.3223	.9466	.3404	2.937	.2
.9	.3239	.9461	.3424	2.921	71.1
19.0	0.3256	0.9455	0.3443	2.904	71.0
.1	.3272	.9449	.3463	2.888	70.9
.2	.3289	.9444	.3482	2.872	.8
.3	.3305	.9438	.3502	2.856	.7
.4	.3322	.9432	.3522	2.840	.6
.5	.3338	.9426	.3541	2.824	.5
.6	.3355	.9421	.3561	2.808	.4
.7	.3371	.9415	.3581	2.793	.3
.8	.3387	.9409	.3600	2.778	.2
.9	.3404	.9403	.3620	2.762	70.1
20.0	0.3420	0.9397	0.3640	2.747	70.0
	cos	sin	cot	tan	←↑

↓→	sin	cos	tan	cot	
20.0	0.3420	0.9397	0.3640	2.747	70.0
.1	.3437	.9391	.3659	2.733	69.9
.2	.3453	.9385	.3679	2.718	.8
.3	.3469	.9379	.3699	2.703	.7
.4	.3486	.9373	.3719	2.689	.6
.5	.3502	.9367	.3739	2.675	.5
.6	.3518	.9361	.3759	2.660	.4
.7	.3535	.9354	.3779	2.646	.3
.8	.3551	.9348	.3799	2.633	.2
.9	.3567	.9342	.3819	2.619	69.1
21.0	0.3584	0.9336	0.3839	2.605	69.0
.1	.3600	.9330	.3859	2.592	68.9
.2	.3616	.9323	.3879	2.578	.8
.3	.3633	.9317	.3899	2.565	.7
.4	.3649	.9311	.3919	2.552	.6
.5	.3665	.9304	.3939	2.539	.5
.6	.3681	.9298	.3959	2.526	.4
.7	.3697	.9291	.3979	2.513	.3
.8	.3714	.9285	.4000	2.500	.2
.9	.3730	.9278	.4020	2.488	68.1
22.0	0.3746	0.9272	0.4040	2.475	68.0
.1	.3762	.9265	.4061	2.463	67.9
.2	.3778	.9259	.4081	2.450	.8
.3	.3795	.9252	.4101	2.438	.7
.4	.3811	.9245	.4122	2.426	.6
.5	.3827	.9239	.4142	2.414	.5
.6	.3843	.9232	.4163	2.402	.4
.7	.3859	.9225	.4183	2.391	.3
.8	.3875	.9219	.4204	2.379	.2
.9	.3891	.9212	.4224	2.367	67.1
23.0	0.3907	0.9205	0.4245	2.356	67.0
.1	.3923	.9198	.4265	2.344	66.9
.2	.3939	.9191	.4286	2.333	.8
.3	.3955	.9184	.4307	2.322	.7
.4	.3971	.9178	.4327	2.311	.6
.5	.3987	.9171	.4348	2.300	.5
.6	.4003	.9164	.4369	2.289	.4
.7	.4019	.9157	.4390	2.278	.3
.8	.4035	.9150	.4411	2.267	.2
.9	.4051	.9143	.4431	2.257	66.1
24.0	0.4067	0.9135	0.4452	2.246	66.0
	cos	sin	cot	tan	←↑

Table II Trigonometric Functions—Decimal Degrees (csc $\theta = 1/\sin\theta$; sec $\theta = 1/\cos\theta$) (*continued*)

↳	sin	cos	tan	cot	
24.0	0.4067	0.9135	0.4452	2.246	66.0
.1	.4083	.9128	.4473	2.236	65.9
.2	.4099	.9121	.4494	2.225	.8
.3	.4115	.9114	.4515	2.215	.7
.4	.4131	.9107	.4536	2.204	.6
.5	.4147	.9100	.4557	2.194	.5
.6	.4163	.9092	.4578	2.184	.4
.7	.4179	.9085	.4599	2.174	.3
.8	.4195	.9078	.4621	2.164	.2
.9	.4210	.9070	.4642	2.154	65.1
25.0	0.4226	0.9063	0.4663	2.145	65.0
.1	.4242	.9056	.4684	2.135	64.9
.2	.4258	.9048	.4706	2.125	.8
.3	.4274	.9041	.4727	2.116	.7
.4	.4289	.9033	.4748	2.106	.6
.5	.4305	.9026	.4770	2.097	.5
.6	.4321	.9018	.4791	2.087	.4
.7	.4337	.9011	.4813	2.078	.3
.8	.4352	.9003	.4834	2.069	.2
.9	.4368	.8996	.4856	2.059	64.1
26.0	0.4384	0.8988	0.4877	2.050	64.0
.1	.4399	.8980	.4899	2.041	63.9
.2	.4415	.8973	.4921	2.032	.8
.3	.4431	.8965	.4942	2.023	.7
.4	.4446	.8957	.4964	2.014	.6
.5	.4462	.8949	.4986	2.006	.5
.6	.4478	.8942	.5008	1.997	.4
.7	.4493	.8934	.5029	1.988	.3
.8	.4509	.8926	.5051	1.980	.2
.9	.4524	.8918	.5073	1.971	63.1
27.0	0.4540	0.8910	0.5095	1.963	63.0
.1	.4555	.8902	.5117	1.954	62.9
.2	.4571	.8894	.5139	1.946	.8
.3	.4586	.8886	.5161	1.937	.7
.4	.4602	.8878	.5184	1.929	.6
.5	.4617	.8870	.5206	1.921	.5
.6	.4633	.8862	.5228	1.913	.4
.7	.4648	.8854	.5250	1.905	.3
.8	.4664	.8846	.5272	1.897	.2
.9	.4679	.8838	.5295	1.889	62.1
28.0	0.4695	0.8829	0.5317	1.881	62.0
	cos	sin	cot	tan	↲

↳	sin	cos	tan	cot	
28.0	0.4695	0.8829	0.5317	1.881	62.0
.1	.4710	.8821	.5340	1.873	61.9
.2	.4726	.8813	.5362	1.865	.8
.3	.4741	.8805	.5384	1.857	.7
.4	.4756	.8796	.5407	1.849	.6
.5	.4772	.8788	.5430	1.842	.5
.6	.4787	.8780	.5452	1.834	.4
.7	.4802	.8771	.5475	1.827	.3
.8	.4818	.8763	.5498	1.819	.2
.9	.4833	.8755	.5520	1.811	61.1
29.0	0.4848	0.8746	0.5543	1.804	61.0
.1	.4863	.8738	.5566	1.797	60.9
.2	.4879	.8729	.5589	1.789	.8
.3	.4894	.8721	.5612	1.782	.7
.4	.4909	.8712	.5635	1.775	.6
.5	.4924	.8704	.5658	1.767	.5
.6	.4939	.8695	.5681	1.760	.4
.7	.4955	.8686	.5704	1.753	.3
.8	.4970	.8678	.5727	2.746	.2
.9	.4985	.8669	.5750	1.739	60.1
30.0	0.5000	0.8660	0.5774	1.7321	60.0
.1	.5015	.8652	.5797	1.7251	50.9
.2	.5030	.8643	.5820	1.7182	.8
.3	.5045	.8634	.5844	1.7113	.7
.4	.5060	.8625	.5867	1.7045	.6
.5	.5075	.8616	.5890	1.6977	.5
.6	.5090	.8607	.5914	1.6909	.4
.7	.5105	.8599	.5938	1.6842	.3
.8	.5120	.8590	.5961	1.6775	.2
.9	.5135	.8581	.5985	1.6709	59.1
31.0	0.5150	0.8572	0.6009	1.6643	59.0
.1	.5165	.8563	.6032	1.6577	58.9
.2	.5180	.8554	.6056	1.6512	.8
.3	.5195	.8545	.6080	1.6447	.7
.4	.5210	.8536	.6104	1.6383	.6
.5	.5225	.8526	.6128	1.6319	.5
.6	.5240	.8517	.6152	1.6255	.4
.7	.5255	.8508	.6176	1.6191	.3
.8	.5270	.8499	.6200	1.6128	.2
.9	.5284	.8490	.6224	1.6066	58.1
32.0	0.5299	0.8480	0.6249	1.6003	58.0
	cos	sin	cot	tan	↲

Table II Trigonometric Functions—Decimal Degrees (csc $\theta = 1/\sin\theta$; sec $\theta = 1/\cos\theta$) (*continued*)

↓→	sin	cos	tan	cot	
32.0	0.5299	0.8480	0.6249	1.6003	58.0
.1	.5314	.8471	.6273	1.5941	57.9
.2	.5329	.8462	.6297	1.5880	.8
.3	.5344	.8453	.6322	1.5818	.7
.4	.5358	.8443	.6346	1.5757	.6
.5	.5373	.8434	.6371	1.5697	.5
.6	.5388	.8425	.6395	1.5637	.4
.7	.5402	.8415	.6420	1.5577	.3
.8	.5417	.8406	.6445	1.5517	.2
.9	.5432	.8396	.6469	1.5458	57.1
33.0	0.5446	0.8387	0.6494	1.5399	57.0
.1	.5461	.8377	.6519	1.5340	56.9
.2	.5476	.8368	.6544	1.5282	.8
.3	.5490	.8358	.6569	1.5224	.7
.4	.5505	.8348	.6594	1.5166	.6
.5	.5519	.8339	.6619	1.5108	.5
.6	.5534	.8329	.6644	1.5051	.4
.7	.5548	.8320	.6669	1.4994	.3
.8	.5563	.8310	.6694	1.4938	.2
.9	.5577	.8300	.6720	1.4882	56.1
34.0	0.5592	0.8290	0.6745	1.4826	56.0
.1	.5606	.8281	.6771	1.4770	55.9
.2	.5621	.8271	.6796	1.4715	.8
.3	.5635	.8261	.6822	1.4659	.7
.4	.5650	.8251	.6847	1.4605	.6
.5	.5664	.8241	.6873	1.4550	.5
.6	.5678	.8231	.6899	1.4496	.4
.7	.5693	.8221	.6924	1.4442	.3
.8	.5707	.8211	.6950	1.4388	.2
.9	.5721	.8202	.6976	1.4335	55.1
35.0	0.5736	0.8192	0.7002	1.4281	55.0
.1	.5750	.8181	.7028	1.4229	54.9
.2	.5764	.8171	.7054	1.4176	.8
.3	.5779	.8161	.7080	1.4124	.7
.4	.5793	.8151	.7107	1.4071	.6
.5	.5807	.8141	.7133	1.4019	.5
.6	.5821	.8131	.7159	1.3968	.4
.7	.5835	.8121	.7186	1.3916	.3
.8	.5850	.8111	.7512	1.3865	.2
.9	.5864	.8100	.7239	1.3814	54.1
36.0	0.5878	0.8090	0.7265	1.3764	54.0
	cos	sin	cot	tan	←↑

↓→	sin	cos	tan	cot	
36.0	0.5878	0.8090	0.7265	1.3764	54.0
.1	.5892	.8080	.7292	1.3713	53.9
.2	.5906	.8070	.7319	1.3663	.8
.3	.5920	.8059	.7346	1.3613	.7
.4	.5934	.8049	.7373	1.3564	.6
.5	.5948	.8039	.7400	1.3514	.5
.6	.5962	.8028	.7427	1.3465	.4
.7	.5976	.8018	.7454	1.3416	.3
.8	.5990	.8007	.7481	1.3367	.2
.9	.6004	.7997	.7508	1.3319	53.1
37.0	0.6018	0.7986	0.7536	1.3270	53.0
.1	.6032	.7976	.7563	1.3222	52.9
.2	.6046	.7965	.7590	1.3175	.8
.3	.6060	.7955	.7618	1.3127	.7
.4	.6074	.7944	.7646	1.3079	.6
.5	.6088	.7934	.7673	1.3032	.5
.6	.6101	.7923	.7701	1.2985	.4
.7	.6115	.7912	.7729	1.2938	.3
.8	.6129	.7902	.7757	1.2892	.2
.9	.6143	.7891	.7785	1.2846	52.1
38.0	0.6157	0.7880	0.7813	1.2799	52.0
.1	.6170	.7869	.7841	1.2753	51.9
.2	.6184	.7859	.7869	1.2708	.8
.3	.6198	.7848	.7898	1.2662	.7
.4	.6211	.7837	.7926	1.2617	.6
.5	.6226	.7826	.7954	1.2572	.5
.6	.6239	.7815	.7983	1.2527	.4
.7	.6252	.7804	.8012	1.2482	.3
.8	.6266	.7793	.8040	1.2437	.2
.9	.6280	.7782	.8069	1.2393	51.1
39.0	0.6293	0.7771	0.8098	1.2349	51.0
.1	.6307	.7760	.8127	1.2305	50.9
.2	.6320	.7749	.8156	1.2261	.8
.3	.6334	.7738	.8185	1.2218	.7
.4	.6347	.7727	.8214	1.2174	.6
.5	.6361	.7716	.8243	1.2131	.5
.6	.6374	.7705	.8273	1.2088	.4
.7	.6388	.7694	.8302	1.2045	.3
.8	.6401	.7683	.8332	1.2002	.2
.9	.6414	.7672	.8361	1.1960	50.1
40.0	0.6428	0.7660	0.8391	1.1918	50.0
	cos	sin	cot	tan	←↑

Table II Trigonometric Functions—Decimal Degrees (csc $\theta = 1/\sin\theta$; sec $\theta = 1/\cos\theta$) (*continued*)

↓→	sin	cos	tan	cot		↓→	sin	cos	tan	cot	
40.0	0.6428	0.7660	0.8391	1.1918	50.0	43.0	0.6820	0.7314	0.9325	1.0724	47.0
.1	.6441	.7649	.8421	1.1875	49.9	.1	.6833	.7302	.9358	1.0686	46.9
.2	.6455	.7638	.8451	1.1833	.8	.2	.6845	.7290	.9391	1.0649	.8
.3	.6468	.7627	.8481	1.1792	.7	.3	.6858	.7278	.9424	1.0612	.7
.4	.6481	.7615	.8511	1.1750	.6	.4	.6871	.7266	.9457	1.0575	.6
.5	.6494	.7604	.8541	1.1708	.5	.5	.6884	.7254	.9490	1.0538	.5
.6	.6508	.7593	.8571	1.1667	.4	.6	.6896	.7242	.9523	1.0501	.4
.7	.6521	.7581	.8601	1.1626	.3	.7	.6909	.7230	.9556	1.0464	.3
.8	.6534	.7570	.8632	1.1585	.2	.8	.6921	.7218	.9590	1.0428	.2
.9	.6547	.7559	.8662	1.1544	49.1	.9	.6934	.7206	.9623	1.0392	46.1
41.0	0.6561	0.7547	0.8693	1.1504	49.0	44.0	0.6947	0.7193	0.9657	1.0355	46.0
.1	.6574	.7536	.8724	1.1463	48.9	.1	.6959	.7181	.9691	1.0319	45.9
.2	.6587	.7524	.8754	1.1423	.8	.2	.6972	.7169	.9725	1.0283	.8
.3	.6600	.7513	.8785	1.1383	.7	.3	.6984	.7157	.9759	1.0247	.7
.4	.6613	.7501	.8816	1.1343	.6	.4	.6997	.7145	.9793	1.0212	.6
.5	.6626	.7490	.8847	1.1303	.5	.5	.7009	.7133	.9827	1.0176	.5
.6	.6639	.7478	.8878	1.1263	.4	.6	.7022	.7120	.9861	1.0141	.4
.7	.6652	.7466	.8910	1.1224	.3	.7	.7034	.7108	.9896	1.0105	.3
.8	.6665	.7455	.8941	1.1184	.2	.8	.7046	.7096	.9930	1.0070	.2
.9	.6678	.7443	.8972	1.1145	48.1	.9	.7059	.7083	.9965	1.0035	45.1
42.0	0.6691	0.7431	0.9004	1.1106	48.0	45.0	0.7071	0.7071	1.0000	1.0000	45.0
.1	.6704	.7420	.9036	1.1067	47.9						
.2	.6717	.7408	.9067	1.1028	.8						
.3	.6730	.7396	.9099	1.0990	.7						
.4	.6743	.7385	.9131	1.0951	.6						
.5	.6756	.7373	.9163	1.0913	.5						
.6	.6769	.7361	.9195	1.0875	.4						
.7	.6782	.7349	.9228	1.0837	.3						
.8	.6794	.7337	.9260	1.0799	.2						
.9	.6807	.7325	.9293	1.0761	47.1						
	cos	sin	cot	tan	←↑		cos	sin	cot	tan	←↑

Table III Common Logarithms

N	0	1	2	3	4	5	6	7	8	9
1.0	.0000	.0043	.0086	.0128	.0170	.0212	.0253	.0294	.0334	.0374
1.1	.0414	.0453	.0492	.0531	.0569	.0607	.0645	.0682	.0719	.0755
1.2	.0792	.0828	.0864	.0899	.0934	.0969	.1004	.1038	.1072	.1106
1.3	.1139	.1173	.1206	.1239	.1271	.1303	.1335	.1367	.1399	.1430
1.4	.1461	.1492	.1523	.1553	.1584	.1614	.1644	.1673	.1703	.1732
1.5	.1761	.1790	.1818	.1847	.1875	.1903	.1931	.1959	.1987	.2014
1.6	.2041	.2068	.2095	.2122	.2148	.2175	.2201	.2227	.2253	.2279
1.7	.2304	.2330	.2355	.2380	.2405	.2430	.2455	.2480	.2504	.2529
1.8	.2553	.2577	.2601	.2625	.2648	.2672	.2695	.2718	.2742	.2765
1.9	.2788	.2810	.2833	.2856	.2878	.2900	.2923	.2945	.2967	.2989
2.0	.3010	.3032	.3054	.3075	.3096	.3118	.3139	.3160	.3181	.3201
2.1	.3222	.3243	.3263	.3284	.3304	.3324	.3345	.3365	.3385	.3404
2.2	.3424	.3444	.3464	.3483	.3502	.3522	.3541	.3560	.3579	.3598
2.3	.3617	.3636	.3655	.3674	.3692	.3711	.3729	.3747	.3766	.3784
2.4	.3802	.3820	.3838	.3856	.3874	.3892	.3909	.3927	.3945	.3962
2.5	.3979	.3997	.4014	.4031	.4048	.4065	.4082	.4099	.4116	.4133
2.6	.4150	.4166	.4183	.4200	.4216	.4232	.4249	.4265	.4281	.4298
2.7	.4314	.4330	.4346	.4362	.4378	.4393	.4409	.4425	.4440	.4456
2.8	.4472	.4487	.4502	.4518	.4533	.4548	.4564	.4579	.4594	.4609
2.9	.4624	.4639	.4654	.4669	.4683	.4698	.4713	.4728	.4742	.4757
3.0	.4771	.4786	.4800	.4814	.4829	.4843	.4857	.4871	.4886	.4900
3.1	.4914	.4928	.4942	.4955	.4969	.4983	.4997	.5011	.5024	.5038
3.2	.5051	.5065	.5079	.5092	.5105	.5119	.5132	.5145	.5159	.5172
3.3	.5185	.5198	.5211	.5224	.5237	.5250	.5263	.5276	.5289	.5302
3.4	.5315	.5328	.5340	.5353	.5366	.5378	.5391	.5403	.5416	.5428
3.5	.5441	.5453	.5465	.5478	.5490	.5502	.5514	.5527	.5539	.5551
3.6	.5563	.5575	.5587	.5599	.5611	.5623	.5635	.5647	.5658	.5670
3.7	.5682	.5694	.5705	.5717	.5729	.5740	.5752	.5763	.5775	.5786
3.8	.5798	.5809	.5821	.5832	.5843	.5855	.5866	.5877	.5888	.5899
3.9	.5911	.5922	.5933	.5944	.5955	.5966	.5977	.5988	.5999	.6010
4.0	.6021	.6031	.6042	.6053	.6064	.6075	.6085	.6096	.6107	.6117
4.1	.6128	.6138	.6149	.6160	.6170	.6180	.6191	.6201	.6212	.6222
4.2	.6232	.6243	.6253	.6263	.6274	.6284	.6294	.6304	.6314	.6325
4.3	.6335	.6345	.6355	.6365	.6375	.6385	.6395	.6405	.6415	.6425
4.4	.6435	.6444	.6454	.6464	.6474	.6484	.6493	.6503	.6513	.6522
4.5	.6532	.6542	.6551	.6561	.6571	.6580	.6590	.6599	.6609	.6618
4.6	.6628	.6637	.6646	.6656	.6665	.6675	.6684	.6693	.6702	.6712
4.7	.6721	.6730	.6739	.6749	.6758	.6767	.6776	.6785	.6794	.6803
4.8	.6812	.6821	.6830	.6839	.6848	.6857	.6866	.6875	.6884	.6893
4.9	.6902	.6911	.6920	.6928	.6937	.6946	.6955	.6964	.6972	.6981
5.0	.6990	.6998	.7007	.7016	.7024	.7033	.7042	.7050	.7059	.7067
5.1	.7076	.7084	.7093	.7101	.7110	.7118	.7126	.7135	.7143	.7152
5.2	.7160	.7168	.7177	.7185	.7193	.7202	.7210	.7218	.7226	.7235
5.3	.7243	.7251	.7259	.7267	.7275	.7284	.7292	.7300	.7308	.7316
5.4	.7324	.7332	.7340	.7348	.7356	.7364	.7372	.7380	.7388	.7396
N	0	1	2	3	4	5	6	7	8	9

Table III Common Logarithms *(continued)*

N	0	1	2	3	4	5	6	7	8	9
5.5	.7404	.7412	.7419	.7427	.7435	.7443	.7451	.7459	.7466	.7474
5.6	.7482	.7490	.7497	.7505	.7513	.7520	.7528	.7536	.7543	.7551
5.7	.7559	.7566	.7574	.7582	.7589	.7597	.7604	.7612	.7619	.7627
5.8	.7634	.7642	.7649	.7657	.7664	.7672	.7679	.7686	.7694	.7701
5.9	.7709	.7716	.7723	.7731	.7738	.7745	.7752	.7760	.7767	.7774
6.0	.7782	.7789	.7796	.7803	.7810	.7818	.7825	.7832	.7839	.7846
6.1	.7853	.7860	.7868	.7875	.7882	.7889	.7896	.7903	.7910	.7917
6.2	.7924	.7931	.7938	.7945	.7952	.7959	.7966	.7973	.7980	.7987
6.3	.7993	.8000	.8007	.8014	.8021	.8028	.8035	.8041	.8048	.8055
6.4	.8062	.8069	.8075	.8082	.8089	.8096	.8102	.8109	.8116	.8122
6.5	.8129	.8136	.8142	.8149	.8156	.8162	.8169	.8176	.8182	.8189
6.6	.8195	.8202	.8209	.8215	.8222	.8228	.8235	.8241	.8248	.8254
6.7	.8261	.8267	.8274	.8280	.8287	.8293	.8299	.8306	.8312	.8319
6.8	.8325	.8331	.8338	.8344	.8351	.8357	.8363	.8370	.8376	.8382
6.9	.8388	.8395	.8401	.8407	.8414	.8420	.8426	.8432	.8439	.8445
7.0	.8451	.8457	.8463	.8470	.8476	.8482	.8488	.8494	.8500	.8506
7.1	.8513	.8519	.8525	.8531	.8537	.8543	.8549	.8555	.8561	.8567
7.2	.8573	.8579	.8585	.8591	.8597	.8603	.8609	.8615	.8621	.8627
7.3	.8633	.8639	.8645	.8651	.8657	.8663	.8669	.8675	.8681	.8686
7.4	.8692	.8698	.8704	.8710	.8716	.8722	.8727	.8733	.8739	.8745
7.5	.8751	.8756	.8762	.8768	.8774	.8779	.8785	.8791	.8797	.8802
7.6	.8808	.8814	.8820	.8825	.8831	.8837	.8842	.8848	.8854	.8859
7.7	.8865	.8871	.8876	.8882	.8887	.8893	.8899	.8904	.8910	.8915
7.8	.8921	.8927	.8932	.8938	.8943	.8949	.8954	.8960	.8965	.8971
7.9	.8976	.8982	.8987	.8993	.8998	.9004	.9009	.9015	.9020	.9025
8.0	.9031	.9036	.9042	.9047	.9053	.9058	.9063	.9069	.9074	.9079
8.1	.9085	.9090	.9096	.9101	.9106	.9112	.9117	.9122	.9128	.9133
8.2	.9138	.9143	.9149	.9154	.9159	.9165	.9170	.9175	.9180	.9186
8.3	.9191	.9196	.9201	.9206	.9212	.9217	.9222	.9227	.9232	.9238
8.4	.9243	.9248	.9253	.9258	.9263	.9269	.9274	.9279	.9284	.9289
8.5	.9294	.9299	.9304	.9309	.9315	.9320	.9325	.9330	.9335	.9340
8.6	.9345	.9350	.9355	.9360	.9365	.9370	.9375	.9380	.9385	.9390
8.7	.9395	.9400	.9405	.9410	.9415	.9420	.9425	.9430	.9435	.9440
8.8	.9445	.9450	.9455	.9460	.9465	.9469	.9474	.9479	.9584	.9489
8.9	.9494	.9499	.9504	.9509	.9513	.9518	.9523	.9528	.9533	.9538
9.0	.9542	.9547	.9552	.9557	.9562	.9566	.9571	.9576	.9581	.9586
9.1	.9590	.9595	.9600	.9605	.9609	.9614	.9619	.9624	.9628	.9633
9.2	.9638	.9643	.9647	.9652	.9657	.9661	.9666	.9671	.9675	.9680
9.3	.9685	.9689	.9694	.9699	.9703	.9708	.9713	.9717	.9722	.9727
9.4	.9731	.9736	.9741	.9745	.9750	.9754	.9759	.9763	.9768	.9773
9.5	.9777	.9782	.9786	.9791	.9795	.9800	.9805	.9809	.9814	.9818
9.6	.9823	.9827	.9832	.9836	.9841	.9845	.9850	.9854	.9859	.9863
9.7	.9868	.9872	.9877	.9881	.9886	.9890	.9894	.9899	.9903	.9908
9.8	.9912	.9917	.9921	.9926	.9930	.9934	.9939	.9943	.9948	.9952
9.9	.9956	.9961	.9965	.9969	.9974	.9978	.9983	.9987	.9991	.9996
N	0	1	2	3	4	5	6	7	8	9

Table IV Natural Logarithms and Powers of e

x	e^x	e^{-x}	$\ln x$
0.00	1.0000	1.0000	
0.01	1.0101	0.9900	−4.6052
0.02	1.0202	0.9802	−3.9120
0.03	1.0305	0.9704	−3.5066
0.04	1.0408	0.9608	−3.2189
0.05	1.0513	0.9512	−2.9957
0.06	1.0618	0.9418	−2.8134
0.07	1.0725	0.9324	−2.6593
0.08	1.0833	0.9231	−2.5257
0.09	1.0942	0.9139	−2.4079
0.10	1.1052	0.9048	−2.3026
0.11	1.1163	0.8958	−2.2073
0.12	1.1275	0.8869	−2.1203
0.13	1.1388	0.8781	−2.0402
0.14	1.1503	0.8694	−1.9661
0.15	1.1618	0.8607	−1.8971
0.16	1.1735	0.8521	−1.8326
0.17	1.1853	0.8437	−1.7720
0.18	1.1972	0.8353	−1.7148
0.19	1.2092	0.8270	−1.6607
0.20	1.2214	0.8187	−1.6094
0.30	1.3499	0.7408	−1.2040
0.40	1.4918	0.6703	−0.9163
0.50	1.6487	0.6065	−0.6931
0.60	1.8221	0.5488	−0.5108
0.70	2.0138	0.4966	−0.3567
0.80	2.2255	0.4493	−0.2231
0.90	2.4596	0.4066	−0.1054
1.00	2.7183	0.3679	0.0000
1.10	3.0042	0.3329	0.0953
1.20	3.3201	0.3012	0.1823
1.30	3.6693	0.2725	0.2624
1.40	4.0552	0.2466	0.3365
1.50	4.4817	0.2231	0.4055

x	e^x	e^{-x}	$\ln x$
1.60	4.9530	0.2019	0.4700
1.70	5.4739	0.1827	0.5306
1.80	6.0496	0.1653	0.5878
1.90	6.6859	0.1496	0.6419
2.00	7.3891	0.1353	0.6931
2.10	8.1662	0.1225	0.7419
2.20	9.0250	0.1108	0.7885
2.30	9.9742	0.1003	0.8329
2.40	11.0232	0.0907	0.8755
2.50	12.1825	0.0821	0.9163
2.60	13.4637	0.0743	0.9555
2.70	14.8797	0.0672	0.9933
2.80	16.4446	0.0608	1.0296
2.90	18.1741	0.0550	1.0647
3.00	20.0855	0.0498	1.0986
3.50	33.1155	0.0302	1.2528
4.00	54.5982	0.0183	1.3863
4.50	90.0171	0.0111	1.5041
5.00	148.4132	0.0067	1.6094
5.50	224.6919	0.0041	1.7047
6.00	403.4288	0.0025	1.7918
6.50	665.1416	0.0015	1.8718
7.00	1096.6332	0.00091	1.9459
7.50	1808.0424	0.00055	2.0149
8.00	2980.9580	0.00034	2.0794
8.50	4914.7688	0.00020	2.1401
9.00	8130.0839	0.00012	2.1972
9.50	13359.7268	0.00007	2.2513
10.00	22026.4658	0.00005	2.3026

Answers to Selected Exercises

Exercises 1.1, page 9

1. $\{0, 2, 3, 4, 5, 6, 9\}$ **3.** $\{3, 4, 6\}$ **5.** $\{0, 2, 3, 4, 5, 6\}$ **7.** $\{0, 2, 3, 4, 6\}$ **9.** $\{3, 4, 6\}$ **11.** $N, W, Z, Q, \mathcal{R}$ **13.** $\mathcal{R}, H$ **15.** $\mathcal{R}, H$ **17.** $Q, \mathcal{R}$ **19.** $Q, \mathcal{R}$ **21.** N **23.** $\mathcal{R}$ **25.** Z **27.** N **29.** Z **31.** Q **33.** Additive inverse **35.** Distributive property **37.** Multiplicative identity **39.** Additive identity **41.** Multiplicative inverse **43.** $7 + 5$ **45.** $(x \cdot y) \cdot z$ **47.** 0 **49.** $1/a$ **51.** ay **53.** 3 **55.** 4 **57.** 5 **59.** 41 **61.** -0.0270 **63.** 2.1087 **65.** $17r - 43$ **67.** $-5a + 6b + 1$ **69.** $26 + 4q$ **71.** $6p - 15$ **73.** $21x - 27$ **75.** $6y - 32$ **77.** $6 + 9a + 12b$ **79.** $2a - 14b$ **81.** $-24 - 16x$ **83.** $45y - 35$ **85.** $60x - 12y + 30$ **87.** $10b - 31a$ **89.** $14 - 60p + 42r$

Exercises 1.2, page 18

1. 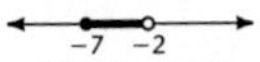 (number line: −7, −2) **3.** 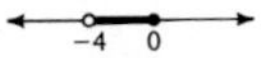 (number line: −4, 0) **5.**  (number line: 1, 3) **7.** 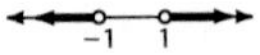 (number line: −1, 1) **9.** 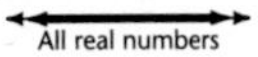 (number line: All real numbers) **11.** 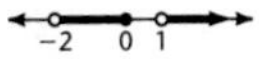 (number line: −2, 0, 1) **13.** 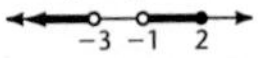 (number line: −3, −1, 2) **15.** 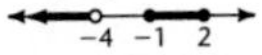 (number line: −4, −1, 2) **17.** $x < -2$ **19.** $2 \le x \le 5$ **21.** $x \le -2$ or $x > 1$ **23.** $1 < x < 3$ or $x \ge 5$ **25.** Addition property **27.** Multiplication property **29.** Addition property **31.** Multiplication property **33.** Multiplication property **35.** $-8 - |3|, -|8 - 3|, 0, |8 - 3|, |8| + 3$ **37.** $-2, -\sqrt{2}, -1, 1, \sqrt{2}, |-2|$ **39.** $-9, -\sqrt{80}, \sqrt{80} - 9, |\sqrt{80} - 9|, \sqrt{80}, |-9|$ **41.** 7 **43.** 4 **45.** 10 **47.** 2 **49.** 16 **51.** -2 **53.** $y - 4$ **55.** $4 - y$ **57.** $9 - \sqrt{80}$ **59.** $10 - 2x$ **61.** $b - a$ **63.** $b - 2a$ **65.** $-x/5$ **67.** 1 **69.** a^2 **71.** $-5 - 2a$ **73.** $3a - 14$ **75.** Multiplication property **77.** Division property **79.** Triangle inequality **81.** (number line: −2, 2) **83.** (number line: −3, 3) **85.** (number line: −4, 4) **87.** (number line: 1, 3) **89.** (number line: −3, 1) **91.** (number line: −10, −2) **93.** (number line: −1, 7) **95.** (number line: −9, 1)

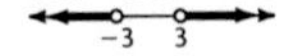

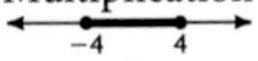

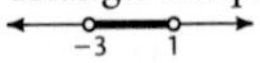

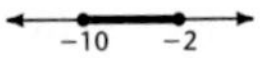

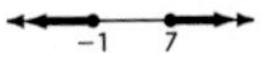

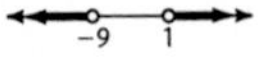

Exercises 1.3, page 29

1. 81 **3.** -25 **5.** $-\frac{1}{16}$ **7.** $\frac{4}{9}$ **9.** $\frac{1}{64}$ **11.** $64p^3q^6$ **13.** $4x^2/49$ **15.** $2/y^3$ **17.** $a/(9b^2)$ **19.** $15r^6$ **21.** $y^3/(2x^2)$ **23.** v^6/u^6 **25.** $1/(64x^7)$ **27.** s^5/r^3 **29.** $16r^3/t^3$ **31.** $125z^2/8$ **33.** $\frac{9}{16}$ **35.** $24/(p^4q^4)$ **37.** $x^{12}/(81y^8)$ **39.** $z^9/(x^6y^3)$ **41.** $-z^3/(64x^6y^6)$ **43.** b^6 **45.** $4z^2/(x - y)^2$ **47.** $9x^{2m}/y^{2n}$ **49.** x^{n-1} **51.** 1.02×10^3 **53.** 4.161×10^7 **55.** 3.5×10^{-3} **57.** 5.6×10^8 **59.** 6.0×10^2 **61.** 6.0×10^{-2} **63.** 8220 **65.** 0.66 **67.** 0.0000287 **69.** 0.0008 **71.** 40 **73.** 1.015×10^{13} **75.** 8.626×10^{-1} **77.** 6.037×10^{-3} **79.** 1.736×10 **81.** 3.705×10^{-5} **83.** 9.111×10^7 **85.** 0.00000501 gram **87.** 10^{-6} **89.** Approximately 8.14 minutes

Exercises 1.4, page 36

1. $5x^4 + 2x^3 - x^2 - 5$ **3.** $4 - r^2 + 6r^4 - 2r^5$ **5.** $9 + 8z + 5z^2 + z^3$ **7.** $20x^5 - 15x^3 + 40x^2$ **9.** $8z^2 - 22z - 21$ **11.** $40q^2 + 19q - 3$ **13.** $10a^2 + 13ab - 3b^2$ **15.** $2x^4 - 5x^3 - 12x^2 - x + 4$ **17.** $2x^4 - x^3 + x^2 + 14x - 4$ **19.** $3y^5 - 10y^4 + 5y^3 - y^2 + 7y + 2$ **21.** $1 + 2x + x^2 - x^3 - x^4$ **23.** $b^2 - 9$ **25.** $4r^2 - s^2$ **27.** $y^4 - 4y^2 + 4$ **29.** $a^2 + 6abc^2 + 9b^2c^4$ **31.** $36m^2 - 96mn + 64n^2$ **33.** $(x^2/9) - (y^2/4)$ **35.** $4x^6 - y^2$ **37.** $p^3 - q^3$ **39.** $8x^3 + 1$ **41.** $n^3 + 3n^2k + 3nk^2 + k^3$ **43.** $x^3 - 6x^2y + 12xy^2 - 8y^3$ **45.** $3x^4 + x^2y^2 - 10y^4$ **47.** $2x^6 - 5x^3y - 12y^2$ **49.** $27a^3 - 8b^3$ **51.** $x^2 + 2xy + y^2 - 1$ **53.** $a^2 + 4b^2 + c^2 + 4ab + 2ac + 4bc$ **55.** $x^2 - y^2 - 2yz - z^2$ **57.** $6x^{2n} + 5x^n - 4$ **59.** $4z^{2r} - 81$ **61.** $25y^{2n} + 30y^n + 9$ **63.** $4b^{2k} - 20b^k + 25$ **65.** (a) 768 cubic inches (b) 1536 cubic inches **67.** $60,000 **69.** 732.4814 **71.** -7738.5160

Exercises 1.5, page 43

1. $3(a-2b)$ **3.** $4x(x^2-4y)$ **5.** $6m(m^2-4m+n)$ **7.** $4xy(3x-1+5y)$ **9.** $4(x+2y)$ **11.** $(x+5)(x-5)$ **13.** $(3y+1)(3y-1)$ **15.** $(4x+5y)(4x-5y)$ **17.** $(x-1)(x+y)(x-y)$ **19.** $-y^3(x+3)(x-3)$ **21.** $(u-4)^2$ **23.** $(5y+1)^2$ **25.** $(2a-5b)^2$ **27.** $3a(a+1)^2$ **29.** $(1-6rs)^2$ **31.** $(x-z)(x^2+xz+z^2)$ **33.** $(2+u)(4-2u+u^2)$ **35.** $(2x+3)(4x^2-6x+9)$ **37.** $8(3x-y)(9x^2+3xy+y^2)$ **39.** $-4a(a-4)(a^2+4a+16)$ **41.** $(x-6)(x+4)$ **43.** $(x+7)(x+3)$ **45.** $(4y+x)(y-x)$ **47.** $(3x-2a)(x+3a)$ **49.** $2x(x+5)(x-3)$ **51.** $4y(2y-3)(y-1)$ **53.** $(a+b)(3c+d)$ **55.** $(a+b)(x^2+d^2)$ **57.** $(3x+y-z)(3x-y+z)$ **59.** $(1+2x)(2-5x^3)$ **61.** $(x-1+y+z)(x-1-y-z)$ **63.** $(2x-7)(x-3)$ **65.** $(4a-3)(a+1)$ **67.** $-(2x+3y)(a+h)$ **69.** $16a^2(a+2)^2$ **71.** $x(x+1)(x-1)$ **73.** $(2y+5)(3y-2)$ **75.** $(6a-5)(a+3)$ **77.** Does not factor **79.** $3(x^2-3)$ **81.** $(5x+1+y)(5x+1-y)$ **83.** $-x^3(x+1)(x^2-x+1)$ **85.** $a(x+3y^2)(x-3y^2)$ **87.** $(4x-3y+5)(4x-3y-5)$ **89.** $(4a-b+2x-z)(4a-b-2x+z)$ **91.** $(x^2-8)^2$ **93.** $(3x^2-5)^2$ **95.** $(2a^2+3b^2)^2$ **97.** $(u+v)^2(u-v)$ **99.** $9w^2(w^6+3w^3-1)$ **101.** $(x^2+2x+2)(x^2-2x+2)$ **103.** $(z^2+z+1)(z^2-z+1)$ **105.** $2(x^n+2)(x^n+3)$ **107.** $(3z^m+w^n)(3z^m-w^n)$

Exercises 1.6, page 52

1. $x \neq -2$ **3.** $x \neq 0$ and $x \neq -3$ **5.** $x \neq y$ **7.** No restriction is necessary. **9.** $x+y$ **11.** $(x+3)/(x+2)$ **13.** $4(a+2)^2/(a-1)$ **15.** $(x^2+3xy+9y^2)(x+2y)$ **17.** $q/[3p(p+q)]$ **19.** $\frac{1}{3}$ **21.** $5a(x+2y)/y$ **23.** $3y^2/(d-c)$ **25.** $3(2x-y)/[2(x-y)]$ **27.** $(5-y)/4$ **29.** $1/(w+5)$ **31.** $(y-4)/(y-5)$ **33.** $(x+y)/(x-y)$ **35.** $x(5+2x)/[(x-1)(3+x)]$ **37.** $(s^2+1)/[s^2(r^2+1)]$ **39.** $(x^2+3x+9)/[3(2x^2+5)]$ **41.** $a(2w+5)/[(w+3)(w^2+4w+16)]$ **43.** $(a^2+axb+x^2b^2)(a+2b)/[2xy(3a-2b)]$ **45.** $1/(2x+3)$ **47.** $2/(2+x)$ **49.** $1/(x-1)$ **51.** $2/(x-1)$ **53.** $(a+b)/(a-b)$ **55.** $(3y-4xz)/(12x^2)$ **57.** $(5x-14)/[2x(x-2)]$ **59.** $-2/(x+3)$ **61.** $(xy-y^2-3)/[(x-y)(x+y)]$ **63.** $6/(x+2)$ **65.** $(x-8)/[(2x-3)(x+2)]$ **67.** $(2w^2-9w-9)/[(2w-3)(w-6)(w-5)]$ **69.** $-(z^2-3z+14)/[4(z-3)(z+3)]$ **71.** $(y^3-3)/[(y+1)(y-1)]$ **73.** $3x(3x-5y)$ **75.** $(x-6)/[(x+2)(x+1)]$ **77.** $x(x+2)/(x+5)$ **79.** $1/(x-y)$ **81.** $(4w+5)/(4w-5)$ **83.** $(2x+3y)(x^2-xy+y^2)/2$ **85.** $(x+y)/(x-y)$ **87.** $(3x+2y)(y-2x)$ **89.** $(y-x)/(xy)$ **91.** $2b(b^2+ab+a^2)/a^2$ **93.** $(2x+1)/(x+1)$ **95.** $x(x-1)/(2x-1)$ **97.** $(x^2-y^2)/(xy)$ **99.** $1/(b^2-a^2)$ **101.** $xy/(y+x)$ **103.** $(y-x)/[x^2y^2(x+y)]$ **105.** -0.1058

Exercises 1.7, page 65

1. 5 **3.** 4 **5.** 73 **7.** $\frac{4}{3}$ **9.** 7 **11.** 2 **13.** $5|x|$ **15.** ab **17.** $|2x|$ **19.** $(r+s)^2$ **21.** $2b^2$ **23.** y^2w^3 **25.** $0.4x^2$ **27.** -1 **29.** $2x/(3y^2)$ **31.** $6x/5$ **33.** x^2/y^3 **35.** $19y^5$ **37.** xy^2 **39.** x^3/z **41.** $5z^3/(9x^2)$ **43.** $\sqrt[5]{a}$ **45.** $3\sqrt[4]{a}$ **47.** $a^{5/4}$ **49.** $x^{10/9}$ **51.** $a^{6/3}$ **53.** $\sqrt[4]{3b}$ **55.** $\sqrt[3]{(2xy^2)^2}$ or $(\sqrt[3]{2xy^2})^2$ **57.** $(bx^2)^{1/7}$ **59.** 2 **61.** Undefined **63.** $\frac{1}{256}$ **65.** 27 **67.** 16 **69.** $4p^4$ **71.** a^3b^6 **73.** $\frac{9}{16}$ **75.** $-8m^3/n^6$ **77.** $-64y^3/x^6$ **79.** x^3 **81.** x^4y^{10} **83.** $a^{2/3}/b^2$ **85.** a^8/b^3 **87.** 1 **89.** $x-1$ **91.** $a-b$ **93.** $a^{1/4}b^{1/3}c^{1/5}$ **95.** 3.61 **97.** 3.74 **99.** 3.74 **101.** 0.26 **103.** 24 square feet **105.** 0.84 second **107.** 45 miles per hour **109.** 1.6 inches **111.** 1.3 centimeters **113.** 2.5 feet

Exercises 1.8, page 77

1. $10\sqrt{5}$ **3.** $4ab^2\sqrt{2a}$ **5.** $4x^2\sqrt[3]{2x}$ **7.** $x^2\sqrt[5]{x^3}$ **9.** $-4a^3\sqrt[3]{2}$ **11.** $3(a+b)\sqrt[4]{2(a+b)}$ **13.** $\sqrt{2}+\sqrt{3}$ **15.** $3\sqrt{6}$ **17.** $9\sqrt{2}$ **19.** $3\sqrt{2}-2\sqrt[3]{2}$ **21.** $(3a+4b)\sqrt[3]{2}$ **23.** $4xy\sqrt[3]{3y}$ **25.** $5\sqrt[3]{2}$ **27.** $\sqrt{5}$ **29.** $x\sqrt[3]{10}$ **31.** $2xy\sqrt[3]{3xy}$ **33.** $\sqrt{3}$ **35.** -27 **37.** $14\sqrt{2}-9\sqrt{5}$ **39.** $\sqrt[6]{5}$ **41.** $\sqrt[6]{u^2}$ **43.** $\sqrt[3]{y^2}$ **45.** $\sqrt[5]{5a^3b^2}$ **47.** $\sqrt{5}/5$ **49.** $\sqrt{10}/4$ **51.** $5\sqrt{7}/7$ **53.** $\sqrt[3]{2}/2$ **55.** $3\sqrt{35}/14$ **57.** $\sqrt[3]{3}$ **59.** $\sqrt{3a}/3$ **61.** $\sqrt{10ab}/(2b)$ **63.** $\sqrt[3]{6cdx}/(3x)$ **65.** $\sqrt{5u}/(5u^3)$ **67.** $y\sqrt[4]{2a^2xy^2z^3}/(xz^2)$ **69.** $(8+\sqrt{7})/3$ **71.** $3(5\sqrt{2}+\sqrt{7})/43$ **73.** $(3\sqrt{15}-\sqrt{10}+\sqrt{6}-9)/25$ **75.** $2v\sqrt[5]{uv^2}$ **77.** $\sqrt[4]{3}/y^2$ **79.** $x^4\sqrt{x}/y$ **81.** $(a+2b)^3$ **83.** $\sqrt[4]{10}/2$ **85.** $\sqrt[3]{25v}/5$ **87.** $x^2/2$ **89.** $3(a+b)^2\sqrt{3(a+b)}$ **91.** $y-6\sqrt{xy}+9x$ **93.** $3-2\sqrt{2}$ **95.** $(2+y^2)\sqrt[3]{4y^2}/y^2$ **97.** $-2x\sqrt[3]{x}$ **99.** x **101.** $1/(2\sqrt{3})$ **103.** $-1/(1+\sqrt{3})$ **105.** $(x-4)/[x(\sqrt{x}+2)]$ **107.** $1/(\sqrt{a}-\sqrt{b})$ **109.** $x\sqrt[12]{x^5}$ **111.** 4.14 **113.** 2.95 **115.** 1.72

Exercises 1.9, page 85

1. $x = 2, y = -6$ **3.** $x = 0, y = -7$ **5.** $x = 4, y = -1$ **7.** $4i$ **9.** $-7i$ **11.** -15 **13.** 3 **15.** $9 - i$
17. $15 + 67i$ **19.** $-4 - 2i$ **21.** $-5 + 3i$ **23.** $-1 - 17i$ **25.** $9 + 7i$ **27.** $-54 - 10i$ **29.** $27 - 36i$
31. $40i$ **33.** $14 + 5i$ **35.** $2 - \frac{7}{3}i$ **37.** $-2i$ **39.** $-\frac{4}{3}i$ **41.** $\frac{1}{3} + 2i$ **43.** $\frac{3}{25} - \frac{4}{25}i$ **45.** $\frac{5}{169} + \frac{12}{169}i$
47. $\frac{9}{5} + \frac{3}{5}i$ **49.** $\frac{27}{37} - \frac{23}{37}i$ **51.** $\frac{18}{5} + \frac{1}{5}i$ **53.** $-i$ **55.** $-\frac{13}{10} + \frac{1}{10}i$ **57.** $-2 - 2i$ **59.** $-\frac{1}{2}i$ **61.** 1 **63.** i
65. $-i$ **67.** i **69.** -1 **71.** $-i$

Critical Thinking: Find the Errors, page 90

1. Errors

$$\begin{aligned} 4[-(2 - x) - (1 + 2x)] &= 4[-2 \overset{\times}{-} x - 1 \overset{\times}{+} 2x] \\ &= 4[-3 + x] \\ &= 4 - 3 + x \quad \times \end{aligned}$$

Correction

$$\begin{aligned} 4[-(2 - x) - (1 + 2x)] &= 4[-2 + x - 1 - 2x] \\ &= 4[-3 - x] \\ &= -12 - 4x \end{aligned}$$

2. Error

$$\begin{aligned} x - y(z + 3) - 2 &= 1 - (-2)(3 + 3) - 2 \\ &= 1 + 2(6) - 2 \\ &\overset{\times}{=} 3(6) - 2 \end{aligned}$$

Correction

$$\begin{aligned} x - y(z + 3) - 2 &= 1 - (-2)(3 + 3) - 2 \\ &= 1 + 2(6) - 2 \\ &= 1 + 12 - 2 \\ &= 11 \end{aligned}$$

3. Error

Figure 1.22

Correction

The inequality $|x| \le 2$ is equivalent to $-2 \le x \le 2$. Its graph is given in the accompanying figure.

4. Errors

$$(x^{-2}y)^{-3} = \frac{1}{\underset{\times}{(x^2y)^3}} = \frac{1}{\underset{\times}{x^5y^3}}$$

Correction

$$(x^{-2}y)^{-3} = \frac{1}{(x^{-2}y)^3} = \frac{1}{x^{-6}y^3} = \frac{x^6}{y^3}$$

5. Error

$$(x^0 - x^{-1})^{-1} \overset{\times}{=} (x^0)^{-1} - (x^{-1})^{-1}$$

Correction

$$\begin{aligned} (x^0 - x^{-1})^{-1} &= \left(1 - \frac{1}{x}\right)^{-1} \\ &= \left(\frac{x - 1}{x}\right)^{-1} \\ &= \frac{x}{x - 1} \end{aligned}$$

6. Error

$$x^3 + 8 = (x + 2)(x^2 \overset{\times}{+} 4x + 4)$$

Correction

$$x^3 + 8 = (x + 2)(x^2 - 2x + 4)$$

7. Error

$$r^2 - s^2 - r - s = (r^2 - s^2) - (r \overset{\times}{-} s)$$

Correction

$$\begin{aligned} r^2 - s^2 - r - s &= (r^2 - s^2) - (r + s) \\ &= (r - s)(r + s) - (r + s) \\ &= (r + s)(r - s - 1) \end{aligned}$$

8. Error

$$\frac{x^3 - y^3}{x - y} \overset{\times}{=} \frac{\cancel{x^3}^{\,x^2} - \cancel{y^3}^{\,y^2}}{\cancel{x} - \cancel{y}}$$

Correction

$$\begin{aligned} \frac{x^3 - y^3}{x - y} &= \frac{(x - y)(x^2 + xy + y^2)}{x - y} \\ &= x^2 + xy + y^2 \end{aligned}$$

9. Error

$$\frac{p-3}{p+1}-\frac{2p-1}{p+2}\overset{\times}{=}\frac{p-3-2p+1}{(p+1)(p+2)}$$

Correction

$$\begin{aligned}\frac{p-3}{p+1}-\frac{2p-1}{p+2}&=\frac{(p-3)(p+2)}{(p+1)(p+2)}-\frac{(2p-1)(p+1)}{(p+2)(p+1)}\\&=\frac{p^2-p-6-(2p^2+p-1)}{(p+1)(p+2)}\\&=\frac{p^2-p-6-2p^2-p+1}{(p+1)(p+2)}\\&=\frac{-p^2-2p-5}{(p+1)(p+2)}\\&=-\frac{p^2+2p+5}{(p+1)(p+2)}\end{aligned}$$

10. Error

$$(-27)^{2/3}=[(-27)^{1/3}]^2\overset{\times}{=}(-9)^2$$

Correction

$$(-27)^{2/3}=[(-27)^{1/3}]^2=(-3)^2=9$$

11. Errors

$$\begin{aligned}2\sqrt{3}-4\sqrt{2}&=(2-4)(\sqrt{3}-\sqrt{2})\quad\times\\&=(-2)(\sqrt{1})\quad\times\end{aligned}$$

Correction

$2\sqrt{3}-4\sqrt{2}$ cannot be combined by using the distributive property.

12. Error

$$\sqrt[3]{8x^2y^3}\overset{\times}{=}\sqrt{2\cdot 4\cdot x^2\cdot y\cdot y^2}$$

Correction

$$\sqrt[3]{8x^2y^3}=\sqrt[3]{2^3y^3x^2}=2y\sqrt[3]{x^2}$$

13. Error

$$\frac{\sqrt{8x^3y}}{\sqrt{16xy^3}}=\sqrt{\frac{8x^3y}{16xy^3}}=\sqrt{\frac{x^2}{2y^2}}=\frac{x}{y\sqrt{2}}=?$$

Correction

$$\frac{x}{y\sqrt{2}}=\frac{x\sqrt{2}}{y\sqrt{2}\sqrt{2}}=\frac{x\sqrt{2}}{2y}$$

14. Error

$$\sqrt{-25}\sqrt{-4}\overset{\times}{=}\sqrt{100}$$

Correction

$$\sqrt{-25}\sqrt{-4}=(5i)(2i)=-10$$

15. Error

$$\begin{aligned}\frac{1}{2+i}&=\frac{1}{2+i}\cdot\frac{2-i}{2-i}\\&=\frac{2-i}{\underset{\times}{4-1}}\end{aligned}$$

Correction

$$\frac{1}{2+i}\cdot\frac{2-i}{2-i}=\frac{2-i}{4+1}=\frac{2}{5}-\frac{1}{5}i$$

Review Problems for Chapter 1, page 92

1. $Z, Q, \mathcal{R}$ **2.** $W, Z, Q, \mathcal{R}$ **3.** $Q, \mathcal{R}$ **4.** $\mathcal{R}, H$ **5.** $Q, \mathcal{R}$ **6.** $N, W, Z, Q, \mathcal{R}$ **7.** $cp + cq$ **8.** ba **9.** $(ab)c$ **10.** $1/x$ **11.** $ay + ax$ **12.** $1 \cdot x$ **13.** Additive identity **14.** Multiplicative identity **15.** Distributive property **16.** Commutative property, $\cdot$ **17.** Associative property, $\cdot$ **18.** Associative property, + **19.** Additive inverse **20.** Commutative property, + **21.** 20 **22.** -10 **23.** -8 **24.** $-\frac{5}{6}$ **25.** $-2x + 4$ **26.** $6z + 30$ **27.** 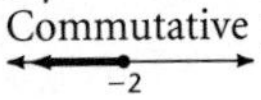

28. 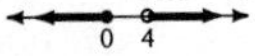

29. 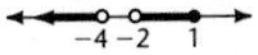

30. 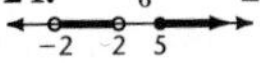

31. $-1 \le x < 4$ **32.** $x < 1$ or $x > 5$ **33.** $x < -1$ or $2 \le x \le 5$ **34.** $-2 \le x < 0$ or $x \ge 3$ **35.** 

36. 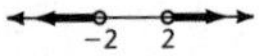

37. 

38. 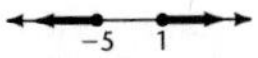

39. 2 **40.** $6 - \sqrt{35}$ **41.** $\sqrt{37} - 6$ **42.** $-x/3$ **43.** $y + 3$ **44.** $14 - 3x$ **45.** False **46.** True **47.** False **48.** True **49.** $\frac{1}{16}$ **50.** $1/(25x^6)$ **51.** x^2/y^2 **52.** $5x/2$ **53.** $4x^6y^8$ **54.** 1 if $a \ne -b$, undefined if $a = -b$ **55.** $(2a^2)/(3b^5)$ **56.** 2.93×10^4 **57.** 5.79×10^6 **58.** 1.6×10^{-4} **59.** 7.06×10^{-2} **60.** 493,000 **61.** 0.00807 **62.** 0.768 **63.** 11.9 **64.** $1 - 2x + 2x^2 - 7x^3$ **65.** $9x^4 - 5x^3 - 7x^2 + 10x + 2$ **66.** $35r^3 - 21r^2 - 15r + 9$ **67.** $30x^5 - 15x^4 - 13x^3 + 9x^2 - 3x$

68. $4p^2 - 9q^2$ **69.** $9x^2 - 6xy + y^2$ **70.** $u^3 - v^3$ **71.** $64x^3 - 96x^2y + 48xy^2 - 8y^3$
72. $8x^3 + 4x^2y - 2xy^2 - y^3$ **73.** $27a^3 + b^3$ **74.** $(a + 4b)(a - 4b)$ **75.** $(7x - 1)^2$ **76.** $4(x - y)^2$
77. $2pq(4q + p - 2)$ **78.** $(m - n)(m - 2n)$ **79.** $(2x - 3)(4x^2 + 6x + 9)$ **80.** $a(a + 4b)(a^2 - 4ab + 16b^2)$
81. $2t(3t + 2)(t - 3)$ **82.** $(z - 4)(z + 3)(z - 3)$ **83.** $(2x + y - 2z)(2x - y + 2z)$ **84.** $4x(3x + 4y)$
85. $5y^2(x + y)/[x(x - y)]$ **86.** $2(x + 2)/(x - 2)$ **87.** $x(3x + 4)/[(x - 2)(x + 1)]$ **88.** $(w + 3)/(w - 5)$
89. $10(x - 2)$ **90.** $(x - y)/(2x + y)$ **91.** $3/(a + 3)$ **92.** $(y^2 - 3x^2)/(6x^2y)$ **93.** $(3x - 5y)/[2(x - y)(x + y)]$
94. $-(x + 3)/[2(x - 1)(x - 3)]$ **95.** $5/(x - 2)$ **96.** $(21x^2 + 3x - 16)/[(2x - 3)(x + 1)(5x - 2)]$ **97.** $5x(5x - 3y)$
98. $1/(x + 2)$ **99.** $4x^2(x - 2y)$ **100.** $2(y - x)(4y + 3x)$ **101.** $(x + y)/(xy)$ **102.** $\frac{11}{6}$ **103.** 79 **104.** $-\frac{4}{5}$
105. $-3xy^2$ **106.** 2 **107.** 3 **108.** $-3y^2/x$ **109.** $2xy^2$ **110.** $25a^2b^6$ **111.** $y^4/(4x^2)$ **112.** $x^{3/7}$
113. $\sqrt[5]{(5ab^2)^2}$ or $(\sqrt[5]{5ab^2})^2$ **114.** -8 **115.** 25 **116.** $\frac{1}{216}$ **117.** $-2a$ **118.** $0.2a^2$ **119.** $1/(25a^4)$
120. $3a\sqrt{3a}$ **121.** $2x\sqrt[6]{y^5}$ **122.** $2ab\sqrt[4]{8a}$ **123.** $-x\sqrt[5]{x^2y^4}$ **124.** $3ab\sqrt[3]{3ab^2}$ **125.** $3x^2/(2y)$ **126.** $6\sqrt{2}$
127. $-16\sqrt{2}$ **128.** $2a(1 - 2ab)\sqrt{2b}$ **129.** $2(b + 2a)\sqrt[3]{ab}$ **130.** $4a\sqrt[3]{a}$ **131.** $3\sqrt[3]{3}$ **132.** $2a\sqrt[4]{4b}$ **133.** $2/a$
134. $8\sqrt{5} - 34$ **135.** $11 - 3\sqrt{15}$ **136.** $\sqrt[4]{5}$ **137.** $x\sqrt[3]{5x}$ **138.** $\sqrt[6]{2x^2y^3}$ **139.** $\sqrt[5]{3x^2y^4}$ **140.** $\sqrt{15}/5$
141. $\sqrt[3]{6}/2$ **142.** $(8 - 5\sqrt{3})/11$ **143.** $3x^2\sqrt{3}/y$ **144.** $\sqrt[3]{ab^2}/(2b)$ **145.** $2\sqrt[3]{5x^2y^2}/y$ **146.** $(6 + 5\sqrt{y} + y)/(4 - y)$
147. $\sqrt{2(x - 2)}/(x - 2)$ **148.** $3\sqrt[3]{2}$ **149.** $y\sqrt[10]{x^9y}$ **150.** $3 + 10i$ **151.** -30 **152.** $10 + 5i$ **153.** $\frac{1}{29} - \frac{17}{29}i$
154. $-i$

Exercises 2.1, page 101

1. $\{3\}$ **3.** $\{6\}$ **5.** $\{\frac{8}{3}\}$ **7.** $\{-3\}$ **9.** $\{2\}$ **11.** $\{3\}$ **13.** $\{\frac{27}{20}\}$ **15.** $\{-3\}$ **17.** $\{\frac{7}{10}\}$ **19.** $\{4\}$
21. $\varnothing$ **23.** $\{15\}$ **25.** $\varnothing$ **27.** $\{\frac{4}{3}\}$ **29.** $\{-1\}$ **31.** $\varnothing$ **33.** $x = (c - by)/a$ **35.** $r = (A - P)/(Pt)$
37. $b = (2A - ha)/h$ **39.** $B = (6V - hb - 4hM)/h$ **41.** $S = pF/(1 - p)$ **43.** $r = (a - S)/(L - S)$ **45.** $\{-3,9\}$
47. $\{2, -9\}$ **49.** $\{\frac{7}{2}\}$ **51.** $\varnothing$ **53.** $\{-\frac{3}{2}, -5\}$ **55.** $\{1, 7\}$ **57.** $\{\frac{9}{4}\}$ **59.** $\{\frac{9}{4}\}$ **61.** $\{-0.64\}$ **63.** $\{-0.60\}$
65. $23\frac{1}{3}°$ Celsius **67.** \$3304 **69.** 17% **71.** \$196

Exercises 2.2, page 110

1. 48, 49, 50 **3.** 92 **5.** 47 **7.** 2 seconds, 3 seconds **9.** \$5400 **11.** \$6000 at 9%, \$8000 at 12% **13.** \$81,000 **15.** 7 dimes, 5 nickels **17.** 3 cartons of milk, 5 cartons of eggs **19.** Length = 36 meters, width = 20 meters **21.** 4 pounds peanuts, 8 pounds cashews **23.** 25 grams of 12% silver, 15 grams of 20% silver **25.** 24 liters **27.** $4\frac{2}{7}$ hours **29.** 21 kilometers per hour, 24 kilometers **31.** 70 kilometers per hour, 75 kilometers per hour **33.** $3\frac{1}{13}$ hours **35.** 45 hours **37.** 60 hours

Exercises 2.3, page 120

1. $(-\infty, 4)$ **3.** $(5, \infty)$ **5.** $(-\infty, 6)$ **7.** $(-\infty, 2)$ **9.** $(\sqrt{2}, \infty)$ **11.** $[-36, \infty)$ **13.** $(-\infty, -3]$ **15.** $[0, \infty)$
17. $(-\infty, -13]$ **19.** $(-7, \infty)$ **21.** $(-2, 4]$ **23.** $[1, 4)$ **25.** $[1, 4]$ **27.** $(-1, 3)$ **29.** $\varnothing$ **31.** $[2, \infty)$
33. $(-\infty, -2] \cup (3, \infty)$ **35.** $(-\infty, 2] \cup (4, \infty)$ **37.** $(-2, \infty)$ **39.** $\mathscr{R}$ **41.** $(1, 4)$
43. $[5, 9]$ **45.** $(-\frac{15}{4}, \frac{5}{2})$ **47.** $(-\infty, \frac{1}{2}] \cup [\frac{7}{2}, \infty)$
49. $(-\infty, \frac{1}{4}) \cup (\frac{5}{4}, \infty)$ **51.** $(-\infty, -2) \cup (\frac{4}{5}, \infty)$ **53.** $\varnothing$ **55.** $\{2\}$
57. $(-\infty, 1) \cup (1, \infty)$ **59.** $\mathscr{R}$ **61.** $(-\infty, 1.31)$ **63.** $(-26.19, 41.32)$ **65.** $(-0.32, 3.15)$
67. $(-\infty, -4.33] \cup [5.10, \infty)$ **69.** No more than 4.3 hours **71.** $3\frac{1}{2}$ years **73.** $6 < d \leq 12$ **75.** $(11, 19)$

Exercises 2.4, page 128

1. $\pm\frac{5}{2}$ **3.** $-3, 0$ **5.** $-5, 2$ **7.** -3 **9.** $\frac{1}{3}, 3$ **11.** $-\frac{2}{7}, 0, \frac{1}{7}$ **13.** $-3, -1$ **15.** $(-1 \pm \sqrt{5})/2$ **17.** $1 \pm i$
19. $2 \pm 2i$ **21.** $(-5 \pm \sqrt{17})/4$ **23.** $-\frac{3}{2}, -\frac{2}{3}$ **25.** $\frac{1}{2}, 7$ **27.** $-1 \pm \sqrt{2}$ **29.** $-\frac{3}{4}, 2$ **31.** $-\frac{2}{9}, \frac{1}{3}$
33. $(-1 \pm \sqrt{11})/2$ **35.** $(2 \pm i)/2$ **37.** $\pm\sqrt{2}$ **39.** $-2, 1 \pm i\sqrt{3}, \frac{1}{3}, (-1 \pm i\sqrt{3})/6$ **41.** $-4, 5$ **43.** $-1, -\frac{1}{5}$
45. $-1, \frac{1}{8}$ **47.** 9, 25 **49.** 67 yards **51.** 30 meters by 60 meters **53.** Yes **55.** 20 meters

Exercises 2.5, page 135

1. $-7, 4$ **3.** $\frac{1}{2}, \frac{3}{2}$ **5.** $-1, 0$ **7.** $-6, 2$ **9.** $\pm\frac{5}{4}$ **11.** $(-7 \pm \sqrt{17})/8$ **13.** $(-5 \pm \sqrt{5})/2$ **15.** $(-4 \pm \sqrt{7})/3$ **17.** $1 \pm i\sqrt{3}$ **19.** $-\frac{9}{7}, \frac{5}{4}$ **21.** $-1.13, -2.79$ **23.** $-0.58, -3.10$ **25.** 16, two distinct real roots **27.** -31, two distinct complex numbers that are conjugates **29.** 37, two distinct real roots **31.** 0, two equal real roots **33.** Approx. 48, two distinct real roots **35.** Approx. -1085, two distinct complex numbers that are conjugates **37.** $-\frac{9}{2}$ **39.** $\pm 4\sqrt{3}$ **41.** $x^2 + x - 2 = 0$ **43.** $x^2 - \frac{11}{2}x + 6 = 0$ **45.** $x^2 + 4x + 4 = 0$ **47.** $x^2 - 2x + 1 = 0$ **49.** $x^2 + 3x - 4 = 0$ **51.** $x^2 - 2x = 0$ **53.** 21, 22, or $-22, -21$ **55.** 5, 6, 7, 8 or $-8, -7, -6, -5$ **57.** 1 second, 2 seconds **59.** 10 **61.** 20 or 80 **63.** 100 **65.** 4 or 6 **67.** \$6000 **69.** 25 **71.** $\frac{1}{2}$ inch by $\frac{1}{2}$ inch

Exercises 2.6, page 143

1. $\{-3, 2\}$ **3.** $\{-4, 1\}$ **5.** $\{-\frac{3}{4}, -3\}$ **7.** $\{(1 \pm \sqrt{5})/2\}$ **9.** $\{-4, 1\}$ **11.** $\{1, \frac{2}{3}\}$ **13.** $\{2\}$ **15.** $\{\frac{3}{2}\}$ **17.** $\{6\}$ **19.** $\{2\}$ **21.** $\{6\}$ **23.** $\{-10, -9\}$ **25.** $\{1\}$ **27.** $\{-7\}$ **29.** $\{11\}$ **31.** $\{7\}$ **33.** $\{28\}$ **35.** $\{2\}$ **37.** $\{-1\}$ **39.** $\{0, 3\}$ **41.** $\varnothing$ **43.** $\{14\}$ **45.** $\{5\}$ **47.** $\{-5\}$ **49.** $\{2\}$ **51.** $\{15\}$ **53.** 55 miles per hour **55.** 27 miles per hour

Exercises 2.7, page 151

1. $(-\infty, -3) \cup (4, \infty)$ 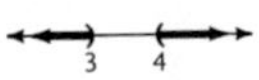 **3.** $[-1, 3]$ 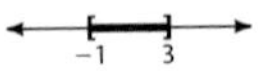 **5.** $(-\infty, 1] \cup [3, \infty)$

7. $(-\infty, -\sqrt{3}) \cup (\sqrt{3}, \infty)$ 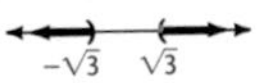 **9.** $(-\infty, 0) \cup (9, \infty)$ 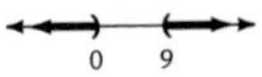 **11.** $[0, 4]$

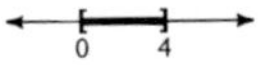

13. $(-\infty, -\sqrt{2}) \cup (\sqrt{2}, \infty)$ 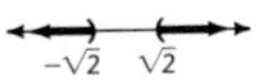 **15.** $(-\infty, 0] \cup [2, \infty)$ **17.** $(-4, -2)$

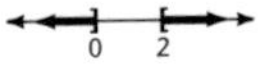

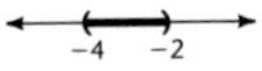

19. $[-5, 3]$ 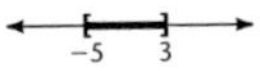 **21.** $(-\infty, -5) \cup (\frac{2}{3}, \infty)$ 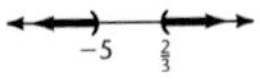 **23.** $(\frac{2}{3}, \frac{6}{5})$

25. $(-\infty, 0) \cup (\frac{4}{3}, \infty)$  **27.** $(-\frac{7}{3}, \frac{5}{2})$ 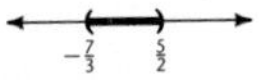 **29.** $(-\frac{7}{4}, 3)$

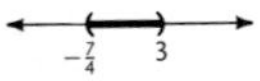

31. $[(-1 - \sqrt{2})/2, (-1 + \sqrt{2})/2]$ 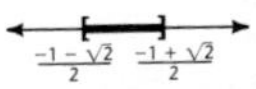 **33.** $(1, 2) \cup (3, \infty)$ 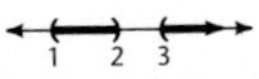 **35.** $(\frac{1}{2}, \frac{4}{5})$

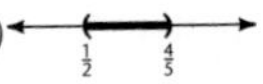

37. $(-2, \frac{3}{2})$ 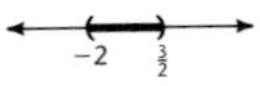 **39.** $[-5, -3)$ **41.** $(2, 3) \cup (7, \infty)$

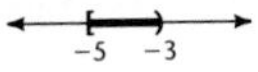

43. $(-2, -\frac{1}{2}) \cup [0, \infty)$ **45.** $(-\frac{3}{2}, -1] \cup (0, \infty)$ 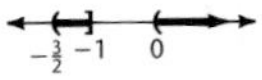 **47.** $(-\infty, -1.97) \cup (0.62, \infty)$

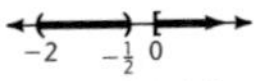

49. $[-0.41, -0.01]$ **51.** Width $= w \geq 30$ feet, Length $= 2w \geq 60$ feet **53.** $2 \leq t \leq 6$

Critical Thinking: Find the Errors, page 154

1. Error

$2x < -4 \qquad 4 - 1 < x$
$x < -2 \qquad 3 < x$

The solution set is $\{x \mid 3 < x < -2\}$. ×

Correction

The solution set is $\{x \mid x < -2\} \cup \{x \mid x > 3\} = (-\infty, -2) \cup (3, \infty)$.

2. Error

$x(6x - 1) = 1$
$x = 1 \quad$ or $\quad 6x - 1 = 1$ ×

Correction

$6x^2 - x - 1 = 0$
$(3x + 1)(2x - 1) = 0$
$3x + 1 = 0 \quad$ or $\quad 2x - 1 = 0$
$3x = -1 \qquad 2x = 1$
$x = -\frac{1}{3} \qquad x = \frac{1}{2}$

The solution set is $\{-\frac{1}{3}, \frac{1}{2}\}$.

3. Error

$$2 - \sqrt{1-x} = 2x$$
$$4 + 1 - x = 4x^2 \quad \times$$

Correction

$$2 - 2x = \sqrt{1-x}$$
$$4 - 8x + 4x^2 = 1 - x$$
$$4x^2 - 7x + 3 = 0$$
$$(4x-3)(x-1) = 0$$
$$4x - 3 = 0 \quad \text{or} \quad x - 1 = 0$$
$$4x = 3 \qquad x = 1$$
$$x = \tfrac{3}{4}$$

With $x = \frac{3}{4}$, $2 - \sqrt{1 - \frac{3}{4}} = \frac{3}{2} = 2(\frac{3}{4})$

With $x = 1$, $2 - \sqrt{1-1} = 2 = 2(1)$

The solution set is $\{\frac{3}{4}, 1\}$.

4. Error

$$(x+4)(x+2) < 3$$
$$x + 4 < 3 \quad \text{or} \quad x + 2 < 3 \quad \times$$

Correction

$$x^2 + 6x + 8 < 3$$
$$x^2 + 6x + 5 < 0$$
$$(x+5)(x+1) < 0$$

$x+5$	−	−	0	+	+	+	+	+	+
$x+1$	−	−	−	−	−	−	0	+	+
$(x+5)(x+1)$	+	+	0	−	−	−	0	+	+

−5 −1

The solution set is $(-5, -1)$.

5. Error

$$\frac{2}{w-1}(w-1)(w+2) > \frac{1}{w+2}(w-1)(w+2) \quad \times$$

Correction

$$\frac{2}{w-1} > \frac{1}{w+2}$$
$$\frac{2}{w-1} - \frac{1}{w+2} > 0$$
$$\frac{2w+4-w+1}{(w-1)(w+2)} > 0$$
$$\frac{w+5}{(w-1)(w+2)} > 0$$

Interval	$(-\infty, -5)$	$(-5, -2)$	$(-2, 1)$	$(1, \infty)$
Test point	−6	−3	0	2
Value	$-\frac{1}{28}$	$\frac{1}{2}$	$-\frac{5}{2}$	$\frac{7}{4}$
Solutions?	No	Yes	No	Yes

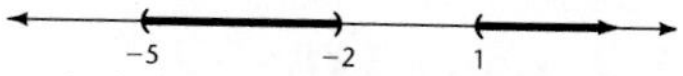

−5 −2 1

The solution set is $(-5, -2) \cup (1, \infty)$

Review Problems for Chapter 2, page 155

1. $\{-\frac{1}{6}\}$ **2.** $\{\frac{7}{2}\}$ **3.** $\{0\}$ **4.** $\{-8\}$ **5.** $\{-1\}$ **6.** $\{\frac{9}{2}\}$ **7.** $\{-\frac{23}{3}, 5\}$ **8.** $\varnothing$ **9.** $\{\frac{4}{3}, -\frac{2}{11}\}$ **10.** $\{1, \frac{5}{3}\}$ **11.** $y = x$ **12.** $z = xy/(xy - x - y)$ **13.** (a) $t = \frac{5}{2}$ and $t = \frac{17}{2}$ (b) $t = \frac{11}{2}$ **14.** 5000 sophomores, 3650 juniors **15.** 50 inches by 70 inches **16.** 11 years **17.** \$15,000 at 8%, \$21,000 at 9% **18.** \$3200 **19.** 146 children's tickets, 132 adult's tickets **20.** $\frac{23}{15} = 1\frac{8}{15}$ liters **21.** 1000 grams **22.** 350 pounds **23.** 3 miles per hour **24.** 10 miles per hour **25.** 1 hour 20 minutes **26.** $(-\infty, 1)$ **27.** $(2, \infty)$ **28.** $(-\infty, -4]$ **29.** $(-7, \infty)$ **30.** $(\frac{7}{9}, \infty)$ **31.** $(-4, -1]$ **32.** $[-3, 4)$ **33.** $(-\infty, \infty)$ **34.** $(-\infty, -5) \cup (-3, \infty)$ −5 −3

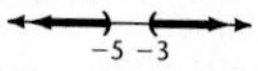

35. $[\frac{3}{5}, \frac{9}{5}]$ $\frac{3}{5}$ $\frac{9}{5}$ 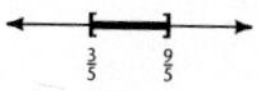 **36.** $(-4, 11)$ 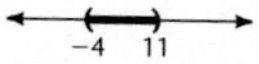 −4 11 **37.** $(-\infty, \frac{4}{3}] \cup [4, \infty)$ $\frac{4}{3}$ 4

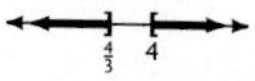

38. $(-\infty, 0] \cup [\frac{5}{3}, \infty)$ 0 $\frac{5}{3}$ **39.** $-3, \frac{2}{3}$ **40.** $-3, \frac{1}{2}$ **41.** $-\frac{3}{5}, 2$ **42.** $\frac{5}{3}, -\frac{5}{6}$ **43.** $9, -5$

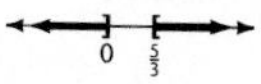

44. $-2 \pm i\sqrt{2}$ **45.** $(3 \pm i)/2$ **46.** $(-6 \pm \sqrt{34})/2$ **47.** $(-2 \pm \sqrt{5})/2$ **48.** $3, -\frac{1}{2}$ **49.** $\pm 1, \pm\frac{1}{3}$

50. $\pm 1, \pm\frac{2}{3}$ **51.** $49, \frac{1}{4}$ **52.** $1 \pm \sqrt{6}$ **53.** $(-1 \pm \sqrt{17})/4$ **54.** $(-1 \pm 3i)/2$

55. $(3 \pm i\sqrt{11})/10$ **56.** Two distinct real numbers **57.** Nonreal complex numbers that are conjugates **58.** Two distinct real numbers **59.** Nonreal complex numbers that are conjugates **60.** Two equal real numbers **61.** 10 or 20 **62.** \$4 or \$8 **63.** 12 **64.** \$12 per pound **65.** 18 inches by 24 inches **66.** 10 feet **67.** $\{1, -2\}$ **68.** $\{0, \frac{7}{2}\}$ **69.** $\{\frac{3}{2}, \frac{9}{2}\}$ **70.** $\{1\}$ **71.** $\{5\}$ **72.** $\{6\}$ **73.** $\{1\}$ **74.** $\{0, 3\}$ **75.** $\{1\}$ **76.** $\{1, 5\}$ **77.** $\{13\}$ **78.** $\{7, -1\}$ **79.** $(-4, -2)$ **80.** $(-\infty, -\frac{5}{2}] \cup [2, \infty)$ **81.** $(-\frac{2}{3}, \frac{1}{4})$ **82.** $(-\infty, -4) \cup (1, \infty)$ **83.** $(-3, 0) \cup (\frac{1}{2}, \infty)$ **84.** $(-3, -2]$ **85.** $(-3, -\frac{3}{2}) \cup [-1, \infty)$

Exercises 3.1, page 168

1. x-intercept $\frac{3}{2}$; y-intercept -3 **3.** x-intercept $\frac{5}{4}$; no y-intercept **5.** x-intercept 0; y-intercept 0 **7.** x-intercept $\frac{7}{2}$; y-intercept $\frac{7}{5}$ **9.** No x-intercept; y-intercept -3

11. $-\frac{4}{3}$ **13.** $\frac{2}{3}$ **15.** Undefined **17.** 0 **19.** $\frac{1}{2}$ **21.** $(-1, -\frac{1}{2})$ **23.** $(-4, 2)$ **25.** $(\frac{3}{2}, 2)$ **27.** $(\pi, 0)$ **29.** $\left(\frac{a+b}{2}, \frac{a+b}{2}\right)$ **31.** $(0, 2)$ **33.** $\frac{3}{4}$ **35.** Undefined **37.** 0 **39.** $\frac{7}{5}$ **41.** $\frac{7}{2}$ **43.** $-\frac{5}{2}$ **45.** $\frac{1}{2}$ **47.** -2 **49.** 6 **51.** Yes **53.** No **55.** **57.** **59.**

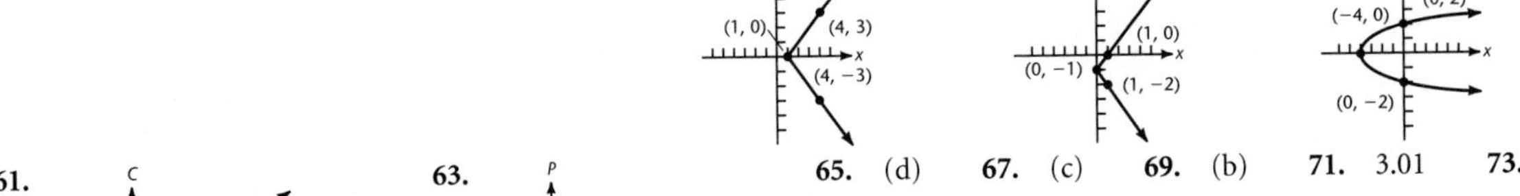

65. (d) **67.** (c) **69.** (b) **71.** 3.01 **73.** -4.41

61. **63.**

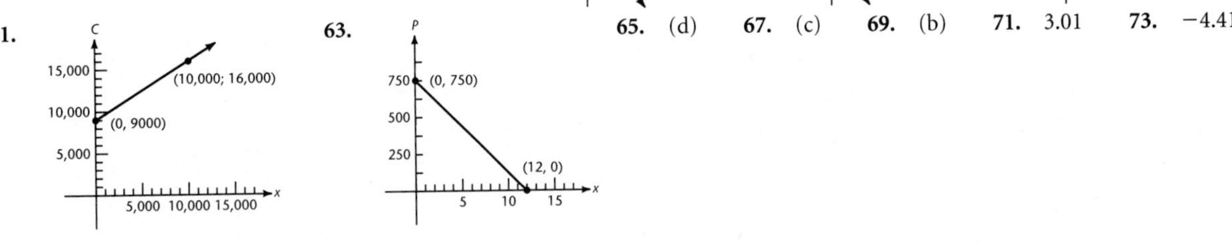

Exercises 3.2, page 178

1. $D = \mathcal{R}, R = \mathcal{R}$ **3.** $D = [4, \infty), R = (-\infty, 0]$ **5.** $D = \mathcal{R}, R = [-3, \infty)$ **7.** $D = (-\infty, 1) \cup (1, \infty), R = (-\infty, 0) \cup (0, \infty)$ **9.** $D = \mathcal{R}, R = [0, \infty)$ **11.** $D = \mathcal{R}, R = [3, \infty)$ **13.** Function **15.** Not a function **17.** Not a function **19.** Function **21.** Function **23.** 10 **25.** 16 **27.** $4b - 2$ **29.** $-4x - 2$ **31.** $(x + 1)^2$ **33.** 6 **35.** 12 **37.** 0 **39.** $-2x$ **41.** $2(x + h)$ **43.** $h(2x + h - 1)$ **45.** 3 **47.** $4x + 2h$ **49.** $6x + 3h - 2$ **51.** $6x^2 + 6xh + 2h^2$ **53.** 721.439 **55.** -4.99 **57.** **59.** **61.** **63.** **65.**

67. **69.** Not a function **71.** Function **73.** Function **75.** Not a function

77.

79. (a) $\ell = 60 - \frac{3}{2}w$ (b) $A = (60 - \frac{3}{2}w)(w)$ **81.** $V = 4x(9 - x)^2$

Exercises 3.3, page 186

1. 8 **3.** 6 **5.** −9 **7.** 4 **9.** −3 **11.** 0 **13.** 5 **15.** 0 **17.** 5 **19.** 9 **21.** 3 **23.** 7
25. $(f+g)(x) = 3x - 1,\ x \in \mathcal{R}$; $(f-g)(x) = -x - 3,\ x \in \mathcal{R}$; $(f\cdot g)(x) = 2x^2 - 3x - 2,\ x \in \mathcal{R}$; $(f/g)(x) = (x-2)/(2x+1),\ x \in \mathcal{R}$ and $x \neq -\frac{1}{2}$ **27.** $(f+g)(x) = x^2 + 3x - 1,\ x \in \mathcal{R}$; $(f-g)(x) = 5x - 1 - x^2,\ x \in \mathcal{R}$; $(f\cdot g)(x) = 4x^3 - 5x^2 + x,\ x \in \mathcal{R}$ $(f/g)(x) = (4x-1)/(x^2 - x),\ x \in \mathcal{R}$ and $x \neq 0,\ x \neq 1$ **29.** $(f+g)(x) = \sqrt{x} + 2x,\ x \in [0, \infty)$; $(f-g)(x) = \sqrt{x} - 2x,\ x \in [0, \infty)$; $(f\cdot g)(x) = 2x\sqrt{x},\ x \in [0, \infty)$; $(f/g)(x) = \sqrt{x}/(2x),\ x \in (0, \infty)$ **31.** $(f+g)(x) = 2x + 1 + \sqrt{x+2},\ x \in [-2, \infty)$; $(f-g)(x) = 2x + 1 - \sqrt{x+2},\ x \in [-2, \infty)$; $(f\cdot g)(x) = (2x+1)\sqrt{x+2},\ x \in [-2, \infty)$; $(f/g)(x) = (2x+1)/\sqrt{x+2},\ x \in (-2, \infty)$
33. $(f+g)(x) = \sqrt{x-2} + \sqrt{3-x},\ x \in [2, 3]$; $(f-g)(x) = \sqrt{x-2} - \sqrt{3-x},\ x \in [2, 3]$; $(f\cdot g)(x) = \sqrt{5x - x^2 - 6},\ x \in [2, 3]$; $(f/g)(x) = \sqrt{x-2}/\sqrt{3-x},\ x \in [2, 3)$ **35.** $(f+g)(x) = \sqrt{x+1} + \sqrt{x+6},\ x \in [-1, \infty)$; $(f-g)(x) = \sqrt{x+1} - \sqrt{x+6},\ x \in [-1, \infty)$; $(f\cdot g)(x) = \sqrt{x^2 + 7x + 6},\ x \in [-1, \infty)$; $(f/g)(x) = \sqrt{x+1}/\sqrt{x+6},\ x \in [-1, \infty)$
37. $(f+g)(x) = x^2 - 4x + 2,\ x \in \mathcal{R}$; $(f-g)(x) = 6x - 10 - x^2,\ x \in \mathcal{R}$; $(f\cdot g)(x) = x^3 - 9x^2 + 26x - 24,\ x \in \mathcal{R}$; $(f/g)(x) = (x-4)/(x^2 - 5x + 6),\ x \in \mathcal{R}$ and $x \neq 2,\ x \neq 3$.
39. $(f+g)(x) = 7x^2 + 5x - 9,\ x \in \mathcal{R}$; $(f-g)(x) = 5x^2 - x - 1,\ x \in \mathcal{R}$; $(f\cdot g)(x) = 6x^4 + 20x^3 - 23x^2 - 23x + 20,\ x \in \mathcal{R}$; $(f/g)(x) = (6x^2 + 2x - 5)/(x^2 + 3x - 4),\ x \in \mathcal{R}$ and $x \neq -4,\ x \neq 1$ **41.** $(f\circ g)(x) = x^2 + 1,\ x \in \mathcal{R}$; $(g\circ f)(x) = (x+1)^2,\ x \in \mathcal{R}$
43. $(f\circ g)(x) = (x-1)^{10},\ x \in \mathcal{R}$; $(g\circ f)(x) = x^{10} - 1,\ x \in \mathcal{R}$ **45.** $(f\circ g)(x) = 2(x+3)^2 + x + 3,\ x \in \mathcal{R}$; $(g\circ f)(x) = 2x^2 + x + 3,\ x \in \mathcal{R}$ **47.** $(f\circ g)(x) = \sqrt{x-3},\ x \in [3, \infty)$; $(g\circ f)(x) = \sqrt{x} - 3;\ x \in [0, \infty)$
49. $(f\circ g)(x) = 1/(1-x),\ x \in \mathcal{R}$ and $x \neq 1$; $(g\circ f)(x) = 1 - (1/x),\ x \in \mathcal{R}$ and $x \neq 0$
51. $(f\circ g)(x) = \sqrt{x-1},\ x \in [1, \infty)$; $(g\circ f)(x) = \sqrt{x+2} - 3,\ x \in [-2, \infty)$ **53.** $(f\circ g)(x) = x,\ x \in \mathcal{R}$; $(g\circ f)(x) = x,\ x \in \mathcal{R}$
55. $(f\circ g)(x) = x,\ x \in \mathcal{R}$ and $x \neq 0$; $(g\circ f)(x) = x,\ x \in \mathcal{R}$ and $x \neq -1$ **57.** $(f\circ g)(x) = |x|,\ x \in \mathcal{R}$; $(g\circ f)(x) = x,\ x \in [1, \infty)$

The answers in Exercises 59–65 are not unique. One correct solution is given.
59. $f(x) = x - 1,\ g(x) = x^3$ **61.** $f(x) = \sqrt{x},\ g(x) = x + 3$ **63.** $f(x) = x^{50},\ g(x) = 2x - 9$
65. $f(x) = x^3,\ g(x) = 1/(x+3)$
67. (*a*) $Y_1 = \sqrt{X - 2}$, Viewing rectangle† $[-10, 10]$ by $[-10, 10]$; $Y_2 = \sqrt{|X - 2|}$, Viewing rectangle $[-10, 10]$ by $[-10, 10]$

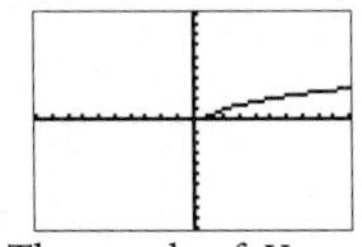

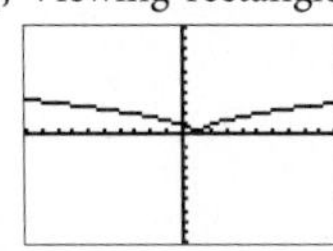

(b) The graph of Y_1 consists of the right half of the graph of Y_2.

Exercises 3.4, page 192

1. $x + y = 0$

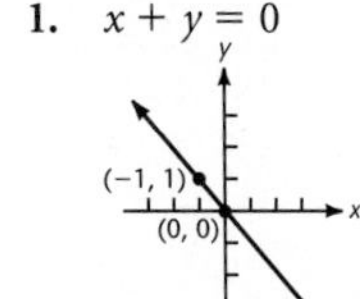

3. $2x - y = -4$

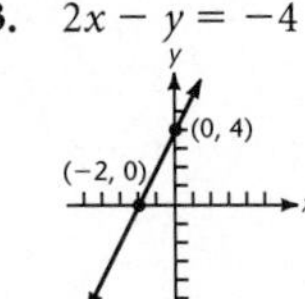

5. $y = 5$

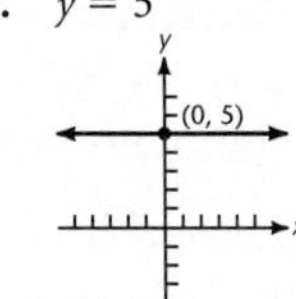

7. $x - 2y = -6$ **9.** $x + y = 7$ **11.** $4x - y = 28$

13. $x - 2y = 2$ **15.** $4x - y = -3$ **17.** $y = 3$ **19.** $x = 5$ **21.** $3x - y = -9$ **23.** $6x + 5y = -45$
25. $x + y = 0$ **27.** $5x - 7y = 7$ **29.** $y = -x + 3$ **31.** $y = 4x$ **33.** $y = 4$

†The graphing device figures were generated using Texas Instruments PC-81 Emulation Software.

35.

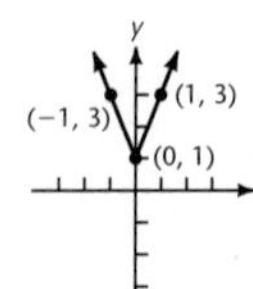

37.

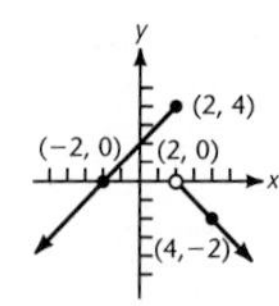

39.

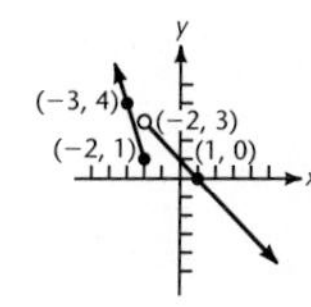

41. Viewing rectangle $[-10, 10]$ by $[-10, 10]$

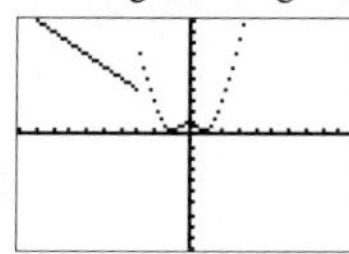

43. $m_1 = \dfrac{7-3}{4-(-1)} = \dfrac{4}{5}$, $m_2 = \dfrac{10-7}{6-4} = \dfrac{3}{2}$. No. The slope of the line between $(-1, 3)$ and $(4, 7)$ is $\frac{4}{5}$, and the slope of the line between $(4, 7)$ and $(6, 10)$ is $\frac{3}{2}$.
45. Yes **47.** No **49.** $f(x) = 30 + 28x$, $0 \le x \le 8$
51. $f(x) = 18{,}000 - 2500x$, $0 \le x \le 6$ **53.** $9\frac{1}{3}$ inches **55.** $53\sqrt{626} \approx 1326$ feet

Exercises 3.5, page 201

1. V: $(0, -4)$; A: $x = 0$; Pts. $(-2, 0)$, $(2, 0)$

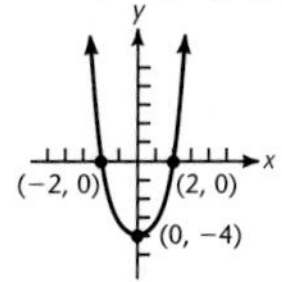

3. V: $(2, -1)$; A: $x = 2$; Pts. $(-1, 5)$, $(5, 5)$

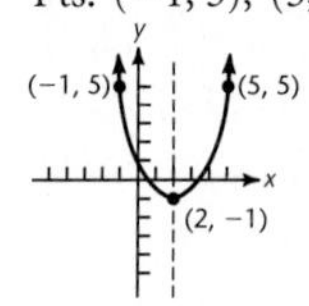

5. V: $(2, 4)$; A: $x = 2$; Pts. $(0, 0)$, $(4, 0)$

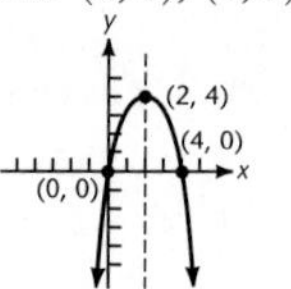

7. V: $(-4, -3)$; A: $x = -4$; Pts. $(-6, 1)$, $(-2, 1)$

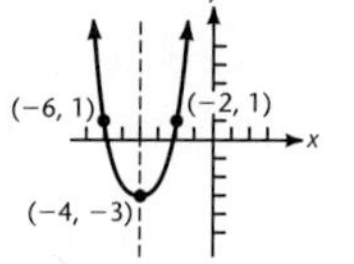

9. V: $(-2, 3)$; A: $x = -2$; Pts. $(0, -1)$, $(-4, -1)$

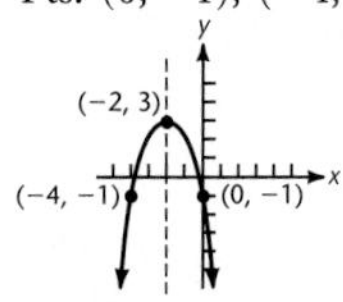

11. V: $(-2, 5)$; A: $x = -2$; Pts. $(-3, 8)$, $(-1, 8)$

(−4, 17) (0, 17) (−3, 8) (−1, 8) (−2, 5)

13. V: $(1, -3)$; A: $x = 1$; Pts. $(0, -5)$, $(2, -5)$

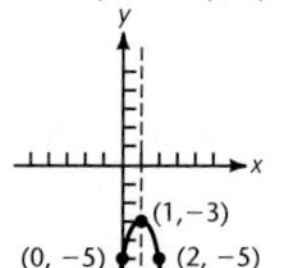

15. $f(4) = -11$ is the minimum. **17.** $f(-5) = -45$ is the minimum. **19.** $f(3) = 12$ is the maximum.
21. $f(1) = -1$ is the maximum. **23.** $f(-3) = 0$ is the maximum. **25.** $f(-\frac{3}{2}) = \frac{29}{4}$ is the minimum. **27.** 27 and 27
29. 12 feet by 12 feet **31.** The minimum cost is $C = \$737.50$ when $x = 250$.
33. The maximum height is 64 feet after 2 seconds. **35.** 10 feet by 20 feet, with the long side parallel to the wall
37. $r = h = \dfrac{20}{\pi + 4}$ feet **39.** x-intercepts, 1.80, 2.20; vertex $(2, -0.04)$ **41.** Viewing rectangle $[-6, 6]$ by $[-6, 6]$

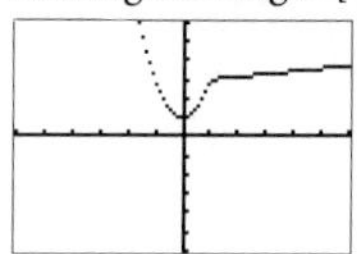

Critical Thinking: Find the Errors, page 206

1. **Error**

$f(g(7)) = (7^2 - 1)(\sqrt{7} - 3)$ ×

Correction

$g(7) = \sqrt{7-3} = \sqrt{4} = 2$
$f(g(7)) = f(2) = 2^2 - 1 = 4 - 1 = 3$

2. **Errors**

$m = \dfrac{-2}{3} = -\dfrac{2}{3}$ ×
$y - (-2) = -\frac{2}{3}(x - 3)$ ×

Correction

$(x_1, y_1) = (3, 0)$ and $(x_2, y_2) = (0, -2)$
$m = \dfrac{-2-0}{0-3} = \dfrac{2}{3}$
$y = \frac{2}{3}x + (-2)$
$3y = 2x - 6$
$2x - 3y = 6$

Review Problems for Chapter 3, page 206

1. No x-intercept; y-intercept 4 **2.** x-intercept -2; No y-intercept **3.** x-intercept 5; y-intercept -2 **4.** 0 **5.** 2 **6.** -4 **7.** Does not exist

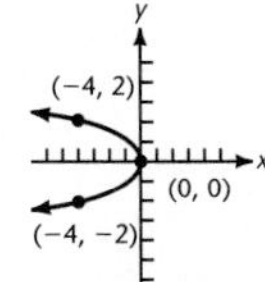

8. $(4, 4)$ **9.** $(2, 5)$ **10.** $(\frac{7}{2}, 1)$ **11.** $(-\frac{5}{2}, \frac{3}{2})$ **12.** 3 **13.** $-\frac{1}{3}$ **14.** $-\frac{2}{7}$ **15.** $\frac{5}{3}$
16. **17.** **18.** **19.** **20.** $D = \mathcal{R}$, $R = [1, \infty)$

21. $D = [9, \infty)$, $R = [0, \infty)$ **22.** $D = \mathcal{R}$, $R = [0, \infty)$ **23.** $D = (-\infty, 4) \cup (4, \infty)$, $R = (-\infty, 0) \cup (0, \infty)$ **24.** 8
25. 24 **26.** $a + 4\sqrt{a} + 3$ **27.** $h(2x + h - 2)$ **28.**

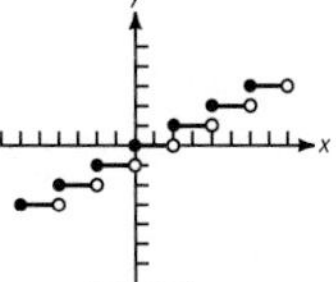

29. (a) $(f + g)(x) = \sqrt{x - 3} + x - 4$, $x \in [3, \infty)$ (b) $(f - g)(x) = \sqrt{x - 3} - x + 4$, $x \in [3, \infty)$
30. (a) $(f \cdot g)(x) = (x - 4)\sqrt{x - 3}$, $x \in [3, \infty)$ (b) $(f/g)(x) = \sqrt{x - 3}/(x - 4)$, $x \in [3, 4) \cup (4, \infty)$ **31.** No **32.** Yes
33. Yes **34.** No **35.** $2x + y = -2$ **36.** $3x + 4y = 7$ **37.** $3x - 4y = 12$ **38.** $5x - y = -9$

39. $x - 4y = 24$ **40.** $x = -2$ **41.** $y = 9$ **42.** $2x + y = 4$ **43.** $2x + y = 5$ **44.** $2x - y = 0$
45. $x - 2y = 0$ **46.** $2x - y = 3$ **47.** **48.** (a) \$40 (b) $V(t) = -40t + 1800$ **49.** $C(x) = 14x + 352$

50. V: $(1, 2)$; A: $x = 1$; Pts. $(0, 5)$, $(2, 5)$

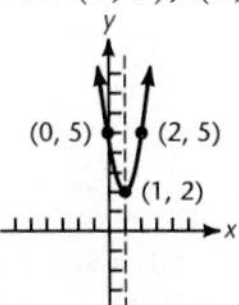

51. V: $(2, 4)$; A: $x = 2$; Pts. $(0, 0)$, $(4, 0)$

52. V: $(-1, -4)$; A: $x = -1$; Pts. $(1, 0)$, $(-3, 0)$

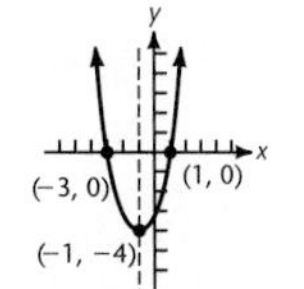

53. V: $(\frac{3}{4}, \frac{17}{8})$; A: $x = \frac{3}{4}$; Pts. $(0, 1)$, $(\frac{3}{2}, 1)$

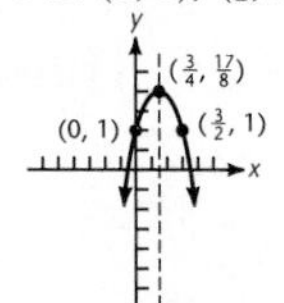

54. $f(3) = 1$ is the minimum. **55.** $f(2) = 1$ is the maximum. **56.** $f(2) = -2$ is the minimum. **57.** $f(-3) = 10$ is the maximum. **58.** $f(-1) = 3$ is the maximum. **59.** $f(2) = 5$ is the minimum. **60.** The maximum profit is \$160 when $x = 400$. **61.** The maximum height is 36 feet after 1.5 seconds. **62.** The minimum cost is \$1200 when $x = 10$. **63.** The maximum revenue is \$250 when $x = 20$. **64.** The maximum revenue is \$80,000 when $n = 2000$.

Cumulative Test Chapters 1–3, page 211

1. (a) $-4x - 10$ (b) $2m + 13$ **2.** (a) (b) **3.** (a) x^2y (b) $-8x^6/y^3$ (c) $(16x^4)/(9y^3)$
4. (a) $2x^3 + 13x^2 + 11x - 6$ (b) $25a^2 - 4b^2$ (c) $12q^4 - 11q^2p - 5p^2$ (d) $4m^2 - 12mn + 9n^2$ **5.** (a) $(x + y)/(2x - y)$

(b) $(z-2)^2/[3(z-1)(2z+5)]$ (c) $1/[(x+3)(2x+5)]$ (d) $-(x+y)$ **6.** (a) $\frac{1}{1296}$ (b) $1/(0.09x^4)$ **7.** (a) $-\frac{5}{2}$ (b) $4z^2/(3x)$ (c) $-3(2+\sqrt{5})$ (d) $(x\sqrt{2y})/(6y^2)$ **8.** (a) $-1+6i$ (b) $\frac{3}{10}+\frac{1}{10}i$ (c) -1 **9.** $\{2\}$ **10.** $\{3, \frac{1}{3}\}$ **11.** $x=(1-3y)/4$ **12.** 7 pounds **13.** 160 miles at 50 miles per hour, 78 miles at 60 miles per hour **14.** (a) $(-\infty, 2)\cup(\frac{10}{3}, \infty)$ (b) $(\frac{1}{2}, 3)$ **15.** (a) $\frac{4}{3}, -\frac{5}{2}$ (b) $0, -2$ (c) $2\pm\sqrt{7}$ **16.** (a) $\{2, -\frac{2}{3}\}$ (b) $\{4\}$ **17.** (a) $(-\infty, -4]\cup[0, 3]$ (b) $(-\infty, -1)\cup(3, \infty)$ **18.**

19. $m=\frac{3}{4}$, midpoint $(1, -5)$ **20.** $D=(2, \infty)$, $R=(0, \infty)$ **21.** (a) $(f-g)(x)=\sqrt{x+1}-x+1$, $x\in[-1, \infty)$ (b) $(f\cdot g)(x)=(x-1)\sqrt{x+1}$, $x\in[-1, \infty)$ (c) $(g\circ f)(x)=\sqrt{x+1}-1$, $x\in[-1, \infty)$ **22.** $x+2y=-1$ **23.** **24.** $V(2, -11)$; A: $x=2$; Pts. $(0, -3)$, $(4, -3)$

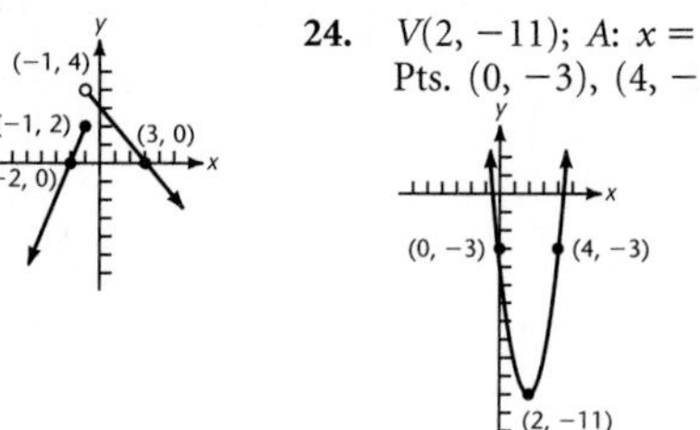

25. The maximum profit is \$5000 when $t=100$.

Exercises 4.1, page 220

1. 5 **3.** 5 **5.** 29 **7.** $2\sqrt{85}$ **9.** 4 **11.** $10-f(x)$ **13.** Circle with center $(0, 0)$ and radius 4 **15.** Circle with center $(0, -4)$ and radius $\sqrt{3}/5$ **17.** Circle with center $(-1, -3)$ and radius $\sqrt{10}$ **19.** Circle with center $(\frac{3}{2}, 2)$ and radius $\frac{3}{2}$ **21.** Not a circle **23.** Circle with center $(2, -8)$ and radius 0 (point-circle) **25.** $x^2+(y-1)^2=4$ **27.** $(x-2)^2+(y+2)^2=18$ **29.** $(x-3)^2+(y-1)^2=18$

31.

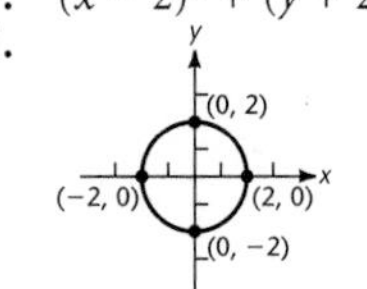

33.

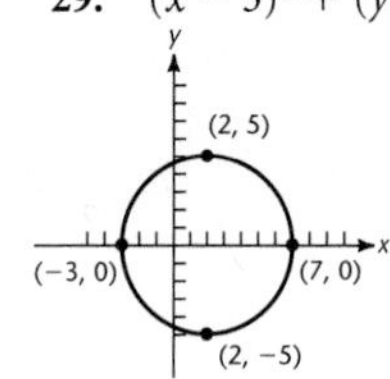

35.

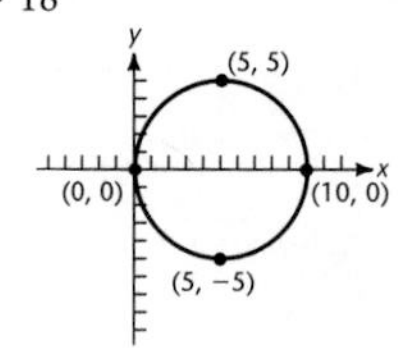

37.

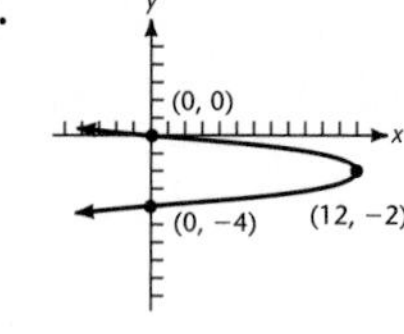

39.

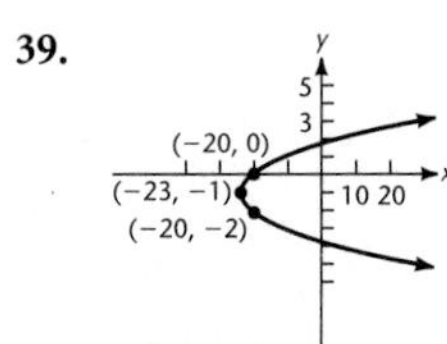

41.

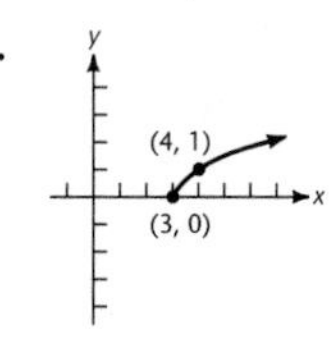

43.

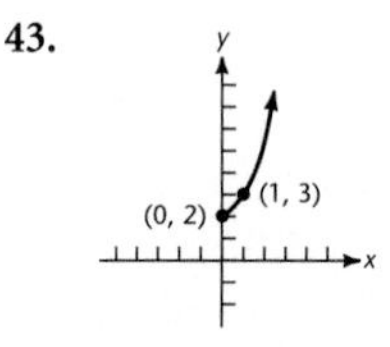

45.

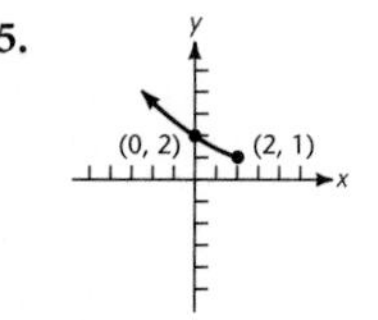

47. $y=4-x^2$ **49.** $x=(y-2)^2-4$ **51.** The lengths of the sides of the triangle formed using the given points as vertices are $3\sqrt{2}$, $4\sqrt{2}$, and $5\sqrt{2}$. Since these lengths satisfy the Pythagorean theorem, the triangle is a right triangle. **53.** All four sides of the quadrilateral formed using the given points have the same length ($\sqrt{109}$), and the two diagonals have the same length ($\sqrt{218}$). Thus the given points are vertices of a square. **55.** 9 meters **57.** $30\sqrt{3}$ centimeters **59.** Viewing rectangle $[-10, 10]$ by $[-7, 7]$ **61.** Viewing rectangle $[-10, 10]$ by $[-10, 10]$

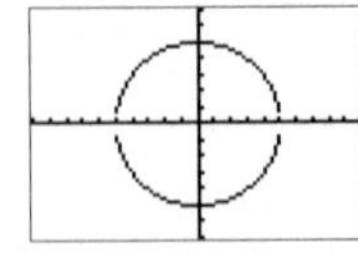

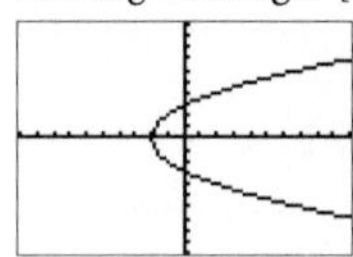

Exercises 4.2, page 228

1. (a) (b) (c)

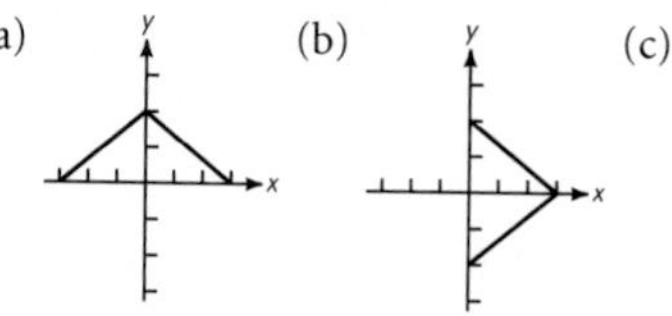

3. (a) (b) (c)

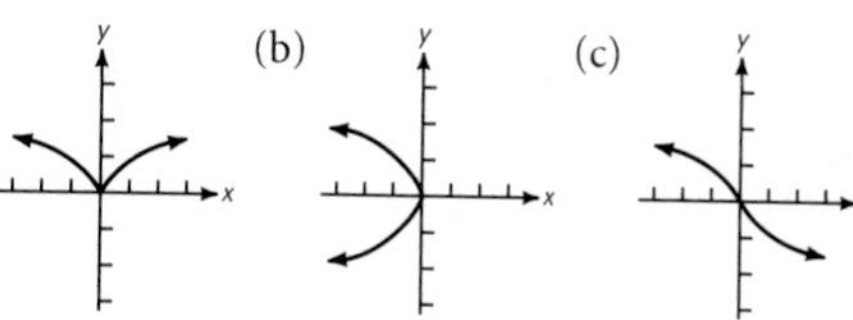

5. (a) (b) (c)

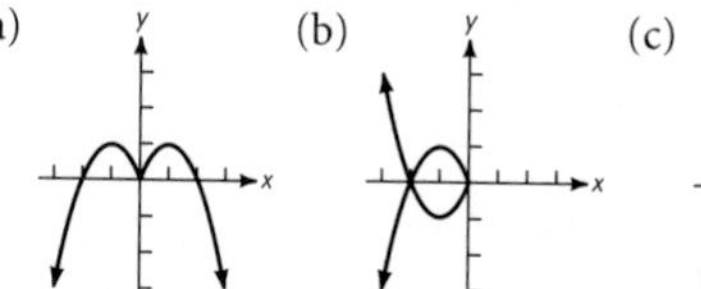

7. (a) (b) (c)

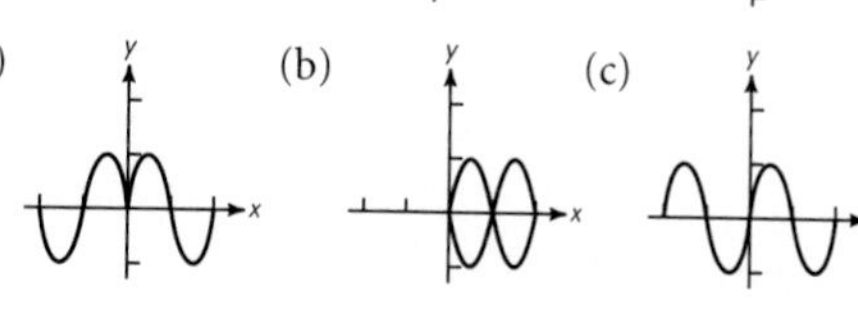

9. Symmetric with respect to the y-axis **11.** Symmetric with respect to the x-axis, the y-axis, and the origin **13.** Symmetric with respect to the x-axis **15.** Symmetric with respect to the origin **17.** Symmetric with respect to the x-axis, the y-axis, and the origin **19.** Symmetric with respect to the origin **21.** Symmetric with respect to the y-axis **23.**

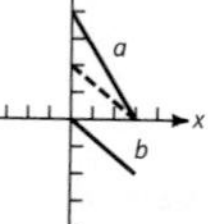

25.

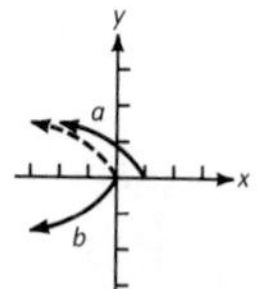

27.

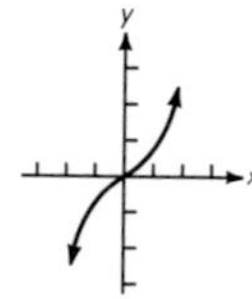

29.

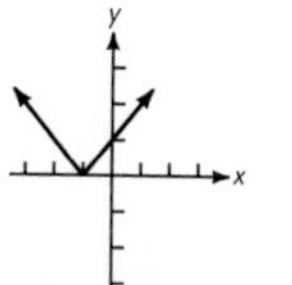

31. **33.**

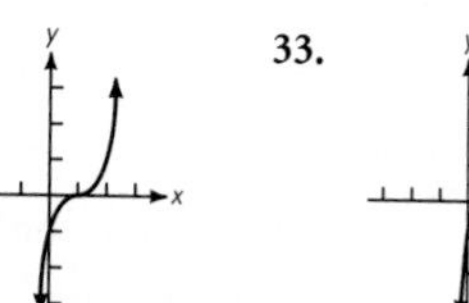

35.

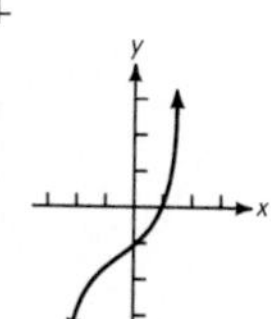

37. Viewing rectangle $[-10, 10]$ by $[-10, 10]$. The original graph has been shifted to the right 5 units.

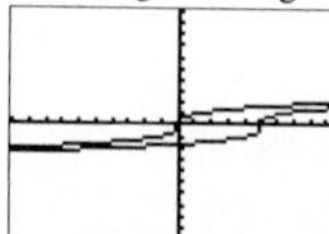

39. Viewing rectangle $[-10, 10]$ by $[-10, 10]$. The original graph has been stretched by a factor of 3; that is, each y-coordinate has been multiplied by 3.

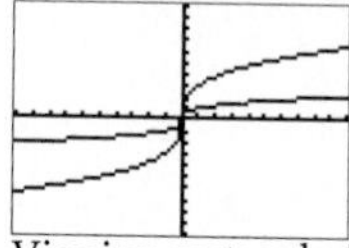

41. Viewing rectangle $[-10, 10]$ by $[-10, 10]$. The original graph has been reflected through the y-axis; that is, each point on the graph has moved to the point that is symmetric with respect to the y-axis.

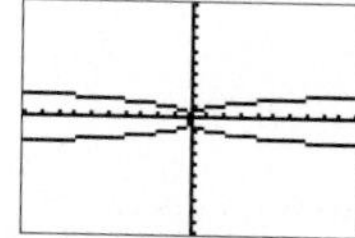

Exercises 4.3, page 234

1.

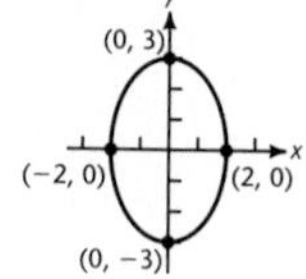

3.

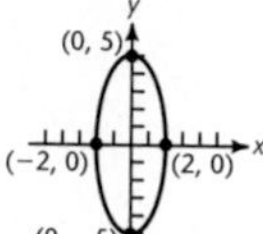

5.

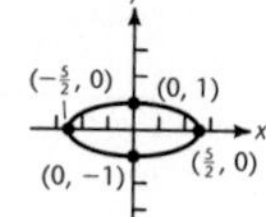

7.

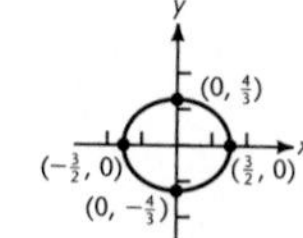

9.

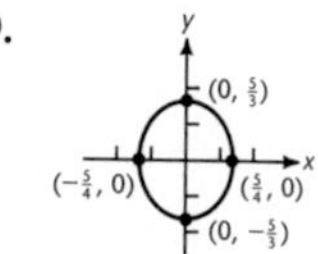

11. 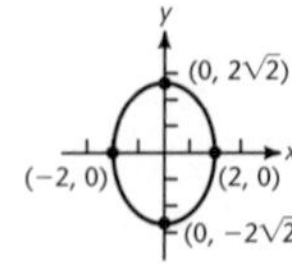

13. 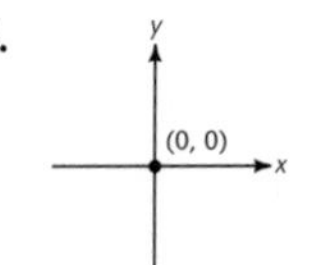

15. 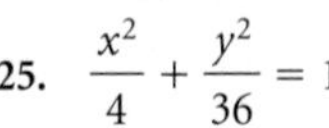

17. 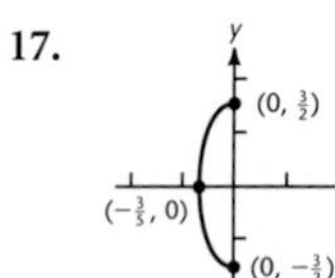

19. 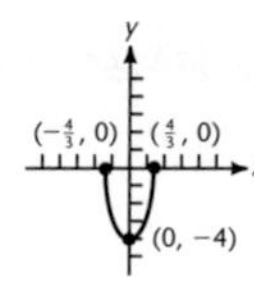

21. 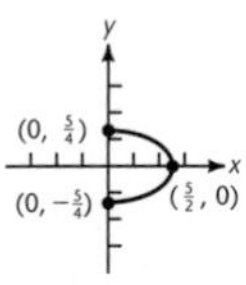

23. $\dfrac{x^2}{16} + \dfrac{y^2}{4} = 1$ **25.** $\dfrac{x^2}{4} + \dfrac{y^2}{36} = 1$ **27.** $9\sqrt{3}/2$ feet **29.** $3\sqrt{3}$ feet

31. Viewing rectangle $[-10, 10]$ by $[-10, 10]$

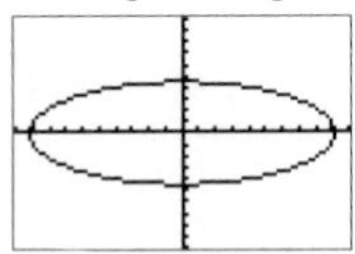

33. Viewing rectangle $[-10, 10]$ by $[-13, 10]$

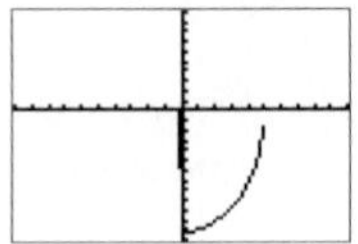

Exercises 4.4, page 240

1. V: $(\pm 3, 0)$; A: $y = \pm 4x/3$

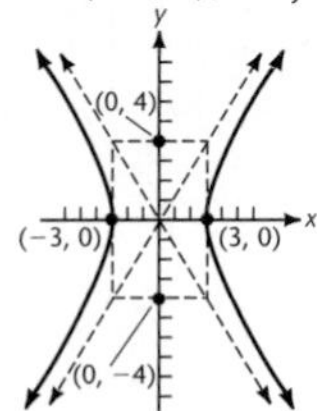

3. V: $(0, \pm 2)$; A: $y = \pm 2x/5$

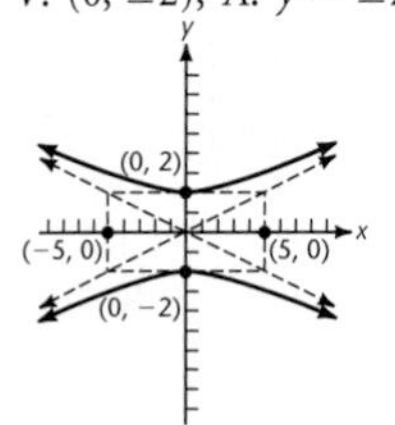

5. V: $(0, \pm 3)$; A: $y = \pm x$

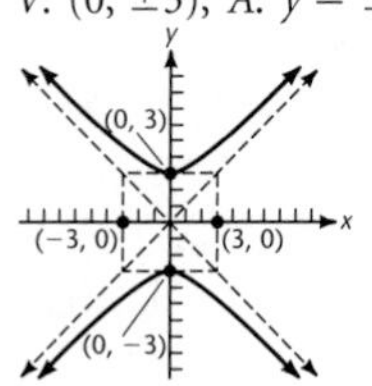

7. V: $(\pm 3, 0)$; A: $y = \pm 2x/3$

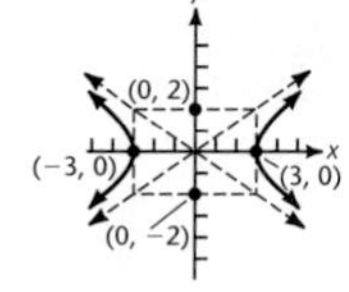

9.

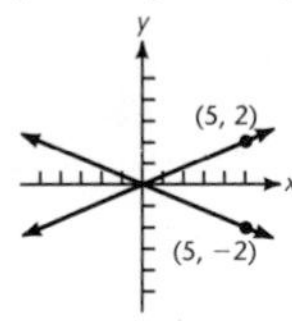

11. V: $(\pm 2, 0)$; A: $y = \pm x/2$

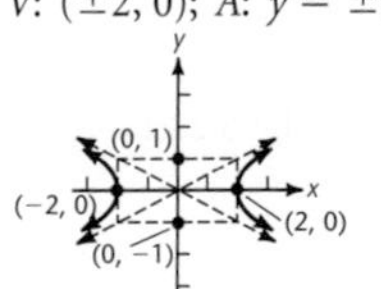

13. V: $(0, \pm \frac{5}{4})$; A: $y = \pm 15x/8$

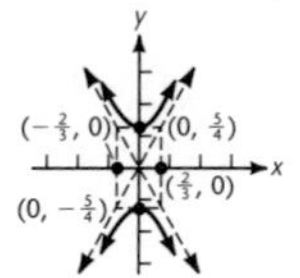

15. V: $(\pm 2, 0)$; A: $y = \pm\sqrt{2}x$

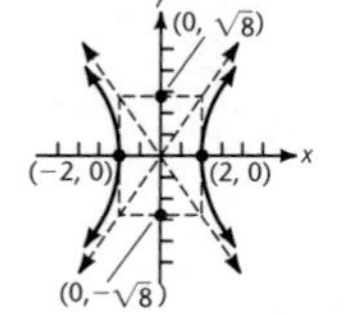

17. V: $(0, \frac{3}{5})$; A: $y = \pm 2x/5$

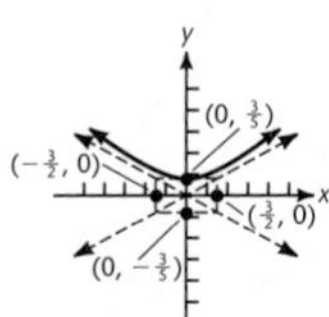

19. V: $(2, 0)$; A: $y = \pm 2x$

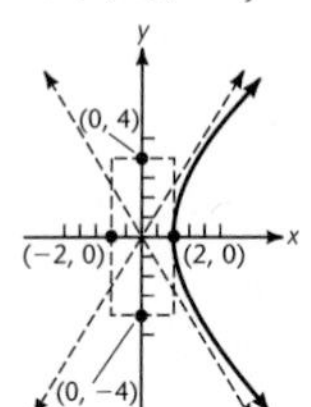
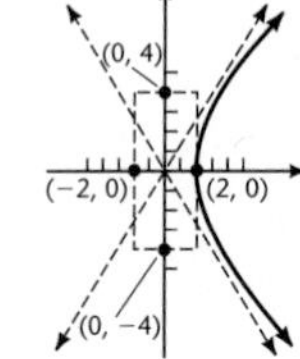

21. $\dfrac{x^2}{4} - \dfrac{y^2}{9} = 1$ **23.** $\dfrac{y^2}{4} - \dfrac{x^2}{4} = 1$

25. Viewing rectangle $[-10, 10]$ by $[-10, 10]$

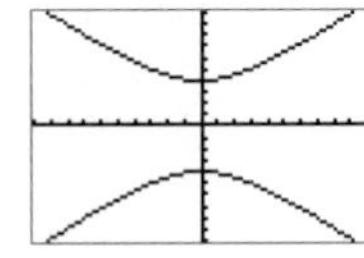

27. Viewing rectangle $[-10, 10]$ by $[-10, 10]$

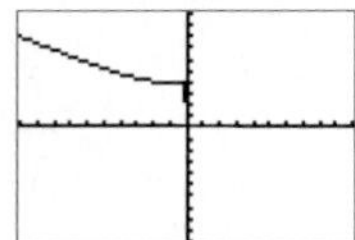

Exercises 4.5, page 247

1. C: $(2, 1)$; V: $(2, 1 \pm 3)$

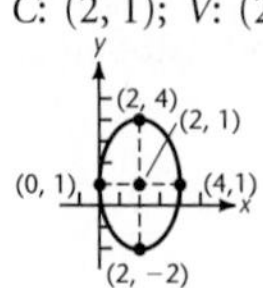

3. C: $(1, 2)$; V: $(1 \pm 3, 2)$

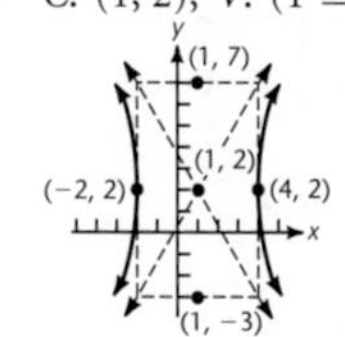

5. C: $(2, -3)$; V: $(2, -3 \pm 2)$

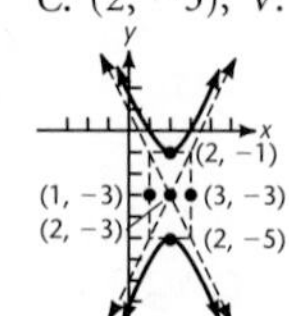

7. C: $(-2, -2)$; V: $(-2 \pm 3, -2)$

9. C: $(-1, 3)$; V: $(-1, 3 \pm 5)$

11. C: $(-2, 3)$; V: $(-2, 3 \pm 9)$

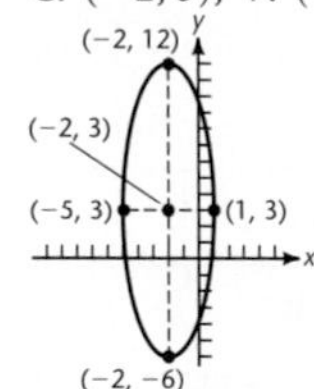

13. Ellipse, C: $(1, 2)$, $a = 3$, $b = 2$

15. Hyperbola, C: $(2, 1)$, $a = 3$, $b = 1$ **17.** Ellipse, C: $(3, -2)$, $a = 4$, $b = 3$ **19.** Hyperbola, C: $(-3, 4)$, $a = 10$, $b = 5$

21. C: $(0, 1)$; V: $(\pm 2, 1)$

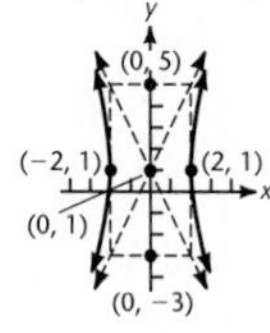

23. C: $(1, -3)$; V: $(1, -3 \pm 4)$

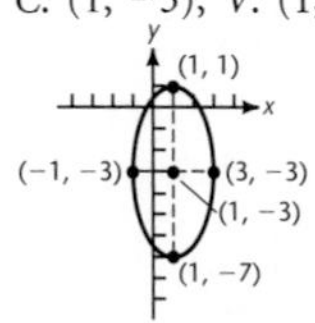

25. C: $(-3, 0)$; V: $(-3 \pm 6, 0)$

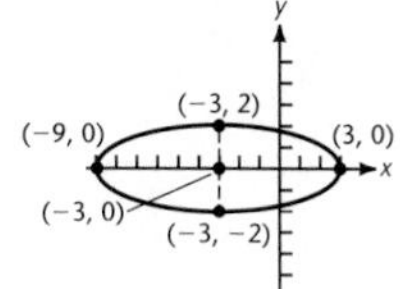

27. C: $(-2, -1)$; V: $(-2, -1 \pm 3)$

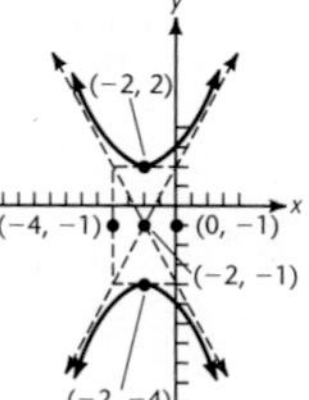

29. $\dfrac{x^2}{9} + \dfrac{y^2}{4} = 1$ **31.** $x^2 + \dfrac{y^2}{16} = 1$ **33.** $\dfrac{(x-1)^2}{4} + y^2 = 1$ **35.** $\dfrac{(x-2)^2}{4} + \dfrac{(y-3)^2}{9} = 1$

37. Viewing rectangle $[-10, 10]$ by $[-10, 10]$

39. Viewing rectangle $[-10, 10]$ by $[-10, 10]$

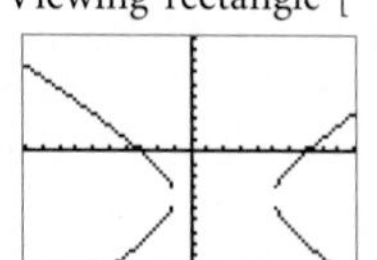

Exercises 4.6, page 254

1. Not the graph of a one-to-one function **3.** Is the graph of a one-to-one function

5. $g^{-1}(x) = x - 3$

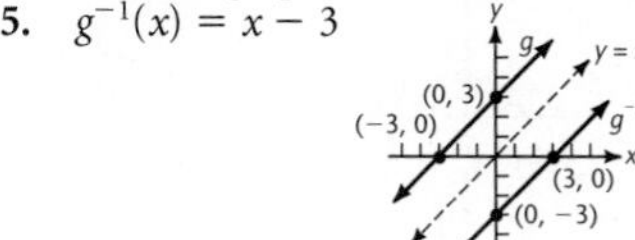

7. $g^{-1}(x) = \frac{3}{4}x + 3$

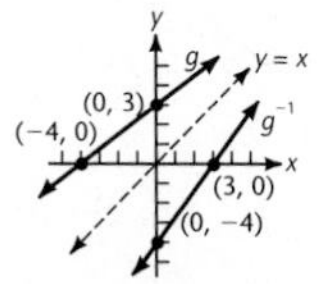

9. $g^{-1}(x) = \sqrt[3]{8 - x}$

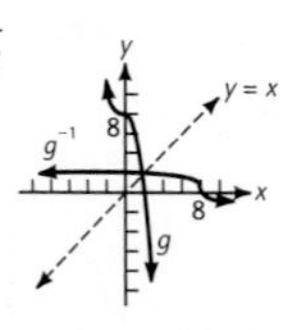

11. $f(x) = 2x - 4$, $f^{-1}(x) = \frac{1}{2}x + 2$; $f^{-1}(f(x)) = \frac{1}{2}(2x - 4) + 2 = x - 2 + 2 = x$; $f(f^{-1}(x)) = 2(\frac{1}{2}x + 2) - 4 = x + 4 - 4 = x$

13. $f(x) = x^2 + 2$, f is not one-to-one **15.** $f(x) = (x - 1)^2$, f is not one-to-one

17. $f(x) = x^3 - 1$, $f^{-1}(x) = \sqrt[3]{x + 1}$; $f^{-1}(f(x)) = \sqrt[3]{x^3 - 1 + 1} = \sqrt[3]{x^3} = x$; $f(f^{-1}(x)) = (\sqrt[3]{x + 1})^3 - 1 = x + 1 - 1 = x$

19. f and g are inverse functions. **21.** f and g are inverse functions. **23.** f and g are inverse functions. **25.** f and g are not inverse functions. **27.** f and g are inverse functions. **29.** f and g are not inverse functions.

Exercises 4.7, page 261

1. $\frac{3}{4}$ **3.** 8 **5.** 3 boys/1 girl **7.** 30 miles per hour **9.** 432 pounds per square foot **11.** 5 miles per quart **13.** 216 miles **15.** $\frac{24}{25}$ pint **17.** \$15, \$25 **19.** 75 **21.** $\frac{1}{2}$ **23.** 9 **25.** $\pm\frac{1}{2}$ **27.** -4 **29.** 40 pounds **31.** 3125 pounds per square foot **33.** \$12,800 **35.** 1.875 seconds **37.** 1600 feet **39.** 2 centimeters **41.** $6/\pi$ inches **43.** 6 workers

Critical Thinking: Find the Errors, page 267

1. **Error**

$$d = \sqrt{(4-1)^2 + (6-2)^2}$$
$$= \sqrt{3^2 + 4^2}$$
×
$$= 3 + 4$$

Correction

$$\sqrt{3^2 + 4^2} = \sqrt{9 + 16} = \sqrt{25} = 5$$

2. **Errors**

$$x^2 + 2x + y^2 + 4y = 11$$
$$x^2 + 2x + 1 + y^2 + 4y + 4 = 11 + 1 + 4$$
$$(x+1)^2 + (y+2)^2 = 16$$

The center is at (1, 2), ×
and the radius is 16. ×

Correction

The center is at $(-1, -2)$, and the radius is 4.

3. **Error**

$$f(x) = \sqrt{x+1}$$
×
$$f^{-1}(x) = \frac{1}{\sqrt{x+1}}$$

Correction

$$x = \sqrt{y+1}$$
$$x^2 = y + 1,\ x \geq 0$$
$$y = x^2 - 1,\ x \geq 0$$
$$f^{-1}(x) = x^2 - 1,\ x \geq 0$$

Review Problems for Chapter 4, page 268

1. 5 **2.** $2\sqrt{2}$ **3.** $3\sqrt{5}$ **4.** 8 **5.** Center $(3, -5)$, radius 4 **6.** Center $(-1, -2)$, radius 3 **7.** Not a circle **8.** $(x-2)^2 + (y+3)^2 = 25$ **9.** $(x+3)^2 + (y+4)^2 = 25$ **10.** $(x+1)^2 + (y+1)^2 = 13$ **11.** $(x-1)^2 + (y+3)^2 = 10$

12. (a)

(−5, 2) (−3, 2) (3, 2) (5, 2) (0, 1)

(b)

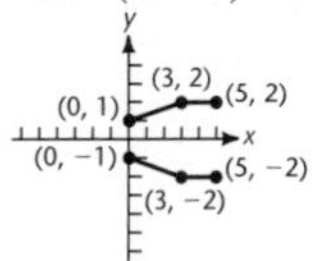

(c)

(3, 2) (5, 2) (0, 1) (0, −1) (−5, −2) (−3, −2)

13. Symmetric with respect to the x-axis, the y-axis, and the origin

(−2, 2) (2, 2) (−2, −2) (2, −2)

14. Symmetric with respect to the y-axis only

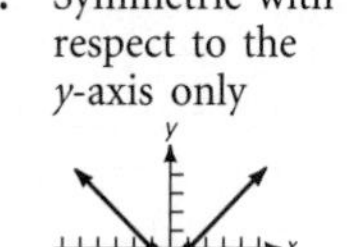

15.

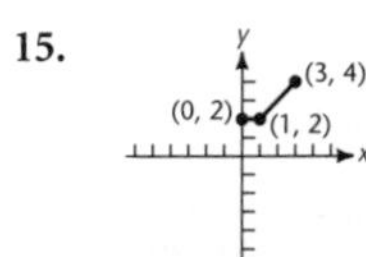

16.

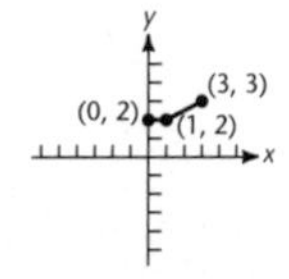

17.

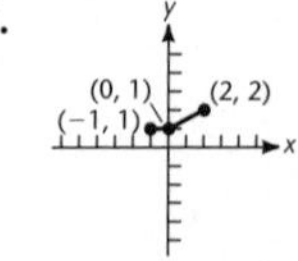

18.

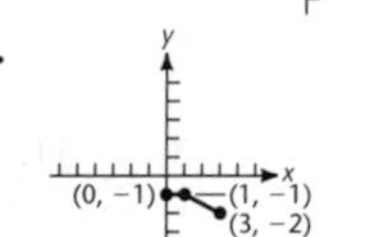

19.

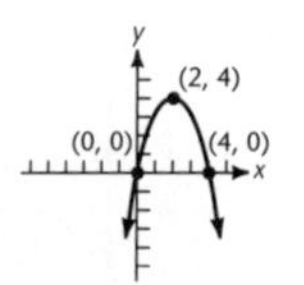

20.

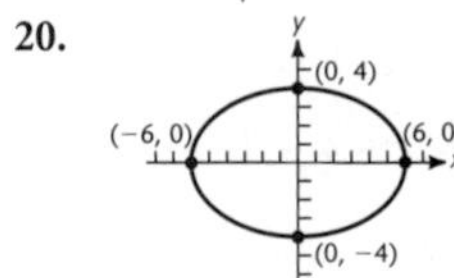

21. V: $(\pm 3, 0)$; A: $y = \pm 2x$

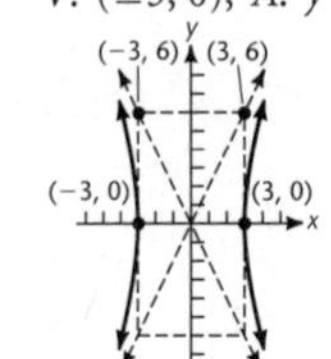

22. V: $(0, \pm 2)$; A: $y = \pm\frac{2}{5}x$

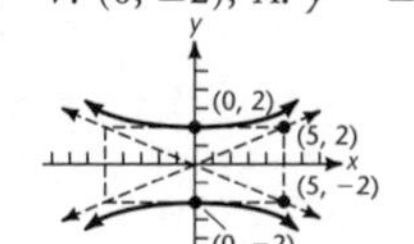

23.

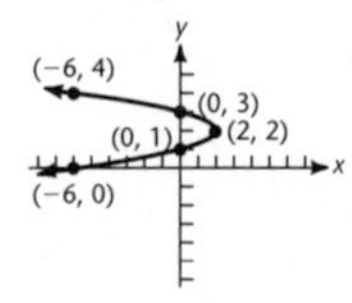

24. **25.**

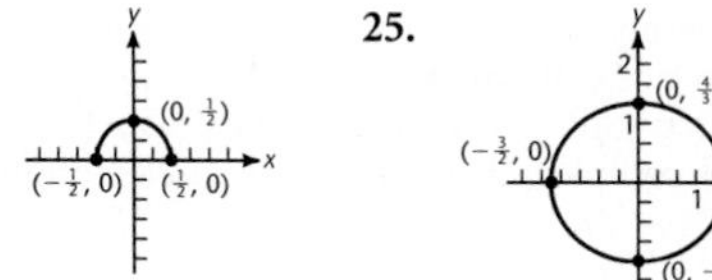

26.

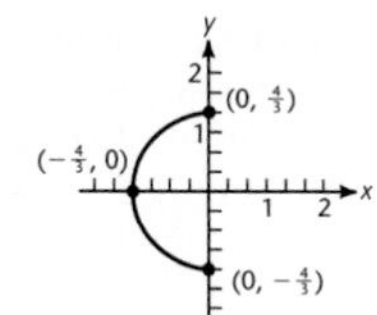

27. V: $(0, -2)$; A: $y = \pm\frac{1}{2}x$

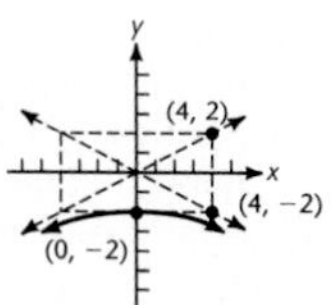

28. C: $(-3, 2)$; V: $(-7, 2)$, $(1, 2)$

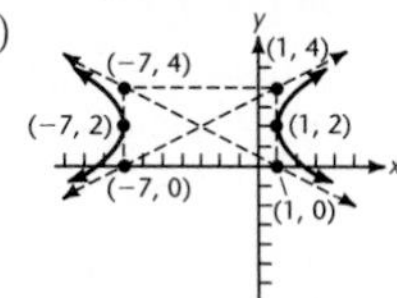

29. The graph of the equation is Ø. **30.** No **31.** It is not the graph of a one-to-one function. **32.** It is not the graph of a one-to-one function. **33.** It is the graph of a one-to-one function. **34.** It is the graph of a one-to-one function. **35.** $f(x) = \frac{2}{3}x - 2$, $f^{-1}(x) = \frac{3}{2}x + 3$ **36.** $f(x) = x^2 - 4$, f is not one-to-one. **37.** $f(x) = x^2 + 1$, f is not one-to-one. **38.** f and g are not inverse functions of each other. **39.** f and g are inverse functions of each other. **40.** **41.** **42.** **43.** 24 cubic meters

44. 66 feet per second **45.** 3 **46.** 8 **47.** 44 amps **48.** 700

Exercises 5.1, page 277

1. $\dfrac{x^3 - 4x^2 + 6x - 3}{x - 1} = x^2 - 3x + 3 + \dfrac{0}{x - 1}$ **3.** $\dfrac{2x^3 - 11x^2 + 19x - 10}{2x - 5} = x^2 - 3x + 2 + \dfrac{0}{2x - 5}$

5. $\dfrac{3x^3 - 5x^2 + 14x + 3}{3x - 2} = x^2 - x + 4 + \dfrac{11}{3x - 2}$ **7.** $\dfrac{2x^4 + 3x^3 - 3x^2 + 5x - 8}{x^2 + 2x - 2} = 2x^2 - x + 3 + \dfrac{-3x - 2}{x^2 + 2x - 2}$

9. $\dfrac{4x^4 + 5x^2 - 7x + 3}{2x^2 + 3x - 2} = 2x^2 - 3x + 9 + \dfrac{-40x + 21}{2x^2 + 3x - 2}$ **11.** $2x^2 + 5 + \dfrac{9}{x - 2}$ **13.** $-3x^2 + 9x - 25 + \dfrac{0}{x + 3}$

15. $x^4 + 2x^3 - x^2 - 3x - 9 - \dfrac{3}{x - 3}$ **17.** $4x^2 - 4x + 5 + \dfrac{2}{x + 1}$ **19.** $4m^3 - 3m - 2 = (m + 1)(4m^2 - 4m + 1) - 3$

21. $3x^5 + 3x^4 - 10x^3 - 10x^2 - 8x - 8 = (x^2 + 3x + 2)(3x^3 - 6x^2 + 2x - 4) + 0$

23. $18k^2 - 3rk - 10r^2 = (6k - 5r)(3k + 2r) + 0$

25. $4x^4 - 13ax^3 + 12a^2x^2 - 5a^3x + 2a^4 = (4x^2 - ax + a^2)(x^2 - 3ax + 2a^2) + 0$ **27.** $x^3 - 27 = (x + 3)(x^2 - 3x + 9) - 54$

29. $Q(x) = x - 4$, $r = 0$ **31.** $Q(x) = 2x^2 + 3x + 4$, $r = 10$ **33.** $Q(x) = x^4 - 2x^3 - 4x^2 - 9x - 18$, $r = -34$

35. $Q(x) = 4x^3 - 2x^2 + 6$, $r = -5$ **37.** $Q(x) = x^2 + (3 - i)x - 3i$, $r = 0$ **39.** $Q(x) = -ix^2 + 7x + 21i$, $r = -58$ **41.** 1

43. 9 **45.** 0 **47.** 0 **49.** 0 **51.** 8 **53.** $k > \frac{1}{2}$ **55.** 33.86 **57.** 0.84

Exercises 5.2, page 285

1. 7 **3.** 5 **5.** 7 **7.** $4\sqrt{2}$ **9.** $-3i$ **11.** 9 **13.** 17 **15.** -90 **17.** 0 **19.** $6 + 6\sqrt{2}$ **21.** $-16 + 3i$ **23.** $-16 - 22i$ **25.** No **27.** Yes **29.** No **31.** Yes **33.** Yes **35.** No **37.** Yes **39.** No **41.** Yes **43.** $x = 3$, $x = \pm i$ **45.** $-2, \frac{1}{2}$ **47.** $k = \frac{14}{3}$ **49.** $P(1) = -8$ and $P(2) = 5$ are of opposite sign. **51.** $P(3) = -39$ and $P(4) = 84$ are of opposite sign. **53.** $P(0) = -3$ and $P(-1) = 1$ are of opposite sign. **55.** Between 0 and 1 **57.** Between 1 and 2, and between -1 and -2

Exercises 5.3, page 293

1. Degree 3; 1 with multiplicity 1, -2 with multiplicity 2 **3.** Degree 5; -3 with multiplicity 2, $\frac{3}{4}$ with multiplicity 3 **5.** Degree 6; 0 with multiplicity 2, each of i, $-i$, 1, and -1 with multiplicity 1 **7.** $P(x) = (x - 3)^2(x + 1)$ **9.** $P(x) = x^5(x + 2)^3$ **11.** $P(x) = 2(x - 1)(x + 1)(x - 2)$ **13.** 2 positive, 1 negative; 1 negative, 2 nonreal complex

15. 2 positive, 2 negative; 2 positive, 2 nonreal complex; 2 negative, 2 nonreal complex; 4 nonreal complex **17.** 3 positive, 1 negative; 1 positive, 1 negative, 2 nonreal complex **19.** 2 negative, 4 nonreal complex; 6 nonreal complex **21.** 4 positive, 1 negative; 2 positive, 1 negative, 2 nonreal complex; 1 negative, 4 nonreal complex **23.** 2 positive, 5 negative; 2 positive, 3 negative, 2 nonreal complex; 2 positive, 1 negative, 4 nonreal complex; 5 negative, 2 nonreal complex; 3 negative, 4 nonreal complex; 1 negative, 6 nonreal complex **25.** 6 nonreal complex **27.** $3i$ **29.** $1 - 2i$ **31.** $2i, -4i$ **33.** $2i, -4$

35. $2 - i, -2$ **37.** $3i, -\dfrac{1}{2} \pm \dfrac{\sqrt{3}}{2}i$ **39.** $1 - i, 1, -1$ **41.** $1 - i$ of multiplicity 2, 2 **43.** $P(x) = Q(x) = x^2 + 2x - 15$

45. $P(x) = x^2 - ix + 2$, $Q(x) = x^4 + 5x^2 + 4$ **47.** $P(x) = x^2 + (-5 + i)x + 6 - 3i$, $Q(x) = x^3 - 7x^2 + 17x - 15$

49. $P(x) = x^3 - (7 + i)x^2 + (17 + 2i)x - 15 + 3i$, $Q(x) = x^5 - 11x^4 + 51x^3 - 119x^2 + 140x - 78$

Exercises 5.4, page 302

3. (a) 3 (b) -3 **5.** (a) 2 (b) -1 **7.** (a) 2 (b) -3 **9.** (a) 2 (b) -3 **11.** $\{\pm 1, \pm 3\}$ **13.** $\{\pm 1, \pm 2, \pm 4, \pm 8\}$ **15.** $\{\pm 1, \pm 2, \pm 3, \pm 4, \pm 6, \pm 12, \pm\frac{1}{2}, \pm\frac{3}{2}\}$ **17.** $\{\pm 1, \pm 2, \pm\frac{1}{2}\}$ **19.** $\{\pm 1, \pm 2, \pm 3, \pm 6, \pm\frac{1}{2}, \pm\frac{3}{2}, \pm\frac{1}{4}, \pm\frac{3}{4}\}$ **21.** $3, 2i, -2i$ **23.** $-3, -1, \frac{1}{2}$ **25.** $2, (-1 \pm i\sqrt{23})/6$ **27.** $-2, 1, 1 \pm i$ **29.** $-1, \frac{3}{2}, \pm\sqrt{5}$ **31.** $-3, \frac{1}{2}, \pm\sqrt{2}$ **33.** No rational zeros **35.** $1, -2$ **37.** $-2, -1, \frac{1}{3}$ **39.** $-3, -5, -\frac{1}{2}$ **41.** $2, -\frac{3}{2}$ **43.** $-\frac{3}{2}$ **45.** $2, -\frac{3}{2}$ **47.** No rational zeros **51.** $\frac{1}{2}$ meter by $\frac{1}{2}$ meter **53.** 2 **55.** $5, (-5 + \sqrt{69})/2$ **57.** 3

59. Possible rational zeros: $\{\pm 1, \pm 3, \pm 5, \pm 15, \pm\frac{1}{2}, \pm\frac{3}{2}, \pm\frac{5}{2}, \pm\frac{15}{2}, \pm\frac{1}{4}, \pm\frac{3}{4}, \pm\frac{5}{4}, \pm\frac{15}{4}, \pm\frac{1}{8}, \pm\frac{3}{8}, \pm\frac{5}{8}, \pm\frac{15}{8}\}$; rational zeros: $\{\frac{5}{4}, -\frac{3}{2}\}$

61. Possible rational zeros: $\{\pm 1, \pm 2, \pm 5, \pm 10, \pm\frac{1}{3}, \pm\frac{2}{3}, \pm\frac{5}{3}, \pm\frac{10}{3}, \pm\frac{1}{5}, \pm\frac{2}{5}, \pm\frac{1}{15}, \pm\frac{2}{15}\}$; rational zeros: $\{\frac{5}{3}, -\frac{1}{5}\}$

Exercises 5.5, page 309

1. $P(0) = 3$ and $P(1) = -1$ are of opposite sign. **3.** $P(1) = -1$ and $P(1.5) = 9.0625$ are of opposite sign. **5.** [1, 1.5] **7.** [−0.5, −0.4] **9.** 2.2 **11.** 0.5 **13.** 2.7 **15.** 0.4 **17.** 1.1 **19.** 1.3 **21.** 0.5 **23.** −0.8 **25.** 1.1 **27.** 1.2 **29.** −1.5, 0.3, 2.2 **31.** −1.5, −0.4, 0.9 **33.** −2.3, −1.5, 4.8 **35.** 0.9, 3.8 **37.** 0.68 **39.** 1.00, −1.22 **41.** −1.00, 0.50, 3.00 **43.** −2.61, −0.58, 2.70

Exercises 5.6, page 316

1.

3.

5.

7.

9.

11.

13.

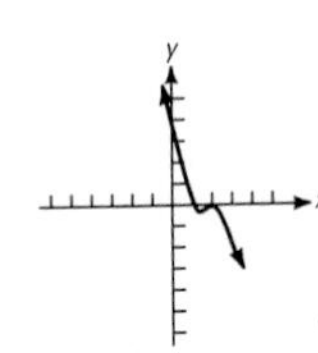

15.

17. $P(x) = (x + 1)(x - 1)(x - 3)$

19. $P(x) = 2(x + 1)(x - 2)^2$ **21.** $P(x) = x(x + 1)(x - 1)(x - 3)$ **23.** $P(x) = -3(x + 1)^2(x - 1)^2$

25.

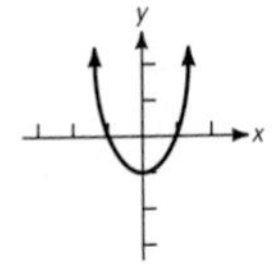

27.

29.

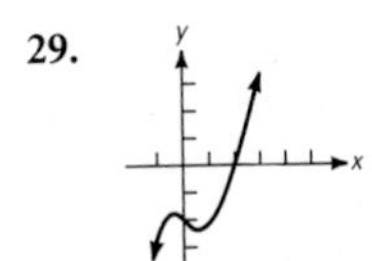

31.

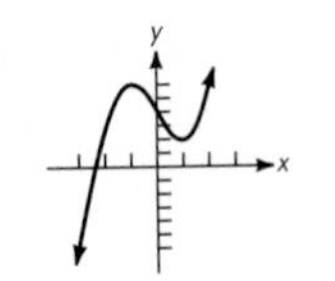

33.

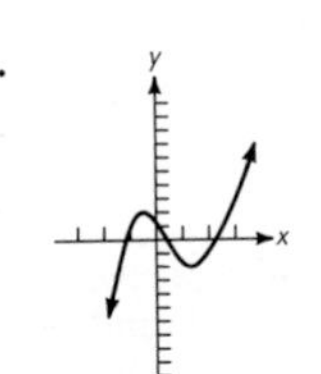

35.

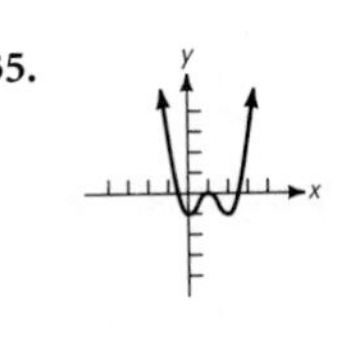

37.

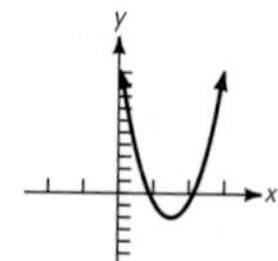

39. Viewing rectangle $[-5, 5]$ by $[-50, 5]$

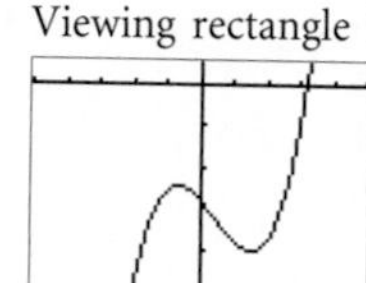

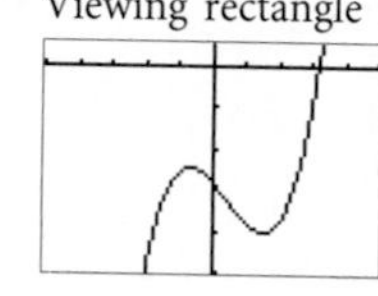

41. Viewing rectangle $[-10, 10]$ by $[-40, 10]$

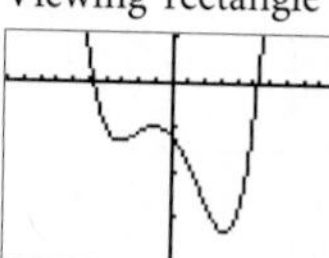

43. (a) $[0, 400]$ (b)

45. $[0, 4]$ **47.** $(-\infty, 1.4)$ **49.** $(-3.3, -0.7) \cup (2.1, \infty)$ **51.** $(1.6, 3.9)$

Exercises 5.7, page 330

1. $y = 0,\ x = -3$ **3.** $y = 0,\ x = 0,\ x = -3$ **5.** $y = 3,\ x = -2$ **7.** $y = \frac{1}{2},\ x = -3,\ x = \frac{3}{2}$ **9.** $y = x - 2,\ x = -2$
11. $y = x,\ x = 0$ **13.** $y = 1,\ x = -1$ **15.** No asymptotes **17.** V: $x = -4$; H: $y = 0$ **19.** V: $x = 0,\ x = 3$; H: $y = 0$

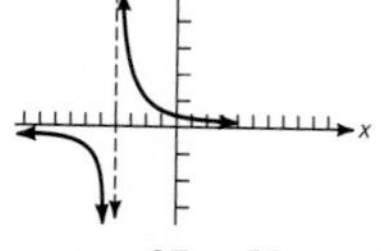

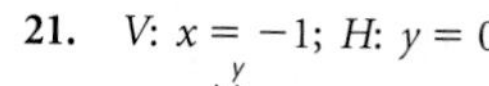

21. V: $x = -1$; H: $y = 0$

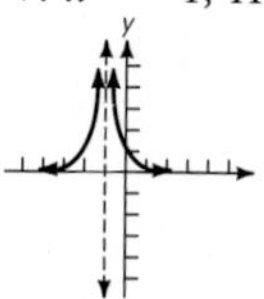

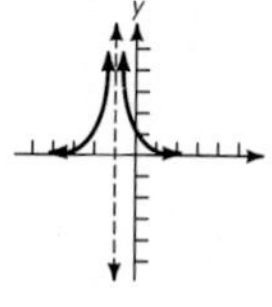

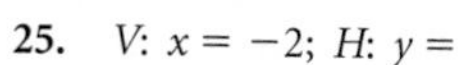

23. H: $y = 0$

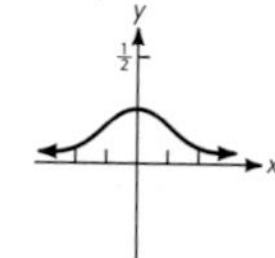

25. V: $x = -2$; H: $y = 0$

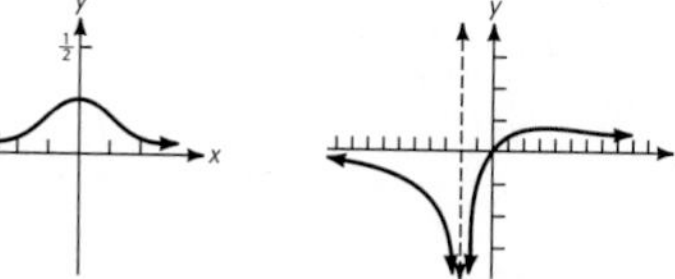

27. V: $x = 0,\ x = -1$; H: $y = 0$

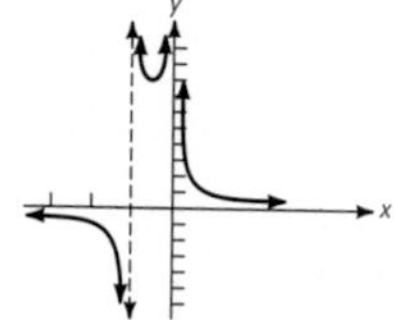

29. V: $x = -3$; H: $y = 2$

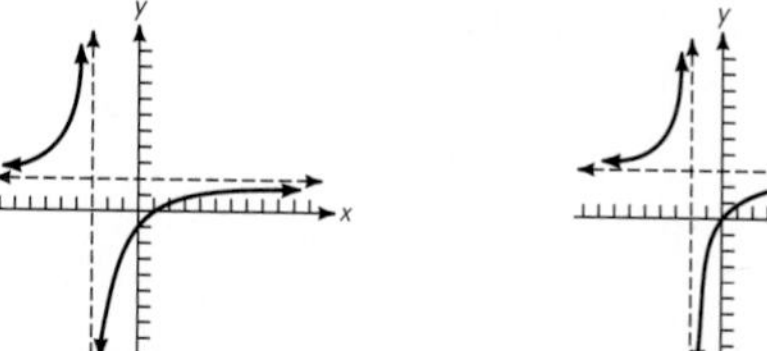

31. V: $x = -2$; H: $y = 3$

33. V: $x = 0,\ x = 1$; H: $y = 1$

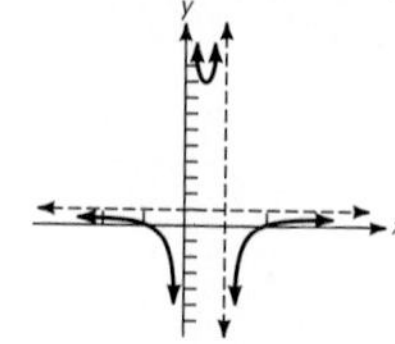

35. V: $x = -1,\ x = \frac{3}{2}$; H: $y = \frac{1}{2}$

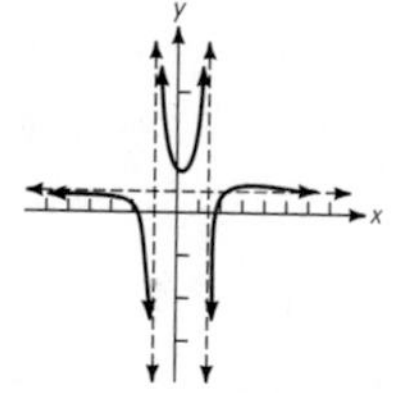

37. V: $x = 2,\ x = -2$; H: $y = 1$

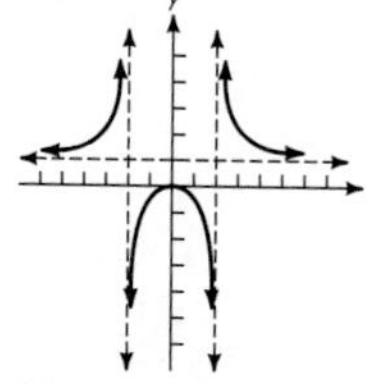

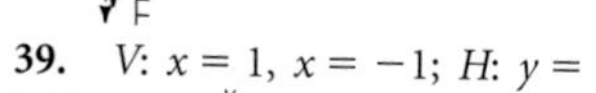

39. V: $x = 1,\ x = -1$; H: $y = 1$

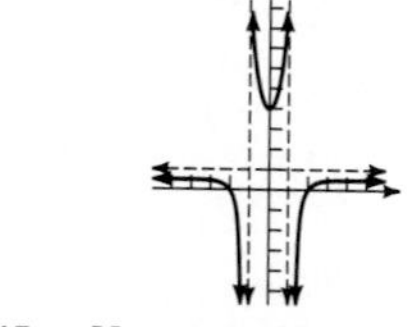

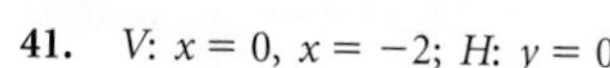

41. V: $x = 0,\ x = -2$; H: $y = 0$

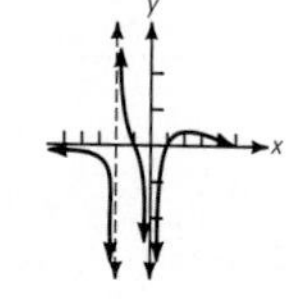

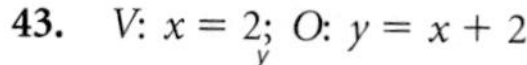

43. V: $x = 2$; O: $y = x + 2$

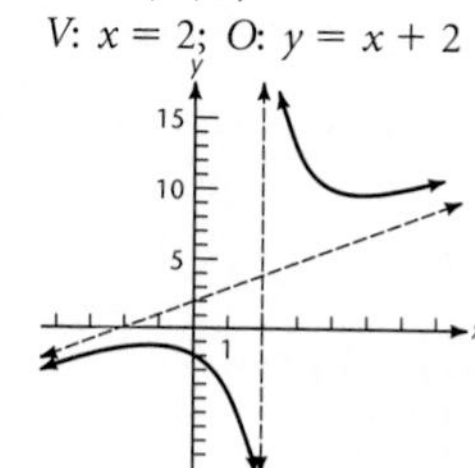

45. V: $x = 0$; H: $y = 0$

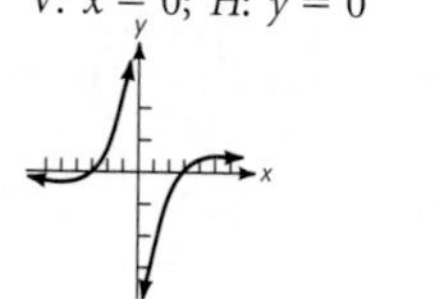

47.

49.

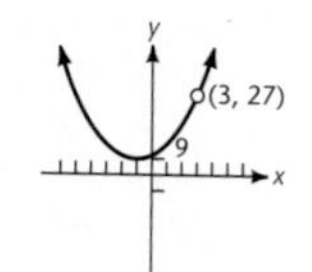

51.

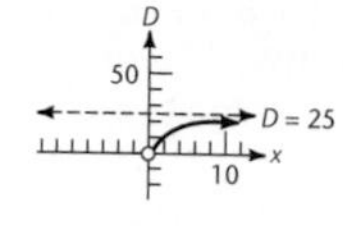

53.

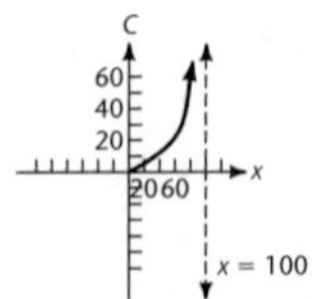

55. Viewing rectangle $[-10, 10]$ by $[-10, 10]$

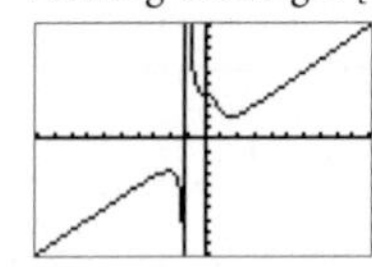

57. Viewing rectangle $[-5, 5]$ by $[0, 10]$

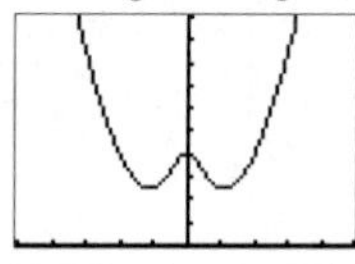

59. (a) Viewing rectangle $[-10, 10]$ by $[-10, 10]$ (b) Viewing rectangle $[-1000, 1000]$ by $[-10, 10]$

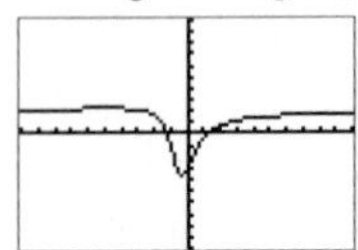

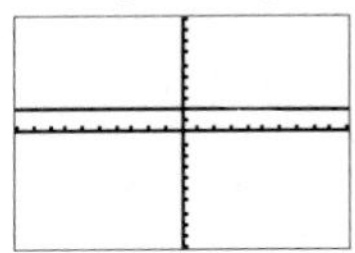

Critical Thinking: Find the Errors, page 336

1. **Error**

$$\begin{array}{r|rrr} 2 & 2 & -7 \times & 6 \\ & & 4 & -6 \\ \hline & 2 & -3 & 0 \end{array}$$

Correction

$$\begin{array}{r|rrrr} 2 & 2 & -7 & 0 & 6 \\ & & 4 & -6 & -12 \\ \hline & 2 & -3 & -6 & -6 \end{array}$$

Since the remainder is -6, $x - 2$ is not a factor of $P(x)$ and 2 is not a zero of $P(x)$.

2. **Error**

$P(x) = (x + 2)(x - 3i)$ ×

Correction

$$\begin{aligned} P(x) &= (x + 2)(x - 3i)(x + 3i) \\ &= x^3 + 2x^2 + 9x + 18 \end{aligned}$$

3. **Error**

There are two variations of sign in $P(x) = x^3 + x^2 - x + 15$ and one variation of sign in $P(-x) = -x^3 + x^2 + x + 15$. Therefore $P(x)$ has two positive zeros and one negative zero. ×

Correction

There are two possibilities. Either **(a)** $P(x)$ has two positive zeros and one negative zero, or **(b)** $P(x)$ has one negative zero and two nonreal complex zeros.

Review Problems for Chapter 5, page 337

1. $\dfrac{2x^3 - x^2 - 4x - 30}{x - 3} = 2x^2 + 5x + 11 + \dfrac{3}{x - 3}$ **2.** $\dfrac{3x^3 + 5x^2 + 7}{x + 2} = 3x^2 - x + 2 + \dfrac{3}{x + 2}$

3. $\dfrac{2x^4 + 6x^2 - 2x + 1}{x + 1} = 2x^3 - 2x^2 + 8x - 10 + \dfrac{11}{x + 1}$ **4.** $Q(x) = 6x^2 - 2x - 4,\ r = 0$

5. $Q(x) = x^3 - (1 + i)x^2 + (2 + i)x - 2i,\ r = 1$ **6.** 6 **7.** $-4 - 2i$ **8.** Yes **9.** Yes **10.** No **11.** No **12.** Yes **13.** $3, -5$ **14.** $P(2) = -12$ and $P(3) = 20$ are of opposite sign. **15.** Between 1 and 2, and between -1 and -2 **16.** $P(x) = (x - 2)^2(x + 1)^3$ **17.** $P(x) = -4(x - 2)(x + 1)(x - 3)$ **18.** 2 positive, 1 negative; 1 negative, 2 nonreal complex **19.** 1 positive, 3 negative; 1 positive, 1 negative, 2 nonreal complex **20.** 2 positive, 2 negative; 2 positive, 2 nonreal complex; 2 negative, 2 nonreal complex; 4 nonreal complex **21.** $1 + i$ **22.** $2 + i, 2 - i$ **23.** $i, 2, -1$ **24.** $x^2 - (4 + 2i)x + 8i$ **25.** $x^4 + 3x^3 + 11x^2 + 27x + 18$ **26.** (a) 2 (b) -3 **27.** -2 **28.** None **29.** $1, \frac{1}{3}, -2$ **30.** $2, 5, -1$ **31.** $2, (-1 \pm i\sqrt{3})/2$ **32.** $1, 4, -2, -2$ **33.** $\frac{1}{3}, 1 \pm i$ **34.** $-2, \frac{2}{3}, -\frac{5}{3}$ **35.** $-2, -4, -\frac{1}{3}$ **36.** (1.1, 1.2) **37.** -0.6 **38.** -0.4 **39.**

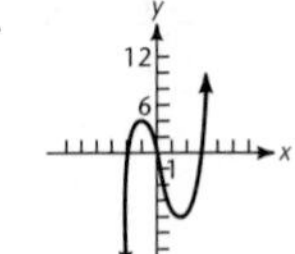

40.

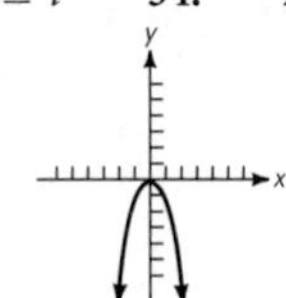

41.

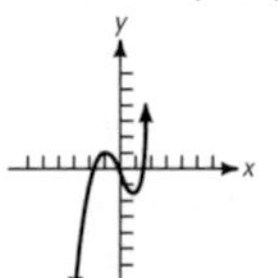

42. $y = 0,\ x = 1$ **43.** $y = 1,\ x = -2$ **44.** $y = 0,\ x = 1,\ x = -2$ **45.** $y = -x - 3,\ x = 3$

46. V: $x = -1$; H: $y = 2$ **47.** Oblique: $y = x$; V: $x = 0$

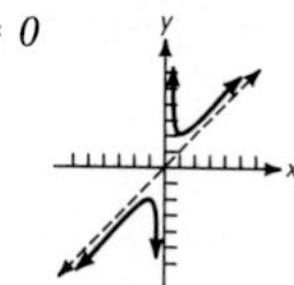

48. No asymptotes

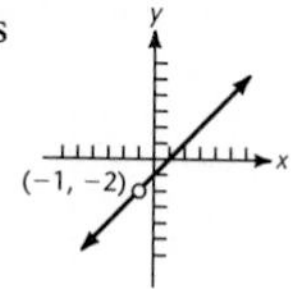

Exercises 6.1, page 346

1. 1 **3.** $\frac{9}{4}$ **5.** $\frac{2}{3}$ **7.** $\frac{\sqrt{6}}{3}$ **9.** 3 **11.** -4 **13.** 0 **15.** -3 **17.** -2 **19.** $-\frac{4}{3}$ **21.** -4

23. $-\frac{3}{2}$ **25.** $-\frac{5}{8}$ **27.** **29.**

31.

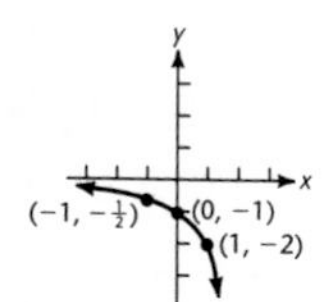

33.

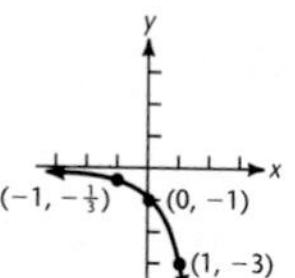

35.

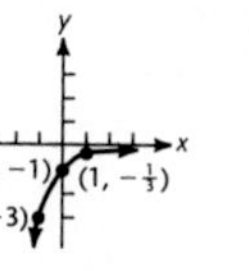

37.

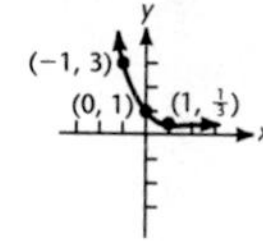

39.

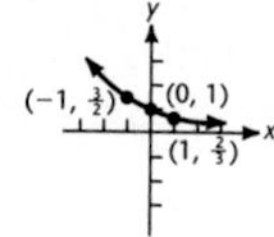

41.

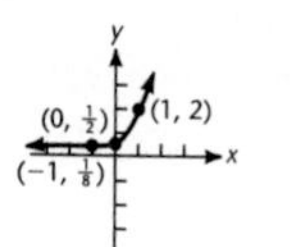

43.

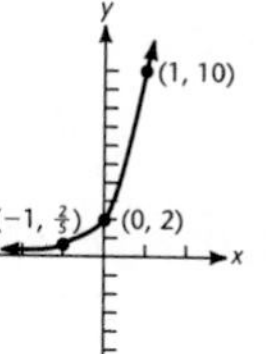

45.

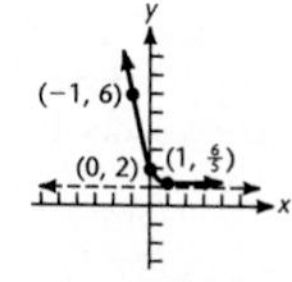

47.

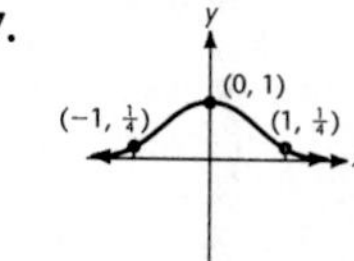

49. 256,000; every 10 years **51.** \$51,501.66 **53.** \$23,758.60 **55.** \$3280.50 **57.** 8,192,000 **59.** $A(t) = 1000(\frac{1}{2})^t$

61. Viewing rectangle $[-10, 10]$ by $[-10, 10]$

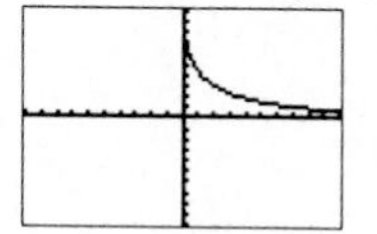

63. Viewing rectangle $[-10, 10]$ by $[-10, 10]$

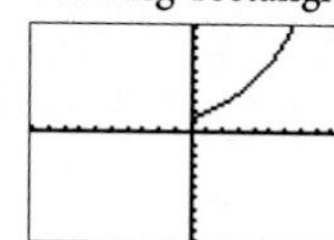

65. -0.8 **67.** $(2.81, \infty)$ **69.** $(0.31, 4.00)$

Exercises 6.2, page 355

1.

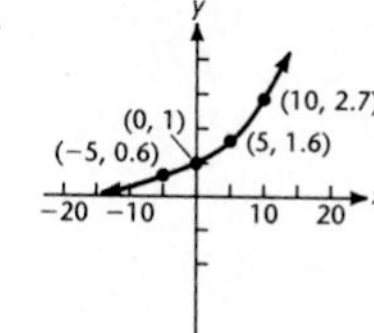

3.

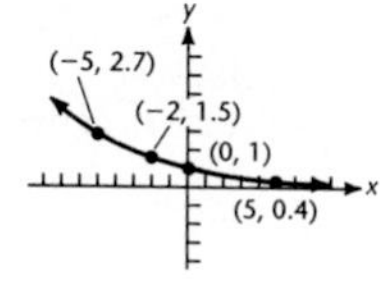

5.

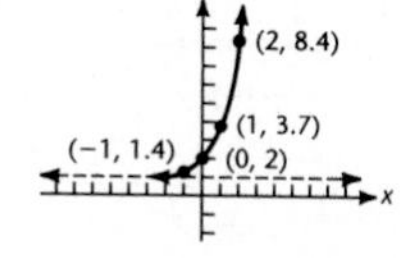

7.

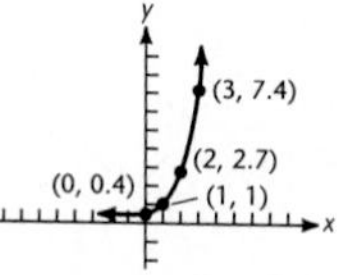

9.

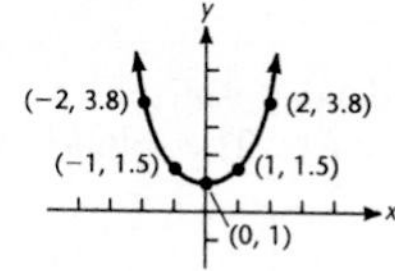

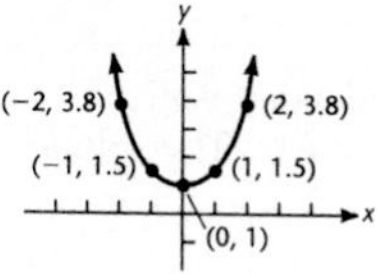

11. \$52,233.93 **13.** \$23,869.11 **15.** 92.85% **17.** 66.30 milligrams **19.** 68.94% **21.** 43
23. Viewing rectangle $[-10, 10]$ by $[-10, 10]$ **25.** Viewing rectangle $[-10, 10]$ by $[-10, 10]$ **27.** 1.9

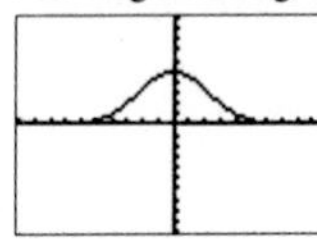

Exercises 6.3, page 363

1. $\log_4 16 = 2$ **3.** $\log_3 81 = 4$ **5.** $\log_3 \frac{1}{9} = -2$ **7.** $\log_{10} \frac{1}{100} = -2$ **9.** $\log_5 1 = 0$ **11.** $\log_4 \frac{1}{2} = -\frac{1}{2}$
13. $2^6 = 64$ **15.** $3^{-3} = \frac{1}{27}$ **17.** $(\frac{3}{4})^2 = \frac{9}{16}$ **19.** $(10)^{-2} = 0.01$
21.

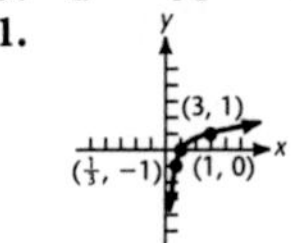

23.

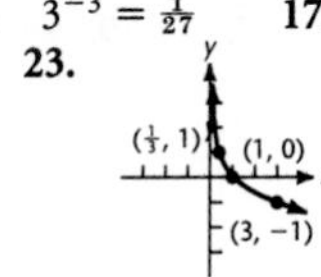

25.

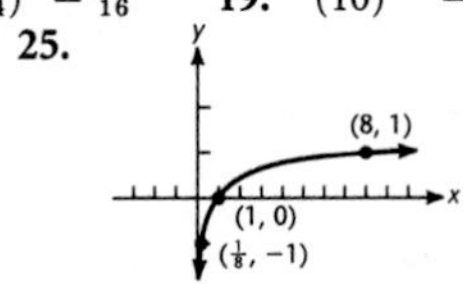

27.

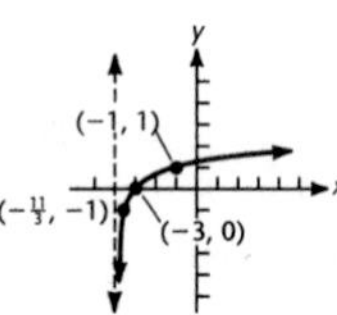

29. $y = 0$ **31.** $y = 3$ **33.** $y = 2$ **35.** $x = 9$ **37.** $x = \frac{1}{81}$ **39.** $y = -2$ **41.** $y = -2$ **43.** $a = 2$
45. $a = \frac{1}{5}$ **47.** $y = -3$ **49.** $y = 2$ **51.** $\log_{10} 3 + \log_{10} 5$ **53.** $3 \log_{10} 3$ **55.** $\frac{1}{2} \log_{10} 5$ **57.** $\log_{10} 3 - \log_{10} 5$
59. $2 \log_{10} 3 - 2 \log_{10} 5$ **61.** $\log_a x + \log_a y$ **63.** $2 \log_a x + 3 \log_a y - 4 \log_a z$ **65.** $3 \log_a x - 5 \log_a y - 2 \log_a z$
67. $(\log_a x)/5 + 2(\log_a z)/5 - \log_a y$ **69.** $(\log_a x)/2 + 2 \log_a y - 3(\log_a z)/2$ **71.** $\log_{10} 2500$ **73.** $\log_{10} 3.2$
75. $\log_{10} 0.16$ **77.** $\log_a (x\sqrt{z}/y^2)$ **79.** $\log_a (5x^2y^5/2)$ **81.** $\log_a(36x^{8/3}y^2z^5)$

Exercises 6.4, page 369

1. $\{\frac{1}{3}\}$ **3.** $\{9\}$ **5.** $\{7\}$ **7.** $\{\frac{8}{7}\}$ **9.** $\{\frac{10}{3}\}$ **11.** $\{\frac{1}{2}\}$ **13.** $\{2\}$ **15.** $\{2\}$ **17.** $\{-2, 2\}$ **19.** $\{2\}$
21. 2.58 **23.** −1.12 **25.** ±1.02 **27.** −1.24 **29.** −23.2 **31.** 1.74 **33.** 0.461 **35.** 19.8 **37.** 1.77
39. 2.10 **41.** 4.84 **43.** 2.51 **45.** 1.44 **47.** −1.67 **49.** 3.21 **51.** 1.87 **53.** \$1806.11 **55.** $t = 5.12$, hence $5\frac{1}{2}$ years **57.** $P = \$5528.75$ **59.** Approximately 6.93 years **61.** Approximately 9.90 days **63.** Approximately 6.10 days **65.** Approximately 11,200 years **67.** Approximately 1700 years **69.** L increases by approximately 5 decibels.
71. Viewing rectangle $[-1, 10]$ by $[-2, 2]$ **73.** Viewing rectangle $[-1, 10]$ by $[-2, 2]$

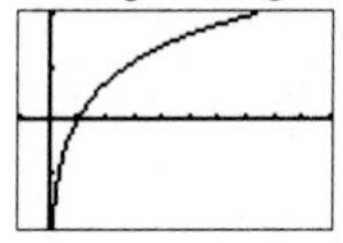

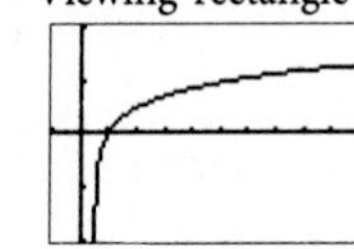

75. Viewing rectangle $[0, 3]$ by $[-1, 3]$ **77.** Viewing rectangle $[-10, 10]$ by $[-10, 10]$

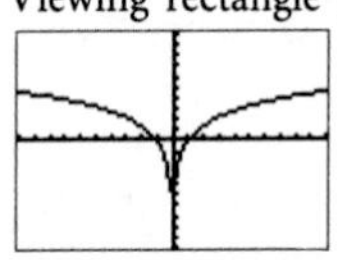

79. Viewing rectangle $[0, 10]$ by $[-5, 5]$ **81.** Viewing rectangle $[0, 10]$ by $[-5, 5]$ **83.** 0.57 **85.** 1.36

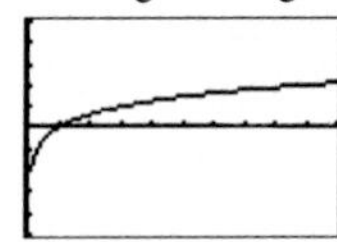

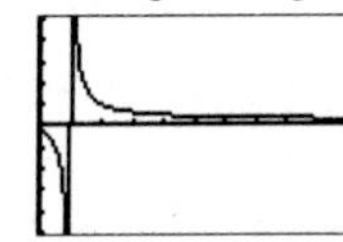

Critical Thinking: Find the Errors, page 373

1. **Error**

The error is the failure to check if any of the functions in the original equation are undefined at $x = -\frac{8}{7}$.

Correction

Both $\log_2 (-\frac{8}{7})$ and $\log_2 (-\frac{8}{7} + 1)$ are undefined, since logarithms are not defined at negative numbers. The solution set is $\varnothing$.

2. Error

$$\log [(2x + 3) + (3x - 1)] = 1 \quad \times$$

Correction

$$\begin{aligned} \log (2x + 3)(3x - 1) &= 1 \\ (2x + 3)(3x - 1) &= 10 \\ 6x^2 + 7x - 13 &= 0 \\ (6x + 13)(x - 1) &= 0 \\ 6x + 13 = 0 \quad \text{or} \quad x - 1 &= 0 \\ x = -\tfrac{13}{6} \quad \text{or} \quad x &= 1 \end{aligned}$$

The value $x = -\frac{13}{6}$ yields logarithms of negative numbers in the original equation. The only solution is $x = 1$.

3. Error

$$x(2x + 3) = 2 \quad \times$$

Correction

$$\begin{aligned} x(2x + 3) &= 3^2 \\ 2x^2 + 3x - 9 &= 0 \\ (2x - 3)(x + 3) &= 0 \\ 2x - 3 = 0 \quad \text{or} \quad x + 3 &= 0 \\ x = 3/2 \quad \text{or} \quad x &= -3 \end{aligned}$$

When $x = -3$, the original equation contains logarithms of negative numbers. The only solution is $x = 3/2$.

Review Problems for Chapter 6, page 374

1. (a) $\frac{9}{4}$ (b) $\sqrt{6}/3$ **2.** $x = -\frac{3}{4}$ **3.** $x = -\frac{5}{3}$ **4.** $x = \frac{31}{8}$ **5.** -9 **6.** **7.**

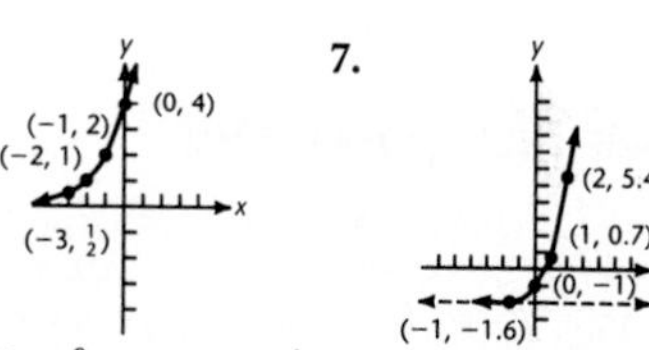

8. **9.**

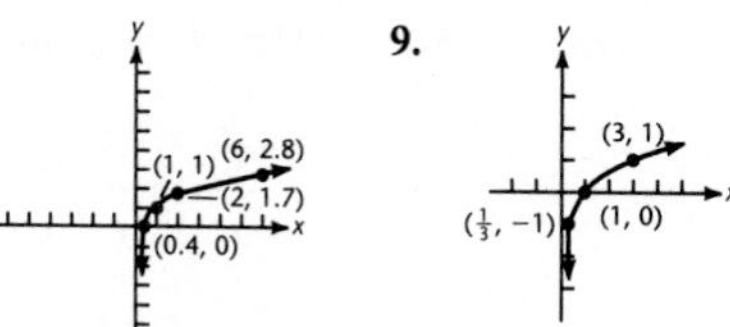

10. 16 **11.** $-\frac{5}{4}$ **12.** $\frac{1}{32}$ **13.** $\frac{9}{2}$ **14.** e^3 **15.** -3

16. $2 \log_a x + \log_a z - 3 \log_a y$ **17.** $\frac{2}{3} \log_a x + \frac{1}{3} \log_a y - 4 \log_a z$ **18.** $\log_a \dfrac{9x^5y}{16z^4}$ **19.** $\log_a 2x^4y^3z^4$ **20.** 2.58
21. 1.55 **22.** 2.27 **23.** 3.24 **24.** $\{\frac{5}{3}\}$ **25.** $\{3\}$ **26.** 33(rounded) **27.** Approximately 96°F
28. Approximately 49 months **29.** Approximately 9.87 hours **30.** Approximately 6.9 months **31.** Approximately 6.74 years **32.** Approximately 2009 days

Cumulative Test Chapters 3–6, page 377

1. $\frac{2}{7}$ **2.** $2x + h$ **3.** V: $(-1, 8)$; A: $x = -1$ **4.** The maximum profit is \$720 when $w = 27$.
5. Center $(-1, 0)$, radius $\frac{1}{2}$ **6.** $(x - 10)^2 + (y + 3)^2 = 25$
7. (a) (b) (c) **8.** V: $(\pm 4, 0)$; A: $y = \pm\frac{1}{2}x$ **9.**

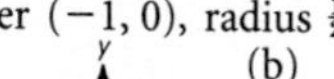

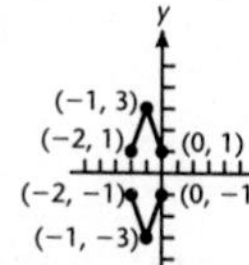

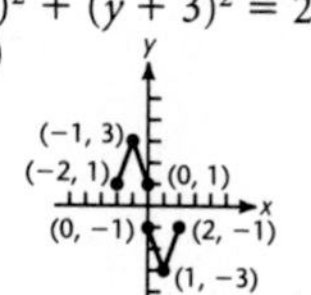

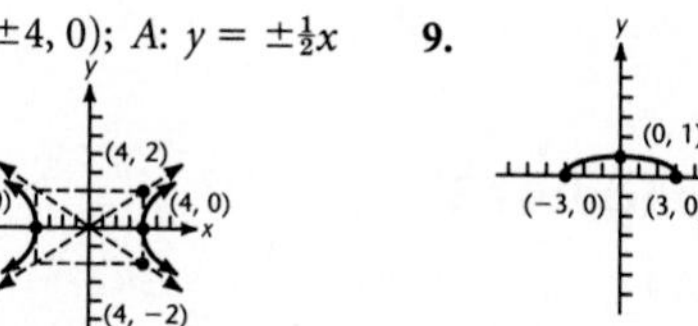

10. $f(x) = -\frac{3}{2}x + 6$, $f^{-1}(x) = -\frac{2}{3}x + 4$ **11.** 10 **12.** 105 **13.** Yes
14. 2 positive, 2 nonreal complex; 4 nonreal complex

15. $\frac{1}{2}, -1, -\frac{4}{3}$ **16.** 0.8 **17.**

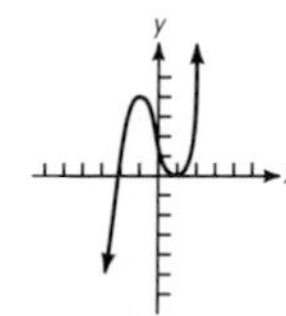

18. V: $x = 2$; H: $y = 0$

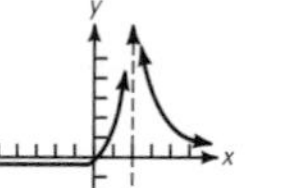

19. $x = \frac{1}{4}$

20. Approximately 23.33 inches of mercury **21.**

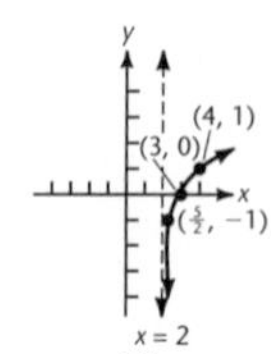

22. (a) $a = 64$ (b) $y = -\frac{3}{2}$

23. (a) $\frac{1}{2}\log_a x - \log_a y - 3\log_a z$ (b) $\log_a xy^3z^7$ **24.** $\{\frac{3}{2}\}$ **25.** Approximately 25 minutes

Exercises 7.1, page 385

1. 48° **3.** 75°35′ **5.** 71°17′46″ **7.** 60°47′17″ **9.** 71.73° **11.** 53.7418° **13.** 136° **15.** 41°39′ **17.** 160°36′46″ **19.** 38°59′57″ **21.** 78.63° **23.** 136.4873° **25.** 14°16′12″ **27.** −63°43′12″ **29.** 27°31′12″ **31.** −45°5′39″ **33.** 40.30° **35.** −18.83° **37.** 53.2406° **39.** −421.0047° **41.** Positive **43.** Negative **45.** Negative **47.** Negative **49.** Quadrant I, 402°, 762°, −318°, −678° **51.** Quadrant II, 510°, 870°, −210°, −570° **53.** Quadrant III, 574°, 934°, −146°, −506° **55.** Quadrant IV, 680°, 1040°, −40°, −400° **57.** Quadrant I, 55°, 775°, −305°, −665° **59.** Quadrant IV, 332°, 692°, −388°, −748° **61.** 80 feet **63.** 37.5 feet **65.** 35 yards **67.** 6 feet

Exercises 7.2, page 393

1. 30° **3.** −15° **5.** −252° **7.** 495° **9.** $-\pi/2$ **11.** $2\pi/3$ **13.** $26\pi/9$ **15.** $7\pi/12$ **17.** 171.887° **19.** 29.393° **21.** 0.3229 **23.** 1.2608 **25.** 4π **27.** 3π **29.** $\frac{1}{72}$ revolution **31.** $\frac{1}{3}$ revolution **33.** 355π meters **35.** 2.1π feet **37.** $\frac{9}{2}$ centimeters **39.** $18\pi/25 \approx 2.3$ feet **41.** $\pi \approx 3.1$ inches **43.** 840 miles **45.** $15\pi/11 \approx 4.3$ miles per hour **47.** $3\pi \approx 9.4$ radians per second **49.** $20/(3\pi) \approx 2.1$ feet **51.** $1000\pi/3 \approx 1050$ miles per hour

Exercises 7.3, page 406

In all these answers, the function values are given in this order: sine, cosine, tangent, cotangent, secant, cosecant.
1. $\frac{4}{5}, \frac{3}{5}, \frac{4}{3}, \frac{3}{4}, \frac{5}{3}, \frac{5}{4}$ **3.** $\frac{24}{25}, -\frac{7}{25}, -\frac{24}{7}, -\frac{7}{24}, -\frac{25}{7}, \frac{25}{24}$ **5.** $-\frac{15}{17}, -\frac{8}{17}, \frac{15}{8}, \frac{8}{15}, -\frac{17}{8}, -\frac{17}{15}$ **7.** $-7\sqrt{51}/51, -\sqrt{102}/51, 7\sqrt{2}/2, \sqrt{2}/7, -\sqrt{102}/2, -\sqrt{51}/7$ **9.** $1/\sqrt{1+x^2}, x/\sqrt{1+x^2}, 1/x, x, \sqrt{1+x^2}/x, \sqrt{1+x^2}$ **11.** $\sqrt{1-x^2}, x, \sqrt{1-x^2}/x, x/\sqrt{1-x^2}, 1/x, 1/\sqrt{1-x^2}$ **13.** 0.926, −0.377, −2.456, −0.407, −2.652, 1.080 **15.** −0.961, 0.276, −3.478, −0.288, 3.619, −1.041 **17.** II **19.** III **21.** II **23.** III, IV **25.** II, IV **27.** $\frac{3}{5}, \frac{4}{5}, \frac{3}{4}, \frac{4}{3}, \frac{5}{4}, \frac{5}{3}$ **29.** $\frac{24}{25}, -\frac{7}{25}, -\frac{24}{7}, -\frac{7}{24}, -\frac{25}{7}, \frac{25}{24}$ **31.** $-\sqrt{10}/10, -3\sqrt{10}/10, 1/3, 3, -\sqrt{10}/3, -\sqrt{10}$ **33.** $-2\sqrt{13}/13, 3\sqrt{13}/13, -2/3, -3/2, \sqrt{13}/3, -\sqrt{13}/2$
35. $-\sqrt{21}/5, -2/5, \sqrt{21}/2, 2\sqrt{21}/21, -5/2, -5\sqrt{21}/21$ **37.** $y, \sqrt{1-y^2}, y/\sqrt{1-y^2}, \sqrt{1-y^2}/y, 1/\sqrt{1-y^2}, 1/y$
39. $\sqrt{1-x^2}, x, \sqrt{1-x^2}/x, x/\sqrt{1-x^2}, 1/x, 1/\sqrt{1-x^2}$ **41.** 0, 1, 0, undefined, 1, undefined **43.** −1, 0, undefined, 0, undefined, −1 **45.** 33° **47.** $\pi/5$ **49.** $\pi/9$ **51.** $-\sin(2\pi/9)$ **53.** cos 73° **55.** $-\tan(2\pi/9)$
57. −csc 59°30′ **59.** $1/2, -\sqrt{3}/2, -\sqrt{3}/3, -\sqrt{3}, -2\sqrt{3}/3, 2$ **61.** $-\sqrt{3}/2, -1/2, \sqrt{3}, \sqrt{3}/3, -2, -2\sqrt{3}/3$
63. $-\sqrt{3}/2, 1/2, -\sqrt{3}, -\sqrt{3}/3, 2, -2\sqrt{3}/3$ **65.** $\sqrt{2}/2, -\sqrt{2}/2, -1, -1, -\sqrt{2}, \sqrt{2}$ **67.** −1/2 **69.** −1
71. $4\sqrt{3}/3$ **73.** False **75.** False

Exercises 7.4, page 413

1. 4.5600 **3.** 0.4679 **5.** −1.0951 **7.** −0.6861 **9.** 0.5357 **11.** −0.7989 **13.** 0.9883 **15.** −0.9926 **17.** $\tan\theta = -\frac{3}{4}$, $\cot\theta = -\frac{4}{3}$, $\sec\theta = \frac{5}{4}$, $\csc\theta = -\frac{5}{3}$ **19.** $\sin\theta = \frac{12}{13}$, $\cos\theta = \frac{5}{13}$, $\cot\theta = \frac{5}{12}$, $\csc\theta = \frac{13}{12}$ **21.** $\sin\alpha = \frac{8}{17}$, $\cos\alpha = -\frac{15}{17}$, $\cot\alpha = -\frac{15}{8}$, $\sec\alpha = -\frac{17}{15}$ **23.** $\cos v = -\frac{24}{25}$, $\tan v = \frac{7}{24}$, $\cot v = \frac{24}{7}$, $\csc v = -\frac{25}{7}$ **25.** $\sin t = 2\sqrt{2}/3$, $\cot t = -\sqrt{2}/4$, $\sec t = -3$, $\csc t = 3\sqrt{2}/4$ **27.** $\sin s = -\sqrt{2}/2$, $\cos s = \sqrt{2}/2$, $\cot s = -1$, $\sec s = \sqrt{2}$

29. $\tan \alpha = \sqrt{14}/2$, $\cot \alpha = \sqrt{14}/7$, $\sec \alpha = -3\sqrt{2}/2$, $\csc \alpha = -3\sqrt{7}/7$ **31.** $\sin \theta = \sqrt{14}/7$, $\cos \theta = -\sqrt{35}/7$, $\tan \theta = -\sqrt{10}/5$, $\sec \theta = -\sqrt{35}/5$ **33.** Impossible **35.** Possible **37.** Possible **39.** Impossible **41.** Impossible **43.** Impossible **45.** Impossible **47.** Impossible **49.** Impossible **51.** Impossible

Exercises 7.5, page 418

1. 0.5195 **3.** 0.5195 **5.** 1.980 **7.** −0.4791 **9.** 0.8450 **11.** −2.960 **13.** −2.577 **15.** −0.6626 **17.** 0.3007 **19.** 1.263 **21.** −0.3378 **23.** 0.4436 **25.** 32.0° **27.** 30.7° **29.** 54.6° **31.** 145.3° **33.** 215.2° **35.** 333.4° **37.** 1.0558 **39.** 0.6226 **41.** 3.6507 **43.** 5.4774 **45.** 2.2020 **47.** 4.0753 **49.** 25°40′ **51.** 67°40′ **53.** 214°10′ **55.** 154°10′ **57.** 222°20′ **59.** 301°40′ **61.** 19°

Exercises 7.6, page 427

1. $b = 9.4$, $c = 36$, $A = 75°$ **3.** $b = 0.80$, $c = 1.7$, $B = 28°$ **5.** $a = 0.34$, $b = 0.51$, $A = 34°$ **7.** $c = 59$, $A = 42°$, $B = 48°$ **9.** $a = 0.11$, $b = 0.15$, $B = 54°$ **11.** $b = 4.0$, $A = 37°$, $B = 53°$ **13.** $b = 207$, $c = 269$, $B = 50°20'$ **15.** $b = 191$, $c = 201$, $A = 18.2°$ **17.** $a = 0.285$, $b = 0.148$, $B = 27°30'$ **19.** $a = 187$, $b = 170$, $A = 47.8°$ **21.** $c = 13.4$, $A = 17.6°$, $B = 72.4°$ **23.** $b = 0.130$, $A = 25.7°$, $B = 64.3°$ **25.** $b = 288.0$, $c = 321.9$, $A = 26°32'$ **27.** $a = 28.01$, $b = 86.67$, $B = 72.09°$ **29.** $c = 13.01$, $A = 27.52°$, $B = 62.48°$ **31.** $a = 96.96$, $A = 79.92°$, $B = 10.08°$ **33.** 16 feet **35.** 26° **37.** 54.8 feet **39.** 67 feet **41.** 350 feet **43.** 102 kilometers **45.** 248 feet **47.** N 70° E **49.** 89 miles **51.** 120 feet, 0 not significant **53.** 73.0 feet

Exercises 7.7, page 434

1. $(-1, 0)$ **3.** $(0, 1)$ **5.** $(\sqrt{3}/2, 1/2)$ **7.** $(\sqrt{2}/2, -\sqrt{2}/2)$ **9.** $(-\sqrt{3}/2, 1/2)$ **11.** $(1/2, -\sqrt{3}/2)$ **13.** $(-\sqrt{2}/2, -\sqrt{2}/2)$ **15.** $(1/2, -\sqrt{3}/2)$ **17.** $(-\sqrt{2}/2, -\sqrt{2}/2)$ **19.** $(\sqrt{3}/2, -1/2)$ **21.** Quadrant III **23.** Quadrant III **25.** Quadrant III **27.** Quadrant IV **29.** 0 **31.** $-\sqrt{2}/2$ **33.** $-\sqrt{3}/3$ **35.** $-\sqrt{3}/2$ **37.** 1 **39.** 0 **41.** False **43.** True

Exercises 7.8, page 452

1. Amp = 4, $P = 2\pi/3$, no phase shift **3.** Amp = 5, $P = \pi/2$, no phase shift **5.** Amp = 1/2, $P = 2\pi$, phase shift is $\pi/4$ units to the right. **7.** Amp = 1/3, $P = 2\pi/3$, phase shift is $\pi/3$ units to the left. **9.** Amp = 4, $P = 2\pi/3$, phase shift is $\pi/12$ units to the right. **11.** Amp = 3, $P = \pi$, phase shift is $\pi/8$ units to the left.

13.
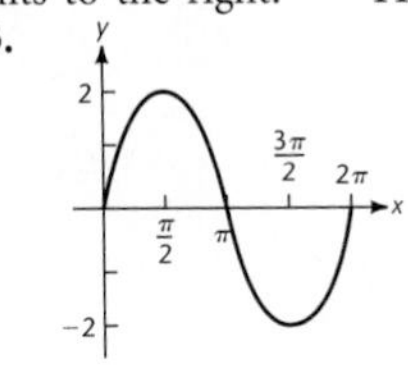

15.
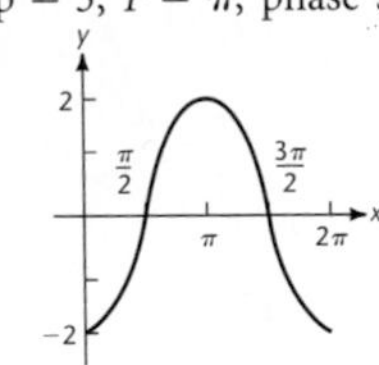

17.
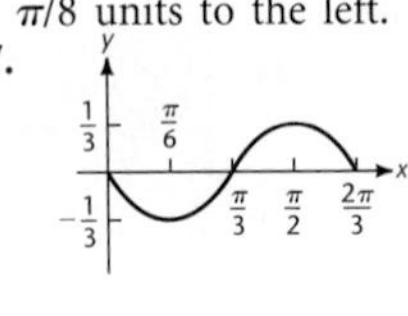

19.
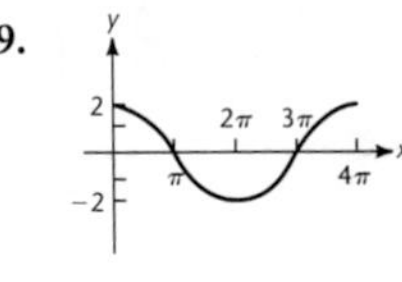

21.
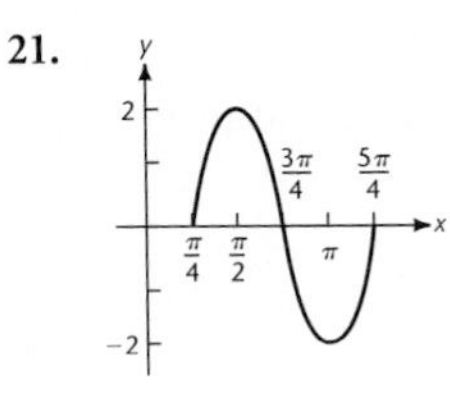

23.
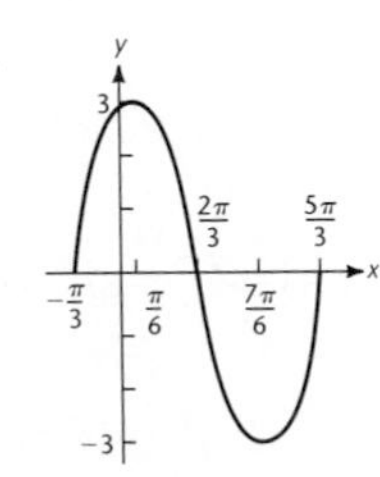

25.
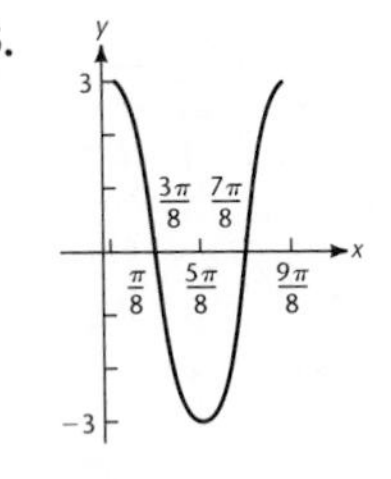

27.
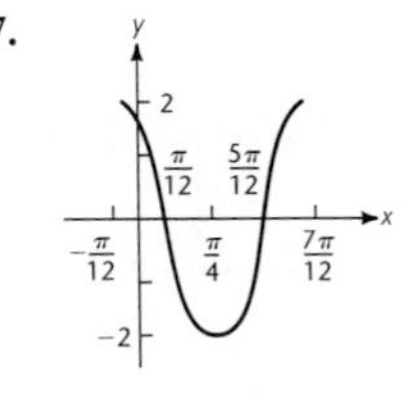

29.
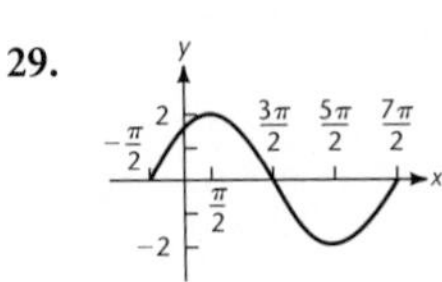

31.
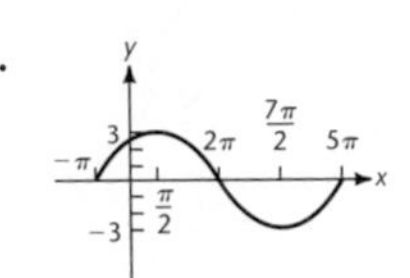

33.
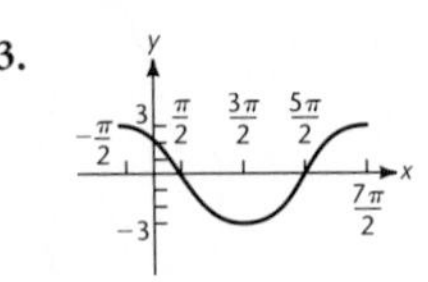

35.
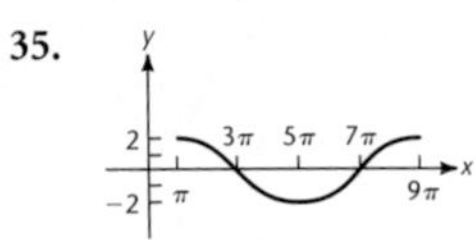

37. 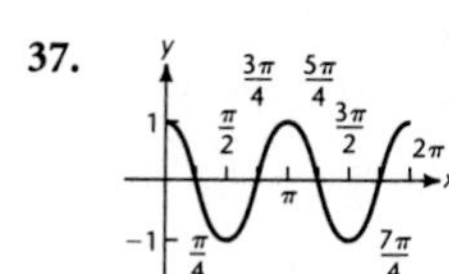

39. 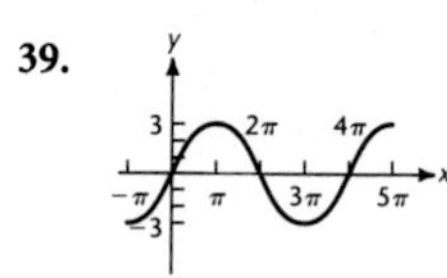

41. $y = 3\sin\left(2x + \dfrac{\pi}{2}\right)$ **43.** $y = 4\cos\left(3x - \dfrac{\pi}{2}\right)$

45. 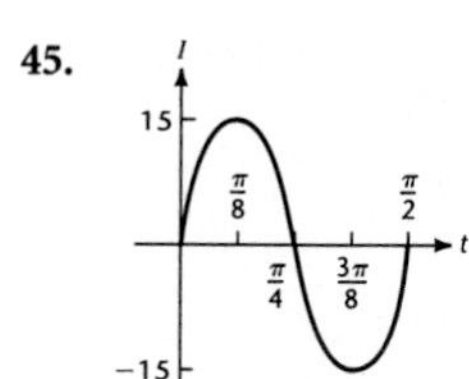

47. Viewing rectangle $[-10, 10]$ by $[-10, 10]$

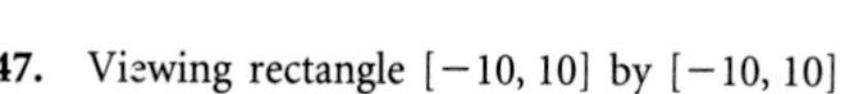

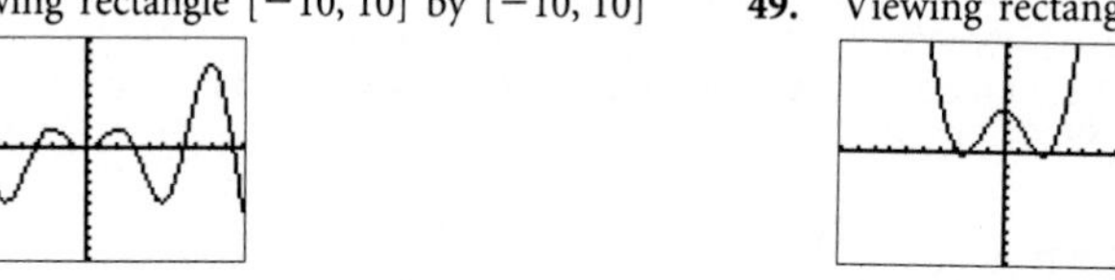

49. Viewing rectangle $[-10, 10]$ by $[-10, 10]$

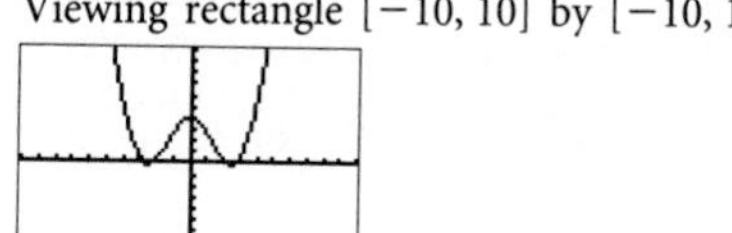

51. Viewing rectangle $[-5, 5]$ by $[-2, 2]$

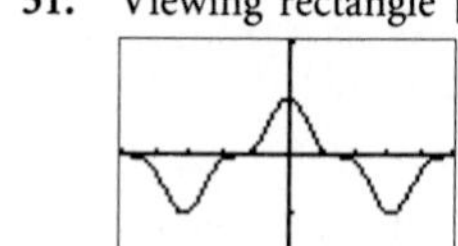

53. Viewing rectangle $[-10, 10]$ by $[-2, 2]$

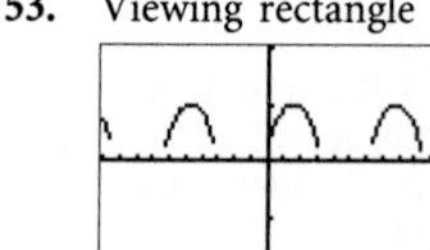

Exercises 7.9, page 465

1. $P = \pi$, phase shift is $\pi/8$ units to the right. **3.** $P = 2\pi/3$, phase shift is $\pi/9$ units to the left. **5.** $P = \pi/2$, phase shift is $\pi/8$ units to the left. **7.** $P = 2\pi$, phase shift is $\pi/2$ units to the right. **9.** $P = 2\pi$, Amp $= 2$, no phase shift, no vertical translation **11.** $P = 4\pi$, Amp $= 2$, phase shift is $\pi/3$ units to the left, vertical translation is 1 unit upward. **13.** $P = 6\pi$, amplitude is not defined, phase shift is $3\pi/2$ units to the right, vertical translation is 2 units upward. **15.** $P = \pi/4$, amplitude is not defined, phase shift is $\pi/4$ units to the right, vertical translation is 1 unit upward.

17. 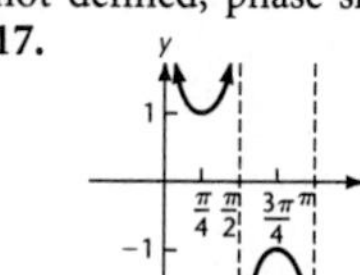

19. 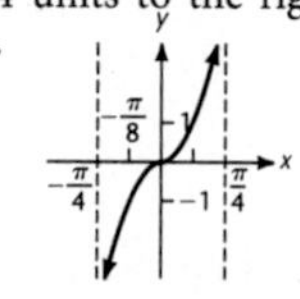

21. 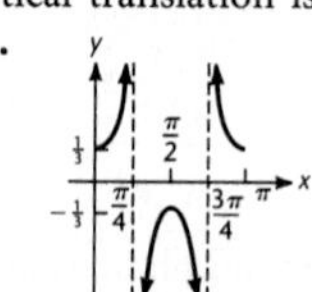

23. 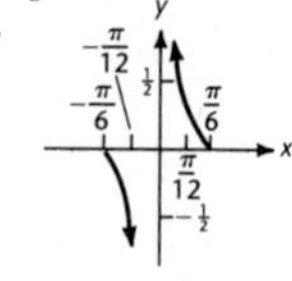

25. 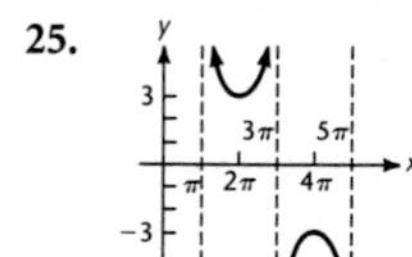

27. 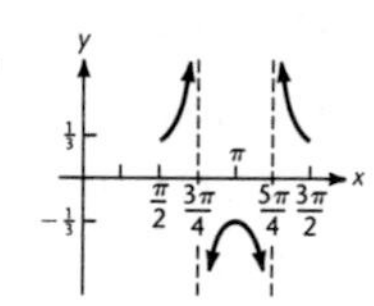

29. 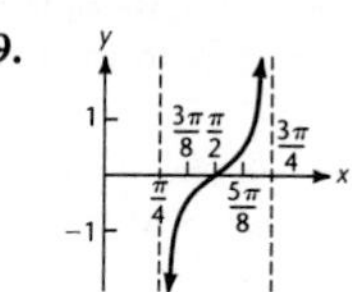

31. 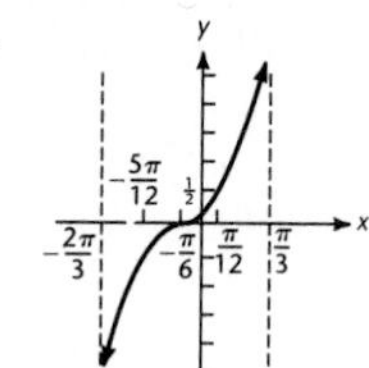

33. 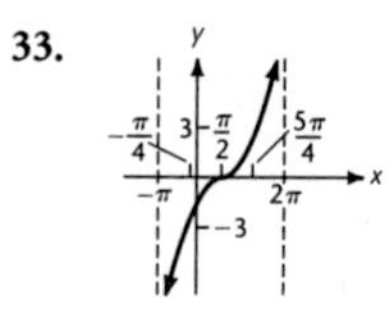

35. 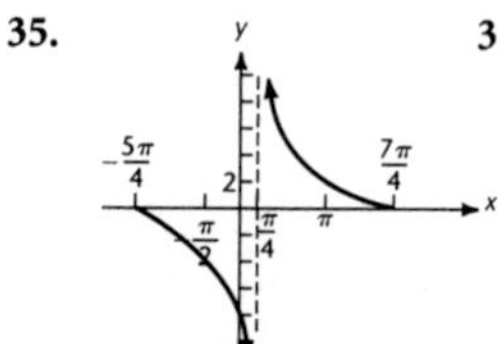

37. 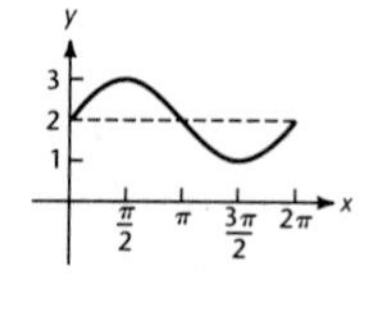

39. 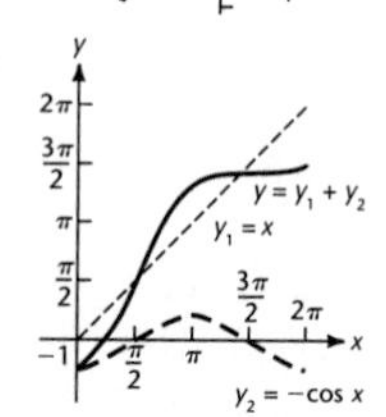

41. 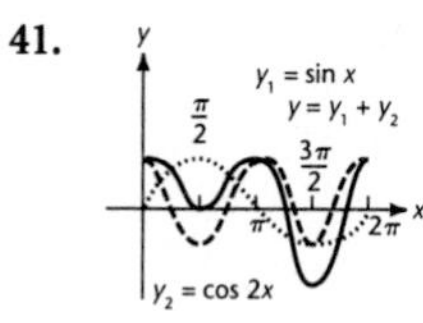

43. 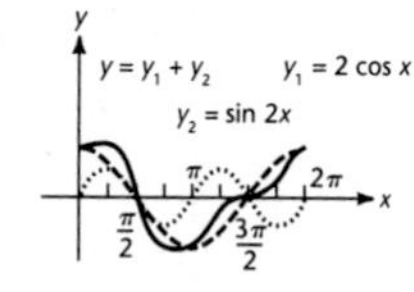

45. 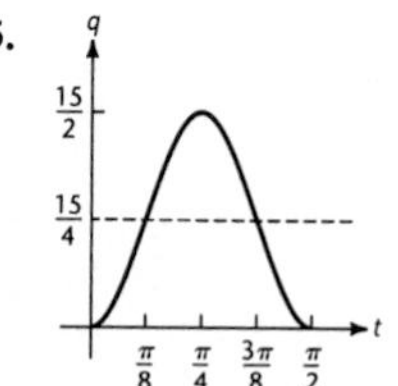

47. 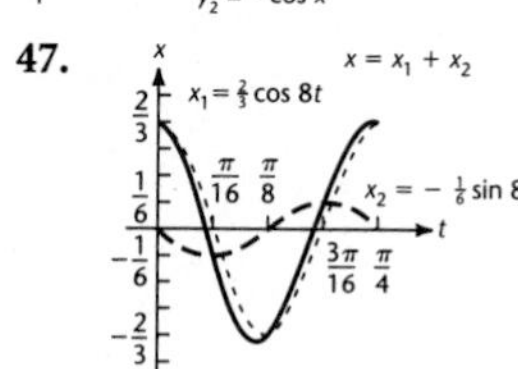

49. Viewing rectangle $[-8, 8]$ by $[-5, 5]$

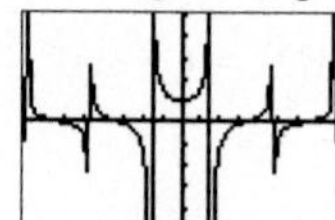

51. Viewing rectangle $[-10, 10]$ by $[-10, 10]$

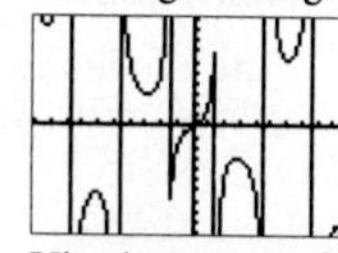

53. Viewing rectangle $[-5, 5]$ by $[-5, 5]$

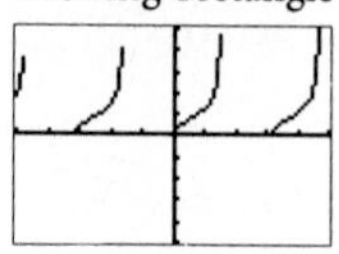

55. Viewing rectangle $[-10, 10]$ by $[-10, 10]$

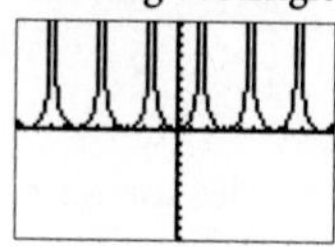

57. 13.9 feet

Critical Thinking: Find the Errors, page 470

1. **Error**

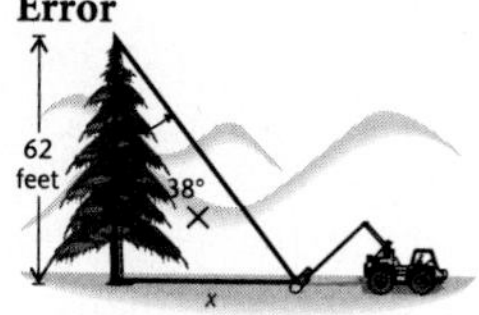

Correction

$$\cot 38° = \frac{x}{62}$$
$$x = 62 \cot 38°$$
$$= 79 \text{ feet}$$

38° = Angle of depression

2. **Error**

$48°24' = 48.24°$ ×

Correction

$$48°24' = 48° + \frac{24°}{60} = 48° + 0.4° = 48.4°$$

3. **Error**

$x = 3$ ×

Correction

$$9 = x^2$$
$$x = -3$$
$$\cos\theta = \frac{x}{r} = -\frac{3}{5}$$

4. **Error**

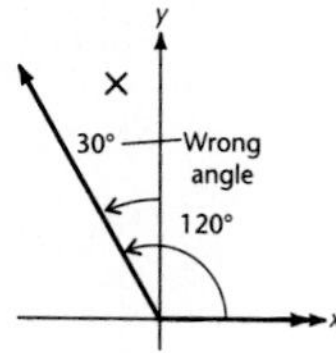

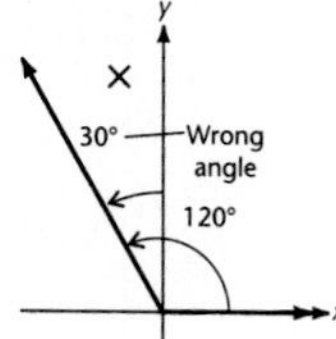

Correction

$$\cos 120° = -\cos 60°$$
$$= -\tfrac{1}{2}$$

5. **Error**

$\omega = \frac{v}{r} = \frac{36}{9}$ ×

Correction

$$\omega = \frac{v}{r} = \frac{36}{\frac{3}{4}} = 48 \text{ radians per second}$$

6. Error

$$y = 2\sin\left(\frac{x}{2} - \frac{\pi}{4}\right)$$
$$= \sin 2\left(\frac{x}{2} - \frac{\pi}{4}\right)$$

Correction

The amplitude is $|a| = 2$, the period is $2\pi/(1/2) = 4\pi$, and a period starts where

$$\frac{x}{2} - \frac{\pi}{4} = 0$$
$$x = \frac{\pi}{2}.$$

The correct graph is shown in the following figure.

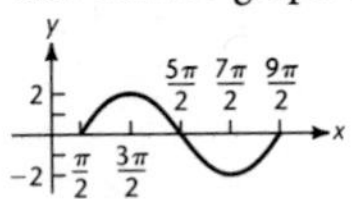

7. Errors

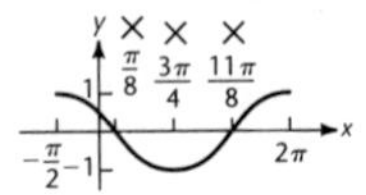

Correction

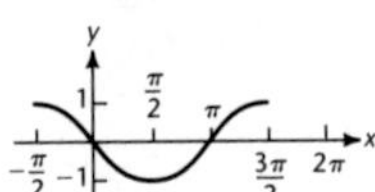

Review Problems for Chapter 7, page 472

1. (a) Quadrant II, 483°, 843°, −237°, −597° (b) Quadrant IV, 316°, 676°, −404°, −746° **2.** $6\frac{2}{3}$ feet **3.** $\sin\theta = 3/7$, $\cos\theta = -2\sqrt{10}/7$, $\tan\theta = -3\sqrt{10}/20$, $\cot\theta = -2\sqrt{10}/3$, $\sec\theta = -7\sqrt{10}/20$, $\csc\theta = 7/3$ **4.** $\sin\alpha = -\frac{24}{25}$, $\cos\alpha = \frac{7}{25}$, $\cot\alpha = -\frac{7}{24}$, $\sec\alpha = \frac{25}{7}$, $\csc\alpha = -\frac{25}{24}$ **5.** $\sin\beta = 2/3$, $\cos\beta = \sqrt{5}/3$, $\tan\beta = 2\sqrt{5}/5$, $\sec\beta = 3\sqrt{5}/5$ **6.** 630 miles **7.** 840 miles **8.** $7\pi/40 \approx 0.55$ feet per second **9.** $7\pi/11 \approx 2.0$ miles per hour **10.** $1056 \approx 1100$ radians per minute, two zeros not significant **11.** (a) −210° (b) 300° **12.** (a) $7\pi/2$ (b) $15\pi/64$ **13.** $3\pi \approx 9.4$ centimeters **14.** The function values are given in this order: sine, cosine, tangent, cotangent, secant, cosecant. (a) 1, 0, undefined, 0, undefined, 1 (b) 1/2, $-\sqrt{3}/2$, $-\sqrt{3}/3$, $-\sqrt{3}$, $-2\sqrt{3}/3$, 2 (c) $-\sqrt{2}/2$, $\sqrt{2}/2$, −1, −1, $\sqrt{2}$, $-\sqrt{2}$ (d) −1, 0, undefined, 0, undefined, −1 (e) −1/2, $-\sqrt{3}/2$, $\sqrt{3}/3$, $\sqrt{3}$, $-2\sqrt{3}/3$, −2 (f) $-\sqrt{3}/2$, −1/2, $\sqrt{3}$, $\sqrt{3}/3$, −2, $-2\sqrt{3}/3$ **15.** (a) Impossible (b) possible (c) impossible (d) possible (e) impossible **16.** (a) 1.029 (b) −0.7490 (c) 1.649 (d) −0.5793 (e) 0.9789 **17.** (a) 0.1363 (b) −0.9217 **18.** (a) 24.7° (b) 64.3° (c)152.0° (d) 228.7° (e) 289.3° **19.** (a) 1.7017 (b) 3.4266 (c) 1.9635 (d) 5.2971 (e) 4.4128 **20.** (a) 0.3200 (b) 1.0996 **21.** (a) $\pi/4$, $5\pi/4$ (b) $7\pi/6$, $11\pi/6$ **22.** $b = 310$ feet, $c = 320$ feet, zeros not significant, $B = 75°$ **23.** 652 yards **24.** 931 meters **25.** (a) (0, 1) (b) $(-1/2, \sqrt{3}/2)$ **26.** (a) $P = \pi$, Amp = 2, phase shift is $\pi/2$ units to the right, vertical translation is 1 unit upward. (b) $P = \pi/2$, Amp = 1/2, phase shift is $\pi/4$ units to the left, vertical translation is 2 units downward. (c) $P = \pi/3$, amplitude is not defined, no phase shift, vertical translation is 4 units upward (d) $P = 2\pi$, amplitude is not defined, phase shift is $\pi/2$ units to the left, no vertical translation (e) $P = 4\pi$, amplitude is not defined, phase shift is $\pi/3$ units to the right, no vertical translation

27.

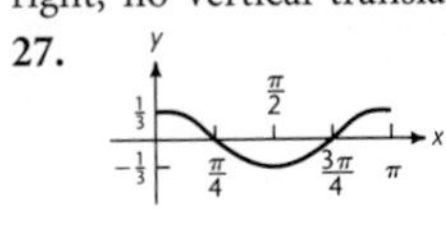

28.

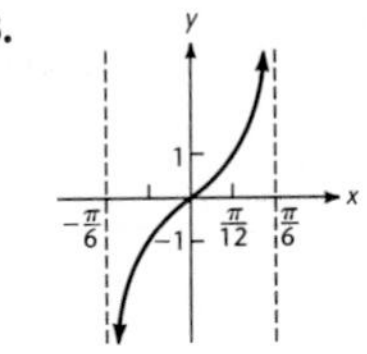

29.

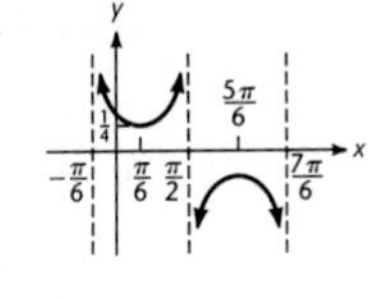

30.

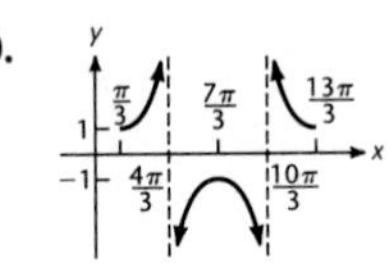

31.

32.

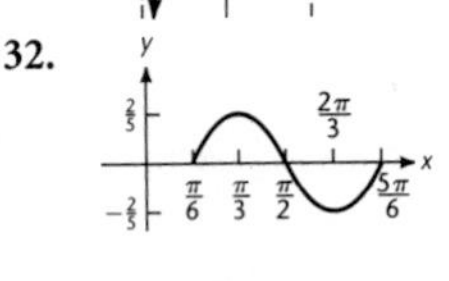

33.

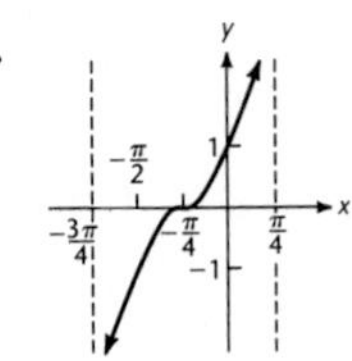

34.

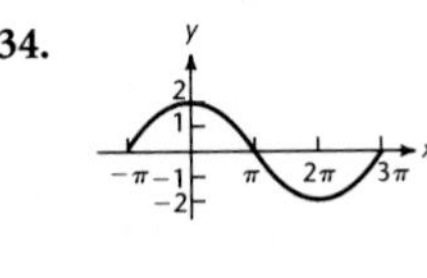

35.

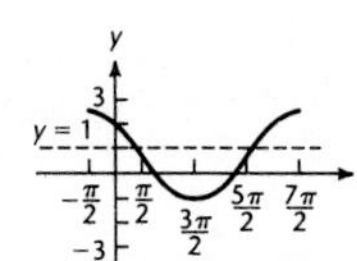

36. $y = y_1 + y_2$, $y_1 = 2\cos x$, $y_2 = \sin x$

37. $y = 3\sin\left(x - \frac{\pi}{4}\right)$

38. $y = \frac{2}{3}\sin\left(3x + \frac{\pi}{2}\right)$ **39.** $y = \frac{1}{2}\cos(2x + \pi)$ **40.** $y = 2\cos\left(x + \frac{\pi}{2}\right)$

41. Maximum value of $H = 14$ at $n = 171$ (rounded from $n = 170.25$); minimum value of $H = \frac{31}{3}$ at $n = 353$.

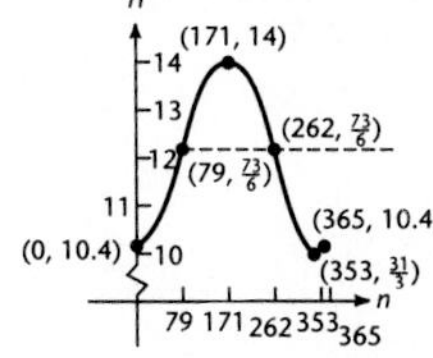

Exercises 8.1, page 484

15. $\sin\theta = \pm\sqrt{1-\cos^2\theta}$, $\tan\theta = \pm\sqrt{1-\cos^2\theta}/\cos\theta$, $\cot\theta = \pm\cos\theta/\sqrt{1-\cos^2\theta}$, $\sec\theta = 1/\cos\theta$, $\csc\theta = \pm 1/\sqrt{1-\cos^2\theta}$
17. $\sin\theta = \pm\tan\theta/\sqrt{1+\tan^2\theta}$, $\cos\theta = \pm 1/\sqrt{1+\tan^2\theta}$, $\cot\theta = 1/\tan\theta$, $\sec\theta = \pm\sqrt{1+\tan^2\theta}$, $\csc\theta = \pm\sqrt{1+\tan^2\theta}/\tan\theta$
19. $\sec\theta$ **21.** $2\cos\theta$ **23.** $1/(4\cos u) = (\sec u)/4$ **25.** $27\sec^3 u$ **27.** $\csc t$ **29.** $1/(a^2 \sec t \tan t)$

Exercises 8.2, page 494

1. 0 **3.** 1 **5.** −1 **7.** $-\cos\theta$ **9.** $-\sin\theta$ **11.** $(\cos\theta - \sqrt{3}\sin\theta)/2$ **13.** $\frac{24}{25}$, 0 **15.** $-\frac{16}{65}$, $\frac{56}{65}$
17. $\frac{7}{25}$, $-\frac{7}{25}$ **19.** $(5\sqrt{3} - 12)/26$, $(5\sqrt{3} + 12)/26$ **21.** $-(\sqrt{77} + \sqrt{10})/12$, $(\sqrt{10} - \sqrt{77})/12$ **23.** $\cos 3x$
25. $\cos(x/12)$
27. $y = \cos x$

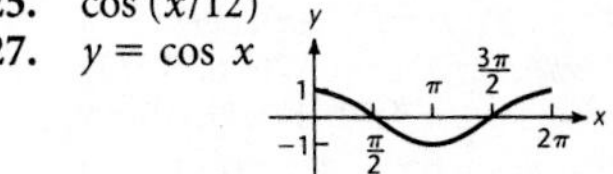

29. $y = -3\sin x$

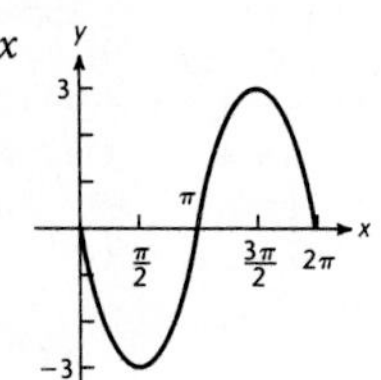

31. $y = -3\cos 2x$

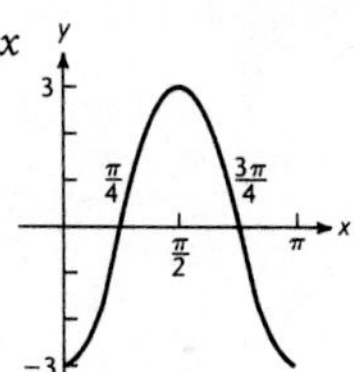

33. $\sqrt{2}(\sqrt{3} - 1)/4$ **35.** $\sqrt{2}(\sqrt{3} + 1)/4$ **37.** $\sqrt{2}(1 - \sqrt{3})/4$ **39.** $\sqrt{2}(1 + \sqrt{3})/4$

Exercises 8.3, page 500

1. $\sqrt{3}/2$ **3.** 0 **5.** $-\sqrt{3}$ **7.** $-\sqrt{3}/3$ **9.** $\frac{180}{163}$ **11.** $\frac{5}{301}$ **13.** (a) 1 (b) $\frac{7}{25}$ (c) undefined (d) $\frac{7}{24}$
15. (a) $\frac{33}{65}$ (b) $-\frac{63}{65}$ (c) $\frac{33}{56}$ (d) $\frac{63}{16}$ **17.** $\sqrt{2}(\sqrt{3} + 1)/4$ **19.** $\sqrt{2}(\sqrt{3} - 1)/4$ **21.** $(1 + \sqrt{3})/(1 - \sqrt{3}) = -(2 + \sqrt{3})$
23. $(1 - \sqrt{3})/(1 + \sqrt{3}) = \sqrt{3} - 2$ **25.** $\sin\theta$ **27.** $(-\sqrt{3}/2)\sin\theta + (1/2)\cos\theta$ **29.** $\tan\theta$
31. $(\tan\theta + 1)/(1 - \tan\theta)$ **33.**

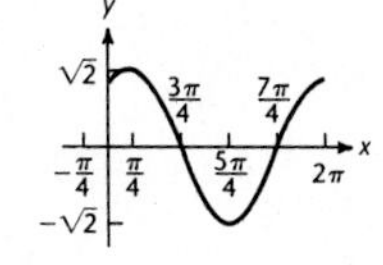

35.

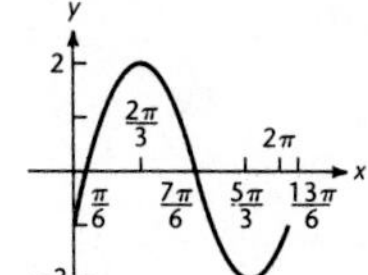

57. $\theta(t) = \frac{1}{2}\sin\left(4t + \frac{\pi}{6}\right)$, Amp $= \frac{1}{2}$, $P = \frac{\pi}{2}$ **59.** Not an identity **61.** Identity **63.** The maximum value of Y is 0.5, agreeing with the earlier answer.

Exercises 8.4, page 509

1. sin 44° **3.** 2 csc 210° **5.** cos (π/9) **7.** sec 2γ **9.** tan (2π/7) **11.** cot 2α **13.** sin 121° **15.** tan 99° **17.** cos 189° **19.** tan 204°

In answers to Exercises 21, 23, 29, and 31, the function values are given in this order: sine, cosine, tangent, cotangent, secant, cosecant.
21. −24/25, −7/25, 24/7, 7/24, −25/7, −25/24 **23.** $4\sqrt{5}/9$, −1/9, $-4\sqrt{5}$, $-\sqrt{5}/20$, −9, $9\sqrt{5}/20$ **25.** $8\cos^4\theta - 8\cos^2\theta + 1$ **27.** $(3\tan\theta - \tan^3\theta)/(1 - 3\tan^2\theta)$ **29.** $2\sqrt{5}/5$, $\sqrt{5}/5$, 2, 1/2, $\sqrt{5}$, $\sqrt{5}/2$ **31.** $\sqrt{30}/6$, $-\sqrt{6}/6$, $-\sqrt{5}$, $-\sqrt{5}/5$, $-\sqrt{6}$, $\sqrt{30}/5$ **33.** $\sqrt{2-\sqrt{2}}/2$, $\sqrt{2+\sqrt{2}}/2$, $\sqrt{2}-1$ **35.** $\sqrt{2+\sqrt{3}}/2$, $-\sqrt{2-\sqrt{3}}/2$, $-(2+\sqrt{3})$ **37.** $\sqrt{2-\sqrt{2}}/2$, $-\sqrt{2+\sqrt{2}}/2$, $1-\sqrt{2}$ **39.** $-\sqrt{2-\sqrt{3}}/2$, $-\sqrt{2+\sqrt{3}}/2$, $2-\sqrt{3}$ **73.** Identity **75.** Not an identity

Exercises 8.5, page 516

1. sin 111° + sin 57° **3.** 4(cos 25° + cos 1°) **5.** (sin 11α + sin 5α)/2 **7.** (cos x − cos 3x)/2 **9.** 2 sin 49° cos 7° **11.** 2 sin 55° sin 38° **13.** −2 cos 10α sin α **15.** 2 cos 2x cos x **33.** 1.3 **35.** 0.4

Exercises 8.6, page 522

In answers to Exercises 1–11, n represents an arbitrary integer.
1. 150° + n180° **3.** 30° + n360°, 150° + n360° **5.** 60° + n120° **7.** 120° + n360°, 240° + n360° **9.** n360°, 60° + n360°, 300° + n360° **11.** 30° + n180°, 150° + n180° **13.** 30°, 210° **15.** 135°, 225° **17.** 17.6°, 162.4° **19.** 66.5°, 127.5°, 246.5°, 307.5° **21.** 0, π, $\pi/4$, $5\pi/4$ **23.** 0, $\pi/4$, $3\pi/4$, π, $5\pi/4$, $7\pi/4$ **25.** $\pi/6$, $5\pi/6$ **27.** $\pi/3$, π, $5\pi/3$ **29.** $7\pi/6$, $3\pi/2$, $11\pi/6$ **31.** 0, $\pi/3$, $5\pi/3$ **33.** 0, π **35.** $\pi/4$, $3\pi/4$, $5\pi/4$, $7\pi/4$ **37.** $\pi/3$, $3\pi/2$, $5\pi/3$ **39.** $\pi/4$, $5\pi/4$ **41.** $3\pi/2$ **43.** $4\pi/3$ **45.** $\pi/9$, $4\pi/9$, $7\pi/9$, $10\pi/9$, $13\pi/9$, $16\pi/9$ **47.** $\pi/6$, $\pi/2$, $5\pi/6$, $7\pi/6$, $3\pi/2$, $11\pi/6$ **49.** $\pi/8$, $5\pi/8$, $9\pi/8$, $13\pi/8$ **51.** $\pi/12$, $5\pi/12$, $3\pi/4$, $13\pi/12$, $17\pi/12$, $7\pi/4$ **53.** $\pi/9$, $5\pi/9$, $7\pi/9$, $11\pi/9$, $13\pi/9$, $17\pi/9$ **55.** 0, $\pi/3$, π, $5\pi/3$ **57.** 0 **59.** $\pi/4$, $\pi/2$, $5\pi/4$, $3\pi/2$ **61.** $\pi/3$, $2\pi/3$, $4\pi/3$, $5\pi/3$ **63.** $\pi/3$, $5\pi/3$ **65.** $2\pi/3$, $4\pi/3$ **67.** 0, $\pi/3$, $2\pi/3$, π, $7\pi/6$, $4\pi/3$, $5\pi/3$, $11\pi/6$ **69.** $\pi/4$ **71.** $\pi/3$ **73.** $t = n\pi/4$; n = 0, 1, 2, . . . **75.** $\pi/2$, $7\pi/6$, $3\pi/2$, $11\pi/6$ **77.** $t = 5\pi/24 + n\pi/4$; n = 0, 1, 2, . . . **79.** ($\pi/2$, 3), ($7\pi/6$, −3/2), ($3\pi/2$, −1), ($11\pi/6$, −3/2) **81.** 0, ±1.9 **83.** 0.5, −1.5 **85.** 0.9

Exercises 8.7, page 534

1. $\pi/2$ **3.** $\pi/4$ **5.** $3\pi/4$ **7.** $-\pi/4$ **9.** $-\pi/3$ **11.** $5\pi/6$ **13.** Not defined **15.** $\sqrt{3}/3$ **17.** −1.3 **19.** $\frac{5}{6}$ **21.** $\frac{4}{5}$ **23.** $-\frac{12}{5}$ **25.** 0.99 **27.** $5\pi/6$ **29.** 0.1978 **31.** 2.2108 **33.** 0.5250 **35.** −0.7046 **37.** $-\frac{17}{49}$ **39.** $-\frac{24}{25}$ **41.** $\frac{119}{169}$ **43.** $\frac{33}{65}$ **45.** $\frac{56}{65}$ **47.** $\sqrt{1-x^2}/x$ **49.** $\sqrt{1+x^2}/x$ **51.** $(1-x^2)/(1+x^2)$ **53.** $xy + \sqrt{(1-x^2)(1-y^2)}$ **55.** $-\sqrt{3}$ **57.** $-\sqrt{2}/2$ **59.** $-\sqrt{2}/2$ **61.** $\sqrt{2}/2$ **63.** $\frac{5}{13}$ **65.** 0 **67.** $\frac{1}{2}$ **69.** $-\frac{1}{2}$ **71.** No solution **73.** $-1 \le x \le 1$

75. **77.** **79.** **81.**

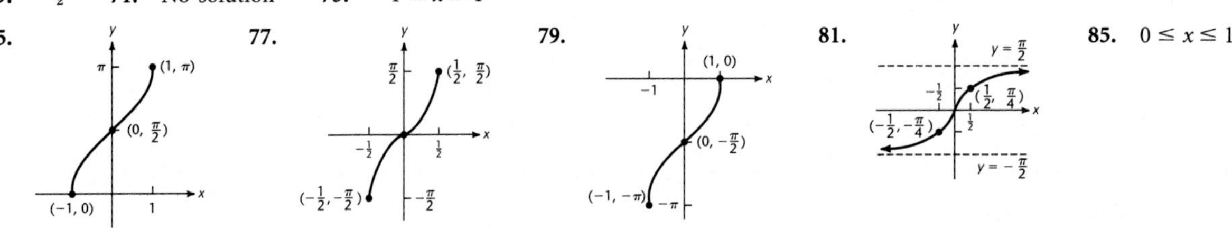

85. $0 \le x \le 1$

Critical Thinking: Find the Errors, page 539

1. Errors

$$\frac{\sin\theta}{\cos\theta}+\frac{\cos\theta}{\sin\theta}=\frac{\overset{\times}{\sin\theta}+\overset{\times}{\cos\theta}}{\cos\theta\sin\theta}$$

$$\frac{1}{\cos\theta}+\frac{1}{\sin\theta}=\frac{1}{\cos\theta\sin\theta}\quad\times$$

Corrections

$$\begin{aligned}\tan\theta+\cot\theta&=\frac{\sin\theta}{\cos\theta}+\frac{\cos\theta}{\sin\theta}\\&=\frac{\sin^2\theta+\cos^2\theta}{\cos\theta\sin\theta}\\&=\frac{1}{\cos\theta\sin\theta}\\&=\frac{1}{\cos\theta}\cdot\frac{1}{\sin\theta}\\&=\sec\theta\csc\theta\end{aligned}$$

2. Errors

$\overset{\times}{\cos\theta=1}$ or $\overset{\times}{2\cos\theta-1=1}$

Correction

$$\begin{aligned}2\cos^2\theta-\cos\theta&=1\\2\cos^2\theta-\cos\theta-1&=0\\(2\cos\theta+1)(\cos\theta-1)&=0\end{aligned}$$

$2\cos\theta+1=0$ or $\cos\theta-1=0$

$\cos\theta=-\frac{1}{2}$ or $\cos\theta=1$

$\theta=120°, 240°$ or $\theta=0°$

The solutions are 0°, 120°, and 240°.

3. Errors

$$\frac{\cos 2\theta}{\cos^2\theta}=\frac{\cos^2\theta\overset{\times}{+}\sin^2\theta}{\cos^2\theta}$$

$$1+\tan^2\theta\overset{\times}{=}1+(1+\sec^2\theta)$$

Correction

$$\begin{aligned}\frac{\cos 2\theta}{\cos^2\theta}&=\frac{\cos^2\theta-\sin^2\theta}{\cos^2\theta}\\&=\frac{\cos^2\theta}{\cos^2\theta}-\frac{\sin^2\theta}{\cos^2\theta}\\&=1-\tan^2\theta\\&=1-(\sec^2\theta-1)\\&=2-\sec^2\theta\end{aligned}$$

4. Errors

$$(\cos\theta-\sin\theta)^2\overset{\times}{=}\cos^2\theta-\sin^2\theta$$

$$\cos 2\theta\overset{\times}{=}1-\sin 2\theta$$

Correction

$$\begin{aligned}(\cos\theta-\sin\theta)^2&=\cos^2\theta-2\cos\theta\sin\theta+\sin^2\theta\\&=1-2\sin\theta\cos\theta\\&=1-\sin 2\theta\end{aligned}$$

5. Error

$$\begin{aligned}2\sin x\cos x&=\cos x\\2\sin x&=1\quad\times\end{aligned}$$

Correction

$$\begin{aligned}2\sin x\cos x&=\cos x\\2\sin x\cos x-\cos x&=0\\(2\sin x-1)\cos x&=0\end{aligned}$$

$2\sin x-1=0$ or $\cos x=0$

$\sin x=\frac{1}{2}$

$x=\frac{\pi}{6},\frac{5\pi}{6}$ or $x=\frac{\pi}{2},\frac{3\pi}{2}$

The solutions are $\pi/6$, $\pi/2$, $5\pi/6$, and $3\pi/2$.

6. Error

$$\operatorname{arccot}(-\sqrt{3})\overset{\times}{=}\arctan\left(-\frac{1}{\sqrt{3}}\right)$$

Correction

$y=\operatorname{arccot}(-\sqrt{3})$ is equivalent to $\cot y=-\sqrt{3}$ and $0<y<\pi$. This is shown in the accompanying figure. From the figure, we see that $\operatorname{arccot}(-\sqrt{3})=5\pi/6$.

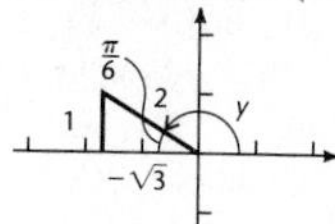

7. **Error**

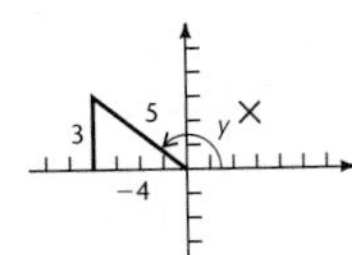

Correction

$y = \tan^{-1}(-\frac{3}{4})$ is equivalent to $\tan y = -\frac{3}{4}$ and $-\pi/2 < y < \pi/2$. From the accompanying figure, we see that $\sin(\tan^{-1}(-\frac{3}{4})) = \sin y = -\frac{3}{5}$.

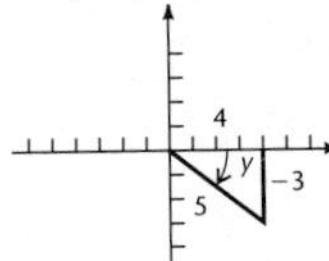

8. **Error**

$$\sin^{-1}\left(\sin\frac{5\pi}{6}\right) = \frac{5\pi}{6}$$

Correction

The range for $\sin^{-1} x$ is $-\pi/2 \le y \le \pi/2$, and $5\pi/6$ is not in this range.

$$\sin^{-1}\left(\sin\frac{5\pi}{6}\right) = \sin^{-1}\frac{1}{2} = \frac{\pi}{6}$$

Review Problems for Chapter 8, page 541

4. $\sin\theta = 1/\csc\theta$, $\cos\theta = \pm\sqrt{\csc^2\theta - 1}/\csc\theta$, $\tan\theta = \pm 1/\sqrt{\csc^2\theta - 1}$, $\cot\theta = \pm\sqrt{\csc^2\theta - 1}$, $\sec\theta = \pm\csc\theta/\sqrt{\csc^2\theta - 1}$ **12.** (a) 2 (b) 0 (c) 1 (d) $\sqrt{3}/2$ (e) $\sqrt{2} + 1$ **13.** $\sqrt{2}(\cos\theta - \sin\theta)/2$ **14.** $-\sin\theta$ **15.** $\tan\theta$

16. $y = 2\sin x$

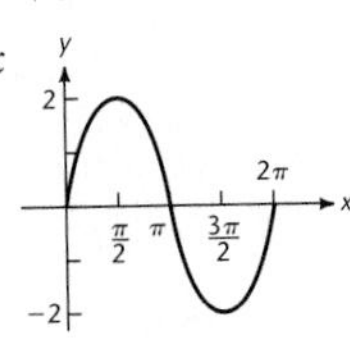

17. $y = 2\sin\left(x + \frac{\pi}{3}\right)$

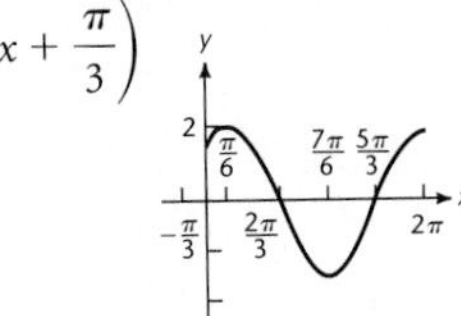

18. $\sin(\alpha - \beta) = \frac{56}{65}$, $\cos(\alpha - \beta) = -\frac{33}{65}$, $\tan(\alpha - \beta) = -\frac{56}{33}$, $\cot(\alpha - \beta) = -\frac{33}{56}$, $\sec(\alpha - \beta) = -\frac{65}{33}$, $\csc(\alpha - \beta) = \frac{65}{56}$ **19.** $\sin 2\theta = -\frac{24}{25}$ **20.** $\cos 2\theta = \frac{7}{25}$ **21.** $\tan 2\theta = -\frac{24}{7}$ **22.** $\sin(\theta/2) = \sqrt{10}/10$ **23.** $\cos(\theta/2) = -3\sqrt{10}/10$ **24.** $\tan(\theta/2) = -1/3$ **25.** $1 - 8\sin^2\theta + 8\sin^4\theta$ **31.** (a) $\cos 8\theta + \cos 2\theta$ (b) $\sin 10x + \sin 2x$ (c) $(\cos 2A + \cos 2B)/2$ **32.** (a) $2\sin 3t\cos t$; (b) $2\cos 6\alpha\cos\alpha$; (c) $2\sin 6t\sin 3t$ **33.** 210°, 330° **34.** 0°, 120°, 240° **35.** 45°, 135° **36.** 45°, 270°, 315° **37.** 0, π **38.** $\pi/6$, $5\pi/6$ **39.** $\pi/2$, $7\pi/6$, $11\pi/6$ **40.** $\pi/2$, $7\pi/6$ **41.** $4\pi/3$ **42.** $\pi/4$, $\pi/2$, $3\pi/4$, $5\pi/4$, $3\pi/2$, $7\pi/4$ **43.** 0 **44.** $3\pi/4$ **45.** $-\pi/4$ **46.** $\pi/4$ **47.** Not defined **48.** 0.2 **49.** $\sqrt{2}$ **50.** 0 **51.** $\frac{1}{4}$ **52.** $-\frac{4}{5}$ **53.** $-\frac{1}{8}$ **54.** $-\frac{24}{25}$ **55.** $\frac{16}{65}$ **56.** 1 **57.** $-\frac{16}{65}$ **58.** $\frac{1}{2}$

59.

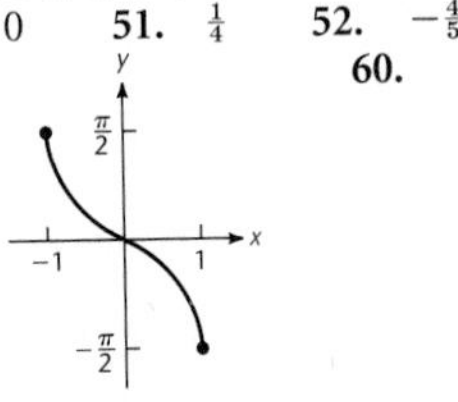

60.

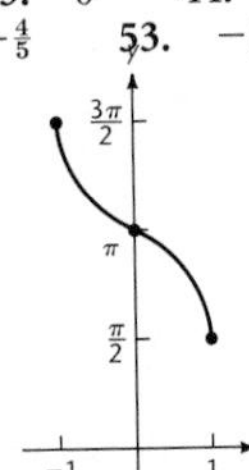

61.

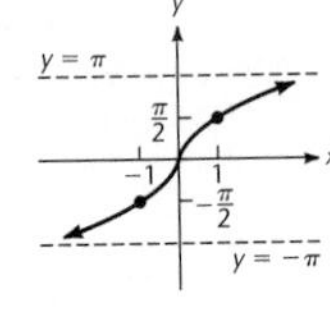

62.

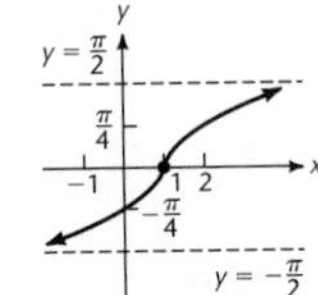

Exercises 9.1, page 553

1. 1.7 **3.** 218 **5.** 19.16 **7.** 7° **9.** 90° **11.** 46° **13.** 59° **15.** 66° or 114° **17.** $C = 50°$, $a = 58$, $b = 55$ **19.** $A = 65.7°$, $b = 12.2$, $c = 7.13$ **21.** No solution **23.** $B = 16°$, $C = 37°$, $b = 14$ **25.** $B = 40°$, $C = 93°$, $c = 110$ (zero not significant) **27.** No solution **29.** $A = 54°$, $C = 79°$, $c = 24$; $A' = 126°$, $C' = 7°$, $c' = 3.0$ **31.** $B = 49.2°$, $C = 68.5°$, $b = 152$; $B' = 6.2°$, $C' = 111.5°$, $b' = 21.7$ **33.** 35 feet, 41 feet **35.** 4.2 feet **37.** 16 feet **39.** 110 miles (zero not significant) **41.** 4.3° **43.** (b) $x = 0.26$ radians

Exercises 9.2, page 562

1. 17 **3.** 8.7 **5.** 8.7 **7.** 5.0 **9.** 120° **11.** 45° **13.** No solution **15.** No solution **17.** $B = 41°$, $C = 83°$, $a = 25$ **19.** $A = 10°$, $C = 58°$, $b = 8.3$ **21.** $A = 26°$, $B = 37°$, $C = 117°$ **23.** $A = 20°$, $B = 149°$, $C = 11°$ **25.** $B = 27°30'$, $C = 125°20'$, $a = 304$ **27.** $A = 18.7°$, $C = 38.7°$, $b = 37.6$ **29.** 210 yards **31.** N 71° E **33.** 7.8 miles **35.** 60 feet **37.** 405 miles **39.** Approximately 14.6 kilometers

Exercises 9.3, page 571

1. 250, zero not significant **3.** 5.3 **5.** 0.73 **7.** 1200, zeros not significant **9.** 6.5 **11.** 600, last zero not significant **13.** 2800, zeros not significant **15.** 140, zero not significant **17.** No solution **19.** No solution **21.** 290, zero not significant **23.** 3.92 or 2.65 (two triangles) **25.** 220, zero not significant **27.** 18.6 square miles **29.** 30°

Exercises 9.4, page 582

1. **3.**

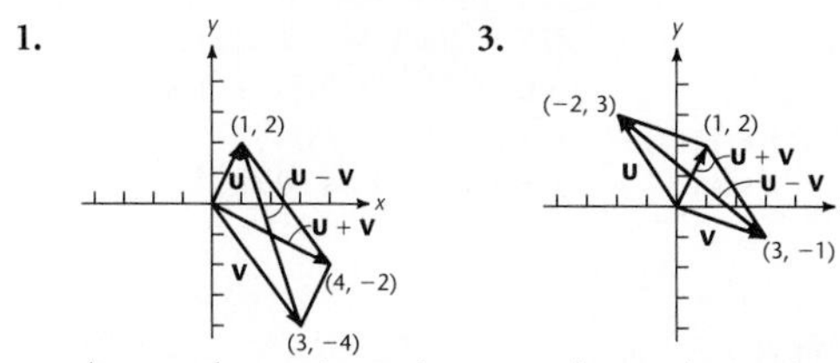

5. $\langle 11, -19\rangle = 11\mathbf{i} - 19\mathbf{j}$ **7.** $\langle 29, -7\rangle = 29\mathbf{i} - 7\mathbf{j}$ **9.** $\langle -1, 4\rangle$, $\sqrt{17}$ **11.** $\langle 1, 4\rangle$, $\sqrt{17}$ **13.** 4, $3\pi/2$ **15.** 8, 0 **17.** $\sqrt{2}$, $7\pi/4$ **19.** 4, $3\pi/4$ **21.** 39 pounds **23.** 35° **25.** 301 pounds **27.** 4.0 miles per hour, S 58° W **29.** 110 miles per hour, zero not significant; N 12° E **31.** 71.4 miles per hour, 173°30′ **33.** 33.7 miles per hour **35.** 5.9° **37.** 10 miles per hour, N 89° E **39.** 0.39 tons **45.** 60°

Exercises 9.5, page 592

1. **3.** **5.** **7.**

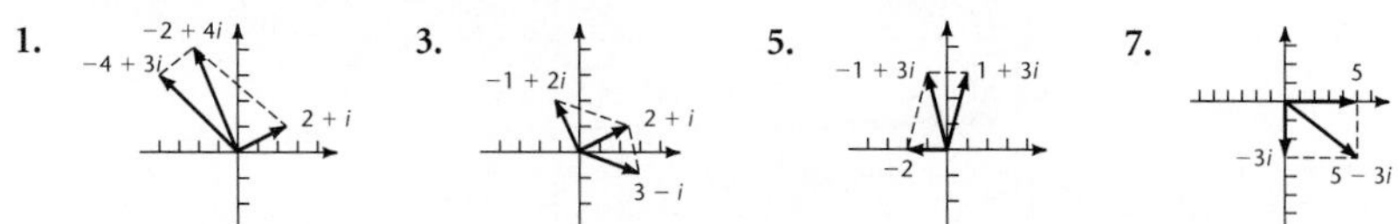

9. $\sqrt{2}(\cos 135° + i \sin 135°)$ **11.** $3\sqrt{2}(\cos 315° + i \sin 315°)$ **13.** $4(\cos 180° + i \sin 180°)$ **15.** $7(\cos 90° + i \sin 90°)$ **17.** $8(\cos 210° + i \sin 210°)$ **19.** $\dfrac{\sqrt{2}}{4}(\cos 45° + i \sin 45°)$ **21.** $2(\cos 300° + i \sin 300°)$ **23.** $6(\cos 150° + i \sin 150°)$ **25.** $\cos 25.4° + i \sin 25.4°$ **27.** $\cos 136.1° + i \sin 136.1°$ **29.** $2\sqrt{2} + 2i\sqrt{2}$ **31.** $(-3/2) - (3\sqrt{3}/2)i$ **33.** $-9i$ **35.** $-1 - i$ **37.** $2.6730 + 1.3620i$ **39.** $4.6839 - 5.2020i$ **41.** $-12 + 12i\sqrt{3}$ **43.** -1 **45.** 6 **47.** $-8 + 8i\sqrt{3}$ **49.** $\sqrt{2} + i\sqrt{2}$ **51.** $2i$ **53.** $(\sqrt{2}/2) - (\sqrt{2}/2)i$ **55.** $(-\sqrt{3}/2) - (1/2)i$ **57.** $4[\cos(3\pi/4) + i \sin(3\pi/4)] = -2\sqrt{2} + 2i\sqrt{2}$, $4[\cos(-\pi/2) + i \sin(-\pi/2)] = -4i$ **59.** $6[\cos(2\pi/3) + i \sin(2\pi/3)] = -3 + 3i\sqrt{3}$, $\frac{2}{3}(\cos \pi + i \sin \pi) = -\frac{2}{3}$ **61.** $4(\cos 150° + i \sin 150°) = -2\sqrt{3} + 2i$, $\cos(-90°) + i \sin(-90°) = -i$ **63.** $4(\cos 180° + i \sin 180°) = -4$, $\frac{1}{2}[\cos(-90°) + i \sin(-90°)] = -\frac{1}{2}i$ **65.** $20(\cos 60° + i \sin 60°) = 10 + 10i\sqrt{3}$, $\dfrac{1}{5}[\cos(-60°) + i \sin(-60°)] = \dfrac{1}{10} - \dfrac{\sqrt{3}}{10}i$ **67.** $6\sqrt{2}(\cos 135° + i \sin 135°) = -6 + 6i$, $\dfrac{3\sqrt{2}}{4}(\cos 45° + i \sin 45°) = \dfrac{3}{4} + \dfrac{3}{4}i$

Exercises 9.6, page 598

1. $\cos 126° + i \sin 126°$ **3.** $512(\cos 108° + i \sin 108°)$ **5.** $\cos(15\pi/8) + i \sin(15\pi/8)$ **7.** $9[\cos(8\pi/5) + i \sin(8\pi/5)]$ **9.** $(-\sqrt{3}/2) - (1/2)i$ **11.** 1 **13.** $-i$ **15.** $-128 - 128i\sqrt{3}$ **17.** $1024i$ **19.** $512 + 512i\sqrt{3}$ **21.** 16 **23.** $-250 + 250i$ **25.** **27.**

29. **31.** **33.** **35.**

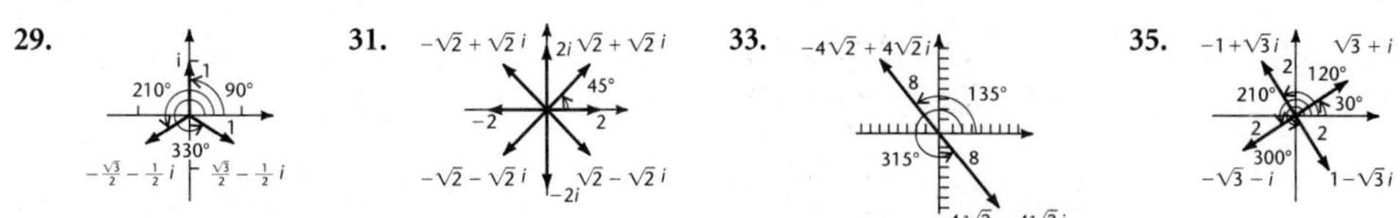

37. $2(\cos 45° + i \sin 45°)$, $2(\cos 105° + i \sin 105°)$, $2(\cos 165° + i \sin 165°)$, $2(\cos 225° + i \sin 225°)$, $2(\cos 285° + i \sin 285°)$, $2(\cos 345° + i \sin 345°)$ **39.** $\cos 10° + i \sin 10°$, $\cos 130° + i \sin 130°$, $\cos 250° + i \sin 250°$ **41.** $\cos 75° + i \sin 75°$, $\cos 165° + i \sin 165°$, $\cos 255° + i \sin 255°$, $\cos 345° + i \sin 345°$ **43.** $\cos 45° + i \sin 45°$, $\cos 117° + i \sin 117°$, $\cos 189° + i \sin 189°$, $\cos 261° + i \sin 261°$, $\cos 333° + i \sin 333°$ **45.** $2(\cos 63° + i \sin 63°)$, $2(\cos 135° + i \sin 135°)$, $2(\cos 207° + i \sin 207°)$, $2(\cos 279° + i \sin 279°)$, $2(\cos 351° + i \sin 351°)$ **47.** $2(\cos 105° + i \sin 105°)$, $2(\cos 225° + i \sin 225°)$, $2(\cos 345° + i \sin 345°)$ **49.** $(3/2) + (3\sqrt{3}/2)i$, -3, $(3/2) - (3\sqrt{3}/2)i$ **51.** $(\sqrt{3}/2) + (1/2)i$, $(-\sqrt{3}/2) + (1/2)i$, $-i$ **53.** $\sqrt{3} + i$, $2i$, $-\sqrt{3} + i$, $-\sqrt{3} - i$, $-2i$, $\sqrt{3} - i$ **55.** $(\sqrt{3}/2) + (1/2)i$, $(-1/2) + (\sqrt{3}/2)i$, $(-\sqrt{3}/2) - (1/2)i$, $(1/2) - (\sqrt{3}/2)i$ **57.** $2(\cos 18° + i \sin 18°)$, $2(\cos 90° + i \sin 90°)$, $2(\cos 162° + i \sin 162°)$, $2(\cos 234° + i \sin 234°)$, $2(\cos 306° + i \sin 306°)$ **59.** $\cos 20° + i \sin 20°$, $\cos 140° + i \sin 140°$, $\cos 260° + i \sin 260°$

Exercises 9.7, page 607

1. $(2\sqrt{3}, 2)$ **3.** $(-\sqrt{2}/2, \sqrt{2}/2)$ **5.** $(-1, \sqrt{3})$ **7.** $(-3\sqrt{3}/2, -3/2)$ **9.** $(-\sqrt{2}, -\sqrt{2})$ **11.** $(0, 5)$ **13.** $(-1, \sqrt{3})$ **15.** $(3, 0)$ **17.** $(2, 135°)$ **19.** $(2, 210°)$ **21.** $(1, 270°)$ **23.** $(4\sqrt{2}, 315°)$ **25.** $r \cos\theta = 3$ **27.** $\theta = 45°$ **29.** $r^2 = 1$, or $r = 1$, or $r = -1$ **31.** $r^2 - 4r\cos\theta = 0$, or $r = 4\cos\theta$ **33.** $y = 2$ **35.** $x^2 + y^2 + x - y = 0$ **37.** $x + y = 1$ **39.** $x^2 + y^2 - x = 1$ **41.** $x^2 + 4y = 4$ **43.** $3x^2 + 4y^2 - 12x = 36$

45. **47.** **49.** **51.**

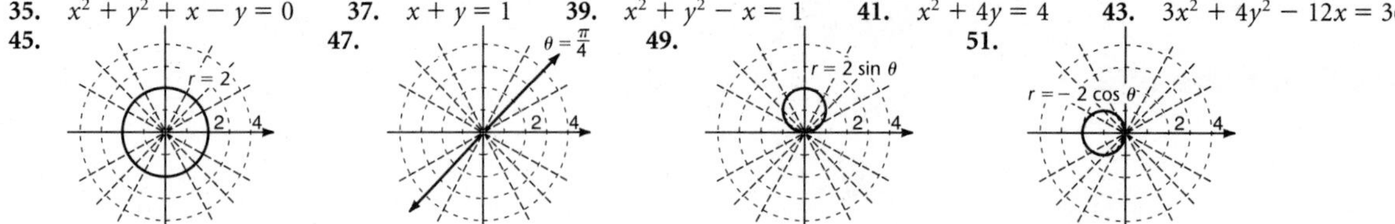

53. **55.** **57.** **59.**

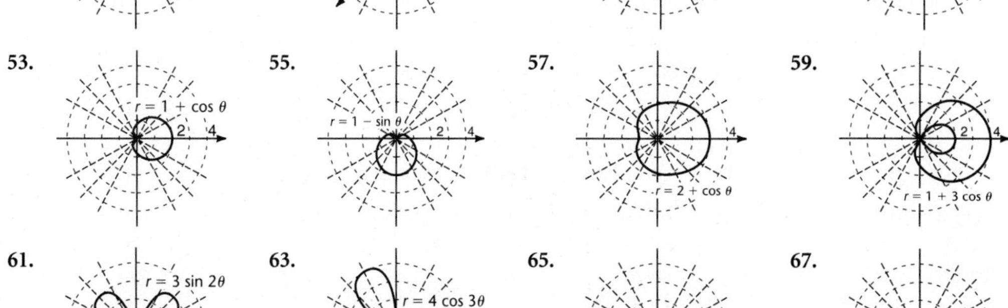

61. **63.** **65.** **67.**

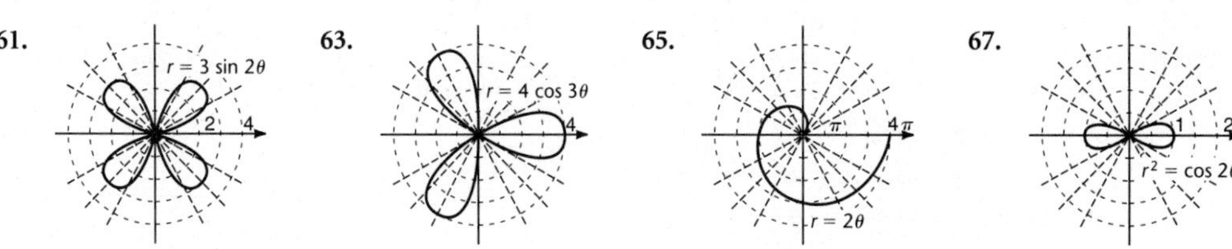

69. Viewing rectangle $[-5, 5]$ by $[-5, 5]$

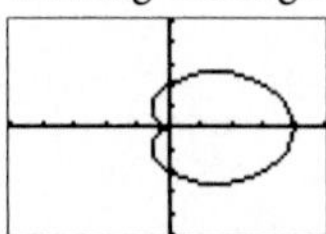

71. Viewing rectangle $[-10, 10]$ by $[-10, 10]$, $0 \le \theta \le 12\pi$

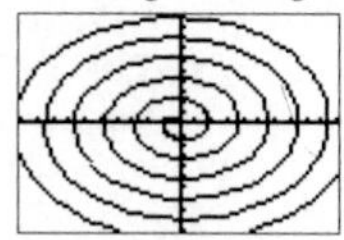

73. Viewing rectangle $[-10, 10]$ by $[-10, 10]$

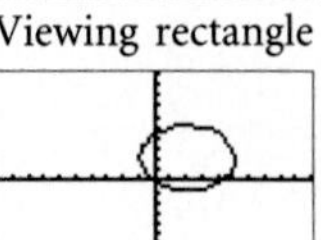

75. Viewing rectangle $[-5, 5]$ by $[-5, 5]$

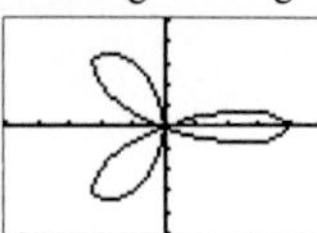

77. Viewing rectangle $[-10, 10]$ by $[-10, 10]$

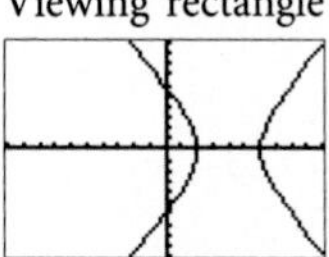

Critical Thinking: Find the Errors, page 610

1. Error

$49 + 9 - 42(\frac{1}{2}) = 16(\frac{1}{2})$

Correction

$$\begin{aligned} 49 + 9 - 42(\tfrac{1}{2}) &= 49 + 9 - 21 \\ &= 37 \\ c &= \sqrt{37} \end{aligned}$$

2. Error

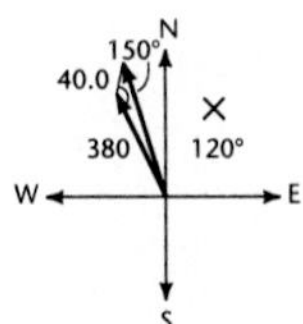

Correction

The heading of 120.0° is an angle to be measured clockwise from north, as shown in the accompanying figure.

$$\begin{aligned} c^2 &= (380)^2 + (40)^2 - 2(380)(40)\cos 60^\circ \\ &= 144{,}400 + 1600 - 30{,}400(\tfrac{1}{2}) \\ &= 130{,}800 \\ c &= 362 \text{ miles per hour} \end{aligned}$$

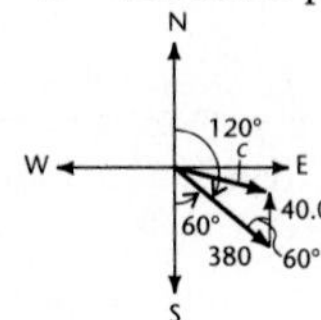

3. Error

$-\sqrt{3} - i = 2(\cos 30^\circ + i \sin 30^\circ)$

Correction

The *related angle* is 30°,
$-\sqrt{3} - i = 2(\cos 210^\circ + i \sin 210^\circ)$.

4. Error

$(1 + i\sqrt{3})^5 = 2(\cos 60^\circ + i \sin 60^\circ)^5$

Correction

$$\begin{aligned} (1 + i\sqrt{3})^5 &= [2(\cos 60^\circ + i \sin 60^\circ)]^5 \\ &= 2^5(\cos 300^\circ + i \sin 300^\circ) \\ &= 32\left(\frac{1}{2} - \frac{\sqrt{3}}{2}i\right) \\ &= 16 - 16i\sqrt{3} \end{aligned}$$

Review Problems for Chapter 9, page 612

1. $A = 50^\circ$, $b = 22$, $c = 15$ **2.** No solution **3.** $A = 117.3^\circ$, $B = 25.5^\circ$, $a = 20.9$ **4.** $A = 84.6^\circ$, $C = 58.2^\circ$, $a = 16.6$; $A' = 21.0^\circ$, $C' = 121.8^\circ$, $a' = 5.99$ **5.** $A = 101.4^\circ$, $B = 59.1^\circ$, $C = 19.5^\circ$ **6.** $B = 23^\circ$, $C = 37^\circ$, $a = 31$ **7.** $K = 130$, zero not significant **8.** $K = 326$ **9.** $K = 130$, zero not significant **10.** $-1 + i\sqrt{3}$ **11.** $-2\sqrt{3} - 2i$

12. $3(\cos 270° + i \sin 270°)$ **13.** $2\sqrt{2}(\cos 315° + i \sin 315°)$ **14.** $-12i$ **15.** $(-2\sqrt{3}/3) + (2/3)i$
16. $16(\cos 150° + i \sin 150°) = -8\sqrt{3} + 8i$, $4(\cos 90° + i \sin 90°) = 4i$ **17.** $-9i$ **18.** $128i$ **19.** -64
20. $(1/2) + (\sqrt{3}/2)i$, $(-\sqrt{3}/2) + (1/2)i$, $(-1/2) - (\sqrt{3}/2)i$, $(\sqrt{3}/2) - (1/2)i$ **21.** 2, $1 + i\sqrt{3}$, $-1 + i\sqrt{3}$, -2, $-1 - i\sqrt{3}$, $1 - i\sqrt{3}$ **22.** 84 feet **23.** 16° **24.** 38 inches **25.** 254 miles per hour, 352.8° **26.** N 67° E, 20 miles per hour
27. 21.1 pounds **28.** 30.6° **29.** 496 pounds **30.** 66.1 pounds **31.** 42.0 miles per hour **32.** $r = 2\cos\theta$
33. $r(3\cos\theta + 4\sin\theta) = 12$ **34.** $y - 2x = 4$ **35.** $xy = 1$ **36.**

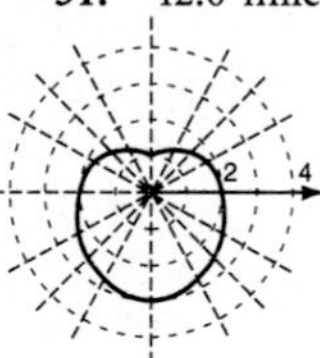

37.

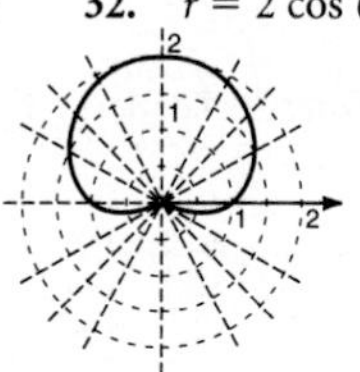

Cumulative Test Chapters 3–9, page 616

1. $2x - 7y = 14$ **2.** (a) $(f/g)(x) = \sqrt{x+2}/(1-x)$, $x \in [-2,1) \cup (1, \infty)$ (b) $(f \circ g)(x) = \sqrt{3-x}$, $x \in [3, \infty)$
3. The maximum demand is 18,000 books when $p = 4$. **4.** $(x+3)^2 + (y-5)^2 = 16$
5. C: $(2, -1)$, V: $(-1, -1)$, $(5, -1)$

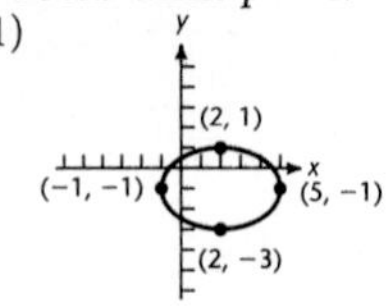

6.

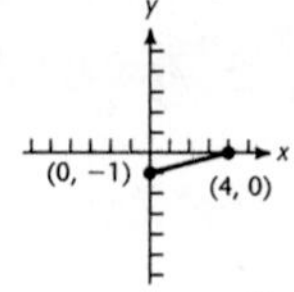

7. $1 + i$, -3 **8.** $3, -2, -4$ **9.** $y = 2$, $x = 1$, $x = -1$ **10.**

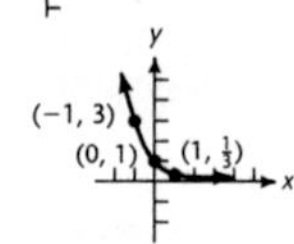

11. $\{-\frac{1}{3}, -\frac{4}{3}\}$ **12.** Approximately 12.6 years **13.** $\sin\alpha = -\frac{15}{17}$, $\cos\alpha = -\frac{8}{17}$, $\tan\alpha = \frac{15}{8}$, $\cot\alpha = \frac{8}{15}$, $\csc\alpha = -\frac{17}{15}$
14. 88 radians per minute **15.** 66 feet **16.** (a) (b)

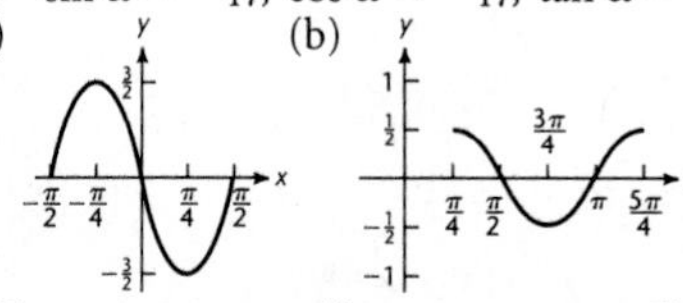

18. $\sin(\alpha + \beta) = \frac{16}{65}$, $\cos(\alpha + \beta) = \frac{63}{65}$, $\tan(\alpha + \beta) = \frac{16}{63}$, $\cot(\alpha + \beta) = \frac{63}{16}$, $\sec(\alpha + \beta) = \frac{65}{63}$, $\csc(\alpha + \beta) = \frac{65}{16}$ **19.** $\pi/12$, $5\pi/12$, $3\pi/4$, $13\pi/12$, $17\pi/12$, $7\pi/4$ **20.** $\frac{56}{65}$ **21.** $B = 49.9°$, $C = 96.4°$, $c = 19.9$; $B' = 130.1°$, $C' = 16.2°$, $c' = 5.58$
22. $2(\cos 15° + i \sin 15°)$, $2(\cos 75° + i \sin 75°)$, $2(\cos 135° + i \sin 135°)$, $2(\cos 195° + i \sin 195°)$, $2(\cos 225° + i \sin 225°)$, $2(\cos 315° + i \sin 315°)$ **23.** 274.9° **24.** 42° **25.**

Exercises 10.1, page 623

1. $\{(-2, 3)\}$

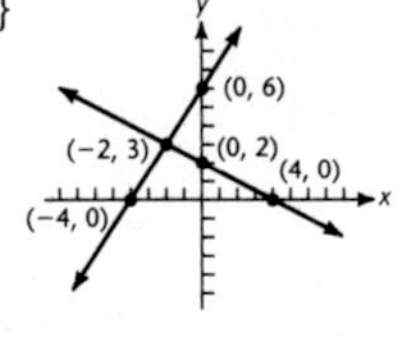

3. $\varnothing$

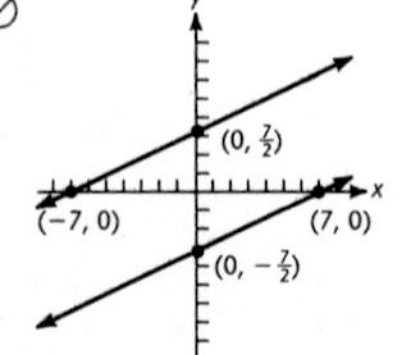

5. $\{(3, -1)\}$ **7.** $\{(3, -4)\}$ **9.** $\{(2, -3)\}$

11. $\{(1, -3)\}$ **13.** $\{(1, 3)\}$ **15.** $\{(10, -3)\}$ **17.** $\{(8, 6)\}$ **19.** $\{(46, -64)\}$ **21.** $\{(3, -2)\}$ **23.** $\varnothing$
25. $\{(0, 4)\}$ **27.** $\varnothing$ **29.** $\{(x, y) \mid 2x - 3y = 4\}$ **31.** $\{(x, y) \mid x + 2y = 4\}$ **33.** $\{(1, \frac{1}{4})\}$ **35.** $\{(\frac{1}{9}, \frac{1}{16})\}$ **37.** \$9500

at 9%, \$5500 at 5% **39.** 4 miles per hour, 20 miles per hour **41.** 5.6 grams of the ingredient that is 50% protein, 22.4 grams of the other ingredient **43.** 40 pounds of peanuts, 60 pounds of cashews **45.** \$7 for regular admission, \$14 for reserved seats **47.** $p = 30$ **49.** 40 desks of the first type, 100 desks of the second type **51.** 3 hours **53.** $(3.01, -1.95)$ **55.** $(1.15, -1.39)$

Exercises 10.2, page 633

1. $\{(-1, 2, 1)\}$ **3.** $\{(3, -2, 2)\}$ **5.** $\{(1, 2, 2)\}$ **7.** $\{(-4, -3, -7)\}$ **9.** $\{(-3, -1, 2)\}$ **11.** $\{(-7, -9, 0)\}$ **13.** $\{(4, 0, -2)\}$ **15.** Inconsistent system **17.** Inconsistent system **19.** $\{(4, 2, -1)\}$ **21.** $\{(\frac{1}{2}, -\frac{1}{2}, 1)\}$ **23.** $\{(\frac{1}{2}, -1, -\frac{1}{3})\}$ **25.** $a = 2$, $b = 4$, $c = 3$ **27.** $x^2 + y^2 - 4x + 2y - 4 = 0$ **29.** 11 **31.** 12 days **33.** 45°, 55°, 80° **35.** 40 pounds of 8-4-8, 120 pounds of 15-30-15, 40 pounds of 12-6-12

Exercises 10.3, page 642

1. $\{(-2, 2), (6, 18)\}$

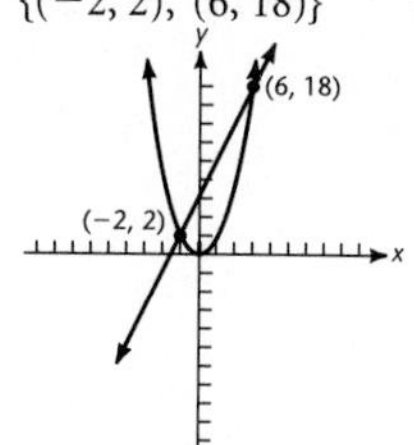

3. $\varnothing$

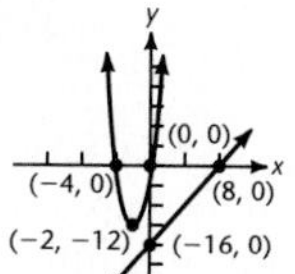

5. $\{(0, -2), (\frac{8}{5}, \frac{6}{5})\}$

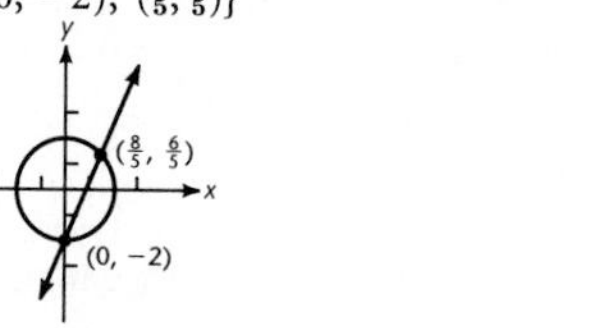

7. $\{(8, 4), (\frac{8}{25}, -\frac{44}{25})\}$

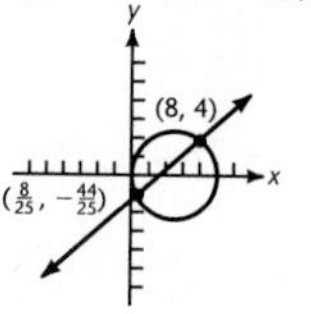

9. $\{(-1, 1), (\frac{3}{5}, -\frac{7}{5})\}$

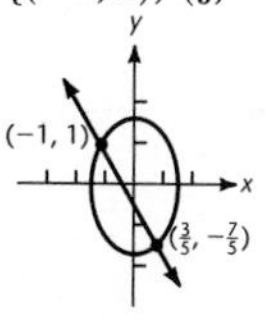

11. $\varnothing$

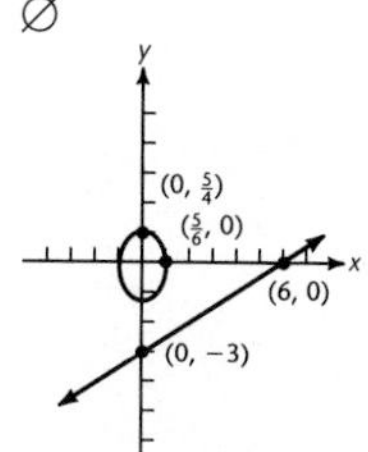

13. $\{(4, 3), (4, -3), (-4, 3), (-4, -3)\}$

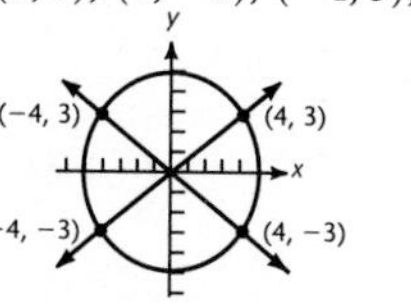

15. $\{(1, 2\sqrt{2}), (1, -2\sqrt{2})\}$

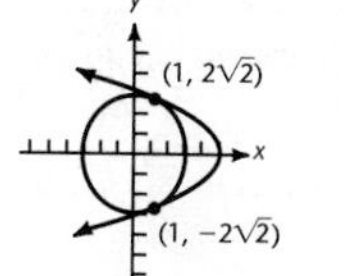

17. $\{(1, 0), (-1, 0)\}$ **19.** $\varnothing$ **21.** $\{(\frac{1}{2}, 2), (7, \frac{1}{7})\}$ **23.** $\{(\frac{1}{3}, -3), (-2, \frac{1}{2})\}$ **25.** $\{(\frac{3}{2}, -\frac{4}{3}), (-\frac{3}{2}, \frac{4}{3})\}$ **27.** $\{(1, 2), (-1, -2), (\sqrt{2}, \sqrt{2}), (-\sqrt{2}, -\sqrt{2})\}$ **29.** $\{(2, 2), (-2, -2)\}$ **31.** $\{(1, 1), (-1, -1), (5, -3), (-5, 3)\}$ **33.** $\{(1, -1), (2, 2)\}$ **35.** $\{(3, 4), (-3, 4)\}$ **37.** $(20, 1000)$ **39.** 4 and 19, or −4 and −19 **41.** 4 and 7, or −4 and 7, or 4 and −7, or −4 and −7 **43.** $\pm\frac{4}{3}$ **45.** For 1200 items, both cost and revenue equal \$264,000. **47.** $s = 4$ feet, $\ell = 2\sqrt{3}$ feet **49.** $\{(1, 0), (0.23, -1.46)\}$ **51.** $\{(1.44, 1.39)\}$

Exercises 10.4, page 650

1.

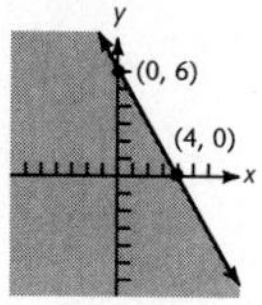

3.

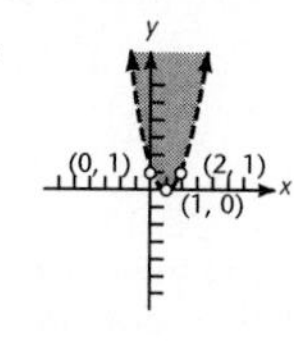

5.

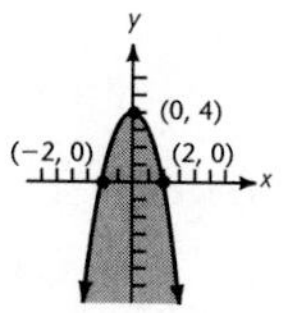

7.

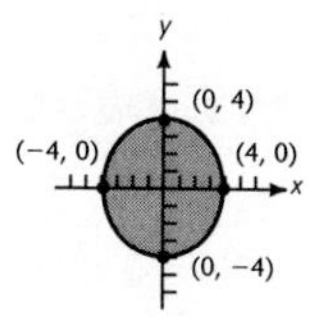

9.

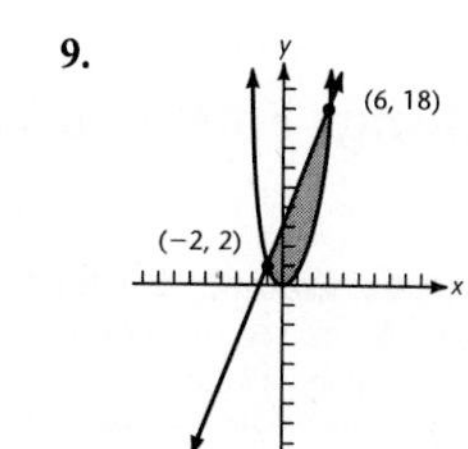

11. No solution

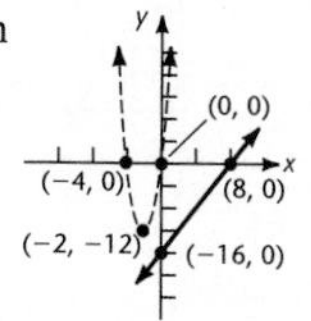

13.

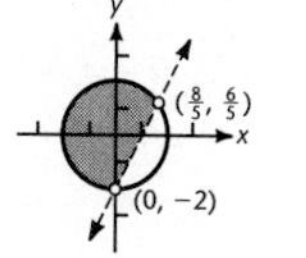

15.

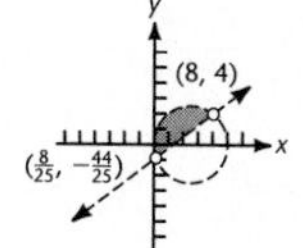

17.

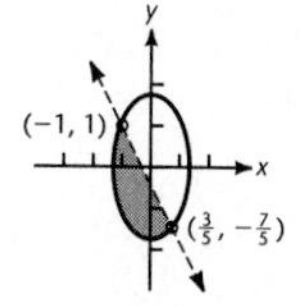

19. $\varnothing$

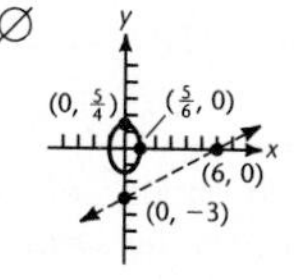

21. 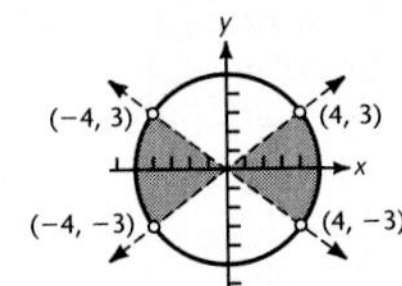

23. 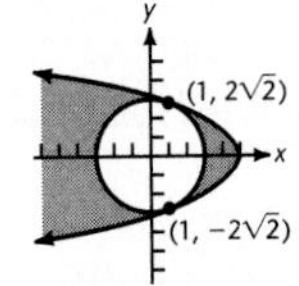

25. 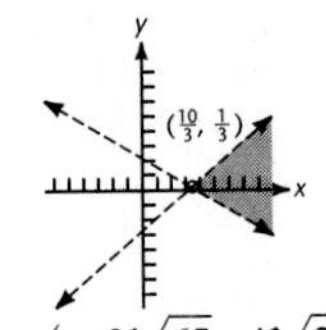

27. 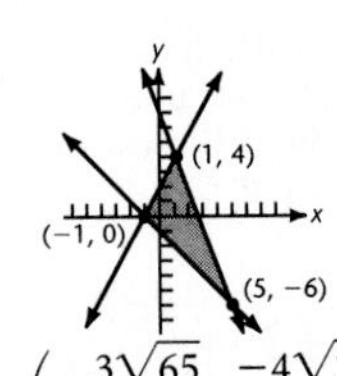

29. 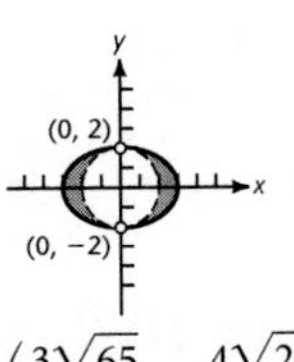

31. 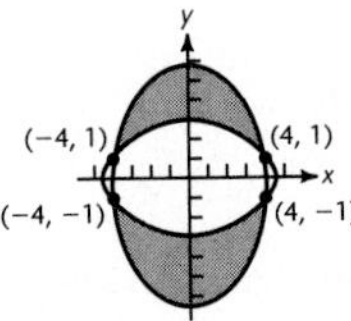

33. 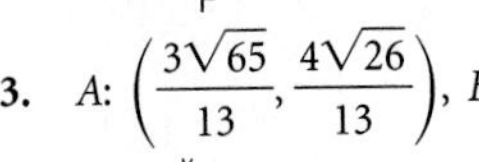 A: $\left(\frac{3\sqrt{65}}{13}, \frac{4\sqrt{26}}{13}\right)$, B: $\left(\frac{-3\sqrt{65}}{13}, \frac{4\sqrt{26}}{13}\right)$, C: $\left(-\frac{3\sqrt{65}}{13}, \frac{-4\sqrt{26}}{13}\right)$, D: $\left(\frac{3\sqrt{65}}{13}, -\frac{4\sqrt{26}}{13}\right)$

35. 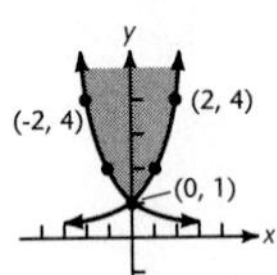

37. $x + 2y \geq 4$
$x - y \leq 1$

39. $y \leq -x^2 - 2x + 3$
$x \leq 0$

41. $y > x^2$
$y - x < 2$

43. $y \geq 2^x$
$x + y \leq 1$
$x \geq -2$

45. Viewing rectangle $[-10, 10]$ by $[-10, 10]$

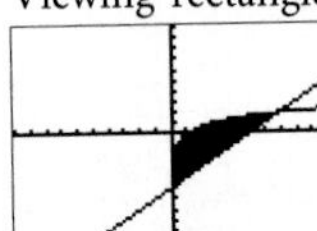

47. Viewing rectangle $[-10, 10]$ by $[-10, 10]$

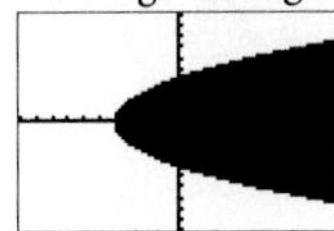

49. Viewing rectangle $[-10, 10]$ by $[-10, 10]$

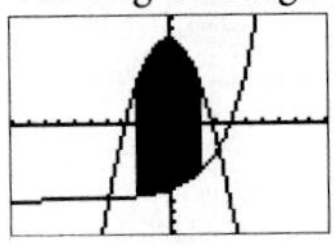

Exercises 10.5, page 657

1. Maximum value of 14 at (6, 2) **3.** Maximum value of 9 at (0, 2) **5.** Minimum value of 11 at (1, 4) **7.** Maximum value of 14 at (6, 2) **9.** Minimum value of 0 at (4, 1) **11.** Maximum value of 4 at (3, 3) **13.** Make 30 cans of the first mixture and 12 cans of the second mixture for a maximum profit of \$180. **15.** Make 40 desks of the first type and 100 desks of the second type for a maximum profit of \$4800. **17.** Manufacture 8 chairs and 14 stools for a maximum profit of \$520. **19.** Production of 14 cold cut trays and 9 seafood trays yields a maximum profit of \$59.50.

Critical Thinking: Find the Errors, page 661

1. Error

To obtain an equation without a y-term, we add -2 times the first equation to the second equation.

$$\begin{array}{rcl} x - y + 2z = 3 & \xrightarrow{\text{Multiply by } -2} & -2x + 2y - 4z = -6 \\ 4x - 2y + 6z = 8 & \longrightarrow & \underline{4x - 2y + 6z = 8} \\ & \text{Add} & 2x \qquad + 2z = 2 \end{array}$$

To obtain another equation without a y-term, we add $-\frac{1}{2}$ times the second equation to the first equation. ×

Correction

It is a mistake to use the *same* pair of equations when eliminating y. A correct choice could be made by using the first and third equations as follows.

$$\begin{array}{rcl} x - y + 2z = 3 & \xrightarrow{\text{Multiply by } 4} & 4x - 4y + 8z = 12 \\ 3x + 4y - 5z = -1 & \longrightarrow & \underline{3x + 4y - 5z = -1} \\ & \text{Add} & 7x \qquad + 3z = 11 \end{array}$$

We solve the x,z subsystem as follows.

$$\begin{array}{lcr} 2x + 2z = \ \ 2 & \xrightarrow{\text{Multiply by } -\frac{3}{2}} & -3x - 3z = -3 \\ 7x + 3z = 11 & \xrightarrow{\hspace{3em}} & \underline{\ \ 7x + 3z = \ \ 11} \\ & \text{Add} & 4x \qquad = \ \ 8 \\ & & x = \ \ 2 \end{array}$$

Substituting $x = 2$ in $7x + 3z = 11$ yields

$14 + 3z = 11$ and $z = -1$.

Using $x = 2$ and $z = -1$ in the original first equation gives

$$\begin{aligned} 2 - y - 2 &= 3 \\ y &= -3. \end{aligned}$$

The solution set is $\{(2, -3, -1)\}$.

2. **Errors**

Testing (0, 0), we find that the origin is in the solution set of each inequality, so the solution set lies to the left of the parabola and to the right of the line as shown in Figure 10.14.

Correction

The origin is not a solution to the inequality $x + y < -1$, so the solution set lies to the left of the straight line as well as the parabola.

3. **Errors**

Vertex	$f = 7x + 2y$
(0, 0)	0
(4, 0)	28
(5, 0)×	35
(3, 2)	25
(0, 5)	10
(0, 8)×	16

Neither (5, 0) nor (0, 8) is a vertex of the solution set.

Correction

The maximum value of f is 28 at the vertex (4, 0).

Review Problems for Chapter 10, page 663

1. $\{(1, -2)\}$ **2.** $\{(2, -1)\}$ **3.** $\{(-3, 5)\}$ **4.** $\{(1, -3)\}$ **5.** $\{(-1, -1)\}$ **6.** $\{(2, 1)\}$ **7.** $\{(4, -1)\}$ **8.** $\varnothing$ **9.** $\{(x, y) \mid x = 2 - 2y\}$ **10.** 3 miles per hour, 15 miles per hour **11.** 25 miles per hour, 125 miles per hour **12.** 120 pounds of the first alloy, 180 pounds of the second alloy **13.** $a = \sqrt{2}/2$, $b = 1$ **14.** $\{(2, -4, 3)\}$ **15.** $\{(0, 6, -2)\}$ **16.** $\{(-1, 2, -3)\}$ **17.** $\varnothing$ **18.** $\varnothing$ **19.** $a = 2$, $b = 1$, $c = -1$ **20.** 11 nickels, 20 dimes, 10 quarters **21.** 4 hours for backhoe A, 6 hours for backhoe B, 12 hours for backhoe C **22.** $\{(5, 2), (2, -1)\}$ **23.** $\{(3, 0), (8, 5)\}$ **24.** $\{(4, 5), (2, 1)\}$ **25.** $\{(\sqrt{6}, \sqrt{6}/6), (-\sqrt{6}, -\sqrt{6}/6)\}$ **26.** $\{(4, 1), (9, -\frac{3}{2})\}$ **27.** $\{(2, 1), (\frac{3}{2}, \frac{5}{4})\}$ **28.** $\{(2, \sqrt{3}), (2, -\sqrt{3}), (2, -\sqrt{3}), (-2, -\sqrt{3})\}$ **29.** **30.**

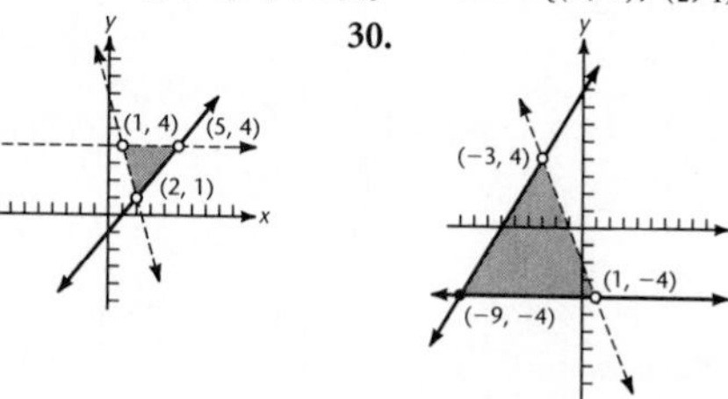

31. 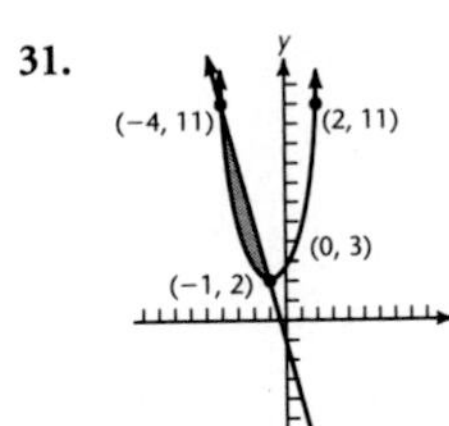

32. 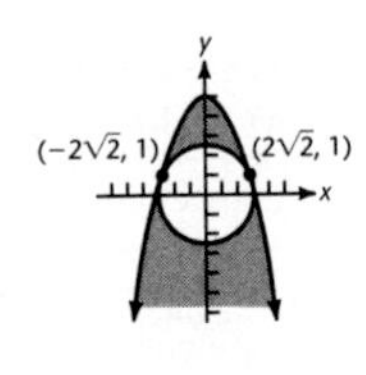

33. 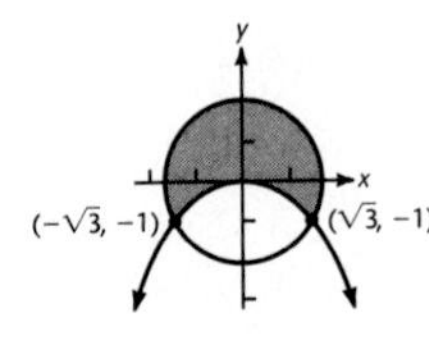

34. Maximum value of 29 at (4, 3)

35. Minimum value of 13 at (2, 1) **36.** Maximum value of 39 at $(3, \frac{3}{2})$ **37.** Preparation of 200 pounds of the \$6 mixture and 400 pounds of the \$5 mixture yields a maximum revenue of \$3200. **38.** 100

Exercises 11.1, page 673

1. 2×3 **3.** 3×1, column matrix **5.** 2×4 **7.** 1×1, column matrix, row matrix, square matrix, diagonal matrix **9.** 1×2, row matrix **11.** 3×3, square matrix, diagonal matrix

13. $\begin{bmatrix} 3 & 4 & 5 & 6 \\ 5 & 6 & 7 & 8 \end{bmatrix}$ **15.** $\begin{bmatrix} -1 & -2 \\ 1 & 2 \\ -1 & -2 \\ 1 & 2 \end{bmatrix}$ **17.** $\begin{bmatrix} 0 & 1 & 1 & 1 \\ 0 & 0 & 1 & 1 \end{bmatrix}$ **19.** $\begin{bmatrix} 1 & 8 & 11 \\ 1 & 1 & 13 \\ 1 & 1 & 1 \\ 1 & 1 & 1 \end{bmatrix}$ **21.** Not equal **23.** Equal

25. Not equal **27.** Not equal **29.** Equal only if $x = 2$ and $y = 3$ **31.** Equal only if $x = 2$ **33.** $\begin{bmatrix} 5 & 0 & 1 \\ 3 & 10 & -5 \end{bmatrix}$

35. Not possible **37.** $\begin{bmatrix} 2 & -8 \\ 4 & 7 \end{bmatrix}$ **39.** $\begin{bmatrix} -11 & 13 \\ 16 & -8 \\ -3 & 9 \end{bmatrix}$ **41.** Not possible **43.** $\begin{bmatrix} -1 \\ -8 \end{bmatrix}$ **45.** $\begin{bmatrix} 8 & 6 \\ 19 & 32 \\ 2 & 13 \end{bmatrix}$

47. $\begin{bmatrix} 50 & 6 & 13 & -23 \\ 34 & -21 & 24 & -14 \end{bmatrix}$

Exercises 11.2, page 681

1. Conformable, 2×7 **3.** Not conformable **5.** Conformable, 1×1 **7.** Conformable, 2×7

9. Conformable, 8×1 **11.** Conformable, 3×3 **13.** Not conformable **15.** Conformable, 1×12 **17.** $\begin{bmatrix} 7 & -1 \\ 7 & -5 \end{bmatrix}$

19. $\begin{bmatrix} -1 & 2 \\ 5 & 20 \\ -11 & -13 \end{bmatrix}$ **21.** Not possible **23.** Not possible **25.** [22] **27.** $\begin{bmatrix} -8 & -22 & 4 \\ 4 & 11 & -2 \\ 4 & 11 & -2 \end{bmatrix}$ **29.** Not possible

31. $\begin{bmatrix} 0 & 0 \\ 0 & 0 \\ 0 & 0 \end{bmatrix}$ **33.** Not possible **35.** $\begin{bmatrix} 1 & 0 \\ 0 & 1 \end{bmatrix}$ **37.** $\begin{bmatrix} 1 & 0 & 0 \\ 0 & 1 & 0 \\ 0 & 0 & 1 \end{bmatrix}$ **39.** $BA = \begin{bmatrix} 9 & -17 \\ -1 & 3 \\ 0 & 10 \end{bmatrix}$, $CB = \begin{bmatrix} -4 & -2 \\ 5 & -2 \\ 1 & 4 \end{bmatrix}$

41. $AC = \begin{bmatrix} -3 & 2 \\ 5 & 0 \\ 1 & -3 \\ 3 & -2 \end{bmatrix}$, $BC = \begin{bmatrix} -1 & 8 \\ 9 & 7 \end{bmatrix}$, $BA = \begin{bmatrix} 0 & -2 & 2 & 8 \\ 1 & 4 & 2 & 29 \end{bmatrix}$, $CB = \begin{bmatrix} 1 & 3 & 22 & 12 \\ 4 & -4 & 8 & 0 \\ 1 & 1 & 12 & 6 \\ 0 & -1 & -5 & -3 \end{bmatrix}$ **43.** $A = \begin{bmatrix} 1 & 2 \\ 0 & 1 \\ 1 & 1 \end{bmatrix}$, $B = \begin{bmatrix} 2 & 1 & 1 \\ -1 & 2 & 0 \end{bmatrix}$ is a possible answer. **45.** $A = \begin{bmatrix} 1 & 2 & -1 \\ 0 & 1 & 2 \\ 1 & 0 & 0 \end{bmatrix}$, $B = \begin{bmatrix} 1 & 0 & 0 \\ 0 & -1 & 2 \\ 1 & 1 & 1 \end{bmatrix}$ is a possible answer. **47.** $A = \begin{bmatrix} 1 & 0 \\ 1 & 2 \end{bmatrix}$,

$B = \begin{bmatrix} 0 & 0 \\ 2 & 2 \end{bmatrix}$ is a possible answer. **49.** $AB + AC = A(B + C) = \begin{bmatrix} -16 & -7 \\ -14 & -8 \\ -10 & -5 \end{bmatrix}$ **51.** $(A - B)(A + B) = \begin{bmatrix} 42 & -15 \\ 2 & 5 \end{bmatrix}$, $A^2 - B^2 = \begin{bmatrix} 39 & -14 \\ -5 & 8 \end{bmatrix}$ **53.** \$3.44 at store A, \$3.24 at store B, \$3.55 at store C.

Exercises 11.3, page 691

1. $\left[\begin{array}{cc|c} 3 & -1 & 0 \\ -1 & 1 & 1 \end{array}\right]$ **3.** $\left[\begin{array}{ccc|c} 3 & -2 & 5 & 0 \\ 4 & 7 & -1 & 0 \\ 1 & 0 & 1 & 2 \end{array}\right]$ **5.** $\left[\begin{array}{cccc|c} 1 & -1 & 0 & 0 & 0 \\ 0 & 0 & 1 & 1 & 0 \\ 3 & 0 & 0 & 2 & 0 \\ 0 & 5 & -1 & 0 & 0 \end{array}\right]$ **7.** $\begin{aligned} x + 2y &= 5 \\ 3x + 4y &= 6 \end{aligned}$ **9.** $\begin{aligned} x &= a \\ z &= b \\ y &= c \end{aligned}$

11. $\begin{aligned} x \quad\quad + z \quad\quad &= 0 \\ 2y + z + 3w &= 7 \\ 3x + y + z + w &= 1 \\ -3x + y - z + 2w &= 5 \end{aligned}$ **13.** $x = 5, y = -4$ **15.** $a = -3, b = -2$ **17.** No solution **19.** $x = -8, y = -5$

21. $x = 0, y = 0$ **23.** $x = 3, y = 1, z = -1$ **25.** $x = 4, y = 4, z = 1$ **27.** $x = 1, y = 3, z = -2$
29. $r = 4, s = -6, t = 3$ **31.** No solution **33.** $x = 1, y = -1, z = 2, t = -2$
35. There are many solutions of the form $a = 5 - 3r$, $b = 10 - 7r$, $c = -4r$, $d = r$, where r is any real number.
37. (a) No solution exists for $c = -2$ (b) exactly one solution for $c \neq 2$ and $c \neq -2$ (c) many solutions for $c = 2$
39. (a) Solutions exist for all values of c (b) exactly one solution for $c \neq 0$ (c) many solutions for $c = 0$

Exercises 11.4, page 699

1. $\begin{bmatrix} \frac{2}{3} & -\frac{1}{2} \\ \frac{1}{3} & 0 \end{bmatrix}$ **3.** $-\frac{1}{8}\begin{bmatrix} 2 & -3 \\ -2 & -1 \end{bmatrix}$ **5.** Does not exist **7.** $-\frac{1}{29}\begin{bmatrix} 7 & -3 \\ -5 & -2 \end{bmatrix}$ **9.** $\begin{bmatrix} -7 & 1 & 1 \\ 1 & 0 & 0 \\ 3 & 0 & -1 \end{bmatrix}$

11. $\begin{bmatrix} 11 & -6 & 2 \\ 3 & -2 & 1 \\ 1 & -1 & 1 \end{bmatrix}$ **13.** $\begin{bmatrix} \frac{5}{4} & \frac{1}{4} & -\frac{1}{2} \\ \frac{3}{4} & \frac{3}{4} & -\frac{1}{2} \\ -\frac{1}{4} & -\frac{1}{4} & \frac{1}{2} \end{bmatrix}$ **15.** Does not exist **17.** $\begin{bmatrix} -23 & 4 & 5 & 2 \\ -97 & 15 & 22 & 8 \\ -50 & 8 & 11 & 4 \\ -13 & 2 & 3 & 1 \end{bmatrix}$ **19.** $x = 5, y = -3$

21. $x_1 = 3, x_2 = 3$ **23.** $x = 6, y = 2$ **25.** $x = -\frac{3}{4}, y = \frac{7}{4}$ **27.** $x_1 = -4, x_2 = 0, x_3 = -4$
29. $x = 2, y = 1, z = -3$ **31.** $x = 5, y = 3, z = 1$ **33.** $x = 3, y = 1, z = -1$ **35.** $x = 1, y = 3, z = -2, w = 0$

37. $\begin{bmatrix} 1/a & 0 & 0 \\ 0 & 1/b & 0 \\ 0 & 0 & 1/c \end{bmatrix}$ **39.** $X = A^{-1}BC^{-1}$

Exercises 11.5, page 706

1. -5 **3.** -1 **5.** -21 **7.** 0 **9.** $-\frac{1}{9}$ **11.** 6 **13.** 26 **15.** $5x$ **17.** 12 **19.** 7 **21.** -3
23. -1 **25.** 4 **27.** -3 **29.** -1 **31.** 0 **33.** -2 **35.** 1 **37.** -1 **39.** 2 **41.** -2 **43.** -3
45. $2, -3$ **47.** $-1, -2$ **49.** $5, -3, 2$ **51.** $0, 4, -3$
53. (b)

$$\begin{aligned} |A| &= a_{11}a_{22}a_{33} + a_{12}a_{23}a_{31} + a_{13}a_{21}a_{32} - a_{11}a_{23}a_{32} - a_{12}a_{21}a_{33} - a_{13}a_{22}a_{31} \\ &= -a_{21}(a_{12}a_{33} - a_{13}a_{32}) + a_{22}(a_{11}a_{33} - a_{13}a_{31}) - a_{23}(a_{11}a_{32} - a_{12}a_{31}) \\ &= -a_{21}\begin{vmatrix} a_{12} & a_{13} \\ a_{32} & a_{33} \end{vmatrix} + a_{22}\begin{vmatrix} a_{11} & a_{13} \\ a_{31} & a_{33} \end{vmatrix} - a_{23}\begin{vmatrix} a_{11} & a_{12} \\ a_{31} & a_{32} \end{vmatrix} \\ &= -a_{21}M_{21} + a_{22}M_{22} - a_{23}M_{23} \\ &= a_{21}C_{21} + a_{22}C_{22} + a_{23}C_{23} \end{aligned}$$

Exercises 11.6, page 713

1. $x = -4$ **3.** $a = -1$ **5.** $x = 1$ **7.** $x = -6$ **9.** $x = 2$ **11.** $x = -2$ **13.** 15 **15.** 10 **17.** -20
19. -1 **21.** 1 **23.** 0 **25.** -1 **27.** 4 **29.** 23 **31.** 4

Exercises 11.7, page 720

1. $x = -3, y = 1$ **3.** $x = 1, y = -2$ **5.** $a = -2, b = 3$ **7.** Dependent system **9.** Inconsistent
11. $x = 2, y = 1, z = -1$ **13.** $x = 5, y = -10, z = 4$ **15.** $x = -21, y = -12, z = 39$ **17.** $a = 1, b = 2, c = 3$
19. $x = 2, y = -1, z = 4$ **21.** Dependent system **23.** $x_1 = \frac{28}{13}, x_2 = -\frac{6}{13}, x_3 = \frac{53}{13}$ **25.** $x = \frac{1}{6}, y = \frac{1}{4}, z = \frac{1}{3}$
27. Inconsistent **29.** $x = -\frac{1}{2}, y = \frac{15}{14}, z = -\frac{25}{14}$ **31.** $x = -1, y = -1, z = 1$ **33.** $x = -2, y = 5, z = 1$
35. $x = 2, y = -1, z = 0, w = 1$ **37.** $x = w = 1, y = z = 0$ **39.** $m = 2, b = -4$ **41.** $a = 2, b = 3, c = -4$

Critical Thinking: Find the Errors, page 723

1. (a) **Error**

$$[2 \quad -5 \quad 0] + [1 \quad 3] = [3 \quad -2 \quad 0]$$

Correction

Not possible to form the sum

(b) **Error**

$$\begin{bmatrix} 0 & 1 \\ 2 & 3 \end{bmatrix}\begin{bmatrix} 4 & 5 \\ -1 & -2 \end{bmatrix} = \begin{bmatrix} 0 & 5 \\ -2 & -6 \end{bmatrix}$$

Correction

$$\begin{bmatrix} 0 & 1 \\ 2 & 3 \end{bmatrix}\begin{bmatrix} 4 & 5 \\ -1 & -2 \end{bmatrix} = \begin{bmatrix} -1 & -2 \\ 5 & 4 \end{bmatrix}$$

2. **Errors**

$$\left[\begin{array}{ccc|c} 1 & 1 & 2 & 2 \\ 0 & 1 & 3 & 1 \\ 1 & 0 & -2 & 1 \end{array}\right] \to \left[\begin{array}{ccc|c} 1 & 1 & 2 & 2 \\ 0 & 1 & 3 & 1 \\ 0 & -1 & -4 & 1 \end{array}\right] \to \left[\begin{array}{ccc|c} 1 & 1 & 2 & 2 \\ 0 & 1 & 3 & 1 \\ 0 & 0 & -1 & 1 \end{array}\right] \to \left[\begin{array}{ccc|c} 1 & 0 & -1 & 2 \\ 0 & 1 & 3 & 1 \\ 0 & 0 & 1 & 1 \end{array}\right] \to \left[\begin{array}{ccc|c} 1 & 0 & 0 & 2 \\ 0 & 1 & 0 & 1 \\ 0 & 0 & 1 & 1 \end{array}\right]$$

Correction

$$\left[\begin{array}{ccc|c} 1 & 1 & 2 & 2 \\ 0 & 1 & 3 & 1 \\ 1 & 0 & -2 & 1 \end{array}\right] \to \left[\begin{array}{ccc|c} 1 & 1 & 2 & 2 \\ 0 & 1 & 3 & 1 \\ 0 & -1 & -4 & -1 \end{array}\right] \to \left[\begin{array}{ccc|c} 1 & 1 & 2 & 2 \\ 0 & 1 & 3 & 1 \\ 0 & 0 & -1 & 0 \end{array}\right] \to \left[\begin{array}{ccc|c} 1 & 0 & -1 & 1 \\ 0 & 1 & 3 & 1 \\ 0 & 0 & 1 & 0 \end{array}\right] \to \left[\begin{array}{ccc|c} 1 & 0 & 0 & 1 \\ 0 & 1 & 0 & 1 \\ 0 & 0 & 1 & 0 \end{array}\right]$$

The solution is $x = 1, y = 1, z = 0$.

3. **Error**

$$\begin{vmatrix} 3 & 1 & -1 \\ -1 & 1 & 2 \\ 1 & 2 & 1 \end{vmatrix} = 3\begin{vmatrix} 1 & 2 \\ 2 & 1 \end{vmatrix} - 1\begin{vmatrix} 1 & -1 \\ 2 & 1 \end{vmatrix} + 1\begin{vmatrix} 1 & -1 \\ 1 & 2 \end{vmatrix}$$

Correction

$$\begin{aligned} \begin{vmatrix} 3 & 1 & -1 \\ -1 & 1 & 2 \\ 1 & 2 & 1 \end{vmatrix} &= 3\begin{vmatrix} 1 & 2 \\ 2 & 1 \end{vmatrix} - (-1)\begin{vmatrix} 1 & -1 \\ 2 & 1 \end{vmatrix} + 1\begin{vmatrix} 1 & -1 \\ 1 & 2 \end{vmatrix} \\ &= 3(1 - 4) + 1(1 + 2) + 1(2 + 1) \\ &= 3(-3) + 1(3) + 1(3) \\ &= -3 \end{aligned}$$

4. **Error**

$$\begin{vmatrix} 0 & 1 & 4 \\ -1 & 2 & 3 \\ 1 & -2 & 0 \end{vmatrix} \overset{R_2 + 1}{=} \begin{vmatrix} 0 & 1 & 4 \\ 0 & 3 & 4 \\ 1 & -2 & 0 \end{vmatrix}$$

Correction

$$\begin{vmatrix} 0 & 1 & 4 \\ -1 & 2 & 3 \\ 1 & -2 & 0 \end{vmatrix} \overset{R_3 + R_2}{=} \begin{vmatrix} 0 & 1 & 4 \\ 0 & 0 & 3 \\ 1 & -2 & 0 \end{vmatrix}$$

$$= 1\begin{vmatrix} 1 & 4 \\ 0 & 3 \end{vmatrix} = 3$$

5. Errors

$$D_y = \begin{vmatrix} 2 & -8 & 2 \\ 0 & 4 & 2 \\ 3 & 2 & 1 \end{vmatrix} \overset{\frac{1}{2}R_2}{=} \begin{vmatrix} -2 & -8 & 2 \\ 0 & 2 & 1 \\ 3 & 2 & 1 \end{vmatrix} \overset{-2C_3 + C_2}{=} \begin{vmatrix} -2 & -12 & 2 \\ 0 & 0 & 1 \\ 3 & 0 & 1 \end{vmatrix} = -1\begin{vmatrix} -2 & -12 \\ 3 & 0 \end{vmatrix} = -36$$

Thus $y = -36$.

Correction

$$D_y = \begin{vmatrix} 2 & -8 & 2 \\ 0 & 4 & 2 \\ 3 & 2 & 1 \end{vmatrix} \overset{-2C_3 + C_2}{=} \begin{vmatrix} 2 & -12 & 2 \\ 0 & 0 & 2 \\ 3 & 0 & 1 \end{vmatrix} = -2\begin{vmatrix} 2 & -12 \\ 3 & 0 \end{vmatrix} = -2(36) = -72$$

$$D = \begin{vmatrix} 2 & -4 & 2 \\ 0 & 3 & 2 \\ 3 & 0 & 1 \end{vmatrix} \overset{-3C_3 + C_1}{=} \begin{vmatrix} -4 & -4 & 2 \\ -6 & 3 & 2 \\ 0 & 0 & 1 \end{vmatrix} = \begin{vmatrix} -4 & -4 \\ -6 & 3 \end{vmatrix} = -12 - 24 = -36$$

Thus $y = D_y/D = -72/(-36) = 2$.

Review Problems for Chapter 11, page 725

1. $\begin{bmatrix} 4 & -3 \\ -10 & 17 \\ -2 & 2 \end{bmatrix}$ **2.** Not possible **3.** $\begin{bmatrix} -7 & -19 & 12 \\ 2 & 7 & 17 \end{bmatrix}$ **4.** Not possible **5.** $\begin{bmatrix} -7 & -8 \\ 9 & 10 \\ -5 & -1 \end{bmatrix}$ **6.** Not possible
7. $\begin{bmatrix} -3 & 23 & 35 \\ 4 & -12 & -20 \end{bmatrix}$ **8.** $x = 4, y = 6$ **9.** No solution **10.** $x = 0, y = 2, z = -5$ **11.** $x = -1, y = 2, z = -2$
12. $x = 4, y = -2, z = 1$ **13.** $x = -1, y = -2, z = 4$ **14.** $\begin{bmatrix} \frac{3}{2} & 2 \\ \frac{1}{2} & 1 \end{bmatrix}$ **15.** Does not exist **16.** $\begin{bmatrix} 1 & 0 & 0 \\ 2 & 2 & -1 \\ -1 & -1 & 1 \end{bmatrix}$
17. $\begin{bmatrix} -3 & 2 & -2 \\ -4 & 2 & -1 \\ 2 & -1 & 1 \end{bmatrix}$ **18.** $\begin{bmatrix} -3 & 2 & 4 \\ 2 & -1 & -1 \\ 2 & -1 & -2 \end{bmatrix}$ **19.** Does not exist **20.** $r = -9, s = -5$ **21.** $x = 1, y = 2$
22. $x = -2, y = 1, z = -4$ **23.** $x = 5, y = 0, z = 4$ **24.** -14 **25.** 73 **26.** 39 **27.** -15 **28.** 0
29. -1 **30.** -1 **31.** 8 **32.** 6 **33.** Dependent **34.** $x = 2, y = -3$ **35.** $x = -2, y = 3, z = 1$
36. Dependent

Cumulative Test Chapters 3–11, page 729

1. (a) $8x + 5y = 17$ (b) $m = \frac{5}{3}$, x-intercept 6, y-intercept -10 **2.** $f(3) = -11$ is the minimum value.
3. (a) Center $(-2, 4)$, radius 7 (b) $\frac{(x+2)^2}{16} + \frac{(y-3)^2}{9} = 1$ **4.** $f(x) = \frac{3}{4}x - 6$, $f^{-1}(x) = \frac{4}{3}x + 8$ **5.** $2 - i, 2 + i, \frac{1}{2}$
6. $-2, \frac{3}{2}, i, -i$ **7.** $x = 13$ **8.** $\{2\}$ **9.** The function values are given in this order: sine, cosine, tangent, cotangent, secant, cosecant. (a) $\sqrt{3}/2, -1/2, -\sqrt{3}, \sqrt{3}/3, -2, 2\sqrt{3}/3$ (b) $-\sqrt{2}/2, -\sqrt{2}/2, 1, 1, -\sqrt{2}, -\sqrt{2}$
10. 440 radians per minute **11.** (a) (b)

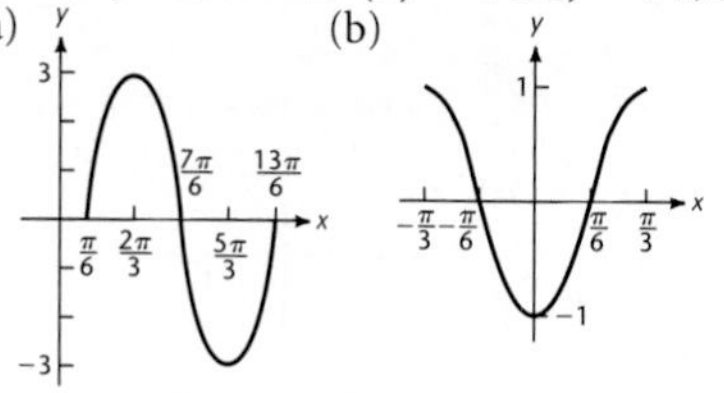

13. 90°, 210°, 330° **14.** $-\frac{1}{2}$ **15.** $A = 54°, B = 18°, c = 98$ **16.** 55.3 miles per hour

17.

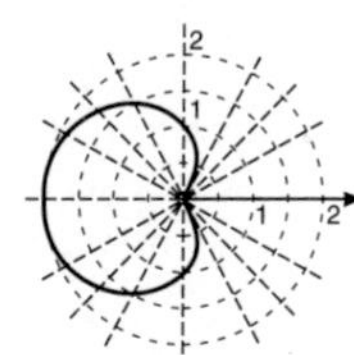

18. (a) $\{(5, -2)\}$ (b) $\varnothing$ (c) $\{(3, 5, 1)\}$

19. $\{(0, 4), (\sqrt{7}, -3), (-\sqrt{7}, -3)\}$ **20.**

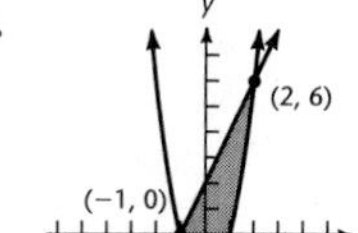

21. Minimum value of -16 at $(4, 1)$

22. (a) $\begin{bmatrix} 1 & -10 \\ 12 & -8 \\ -5 & 17 \end{bmatrix}$ (b) not possible (c) not possible (d) $\begin{bmatrix} 3 & 2 & -3 \\ 15 & 25 & 1 \end{bmatrix}$

23. $\{(0, 2, -1)\}$ **24.** $\{(2, -1, 3)\}$ **25.** $\{(4, -2, 1)\}$

Exercises 12.1, page 735

1. $\frac{1}{2}, \frac{1}{3}, \frac{1}{4}, \frac{1}{5}, \frac{1}{6}$ **3.** $0, \frac{1}{2}, \frac{2}{3}, \frac{3}{4}, \frac{4}{5}$ **5.** $-2, 4, -8, 16, -32$ **7.** $2, 2, 2, 2, 2$ **9.** $-1, \frac{1}{4}, -\frac{1}{9}, \frac{1}{16}, -\frac{1}{25}$ **11.** $1, \frac{2}{3}, \frac{3}{5}, \frac{4}{7}, \frac{5}{9}$
13. $1, 4, 7, 10, 13$ **15.** $x - 1, (x - 1)^2, (x - 1)^3, (x - 1)^4, (x - 1)^5$ **17.** $-x^2, x^4, -x^6, x^8, -x^{10}$ **19.**

(4, 9), (3, 7), (2, 5), (1, 3)

21.

(4, 10), (3, 3), (2, −2), (1, −5)

23. $1, 2, 4, 8, 16, 32$ **25.** $-5, 5, 5, -5, -5, 5$ **27.** $2, 3, 8, 19, 46, 111$

29. $2 + 4 + 8 + 16 + 32 = 62$ **31.** $\frac{1}{4} + \frac{2}{5} + \frac{3}{6} + \frac{4}{7} + \frac{5}{8} + \frac{6}{9} + \frac{7}{10} = \frac{3119}{840}$ **33.** $0 + 1 + 3 + 6 + 10 + 15 + 21 = 56$
35. $-1 + 1 - 1 + 1 - 1 + 1 - 1 = -1$ **37.** $2 + 2 + 2 + 2 + 2 + 2 = 12$ **39.** $4 - 9 + 16 - 25 + 36 - 49 = -27$
41. $-1 + 0 + 1 + 8 + 27 + 64 + 125 = 224$ **43.** $(1 - \frac{1}{2}) + (\frac{1}{2} - \frac{1}{3}) + (\frac{1}{3} - \frac{1}{4}) + (\frac{1}{4} - \frac{1}{5}) + (\frac{1}{5} - \frac{1}{6}) + (\frac{1}{6} - \frac{1}{7}) = \frac{6}{7}$

45. $\frac{1}{4} - \frac{1}{8} + \frac{1}{16} - \frac{1}{32} = \frac{5}{32}$ **47.** $\sum_{k=1}^{17} k$

49. $\sum_{j=1}^{7} 2^j$ **51.** $\sum_{n=1}^{7} (-1)^{n+1}(2n - 1)$ **53.** $\sum_{i=1}^{5} \frac{i}{i + 1}$ **55.** $\sum_{j=0}^{5} (-1)^j \frac{2j + 1}{2^{j+1}}$ **57.** $\sum_{k=1}^{43} \frac{1}{(2k - 1)(2k + 1)}$

59. $\sum_{i=0}^{n} \frac{x^i}{i + 1}$ **61.** Viewing rectangle $[0, 10]$ by $[0, 15]$ **63.** Viewing rectangle $[-10, 10]$ by $[-10, 10]$

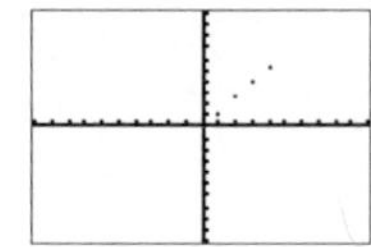

Exercises 12.2, page 743

1. 1, 2, 3, 4, 5, 6 **3.** 13, 11, 9, 7, 5, 3 **5.** 3, -1, -5, -9, -13, -17 **7.** 2, 6, 10, 14, 18, 22 **9.** $d = 5$, $a_n = 5n - 22$ **11.** $d = -3$, $a_n = -3(n + 1)$ **13.** Not arithmetic **15.** $d = -14$, $a_n = 21 - 14n$ **17.** $d = y$, $a_n = x - y + ny$ **19.** Not arithmetic unless $x = 0$ or $x = 1$. If $x = 0$, $d = 0$ and $a_n = 0$. If $x = 1$, $d = 0$ and $a_n = 1$. **21.** $d = x$, $a_n = nx$ **23.** $a_{11} = -25$, $S_{11} = -165$ **25.** $a_{11} = 7$, $S_{11} = -88$ **27.** $a_{11} = -13$, $S_{11} = 22$ **29.** $a_{11} = m - 10x$, $S_{11} = 11(m - 5x)$ **31.** $a_{11} = x - 10k$, $S_{11} = 11x$ **33.** 53 **35.** -60 **37.** 23 **39.** $-\frac{3933}{4}$ **41.** $a_1 = -7$, $d = \frac{1}{2}$ **43.** $a_1 = 17$, $d = -5$ **45.** 91 **47.** $\frac{110}{3}$ **49.** 2550 **51.** -117 **55.** $-20, -25, -30, -35$ **57.** 4510 **59.** $n = 36$ **61.** 39 **63.** 9 **65.** 1065 **67.** 1440

Exercises 12.3, page 751

1. $-1, 2, -4, 8, -16$ **3.** $4, 1, \frac{1}{4}, \frac{1}{16}$ **5.** $\frac{3}{4}, 3, 12, 48$ **7.** $\frac{1}{4}, -\frac{1}{2}, 1, -2, 4$ **9.** $-3, -6, -12$ or $-3, 6, -12$ **11.** $a_5 = 4$, $a_n = 2^n/8$, $S_5 = \frac{31}{4}$ **13.** $a_5 = 27$, $a_n = -(-3)^n/9$, $S_5 = \frac{61}{3}$ **15.** $a_5 = -8$, $a_n = (-2)^n/4$, $S_5 = -\frac{11}{2}$ **17.** $r = \frac{1}{2}$, $a_n = 14/2^n$ **19.** $r = \sqrt{2}/2$, $a_n = (\sqrt{2})^{5-n}$ **21.** Not geometric **23.** Not geometric **25.** $r = -\frac{1}{2}$, $a_n = (-4)(-\frac{1}{2})^{n-1}$ **27.** $r = -\frac{1}{7}$, $a_n = (-1)^{n-1}/7^{n-4}$ **29.** 31 **31.** $\frac{4323}{3125}$ **33.** $\frac{275}{81}$ **35.** $\frac{2101}{4}$ **37.** $\frac{63}{4}$ **39.** $r = \pm\sqrt{3}$, $a_1 = -\frac{1}{3}$ **41.** $r = 1.01$, $a_1 = 4/1.01$, or $r = -1.01$, $a_1 = -4/1.01$ **43.** \$511 **45.** \$2,688,000

Exercises 12.4, page 757

1. -2 **3.** $\frac{27}{8}$ **5.** Sum does not exist **7.** Sum does not exist **9.** $2(2 + \sqrt{2})$ **11.** $\frac{50}{9}$ **13.** 5 **15.** 30 **17.** 3 **19.** Sum does not exist; geometric series with $|r| = |-1.02| > 1$. **21.** Sum does not exist; geometric series with $|r| = |-\frac{100}{99}| > 1$. **23.** $\frac{4096}{7}$ **25.** Sum does not exist; geometric series with $|r| = |6| > 1$. **27.** 10 **29.** Sum does not exist; geometric series with $|r| = |-3| > 1$. **31.** $\frac{13}{3}$ **33.** $-\frac{4}{3}$ **35.** 1 **37.** 1 **39.** $\frac{1}{99}$ **41.** $\frac{131}{9999}$ **43.** $\frac{28}{9}$ **45.** $-\frac{4579}{1998}$ **47.** $\frac{10{,}033}{990}$ **49.** 10 feet **51.** 256,000

Exercises 12.5, page 764

1. $x^7 + 7x^6y + 21x^5y^2 + 35x^4y^3 + 35x^3y^4 + 21x^2y^5 + 7xy^6 + y^7$ **3.** $16x^4 + 32x^3y + 24x^2y^2 + 8xy^3 + y^4$ **5.** $x^{10} + 10x^8y + 40x^6y^2 + 80x^4y^3 + 80x^2y^4 + 32y^5$ **7.** $x^8 - 4x^6y^2 + 6x^4y^4 - 4x^2y^6 + y^8$ **9.** $16x^4 - 32x^3/y + 24x^2/y^2 - 8x/y^3 + 1/y^4$ **11.** $81x^4 + 36x^3y + 6x^2y^2 + 4xy^3/9 + y^4/81$ **13.** $x^{18} - 3x^{17} + 15x^{16}/4 - 5x^{15}/2 + 15x^{14}/16 - 3x^{13}/16 + x^{12}/64$ **15.** $a^5/32 - 15a^4b^2/16 + 45a^3b^4/4 - 135a^2b^6/2 + 405ab^8/2 - 243b^{10}$ **17.** 45 **19.** 80 **21.** $1760x^3y^9$ **23.** -29, $120x^{12}y^6$ **25.** $3003p^{12}q^{24}$ **27.** $8192y^{13}$ **29.** $1 + x/4 - 3x^2/32 + 7x^3/128 - 77x^4/2048 + \cdots$, $-1 < x < 1$ **31.** $1 - 2y/3 - y^2/9 - 4y^3/81 - 7y^4/243 + \cdots$, $-1 < y < 1$ **33.** $1 - x + 3x^2/2 - 5x^3/2 + 35x^4/8 - \cdots$, $-\frac{1}{2} < x < \frac{1}{2}$ **35.** $1 - 2x^2 + 3x^4 - 4x^6 + 5x^8 - \cdots$, $-1 < x < 1$ **37.** $1 + a + 3a^2/4 + a^3/2 + 5a^4/16 + \cdots$, $-2 < a < 2$ **39.** $x^4 + 4x^3y + 6x^2y^2 + 4xy^3 + y^4 + 4x^3 + 12x^2y + 12xy^2 + 4y^3 + 6x^2 + 12xy + 6y^2 + 4x + 4y + 1$ **41.** $x^3 - y^3 + z^3 - w^3 - 3x^2y + 3xy^2 + 3x^2z + 3xz^2 + 3y^2z - 3yz^2 - 3x^2w + 3xw^2 - 3y^2w - 3yw^2 - 3z^2w + 3zw^2 - 6xyz + 6xyw - 6xzw + 6yzw$ **43.** 1.061520 **45.** 0.922744 **47.** 27.270901 **49.** 1.0099505 **51.** 4,280,561,376 **53.** 1,609,344,100

Exercises 12.7, page 779

1. 12 different T-shirts are available.

- S
  - Red
  - White
  - Blue
- M
  - Red
  - White
  - Blue
- L
  - Red
  - White
  - Blue
- XL
  - Red
  - White
  - Blue

3. 0; 1; 0, 1; 1, 0

5. 1, 2; 2, 1; 3, 1; 4, 1; 5, 1; 1, 3; 2, 3; 3, 2; 4, 2; 5, 2; 1, 4; 2, 4; 3, 4; 4, 3; 5, 3; 1, 5; 2, 5; 3, 5; 4, 5; 5, 4 **7.** 24 **9.** 210 **11.** 1 **13.** $n(n-1)$ **15.** $n!/2$ **17.** $n(n-1)$ **19.** $1/[(n+2)(n+1)]$ **21.** 12 **23.** 120 **25.** 1024 **27.** 16 **29.** 120 **31.** 1,000,000,000 **33.** 7999 **35.** 108 **37.** (a) 720 (b) 5040 **39.** 39,916,800 **41.** 144 **43.** 1440 **45.** 120 **47.** (a) 2520 (b) 5040 (c) 3 (d) 180 **49.** 630,630 **51.** 3150 **53.** 144 **55.** 24 **57.** 144, 576 **59.** $7.8663312 \times 10^9 = 7{,}866{,}331{,}200$ **61.** $1.708663123 \times 10^{17} = 170{,}866{,}312{,}300{,}000{,}000$

Exercises 12.8, page 785

1. {0}, {1}, {0, 1} **3.** {1, 2}, {1, 3}, {1, 4}, {1, 5}, {2, 3}, {2, 4}, {2, 5}, {3, 4}, {3, 5}, {4, 5} **5.** 6 **7.** 1 **9.** 1 **11.** 15 **13.** n **15.** 1 **17.** 495 **19.** 5,005; 3,003 **21.** 280 **23.** 3,300 **25.** 44,100 **27.** 10,090,080 **29.** (a) 70 (b) 10 (c) 70 (d) 0 **31.** 36 **33.** (a) 8,211,173,256 (b) 169 (c) 2,944,656 **37.** 75,287,520 **39.** 635,013,559,600

Exercises 12.9, page 795

1. (a) $\{(H,1), (H,2), (H,3), (H,4), (H,5), (H,6), (T,1), (T,2), (T,3), (T,4), (T,5), (T,6)\}$ (b) $\{(H,2), (H,4), (H,6)\}$ **3.** (a) $\frac{4}{25}$ (b) $\frac{1}{5}$ (c) $\frac{2}{5}$ **5.** (a) $\frac{1}{4}$ (b) $\frac{1}{4}$ (c) $\frac{3}{4}$ **7.** $\frac{4}{7}$ **9.** $\frac{1}{4}$ **11.** (a) $\frac{1}{13}$ (b) $\frac{3}{26}$ (c) $\frac{1}{2}$ (d) $\frac{11}{26}$ **13.** (a) $\frac{5}{143}$ (b) $\frac{15}{26}$ (c) $\frac{30}{143}$ (d) $\frac{11}{26}$ (e) 0 **15.** (a) $\frac{1}{17}$ (b) $\frac{1}{221}$ (c) $\frac{11}{221}$ **17.** (a) $\frac{33}{16660}$ (b) $\frac{1}{108290}$ **19.** (a) $\frac{1}{22}$ (b) $\frac{5}{11}$ (c) $\frac{2}{11}$ **21.** (a) $\frac{4}{9}$ (b) $\frac{7}{9}$ **23.** 0.3

Critical Thinking: Find the Errors, page 799

1. Error

$$\sum_{n=2}^{7} (2n-1) = 1 + 3 + 5 + 7$$

Correction

$$\sum_{n=2}^{7} (2n-1) = 3 + 5 + 7 + 9 + 11 + 13 = 48$$

2. Error

$a_7 = a_1 r^6$

Correction

$a_7 = a_1 + (7-1)d = 2 + 6(-3) = -16$

3. Error

Since $a = \frac{2}{3}$ and $r = -\frac{3}{2}$ then the value of the sum is

$$\frac{a}{1-r} = \frac{\frac{2}{3}}{1-(-\frac{3}{2})} = \frac{\frac{2}{3}}{\frac{5}{2}} = \frac{4}{15}.$$

Correction

Since $|r| > 1$, the sum does not exist.

4. Error

Since the president must be a senior there are 10 choices for the president and 8 choices of a junior for the secretary. Hence there are $10 \cdot 8 = 80$ ways of electing a president and a secretary.

Correction

Since the president must be a senior, there are 10 choices for president. There are 17 choices for secretary since the secretary can be either a senior or a junior. Hence there are $10 \cdot 17 = 170$ ways of electing a president and a secretary.

5. Error

We count the number of ways that 3 of the 8 horses will finish the race. That is

$$\binom{8}{3} = \frac{8!}{3!5!} = \frac{8 \cdot 7 \cdot 6 \cdot 5!}{3 \cdot 2 \cdot 1 \cdot 5!} = 56$$

possible outcomes.

Correction

The number of ways that 3 of the 8 horses will finish the race is $P(8, 3) = 8 \cdot 7 \cdot 6 = 336$.

6. Error

The third term is

$$\binom{9}{3} x^6 (2y)^3 = \frac{9 \cdot 8 \cdot 7 \cdot 6!}{3 \cdot 2 \cdot 1 \cdot 6!} x^6 8y^3 = 672x^6y^3.$$

Correction

The third term is

$$\binom{9}{2} x^7 (-2y)^2 = \frac{9 \cdot 8 \cdot 7!}{2! \cdot 7!} x^7 4y^2 = 144x^7y^2.$$

Review Problems for Chapter 12, page 800

1. $0, \frac{1}{3}, \frac{1}{2}, \frac{3}{5}, \frac{2}{3}$ **2.** $-2, 2, -\frac{4}{3}, \frac{2}{3}, -\frac{4}{15}$ **3.** 1, 3, 5, 7, 9, 11 **4.** 2, 1, −1, −5, −13, −29
5. $(-2) + (-2)^2 + (-2)^3 + (-2)^4 + (-2)^5 = -22$ **6.** $1 + \frac{1}{2} + \frac{1}{2^2} + \frac{1}{2^3} + \frac{1}{2^4} + \frac{1}{2^5} = \frac{63}{32}$
7. $5 + 5 + 5 + 5 + 5 + 5 + 5 = 35$ **8.** $\frac{2!}{2} + \frac{4!}{2^2} + \frac{6!}{2^3} + \frac{8!}{2^4} = 2617$
9. $\frac{1!}{-1} + \frac{2!}{(-1)^2} + \frac{3!}{(-1)^3} + \frac{4!}{(-1)^4} + \frac{5!}{(-1)^5} = -101$ **10.** $\sum_{k=1}^{7} 2k$ **11.** $\sum_{j=1}^{5} 2^{j-1}(2j)$ **12.** $\sum_{n=1}^{6} (-1)^{n+1}(2n-1)$
13. $\sum_{i=1}^{7} \frac{i^2}{i+1}$ **14.** $\sum_{k=1}^{6} \frac{k!}{2^{k-1}}$ **15.** $a_{11} = -8, S_{11} = 22$ **16.** $a_1 = -2, d = 2$ **17.** 5, 3, 1, −1, −3
18. $a_{19} = -33, a_n = 5 - 2n$ **19.** (a) Geometric sequence, $r = -\frac{3}{2}$, $a_n = -16(-\frac{3}{2})^n$ (b) Not a geometric sequence
20. $r = -\frac{2}{3}, a_1 = \frac{729}{4}$ **21.** $a_6 = -\frac{1}{8}, a_n = 4(-\frac{1}{2})^{n-1} = (-1)^{n-1}/2^{n-3}$ **22.** (a) The sum does not exist. (b) $\frac{27}{5}$ **23.** 50
24. $\frac{106}{33}$ **25.** \$163.84, \$327.67 **26.** $4\frac{2}{3}$ meters **27.** $x^6 - 12x^5y + 60x^4y^2 - 160x^3y^3 + 240x^2y^4 - 192xy^5 + 64y^6$
28. $-2016x^4y^{10}$ **29.** 1.14868 **34.** 9,000,000 **35.** (a) 15,600 (b) 8400 **36.** (a) 18,720 (b) 13,000 **37.** 336
38. 120 **39.** 20,160 **40.** 147 **41.** (a) 1 (b) 360 (c) $n + 1$ (d) $n(n + 1)$ **42.** (a) 12 (b) 60 (c) 56 (d) 35 (e) n
(f) $n(n + 1)/2$ **43.** 1260 **44.** 6720 **45.** 3360 **46.** 10,080 **47.** 495 **48.** 672 **49.** 792 **50.** 0.4
51. (a) $\frac{15}{28}$ (b) $\frac{5}{28}$ **52.** (a) $\frac{4}{11}$ (b) $\frac{4}{33}$

Exercises A.1, page A-5

1. 1.6191 − 10 **3.** 1.4871 **5.** 0.6542 **7.** 1 **9.** 6.0294 **11.** 7.0294 − 10 **13.** 7.18 **15.** 283,000,000
17. 0.00476 **19.** 0.001 **21.** 1.0047 **23.** 0.6202 **25.** 5.6210 **27.** 7.8875 − 10 **29.** 1.292 **31.** 87,760
33. 0.004954 **35.** 0.2444

Index

Index

Abscissa, 162
Absolute value, 16–17, 86, 588
  equations, 98–100
  inequalities, 118
Acute angle, 381
Addition, 4
  of complex numbers, 81
  of fractions, 8
  of functions, 184
  of matrices, 671
  of ordinates, 463
  of polynomials, 33
  of rational expressions, 49
  of vectors, 574
Addition property of equality, 4, 97
Addition property of inequalities, 15, 115
Additive identity, 5
  for matrices, 672
Additive inverse, 5
  for matrices, 672
Adjacent side, 420
Airspeed, 580
Algebraic expression, 32
Algebraic method of solution of an inequality, 148
Ambiguous case, 548–549
Amount, 103
  compound, 345
Amplitude of a complex number, 588
Amplitude of a function, 442
Angle, 380
  acute, 381
  central, 388
  complementary, 381
  composite 489
  coterminal, 383
  degree measure, 381
  of depression, 424
  of elevation, 424
  of incidence, 417
  initial side, 380
  measure, 380
  negative, 380
  obtuse, 381
  positive, 380
  quadrantal, 401
  radian measure, 389
  reference, 403
  of refraction, 417
  related, 403
  right, 381
  in standard position, 382
  straight, 381
  supplementary, 381
  terminal side, 380
  vertex, 380
Angular speed, 392
Antilogarithm, A-3
Approximate equality, 12
Approximation of real zeros, 307
Arc length formula, 391
Arcsine function, 529
Area of a triangle, 567–569
Argument of a complex number, 588
Argument of a function, 443
Arithmetic sequence, 738
Associative property, 4, 5
  of addition of matrices, 672
  of multiplication of matrices, 679
Asymptotes:
  horizontal, 321, 324
  of a hyperbola, 238
  oblique, 328
  parabolic, 333
  vertical, 321, 324
Augmented matrix, 685
Axes of an ellipse, 233
Axis, 161
  imaginary, 587
  major, 233
  minor, 233
  polar, 601
  real, 587
  of symmetry, 198, 200

Back substitution, 686
Base, 21, 341, 357
Base vectors, 578
Basic:
  formulas for fractions, 8
  properties of radicals, 60
  properties for real numbers, 4
  properties of vector operations, 577
Bearing, 425
Bermuda triangle, 614
Binomial, 33
  coefficients, 760, 762–763
  theorem, 761
Biorhythm, 443
Boundary, 646
Bounds theorem, 299
Boyle's law, 265
Break-even point, 643
Broken-line graphs, 191

Calculation rules for approximate numbers, 421–422
Carbon-14 dating, 346
Cardano, Girolamo, 288
Cardioid, 605
Cartesian coordinate system, 161
Center:
  of a circle, 217
  of an ellipse, 233, 242
  of a hyperbola, 245
Central angle, 388
Change-of-base formula, 369
Characteristic of a logarithm, A-2
Characteristic values, 708
Circle, 214, 217–218
Circular function, 437
Closure properties, 4, 5
Coefficient, 7
  matrix, 685, 698
Cofactor, 703
  expansion, 705
Cofunction, 421, A-8
Cofunction identities, 491
Column matrix, 670
Column operations, 709
Combinations, 765, 783, 784
Combined variation, 259
Common:
  difference, 738
  logarithm, 365
  ratio, 747
Commutative property, 5
  of matrix addition, 672
Compact notation for inequalities, 14
Complement, 792
Complementary angles, 381, A-7
Complementary function, 421, A-8
Completely factored, 39
Completing the square, 126
Complex fraction, 48
Complex numbers, 80
  absolute value, 588
  addition, 81
  amplitude, 588
  argument, 588
  conjugate, 83
  difference, 81
  division, 83
  equality, 81
  imaginary part, 80
  modulus, 588
  multiplication, 82, 591
  polar form, 588
  powers, 596
Complex numbers *(Cont.)*
  products, 82, 591
  quotients, 83, 591
  real part, 80
  roots, 597
  standard form, 82
  subtraction, 81
  trigonometric form, 588
Components of an ordered pair, 172, 575
Components of a vector, 575
Composition function, 185
Compound amount, 345
Compound inequality, 15, 115
Compound interest, 345
  continuous, 352
Conditional equation, 483
Cone, 214
Conformable matrices, 676
Conic sections, 214
Conjugate:
  of a binomial, 74, 480
  of a complex number, 83
  zeros theorem, 292
Constant, 7
  matrix, 685, 698
  of variation, 259
Constraints, 653
Conversion of degrees to radians, 389
Conversion of radians to degrees, 389
Convex region, 653
Coordinates, 162, 601
Cosecant function, 399, 421, 437
Cosine:
  of the difference of two angles, 490

Cosine *(Cont.)*
of half an angle, 507
of the sum of two angles, 493
of twice an angle, 504
Cosine function, 399, 421, 437
Cosines, law of, 559, 560
Cotangent function, 399, 421, 437
Coterminal angles, 383
Counterexample, 483
Counting principle, 773
Course, 580
Cramer's Rule, 717, 719
Cube root, 58

Decibels, 367
Degenerate circle, 218
Degree, 381
of a polynomial, 33
of a term, 33
De Moivre's theorem, 596
Dependent system, 619, 631, 717, 719
Dependent variable, 172
Depreciation, 192, 616, 703
linear, 192
Descartes' rule of signs, 290
Determinant, 702, 705
cofactors of, 703
expansion of, 705
order of, 702
properties of, 710–711
Diagonal matrix, 670
Difference, 5
common, 738
of complex numbers, 81
of matrices, 672
quotient, 174
Dimension of a matrix, 669
Direction angle, 579
Direct variation, 258
Discount rate, 103
Discriminant, 134
Distance formula, 217
Distributive property, 5, 72, 679
Dividend, 274
Division, 6
by detached coefficients, 275
of fractions, 8
of functions, 184
of polynomials, 274
of rational expressions, 47
synthetic, 275
Divisor, 274
Domain, 171, 172
of an exponential function, 343
of a logarithmic function, 360
Dot product, 586
Double-angle identities, 504

Eccentricity of an ellipse, 271
Eigenvalues, 708
Element of a matrix, 669
Element of a set, 2
Elimination method, 620, 630, 638
Ellipse, 214, 231, 242
center of, 233, 242
foci of, 231
major axis of, 233
minor axis of, 233
standard equations for, 232, 242
vertices of, 233
Empty set, 3
Equality, 3
of complex numbers, 81
of fractions, 8
of matrices, 670
Equality *(Cont.)*
of ordered pairs, 172
properties of, 4
of sets, 2
of vectors, 574
Equations, 3
absolute value, 98–100
conditional, 483
equivalent, 96
exponential, 366
inverse trigonometric, 529
involving radicals, 143
linear, 96, 163, 191, 618
logarithmic, 365
polar, 603
quadratic, 123
trigonometric, 517
Equilateral triangle, 383
Equilibrium price, 626
Equivalent:
equations, 96
inequalities, 115
systems, 620
Event, 789
Experiment, 789
Exponent, 21
definition of, 21, 24, 25, 62
laws of, 25, 341
Exponential:
decay, 353
equation, 366
form, 63, 357–358
function, 342, 353
growth, 353
Expressions:
algebraic, 32
rational, 46
trigonometric, 478

Extraneous solutions, 98, 139, 142
Extreme values, 655

Factor, 38
  of multiplicity k, 289
Factorial, 759
Factoring by grouping, 42
Factorization formulas, 39
Factor theorem, 282
Feasible solutions, 653
Ferrari, Ludovico, 288
Fibonacci sequence, 734, 737–738
Finite sequence, 732
Focus of an ellipse, 231
Focus of a hyperbola, 238–239
Four-leaved rose, 607
Fractions:
  complex, 48
  operations with, 8
Function, 171, 172
  addition, 184
  circular, 437
  complementary, 421, A-8
  composition, 185
  defined piecewise, 191
  division, 184
  exponential, 342, 353
  greatest integer, 177
  inverse, 250, 529
  linear, 187
  logarithmic, 357
  multiplication, 184
  one-to-one, 249
  periodic, 442
  quadratic, 197
  rational, 321
  sinusoidal, 443
  step, 178
  subtraction, 184
Function *(Cont.)*
  trigonometric, 399, 421, 437
  wrapping, 436
Function value notation, 171
Fundamental identities, 479
Fundamental operations, 6
  on functions, 184
Fundamental principle of fractions, 8, 46
Fundamental theorem of algebra, 288

Gauss, C.F., 288
Gaussian elimination, 687
Gauss-Jordan elimination to find an inverse, 695
Gauss-Jordan elimination to solve a system, 688
General term:
  of an arithmetic sequence, 741
  of a geometric sequence, 750
  of a sequence, 732
Geometric sequence, 747
Graeco-Latin square, 726, 728
Graph:
  broken-line, 191
  of a cosecant function, 456
  of a cosine function, 449
  of a cotangent function, 462
  of an equation, 162
  of an exponential function, 343
  of a function, 175
  of an inequality, 13–15, 646
  of a logarithmic function, 359
  of a polynomial function, 311
  of a rational function, 321
  of a relation, 175
  of a secant function, 458
  of a sine function, 444, 447
  of a tangent function, 460, 461
Great circle, 613
Greater than, 13
Greatest integer function, 177
Ground speed, 580

Half-angle identities, 507
Half-life, 355–356
Half-line, 380
Half-plane, 646
Head of a vector, 573
Heading, 580
Heron's formula, 61, 569
Hooke's law, 262
Horizontal:
  asymptote, 321, 324
  line, 165
  line test, 249
  translation, 228, 446
Hubble space telescope, 27–28
Hyperbola, 214, 237
  asymptotes of, 238
  center of, 245
  foci of, 238–239
  standard equations for, 238–239
  vertices of, 237

Identities, 410
  cofunction, 491
  double-angle, 504
  fundamental, 479
  half-angle, 507
  negative angle, 492
  product, 513
  Pythagorean, 479
  quotient, 411, 479
  reciprocal, 410, 479
  reduction, 500
  sum, 514
Identity elements, 5

Identity matrix, 679
Imaginary:
  axis, 587
  number, 80
  part of a complex number, 80
Inconsistent system, 618, 631, 717, 719
Independent variable, 171
Index:
  of a radical, 58
  of refraction, 417
  of summation, 734
Induction hypothesis, 767
Induction principle, 767
Inequalities, 14
  absolute value, 118
  compound, 15, 115
  equivalent, 115
  linear, 115, 646
  nonlinear, 145
  quadratic, 145
  systems of, 646
Infinite sequence, 732
Infinity, 117
Initial point of a vector, 573
Initial side, 380
Inner product, 586
Integers, 4
Intercepts of a graph, 162
Intermediate value theorem for polynomials, 283
Interpolation, A-4
Intersection, 3, 618
Interval notation, 117
Inverse, 5
  of a function, 250
  of a matrix, 694
  sine function, 529
  trigonometric equations, 533
Inverse *(Cont.)*
  trigonometric functions, 529
  variation, 259
Irrational numbers, 4
Isosceles triangle, 383

Joint variation, 259

Key points on a graph, 443, 448, 460

Latin square, 727
Law of cosines, 559, 560
Law of sines, 547
Laws of exponents, 25, 341
Leading coefficient, 33
Least common denominator, 49
Length of an arc, 391
Length of a vector, 574
Less than, 13
Light year, 32
Like terms, 7
Limit, 753–755
Linear depreciation, 192
Linear equation, 96, 163
  point-slope form of, 191
  slope-intercept form of, 191
  standard form of, 191
Linear function, 187, 653
Linear inequality, 115, 646
Linear interpolation, A-4
Linear programming problem, 653
Linear speed, 392
Lines:
  equations of, 191
  parallel, 190
  perpendicular, 190
  straight, 188
Logarithm, 357, A-7, A-8
  characteristic, A-2
  common, 365
  mantissa, A-2
  natural, 368, A-9
  properties of, 361
Logarithmic:
  equation, 365
  form, 357
  function, 357
Lower bound, 299
Lowest terms, 8, 46

Magnitude of a vector, 574
Main diagonal elements, 669
Major axis of an ellipse, 233
Mantissa, A-2
Mathematical induction, 767
Mathematical model, 1
Matrix, 669
  addition, 671
  augmented, 685
  coefficient, 685, 698
  column, 670
  constant, 685, 698
  determinant of, 702, 705
  diagonal, 670
  dimension of, 669
  equality of, 670
  identity, 679
  multiplication, 676
  multiplicative inverse of, 694
  row, 670
  square, 670
  sum, 671
  unknown, 698
  zero, 672
Maximize, 655
Mean, 69

Measure of an angle, 380, 381, 389
Members of a set, 2
Method:
 of completing the square, 126
 of elimination, 620, 630, 638
 of substitution, 619, 637
Midpoint formula, 167
Minimize, 655
Minor, 703
 axis of an ellipse, 233
Minute, 381
Mixture problems, 106
Modulus, 588
Monomial, 33
Mount Everest, 475
Multiplication, 4
 of complex numbers, 82
 of fractions, 8
 of functions, 184
 of matrices, 676
 of polynomials, 33
 properties of inequalities, 15, 115
 property of equality, 4, 97
 of radicals, 73
 of rational expressions, 47
Multiplicative identity, 5, 679
Multiplicative inverse, 5
 of a matrix, 694
Multiplicity of zeros, 289

Natural logarithm, 368
Natural number, 4
Natural number *e*, 350
Negative:
 angle, 380
 angle identities, 492
 infinity, 117
 number, 13
Newton's law of cooling, 356

n^{th} root, 58
n^{th} term:
 of an arithmetic sequence, 741
 of a geometric sequence, 750
 of a sequence, 732
Number line, 13
Numbers:
 complex, 80
 irrational, 4
 natural, 4
 negative, 13
 positive, 13
 rational, 4
 real, 3, 4

Oblique asymptote, 328
Oblique triangle, 383, 546
Obtuse angle, 381
Ohm's law, 270
One-to-one function, 249
One-to-one property, 344
Operations with fractions, 8
Operations with radicals, 71–74
Opposite side, 420
Orbits of the planets, 271–272
Order:
 of a determinant, 702
 of a matrix, 670
 of operations, 6
 of a radical, 58
Ordered pair, 172
Ordering of real numbers, 13
Ordinate, 162
Origin, 12, 161

Parabola, 197, 214, 215
Parallel lines, 190
Parallelogram rule, 574, 588
Pascal's triangle, 764

Period, 442
Periodic function, 442
Permutations, 775, 777, 778
Perpendicular lines, 190
Phase shift, 446
pH scale, 31
Point-circle, 218
Points of intersection, 618
Point-slope form, 191
Polar:
 axis, 601
 coordinates, 601, 603
 equations, 603
 form, 588
Pole, 601
Polynomial, 32
 addition of, 33
 completely factored, 39
 degree of, 33
 division of, 274
 irreducible, 40
 monic, 33
 multiplication of, 33
 prime, 40
 rational zeros of, 296
 simplest form, 33
 zeros of, 86, 284, 288, 289
Positive angle, 380
Positive number, 13
Power, 21
Powers of a complex number, 596
Powers of *i*, 84–85
Principal, 103, 345
 n^{th} root, 58, 62
 square root, 58, 84
Principle of mathematical induction, 767
Prism, 476
Probability, 789

Product:
- of complex numbers, 82, 591
- of matrices, 676
- of polynomials, 33
- of a scalar and a matrix, 672
- of vectors, 586

Product identities, 513
Projectiles, 543
Properties:
- of absolute value, 16
- of equality, 4
- of inequalities, 15
- of logarithms, 361
- of probability, 793
- of radicals, 60

Proportion, 258
Proportionality constant, 259
Pythagorean identities, 479
Pythagorean theorem, 124–125, 216

Quadrantal angles, 401
Quadrants, 162
Quadratic:
- equation, 123
- formula, 131
- function, 197
- inequality, 145

Quotient identities, 411, 479
Quotients, 6
- of complex numbers, 83, 591
- of polynomials, 274

Radian, 389
Radical, 58
- form, 63
- index of, 58
- order of, 58

Radicand, 58
Radius of a circle, 217
Range, 171, 172
- of an exponential function, 343
- of a logarithmic function, 360
- of a projectile, 544
- of the trigonometric functions, 412

Ratio, 256, 747, 752
Rational:
- exponents, 62
- expression, 46
- function, 321

Rationalizing the denominator, 74
Rationalizing the numerator, 75, 78
Rational number, 4
Rational zeros, 296
Ray, 380
Real axis, 587
Real part of a complex number, 80
Reciprocal identities, 410, 479
Rectangular coordinate system, 161, 603
Reduction identity, 500
Reduction of radical index, 73
Reference angle, 403
Reference triangle, 404
Reflection, 227, 417
Related:
- angle, 403
- angle theorem, 405
- number, A-9
- triangle, 404

Relation, 172
Remainder theorem, 280
Resultant, 574
Revolution, 381
Richter scale, 376
Right angle, 381
Root, n^{th}, 58
Root, principal square, 58, 84
Roots of complex numbers, 597
Roots of an equation, 123, 134
Rose, 607
Row:
- matrix, 670
- operations on a determinant, 709
- operations on a matrix, 687

r^{th} term of a binomial expansion, 762
Rules for calculations with approximate numbers, 421–422

Sample space, 789
Scalar, 573, 576, 672
- multiplication, 573, 576, 672
- quantity, 573

Scientific notation, 27, A-1
Secant function, 399, 421, 437
Second, 381
Sequence, 732
- arithmetic, 738
- Fibonacci, 734, 737–738
- finite, 732
- geometric, 747
- infinite, 732
- terms of, 732

Series, 734
Set-builder notation, 2
Sets, 2
- empty, 3
- equal, 2
- intersection of, 3
- union of, 2

Shroud of Turin, 591
Side adjacent, 420
Side opposite, 420
Sides of an angle, 380
Sigma notation, 734
Sign graph, 146

Significant digits, 421
Signs of the trigonometric functions, 400
Similar triangles, 383
Simple event, 789
Simple interest, 103
Simplest form of a polynomial, 33
Simplest radical form, 75
Sine, 399, 421, 437
  of half an angle, 507
  of the sum and difference of two angles, 497
  of twice an angle, 504
Sines, law of, 547
Sinusoidal function, 443
Slope of a line, 166
Slope-intercept form, 191
Snell's law, 417
Solutions, 96
  extraneous, 98
  feasible, 653
  of inequalities, 115, 145, 646
  of linear equations, 96
  of linear inequalities, 115
  of nonlinear inequalities, 145
  of quadratic equations, 123, 131
  of quadratic inequalities, 145
  of systems of equations, 618
  of systems of inequalities, 647
  of systems of linear equations, 618, 629, 687, 688, 698
Special products, 35
Speed, 392, 580
Square matrices, 670
Square root, 57
  property, 124
Standard form:
  for a complex number, 82
  for the equation of a circle, 218

Standard form *(Cont.)*
  for the equation of an ellipse, 232, 242
  for the equation of a hyperbola, 238–239, 245
  for the equation of a line, 191, 196
Standard position, 382
Step function, 178
Straight angle, 381
Straight line, 163, 188
  depreciation, 192
Standard deviation, 69
Stretching, 227
Subset, 2
Substitution method, 619, 637
Substitution property, 4
Subtraction, 5
  of complex numbers, 81
  of functions, 184
  of matrices, 672
Sum:
  of an arithmetic sequence, 741
  of complex numbers, 81
  of an infinite geometric sequence, 755
  of matrices, 671
  of n terms of a geometric sequence, 750
Sum identities, 514
Summation notation, 734
Supplementary angles, 381
Symbols of grouping, 6
Symmetry, 224
Synthetic division, 275
Systems of equations, 618, 637
Systems of inequalities, 646
Systems of linear equations, 618, 629
  consistent, 618, 631

Systems of linear equations *(Cont.)*
  dependent, 619, 631, 717, 719
  inconsistent, 618, 631, 717, 719
  independent, 618, 631

Tail of a vector, 573
Tangent:
  function, 399, 421, 437
  of half an angle, 507
  of the sum and difference of two angles, 497
  of twice an angle, 504
Tartaglia, Niccolò, 288
Terminal point of a vector, 573
Terminal side of an angle, 380
Terms:
  of a polynomial, 33
  of a sequence, 732
  of a sum, 7
Tests for symmetry, 226
Translation, 228
Tree diagram, 772
Trial, 789
Triangle:
  equilateral, 383
  inequality, 16
  isosceles, 383
  oblique, 383, 546
  similar, 383
Trigonometric:
  equations, 517
  expressions, 478
  form, 588
  functions of an angle, 399, 421
  functions of a real number, 437
Trinomial, 33
  theorem, 766
Typical cycle:
  for a cosine curve, 449

Typical cycle *(Cont.)*
for a cotangent curve, 462
for a sine curve, 444
for a tangent curve, 460

Undamped motion, 451
Uniform motion, 108, 140
Union of sets, 2
Unit circle, 434
Unit vector, 579
Unknown matrix, 698
Upper bound, 299

Variable, 7
dependent, 172
dummy, 735
independent, 171
Variation:
combined, 259
constant of, 259
Variation *(Cont.)*
direct, 258
inverse, 259
joint, 259
of sign, 290
Vector, 573
addition, 574
quantity, 573
Velocity, 580
Vertex, 416
of an angle, 380
of a parabola, 198, 200, 215
Vertical:
asymptote, 321, 324
line, 165
line test, 177
translation, 228
Vertices:
of an ellipse, 233
of a hyperbola, 237
of parabolas, 198, 200, 215

Whole numbers, 4
Work formula, 110
Wrapping function, 436

x-axis, 161
x-component, 575
x-coordinate, 162
x-intercept, 162

y-axis, 161
y-component, 575
y-coordinate, 162
y-intercept, 162

Zero:
expansion property, 706
exponent, 24
matrix, 672
product property, 123
Zeros of a polynomial, 86, 284, 288, 289
rational, 296

The Trigonometric Functions

$$\sin\theta = \frac{y}{r} \qquad \cot\theta = \frac{x}{y}$$
$$\cos\theta = \frac{x}{r} \qquad \sec\theta = \frac{r}{x}$$
$$\tan\theta = \frac{y}{x} \qquad \csc\theta = \frac{r}{y}$$

Fundamental Identities

$$\sin\theta\csc\theta = 1 \qquad \sin^2\theta + \cos^2\theta = 1$$
$$\cos\theta\sec\theta = 1 \qquad 1 + \tan^2\theta = \sec^2\theta$$
$$\tan\theta\cot\theta = 1 \qquad 1 + \cot^2\theta = \csc^2\theta$$
$$\tan\theta = \frac{\sin\theta}{\cos\theta} \qquad \cot\theta = \frac{\cos\theta}{\sin\theta}$$

Cofunction Identities

$$\cos\left(\frac{\pi}{2} - \theta\right) = \sin\theta \qquad \sin\left(\frac{\pi}{2} - \theta\right) = \cos\theta$$
$$\cot\left(\frac{\pi}{2} - \theta\right) = \tan\theta \qquad \tan\left(\frac{\pi}{2} - \theta\right) = \cot\theta$$
$$\csc\left(\frac{\pi}{2} - \theta\right) = \sec\theta \qquad \sec\left(\frac{\pi}{2} - \theta\right) = \csc\theta$$

Negative-Angle Identities

$$\sin(-\theta) = -\sin\theta \qquad \csc(-\theta) = -\csc\theta$$
$$\cos(-\theta) = \cos\theta \qquad \sec(-\theta) = \sec\theta$$
$$\tan(-\theta) = -\tan\theta \qquad \cot(-\theta) = -\cot\theta$$

Identities for Sum and Difference of Two Angles

$$\cos(\alpha - \beta) = \cos\alpha\cos\beta + \sin\alpha\sin\beta$$
$$\cos(\alpha + \beta) = \cos\alpha\cos\beta - \sin\alpha\sin\beta$$
$$\sin(\alpha + \beta) = \sin\alpha\cos\beta + \cos\alpha\sin\beta$$
$$\sin(\alpha - \beta) = \sin\alpha\cos\beta - \cos\alpha\sin\beta$$
$$\tan(\alpha + \beta) = \frac{\tan\alpha + \tan\beta}{1 - \tan\alpha\tan\beta}$$
$$\tan(\alpha - \beta) = \frac{\tan\alpha - \tan\beta}{1 + \tan\alpha\tan\beta}$$

Greek Alphabet

Alpha	Α, α	Nu	Ν, ν
Beta	Β, β	Xi	Ξ, ξ
Gamma	Γ, γ	Omicron	Ο, o
Delta	Δ, δ	Pi	Π, π
Epsilon	Ε, ϵ	Rho	Ρ, ρ
Zeta	Ζ, ζ	Sigma	Σ, σ
Eta	Η, η	Tau	Τ, τ
Theta	Θ, θ	Upsilon	Υ, υ
Iota	Ι, ι	Phi	Φ, ϕ
Kappa	Κ, κ	Chi	Χ, χ
Lambda	Λ, λ	Psi	Ψ, ψ
Mu	Μ, μ	Omega	Ω, ω

Double-Angle Identities

$$\sin 2\theta = 2\sin\theta\cos\theta$$
$$\cos 2\theta = \cos^2\theta - \sin^2\theta = 2\cos^2\theta - 1 = 1 - 2\sin^2\theta$$
$$\tan 2\theta = \frac{2\tan\theta}{1 - \tan^2\theta}$$

Half-Angle Identities

$$\sin\frac{\theta}{2} = \pm\sqrt{\frac{1 - \cos\theta}{2}} \qquad \tan\frac{\theta}{2} = \pm\sqrt{\frac{1 - \cos\theta}{1 + \cos\theta}}$$
$$\cos\frac{\theta}{2} = \pm\sqrt{\frac{1 + \cos\theta}{2}} \qquad = \frac{\sin\theta}{1 + \cos\theta}$$
$$= \frac{1 - \cos\theta}{\sin\theta}$$

Reduction Identity

$$a\sin\theta + b\cos\theta = \sqrt{a^2 + b^2}\,\sin(\theta + \alpha),$$

where

$$\cos\alpha = \frac{a}{\sqrt{a^2 + b^2}} \quad \text{and} \quad \sin\alpha = \frac{b}{\sqrt{a^2 + b^2}}$$